Performance, Stability, Dynamics, and Control of Airplanes

Third Edition

Bandu N. Pamadi
NASA Langley Research Center
Hampton, Virginia

AIAA EDUCATION SERIES

Joseph A. Schetz, Editor-in-Chief
Virginia Polytechnic Institute and State University
Blacksburg, Virginia

Published by the
American Institute of Aeronautics and Astronautics, Inc.
1801 Alexander Bell Drive, Reston, Virginia 20191-4344

American Institute of Aeronautics and Astronautics, Inc., Reston, Virginia

5 4 3 2 1

Library of Congress Cataloging-in-Publication Data

Pamadi, Bandu N., 1945-
Performance, stability, dynamics, and control of airplanes / Bandu N. Pamadi, NASA Langley Research Center, Hampton, Virginia. – Third edition.
pages cm. – (AIAA education series)
ISBN 978-1-62410-274-5
1. Airplanes–Performance. 2. Stability of airplanes. 3. Airplanes–Control systems. 4. Aerodynamics. I. Title.
TL671.4.P28 2015
629.132–dc23

2014040138

AIAA EDUCATION SERIES

FOREWORD

This third edition of the textbook *Performance, Stability, Dynamics, and Control of Airplanes* by Bandu N. Pamadi is a comprehensive and integrated treatment of important subject areas in the aerospace field, including worked examples. The first two editions have proven to be a valuable part of the AIAA Education Series, and we are pleased to welcome this new edition to the series. New to the third edition is a chapter on stability and control issues/challenges of unmanned vehicles.

The AIAA Education Series aims to cover a broad range of topics in the general aerospace field, including basic theory, applications, and design. The philosophy of the series is to develop textbooks that can be used in a college or university setting, instructional materials for intensive continuing education and professional development courses, and also books that can serve as the basis for independent self-study for working professionals in the aerospace field. Suggestions for new topics and authors for the series are always welcome.

Joseph A. Schetz
Editor-in-Chief
AIAA Education Series

FOREWORD TO THE FIRST EDITION

Performance, Stability, Dynamics, and Control of Aircraft by Bandu N. Pamadi represents a novel approach of teaching an aeronautical engineering course as a comprehensive and integrated exposure to several interrelated disciplines. The text reflects many years of teaching experience of the author and his consulting work as a senior research engineer with NASA Langley Research Center. It contains all the necessary background material on aerodynamics, dynamics, and control systems, discussing the fundamental principles with the use of sketches, solved examples, and design exercises. It takes the reader from the early days of the Wright Brothers to the modern era of combat aircraft flying at poststall angles of attack.

Chapters 1 through 3 present basic principles of aerodynamics, aircraft performance, and static equilibrium and control in steady flight. Chapter 4 deals with aircraft dynamics and decoupled equations for longitudinal and lateral motion, and it introduces the concept of stability derivatives. Chapter 5 discusses design of stability augmentation systems and autopilots. Chapter 6 discusses aircraft response and methods of closed-loop control of aircraft, while Chapters 7 and 8 discuss problems of inertial coupling, aircraft spin, and high angle of attack. A useful feature of the text is the inclusion of solved examples presented to illustrate the theory and basic principles involved.

The Education Series of textbooks and monographs published by the American Institute of Aeronautics and Astronautics embraces a broad spectrum of theory and application of different disciplines in aeronautics and astronautics, including aerospace design practice. The series also includes texts on defense science, engineering, and management. The series serves as teaching texts as well as reference materials for practicing engineers, scientists, and managers.

J. S. Przemieniecki
Editor-in-Chief
AIAA Education Series

CONTENTS

PREFACE TO THE THIRD EDITION

All the material presented in the second edition has been retained in this new edition, and a new chapter on stability and control issues/challenges of unmanned vehicles has been added. However, several figures have been redrawn, and these new figures replace the old ones. Typographical mistakes and numerical errors detected in the second edition have been fixed.

Some of the major corrections are as follows:

- Section 4.2.5: The transformation matrix from wind axis to body axis was in error. We had a third rotation (ϕ), which was not needed. This error propagated to solved example 4.4 and exercise problems 4.5, 4.6, and 4.7. All of these errors have been corrected.
- Section 4.26: The part entitled "Relation Between Angular Velocities in Wind and Body Axes Systems" has been revised and updated to reflect the correction made in Section 4.2.5.
- Section 6.3: The signs of the numerical data used for the aileron control derivatives of the general aviation airplane were opposite to the sign convention used in this book. This numerical error has been fixed and the revised solution is presented. Also, the numerical data given in exercise problem 6.1 for a business jet has been corrected for the same sign error and the solution manual will be updated.

My special appreciation and thanks go to Anne Costa Rhodes for her patient and meticulous redrawing of numerous figures to make them look sharper. I also thank James R. Beaty for numerous helpful discussions, and John W. Paulson Jr., Jing Pei, and Jay M. Brandon for reviewing the new chapter.

Finally, I would like to record my appreciation to AIAA for bringing out the third edition. I want to express my sincere thanks to David Arthur, Toni Z. Ackley, and all other AIAA publishing staff, whose names I do not know.

Bandu N. Pamadi
June 5, 2015

PREFACE TO THE SECOND EDITION

All the material presented in the first edition is retained and no new material has been introduced in the second edition. Typographical mistakes, missing numerical data, numerical errors in solved examples, and answers to exercise problems detected after printing of the first edition have been fixed. I wish to express my sincere thanks to Professors Bong Wie, Vicki Johnson, Chuck Eastlake, Tracy Doryland, Robert P. Annex, Robert F. Stengel, V. R. Murthy, and K. Sudhakar for their comments and help in locating and fixing these typographical and numerical errors. Also, I would like to record my appreciation to AIAA for publishing the second edition and express my sincere thanks to all the AIAA publications staff. The help and cooperation I received from Rodger Williams and Meredith Cawley is gratefully acknowledged.

Bandu N. Pamadi
October 2003

PREFACE TO THE FIRST EDITION

The objective of this book is to provide a comprehensive and integrated exposure to airplane performance, stability, dynamics, and flight control. This book is intended as a text for senior undergraduate or first-year graduate students in aerospace engineering. The material presented in this book has mainly evolved from the lecture notes I used to prepare for teaching while I was at the Aerospace Engineering Department of the Indian Institute of Technology, Bombay.

Ideally, the material presented in this text could be covered in two semesters. The text includes adequate background material on basic aerodynamics, dynamics, and linear control systems to help the reader grasp the main subject matter. In this text, the airplane is assumed to be a rigid body, and elastic deformations and their effects on airplane motion are not considered.

Chapter 1 presents a brief review of basic principles of aerodynamics. The reader is also exposed to various modern concepts like supercritical airfoils, swept-forward wings, and slender delta wings. The slender delta wing represents a departure from the long-held doctrine of attached flow to the concept of controlled separated flow at high angles of attack.

Chapter 2 discusses the subject of airplane performance starting with the simple power-off gliding flight and covers various categories of flight in vertical and horizontal planes with a discussion of the conditions that optimize the local or the point performance. The limitations of the method of point performance optimization are brought to the attention of the reader.

The airplane is a dynamic system with all six degrees of freedom. However, when it is in a steady flight with uniform speed, the principles of static equilibrium can be applied, and this forms the subject matter of Chapter 3. The discussion is focused on concepts such as longitudinal and lateral-directional stability, determination of control surface deflections for trim, hinge moments, and stick force gradients for various flight conditions. A brief discussion is included on the concept of relaxed static stability.

Chapter 4 deals with airplane dynamics, starting with the derivation of equations of motion and a discussion on various coordinate systems. The equations of motion are derived using the moving axes theorem, and then simplifications are introduced to obtain the well-known small disturbance, decoupled equations for the longitudinal and lateral-directional motions. The concept of stability and control derivatives is introduced. Simple methods based on strip theory are presented for the evaluation of stability and control derivatives that should be helpful for understanding the physical principles involved in airplane motion. Also included is a discussion of engineering methods based on Datcom (Data Compendium) for the estimation of the stability and control derivatives.

Chapter 5 presents a brief review of the linear system theory and design with an objective to provide the background material for understanding airplane response and design of stability augmentation systems and autopilots. The discussion covers the frequency domain methods like Nyquist and Bode plots, time domain methods like the root-locus, modern state-space methods, and the design of various types of compensators.

Chapter 6 discusses airplane response and methods of closed-loop control of the airplane. The longitudinal and lateral-directional stability and response to various control inputs are discussed. Also discussed briefly are the concept of handling qualities, design of various longitudinal and lateral-directional stability augmentation systems, and autopilots to meet the desired level of handling qualities.

The problems of inertial coupling and airplane spin are discussed in Chapter 7. These are the typical examples of flight conditions where the longitudinal and lateral-directional motions of the airplane are coupled due to inertia terms in the equations of motion. The discussion includes the basic principles of inertia coupling, divergence in pitch or yaw, and methods of preventing inertia coupling. The discussion on spinning motion includes kinematics of spin, steady-state spin, recovery, and methods of improving spin resistance.

The problems of stability and control at high angles of attack are discussed in Chapter 8. The discussion is focused on the high angle-of-attack aerodynamics of slender fuselages and delta wings, which are characteristic of modern combat aircraft. The discussion includes phenomena such as wing rock caused by wings and slender forebodies, methods of suppressing wing rock, directional divergence/departure, roll reversal, and modern concepts of forebody flow control and thrust vectoring.

Effort is made to present a number of solved examples to illustrate the theory and basic principles. Also, several exercise problems are included to help the reader develop problem-solving skills. It is hoped that this text will be useful to the students of aerospace engineering and serve as a useful reference to the practicing aerospace engineer.

I would like to express my sincere thanks to several of my friends and colleagues whose help was of crucial importance in the preparation of the material presented in this text. First, I am extremely thankful to all my students, who have actually taught me this subject. I am grateful to Prof. Colin

P. Britcher and Dr. Atul Kelkar for reviewing various parts of the manuscript and making helpful suggestions. I greatly appreciate the discussions I had at various times with Walt Engelund, Richard Powell, Dr. Eric Queen, Dr. Suresh Joshi, Dr. D. M. Rao, Fred Lalman, and H. Paul Stough. I am thankful to Dr. Sudhir Mehrotra for providing me with the facilities to work on this project and to Charles Eldred and Larry Rowell for their support and encouragement to bring this project to a successful completion. I appreciate the help of Dr. K. Sudhakar in the preparation of the manuscript. I am also thankful to John Aguire for his help with computers.

I am especially grateful to my wife, Vijaya, whose constant love, affection, and untiring support has sustained me all the time during this project. I cannot thank her enough for her patience to put up with me on countless weeknights, weekends, and holidays when I was working on this book. I am deeply indebted to my son, Kishore, and daughter, Chitra, whose love and affection have given me the inspiration to work on this text.

I certainly owe a sense of gratitude to various authors on this subject, including C. D. Perkins, R. E. Hage, B. Etkins, A. W. Babister, B. W. McCormick, Angelo Miele, J. H. Blakelock, and many others whose works have made a significant contribution to this subject and have been useful to me in the preparation of the material presented in this text.

Finally, I would like to record my appreciation to AIAA for publishing this as an AIAA textbook. The help and encouragement of all the editorial staff of AIAA and in particular that of Rodger Williams and John Calderone is gratefully acknowledged.

Bandu N. Pamadi
July 1998

Chapter 1 Review of Basic Aerodynamic Principles

1.1 Introduction

The performance, stability, and control characteristics of an airplane depend on the aerodynamic forces and moments acting on it that, in turn, depend on the shape and size of the body, the attitude of the body relative to the airstream, the density ρ and viscosity μ of the air, velocity of the body V_∞, and the speed of sound a in air. By a dimensional analysis, it can be shown that a number of these variables can be grouped into two important nondimensional parameters, the Reynolds number Re and the Mach number M.

The Reynolds number is the ratio of inertial forces to viscous forces.

$$Re = \frac{\text{Inertia force}}{\text{Viscous force}} \tag{1.1}$$

$$= \frac{\text{Change in momentum/time}}{\text{Shear stress} \times \text{Area}} \tag{1.2}$$

$$= \frac{\frac{\mathrm{d}}{\mathrm{d}t}(mV_\infty)}{\mu \frac{\partial V}{\partial y} A} \tag{1.3}$$

We assume

$$m = \rho L^3 \tag{1.4}$$

$$\frac{\mathrm{d}}{\mathrm{d}t} = \frac{1}{\text{time}} \tag{1.5}$$

$$= \frac{V_\infty}{L} \tag{1.6}$$

$$\frac{\partial V}{\partial y} = \frac{V_\infty}{L} \tag{1.7}$$

where L is the characteristic dimension (length) of the body. Then

$$Re = \frac{\rho V_\infty L}{\mu} \tag{1.8}$$

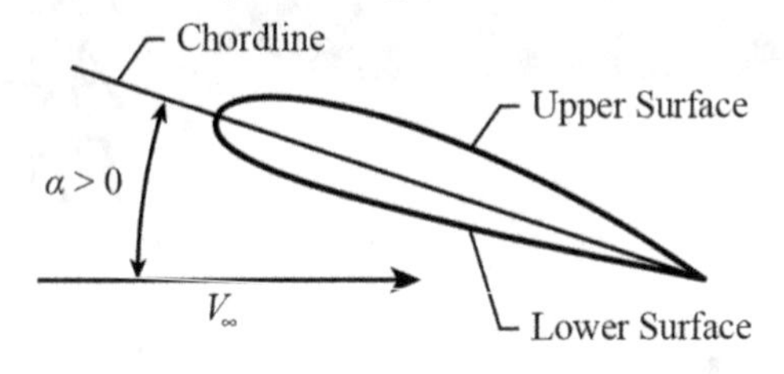

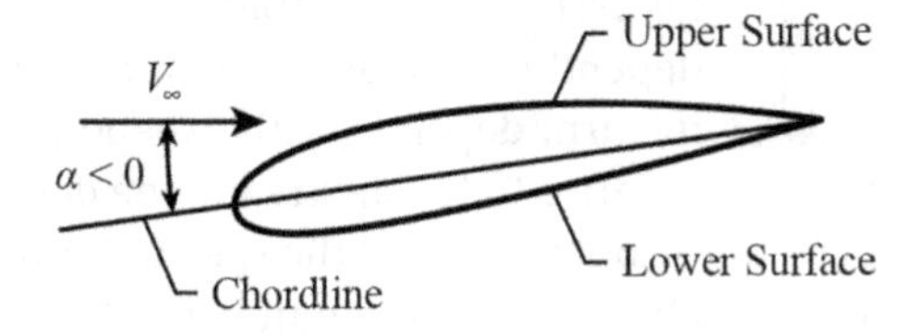

Fig. 1.1 Definition of angle of attack.

The Mach number is defined as the ratio of velocity of the body V_∞ to the speed of sound a.

$$M = \frac{V_\infty}{a} \tag{1.9}$$

The attitude of the body relative to the airstream is also known as the angle of attack α, which is defined as the angle between the airstream and a reference line fixed to the body as shown in Fig. 1.1. For airplane wings and horizontal tail, the reference line is typically the chordline and, for fuselages, it is the centerline.

1.2 Fluid Flow over Wings and Bodies

The hydrodynamic theory of fluids deals with inviscid or ideal fluid flows. This theory predicts that the fluid flow always closes behind the body no matter what the body shape is. The ideal fluid flow pattern for a two-dimensional wing and a circular cylinder are schematically shown in Fig. 1.2. This theory also states that there is no loss of energy in the flow. However, all the real fluids have viscosity to varying degrees. Because of viscosity, a real fluid always sticks to the body surface. In other words, the fluid velocity at any point on the body surface is always zero. However, the ideal fluid theory predicts a finite, nonzero velocity on the body surface. In fact, in the ideal fluid theory, the body surface is a part of the stagnation streamline. At the front and rear stagnation points (S_1 and S_2 in Fig. 1.2), the fluid velocity is zero. Downstream of the front stagnation point S_1, the velocity increases and may even exceed the freestream value and then drop back to zero at the rear stagnation point S_2. A typical variation of the velocity on the surface of an airfoil and a circular cylinder in

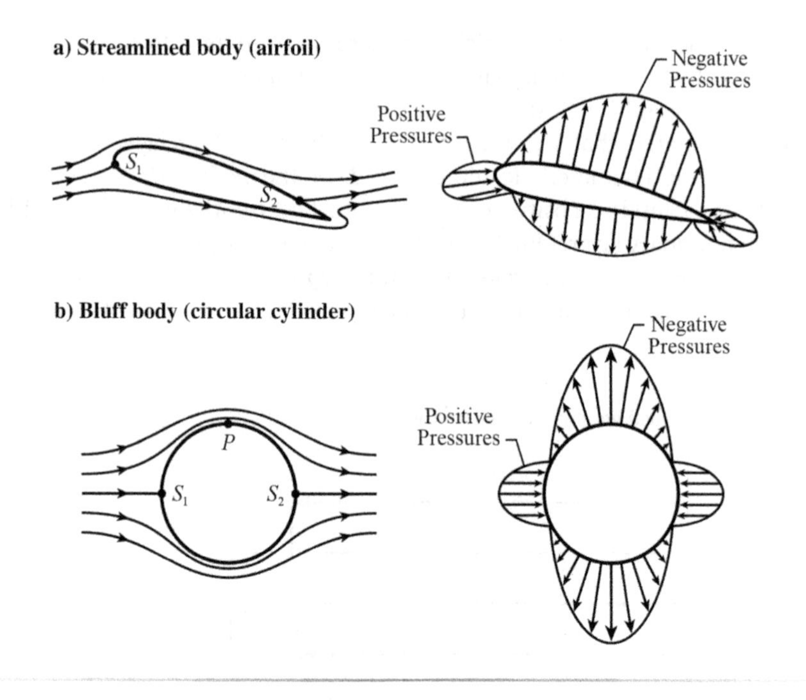

Fig. 1.2 Ideal fluid flow over bodies.

crossflow are shown in Fig. 1.3. For a circular cylinder, the peak velocity occurs at the maximum thickness point and is equal to $2V_\infty$, where V_∞ is the freestream velocity.

In real fluid flow, the velocity on the surface of the body is equal to zero and rises rapidly to the local value V_1 within a small distance δ as shown in

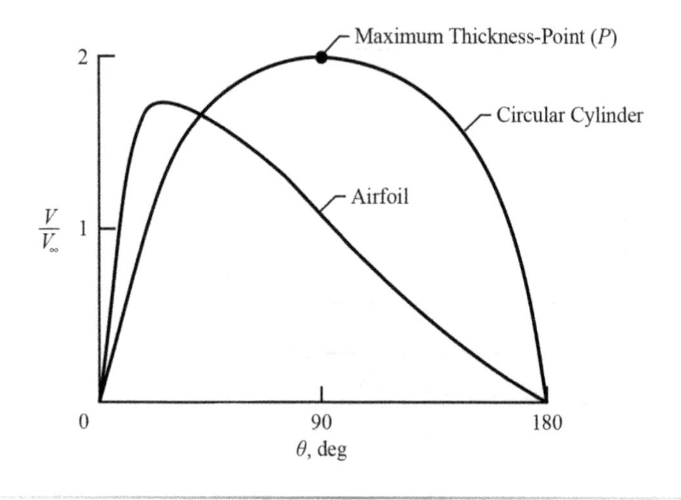

Fig. 1.3 Velocity distribution over bodies in ideal fluid flow.

Fig. 1.4a. The thin layer in which the increase from zero to local value V_1 occurs is called the boundary layer. The concept of boundary layer was introduced by L. Prandtl in 1904 [1].

He postulated that the thickness δ of the boundary layer is small in relation to the characteristic dimension L of the body ($\delta/L \ll 1$) so that the effects of fluid viscosity are assumed to be essentially confined within this thin boundary layer. Outside the boundary layer, the fluid flow practically behaves as though it is inviscid. With this hypothesis, one can consider the effect of fluid viscosity or the boundary-layer effect as equivalent to shifting

a) Typical boundary-layer profile

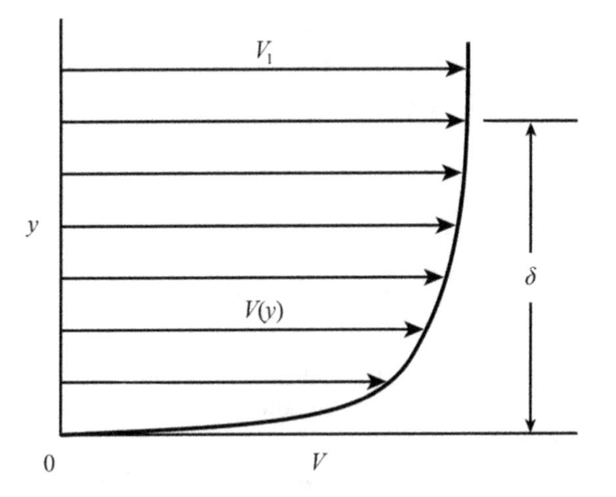

b) Concept of displacement thickness

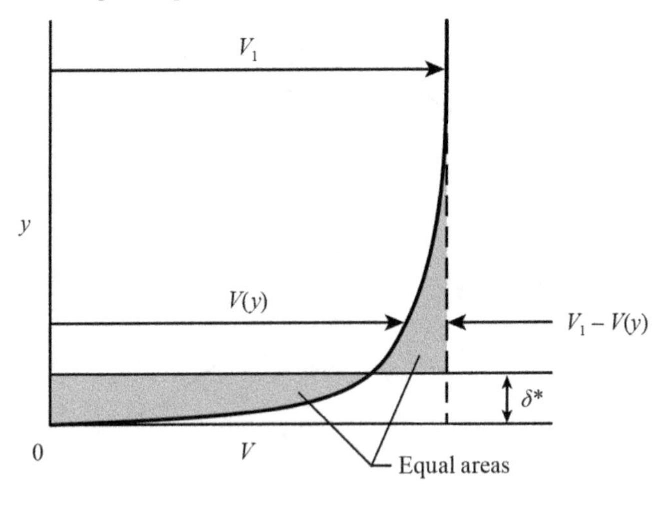

Fig. 1.4 Velocity distribution in boundary layer.

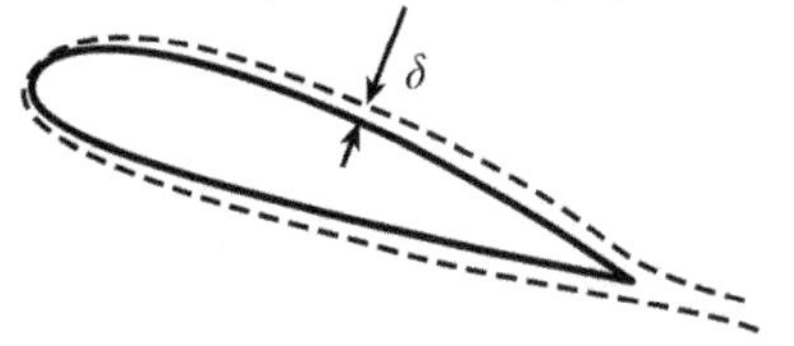

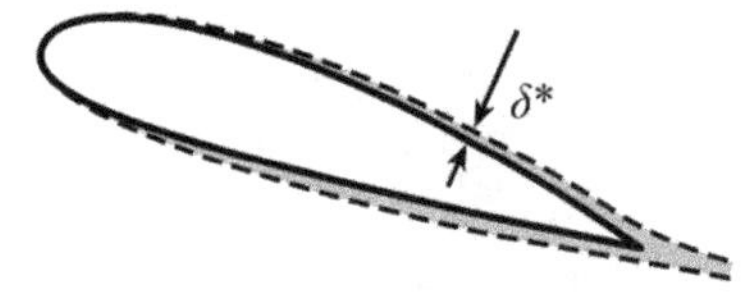

Fig. 1.5 Concepts of equivalent body based on boundary-layer displacement thickness.

the inviscid (potential) flow by a small amount equal to boundary-layer displacement thickness δ^*, which is defined as

$$\delta^* = \frac{1}{V_1}\int_0^\infty [V_1 - V(y)]\mathrm{d}y \tag{1.10}$$

The concept of δ^* is illustrated in Fig. 1.4b.

The actual body shape now consists of the given shape plus a displacement thickness δ^* as shown schematically in Fig. 1.5. The concept of boundary layer leads to a simple and practical method of finding the pressure distribution over a given body surface as follows: 1) using the ideal fluid theory, obtain the inviscid velocity and pressure distribution, and 2) with this velocity distribution, obtain the variation of the boundary-layer displacement thickness using the boundary-layer theory. The new surface now consists of the given shape plus the displacement thickness. For this new body shape, use the ideal fluid theory once again and predict the velocity and pressure distribution. Repeat these steps until there is a convergence within a certain specified tolerance.

This approach is generally applicable as long as the body is streamlined and the fluid flow is attached to the body along its surface. If the flow is separated from the surface of the body, this approach cannot be used.

1.2.1 Flow Separation

The ideal fluid theory predicts that the fluid flow closes behind any body no matter what the body shape is. We have two stagnation points S_1 and S_2 as shown in Fig. 1.2. Positive pressures acting in front and rear of the body

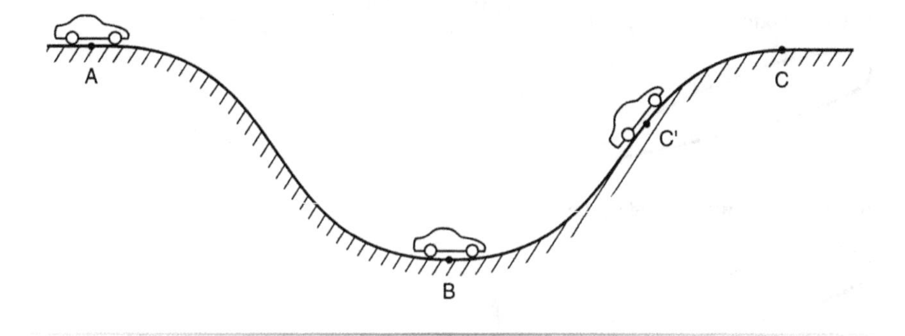

Fig. 1.6 An example of total energy.

balance out each other so that the net drag force is zero. Similarly, negative pressures on the top and bottom surfaces balance out, resulting in a zero net lift force. However, in real fluid flow, the flow pattern will be different. To understand this, let us consider a mechanical analogy forwarded by Prandtl.

Consider a roller coaster starting from rest at an elevation A and rolling down along a track as shown in Fig. 1.6. During this motion, the potential energy at A is transferred to kinetic energy at B and back to potential energy while ascending the hill towards C. The roller coaster would regain the same elevation at C as that at point A if there is no loss of energy during its motion. Because there is friction between the wheels of the roller coaster and the track, the roller coaster can only make it to point C' and not to point C.

Now let us consider the flow in the boundary layer as schematically shown in Fig. 1.7. The innermost fluid particles traveling within the boundary

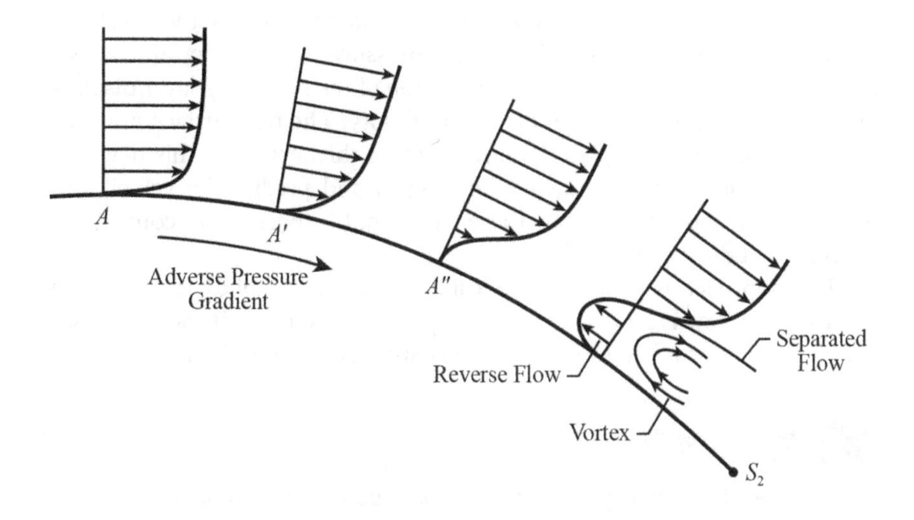

Fig. 1.7 Boundary-layer separation.

layer experience a retardation and come to a halt before the rear stagnation point S_2 is reached. Assume that at point A the velocity is maximum and falls gradually to zero at the rear stagnation point S_2. According to Bernoulli's theorem, when the velocity decreases, the pressure must increase so that the total pressure, which is the sum of static and dynamic pressures, remains constant. Thus, moving from A towards S_2, the pressure increases in the streamwise direction. Such a pressure rise in the streamwise direction is called adverse pressure gradient. If the pressure decreases in the streamwise direction, it is called a favorable pressure gradient. At point A', the boundary-layer velocity profile has a considerably different shape than that at point A because of retardation.

To overcome the adverse pressure gradient, fluid particles need to have sufficient energy in the reserve. However, because of friction, fluid particles will have lost part of their energy and hence cannot reach the stagnation point S_2, which would have been the case if the fluid flow were frictionless. As a result, fluid particles momentarily come to rest at point A''. Downstream of A'', the flow is reversed in direction. Thus, downstream of A'', the flow is separated.

Once the flow separates from the body, the pressure distribution is altered, particularly in the region of separated flow. It is different from that predicted by the ideal fluid theory. The adverse pressure gradient is no longer there. In other words, the picture presented in Fig. 1.7, based on the ideal fluid theory, existed only during the initial moments of the flow over the body. To understand this concept, let us assume that at $t < 0$ there is no flow over the body and let the flow start impulsively at $t = 0$. For the "first batch" of fluid particles that arrive at the surface of the body, there is no adverse pressure gradient to overcome. So the flow is smooth over the body and closes behind the body, forming front and rear stagnation points S_1 and S_2 (Fig. 1.2) as postulated by the ideal fluid theory. Once this happens, adverse pressure gradients are established. The immediate "next batch" of fluid particles faces the adverse pressure gradient and separates from the body surface in a manner discussed earlier.

1.2.2 Flow Past Circular Cylinder

The fluid flow past a circular cylinder in crossflow has attracted the attention of several researchers, including Theodore von Kármán. According to the ideal fluid theory, the maximum velocity is equal to twice the freestream velocity and occurs at the maximum thickness point as shown in Fig. 1.3.

The flow pattern over a circular cylinder in a real fluid flow depends on the Reynolds number. At a low Reynolds number, that is, when the Reynolds number is below the critical value (subcritical), the flow within the boundary layer is laminar and the separation point is located ahead of the maximum thickness point as shown schematically in Fig. 1.8a. This separation is of

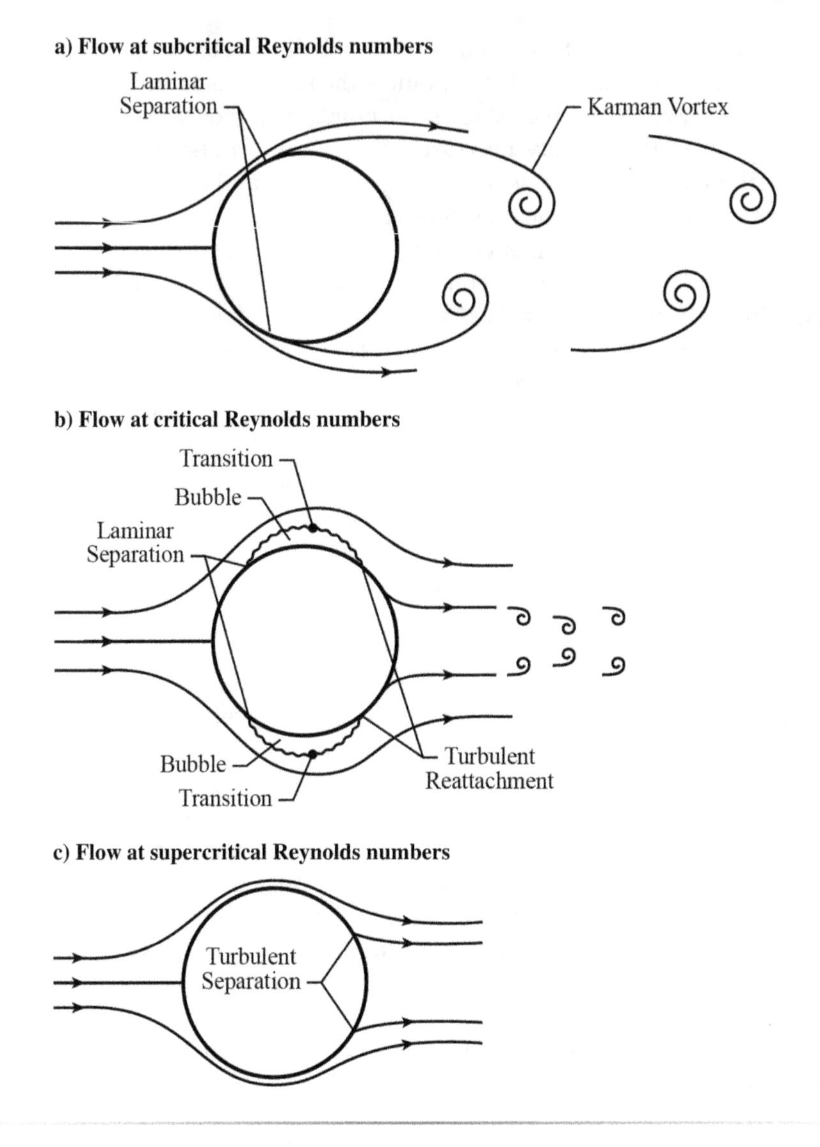

Fig. 1.8 Flow over circular cylinder.

laminar type and is of permanent nature. The separated flow rolls into a pair of alternating vortices, which are known as Kármán vortices. The pattern of alternating vortices is also known as Kármán vortex street. The wake flow is oscillatory and the drag coefficient is a function of time. The frequency of the Kármán vortex shedding is usually given in terms of a nondimensional number known as Strouhal number *St*, which is defined as

$$St = \frac{fd}{V_\infty} \tag{1.11}$$

where f is the frequency of the vortex shedding, d is the diameter of the circular cylinder, and V_∞ is the freestream velocity. For a circular cylinder, the Strouhal number is approximately 0.21 for a Reynolds number range of 10^3 to 10^4 [1].

As the freestream Reynolds number increases towards the critical value, there is an increasing tendency for the transition to occur in the separated boundary layer. For a smooth circular cylinder, the critical Reynolds number, based on the diameter, is in the range of 4×10^5 to 8×10^5. As a result of flow transition, the separated boundary layer reattaches to the surface of the circular cylinder, forming a bubble as shown in Fig. 1.8b. Because the reattached turbulent boundary layer is more resistant to the adverse pressure gradient, it sticks to the surface to a greater extent before eventually separating again. As a result, the drag coefficient drops from its subcritical value to a minimum value $C_{d,\min}$ in the critical Reynolds number range as shown in Fig. 1.9. For a smooth circular cylinder, the subcritical value of drag coefficient is approximately 1.18 and $C_{d,\min} = 0.3$ [1]. In the critical range of Reynolds number, the organized Kármán vortex pattern is suppressed and the wake flow consists of random, large eddy fluid motion.

As the freestream Reynolds number increases further and exceeds the critical Reynolds number (supercritical flow), the transition point in the bubble moves forward and eventually moves ahead of the separation point, wiping out the bubble. This process leads to the supercritical flow pattern as shown in Fig. 1.8c. However, the flow eventually separates, and the separation point is usually located upstream of that in the critical Reynolds

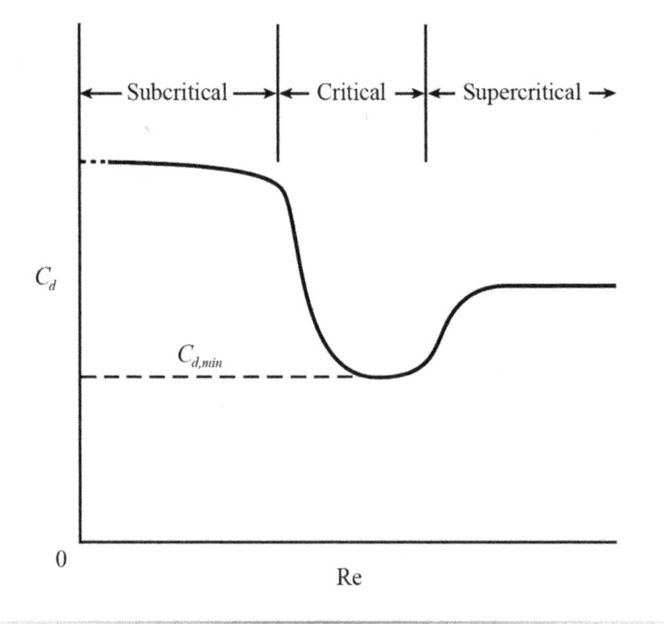

Fig. 1.9 Drag coefficient of circular cylinder with Reynolds number.

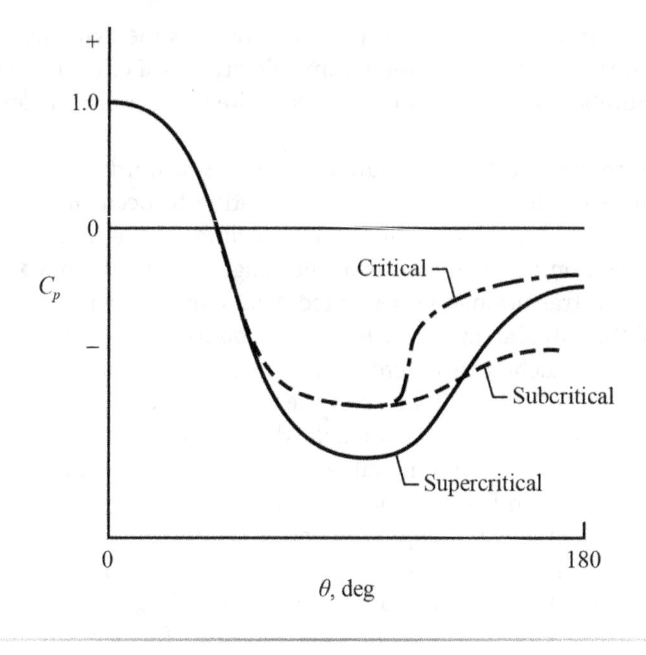

Fig. 1.10 Pressure distribution over circular cylinder.

number range. Because of this, the drag coefficient starts rising slowly in the supercritical Reynolds number range as indicated in Fig. 1.9.

A schematic variation of the pressure distribution over a circular cylinder in crossflow is shown in Fig. 1.10.

The discussed phenomenon of laminar and turbulent flow separations can be used to explain the swing of a cricket ball. It is common knowledge that the swing occurs when the ball is smooth (new) and is thrown by a fast bowler. As shown in Fig. 1.11, on the top surface, the flow separation is turbulent because of the transition caused by the "seam," which trips the local laminar boundary layer, causing a transition to turbulent flow whereas, on the bottom surface,

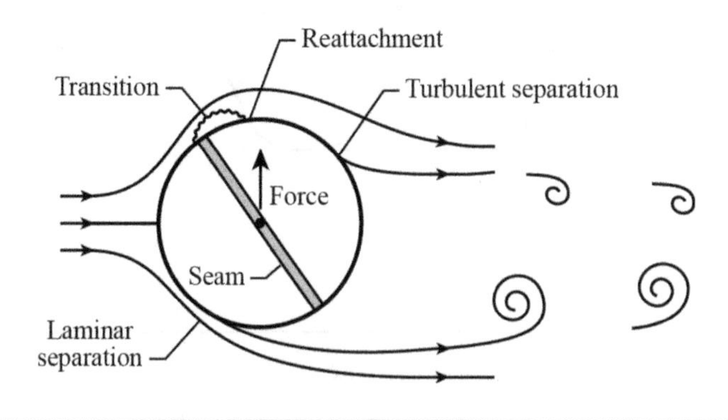

Fig. 1.11 The swing of a cricket ball.

the flow separation is laminar. As a result, the pressure distribution is asymmetric, and the cricket ball develops a net upward force as shown. If the ball is so oriented that the whole picture is rotated by 90 deg, the ball develops a cross force, which will make the ball swing inward or outward with respect to the batsman, depending on the orientation of the seam.

1.3 Drag of Bodies

The total drag of bodies exposed to airstream at low speeds consists of 1) skin friction, 2) pressure drag, and 3) induced drag. Induced drag is also called drag due to lift. Whereas the contributions to skin friction and pressure drag come from all parts exposed to the airstream, contributions to induced drag come only from lifting surfaces. At low speeds, contributions to lift come mainly from wing and horizontal tail. The lift caused by fuselage is small and hence is negligible.

The skin-friction drag is the drag caused by shear stresses present at the body surface where the flow is attached. The region of the body surface where the flow is attached is called the wetted surface, and the corresponding surface area is called the wetted area.

The skin-friction coefficient is defined as

$$C_f = \frac{D_f}{\frac{1}{2}\rho V_\infty^2 S} \tag{1.12}$$

Here, S is the wetted area. The schematic variation of skin-friction coefficient with Reynolds number for a flat plate is shown in Fig. 1.12. For approximate calculation of skin friction of streamlined bodies for attached flow conditions, the following flat plate results may be used [7].

For laminar flow

$$C_f = \frac{1.328}{\sqrt{Re}} \tag{1.13}$$

For turbulent flow

$$C_f = \frac{0.455}{(\log_{10} Re)^{2.58}} \tag{1.14}$$

where Re is the Reynolds number based on the length of the flat plate.

The pressure drag arises because of flow separation. If the flow does not close behind the body, forming a rear stagnation point, but separates, positive pressures acting in front of the body do not balance those acting on the rear of the body, and pressure drag results. For streamlined bodies, skin friction is the primary component of the drag, and pressure drag is relatively small. However, for bluff bodies like circular cylinder, the skin friction is small, and it is the pressure drag that forms the primary component of the total drag.

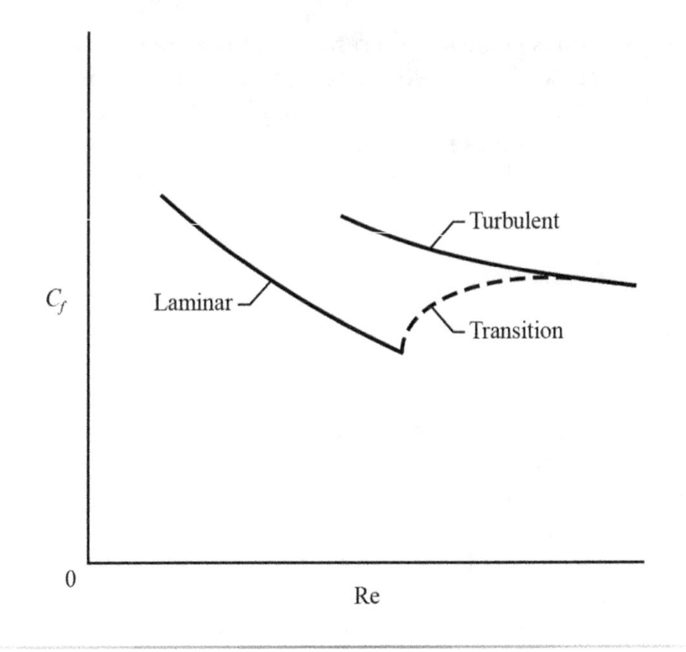

Fig. 1.12 Skin-friction coefficient of a flat plate with Reynolds number.

Induced drag is a component of the normal force in the freestream direction. Whereas the skin-friction and pressure drag are caused by fluid viscosity, induced drag is not directly caused by the effects of viscosity.

1.4 Wing Parameters

The wing section and planform parameters that are useful in estimating the aerodynamic characteristics of aircraft are discussed in the following sections.

1.4.1 Section Parameters

The cross section *AA* of an airplane wing, as shown in Fig. 1.13, is known as an airfoil section. A two-dimensional wing (a wing that extends to infinity in both directions), having an identical airfoil section at all cross sections, is called an airfoil. In other words, an airfoil is a two-dimensional wing with a constant airfoil section.

The objective of analytical, computational, or experimental investigations is to design an efficient airfoil section that has a low drag coefficient, a high lift-to-drag ratio, a high value of maximum lift coefficient, a small value of pitching-moment coefficient, and a smooth gradual stall.

Since the early days of aviation, several efforts have been made to design and develop series of efficient airfoil sections. Notable among these are the works at Goettingen, Germany; Royal Aircraft Establishment in the United Kingdom; and National Advisory Committee for Aeronautics (NACA) in the United States. The NACA airfoils have gained wide acceptance throughout the world and have influenced the airfoil design of others. The NACA airfoils have been used on several commercial and military airplanes that have been built over the years. The following is a brief description of NACA airfoil sections. For more information, see Ref. 2.

The typical geometry of an airfoil is shown in Fig. 1.14. The line drawn midway between the upper and lower surfaces is called the meanline. The leading and trailing edges are defined as the forward and rearward extremities, respectively, of the meanline. The straight line joining the leading and trailing edges is called the chordline, and the length of the chordline is usually known as the chord of the airfoil. The distance between the meanline and the chordline measured normal to the chordline is denoted by y_c. The variation of y_c along the chord defines the camber of the airfoil. In view of this, the meanline is also known as the camberline. The distance between the upper and lower surfaces measured perpendicular to the meanline is called the thickness of the airfoil. The abscissas, ordinates, and slopes of the meanline are designated as x_c, y_c, and tan θ, respectively. If x_u and y_u represent, respectively, the abscissa and ordinates of a point on the upper and lower surfaces of the airfoil and y_t is the ordinate of the symmetrical thickness distribution at chordwise position x, then the upper and lower surface coordinates are given by the following relations:

$$x_u = x_c - y_t \sin\theta \tag{1.15}$$

$$y_u = y_c + y_t \cos\theta \tag{1.16}$$

$$x_l = x_c + y_t \sin\theta \tag{1.17}$$

$$y_l = y_c - y_t \cos\theta \tag{1.18}$$

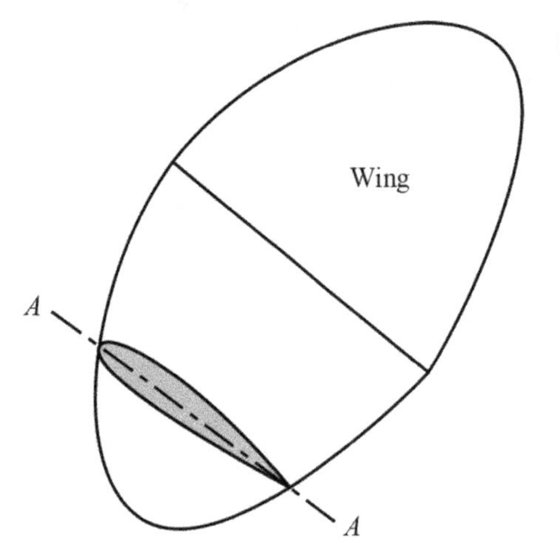

Fig. 1.13 Wing section geometry.

To find the center of the leading-edge radius, we draw a line through the end of the chordline at the leading edge with a slope equal to the slope of the meanline at that point and mark a distance equal to the leading-edge radius along that line.

Thus, the geometrical shape of the airfoil is specified by three parameters, the leading-edge radius ρ, the variation of y_c along the chordline, and the variation of y_t along the chordline for the symmetric thickness distribution.

Basically there are two types of airfoil sections, symmetrical and cambered sections. A symmetrical airfoil is one whose lower surface is a mirror image of the upper surface about the chordline. In other words, a symmetrical airfoil has zero camber, or the meanline coincides with the chordline as shown in Fig. 1.15a. If the meanline is convex up, there is a positively cambered airfoil as shown in Fig. 1.15b. If the meanline is concave up, then it is a negatively cambered airfoil as shown in Fig. 1.15c.

For a symmetrical airfoil, $C_l = C_m = 0$ at $\alpha = 0$. Here, C_l and C_m are the sectional lift and pitching-moment coefficients of the airfoil. If α_{0L} denotes the angle of attack when $C_l = 0$, then, for a symmetrical airfoil, $\alpha_{0L} = 0$. For a positively cambered airfoil, at $\alpha = 0$, $C_l > 0$ and, therefore, $\alpha_{0L} < 0$ as shown in Fig. 1.15d. If C_{mo} is denoted as the value of the pitching-moment coefficient at $\alpha = \alpha_{0L}$, then, for a positively cambered airfoil, $C_{mo} < 0$. Similarly, for a negatively cambered airfoil, $\alpha_{0L} > 0$ and $C_{mo} > 0$. The effective angle of attack is given by $\alpha_{\text{eff}} = \alpha - \alpha_{0L}$ as shown in Fig. 1.15e for a positively cambered airfoil.

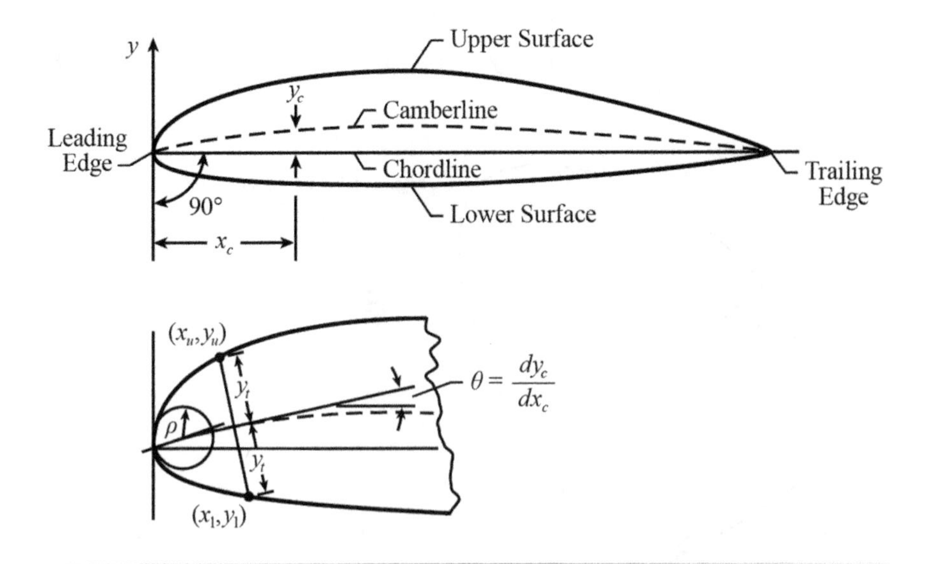

Fig. 1.14 Airfoil geometry.

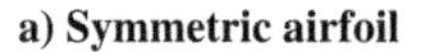

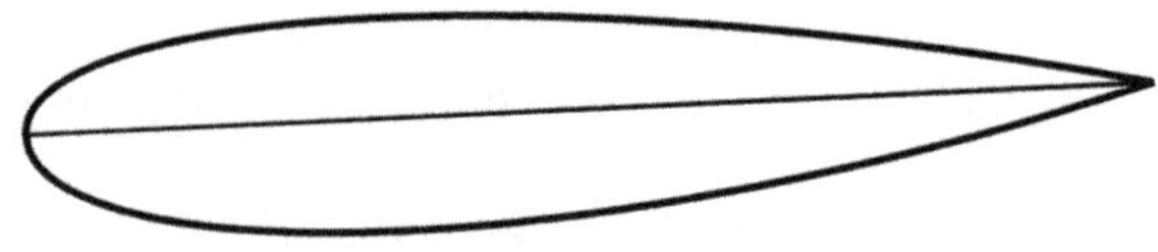

b) Positively cambered airfoil

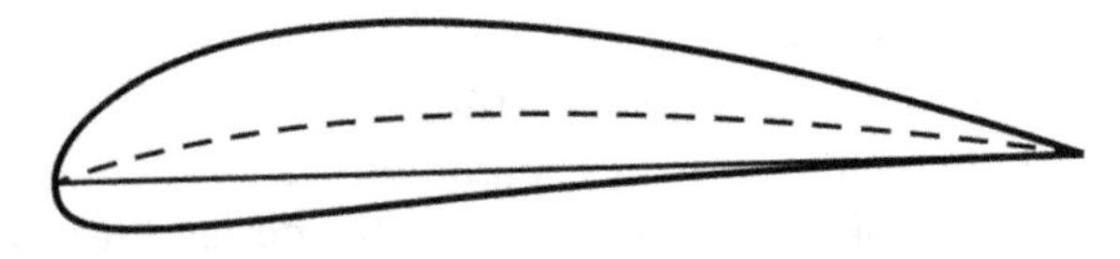

c) Negatively cambered airfoil

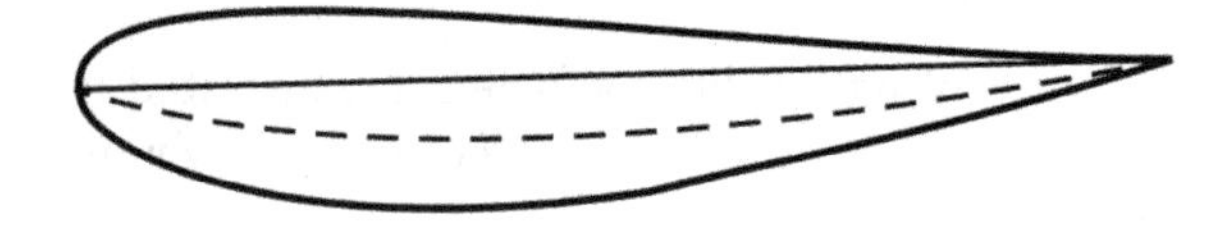

d) Orientation of zero-lift

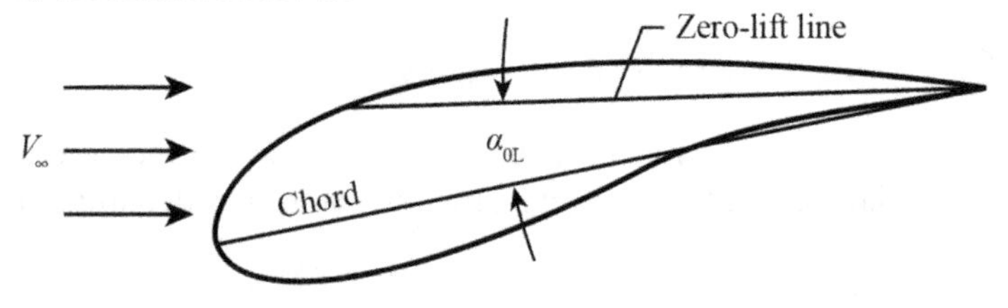

e) Effective angle of attack

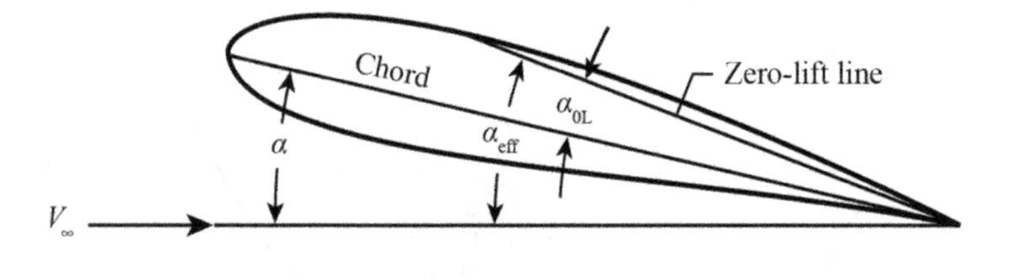

Fig. 1.15 Typical airfoil shapes.

Most of the NACA airfoils are classified among three types of airfoils: the four digit, the five digit, and the series 6 sections. The nomenclature and meaning of various digits are explained with the help of following examples:

1. Four-digit series, example NACA 2412
   2: The maximum camber of the meanline is 0.02c.
   4: The position of the maximum camber is at 0.4c.
   12: The maximum thickness is 0.12c.

2. Five-digit series, example NACA 23012
   2: The maximum camber of the meanline is approximately equal to 0.02c. The design lift coefficient is 0.15 times the first digit of the series, which in this case is 2.
   30: The position of maximum camber is $0.30/2 = 0.15c$.
   12: The maximum thickness is 0.12c.
3. Series 6 sections, example NACA 65_3-418
   6: Series designation.
   5: The minimum pressure is at 0.5c.
   3: The drag coefficient is near minimum value over a range of lift coefficients of 0.3 above and below the design lift coefficient.
   4: The design lift coefficient is 0.4.
   18: The maximum thickness is 0.18c.

The concept of design lift coefficient is illustrated in Fig. 1.16. It is the mean value of the range of lift coefficients for which the sectional drag coefficient is minimum.

For additional information on airfoil data, refer to Refs. 2 and 3.

1.4.2 Wing Planform Parameters

General planform parameters that are useful in estimating aerodynamic data are the planform area S, aspect ratio A, mean aerodynamic chord $\bar{c}$,

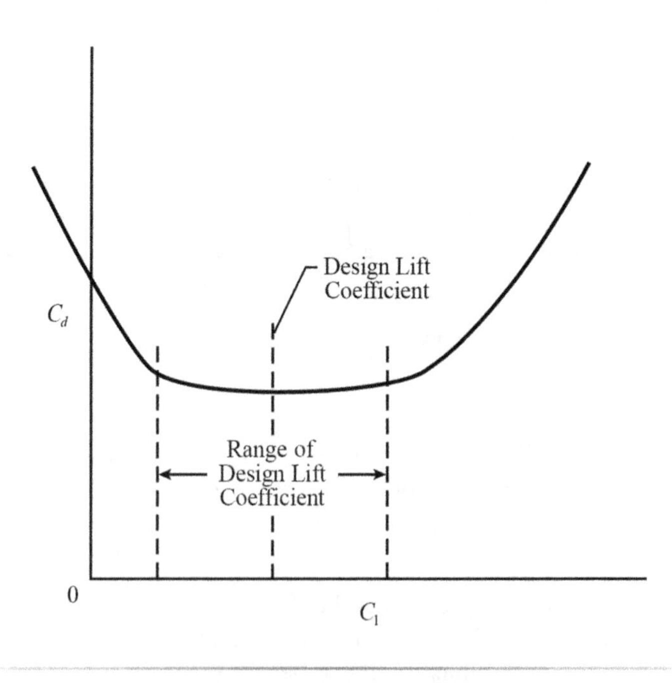

Fig. 1.16 Design lift coefficient.

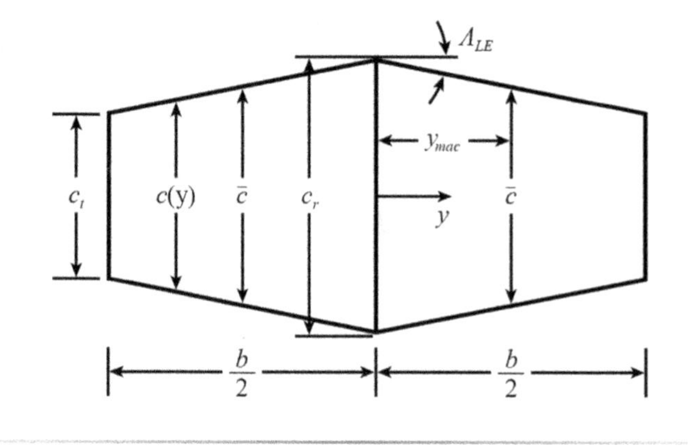

Fig. 1.17 Wing planform parameters.

and the spanwise location of the mean aerodynamic chord y_{mac}. These parameters are given by the following expressions:

$$S = 2\int_0^{\frac{b}{2}} c(y)\,\mathrm{d}y \tag{1.19}$$

$$A = \frac{b^2}{S} \tag{1.20}$$

$$\bar{c} = \frac{2}{S}\int_0^{\frac{b}{2}} c^2(y)\,\mathrm{d}y \tag{1.21}$$

$$y_{\text{mac}} = \frac{2}{S}\int_0^{\frac{b}{2}} c(y)y\,\mathrm{d}y \tag{1.22}$$

where $c(y)$ is the local chord, y is the spanwise coordinate, and b is the wing span.

For a conventional straight wing (Fig. 1.17), Eqs. (1.19–1.22) reduce to

$$S = \frac{b}{2}c_r(1+\lambda) \tag{1.23}$$

$$A = \frac{2b}{c_r(1+\lambda)} \tag{1.24}$$

$$\bar{c} = \frac{2}{3}c_r\left(\frac{1+\lambda+\lambda^2}{1+\lambda}\right) \tag{1.25}$$

$$y_{\text{mac}} = \frac{b}{6}\left(\frac{1+2\lambda}{1+\lambda}\right) \tag{1.26}$$

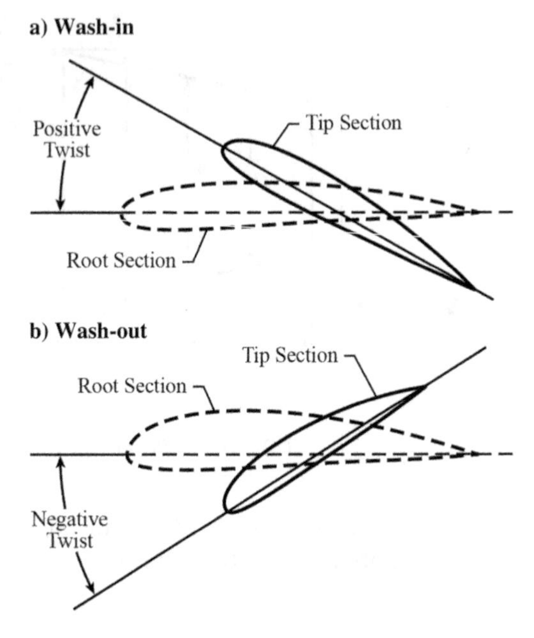

Fig. 1.18 Wing twist.

where λ is the taper ratio given by

$$\lambda = \frac{c_t}{c_r} \tag{1.27}$$

Here, c_t is the tip chord and c_r is the root chord.

To have the desired variation of aerodynamic loading and stall pattern along the span, wing sections are sometimes given what is called the twist. If a wing is so twisted that the angle of attack increases from root to the tip, such a twist is called wash-in. If it is the other way, that is, the angle of attack decreases from the root to the tip, then that type of twist is called wash-out. The concepts of wash-in and wash-out are illustrated in Fig. 1.18.

1.5 Aerodynamic Characteristics of Wing Sections

Typical pressure distribution over an airfoil at an angle of attack is schematically shown in Fig. 1.19. Arrows toward the surface indicate positive pressures, and those that point away from the surface represent negative pressures or suction. An integration of the pressure distribution over the body gives a net or resultant force that can be resolved into two components called lift and drag. The component of the force perpendicular to the freestream direction is called the lift, and that along the freestream direction is called the drag. The point along the chordline where the resultant force acts is known as the center of pressure. Thus, if the moment reference point coincides with the center of pressure, the pitching moment is zero. For any other moment reference point along the chord, the pitching moment is in general nonzero. Therefore, to specify forces acting on the airfoil, one has to specify the lift, drag, and pitching moments and the location of moment reference point on the chord. Because values of the lift, drag, and pitching moments vary with the speed, it is usual to specify the aerodynamic characteristics of an airfoil in terms of nondimensional lift, drag, and pitching moment coefficients, which are defined as follows:

$$C_l = \frac{\text{Lift}}{\frac{1}{2}\rho_\infty V_\infty^2 S} \tag{1.28}$$

$$C_d = \frac{\text{Drag}}{\frac{1}{2}\rho_\infty V_\infty^2 S} \tag{1.29}$$

$$C_m = \frac{\text{Moment}}{\frac{1}{2}\rho_\infty V_\infty^2 S\bar{c}} \tag{1.30}$$

where C_l, C_d, and C_m are known as lift, drag, and pitching-moment coefficients, respectively; S is the reference area, usually the wing planform area; and $\bar{c}$ is the mean aerodynamic chord used as the reference length. For the airfoil that is a wing of infinite aspect ratio, the lift, drag, and pitching moments are per unit span. Also, S is the wing planform area per unit span, which will be numerically equal to $\bar{c}$.

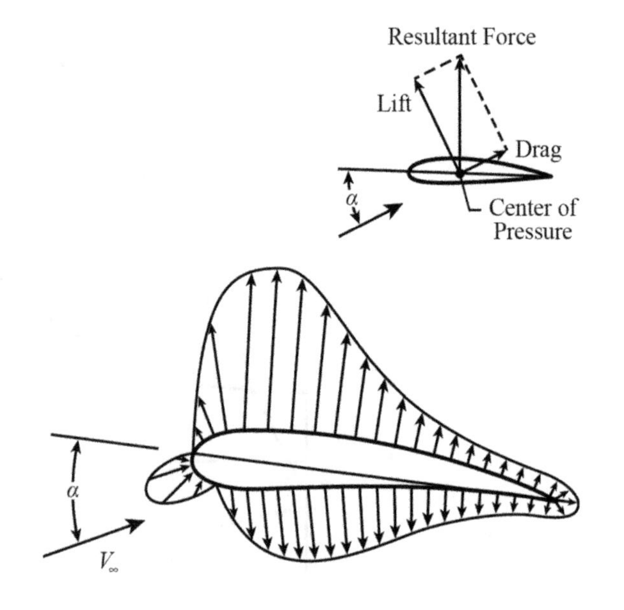

Fig. 1.19 Pressure distribution over an airfoil.

It may be noted that the notations C_l and C_d will be used to denote the lift and drag coefficients of an airfoil (two-dimensional wing) and C_L and C_D will be used to denote those of a finite wing.

As said before, the lift and drag coefficient depend on the body shape, angle of attack, Reynolds number, and Mach number. This type of dependence can be expressed in functional form as follows:

$$C_l = C_l(\alpha, Re, M) \tag{1.31}$$

$$C_d = C_d(\alpha, Re, M) \tag{1.32}$$

$$C_m = C_m(\alpha, Re, M) \tag{1.33}$$

The dependence of aerodynamic coefficients on Reynolds number is usually known as "scale effect" and that on the Mach number is called compressibility effect. Scale effects become important in the problem of extrapolating the aerodynamic coefficients obtained from model wind-tunnel tests to full-scale aircraft flight conditions [4]. A schematic variation of the aerodynamic coefficients is shown in Fig. 1.20.

For low values of angle of attack, the lift and moment coefficients vary linearly with angle of attack. This range of angle of attack, where the lift and pitching-moment coefficients have a linear relationship with angle of attack, is called the linear range of angle of attack.

The lift coefficient assumes a maximum value at some angle of attack and then falls. At this point, the airfoil is said to be stalled. The maximum value of

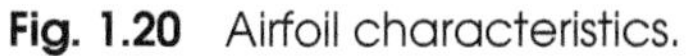
Fig. 1.20 Airfoil characteristics.

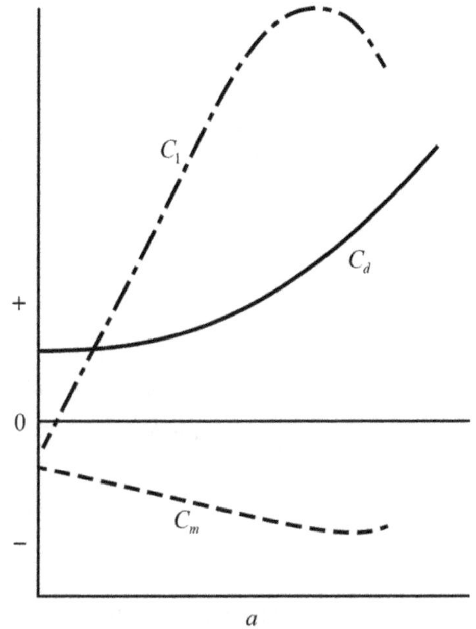

the lift coefficient of the airfoil is denoted by $C_{l,\max}$. For simplicity, assume that the linear range of angle of attack extends up to $C_{l,\max}$. Typical values of $C_{l,\max}$ for airfoils are in the range of 1.1 to 1.7.

The drag coefficient varies nonlinearly with angle of attack. This rise in drag coefficient is primarily because of the increase in pressure-drag coefficient with angle of attack. Note that for an airfoil, the drag consists of only skin friction and pressure drag. Finite wings will have another component of drag known as induced drag. For airfoils, the induced drag is zero.

1.5.1 Aerodynamic Center

In general, for the linear range of angle of attack, the wing pitching-moment coefficient can be expressed as follows:

$$C_m = C_{mo} + \xi C_l \tag{1.34}$$

where C_{mo} is the pitching-moment coefficient at zero-lift coefficient and ξ is an empirical constant that depends on the location of the moment reference point.

At zero-lift condition, the lift on the upper and lower surfaces of the airfoil are equal and opposite and hence cancel each other. For a symmetric airfoil, these two forces act at the same point on the chord and $C_{mo} = 0$. However, for a cambered airfoil, the two forces act at different points along the chord and, as a result, a pure couple is acting on the airfoil. Because C_{mo} is a pure couple, it is independent of the moment reference point. The sign of ξ depends on the location of the moment reference point. If this point is located close to the leading edge, the incremental lift caused by angle of attack acts aft of the moment reference point, and the associated increment in pitching moment will be negative, making $\xi < 0$. On the other hand, if the moment reference point is located close to the trailing edge, then the incremental lift acts ahead of the moment reference point, and the corresponding incremental pitching moment is positive; hence $\xi > 0$. Therefore, there must be some point on the chordline where $\xi = 0$ so that the incremental pitching moment caused by angle of attack is zero and the pitching-moment coefficient remains constant with respect to angle of attack and equal to C_{mo}. This point is called aerodynamic center. In other words, all the incremental lift caused by angle of attack acts at the aerodynamic center. The pitching-moment coefficient with aerodynamic center as the moment reference point is denoted by C_{mac}. Note that

$$C_{mac} = C_{mo} \tag{1.35}$$

For symmetric airfoils, $C_{mac} = C_{mo} = 0$.

The concept of aerodynamic center is helpful in the study of airplane stability and control because it is sufficient to specify the pitching-moment coefficient at any one angle of attack in the entire linear range of angle of attack.

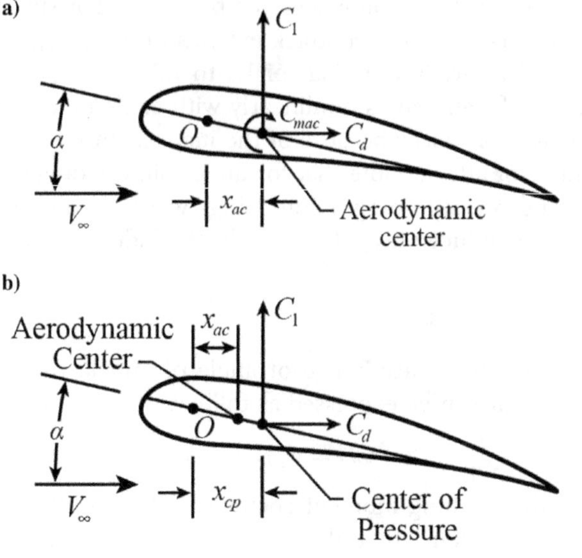

Fig. 1.21 Aerodynamic centerand center of pressure.

At low speeds, the aerodynamic center usually lies close to the chordline and is located at approximately 22 to 26% chord from the leading edge. For approximate purposes, it can be assumed that it is located on the chordline at the quarter chord point. For high subsonic and supersonic speeds, both the aerodynamic center and center of pressure move aft.

The exact location of the aerodynamic center can be determined if given the lift, drag, and pitching-moment coefficients about any reference point along the chordline. Let O be the moment reference point (see Fig. 1.21a) about which we are given C_l, C_d, and C_m. Let x_{ac} be the distance of the aerodynamic center from O, positive aft. Let C_{mac} denote the pitching moment coefficient about the aerodynamic center, which by definition is invariant with angle of attack. Then, the pitching moment about the given moment reference point can be written as

$$M = M_{ac} - x_{ac}L \tag{1.36}$$

In terms of coefficients

$$C_m = C_{\text{mac}} - \overline{x}_{ac}C_l \tag{1.37}$$

where $\overline{x}_{ac} = x_{ac}/c$. Differentiate with respect to C_l and note that C_{mac} (by definition) is constant. Then

$$\frac{\mathrm{d}C_m}{\mathrm{d}C_l} = -\overline{x}_{ac} \tag{1.38}$$

or

$$\bar{x}_{ac} = -\frac{dC_m}{dC_l} \tag{1.39}$$

Thus, for linear range of attack, $\bar{x}_{ac}$ depends on the slope of the pitching moment curve. If $dC_m/dC_l > 0$, then $\bar{x}_{ac}$ is negative, implying that the aerodynamic center is located ahead of the moment reference point. On the other hand, if $dC_m/dC_l < 0$, then $\bar{x}_{ac}$ is positive and the aerodynamic center is aft of the moment reference point as assumed. If, on the other hand, the lift and moment coefficients vary nonlinearly with angle of attack, then dC_m/dC_l is not a constant, and the concept of aerodynamic center is not valid.

1.5.2 Relation Between Center of Pressure and Aerodynamic Center

For linear range of angle of attack,

$$C_m = C_{mo} + \frac{dC_m}{dC_l} C_l \tag{1.40}$$

$$= -\bar{x}_{cp} C_l \tag{1.41}$$

where $\bar{x}_{cp} = x_{cp}/c$ and C_{mo} is the pitching-moment coefficient at that angle of attack when the lift-coefficient is zero. Let α_{0L} denote this angle of attack. As said before, for symmetric airfoil sections, $\alpha_{0L} = 0$; for sections with positive camber, α_{0L} is negative; and for sections with negative camber, α_{0L} is positive.

Then,

$$\bar{x}_{cp} = -\frac{C_{mo}}{C_l} - \frac{dC_m}{dC_l} \tag{1.42}$$

Using Eq. (1.39), we get

$$\bar{x}_{cp} = \bar{x}_{ac} - \frac{C_{mo}}{C_l} \tag{1.43}$$

Thus, the location of the aerodynamic center relative to the center of pressure depends on the sign of C_{mo}. For positively cambered airfoils, $C_{mo} < 0$ so that the center of pressure is aft of the aerodynamic center as shown in Fig. 1.21b. However, as the angle of attack increases ($0 \leq C_l \leq C_{l,\max}$), the center of pressure moves towards the aerodynamic center as shown in Fig. 1.22.

1.5.3 Stall of Wing Sections

At low angles of attack, the flow separation occurs at or close to the trailing edge. As the angle of attack increases, the separation point gradually

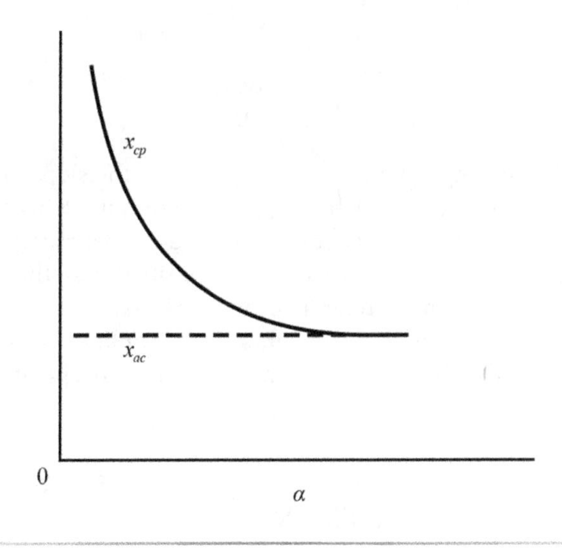

Fig. 1.22 Center of pressure location with angle of attack.

moves towards the leading edge. In this process, the lift coefficient continues to increase and, at some point, attains a maximum value. Beyond this value of angle of attack, the lift coefficient drops, and the airfoil is said to have stalled. This type of stall generally occurs on thick airfoils and is characterized by a gradual loss of lift beyond the stall as shown by the curve a in Fig. 1.23.

For thin airfoils that have sharply curved leading edges, a slightly different type of stall occurs. From the leading edge and up to the minimum pressure

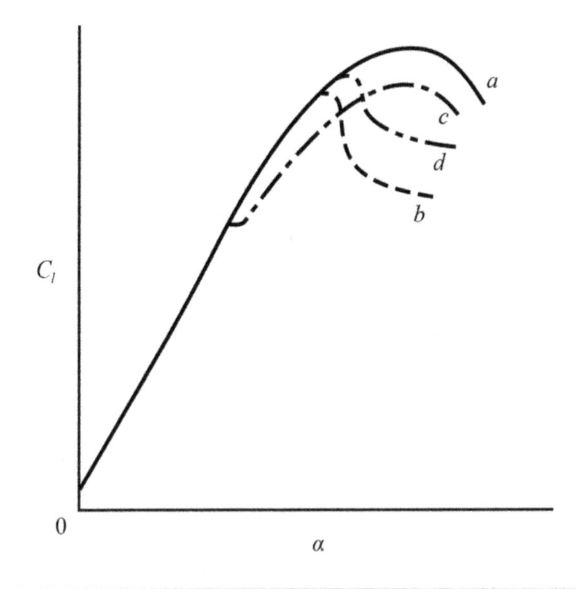

Fig. 1.23 Stall of airfoil sections.

point, a large favorable pressure gradient exists that tends to promote the existence of a laminar boundary layer. Beyond the minimum pressure point, an adverse pressure gradient exists towards the trailing edge. Because of the sharp curvature of the leading edge of a thin airfoil, the adverse pressure gradient is sufficiently strong to cause flow separation. If the Reynolds number is low, this type of flow separation is of a permanent nature (no subsequent reattachment) and, as a result, the airfoil stalls. Because the flow separation occurs close to the leading edge, this type of stall is usually abrupt and is marked by a sudden loss of lift beyond the maximum lift as shown by curve *b* in Fig. 1.23.

If the freestream Reynolds number is sufficiently high, then there is an increasing tendency for the transition to occur in the separated boundary layer and cause the separated boundary-layer flow to reattach back to the upper surface, forming a bubble in a similar fashion to that discussed earlier for a circular cylinder. The reattached turbulent boundary layer is more capable of resisting the adverse pressure gradient than the laminar boundary layer. The reattached turbulent flow eventually separates from a point closer to the trailing edge.

As the angle of attack increases, two types of flow patterns are found to occur [5], 1) the bubble grows in size so that the reattachment point moves downstream towards the trailing edge until at some point there is no more reattachment and the airfoil stalls, and 2) the bubble shortens so that the reattachment point moves upstream towards the leading edge until at some point the bubble suddenly bursts causing the airfoil to stall. The first type of stall caused by the long bubble is called the thin airfoil stall (curve *c* of Fig. 1.23), and the second type of stall caused by the bursting of the short bubble is called the leading-edge stall (curve *d* of Fig. 1.23). The leading-edge stall is accompanied by a sudden loss of lift and is highly undesirable.

1.5.4 High-Lift Devices

As said before, the conventional airfoils have a maximum lift coefficient $C_{l,\max}$ in the range of 1.1 to 1.7. This value of $C_{l,\max}$ depends mainly on the thickness ratio, camber, and the flow Reynolds number. As stated later in the text, several of the airplane performance characteristics depend on the value of $C_{l,\max}$. Especially for better landing and takeoff performance, it is desirable to have a high value of maximum lift coefficient. Some of the commonly used methods to improve the maximum lift coefficient capability of the aircraft are discussed in the following sections [6].

Flaps

Flaps are mechanical devices deflected either from the trailing edge (or close to the trailing edge) or the leading edge. Several different designs of the flaps have come into existence, and some of the commonly used configurations are shown in Fig. 1.24. Flap configurations depicted in Figs. 1.24a–1.24g are mechanical flaps, and that shown in Fig. 1.24h is the so-called jet flap.

A deflection of a hinged mechanical flap basically alters the effective camber of the airfoil section. The deflection of the flap increases the value of $C_{l,\max}$ without essentially altering the lift-curve slope or the stall angle. Therefore, at the stall, an airfoil with a deflected flap will have an incremental lift coefficient $\Delta C_{l,\max}$ as shown in Fig. 1.25. The plain flap, split flap, Kruger flap, and leading-edge flaps belong to this category of plain mechanical flaps.

In addition to changing the camber, the slotted flap and the Fowler-flap also act as boundary-layer control devices. A schematic diagram of the operation of a trailing-edge slotted flap is shown in Fig. 1.26. High-pressure air from the lower surface leaks through the gap and energizes the retarded upper surface boundary layer. This process helps to delay the flow separation on the upper surface and hence improve the value of $C_{l,\max}$.

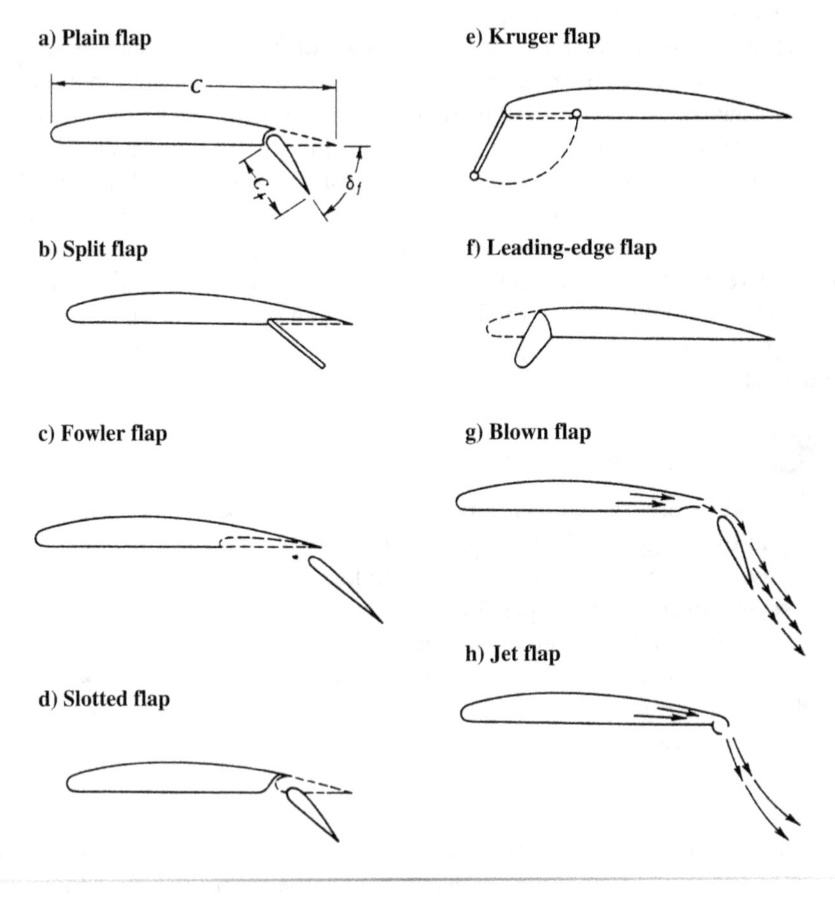

Fig. 1.24 Typical flap configurations.

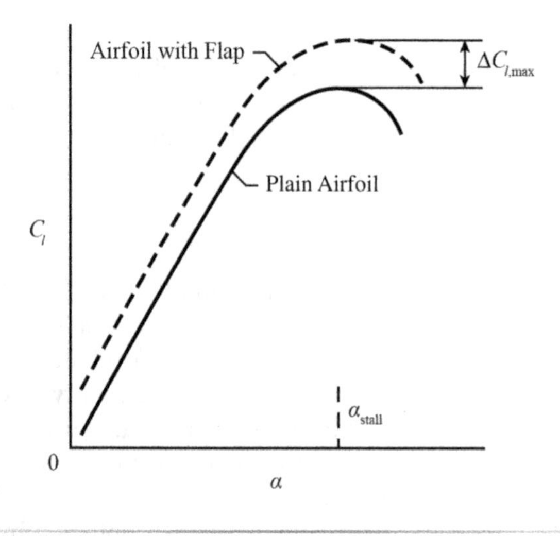

Fig. 1.25 Lift coefficient for airfoils with and without flaps.

For the blown flap, high-pressure air is blown over the upper surface of the flap as shown in Fig. 1.24g. In this way, considerable energy is added to stabilize the upper surface boundary layer that delays flow separation. The magnitude of $C_{l,\max}$ depends on the mass of the added air and the blowing velocity (momentum of the blown air). A jet flap (Fig. 1.24h) also functions in a similar way.

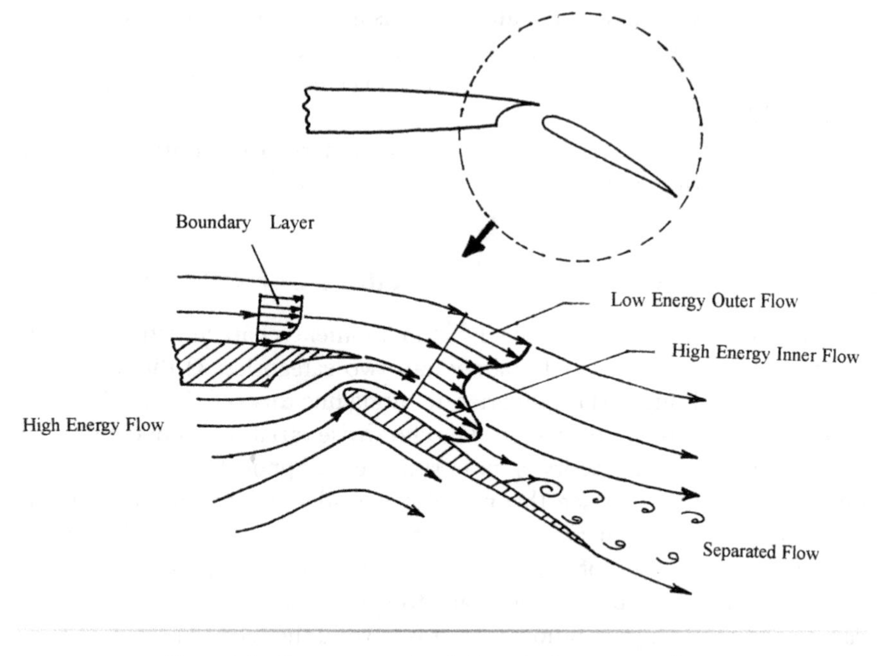

Fig. 1.26 Flow around a slotted trailing-edge flap.

The flap effectiveness depends on many factors. The most important ones are the flap-chord-to-wing-chord ratio, shape of the flap leading edge, and width of the gap. For a finite wing, value of the increment in maximum lift coefficient also depends on the ratio of flap span to the main wing span, sweep angle of the flap hingeline, etc.

Boundary-Layer Control

Another method of increasing the maximum lift coefficient of the wing is the application of boundary-layer control methods. The slotted flap and the Fowler-flap are some forms of boundary-layer control although this term is usually used for actual flow control using boundary-layer suction or blowing methods. In the boundary-layer suction method, low energy fluid from the upper surface is removed by the application of suction. This process helps to delay the flow separation to higher angles of attack. Significant increases in maximum lift coefficients have been obtained by this method. Another advantage of this method is that the application of suction also stabilizes the laminar boundary layer by delaying the flow transition. This results in a significant reduction in the skin-friction drag. This method has been flight tested on Northrop X-21A as well as by NASA on the F-16XL aircraft.

In boundary-layer control using the blowing methods, high-energy air is blown tangentially on the upper surface of the wing and the flap as shown in Fig. 1.24g. This addition of energy helps to delay boundary-layer flow separation on the upper surface. As a result, the maximum lift coefficient increases. This method of boundary layer control is sometimes called "circulation control." In this method, the main emphasis is to enhance the lift coefficient without worrying too much about the drag coefficient. Therefore, the drag coefficient of an airfoil with blown flap may be higher than that of the basic airfoil.

However, the main disadvantage of boundary-layer control based on suction or blowing is the mechanical complexity and additional weight.

1.6 Aerodynamic Characteristics of Finite Wings

Theories for calculating the lift and moment characteristics of finite wings at subsonic speeds fall mainly into two categories: 1) the lifting line theory, and 2) lifting surface theories. In lifting line theory [7], only the spanwise lift distribution is considered. The chordwise variation is not considered. Hence, the lifting line theory is more suitable for application to high-aspect ratio wings. Lifting surface theories also consider the chordwise variation of lift distribution and hence give more accurate results for lift and pitching-moment curve slopes of finite wings. In general, lifting surface theories [8–10] are more difficult to apply than lifting line theories and hence are typically used for low-aspect ratio wings. It is beyond the scope of this text to go into the details of these two types of wing theories. In the following, the lifting line theory will be used to obtain expressions for the lift-curve slope and induced-drag coefficient of finite wings.

In lifting line theory, the lifting wing is modeled as a horseshoe vortex as depicted in Fig. 1.27. The part of the vortex sheet attached to the wing surface is called the bound vortex. The bound vortex continues beyond the wing tips in the downstream direction, and these parts of the horseshoe vortex are called the trailing vortices or tip vortices. According to Helmholtz theorem, a vortex system cannot end abruptly in a fluid medium. Therefore, the system of bound vortex and trailing vortices must be closed in some manner. This closure is provided by the so-called starting vortex as shown in Fig. 1.27.

To understand how the starting vortex is a physical reality, let us consider the motion of a finite wing, which starts to move forward impulsively, i.e., it is imparted a forward velocity instantaneously. As explained earlier, the first batch of fluid particles goes around the wing sections smoothly, forming the front and rear stagnation points as shown in Fig. 1.28a. Once this flow pattern is formed, pressure gradients come into existence. Of particular interest is the adverse pressure gradient on the upper surface. As a result, the flow separates on the upper surface upstream of the trailing edge, and the flow coming from the lower surface cannot go around the sharp trailing edge as it did at $t = 0$. Thus, the curved or vortex flow formed at $t = 0$ (Fig. 1.28a) is no more formed for $t > 0$ (Fig. 1.28b), and that formed at $t = 0$ is swept away in the downstream direction.

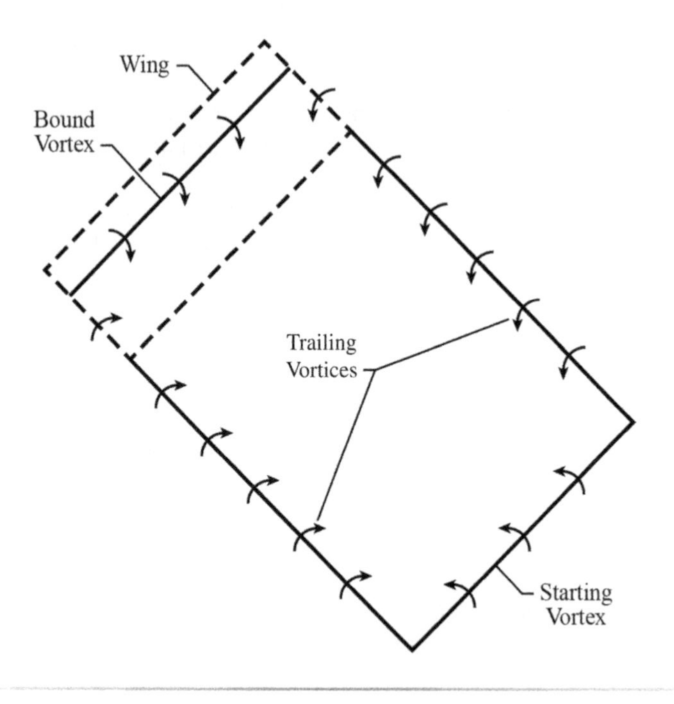

Fig. 1.27 Horseshoe vortex model of a finite wing.

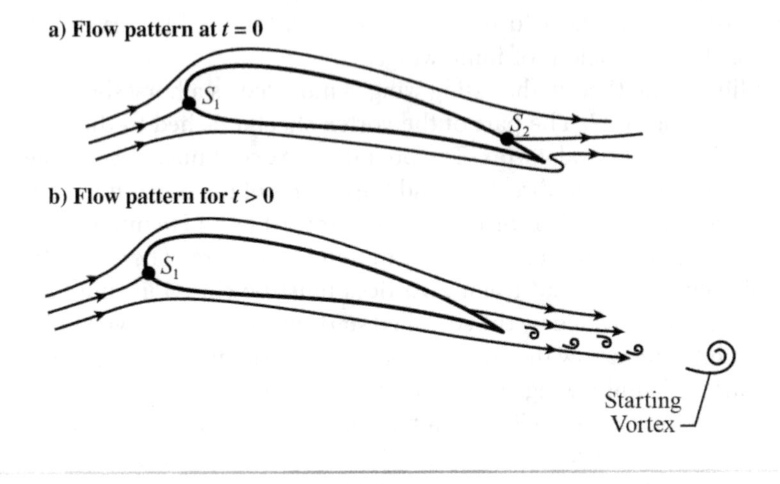

Fig. 1.28 Flow over a finite wing section.

The bound vortex induces upwash in front of the wing and downwash behind. The trailing vortices induce downwash everywhere on the wing surface. The downwash effect caused by the trailing vortex is negligible in the vicinity of the wing. A schematic variation of induced flow field around the wing is shown in Fig. 1.29. As a result of these induced upwash/downwash effects, the effective angle of attack varies along the wing span.

One direct consequence of the induced flow field around a finite wing is that the lift vector is now perpendicular to the local velocity vector and not to the freestream velocity vector as shown in Fig. 1.30. By definition, the component of the force perpendicular to the freestream is the lift and that along the freestream direction is the drag. As shown in Fig. 1.30, we now have a component of the local lift in the freestream direction, which is a form of drag. This component of drag, which is caused by lift, is called the induced drag. It is important to bear in mind that induced drag is not caused by fluid viscosity but represents a kind of penalty to be paid for developing the lift.

Referring to Fig. 1.30,

$$\alpha_l = \alpha - \alpha_i \tag{1.44}$$

The basic contribution of the lifting line theory is the following expression for the induced angle of attack,

$$\alpha_i = \frac{C_L}{\pi A} \tag{1.45}$$

This formula is applicable only to wings of elliptical planform. For all other wing planforms, this formula is modified as follows:

$$\alpha_i = \frac{C_L}{\pi A e} \tag{1.46}$$

where e is called the planform efficiency factor. We note that, for elliptical wings, $e = 1$. With this, the induced-drag coefficient of a lifting wing is given by

$$C_{Di} = C_L \alpha_i \tag{1.47}$$

$$= \frac{C_L^2}{\pi A e} \tag{1.48}$$

This is an important result of the lifting line theory. It shows that for a given aspect ratio, the induced drag coefficient varies to the square of the lift coefficient and, for a given lift coefficient, the induced drag varies inversely to the aspect ratio. In other words, induced drag is high for low-aspect ratio wings and insignificant for high-aspect ratio wings.

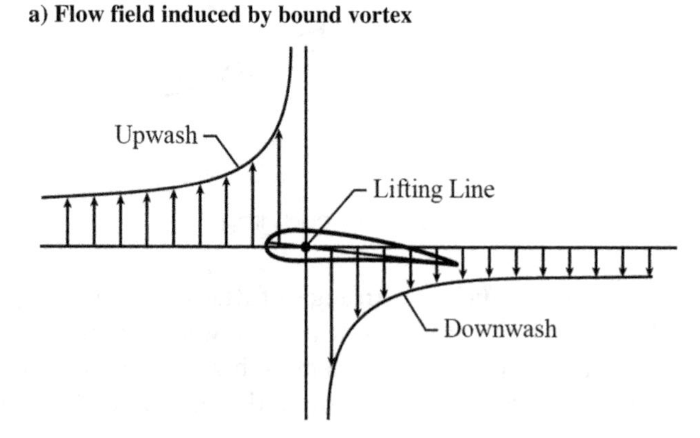

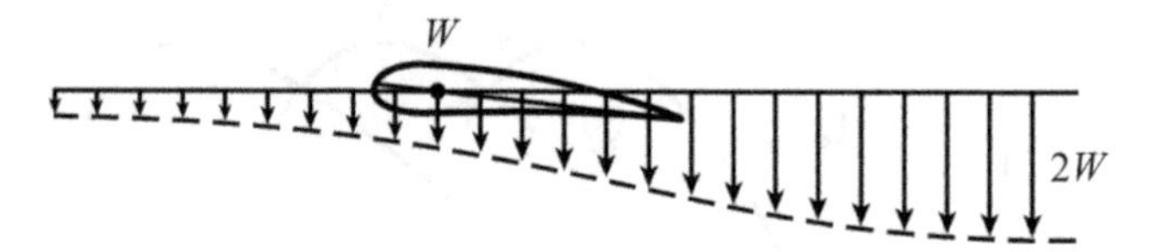

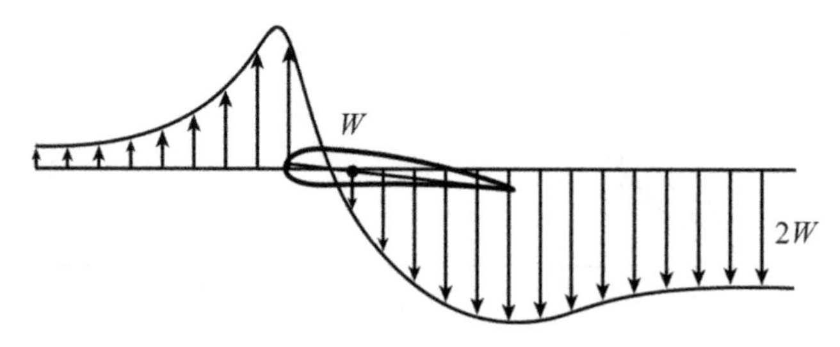

Fig. 1.29 Induced flow field around a finite wing.

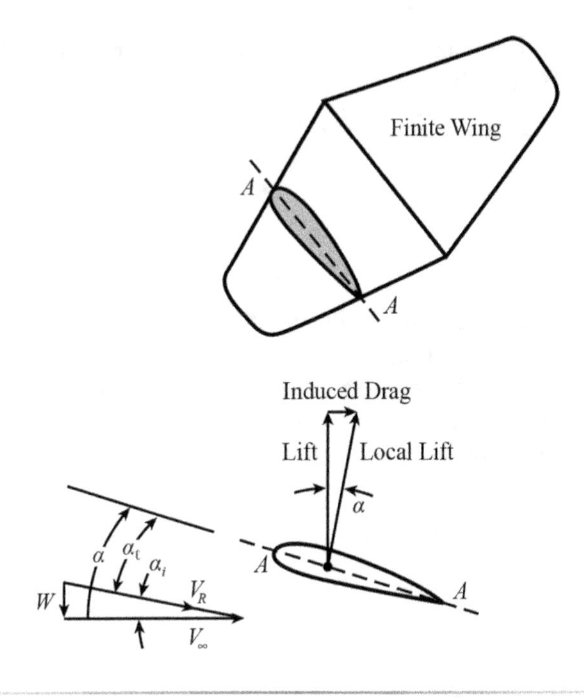

Fig. 1.30 Induced drag.

The variation of lift coefficient with angle of attack for various aspect ratios is schematically shown in Fig. 1.31. For wings of low aspect ratio, the effective angle of attack for a given lift coefficient is higher than that for a two-dimensional wing ($A = \infty$). In other words, the lift-curve slope effectively

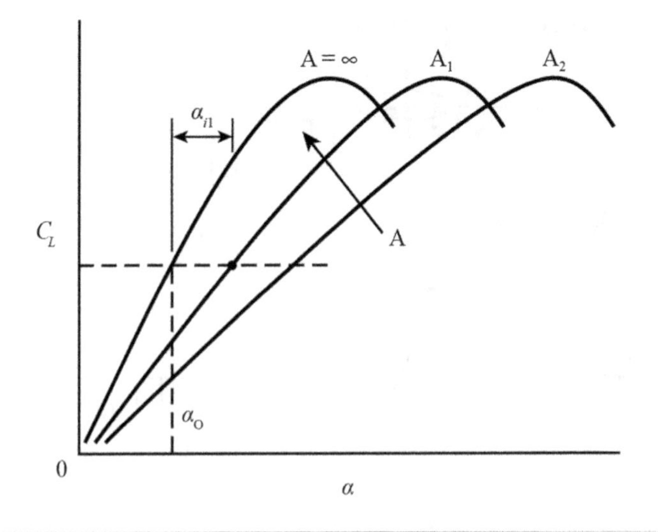

Fig. 1.31 Lift coefficient with angle of attack for finite wings.

reduces with a decrease in aspect ratio. Based on the lifting line theory, we can obtain an expression for the lift-curve slope of the finite wings as follows.

For a given lift coefficient,

$$C_L = a_o \alpha_o \tag{1.49}$$

$$= a(\alpha_o + \alpha_i) \tag{1.50}$$

$$= a\alpha_o \left(1 + \frac{\alpha_i}{\alpha_o}\right) \tag{1.51}$$

or

$$a = \frac{a_o}{1 + \frac{a_o}{\pi A e}} \tag{1.52}$$

where a_o is the sectional or two-dimensional lift-curve slope and a is the lift-curve slope of a finite wing of aspect ratio A. In the formula given, the values of a and a_o are per rad.

1.7 Methods of Reducing Induced Drag

An obvious method of reducing induced drag is to increase the aspect ratios of a lifting surface or install end plates (at the wing tips) to prevent the crossflow around wing tips. However, either of these two approaches is not always the best option from structural considerations. The weight of the wing increases considerably because of the additional structure that is needed to resist the large bending loads caused by increased span or the added weight of the end plates. Another method that is simpler and does not encounter this problem is the use of winglets as discussed in the following.

One effective method of reducing induced drag of a lifting surface is the application of winglets [12]. A winglet is a small "wing" placed near the wing tips and almost normal to the main wing surface as shown in Fig. 1.32. Placing a winglet on an existing wing alters the spanwise distribution of the circulation, hence the structure of the wing tip vortices. The key to achieving induced-drag reduction is the efficient generation of a side force. The side force is generated due to the lift-induced inflow above the wing tip or due to the outflow below the wing tip. The side force dy_N generated by the winglet because of flow angle α' is directed inboard as shown in Fig. 1.32b. This side force provides a component $-dD$ in the flight direction that contributes to a reduction in induced drag. This effect is similar to that on a sail of a sailboat tacking upwind.

The application of winglets to KC135 aerial-tanker aircraft resulted in approximately 9% reduction in drag at cruise conditions [12]. Winglets are

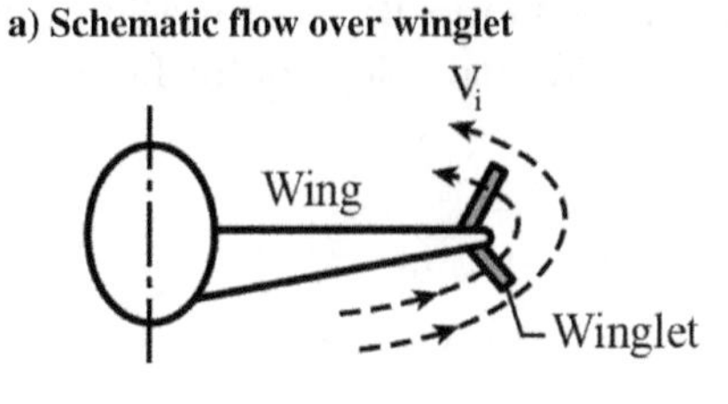

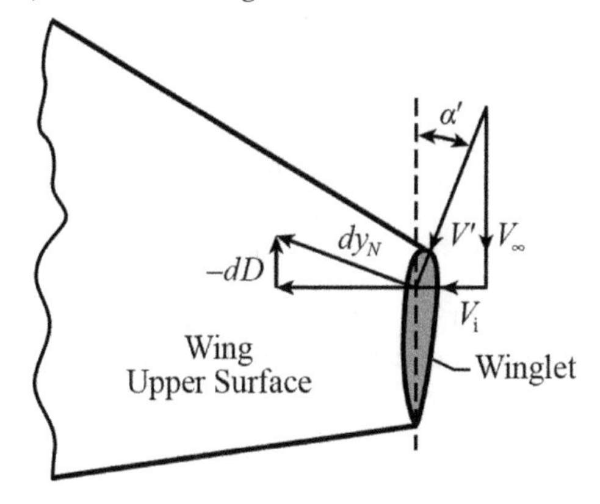

Fig. 1.32 Winglets.

also used on the Boeing 747-400 aircraft, which is a long-range derivative of the Boeing 747 wide-body transport family.

1.8 Tip Vortices: Formation and Hazards

A salient feature of finite wings is the existence of tip vortices and their influence on the aircraft that fly in their vicinity. Note that for a two-dimensional wing, which is supposed to extend from $-\infty$ to ∞, there are no tip effects and tip vortices do not exist.

Tip vortices on a finite wing are mainly caused by the pressure differences existing on the upper and lower surfaces of the wing. For a lifting wing operating at positive angles of attack, the pressure on the upper surface is lower compared to that on the lower surface. Because of this pressure differential, there is a tendency for the fluid to "leak" from lower surfaces to higher surfaces, and wing tips provide a path to this "leakage." Thus, fluid particles from the lower surface tend to go around the wing tips and move on to the upper surface, forming a curved, vortex type of flow as shown in Fig. 1.33a. This curved flow, combined with the freestream airflow, when viewed from top, appears to be moving in a helical path as shown in Fig. 1.33b. A short distance downstream of the wing, the vortices roll up tightly into so-called tip vortices.

These tip vortices remain in fluid medium for some time before getting dissipated because of viscosity and ambient turbulence. For the time duration when these tip vortices exist, they can create significant hazards to the aircraft that fly into this area unaware of their existence.

Consider an aircraft flying perpendicular to the flight path of the aircraft that generated the tip vortices and left them behind as shown in Fig. 1.34. This aircraft experiences an upwash followed by a relatively long period of downwash and then an upwash again. On encountering the first upwash, an obvious thing for a pilot is to reduce the angle of attack to prevent the airplane from climbing. After he will have done this, he will encounter the

a) Crossflow around wing tips

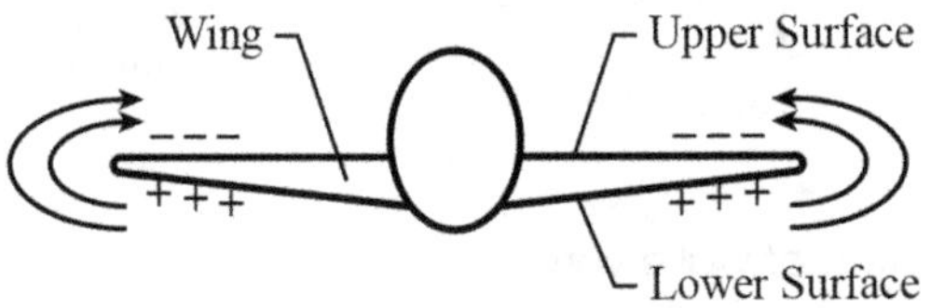

b) Roll-up of trailing vortices

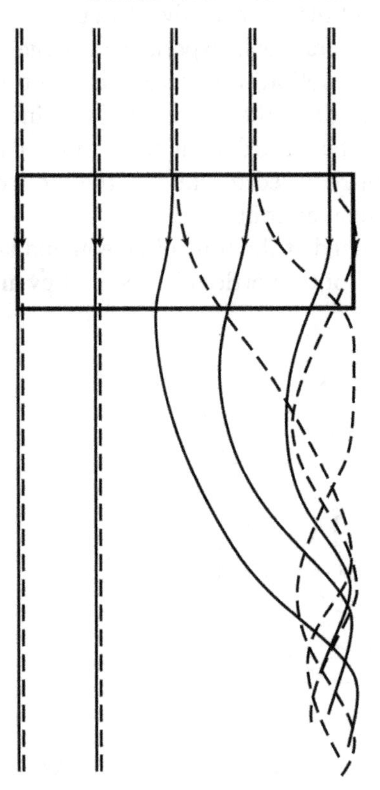

Fig. 1.33 Development of trailing vortices.

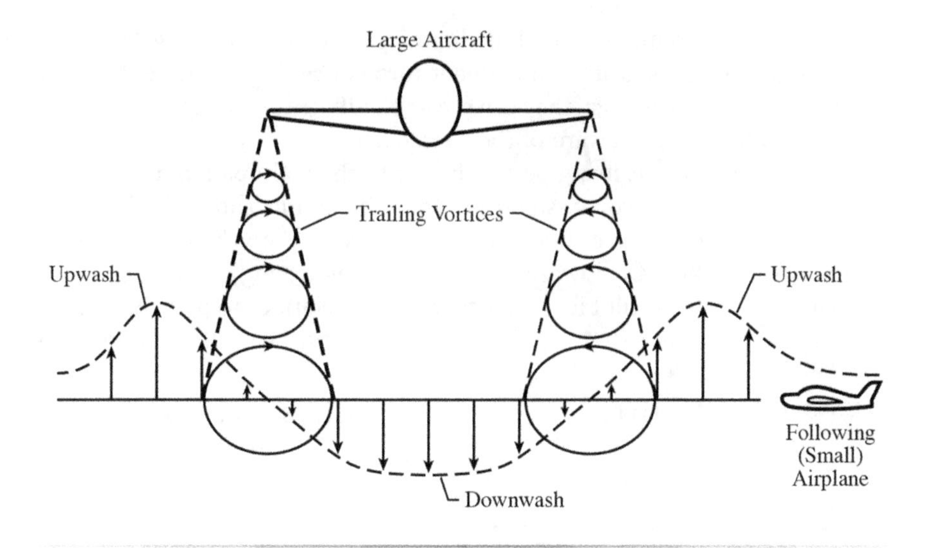

Fig. 1.34 Effect of wake vortices.

downwash field that reduces the angle of attack further. This sequence of events can be of catastrophic consequence if the aircraft is close to the ground.

Another possibility is that an aircraft may directly fly into a trailing vortex system as shown in Fig. 1.35. This aircraft will experience a large rolling motion as shown. This induced rolling motion will be significant if the trailing vortices are caused by a large aircraft and the aircraft that flies into them is a small aircraft. This condition can be hazardous if an encounter occurs close to the ground and the roll rate required to counter the induced roll rate is beyond the capability of the small aircraft.

A number of accidents have happened at the time of takeoff and landing when one aircraft enters into the field of tip vortices left behind by another

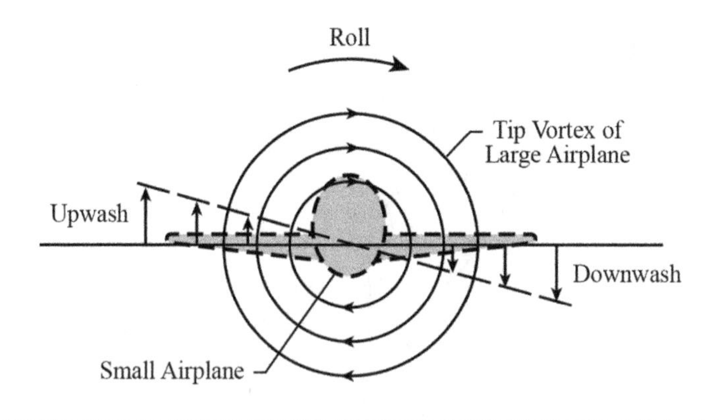

Fig. 1.35 Induced rolling motion because of wake vortices.

aircraft. To minimize such mishaps, a safe separation distance or time between a leading and a following aircraft is specified by aviation regulatory authorities. This separation distance or time depends on the gross weight of the aircraft. Generally, air traffic controllers typically maintain at least three minutes separation for an aircraft like a Boeing 757, which has an approximate gross weight of 240,000 lb.

1.9 Flow of a Compressible Fluid

Infinitesimal disturbances in fluid propagate at a velocity equal to the speed of sound in that medium. The speed of sound depends on the nature of the fluid and its temperature. In a stationary fluid, the pressure disturbances travel in concentric circles as shown in Fig. 1.36a. For a two-dimensional flow, these circles are cross sections of a cylinder and, for a three-dimensional flow,

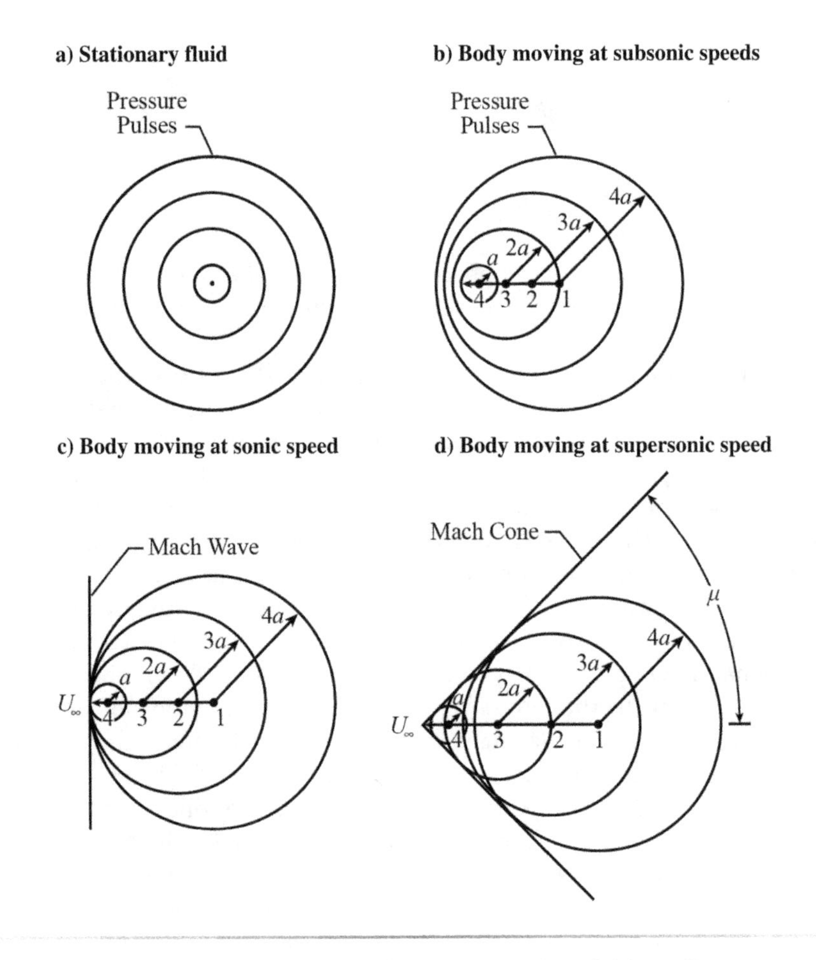

Fig. 1.36 Propagation of disturbances in a fluid medium.

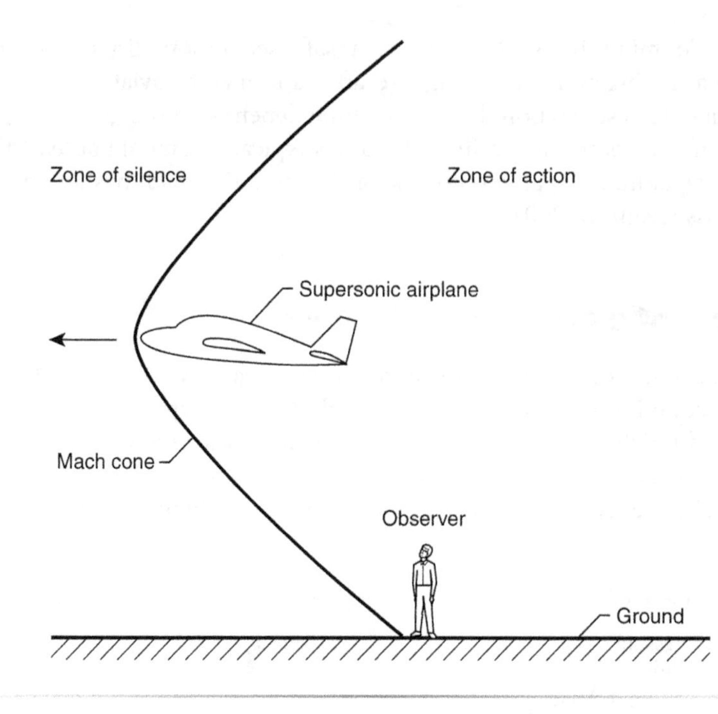

Fig. 1.37 Zone of action and zone of silence.

they are cross sections of a sphere. However, if the object that creates these pressure disturbances moves, the picture changes as shown in Figs. 1.36b–1.36d.

When the speed of the body is subsonic ($U_\infty < a$), pressure pulses traveling at the speed of sound a will always be ahead of the body. These pulses would crowd in front of the body and spread out behind it. As the speed of the body U_∞ approaches the speed of sound a, the crowding in front of the body becomes intense until at $U_\infty = a$; the pressure pulses merge to form a moving front, which is called a Mach wave. When the speed of the body U_∞ exceeds the speed of the sound a, the body will be ahead of the moving wave front as shown in Fig. 1.36d. All the disturbances emitted by the body are confined into a narrow region, which is called the Mach wedge (two-dimensional flow) or the Mach cone (three-dimensional flow), obtained by drawing a tangent to the circles representing the propagation of pressure pulses. The zone inside the Mach cone (or wedge) is called the zone of action, and that outside the Mach cone (or wedge) is called the zone of silence.

An interesting example of the zone of silence and the zone of action is the flight of a supersonic aircraft past an observer stationed on the ground. The observer looking at the aircraft will not hear any sound emitted by the aircraft until the aircraft flies past and the observer comes within the Mach cone as schematically shown in Fig. 1.37. If this aircraft were subsonic, the observer would have heard the sound long before the aircraft flies past him.

The semi-included angle of the Mach cone is called the Mach angle μ and is given by

$$\sin \mu = \frac{a}{U_\infty} \tag{1.53}$$

$$= \frac{1}{M} \tag{1.54}$$

1.10 Aerodynamic Forces in Supersonic Flow

In supersonic flow, fluid particles are not aware of the existence of the body downstream because this information is confined only to that region that is within the Mach cone of the body. Therefore, a fluid particle comes to know of the existence of the body only when it strikes it abruptly and comes within its Mach cone. As a result, a fluid particle cannot adjust itself to flow smoothly over the body as it would have done if the flow were subsonic. In subsonic flow, the streamlines bend well ahead of the body and smoothly flow past it as shown in Fig. 1.38a. However, in supersonic flow, a shock wave is formed at the leading edge of the body as shown in Fig. 1.38b, and the fluid particles experience a sudden rise in pressure, temperature, and density on crossing the shock wave. The Mach number downstream of the shock wave is always smaller than the upstream Mach number. The shock wave is attached to the leading edge if the leading edge

a) Subsonic flow

b) Supersonic flow over a sharp leading-edge body

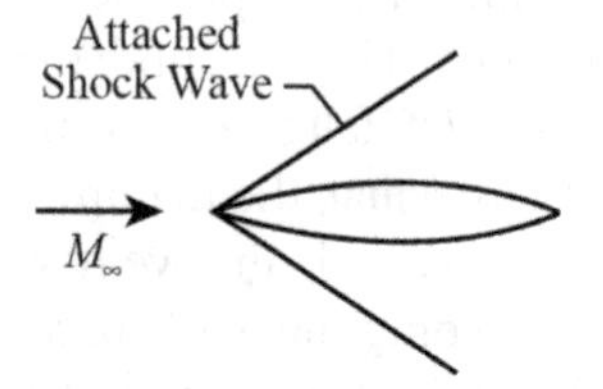

c) Supersonic flow over a blunt leading-edge body

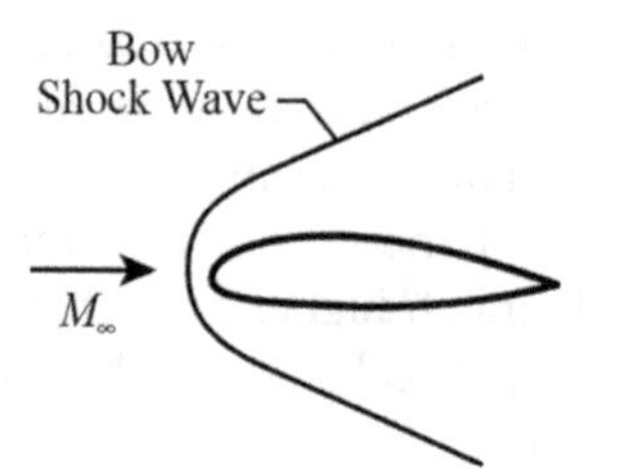

Fig. 1.38 High-speed flow over streamlined bodies.

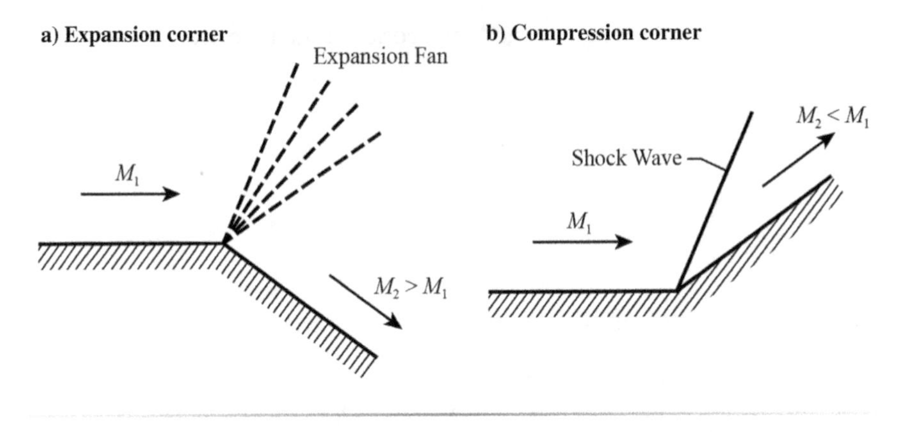

Fig. 1.39 Supersonic flow around corners.

is sufficiently sharp and the Mach number is much greater than unity. If the leading edge is blunt or the Mach number is close to unity or both, then a detached bow shock wave is formed ahead of the body as shown in Fig. 1.38c.

When a supersonic flow goes around a corner that turns away from the flow direction, then an expansion fan originates at the corner as shown in Fig. 1.39a. In contrast to a shock wave, the temperature, pressure, and density change gradually across an expansion fan. On crossing an expansion wave, the Mach number increases, and pressure and temperature decrease. Such a corner is called an expansion corner. On the other hand, if the corner turns into the flow (Fig. 1.39b), a shock wave is formed at the corner, and the pressure, temperature, and density increase on crossing the shock wave. Such a corner is called a compression corner.

Consider supersonic flow past a sharp symmetrical double wedge airfoil at zero angle of attack as shown in Fig. 1.40a. At the leading edge, an attached shock wave is formed. As a result, pressures on surfaces *AB* and *AD* are higher than freestream pressure p_∞ because of a compression produced by the shock wave. At *B* and *D*, expansion fans are formed. The flow undergoes a gradual expansion and pressures on surfaces *BC* and *DC* fall below the freestream value p_∞. Because pressures on *AB* and *AD* are equal, the net upward force or lift is zero. But the axial components do not cancel, and we have a nonzero force in the axial direction. Similarly, the component of force in the direction normal to the flow on surfaces *BC* and *DC* cancel, but the axial components add up to give a nonzero force in the axial direction. The two nonzero axial force components add up to a net nonzero force in the flow direction, which is called wave drag. The wave drag is solely caused by the compressible nature of the fluid. Therefore, in supersonic flow, the wave drag is an additional component of the drag. Thus, for a given body, the drag in supersonic flow is usually much higher than that in subsonic flow. The wave drag depends on the geometrical shape and thickness ratio of the body and freestream Mach number.

Now consider the sharp double wedge airfoil to be at a small angle of attack ($\alpha \leq \theta$) as shown in Fig. 1.40b. The oblique shock waves AA' and AA'', formed respectively on the upper and lower surfaces, are not of equal strength. The reason for this difference is that the flow turning angles on the top and bottom surfaces are not equal. On the top side AB, it is equal to $\theta - \alpha$, whereas on the bottom side AD it is equal to $\alpha + \theta$. Thus, the pressure on AD is higher than that on AB. Similarly, the strengths of the expansion fans originating at B and D are different because Mach numbers M_u and M_l are different. Actually, $M_u > M_l$. Therefore, the pressure on BC is lower compared to that on DC. The net result is that we have an upward force (F_N) acting on the airfoil. The component of this upward force in the direction perpendicular to the freestream is the lift L, and that along the freestream direction D is the wave drag.

For a double wedge airfoil,

$$a = \frac{4}{\sqrt{M_\infty^2 - 1}} \tag{1.55}$$

$$C_l = \frac{4\alpha}{\sqrt{M_\infty^2 - 1}} \tag{1.56}$$

$$C_{dw} = \frac{4}{\sqrt{M_\infty^2 - 1}}\left[\alpha^2 + \left(\frac{t}{c}\right)^2\right] \tag{1.57}$$

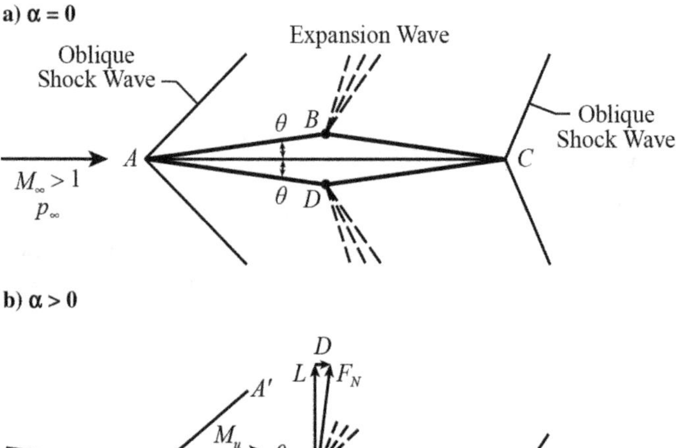

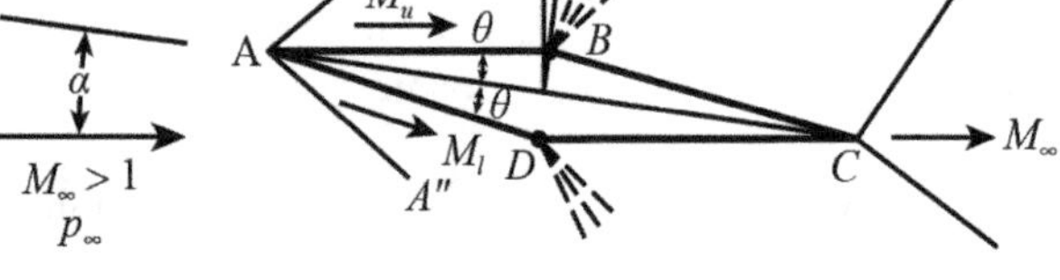

Fig. 1.40 Supersonic flow over double wedge airfoil.

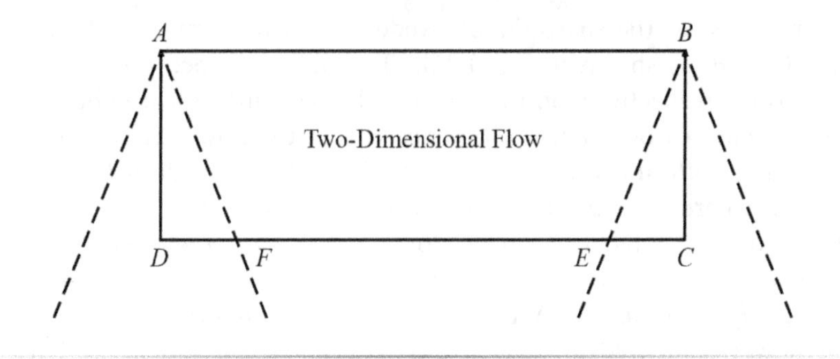

Fig. 1.41 Finite wings in supersonic flow.

Here, the wave drag has two components, one caused by thickness distribution and the other caused by angle of attack. For a thin flat plate, $t/c = 0$ and hence $C_{dw} = 0$ at $\alpha = 0$.

A finite wing in supersonic flow behaves in a different manner compared to that in a subsonic flow. For subsonic flow, the effects of the wing tips are felt all over the wing surface. However, in supersonic flow, the effect of the wing tips are confined to the Mach cones emanating from the leading edges of the tip chord as shown in Fig. 1.41. Influences of the wing tips *AD* and *BC* are confined to the regions *ADF* and *BCE*. The rest of the wing *ABEF* is not aware of the wing tips and functions as though it were part of a two-dimensional wing. If the aspect ratio is sufficiently high, then the region of the wing falling within these Mach cones is quite small, and the entire wing can be assumed to behave like a two-dimensional wing.

1.11 Critical Mach Number

So far we have discussed the flow over an airfoil at low subsonic or supersonic speeds, assuming that the flow is either completely subsonic or completely supersonic. However, if the freestream Mach number is in the high subsonic or low supersonic range, then flow field around the body may consist of mixed subsonic and supersonic flow regions. When the freestream Mach number is in the range of 0.8–1.2, the flow is said to be transonic. To understand this type of complex flow field, let us study the flow over an airfoil held at a constant positive angle of attack when the freestream Mach number increases from a low subsonic to a high subsonic or transonic value.

For a given freestream Mach number, we will have some point like *P* on the top surface of the airfoil section, where the local velocity is maximum (Fig. 1.42a). At this point, the local velocity will continuously increase as the freestream Mach number increases. No drastic changes in the nature of the flow take place as long as the local flow everywhere on the body surface is subsonic. For this range of Mach numbers, the pressure coefficient and lift-curve slopes are given by the well-known Prandtl–Glauert rule,

$$C_{p,c} = \frac{C_{p,i}}{\sqrt{1 - M_\infty^2}} \tag{1.58}$$

$$a_{o,c} = \frac{a_{o,i}}{\sqrt{1 - M_\infty^2}} \tag{1.59}$$

a) Subsonic flow

$M < 1$

P

$M_\infty < M_{cr}$

b) Local sonic flow

$M < 1$

$P\ (M = 1)$

$M_\infty = M_{cr}$

c) Supersonic flow on upper surface

$M < 1$

Shock Wave

$M \geq 1$

$M_\infty > M_{cr}$

d) Supersonic flow on upper and lower surfaces

$M < 1$

Shock Wave

$M \geq 1$

Wake

$M \geq 1$

$M_\infty > M_{cr}$

e) Airfoil in supersonic flow

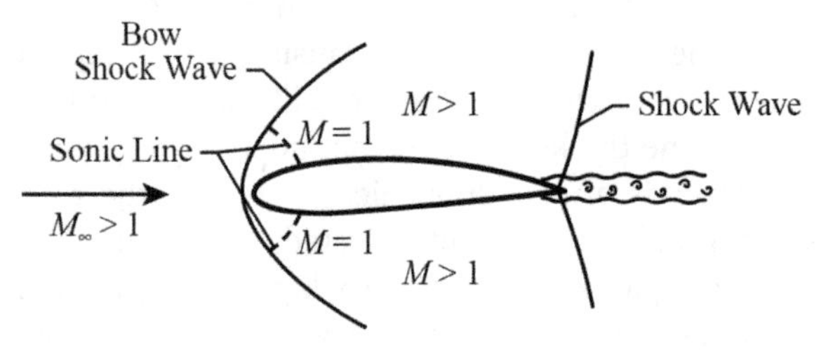

Fig. 1.42 Flow over airfoil at high speeds.

where C_p is the pressure coefficient and a_o is the lift-curve slope of the airfoil. The suffixes i and c denote incompressible and compressible values, respectively. It may be noted that the pressure coefficient is defined as

$$C_p = \frac{p - p_\infty}{1/2\rho_\infty V_\infty^2} \tag{1.60}$$

According to Eqs. (1.58) and (1.59), which are based on potential flow theory, the magnitude of the pressure coefficient at any point on the airfoil surface and the section lift-curve slope increases steadily with Mach number above their incompressible values. These formulas are applicable as long as the flow everywhere is subsonic and shock free.

When the local Mach number at point P on the airfoil surface reaches the value of unity, the corresponding freestream Mach number is called the critical Mach number and is denoted by M_{cr} as schematically illustrated in Fig. 1.42b. Elsewhere on the surface of the airfoil, the local Mach number is below unity, and the flow is subsonic. The value of M_{cr} depends on the geometrical shape of the body and the angle of attack. For a given airfoil, the critical Mach number usually decreases with increase in angle of attack.

The typical values of M_{cr} for airfoils at zero-lift lie in the range 0.6–0.85. For example, at $C_l = 0$, the values of M_{cr} for NACA 2412, NACA 23012, and NACA 65_3-418 airfoils are, respectively, equal to 0.69, 0.672, and 0.656. For a thin airfoil like NACA 0006, the critical Mach number is 0.805 at $C_l = 0$. The values of M_{cr} for other airfoils may be found elsewhere [3, 11].

As the freestream Mach number increases beyond M_{cr}, the local velocity on the surface of the airfoil exceeds the sonic velocity at more than one point. In fact, a small region appears where the local Mach number is either equal to or greater than unity as shown in Fig. 1.42c. Because the flow Mach number behind the airfoil has to be equal to the freestream value, which is subsonic, the region of supersonic flow is terminated by a shock wave as shown. As M_∞ increases further but still below unity, the region of supersonic flow expands on the upper surface, and a region of supersonic flow may even appear on the lower surface of the airfoil as shown in Fig. 1.42d.

The formation of shock waves on the surface of the airfoil leads to flow separation, loss of lift, and increase in drag. Downstream of a shock wave, the pressure is always higher. As a result, an adverse pressure gradient is impressed upon the boundary layer, causing it to separate from the surface. Because the velocity rises from zero at the body surface to the freestream value across the thickness of the boundary layer, a part of the supersonic boundary layer is always subsonic. Therefore, the formation of the adverse pressure gradient is communicated upstream of a shock wave through the subsonic part of the boundary layer. As a result, the boundary-layer flow upstream of the shock wave is aware of the pressure gradient and may separate even ahead of the shock wave, causing a modification in the effective body shape, hence a change in the shock wave structure. This process of mutual interaction between the shock wave and the boundary layer is called shock–boundary layer interaction.

As a result of the shock-induced flow separation, the drag coefficient rises very rapidly, and the lift coefficient drops. Another consequence of the shock-induced flow separation is the buffeting of horizontal tail. Buffeting is said to occur when the separated, unsteady, turbulent flow from the wing surface passes over the horizontal tail and causes the tail loads (lift, drag, and pitching moment) to fluctuate. The fluctuating tail loads create problems of stability and control.

In Figs. 1.43 and 1.44, the schematic variations of lift and drag coefficients with Mach number at various angles of attack are shown. The lift coefficient increases in the high subsonic Mach numbers as indicated by Eq. (1.59). This increase continues slightly beyond the critical Mach number because even though shock waves are formed on the airfoil surface they may not be strong enough to cause flow separation. However, as the freestream Mach number increases further, the shock waves become stronger and eventually cause flow separation. Around this Mach number, the lift coefficient reaches its maximum value and then falls off as shown in Fig. 1.43. This type of stall is sometimes called "shock stall." The nature and severity of the shock stall depends on the camber and thickness ratio of the airfoil. The drop in the lift coefficient is more severe for highly cambered and thicker airfoils because these airfoils have relatively higher local velocities that lead to stronger shock waves and more severe adverse pressure gradients.

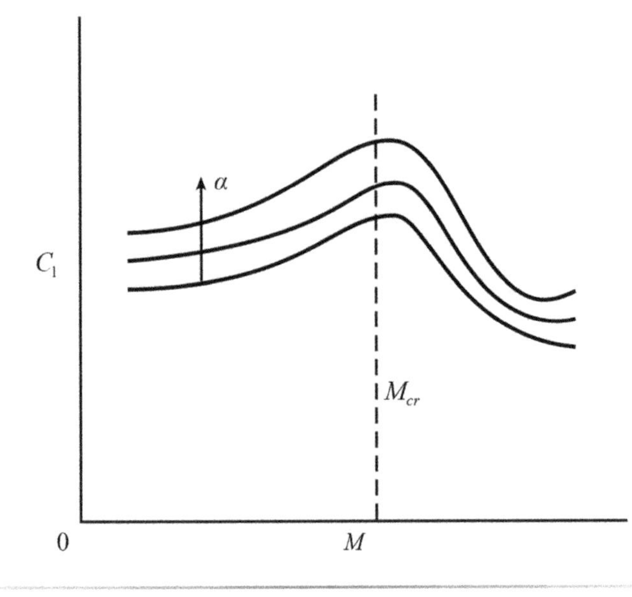

Fig. 1.43 Sectional lift coefficient at high speeds.

The fall in the lift coefficient is usually accompanied by a steep rise in the drag coefficient as shown in Fig. 1.44. The drag coefficient reaches a peak value around Mach 1 and then starts dropping off when a clear supersonic flow is established on the entire surface of the airfoil as shown in Fig. 1.42e. When this happens, the flow is smooth and attached because shock waves that were formed on the upper or lower surface of the airfoil and caused flow separation are now pushed towards the trailing edge.

To characterize the drag rise in high subsonic/transonic flow, it is usual to define what is called the drag divergence Mach number M_d as that value of freestream Mach number when the drag coefficient begins to rise sharply as shown schematically in Fig. 1.44.

The typical drag polars of an airfoil section in the transonic Mach number range are schematically shown in Fig. 1.45. The graph in which the lift coefficient is plotted against the drag coefficient is called the drag polar. As the Mach number increases, the drag polars shift to the right and bend forward because at a given lift coefficient the drag coefficient will be higher as the Mach number increases.

For a conventional low-speed airfoil with a well-rounded leading edge, a detached bow shock wave is formed when the freestream Mach number is in the supersonic range (Fig. 1.42e). Because of symmetry, a detached shock wave is normal to the flow direction on the line of symmetry. Because the flow behind the normal shock wave is always subsonic, there is a small patch of subsonic flow around the leading edge region. The subsonic flow

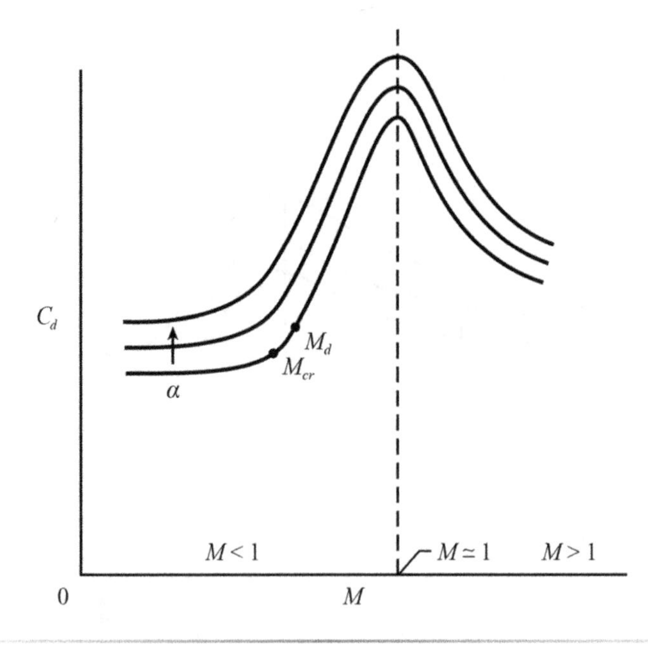

Fig. 1.44 Sectional drag coefficient at high speeds.

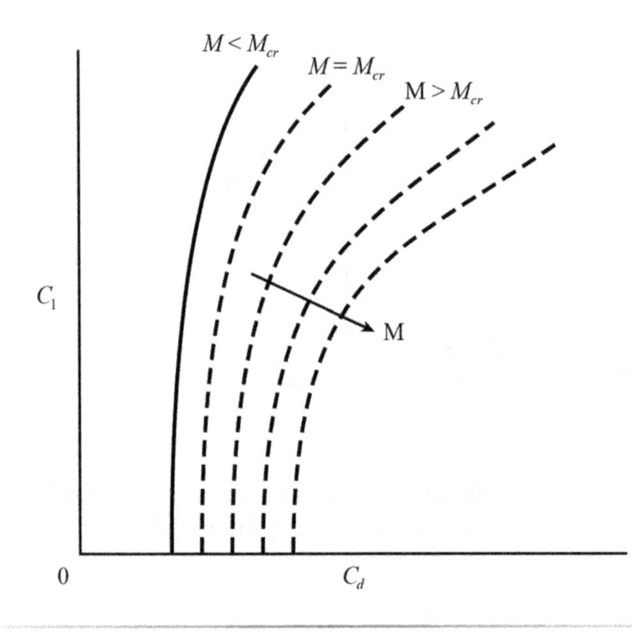

Fig. 1.45 Sectional drag polars at transonic speeds.

quickly accelerates to supersonic flow over the airfoil surface. Somewhere on the upper and lower surfaces, we have sonic lines as shown. The flow over this type of airfoil is one of mixed subsonic and supersonic flow. Furthermore, the wave drag is high because a detached bow shock wave is formed.

One way of reducing the wave drag of the wing in supersonic flow and avoiding the formation of a detached bow shock wave ahead of the airfoil is to use a sharp leading-edge airfoil like the diamond section of Fig. 1.40. An attached shock wave is formed at the leading edge and the flow everywhere is supersonic on the airfoil surface as we have discussed earlier.

There are a number of ways of delaying the adverse effects of compressibility to higher Mach numbers. This is of considerable importance to the airline industry because even a modest rise in the cruise Mach number in the high subsonic/transonic regime helps to cut down operating costs considerably. Some of the most commonly used methods are discussed in the following sections.

Thin Airfoils

The thinner the airfoil, the smaller the peak local velocity on its surface will be and the higher the critical Mach number will be as schematically shown in Fig. 1.46. The use of thin airfoils pushes the critical Mach number and the drag divergence number further towards unity. However, a disadvantage of using very thin airfoils is that their low-speed characteristics, especially the stalling characteristics, are poor. As we have discussed

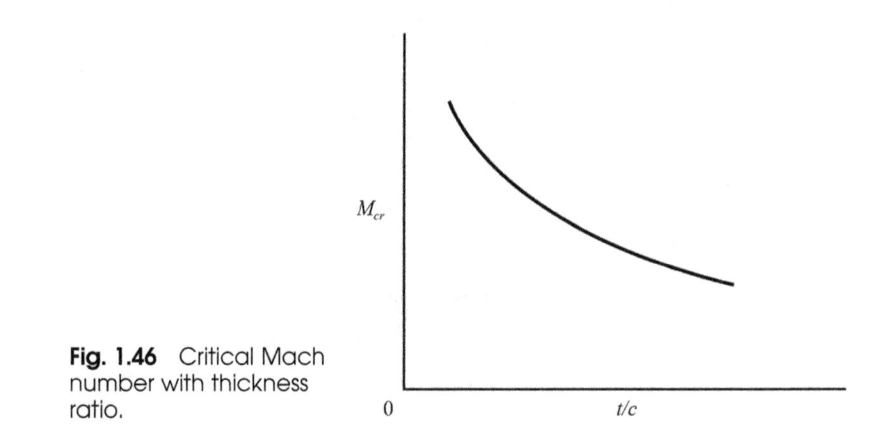

Fig. 1.46 Critical Mach number with thickness ratio.

earlier, thin airfoils exhibit an abrupt loss of lift at stall, which leads to stability and control problems as we shall study later in the text.

Low-Aspect Ratio Wings

As we know, the lower the aspect ratio, the more pronounced the induced flow effects will be and the lower the peak velocities on the wing surface will be. As a result, the critical Mach number will increase with a decrease in aspect ratio as shown schematically in Fig. 1.47. However, the disadvantage of low-aspect ratio wings is that they have higher induced drag and lower lift-curve slopes.

Supercritical Airfoils

For a conventional airfoil at Mach numbers above the critical Mach number, the local region of supersonic flow terminates in a strong shock wave (Figs. 1.42c and 1.42d) that induces significant flow separation and a steep rise in the drag coefficient. For a supercritical airfoil [12, 13] shown

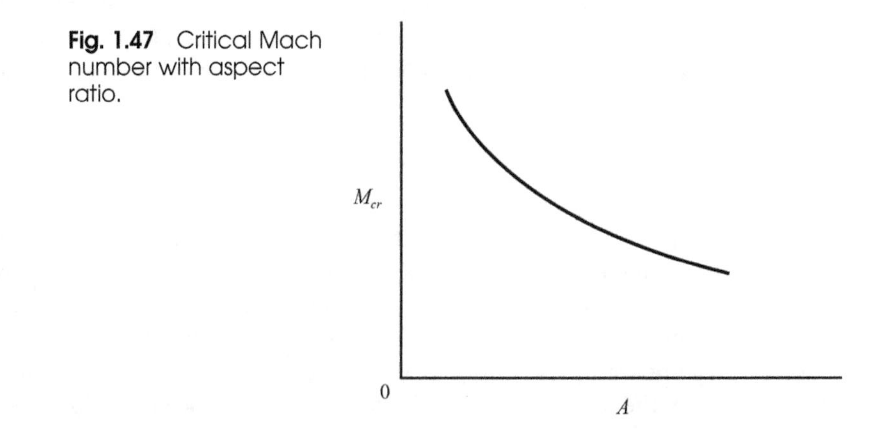

Fig. 1.47 Critical Mach number with aspect ratio.

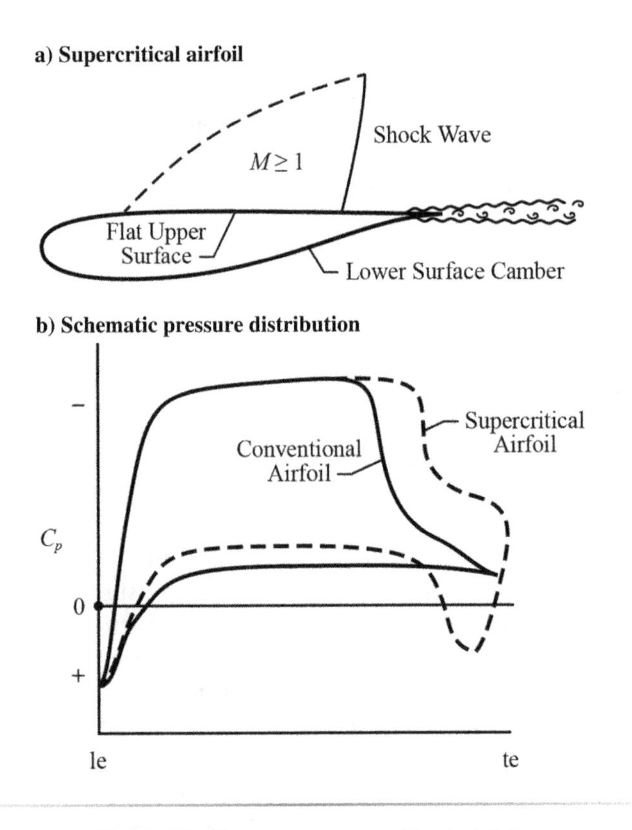

Fig. 1.48 Flow over supercritical airfoil.

in Fig. 1.48a, the curvature of the middle region of the upper surface is substantially reduced with a resulting decrease in the strength and extent of the shock wave. As a result, the drag associated with the shock wave is reduced and, more importantly, the onset of separation is substantially delayed. The lift lost by reducing the upper surface curvature is regained by substantial camber of the rear portion of the supercritical airfoil. The supercritical airfoil gives considerable increase in the critical Mach number and the drag divergence Mach number. For example, at a lift coefficient of 0.7 for the NACA 64A-410, the drag divergence Mach number is around 0.66. For a supercritical airfoil of the same 10% thickness ratio, the drag divergence Mach number is 0.80 [6, 7]. This increase in drag divergence Mach number of 0.14 is of great significance to the airline industry.

A comparison of the pressure distribution for conventional and supercritical airfoils is shown schematically in Fig. 1.48b.

Wing Sweep

Research on swept-wing concept for high-speed flight originated in Germany in the late 1930s or 1940s but was not known outside of

Germany because of World War II. The Allied pilots first encountered the German swept-wing Messerschmitt Me 262 jet fighter during 1944, which had speed advantage because of a delay in the onset of compressibility effect and the attendant drag rise. The choice of 18 deg of sweep-back was fortuitous because this was done primarily to resolve the center of gravity problem. However, at approximately the same time, researchers in Germany demonstrated through wind-tunnel tests that Busemann's supersonic swept-wing theory also applied to subsonic speeds and helped to delay compressibility effects. This effort led to the design of more highly swept wings for the Me 262 aircraft. Thus, the Me 262 program was the first systematic effort towards the design of modern swept-wing aircraft.

The concept of wing sweep is a powerful method of delaying the adverse effects of compressibility and pushing the critical Mach number to much higher values and even beyond Mach 1. The fundamental idea is that only the component of the freestream velocity normal to the wing leading edge $V_\infty \cos \Lambda$ affects the pressure distribution. The spanwise component $V_\infty \sin \Lambda$ (Fig. 1.49) does not alter the pressure distribution. The only effect the spanwise component has is on skin-friction drag. Therefore, the freestream Mach number at which the critical condition (local sonic velocity) occurs on the wing is increased by a factor $1/\cos \Lambda$ compared to a straight-unswept wing. In other words, if M_{cr} is the critical Mach number for a straight wing, then the critical Mach number for a swept wing having identical airfoil section is equal to $M_{cr}/\cos \Lambda$. This is true whether the wing is swept-back or swept-forward as shown in Figs. 1.49a and 1.49b.

The leading edge of a swept wing is said to be subsonic if the component of velocity normal to the leading edge is below the sonic velocity or the corresponding Mach number is less than unity. Therefore, the leading edge is subsonic if it is swept behind the Mach cone as shown in Fig. 1.50a. Similarly, if the component of velocity normal to the leading edge is above the sonic velocity, the wing leading edge is said to be supersonic. Therefore, the wing leading edge is supersonic if the leading edge sweep angle is smaller than the Mach angle μ as shown in Fig. 1.50b. In Chapter 3 we make use of these definitions when we discuss the determination of wing lift and pitching-moment coefficients. The wave drag is usually much smaller for wings whose leading edges are swept behind the Mach cone.

To demonstrate the advantage of sweep, consider two aircraft, one straight wing and the other swept-back wing. We assume that the gross weight, wing area, and airfoil section of both aircraft are identical. Furthermore, assume that both aircraft are required to operate at the same values of normal dynamic pressure, lift, and drag forces.

Let V_N be the flight velocity of the straight-wing aircraft. Then for equal normal dynamic pressure, the flight velocity of the swept-wing aircraft will be equal to $V_N/\cos \Lambda$.

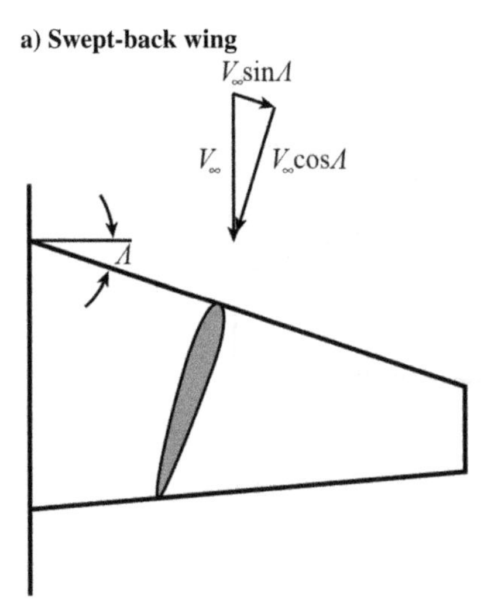

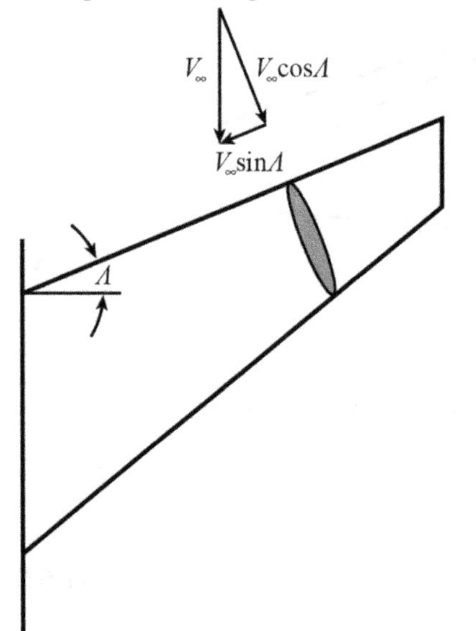

Fig. 1.49 Effect of wing sweep.

For the straight-wing aircraft,

$$L = \frac{1}{2}\rho V_N^2 S C_l \tag{1.61}$$

and for the swept-wing aircraft,

$$L_s = \frac{1}{2}\rho\left(\frac{V_N}{\cos\Lambda}\right)^2 S C_{ls} \tag{1.62}$$

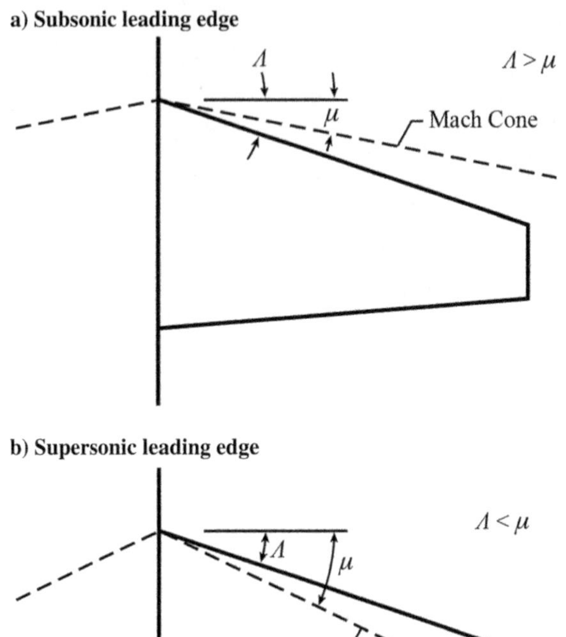

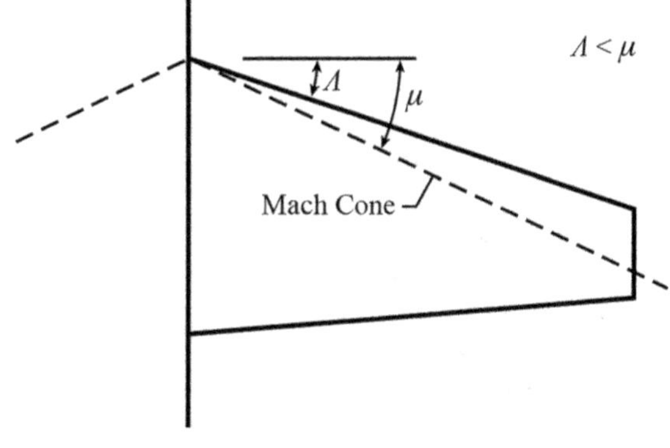

Fig. 1.50 Subsonic and supersonic leading edges.

Equating the two,

$$C_{ls} = C_l \cos^2 \Lambda \tag{1.63}$$

Similarly, we can show that

$$C_{ds} = C_d \cos^2 \Lambda \tag{1.64}$$

Thus, for equal lift and drag forces (which means equal engine power), an aircraft with swept wings can fly at a higher Mach number by a factor of $1/\cos \Lambda$ compared to an aircraft with straight unswept wings, and the use of wing sweep considerably softens the amount of drag-coefficient rise in the transonic range.

It may be noted these observations are mainly based on considering two-dimensional wings. However, for finite or three-dimensional wings, the benefits appear to be smaller and closer to the factor of $1/\sqrt{\cos \Lambda}$ instead of $1/\cos \Lambda$ as assumed here. A schematic variation of drag coefficient for various values of wing sweep is shown in Fig. 1.51.

While the application of wing sweep offers significant benefits for high-speed flight, it is accompanied by poor subsonic capabilities because of its high induced drag and low lift-curve slope. Because of the low value of lift-curve slope, the angle of attack needs to be considerably high during takeoff and landing, which may create problems of tail scraping and pilot visibility. Furthermore, conventional high-lift devices like the trailing-edge flaps perform poorly on swept wings. For those aircraft missions requiring very high levels of both subsonic and supersonic performance, it is advantageous to use variable sweep. Examples of variable-sweep designs are the F-111, F-14, and B-1 aircraft. Variable-sweep aircraft position the wings at zero or small sweep angle for low-speed operations such as landing and takeoff and then sweep the wings as required for high-speed operation.

To understand why the lift-curve slope of a swept wing is smaller compared to a straight wing, consider straight and swept wings, both operating at an angle of attack α (Fig. 1.52). Then the effective angle of attack of the swept wing is given by

$$\tan \alpha_{\text{eff}} = \frac{V_\infty \sin \alpha}{V_\infty \cos \Lambda \cos \alpha} \tag{1.65}$$

or

$$\alpha_{\text{eff}} \simeq \alpha \sec \Lambda \tag{1.66}$$

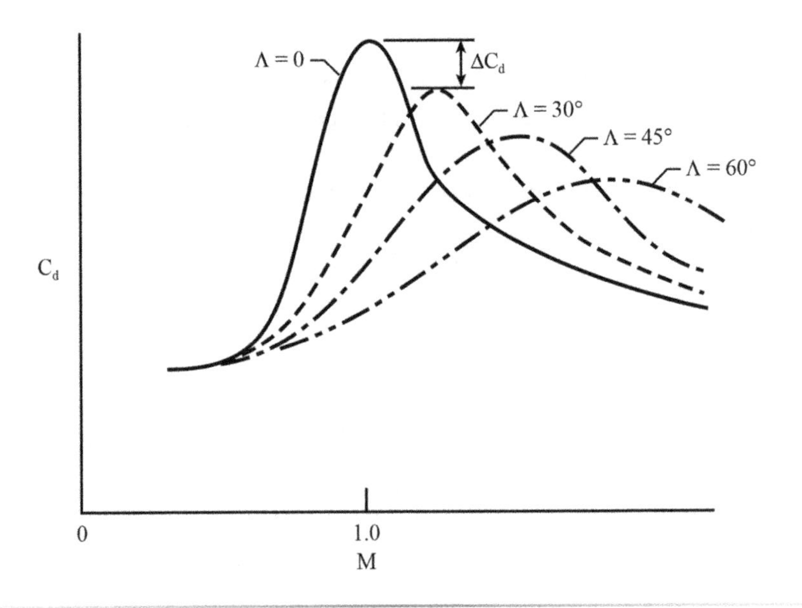

Fig. 1.51 Effect of wing sweep on drag coefficient at high speeds.

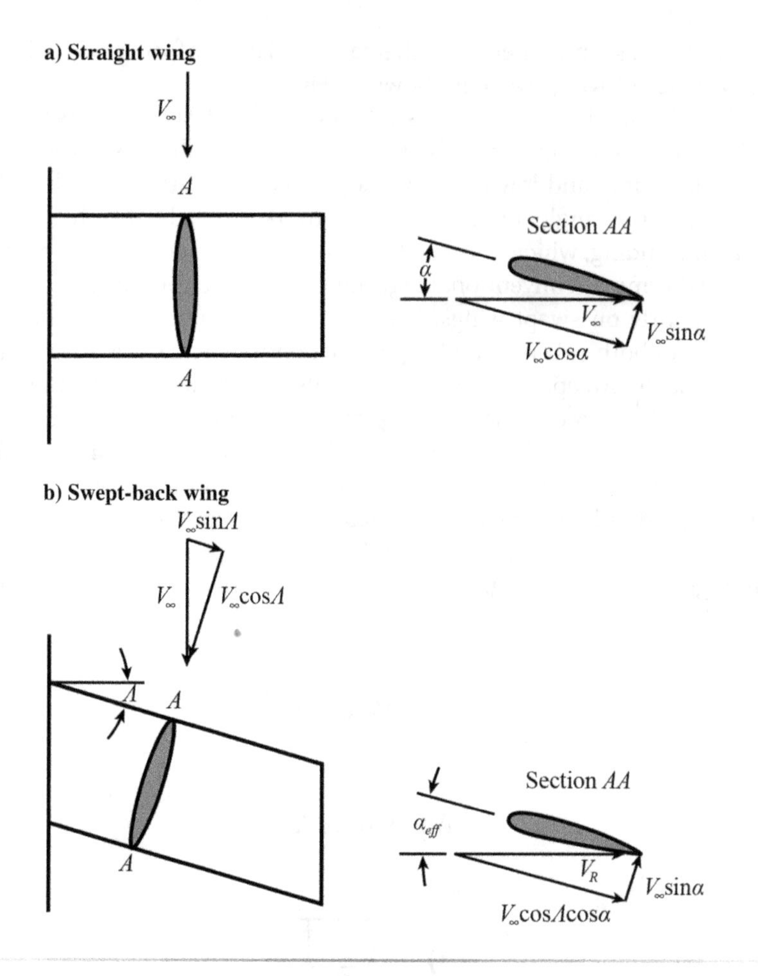

Fig. 1.52 Straight and swept wings at angle of attack.

Let a be the lift-curve slope of the wing. Then,

$$L = \left(\frac{1}{2}\rho V_\infty^2 \cos^2\Lambda\right) Sa(\alpha \sec\Lambda) \tag{1.67}$$

$$= \frac{1}{2}\rho V_\infty^2 \cos\Lambda Sa\alpha \tag{1.68}$$

According to the conventional definition of lift, we have

$$L = \frac{1}{2}\rho V_\infty^2 SC_{Ls} \tag{1.69}$$

Equating the two, we get

$$C_{Ls} = a\alpha\cos\Lambda \tag{1.70}$$

or

$$a_s = a \cos \Lambda \tag{1.71}$$

where

$$a_s = \left(\frac{\partial C_L}{\partial \alpha} \right)_s \tag{1.72}$$

From the consideration of delaying the adverse effects of compressibility to higher Mach numbers, one can either use a swept-back or a swept-forward wing. However, in many other ways, the two wings do not function in the same manner as discussed in the following section.

Swept-Back Wings

On a finite straight-rectangular wing, the root sections operate at relatively higher angles of attack compared to the tip sections. As a result, the stall on a straight-rectangular wing of finite aspect ratio originates at the root and progresses outboard as shown in Fig. 1.53a. As angle of attack increases, the stall spreads outward, and eventually the entire wing will

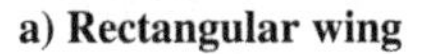

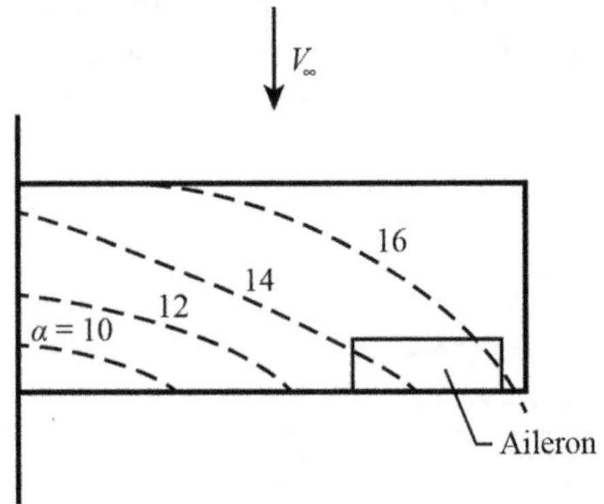

b) Swept-back wing

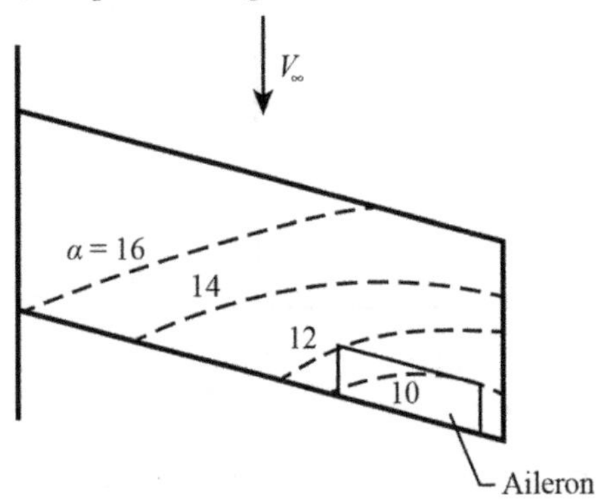

Fig. 1.53 Stall progression on rectangular and swept-back wings.

stall at some angle of attack. A finite wing is said to stall when any of the sections stall. Thus, at stall, the ailerons located outboard are still effective, and the aircraft has good lateral or roll control authority.

On the other hand, for the swept-back wing, the tip sections stall first as shown in Fig. 1.53b. To understand why this happens, let us refer to the schematic variation of the lift coefficient along the span as shown in Fig. 1.54. We observe that the tip sections on a swept-back wing develop relatively higher lift coefficients, which means that they are operating at higher angles of attack compared to the root sections. Thus, the stall originates from the tip sections. In addition, the existence of spanwise pressure gradient causes the wing tips to be loaded with thick boundary layers, which further aggravate the tip stall tendency of swept-back wings. As a result, an aircraft with swept-back wings will lose roll control at stall. As we shall discuss later in the text, a loss of roll control at stall makes an aircraft susceptible to problems of stability and control at high angles of attack.

To understand why a spanwise pressure gradient exists on a swept-back wing, consider the pressure distribution along chordwise sections I, II, and III as shown in Fig. 1.55. Now let us pick three points such as A_1, B_1, C_1 on section I as shown. Because point C_1 is closest to the leading edge compared to B_1 and A_1, the pressure at C_1 is the lowest of all the three points on section I. Similarly, the pressure at point B_1 is lower that at A_1. Thus, we have a spanwise pressure gradient along the section I. Similarly, we can show that spanwise pressure gradient exists along other chordwise sections such as II and III. Because of this spanwise pressure gradient, the inboard fluid particles

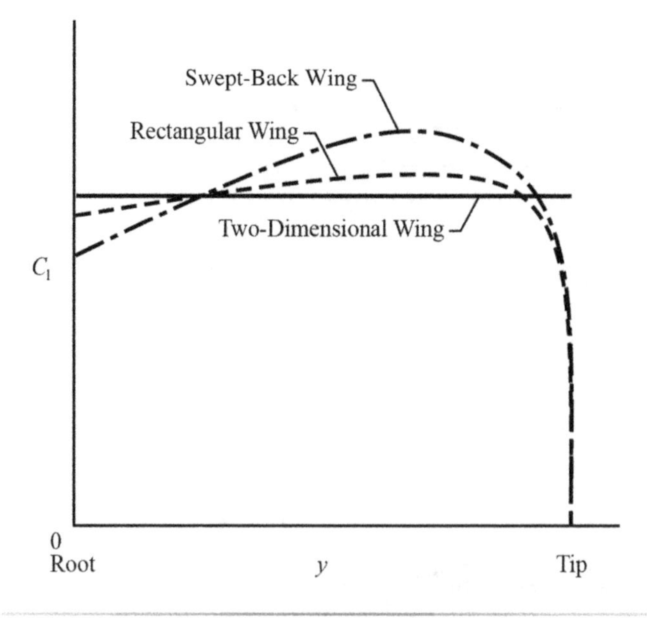

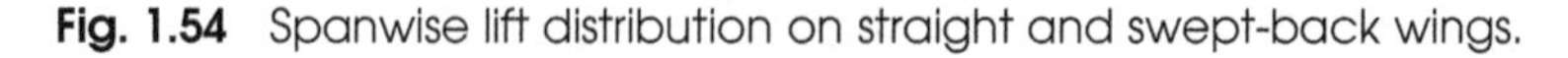

Fig. 1.54 Spanwise lift distribution on straight and swept-back wings.

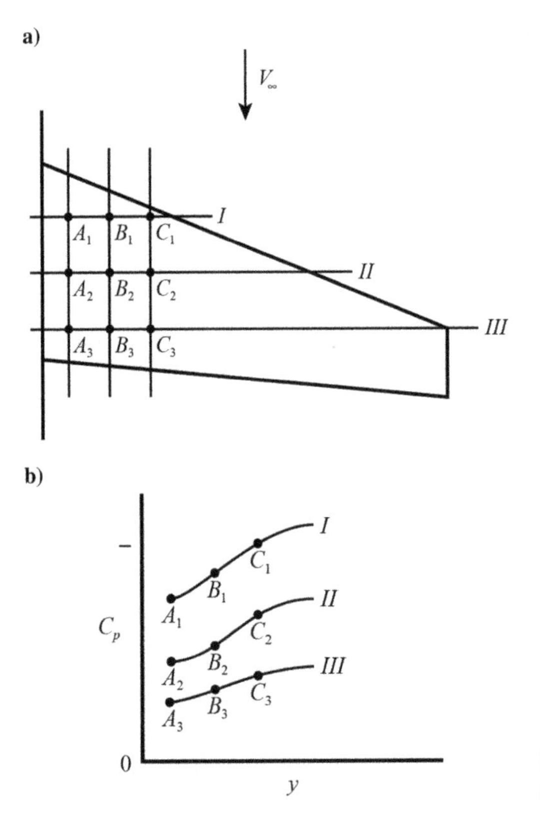

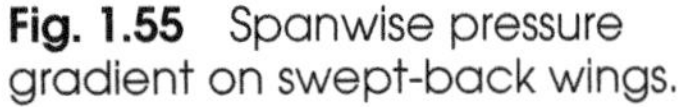

Fig. 1.55 Spanwise pressure gradient on swept-back wings.

will drift towards the tip. As a result, the wing tips will be loaded with thicker boundary layers that are more susceptible to flow separation.

Another consequence of the stall progression from tips to the root is the pitch-up tendency exhibited by the swept-back wings. This pitch-up occurs because tip regions that have the largest moment arm lose lift. The only contribution comes from unstalled sections closer to the root that have a smaller moment arm. As a result, the stabilizing contribution of the tip sections decreases and the swept wing experiences the pitch-up as shown schematically in Fig. 1.56.

One way of controlling the tip stall of swept-back wings is the application of boundary-layer fences or vortex generators [14]. Boundary-layer fences are obstacles positioned at various spanwise stations to prevent the spanwise flow within the boundary layer. The height of these fences is sufficiently small so that they do not disturb the external flow. Vortex generators are devices that generate stream-wise vorticity to energize the boundary layer and make it more resistant to flow separation. Vortex generators may be either submerged within the boundary layer or may protrude outside of it. The vortex generators that protrude outside the boundary layer may produce considerable skin friction, whereas the submerged type alleviate this problem because they are not exposed to the external airstream.

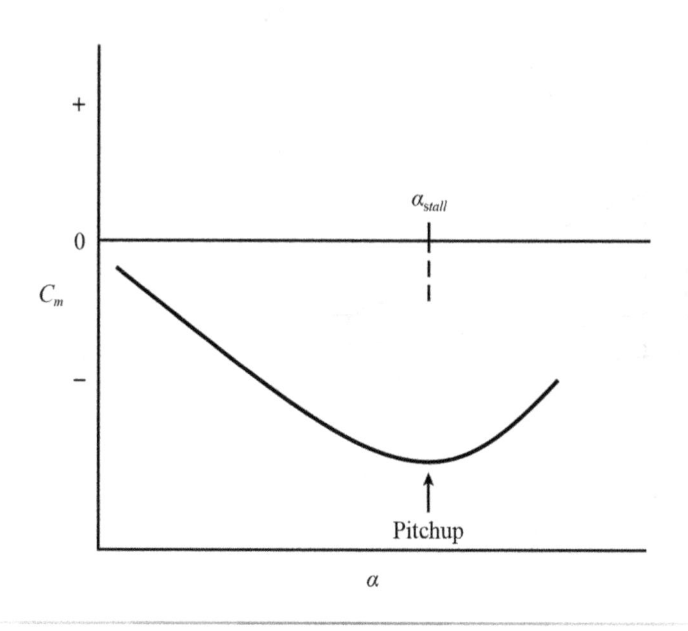

Fig. 1.56 Pitch-up of swept-back wings.

Delta Wings

In the late 1950s or the early 1960s, the use of sweep-back at supersonic speeds, wherein the wing leading edge was swept behind the Mach cone, provided an efficient method of reducing the wave drag. However, at low speeds, the increased sweep caused difficulties in maintaining attached flow even at moderate angles of attack. Flow separations occurred at quite modest values of angles of attack and tended to spread in an unpredictable way. This often caused serious stability and control problems. The most troublesome of these separations originated at the leading edge and rolled up into a vortex sheet. The point of origin of these leading-edge separations were difficult to tie down because the swept-back wing had a round leading edge. Designers tried various fixes to alleviate the problem but not with much success. This led aerodynamicists to believe that the requirements for high- and low-speed flights were in direct conflict.

The emergence of the thin slender, sharp-edged delta wing solved this problem [15, 16]. The flow over a sharp-edged delta wing differs significantly from that over a swept-back wing with round leading edge. The most important difference is that the flow separation occurs right at the sharp leading edges and, in this way, flow separation points are fixed for all angles of attack. With fixed flow separation points, the problems associated with unpredictable stall progression of the swept-back wings were eliminated. Along with this, many of the stability and control problems associated with the uncertain flow separation pattern of the swept-back wing also disappeared.

The separated flow on the lee side rolls up to form a spiral vortex over the lee side of the wing, as shown in Fig. 1.57a. The pressure in the vortex core is considerably lower and, as a result, substantial lift increment is obtained as shown in Fig. 1.57b. This incremental lift is called vortex lift and is associated with the large mass of air accelerated downward by the vortex sheet. This solved the problem of lift deficiency associated with swept-back wings. Thus, a new concept of controlled flow separation came into existence that represented a departure from the time-honored concept of attached flow. Under this new concept, a wing is designed for attached flow at supersonic cruise conditions and at low speeds the flow is allowed to separate at the

a) Delta wing at angle of attack

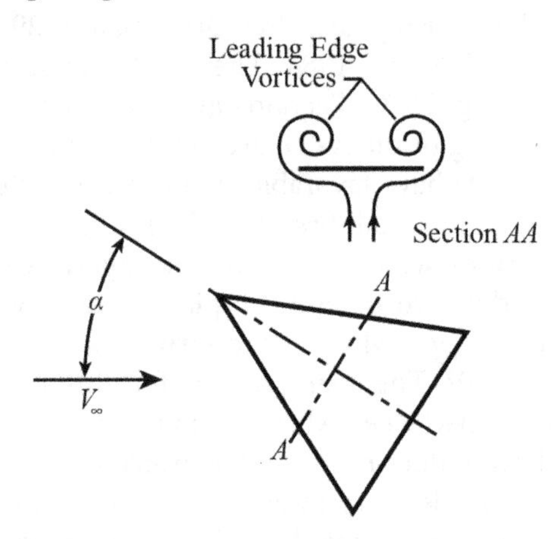

b) Variation of lift coefficient with angle of attack

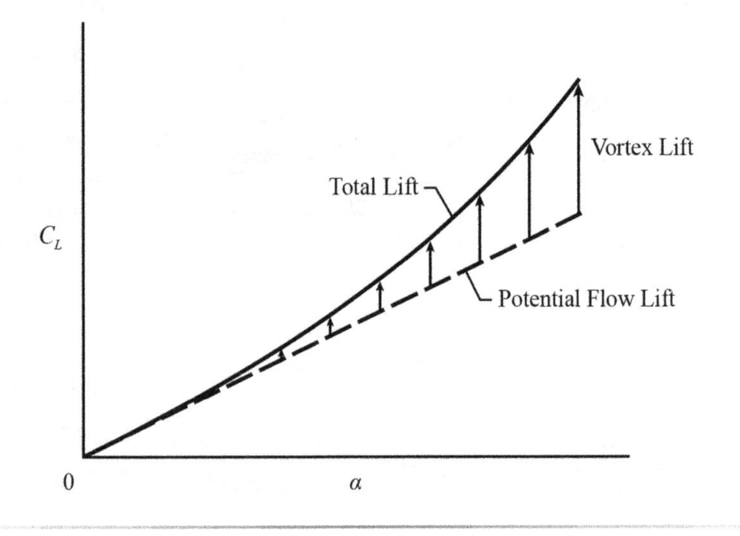

Fig. 1.57 Delta wing in low-speed flow at angle of attack.

leading edge and generate the beneficial vortex lift. We study more about slender delta wings in Chapter 8.

An example of the aircraft design based on the controlled flow separation concept is the Anglo–French Concorde, which was the first supersonic commercial aircraft with a slender ogee-delta wing planform [17].

Swept-Forward Wings

Historically, the first airplane to use the forward-swept wing [18] was the Junkers Ju 287, which was a jet-powered German bomber aircraft during World War II. The wing had a forward sweep of 25 deg. Then in 1961, Messerschmitt of Germany produced the HFB 320 Hansa aircraft, which also had forward-swept wings. After these brief appearances, the forward-swept wing disappeared from the horizon until the Grumman X-29 research aircraft made its first flight on Dec. 14, 1984. The X-29 aircraft was designated as the Forward Swept Wing (FSW) Technology Demonstrator aircraft. A schematic three-view drawing of the X-29 aircraft is presented in Fig. 1.58 [19].

Forward-swept wings have favorable stall characteristics. Unlike swept-back wings, the root regions stall first, and the stall progresses from root to tips. Therefore, forward-swept wings do not experience the loss of roll control at stall and thus are inherently spin resistant (we study spinning motion in Chapter 7). Then why did the forward-swept wing vanish after the German Hansa 320? The main reason for this is the fact that the forward-swept wing experiences what is known as aeroelastic divergence, and the swept-back wing does not face this problem.

Because wing tips of a forward-swept wing are effective in producing lift, the swept-forward wing experiences larger twisting moments, which tend to

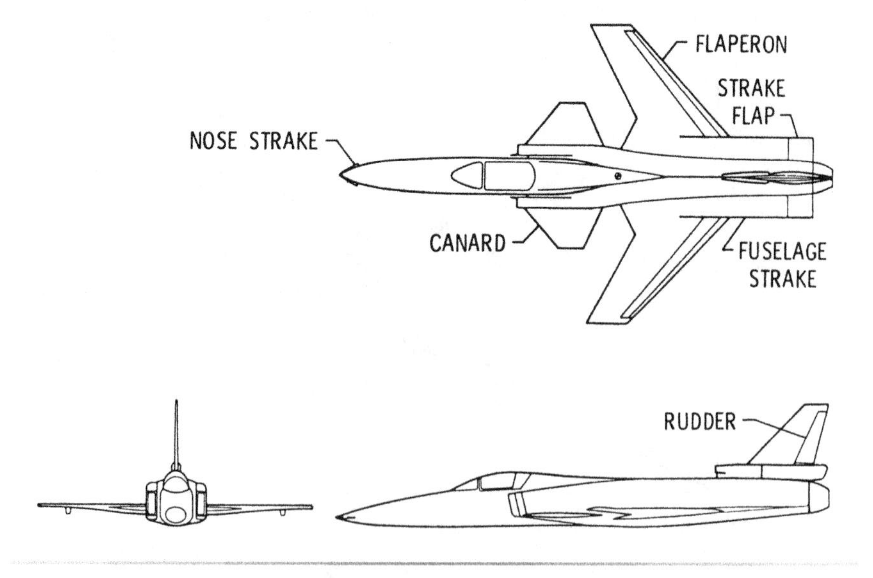

Fig. 1.58 X-29 forward-swept wing airplane (18).

twist the wing sections in a direction that increases their angle of attack further, hence their lift. As a result, the twisting moment increases further. The lift and twisting moments are proportional to the square of the flight speed. Therefore, the problem becomes more severe as flight speed increases. This phenomenon is known as aeroelastic divergence. When the twisting moments exceed the structural load limit, the wing fails. The traditional aluminum structure could not resist such aeroelastic deformations of the forward-swept wing. What made a difference in the X-29 aircraft was the use of an aeroelastically tailored composite wing. Composite materials forming the wings of X-29 are so designed that the wing actually twists in the opposite direction, i.e., leading edge down under the action of twisting loads, thereby reducing the angle of attack and effectively preventing aeroelastic divergence. This advance in aircraft materials paved the way for the resurgence of interest in forward-swept wing technology.

Forward-swept wings also experience pitch-up like swept-back wings, and the pitch-up tendency is more severe. Because of this problem, the X-29 is statically unstable and has about 35% negative static margin [23]. (We study the concept of static stability and stability margins in Chapter 3.) Therefore, this inherent instability requires the use of an automatic flight control system that has to make the aircraft safely flyable. The X-29 aircraft is stabilized by a highly augmented triplex digital-analog, fly-by-wire flight control system. The flight control system has three modes: normal (primary) mode, digital reversion mode, and analog reversion mode [23].

Oblique Wings

As said earlier, if we need good low-speed characteristics combined with good high-speed characteristics such as low-wave drag, then we have to use a variable-sweep wing. However, one of the main disadvantages of the variable-sweep wing is the large weight penalty associated with the additional structure required to handle structural loads. With oblique wings [20], a straight wing is rotated in flight so that, at low speeds, it is essentially at zero sweep and, at high speeds, one wing is swept forward and the other is swept back as shown in Fig. 1.59. An oblique wing is continuous from tip to tip and is attached to the fuselage at one point only—the pivot. The bending moment on one half of the wing is reacted by the other half. As a result, the pivot carries only the lift load; hence, the wing structure is much lighter.

Another significant advantage of the oblique wing is the absence of aerodynamic center shift at transonic speeds. On a conventional variable geometry wing at transonic speeds, the aerodynamic center moves aft as the wing sweep increases to reduce the wave drag. This creates more nose-down moment and requires a large horizontal tail surface to trim the aircraft. This increases the trim drag and hence affects the performance. With the oblique wing, this problem is considerably alleviated as one half of the wing sweeps aft and the other half sweeps forward and the overall

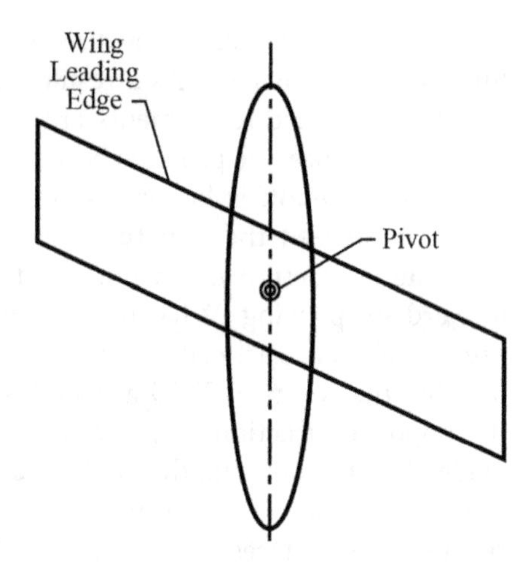

Fig. 1.59 Oblique wing.

aerodynamic center hardly moves. Therefore, a large horizontal tail is not needed to trim the aircraft in transonic speeds.

The benefits of the oblique wing were demonstrated in flight using the F-8 oblique wing research aircraft [24] during the mid-1980s.

1.12 Area Rule

Area ruling [21, 22] is a systematic method of minimizing the transonic/supersonic wave drag of airplane configurations. Fundamental to this method is the assumption that, at Mach numbers close to unity and at large distances from the body, disturbances and shock waves are independent of the arrangement of the components and are only functions of the longitudinal variation of the cross-sectional area. In other words, the wave drags of a given wing–body and an equivalent body having an identical longitudinal cross-sectional area variation (Fig. 1.60) are essentially the same.

For most airplane configurations, adding cross-sectional areas of wing to that of the fuselage results in a bump in the overall area distribution as shown in Fig. 1.61b. To obtain the minimum wave drag, the overall distribution

Fig. 1.60 Equivalent body for area rule (21, 22).

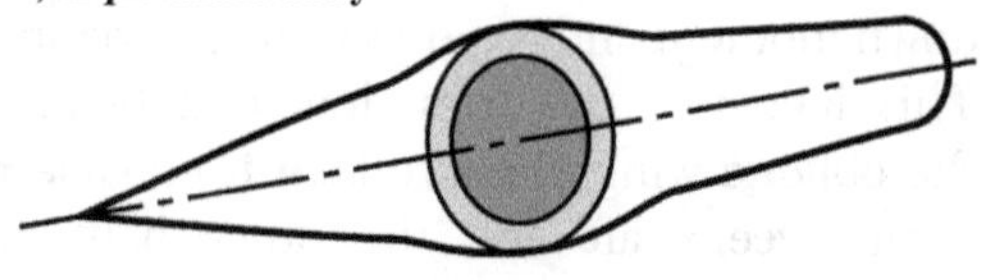

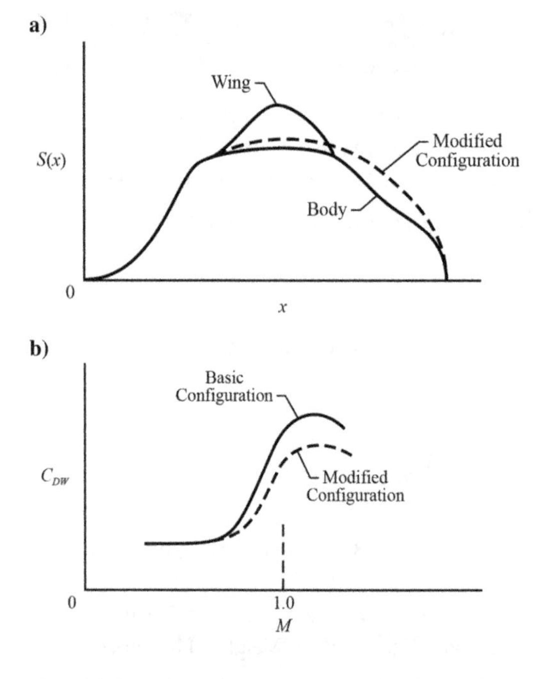

Fig. 1.61 Area-rule concept.

should be that for a smooth body with minimum wave drag. The most obvious way to achieve this is to remove the cross-sectional area of the wing from the fuselage in that region where the fuselage joins the wing. This results in a modified shape that looks like a coke bottle. The cross-sectional areas of other components like engines, nacelles, and tail surfaces can be treated in a similar manner. The wave drag of the modified body is lower than that of the basic configuration as shown in Fig. 1.61b.

For supersonic Mach numbers, cross-sectional areas are to be obtained by taking planes inclined at an angle $\mu = \sin^{-1}(1/M_\infty)$ to the axis of the given body.

The application of area rule resulted in significant transonic/supersonic drag reductions of several aircraft. One particular example was the Convair F-102 interceptor aircraft [21, 22]. Wind-tunnel tests on the original design indicated that the transonic drag was so high that the aircraft would not be able to fly past Mach 1. This prediction was found to be true when the prototype aircraft flew in 1953. The concept of area rule was then applied, and the fuselage of the F-102 was redesigned. In late 1954, the modified F-102 flew past Mach 1 successfully. After this flight validation, several aircraft were redesigned on the basis of the area rule concept. These include the Convair F-106, the Convair B-58, and the Vought F8V. Some recent examples are the Air Force B-1 and the Boeing 747 aircraft.

It should be noted that with the application of the area rule concept, an aircraft configuration can be derived to give a minimum wave drag only at one flight Mach number. At any other flight Mach number, the benefits of area rule may be lost and the wave drag may be substantially higher.

Example 1.1

A flying wing with an area of 27.75 m^2 has a NACA 2412 airfoil section. The weight of the flying wing is 2270.663 kg and the aspect ratio is 6. For level flight at an altitude of 1500 m ($\sigma = \frac{\rho}{\rho_0} = 0.864$) and a velocity of 160 km/h, determine angle of attack, induced-drag coefficient, lift-to-drag ratio, and drag. Assume $e = 0.95$.

Solution:

For the NACA 2412 airfoil [3], we have $a_o = 0.104/\text{deg}$ and $C_{do} = 0.0060$. We have $V = 160\ \text{km/h} = 44.44\ \text{m/s}$.

From Eq. (1.52), we have

$$a = \frac{a_o}{1 + \dfrac{57.3 a_o}{\pi A e}}$$

Here, we have to multiply the value of a_o by 57.3 in the denominator because we are given a_o/deg.

Substituting $a_o = 0.104$, $A = 6$, and $e = 0.95$, we get $a = 0.079/\text{deg}$.

As we discuss in Chapter 2, for level flight, Lift = Weight. Therefore

$$\begin{aligned} C_L &= \frac{\text{Lift}}{\frac{1}{2}\rho V^2 S} \\ &= \frac{2270.663 * 9.81}{\frac{1}{2} 1.225 * 0.864 * 44.44^2 * 27.75} \\ &= 0.768 \end{aligned}$$

Then,

$$\begin{aligned} \alpha &= \frac{C_L}{a} \\ &= \frac{0.768}{0.079} \\ &= 9.7203\ \text{deg} \end{aligned}$$

and

$$\begin{aligned} C_{Di} &= k C_L^2 \\ &= \left(\frac{1}{\pi A e}\right) C_L^2 \\ &= \frac{0.768^2}{\pi 6 * 0.95} \\ &= 0.03294 \end{aligned}$$

(*Continued*)

Example 1.1 (*Continued*)

$$\begin{aligned} C_D &= C_{DO} + C_{Di} \\ &= 0.006 + 0.03294 \\ &= 0.03894 \end{aligned}$$

The drag is given by

$$\begin{aligned} D &= \frac{1}{2}\rho V^2 S C_D \\ &= \frac{1}{2}(1.225 * 0.864)44.44^2 * 27.75 * 0.03294 \\ &= 1129.3456\,\text{N} \end{aligned}$$

With this, we get the lift-to-drag ratio $\frac{C_L}{C_D} = \frac{L}{D} = 19.7227$.

Example 1.2

A two-dimensional wing was tested in a wind tunnel, and the pitching-moment coefficient about the leading edge at zero-lift was found to be equal to -0.02. At $\alpha = 8$ deg, $C_l = 0.7$, $C_d = 0.04$, and $C_{m,le} = -0.20$, determine the location of the aerodynamic center.

Solution:

From Eq. (1.37),

$$C_m = C_{\text{mac}} - \bar{x}_{ac} C_l$$

We have $C_{\text{mac}} = C_{mo} = -0.02$, and $C_l = 0.7$, $C_m = -0.2$ at $\alpha = 8$ deg. Then,

$$\begin{aligned} \bar{x}_{ac} &= \frac{C_{mo} - C_m}{C_l} \\ &= \frac{-0.02 - (-0.20)}{0.7} \\ &= 0.2571 \end{aligned}$$

Thus, the aerodynamic center is located at 25.71% chord from the leading-edge.

Example 1.3

Determine the mean aerodynamic chord and its spanwise location for an aircraft wing having a root chord of 6.1 m, tip chord of 3.28 m, leading-edge sweep of 30 deg, and a semispan of 15.25 m.

Solution:

From Eqs. (1.25) and (1.26), we have

$$\bar{c} = \frac{2}{3}c_r\left(\frac{1+\lambda+\lambda^2}{1+\lambda}\right)$$

$$y_{\text{mac}} = \frac{b}{6}\left(\frac{1+2\lambda}{1+\lambda}\right)$$

We have $c_r = 6.1$, $b = 2 * 15.25 = 30.5$, and the taper ratio $\lambda = 3.28/6.1 = 0.5377$. Substituting, we get

$$\bar{c} = 4.8313\,\text{m}$$

$$y_{\text{mac}} = 6.8608\,\text{m}$$

Example 1.4

For a straight rectangular wing having a chord of 3 m and a span of 15 m, determine the skin-friction coefficient and skin-friction drag at a velocity of 50 m/s, assuming the boundary layer to be laminar and the boundary layer to be fully turbulent. Assume $\rho = 1.225\ \text{kg/m}^3$ and $\nu = 1.4607 \times 10^{-5}\ \text{m}^2/\text{s}$.

Solution:

The Reynolds number is given by

$$\begin{aligned} Re &= \frac{VL}{\nu} \\ &= \frac{50 * 3}{1.4067 \times 10^{-5}} \\ &= 1.0663 \times 10^{7} \end{aligned}$$

Now let us consider the two cases as follows.

Laminar flow: from Eq. (1.13), we have

$$\begin{aligned} C_f &= \frac{1.328}{\sqrt{Re}} \\ &= \frac{1.328}{\sqrt{1.0663 \times 10^{7}}} \\ &= 0.0004 \end{aligned}$$

(Continued)

Example 1.4 (*Continued*)

Assuming that the flow is fully attached on the top and bottom sides of the airfoil, we note that the wetted area S is equal to twice the wing planform area. Thus, $S = 2 * 3 * 15 = 90\,\text{m}^2$.

$$\begin{aligned} D_f &= \frac{1}{2}\rho V^2 S C_f \\ &= \frac{1}{2} * 1.225 * 50^2 * 90 * 0.0004 \\ &= 55.1250\,\text{N} \end{aligned}$$

Turbulent flow: from Eq. (1.14), we have

$$\begin{aligned} C_f &= \frac{0.455}{(\log_{10} Re)^{2.58}} \\ &= \frac{0.455}{(\log_{10} 1.0663 \times 10^7)^{2.58}} \\ &= 0.0030 \end{aligned}$$

$$\begin{aligned} D_f &= \frac{1}{2}\rho V^2 S C_f \\ &= \frac{1}{2} * 1.225 * 50^2 * 90 * 0.0030 \\ &= 413.4375\,\text{N} \end{aligned}$$

Example 1.5

For a double wedge airfoil of 2 m chord and 4% thickness ratio, determine the lift and wave-drag coefficients per unit span at a Mach number of 2.0 and $\alpha = 5$ deg.

Solution:

We have

$$\alpha = 5\,\text{deg} = \frac{5}{57.3} = 0.08726\,\text{rad}$$

$$\begin{aligned} C_l &= \frac{4\alpha}{\sqrt{M^2 - 1}} \\ &= \frac{4 * 0.08726}{\sqrt{2^2 - 1}} \\ &= 0.2015 \end{aligned}$$

$$\begin{aligned} C_{dw} &= \frac{4}{\sqrt{M^2 - 1}}\left[\alpha^2 + \left(\frac{t}{c}\right)^2\right] \\ &= 0.02128 \end{aligned}$$

Example 1.6

For a two-dimensional rectangular wing having a NACA 23012 airfoil, the critical Mach number is 0.672 and the lift-curve slope is 0.104/deg. What should be the leading-edge sweep angle to have a critical Mach number of 1.5? What is the lift-curve slope of this swept wing?

Solution:

We have

$$M_{cr,s} = \frac{M_{cr}}{\cos\Lambda}$$

Therefore,

$$\begin{aligned}\cos\Lambda &= \frac{M_{cr}}{M_{cr,s}} \\ &= \frac{0.672}{1.5} \\ &= 0.448 \\ \Lambda &= 63.3845\ \text{deg}\end{aligned}$$

The lift-curve slope is given by

$$\begin{aligned}a_{os} &= a\cos\Lambda \\ &= 0.104 * 0.448 \\ &= 0.0466/\text{deg}\end{aligned}$$

1.13 Summary

In this chapter, we reviewed the basic principles of aerodynamics applicable to aircraft. We studied elements of the fluid flow over wings and bodies at subsonic, transonic, and supersonic speeds and understood how lift, drag, and pitching moments vary with angle of attack and Mach number. This aerodynamic background should be useful in the study of performance, stability, and control of the aircraft, which will be discussed in the next chapters of this text.

References

[1] Schlichting, H., *Boundary Layer Theory*, 6th ed., McGraw-Hill, New York, 1968.
[2] Abbott, I. H., and Von Doenhoff, A. E., *Theory of Wing Sections*, Dover, New York, 1959.
[3] Perkins, C. D., and Hage, R. E., *Airplane Performance, Stability and Control*, 10th ed., Wiley, New York, 1965.
[4] Jacobs, E. W. and Sherman, A., "Airfoil Section Characteristics as Affected by Variations of the Reynolds Number," Rept. No. 586, 1936.

[5] Cooke, J. C., and Brebner, C. G., "The Nature of Separation and its Prevention by Geometric Design in a Wholly Subsonic Flow," *Boundary Layer and Flow Control,* Vol. I, edited by G. V., Lachmann, Pergamon, New York, 1961, pp. 145–185.
[6] McCormick, B.W., *Aerodynamics, Aeronautics and Flight Mechanics,* Wiley, New York, 1979.
[7] Kuethe, A. M., and Schetzer, J. D., *Foundations of Aerodynamics,* 2nd ed., Wiley, New York, 1961.
[8] Multhopp, H., "Methods for Calculating the Lift Distribution of Wings (Subsonic Lifting Surface Theory)," ARC R and M 2884, 1950(U).
[9] Falkner, V. M., "The Calculation of the Aerodynamic Loading on Surface of Any Shape," ARC R and M 1910, 1943(U).
[10] De Young, J., and Harper, C. W., "Theoretical Symmetric Spanwise Loading at Subsonic Speeds for Wings Having Arbitrary Planform," NACA TR 921, 1948(U).
[11] Neumark, S., "Critical Mach Numbers for Thin Untapered Swept Wings at Zero Incidence," ARC R and M 2821, 1954(U).
[12] Whitcomb, R. E., "A Design Approach and Selected Wind Tunnel Results at High Subsonic Speed for Wing-Tip Mounted Winglets," NASA TND-8260, 1976.
[13] Whitcomb, R. E., "Review of NASA Supercritical Airfoils," *Ninth Congress of ICAS,* Haifa, Israel, 1974.
[14] Rao, D. M., and Kariya, T. T., "Boundary-Layer Submerged Vortex Generators for Separation Control—An Exploratory Study," *Proceedings of 1st National Fluid Dynamic Conference,* Cincinnatti, OH, 1988, pp. 839–846.
[15] Maltby, R. L., "The Development of the Slender Delta Concept," *Aircraft Engineering,* March 1968, pp. 12–17.
[16] Polhamus, E. C., "Vortex Lift Research, Early Contributions and Some Current Challenges," NASA CP-2416, Vol. 1, 1992.
[17] Birtles, P., *Concorde,* Ian Allen, London, 1984.
[18] Moore, M., and Frei, D., "X-29 Forward-Swept Wing Aerodynamic Overview," AIAA Paper 83-1834, 1983.
[19] Murri, D. G., Nguyen, L. T., and Grafton, S. B., "Wind Tunnel Free-Flight Investigations of a Model of Forward-Swept Wing Fighter Configuration," NASA TP 2230, Feb. 1984.
[20] Jones, R. T., "Flying-Wing SST for the Pacific," *Aerospace America,* Nov. 1986, pp. 32, 33.
[21] Whitcomb, R. E., "Research Methods for Reducing Aerodynamic Drag at Transonic Speeds," *Inagural Eastman Jacobs Lecture,* NASA Langley Research Center, Hampton, VA, 14 Nov. 1994.
[22] Whitcomb, R. E., and Silver, J. R. Jr., "A Supersonic Area Rule and an Application to the Design of a Wing-Body Combination with High Lift-to-Drag Ratio," NASA TR R-72, 1960.
[23] Jackson, B. E., "Assurance of the X-29A Advanced Technology Demonstrator Flight Control Software," AIAA Paper 83-2724.
[24] Enns, D. F., Bugajski, D., and Klepl, M. J., "Flight Control for the F-8 Oblique Wing Research Aircraft," *Proceedings of the American Control Conference,* Minneapolis, MN, June 10–12, 1987, pp. 81–86.

Problems

1.1 A flying wing weighing 20,000 N has a NACA 65_3-418 airfoil section, area of 30 m^2, and an aspect ratio 5.0. Determine the angle of attack for a sea-level flight at 60 m/s. What is the lift-to-drag ratio? Assume $a_o = 0.106/\text{deg}$, $C_{do} = 0.0043$, and $e = 0.92$.

1.2 For a certain airfoil section, the pitching-moment coefficient about 0.33 chord behind the leading edge varies with C_L as shown in Table P1.2.

Table P1.2

C_l	C_m
0.20	−0.02
0.40	0
0.60	0.02
0.80	0.04

Determine the locations of the aerodynamic center and the center of pressure for $C_l = 0.5$.

1.3 For an aircraft wing with a leading-edge sweep of 45 deg, root chord of 5 m, tip chord of 2 m, and semispan of 10 m, determine the aspect ratio, mean aerodynamic chord, and spanwise location of the mean aerodynamic chord.

1.4 Estimate the skin-friction coefficient for a flat plate wing of 4 m chord and 10 m span exposed to an airstream of 75 m/s. Assume that the boundary-layer transition occurs at 50% chord and $\nu = 1.5 \times 10^{-5}\ \text{m}^2/\text{s}$.

1.5 Determine the wave-drag coefficient of a double wedge airfoil of 6% thickness ratio held at an angle of attack of 4 deg and a Mach number of 3.

1.6 The critical Mach number of a two-dimensional rectangular wing is 0.75. What will be the critical Mach number of this wing if the leading edge is swept back at 45 deg?

Chapter 2 Aircraft Performance

2.1 Introduction

An airplane is a flying machine. Like any other machine, it is judged by its performance. Some of the questions that usually come to mind are *How fast can it fly? How high can it fly? How fast and how steep can it climb? How far can it go with a tank-load of fuel? What length of runway does it need for takeoff and landing? How sharp and how fast can it turn?* Answers to these and many other questions form the subject matter of aircraft performance.

Performance characteristics depend on the weight of the airplane, aerodynamic characteristics of the airframe, and the thrust or the power developed by the powerplant. For a given airplane configuration, aerodynamic characteristics depend on angle of attack/sideslip, Mach number, and Reynolds number. The thrust/power characteristics of the powerplant depend on altitude, flight velocity, and engine operating conditions. Therefore, in general, it is not possible to analytically estimate airplane performance considering the arbitrary variations of aerodynamic and propulsive characteristics. Prior to the arrival of modern digital computers, it was common practice to use graphical methods for performance evaluations. Using high-speed computers, these calculations, which once took several hours or days, can be performed now in a matter of a few minutes with much more precision and accuracy. However, we are not going to discuss these computational methods of performance estimation. Instead, we will introduce some simplifying assumptions so that performance calculations become amenable to the methods of ordinary calculus.

We assume that the aerodynamic forces acting on the airplane are given in the coefficient form as follows:

$$C_L = a\alpha \tag{2.1}$$

$$C_D = C_{DO} + kC_L^2 \tag{2.2}$$

Here, C_L is the lift coefficient, a is the lift-curve slope, C_D is the drag coefficient, C_{DO} is the zero-lift drag coefficient, and kC_L^2 is the induced-drag term. Usually, C_{Di} is used to denote the induced-drag coefficient so that we have

$$C_{Di} = kC_L^2 \tag{2.3}$$

The variation of the drag coefficient as given by Eq. (2.2) is often called the parabolic drag polar of the airplane.

At low speeds, C_{DO} includes skin-friction and pressure drag of all wetted (exposed to airflow) components of the airplane. At high speeds, another form of zero-lift drag arises, which is known as the wave drag. Therefore, for high subsonic and supersonic airplanes, the term C_{DO} will also include the wave drag. The induced-drag term C_{Di} is primarily the drag due to lift. However, it also includes those parts of the skin-friction and pressure drag that vary with angle of attack. The induced-drag parameter k in Eq. (2.2) is given by

$$k = \frac{1}{\pi e A} \tag{2.4}$$

For finite wings, e is the wing planform efficiency factor. For elliptical wing planform, $e = 1$. In other words, for a given aspect ratio A, the induced drag is a minimum for an elliptical wing. However, for the complete airplane, the definition of e is modified to include the variation of skin-friction and pressure drag coefficients with angle of attack and is called the Oswald's efficiency factor.

Generally, two types of powerplants are commonly used on modern airplanes, 1) the piston engine–propeller combination or the piston-prop and 2) the turbojet. Piston-props are widely used on light general aviation airplanes, whereas jet engines are used for commercial transport and military aircraft. In view of this, the performance analyses presented in this text will be based on these two types of powerplants. We will be referring to an airplane with a piston-prop powerplant as a propeller airplane and that with a turbojet powerplant as a jet airplane.

The propulsive characteristics of the propeller airplane are normally given in terms of the power developed by the reciprocating (piston) engine and the propulsive efficiency of the engine–propeller combination. The power available for propulsion is the product of the power developed by the reciprocating (piston) engine and the propulsive efficiency of the engine–propeller combination. Because of this, it is common practice to analyze the performance problems of a propeller airplane in terms of the power available and the power required. The power required is the power necessary to overcome the aerodynamic drag of the airplane. In this text, we will use kW as the unit of power and denote the power developed by the engine in kW using the notation P(kW).

The propulsive characteristics of the jet airplane are normally specified in terms of the thrust produced by the jet engine, which is referred to as the thrust available. Hence, the performance characteristics of the jet airplane are usually discussed in terms of the thrust available and the thrust required. The thrust required is the thrust necessary to balance the aerodynamic drag of the airplane.

The specific fuel consumption of a propeller airplane is the weight of fuel consumed per unit power per unit time. In this text, we will use the units of N/kWh to specify the specific fuel consumption of propeller airplanes. For jet aircraft, specific fuel consumption is the amount of fuel consumed per unit thrust per unit time and will have the units of N/Nh.

For a given altitude, the power developed by a piston (reciprocating) engine is virtually constant with flight velocity. The only variation of power arises due to the variation of the ram pressure in the intake manifold with flight velocity. However, the power developed by a reciprocating engine decreases with an increase in altitude because of a fall in air density. With turbosupercharging, this decrease in engine power can be minimized up to a certain altitude. The propulsive efficiency, in general, varies with flight velocity. However, if the aircraft is equipped with a variable-pitch, constant-speed propeller, the propulsive efficiency can be assumed constant over the design operating range. For a given altitude, the thrust developed by a jet engine is roughly constant with flight velocity. However, the thrust produced by the jet engine decreases with altitude because of a fall in air density. These variations are schematically shown in Fig. 2.1.

In the performance analyses presented in this text, we assume that, at a given altitude, the power developed by the reciprocating engine and the propulsive efficiency are constants so that the power available is independent of

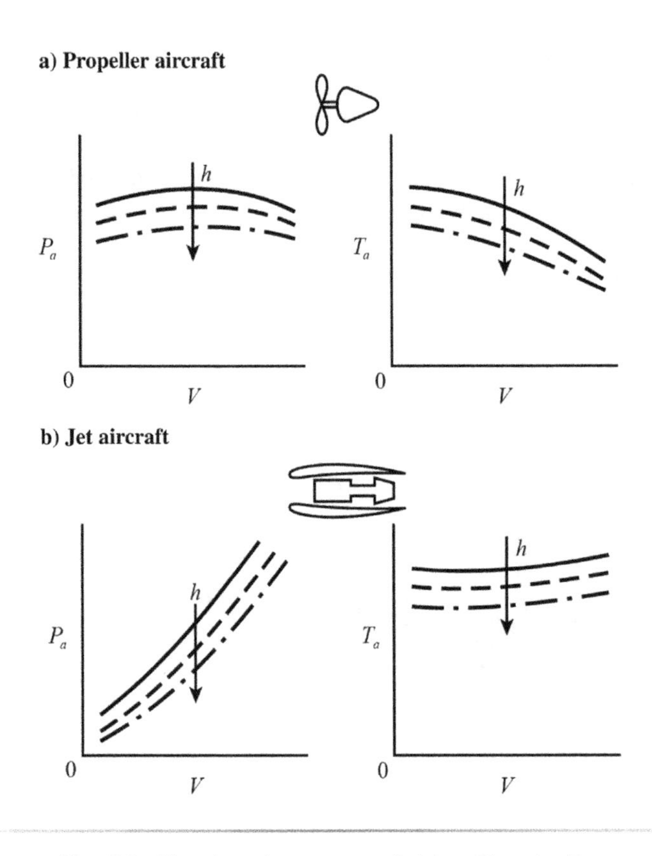

Fig. 2.1 Thrust and power available with velocity.

flight velocity. Similarly, we assume that the thrust available for jet aircraft is independent of flight velocity. However, both power available and thrust available will be assumed to vary with altitude as specified in each case.

Basically there are two types of problems in aircraft performance: 1) point performance problems and 2) path performance problems. Point performance problems are concerned with the investigation of the local characteristics along the flight path, whereas the path performance problems deal with the entire flight path, i.e., with the overall behavior of the airplane all along the flight path starting from a given initial condition to the specified final condition. In this chapter, we will study the basic methods of estimating point performance characteristics of the airplane in categories of flight such as the power-off glide, level flight, climbing flight, range and endurance, takeoff, landing, and turning flights. We will also determine the flight variables that optimize the point performance of the airplane in such flights. We will not be dealing with path performance problems that are essentially problems in variational calculus. Interested readers may refer elsewhere [1] for more information on path performance problems.

2.2 Equations of Motion for Flight in Vertical Plane

Let us consider an airplane whose flight path is contained in a vertical plane as shown in Fig. 2.2. Let V be the flight velocity that is directed along the tangent to the flight path. Let the tangent to the flight path at a given instant of time make an angle γ with respect to the local horizontal. The angle γ is usually called the flight path angle. Let θ be the inclination of the airplane reference line or the zero-lift line with respect to the local

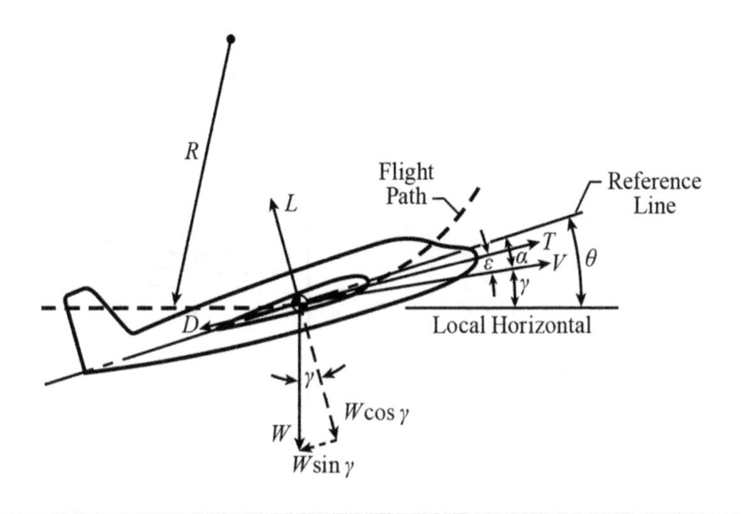

Fig. 2.2 Forces acting on an airplane in flight in a vertical plane.

horizontal so that the angle of attack $\alpha = \theta - \gamma$. Let R be the radius of curvature of the flight path in the vertical plane.

The external forces acting on the airplane are lift L acting normal to the flight velocity, the drag D parallel to the flight path and in opposite direction to the flight velocity V, the thrust T in the forward direction making an angle ϵ with respect to the flight path, and the weight W in the direction of gravity directed towards the center of the Earth.

Resolving the forces along and normal to the flight path, we have

$$T \cos \epsilon - D - W \sin \gamma = \frac{W}{g} \frac{\mathrm{d}V}{\mathrm{d}t} \tag{2.5}$$

$$T \sin \epsilon + L - W \cos \gamma = \frac{W}{g} \frac{V^2}{R} \tag{2.6}$$

Equations (2.5) and (2.6) describe the accelerated motion of the airplane in a vertical plane. Generally, the thrust inclination ϵ is very small. Therefore, it is usual in performance analyses to assume that the thrust is aligned with the flight path ($\epsilon = 0$).

For the special case of an airplane whose flight path is a straight line in the vertical plane and whose flight velocity is constant, acceleration terms on the right-hand sides of Eqs. (2.5) and (2.6) vanish. The performance of an airplane based on this assumption is called static performance. Examples of static performance are steady level flight, steady climb, range, and endurance in constant-velocity cruise. Performance problems that involve acceleration terms are accelerated climbs, takeoff, and landing.

Assuming that the thrust vector is aligned with the flight path, equations for static performance are given by

$$T - D - W \sin \gamma = 0 \tag{2.7}$$

$$L - W \cos \gamma = 0 \tag{2.8}$$

In addition to these equations of motion, we also need the following kinematic relations to calculate distances with respect to the ground:

$$\dot{x} = V \cos \gamma \tag{2.9}$$

$$\dot{h} = V \sin \gamma \tag{2.10}$$

where x and h are the horizontal and vertical distances measured with respect to the origin of a suitable coordinate system fixed on the ground. The "·" over a symbol denotes differentiation with respect to time. Thus, $\dot{x} = \mathrm{d}x/\mathrm{d}t$ is the horizontal velocity with respect to the ground, and $\dot{h} = \mathrm{d}h/\mathrm{d}t$ is the rate of increase of altitude or the rate of climb. In calm air, the velocity with respect to the ground and that with respect to air are equal. In the presence of the wind, $\dot{x} = V \mp V_w$, where V_w is the wind velocity and "–" applies to the headwind and "+" for the tailwind.

Let us introduce some nondimensional parameters so that we can express the previous equations in a more compact form. Let

$$E = \frac{C_L}{C_D} \tag{2.11}$$

$$n = \frac{L}{W} \tag{2.12}$$

$$z = \frac{TE_m}{W} \tag{2.13}$$

$$u = \frac{V}{V_R} \tag{2.14}$$

$$V_R = \sqrt{\frac{2W}{\rho S}}\sqrt[4]{\frac{k}{C_{DO}}} \tag{2.15}$$

Here, E is known as the lift-to-drag ratio and the maximum value of E is denoted by E_m. The parameter n, which is the ratio of lift to weight, is called the load factor. The parameter z is the nondimensional thrust, and u is the nondimensional flight velocity. We will show later that the reference velocity V_R happens to be the flight velocity in level flight when the drag is minimum. The lift-to-drag ratio E is a measure of the aerodynamic efficiency of the aircraft. Using the drag polar as in Eq. (2.2), we can determine E_m, the maximum value of aerodynamic efficiency as follows:

$$\frac{C_D}{C_L} = \frac{C_{DO} + kC_L^2}{C_L} \tag{2.16}$$

Then,

$$\begin{aligned} \frac{\mathrm{d}}{\mathrm{d}C_L}\left(\frac{C_D}{C_L}\right) &= \frac{-C_{DO} + kC_L^2}{C_L^2} \\ &= 0 \end{aligned} \tag{2.17}$$

which gives

$$C_L = \sqrt{\frac{C_{DO}}{k}} \tag{2.18}$$

$$\left(\frac{C_D}{C_L}\right)_{\min} = 2\sqrt{kC_{DO}} \tag{2.19}$$

Inverting Eq. (2.19) we obtain

$$E_m = \left(\frac{C_L}{C_D}\right)_{\max} = \frac{1}{2\sqrt{kC_{DO}}} \tag{2.20}$$

Let C_L^* denote the value of lift coefficient when $E = E_m$. Then,

$$C_L^* = \sqrt{\frac{C_{DO}}{k}} \tag{2.21}$$

Schematic variations of C_L, C_D, and E with angle of attack are shown in Fig. 2.3.

The drag of the aircraft is given by

$$D = \frac{1}{2}\rho V^2 S\left(C_{DO} + kC_L^2\right) \tag{2.22}$$

With $L = nW$, we have

$$C_L = \frac{2nW}{\rho V^2 S} \tag{2.23}$$

Then,

$$\begin{aligned} D &= \frac{1}{2}\rho V^2 S\left[C_{DO} + \left(\frac{4kn^2W^2}{\rho^2 V^4 S^2}\right)\right] \\ &= \frac{1}{2}\rho S\left[C_{DO}V^2 + \left(\frac{4kn^2W^2}{\rho^2 V^2 S^2}\right)\right] \end{aligned} \tag{2.24}$$

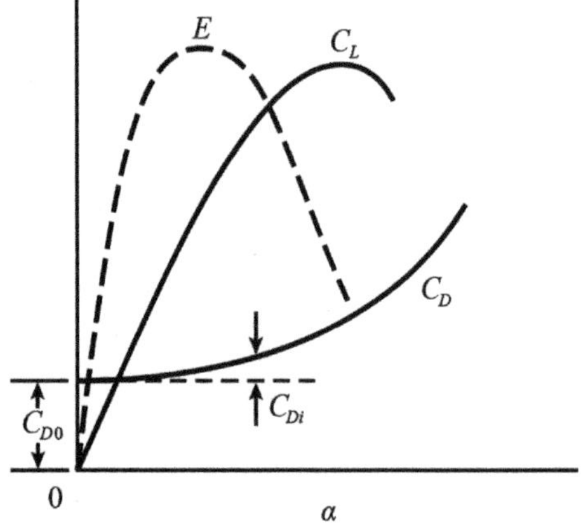

Fig. 2.3 Aerodynamic parameters with angle of attack.

Substituting $V = uV_R$, where V_R is given by Eq. (2.15), we obtain

$$D = \frac{W}{2E_m}\left(u^2 + \frac{n^2}{u^2}\right) \tag{2.25}$$

Equations (2.7) and (2.8) for static performance can be expressed now as

$$\frac{zW}{E_m} - \frac{W}{2E_m}\left(u^2 + \frac{n^2}{u^2}\right) - W\sin\gamma = 0 \tag{2.26}$$

or

$$2zu^2 - u^4 - n^2 - 2E_m u^2 \sin\gamma = 0 \tag{2.27}$$

and

$$n - \cos\gamma = 0 \tag{2.28}$$

Equations (2.27) and (2.28) describe the static performance of the airplane for such flight conditions as steady level flight, steady climb, range, and endurance. Before we discuss such problems of powered flights, let us consider the simple power-off gliding flight.

2.3 Gliding Flight

A glider is an unpowered light airplane. The power to overcome the aerodynamic drag comes at the expense of its potential energy or its height above the ground. Because a glider cannot attain the height on its own, it must be towed by another powered airplane to the desired height and then launched to fly on its own.

An interesting example of power-off gliding flight is the return of the Space Shuttle orbiter from space. The Space Shuttle orbiter glides back to the Earth and lands on a runway like a conventional airplane.

Consider a glider flying in a vertical plane as shown in Fig. 2.4. Note that, for gliding flight, $\gamma < 0$. With $T = 0$, Eqs. (2.7) and (2.8) reduce to the following form:

$$D + W\sin\gamma = 0 \tag{2.29}$$

$$L - W\cos\gamma = 0 \tag{2.30}$$

Kinematic equations as given by Eqs. (2.9) and (2.10) are

$$\dot{x} = V\cos\gamma \tag{2.31}$$

$$\dot{h} = V\sin\gamma \tag{2.32}$$

Generally, the glide angle γ is small so that we have

$$D + W\gamma = 0 \tag{2.33}$$

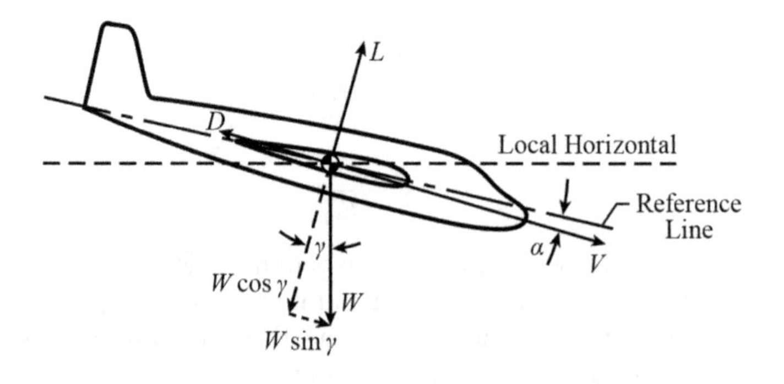

Fig. 2.4 Forces acting on a glider.

$$L = W \tag{2.34}$$

$$\dot{x} = V \tag{2.35}$$

$$\dot{h} = V\gamma \tag{2.36}$$

Lift and drag forces are given by

$$L = \frac{1}{2}\rho V^2 S C_L \tag{2.37}$$

$$\begin{aligned} D &= \frac{1}{2}\rho V^2 S C_D \\ &= \frac{1}{2}\rho V^2 S (C_{DO} + kC_L^2) \end{aligned} \tag{2.38}$$

The velocity is given by

$$V = \sqrt{\frac{2W}{\rho S C_L}} \tag{2.39}$$

The glide angle can be obtained from Eqs. (2.33) and (2.34) as

$$\begin{aligned} \gamma &= -\frac{D}{W} \\ &= -\frac{D}{L} \\ &= -\frac{C_D}{C_L} \\ &= -\frac{1}{E} \end{aligned} \tag{2.40}$$

so that

$$\gamma_{\min} = -\frac{D_{\min}}{W} = -\frac{1}{E_m} \tag{2.41}$$

Thus, the flattest glide ($\gamma = \gamma_{\min}$) occurs when the glider flies at that angle of attack when the drag per unit weight is minimum or $E = E_m$. Because the weight of a glider is constant, we can say that the flattest glide occurs when the drag is at minimum. We know that when $E = E_m$, $C_L = C_L^* = \sqrt{C_{DO}/k}$. The flight velocity for the flattest glide is given by

$$V = \sqrt{\frac{2W}{\rho S C_L^*}} = \sqrt{\frac{2W}{\rho S}}\sqrt[4]{\frac{k}{C_{DO}}} \tag{2.42}$$

In other words, the velocity for the flattest glide is equal to the reference velocity V_R and $u = 1$.

For a given height, the distance covered with respect to the ground can be obtained as follows:

$$\frac{\mathrm{d}x}{\mathrm{d}t} = V \tag{2.43}$$

$$\frac{\mathrm{d}x}{\mathrm{d}h}\frac{\mathrm{d}h}{\mathrm{d}t} = V \tag{2.44}$$

$$\frac{\mathrm{d}x}{\mathrm{d}h} = \frac{V}{V\gamma} = \frac{1}{\gamma} = -E \tag{2.45}$$

Let $R = x_f - x_i$ denote the range, which is equal to the horizontal distance covered with respect to the ground. Then,

$$R = -\int_{h_i}^{h_f} E\mathrm{d}h \tag{2.46}$$

where h_i and h_f are the initial and final altitudes, respectively. Assuming that the angle of attack α is held constant so that E is constant during the glide, we have

$$R = E\Delta h \tag{2.47}$$

where $\Delta h = h_i - h_f$ is the height lost during the glide. From Eq. (2.47), we observe that, for a given height difference, the range is maximum when $E = E_m$, which is also the condition for the flattest glide. In other words, maximum range occurs when the glide angle is minimum. Using Eq. (2.20), we get

$$R_{\max} = \frac{\Delta h}{2\sqrt{kC_{DO}}} \tag{2.48}$$

Here, we have ignored the effect of wind on the range. Actually, the range depends on wind conditions. A headwind reduces the range, whereas a tailwind increases it.

In sailplane terminology, glide ratio is the ratio between the ground distance traversed and the height lost. The glide ratio is also equal to E. A high-performance sail plane with a glide ratio of 40 can cover 4 km with respect to the ground for every 100 m of height lost.

Let the rate of sink, the speed with which the glider is heading towards the Earth, be denoted by $\dot{h}_s$. Usually, $\dot{h}$ denotes the rate of climb. Therefore, $\dot{h}_s = -\dot{h}$. With this, the rate of sink is given by

$$\begin{aligned} \dot{h}_s &= -V\gamma \\ &= \frac{DV}{W} \\ &= \sqrt{\frac{2W}{\rho S C_L}}\left(\frac{C_D}{C_L}\right) \\ &= \sqrt{\frac{2W}{\rho S}}\left(\frac{C_D}{C_L^{\frac{3}{2}}}\right) \end{aligned} \tag{2.49}$$

The term DV in Eq. (2.49) represents the power required P_R to sustain the gliding flight. Thus, we observe that the rate of sink is minimum when the power required per-unit-weight is minimum, which also corresponds to the case when the parameter $(C_D/C_L^{3/2})$ is a minimum or $(C_L^{3/2}/C_D)$ is a maximum. It can be shown that this happens when

$$C_{Di} = 3C_{DO} \tag{2.50}$$

or

$$C_L = \sqrt{\frac{3C_{DO}}{k}} \tag{2.51}$$

Let the value of lift coefficient when $(C_L^{3/2}/C_D)$ is maximum be denoted by $C_{L,m}$. Then,

$$C_{L,m} = \sqrt{\frac{3C_{DO}}{k}}$$
$$= \sqrt{3}C_L^* \tag{2.52}$$

With this, it can be shown that

$$\left(\frac{C_L^{\frac{3}{2}}}{C_D}\right)_{\max} = \frac{1}{4}\sqrt[4]{\frac{27}{k^3 C_{DO}}} \tag{2.53}$$

We observe that for glide with minimum sink rate, the induced drag is three times the zero-lift drag. The schematic variations of E and $(C_L^{3/2}/C_D)$ with angle of attack are shown in Fig. 2.5.

The velocity V_m for glide with minimum sink rate is given by

$$V_m = \sqrt{\frac{2W}{\rho S}}\sqrt[4]{\frac{k}{3C_{DO}}}$$
$$\simeq 0.76\, V_R \tag{2.54}$$

From this relation, we observe that the velocity for the glide with minimum sink rate is about 0.76 V_R or 0.76 times the velocity for flattest glide. Whereas the flattest glide occurs when the drag is minimum, the glide with minimum sink rate occurs when the power required is a minimum. These velocities are shown in Fig. 2.6.

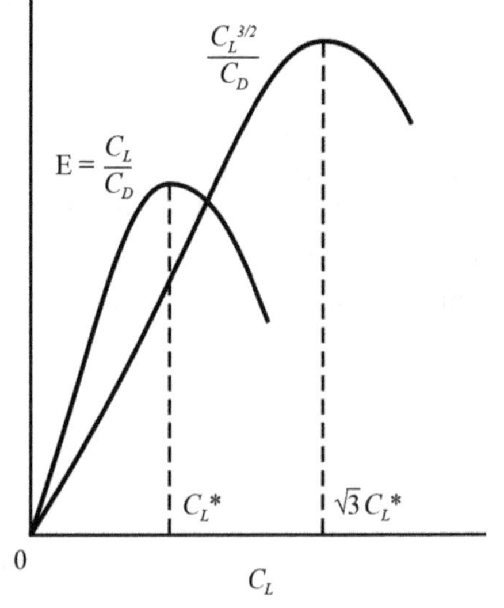

Fig. 2.5 Aerodynamic parameters with lift coefficient.

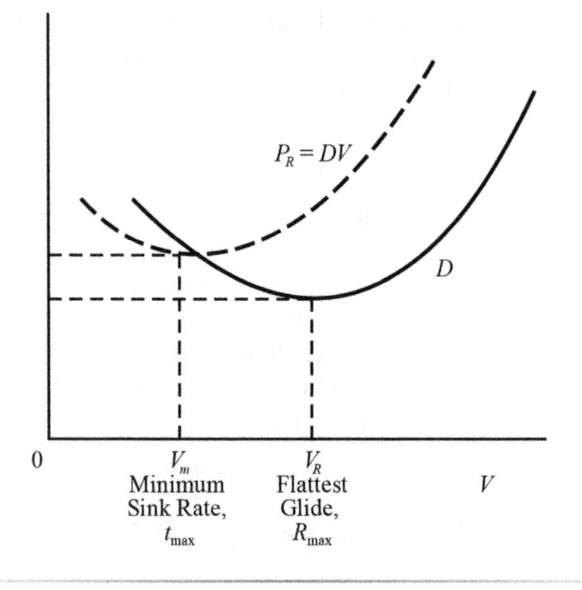

Fig. 2.6 Optimal velocities for gliding flight.

The minimum sink rate is given by

$$\dot{h}_{s,\,\min} = \sqrt{\frac{2W}{\rho S}}\sqrt[4]{\frac{k^3 C_{DO}}{27}} \tag{2.55}$$

Sailplane pilots fly at $V = V_m$ when they are in the "lift" mode, i.e., when they encounter an upward gust. When the "lift" dies, they accelerate to V_R, the velocity for flattest glide to cover the most ground while searching for another "lift." An instrument called "variometer" tells sailplane pilots whether they are in "lift" mode or not.

The endurance is the total time the glider remains in the air and can be determined as follows:

$$dt = \frac{dh}{V\gamma} \tag{2.56}$$

so that

$$t = -\int_{h_i}^{h_f} \sqrt{\frac{\rho S}{2W}} \left(\frac{C_L^{\frac{3}{2}}}{C_D} \right) dh \tag{2.57}$$

Assuming that the angle of attack is held constant during the glide and ignoring the variations in density because of changes in altitude, we obtain

$$t = \sqrt{\frac{\rho S}{2W}} \left(\frac{C_L^{\frac{3}{2}}}{C_D} \right) (h_i - h_f) \tag{2.58}$$

If the difference between the initial and final altitude is significant, then the variation in density may have to be considered. In such cases, the following approximate equation may be used.

$$\rho = \rho_o e^{-0.000114h} \tag{2.59}$$

where ρ_o is the density at sea level and h is the altitude in meters. Usually, $\rho_o = 1.225\ \mathrm{kg/m^3}$.

For maximum endurance, the glider has to fly at that angle of attack or lift coefficient when the parameter $(C_L^{3/2}/C_D)$ is maximum, which occurs when $C_L = \sqrt{3}C_L^*$ and $V = 0.76\ V_R$. Note that this is also the condition for minimum sink rate. Thus, the endurance is maximum when the sink rate is minimum. Using Eq. (2.53), we obtain

$$t_{\max} = \sqrt{\frac{\rho S}{2W}}\sqrt[4]{\frac{27}{k^3 C_{DO}}}\left(\frac{h_i - h_f}{4}\right) \tag{2.60}$$

The range and endurance are important measures of a glider's performance. Whereas the maximum range was independent of the weight, the maximum endurance depends on the weight. This calls for the designer to make the glider as light as possible. Furthermore, both the maximum range and endurance improve if the aerodynamic parameters k and C_{DO} are kept to their lowest possible values. Because of this, gliders tend to have an elliptical wing with a high aspect ratio and an efficient low-drag, laminar-flow airfoil section.

Example 2.1

A glider having $W = 2000$ N, $S = 8.0\ \mathrm{m^2}$, $A = 16.0$, $e = 0.95$, and $C_{DO} = 0.015$ is launched from a height of 300 m. Determine the maximum range, corresponding glide angle, forward velocity, and lift coefficient at sea level.

Solution:

At sea level, we have $\rho_o = 1.225\ \mathrm{kg/m^3}$. Furthermore,

$$\begin{aligned} k &= \frac{1}{\pi A e} \\ &= \frac{1}{\pi * 16 * 0.95} \\ &= 0.02 \end{aligned}$$

(Continued)

Example 2.1 (Continued)

$$E_m = \frac{1}{2\sqrt{KC_{DO}}}$$
$$= \frac{1}{2\sqrt{0.02 * 0.015}}$$
$$= 28.86$$
$$R_{\max} = (h_f - h_i)E_m$$
$$= 28.86 * 300\,\text{m}$$
$$= 8.658\,\text{km}$$

The maximum range occurs when the glide angle is minimum

$$\gamma_{\min} = \frac{1}{E_m}$$
$$= \frac{1}{28.86}$$
$$= 0.0346\,\text{rad}$$
$$= 1.985\,\text{deg}$$

The velocity and lift coefficient for the flattest glide are given by

$$V = V_R$$
$$= \sqrt{\frac{2W}{\rho S}}\sqrt[4]{\frac{k}{C_{DO}}}$$
$$= \sqrt{\frac{2 * 2000.0}{1.225 * 8.0}}\sqrt[4]{\frac{0.02}{0.015}}$$
$$= 23.328\,\text{m/s}$$
$$C_L = C_L^*$$
$$= \sqrt{\frac{C_{DO}}{k}}$$
$$= \sqrt{\frac{0.015}{0.02}}$$
$$= 0.866$$

Example 2.2

A glider weighing 5000 N and having an elliptical wing with an area of 10 m^2 is required to maintain a glide angle of 3 deg at a forward speed of 50 m/s. Assuming $C_{DO} = 0.015$, find the aspect ratio of the glider.

Solution:

Because the glider has an elliptical wing, $e = 1$. Then,

$$\gamma = \frac{C_D}{C_L} = \frac{3}{57.30} = 0.0524 \text{ rad}$$

$$C_L = \frac{2W}{\rho S V^2} = \frac{2 * 5000}{1.25 * 10 * 50^2} = 0.3260$$

$$C_D = C_L \gamma = 0.326 * 0.0524 = 0.01708$$

$$kC_L^2 = C_D - C_{DO}$$

$$k = \frac{0.01708 - 0.015}{0.326^2} = 0.0196$$

$$A = \frac{1}{k\pi e} = \frac{1}{0.0196 * \pi * 1} = 16.20$$

2.4 Level Flight

The forces acting on an airplane in level flight are shown in Fig. 2.7. In level flight, the airplane is at a constant altitude so that $\dot{h} = \gamma = 0$. With this, Eqs. (2.7) and (2.8) take the following form:

$$L - W = 0 \tag{2.61}$$

$$T - D = 0 \tag{2.62}$$

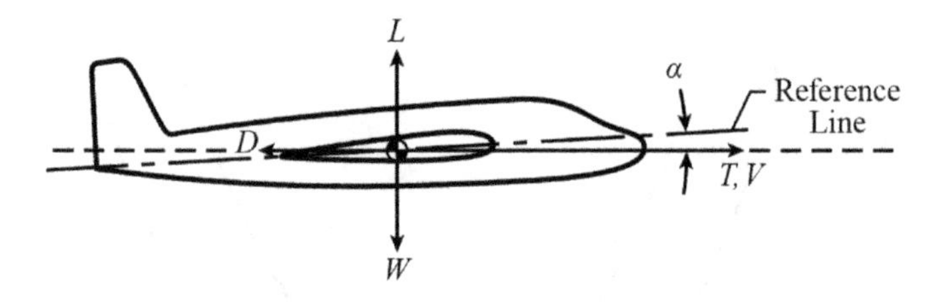

Fig. 2.7 Airplane in level flight.

Here, the thrust T is the thrust produced by the powerplant of the airplane or the thrust available; D is the drag of the airplane or, essentially, the thrust required to sustain steady level flight. Thus, for level flight, the condition for force balance can be stated in simple terms as the lift is equal to weight and thrust available is equal to thrust required. With $L = W$, the load factor $n = 1$ for level flight.

The kinematic equations as given by Eqs. (2.9) and (2.10) assume the form

$$\dot{x} = V \tag{2.63}$$

$$\dot{h} = 0 \tag{2.64}$$

A solution of Eqs. (2.61) and (2.62) gives the velocity V and angle of attack or the lift coefficient C_L for steady, unaccelerated level flight.

The drag of an airplane in level flight has an interesting variation with flight velocity. To understand this, let us proceed as follows:

$$D = \frac{1}{2}\rho V^2 S\left(C_{DO} + kC_L^2\right) \tag{2.65}$$

With $L = W$, we have

$$C_L = \frac{2W}{\rho V^2 S} \tag{2.66}$$

Then,

$$\begin{aligned} D &= \frac{1}{2}\rho V^2 S\left(C_{DO} + \frac{4kW^2}{\rho^2 V^4 S^2}\right) \\ &= \frac{1}{2}\rho V^2 S C_{DO} + \left(\frac{2kW^2}{\rho V^2 S}\right) \end{aligned} \tag{2.67}$$

The first term on the right-hand side of Eq. (2.67) is the zero-lift drag D_O and the second term is the induced drag D_i. The zero-lift drag varies directly as the square of the velocity, whereas the induced drag varies inversely with the square of the velocity. The variation of these two drag components and the total drag with velocity are shown in Fig. 2.8. At low speeds, the induced drag is dominant because the lift coefficient needed to sustain

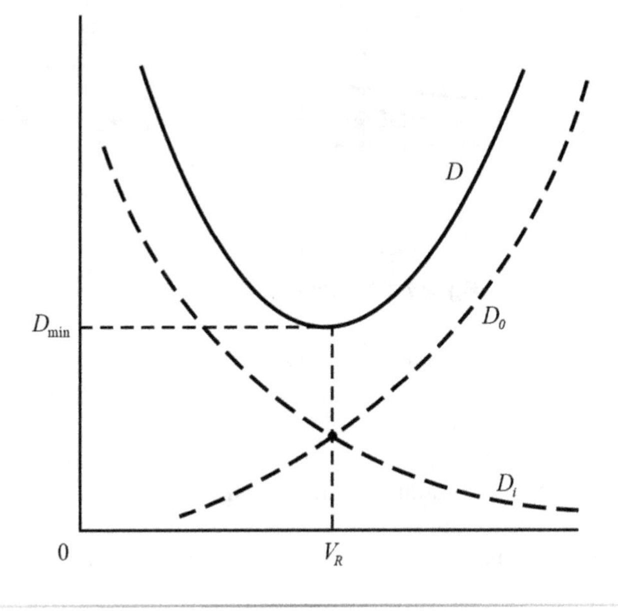

Fig. 2.8 Drag components with velocity.

level flight is quite high. As the velocity increases, the zero-lift drag rises very rapidly, but the induced drag becomes insignificant.

At low subsonic speeds, the zero-lift drag coefficient C_{DO} can be assumed constant with respect to the velocity. However, at high subsonic, transonic, and supersonic speeds, the zero-lift drag coefficient varies with flight speed (Mach number) because it includes the wave drag component. This variation must be taken into account.

Because of opposite trends of zero-lift drag and induced drag, the total drag assumes a minimum value at a certain velocity. The speed at which the total drag is a minimum can be obtained by differentiating Eq. (2.67) with respect to the velocity and equating the resulting expression to zero to obtain

$$\rho VSC_{DO} - \frac{4kW^2}{\rho SV^3} = 0 \tag{2.68}$$

or

$$V = \sqrt{\frac{2W}{\rho S}}\sqrt[4]{\frac{k}{C_{DO}}} \tag{2.69}$$

It may be observed that this velocity is equal to the reference velocity V_R introduced earlier in Eq. (2.15) for the definition of nondimensional flight speed u. Thus, the reference velocity happens to be the velocity at which the drag is minimum in a steady, unaccelerated level flight.

Substituting the expression for velocity from Eq. (2.69) in Eq. (2.67), we obtain

$$D_O = D_i = W\sqrt{kC_{DO}} \tag{2.70}$$

$$D_{\min} = 2D_O = 2D_i = 2W\sqrt{kC_{DO}} \tag{2.71}$$

Thus, we observe that when the total drag in level flight is minimum, the zero-lift drag equals the induced drag.

The power required in level flight is given by

$$P_R = DV = \frac{1}{2}\rho V^3 S C_{DO} + \left(\frac{2kW^2}{\rho VS}\right) \tag{2.72}$$

The first term represents the power required to overcome the zero-lift drag and is denoted by P_{Ro}. Similarly, the second term represents that required to overcome induced drag and is denoted by P_{Ri}. Typical variations of P_{Ro} and P_{Ri} are shown in Fig. 2.9. We observe that, at low speeds, the power required to overcome induced drag is dominating, whereas, at high speeds, it is the power required to overcome the zero-lift drag that assumes significance. Because of the opposite nature of variations of these terms, the total power

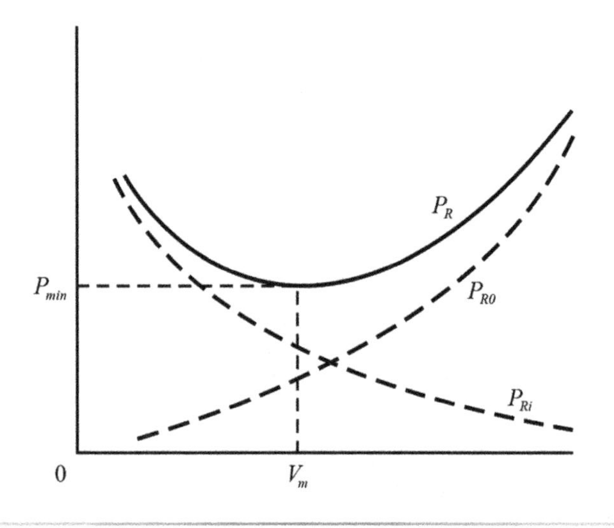

Fig. 2.9 Power-required curves for propeller aircraft.

required for level flight assumes a minimum value at some velocity, which will be denoted by V_{mp} and can be obtained as follows.

For $P_{R,\min}$,

$$\frac{dP_R}{dV} = \frac{3}{2}\rho V^2 S C_{DO} - \left(\frac{2kW^2}{\rho V^2 S}\right) = 0 \tag{2.73}$$

or

$$V_{mp}^4 = \frac{4kW^2}{3\rho^2 S^2 C_{DO}} \tag{2.74}$$

$$V_{mp} = \sqrt{\frac{2W}{\rho S}}\sqrt[4]{\frac{k}{3C_{DO}}} = \frac{1}{\sqrt[4]{3}} V_R \tag{2.75}$$

The lift coefficient in level flight when the power required in level flight is minimum is as follows.

$$C_{L,mp} = \frac{2W}{\rho S V_{mp}^2} = \sqrt{\frac{3C_{DO}}{k}} = \sqrt{3} C_L^* \tag{2.76}$$

Similarly, we can show that

$$E_{mp} = 0.866\, E_m \tag{2.77}$$

Then, we get

$$P_{R,\min} = \frac{8}{3}\frac{kW^2}{\rho S V_{mp}} = \frac{W V_{mp}}{0.866\, E_m} \tag{2.78}$$

Thus, we observe that both thrust- and power-required curves assume their minimum values at some velocities. In view of this, we have two intersection points between power-available and power-required curves as shown in Fig. 2.10a for the propeller airplane and between thrust-available and thrust-required curves for a jet airplane as shown in Fig. 2.10b. These two intersection points give us two solutions of Eqs. (2.61) and (2.62). Thus, spending the same amount of fuel and developing the same magnitude of

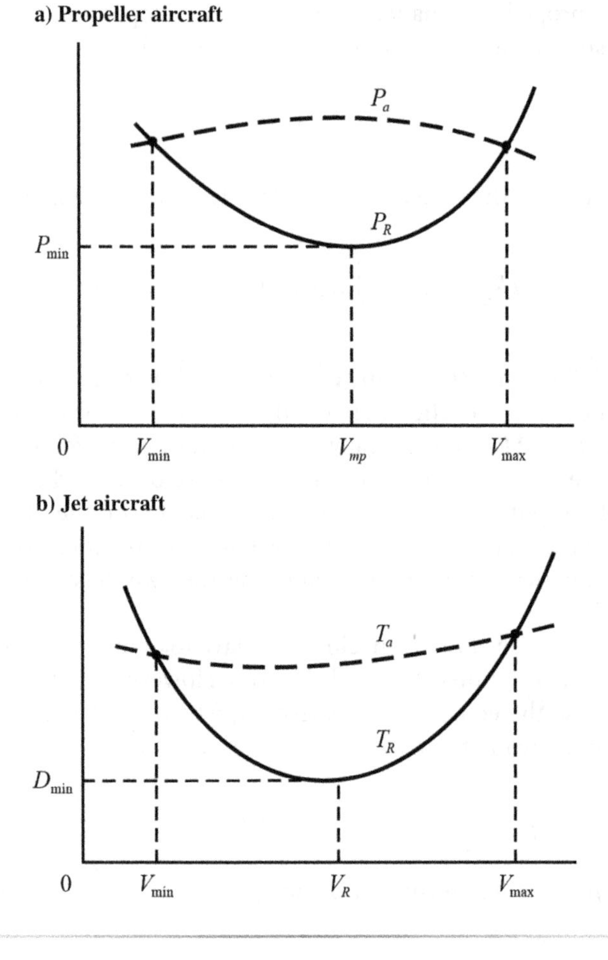

Fig. 2.10 Level flight solutions.

thrust or the power, an airplane can fly at either of these two velocities. At the low or minimum velocity V_{min}, the thrust or power available is essentially used to overcome induced drag, whereas, at high or maximum speed V_{max}, it is mostly used to balance the zero-lift drag.

2.4.1 Analytical Solutions of Level Flight for Propeller Airplanes

Because the power available and power required are the basic quantities for propeller aircraft, we rewrite Eq. (2.62) as follows:

$$TV - DV = 0 \tag{2.79}$$

$$P_a - P_R = 0 \tag{2.80}$$

For an ideal propeller airplane, whose power developed in kilowatts P(kW) and propulsive efficiency η_p are independent of flight velocity,

$$P_a = k'\eta_p P(\mathrm{kW}) \tag{2.81}$$

where $k' = 1000$ is the conversion constant from kW to Nm/s. Then,

$$k'\eta_p P(\mathrm{kW}) - \frac{1}{2}\rho S C_{DO} V^3 - \frac{2kW^2}{\rho S V} = 0 \tag{2.82}$$

Equation (2.82) is of the form $V^4 + aV + b = 0$, which has no closed-form analytical solution. In other words, we have to solve the level flight equation in Eq. (2.82) numerically or graphically even for this simplified case where we have assumed that the power developed by the reciprocating engine and propulsive efficiency are independent of the flight velocity. Therefore, for propeller aircraft, it is a common practice to obtain graphical or numerical solutions to determine the two speeds in level flight V_{min} and V_{max}.

To construct a level flight envelope, we have to obtain this type of graphical or numerical solutions at several altitudes. However, this task can be simplified if we use the equivalent air speed V_e, which is related to the true air speed by the following relation:

$$V_e = V\sqrt{\sigma} \tag{2.83}$$

where $\sigma = \rho/\rho_0$ is the density ratio and ρ_0 is the air density at sea level.

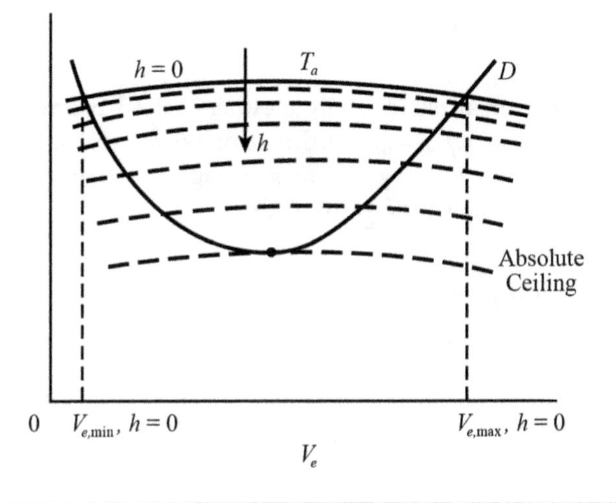

Fig. 2.11 Maximum and minimum speeds in level flight.

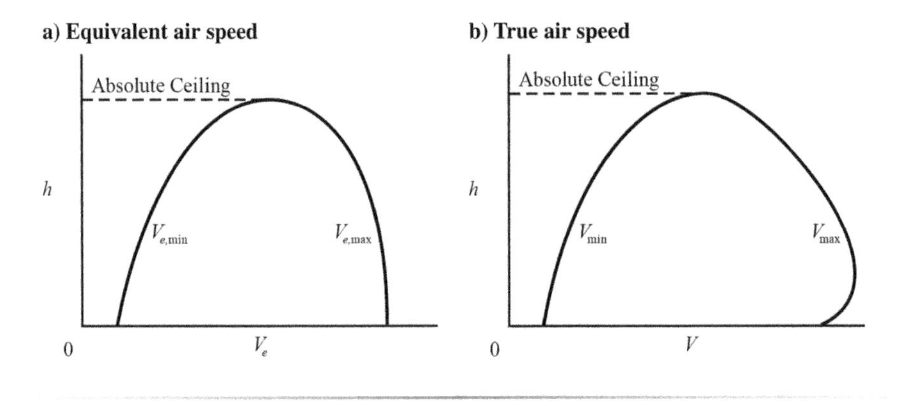

Fig. 2.12 Level flight envelope.

With this we can rewrite Eq. (2.67) as follows:

$$D = \frac{1}{2}\rho_0 V_e^2 SC_{DO} + \left(\frac{2kW^2}{\rho_0 V_e^2 S}\right) \tag{2.84}$$

The advantage of introducing the equivalent air speed is obvious. We have only one drag curve given by Eq. (2.84) that holds for all altitudes. Now, on this drag curve, let us superpose the thrust-available curves at various altitudes as shown in Fig. 2.11. The thrust available drops as the altitude increases because of a fall in air density. The thrust-available curve intersects the drag curve at two points designated as $V_{e,\min}$ and $V_{e,\max}$. At these two intersection points, the level flight Eqs. (2.61) and (2.62) are identically satisfied, and, therefore, each point is a level flight solution.

The schematic variation of $V_{e,\min}$ and $V_{e,\max}$ with altitude is shown in Fig. 2.12a. If we plot the true air speeds corresponding to $V_{e,\min}$ and $V_{e,\max}$, we obtain the level flight envelope as shown in Fig. 2.12b. It is interesting to observe that the true air speed corresponding to high-speed solution increases initially with altitude and then begins to decrease, whereas the true air speed corresponding to the low-speed solution increases monotonically with altitude. At a certain altitude, the two solutions merge, and we have only one level flight solution. This altitude is called the absolute ceiling of the airplane. Note that the absolute ceiling is also the altitude where the thrust-available curve is tangential to the thrust-required (drag) curve. In other words, at absolute ceiling, the thrust available has dropped so much that the level flight is possible only at one speed. This speed happens to be the speed at which the drag is minimum. Also, at absolute ceiling, the rate of climb will be zero as we will see later. Beyond the absolute ceiling, steady level flight is not possible because thrust available is not sufficient to balance the aerodynamic drag.

The velocity corresponding to the high-speed solution increases initially with altitude because at altitudes close to the sea level, the drop in thrust available is very gradual. As a result, $V_{e,\max}$, even though decreasing, is still quite close to its value at sea level. On the other hand, the density ratio σ drops at a higher rate so that the true air speed, which is the ratio of two decreasing quantities with the denominator decreasing faster than the numerator, increases initially with altitude. At some altitude, the true air speed attains a maximum value and then starts dropping when this trend reverses.

2.4.2 Analytical Solutions for Jet Aircraft

If we assume that at a given altitude, the thrust developed by a jet engine is independent of flight velocity, then we can obtain analytical solution for level flight. We have

$$
\begin{aligned}
T &= D \\
&= \frac{1}{2}\rho V^2 S\left(C_{DO} + kC_L^2\right) \\
&= \frac{1}{2}\rho V^2 S C_{DO} + \left(\frac{2kW^2}{\rho V^2 S}\right)
\end{aligned}
\tag{2.85}
$$

Equation (2.85) can be written as follows:

$$AV^4 - TV^2 + B = 0 \tag{2.86}$$

where

$$A = \frac{1}{2}\rho S C_{DO} \tag{2.87}$$

$$B = \left(\frac{2kW^2}{\rho S}\right) \tag{2.88}$$

Solving Eq. (2.86), we obtain

$$V_{\max} = \sqrt{\frac{T + \sqrt{T^2 - 4AB}}{2A}} \tag{2.89}$$

$$V_{\min} = \sqrt{\frac{T - \sqrt{T^2 - 4AB}}{2A}} \tag{2.90}$$

However, a more elegant form of analytical solution can be obtained using the nondimensional form of Eqs. (2.27) and (2.28). With $\gamma = 0$, Eqs. (2.27) and (2.28) take the following form,

$$u^4 - 2zu^2 + 1 = 0 \tag{2.91}$$

$$n = 1 \tag{2.92}$$

Then,

$$u^2 = z \pm \sqrt{z^2 - 1} \tag{2.93}$$

$$u_{\max} = \sqrt{z + \sqrt{z^2 - 1}} \tag{2.94}$$

$$u_{\min} = \sqrt{z - \sqrt{z^2 - 1}} \tag{2.95}$$

$$V_{\max} = u_{\max} V_R \tag{2.96}$$

$$V_{\min} = u_{\min} V_R \tag{2.97}$$

Given the variation of thrust available with altitude, using this approach, we can construct the level flight envelope for a jet aircraft whose thrust is independent of flight velocity. At the absolute ceiling, $u_{\min} = u_{\max}$, which gives $z = 1$. Physically, when $z = 1$, $T = D_{\min}$. As said before, at absolute ceiling, level flight is possible only at only one speed where the drag is minimum.

Whether the flight is possible at $V_{\min}$ depends on the magnitude of the level flight stalling velocity, which is given by

$$V_{\text{stall}} = \sqrt{\frac{2W}{\rho S C_{L,\max}}} \tag{2.98}$$

If $V_{\min} < V_{\text{stall}}$, then the airplane cannot fly at $V_{\min}$ in steady level flight because it asks for a lift coefficient in excess of $C_{L,\max}$. For such cases, the level flight solution $V_{\min}$ has no physical significance, and V_{stall} effectively becomes the low-speed solution.

To fly in level flight at any speed at a speed V such that $V_{\min} < V < V_{\max}$, the pilot has to throttle the engine so that the drag at this speed equals the thrust available as schematically shown in Fig. 2.13.

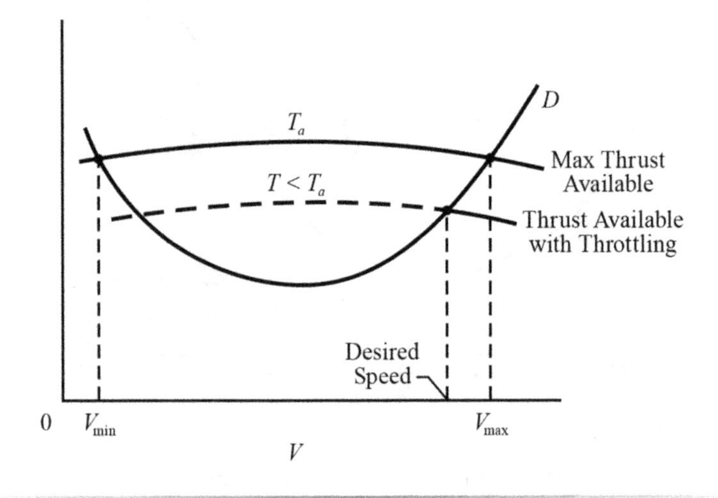

Fig. 2.13 Level flight at any desired speed.

Example 2.3

A propeller aircraft weighs 50,000 N and has a wing area of 30 m^2. The reciprocating engine develops 840 kW at sea level and the adjustable pitch constant-speed propeller has an efficiency of 0.85. The drag polar of the aircraft is given by $C_D = 0.025 + 0.05C_L^2$, and the maximum lift coefficient is 1.75. Determine the maximum and minimum speeds in level flight at sea level.

Solution:

We have

$$P_a = k'\eta_p P(kW)$$

$$P_R = \frac{1}{2}\rho S C_{DO} V^3 + \frac{2kW^2}{\rho S V}$$

We have $k' = 1000$, $\eta_p = 0.85$, $P(\text{kW}) = 840$, $W = 50{,}000$ N, $S = 30$ m^2, $\rho = \rho_o = 1.225$ kg/m^3, $C_{DO} = 0.025$, $k = 0.05$, and $C_{L,\max} = 1.75$. Because the analytical solution is not possible, we have to use a graphical method for the given propeller airplane.

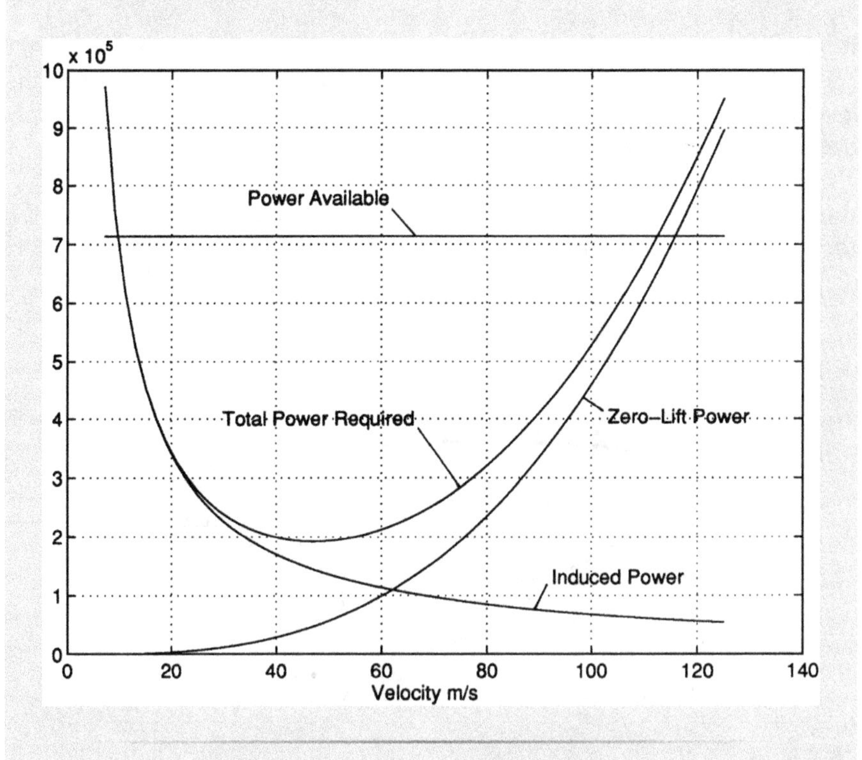

Fig. 2.14 Power curves for aircraft.

(Continued)

Example 2.3 (*Continued*)

The power-required and power-available curves are plotted as shown in Fig. 2.14. The stalling velocity is given by

$$V_{stall} = \sqrt{\frac{2W}{\rho S C_{L,max}}}$$
$$= \sqrt{\frac{2 * 50{,}000}{1.225 * 30 * 1.75}}$$
$$= 39.4323 \, m/s$$

From Fig. 2.14, we find that the minimum and maximum velocities are approximately 10 and 113.0 m/s, respectively. The flight at V_{min} is not possible because it is lower than V_{stall}. Hence, V_{stall} is effectively the minimum speed in level flight at sea level.

Example 2.4

A turbojet airplane weighs 45,000 N and has a wing loading W/S of 1450 N/m^2. The drag polar is given by $C_D = 0.014 + 0.038C_L^2$, $C_{L,max} = 1.5$, and the thrust $T = 20{,}000\ \sigma$. Determine the maximum and minimum speeds in level flight at sea level and at an altitude of 9000 m ($\sigma = 0.3813$). What is the absolute ceiling of this airplane?

Solution:

We have

$$E_{max} = \frac{1}{2\sqrt{kC_{DO}}}$$
$$= \frac{1}{2\sqrt{0.014 * 0.038}}$$
$$= 21.6777$$

For level flight at sea level ($\sigma = 1$),

$$T = 20{,}000 \, N$$

$$z = \frac{TE_m}{W}$$
$$= \frac{20{,}000 * 21.6777}{45{,}000}$$
$$= 9.6345$$

(*Continued*)

Example 2.4 (Continued)

$$V_R = \sqrt{\frac{2W}{\rho S}}\sqrt[4]{\frac{k}{C_{DO}}}$$

$$= \sqrt{\frac{2 * 1450}{1.225}}\sqrt[4]{\frac{0.038}{0.014}}$$

$$= 62.4518\,\mathrm{m/s}$$

$$u_{\max} = \sqrt{z + \sqrt{z^2 - 1}}$$

$$= \sqrt{9.6345 + \sqrt{9.6345^2 - 1}}$$

$$= 4.3837$$

$$V_{\max} = u_{\max} V_R$$

$$= 273.77\,\mathrm{m/s}$$

$$u_{\min} = \sqrt{z - \sqrt{z^2 - 1}}$$

$$= \sqrt{9.6345 - \sqrt{9.6345^2 - 1}}$$

$$= 0.2281$$

$$V_{\min} = u_{\min} V_R$$

$$= 14.2459\,\mathrm{m/s}$$

$$V_{\mathrm{stall}} = \sqrt{\frac{2W}{\rho S C_{L,\max}}}$$

$$= \sqrt{\frac{2 * 1450}{1.225 * 1.50}}$$

$$= 39.7270\,\mathrm{m/s}$$

Because $V_{\min} < V_{\mathrm{stall}}$, level flight is not possible at $V_{\min}$. Therefore, the minimum level flight speed at sea level is $V_{\mathrm{stall}} = 39.7270\,\mathrm{m/s}$.

At an altitude of 8000 m ($\sigma = 0.383$),

$$T = T_0 \sigma$$

$$= 20{,}000 * 0.3813\,\mathrm{N}$$

$$= 7626.0\,\mathrm{N}$$

$$z = \frac{T E_m}{W}$$

$$= \frac{7626.0 * 21.6777}{45{,}000}$$

$$= 3.6736$$

(Continued)

Example 2.4 (*Continued*)

$$V_R = \sqrt{\frac{2W}{\rho S}} \sqrt[4]{\frac{k}{C_{DO}}}$$

$$= \sqrt{\frac{2 * 1450}{1.225 * 0.3813}} \sqrt[4]{\frac{0.038}{0.014}}$$

$$= 101.1373 \text{ m/s}$$

$$u_{\text{max}} = \sqrt{z + \sqrt{z^2 - 1}}$$

$$= \sqrt{3.6736 + \sqrt{3.6736^2 - 1}}$$

$$= 2.6849$$

$$V_{\text{max}} = u_{\text{max}} V_R$$

$$= 271.5436 \text{ m/s}$$

$$u_{\text{min}} = \sqrt{z - \sqrt{z^2 - 1}}$$

$$= \sqrt{3.6736 - \sqrt{3.6736^2 - 1}}$$

$$= 0.1137$$

$$V_{\text{min}} = u_{\text{min}} V_R$$

$$= 11.4993 \text{ m/s}$$

$$V_{\text{stall}} = \sqrt{\frac{2W}{\rho S C_{L,\text{max}}}}$$

$$= \sqrt{\frac{2 * 1450}{1.225 * 0.3813 * 1.50}}$$

$$= 64.3357 \text{ m/s}$$

Because $V_{\text{min}} < V_{\text{stall}}$, level flight is not possible at V_{min}. Therefore, the minimum level flight speed at 8000 m altitude is $V_{\text{stall}} = 64.3357$ m/s.

At absolute ceiling, $z = 1$. Therefore,

$$T = \frac{zW}{E_m}$$

$$= \frac{1.0 * 45{,}000}{21.6777}$$

$$= 2075.866 \text{ N}$$

(Continued)

Example 2.4 (Continued)

$$\sigma = \frac{T}{T_0} = \frac{2075.866}{20{,}000} = 0.1038$$

$$\rho = \rho_0 \sigma = 1.225 * 0.1038 = 0.1271 \text{ kg/m}^3$$

From atmospheric tables (see Appendix A), the corresponding altitude, which in this case is the absolute ceiling of the airplane, is found to be approximately equal to 17.7 km.

2.5 Climbing Flight

For all altitudes below the absolute ceiling, we have two solutions for steady level flight, a low-speed solution V_{min} and a high-speed solution V_{max}. What happens if we try to fly at any speed V, $V_{min} < V < V_{max}$ using full thrust or power? Now we have excess thrust or excess power, and steady level flight is not possible. What may happen is that either 1) the airplane will accelerate to V_{max} if the altitude is forced to remain constant or 2) the airplane will climb. In this section, we discuss climbing flight. To begin with, let us consider the steady or constant velocity climb and then the more general case of accelerated climb. For simplicity, let us assume that the weight of the airplane remains constant during the climb.

2.5.1 Steady Climb

Consider an airplane in a steady climbing flight. The flight path of such an airplane is a straight line in vertical plane. From Eqs. (2.7) and (2.8), we have

$$L - W \cos \gamma = 0 \tag{2.99}$$

$$T - D - W \sin \gamma = 0 \tag{2.100}$$

and the kinematic equations are

$$\dot{x} = V \cos \gamma \tag{2.101}$$

$$\dot{h} = V \sin \gamma \tag{2.102}$$

From Eq. (2.100), we obtain

$$\sin\gamma = \frac{T-D}{W} = \frac{\text{Excess thrust}}{\text{Weight}} \tag{2.103}$$

We observe that the climb angle γ is directly proportional to the excess thrust per unit weight. Therefore, steepest climb $\gamma_{\max}$ occurs when the excess thrust per unit weight is maximum.

Let R/C denote the rate of climb. Then

$$\begin{aligned} R/C &= \dot{h} \\ &= V\sin\gamma \\ &= \frac{V(T-D)}{W} \\ &= \frac{P_a - P_R}{W} \\ &= P_s \end{aligned} \tag{2.104}$$

where P_s is the excess power per unit weight. Usually, P_s is called specific excess power. We observe that the R/C is maximum when the specific excess power is maximum.

Analytical Solution for Climbing Flight of Propeller Aircraft

For propeller aircraft whose power developed by the engine P(kW) and propulsive efficiency η_p are independent of flight velocity, Eq. (2.104) takes the form

$$(R/C) = \frac{1}{W}\left(k'\eta_p P(\text{kW}) - \frac{1}{2}\rho S C_{DO} V^3 - \frac{2kW^2}{\rho S V}\right) \tag{2.105}$$

For maximum R/C,

$$\begin{aligned} \frac{\mathrm{d}(R/C)}{\mathrm{d}V} &= -\frac{3}{2}\rho S C_{DO} V^2 + \frac{2kW^2}{\rho S V^2} \\ &= 0 \end{aligned} \tag{2.106}$$

or

$$\begin{aligned} V_{R/C,\max} &= \frac{1}{\sqrt[4]{3}}\sqrt{\frac{2W}{\rho S}}\sqrt[4]{\frac{k}{C_{DO}}} \\ &= \frac{1}{\sqrt[4]{3}} V_R \end{aligned} \tag{2.107}$$

It may be observed that the speed at which the rate of climb is maximum is the same as the speed at which the power required in level flight is minimum [see Eq. (2.75)]. The two speeds are the same because, for the propeller aircraft, we have assumed that the power available is independent of forward speed. Thus, essentially, the speed at which excess power is maximum becomes the same as that when the power required is minimum, i.e., $V_{R/C,\max} = V_{mp}$.

Using Eq. (2.78) in Eq. (2.105), we get

$$(R/C)_{\max} = k'\eta_p P'(\mathrm{kW}) - \frac{V_{mp}}{0.866\,E_m} \tag{2.108}$$

where

$$P'(\mathrm{kW}) = \frac{P(\mathrm{kW})}{W} \tag{2.109}$$

The climb angle is given by

$$\begin{aligned} \gamma &= \frac{\dot{h}}{V} \\ &= \frac{R/C}{V} \\ &= \frac{k'\eta_p P'(\mathrm{kW})}{V} - \frac{1}{2W}\rho S C_{DO} V^2 - \frac{2kW}{\rho S V^2} \end{aligned} \tag{2.110}$$

For steepest climb, i.e., when ($\gamma = \gamma_{\max}$),

$$\begin{aligned} \frac{d\gamma}{dV} &= -\frac{k'\eta_p P'(\mathrm{kW})}{V^2} - \frac{\rho S C_{DO} V}{W} + \frac{4kW}{\rho S V^3} \\ &= 0 \end{aligned} \tag{2.111}$$

This equation is of the form $V^4 + aV + b = 0$, where

$$a = \frac{k'\eta_p P(\mathrm{kW})}{\rho S C_{DO}} \tag{2.112}$$

$$b = \frac{-4k}{\rho^2 C_{DO}}\left(\frac{W}{S}\right)^2 \tag{2.113}$$

The equation $V^4 + aV + b = 0$ has no closed-form analytical solution. We have to obtain either a numerical or graphical solution to determine $\gamma_{\max}$. For this purpose, let us plot R/C against velocity as shown in Fig. 2.15. This plot is called the hodograph of climbing flight. A hodograph is a graph in which the variation of one velocity component is plotted against the other.

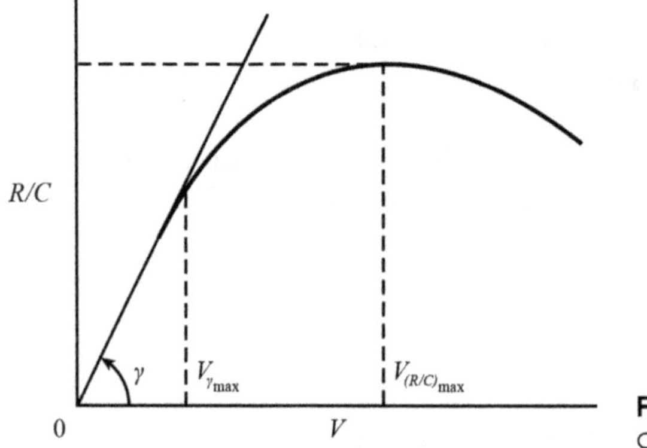

Fig. 2.15 Hodograph of climbing flight.

The maximum climb angle occurs at that velocity when the excess thrust per unit weight is maximum. This corresponds to the point where a line drawn from the origin is tangent to the hodograph as shown in Fig. 2.15.

The time to climb from a given initial altitude h_i to final altitude h_f is given by

$$t = \int_{h_i}^{h_f} \frac{\mathrm{d}h}{P_s} \tag{2.114}$$

which is equal to the area under the curve obtained by plotting inverse specific excess power against altitude as shown in Fig. 2.16. A special case of interest in climbing flight is the minimum time to climb. We will discuss this in more detail when we address the same issue for a jet aircraft as the approach will be essentially similar for the propeller aircraft.

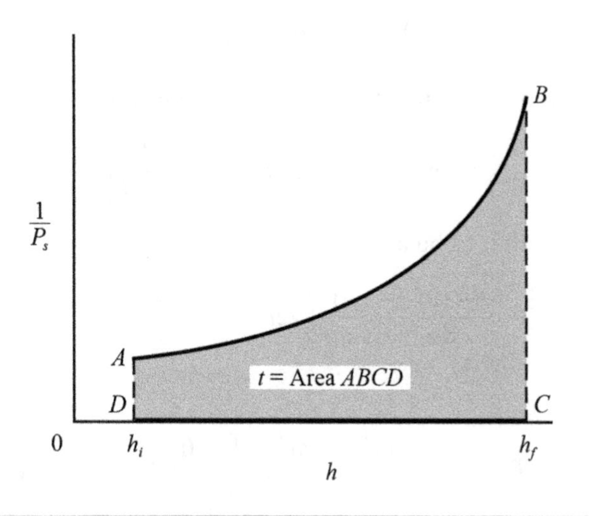

Fig. 2.16 Determination of time to climb.

Analytical Solutions for Climbing Flight of Jet Aircraft

If we assume that the thrust developed by a jet aircraft is independent of flight velocity, then it is possible to obtain some analytical solutions for climbing flight as follows.

Assuming that the climb angle γ is small, Eq. (2.99) reduces to $L = W$ or $n = 1$, and, using Eq. (2.26), Eq. (2.100) takes the following form:

$$\frac{zW}{E_m} - \frac{W}{2E_m}\left(u^2 + \frac{1}{u^2}\right) - W \sin\gamma = 0 \tag{2.115}$$

or

$$\sin\gamma = \frac{1}{2E_m}\left[2z - \left(u^2 + \frac{1}{u^2}\right)\right] \tag{2.116}$$

For γ to be a maximum,

$$\frac{\mathrm{d}\sin\gamma}{\mathrm{d}u} = -\frac{1}{2E_m}\left(2u - \frac{2}{u^3}\right)$$
$$= 0 \tag{2.117}$$

or

$$u = 1 \tag{2.118}$$

Thus, the climb angle γ is maximum when $u = 1$, which we know is also the condition for minimum drag in level flight. Also, when $u = 1$, we have $V = V_R$, $C_L = C_L^* = \sqrt{C_{DO}/k}$, $C_{DO} = C_{Di}$, and $C_D = 2C_{DO}$.

Substituting $u = 1$ in Eq. (2.116), we get

$$\gamma_{\max} = \sin^{-1}\left(\frac{z-1}{E_m}\right) \tag{2.119}$$

The rate of climb in a nondimensional form is given by

$$u\sin\gamma = \frac{1}{2E_m}\left[2zu - \left(u^3 + \frac{1}{u}\right)\right] \tag{2.120}$$

For the rate of climb to be a maximum,

$$\frac{\mathrm{d}(u\sin\gamma)}{\mathrm{d}u} = \frac{\mathrm{d}}{\mathrm{d}u}\left(2zu - u^3 - \frac{1}{u}\right) = 0 \tag{2.121}$$

or

$$3u^4 - 2zu^2 - 1 = 0 \tag{2.122}$$

$$u = \sqrt{\frac{z \pm \sqrt{z^2+3}}{3}} \tag{2.123}$$

Then, the maximum rate of climb in nondimensional form is given by

$$(u \sin \gamma)_{max} = \frac{1}{2E_m}\left[2zu_m - \left(u_m^3 + \frac{1}{u_m}\right)\right] \tag{2.124}$$

where u_m is the nondimensional flight velocity at which the rate of climb is maximum. The value of u_m is obtained by using the positive sign before the term $\sqrt{z^2+3}$ in Eq. (2.123). The corresponding dimensional speed is equal to $u_m\ V_R$, and the dimensional maximum rate of climb is given by

$$(R/C)_{max} = (V \sin \gamma)_{max} = (u \sin \gamma)_{max} V_R \tag{2.125}$$

As in the case of propeller aircraft, the time to climb from a given initial altitude h_i to the desired final altitude h_f is given by

$$t = \int_{h_i}^{h_f} \frac{dh}{P_s} \tag{2.126}$$

For minimum time to climb, the aircraft has to follow the path along which the specific excess power is maximum at each altitude. For a propeller airplane, this has to be done graphically or numerically whereas, for a jet airplane whose thrust is independent of the flight velocity, it is possible to find an analytical solution. We will discuss this problem in more detail when we consider energy climb later in this section.

2.5.2 Absolute and Service Ceilings

Previously, we defined the absolute ceiling as that altitude where the two level flight solutions merge into one. Based on the rate of climb, we can have an alternate definition of the absolute ceiling.

The $(R/C)_{max}$ decreases with altitude as shown in Fig. 2.17 because of the drop in thrust available or power available as the altitude increases. The altitude where the maximum rate of climb drops to 100 ft/min (30.5 m/min) is called service ceiling. Extrapolating further to $(R/C)_{max} = 0$, we get the absolute ceiling. For jet aircraft whose thrust is independent of flight velocity, $z = 1$ when the $(R/C)_{max} = 0$. Thus, both definitions of absolute ceiling are essentially identical.

For propeller airplanes, the service ceiling is in the range of 4–6 km, whereas for commercial jet airplanes it is 11–17 km.

2.5.3 Energy Climb Method

The steady or constant-velocity climb just discussed is based on the assumption that accelerations along and normal to the flight path are both equal to zero. In this section, we study a more general case of climbing

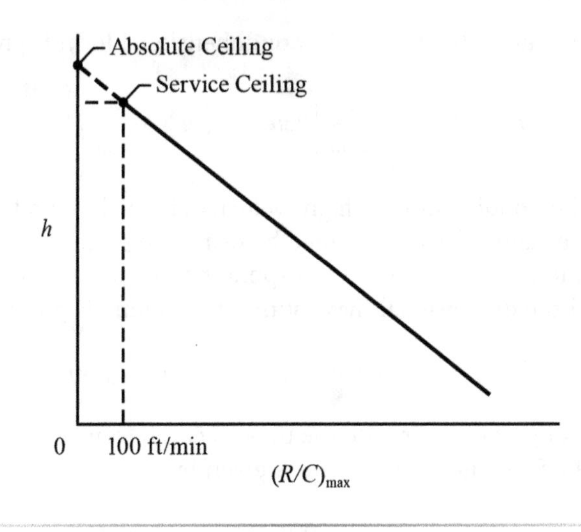

Fig. 2.17 Service and absolute ceilings.

flight that involves nonzero accelerations and is called energy climb method [8]. The equations of motion for this class of flight are given by Eqs. (2.5) and (2.6). With $\epsilon = 0$ and $\dot{\gamma} = V/R$, we have

$$T - D - W \sin \gamma = \frac{W}{g}\frac{dV}{dt} \tag{2.127}$$

$$L - W \cos \gamma = \frac{W}{g} V \frac{d\gamma}{dt} \tag{2.128}$$

The kinematic equations are

$$\dot{x} = V \cos \gamma \tag{2.129}$$

$$\dot{h} = V \sin \gamma \tag{2.130}$$

Equation (2.127) can be expressed as

$$\frac{T - D}{W} - \sin \gamma = \frac{1}{g}\frac{dV}{dt} \tag{2.131}$$

Multiply both sides of Eq. (2.131) by V to obtain

$$V\left(\frac{T - D}{W}\right) - V \sin \gamma = \frac{1}{2g}\left(\frac{dV^2}{dt}\right) \tag{2.132}$$

Define

$$h_e = h + \frac{V^2}{2g} \tag{2.133}$$

where h_e is called the energy height. Using Eqs. (2.130) and (2.132), we obtain

$$\frac{\mathrm{d}h_e}{\mathrm{d}t} = V\left(\frac{T-D}{W}\right) \tag{2.134}$$

or

$$\frac{\mathrm{d}h_e}{\mathrm{d}t} = P_s \tag{2.135}$$

Thus, the specific excess-power is more realistically equal to energy rate of climb rather than the kinematic rate of climb $\mathrm{d}h/\mathrm{d}t$ as assumed for steady-state climb. The time to climb from a given initial energy height h_{ei} to a given final energy height h_{ef} is given by

$$t = \int_{h_{ei}}^{h_{ef}} \frac{\mathrm{d}h_e}{P_s} \tag{2.136}$$

To evaluate the integral in Eq. (2.136), we need to draw the P_s curves at various altitudes. The procedure of drawing P_s curves is as follows:

1. Given the data on P_a and P_R at a number of altitudes, plot the specific excess-power P_s curves at selected altitudes to cover the range h_i to h_f, say $h_1 - h_5$ as shown in the upper sketch of Fig. 2.18. Note that $h =$ const for each of these curves. Then, draw horizontal lines along which $P_s =$ const, say P_{s1} and P_{s2} as shown, and read the values of altitude and velocity corresponding to every point of intersection.
2. Cross plot the altitude vs velocity with P_s a parameter as shown in the lower sketch of Fig. 2.18. Note that for each of these curves, $P_s =$ const. Draw as many curves as possible to cover the range 0 to $P_{s,\max}$. Note that the curve for which $P_s = 0$ represents the level flight envelope.

The problem of minimum time to climb from a given initial altitude to the desired final altitude is of interest to both commercial and military aircraft. A commercial airliner needs to clear the terminal airspace quickly to reduce terminal area congestion and reach the cruise altitude as early as possible so that the fuel consumed during climb is kept to a minimum. Also, this problem is of special interest to interceptor/fighter aircraft because minimum time to climb is an important measure of air superiority. In the following, we will discuss the application of energy climb method to the problem of minimum time to climb for typical subsonic and supersonic aircraft.

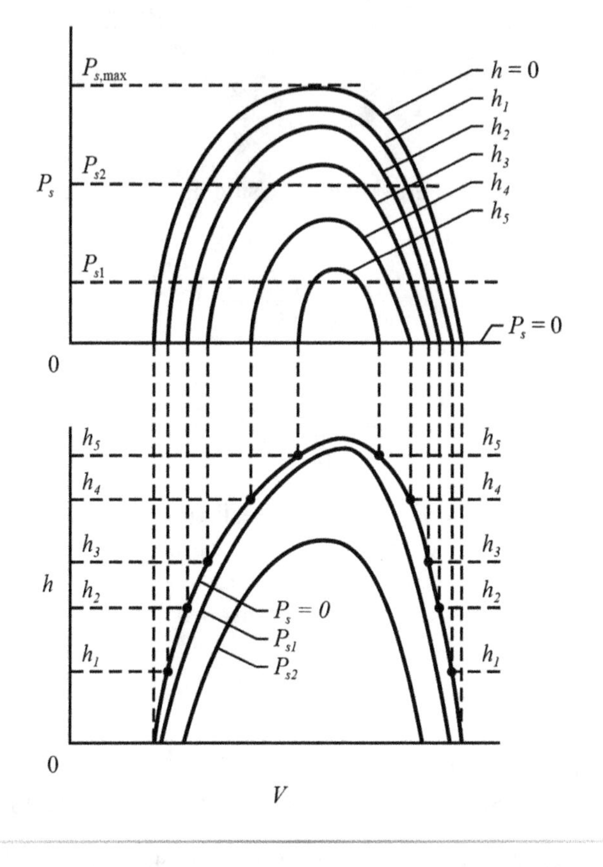

Fig. 2.18 Specific excess power curves at various altitudes.

Minimum Time to Climb for Subsonic Aircraft

From Eq. (2.136), we observe that the time to climb is minimum when the aircraft, at each energy state h_e, follows the path along which the specific excess power is locally maximum.

Typically, the P_s curves for a subsonic aircraft are continuous and smooth as shown in Fig. 2.19. According to the steady (static) climb method, the aircraft will fly such that R/C is a local maximum at each altitude. The locus of this path is given by the curve AB. However, the minimum time obtained by this method will be in error if the path AB is not vertical in $h-V$ plane because it would violate the basic assumption that the flight velocity is constant all along the flight path AB.

The minimum time according to the energy height method is obtained when the aircraft follows the path AB', which is the locus of all those points on P_s curves that are locally tangential to $h_e = \text{const}$ curves. Along AB', the value of P_s is a local maximum for each $h_e = \text{const}$ curve. For subsonic aircraft, the curve AB' is usually quite close to AB so that the minimum time given by the steady-state method is nearly equal to that given by the more accurate energy height method.

Minimum Time to Climb for Supersonic Aircraft

Here, we will discuss an application of the energy climb method for a supersonic aircraft of the 1960s and 1970s because these aircraft had limited excess-thrust capability in the transonic and supersonic regions. Because of this, the flight path for minimum time to climb assumes some interesting variations as discussed in the following.

The excess-power curves for a typical supersonic aircraft of the 1960s and 1970s are schematically shown in Fig. 2.20. In the subsonic region, P_S curves are well defined, smooth, and open. However, because of limited thrust capability, discontinuities exist in the transonic and supersonic regions where P_S curves form closed contours. Also, P_S curves may not exist at lower altitudes for transonic and supersonic Mach numbers.

The basic principle in the energy height method is that the aircraft always follows the locus of highest possible values of P_S without decreasing its energy height. With this criterion, the minimum time to climb starts in the subsonic region as in the case of a steady climb and follows the path AB until point B, where a certain h_e = const curve is tangential to two equal-valued P_S curves. At this point, the aircraft follows a small segment BC of a constant energy height until it finds another curve of equal value of P_S, which it does at point C. In this process of tracking the segment BC, the aircraft is essentially in a dive and is accelerating to supersonic speeds. On reaching point C, the aircraft resumes its climb in supersonic region along the path CD. At point D, another change occurs and the aircraft follows the path DE to its final

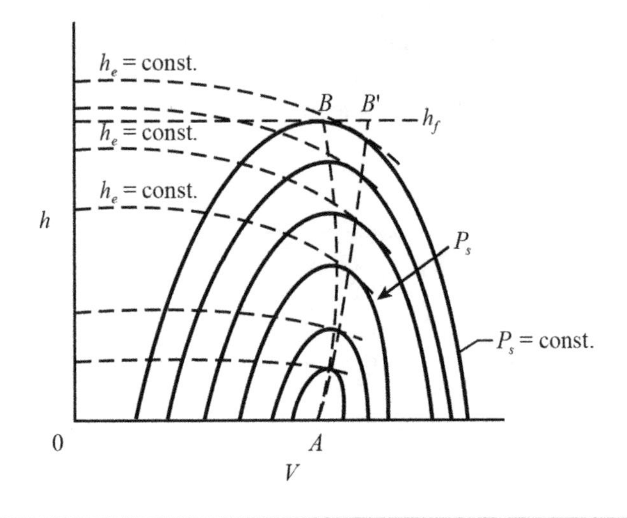

Fig. 2.19 Minimum time to climb for a subsonic aircraft (8).

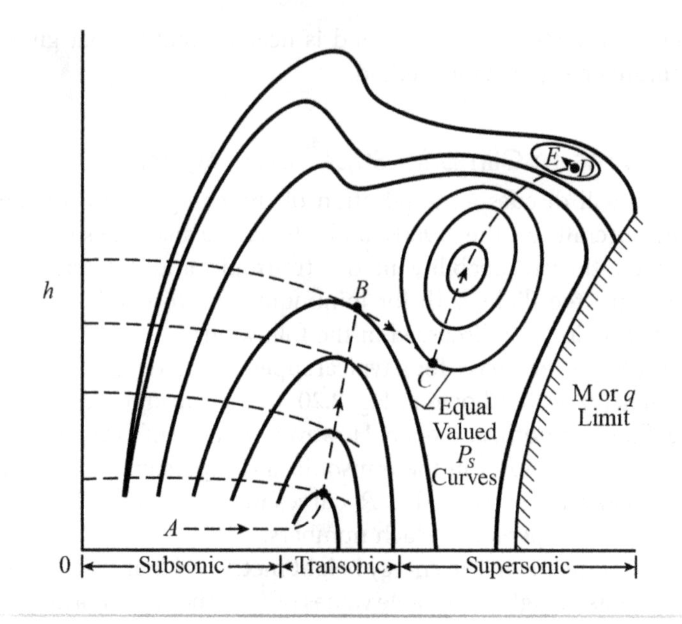

Fig. 2.20 Minimum time to climb for a supersonic aircraft with limited excess-thrust capability (8).

destination, represented by point E, during which it is in a decelerating climb. Even though such a flight path is an unusual one to fly, it is still the optimal compared to any other possible flight path connecting energy states A and E.

For a modern (supersonic) combat aircraft with sufficiently high excess-thrust capability, transonic discontinuities and closed contours in supersonic regions disappear, and the optimal climb schedule will be quite similar to that of the subsonic aircraft as shown in Fig. 2.19.

The $(R/C)_{\max}$ and the minimum time to climb to a given altitude are important performance metrics for interceptor/fighter aircraft. Table 2.1 gives these two parameters for some modern combat aircraft.

Table 2.1 Climb Performance of Some Aircraft (2)

Aircraft	Type	Sea Level, $(R/C)_{\max}$	Minimum Time to Climb
Mirage III	Interceptor	5.0 km/min	3.0 mt to 11 km
Mirage 2000	Interceptor	14.94 km/min	2.4 mt to 14.94 km
F-111	Fighter	10.94 km/min	N/A
F-4	Fighter	8.534 km/min	N/A
F-14	Interceptor	9.14 km/min	N/A
F-15	Fighter	15.24 km/min	N/A
F-16	Fighter	15.24 km/min	N/A

Example 2.5

For the propeller airplane of Example 2.3, determine the maximum climb angle, maximum rate of climb, and corresponding speed and lift coefficients.

Solution:

Using the power-available and power-required curves of Fig. 2.14, the rate of climb R/C is plotted against velocity V in Fig. 2.21. Also plotted in Fig. 2.21 is the climb angle γ obtained by dividing R/C by V.

From this graph, we find that the maximum rate of climb is 10.5 m/s and occurs at 50 m/s. The corresponding lift coefficient is found to be 1.084.

The steepest climb angle is 21.8 deg and occurs at 19.0 m/s but needs a lift coefficient of 7.54, which is in excess of the maximum lift coefficient of 1.75. Therefore, the permissible steepest climb angle for this aircraft occurs at V_{stall} of 39.4323 m/s and is equal to 14.5 deg.

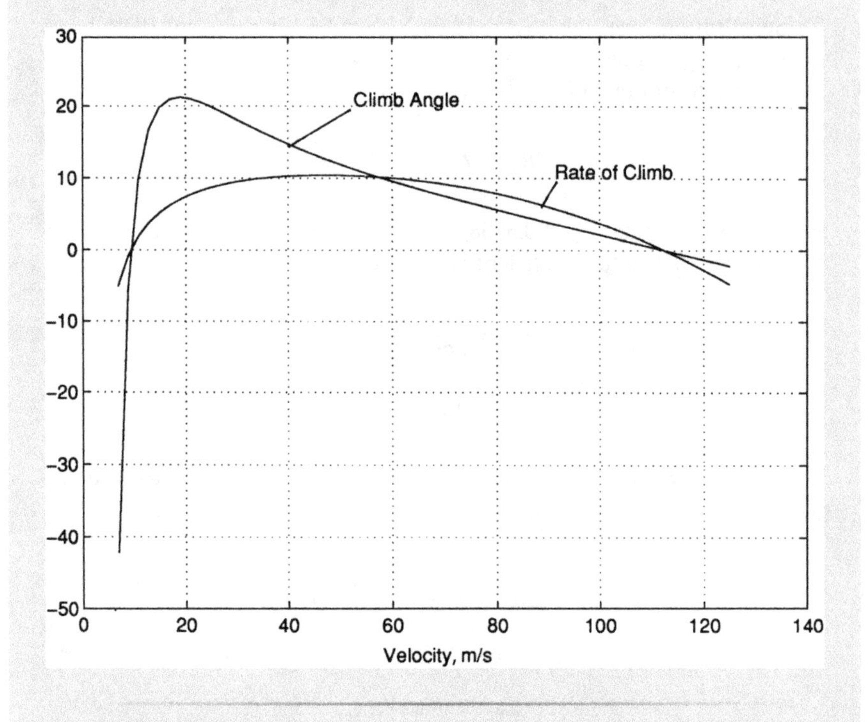

Fig. 2.21 Climb angle and rate of climb for propeller aircraft.

Example 2.6

An aircraft weighing 156,960 N is powered by a jet engine whose thrust is independent of flight speed. The maximum rate of climb at sea level occurs at 152.5 m/s. The wing area is 46 m^2. The drag polar is given by $C_D = 0.016 + 0.045C_L^2$ and $C_{L,\max} = 1.5$. Determine 1) the sea-level thrust T_0 developed by the engine, 2) lift coefficient, 3) maximum and minimum speeds in level flight at sea level, and 4) the absolute ceiling assuming that the thrust varies with altitude as $T = T_0\sigma^{0.6}$.

Solution:

We are given $W = 156{,}960.0$ N, $S = 46$ m^2, $C_{DO} = 0.016$, $k = 0.045$, $C_{L,\max} = 1.5$, $\rho = 1.225$ kg/m^3, and $V_{(R/C)\max} = 152.5$ m/s.

We have

$$E_m = \frac{1}{2\sqrt{kC_{DO}}}$$

$$V_R = \sqrt{\frac{2W}{\rho S}}\sqrt[4]{\frac{k}{C_{DO}}}$$

Substituting, we get $E_m = 18.64$, $V_R = 96.136$ m/s so that $u_m = V_{(R/C)\max}/V_R = 1.5863$.

For maximum rate of climb at sea level,

$$3u_m^4 - 2z_o u_m^2 - 1 = 0$$

Solving, we obtain $z_o = 3.5758$, which gives sea-level thrust $T_0 = 30{,}110.6790$ N. The lift coefficient is given by

$$\begin{aligned} C_L &= \frac{2W}{\rho S V^2_{(R/C)\max}} \\ &= 0.2370 \end{aligned}$$

The maximum and minimum velocities in level flight at sea level are given by

$$\begin{aligned} u_{\max} &= \sqrt{z_o + \sqrt{z_o^2 - 1}} \\ &= 2.6474 \end{aligned}$$

$$\begin{aligned} V_{\max} &= u_{\max} V_R \\ &= 254.51 \text{ m/s} \end{aligned}$$

(Continued)

Example 2.6 (Continued)

$$u_{\min} = \sqrt{z_o - \sqrt{z_o^2 - 1}}$$
$$= 0.3770$$
$$V_{\min} = u_{\min} V_R$$
$$= 36.31\,\text{m/s}$$
$$V_{\text{stall}} = \sqrt{\frac{2W}{\rho S C_{L,\max}}}$$
$$= 60.6135\,\text{m/s}$$

The flight at $V_{\min}$ is not possible because $V_{\min} < V_{\text{stall}}$.

We have

$$z = \frac{T_0 \sigma^{0.6} E_m}{W}$$
$$\sigma = \left(\frac{zW}{T_0 E_m}\right)^{\frac{1}{0.6}}$$

With $z = 1$ at absolute ceiling, we get $\sigma = 0.1196$. From standard atmospheric tables (see Appendix A), we get the altitude as approximately equal to 16.8 km.

2.6 Range and Endurance

Cruise flight begins at the end of the climb phase when the airplane has reached the desired altitude, and cruise flight ends when the descent phase begins. For a given amount of fuel load, the horizontal distance covered with respect to the ground during the cruise flight is called cruise range and does not include any distance covered during the climb or descent. In this text, the term range refers to the cruise range and will be expressed in kilometers.

Range is an important performance measure for commercial transport airplanes like the Boeing 747 or Airbus 320. Range is also important for military bombers like the B-52 or the B-1.

From physical considerations, the range will be maximum when the airplane cruises at that velocity when the ratio of velocity to fuel consumed per unit time is maximum.

Endurance is the total time that an airplane can remain in the air for a given fuel load and is usually expressed in hours. The endurance is an important performance measure for reconnaissance or surveillance airplanes. It is also of interest to all aircraft during the loiter, which is defined as that phase of flight where the primary aim is to remain airborne and not worry about the ground distance covered. Long-range bombers like the B-52 or the B-1 may loiter while on an airborne alert. A fighter aircraft on a combat air patrol mission loiters while guarding the airspace from intruding aircraft. When the primary aim is to be airborne, the aircraft is essentially in level flight, and it will fly at that speed when the fuel consumed per unit time is a minimum.

In the following sections, we will study the range and endurance of propeller and jet aircraft and determine the conditions when these quantities assume their maximum values.

2.6.1 Range of a Jet Aircraft

Because the airplane during the cruise is essentially in level flight, we have

$$L - W = 0 \tag{2.137}$$

$$T - D = 0 \tag{2.138}$$

Kinematic equations are

$$\dot{x} = V \tag{2.139}$$

$$\dot{h} = 0 \tag{2.140}$$

For range and endurance problems, we need an additional equation that describes the variation of weight. This is obtained by considering fuel consumption. For a jet airplane, the specific fuel consumption is the amount of fuel consumed per unit thrust per unit time so that the variation of weight is given by

$$\dot{W} = -cT \tag{2.141}$$

where c is the specific fuel consumption. For the specific fuel consumption of jet engines, we will use the units of N/Nh. Here, we assume that for a given altitude, specific fuel consumption is a constant. In other words, variation of specific fuel consumption with velocity is ignored. A negative sign is chosen in Eq. (2.141) because weight decreases as fuel is consumed.

Equation (2.139) can be written as

$$\frac{dx}{dW}\frac{dW}{dt} = V \tag{2.142}$$

$$\begin{aligned}\frac{dx}{dW} &= -\frac{V}{cT}\\ &= -\frac{V}{cD}\end{aligned} \tag{2.143}$$

The quantity dx/dW is often called the instantaneous range and is equal to the horizontal distance traversed per unit load of fuel or the specific range. It is analogous to the gas mileage of an automobile.

Multiply and divide the right-hand side of Eq. (2.143) by W and use the relation $L = W$ to obtain

$$\frac{dx}{dW} = -\left(\frac{L}{D}\right)\left(\frac{V}{cW}\right)$$
$$= -\left(\frac{C_L}{C_D}\right)\frac{V}{cW} \tag{2.144}$$

Let $R = x_f - x_i$ denote the horizontal distance covered during the cruise. Then,

$$R = -\int_{W_0}^{W_1}\left(\frac{C_L}{C_D}\right)\left(\frac{1}{c}\right)V\frac{dW}{W} \tag{2.145}$$

where W_0 is the initial weight and W_1 is the final weight. Let

$$W_f = W_0 - W_1 \tag{2.146}$$

denote the weight of fuel consumed during the cruise.

To integrate Eq. (2.145), we have to specify the variation of lift and drag coefficients and the velocity during the cruise. We assume that the angle of attack is held constant at some value throughout the cruise so that the lift and drag coefficients are constants. To determine this value of angle of attack and the corresponding flight velocity, we consider two special cases as discussed in the following subsections.

Range at Constant Altitude

To hold the altitude at a constant value, the velocity must vary continuously during the cruise to compensate for the variation in weight (because of fuel consumption) according to the relation,

$$V = \sqrt{\frac{2W}{\rho S C_L}} \tag{2.147}$$

where W is the instantaneous weight of the airplane. Then, substitution in Eq. (2.145) gives

$$R = -\int_{W_0}^{W_1}\left(\frac{\sqrt{C_L}}{C_D}\right)\left(\frac{1}{c}\right)\sqrt{\frac{2}{\rho S}}\frac{dW}{\sqrt{W}}$$
$$= \left(\frac{2}{c}\right)\left(\frac{\sqrt{C_L}}{C_D}\right)\sqrt{\frac{2}{\rho S}}\left[\sqrt{W_0} - \sqrt{W_1}\right] \tag{2.148}$$

Thus, for a given amount of fuel load, the range is maximum when the aerodynamic parameter $(\sqrt{C_L}/C_D)$ is a maximum. It can be shown that this happens when $C_L = \sqrt{C_{DO}/3k}$ or $C_L = (1/\sqrt{3})C_L^*$ and $V = \sqrt[4]{3}V_R$. Thus, the velocity for best constant-altitude range is equal to $\sqrt[4]{3}$ times the velocity for minimum drag.

According to Eq. (2.148), the range improves as altitude increases because the air density appears in the denominator. In general, the thrust available drops with altitude so that there is an altitude where the overall range is maximum. This altitude is called the most economical or cruise altitude. We will discuss an approximate method to determine this altitude later in this section.

Constant-Velocity Range

When the cruise velocity is held constant, Eq. (2.145) can be integrated to obtain

$$R = \left(\frac{V}{c}\right) E \ln\left(\frac{W_0}{W_1}\right) \tag{2.149}$$

Equation (2.149) is known as the Breguet range formula. The Breguet range is maximum when the aircraft flies at that velocity for which $E = E_m$. Recall that when $E = E_m$, $C_L = C_L^*$, the flight velocity is equal to V_R, and the drag in level flight is minimum.

The maximum range is given by

$$R_{\max} = \left(\frac{V_R}{c}\right) E_m \ln\left(\frac{W_0}{W_1}\right) \tag{2.150}$$

where V_R is based on initial weight W_0 and is given by

$$V_R = \sqrt{\frac{2W_0}{\rho S}} \sqrt[4]{\frac{k}{C_{DO}}} \tag{2.151}$$

During this constant-velocity cruise, the airplane will steadily gain the altitude because of continuous decrease in weight. This phenomenon is called cruise climb. The altitude gained could be substantial for long cruises. Later in this section, we will discuss an approximate method to estimate the height gained during the cruise climb.

2.6.2 Range of Propeller Aircraft

For propeller airplanes, the specific fuel consumption is the amount of fuel consumed per unit power developed by the engine per unit time so that

$$\dot{W} = -cP \tag{2.152}$$

where P is the power developed by the engine. The Système International (SI) unit for power is kW. Here, we assume that for a given altitude the specific fuel consumption is a constant. Then,

$$\frac{dx}{dW} = -\frac{V}{cP} \tag{2.153}$$

The power required is equal to the product of drag D and velocity V. The power available is the product of the power developed by the engine P and the propulsive efficiency η_p. For steady level flight condition, the power available must be equal to power required, so that

$$P = \frac{DV}{\eta_p} \tag{2.154}$$

Then

$$\frac{dx}{dW} = -\frac{\eta_p}{cD} \tag{2.155}$$

Multiply and divide the right-hand side of Eq. (2.155) by W and use the relation $L = W$ to obtain

$$\frac{dx}{dW} = -\left(\frac{C_L}{C_D}\right)\frac{\eta_p}{cW} \tag{2.156}$$

With $R = x_f - x_i$, we get

$$R = -\int_{W_0}^{W_1} E\left(\frac{\eta_p}{c}\right)\frac{dW}{W} \tag{2.157}$$

We assume that the angle of attack is held constant at some value throughout the cruise so that the lift-to-drag ratio E remains constant. Furthermore, we assume that for a given altitude, $\eta_p = \text{const}$. In other words, we assume that the propulsive efficiency is independent of flight velocity at any given altitude. With these assumptions, Eq. (2.157) can be integrated to obtain

$$R = \left(\frac{\eta_p}{c}\right)E\,\ell n\left(\frac{W_0}{W_1}\right) \tag{2.158}$$

It is interesting to observe that both altitude and flight velocity do not explicitly appear in the equation for the range of a propeller aircraft. Because both power developed and specific fuel consumption vary with altitude, the range will actually depend on altitude.

The range of a propeller aircraft assumes a maximum value when flying at $E = E_m$ and is given by

$$R_{\max} = \left(\frac{\eta_p}{c}\right)E_m\,\ell n\left(\frac{W_0}{W_1}\right) \tag{2.159}$$

We know that when $E = E_m$, $C_L = C_L^*$, this gives us the value of lift coefficient or the angle of attack for $R_{\max}$. What about the flight velocity for $R_{\max}$? For this purpose, we consider two options: 1) constant-velocity cruise, and 2) constant-altitude cruise. In constant-velocity cruise, the cruise velocity is taken equal to the velocity V_R, i.e., based on the initial weight W_0 as given by

$$V = V_R = \sqrt{\frac{2W_0}{\rho S}}\sqrt[4]{\frac{k}{C_{DO}}} \tag{2.160}$$

In constant-velocity cruise, the propeller aircraft will gain altitude in a similar manner to that of the jet aircraft.

In constant-altitude cruise, the flight velocity varies continuously and is equal to the instantaneous value of V_R as given by

$$V = V_R = \sqrt{\frac{2W}{\rho S}}\sqrt[4]{\frac{k}{C_{DO}}} \tag{2.161}$$

where W is the instantaneous weight.

2.6.3 Effect of Wind on Range

The expressions just derived for range are for calm air, i.e., we assumed that the Earth's atmosphere is stationary. However, in practice this is not always true. It is a common knowledge that tailwind has a beneficial effect and headwind has an adverse effect on the range. In the following, we will derive approximate expressions for the range of a jet airplane in the presence of the wind. A similar approach can be used for the propeller airplane, and this is left as an exercise for the reader.

The velocity with respect to the ground in presence of the wind is given by

$$\frac{dx}{dt} = V \mp V_w \tag{2.162}$$

where V_w is the wind velocity. The "−" and "+" signs apply for headwind and tailwind, respectively. For a given air speed V, the headwind reduces the Earth-related velocity, whereas tailwind increases it.

Then,

$$\begin{aligned} dx &= -(V \mp V_w)\frac{dW}{cT} \\ &= -(V \mp V_w)E\left(\frac{1}{c}\right)\frac{dW}{W} \end{aligned} \tag{2.163}$$

so that the range in the presence of the wind is given by

$$R_w = -\int_{W_0}^{W_1} (V \mp V_w)E\left(\frac{1}{c}\right)\frac{dW}{W} \tag{2.164}$$

As before, we assume that the angle of attack is held constant. Furthermore, assuming that the wind velocity is constant, we have

$$R_W = R \mp E\left(\frac{1}{c}\right) V_W \, \ell n \frac{W_0}{W_1} \tag{2.165}$$

In Eq. (2.165), the "$-$" sign applies for the headwind and "$+$" sign for the tailwind. Note that R_W is the range in presence of the wind and R is the still air range. Equation (2.165) applies for both constant-altitude cruise and constant-velocity cruise. As expected, the headwind reduces the range, and tailwind improves it.

2.6.4 Estimation of Cruise Altitude

The cruise, or the most economical altitude, is that altitude where the overall range is maximum. For propeller airplanes, it is in the range of 4–7 km and, for jet airplanes, it is 11–17 km.

An approximate estimation of the cruise altitude can be made if we ignore the variation of specific fuel consumption with altitude. The basic concept of this method is illustrated in Fig. 2.22 for propeller airplanes and in Fig. 2.23 for jet airplanes. For propeller airplanes, the power required P_R based on W_0 is plotted against the equivalent air speed V_e and, on this graph, the power-available P_a curves at various altitudes are superposed as shown. The cruise altitude corresponds to the altitude of that P_a curve, which intersects the P_R curve at the point where $V_e = V_{e,R_{\max}}$. Here, $V_{e,R_{\max}}$ is the equivalent air speed for the maximum range and is obtained by drawing a tangent from the origin to the P_R curve as shown in Fig. 2.22.

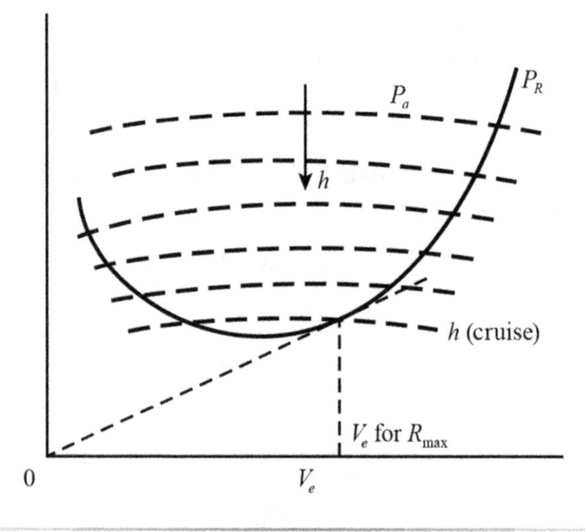

Fig. 2.22 Cruise altitude for propeller aircraft.

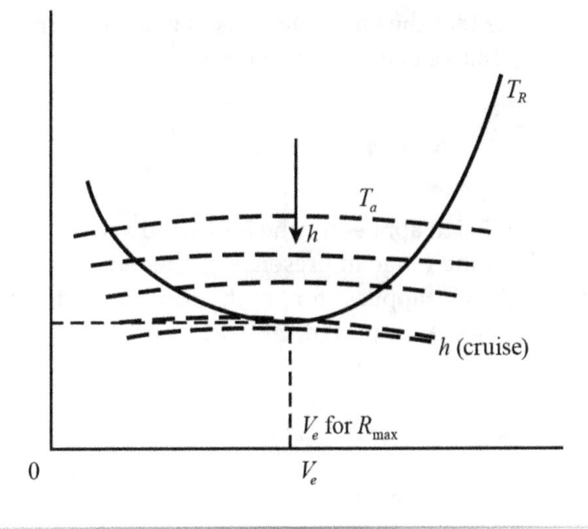

Fig. 2.23 Cruise altitude for jet aircraft.

For the jet aircraft, the thrust required T_R based on initial weight W_0 is plotted against the equivalent air speed V_e, and the thrust-available T_a curves at various altitudes are superposed on this graph as shown in Fig. 2.23. The cruise altitude corresponds to the altitude of that T_a curve, which is tangential to the T_R curve. The velocity at this tangential point is the velocity for maximum Breguet range. Note that this altitude is not the absolute ceiling because T_R is based on the initial weight W_0 and not the instantaneous weight.

A similar approach can be used for the constant-altitude range program.

For jet airplanes, the cruise altitude can also be determined analytically if we assume: 1) the thrust varies with altitude as $T = T_0\sigma^\beta$, where β is a constant smaller than unity, and 2) we ignore the variation of specific fuel consumption with altitude. We have

$$T = D \tag{2.166}$$

Multiply and divide the right-hand side of Eq. (2.166) by W_0 and use the condition $L = W_0$ to obtain

$$\begin{aligned} T_0\sigma^\beta &= \frac{W_0 D}{L} \\ &= \frac{W_0 C_D}{C_L} \\ &= \frac{W_0}{E} \end{aligned} \tag{2.167}$$

$$\sigma = \left(\frac{W_0}{ET_0}\right)^{\frac{1}{\beta}} \tag{2.168}$$

With this value of σ, the most economical cruise altitude h in kilometers can be approximately obtained using

$$h = 44.3(1 - \sigma^{0.235}) \tag{2.169}$$

2.6.5 Approximate Estimation of Height Gained During Cruise Climb

This method is based on the assumption that the relations $L = W$ and $T = D$ hold at the beginning and at the end of the constant-velocity cruise. In other words, we assume that, at the endpoint of the constant-velocity cruise, the decrease in thrust and drag caused by increase in altitude compensate each other, and similarly the decrease in lift (caused by increase in altitude) and weight (caused by fuel consumption) also balance each other. Furthermore, we ignore any variations in specific fuel consumption during the cruise.

Let ρ_i and ρ_f denote densities at the initial and final altitudes. Then

$$W_0 = \frac{1}{2}\rho_i V^2 S C_L \tag{2.170}$$

$$W_1 = \frac{1}{2}\rho_f V^2 S C_L \tag{2.171}$$

so that

$$\rho_f = \frac{W_1}{W_0}\rho_i \tag{2.172}$$

or

$$\sigma_f = \frac{W_1}{W_0}\sigma_i \tag{2.173}$$

The corresponding altitude h_f in kilometers can be approximately estimated using the following relation:

$$h_f = 44.3\left(1 - \sigma_f^{0.235}\right) \tag{2.174}$$

so that the height gained is given by

$$\Delta h = h_f - h_i \tag{2.175}$$

2.7 Endurance

The endurance of an airplane is the total time an airplane remains in the air and is usually expressed in hours. Assume that most of the time the airplane is in level flight; we can determine the endurance of jet and propeller airplanes as follows.

2.7.1 Endurance of Jet Aircraft

We have

$$\frac{dW}{dt} = -cT \tag{2.176}$$

$$dt = -\frac{dW}{cT}$$

$$= -\frac{dW}{cT}\frac{W}{W} \tag{2.177}$$

With $T = D$ and $L = W$,

$$dt = -\left(\frac{1}{c}\right)E\left(\frac{dW}{W}\right) \tag{2.178}$$

or

$$t = -\int_{W_0}^{W_1}\left(\frac{1}{c}\right)E\left(\frac{dW}{W}\right) \tag{2.179}$$

Assuming that the angle of attack is held constant throughout the flight,

$$t = \left(\frac{1}{c}\right)E\,\ell n\left(\frac{W_0}{W_1}\right) \tag{2.180}$$

$$t_{\max} = \left(\frac{1}{c}\right)E_m\,\ell n\left(\frac{W_0}{W_1}\right) \tag{2.181}$$

We observe that the altitude and flight velcocity do not directly affect the endurance. However, the altitude has an indirect influence through the specific fuel consumption, which generally varies with altitude. Therefore, the endurance improves when the jet airplane flies at that altitude where the specific fuel consumption is low.

For a given altitude, the endurance is maximum when the airplane angle of attack is such that $E = E_m$. We know that when $E = E_m$, $C_L = C_L^*$, $V = V_R$, and $D = D_{\min}$.

2.7.2 Endurance of Propeller Aircraft

For propeller aircraft, we have

$$\frac{dW}{dt} = -cP \tag{2.182}$$

or

$$\begin{aligned} dt &= -\frac{dW}{cP} \\ &= -\frac{\eta_p dW}{cDV} \\ &= -\left(\frac{\eta_p}{c}\right)\left(\frac{L}{D}\right)\left(\frac{1}{V}\right)\frac{dW}{W} \end{aligned} \tag{2.183}$$

or

$$t = -\left(\frac{\eta_p}{c}\right)\int_{W_0}^{W_1} E\left(\frac{1}{V}\right)\frac{dW}{W} \tag{2.184}$$

As said before, we assume that the angle of attack is held constant during the flight so that E is constant. The endurance of a propeller aircraft depends on velocity. Here, we consider two options: 1) flight at constant altitude or 2) flight with constant velocity.

For flight with constant altitude, the velocity is allowed to vary to compensate for the decrease in weight as given by

$$V = \sqrt{\frac{2W}{\rho S C_L}} \tag{2.185}$$

where W is the instantaneous weight. With this, we obtain

$$t = \left(\frac{2\eta_p}{c}\right)\left(\frac{C_L^{\frac{3}{2}}}{C_D}\right)\sqrt{\frac{\rho S}{2}}\left(\sqrt{\frac{1}{W_1}} - \sqrt{\frac{1}{W_0}}\right) \tag{2.186}$$

From Eq. (2.186) we observe that the endurance of the propeller aircraft depends explicitly on the altitude. The maximum endurance occurs at sea level when the aerodynamic parameter $(C_L^{3/2}/C_D)$ is maximum. From Eq. (2.53), we have

$$\left(\frac{C_L^{\frac{3}{2}}}{C_D}\right)_{\max} = \frac{1}{4}\sqrt[4]{\frac{27}{k^3 C_{DO}}} \tag{2.187}$$

and this happens when $C_L = \sqrt{3C_{DO}/k}$ or $C_L = \sqrt{3}C_L^*$, $C_{D,i} = 3C_{DO}$, $C_D = 4C_{DO}$, and $V = V_R/\sqrt[4]{3} = V_{mp}$, the velocity for $P_{R,\min}$ in level flight.

For flight at constant velocity, the endurance is given by

$$t = \left(\frac{\eta_p}{c}\right)\left(\frac{E}{V_R}\right)\ell n\left(\frac{W_0}{W_1}\right) \tag{2.188}$$

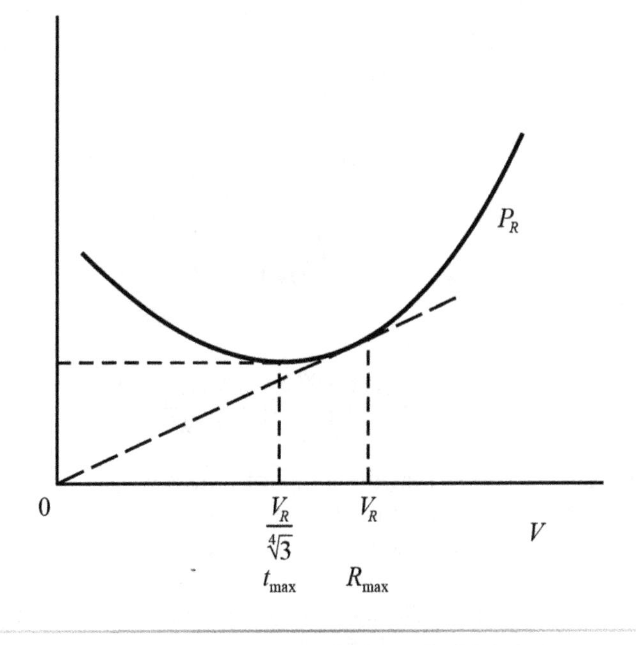

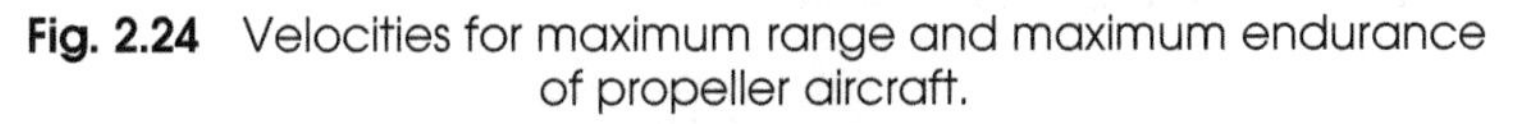

Fig. 2.24 Velocities for maximum range and maximum endurance of propeller aircraft.

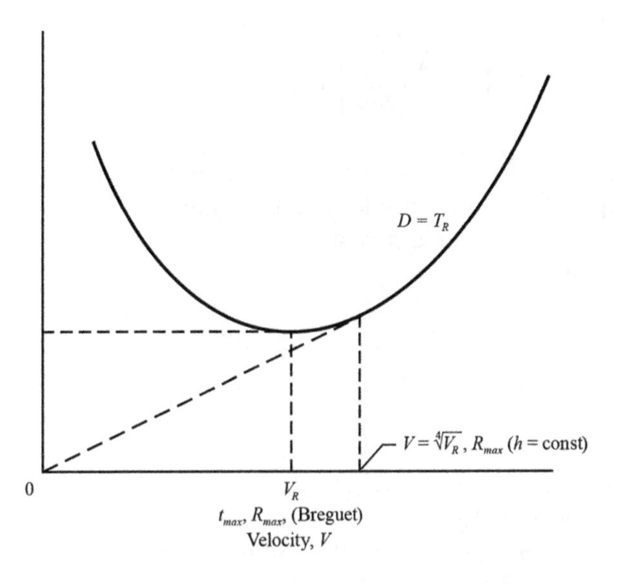

Fig. 2.25 Velocities for maximum range and maximum endurance of jet aircraft.

The maximum endurance occurs when $V = V_R$ and $E = E_m$. Here, the velocity V_R is based on initial weight W_0 as given by

$$V_R = \sqrt{\frac{2W_0}{\rho S}} \sqrt[4]{\frac{C_{DO}}{k}} \tag{2.189}$$

Note that the flight velocity is held constant at this value throughout the flight.

The velocities for maximum range and endurance are schematically shown for propeller and jet aircraft in Figs. 2.24 and 2.25.

Example 2.7

A turbojet airplane weighs 80,343.9 N, has a wing area of 37.2 m^2, and has a specific fuel consumption equal to 1.3 N/Nh at an altitude of 10,000 m ($\sigma = 0.374$). The airplane drag polar is given by $C_D = 0.02 + 0.06C_L^2$. Determine the fuel loads for 1) range of 2400 km and 2) endurance of 5 h. With fuel load as obtained in (1), find the change in the range if the airplane encounters a steady headwind of 10 m/s throughout the flight.

Solution:

We will solve this range problem using both constant-altitude and constant-velocity (Breguet) methods.

The maximum Breguet range is given by Eq. (2.150).

$$R_{max} = \left(\frac{V_R}{c}\right) E_m \, \ell n \frac{W_0}{W_1}$$

We have $W_0 = 80{,}343.9$ N, $S = 37.2$ m^2, $c = 1.3$ N/Nh, $C_{DO} = 0.02$, $k = 0.06$, $h_i = 10$ km, $t_{max} = 5$ h, $R_{max} = 2400$ km, and $\sigma_i = 0.374$.

The specific fuel consumption has the units of N/Nh, and we have to express it in terms of N/Ns as follows.

$$c = 1.3 \frac{\text{N}}{\text{Nh}}$$
$$= \frac{1.3}{3600} \frac{\text{N}}{\text{Ns}}$$

We have

$$E_m = \frac{1}{2\sqrt{kC_{DO}}}$$
$$= \frac{1}{2\sqrt{0.02 * 0.06}}$$
$$= 14.433$$

(Continued)

Example 2.7 (Continued)

$$V_R = \sqrt{\frac{2W_0}{\rho S}} \sqrt[4]{\frac{k}{C_{DO}}}$$

$$= \sqrt{\frac{2 * 80{,}343.9}{1.225 * 0.374 * 37.2}} \sqrt[4]{\frac{0.06}{0.02}}$$

$$= 127.8108 \text{ m/s}$$

For $R_m = 2400 * 10^3$ m,

$$2400 * 10^3 = \left(\frac{127.8108 * 3600}{1.3}\right) * 14.433 * \ell n \frac{W_0}{W_1}$$

Solving, $W_1 = 50{,}227.5$ N and $W_f = W_0 - W_1 = 30{,}116.4$ N.

The approximate height gained in cruise climb is obtained as follows.

$$\sigma_f = \frac{W_1}{W_0} \sigma_i$$

$$= \frac{50{,}227.5}{80{,}343.9} * 0.374$$

$$= 0.23378$$

Then,

$$h_f = 44.3\left(1 - \sigma_f^{0.235}\right)$$

$$= 44.3\left(1 - 0.23378^{0.235}\right)$$

$$= 12.8172 \text{ km}$$

$$\Delta h = h_f - h_i$$

$$= 2.8172 \text{ km}$$

Approximate climb angle, $\tan \gamma = 2.8172/2400 = 0.0012$ rad $= 0.0673$ deg. Thus, we observe that γ is very small as assumed.

The maximum constant-altitude range is given by Eq. (2.148),

$$R_{\max} = \left(\frac{2}{c}\right)\left(\frac{\sqrt{C_L}}{C_D}\right)_{\max} \sqrt{\frac{2}{\rho S}}\left[\sqrt{W_0} - \sqrt{W_1}\right]$$

For $(\sqrt{C_L}/C_D)_{\max}$, we have

$$C_L = \sqrt{\frac{C_{DO}}{3k}}$$

$$= \sqrt{\frac{0.02}{3 * 0.06}}$$

$$= 0.3333$$

(Continued)

Example 2.7 (*Continued*)

$$C_D = 0.02 + 0.06 * 0.3333^2$$
$$= 0.0266$$

so that

$$\left(\frac{\sqrt{C_L}}{C_D}\right)_{max} = \frac{\sqrt{0.3333}}{0.0266}$$
$$= 21.705$$

Substituting,

$$2400 * 10^3 = \left(\frac{2 * 3600}{1.3}\right)(21.705)\sqrt{\frac{2}{0.374 * 1.225 * 37.2}}\left[\sqrt{W_0} - \sqrt{W_1}\right]$$

Solving, $W_1 = 50{,}697.024$ N, and $W_f = W_0 - W_1 = 29{,}646.87$ N. Thus, the fuel consumption for constant-altitude range is slightly smaller compared to the constant-velocity (Breguet) range. However, the disadvantage is that the flight velocity continuously varies during the cruise.

The decrease in the range because of headwind for the constant-altitude program is given by Eq. (2.165),

$$\Delta R = -E\left(\frac{1}{c}\right)V_w \ell n \frac{W_0}{W_1}$$
$$= -\frac{0.3333}{0.0266}\left(\frac{3600}{1.3}\right) * 10.0 * \ell n \frac{80{,}343.9}{50{,}227.5}$$
$$= -162.9982\,\text{km}$$

Similarly, the decrease in range for constant-velocity (Breguet) case is given by

$$\Delta R = -E_m\left(\frac{1}{c}\right)V_w \ell n \frac{W_0}{W_1}$$

Substituting, we get

$$\Delta R = -187.7525\,\text{km}$$

The maximum endurance is given by

$$t_{max} = \left(\frac{1}{c}\right)E_m \ell n \frac{W_0}{W_1}$$

Substituting,

$$5 * 3600 = \frac{3600}{1.3} * 14.433 * \ell n \frac{80{,}343.9}{W_1}$$
$$W_1 = 51{,}211.256\,\text{N}$$

(*Continued*)

Example 2.7 (Continued)

Then, the corresponding fuel load is given by

$$W_f = W_0 - W_1$$
$$= 29{,}132.644\,\text{N}$$

Example 2.8

A propeller aircraft weighs 50,000 N and has a wing area of 30 m^2. Its reciprocating engine produces a thrust of 840 kW, and the propulsive efficiency is 0.85. The drag polar of the aircraft is given by $C_D = 0.025 + 0.05C_L^2$ and $C_{L,\max} = 1.60$. The specific fuel consumption is 3.0 N per kW/h. For a fuel load of 10,000 N, determine the best range and best endurance if the aircraft flies at an altitude of 3000 m ($\sigma = 0.7423$).

Solution:
For range,

$$R = \left(\frac{\eta_p}{c}\right) E \, \ell n\left(\frac{W_0}{W_1}\right)$$

For maximum range, we have

$$E = E_m$$
$$= \frac{1}{\sqrt{2kC_{DO}}}$$
$$= \frac{1}{\sqrt{2 * 0.025 * 0.05}}$$
$$= 14.1421$$

$$C_L = \sqrt{\frac{C_{DO}}{k}}$$
$$= \sqrt{\frac{0.025}{0.05}}$$
$$= 0.7071$$

(*Continued*)

Example 2.8 (*Continued*)

The specific fuel consumption c is given in N/kW/h. We have to convert it in terms of N per Nm/s. We have 1 kW = 1000 Nm/s. With this,

$$c = \frac{3.0\,\mathrm{N}}{1000\,\mathrm{Nm/s}\;3600\,\mathrm{s}}$$
$$= \frac{3.0}{1000 * 3600}\frac{1}{\mathrm{m}}$$

The range is given by

$$R = \left(\frac{\eta_p}{c}\right) E\,\ell n\left(\frac{W_0}{W_1}\right)$$

Substituting,

$$R = \left(\frac{3600 * 1000 * 0.85}{3.0}\right) 14.1421\,\ell n\left(\frac{50{,}000}{40{,}000}\right)$$
$$= 3{,}218{,}840.9\,\mathrm{m}$$
$$= 3218.8409\,\mathrm{km}$$

To cover this range, we have two options: 1) cruise at constant altitude or 2) cruise at constant velocity.

For the constant-altitude cruise, the flight velocity varies continuously from V_0 to V_1 as the weight decreases from $W_0 = 50{,}000$ N to $W_1 = 50{,}000 - 10{,}000 = 40{,}000$ N, where

$$V_0 = \sqrt{\frac{2W_0}{\rho S C_L}}$$
$$= \sqrt{\frac{2 * 50{,}000}{(1.225 * 0.7423) * 30.0 * 0.7071}}$$
$$= 72.0021\,\mathrm{m/s}$$

and

$$V_1 = \sqrt{\frac{2W_1}{\rho S C_L}}$$
$$= \sqrt{\frac{2 * 40{,}000}{(1.225 * 0.7423) * 30.0 * 0.7071}}$$
$$= 64.4007\,\mathrm{m/s}$$

(*Continued*)

Example 2.8 (Continued)

For constant-velocity range, $V = V_0$, and the airplane gains altitude during the cruise. The height gained is obtained as follows.

$$\begin{aligned}\sigma_f &= \frac{W_1}{W_0}\sigma_i \\ &= \frac{40{,}000}{50{,}000} * 0.7423 \\ &= 0.5938\end{aligned}$$

Then,

$$\begin{aligned}h_f &= 44.3\left(1 - \sigma_f^{0.235}\right) \\ &= 44.3\left(1 - 0.5938^{0.235}\right) \\ &= 5.1066\,\text{km} \\ \Delta h &= h_f - h_i \\ &= 2.1066\,\text{km}\end{aligned}$$

The maximum endurance is given by

$$t_{max} = \left(\frac{2\eta_p}{c}\right)\left(\frac{C_L^{\frac{3}{2}}}{C_D}\right)_{max}\sqrt{\frac{\rho S}{2}}\left(\sqrt{\frac{1}{W_1}} - \sqrt{\frac{1}{W_0}}\right)$$

Using Eq. (2.53)

$$\begin{aligned}\left(\frac{C_L^{\frac{3}{2}}}{C_D}\right)_{max} &= \frac{1}{4}\sqrt{\frac{27}{k^3 C_{DO}}} \\ &= \frac{1}{4}\sqrt{\frac{27}{0.05^3 * 0.025}} \\ &= 13.5540\end{aligned}$$

Substituting, we obtain $t_{max} = 14.9733$ h.

2.8 Turning Flight

In this section we will study the aircraft performance in turning flight. Whereas the turning flight is one of the routine flights for the majority of aircraft, it is a key tactical maneuver for the combat aircraft. In addition, the turning performance is one important metric of air superiority of the fighter aircraft.

The turning flight can be broadly classified in two categories: 1) steady turning flight in a horizontal plane, and 2) general turning flight. In a steady or constant-velocity turning flight in a horizontal plane, the airplane is at a constant altitude, whereas the general turning flight may involve a gain or loss of altitude. The routine flights of commercial transport and general aviation airplanes usually belong to the first category. The examples of the second category are the turning flight of a glider and that of fighter aircraft performing limiting turns exploiting the full aerodynamic and structural capabilities. We will also discuss a new concept in turning known as the Herbst maneuver, which is said to exceed the best performance attainable by the conventional turning maneuvers of categories 1 and 2.

2.8.1 Equations of Motion for Turning Flight

The forces acting on an airplane in a steady, constant-velocity turn with bank angle μ and sideslip β are shown in Fig. 2.26. The aircraft has to bank or sideslip to generate the necessary centripetal force. The following equations govern the steady turning flight.

Along the flight path,

$$T\cos\beta - D - W\sin\gamma = 0 \tag{2.190}$$

Along the principal normal,

$$L\cos\mu - W\cos\gamma = 0 \tag{2.191}$$

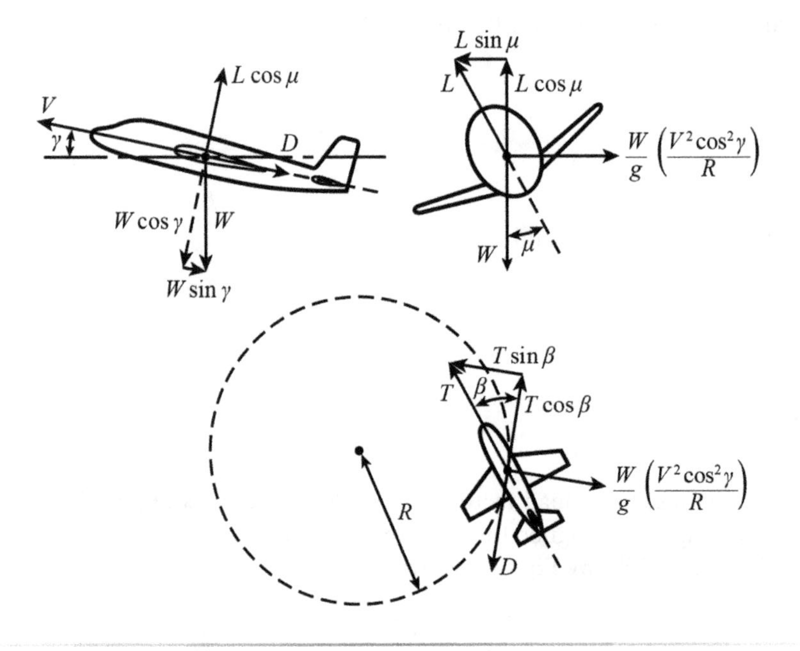

Fig. 2.26 Aircraft in turning flight in a horizontal plane.

Along the binormal,

$$T \sin \beta + L \sin \mu - \left(\frac{W V^2 \cos^2 \gamma}{g R} \right) = 0 \tag{2.192}$$

where μ is the bank angle, β is the sideslip angle, γ is the flight path angle, and R is the radius of turn.

The kinematic equations are

$$\dot{x} = V \cos \gamma \tag{2.193}$$

$$\dot{h} = V \sin \gamma \tag{2.194}$$

2.8.2 Turning Flight in a Horizontal Plane

Let us consider a steady turning flight in horizontal plane so that $\gamma = 0$. For simplicity, let us ignore the variation of the weight because of fuel consumption during the turn. Then, Eqs. (2.190–2.192) take the following form:

$$T \cos \beta - D = 0 \tag{2.195}$$

$$L \cos \mu - W = 0 \tag{2.196}$$

$$T \sin \beta + L \sin \mu - \frac{W V^2}{g R} = 0 \tag{2.197}$$

From Eq. (2.196), we obtain the load factor in turn as

$$\begin{aligned} n &= \frac{L}{W} \\ &= \frac{1}{\cos \mu} \end{aligned} \tag{2.198}$$

The flight velocity is given by

$$\begin{aligned} V &= \sqrt{\frac{2n(W/S)}{\rho C_L}} \\ &= \sqrt{\frac{2(W/S)}{\rho C_L \cos \mu}} \end{aligned} \tag{2.199}$$

For turning flight, the load factor n depends on the bank angle. The higher the bank angle, the higher the load factor that will be developed in turn. Dividing Eq. (2.197) by Eq. (2.196) we get

$$\tan \mu = \frac{V^2}{R g} - \frac{T \sin \beta}{W} \tag{2.200}$$

The radius of curvature R and rate of turn ω are given by

$$R = \frac{V^2}{g\left(\tan\mu + \dfrac{T\sin\beta}{W}\right)} = \frac{2n(W/S)}{\rho g C_L\left(\tan\mu + \dfrac{T\sin\beta}{W}\right)} \tag{2.201}$$

$$\omega = \frac{V}{R} = \frac{g\left(\tan\mu + \dfrac{T\sin\beta}{W}\right)}{V} = g\left(\tan\mu + \frac{T\sin\beta}{W}\right)\sqrt{\frac{\rho C_L}{2n(W/S)}} \tag{2.202}$$

The time for one complete turn of 2π radians is given by

$$t_{2\pi} = \frac{2\pi}{\omega} = \frac{2\pi}{g\left(\tan\mu + \dfrac{T\sin\beta}{W}\right)}\sqrt{\frac{2n(W/S)}{\rho C_L}} \tag{2.203}$$

The important metrics of turning performance are the turn rate or the angular velocity in turn ω and the radius of turn R. From the equations given, we observe that the aircraft with smaller values of wing loading W/S will have higher turn rates and a lower radius of turn, everything else being equal. Notice that the expressions for load factor n, rate of turn ω, and radius of turn R do not explicitly contain the drag term. It is implied that for these relations to be true, the thrust available balances the drag as given by Eq. (2.195).

We also observe that sideslip helps the turning flight because of the availability of a thrust component $T\sin\beta$, which provides part of the centripetal force as shown in Fig. 2.26. However, this benefit may be slightly offset because of the additional drag experienced by the aircraft in sideslip.

2.8.3 Coordinated Turning Flight in Horizontal Plane

Let us consider a steady, coordinated turning flight in a horizontal plane. By coordinated turn, we mean that the sideslip is zero and the aircraft is correctly banked so that the lift vector is tilted away from the plane of symmetry to provide all the centripetal force.

With $\beta = 0$, Eqs. (2.195–2.197) take the following form:

$$T - D = 0 \tag{2.204}$$

$$L \cos \mu - W = 0 \tag{2.205}$$

$$L \sin \mu - \frac{WV^2}{gR} = 0 \tag{2.206}$$

Then,

$$n = \frac{1}{\cos \mu} \tag{2.207}$$

$$V = \sqrt{\frac{2nW}{\rho SC_L}} = \sqrt{\frac{2W}{\rho SC_L \cos \mu}} \tag{2.208}$$

$$\tan \mu = \frac{V^2}{Rg} \tag{2.209}$$

$$R = \frac{V^2}{g \tan \mu} = \frac{V^2}{g\sqrt{n^2 - 1}} \tag{2.210}$$

$$\omega = \frac{g \tan \mu}{V} = \frac{g\sqrt{n^2 - 1}}{V} \tag{2.211}$$

$$t_{2\pi} = \frac{2\pi}{\omega} = \frac{2\pi V}{g \tan \mu} \tag{2.212}$$

As said earlier, the parameters that are of special interest in the turning performance are the rate of turn ω and the radius of turn R. From Eqs. (2.204–2.212), we observe that the turning performance improves with increase in load factor, all the other things being equal.

The load factor is a measure of the stress to which both the aircraft and the pilot are subjected. A load factor of 2 means the aircraft structure and the pilot are stressed twice as much as in steady level flight with a load factor of unity. For transport airplanes, the value of limiting load factor n_{lim} is about 2.5 whereas, for fighter aircraft, it can be as high as 9. However, in most of the cases, this load factor limit of a fighter aircraft becomes more of a limitation to the pilot rather than to the machine he is flying [2]. As load factor increases, the pilot experiences what is called "blackout" or "greyout" caused by the blood draining away from his brain; then he loses color perception. With further increase in load factor, he will experience loss of peripheral vision and tunnel vision. Eventually, he will experience a complete loss of sight. Because an attacking pilot must be in total control of his physical and mental faculties, he may not be actually able to stress the machine to its limiting value if it is as high as 9. He could be unconscious by the time he is pulling 6 g or 7 g. However, a defending pilot is more likely to use all the limiting load factor of his machine.

A fighter aircraft should be designed to have a high structural limit load factor. For air superiority, a fighter aircraft should have a highest possible rate of turn and a lowest possible radius of turn. In general, these two conditions do not occur at the same time. The capability to turn as fast as possible is often preferred compared to the capability to make the sharpest turn because it is generally known that a fighter aircraft having a higher rate of turn but not making as sharp turn as its adversary usually performs better in close combat. Some of the modern high-performance combat aircraft such as the F-15 or the F-16 are capable of producing turn rates as high as 20 deg/s.

For given values of engine thrust or power, wing loading, and aircraft drag polar, the problem of determining optimal values of turn rate, the radius of turn subject to the restrictions imposed by the structural limit load factor n_{lim}, and aerodynamic limitation $C_{L,\text{max}}$ is a nonlinear programming problem and is beyond the scope of this text. Interested readers may refer elsewhere [1]. Instead, we will present a simple analysis that gives a fair estimation of optimum turning performance.

Amongst various possible types of turning flights in a horizontal plane, we will study three cases that are of special interest: 1) turning flight with maximum turn rate, 2) sharpest turn or minimum radius turn, and 3) turning flight with maximum possible load factor.

Turning Flight of Propeller Aircraft

Consider the constant-velocity, coordinated turning flight of a propeller aircraft in a horizontal plane. In general, for propeller airplanes graphical or numerical methods have to be used to determine the turning performance. However, for the propeller airplane whose power developed by the engine and the propulsive efficiency are independent of flight velocity, the problem of turning performance can be formulated in analytical form, but the resulting equations often need numerical or graphical solutions.

For the propeller aircraft, power available and power required are the basic quantities. In view of this, multiply Eq. (2.204) by velocity to obtain

$$TV - DV = 0 \tag{2.213}$$

or

$$k'\eta_p P(\text{kW}) - \frac{1}{2}\rho S C_{DO} V^3 - \left(\frac{2kn^2W^2}{\rho S V}\right) = 0 \tag{2.214}$$

where $TV = k'\eta_p P(\mathrm{kW})$ is the power available P_a in Nm/s, $P(\mathrm{kW})$ is the power developed by the reciprocating piston engine in kW, $k' = 1000$ is the constant to convert the kW into Nm/s, and η_p is the propulsive efficiency. As said before, we assume that for a given altitude, both $P(\mathrm{kW})$ and η_p are independent of flight velocity. Solving for load factor n, we get

$$n = \frac{1}{W/S}\left[\frac{k'\eta_p P'(\mathrm{kW})\rho(W/S)V}{2k} - \frac{\rho^2 C_{DO} V^4}{4k}\right]^{0.5} \tag{2.215}$$

where $P'(\mathrm{kW})$ is the power per unit weight or the specific power in kW/N.

Maximum Sustained Turn Rate

Here we consider the case when the aircraft attempts to make a constant-velocity turn in a horizontal plane with maximum rate of turn or angular velocity in turn. The maximum turn rate so generated while holding a constant altitude is called the maximum sustained turn rate (MSTR).

From Eq. (2.211), we have

$$\omega = \frac{g\sqrt{n^2 - 1}}{V} \tag{2.216}$$

For the MSTR,

$$\frac{\mathrm{d}\omega}{\mathrm{d}V} = g\left(\frac{Vn\dfrac{\mathrm{d}n}{\mathrm{d}V} - (n^2 - 1)}{V^2\sqrt{n^2 - 1}}\right) = 0 \tag{2.217}$$

or

$$\frac{V}{2}\frac{\mathrm{d}n^2}{\mathrm{d}V} - (n^2 - 1) = 0 \tag{2.218}$$

Using Eq. (2.215)

$$\frac{\mathrm{d}n^2}{\mathrm{d}V} = \frac{1}{(W/S)^2}\left(\frac{k'\eta_p P'(\mathrm{kW})(W/S)}{2k} - \frac{\rho^2 C_{DO} V^3}{k}\right) \tag{2.219}$$

Substitution in Eq. (2.218) gives an expression of the form

$$aV^4 + bV + c = 0 \tag{2.220}$$

A solution of this equation gives the velocity that maximizes the turn rate. However, this equation has no analytical solution. Therefore, either a graphical or numerical solution has to be obtained. Let us assume that V_{ft} is such a solution. Then,

$$n = \frac{1}{W/S}\left(\frac{k'\eta_p \rho P'(\mathrm{kW})(W/S)V_{ft}}{2k} - \frac{\rho^2 C_{DO} V_{ft}^4}{4k}\right)^{0.5} \tag{2.221}$$

$$\omega_{\max} = \frac{g\sqrt{n^2-1}}{V_{ft}} \tag{2.222}$$

$$R = \frac{V_{ft}^2}{g\sqrt{n^2-1}} \tag{2.223}$$

$$\cos\mu = \frac{1}{n} \tag{2.224}$$

$$C_L = \frac{2nW}{\rho S V_{ft}^2} \tag{2.225}$$

Here, we assume that the load factor $n \le n_{\lim}$ where $n_{\lim}$ is the structural limit load factor and $C_L \le C_{L,\max}$. If either of these two conditions are not satisfied, then the aircraft cannot generate the maximum sustained turn rate predicted by Eq. (2.222).

Sharpest Sustained Turn

Here, we consider the case when the aircraft attempts to make a sharpest turn or the turn with minimum radius of curvature. Such a turn while holding a constant altitude is called the sharpest sustained turn (SST).

We have

$$R = \frac{V^2}{g\sqrt{n^2-1}} \tag{2.226}$$

For $R_{\min}$,

$$\frac{\mathrm{d}R}{\mathrm{d}V} = \left(\frac{2(n^2-1)V - V^2 n \frac{\mathrm{d}n}{\mathrm{d}V}}{g(n^2-1)^{\frac{3}{2}}}\right) = 0 \tag{2.227}$$

or

$$\frac{V}{4}\frac{\mathrm{d}n^2}{\mathrm{d}V} - (n^2-1) = 0 \tag{2.228}$$

As before, substitution for $\mathrm{d}n^2/\mathrm{d}V$ using Eq. (2.219) gives an equation of the same form as that of Eq. (2.220), which has no analytical solution. We have to use graphical or numerical methods to solve this equation. Let V_{st} be such a solution that minimizes the radius of turn. Then,

$$n = \frac{1}{W/S}\left(\frac{k'\eta_p \rho P'(\mathrm{kW})(W/S)V_{st}}{2k} - \frac{\rho^2 C_{DO} V_{st}^4}{4k}\right)^{0.5} \tag{2.229}$$

$$\omega = \frac{g\sqrt{n^2-1}}{V_{st}} \tag{2.230}$$

$$R_{\min} = \frac{V_{st}^2}{g\sqrt{n^2-1}} \tag{2.231}$$

$$\cos\mu = \frac{1}{n} \tag{2.232}$$

$$C_L = \frac{2nW}{\rho S V_{st}^2} \tag{2.233}$$

Here, we assume that the load factor $n \leq n_{\lim}$ and $C_L \leq C_{L,\max}$. If either of these two conditions are not satisfied, then the aircraft cannot perform the sharpest turn with $R_{\min}$ as predicted by Eq. (2.231).

Turn with Maximum Load Factor

Because the bank angle and load factor are uniquely related by Eq. (2.207), the turn for maximum load factor is same as the turn for maximum bank angle. From Eq. (2.215), we have

$$n^2 = \frac{1}{(W/S)^2}\left(\frac{k'\eta_p \rho P'(\mathrm{kW})(W/S)V}{2k} - \frac{\rho^2 C_{DO} V^4}{4k}\right) \tag{2.234}$$

or

$$n^2 = AV - BV^4 \tag{2.235}$$

where

$$A = \frac{k'\eta_p P'(\mathrm{kW})\rho}{2k(W/S)} \tag{2.236}$$

$$B = \frac{\rho^2 C_{DO}}{4k(W/S)^2} \tag{2.237}$$

For turning in horizontal plane with maximum load factor,

$$2n\frac{\mathrm{d}n}{\mathrm{d}V} = A - 4BV^3 = 0 \tag{2.238}$$

Because $n \neq 0$, the velocity that maximizes the load factor is given by

$$V_{n,\max} = \left[\frac{k'\eta_p P'(\mathrm{kW})(W/S)}{2\rho C_{DO}}\right]^{\frac{1}{3}} \tag{2.239}$$

With this, the maximum load factor in turn is given by

$$n_{\max} = 0.6874\left(\frac{[k'\eta_p P'(\mathrm{kW})]^2 \rho E_m}{k(W/S)}\right)^{\frac{1}{3}} \tag{2.240}$$

Then,

$$\omega = \frac{g\sqrt{n_{\max}^2 - 1}}{V_{n,\max}} \tag{2.241}$$

$$R = \frac{V_{n,\max}^2}{g\sqrt{n_{\max}^2 - 1}} \tag{2.242}$$

$$C_L = \frac{2n_{\max} W}{\rho S V_{n,\max}^2} \tag{2.243}$$

Time for one complete turn of 2π radians is given by

$$t_{2\pi} = \frac{2\pi V_{n,\max}}{g\sqrt{n_{\max}^2 - 1}} \tag{2.244}$$

Turning Performance of Jet Aircraft

For jet aircraft whose thrust is independent of flight velocity, analytical solution can be obtained as follows.

We have

$$z = \frac{TE_m}{W} \tag{2.245}$$

$$D = \frac{W}{2E_m}\left(u^2 + \frac{n^2}{u^2}\right) \tag{2.246}$$

Then, Eq. (2.204) can be expressed in nondimensional form as

$$\frac{zW}{E_m} - \frac{W}{2E_m}\left(u^2 + \frac{n^2}{u^2}\right) = 0 \tag{2.247}$$

or

$$u^4 - 2zu^2 + n^2 = 0 \tag{2.248}$$

$$n = \sqrt{2zu^2 - u^4} \tag{2.249}$$

Maximum Sustained Turn Rate

As said before, the maximum turn rate generated while holding a constant altitude is called the MSTR. Using Eq. (2.249), Eq. (2.211) can be written as

$$\omega = \frac{g\sqrt{2zu^2 - u^4 - 1}}{uV_R} \tag{2.250}$$

$$\left(\frac{\omega V_R}{g}\right)^2 = 2z - u^2 - \frac{1}{u^2} \tag{2.251}$$

where V_R is the reference velocity given by

$$V_R = \sqrt{\frac{2W}{\rho S}}\sqrt[4]{\frac{k}{C_{DO}}} \tag{2.252}$$

For MSTR or $\omega_{\max}$,

$$2\omega\left(\frac{V_R}{g}\right)^2 \frac{d\omega}{du} = -2u + \frac{2}{u^3} \tag{2.253}$$

$$= 0 \tag{2.254}$$

because $\omega \neq 0$, we get

$$u = 1 \tag{2.255}$$

Thus, the maximum turn rate occurs when $V = V_R$. Substituting $u = 1$ in Eqs. (2.249) and (2.250), we obtain

$$n = \sqrt{2z - 1} \tag{2.256}$$

$$\omega_{\max} = \frac{g\sqrt{2z - 2}}{V_R} \tag{2.257}$$

The lift coefficient C_L can be obtained using the nondimensional expression for drag in Eq. (2.246) to express the ratio of induced drag to zero-lift drag as follows:

$$\frac{kC_L^2}{C_{DO}} = \frac{n^2}{u^4} = n^2 = 2z - 1 \tag{2.258}$$

so that

$$C_L = \sqrt{\frac{(2z - 1)C_{DO}}{k}}$$

$$= \sqrt{2z - 1}C_L^* \tag{2.259}$$

Here, we assume that the load factor n given by Eq. (2.256) is smaller than or equal to $n_{\lim}$, and the lift coefficient given by the Eq. (2.259) is smaller than or equal to $C_{L,\max}$. If these two conditions are not satisfied, the aircraft will not be capable of producing the maximum turn rate predicted by Eq. (2.257), and the analysis of this section is not valid.

With $u = 1$, the dimensional flight velocity and radius of turn are given by

$$V = \sqrt{\frac{2W}{\rho S}} \sqrt[4]{\frac{k}{C_{DO}}} \tag{2.260}$$

$$R = \left(\frac{2W}{\rho S g}\right) \sqrt{\frac{k}{2(z-1)C_{DO}}} \tag{2.261}$$

Sharpest Sustained Turn

Here, we consider the case when the aircraft attempts to make the sharpest possible turn ($R = R_{\min}$) in a horizontal plane with constant speed. As said before, such a turn holding a constant altitude is called the SST. From Eq. (2.210), we have

$$R = \frac{V^2}{g\sqrt{n^2 - 1}} \tag{2.262}$$

Substituting for load factor from Eq. (2.249) and using $V = uV_R$, we have

$$R = \frac{u^2 V_R^2}{g\sqrt{u^2(2z - u^2) - 1}} \tag{2.263}$$

For $R_{\min}$,

$$\frac{dR}{du} = \left(\frac{V_R^2}{g}\right) \left(\frac{2u\sqrt{u^2(2z-u^2)-1} - \dfrac{u^2(4zu - 4u^3)}{2\sqrt{u^2(2z-u^2)-1}}}{u^2(2z-u^2)-1} \right)$$

$$= 0 \tag{2.264}$$

Solving, we get

$$u = \frac{1}{\sqrt{z}} \tag{2.265}$$

$$V = \frac{V_R}{\sqrt{z}} \tag{2.266}$$

Then,

$$R_{\min} = \frac{V_R^2}{g\sqrt{z^2 - 1}} \tag{2.267}$$

The rate of turn is

$$\omega = \frac{g}{V_R}\sqrt{\frac{z^2 - 1}{z}} \tag{2.268}$$

The corresponding load factor and lift coefficient are given by

$$n = \frac{\sqrt{2z^2 - 1}}{z} \tag{2.269}$$

$$C_L = \sqrt{\frac{(2z^2 - 1)C_{DO}}{k}} \tag{2.270}$$

Here, it is presumed that $n \leq n_{lim}$ and $C_L \leq C_{L,max}$. If not, the aircraft cannot achieve the R_{min} predicted by Eq. (2.267), and the analysis presented in this section is not valid.

Turn for Maximum Load Factor

We have

$$n^2 = 2zu^2 - u^4 \tag{2.271}$$

For the maximum load factor n_{max},

$$\begin{aligned} 2n\frac{dn}{du} &= 4u(z - u^2) \\ &= 0 \end{aligned} \tag{2.272}$$

or

$$u = \sqrt{z} \tag{2.273}$$

$$n_{max} = z \tag{2.274}$$

Once again, it is presumed that $n_{max} \leq n_{lim}$. Furthermore, Eq. (2.274) implies that

$$T \leq \frac{n_{lim} W}{E_m} \tag{2.275}$$

In other words, the thrust should be limited as shown. If not, the load factor will exceed the structural limit factor, causing a structural failure during such a turn.

The flight velocity is given by

$$\begin{aligned} V &= \sqrt{z}V_R \\ &= \sqrt{\left(\frac{TE_m}{W}\right)\left(\frac{2W}{\rho S}\right)}\sqrt[4]{\frac{k}{C_{DO}}} \end{aligned} \tag{2.276}$$

With $E_m = 1/2\sqrt{kC_{DO}}$, Eq. (2.276) simplifies to

$$V = \sqrt{\frac{T}{\rho S C_{DO}}} \tag{2.277}$$

The lift coefficient is given by

$$C_L = \frac{2n_{\max} W}{\rho S V^2} \tag{2.278}$$

Substituting $n_{\max} = z$ and $V = \sqrt{z}V_R$, Eq. (2.278) reduces to

$$\begin{aligned} C_L &= \sqrt{\frac{C_{DO}}{k}} \\ &= C_L^* \end{aligned} \tag{2.279}$$

Recall that when $C_L = C_L^*$, $E = E_m$, and $D = D_{\min}$. In other words, for a coordinated turn in a horizontal plane with maximum load factor, the aircraft is operating at maximum aerodynamic efficiency or minimum drag condition. The radius of turn and the rate of turn for this case are

$$R = \frac{zV_R^2}{g\sqrt{z^2 - 1}} \tag{2.280}$$

$$\omega = \frac{g}{V_R}\sqrt{\frac{z^2 - 1}{z}} \tag{2.281}$$

Table 2.2 summarizes important performance parameters for three special cases of constant-velocity turn in a horizontal plane for jet aircraft whose thrust is independent of velocity. The variation of these parameters with nondimensional thrust is shown in Fig. 2.27. From this analysis, we observe that the turning performance improves with an increase in the thrust-to-weight ratio T/W, the aerodynamic efficiency E_m, and a decrease in wing loading W/S.

Table 2.2 Summary of Turning Performance of Jet Aircraft

Special Case	u	n	ω	R
MSTR	1	$\sqrt{2z - 1}$	$\frac{g\sqrt{2z - 2}}{V_R}$	$\frac{V_R^2}{g\sqrt{2z - 2}}$
SST	$\frac{1}{\sqrt{z}}$	$\frac{\sqrt{2z^2 - 1}}{z}$	$\frac{g}{V_R}\sqrt{\frac{z^2 - 1}{z}}$	$\frac{V_R^2}{g\sqrt{z^2 - 1}}$
$n_{\max}$	$\sqrt{z}$	z	$\frac{g}{V_R}\sqrt{\frac{z^2 - 1}{z}}$	$\frac{zV_R^2}{g\sqrt{z^2 - 1}}$

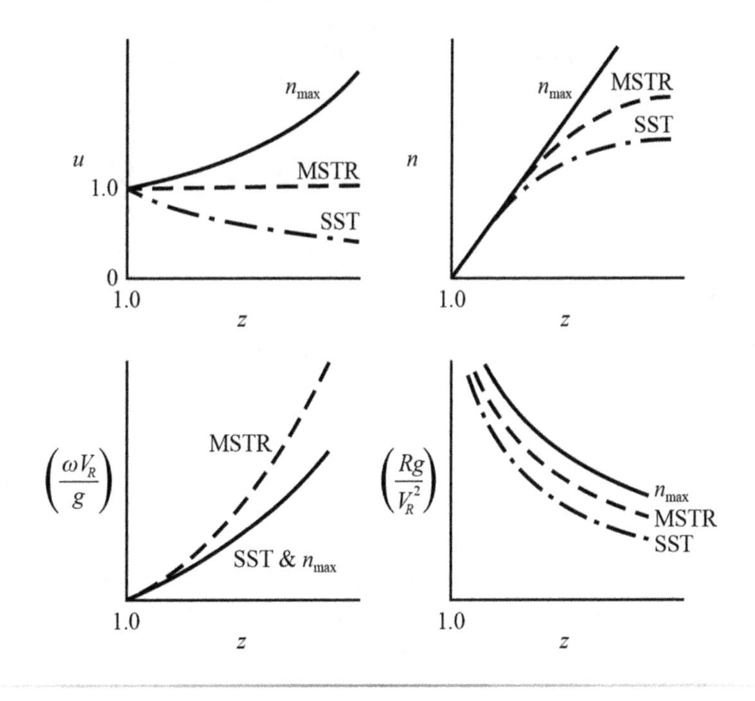

Fig. 2.27 Turning performance parameters with nondimensional thrust.

2.8.4 General Turning Flight

Let us consider a more general case of turning flight wherein the aircraft may gain or lose altitude while turning. We assume that the velocity vector is still contained in the plane of symmetry so that sideslip is zero. To begin with, let us consider the turning flight of a glider.

Turning Flight of a Glider

The equations of motion of a glider executing a constant-velocity turn are obtained by substituting $T = 0$ (assuming $\gamma < 0$) in Eqs. (2.190–2.192) as follows:

$$D = W \sin \gamma \tag{2.282}$$

$$L \cos \mu = W \cos \gamma \tag{2.283}$$

$$L \sin \mu = \frac{WV^2 \cos^2 \gamma}{gR} \tag{2.284}$$

Then,

$$\tan \gamma = \frac{1}{E \cos \mu} \tag{2.285}$$

$$n = \frac{\cos \gamma}{\cos \mu} \tag{2.286}$$

$$V = \sqrt{\frac{2nW}{\rho S C_L}} \tag{2.287}$$

$$\tan\mu = \frac{V^2 \cos\gamma}{Rg} \tag{2.288}$$

$$R = \frac{V^2 \cos\gamma}{g \tan\mu} \tag{2.289}$$

$$\omega = \frac{V \cos\gamma}{R} = \frac{g \tan\mu}{V} \tag{2.290}$$

For shallow glide angles, $\sin\gamma \simeq \gamma$ and $\cos\gamma \simeq 1$, so that

$$\gamma = \frac{1}{E\cos\mu} = \frac{n}{E} = \frac{nC_D}{C_L} \tag{2.291}$$

$$V = \sqrt{\frac{2nW}{\rho S C_L}} \tag{2.292}$$

$$n = \frac{1}{\cos\mu} \tag{2.293}$$

$$\tan\mu = \frac{V^2}{Rg} \tag{2.294}$$

$$R = \frac{V^2}{g\tan\mu} \tag{2.295}$$

$$\omega = \frac{V}{R} = \frac{g\tan\mu}{V} = \frac{g\sqrt{n^2-1}}{V} \tag{2.296}$$

The time for one complete turn of 2π radians is given by

$$t_{2\pi} = \frac{2\pi}{\omega} = \frac{2\pi V}{g\sqrt{n^2-1}} \tag{2.297}$$

Height lost in one complete turn is given by

$$\begin{aligned}\Delta h &= \left(\frac{\mathrm{d}h}{\mathrm{d}t}\right) t_{2\pi} \\ &= (V\gamma) t_{2\pi} \\ &= \frac{2\pi V^2 \gamma}{g\sqrt{n^2-1}} \\ &= \frac{4\pi n^2 (W/S)}{\rho g\sqrt{n^2-1}} \left(\frac{C_D}{C_L^2}\right)\end{aligned} \tag{2.298}$$

For a given load factor n, we observe that the height lost per turn is minimum when the glider flies at an angle of attack when C_D/C_L^2 is a minimum, which happens when $C_L = C_{L,\max}$.

General Turning Flight of Aircraft

The maximum rate of turn generated by an airplane in a coordinated, constant-altitude turn is called the MSTR because the aircraft can produce such a turn rate continuously for some time. However, the aircraft can generate still higher rates of turn if it is permitted to lose altitude so that it can make use of its height or the potential energy in addition to the aerodynamic and propulsive forces. The turn rate so generated can be much higher than MSTR and is called the maximum instantaneous turn rate or the maximum attainable turn rate (MATR). A high value of MATR is a good measure of the air superiority of combat aircraft because it is of crucial importance in close encounters where a pilot getting first point-and-shoot capability has a better chance to win. Because the aircraft will rapidly lose altitude during this maneuver, it is necessary for a pilot to have enough height margin before he initiates this maneuver. In the following, we will study this type of turning flight, assuming that the sideslip is zero and the flight velocity is approximately constant. We will consider only the jet aircraft here. A similar approach can be used for propeller aircraft.

Equations (2.190–2.192) take the following form:

$$T - D - W \sin \gamma = 0 \tag{2.299}$$

$$L \cos \mu - W \cos \gamma = 0 \tag{2.300}$$

$$L \sin \mu - \frac{WV^2 \cos^2 \gamma}{gR} = 0 \tag{2.301}$$

We have

$$D = \frac{W}{2E_m}\left(u^2 + \frac{n^2}{u^2}\right) \tag{2.302}$$

$$\omega = \frac{V \cos \gamma}{R} \tag{2.303}$$

Then,

$$\sin \gamma = \frac{T - D}{W} \tag{2.304}$$

$$n = \frac{\cos \gamma}{\cos \mu} \tag{2.305}$$

$$\tan\mu = \frac{V^2 \cos\gamma}{Rg}$$

$$= \frac{\sqrt{n^2 - \cos^2\gamma}}{\cos\gamma} \tag{2.306}$$

$$\omega = \frac{g\sqrt{n^2 - \cos^2\gamma}}{V\cos^2\gamma} \tag{2.307}$$

$$R = \frac{V^2\cos^2\gamma}{g\sqrt{n^2 - \cos^2\gamma}} \tag{2.308}$$

$$V = \sqrt{\frac{2nW}{\rho S C_L}}$$

$$= \sqrt{\frac{2W\cos\gamma}{\rho S C_L \cos\mu}} \tag{2.309}$$

From the relations given, we observe that, for a given value of thrust T and aerodynamic efficiency E_m, the turning performance improves with increase in load factor n and lift coefficient C_L only to be limited by the aerodynamic limit $C_{L,\max}$ and the structural limit $n_{\lim}$.

Assuming that the flight path angle γ is small, let us plot the variation of turn rate ω with velocity V for various values of load factor n as shown in Fig. 2.28. The curve corresponding to $n = n_{\lim}$ represents the structural boundary and that corresponding to $C_{L,\max}$ represents the aerodynamic boundary. The velocity at the intersection of these two boundaries is called the corner velocity. This is the velocity for maximum instantaneous turn rate or MATR. For a typical fighter aircraft, the corner velocity is about 300–350 kn (550–650 km/h). In a classical turning dogfight situation, the pilot who gets to his corner velocity first usually gains an upper hand and has a better chance for first point-and-shoot capability.

In the following, we will consider the turning flight at corner velocity.

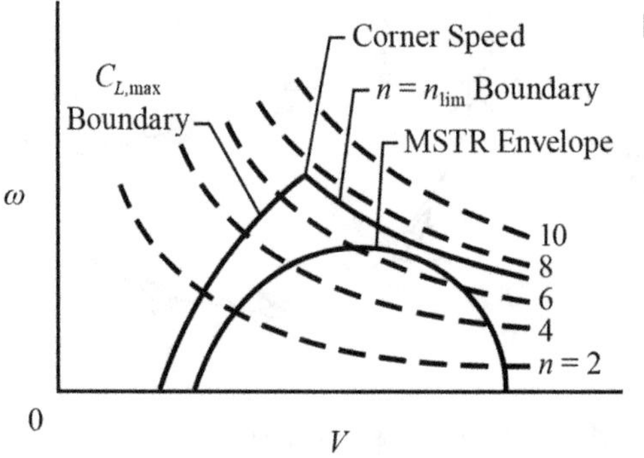

Fig. 2.28 Corner velocity turn.

Corner Velocity Turn

Here, the aircraft touches both the structural and aerodynamic limits as represented by $n = n_{\text{lim}}$ and $C_L = C_{L,\text{max}}$. We have

$$C_D = C_{DO} + kC_{L,\text{max}}^2 \tag{2.310}$$

$$V = \sqrt{\frac{2n_{\text{lim}} W}{\rho S C_{L_{\text{max}}}}} \tag{2.311}$$

$$\sin\gamma = \frac{T - D}{W} \tag{2.312}$$

$$R = \frac{V^2 \cos^2\gamma}{g\sqrt{n_{\text{lim}}^2 - \cos^2\gamma}} \tag{2.313}$$

$$\omega = \frac{g\sqrt{n_{\text{lim}}^2 - \cos^2\gamma}}{V\cos\gamma} \tag{2.314}$$

The time for one complete turn of 2π radians and height lost per turn is given by

$$t_{2\pi} = \frac{2\pi V\cos\gamma}{g\sqrt{n_{\text{lim}}^2 - \cos^2\gamma}} \tag{2.315}$$

$$\Delta h = (V\sin\gamma)t_{2\pi} \tag{2.316}$$

Herbst Maneuver

In the conventional turning flight (Fig. 2.29a), the aircraft that is initially in a level flight (A) banks in to the turn (B) to have a component of lift force $L \sin\mu$ to generate the required centripetal acceleration. The aircraft will

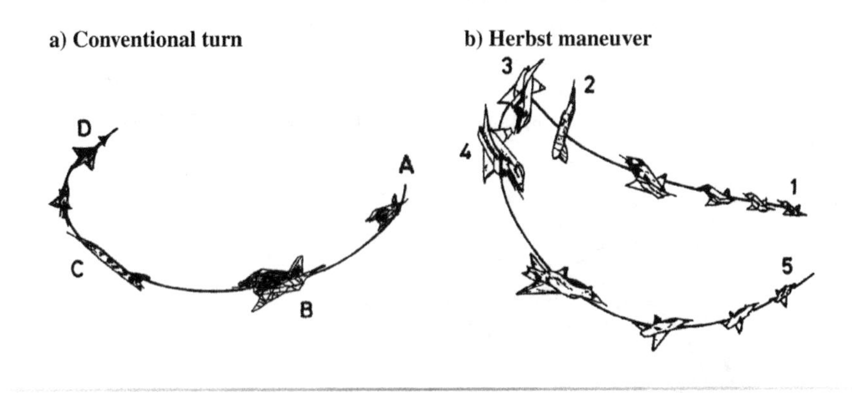

Fig. 2.29 Conventional turn and Herbst maneuver.

continuously turn (C) until the nose points in the right direction (D), at which point the pilot will level the wings and return to the level flight condition. The scenario is quite different in the Herbst maneuver, named after the late German pilot Wolfgang Herbst, who conceived this poststall maneuver as a quick way of pointing the aircraft's guns at a target during a close encounter. The Herbst maneuver was explored using the X-31 aircraft. Schematics of the Herbst maneuver are shown in Fig. 2.29b.

To start the Herbst maneuver from normal level flight condition (1), the pilot decelerates the aircraft by pitching the nose up till the angle of attack goes well beyond the stall (2), and the aircraft is nearly perpendicular to the flight path (3), using the entire airframe as a speed brake. The pilot rolls the aircraft about the velocity vector until the nose points in the right direction (4), lowers the nose, and accelerates back to the desired speed in level flight in a new direction (5).

The rolling motion about the velocity vector is often referred to as "coning" motion. This maneuver is used to minimize the sideslip excursions during the roll. If the aircraft rolls about the body axis while operating at high angles of attack, it will experience significant sideslip buildup. The development of sidesip is not desirable because it increases drag and affects the turning performance.

Example 2.9

A propeller airplane weighs 50,000 N and has a wing area of 30 m^2, structural limit load factor of 2.5, and a maximum lift coefficient of 1.75. The drag polar of the aircraft is given by $C_D = 0.025 + 0.05C_L^2$. The reciprocating engine produces 840 kW, and the propulsive efficiency of the engine–propeller combination is 0.85. Determine the lift coefficient, velocity, load factor, turn rate, and radius of turn for 1) fastest sustained turn rate, 2) sharpest sustained turn, and 3) maximum load factor.

Solution:

Because we cannot obtain an analytical solution to the turning performance of the propeller aircraft for fastest and sharpest sustained turns, we have to use a graphical method. The calculated values of load factor n, lift coefficient C_L, rate of turn ω, and radius of turn are plotted against the velocity as shown in Fig. 2.30.

From Fig. 2.30b, we observe that for $V \leq 60$ m/s, the required lift coefficient exceeds $C_{L,\max}$ of 1.75. Hence, the aircraft cannot achieve the predicted load factor, rate of turn, and radius of turn for velocities lower than 60 m/s. Therefore, we have $\omega_{\max} = 0.344$ rad/s or 19.7112 deg/s, $R_{\min} = 174.5$ m, which occur at $V = 60$ m/s.

(*Continued*)

Example 2.9 (Continued)

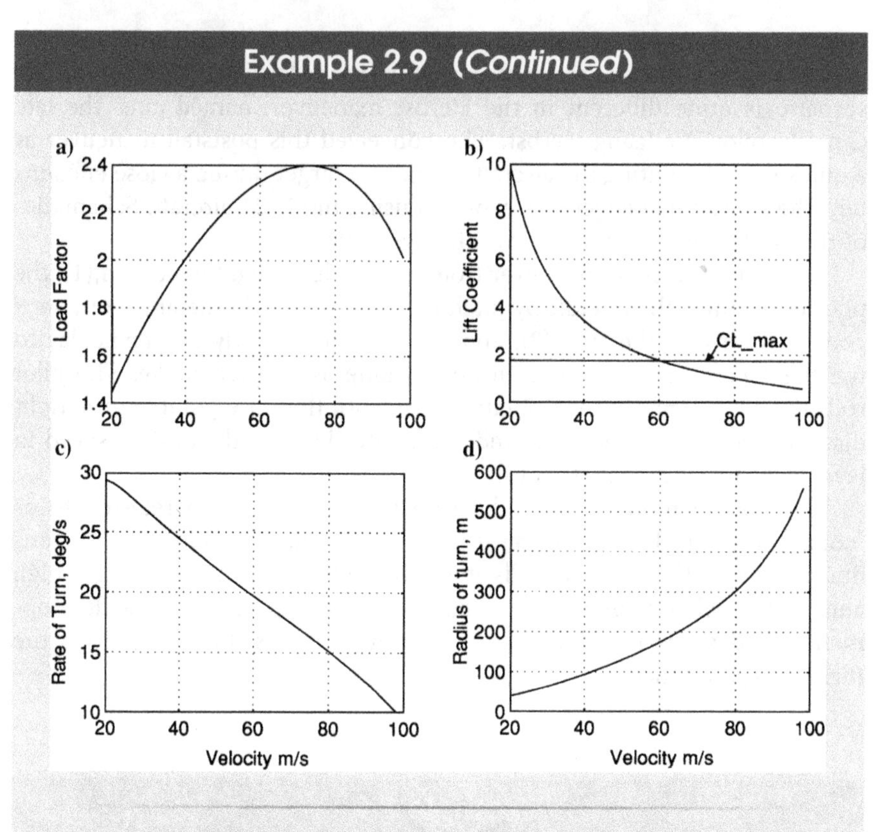

Fig. 2.30 Turning performance of propeller aircraft.

The turning performance for maximum load factor has an analytical solution. Using Eq. (2.240), we find $n_{\max} = 2.3968$. The corresponding velocity $V_{n,\max}$ as given by Eq. (2.239) is equal to 72.969 m/s. The corresponding rate of turn and radius of turn were found to be equal to 16.7798 deg/s and 249.1757 m. These values agree with the graphical solution given in Fig. 2.30.

Example 2.10

A jet aircraft weighs 78,480 N and has a wing area of 30 m^2, $C_{L,\max}$ of 1.2, sea-level thrust of 39,240 N, drag polar $C_D = 0.012 + 0.12C_L^2$, and a limit load factor of 8.0. For turning flight at sea level, determine the turn rate, radius of turn, load factor, lift coefficient, and lift-to-drag ratio when flying for the following cases: 1) MSTR, 2) SST, 3) maximum load factor, and 4) turning at corner velocity. Cases 1–3 correspond to coordinated turns in a horizontal plane.

(Continued)

Example 2.10 (*Continued*)

Solution:

Let us first consider cases 1–3 of constant-velocity turns in a horizontal plane.

Case 1, MSTR:

We have

$$E_m = \frac{1}{2\sqrt{KC_{DO}}} = \frac{1}{2\sqrt{0.012*0.12}} = 13.1762$$

$$V_R = \sqrt{\frac{2W}{\rho S}}\sqrt[4]{\frac{k}{C_{DO}}} = \sqrt{\frac{2*78480}{1.225*30}}\sqrt[4]{\frac{0.12}{0.012}} = 116.2160\,\text{m/s}$$

$$V = V_R = 116.2160\,\text{m/s}$$

$$z = \frac{TE_m}{W} = \frac{39{,}240*13.1762}{78{,}480} = 6.5881$$

$$n = \sqrt{2z-1} = \sqrt{2*6.5881-1} = 3.4894$$

$$\mu = \cos^{-1}\left(\frac{1}{3.4894}\right) = 73.3466\,\text{deg}$$

$$C_L = n\sqrt{\frac{C_{DO}}{k}} = 3.4894*\sqrt{\frac{0.012}{0.12}} = 1.1034$$

(*Continued*)

Example 2.10 (*Continued*)

$$\begin{aligned}\omega &= \omega_{\max} \\ &= \frac{g\sqrt{2z-2}}{V_R} \\ &= \frac{9.81\sqrt{2*6.5881-2}}{116.2160} \\ &= 0.2822 \text{ rad/s} \\ &= 16.1693 \text{ deg/s}\end{aligned}$$

$$\begin{aligned}t &= \frac{2\pi}{\omega} \\ &= 22.2650 \text{ s}\end{aligned}$$

$$\begin{aligned}R &= \frac{V}{\omega} \\ &= \frac{116.216}{0.2822} \\ &= 411.8214 \text{ m}\end{aligned}$$

The lift-to-drag ratio is given by [refer to Problem 2.19(b)]

$$\begin{aligned}E &= \left(\frac{n}{1+n^2}\right)\frac{1}{\sqrt{KC_{DO}}} \\ &= \left(\frac{3.4894}{1+3.4894^2}\right)\frac{1}{\sqrt{0.012*0.12}} \\ &= 6.9789\end{aligned}$$

Case 2, SST:

The lift-to-drag ratio is given by

$$\begin{aligned}E &= \frac{2znE_m}{1+n^2z^2} \\ &= \frac{2*6.5881*1.406*13.1762}{1+6.5881^2*1.406^2} \\ &= 2.8122\end{aligned}$$

We have

$$\begin{aligned}n &= \frac{\sqrt{2z^2-1}}{z} \\ &= 1.406\end{aligned}$$

(*Continued*)

Example 2.10 (*Continued*)

$$\mu = \cos^{-1}\left(\frac{1}{1.406}\right)$$
$$= 44.6643 \text{ deg}$$

$$V = \frac{V_R}{\sqrt{z}}$$
$$= 45.2778 \text{ m/s}$$

$$C_L = \frac{2nW}{\rho S V^2}$$
$$= 2.9292$$

$$R = R_{\min}$$
$$= \frac{V_R^2}{g\sqrt{z^2 - 1}}$$
$$= 211.4288 \text{ m}$$

$$\omega = \frac{V}{R}$$
$$= 0.2142 \text{ rad/s}$$
$$= 12.2709 \text{ deg/s}$$

$$t = \frac{2\pi}{\omega}$$
$$= 29.3333 \text{ s}$$

Because $C_L > C_{L,\max}$, the aircraft cannot perform a steady SST at constant altitude at sea level. The reader may verify that $V < V_{\text{stall}}$.

Case 3, Turn with $n_{\max}$:

We have

$$E = E_m$$
$$= 13.1762$$

$$n = n_{\max}$$
$$= z$$
$$= 6.5881$$

$$\mu = \cos^{-1}\frac{1}{6.5881}$$
$$= 81.2694 \text{ deg}$$

(*Continued*)

Example 2.10 (*Continued*)

$$V = \sqrt{z}V_R$$
$$= 298.2950\,\text{m/s}$$

$$C_L = \sqrt{\frac{C_{DO}}{k}}$$
$$= 0.3162$$

$$\omega = \frac{g\sqrt{z^2 - 1}}{V}$$
$$= 0.2142\,\text{rad/s}$$
$$= 12.2709\,\text{deg/s}$$

$$t = \frac{2\pi}{\omega}$$
$$= 29.3333\,\text{s}$$

$$R = \frac{V}{\omega}$$
$$= 1392.60\,\text{m}$$

Case 4, Corner Velocity Turn:

For this case, we have

$$C_L = C_{L,\max}$$
$$= 1.2$$

$$C_D = C_{DO} + kC_{L,\max}^2$$
$$= 0.012 + 0.120 * 1.2^2$$
$$= 0.1848$$

$$n = n_{\text{lim}}$$
$$= 8.0$$

$$\mu = \cos^{-1}\left(\frac{1}{8.0}\right)$$
$$= 82.8192\,\text{deg}$$

$$V = \sqrt{\frac{2n_{\text{lim}}W}{\rho S C_{L,\max}}}$$
$$= \sqrt{\frac{2 * 8 * 78{,}480}{1.225 * 30 * 1.2}}$$
$$= 168.7408\,\text{m/s}$$

(Continued)

Example 2.10 (*Continued*)

$$D = \frac{1}{2}\rho V^2 S C_D$$
$$= \frac{1}{2} 1.225 * 168.7408^2 * 30 * 0.1848$$
$$= 96{,}687.320 \text{ N}$$

$$\gamma = \sin^{-1}\left(\frac{T - D}{W}\right)$$
$$= -0.7320 \text{ rad}$$
$$= -47.0543 \text{ deg}$$

$$\cos\gamma = 0.6813$$

$$R = \frac{V^2 \cos\gamma}{g \tan\mu}$$
$$= \frac{168.7408^2 * 0.6813}{9.81 * 7.9373}$$
$$= 249.1362 \text{ m}$$

$$\omega = \frac{V\cos\gamma}{R}$$
$$= 0.4614 \text{ rad/s}$$
$$= 26.4409 \text{ deg/s}$$

$$t = \frac{2\pi}{\omega}$$
$$= 13.6231 \text{ s}$$

The height lost per turn is given by

$$\Delta h = (V \sin\gamma)t$$
$$= 1682.6 \text{ m}$$

The results of the calculation are summarized in Table 2.3. From this exercise we note that, for the given aircraft, the MSTR program gives optimum steady turning performance in horizontal plane at sea level. The given aircraft cannot execute the predicted SST at sea level because the required lift coefficient of 2.9292 exceeds $C_{L,\max}$ of 1.2. However, the best possible turning performance occurs when flying at corner velocity but is accompanied by a significant loss of height as much as 1682.6 m per turn.

(*Continued*)

Example 2.10 (Continued)

Table 2.3 Summary of Results

Parameter	MSTR	SST	n_{max}	Corner Velocity
μ, deg	73.3466	44.6643	81.2694	82.8182
R, m	411.8214	211.4288	1392.60	249.1362
ω, deg/s	16.1693	12.2709	12.2709	26.4409
n	3.4894	1.4060	6.5881	8.0
t, s	22.2650	29.3333	29.3333	13.6231
C_L	1.1034	2.9292	0.3162	1.20
E	6.9789	—	13.1762	6.4935
V, m/s	116.2160	45.2778	298.295	168.7408
Δh, m	0	0	0	1682.6

2.9 Takeoff and Landing

Takeoff and landing, although not exactly free-flight conditions, are critical phases of aircraft operation. A safe operation of takeoff and landing not only depends on the aerodynamic and propulsion characteristics of the aircraft but also on the piloting technique.

2.9.1 Takeoff

A typical takeoff phase consists of the following steps: 1) with engines producing maximum power/thrust, the aircraft is accelerated from rest to the takeoff speed; 2) on reaching the takeoff speed, the aircraft is rotated noseup so that the angle of attack increases to generate sufficient lift for liftoff; and 3) the aircraft starts climbing to clear the specified obstacle height (h_{obst}). FAR Part II on takeoff specifies an obstacle height of 35 ft (11.5 m) whereas RAE, U.K., specifies 50 ft (approximately 15 m). The takeoff phase is said to be complete once the aircraft clears this prescribed obstacle height. The pilot then retracts the landing gear to reduce the airplane drag and increase the rate of climb. The aircraft may continue to climb until it reaches the desired cruise altitude or may proceed to perform other missions as may be assigned to it.

The total takeoff distance is the sum of the ground run s_1 and airborne distance s_2 as shown in Fig. 2.31. Ideally, the takeoff distance should be as small as possible for quick and efficient operation of the aircraft.

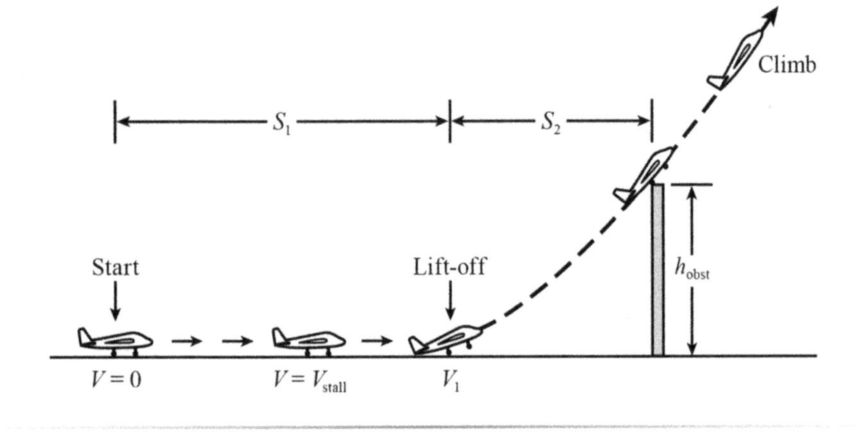

Fig. 2.31 Airplane in takeoff.

Estimation of Ground Run

The forces acting on the aircraft during the ground run are shown in Fig. 2.32. Let

$$F_a = T - D - \mu(W - L) \tag{2.317}$$

denote the net accelerating force during the ground run. Here, T is the thrust, D is the aerodynamic drag, L is the aerodynamic lift, W is the gross takeoff weight, and μ is the coefficient of friction between the wheels and the runway. For concrete runways, μ typically ranges from 0.02 to 0.05. The schematic variation of these forces during the ground run is shown in Fig. 2.33.

The lift and drag forces are to be estimated for the takeoff configuration with landing gear, flaps, and high-lift devices deployed and aircraft operating in the proximity of the ground.

Then,

$$\frac{W}{g}\left(\frac{\mathrm{d}V}{\mathrm{d}t}\right) = \frac{W}{g}\left(V\frac{\mathrm{d}V}{\mathrm{d}s}\right) = F_a \tag{2.318}$$

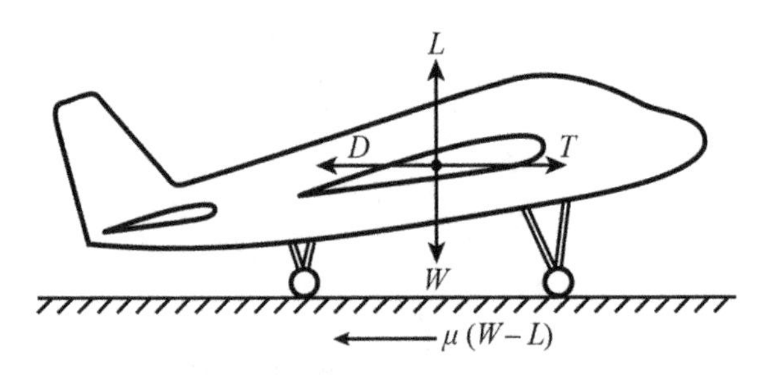

Fig. 2.32 Forces acting on an airplane during takeoff ground run.

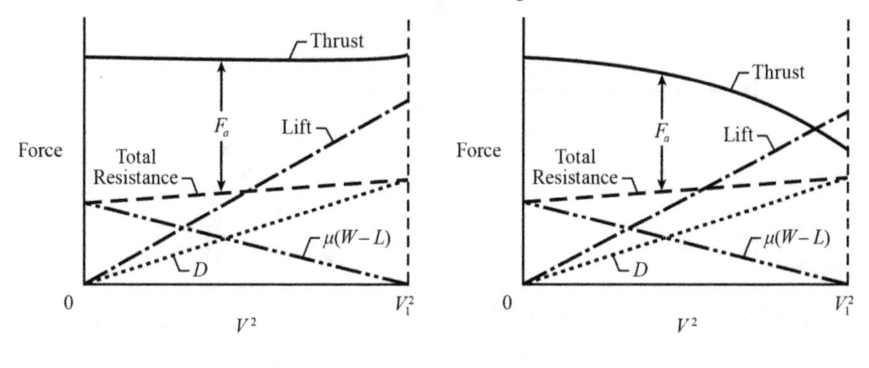

Fig. 2.33 Forces with velocity during takeoff ground run.

or

$$ds = \frac{W\,d(V^2)}{2gF_a} \tag{2.319}$$

During the ground run, the angle of attack is normally held constant at a selected value. Assuming that the net accelerating force F_a varies as the square of the velocity, we have

$$F_a = F_0 + \left(\frac{F_1 - F_0}{V_1^2}\right)V^2 \tag{2.320}$$

where V_1 is the velocity at liftoff and

$$F_0 = T - \mu W \tag{2.321}$$

$$F_1 = T - D \tag{2.322}$$

Then,

$$\int_0^{s_1} ds = \frac{W}{2g}\int_0^{V_1} \frac{dV^2}{F_0 + \left(\frac{F_1 - F_0}{V_1^2}\right)V^2} \tag{2.323}$$

$$s_1 = \frac{W}{2g}\left(\frac{V_1^2}{F_0 - F_1}\right)\ell n\frac{F_0}{F_1} \tag{2.324}$$

The liftoff or the takeoff velocity V_1 is usually equal to 1.2 V_{stall}.

To minimize the ground run s_1, we have to maximize the value of the net accelerating force F_a. One of the parameters that can be adjusted to optimize the takeoff ground run is the angle of attack. Differentiating Eq. (2.317),

$$\begin{aligned}\frac{dF_a}{dC_L} &= -\frac{dD}{dC_L} + \mu\frac{dL}{dC_L}\\ &= -2qSkC_L + \mu qS\\ &= 0\end{aligned} \tag{2.325}$$

where $q = 1/2\rho V^2$. Then,

$$C_L = C_L^* = \frac{\mu}{2k} \tag{2.326}$$

where C_L^* denotes the value of lift coefficient when F_a is maximum. Then,

$$s_{1,\min} = \frac{W}{2g}\left(\frac{V_1^2}{F_0 - F_1^*}\right)\ell n\frac{F_0}{F_1^*} \tag{2.327}$$

where

$$F_1^* = T - qS\left(C_{DO} + kC_L^{*2}\right) \tag{2.328}$$

An alternative way of minimizing the ground run is to deploy high-lift devices for takeoff. The use of high-lift devices increases the value of $C_{L,\max}$ and reduces the value of takeoff speed V_1, but the benefit is partially offset because of an increase in drag. Whether the ground distance is smaller for operation at lift coeficient equal to C_L^* or with the deployment of the high-lift devices depends on the aircraft configuration.

Time for Takeoff

$$\frac{W}{g}\frac{dV}{dt} = F_a \tag{2.329}$$

$$dt = \frac{WdV}{gF_a} \tag{2.330}$$

Let

$$a = F_0 \tag{2.331}$$

$$b = \frac{F_1 - F_0}{V_1^2} \tag{2.332}$$

so that

$$F_a = a + bV^2 \tag{2.333}$$

and

$$dt = \frac{WdV}{g(a + bV^2)} \tag{2.334}$$

$$t_1 = \int_0^{V_1} \frac{W dV}{g(a + bV^2)} \tag{2.335}$$

Then,

$$\begin{aligned} t_1 &= \left(\frac{W}{g\sqrt{ab}}\right) \tan^{-1}\sqrt{\frac{b}{a}}V_1 \quad (a > 0, b > 0) \\ &= \left(\frac{W}{2g\sqrt{ab}}\right) \ell n \left(\frac{\sqrt{a} + \sqrt{b_1}V_1}{\sqrt{a} - \sqrt{b_1}V_1}\right) \quad (a > 0, b < 0) \end{aligned} \tag{2.336}$$

where $b_1 = |b|$ is the absolute value of b. Here, we have assumed that there is no wind during the ground run. If the aircraft operates in the presence of wind, the ground run s_1 and takeoff time t_1 will be quite different. The headwind improves takeoff performance, whereas the tailwind degrades it.

If the airplane is to takeoff under crosswind conditions, the airplane drag has to be estimated for sideslip conditions. With sideslip, the drag will be higher, and, as a result, the ground run s_1 and corresponding time t_1 will be higher compared to the case of zero sideslip. Furthermore, the pilot has to operate the rudder to prevent the airplane from aligning itself along the resultant velocity vector as schematically shown in Fig. 2.34. The aircraft attempts to align along the resultant velocity on account of its inherent directional stability, which we study in Chapter 3.

Aborted Takeoff

Sometimes, takeoff may have to be aborted because an engine failure occurs during the ground run. An immediate detection of an engine failure may not always be possible, and the pilot may not notice it until the aircraft has gained some ground speed. To account for such unforeseen situations, a

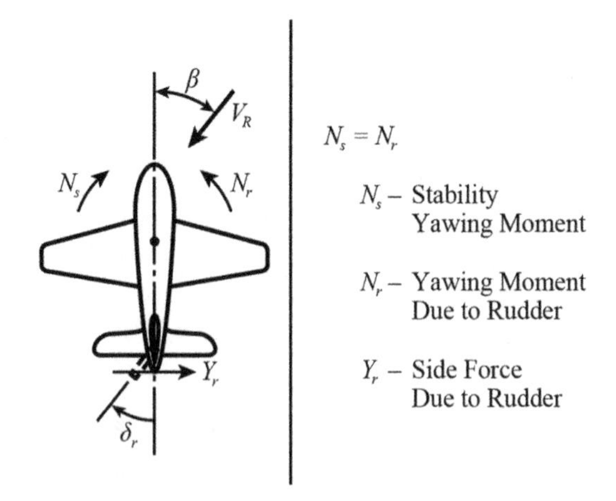

Fig. 2.34 Crosswind takeoff.

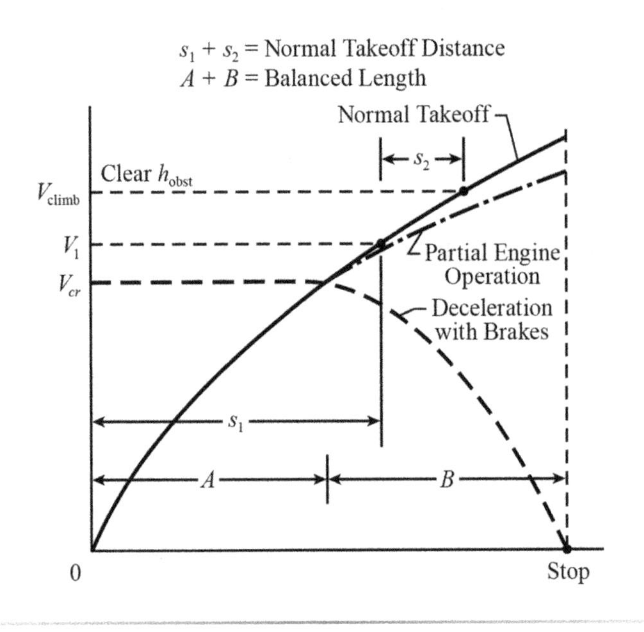

Fig. 2.35 Normal takeoff distance and balanced field length.

certain speed called critical speed V_{cr} is defined for each aircraft to assist the pilot in making a decision whether to abort or continue the takeoff. If an engine failure is detected at speeds below the critical speed, the pilot is advised to abort the takeoff because he still has enough runway length left to bring the aircraft to a complete stop. If engine failure is detected at speeds exceeding the critical speed, then he is advised to continue the takeoff and get airborne because the available runway length may not be sufficient to bring the aircraft to a safe halt. Once airborne, he will have to assess the nature and severity of the situation and take appropriate action. If an engine failure is detected at $V = V_{cr}$, he has both options. He can either abort the takeoff or continue and get airborne.

The field length determined by the considerations given is known as "balanced" field length [3, 9] and is schematically shown in Fig. 2.35.

Estimation of Airborne Distance and Time

During the airborne phase, the aircraft is in an accelerated climb. The equations of motion for this phase are

$$T - D - W \sin\gamma = \frac{W}{g}\frac{\mathrm{d}V}{\mathrm{d}t} \tag{2.337}$$

$$L - W\cos\gamma = \frac{W}{g}V\frac{\mathrm{d}\gamma}{\mathrm{d}t}$$

$$= \frac{WV^2}{Rg} \tag{2.338}$$

and the kinematic equations are given by

$$\dot{x} = V \cos \gamma \tag{2.339}$$

$$\dot{h} = V \sin \gamma \tag{2.340}$$

For accurate estimation of the airborne distance s_2, we have to solve the differential equations. However, a simple and approximate estimation of the airborne distance s_2 and the corresponding time t_2 can be done by assuming that the flight path angle γ is constant during the airborne phase. With these assumptions, we have

$$s_2 = \frac{h_{\text{obst}}}{\tan \gamma} \tag{2.341}$$

$$t_2 = \frac{s_2}{V_1 \cos \gamma} \tag{2.342}$$

where

$$\sin \gamma = \left(\frac{T - D}{W} \right)_{V=V_1} \tag{2.343}$$

2.9.2 Landing

A typical landing phase consists of the following sequences (Fig. 2.36): 1) a steady shallow glide approach to the runway holding a constant glide angle γ; the glide angle is held as small as possible so that the rate of sink is kept down to a minimum (in this way, it is possible to reduce the energy that has to be dissipated at impact with the ground); 2) a flare during which the aircraft is rotated noseup so as to momentarily achieve a level flight condition and to further minimize the rate of sink; 3) a touchdown

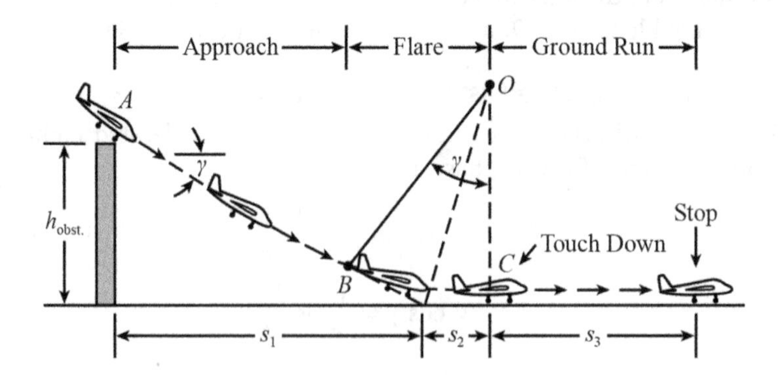

Fig. 2.36 Airplane landing.

prior to which the pilot may momentarily stall the airplane; and 4) the ground run during which full reverse thrust (if available), spoilers or drag parachutes (like the F-15 or the MiG-21 aircraft), and brakes are applied to produce maximum retardation and bring the aircraft to a complete halt.

Estimation of Airborne Distance and Time

Assuming that the glide angle and velocity are constant during the final approach,

$$s_1 = \frac{h_{\text{obst}}}{\tan \gamma} \tag{2.344}$$

$$t_1 = \frac{s_1}{V_A \cos \gamma} \tag{2.345}$$

The approach speed V_A is usually equal to 1.3 V_{stall}. Furthermore, we assume that the segment BC (Fig. 2.36) of the flare can be approximated by a circular arc. Let R be the radius of this arc. Note that $\angle BOC = \gamma$. The appropriate equations of motion for flight along the arc BC are

$$L - W \cos \gamma = \frac{WV^2}{Rg} \tag{2.346}$$

$$T - D - W \sin \gamma = 0 \tag{2.347}$$

Then,

$$\sin \gamma = \frac{T - D}{W} \tag{2.348}$$

Assuming γ to be small, we obtain

$$R = \frac{V^2 W}{g(L - W)} \tag{2.349}$$

Assuming that the velocity remains constant on the arc BC and is equal to V_A and $C_L = C_{L,\max}$,

$$\begin{aligned} L &= \frac{1}{2}\rho V_A^2 S C_{L,\max} \\ &= \frac{1}{2}\rho (1.3 V_{\text{stall}})^2 S C_{L,\max} \\ &= 1.69\, W \end{aligned} \tag{2.350}$$

$$R = \frac{V_A^2}{0.699\, g} \tag{2.351}$$

$$s_2 \simeq \frac{1}{2} R\gamma \tag{2.352}$$

Estimation of Landing Ground Run

The forces acting on the aircraft during the ground run are shown in Fig. 2.37.

Let

$$F_a = T_R + D + \mu(W - L) \tag{2.353}$$

be the net retarding force during the ground run. Here, T_R is the reverse thrust. Then,

$$\frac{W}{g} V \frac{\mathrm{d}V}{\mathrm{d}s} = -F_a \tag{2.354}$$

$$\mathrm{d}s = -\frac{W\mathrm{d}(V^2)}{2gF_a} \tag{2.355}$$

As before, we assume that the angle of attack during the ground run is held constant and the net retarding force F_a varies as the square of the velocity as given by

$$F_a = F_0 + \frac{F_1 - F_0}{V_1^2} V^2 \tag{2.356}$$

where V_1 is the touchdown velocity, approximately equal to V_A. Then,

$$\int_0^{s_3} \mathrm{d}s = -\frac{W}{2g} \int_{V_1}^{0} \frac{\mathrm{d}V^2}{F_0 + \frac{F_1 - F_0}{V_1^2} V^2} \tag{2.357}$$

$$s_3 = \frac{W}{2g} \left(\frac{V_1^2}{F_1 - F_0} \right) \ell n \frac{F_1}{F_0} \tag{2.358}$$

where $F_1 = T_R + D$ and $F_0 = T_R + \mu W$.

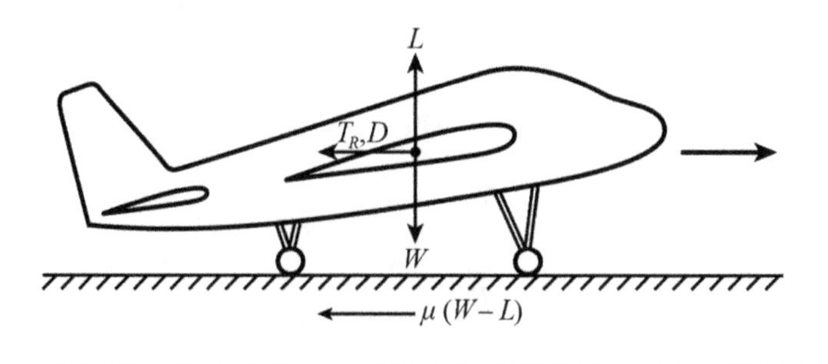

Fig. 2.37 Forces acting on an airplane during landing run.

Example 2.11

A certain jet aircraft has the following data: $W = 50{,}000$ N, $T = 14{,}500$ N, $C_D = 0.02 + 0.04C_L^2$, $C_{L,\max} = 1.2$, and $S = 30$ m^2. Assuming an obstacle height of 15 m and $\mu = 0.05$, calculate the total takeoff distance and time at sea level.

Solution:

$$\begin{aligned} C_L^* &= \frac{\mu}{2k} \\ &= \frac{0.05}{(2)(0.04)} \\ &= 0.625 \end{aligned}$$

$$\begin{aligned} V_{\text{stall}} &= \sqrt{\frac{(2)(50{,}000)}{(1.225)(30)(1.2)}} \\ &= 47.6290\,\text{m/s} \\ V_1 &= 1.2V_{\text{stall}} \\ &= 57.143\,\text{m/s} \end{aligned}$$

$$\begin{aligned} F_0 &= T - \mu W \\ &= 14{,}500 - 0.05(50{,}000) \\ &= 12{,}000\,\text{N} \end{aligned}$$

$$\begin{aligned} F_1 &= T - D \\ &= 14{,}500 - 0.5(1.225)(57.143^2)(30)[0.02 + 0.04(0.625^2)] \\ &= 12{,}362.489\,\text{N} \end{aligned}$$

For the ground run,

$$\begin{aligned} s_1 &= \frac{W}{2g}\left(\frac{V_1^2}{F_0 - F_1}\right) \ell n \frac{F_0}{F_1} \\ &= \frac{50{,}000(57.13^2)}{2(9.81)(12{,}000 - 12{,}362.489)} \ell n \frac{12{,}000}{12{,}362.489} \\ &= 684.1065\,\text{m} \end{aligned}$$

Let us calculate the time for ground run. We have

$$\begin{aligned} a &= F_0 \\ &= 12{,}000 \end{aligned}$$

(*Continued*)

Example 2.11 (*Continued*)

$$b = \frac{F_1 - F_0}{V_1^2}$$
$$= \frac{12{,}362.489 - 12{,}000}{57.143^2}$$
$$= 0.111$$

Because both a and b are positive,

$$t_1 = \frac{W}{g\sqrt{ab}} \tan^{-1}\left[\sqrt{\frac{b}{a}} V_1\right]$$
$$\sqrt{ab} = 36.49$$
$$\sqrt{\frac{b}{a}} V_1 = 0.1733$$
$$t_1 = \frac{50{,}000}{(9.81)(36.49)} \tan^{-1}(0.1733)$$
$$= 23.9681 \text{ s}$$

We have

$$\sin\gamma = \left(\frac{T - D}{W}\right)_{V=V_1}$$
$$= 0.2472$$
$$\gamma = 14.3153 \text{ deg}$$

The airborne distance s_2 and time t_2 are given by

$$s_2 = \frac{15}{\tan\gamma}$$
$$= 58.7863 \text{ m}$$
$$t_2 = \frac{s_2}{V_1 \cos\gamma}$$
$$= 1.0617 \text{ s}$$

Total takeoff distance $= s_1 + s_2 = 742.8928$ m, and the total takeoff time $= t_1 + t_2 = 25.0298$ s.

2.10 Hazards During Takeoff and Landing: Windshear and Microburst

Windshear [4, 5] is the change in wind direction and/or speed in a very short distance with respect to the ground. Whereas a change in the horizontal wind velocity affects the airspeed, the vertical component of windshear affects the flight path angle. Windshears are normally caused by thunderstorm activity, weather fronts, and jet streams. While windshear can occur at any altitude, it is most hazardous if encountered during landing and takeoff. The windshear occurring at low altitudes close to the ground is called low-level windshear.

Another dangerous form of windshear, which has been recognized in recent years, is the microburst. A microburst is formed when a column of air at high altitudes quickly cools because of evaporation of ice, snow, or rain and, becoming denser than the surrounding air, falls rapidly to the ground. On hitting the ground, this mass of air spreads radially outward in all directions. A core of such a downburst can be of about 2–3 km in width, and the winds generated can be as high as 150–200 km/h. The life of a microburst is typically 5–15 min.

A pilot flying through a microburst will experience within a minute or less, a bewildering combination of events during which he will encounter headwind, downburst, and tailwind as shown schematically in Fig. 2.38. The headwind will cause the air speed to rise and make the aircraft start climbing. The pilot, if unaware of what may happen next, is likely to throttle back the engine and reduce power to maintain his original flight path angle. However, the headwind quickly disappears and is followed by the downdraft, which will induce a high rate of descent. At this point, the pilot will attempt to restore the full thrust by opening the throttle. While the engine takes some time to react, the pilot encounters strong tailwind that reduces air speed

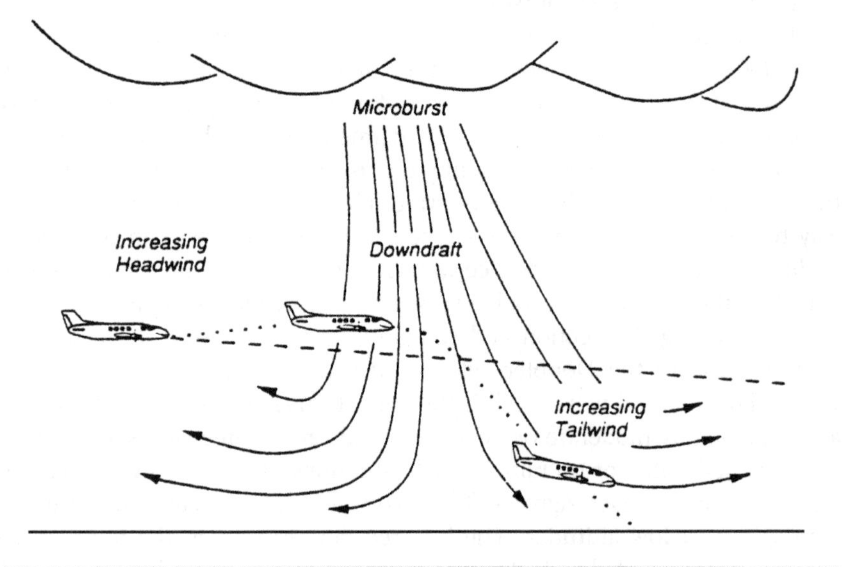

Fig. 2.38 Microburst during takeoff/landing (6).

further with a potential risk of an imminent stall. Depending on the proximity of the ground and the magnitude of windshear velocities, the pilot is likely to run out of space to maneuver the aircraft and may stall and impact the ground. In a modern transport aircraft with the autothrottle connected during the approach, the situation will be made even worse because the throttle will have been closed earlier on encountering the headwind.

During 1964–1985, the windshear/microburst has caused at least 26 civil aviation accidents in the United States, involving as many as 500 fatalities and more than 200 injuries. The July 1982 crash of the Pan American Flight 759 on the outskirts of New Orleans, which killed 153 people, drew the attention of the aviation community on the hazards of flying in low-level windshear. A microburst compounded by heavy rain is believed to be the cause of this crash. Since then, the Federal Aviation Administration (FAA) and NASA in the United States and RAE in the United Kingdom have extensively studied the problems associated with windshear/microburst. These efforts have resulted in the development of airborne and ground-based windshear detection systems, crew alerting, flight management, simulaton, and evaluation of piloting techniques for successful flight through windshear/microburst.

The best defense against the windshear or microburst is complete avoidance. However, if unavoidable, it is now recommended that on recognizing a possible encounter with severe windshear/microburst, the pilot should increase the thrust to its maximum and rotate the aircraft to an initial attitude, called the target pitch angle, which depends on the type of the airplane [6, 7]. The recommended target pitch angle for transport aircraft is about 15 deg. The pilot is advised to hold these settings until he is out of the low-level windshear or microburst region.

The onboard windshear warning/detection systems introduced in the late 1980s were reactive type, i.e., they detected the windshear only after the aircraft has actually encountered the windshear, not before. These systems relied on standard aircraft instrumentation such as accelerometers and air data systems that gave inputs to onboard algorithms that determined the windshear surrounding the aircraft. These reactive systems have given way to more advanced predictive type sensor systems that can provide the airline crew up to 30 s of advance warning of windshear/microburst activity ahead of the aircraft. The onboard warning/detection systems may be of radar, Lidar (light detection and ranging), or infrared type devices. Radar devices measure the Doppler velocity of water droplets moving with the winds. This method works well if the windshear/microburst conditions are accompanied by moderate or heavy rain. However, some windshear/microbursts contain little or no rain. In such situations, the Lidar devices perform well because the laser signal reflects from aerosol particles carried in the atmosphere at low altitudes. The infrared devices rely on the temperature changes in the farfield (1–3 km from the aircraft) caused by the windshear/microburst activity.

The FAA has made it mandatory for the airlines to install onboard warning/detection systems and pilot training to prevent any further mishaps because of windshear/microburst [10].

2.11 Summary

In this chapter, we have studied the basic methods of estimating the performance of piston-prop (propeller) and jet aircraft in power-off glide, level flight, climbing flight, range and endurance, turning flight, takeoff, and landing. These performance characteristics depend on the aerodynamic forces acting on the airframe, the thrust or power produced by the propulsion system, the rate of fuel consumption, and weight of the aircraft. In general, the thrust or power produced by the engine varies with altitude and flight velocity. The aerodynamic forces depend on the angle of attack, Mach number, and Reynolds number. The weight of the airplane continuously varies as the fuel is consumed. Furthermore, we have aerodynamic and structural limitations as expressed by the maximum lift coefficient and limit load factors. Therefore, estimation of the performance characteristics considering all these factors while optimizing the flight path performance is a complex mathematical problem.

However, the problem was simplified by assuming that 1) the airplane has a parabolic drag polar and a linear lift coefficient and 2) the power developed by the piston engine, the propulsive efficiency of the piston-props, and the thrust produced by the jet aircraft are independent of flight velocity. With these assumptions, the performance problems of propeller and jet aircraft become amenable to the methods of ordinary calculus. Although the equations for the performance of jet aircraft have analytical closed-form solutions, most of those for propeller aircraft do not have analytical solutions. One has to solve them graphically or numerically.

This approach gives us so-called point performance characteristics. We have obtained conditions that optimize the point performance in various flight conditions. Even though this method does not deal with the path performance of the aircraft, it gives a very useful understanding of the physical parameters of the airplane that play key roles in improving its path performance.

References

[1] Vinh, N. X., *Optimal Trajectories in Atmospheric Flight,* Elsevier, New York, 1981.
[2] Guntson, B., and Spick, M., *Modern Aircombat,* Crescent, New York, 1983.
[3] McCormick, B. W., *Aerodynamics, Aeronautics, and Flight Mechanics,* Wiley, New York, 1979.
[4] Boyle, D., "Windshear, Taming the Killer," *Interavia,* Jan. 1985, pp. 65–66.
[5] Lewis, M. S., "Sensing a Change in the Wind," *Aerospace America,* Jan. 1993, p. 20.

[6] Mulgund, S. S., and Stengel, R. F., "Target Pitch Angle for the Microburst Escape Maneuver," *Journal of Aircraft*, Vol. 30, No. 6, 1993, pp. 826–832.
[7] Stengel, R. F., "Solving the Pilot's Wind-Shear Problem," *Aerospace America*, Mar. 1985, pp. 82–85.
[8] Shelvel, R. S., *Fundamentals of Flight*, Prentice Hall, 1983.
[9] Loftin, L. K., "Quest for Performance," NASA SP-468, 1985.
[10] "Low Altitude Windshear Equipment Requirement," Federal Aviation Regulations (FAR), 14 CFR, Section 358, Part 121, Vol. 3, 2012.

Problems

2.1 Show that the maximum endurance of a glider is given by

$$t_{\max} = \frac{3E_m}{2\sqrt[4]{3}V_R}(h_i - h_f)$$

2.2 A glider weighs 4500 N and has a wing loading of 600 N/m^2, and its drag polar is given by $C_D = 0.01 + 0.022C_L^2$. It is launched from a height of 400 m in still air. Find (a) greatest possible ground distance it can cover, (b) maximum time it can remain in air, and (c) effect of 10 m/s tailwind for the first two cases (a and b). [Answer: (a) 13.484 km, (b) 6.72 min, and (c) $\Delta R = 3.45$ km and $\Delta t = 0$.]

2.3 A certain airplane weighs 44,440 N and has a wing loading of 1433.55 N/m^2. The drag polar is given by $C_D = 0.02 + 0.04C_L^2$ and $C_{L,\max} = 1.2$. For a power-off glide from 600 m, determine (a) the maximum distance it can cover and (b) the maximum time it can remain in the air. [Answer: (a) 10.6066 km and (b) 210.125 s.]

2.4 A piston-prop aircraft has a wing loading of 1600 N/m^2, wing area of 25 m^2 and its drag polar is given by $C_D = 0.025 + 0.05C_L^2$. The maximum lift coefficient is 1.5. The reciprocating engine develops 750 kW at sea level, and the propulsive efficiency of the engine–propeller combination is 0.85. Draw the power-available and power-required curves at sea level. Determine the maximum and minimum speeds for level flight at sea level.

What is the minimum power required for level flight at sea level? Determine the corresponding velocity and lift coefficient.

2.5 A turbojet airplane weighs 30,000 N, has a wing loading of 1000 N/m^2, and produces a sea-level thrust of 4000 N. The thrust varies with altitude as $T = T_0\sigma^{0.8}$. Assuming $C_D = 0.015 + 0.024C_L^2$ and $C_{L,\max} = 1.4$, find (a) maximum and minimum speeds in level flight at sea level and (b) the absolute ceiling of the airplane. [Answer: (a) 119.2132 m/s and 34.1494 m/s and (b) 13.30 km.]

2.6 What is the thrust required for a turbojet airplane weighing 50,000 N, with a wing loading of 1800 N/m^2 and a maximum sea level flight speed of 241.83 m/s? Assume $C_D = 0.02 + 0.04C_L^2$, and $C_{L,\max} = 1.5$. [Answer: 20,000 N.]

2.7 For the propeller airplane of Problem 2.4, determine, the maximum climb angle and the maximum rate of climb at sea level as well as the velocities and lift coefficients at which they occur.

2.8 A propeller airplane has a wing loading of 1750 N/m^2, wing area of 30 m^2, $C_D = 0.02 + 0.04C_L^2$, and $C_{L,\max} = 1.5$. The propulsive efficiency is 0.85. What should be the power developed by the reciprocating engine to achieve a maximum rate of climb of 12 m/s at sea level?

2.9 The drag polar of a turbojet airplane is given by

$$C_D = C_{DO} + \frac{C_L^2}{\pi Ae}$$

Assuming that the thrust is independent of flight speed, show that the dynamic pressure when the rate of climb is maximum is given by

$$q = \frac{T}{6SC_{DO}} + \sqrt{\left(\frac{T}{6SC_{DO}}\right)^2 + \frac{W^2}{3\pi AeS^2C_{DO}}}$$

2.10 An aircraft is powered by a turbojet engine and has a maximum speed of 790 km/h at sea level. The gross weight of the vehicle is 160,000 N, wing area is 50 m^2, and $C_D = 0.02 + 0.04C_L^2$. Find (a) the thrust developed by the engine, (b) climb angle and rate of climb when flying at 75% maximum aerodynamic efficiency, and (c) maximum rate of climb and velocity at which it occurs. [Answer: (a) 30,190.35 N, (b) 6.5036 deg and 14.4951 m/s, and (c) 14.5195 m/s at 132.1841 m/s.]

2.11 At a sea level speed of 200 m/s, the pilot of a certain jet aircraft can achieve a rate of climb of 20 m/s. Instead of climbing, the pilot chooses to accelerate, holding the altitude constant. Determine the maximum speed achieved at sea level. Assume $W = 200{,}000$ N, $S = 60$ m^2, and $C_D = 0.021 + 0.042C_L^2$. [Answer: 257.8538 m/s.]

2.12 A jet airplane weighs 160,000 N and has a zero-lift drag coefficient of 0.008 and a wing area of 42 m^2. At 100 m/s at sea level, the rate of climb is 11.5 m/s. The thrust developed by the engines is equal to 27,000 N. Determine the maximum rate of climb and the corresponding flight speed at sea level. [Answer: 21.6013 m/s and 214.5577 m/s.]

2.13 A jet aircraft weighs 150,000 N and has a wing area of 30 m^2. Its drag polar is given by $C_D = 0.015 + 0.025C_L^2$. The sea-level thrust T_0 is equal to 23,200 N, and the thrust at altitude is given by $T = T_0\sigma$. The specific fuel consumption at sea level c_0 is 1.2 N/Nh and, at altitude, $c = c_0\sigma$. Determine (a) the most economical cruise altitude and (b) fuel load and cruise velocity for a Breguet range of 2500 km. [Answer: (a) 12.31 km and (b) 5796.96 N and 205.166 m/s.]

2.14 If the aircraft in Problem 2.13 encountered a steady tailwind of 35 km/h during the cruise, how much fuel is saved? [Answer: 263.88 N.]

2.15 If the aircraft in Problem 2.13 followed a constant-altitude range program, find (a) cruise altitude and (b) fuel required to cover a still-air range of 2500 km. [Answer: (a) 11.2032 km and (b) 6365.37 N.]

2.16 A jet aircraft has the following data: $W = 50{,}000.0$ N, $C_D = 0.02 + 0.04C_L^2$, $C_{L,\max} = 1.2$, $S = 30$ m^2, $T = 14{,}500\sigma^{0.65}$ N, $c = 1.5\sigma^{0.25}$ N/Nh, oil consumption = 1 N for every 25 N of fuel, and $W_f = 20{,}000.0$ N.

Determine (a) still-air range and endurance at an altitude of 5000 m ($\sigma = 0.60$) and (b) reduction in the range if the aircraft encounters a steady headwind of 20 m/s during the cruise. [Answer: (a) 2075.05 km and (b) 448.49 km.]

2.17 A jet aircraft has the following data: $W = 50{,}000$ N, $C_D = 0.025 + 0.03C_L^2$, $C_L = 0.08\alpha$, $S = 30$ m^2, $T = 15{,}000\sigma^{0.6}$ N, and $c = 1.0\sigma^{0.2}$ N/Nh.

If the airplane cruises at an angle of attack of 5 deg, determine the percent change in the fuel consumption compared to the minimum possible fuel consumption to cover a range of 1500 km at an altitude of 11 km ($\sigma = 0.293$). Assume that the pilot constantly monitors the speed to prevent the aircraft from gaining altitude. [Answer: 2.6102]

2.18 A propeller airplane weighs 60,000 N and has a wing loading of 2000 N/m^2. The aspect ratio of the wings is 6, zero-lift drag coefficient is equal to 0.021, and Oswald's efficiency factor is equal to 0.920. The airplane produces a thrust of 750 kW at a propulsive efficiency of 0.82 and a specific fuel consumption of 3.5 N/kW/h. What fuel load should the airplane carry (a) to cover a still-air range of 1500 km and (b) to remain in air for 8 h? Assume that the airplane operates at an altitude of 2.4 km ($\sigma = 0.7892$). [Answer: (a) 6985.3523 N and (b) 9156.86 N.]

2.19 a) For a coordinated constant-velocity turn in a horizontal plane, derive the expressions for the turn rate and radius of turn for flight at limit load factor for jet aircraft.

b) Show that for MSTR and SST, the lift-to-drag ratios are respectively given by

1. $E = \dfrac{2nE_m}{(1+n^2)}$

2. $E = \dfrac{2znE_m}{(1+n^2z^2)}$

2.20 A propeller airplane weighs 45,000 N and has a wing area of 28 m^2. The drag polar is given by $C_D = 0.021 + 0.045C_L^2$, and the maximum lift coefficient is 1.5. The power developed by the engine is 900 kW with a propulsive efficiency of 0.82. The structural limit load factor is 3.0.

Determine the performance characteristics for (a) fastest sustained rate of turn, (b) sharpest sustained rate of turn, and (c) maximum load factor turn.

2.21 A jet aircraft weighs 80,000 N and has a maximum thrust of 30,000 N. The lift-curve slope of the wings is 5.0 per radian, the zero-lift incidence is −2.0 deg, and the maximum lift coefficient is 1.5. The wing area is 25 m^2, and the drag polar is given by $C_D = 0.018 + 0.08C_L^2$. The structural limit load factor is 6.0. For flight at an altitude of 1500 m ($\rho = 1.058$ kg/m^3), determine (a) the fastest rate of turn, (b) corresponding bank angle, (c) angle of attack, and (d) lift-to-drag ratio for a properly banked turn in a horizontal plane. [Answer: (a) $\omega = 13.9748$ deg/s, (b) $\mu = 70.3948$ deg, (c) $\alpha = 14.2$ deg, and (d) $E = 7.9477$.]

2.22 A jet aircraft has the following data: $W = 45{,}000$ N, thrust-to-weight ratio of 0.49, $C_{L\alpha} = 4.6$ per radian, $\alpha_{0L} = -2.2$ deg, and $S = 25$ m^2. Determine (a) the radius of a properly banked level turn at an altitude of 2250 m ($\sigma = 0.80$), with $\alpha = 8$ deg and $n = 4$; (b) additional thrust for executing a 5-g turn, with $C_L = 1.5$ and $E = 6.0$; and (c) reduction in the radius of turn, with 25% additional thrust and a sideslip of 20 deg. [Answer: (a) 472.18 m, (b) 15450 N, and (c) 24.23 m.]

2.23 A combat aircraft powered by a jet engine weighs 70,000 N and has a wing area of 25 m^2, $C_D = 0.015 + 0.06C_L^2$, and $C_{L,\max} = 1.4$. The structural limit load factor is 8.0. Determine (a) the bank angle, (b) lift coefficient, (c) lift-to-drag ratio, (d) radius of turn, and (e) load factor developed when the aircraft flying at 300 km/h completes a 180-deg turn at sea level in 15 s. (f) What is the thrust required? [Answer: (a) $\mu = 60.656$ deg, (b) $C_L = 1.3433$, (c) $E = 10.8975$, (d) $R = 397.9615$ m, (e) $n = 2.0406$, and (f) $T = 13{,}111.32$ N.]

2.24 A turbojet aircraft powered by a jet engine weighs 50,000 N and has a maximum thrust of 6000 N at sea level. The wing loading is 1500 N/m^2, $C_D = 0.020 + 0.08C_L^2$, and $C_{L,\max} = 1.8$. The maximum permissible load factor is 2.5. Determine (a) the bank angle, (b) lift coefficient, (c) lift-to-drag ratio, (d) rate of turn, and (e) load factor developed when the aircraft makes a coordinated sea-level turn of 1500 m radius using full thrust. [Answer: This problem has two solutions: 1) (a) $\mu = 37.5726$ deg, (b) $C_L = 0.2729$, (c) $E = 10.4962$, (d) $\omega = 4.0644$ deg/s, and (e) $n = 1.2617$, and 2) (a) $\mu = 7.4005$ deg, (b) $C_L = 1.2945$, (c) $E = 8.4004$, (d) $\omega = 1.6685$ deg/s, and (e) $n = 1.0084$.]

2.25 A combat aircraft weighs 75,000 N and has a wing area of 27 m^2. The maximum lift coefficient with high-lift devices is 1.8, and the structural limit load factor is 6.0. While flying at 250 km/h, the aircraft makes a 90-deg turn in 8 s at sea level holding a constant altitude and at an angle of attack such that the lift-to-drag ratio is 8.0. Find (a) the bank angle, (b) load factor, (c) radius of turn, and (d) the thrust required. [Answer: (a) $\mu = 54.26$ deg, (b) $n = 1.7120$, (c) $R = 353.7665$ m, and (d) $T = 16{,}050$ N.]

2.26 Determine the thrust required for a turbojet aircraft weighing 85,000 N with a wing area of 32 m^2, $C_{L,\max} = 1.70$, $C_D = 0.04 + 0.0833C_L^2$, and $n_{\lim} = 6.0$, so that it can execute a sea-level coordinated 90-deg turn in 6 s. [Answer: 31,681.9 N.]

2.27 A turbojet aircraft weighs 68,000 N and has a thrust-to-weight ratio of 0.6. Assuming $S = 24$ m^2, $C_D = 0.025 + 0.07C_L^2$, $C_{L,\max} = 1.20$, and $n_{\lim} = 8.0$, determine the minimum time to make a 180-deg turn while operating at an altitude of 2250 m ($\sigma = 0.80$) for (a) holding constant altitude and (b) permitting loss of altitude. (c) Determine the height lost in case (b). [Answer: (a) 9.4366 s, (b) 7.6886 s, and (c) 360.3337 m.]

2.28 Show that the reduction in takeoff ground run in the presence of a headwind is approximately given by $\Delta s_1 = (2V_w/V_1)s_1$.

2.29 An aircraft weighs 294,300 N and has a wing area of 100 m^2 and a thrust of 73,575 N. Its drag polar is given by $C_D = 0.02 + 0.05C_L^2$, and the maximum lift coefficient is 1.80 with flaps. Assuming that the coefficient of friction between the aircraft tires and the concrete runway is 0.05, find the minimum ground run and corresponding time at sea level. [Answer: $s_1 = 925.1527$ m.]

2.30 Assuming an obstacle height of 15 m, find total takeoff distance and time for the aircraft in Problem 2.29. [Answer: 990.4157 m, 31.50 s]

2.31 A combat aircraft weighs 78,480 N and has a wing area of 25 m^2, lift-curve slope of 0.06 per deg, $C_{L,\max} = 0.95$, and $C_D = 0.0254 + 0.178C_L^2$. This aircraft is required to land at an airstrip located at an altitude of 1000 m ($\sigma = 0.9074$). Assuming that the coefficient of friction between the tires and the runway is equal to 0.02 and approach glide angle is 3.5 deg, estimate (a) airborne distance (including flare) and (b) ground run. Assume an obstacle height of 15 m.

Assume that flaps are lowered at touchdown and give an increase in $C_{L,\max}$ of 0.45 and an increase in C_D of 0.05. Further assume that the brakes are applied simultaneously giving an increment in frictional coefficient of 0.4. [Answer: (a) 290 m and (b) 1038 m.]

Chapter 3 Static Stability and Control

3.1 Introduction

The airplane is a dynamic system with six degrees of freedom, three in translation and three in rotation. In general, airplane motion is characterized by three linear and three angular accelerations under the action of gravitational, inertial, and aerodynamic and propulsive forces and moments. We will study such a dynamic system in Chapters 4 and 6. However, a significant portion of the airplane's flight envelope consists of steady flight conditions such as cruise, climb, or glide. For such flight conditions, the principles of static equilibrium can be applied. This approach can be extended to some class of maneuvers like pull-up from a dive in vertical plane or a steady turn in horizontal plane. This background will also be helpful to understand the dynamic motion of the airplane. In this chapter, we will study the basic principles of static stability and control and discuss methods that are suitable for preliminary estimation of the static aerodynamic and stability characteristics of typical airplane configurations.

3.2 Concept of Equilibrium and Stability

Consider a ball resting on three different types of surfaces as shown in Fig. 3.1. If the ball is disturbed, it will return to its original equilibrium state in Fig. 3.1a, move away in Fig. 3.1b, and assume a new equilibrium state in Fig. 3.1c. In other words, the forces and moments acting on the ball in its disturbed state are such that they cause it to move towards its original equilibrium state in Fig. 3.1a and away from it in Figs. 3.1b and 3.1c. The ball never attains any equilibrium state in Fig. 3.1b but attains a new equilibrium state in Fig. 3.1c. Thus, the equilibrium states of the ball can be classified as stable, unstable, and neutrally stable.

Let us see how these simple concepts can be applied to an airplane in steady flight as shown in Fig. 3.2. For simplicity, let us restrict ourselves to angles of attack below the stall. To begin with, let us consider steady flights in vertical (longitudinal) plane. The steady flights in horizontal (lateral-directional) plane will be considered later. For a steady level flight in vertical plane, $L = W$ and $T = D$. Now if this airplane is disturbed in angle of attack, the response analogous to that of the ball could be one of three types. The disturbance could be in the form of a vertical gust, air turbulence, or a rapid movement of the elevator. In Fig. 3.2a, the induced nose-down pitching moment M is stabilizing because it tends to restore the airplane to its original angle of attack. In Fig. 3.2b, the induced noseup

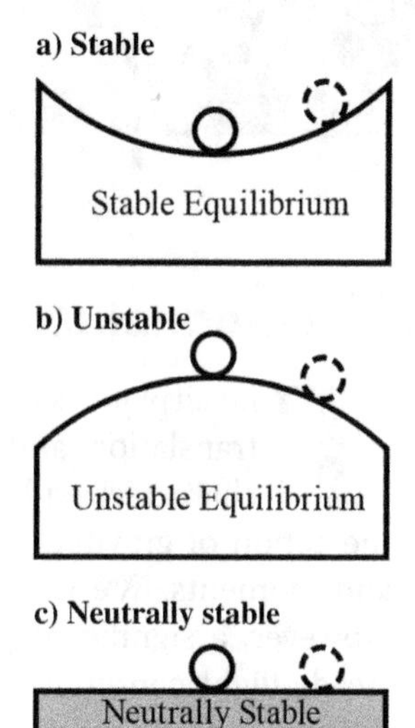

Fig. 3.1 Various forms of equilibrium.

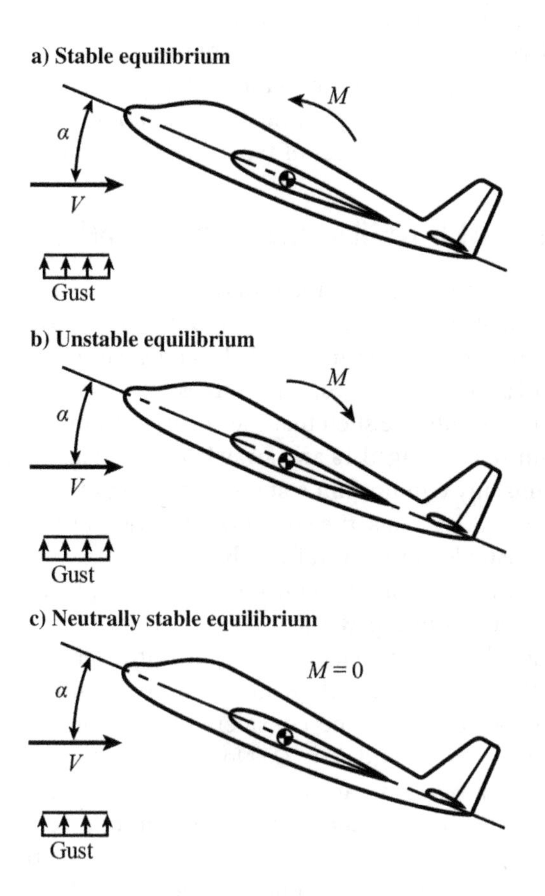

Fig. 3.2 Various forms of equilibrium in pitch.

pitching moment is destabilizing because it tends to increase the angle of attack and stall the airplane. In Fig. 3.2c, the induced pitching moment is zero, and the airplane will assume a new angle of attack depending on the magnitude of disturbance like the ball in Fig. 3.1c. According to the usual sign convention, noseup pitching moment is assumed to be positive and nosedown pitching moment is assumed to be negative. Thus, the criterion for longitudinal or pitch stability can be expressed mathematically as

$$\frac{\mathrm{d}M}{\mathrm{d}\alpha} < 0 \tag{3.1}$$

or in coefficient form

$$\frac{\mathrm{d}C_m}{\mathrm{d}\alpha} < 0 \tag{3.2}$$

where $C_m = M/qS\bar{c}$, q is the dynamic pressure, S is the reference area, and $\bar{c}$ is the reference length. Usually, wing area is used as reference area, and the mean aerodynamic chord (mac) of the wing is used as reference length. Thus, an airplane with $\mathrm{d}C_m/\mathrm{d}\alpha < 0$ is statically stable; that with $\mathrm{d}C_m/\mathrm{d}\alpha > 0$ is statically unstable. If $\mathrm{d}C_m/\mathrm{d}\alpha = 0$, the airplane is neutrally stable in pitch. In control system terminology, this type of system stability is known as open-loop stability.

The equilibrium condition in pitch is usually called trim condition. For pitch trim, the net pitching moment about the center of gravity is zero. For an airplane to be flyable, it must be capable of trimming at all values of angles of attack within the permissible range of angles of attack. Typical variations of the pitching moment with angle of attack are shown in Fig. 3.3. To establish a stable pitch trim the necessary and sufficient conditions are

$$C_{mo} > 0 \tag{3.3}$$

$$\frac{\mathrm{d}C_m}{\mathrm{d}\alpha} < 0 \tag{3.4}$$

If $C_{m\alpha} < 0$ but $C_{mo} < 0$, the airplane cannot be trimmed. In other words, the airplane is not flyable even though it has static stability. On the contrary, if $C_{m\alpha} > 0$ and $C_{mo} < 0$, the airplane is flyable, but it is statically unstable, i.e., the equilibrium condition is unstable like the ball in Fig. 3.1b. With proper feedback control, however, such an airplane can be made closed-loop stable (Chapter 6). In this chapter, we study the open-loop stability of the airplane with following assumptions:

1. The airplane has a vertical plane of symmetry, i.e., it has a symmetric geometry and mass distribution with respect to this plane.
2. Deflection of longitudinal controls like elevators do not generate side force, rolling, or yawing moments. Similarly, deflection of the

lateral-directional controls like ailerons or rudders do not produce lift or pitching moments.
3. Aerodynamic forces and moments vary linearly with aerodynamic/control variables.
4. Total forces and moments acting on the airplane are equal to the sum of forces/moments on individual components.

With these assumptions, longitudinal and lateral-directional motions of the airplane can be decoupled and studied separately. We will evaluate the contributions of fuselage, wing, and tail surfaces to static longitudinal and lateral-directional stabilities. The analysis presented here, in general, applies to airplane-type configurations at all speed ranges. However, the methods for evaluation of aerodynamic coefficients are presented only for subsonic ($0 \leq M \leq 0.8$) and supersonic speeds ($1.2 \leq M \leq 5$). At transonic speeds ($0.8 \leq M \leq 1.2$), satisfactory methods suitable for preliminary estimation of aerodynamic characteristics are not available. In view of this, one may have to carefully interpolate subsonic and supersonic aerodynamic

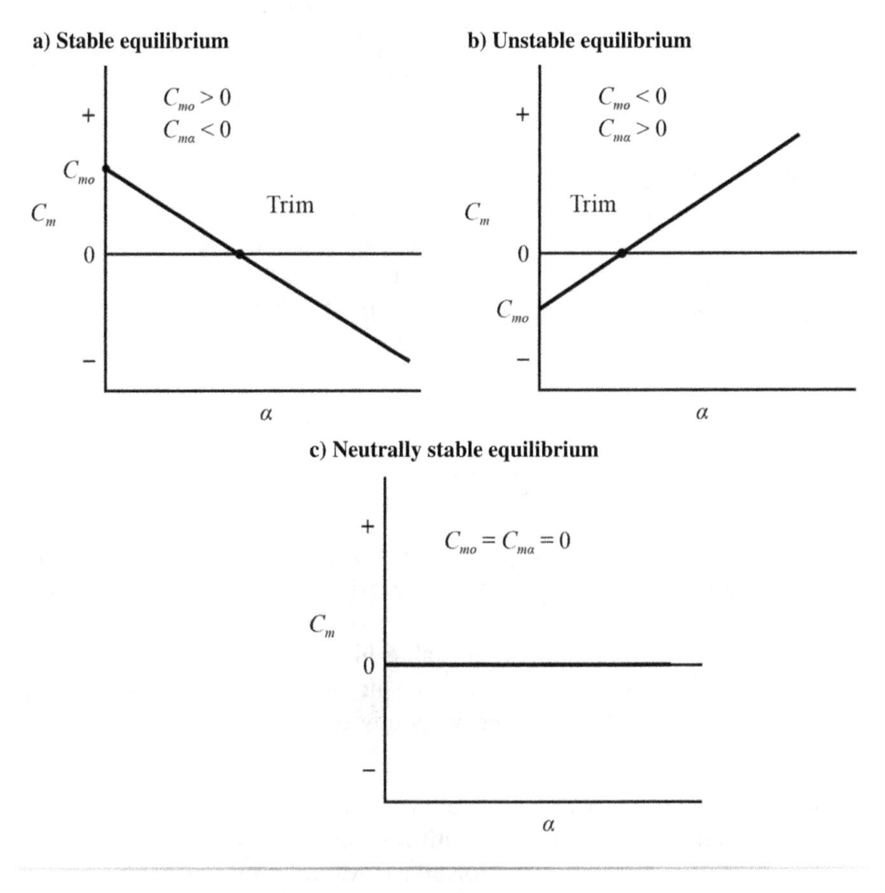

Fig. 3.3 Concept of stability and trim.

data to obtain some crude estimates of aerodynamic coefficients in the transonic Mach number range. The estimation of aerodynamic coefficients at hypersonic speeds ($M > 5$) is not discussed here.

We will be referring quite often to Datcom [1], short for Data Compendium. Datcom is a collection of empirical design methods for estimating stability and control derivatives for subsonic, supersonic, and hypersonic speed regimes.

3.3 Static Longitudinal Stability

The forces and moments acting on wing, fuselage, and horizontal tail surfaces in a steady, unaccelerated level flight are shown schematically in Fig. 3.4. We will be considering only such class of flights unless otherwise stated. In Fig. 3.4, L_w is the wing lift, x_a is the distance between wing aerodynamic center and center of gravity, $M_{ac,w}$ is the wing pitching moment about its aerodynamic center, M_f is the pitching moment due to fuselage about the center of gravity (c.g.), L_t is the horizontal tail lift, l_t is the distance between center of gravity and the horizontal tail aerodynamic center, and $M_{ac,t}$ is the horizontal tail pitching moment about its aerodynamic center. We ignore the lift due to fuselage and assume that the fuselage contributes only to pitching moment. As said before, we assume that the net or total pitching moment is equal to the sum of the individual contributions from fuselage, wing, and tail surfaces. Here, we ignore the contribution of propulsive unit to pitching moment. However, a brief qualitative discussion on power effects is presented later in this chapter. With these assumptions, the pitching moment about the center of gravity is given by

$$M_{cg} = \underbrace{M_f}_{\text{Fuselage Contribution}} + \underbrace{M_{ac,w} + L_w x_a}_{\text{Wing Contribution}} + \underbrace{M_{ac,t} - L_t l_t}_{\text{Tail Contribution}}$$

Here, M_f is the fuselage contribution. The wing contribution consists of two terms, $M_{ac,w}$ and $L_w x_a$. Similarly, the horizontal tail contribution

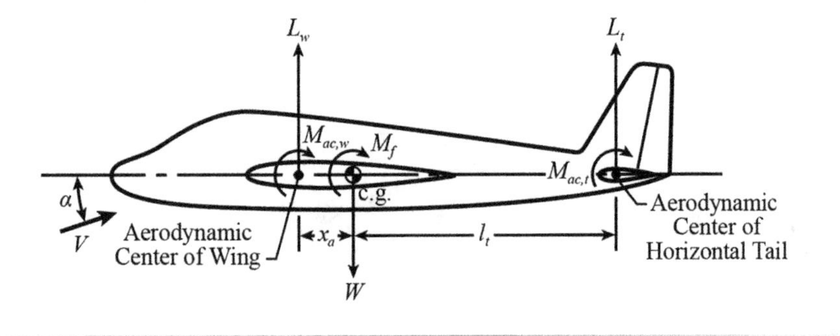

Fig. 3.4 Forces and moments acting on an airplane in level flight.

also consists of two terms, $M_{ac,t}$ and $L_t l_t$. In the following subsections, we discuss the methods that are suitable for preliminary estimation of these contributions.

3.3.1 Fuselage Contribution

The aerodynamics of streamlined bodies, typical of airplane fuselages, was studied by Munk [2] in the early 1920s. He ignored fluid viscosity and assumed ideal fluid flow. According to this theory, the pressure distribution over a streamlined body at an angle of attack yields a zero net force accompanied by a pure couple that is of a destabilizing nature as shown in Fig. 3.5. In other words, both lift and drag are equal to zero, but the pitching moment is nonzero. Mathematically, this is equivalent to a zero lift (normal force) acting at an infinite distance from the body so that the product (pitching moment) is finite. Generally, the fuselage contribution to static longitudinal stability is quite significant and is of a destabilizing nature. The destabilizing pitching moment varies linearly with angle of attack α as given by

$$\left(\frac{\partial M}{\partial \alpha}\right)_f = 2(k_2 - k_1)qV_f \tag{3.5}$$

where V_f is volume of the fuselage and $(k_2 - k_1)$ is the apparent mass constant, which depends on body fineness ratio $(l_f/b_{f,\max})$ as shown in Fig. 3.6. Assuming that the fuselage is a streamlined body with varying width/diameter, we can rewrite Eq. (3.5) in a coefficient form as

$$\left(\frac{\partial C_m}{\partial \alpha}\right)_f = \frac{\pi(k_2 - k_1)}{2S\bar{c}} \int_0^{l_f} b_f^2 \mathrm{d}x \tag{3.6}$$

where b_f is the local width/diameter, l_f is the fuselage length, S is the reference (wing) area, and $\bar{c}$ is the reference length (wing mean aerodynamic chord). Note that the values of slope of the pitching moment and that of the pitching-moment coefficient given by Eqs. (3.5) and (3.6) are per radian.

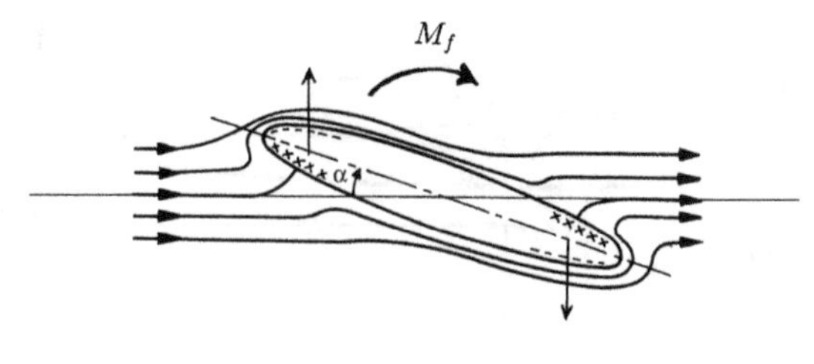

Fig. 3.5 Ideal fluid flow over an airplane fuselage.

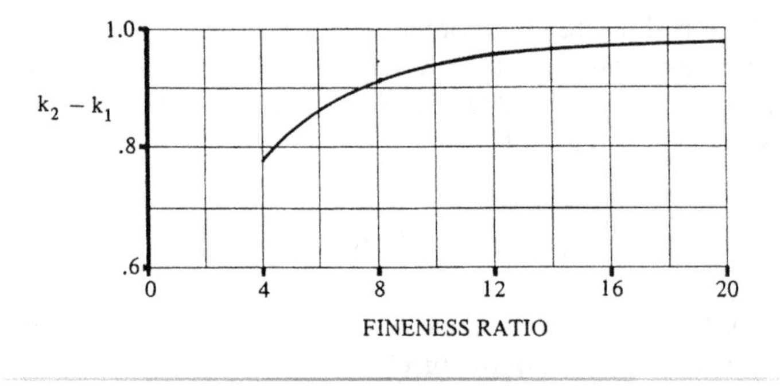

Fig. 3.6 Fuselage apparent mass coefficient (1).

For an isolated fuselage (no wing), the local angle of attack α_f would be a constant along the entire length and equal to α. However, in presence of the wing, the local fuselage angle of attack varies as shown in Fig. 3.7 with upwash in front of the wing leading edge and downwash behind the wing trailing edge. For that part of the fuselage extending from the wing leading edge to the wing trailing edge, the local flow is essentially parallel to the wing chord so that $\alpha_f = 0$. To account for these induced effects, Multhopp [3] modified Munk's theory as discussed in the following. Note that this theory is applicable only for low subsonic speeds.

According to Multhopp,

$$\left(\frac{\partial C_m}{\partial \alpha}\right)_f = \frac{\pi}{2S\bar{c}} \int_0^{l_f} b_f^2 \left(1 + \frac{\partial \epsilon_u}{\partial \alpha}\right) \mathrm{d}x \tag{3.7}$$

where ϵ_u denotes induced upwash or downwash at the axial location x. The slope of the pitching-moment coefficient given by Eq. (3.7) is per

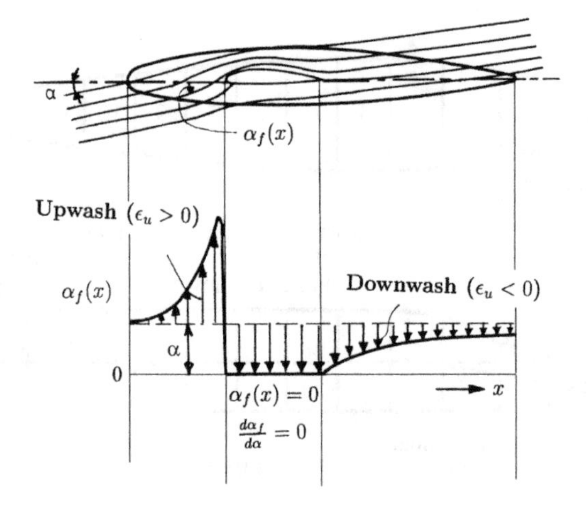

Fig. 3.7 Fuselage flow field in the presence of the wing.

radian. The zero-lift pitching moment $C_{mo,f}$ can be assumed to be equal to zero for uncambered (symmetric) fuselages. However, for cambered fuselages such as those with leading-edge droop or aft upsweep, $C_{mo,f}$ is nonzero and can be estimated as follows [1]:

$$C_{mo,f} = \frac{k_2 - k_1}{36.5S\bar{c}} \int_0^{l_f} b_f^2(\alpha_{ow} + i_{cl,B})\mathrm{d}x \tag{3.8}$$

where α_{ow} is the wing zero-lift angle relative to the fuselage reference line and $i_{cl,B}$ is the incidence angle of the fuselage camberline relative to the fuselage reference line. The parameter $i_{cl,B}$ is assumed to be negative for nose droop or aft upsweep as shown in Fig. 3.8. Note that in Eq. (3.8), both α_{ow} and $i_{cl,B}$ are in degrees.

Thus, in general, the pitching-moment coefficient of a fuselage can be expressed as

$$C_{m,f} = C_{mo,f} + \left(\frac{\partial C_m}{\partial \alpha}\right)_f \alpha \tag{3.9}$$

The variation of $\partial \epsilon_u/\partial \alpha$ for sections ahead of the wing leading edge is shown in Fig. 3.9. The curve in Fig. 3.9a is to be used for those sections that are very close to the wing leading edge and where the upwash gradients are high. The curve in Fig. 3.9b applies for those sections further away from the wing leading edge where the upwash gradients are low or moderate. For that part of the fuselage that is along the wing root chord, both α_f and $\partial \epsilon_u/\partial \alpha$ are assumed to be zero. As a result, this portion of the fuselage does not contribute to the pitching-moment integral of Eq. (3.7). In other words, we can effectively assume that $(1 + [\partial \epsilon_u/\partial \alpha]) = 0$ for this part of the fuselage.

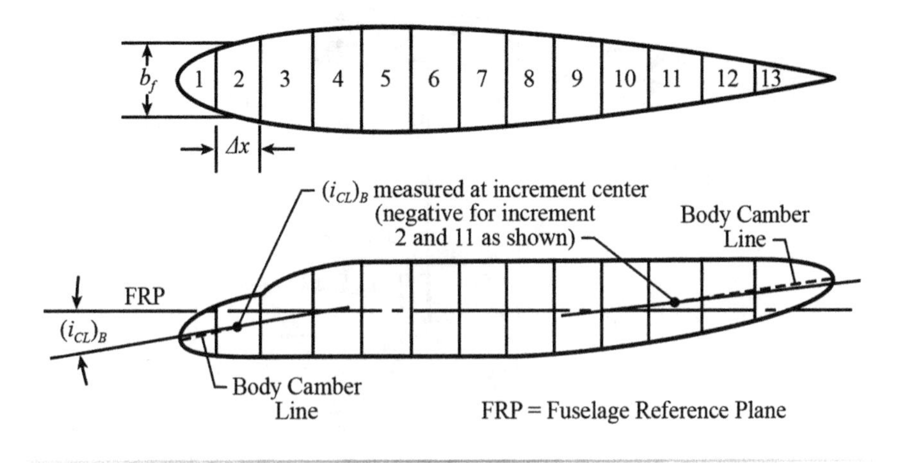

Fig. 3.8 Fuselage nose droop and aft upsweep [1].

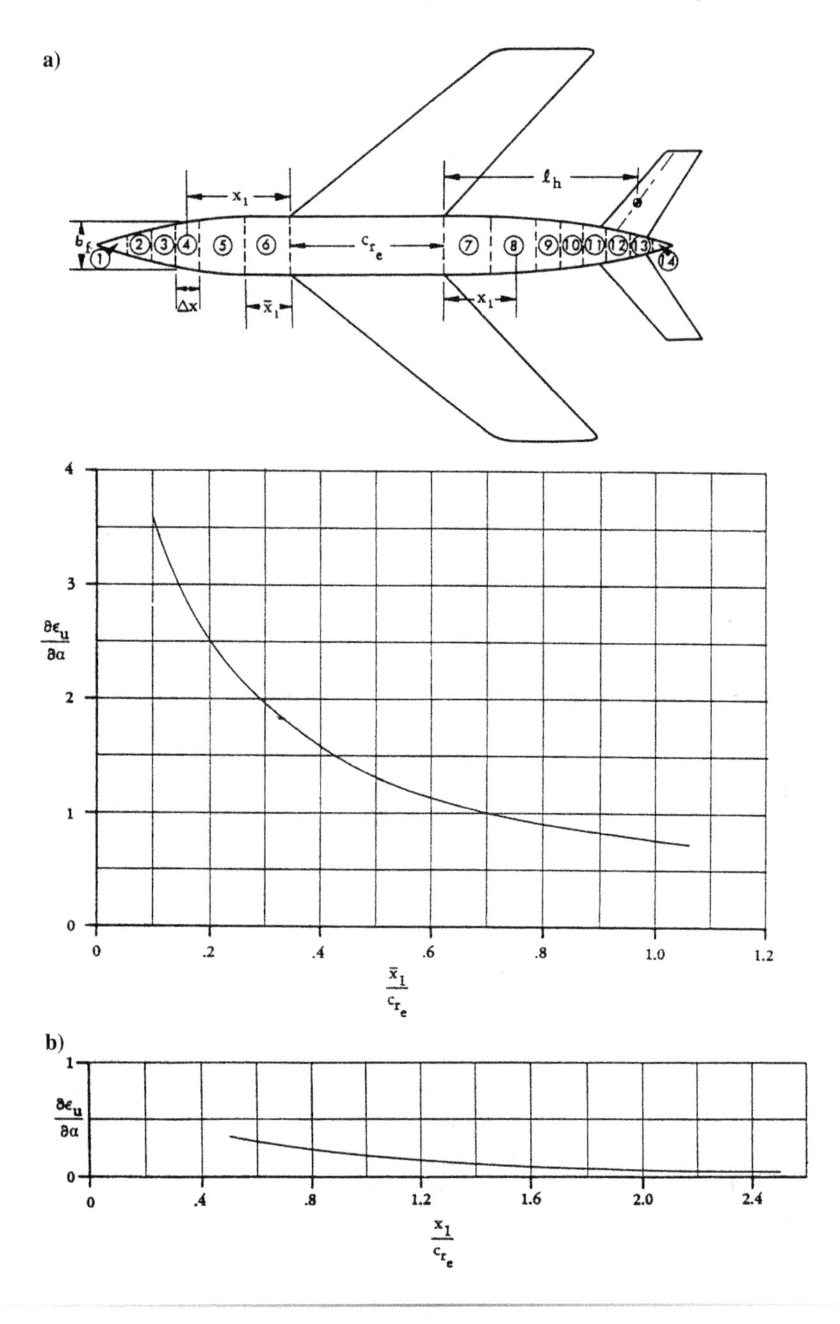

Fig. 3.9 Fuselage upwash ahead of the wing (1).

Both the curves in Figs. 3.9a and 3.9b are based on the wing–body lift-curve slope of 0.0785/deg. For any other values of wing–body lift-curve slope, multiply the values of $\partial\epsilon_u/\partial\alpha$ obtained from Fig. 3.9a or 3.9b by the factor $C_{L\alpha,WB}/0.0785$.

For fuselage sections behind the wing trailing edge, the parameter $(1 + [\partial\epsilon_u/\partial\alpha])$ varies linearly from zero at the wing trailing edge to the value $(1 - [\partial\epsilon/\partial\alpha])$ at the horizontal tail aerodynamic center.

The integral in Eq. (3.7) can be conveniently evaluated by dividing the fuselage into a number of segments as shown in the upper part of Fig. 3.9a. For segments like 1–5, the value of $\partial\epsilon_u/\partial\alpha$ can be computed at section midpoints using the curve in Fig. 3.9b. For the segments such as 6 that are very close to the wing leading edge and where upwash gradients are high, the value of $\partial\epsilon_u/\partial\alpha$ is to be computed using the curve in Fig. 3.9a. For aft segments such as 7–14, which are influenced by the wing downwash, the following linear relation is used:

$$1 + \frac{\partial\epsilon_u}{\partial\alpha} = \frac{x_1}{l_h}\left(1 - \frac{d\epsilon}{d\alpha}\right) \tag{3.10}$$

where l_h is the distance (measured parallel to the root chord) between the trailing edge of the root chord and the horizontal tail aerodynamic center. We will discuss the evaluation of the parameter $(1 - [d\epsilon/d\alpha])$ a little later.

For supersonic speeds, such simple and generalized methods applicable to arbitrary fuselage shapes and account for the wing interference as discussed herein are not available. Available methods that are suitable for preliminary design/analysis are limited to specific configurations such as cylindrical bodies. For ogive-cylinders and cone-cylinders, Datcom [1] gives the following expression:

$$C_{m\alpha,f} = \left(\frac{x_{cg}}{l_f} - \frac{x_{cp}}{l_f}\right) C_{N\alpha,f} \tag{3.11}$$

where $C_{N\alpha,f}$ is the normal-force-curve slope given in Fig. 3.10, x_{cp} is the center of pressure location (measured from nose of the body) given in Fig. 3.11, and l_f is the total length of the fuselage. In Figs. 3.10 and 3.11, $l_f = l_n + l_A$, $f_A = l_A/b_{f,\max}$, $\beta = \sqrt{M^2 - 1}$, and $f_N = l_n/b_{f,\max}$. Here, $b_{f,\max}$ is the maximum width/diameter of the fuselage. Note that the values of $C_{N\alpha}$ given in Fig. 3.10 are based on maximum cross-sectional area $S_{B,\max}$ of the body and the slope of the pitching-moment coefficient given by Eq. (3.11) is based on the product $S_{B,\max}l_f$. Therefore, we convert this result in terms of the standard reference area S (wing area) and reference length $\bar{c}$ (wing mac) as follows:

$$C_{m\alpha,f} = \left(\frac{x_{cg}}{l_f} - \frac{x_{cp}}{l_f}\right) C_{N\alpha,f}\left(\frac{S_{B,\max}l_f}{S\bar{c}}\right) \tag{3.12}$$

3.3.2 Wing Contribution

The nature of wing contribution depends on the relative distance between the wing aerodynamic center and the center of gravity x_a

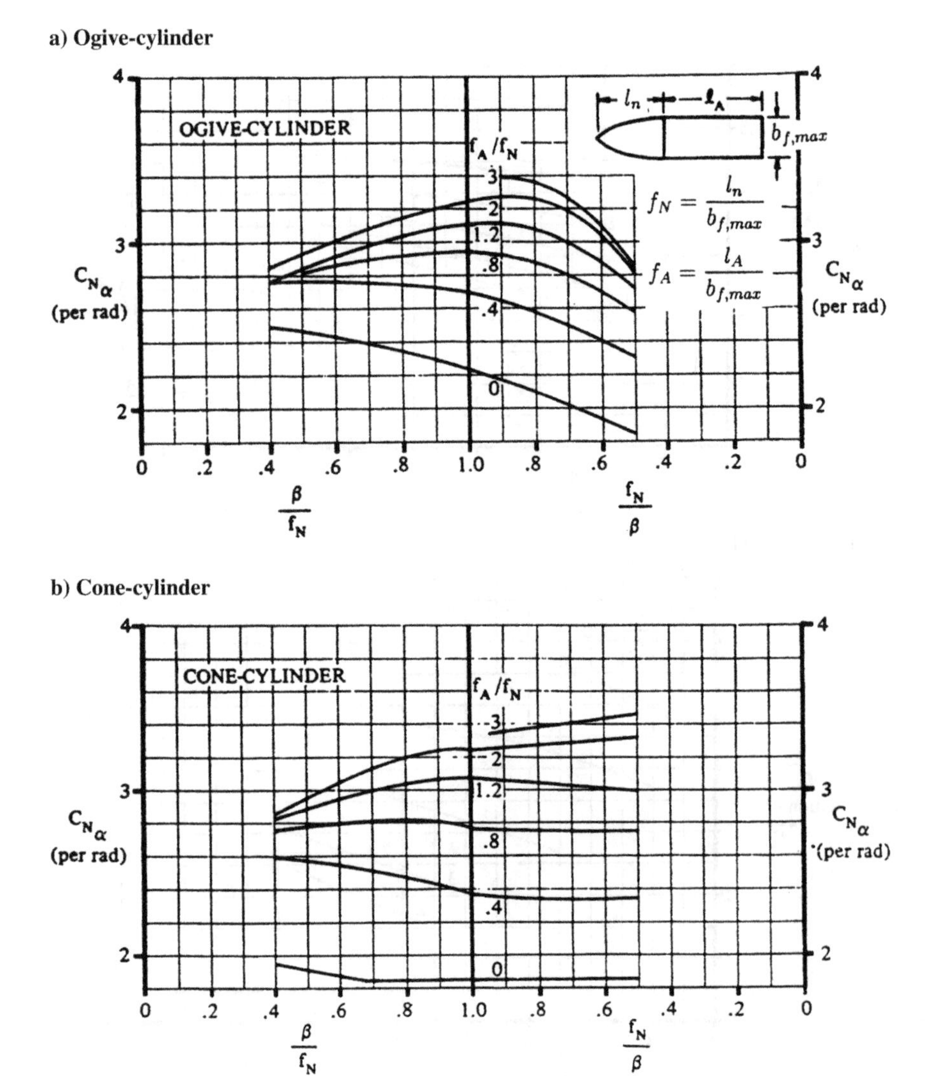

Fig. 3.10 Supersonic normal-force-curve slope (1).

(see Fig. 3.4). If the aerodynamic center is ahead of the center of gravity, which is generally the case for low-speed general aviation airplanes, the wing contribution is destabilizing. For modern high-speed airplanes with swept-back wings, the wing aerodynamic center is usually aft of the center of gravity, and wing contribution is stabilizing. For the low-speed general aviation-type configuration shown in Fig. 3.4, the wing contribution is given by

$$M_W = M_{ac,w} + L_W x_a$$

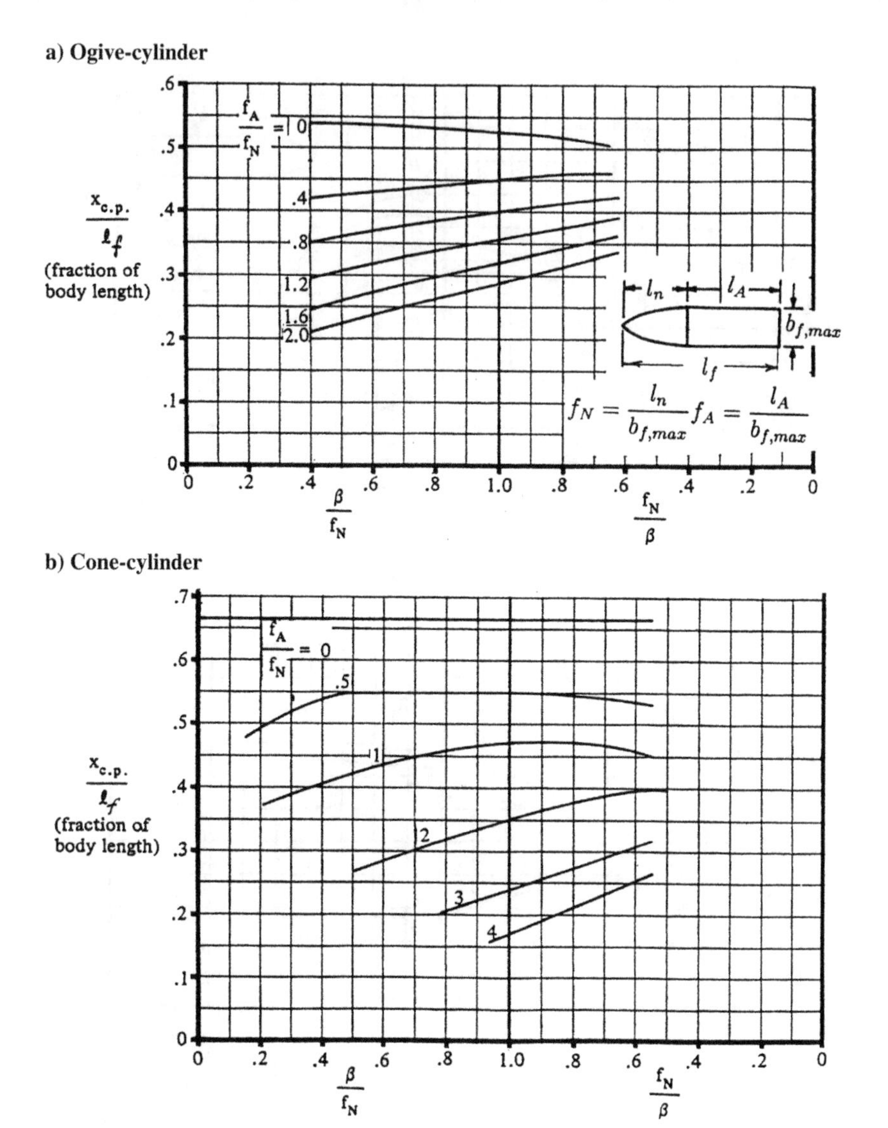

Fig. 3.11 Supersonic center of pressure locations (1).

Dividing throughout by $qS\bar{c}$, we obtain

$$C_{m,w} = C_{mac,w} + C_{L,w}\bar{x}_a \tag{3.13}$$

where

$$C_{L,w} = a_w \alpha_w \tag{3.14}$$

$$\bar{x}_a = \frac{x_a}{\bar{c}} \tag{3.15}$$

Here x_a will be considered positive if the center of gravity is aft of the aerodynamic center and vice versa. The wing angle of attack α_w is usually defined as

$$\alpha_w = \alpha + i_w - \alpha_{oL,w}$$

Here, α is the angle of attack of the airplane. It is defined as the angle between the flight velocity vector and the airplane reference line. Usually, the airplane reference line is assumed to coincide with the fuselage centerline. Further, i_w is the wing incidence with respect to the airplane reference line and $\alpha_{oL,w}$ is the zero-lift angle of the wing.

In Eq. (3.14), a_w is lift-curve slope of the wing in the presence of the fuselage. Just as wing affects the fuselage flow field, fuselage induces a crossflow that affects the spanwise lift distribution over the wing. The fuselage interference effects on the wing are generally small for configurations with large wing-span-to-body-diameter ratios typified by conventional long-range subsonic airplanes. For such configurations, lift-curve slope of the wing in the presence of the fuselage can be assumed to be equal to that of an isolated wing.

For subsonic speeds (below critical Mach number), the lift-curve slope of straight tapered wings can be determined using the data given in Fig. 3.12 or the following expression [1]:

$$a_w = \frac{2\pi A}{2 + \sqrt{\frac{A^2\beta^2}{k^2}\left(1 + \frac{\tan^2\Lambda_{c/2}}{\beta^2}\right) + 4}} \tag{3.16}$$

where A is the aspect ratio, $\beta = \sqrt{1 - M^2}$, $k = a_o/2\pi$, and $\Lambda_{c/2}$ is the mid-chord sweep. The value of a_w given by Eq. (3.16) is per radian.

The sectional (two-dimensional) lift-curve slope a_o is given by

$$a_o = \frac{1.05}{\sqrt{1 - M^2}}\left[\frac{a_o}{(a_o)_{\text{theory}}}\right](a_o)_{\text{theory}} \tag{3.17}$$

Here, $(a_o)_{\text{theory}}$ is the theoretical sectional lift-curve slope in incompressible flow and is given in Fig. 3.13a. The parameter $a_o/(a_o)_{\text{theory}}$ is the empirical correction factor shown in Fig. 3.13b to account for the viscous effects that depend on the Reynolds number and trailing-edge included angle. The Reynolds number R_l is based on the section chord. The trailing-edge included angle ϕ'_{TE} is defined as the angle between straight lines passing through points at 90 and 99% of the chord on the upper and lower surfaces of the airfoil and is given by

$$\tan\frac{\phi'_{TE}}{2} = \frac{0.5y_{90} - 0.5y_{99}}{9} \tag{3.18}$$

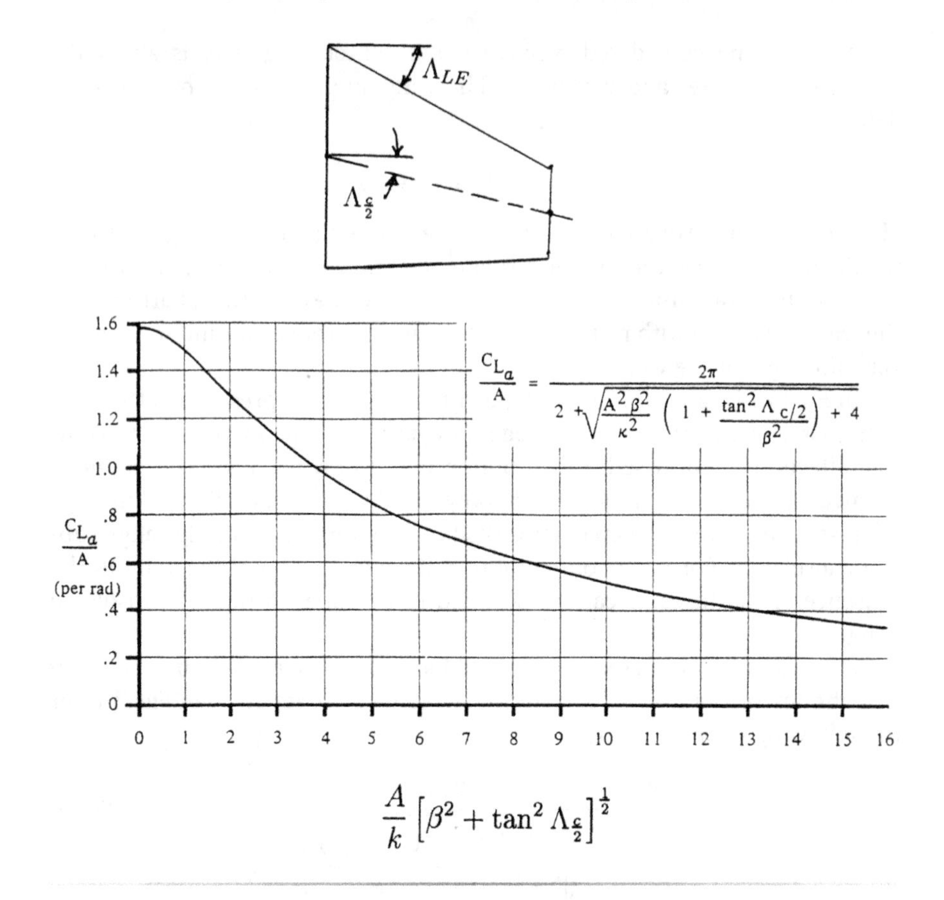

Fig. 3.12 Subsonic wing lift-curve slope (1).

where y_{90} and y_{99} are the y coordinates (in percent of chord) at $x/c = 0.09$ and 0.99, respectively. For a given airfoil, this information may be obtained from the given section geometrical data or from Abbott and Von Doenhoff [4] for most of the NACA family airfoils. Figure 3.13c gives the variation of trailing-edge angle ϕ_{TE} with thickness ratio for several NACA family airfoils. In general, $\phi_{TE} \neq \phi'_{TE}$. However, in the absence of actual geometrical data, we may assume $\phi'_{TE} = \phi_{TE}$ and approximately determine the value of ϕ_{TE} using the data given in Fig. 3.13c.

The midchord sweep $\Lambda_{c/2}$ is given by the following expression:

$$\tan\Lambda_{c/2} = \tan\Lambda_{LE} - \left(\frac{c_r - c_t}{b}\right) \tag{3.19}$$

where Λ_{LE} is the leading-edge sweep angle, c_r is the root chord, c_t is the tip chord, and b is the wing span.

a) Theoretical sectional lift-curve slope

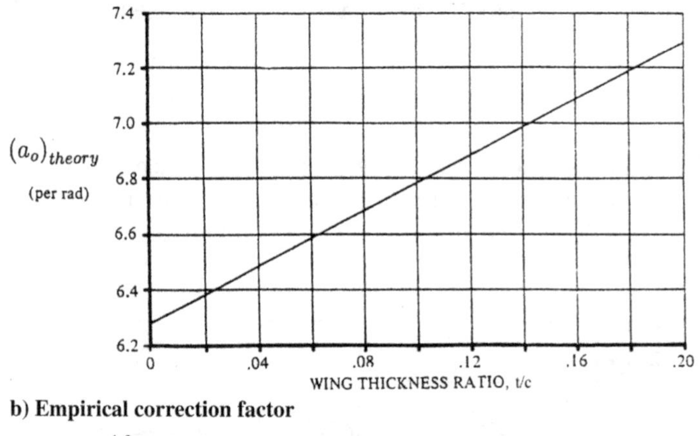

b) Empirical correction factor

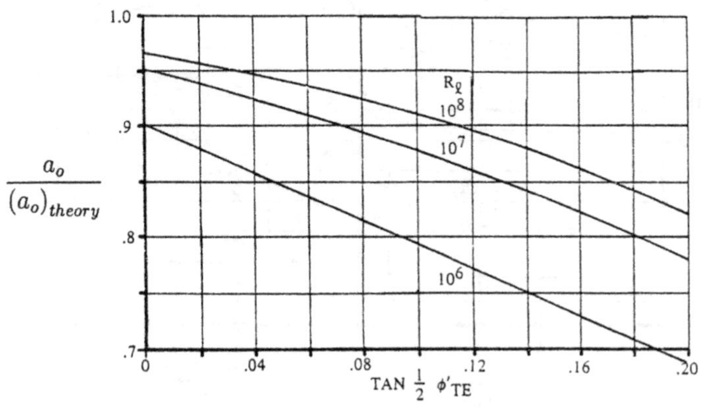

c) Variation of trailing-edge angle

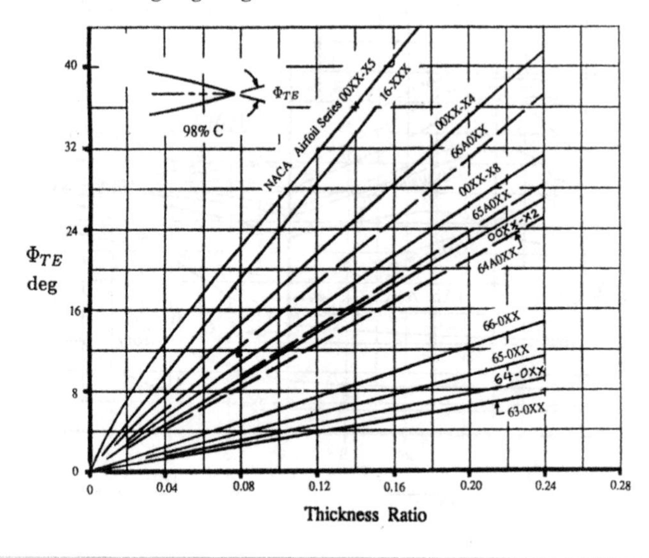

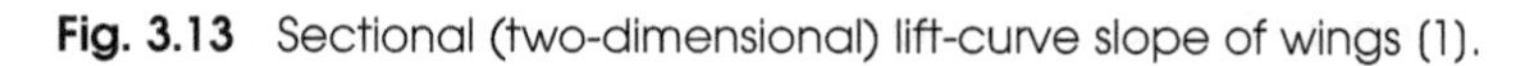

Fig. 3.13 Sectional (two-dimensional) lift-curve slope of wings (1).

For high-aspect ratio rectangular wings at very low speeds (incompressible flow), the following simple formula is quite often used:

$$a_W = \frac{a_o}{1 + \frac{a_o}{\pi A}} \tag{3.20}$$

It may be observed that for such cases, Eq. (3.16) reduces to Eq. (3.20).

The estimation of lift-curve slope of straight tapered wings at supersonic speeds can be done using the Datcom data [1] presented in

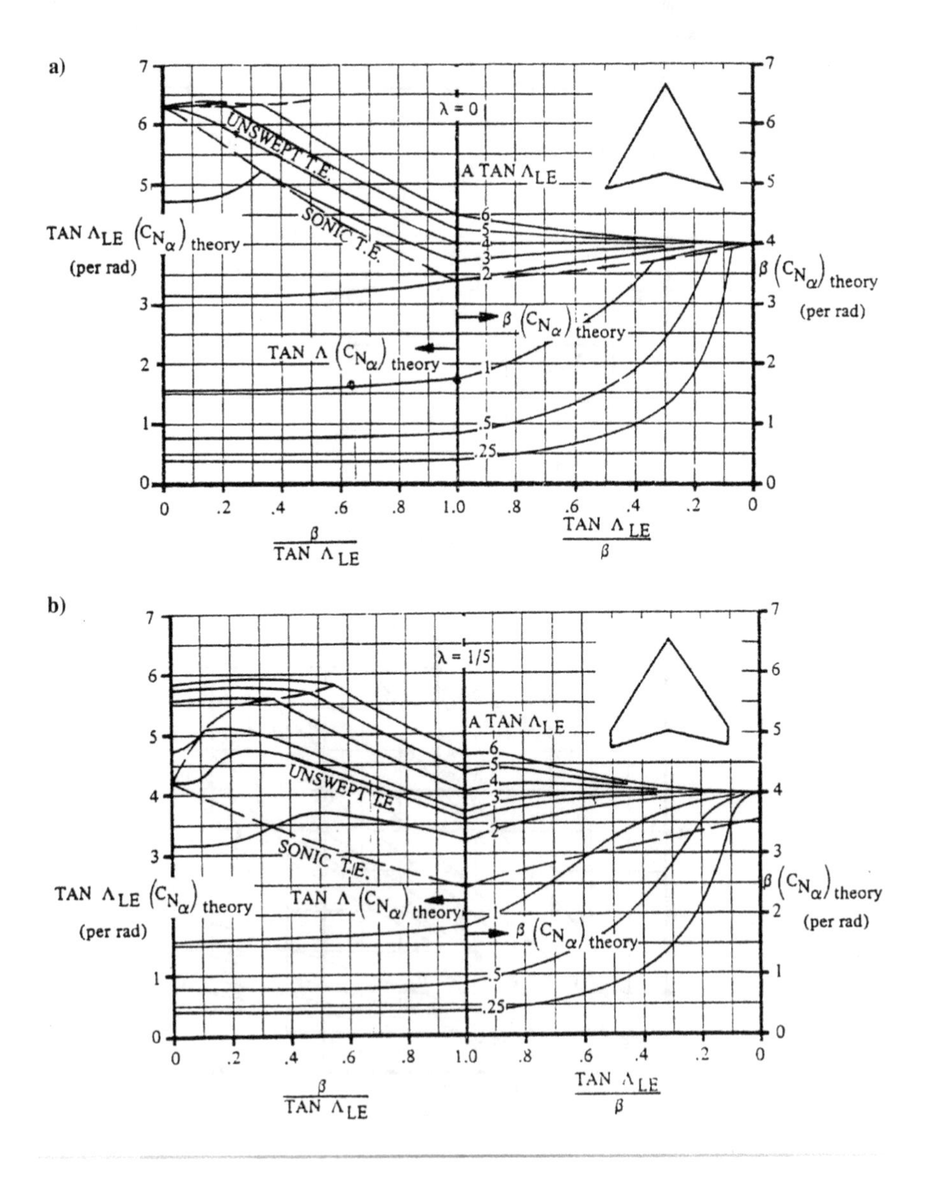

Fig. 3.14 Supersonic normal-force-curve slope of wings (1).

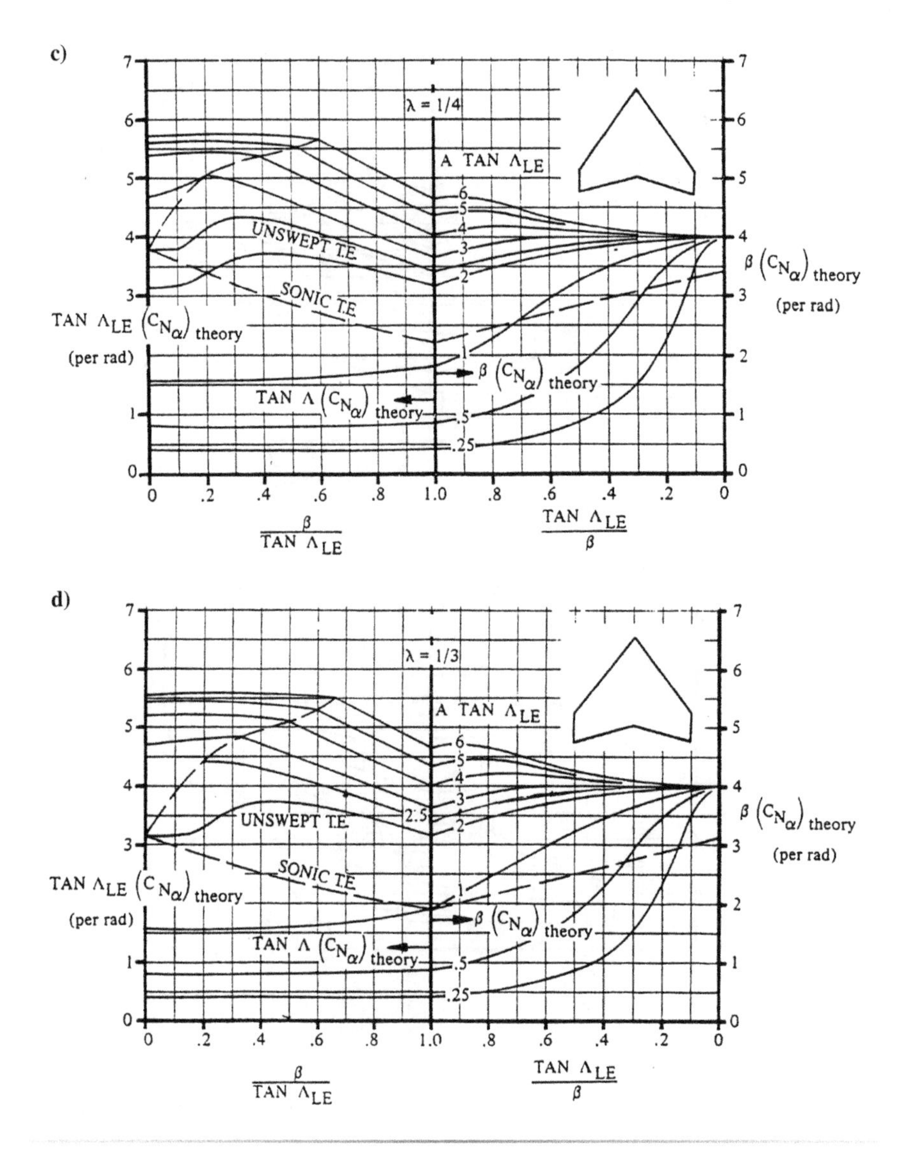

Fig. 3.14 Supersonic normal-force-curve slope of wings (1) (*continued*).

Figs. 3.14 and 3.15. The parameter β used in Figs. 3.14 and 3.15 is defined as follows:

For subsonic speeds

$$\beta = \sqrt{1 - M^2}$$

For supersonic speeds

$$\beta = \sqrt{M^2 - 1}$$

where M is the flight Mach number.

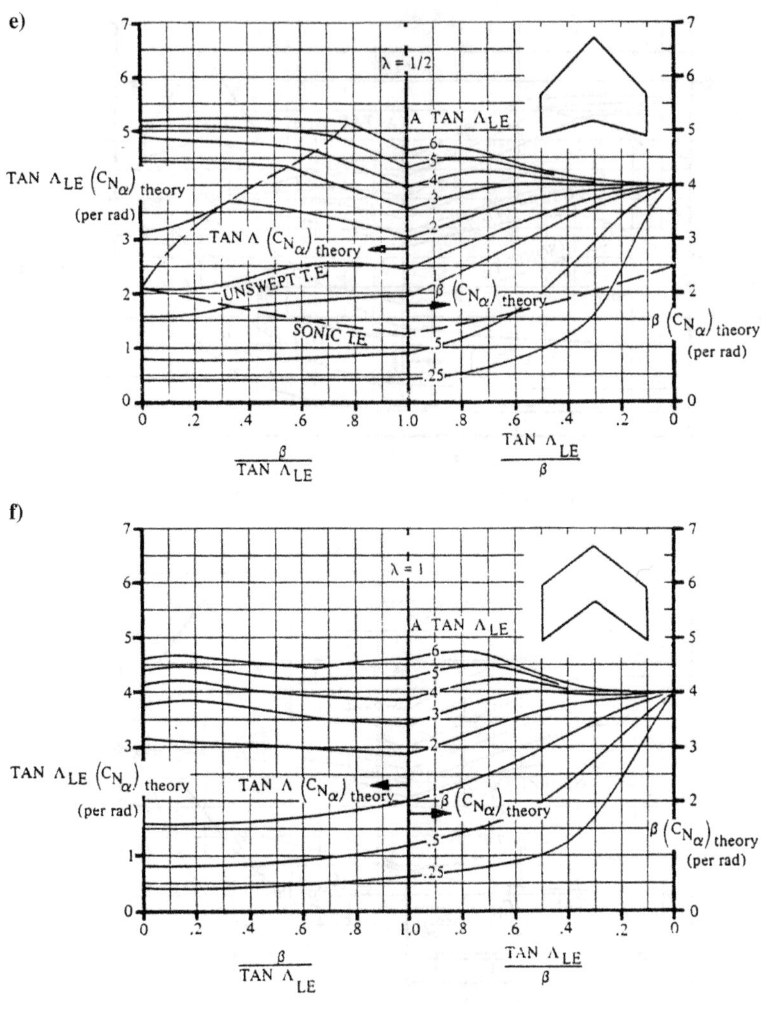

Fig. 3.14 Supersonic normal-force-curve slope of wings (1) *(continued)*.

At high speeds, the slope of normal-force coefficient is usually given instead of the lift-curve slope. Approximately, $C_{N\alpha} = C_{L\alpha}$. The procedure of estimating $C_{N\alpha}$ is as follows:

1. For the given value of taper ratio λ, find $(C_{N\alpha})_{\text{theory}}$ from Figs. 3.14a–3.14f. Use interpolation if the given value of λ is not covered in Figs. 3.14a–3.14f.
2. Apply the correction for sonic leading-edge effects using the data given in Fig. 3.15a and the following relation:

$$C_{N\alpha} = \frac{C_{N\alpha}}{(C_{N\alpha})_{\text{theroy}}}(C_{N\alpha})_{\text{theroy}} \tag{3.21}$$

Data in Fig. 3.15a are plotted as functions of the parameter $\Delta y_\perp$, which is defined as

$$\Delta y_\perp = \frac{\Delta y}{\cos \Lambda_{LE}} \tag{3.22}$$

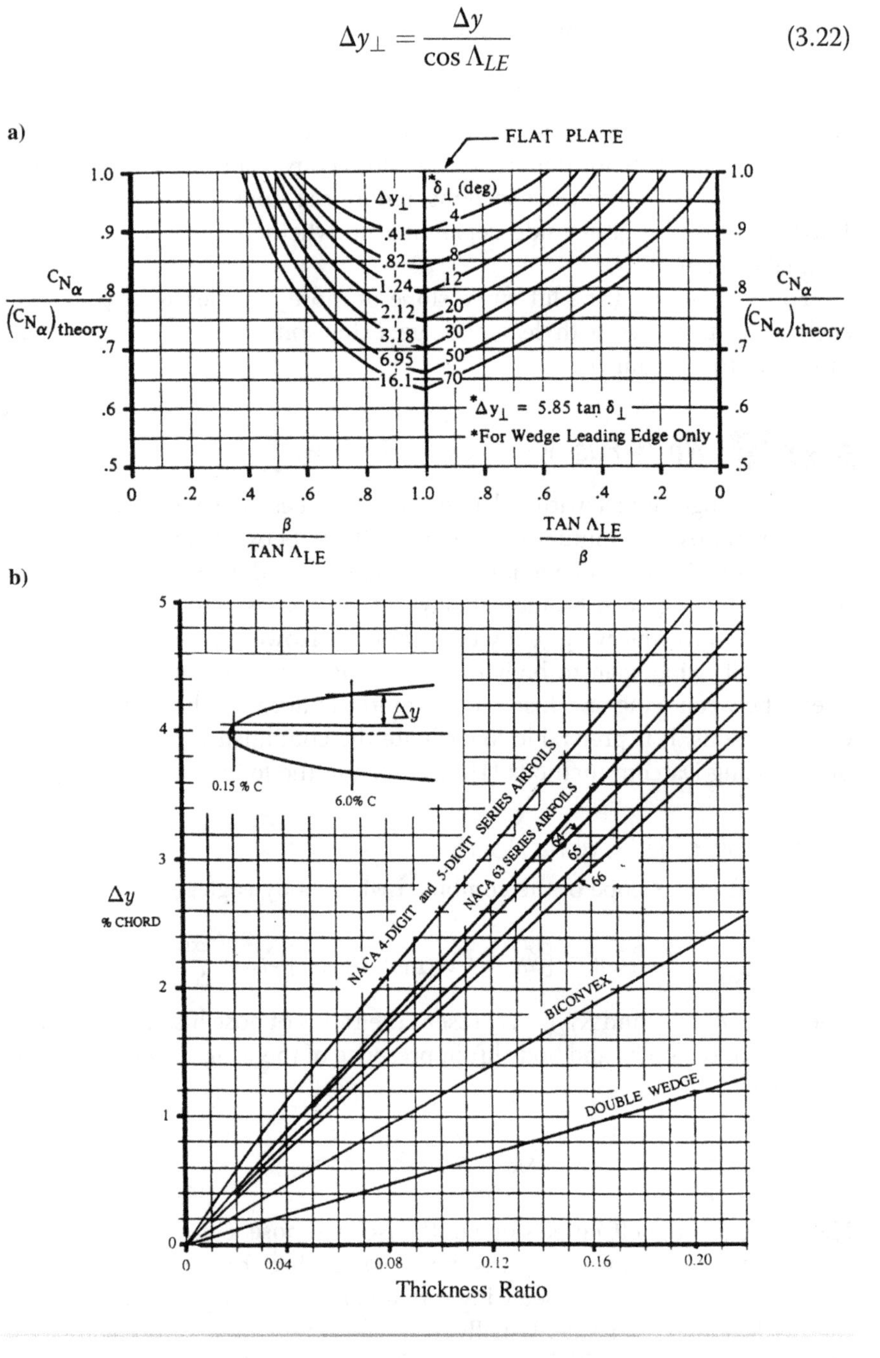

Fig. 3.15 Sonic leading-edge correction factor for supersonic lift-curve slope of wings (1).

where Δy is the difference between the upper-surface coordinates expressed in percent chord at the 6% and 0.15% chord stations as given in Fig. 3.15b for various types of airfoil sections. For double wedge and biconvex airfoils, $\Delta y_{\perp}$ can also be obtained using the following relation:

$$\Delta y_{\perp} = 5.85 \tan \delta_{\perp} \tag{3.23}$$

where $\delta_{\perp}$ is the leading-edge, semiwedge angle. The $\delta_{\perp}$ values superposed in Fig. 3.15a apply only for double wedge and biconvex airfoils and are related by Eq. (3.23). For other wing sections, ignore the values of $\delta_{\perp}$ in Fig. 3.15a. Use the curves on either side depending on whether $\beta/\tan \Lambda_{LE}$ is greater or smaller than unity.

For estimation of the lift-curve slope of more complex wing planforms such as delta wings with leading-edge extensions or strakes, the reader may refer to Datcom [1].

3.3.3 Wing–Fuselage Contribution

For configurations with relatively large values of wing-span-to-body-diameter ratios, the mutual interference effects between the wing and the body are small and can be ignored. For such cases, the contributions of the wing and fuselage can be individually determined as discussed and added together to give the wing–body contribution. However, for configurations with small wing-span-to-body-diameter ratio, the mutual interference effects between wing and fuselage are quite significant. Therefore, for such configurations, it is preferable to evaluate the combined wing–body contribution using Datcom methods [1] discussed in the following sections.

Estimation of Lift-Curve Slope

The lift-curve slope of the combined wing–body is given by

$$C_{L\alpha,WB} = \left[K_N + K_{W(B)} + K_{B(W)}\right] C_{L\alpha,e} \frac{S_{\text{exp}}}{S} \tag{3.24}$$

where K_N, $K_{W(B)}$, and $K_{B(W)}$ represent the ratios of nose lift, the wing lift in presence of the body, and body lift in presence of the wing to wing-alone lift. We have

$$K_N = \left(\frac{C_{L\alpha,N}}{C_{L\alpha,e}}\right) \frac{S}{S_{\text{exp}}} \tag{3.25}$$

Here, $C_{L\alpha,N}$ is the lift-curve slope of the isolated nose, $C_{L\alpha,e}$ is the lift-curve slope of the exposed wing, S_{exp} is the exposed wing area, and S is the reference (wing) area. The exposed wing is that part of the wing that lies outboard the fuselage on either side. Usually, the reference wing area is equal to the theoretical wing area, which is obtained by extending the leading and trailing edges of both the right and left wings to meet on the centerline. Thus, the theoretical wing area is the sum of exposed wing area and the area of that part of the fuselage obtained by an extension of the leading and trailing edges of the wing up to the fuselage centerline as shown in Fig. 3.16.

For subsonic speeds,

$$C_{L\alpha,N} = \frac{2(k_2 - k_1)S_{B,\max}}{S} \tag{3.26}$$

where $k_2 - k_1$ is the apparent mass constant (Fig. 3.6) and $S_{B,\max}$ is the maximum cross-sectional area of the fuselage. The value of $C_{L\alpha,N}$ given by Eq. (3.26) is per radian. For supersonic speeds, as said before, such simple and generalized methods are not available. For cone-cylinders and ogive-cylinders, the Datcom [1] data given in Fig. 3.10 may be used. Values of $C_{N\alpha}$ given in Fig. 3.10 are based on maximum frontal area of the body $S_{B,\max}$. To obtain $C_{L\alpha,N}$ based on wing area S, multiply that value by $S_{B,\max}/S$.

The coefficients $K_{W(B)}$ and $K_{B(W)}$ are given in Fig. 3.17 as functions of the ratio of body diameter to wing span. Data presented in Fig. 3.17 apply for all Mach numbers because they are based on slender body theory.

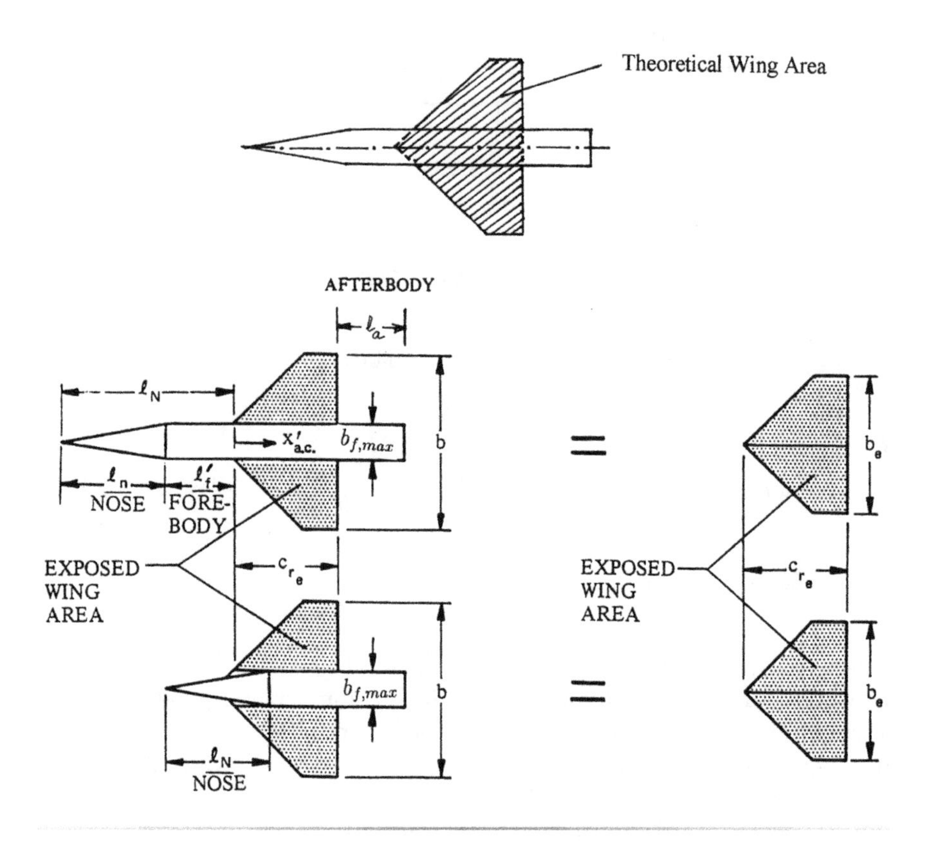

Fig. 3.16 Body and wing nomenclature (1).

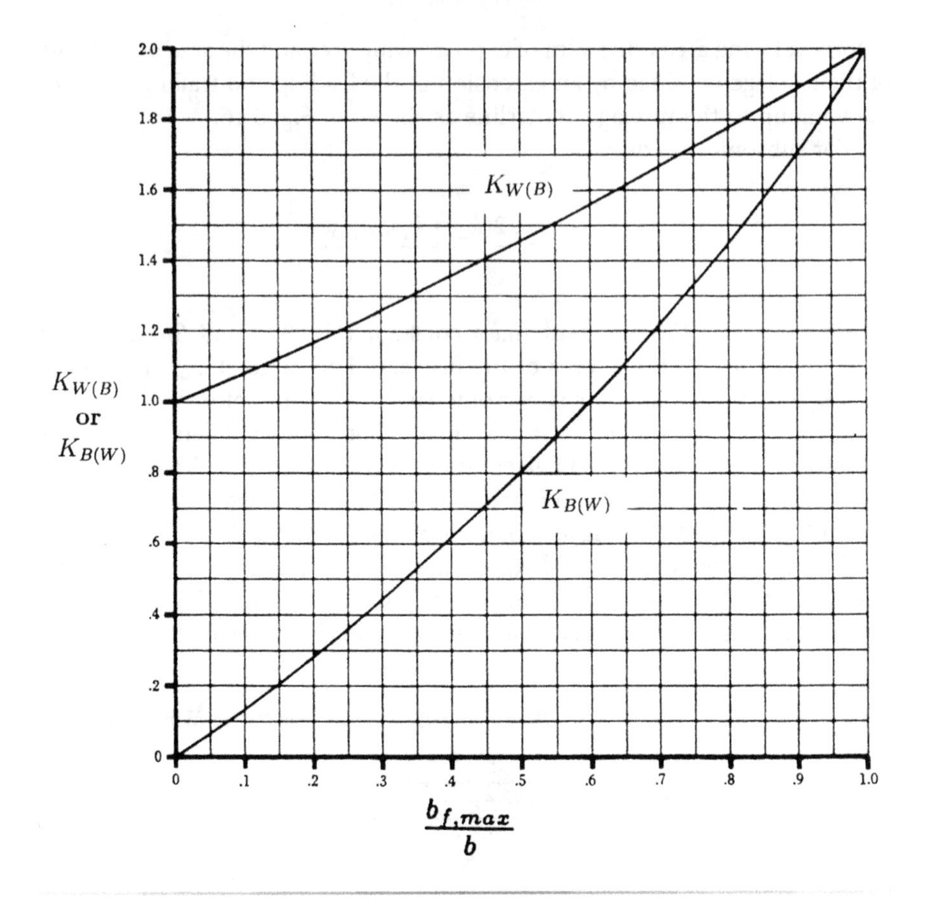

Fig. 3.17 Lift ratios $K_{B(W)}$ and $K_{W(B)}$ (1).

Data for $K_{W(B)}$ and $K_{B(W)}$ of Fig. 3.17 can be approximated by the following expressions:

$$K_{W(B)} = 0.1714\left(\frac{b_{f,\max}}{b}\right)^2 + 0.8326\left(\frac{b_{f,\max}}{b}\right) + 0.9974 \tag{3.27}$$

$$K_{B(W)} = 0.7810\left(\frac{b_{f,\max}}{b}\right)^2 + 1.1976\left(\frac{b_{f,\max}}{b}\right) + 0.0088 \tag{3.28}$$

where $b_{f,\max}$ is the maximum width of the fuselage and b is the wing span. However, for certain combinations of Mach number and geometric parameters, the slender body theory may not strictly hold. Such a situation may occur at supersonic speeds for the parameter $K_{B(W)}$ when 1) the parameter $\beta A_e(1+\lambda_e)[(\tan\Lambda_{LE}/\beta)+1] \geq 4$ for rectangular planforms, or 2) the parameter $\beta A_e \geq 1$ for triangular wing plan-forms. Here, A_e is the exposed wing aspect ratio and $\beta = \sqrt{M^2-1}$. For such cases, $K_{B(W)}$ can be determined using the data given in Fig. 3.18a or 3.18b depending on whether the fuselage extends beyond the wing trailing edge or not, i.e., with or without the afterbody.

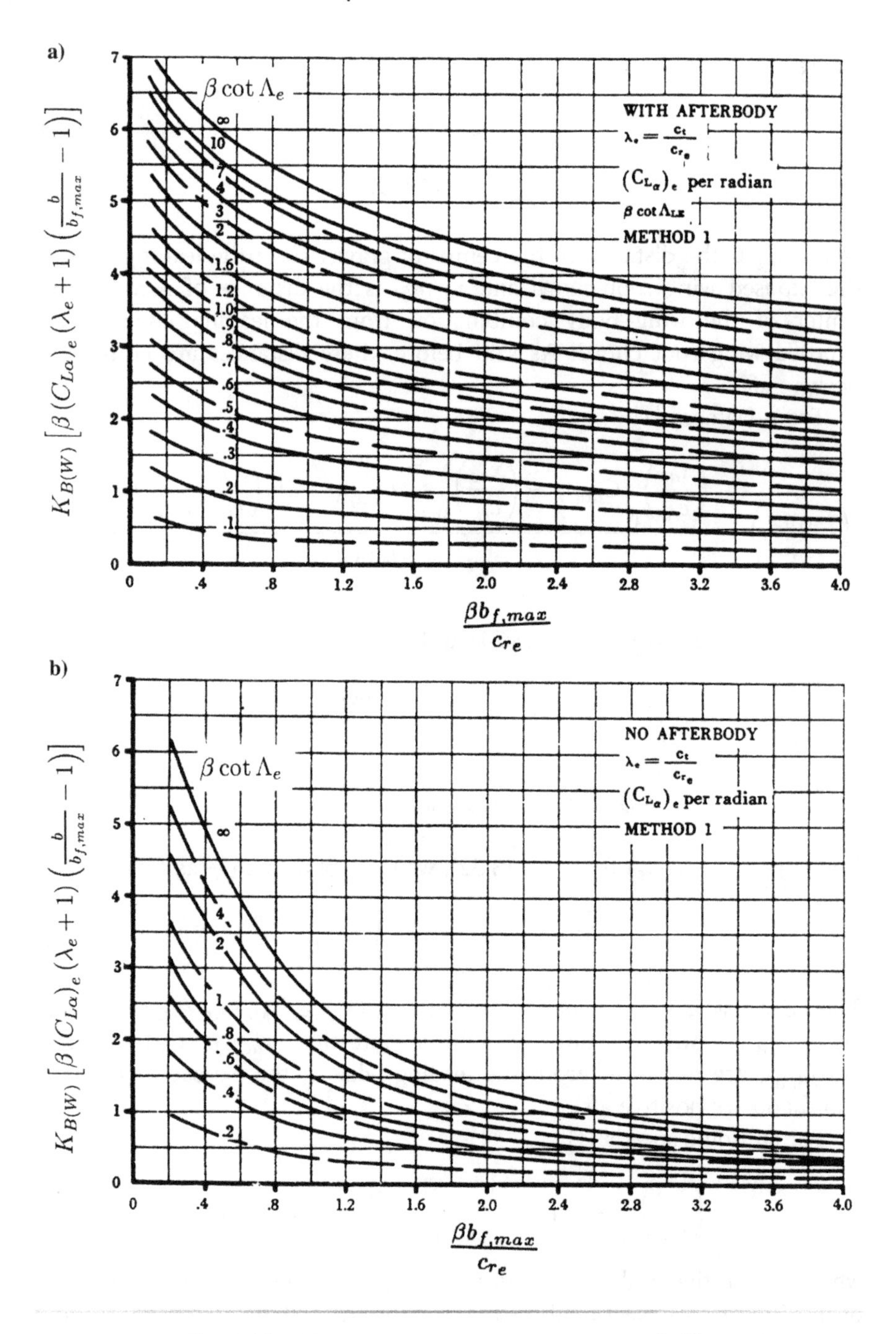

Fig. 3.18 Lift ratio $K_{B(W)}$ at supersonic speeds (1).

Slope of Pitching-Moment Coefficient

The slope of wing–body pitching-moment coefficient is given by

$$C_{m\alpha,WB} = (\bar{x}_{cg} - \bar{x}_{ac,WB})C_{L\alpha,WB} \tag{3.29}$$

where

$$\bar{x}_{cg} = \frac{x_{cg}}{\bar{c}} \tag{3.30}$$

$$\bar{x}_{ac,\,WB} = \frac{x_{ac,WB}}{\bar{c}} \tag{3.31}$$

where x_{cg} is the distance of the center of gravity from the leading edge of the exposed wing root chord and $x_{ac,WB}$ is the distance of aerodynamic center of the wing–body combination from the leading edge of the exposed wing root chord. As said before, these distances are considered positive aft.

From Datcom [1],

$$\left(\frac{x_{ac,WB}}{c_{re}}\right) = \frac{\left(\frac{x_{ac}}{c_{re}}\right)_N C_{L\alpha,N} + \left(\frac{x_{ac}}{c_{re}}\right)_{W(B)} C_{L\alpha,W(B)} + \left(\frac{x_{ac}}{c_{re}}\right)_{B(W)} C_{L\alpha,B(W)}}{C_{L\alpha,WB}} \tag{3.32}$$

where $C_{L\alpha,WB}$ is given by Eq. (3.24) and

$$C_{L\alpha,W(B)} = C_{L\alpha,e}K_{W(B)}\left(\frac{S_{\exp}}{S}\right) \tag{3.33}$$

$$C_{L\alpha,B(W)} = C_{L\alpha,e}K_{B(W)}\left(\frac{S_{\exp}}{S}\right) \tag{3.34}$$

Knowing $(x_{ac,WB}/c_{re})$ from Eq. (3.32), we can obtain $\bar{x}_{ac,WB}$ as follows:

$$\bar{x}_{ac,WB} = \left(\frac{x_{ac,WB}}{c_{re}}\right)\left(\frac{c_{re}}{\bar{c}}\right) \tag{3.35}$$

The next task is to evaluate the parameters $(x_{ac}/c_{re})_N$, $(x_{ac}/c_{re})_{W(B)}$, and $(x_{ac}/c_{re})_{B(W)}$, which is done as follows [note that $(x_{ac})_N$, $(x_{ac})_{W(B)}$, and $(x_{ac})_{B(W)}$ are measured from the leading edge of the exposed wing root chord and are positive aft].

For subsonic speeds [1],

$$\left(\frac{x_{ac}}{c_{re}}\right)_N = -\left(\frac{1}{c_{re}S_{B,\max}}\right)\int_0^{x_o}\frac{\mathrm{d}S_b(x)}{\mathrm{d}x}(l_N - x)\mathrm{d}x \tag{3.36}$$

where $S_b(x)$ is the local cross-sectional area of the nose, x_o is the axial location of maximum cross-sectional area, and l_N is the distance from the leading edge of the fuselage to the leading edge of the exposed wing root chord as shown in Fig. 3.16. If the integral in Eq. (3.36) is positive, then $(x_{ac}/c_{re})_N$ will be negative, which implies that the aerodynamic center of the fuselage nose is ahead of the leading edge of the exposed wing root chord. On the other hand, if the integral is negative, it will be aft of leading edge of the exposed wing root chord.

For supersonic speeds [1],

$$\left(\frac{x_{ac}}{c_{re}}\right)_N = \frac{l_N}{c_{re}}\left(\frac{x_{cp}}{l_N} - 1\right) \tag{3.37}$$

As suggested in Datcom [1], the parameter (x_{cp}/l_N) can be evaluated using the data given in Fig. 3.11 for cone-cylinders and ogive-cylinders. The definition of the nose is given in Fig. 3.16. To use the data of Fig. 3.11, note that l_A of Fig. 3.11 now equals l'_f of Fig. 3.16 so that we have $f_A/f_N = l'_f/l_n$.

According to Datcom [1],

$$\left(\frac{x_{ac}}{c_{re}}\right)_{W(B)} = \left(\frac{x_{ac}}{c_{re}}\right)_W \tag{3.38}$$

that is, the influence of the body on the location of wing aerodynamic center is small and ignored.

The data for estimating $(x_{ac}/c_{re})_W$ at subsonic and supersonic speeds are presented in Figs. 3.19a–3.19f.

The procedure for estimation of $(x_{ac}/c_{re})_{B(W)}$ at subsonic and supersonic speeds is as follows:

1. Subsonic:
   For $\beta A_e \geq 4.0$, where $\beta = \sqrt{1 - M^2}$ and A_e is the exposed aspect ratio of the wing,

$$\left(\frac{x_{ac}}{c_{re}}\right)_{B(W)} = \frac{1}{4} + \left(\frac{b - b_{f,\max}}{2c_{re}}\right)\chi \tan \Lambda_{c/4} \tag{3.39}$$

   where the parameter χ is given in Fig. 3.20 as a function of $b_{f,\max}/b$. For $\beta A_e < 4.0$, do linear interpolation as follows: obtain the value of $(x_{ac}/c_{re})_{B(W)}$ for $\beta A_e = 4.0$ as above using Fig. 3.20 and the value of $(x_{ac}/c_{re})_{B(W)}$ for $\beta A_e = 0$ using Fig. 3.21. Using these two values, do linear interpolation to obtain the value of $(x_{ac}/c_{re})_{B(W)}$ at the desired value of βA_e that lies between 0 and 4.
2. Supersonic:
   For

$$\beta A_e(1 + \lambda_e)\left(1 + \frac{1}{\beta \cot \Lambda_{LE}}\right) \geq 4.0 \tag{3.40}$$

   use the data given in Fig. 3.22a or Fig. 3.22b depending on whether the fuselage has an afterbody or not.

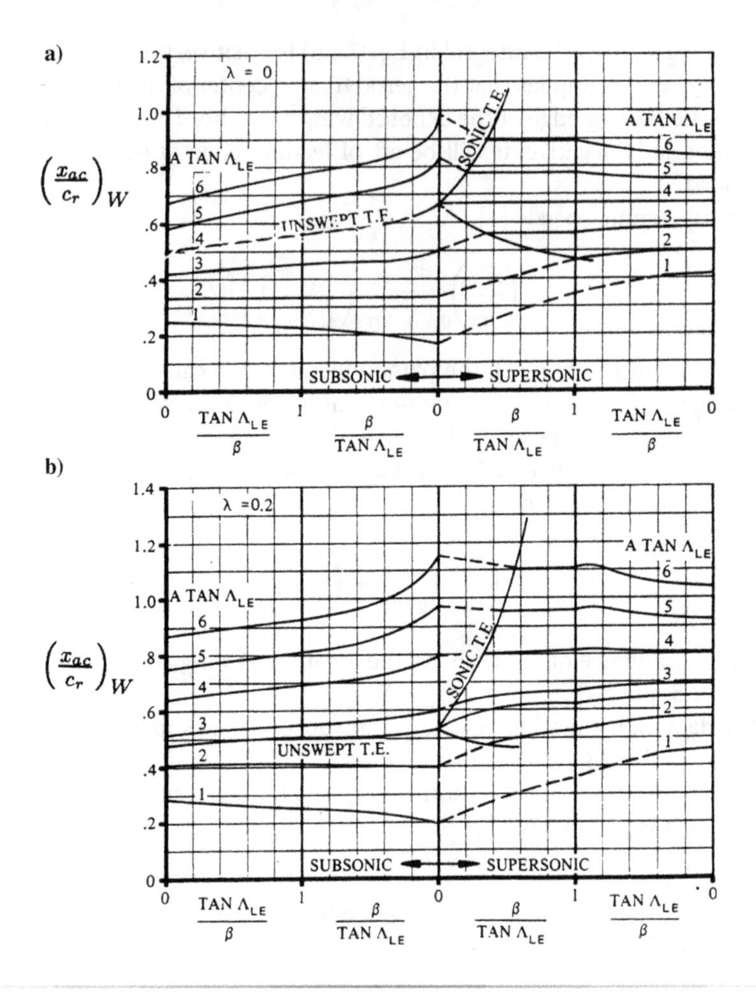

Fig. 3.19 Wing aerodynamic center position at supersonic speeds (1).

For

$$\beta A_e(1+\lambda_e)\left(1+\frac{1}{\beta \cot \Lambda_{LE}}\right) \leq 4.0 \tag{3.41}$$

use the method of interpolation as described for the subsonic case.

It may be noted that a positive value of $\bar{x}_{ac,WB}$ indicates that the aerodynamic center of the wing–body combination is aft of the leading edge of the exposed wing root chord and vice versa.

3.3.4 Tail Contribution

The local angle of attack of the horizontal tail is affected by the wing downwash as shown in Fig. 3.23. In this figure, α is the angle of attack of

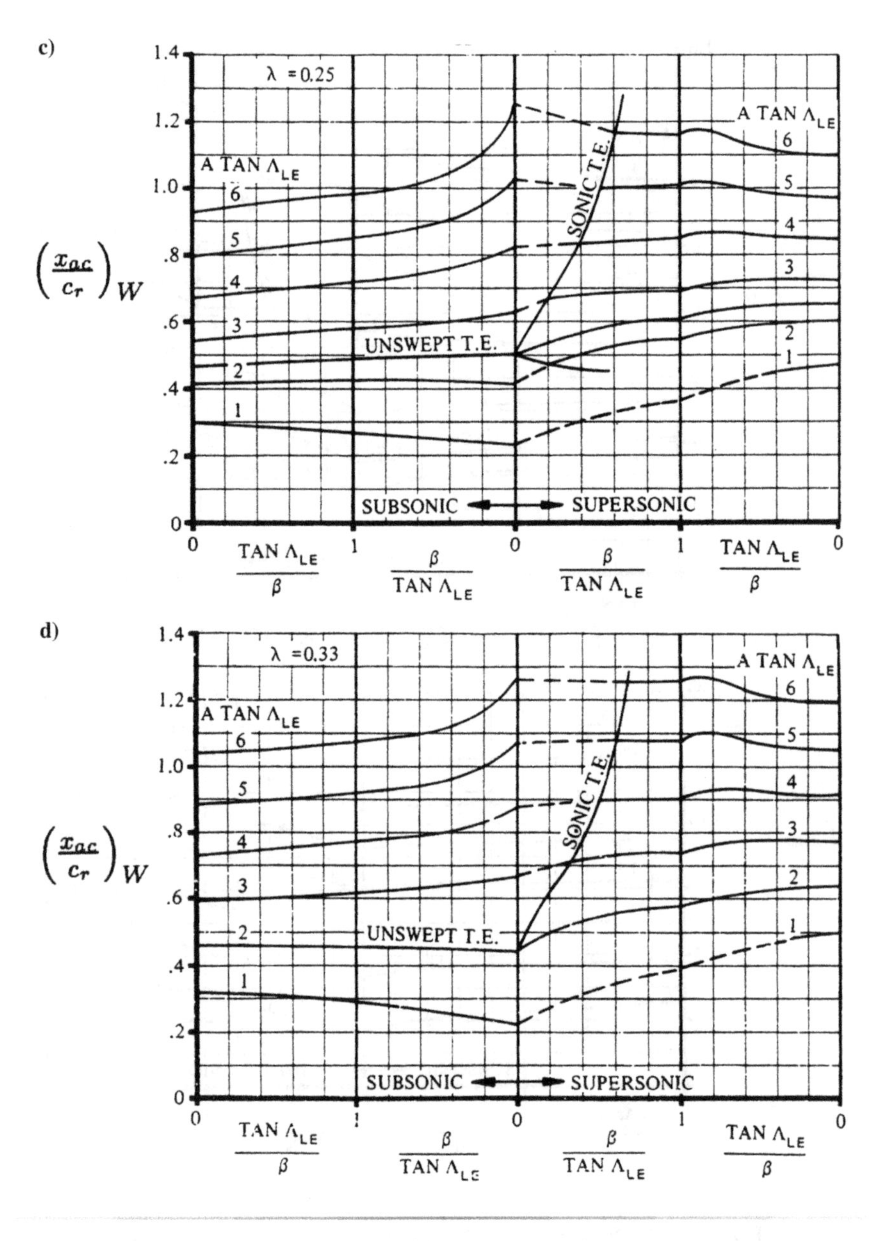

Fig. 3.19 Wing aerodynamic center position at supersonic speeds (1) (*continued*).

the airplane, i_w and i_t are respectively the incidences of wing and horizontal tail surfaces with respect to the reference line of the airplane. Assuming $\alpha_{oL,w} = \alpha_{oL,t} = 0$, the wing and tail angles of attack are given by

$$\alpha_w = \alpha + i_w$$

$$\alpha_t = \alpha + i_t - \epsilon$$

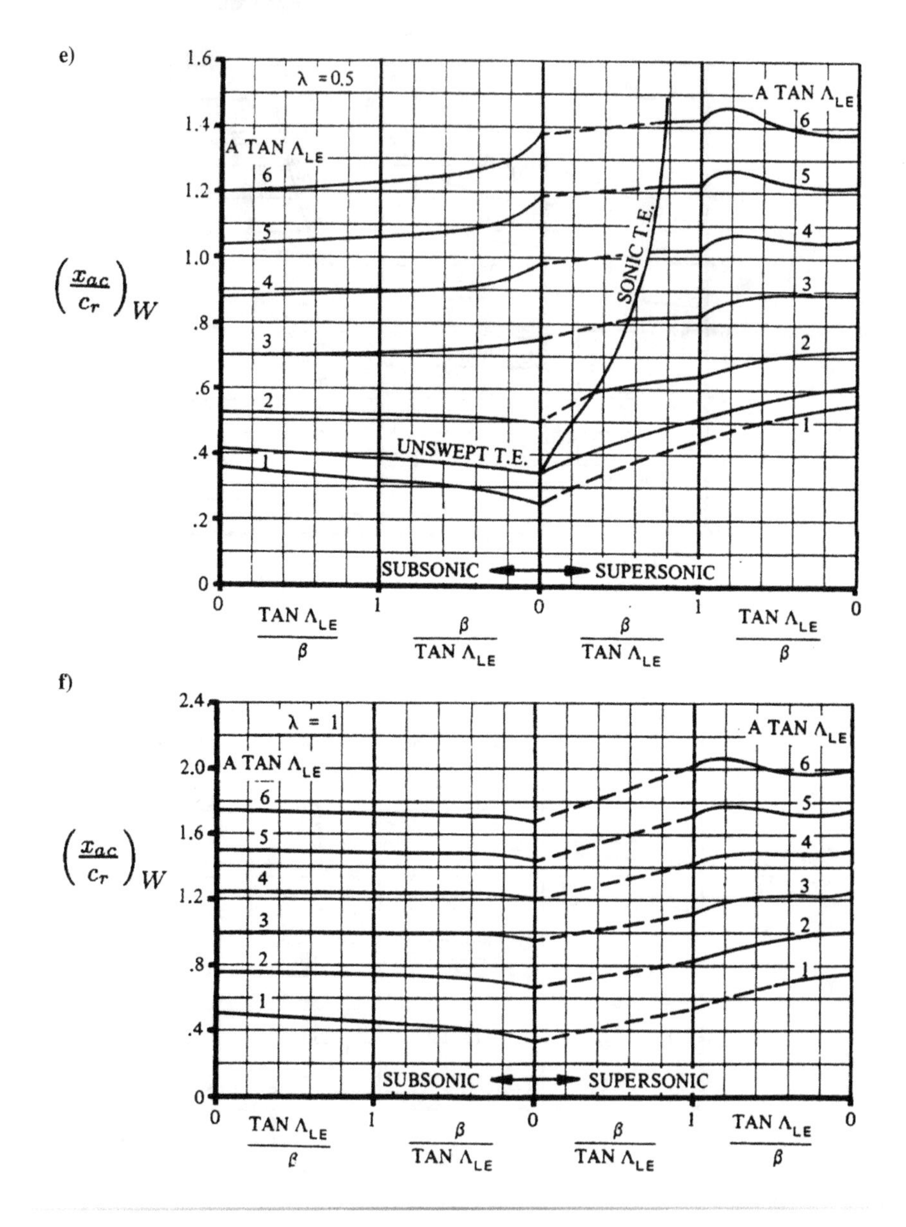

Fig. 3.19 Wing aerodynamic center position at supersonic speeds (1) (*continued*).

The tail angle of attack can be expressed in terms of wing angle attack as

$$\alpha_t = \alpha_w - i_w + i_t - \epsilon \tag{3.42}$$

where ϵ is the downwash angle at the horizontal tail aerodynamic center. The downwash is induced mainly because of wing-trailing vortices, and the magnitude of downwash depends on wing planform, aspect ratio, and

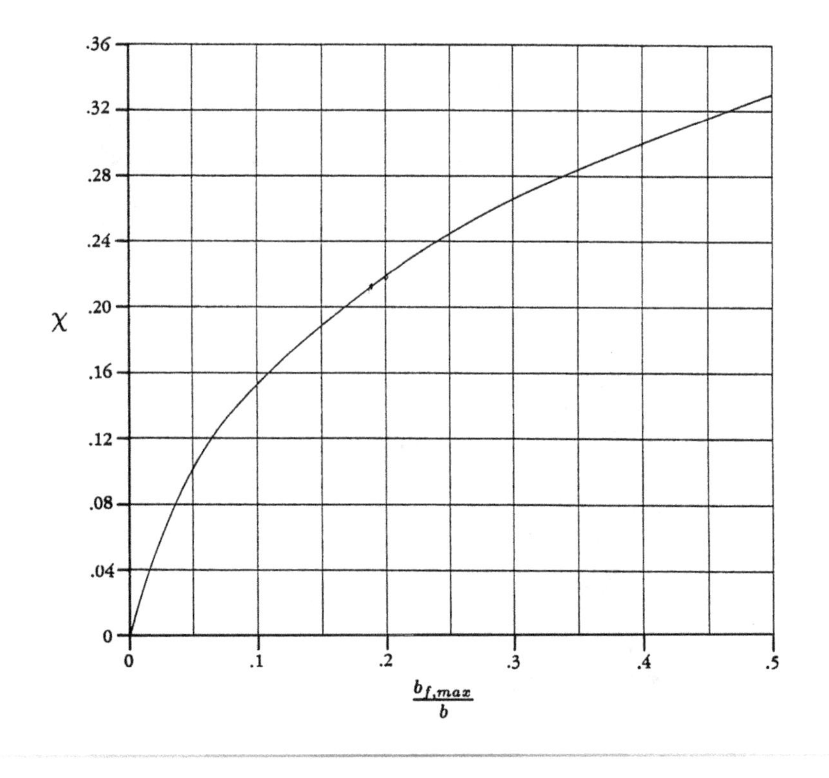

Fig. 3.20 Subsonic wing-lift carryover parameter (1).

the distance between the aerodynamic centers of wing and horizontal tail. The downwash caused by the fuselage is generally small in comparison with that caused by the wing and can be ignored.

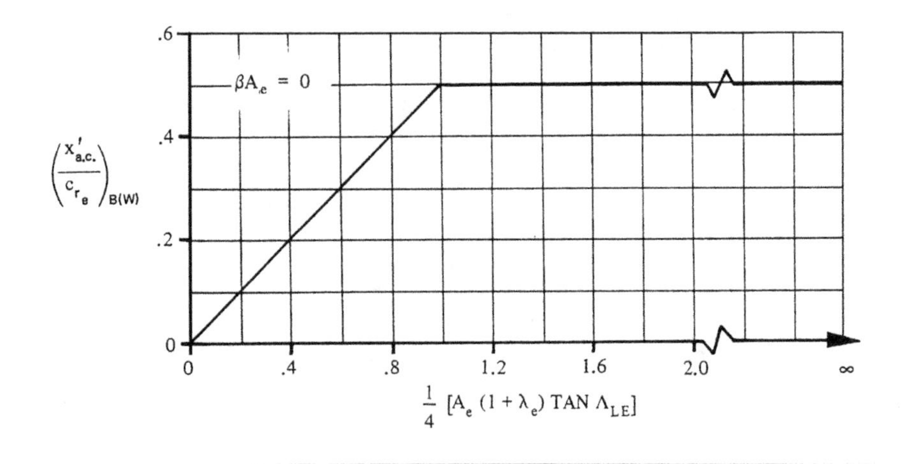

Fig. 3.21 Aerodynamic center location for wing-lift carryover body (1).

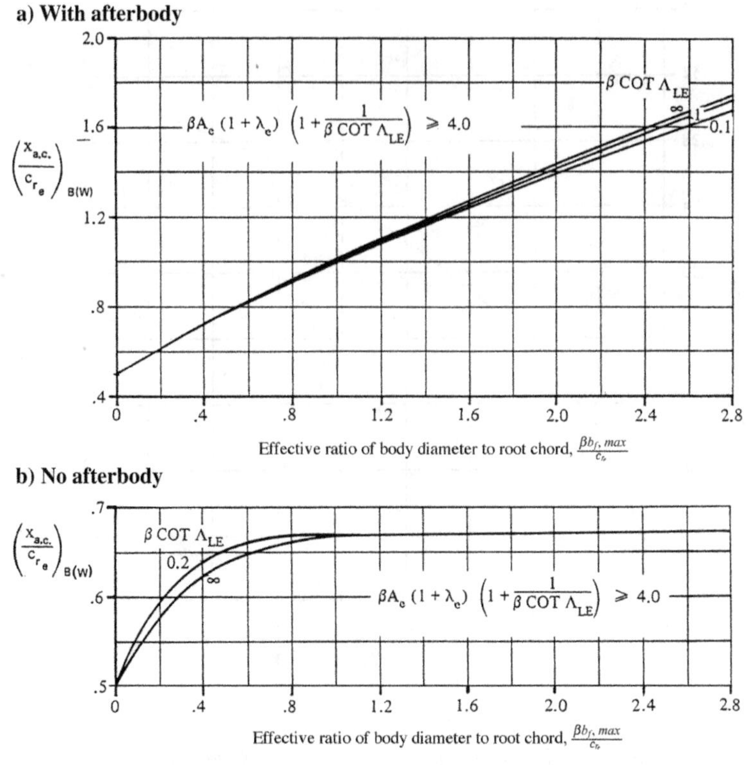

Fig. 3.22 Aerodynamic center location for wing-lift carryover body (1).

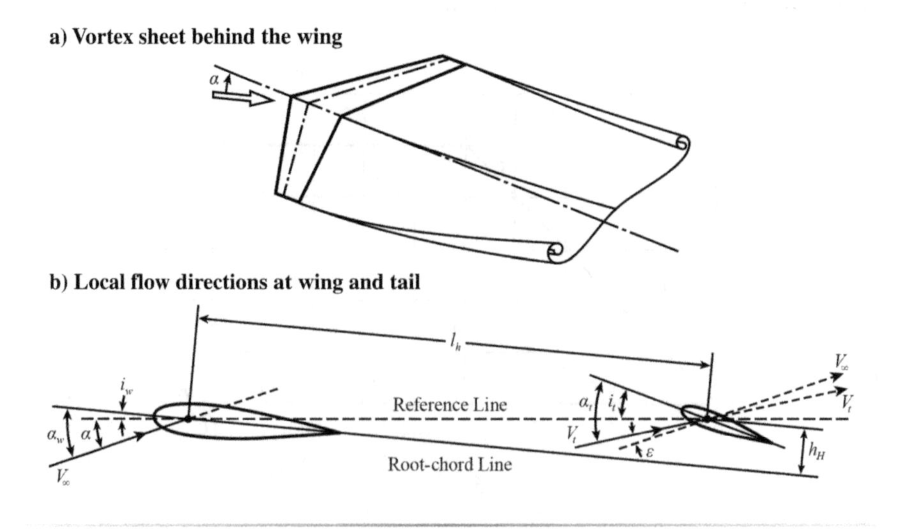

Fig. 3.23 Flow directions at wing and horizontal tail.

Analytical determination of downwash is quite complex and is beyond the scope of this text. Here, we will present empirical relations suitable for preliminary calculations [1].

For subsonic speeds,

$$\frac{d\epsilon}{d\alpha} = 4.44\left[K_A K_\lambda K_H \left(\cos\Lambda_{c/4}\right)^{\frac{1}{2}}\right]^{1.19} \tag{3.43}$$

Here, K_A, K_λ, and K_H are wing aspect ratio, wing taper ratio, and horizontal tail location factors as given by

$$K_A = \frac{1}{A} - \frac{1}{1 + A^{1.7}} \tag{3.44}$$

$$K_\lambda = \frac{10 - 3\lambda}{7} \tag{3.45}$$

$$K_H = \frac{1 - \frac{h_H}{b}}{\sqrt[3]{\frac{2l_h}{b}}} \tag{3.46}$$

where l_h is distance measured parallel to the wing root chord, between wing mac quarter chord point and the quarter chord point of the mac of horizontal tail, and h_H is the height of the horizontal tail mac above or below the plane of wing root chord, measured in the plane of symmetry and normal to the extended wing root chord and positive for horizontal tail mac above the plane of the wing root chord (see Fig. 3.23).

The wing quarter chord sweep $\Lambda_{c/4}$ can be determined using the following relation:

$$\tan\Lambda_{c/4} = \tan\Lambda_{LE} - \frac{c_r - c_t}{2b} \tag{3.47}$$

For straight tapered wings, the exposed and theoretical values of $\Lambda_{c/4}$ are identical. For more complex wing planforms, the two parameters are different.

The following simple formula, which gives the variation of downwash with angle of attack at infinity (the downwash at infinity is twice of that at the wing) and is based on lifting line theory (see Chapter 1) for elliptical wings, can be used for crude estimation of downwash at the horizontal tail at low speeds.

$$\frac{d\epsilon}{d\alpha} = \frac{2a_w}{\pi A} \tag{3.48}$$

Note that the values given by Eqs. (3.43) and (3.48) are per radian. The downwash angle ϵ is then given by

$$\epsilon = \frac{d\epsilon}{d\alpha}\alpha_w \tag{3.49}$$

For supersonic speeds, the estimation of downwash is much more involved because it depends on many factors such as whether the wing leading- and trailing-edge conditions are subsonic or supersonic. However, for very approximate estimation, the following formula may be used [1]:

$$\epsilon = \frac{1.62C_L}{\pi A} \tag{3.50}$$

where ϵ is in radians. It is suggested that adequate care and caution be exercised in using Eq. (3.50).

The tail lift is given by

$$L_t = q_t S_t C_{L,t} \tag{3.51}$$

With $C_{L,t} = a_t \alpha_t$ and using Eq. (3.42) we obtain

$$C_{L,t} = a_t(\alpha_w - i_w + i_t - \epsilon) \tag{3.52}$$

We assume $C_L \approx a_w \alpha_w$. In other words, we assume that the airplane lift coefficient is approximately equal to the wing lift coefficient and the tail lift coefficient is small in comparison with wing lift coefficient. With this assumption

$$\frac{dC_{L,t}}{dC_L} = \left(\frac{dC_{L,t}}{d\alpha_t}\right)\left(\frac{d\alpha_t}{d\alpha_w}\right)\left(\frac{d\alpha_w}{dC_L}\right)$$

or

$$\frac{dC_{L,t}}{dC_L} = \frac{a_t}{a_w}\left(1 - \frac{d\epsilon}{d\alpha}\right) \tag{3.53}$$

The horizontal tail lift-curve slope a_t is to be evaluated in the presence of the fuselage. Just like the wing, the horizontal tail also experiences the cross-flow effect because of aft fuselage. If the ratio of horizontal tail span to body diameter (at tail location) is sufficiently high, then the aft fuselage interference can be ignored, and a_t equals the lift-curve slope of an isolated horizontal tail. For such cases, Eq. (3.16) can be used to determine a_t. On the other hand, if this ratio is small, then the approach similar to that used for wing–body combination may have to be used. Additional information, if necessary, may be obtained from Datcom [1].

Let $\eta_t = q_t/q$ be the dynamic pressure ratio at the horizontal tail. The dynamic pressure ratio η_t at the horizontal tail is less than unity because the horizontal tail comes under the influence of the wing wake, which is a region of reduced dynamic pressure. For subsonic speeds, Datcom [1] gives

$$\eta_t = 1 - \frac{\Delta q}{q} \tag{3.54}$$

where

$$\frac{\Delta q}{q} = \frac{2.42\sqrt{C_{Do,w}}}{\frac{l_{h1}}{\bar{c}} + 0.30} \tag{3.55}$$

where l_{h1} is the distance between the trailing edge of the wing root chord and the aerodynamic center of the horizontal tail. $C_{Do,w}$ is the zero-lift drag coefficient of the wing as given by Datcom [1],

$$C_{Do,w} = C_{f,w}\left[1 + L(t/c) + 100(t/c)^4\right]R_{L,S}\frac{S_{\text{wet}}}{S} \tag{3.56}$$

where $C_{f,w}$ is the turbulent flat plate skin-friction coefficient. The variation of $C_{f,w}$ with Mach number and Reynolds number (based on the mean aerodynamic chord $\bar{c}$) is given in Fig. 3.24. L is the airfoil thickness location parameter. For airfoils where maximum thickness ratio is located within $0.3c$, $L = 2.0$ and, for those beyond $0.3c$, $L = 1.2$. The parameter $R_{L,S}$ depends on the sweep angle at that chord location where t/c is maximum and is given in Fig. 3.25. S_{wet} is the wetted area of the wing. For low/moderate angles of attack assuming fully attached flow on both top and bottom surfaces, $S_{\text{wet}} \approx 2S$.

For supersonic speeds, the determination of η_t is much more involved. The interested reader may refer to Datcom [1].

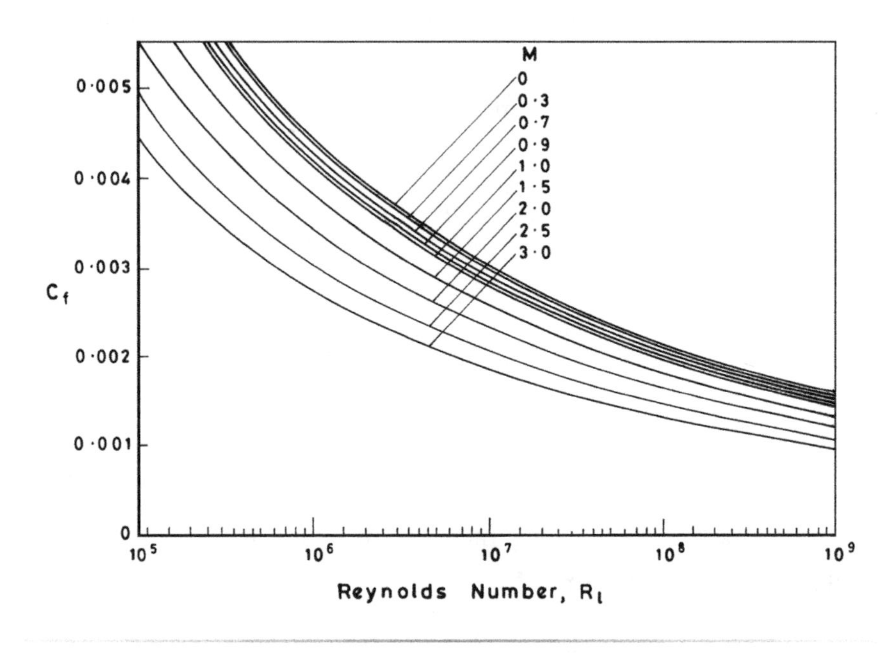

Fig. 3.24 Flat plate skin-friction coefficient (1).

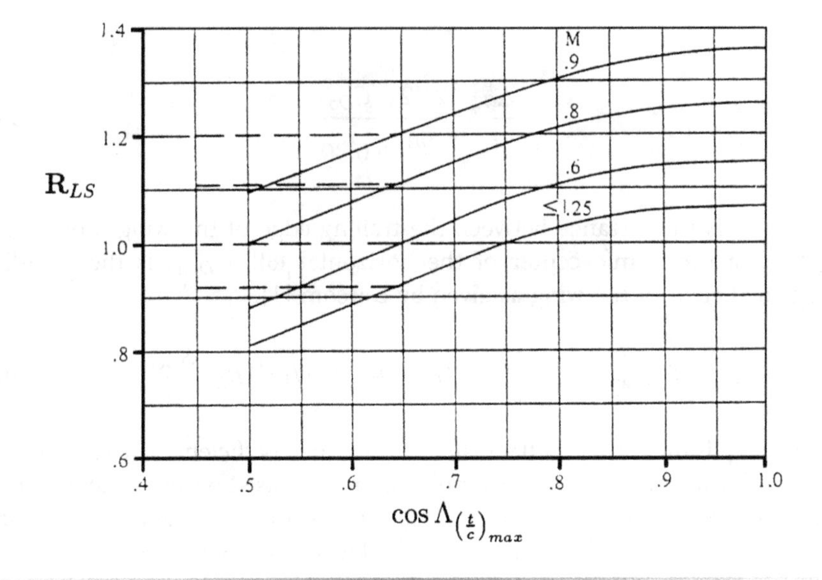

Fig. 3.25 Parameter $R_{L,S}$ (1).

The horizontal tail contribution to pitching moment is given by

$$M_t = M_{ac,t} - L_t l_t \tag{3.57}$$

We ignore $M_{ac,t}$, the zero-lift pitching moment of the horizontal tail, because this term is usually quite small. Then,

$$M_t = -L_t l_t \tag{3.58}$$

or, in coefficient form,

$$C_{m,t} = -C_{L,t} \frac{q_t}{q} \frac{S_t l_t}{S\bar{c}} \tag{3.59}$$

$$= -C_{L,t} \overline{V}_1 \eta_t \tag{3.60}$$

$$= -a_t(\alpha_w - i_w + i_t - \epsilon)\eta_t \overline{V}_1 \tag{3.61}$$

where

$$\overline{V}_1 = \frac{S_t l_t}{S\bar{c}} \tag{3.62}$$

The parameter $\overline{V}_1$ is often called the horizontal tail volume ratio. Differentiating with respect to C_L, we get

$$\frac{dC_{m,t}}{dC_L} = -\frac{a_t}{a_w}\left(1 - \frac{d\epsilon}{d\alpha}\right)\eta_t \overline{V}_1 \tag{3.63}$$

3.3.5 Effect of Power

The effect of the propulsive unit on longitudinal stability and trim can be both significant and difficult to evaluate. These effects also depend on the mode of propulsion such as turbo props or piston props, turbo fans, or turbo jets. Owing to these complexities, it is difficult to make a comprehensive evaluation of the power effects on longitudinal stability and trim. However, to give the reader some idea, a qualitative discussion of the major effects is presented in the following paragraphs. For simplicity, we restrict this discussion to propeller and jet aircraft.

For propeller-driven aircraft, the effect of power on static longitudinal stability consists of two parts: 1) direct effect caused by forces developed by the propulsion unit, and 2) indirect effect caused by propeller slipstream passing over wing or tail surfaces.

The direct effect caused by thrust depends on the vertical location of the thrust line with respect to the center of gravity. A high thrust line is stabilizing, whereas a low thrust line is destabilizing. The direct effect also includes a normal force N_p acting on the propulsion unit, which for a propeller airplane is destabilizing and for a pusher airplane is stabilizing as shown in Fig 3.26.

The indirect effects arise because of the influence of propeller slipstream passing over wing and tail surfaces as shown in Fig. 3.27. The wing sections exposed to the propeller slipstream experience a higher dynamic pressure and hence develop higher local lift and drag forces. The effect of this change in local lift and drag forces on pitching moment is usually small and can be ignored.

For a jet aircraft, the direct effects caused by thrust and intake normal force are similar to those of propeller aircraft. The indirect effects due to the jet-induced flow field may affect the horizontal tail as schematically shown in Fig. 3.28.

In the following analysis, we ignore the contribution of power effects to static longitudinal stability. The interested reader may refer to Datcom [1] for additional information.

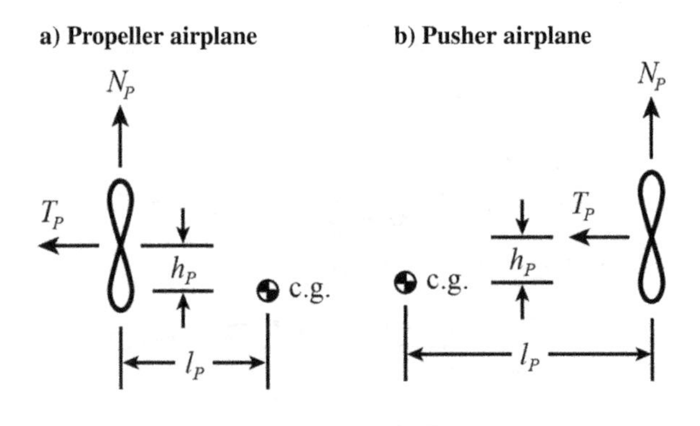

Fig. 3.26 Power effects.

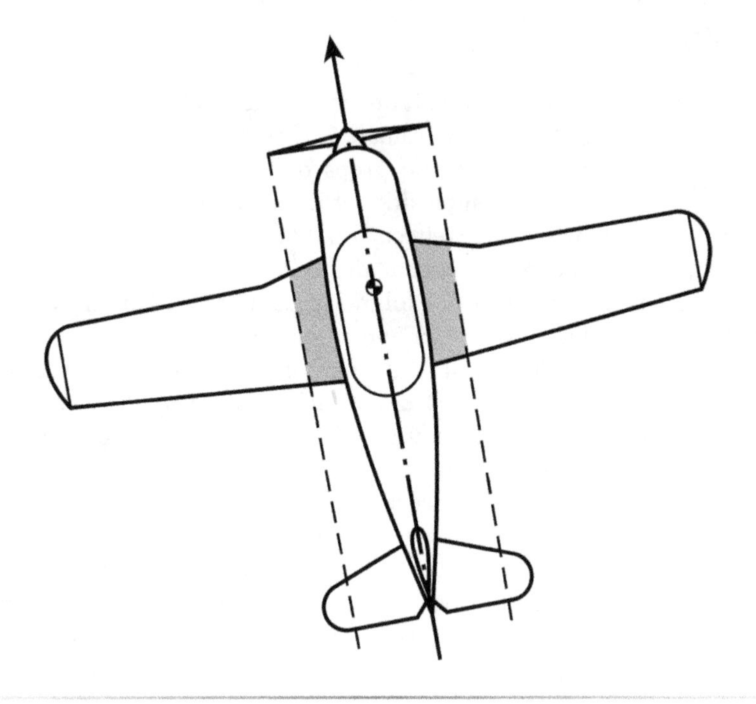

Fig. 3.27 Propeller slipstream effects.

3.3.6 Total Longitudinal Stability

Having evaluated the contribution of each of the components, we are now in a position to analyze the static longitudinal stability of the airplane. In the following analysis, we have assumed that the contribution of the fuselage, wing, and horizontal tail can be separately represented. This approach, as mentioned earlier, is justified for general aviation-type aircraft configurations with high values of wing span to body diameter. For configurations with

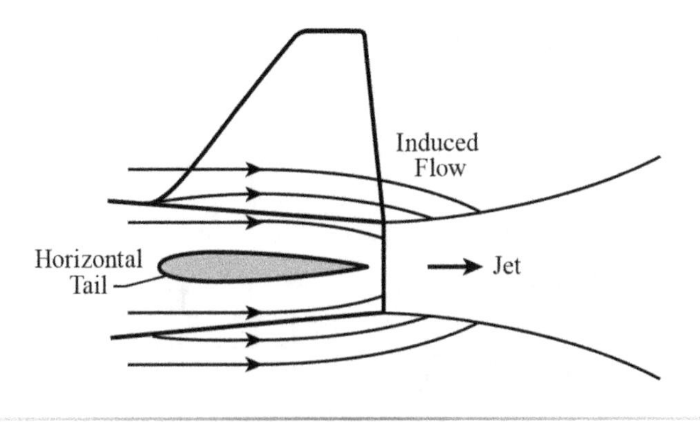

Fig. 3.28 Jet-induced flow field at horizontal tail.

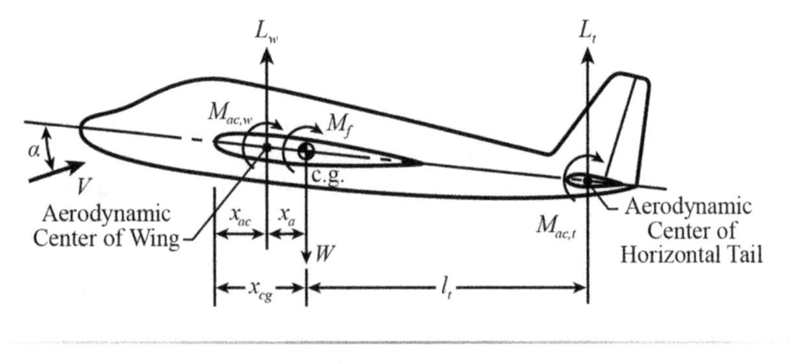

Fig. 3.29 Aerodynamic forces and moment acting on an airplane in steady level flight.

lower values of these parameters, the contributions of wing and fuselage have to be determined as wing–body contribution as discussed in Sec. 3.3.3. For such cases, the terms representing the fuselage and wing contributions should be replaced by the combined wing–body contribution. This exercise, however, is left to the reader.

Consider the forces and moments acting on a typical low-speed general aviation-type airplane as shown in Fig. 3.29. Such an airplane usually has the wing aerodynamic center ahead of the center of gravity. In this chapter, we will refer to such a configuration as a conventional airplane.

We assume that the thrust vector passes through the center of gravity. Furthermore, we ignore power effects. Summing up the pitching-moment contributions caused by the wing, fuselage, and horizontal tail, we have

$$C_{mcg} = C_{mw} + C_{mf} + C_{mt} \tag{3.64}$$

$$= C_{L,w}\bar{x}_a + C_{mac,w} + C_{mf} - C_{L,t}\eta_t\overline{V}_1 \tag{3.65}$$

Let us drop the suffix *cg* with an understanding that $C_m$denotes the pitching-moment coefficient about the center of gravity. Then,

$$C_m = C_{L,w}\bar{x}_a + C_{mac,w} + C_{mf} - C_{L,t}\overline{V}_1\eta_t \tag{3.66}$$

Then,

$$\frac{dC_m}{dC_L} = \bar{x}_a + \left(\frac{dC_m}{dC_L}\right)_f - \frac{a_t}{a_w}\left(1 - \frac{d\epsilon}{d\alpha}\right)\overline{V}_1\eta_t \tag{3.67}$$

As said earlier, we assume that the lift coefficient of the aircraft is more or less equal to the wing-lift coefficient or $C_L \approx C_{L,w}$.

With $\bar{x}_a = \bar{x}_{cg} - \bar{x}_{ac}$, we have

$$\frac{dC_m}{dC_L} = \bar{x}_{cg} - \bar{x}_{ac} + \left(\frac{dC_m}{dC_L}\right)_f - \frac{a_t}{a_w}\left(1 - \frac{d\epsilon}{d\alpha}\right)\overline{V}_1\eta_t \tag{3.68}$$

3.3.7 Stick-Fixed Neutral Point

We observe from Eq. (3.68) that the level of longitudinal static stability depends on the location of the center of gravity. In other words, movement of the center of gravity due to loading changes has a strong influence on the longitudinal static stability of the airplane. The schematic variations of C_m vs C_L and dC_m/dC_L vs center of gravity location are shown in Fig. 3.30. We observe that the level of static stability increases as the center of gravity moves forward and vice versa. Let N_o denote that center of gravity location when $dC_m/dC_L = 0$ or when the airplane becomes neutrally stable. N_o is called the stick-fixed neutral point. The term stick-fixed implies that the longitudinal control (elevator) was held fixed during the time when the airplane was subject to a disturbance in angle of attack. For neutral longitudinal stability, Eq. (3.68) assumes the following form:

$$0 = N_o - \bar{x}_{ac} + \left(\frac{dC_m}{dC_L}\right)_f - \frac{a_t}{a_w}\left(1 - \frac{d\epsilon}{d\alpha}\right)\overline{V}_1\eta_t \tag{3.69}$$

or

$$N_o = \bar{x}_{ac} + \left(\frac{dC_m}{dC_L}\right)_f - \frac{a_t}{a_w}\left(1 - \frac{d\epsilon}{d\alpha}\right)\overline{V}_1\eta_t \tag{3.70}$$

Then, the static longitudinal stability at any other center of gravity location can be expressed as

$$\frac{dC_m}{dC_L} = \bar{x}_{cg} - N_o \tag{3.71}$$

The concept of neutral point has a close similarity with that of the aerodynamic center. As we know, with respect to the aerodynamic center, wing-pitching moment remains constant with angle of attack. In other words, all the incremental wing lift due to a change in the angle of attack acts at the wing aerodynamic center. Remember that the contribution of drag force to pitching moment is ignored. At the neutral point, we have $dC_m/dC_L = 0$, which also implies that the incremental lift (of the complete airplane) due to a change in the angle of attack acts through the neutral point. Thus, the neutral point is, in essence, the aerodynamic center of the complete aircraft.

The static margin is another term often used to describe the level of stick-fixed longitudinal static stability of the airplane. The static margin is defined as

$$H_n = N_o - \bar{x}_{cg} \tag{3.72}$$

$$= -\left(\frac{dC_m}{dC_L}\right)_{\text{fix}} \tag{3.73}$$

For a stable airplane, the static margin is positive.

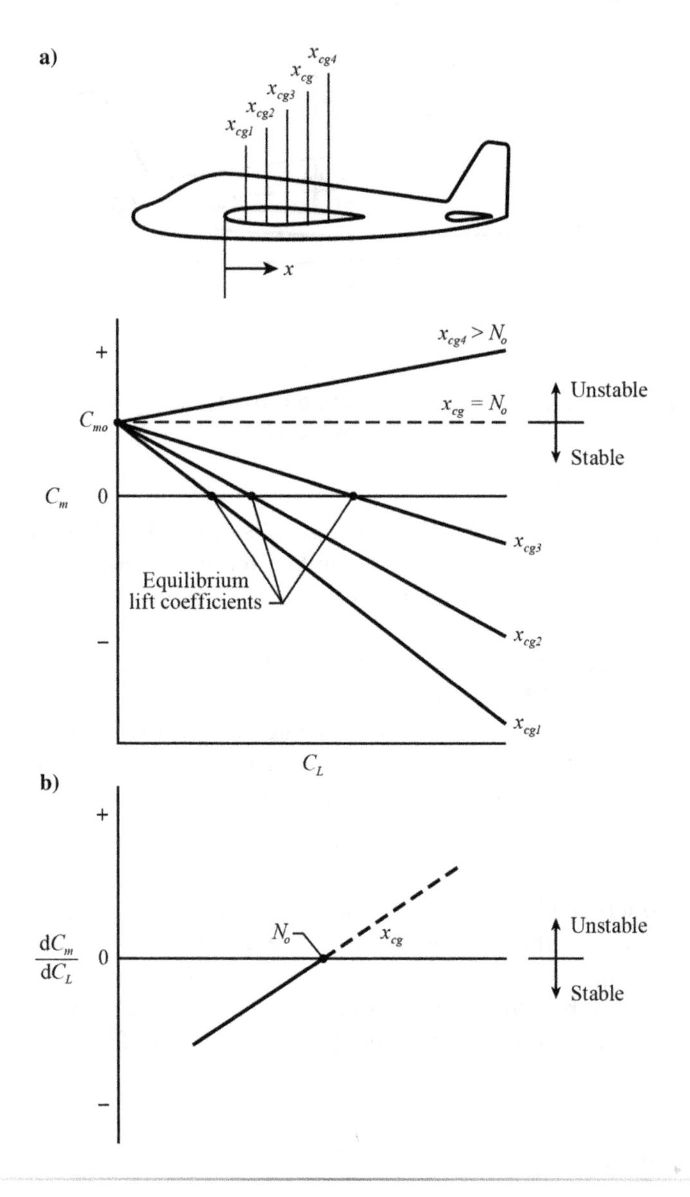

Fig. 3.30 Static stability level with center of gravity position.

3.3.8 Tail Lift for Trim

For the conventional airplane, the tail lift for pitch trim ($C_m = 0$) or equilibrium is given by

$$C_{L,t} = \frac{C_{L,w}\bar{x}_a + C_{mac,w} + C_{mf}}{\overline{V}_1 \eta_t} \tag{3.74}$$

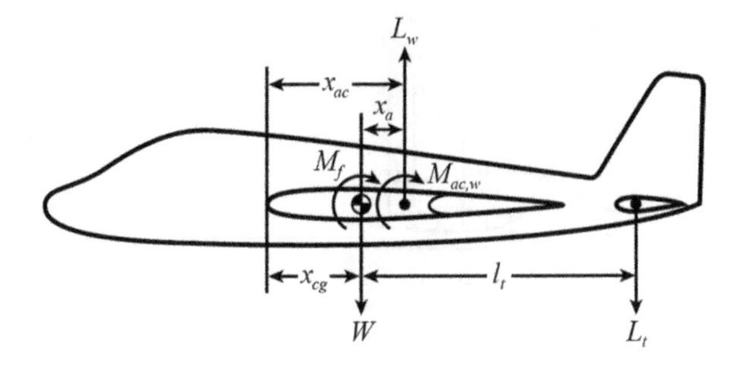

Fig. 3.31 Aerodynamic forces and moments acting on a modern swept-back wing airplane.

Generally, $C_{m,f}$ is positive. Whether $C_{mac,w}$ is positive or negative depends on the wing camber. For positive camber (convex upward), $C_{mac,w} < 0$ and vice versa. Usually, the magnitude of $C_{mac,w}$ is small compared to that of C_{fm}.

The dominant term on the right-hand side of Eq. (3.74) is the wing-lift term, which is positive. Furthermore, $\overline{x}_a > 0$ for a conventional airplane with aerodynamic center ahead of the center of gravity. Therefore, the tail lift is positive. In other words, the tail carries a portion of the aircraft weight, and the wing lift is correspondingly reduced. We have

$$L_w = W - L_t \tag{3.75}$$

For modern high-speed aircraft with long slender fuselages and highly swept-back wings, the wing aerodynamic center is usually aft of the center of gravity or $\overline{x}_a < 0$. For such aircraft, the tail load for pitch trim in steady level flight may be downward as shown in Fig. 3.31. To understand how this happens, let us consider the equilibrium in pitch as follows:

$$C_m = -C_{L,w}\overline{x}_a + C_{mac,w} + C_{mf} - C_{L,t}\overline{V}_1\eta_t \tag{3.76}$$

For equilibrium, $C_m = 0$ and

$$C_{L,t} = \frac{-C_{L,w}\overline{x}_a + C_{mac,w} + C_{mf}}{\overline{V}_1\eta_t} \tag{3.77}$$

As said before, the wing-lift term is the most dominant one in Eq. (3.77).

As a result, the tail-lift coefficient is negative, hence

$$L_w = W + L_t \tag{3.78}$$

This leads to a higher wing structural weight, a larger trim drag, and a reduction in overall vehicle performance. The trim drag is that component of the total drag that is associated with control surface deflections necessary to achieve pitch trim. The part of trim drag that is associated with the downward tail load can be approximately determined as follows:

$$\Delta D = \frac{1}{2}\rho V^2 S \Delta C_D \tag{3.79}$$

where

$$\Delta C_D = k\left(C_{L,w}^2 - C_L^2\right) \tag{3.80}$$

$$C_L = \frac{2W}{\rho V^2 S} \tag{3.81}$$

$$C_{L,w} = \frac{2(W + L_t)}{\rho V^2 S} \tag{3.82}$$

Here, we shall refer to ΔD as the trim drag penalty. One way of avoiding the trim drag penalty and improving the performance of such high-speed aircraft is to use a forward-placed canard as shown in Fig. 3.32. Because the canard is always exposed to the freestream, $\eta_c = 1$. The canard-lift coefficient for pitch trim is given by

$$C_{L,c} = \frac{C_{L,w}\bar{x}_a - C_{mac,w} - C_{mf}}{\overline{V}_c} \tag{3.83}$$

where $\overline{V}_c = S_c l_c / S\bar{c}$ is the canard volume ratio. As said before, the wing lift is the dominant term. Therefore, we see that the canard lift is positive. Then, the wing lift is given by

$$L_w = W - L_c \tag{3.84}$$

This leads to a smaller wing, a lower structural weight, and a smaller trim drag. Another advantage is that the canard aircraft is highly resistant to stall. Because the canard is ahead of the wing, it is the wing that is exposed to the downwash field of the canard, Therefore, when the canard stalls, the

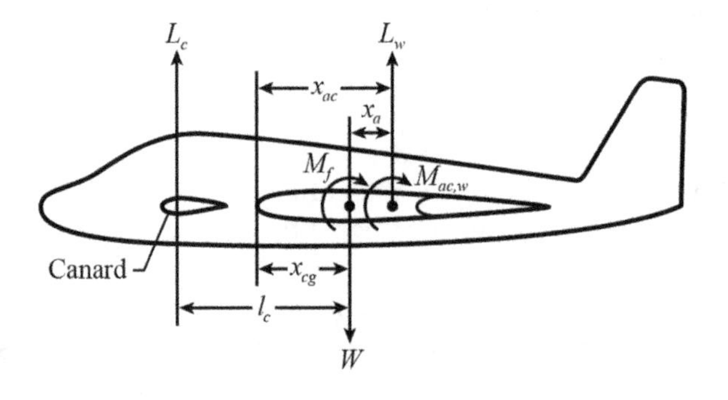

Fig. 3.32 Aerodynamic forces and moments acting on a canard airplane.

wing is still effective and produces a nosedown pitching moment to unstall the canard, thereby making the configuration stall resistant. However, it is possible that, in very strong upward gusts, the canard airplane may not be able to recover from the stall when both canard and wing are accidentally stalled. In such situations, an aircraft with conventional aft horizontal tail may have a better chance of recovering from the stall. Another disadvantage of the canard configuration is that a possible stall of the canard during the landing flare will bring the nose abruptly down resulting in a hard impact on the nose landing gear.

Another method of reducing the trim-drag penalty is the application of so-called relaxed static stability (RSS) concept. In other words, the static stability is compromised in favor of the vehicle performance. The airplane is intentionally made unstable so that the tail lift acts upward.

With relaxed static stability, we have

$$\left(\frac{\mathrm{d}C_m}{\mathrm{d}C_L}\right)_{\text{fix}} = \bar{x}_{cg} - N_o \tag{3.85}$$

$$> 0 \tag{3.86}$$

Then

$$C_m = C_{mo} + \left(\frac{\mathrm{d}C_m}{\mathrm{d}C_L}\right)_{\text{fix}} C_L \tag{3.87}$$

$$= C_{mo} + (\bar{x}_{cg} - N_o)C_L \tag{3.88}$$

Assuming that the tail lift is positive so that $C_L = C_{L,w} + C_{L,t}$, we obtain the tail-lift coefficient for trim as

$$C_{L,t} = \frac{-C_{mo}}{\bar{x}_{cg} - N_o} - C_{L,w} \tag{3.89}$$

As said before, to trim a statically unstable airplane, we must have $C_{mo} < 0$. Therefore, for positive tail lift as assumed,

$$\left|\frac{-C_{mo}}{\bar{x}_{cg} - N_o}\right| > |C_{L,w}| \tag{3.90}$$

Therefore, by suitably choosing the magnitude of C_{mo} and the level of relaxed static stability, the designer can ensure that the tail lift is positive (upward) for the desired operating conditions.

With application of the RSS concept, we note that the airplane is in an unstable trim. In other words, it has no inherent or open-loop stability to counter any external disturbance in angle of attack because of wind gusts or air turbulence. A pilot has to continuously monitor the airplane by operating the control surfaces to maintain a trim condition. To fly such an airplane safely, the airplane must be made closed-loop stable by a suitable design of the longitudinal control system, which we will study later.

For a conventional airplane with wing aerodynamic center ahead of the center of gravity, the tail lift is upward for trim. The wing lift acting ahead of the center of gravity develops a destabilizing moment. However, for a modern highly swept-back-wing airplane, the wing lift acting behind the center of gravity is stabilizing, and the tail lift acting downward generates a destabilizing pitching moment. To understand how both these vehicles are in stable trim, consider what happens when the lift coefficient increases by (ΔC_L) because of a disturbance in angle of attack. For the conventional airplane, the stable pitching moment induced by the horizontal tail exceeds the destabilizing pitching moment caused by the wing and the fuselage and, as a result, the net induced pitching moment is stabilizing. For a modern highly swept-back-wing airplane, the stabilizing moment induced due to wing exceeds the destabilizing pitching moments caused by the horizontal tail and the fuselage. These concepts are schematically illustrated in Fig. 3.33. Thus, for a conventional airplane, the tail plays the dual role of providing both stability and trim, whereas, for a modern highly swept-back-wing airplane with aerodynamic center aft of the center of gravity, the wing provides stability, and the horizontal tail functions as a trimming device.

3.3.9 Longitudinal Control

By longitudinal control, we mean the capability to change the equilibrium or trim-lift coefficient within the permissible range ($0 < C_L \leq C_{L,\max}$). In other words, the airplane should be capable of flying at any desired angle of attack within its aerodynamic limits. To explore the methods of longitudinal control, let us take a look at the expression for the equilibrium lift coefficient, which is obtained from Eq. (3.65) and the condition $C_{m,cg} = 0$.

$$C_L = \frac{-C_{mac,w} - C_{mf} + a_t(\alpha_w - i_w + i_t - \epsilon)\overline{V}_1 \eta_t}{\overline{x}_{cg} - \overline{x}_{ac}} \tag{3.91}$$

Here, we have assumed $C_L \approx C_{L,w}$. In other words, the tail lift, even though contributing to trim drag penalty and structural weight considerations, is assumed small in comparison with wing lift for longitudinal control considerations.

From this equation, we observe that we have the following options for longitudinal control.

Control of Wing Camber

The $C_{mac,w}$ is a function of the wing camber, which can be altered in a relatively simple manner during flight by deploying either leading- or

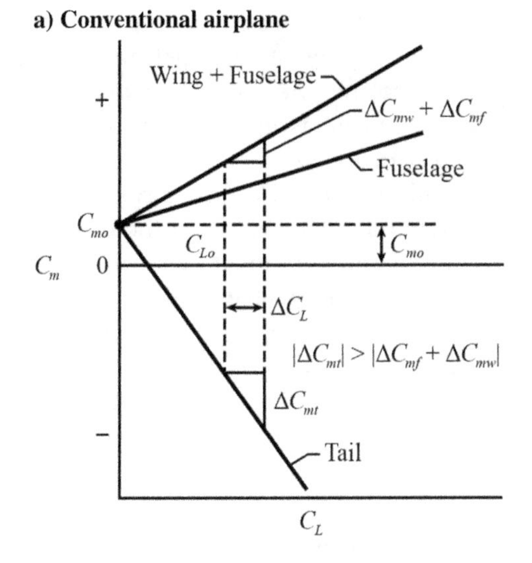

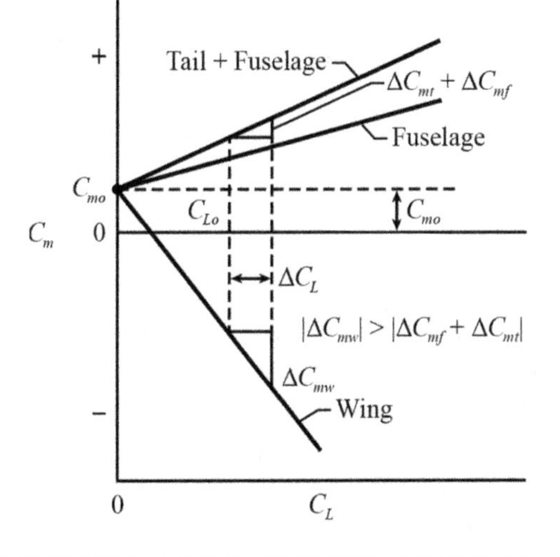

Fig. 3.33 Stability and trim concepts.

trailing-edge flaps. On tailless aircraft, this type of control is usually used. However, a disadvantage of using wing flaps for longitudinal control on aircraft with horizontal tails is that the flap deflection alters the downwash field at the horizontal tail, hence the longitudinal stability level of the aircraft. In other words, the use of flaps for longitudinal control alters the longitudinal stability level from one trim-lift coefficient to another. This is not desirable. Ideally, the method used for control should not affect the stability level of the airplane. As we will see later, any change in the static stability level affects the pilot's feel of the airplane.

Center of Gravity Shift

The forward or backward movement of the center of gravity during flight has a strong effect on the equilibrium or trim-lift coefficient as we have seen in Fig. 3.30. We observe that the equilibrium lift coefficient decreases and the level of static stability increases as the center of gravity moves forward and vice versa.

The Anglo–French supersonic Concorde airliner used this method of longitudinal control to partially offset the rearward movement of the aerodynamic center in the transonic/supersonic region. The center of gravity shift on the Concorde was affected by transferring the fuel from one wing tank to another, which was favorably located. Although this is a powerful method of longitudinal control, it is mechanically complex and is accompanied by a significant change in the level of static stability from one trim-lift coefficient to another. As we will discuss later, it affects the pilot's feel of the airplane.

Elevator Control

This is a powerful and perhaps the most widely used method of longitudinal control. The elevator is a small flap attached at the trailing edge of the horizontal tail as shown in Fig. 3.34a. A deflection of the elevator

a) Elevator geometry

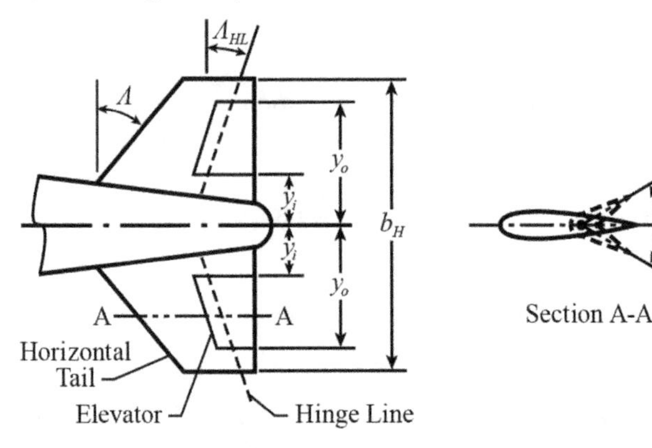

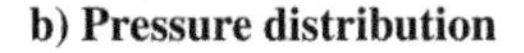

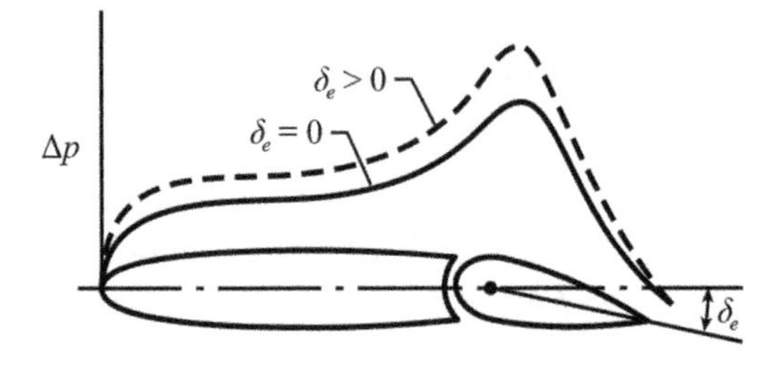

Fig. 3.34 Elevator geometry and pressure distribution.

alters the pressure distribution as shown in Fig. 3.34b, which in turn changes the tail lift. Some high-speed aircraft needing significant pitch control authority to counter the rearward movement of the center of pressure at transonic/supersonic speeds employ a slab tail or an all-movable horizontal tail.

To understand how the elevator control works, let us return to the pitching moment equation:

$$C_m = C_{L,w}\bar{x}_a + C_{mac,w} + C_{mf} - C_{L,t}\overline{V}_1\eta_t \tag{3.92}$$

The sign convention for elevator deflection is as follows. We assume that the elevator deflection is positive if it is deflected downward or trailing edge down and negative if it is deflected upward or the trailing edge up.

The effect of elevator deflection on the tail-lift coefficient can be expressed as

$$C_{L,t} = a_t\alpha_t \tag{3.93}$$

$$= a_t(\alpha_w - i_w + i_t - \epsilon + \tau\delta_e) \tag{3.94}$$

where

$$\tau = \frac{d\alpha_t}{d\delta_e} \tag{3.95}$$

Here, τ denotes the elevator effectiveness, which is the change in tail angle of attack per unit deflection of the elevator. The parameter τ can be estimated as follows.

The lift increment due to a control surface deflection of $\Delta\delta e$ is given by Datcom [1]:

$$\Delta C_L = \Delta C_l\left(\frac{a_t}{a_o}\right)\left[\frac{(\alpha_\delta)_{CL}}{(\alpha_\delta)_{Cl}}\right]K_b \tag{3.96}$$

where a_o is the sectional (two-dimensional) lift-curve slope and ΔC_l is the section-lift increment caused by $\Delta\delta e$. The parameter $(\alpha_\delta)_{CL}/(\alpha_\delta)_{Cl}$ is the ratio of the three-dimensional control effectiveness parameter to the two-dimensional control effectiveness parameter as given by Fig. 3.35, and K_b is the control surface span factor. Using Fig. 3.36, K_b can be evaluated as

$$K_b = (K_b)_{\eta=\eta_o} - (K_b)_{\eta=\eta_i} \tag{3.97}$$

Here, $\eta = 2y/b_H$ and b_H is the span of the horizontal tail. Differentiating Eq. (3.96) with respect to δe and rearranging, we obtain

$$\tau = \left(\frac{Cl_\delta}{a_o}\right)\left[\frac{(\alpha_\delta)_{CL}}{(\alpha_\delta)_{Cl}}\right]K_b \tag{3.98}$$

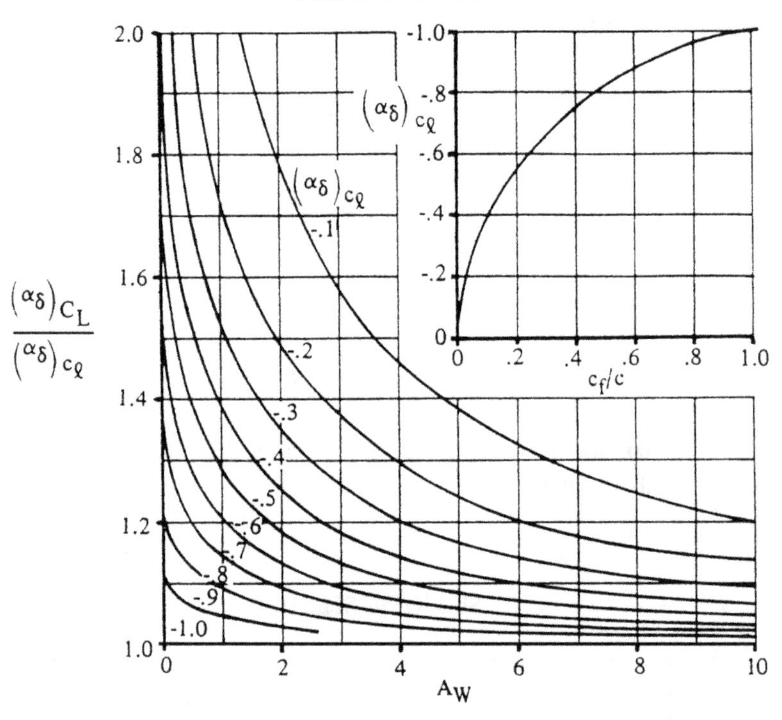

Fig. 3.35 Flap chord factor (1).

where $Cl_\delta = \partial C_l/\partial \delta_e$ can be obtained using the data given in Figs. 3.37a and 3.37b.

It is important to note that the elevator deflection does not affect the stick-fixed static stability level of the airplane. To understand this concept, differentiate Eq. (3.92) with respect to C_L and consider only the horizontal tail contribution, which is given by

$$\frac{dC_{m,t}}{dC_L} = \frac{a_t}{a_w}\left(1 - \frac{d\epsilon}{d\alpha}\right)\eta_t \overline{V}_1 \tag{3.99}$$

We observe that this term is exactly the same as that given in Eq. (3.63), which should not be surprising because, during a disturbance in angle of attack, the elevator is supposed to be held constant (stick-fixed).

Using Eqs. (3.92) and (3.94), the elevator deflection for longitudinal trim ($C_m = 0$) is given by

$$\delta_e = \frac{C_{L,w}\overline{x}_a + C_{mac,w} + C_{mf} - a_t(\alpha_w - i_w + i_t - \epsilon)\overline{V}_1\eta_t}{a_t\overline{V}_1\eta_t\tau} \tag{3.100}$$

Then, the variation of elevator deflection with equilibrium or trim-lift coefficient can be expressed as

$$\frac{d\delta_e}{dC_L} = -\frac{\left(\frac{dC_m}{dC_L}\right)_{\text{fix}}}{C_{m\delta}} \tag{3.101}$$

$$= -\frac{\bar{x}_{cg} - N_o}{C_{m\delta}} \tag{3.102}$$

where

$$C_{m\delta} = -a_t \overline{V}_1 \eta_t \tau \tag{3.103}$$

We observe that, for a statically stable airplane, $d\delta_e/dC_L < 0$. The term $C_{m\delta}$ is often called the elevator control power.

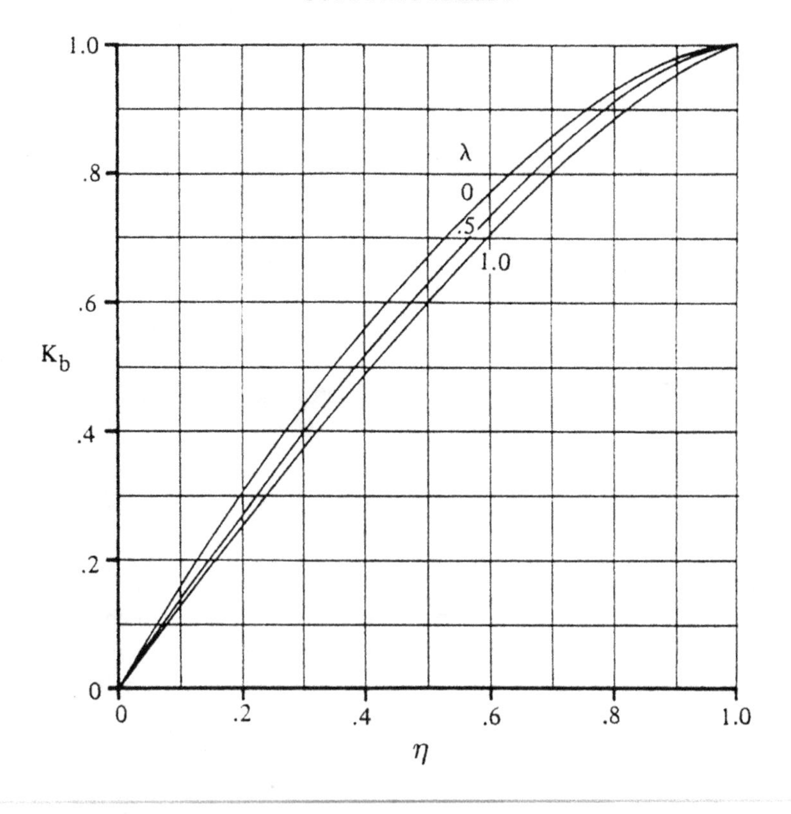

Fig. 3.36 Span factor for flaps (1).

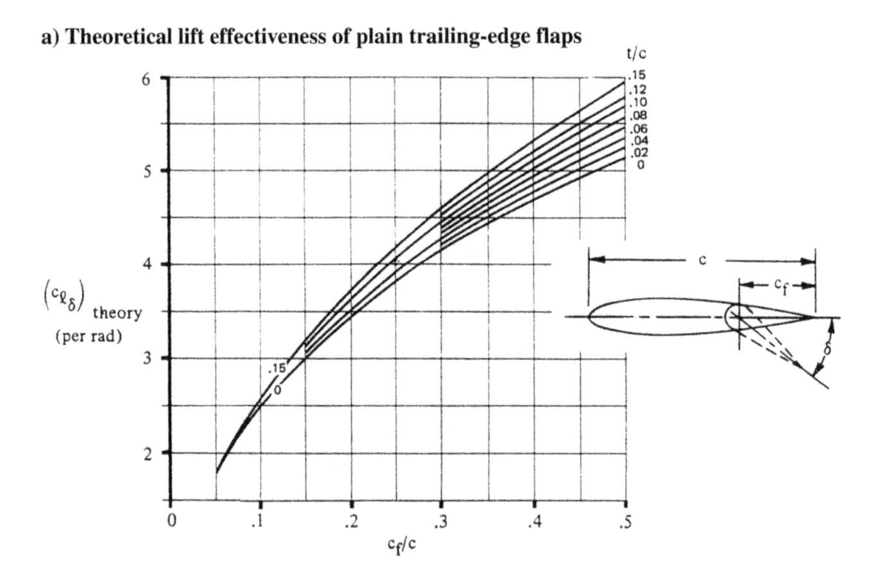

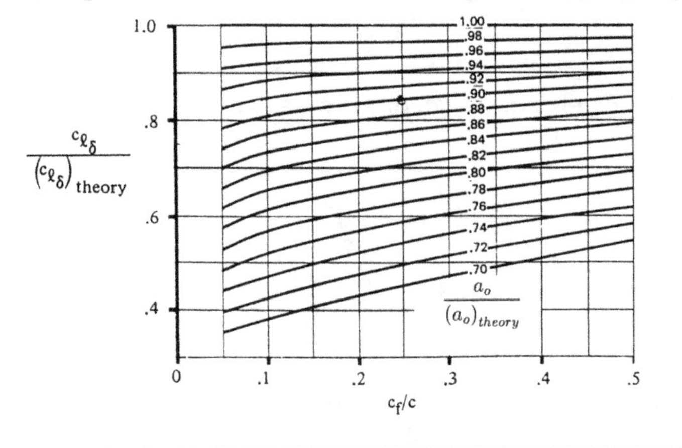

Fig. 3.37 Empirical correction factor for lift effectiveness of plain trailing-edge flaps (1).

We can express the variation of the elevator deflection with trim-lift coefficient as

$$\delta_e = \delta_{e,o} + \frac{d\delta_e}{dC_L} C_L \tag{3.104}$$

$$= \delta_{e,o} - \frac{\left(\dfrac{dC_m}{dC_L}\right)_{\text{fix}}}{C_{m\delta}} C_L \tag{3.105}$$

$$= \delta_{e,o} - \frac{\overline{x}_{cg} - N_o}{C_{m\delta}} C_L \tag{3.106}$$

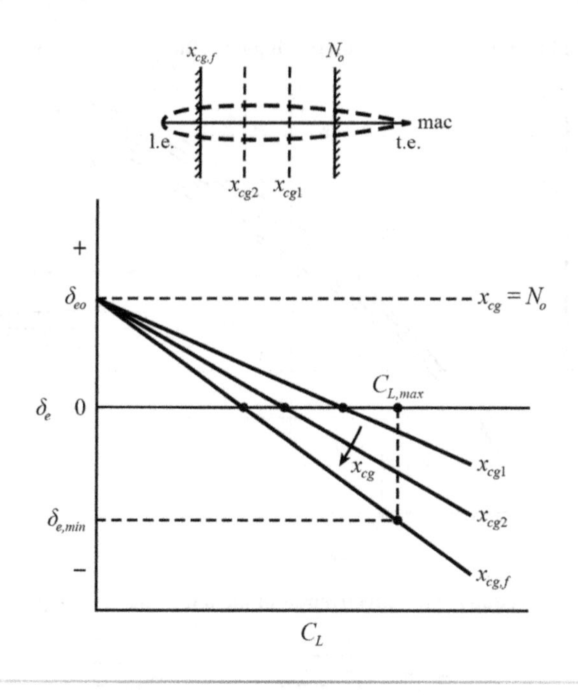

Fig. 3.38 Effect of center of gravity location on elevator deflection for trim.

where

$$\delta_{e,o} = \frac{C_{mac,w} + C_{mf,0} - a_t(\alpha_{w,oL} - i_w + i_t)\overline{V}_1\eta_t}{-C_{m\delta}} \tag{3.107}$$

Here, $\alpha_{w,oL}$ is the wing zero-lift angle of attack, which depends on wing camber and wing twist. Usually, the tail setting angle i_t is adjusted to make $\delta_{e,o}$ positive.

From Eq. (3.105) or (3.106), we observe that the elevator deflection required to trim a given lift coefficient directly depends on the level of stick-fixed static stability. It is through this relation that the pilot gets a feel of the level of stick-fixed stability of the airplane. This level changes from one trim point to another if the center of gravity shift or the wing camber manipulation is employed for longitudinal control. In such cases, the pilot will immediately sense it by observing differing amounts of elevator deflections needed for trim.

Typical variations of the elevator deflection with trim-lift coefficient for various center of gravity locations are shown in Fig. 3.38. For a given center of gravity location, we observe that a statically stable aircraft needs an upward (negative) elevator deflection to trim the aircraft at a higher lift coefficient (lower speed) and a downward (positive) elevator deflection to trim at a lower lift coefficient (higher speed) compared to an existing trim condition.

3.3.10 Most Forward Location of the Center of Gravity

We observe from Eq. (3.105) or (3.106) that the elevator required to trim a given lift coefficient depends directly on the level of static stability of the airplane. As the center of gravity moves forward, more and more upward elevator deflection is required. At some point, the required upward elevator deflection may exceed the maximum allowable upward elevator deflection. The limits on upward or downward deflections of the elevator arise because of flow separation and stall of the horizontal tail. Beyond these limiting values, the horizontal tail effectiveness diminishes very rapidly.

The center of gravity location for which the maximum upward elevator deflection is just capable of trimming the maximum lift coefficient in level flight represents the most forward permissible location of the center of gravity (Fig. 3.39). For any center of gravity location forward of this, the airplane cannot be trimmed at $C_{L,\max}$. Recall that many performance parameters directly depend on the capability of the airplane to fly at $C_{L,\max}$.

The most forward location of the center of gravity $x_{cg,f}$ can be obtained as follows:

$$\delta_e = \delta_{e,o} + \frac{\mathrm{d}\delta_e}{\mathrm{d}C_L} C_L \tag{3.108}$$

$$= \delta_{e,o} - \left(\frac{\overline{x}_{cg} - N_o}{C_{m\delta}}\right) C_L \tag{3.109}$$

$$\overline{x}_{cg,f} = N_o - (\delta_{e,\max} - \delta_{e,o}) \frac{C_{m\delta}}{C_{L,\max}} \tag{3.110}$$

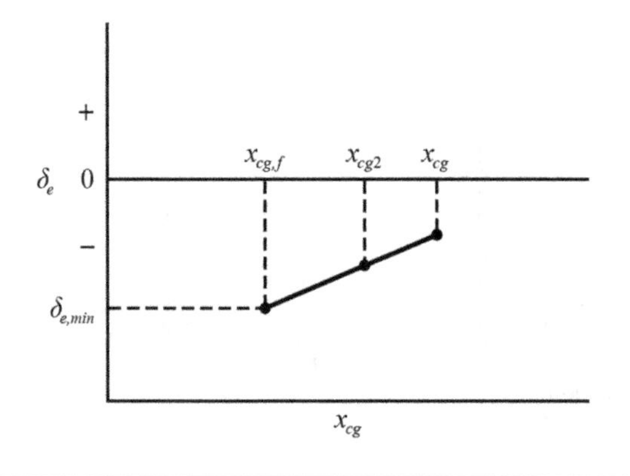

Fig. 3.39 Elevator required for trim at $C_{L,\max}$.

where $\delta_{e,\max}$ denotes the maximum permissible upward elevator deflection. Note that, according to our sign convention, downward deflection is positive and upward deflection is negative. Hence, $\delta_{e,\max}$ happens to be a negative quantity. Beyond this value, the elevator effectiveness falls off rapidly because of stall/flow separation.

3.3.11 Permissible Center of Gravity Travel

For an airplane to be statically stable in pitch, the center of gravity must lie ahead of the stick-fixed neutral point. From a stability point of view, it is desirable that the center of gravity be located as much forward of the neutral point N_o as possible. On the other hand, from the longitudinal control point of view, it is preferable that the center of gravity be located as much aft of $x_{cg,f}$ as possible. Therefore, to satisfy both these requirements, the center of gravity must always lie within these two limits as schematically shown in Fig. 3.40. The parameter $\Delta\bar{x}$ denotes the stick-fixed permissible center of gravity travel, expressed in terms of the mean aerodynamic chord

$$\Delta\bar{x} = N_o - \bar{x}_{cg,f} \qquad (3.111)$$

where

$$\Delta\bar{x} = \frac{\Delta x}{\bar{c}} \qquad (3.112)$$

3.3.12 Ground Effect

The demand on the elevator to trim the $C_{L,\max}$ becomes more severe when the aircraft is operating in close proximity of the ground, which is the case during landing or takeoff. The most forward limit on the center of gravity location given by Eq. (3.110) essentially corresponds to an airplane in free flight where the ground effects are insignificant.

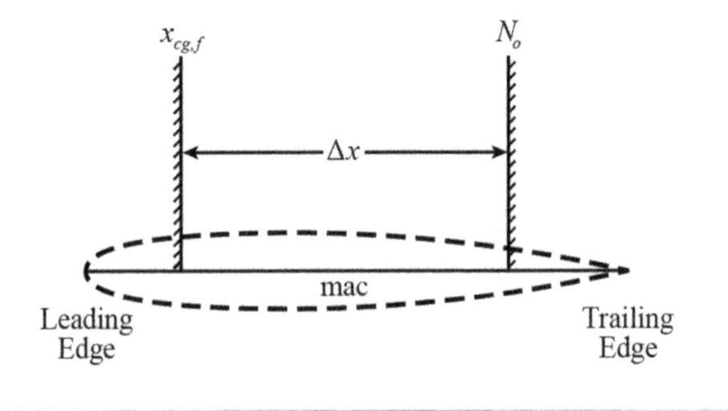

Fig. 3.40 Permissible center of gravity travel.

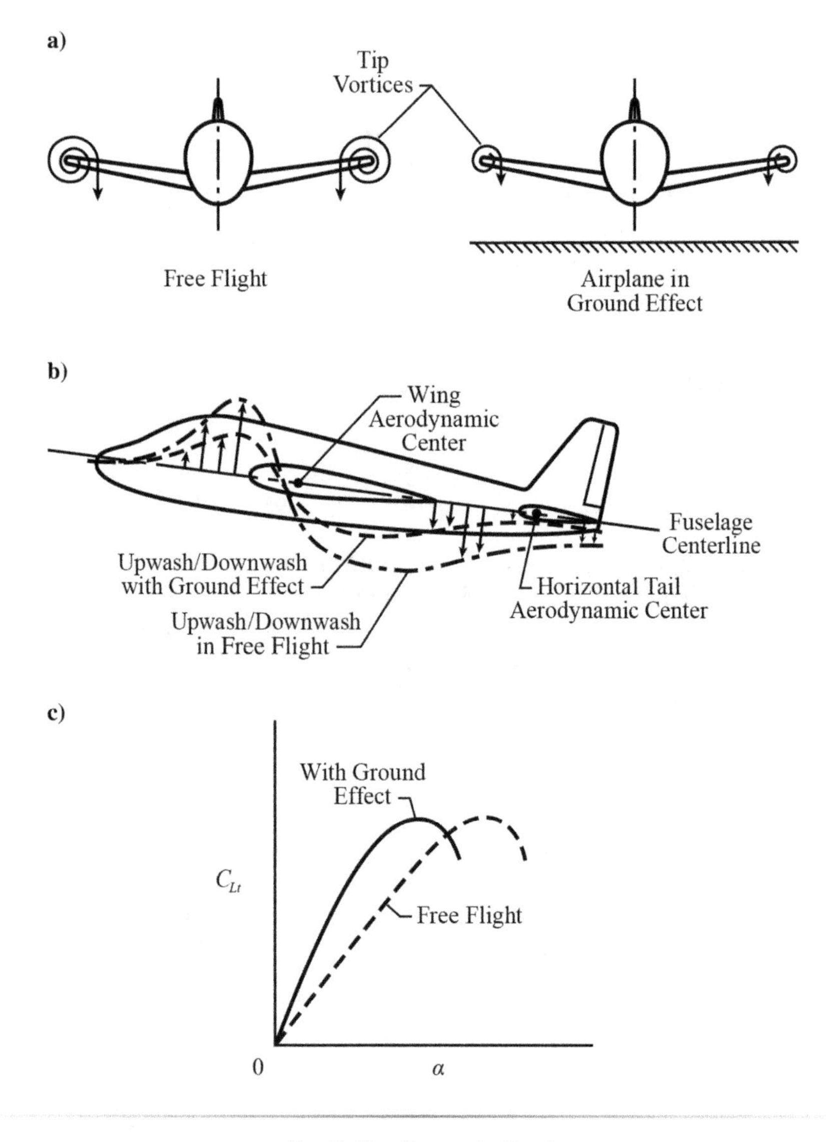

Fig. 3.41 Ground effect.

The proximity of the ground reduces the strength of the wingtip vortices (Fig. 3.41a). As a result, the induced downwash velocities are small everywhere in comparison with those in free flight. In particular, the downwash angle ϵ at the tail is much smaller in the presence of the ground compared to that in free flight (Fig. 3.41b). This leads to an increase in the lift-curve slope of the horizontal tail (Fig. 3.41c) and a corresponding increase in tail effectiveness. As a result, the aircraft becomes more stable, leading to a requirement that the most forward center of gravity location with ground effect lie aft of that in free flight as shown in Fig. 3.42.

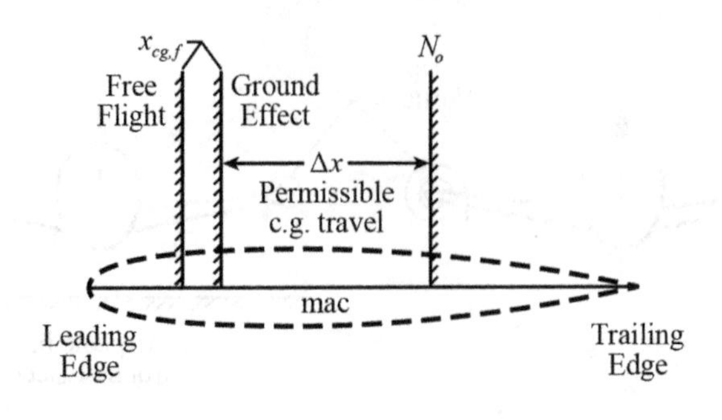

Fig. 3.42 Permissible center of gravity travel with ground effect.

3.3.13 Stick-Free Longitudinal Stability

Let us assume that the elevator is mounted on a frictionless hinge so that it can float freely under the action of aerodynamic forces when allowed to do so. The elevator will either float up or down depending on location of the hingeline relative to its center of pressure. If the hingeline is ahead of the center of pressure, the elevator tends to float up for positive angles of attack and float down at negative angles of attack as shown in Fig. 3.43.

The horizontal tail angle of attack is given by

$$\alpha_t = \alpha_w - i_w + i_t - \epsilon + \tau\delta_e \tag{3.113}$$

Let

$$\alpha_s = \alpha_w - i_w + i_t - \epsilon \tag{3.114}$$

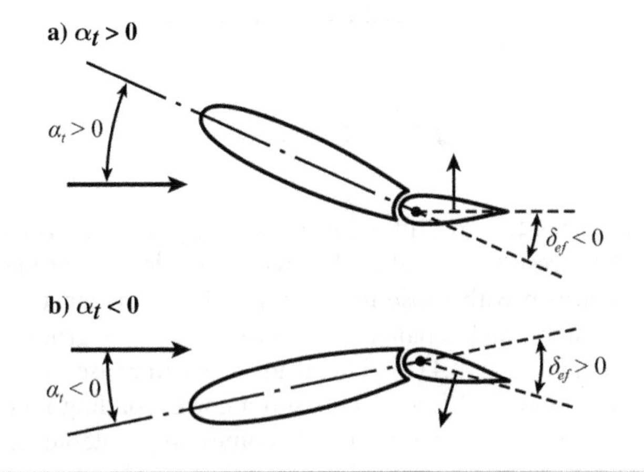

Fig. 3.43 Elevator floating angle.

so that

$$\alpha_t = \alpha_s + \tau\delta_e \tag{3.115}$$

Thus, when $\delta_e = 0$, $\alpha_t = \alpha_s$. Often, α_s is also called the stabilizer angle of attack.

The moment caused by the forces acting on the elevator about its hingeline is called hinge moment. The sign convention for the elevator hinge moment is as follows. The hinge moment is considered positive if it rotates the elevator so that the trailing edge goes down and is negative if the trailing edge goes up.

Assuming that the hinge moment coefficient depends linearly on horizontal tail or stabilizer angle of attack α_s and elevator deflection δ_e, we have

$$C_h = C_{ho} + C_{h\alpha}\alpha_s + C_{h\delta,e}\delta_e \tag{3.116}$$

where C_{ho} is the hinge moment coefficient for $\alpha_s = \delta_e = 0$ and

$$C_h = \frac{M_h}{q_t S_e \bar{c}_e} \tag{3.117}$$

$$C_{h\alpha} = \frac{\partial C_h}{\partial \alpha} \tag{3.118}$$

$$C_{h\delta,e} = \frac{\partial C_h}{\partial \delta_e} \tag{3.119}$$

Here, M_h is the hinge moment. Note that the hinge moment coefficient C_h is based on elevator area S_e, elevator mean aerodynamic chord $\bar{c}_e$. Usually, both $C_{h\alpha}$ and $C_{h\delta,e}$ are negative, and their magnitudes depend on the location of the hingeline with respect to the leading edge.

When the elevator is freely floating, the hinge moment is zero. This condition gives the floating angle of the elevator δ_{ef} as

$$\delta_{ef} = -\frac{C_{ho} + C_{h\alpha}\alpha_s}{C_{h\delta,e}} \tag{3.120}$$

We know that the elevator deflection required to trim the aircraft at a given lift coefficient is directly proportional to the level of stick-fixed longitudinal stability of the aircraft. To maintain a pitch trim during flight, the pilot has to continuously hold the elevator at that position by constantly applying a steady force on the stick (control column). This can be quite strenuous if he has to do this over an extended period of time.

Suppose that the pilot takes his hands off the stick and leaves it free. Then the elevator will float freely and assume the floating angle given by

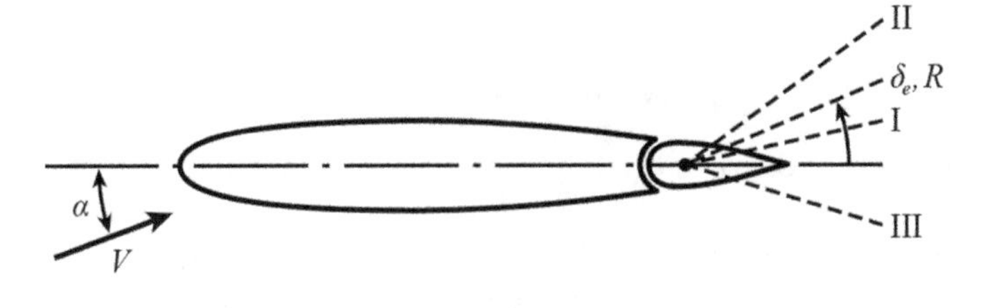

Fig. 3.44 Relation between stick-free stability and elevator floating characteristics.

Eq. (3.120). In principle, this floating angle can be any one of the four possibilities shown in Fig. 3.44.

Let $\delta_{e,R}$ denote the required elevator deflection to trim the aircraft at a particular speed or lift coefficient. Suppose it so happens that the floating elevator assumes exactly this position, i.e., $\delta_{e,f} = \delta_{e,R}$. Then the pilot does not have to move the elevator or the stick to trim the aircraft at that lift coefficient or speed. The aircraft does it by itself. In other words, both the required elevator deflection (stick movement) and stick force are zero. If this is the case, then the airplane is said to be neutrally stable stick-free. On the other hand, if the elevator floats in the right direction (upward) but stops a little short of $\delta_{e,R}$ assuming position I, then the pilot has to move the elevator only by the amount $\delta_{e,R} - \delta_{e,\mathrm{I}}$. Thus, in stick-free condition, the airplane is still stable, but the level of stability has reduced. Suppose the elevator overshoots and assumes the position II; then the pilot has to move the elevator in the opposite (downward) direction by the amount $\delta_{e,\mathrm{II}} - \delta_{e,R}$, which implies that the airplane has become statically unstable in stick-free condition. On the other hand, if the elevator floats down to position III, the pilot has to move the elevator in the right direction (upward) but by an increased amount $\delta_{e,\mathrm{III}} + \delta_{e,R}$. This implies that the airplane has become more stable in stick-free condition. Thus, the floating characteristics of the elevator have a direct bearing on the stick-free stability.

3.3.14 Aerodynamic Balancing

The floating characteristics of a control surface depend on the hinge-moment characteristics and, as we will see later, the stick force also depends on the hinge-moment characteristics. Too low values of the hinge-moment would make the control highly sensitive to small disturbances, whereas too high values would make the controls sluggish to operate. Therefore, a careful design of the hinge-moment parameters $C_{h\alpha}$ and $C_{h\delta,e}$ is necessary to achieve a proper balance between these two conflicting requirements. The control of hinge-moment parameters is called aerodynamic balancing.

When the hingeline is at the control surface leading edge, both $C_{h\alpha}$ and $C_{h\delta,e}$ are negative. If the hingeline is moved further aft, both $C_{h\alpha}$ and $C_{h\delta,e}$ become more positive because the control surface forward of the hingeline produces an opposing moment to that produced by the surface aft of the hingeline. The net hinge moment, which is the algebraic sum of these two moments, is greatly reduced. This type of aerodynamic balance is called set-back hinge balance (Fig. 3.45a). However, one has to be careful because too much area forward of the hingeline may lead to an overbalance of the control at some flight conditions that may affect the pilot's feel of the aircraft.

Horn balance (Fig. 3.45b) is similar to the set-back hinge, except that all the area ahead of the hingeline is concentrated on one part of the surface. The horn balance makes both $C_{h\alpha}$ and $C_{h\delta,e}$ less negative though the effect on $C_{h\delta,e}$ is more pronounced than in the case of set-back hinge method.

The internal balance (Fig. 3.45c) is a modification of the set-back hinge method. The inside of the airfoil has to be vented to the external pressures so that the pressures acting on the balancing area provide the necessary balancing effect. The effectiveness of this balance can be increased by sealing the gap between the leading edge of the control surface and the structure of the airfoil as shown in Fig. 3.45c. The amount of balance can be

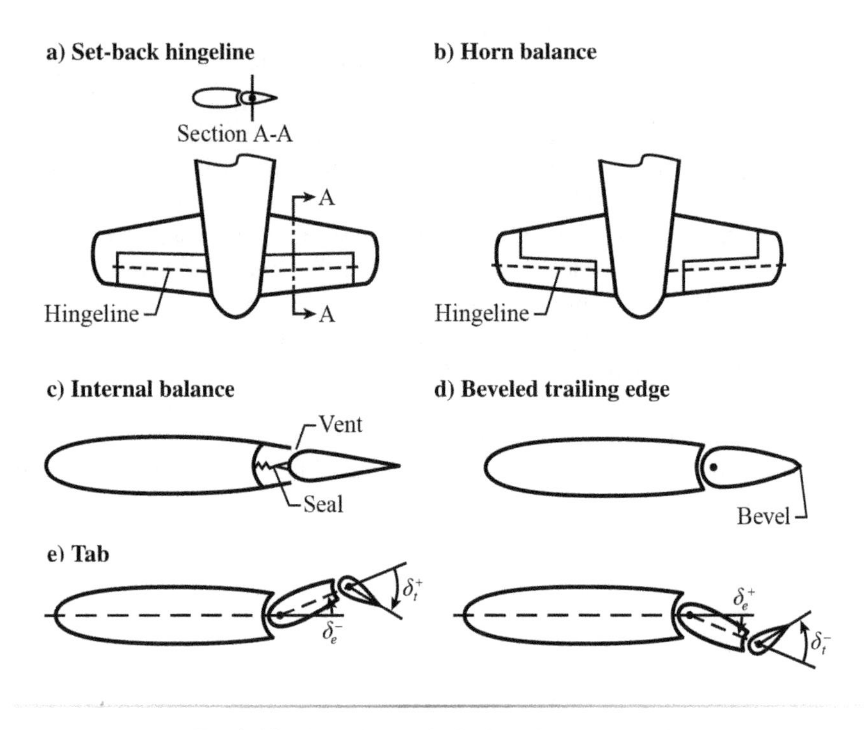

Fig. 3.45 Aerodynamic balancing concepts.

adjusted by properly venting the seal. This type of balance has a greater effect on $C_{h\delta,e}$ than on $C_{h\alpha}$.

The shapes of the leading and trailing edges of the control surface also have an effect on the hinge moment parameters $C_{h\alpha}$ and $C_{h\delta,e}$. For example, beveling the trailing edge (Fig. 3.45d) alters the parameter $C_{h\delta,e}$ because the pressure distribution over the control surface is altered.

The described methods of balancing (except the beveled trailing edge) are based on pressure changes on the control surface ahead of the hingeline. An alternative way of balancing the control surface is by deploying an additional control surface called tab. The tab is much smaller in size compared to the elevator and is usually deflected in the opposite direction as shown in Fig. 3.45e. Even though the tab is small in size, the pressure changes caused by its deflection produce appreciable moments about the elevator hingeline. Because the tab moves only when the elevator moves, it has little or no effect on $C_{h\alpha}$. The tab has the added advantage that it can always provide just the right amount of balancing as needed. Hence, the possibility of control surface overbalance is almost ruled out in this case. However, it has a minor disadvantage that an opposite deflection of the tab results in a small lift loss, hence a reduction in overall control effectiveness. This type of tab is also known as a trim tab because it is used to trim the stick forces as we will study later.

Estimation of Hinge-Moment Characteristics

Here, we consider sealed (no gap), plain trailing-edge flaps. The following procedures for estimating the hinge-moment coefficients $C_{h\alpha}$ and $C_{h\delta,e}$ are based on Datcom [1] methods. These methods are sufficiently general in nature and can be used for estimation of hinge-moment coefficients of wing flaps, ailerons, rudder, etc. Hinge-moment coefficients are based on control surface area S_e and control surface chord c_e. Furthermore, we assume that the included angle (ϕ'_{TE}) between the upper and lower surfaces from 90 to 99% of the chord is constant. If this condition on trailing-edge included angle is not satisfied, an additional correction may be necessary, which is not given here. The reader may refer to Datcom [1] for determining this correction. Furthermore, the methods given here are applicable only for subsonic speeds. Estimation of hinge-moment characteristics at supersonic speeds is much more involved. The interested reader may refer to Datcom [1].

Estimation of $C_{h\alpha}$ at Subsonic Speeds

We have [1]

$$C_{h\alpha} = \left(\frac{A_t \cos\Lambda_{c/4}}{A_t + 2\cos\Lambda_{c/4}}\right) C_{h\alpha_{\text{bal}}} + \Delta C_{h\alpha} \tag{3.121}$$

where A_t is the aspect ratio of the main control surface and, in this case, it is the aspect ratio of the horizontal tail. The estimation of the parameter $C_{h\alpha_{\text{bal}}}$ is done as follows [1]:

$$C_{h\alpha_{\text{bal}}} = \left[\frac{C_{h\alpha_{\text{bal}}}}{C'_{h\alpha}}\right] C'_{h\alpha} \tag{3.122}$$

$$C'_{h\alpha} = \left[\frac{C'_{h\alpha}}{(C_{h\alpha})_{\text{theory}}}\right] (C_{h\alpha})_{\text{theory}} \tag{3.123}$$

Data to evaluate the parameters $C_{h\alpha_{\text{bal}}}/C'_{h\alpha}$, $C'_{h\alpha}/(C_{h\alpha})_{\text{theory}}$, and $(C_{h\alpha})_{\text{theory}}$ are presented in Figs. 3.46 and 3.47 for typical aerodynamic balancing methods. Note that in Fig. 3.46, c_b and c_f are elevator chords forward and

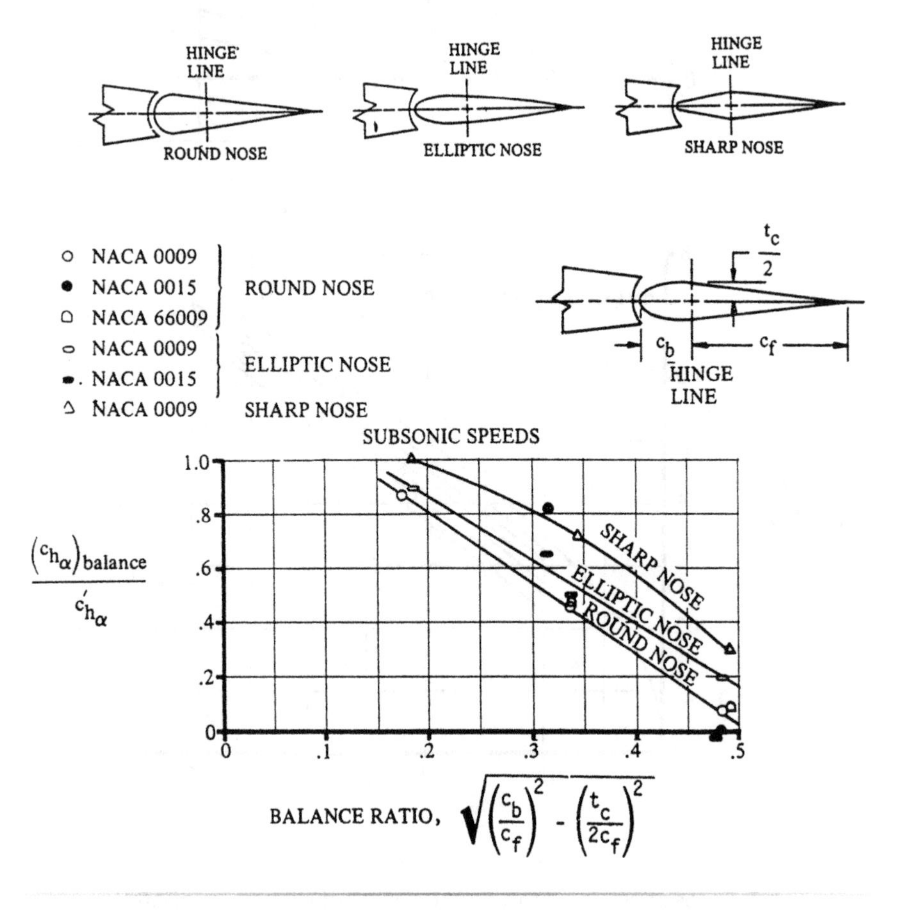

Fig. 3.46 Effect of nose balance on control hinge moment (1).

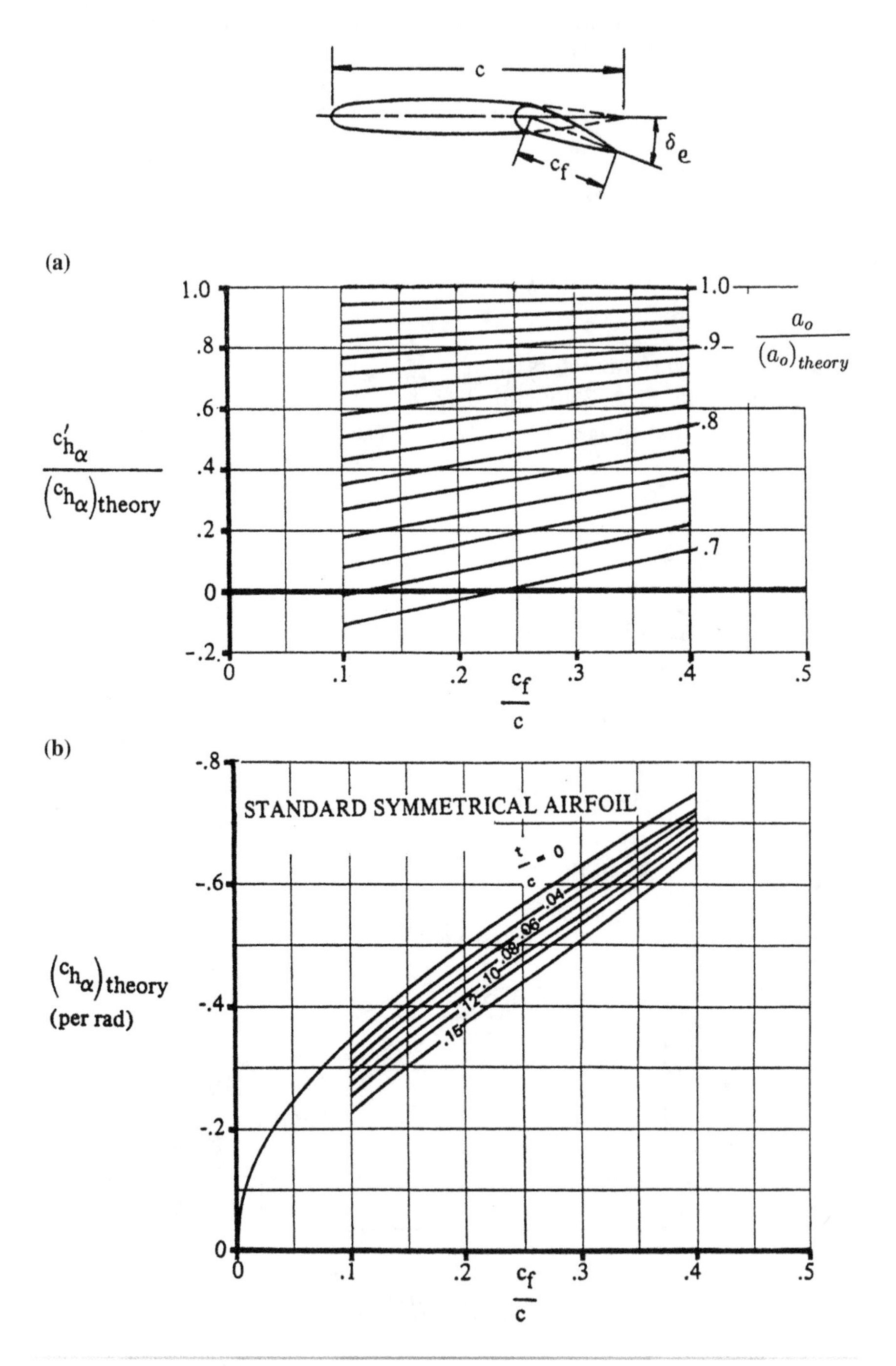

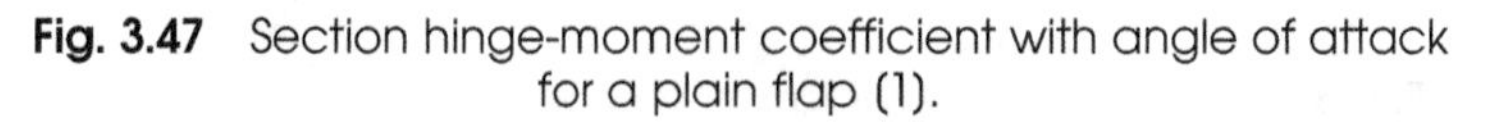
Fig. 3.47 Section hinge-moment coefficient with angle of attack for a plain flap (1).

aft of the hingeline and t_c is the elevator thickness at the hingeline location. The balance ratio used in Fig. 3.46 is defined as follows:

$$BR = \sqrt{\left(\frac{c_b}{c_f}\right)^2 - \left(\frac{t_c}{2c_f}\right)^2} \tag{3.124}$$

The variable $a_o/(a_o)_{\text{theory}}$ used in Fig. 3.47 can be obtained from Fig. 3.13. Finally, the parameter $(C_{h\alpha})_{\text{theory}}$ presented in Fig. 3.47b refers to typical airfoil sections with thickness ratios up to 0.15.

Data to evaluate the parameter $\Delta C_{h\alpha}$ of Eq. (3.121) are presented in Figs. 3.48a–3.48c. Note that c_b', c_f', and c' correspond to the values of c_b, c_f, and c resolved normal to the quarter chordline. The parameter $\Delta C_{h\alpha}/a_o B_2 K_\alpha$ $\cos \Lambda_{c/4}$ is given in Fig. 3.48a. Here, B_2 is a factor to account for the control surface and the balance chord ratios as given in Fig. 3.48c, and K_α

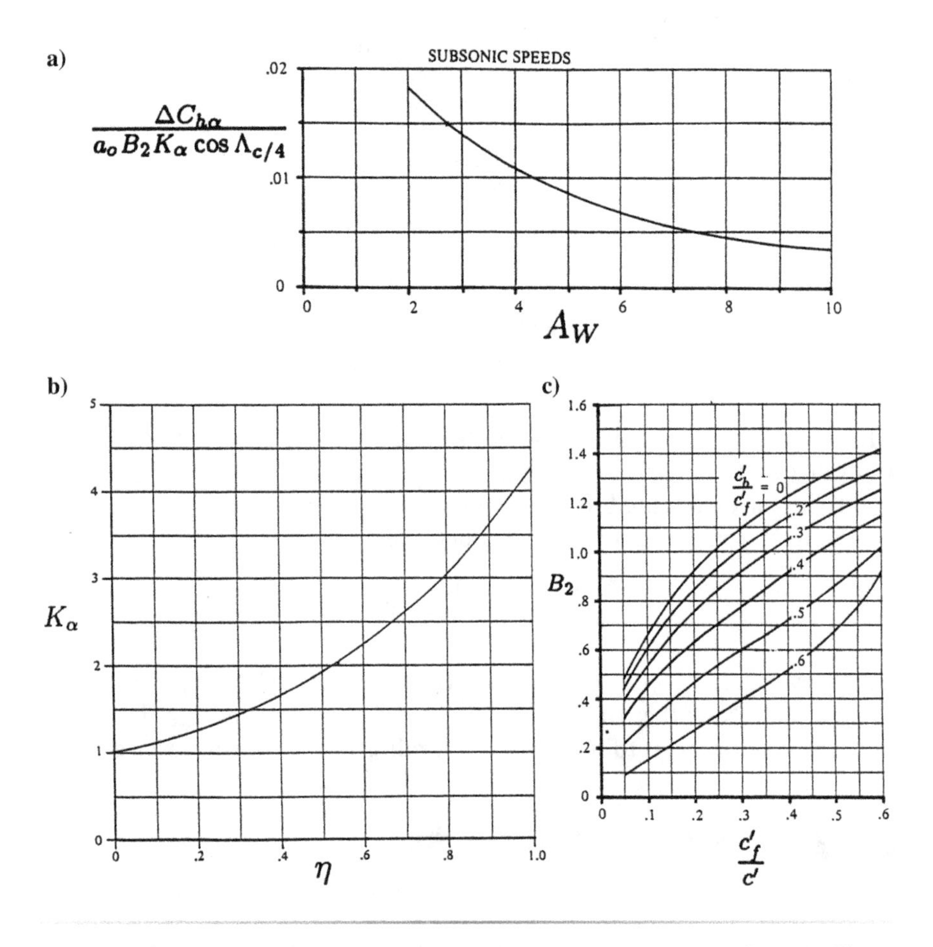

Fig. 3.48 Induced camber correction for hinge-moment coefficients (1).

is a factor to account for the control surface span and is defined as

$$K_\alpha = \frac{(K_\alpha)_{\eta_i}(1-\eta_i) - (K_\alpha)_{\eta_o}(1-\eta_o)}{\eta_o - \eta_i} \tag{3.125}$$

where $\eta = 2y/b$, $\eta_i = 2y_i/b$, and $\eta_o = 2y_o/b$. Here, η_i and η_o are the inboard and outboard span locations of the control surface, which in this case is the elevator. Note that the y coordinate is measured from the fuselage centerline and is positive towards the right wing. Data presented in Fig. 3.48b can be used to estimate the parameters $(K_\alpha)_{\eta_i}$ and $(K_\alpha)_{\eta_o}$.

Estimation of $C_{h\delta,e}$ at Subsonic Speeds

We have [1]

$$C_{h\delta,e} = \cos\Lambda_{c/4}\cos\Lambda_{HL} \times \left[(C_{h\delta,e})_{\text{bal}} + \alpha_\delta (C_{ha})_{\text{bal}}\left(\frac{2\cos\Lambda_{c/4}}{A_t + 2\cos\Lambda_{c/4}}\right)\right] + \Delta C_{h\delta,e} \tag{3.126}$$

where Λ_{HL} is the hingeline sweep and

$$(C_{h\delta,e})_{\text{bal}} = C'_{h\delta,e}\frac{(C_{h\delta,e})_{\text{bal}}}{C'_{h\delta,e}} \tag{3.127}$$

$$\alpha_\delta = -\frac{C_{l\delta}}{a_o} \tag{3.128}$$

$$C'_{h\delta,e} = \frac{C'_{h\delta,e}}{(C_{h\delta,e})_{\text{theory}}}(C_{h\delta,e})_{\text{theory}} \tag{3.129}$$

The parameter $C_{l\delta}$ can be obtained from the data given in Fig. 3.37. Data to evaluate the parameters $(C_{h\delta,e})_{\text{bal}}/C'_{h\delta,e}$, $C'_{h\delta,e}/(C_{h\delta,e})_{\text{theory}}$, $(C_{h\delta,e})_{\text{theory}}$, and $\Delta C_{h\delta,e}$ are presented in Figs. 3.49–3.51. The variable $a_o/(a_o)_{\text{theory}}$ can be obtained from Fig. 3.13. The parameter K_δ is defined as

$$K_\delta = \frac{(K_\delta)_{\eta i}(1-\eta_i) - (K_\delta)_{\eta_o}(1-\eta_o)}{\eta_o - \eta_i} \tag{3.130}$$

The parameters $(K_\delta)_{\eta_i}$ and $(K_\delta)_{\eta_o}$ can be obtained using data given in Fig. 3.48b. Note that the parameters c'_f and c' correspond to values of c_f and c resolved normal to the quarter chordline of the control surface.

3.3.15 Stick-Free Neutral Point

We have

$$C_m = C_L\bar{x}_a + C_{\text{mac},w} + C_{mf} - C_{L,t}\overline{V}_1\eta_t \tag{3.131}$$

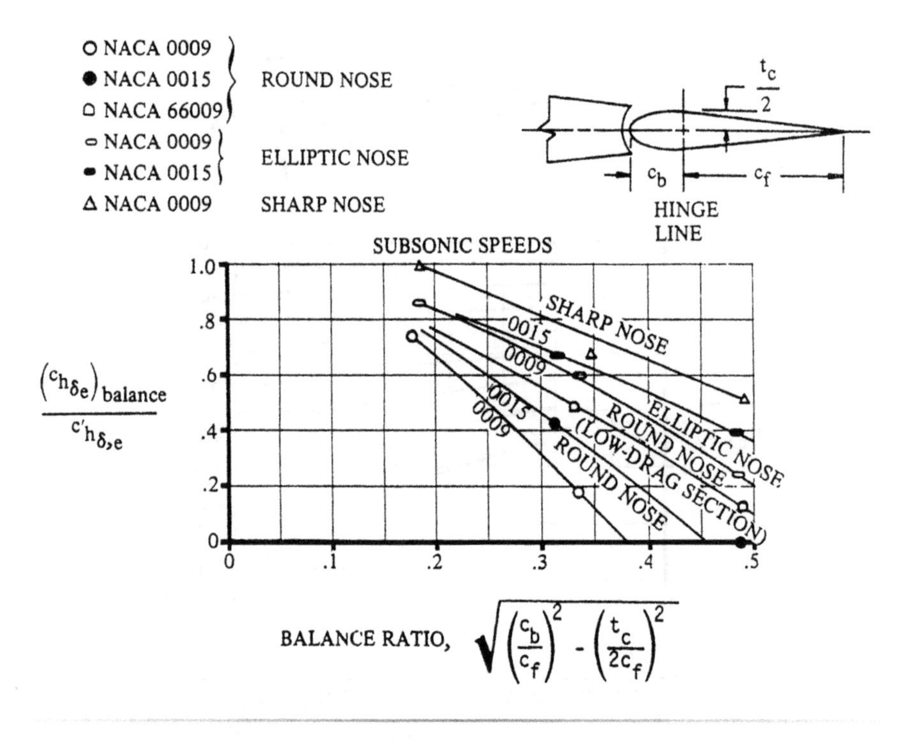

Fig. 3.49 Effect of nose balance on control hinge moment (1).

The floating elevator affects the tail contribution as follows:

$$C_{L,t} = a_t(\alpha_s + \tau\delta_{ef}) \tag{3.132}$$

$$= a_t\alpha_s\left(1 - \tau\frac{C_{h\alpha}}{C_{h\delta,e}}\right) \tag{3.133}$$

Here, we have ignored C_{ho} as it is small. Substituting for $C_{L,t}$ in Eq. (3.131) and differentiating with respect to C_L, we get

$$\left(\frac{\mathrm{d}C_m}{\mathrm{d}C_L}\right)_{\text{free}} = \bar{x}_a + \left(\frac{\mathrm{d}C_m}{\mathrm{d}C_L}\right)_f - \frac{a_t}{a_w}\overline{V}_1\eta_t\left(1 - \frac{\mathrm{d}\epsilon}{\mathrm{d}\alpha}\right)\left(1 - \tau\frac{C_{h\alpha}}{C_{h\delta,e}}\right) \tag{3.134}$$

It can be easily shown that the relation between stick-free and stick-fixed stabilities is given by

$$\left(\frac{\mathrm{d}C_m}{\mathrm{d}C_L}\right)_{\text{free}} = \left(\frac{\mathrm{d}C_m}{\mathrm{d}C_L}\right)_{\text{fix}} - \frac{C_{m\delta}}{a_w}\frac{C_{h\alpha}}{C_{h\delta,e}}\left(1 - \frac{\mathrm{d}\epsilon}{\mathrm{d}\alpha}\right) \tag{3.135}$$

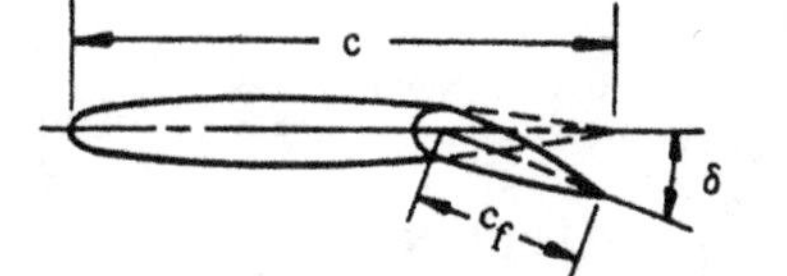

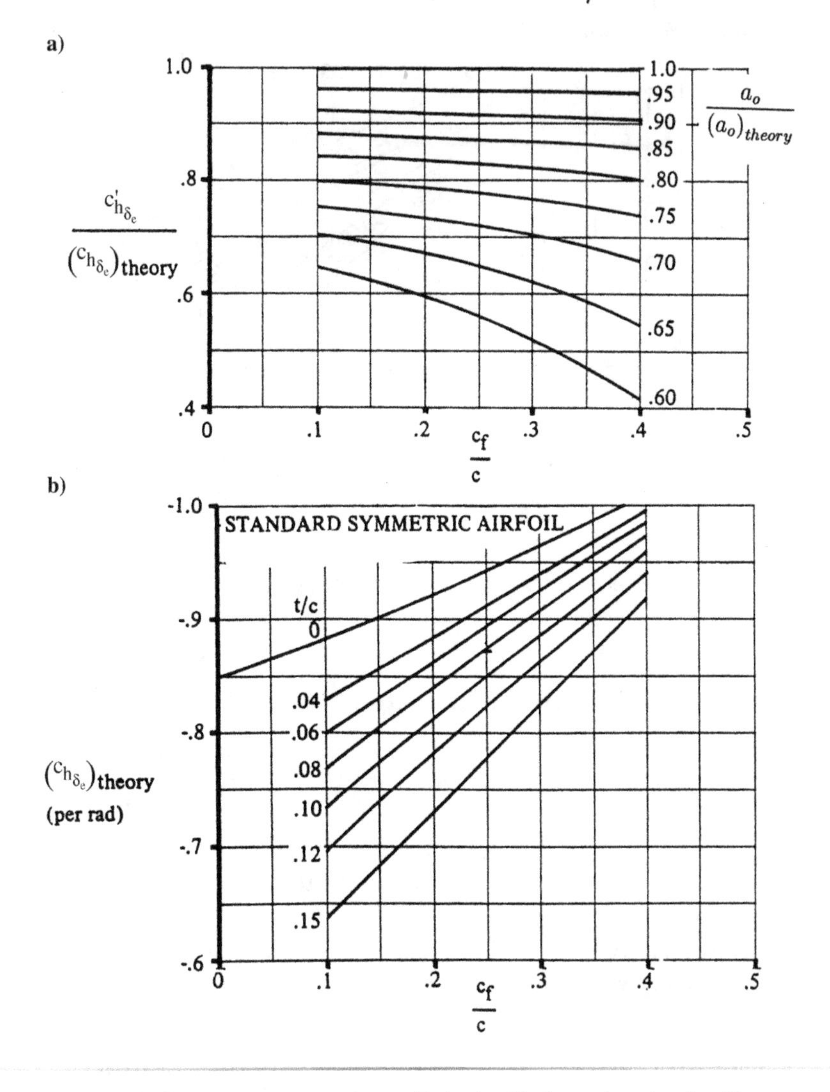

Fig. 3.50 Hinge-moment coefficient with flap chord ratio (1).

The stick-free neutral point is that location of the center of gravity when $(dC_m/dC_L)_{free} = 0$ and is given by

$$N_0' = \bar{x}_{ac,w} - \left(\frac{dC_m}{dC_L}\right)_f + \frac{a_t}{a_w}\overline{V}_1\eta_t\left(1 - \frac{d\epsilon}{d\alpha}\right)\left(1 - \tau\frac{C_{h\alpha}}{C_{h\delta,e}}\right) \tag{3.136}$$

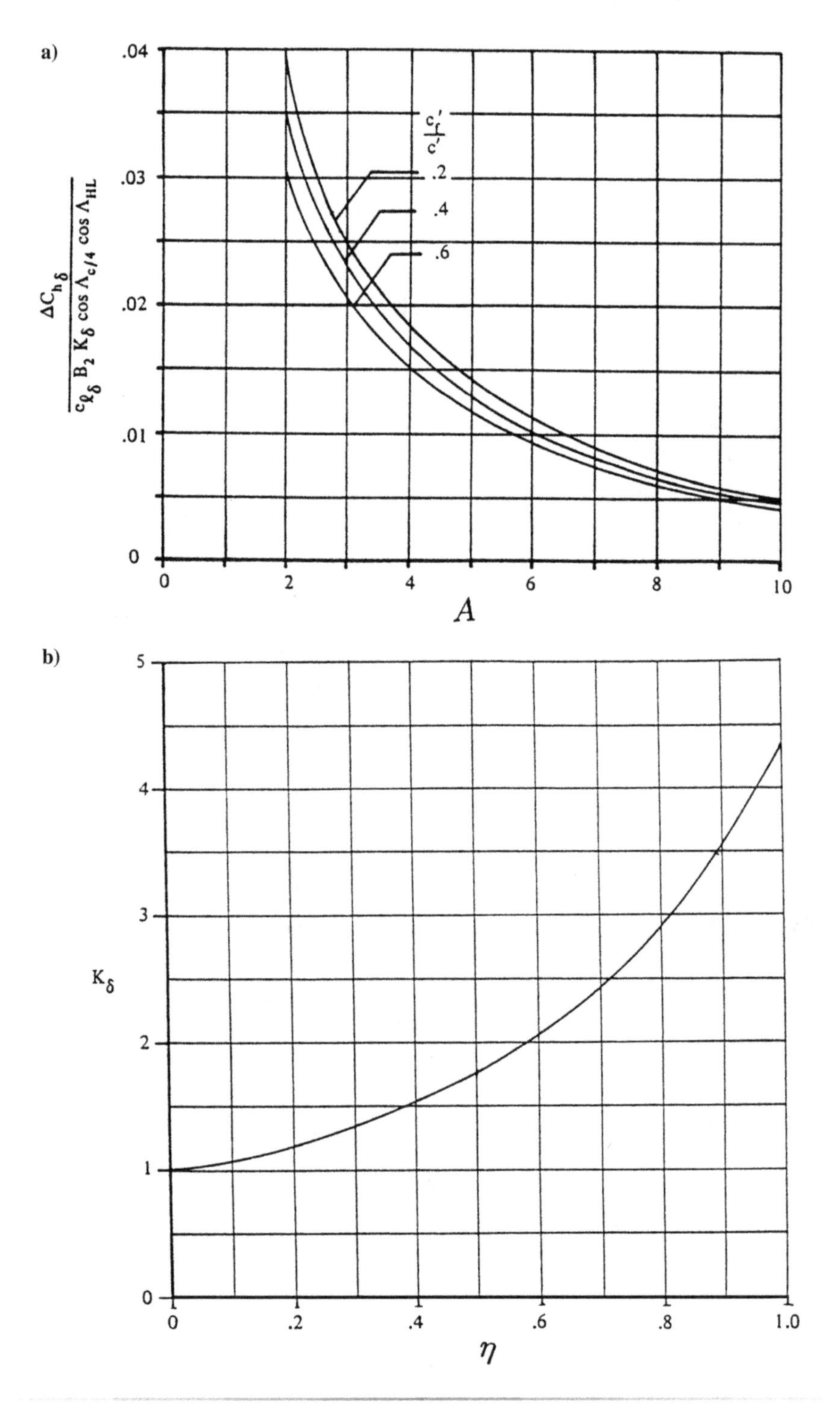

Fig. 3.51 Induced camber corrections for hinge-moment coefficient (1).

3.3.16 Stick-Free Margin

The stick-free margin is defined as

$$H_n' = N_o' - \bar{x}_{cg} \tag{3.137}$$

$$= -\left(\frac{dC_m}{dC_L}\right)_{\text{free}} \tag{3.138}$$

For a stick-free stable airplane, the stick-free margin (H_n') is positive.

With stick-free, the permissible center of gravity travel ($\Delta x'$) is modified as shown schematically in Fig. 3.52.

3.3.17 Trim Tab

A tab is a small flap hinged at the trailing edge of the elevator (see Fig. 3.53). A deflection of the tab alters the pressure distribution over the entire horizontal tail surface and hence gives a control over the hinge-moment coefficient. Thus, a deflection of the tab also affects the floating characteristics of the elevator. As we will discuss soon, the stick force directly depends on the floating characteristics of the elevator, hence on the hinge-moment coefficient. Thus, a tab deflection affects the stick force. The condition $C_h = 0$ corresponds to zero stick force, and a tab used to achieve this condition is called trim tab.

Assuming that the hinge moment depends linearly on tail angle of attack, elevator, and tab deflections,

$$C_h = C_{ho} + C_{h\alpha}\alpha_s + C_{h\delta,e}\delta_e + C_{h\delta,t}\delta_t \tag{3.139}$$

The floating angle of the elevator in the presence of the tab is given by

$$\delta_{ef} = -\frac{C_{ho} + C_{h\alpha}\alpha_s + C_{h\delta,t}\delta_t}{C_{h\delta,e}} \tag{3.140}$$

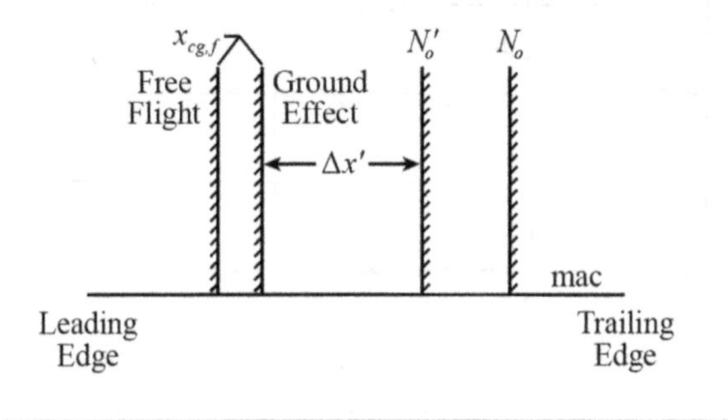

Fig. 3.52 Stick-free permissible center of gravity travel.

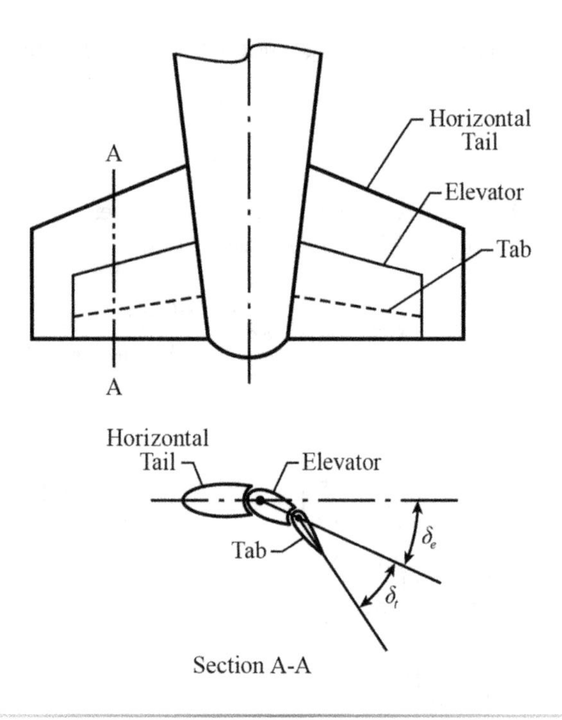

Fig. 3.53 Trim tab.

Usually, C_{ho} is small and can be ignored. Then,

$$\delta_{ef} = -\frac{C_{h\alpha}\alpha_s + C_{h\delta,t}\delta_t}{C_{h\delta,e}} \tag{3.141}$$

From this relation, we note that the deflection of the tab modifies the floating characteristics of the elevator. To see whether the deflection of the tab affects the stick-free stability, consider the horizontal-tail term,

$$C_{m,t} = -C_{Lt}\overline{V}_1\eta_t \tag{3.142}$$

$$= -a_t(\alpha_s + \tau\delta_{ef})\overline{V}_1\eta_t \tag{3.143}$$

$$= -a_t\left[\alpha_s - \frac{\tau(C_{h\alpha}\alpha_s + C_{h\delta,t}\delta_t)}{Ch_{\delta,e}}\right]\overline{V}_1\eta_t \tag{3.144}$$

$$\frac{dC_{m,t}}{dC_L} = -\frac{a_t}{a_w}\left(1 - \frac{d\epsilon}{d\alpha}\right)\left(1 - \tau\frac{C_{h\alpha}}{C_{h\delta,e}}\right)\overline{V}_1\eta_t \tag{3.145}$$

Comparing this term with that in Eq. (3.134), we find that the deflection of the tab does not alter the location of the stick-free neutral point. It only affects the hinge moment as we have seen earlier.

The tab deflection for zero hinge moment ($C_h = 0$) is given by

$$\delta_t = -\frac{C_{h\alpha}\alpha_s + C_{h\delta,e}\delta_e}{C_{h\delta,t}} \tag{3.146}$$

Note that $C_h = 0$ also corresponds to zero stick force.

Substituting for δ_e from Eq. (3.105), we obtain

$$\delta_t = -\frac{1}{C_{h\delta,t}}\left(C_{h\alpha}\alpha_s + C_{h\delta,e}\left[\delta_{e,o} - \left(\frac{dC_m}{dC_L}\right)_{\text{fix}}\frac{C_L}{C_{m\delta}}\right]\right) \tag{3.147}$$

Differentiating with respect to C_L,

$$\frac{d\delta_t}{dC_L} = -\left(\frac{C_{h\alpha}}{C_{h\delta,t}}\right)\left(\frac{1}{a_w}\right)\left(1 - \frac{d\epsilon}{d\alpha}\right) + \left(\frac{C_{h\delta,e}}{C_{m\delta}}\right)\left(\frac{1}{C_{h\delta,t}}\right)\left(\frac{dC_m}{dC_L}\right)_{\text{fix}} \tag{3.148}$$

$$= \left(\frac{C_{h\delta,e}}{C_{h\delta,t}}\right)\left(\frac{1}{C_{m\delta}}\right)\left(\frac{dC_m}{dC_L}\right)_{\text{free}} \tag{3.149}$$

$$= \left(\frac{C_{h\delta,e}}{C_{h\delta,t}}\right)\left(\frac{1}{C_{m\delta}}\right)(\bar{x}_{cg} - N_o') \tag{3.150}$$

$$= -\left(\frac{C_{h\delta,e}}{C_{h\delta,t}}\right)\left(\frac{1}{C_{m\delta}}\right)H_n' \tag{3.151}$$

Thus, we observe that $d\delta_t/dC_L$ directly depends on the level of stick-free stability or the stick-free margin. Usually, $C_{m\delta}$, $C_{h\alpha}$, $C_{h\delta,e}$, and $C_{h\delta,t}$ are all negative. $H_n' > 0$ for a stick-free stable aircraft. Therefore, for a stick-free stable aircraft, $d\delta_t/dC_L > 0$.

3.3.18 Determination of Neutral Points in Flight

The direct dependence of a stick-fixed neutral point on the elevator deflection for pitch trim and that of stick-free neutral point on the trim-tab deflection enable us to determine the stick-fixed and stick-free neutral points by actual flight tests. The flight test procedure is as follows.

The test aircraft is taken to the desired altitude and flown in level flight at various speeds within the permissible speed range ($V_{\min} \leq V \leq V_{\max}$, see Chapter 2 for more information). At each selected speed, the pilot deflects the elevator for pitch trim and rolls the trim tab for stick-force trim (zero stick force). This experiment is repeated for various locations of the center of gravity, which is usually accomplished by carefully moving a ballast weight in a limited range. Test data are then plotted and extrapolated to obtain the locations of stick-fixed and stick-free neutral points as shown in Fig. 3.54.

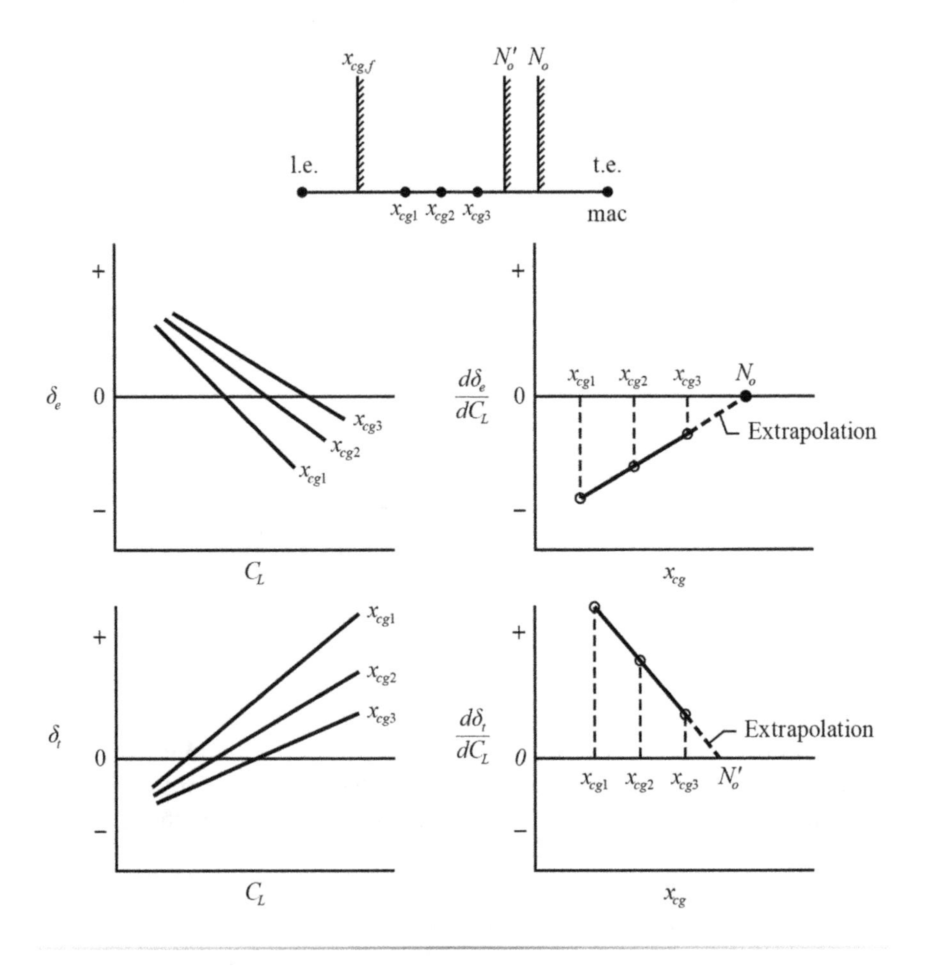

Fig. 3.54 Determination of neutral points in flight.

3.3.19 Stick Force in Steady Level Flight

The elevator is connected to the pilot's control stick through some kind of a linkage as shown schematically in Fig. 3.55. The force applied by the pilot on the stick (control column) to deflect the elevator is known as stick force. The magnitude of the stick force depends on the floating characteristics of the elevator, hence on the hinge-moment coefficients. The pilot needs to deflect the elevator by the amount $\delta_{e,R} - \delta_{e,f}$, which is related to the level of stick-free stability. In the following, we will derive the expressions for stick force and stick-force gradient in level flight and show that these two quantities depend directly on the level of stick-free stability.

Let the stick-force F_s be given by

$$F_s = -G_1 M_h \tag{3.152}$$

$$= -G_1 C_h q_t S_e \bar{c}_e \tag{3.153}$$

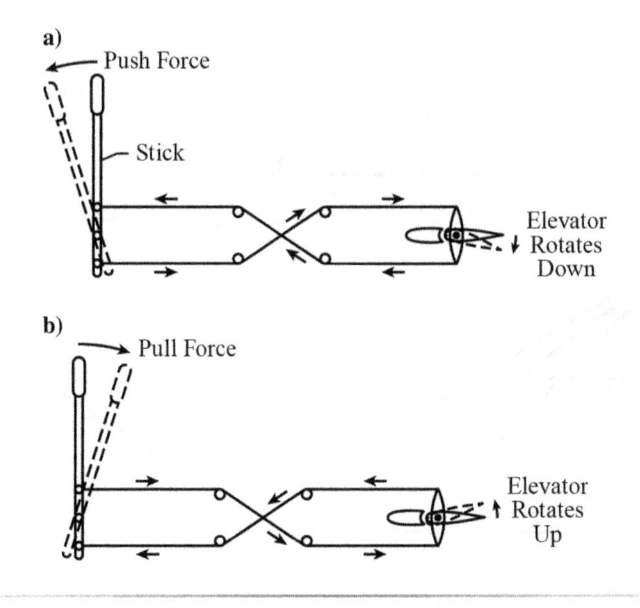

Fig. 3.55 Stick–elevator linkage.

where G_1 is the constant of proportionality between the elevator hinge moment and the stick force or the gearing ratio, M_h is the hinge moment, S_e is the elevator area, and $\bar{c}_e$ is the mean aerodynamic chord of the elevator. The units of G_1 are N/Nm or rad/m. The hinge-moment coefficient C_h in the presence of a trim tab is given by

$$C_h = C_{ho} + C_{h\alpha}\alpha_s + C_{h\delta,e}\delta_e + C_{h\delta,t}\delta_t \tag{3.154}$$

The negative sign is chosen in Eq. (3.152) so that, for a stable aircraft, the push force has a positive sign and the pull force has a negative sign. For a stable aircraft, the pilot usually applies a pull force to deflect the elevator upward (trailing edge up). Note that, according to our sign convention, downward elevator deflection is positive and upward deflection is negative. With elevator deflected upwards, the hinge moment is usually positive and Eq. (3.154) gives a negative (pull) force as desired.

Substituting for δ_e from Eq. (3.105), we have

$$C_h = C_{ho} + C_{h\alpha}\alpha_s + C_{h\delta,e}\left[\delta_{e,o} - \frac{\left(\dfrac{\mathrm{d}C_m}{\mathrm{d}C_L}\right)_{\text{fix}} C_L}{C_{m\delta}}\right] + C_{h\delta t}\delta_t \tag{3.155}$$

where

$$\alpha_s = \alpha_w - i_w + i_t - \epsilon \tag{3.156}$$

Let

$$K_1 = -G_1 S_e c_e \eta_t \tag{3.157}$$

and

$$K_2 = C_{ho} + C_{h\alpha}(\alpha_{w,oL} - i_w + i_t) + C_{h\delta,e}\delta_{e,o} \tag{3.158}$$

Then,

$$F_s = \frac{1}{2}\rho V^2 K_1 \left[K_2 + \frac{C_{h\alpha}C_L}{a_w}\left(1 - \frac{\mathrm{d}\epsilon}{\mathrm{d}\alpha}\right) - \frac{C_{h\delta,e}C_L}{C_{m\delta}}\left(\frac{\mathrm{d}C_m}{\mathrm{d}C_L}\right)_{\text{fix}} + C_{h\delta,t}\delta_t\right] \tag{3.159}$$

$$= \frac{1}{2}\rho V^2 K_1 \left[K_2 + \frac{C_{h\delta,e}C_L}{C_{m\delta}}\left(\frac{\mathrm{d}C_m}{\mathrm{d}C_L}\right)_{\text{free}} + C_{h\delta,t}\delta_t\right] \tag{3.160}$$

$$= \frac{1}{2}\rho V^2 K_1 (K_2 + C_{h\delta,t}\delta_t) - K_1 \frac{C_{h\delta,e}}{C_{m\delta}}\frac{W}{S}\left(\frac{\mathrm{d}C_m}{\mathrm{d}C_L}\right)_{\text{free}} \tag{3.361}$$

Thus, the stick-force variation can be represented as

$$F_s = K' V^2 (K_2 + C_{h\delta,t}\delta_t) + K'' \tag{3.162}$$

where

$$K' = \frac{1}{2}\rho K_1 \tag{3.163}$$

$$K'' = -K_1 \frac{C_{h\delta,e}}{C_{m\delta}}\frac{W}{S}\left(\frac{\mathrm{d}C_m}{\mathrm{d}C_L}\right)_{\text{free}} \tag{3.164}$$

Note that, for a given center of gravity location, K'' is a constant. Furthermore, if the aircraft is operating at a constant altitude as it does in a steady level flight, K' is also a constant. Note that $K_1 < 0$ and $K' < 0$. For a stick-free stable aircraft, $K'' < 0$ and, for a stick-free unstable aircraft, $K'' > 0$.

The stick-force gradient is given by

$$\frac{\mathrm{d}F_s}{\mathrm{d}V} = 2K' V (K_2 + C_{h\delta,t}\delta_t) \tag{3.165}$$

The tab deflection for stick-force trim ($F_s = 0$) is given by

$$\delta_{t,\text{trim}} = -\frac{1}{C_{h\delta,t}}\left(\frac{K''}{K' V_{\text{trim}}^2} + K_2\right) \tag{3.166}$$

When the stick force is trimmed by tab deflection as given by Eq. (3.166), the stick force and stick-force gradient at any other speed can be expressed as

$$F_s = K_1 \frac{W}{S} \frac{C_{h\delta,e}}{C_{m\delta}} \left(\frac{\mathrm{d}C_m}{\mathrm{d}C_L}\right)_{\text{free}} \left[\left(\frac{V}{V_{\text{trim}}}\right)^2 - 1\right] \tag{3.167}$$

$$\frac{\mathrm{d}F_s}{\mathrm{d}V} = 2K_1 \frac{W}{S} \frac{C_{h\delta,e}}{C_{m\delta}} \left(\frac{\mathrm{d}C_m}{\mathrm{d}C_L}\right)_{\text{free}} \frac{V}{V_{\text{trim}}^2} \tag{3.168}$$

$$\left(\frac{\mathrm{d}F_s}{\mathrm{d}V}\right)_{\text{trim}} = 2K_1 \frac{W}{S} \frac{C_{h\delta,e}}{C_{m\delta}} \left(\frac{\mathrm{d}C_m}{\mathrm{d}C_L}\right)_{\text{free}} \frac{1}{V_{\text{trim}}} \tag{3.169}$$

For proper operation by the pilot, the airplane designer has to ensure that the stick-force gradient is positive, i.e.,

$$\frac{\mathrm{d}F_s}{\mathrm{d}V} > 0 \tag{3.170}$$

The schematic variation of the stick force is shown in Fig. 3.56. It is observed that the stick-free stability provides a constant pull force on the stick and ensures a proper sign of the stick-force gradient ($\mathrm{d}F_s/\mathrm{d}V > 0$) at the trim speed.

Airworthiness requirements [5, 6] state that the stick-force gradient must be positive and greater than a certain minimum value. A typical minimum value suggested is 1 lb/kn or 8.72 Ns/m [5, 7]. For short-duration applications, the recommended maximum allowable stick force to operate the pitch control (elevator) is 60 lb or 267.3 N. It is further recommended that for sustained application, the stick force for pitch control should not exceed 10 lb or 44.60 N [5, 7].

Historically, as the flight speed increased, stick forces increased significantly and exceeded normal pilot capabilities. This led to the development of powered controls. Initially, power was supplied only to assist the pilot by means of hydraulic boosters. With such a system, the pilot still had to apply some stick force. If a booster failed, the system would revert to full manual control. Aircraft using such semimanual systems include the Boeing 707 series and the B-52 Stratofortress Bomber.

As flight speed increased further, it became more and more difficult to keep the stick forces within the capability of an average human pilot, and this led to the development of fully powered control systems. With such systems, the pilot hardly does anything more than operate a small hydraulic valve, requiring a small nominal force. The stick force virtually remains constant irrespective of the speed or the deflection of the control surface.

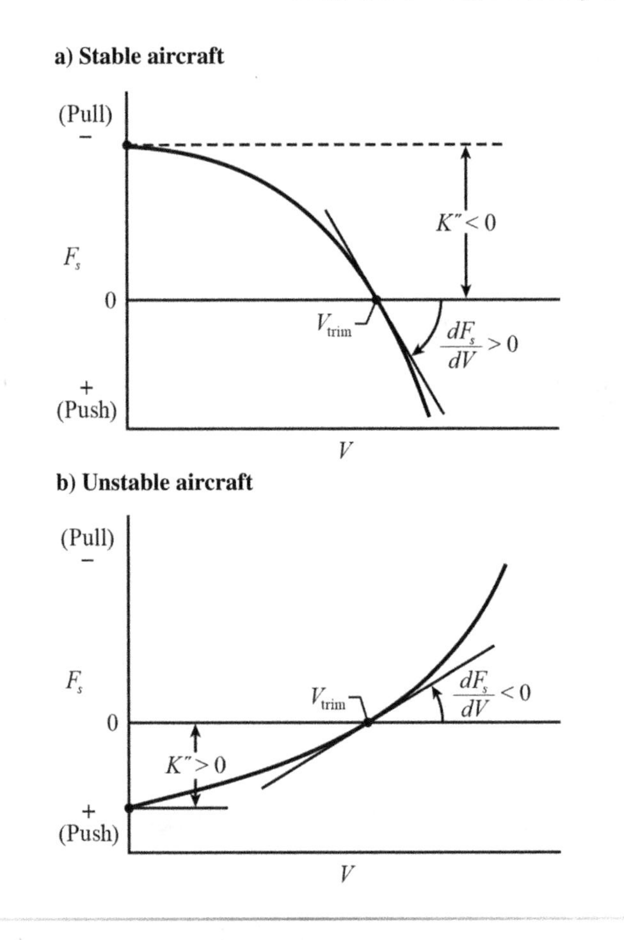

Fig. 3.56 Typical variation of stick force with velocity.

However, to give the pilot some idea of the stick forces involved, an artificial feel is introduced to the system. This may take the form of a simple spring providing an increased resistance with increase of control column movement, or it can be a complex simulator providing forces proportional to the airspeed or Mach number.

Example 3.1

For the generic vehicle configuration shown in Fig. 3.57, length of the fuselage = 13.0264 m, $S = 25.7167\ \text{m}^2$, $c_{re} = 3.48$ m, $\bar{c} = 3.0457$ m, $l_h = 4.287$ m, $d\epsilon/d\alpha = 0.5$, and $C_{L\alpha,WB} = 0.047/\text{deg}$.

Using the following geometrical data of the fuselage, estimate the slope of the fuselage pitching-moment-coefficient curve.

(Continued)

Example 3.1 (*Continued*)

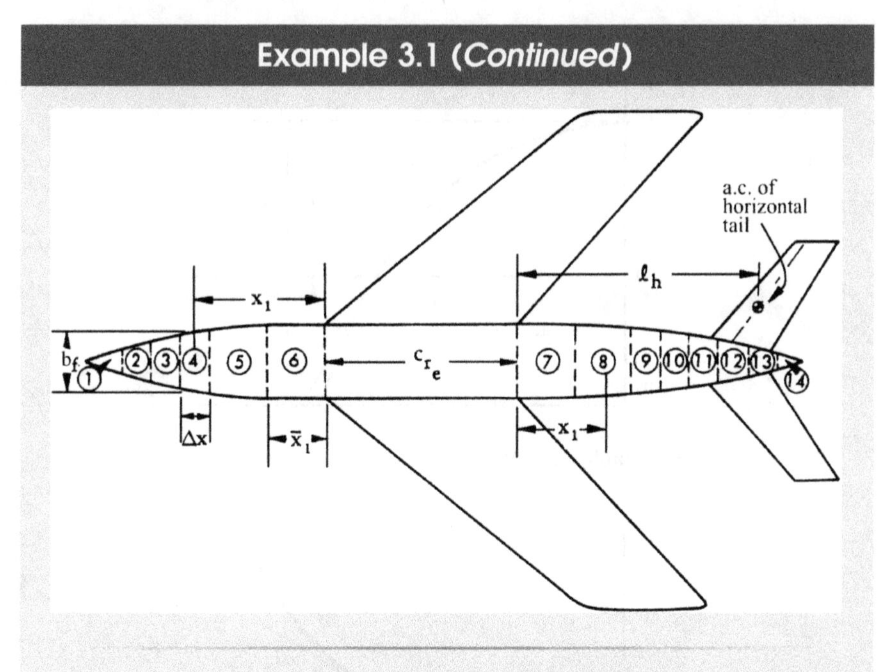

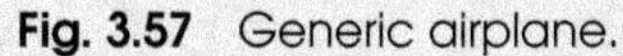

Fig. 3.57 Generic airplane.

In Table 3.1, Δx is the section length, b_f is the section width, and x_1 is the distance measured from leading edge of the root chord to the centroid of the section with one exception for Section 6. For Section 6, $\bar{x}_1 = \Delta x = x_1$.

Table 3.1 Geometrical Fuselage Data

Section	Δx, m	b_f, m	x_1, m
1	0.6150	0.2057	4.0386
2	0.5334	0.5258	3.4671
3	0.5334	0.7849	2.9337
4	0.5334	1.0211	2.4003
5	1.0668	1.2268	1.6002
6	1.0668	1.2802	1.0668
7	1.0668	1.2802	0.5334
8	1.0668	1.2268	1.6002
9	0.5334	1.0744	2.4003
10	0.5334	0.9677	2.9337
11	0.5334	0.8458	3.4671
12	0.5334	0.6248	4.0005
13	0.5334	0.3886	4.5339
14	0.3970	0.1295	4.9987

(*Continued*)

Example 3.1 (*Continued*)

Solution:

The first step is to determine $d\epsilon_u/d\alpha$ at each body section as shown in Tables 3.2 and 3.3. Note that the values of $d\epsilon_u/d\alpha$ in column 3 of Table 3.2 are obtained from Fig. 3.9b for Sections 1–5 and from Fig. 3.9a for Section 6. Also observe that these values are based on x_1 measured to their centroid for Sections 1–5 and 7–14. For Section 6, $\bar{x}_1$ is used to estimate $d\epsilon_u/d\alpha$. Furthermore, these values obtained from Figs. 3.9a and 3.9b correspond to $C_{L\alpha,WB} = 0.0785$. We have $C_{L\alpha,WB} = 0.0047$. Therefore, we have to apply the correction as follows:

$$\left(\frac{d\epsilon_u}{d\alpha}\right)_c = \left(\frac{d\epsilon_u}{d\alpha}\right)\left(\frac{0.047}{0.0785}\right)$$

These corrected values are given in column 4 of Table 3.2.

Table 3.2 Corrected Upwash Values for Sections 1–6

Section	$\frac{x_1 \text{ or } \bar{x}_1}{c_{re}}$	$\frac{d\epsilon_u}{d\alpha}$	$\left(\frac{d\epsilon_u}{d\alpha}\right)_c$
1	1.16	0.14	0.084
2	1.00	0.18	0.108
3	0.84	0.22	0.132
4	0.69	0.27	0.162
5	0.46	0.37	0.222
6	0.31	1.95	1.168

Table 3.3 Downwash Values for Sections 7–14

Section	$\frac{x_1}{l_h}$	$\left(1 + \frac{d\epsilon_u}{d\alpha}\right)$	$\frac{d\epsilon_u}{d\alpha}$
7	0.124	0.062	−0.928
8	0.373	0.187	−0.813
9	0.560	0.280	−0.720
10	0.684	0.342	−0.658
11	0.809	0.404	−0.596
12	0.933	0.467	−0.533
13	1.058	0.529	−0.471
14	1.166	0.583	−0.417

(*Continued*)

Example 3.1 (Continued)

For Sections 7–14 (Table 3.3), we have from Eq. (3.10)

$$\left(1+\frac{\mathrm{d}\epsilon_u}{\mathrm{d}\alpha}\right)=\frac{x_1}{l_h}\left(1-\frac{\mathrm{d}\epsilon}{\mathrm{d}\alpha}\right)$$

With $\mathrm{d}\epsilon/\mathrm{d}\alpha = 0.5$,

$$1+\frac{\mathrm{d}\epsilon_u}{\mathrm{d}_\alpha}=0.5\frac{x_1}{l_h}$$

The calculated values of $\mathrm{d}\epsilon_u/\mathrm{d}\alpha$ are presented in Table 3.3.
Next, we have to evaluate the integral in Eq. (3.7). We assume

$$\int_0^{l_f} b_f^2\left(1+\frac{\partial\epsilon_u}{\partial\alpha}\right)\mathrm{d}x=\sum_{i=1}^{N} b_{f,i}^2\left(1+\frac{\partial\epsilon_u}{\partial\alpha}\right)_i\Delta x_i$$

where N is the number of body sections. In this case, $N = 14$. Using Table 3.4, the sum is evaluated as follows.

$$\sum_{i=1}^{N} b_{f,i}^2\left(1+\frac{\partial\epsilon_u}{\partial\alpha}\right)_i\Delta x_i = 8.0114\,\mathrm{m}^3$$

Table 3.4 Evaluation of Fuselage Integral

Section	b_f^2, m^2	$\left(1+\frac{\partial\epsilon_u}{\partial\alpha}\right)$	Δx, m	$b_f^2\left(1+\frac{\partial\epsilon_u}{\partial\alpha}\right)\Delta x$, m^3
1	0.0423	1.084	0.6149	0.0282
2	0.2764	1.108	0.5334	0.1633
3	0.6161	1.132	0.5334	0.3720
4	1.0428	1.162	0.5334	0.6464
5	1.5050	1.222	1.0668	1.9620
6	1.6386	2.168	1.0668	3.7897
7	1.6386	0.062	1.0668	0.1084
8	1.5050	0.187	1.0668	0.3002
9	1.1543	0.280	0.5334	0.1724
10	0.9366	0.342	0.5334	0.1708
11	0.7154	0.404	0.3334	0.1541
12	0.3902	0.467	0.5334	0.0972
13	0.1510	0.529	0.5334	0.0426
14	0.0168	0.583	0.3970	0.0039

(Continued)

Example 3.1 (Continued)

Then, using Eq. (3.7),

$$C_{m\alpha,f} = \frac{\pi}{2S\bar{c}} \int_0^{l_f} b_f^2 \left(1 + \frac{\partial \epsilon_u}{\partial \alpha}\right) \mathrm{d}x$$

$$= \frac{\pi}{2S\bar{c}} \sum_{i=1}^{14} b_{f,i}^2 \left(1 + \frac{\partial \epsilon_u}{\partial \alpha}\right)_i \Delta x_i$$

$$= \left(\frac{\pi}{2 * 25.7167 * 3.0457}\right) 8.0114$$

$$= 0.1607/\mathrm{rad}$$

$$= 0.0028/\mathrm{deg}$$

Example 3.2

Consider the generic tailless vehicle shown in Fig. 3.58. We will be using this vehicle as an example to illustrate the application of the methods of estimating stability and control characteristics discussed in this text.

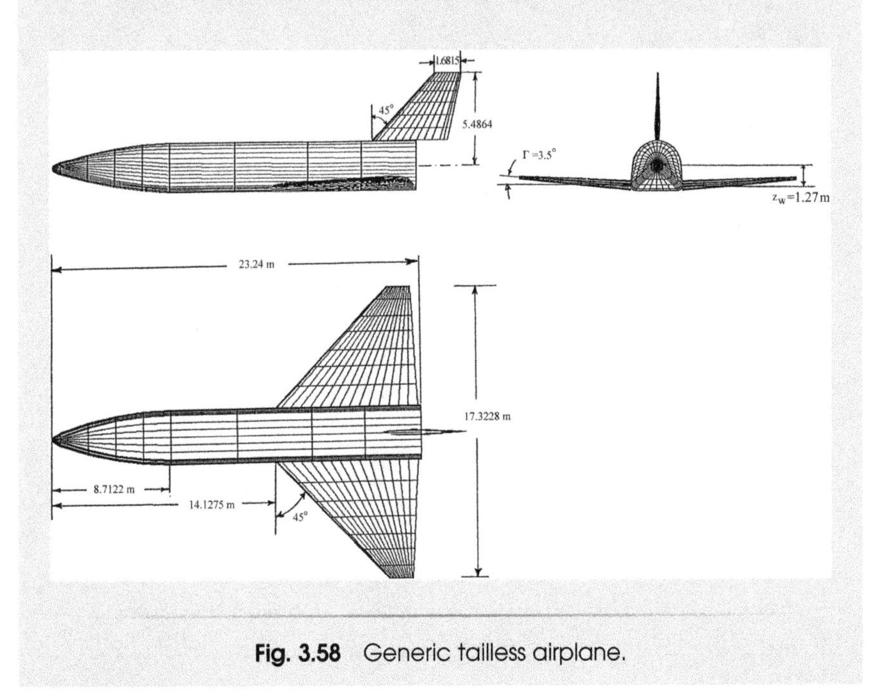

Fig. 3.58 Generic tailless airplane.

(Continued)

Example 3.2 (Continued)

We will evaluate the longitudinal stability parameters here. In Sec. 3.5, we will obtain the directional stability and, in Sec. 3.6, the lateral stability characteristics for this configuration.

The variations of fuselage cross-sectional area S_B and its first derivative dS_B/dx are shown in Fig. 3.59. The leading dimensional characteristics of this vehicle are as follows.

Wing:

Span $b = 17.3228$ m leading sweep $\Lambda_{LE} = 45$ deg, dihedral angle $\Gamma = 3.5$ deg, theoretical area $S = 106.0114$ m^2, exposed area $S_{\text{exp}} = 73.6282$ m^2, theoretical aspect ratio $A = 2.8306$, exposed aspect ratio $A_e = 2.6893$, theoretical taper ratio $\lambda = 0.1427$, exposed taper ratio $\lambda_e = 0.1705$, exposed root chord $c_{re} = 8.94$ m, theoretical root chord $c_r = 10.6766$ m, $c_t = 1.5236$ m, sectional (two-dimensional) lift-curve slope $a_o = 0.0877$/deg, mean aerodynamic chord $\bar{c} = 6.8072$ m, vertical distance between the center of gravity and the quarter chordline of the wing root chord $z_w = 1.27$ m, and the airfoil section geometry parameter $\Delta y = 2.5$.

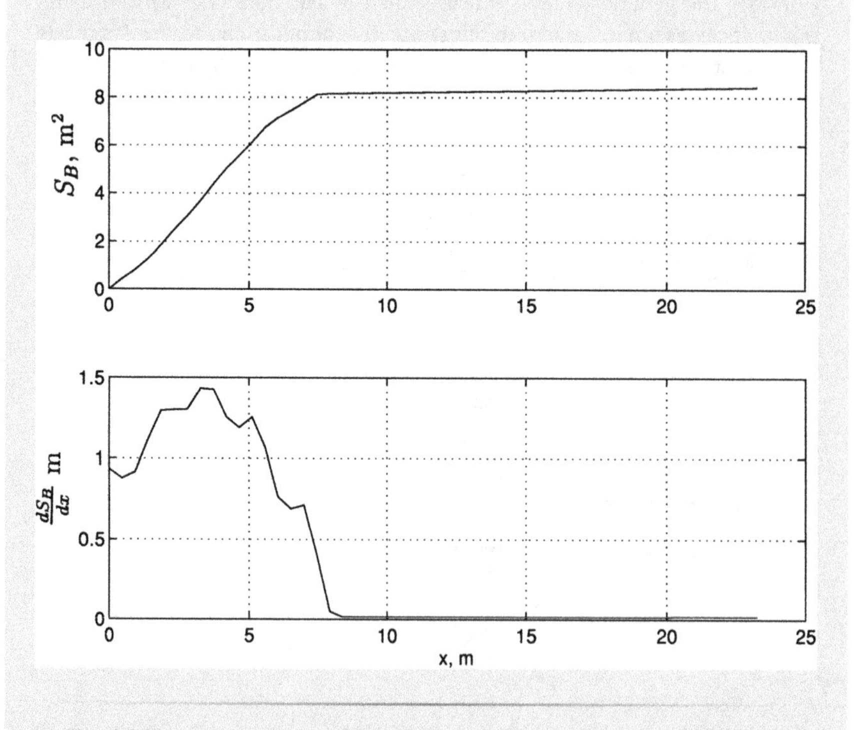

Fig. 3.59 Fuselage cross-sectional area of generic tailless airplane.

(Continued)

Example 3.2 (*Continued*)

Fuselage:
Overall length $l_f = 23.2410$ m, length of the nose $l_n = 8.7122$ m, distance between the fuselage leading edge and the leading edge of the exposed wing root chord $l_N = 14.1275$ m, fuselage height at a distance of $l_f/4$ from the leading edge is 2.7838 m, fuselage height at a distance of $3l_f/4$ from leading edge is 3.048 m, maximum width $b_{f,\max} = 3.2715$ m, maximum height $d_{f,\max} = 3.048$ m, maximum cross-sectional area $S_{B,\max} = 8.3193$ m^2, and projected side area $S_{B,S} = 60.75$ m^2.

Vertical Tail:
Leading-edge sweep $\Lambda_{LE,v} = 45$ deg, theoretical area $S_v = 20.2426$ m^2, root chord (exposed) = 4.5826 m, root chord (theoretical) = 5.6977 m, span (theoretical) $b_v = 5.4864$ m, tip chord $c_t = 1.6815$ m, taper ratio (theoretical) = 0.2951, theoretical aspect ratio $A_v = 1.4869$, horizontal distance between the center of gravity and the vertical tail aerodynamic center $l_v = 7.7561$ m, vertical distance between the fuselage centerline and the aerodynamic center $z_v = 3.8290$ m, and the vertical tail airfoil section geometry parameter $\Delta y = 2.50$.

For a center of gravity location at 15.9334 m from the fuselage leading edge, calculate the wing–body lift-curve slope and pitching-moment-curve slope for subsonic/supersonic speeds.

Solution:
The given configuration has a low-aspect ratio wing and the wing-span-to-body-diameter/width ratio $b/b_{f,\max}$ is 5.2951, which is small. Therefore, it is preferable to use the combined wing–body approach. From the given data, we obtain the following: $\Lambda_{c/4} = 36.3487$ deg, $\Lambda_{c/2} = 25.2485$ deg, $A_e = 2.6893$, $k = 0.8$, $S_{\exp}/S = 0.6945$, $k_2 - k_1 = 0.89$ (see Fig. 3.6), and $\Delta y_\perp = 3.5355$ [Eq. (3.22)].

Calculation of $C_{L\alpha,WB}$:
To begin with, let us calculate the lift-curve slope of the fuselage nose. Using Eq. (3.26), we obtain $C_{L\alpha,N} = 0.0024$/deg for subsonic Mach numbers. For supersonic Mach numbers, we use the data given in Fig. 3.10a for ogive-cylinders. The calculated values of $C_{L\alpha,N}$ (referenced to wing area S) were curve-fitted to obtain the following expression for $1.2 \le M \le 4.0$,

$$C_{L\alpha,N} = 0.0033 + 0.00035\,M/\mathrm{deg}$$

The lift-curve slope of the exposed wing was calculated using Eq. (3.16) for subsonic speeds based on exposed aspect ratio A_e. This equation gives an analytical expression for $C_{L\alpha,e}$ in terms of Mach number.

The lift-curve slope of the exposed wing at supersonic speeds was calculated using the data of Fig. 3.14 and applying the correction obtained from

(*Continued*)

Example 3.2 (*Continued*)

Fig. 3.15a using $\Delta y_{\perp} = \Delta y / \cos \Lambda_{LE} = 3.536$. The calculated values of $C_{L\alpha,e}$ were curve-fitted to obtain the following expression for $1.2 \leq M \leq 4.0$,

$$C_{L\alpha,e} = 0.0038\,M^2 - 0.03088\,M + 0.0791/\text{deg}$$

Knowing $C_{L\alpha,N}$ and $C_{L\alpha,e}$ at subsonic and supersonic speeds, we can now find K_N using Eq. (3.25).

Using Eq. (3.27), we obtain $K_{W(B)} = 1.1607$, which is applicable for both subsonic and supersonic Mach numbers. We get $K_{B(W)} = 0.2627$ using Eq. (3.28) for subsonic speeds. For supersonic speeds, we have to use the data of Fig. 3.18b for $K_{B(W)}$ because $\beta A_e(1 + \lambda_e)[(\tan \Lambda_{LE}/\beta) + 1] \geq 4$. These values of $K_{B(W)}$ were curve-fitted to obtain the following expression applicable for $1.2 \leq M \leq 4.0$:

$$K_{B(W)} = 0.0011\,M^2 - 0.1044\,M = 0.3274$$

With these values, we are now in a position to calculate the wing–body lift coefficient using the following equation:

$$C_{L\alpha,WB} = \left[K_N + K_{W(B)} + K_{B(W)}\right] C_{L\alpha,e} \frac{S_{\text{exp}}}{S}$$

Calculation of $C_{m\alpha,WB}$:

We have

$$C_{m\alpha,WB} = \left(\bar{x}_{cg} - \bar{x}_{ac,WB}\right) C_{L\alpha,WB}$$

where

$$\bar{x}_{ac,WB} = \frac{x_{ac,WB}}{\bar{c}}$$

and

$$\left(\frac{x_{ac,WB}}{c_{re}}\right) = \frac{\left(\frac{x_{ac}}{c_{re}}\right)_N C_{L\alpha,N} = \left(\frac{x_{ac}}{c_{re}}\right)_{W(B)} C_{L\alpha,W(B)} + \left(\frac{x_{ac}}{c_{re}}\right)_{B(W)} C_{L\alpha,B(W)}}{C_{L\alpha,WB}}$$

Note that $x_{ac,WB}$ is the distance measured from the leading edge of the exposed wing root chord to the aerodynamic center of the wing–body combination. Further, we have

$$C_{L\alpha,W(B)} = K_{W(B)} C_{L\alpha,e} \frac{S_e}{S}$$

$$C_{L\alpha,B(W)} = K_{B(W)} C_{L\alpha,e} \frac{S_e}{S}$$

(*Continued*)

Example 3.2 (*Continued*)

Because we have calculated $K_{W(B)}$, $K_{B(W)}$, and $C_{L\alpha,e}$, we can evaluate $C_{L\alpha,B(W)}$ and $C_{L\alpha,W(B)}$ using the given relations.

For subsonic speeds ($0 \leq M \leq 0.80$), using Eq. (3.36), we obtain

$$\left(\frac{x_{ac}}{c_{re}}\right)_N = -1.1379$$

For supersonic speeds ($1.2 \leq M \leq 4.0$), the calculated values using data of Fig. 3.11a and Eq. (3.37) were curve-fitted to obtain the following expression:

$$\left(\frac{x_{ac}}{c_{re}}\right)_N = -0.0171\,M^2 + 0.1295\,M - 1.1364$$

Furthermore, we note that

$$\left(\frac{x_{ac}}{c_{re}}\right)_{W(B)} = \left(\frac{x_{ac}}{c_{re}}\right)_W$$

Then, using the data of Fig. 3.19b and curve-fitting the values, we obtain

$$\left(\frac{x_{ac}}{c_{re}}\right)_W = 0.3367\,M^3 - 0.4810\,M^2 + 0.2214\,M + 0.4918 \quad (0 \leq M \leq 0.80)$$

$$= -0.0121\,M^2 + 0.0837\,M + 0.5342 \quad (1.4 \leq M \leq 4.0)$$

Now we have to evaluate $(x_{ac}/c_{re})_{B(W)}$. We have $A_e = 2.6893$. Therefore, for all subsonic Mach numbers, we find that $\beta A_e < 4$. Therefore, we need to use interpolation. For $\beta A_e = 4$, using the data of Fig. 3.20 and Eq. (3.39), we obtain $(x_{ac}/c_{re})_{B(W)} = 0.3748$. For $\beta A_e = 0$, using Fig. 3.21, we obtain $(x_{ac}/c_{re})_{B(W)} = 0.40$. With this, we get the following interpolation formula for subsonic speeds and $0 \leq \beta A_e \leq 4$,

$$\left(\frac{x_{ac}}{c_{re}}\right)_{B(W)} = 0.4 - 0.0063\beta A_e \quad (0 \leq M \leq 0.8)$$

where $\beta = \sqrt{1 - M^2}$.

For supersonic speeds, with $\Lambda_{LE} = 45$ deg, $\lambda_e = 0.1705$, and $A_e = 2.6893$, we find that $\beta A_e(1 + \lambda_e)[1 + 1/\beta \tan \Lambda_{LE}] \geq 4$. Therefore, for supersonic speeds, we use Fig. 3.22b because the given configuration has no afterbody. The calculated values were curve-fitted (least square) to obtain the following expression:

$$\left(\frac{x_{ac}}{c_{re}}\right)_{B(W)} = -0.0110\,M^2 + 0.0760\,M + 0.5410 \quad (1.2 \leq M \leq 4.0)$$

Having calculated all the required parameters, we can now obtain the slope of the pitching-moment coefficient from subsonic to supersonic

(*Continued*)

Example 3.2 (*Continued*)

speeds. Note that some of the calculated parameters have a discontinuity across the transonic Mach numbers from 0.8 to 1.2. For such cases, a smooth interpolation was used.

The calculated values of $C_{L\alpha,WB}$, $x_{ac,W(B)}$ and $C_{m\alpha,WB}$ at various subsonic and supersonic Mach numbers are shown in Figs. 3.60 and 3.61. For the purpose of comparison, these calculations were also done using the Aerodynamic Preliminary Analysis System (APAS) [8] and the APAS results are included in Figs. 3.60 and 3.61. We observe that $C_{L\alpha}$ increases in high subsonic/transonic range and then drops steadily as Mach number increases further. The aerodynamic center moves aft in transonic/low supersonic range and starts moving forward again as the Mach number increases further.

The APAS is an interactive, graphic user interface program for preliminary aerodynamic analyses of airplane-type configurations. It uses slender body theory for subsonic/supersonic modeling of the fuselage. The surface singularity/panel methods are used for wing and tail surfaces at subsonic/supersonic speeds. The interference effects between fuselage and wing and between fuselage and tail surfaces are modeled using a cylindrical interference shell enveloping the fuselage.

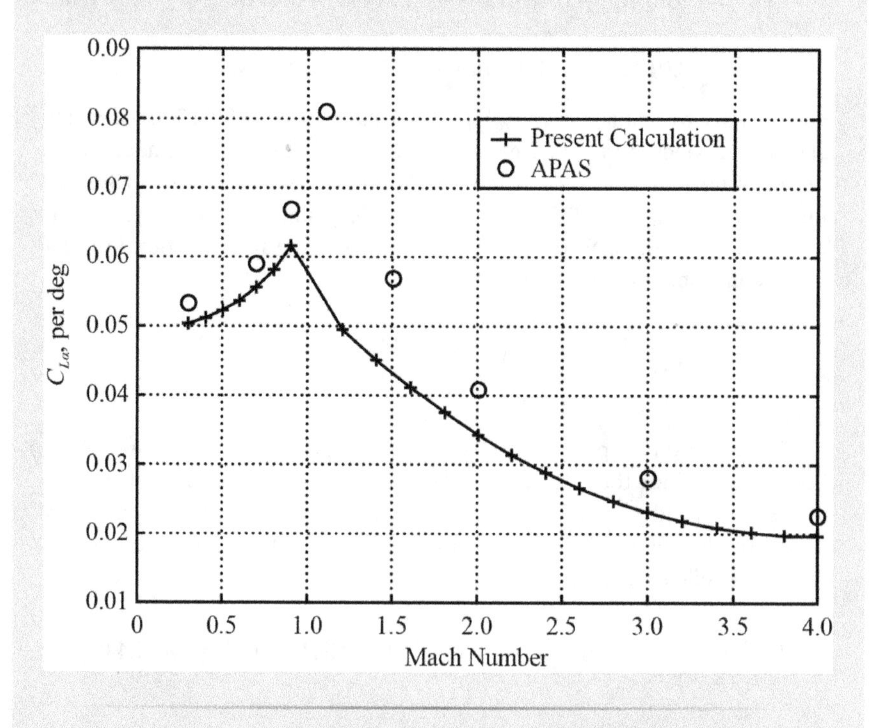

Fig. 3.60 Lift-curve slope of the generic tailless airplane.

(Continued)

Example 3.2 (*Continued*)

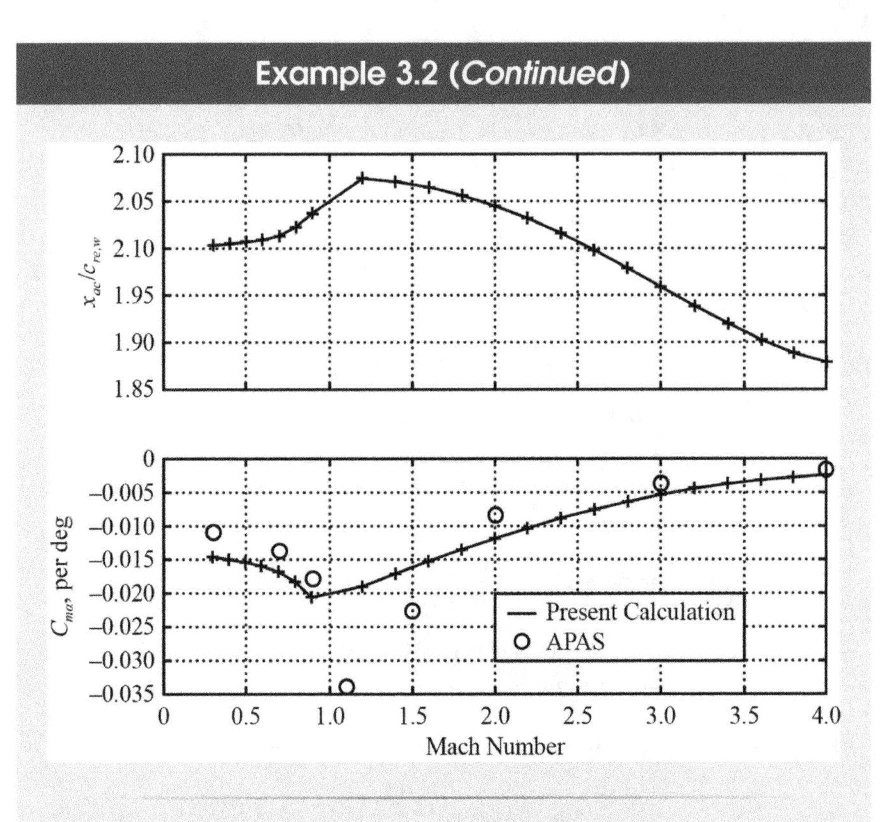

Fig. 3.61 X_{ac} and $C_{m\alpha}$ with Mach number.

The analysis for subsonic/supersonic Mach numbers is performed by the unified distributed panel (UDP) program. For hypersonic Mach numbers, APAS uses the hypersonic arbitrary body program (HABP).

From Figs. 3.60 and 3.61 we observe that present calculations are in quite good agreement with APAS results except around $M = 1$, where both methods are approximate.

Example 3.3

An aircraft has the following data: $\bar{x}_{cg} = 0.3$, $\bar{x}_{ac} = 0.24$, $C_{L,w} = 0.10(\alpha^o + 2.5)$, $C_{L,\max} = 1.2$, $C_{\mathrm{mac},w} = 0.06$, $\epsilon = 0.3\alpha$, $C_{mf} = 0.05 + 0.1C_L$, $a_t = 0.08/\mathrm{deg}$, $\eta_t = 0.9$, $\overline{V}_1 = 0.6$, $i_w = 0$, $i_t = 2\,\mathrm{deg}$, $C_{h\alpha} = -0.002/\mathrm{deg}$, $C_{h\delta_e} = -0.003/\mathrm{deg}$, and $\tau = 0.20$.

Determine 1) the angle of attack in steady level flight if the elevator is locked in neutral position, 2) the permissible most forward position of the center of gravity if the maximum up elevator deflection is limited to 30 deg, and 3) stick-free neutral point and stick-free margin.

(*Continued*)

Example 3.3 (*Continued*)

Solution:
We have

$$C_m = C_{mo} + \left(\frac{dC_m}{dC_L}\right)_{fix} C_L$$

where

$$\begin{aligned} C_{mo} &= C_{m,ac} + C_{mf,o} - a_t \overline{V}_1 \eta_t(\alpha_{oL,w} - i_w + i_t) \\ &= 0.06 + 0.05 - 0.08 * 0.6 * 0.9 * (-2.5 - 0 + 2) \\ &= 0.1316 \end{aligned}$$

and

$$\begin{aligned} \left(\frac{dC_m}{dC_L}\right)_{fix} &= \overline{x}_{cg} - \overline{x}_{ac} + \left(\frac{dC_m}{dC_L}\right)_f - \frac{a_t}{a_w}\left(1 - \frac{d\epsilon}{d\alpha}\right)\overline{V}_1 \eta_t \\ &= 0.30 - 0.24 + 0.1 - (0.08/0.10)(1 - 0.3)0.6 * 0.9 \\ &= -0.1424 \end{aligned}$$

For trim with elevator in neutral or stick-fixed position,

$$C_m = C_{mo} + \left(\frac{dC_m}{dC_L}\right)_{fix} C_L$$

$$0 = 0.1316 - 0.1424 C_L$$

Then,

$$\begin{aligned} C_L &= 0.9242 \\ \alpha &= \frac{C_L}{C_{L\alpha}} + \alpha_{oL,w} \\ &= \frac{0.9242}{0.1} - 2.5 \\ &= 6.7416 \text{ deg} \end{aligned}$$

We have

$$\overline{x}_{cg,f} = N_o - (\delta_{e,\max} - \delta_{e,o})\frac{C_{m\delta}}{C_{L,\max}}$$

$$\begin{aligned} N_o &= \overline{x}_{cg} - \left(\frac{dC_m}{dC_L}\right)_{fix} \\ &= 0.3 + 0.1424 \\ &= 0.4424 \end{aligned}$$

(Continued)

Example 3.3 (Continued)

$$C_{m,\delta} = -a_t \overline{V}_1 \eta_t \tau$$
$$= -0.08 * 0.6 * 0.9 * 0.2$$
$$= -0.00864/\text{deg}$$

$$\delta_{e,o} = \frac{C_{mf,o} + C_{mac,w} - a_t \overline{V}_1 \eta_t (\alpha_o - i_w + i_t)}{-C_{m\delta}}$$
$$= \frac{0.05 + 0.06 - 0.08 * 0.6 * 0.9 * (-2.5 - 0 + 2)}{0.00864}$$
$$= 15.2315\ \text{deg}$$

The permissible most forward center of gravity location is given by

$$\overline{x}_{cg,f} = 0.4424 - (-30 - 15.23215)\frac{-0.00864}{1.2}$$
$$= 0.1167$$

The stick-free neutral point is given by

$$\left(\frac{\mathrm{d}C_m}{\mathrm{d}C_L}\right)_{\text{free}} = \left(\frac{\mathrm{d}C_m}{\mathrm{d}C_L}\right)_{\text{fix}} - \frac{C_{h\alpha}}{C_{h\delta,e}}\frac{C_{m\delta}}{a_w}\left(1 - \frac{\mathrm{d}\epsilon}{\mathrm{d}\alpha}\right)$$
$$= -0.1424 - \left(\frac{-0.002}{-0.003}\right)\left(\frac{-0.00864}{0.10}\right)(1 - 0.3)$$
$$= -0.1021$$

Then,

$$N_o' = \overline{x}_{cg} - \left(\frac{\mathrm{d}C_m}{\mathrm{d}C_L}\right)_{\text{free}}$$
$$= 0.4021$$
$$H_n' = -\left(\frac{\mathrm{d}C_m}{\mathrm{d}C_L}\right)_{\text{free}}$$
$$= 0.1021$$

Example 3.4

An aircraft has a wing loading of 1450 N/m^2, wing aspect ratio of 8, and tail aspect ratio of 4. The aerodynamic center of the wing–fuselage combination is at 0.24 mac from the wing leading edge. The static margin is −0.05. Using the following data, determine the tab deflection to trim the stick force at a

(Continued)

Example 3.4 (*Continued*)

speed of 160 km/h in level flight at sea level. What is the stick-force gradient at this speed?

$a_w = a_t = 0.1/\text{deg}$, $\alpha_{oL,w} = i_t = i_w = 0$, $\delta_{e,o} = -2\,\text{deg}$, $\overline{V}_1 = 0.6$, $\epsilon = 0.5\alpha$, $\eta_t = 0.9$, $C_{m,f} = 0.1C_L$, $\tau = 0.5$, $S_e = 1.85\,\text{m}^2$, $c_e = 0.608\,\text{m}$, $C_{h\alpha} = -0.003/\text{deg}$, $C_{h\delta_e} = -0.006/\text{deg}$, $C_{h,o} = 0$, $C_{h\delta_t} = -0.003/\text{deg}$, and $G_1 = 5.0\,\text{rad/m}$.

Solution:

This aircraft is statically unstable because it has negative static margin ($H_n = -0.05$). We have

$$\delta_{\text{trim}} = -\frac{1}{C_{h\delta,t}}\left(\frac{K''}{K'V_{\text{trim}}^2} + K_2\right)$$

where

$$K'' = -K_1 \frac{C_{h\delta,e}}{C_{m\delta}} \frac{W}{S} \left(\frac{\text{d}C_m}{\text{d}C_L}\right)_{\text{free}}$$

$$K' = \frac{1}{2}\rho K_1$$

and

$$K_2 = C_{ho} + C_{h\alpha}(\alpha_{oL,w} - i_w + i_t) + C_{h\delta,e}\delta_{e,o}$$

Now

$$\begin{aligned} K_1 &= -G_1 S_e c_e \eta_t \\ &= -5.0 * 1.85 * 0.608 * 0.9 \\ &= -5.0616 \end{aligned}$$

$$\begin{aligned} C_{m\delta} &= -a_t \overline{V}_1 \eta_t \tau \\ &= -0.1 * 0.6 * 0.9 * 0.5 \\ &= -0.027 \end{aligned}$$

$$\begin{aligned} \left(\frac{\text{d}C_m}{\text{d}C_L}\right)_{\text{free}} &= \left(\frac{\text{d}C_m}{\text{d}C_L}\right)_{\text{fix}} - \frac{C_{m\delta}}{a_w}\frac{C_{h\alpha}}{C_{h\delta,e}}\left(1 - \frac{\text{d}\epsilon}{\text{d}\alpha}\right) \\ &= 0.05 + \left(\frac{-0.027}{0.1}\right)\left(\frac{-0.003}{-0.006}\right)(1 - 0.5) \\ &= 0.1175 \end{aligned}$$

$$\begin{aligned} V_{\text{trim}} &= \frac{160 * 1000}{3600} \\ &= 44.44\,\text{m/s} \end{aligned}$$

(*Continued*)

Example 3.4 (*Continued*)

$$K'' = -(-5.0616)\left(\frac{-0.006}{-0.027}\right)1450 * 0.1175$$
$$= 191.6378$$
$$K' = (1/2) * 1.225(-5.0616)$$
$$= -3.1002$$

$$K_2 = 0 - 0.003(0 - 0 + 0) - 0.006(-2.0)$$
$$= 0.0120$$
$$\delta_{\text{trim}} = -\left(\frac{1}{-0.003}\right)\left(\frac{191.6378}{-3.1002 * 44.44^2} + 0.012\right)$$
$$= -6.4460 \text{ deg}$$

The stick-force gradient is given by

$$\frac{dF_s}{dV} = 2K'V\left(K_2 + C_{h\delta,t}\delta_t\right)$$
$$= 2(-3.1002) * 44.44(0.0120 - 0.003[-6.4460])$$
$$= -8.6351 \text{ Ns/m}$$

The stick-force gradient is negative because the aircraft is unstable stick-free.

Example 3.5

Estimate the elevator effectiveness for a horizontal tail fitted with an elevator and having the following characteristics: aspect ratio $= 3.78$, $\Lambda_{c/2} =$ 45.5 deg, $c_e/c = 0.224$, $\lambda = 0.586$, $\eta_i = 0.141$, and $\eta_o = 0.61$. The airfoil section is a NACA 65A006. Assume that the Reynolds number $R_l = 6.1 \times 10^6$.

Solution:

We have to determine the elevator effectiveness parameter τ. For NACA 65A006 airfoil section, we obtain $\tan(\phi_{TE}/2) = 0.0616$ using Fig. 3.13c.

From Figs. 3.13a and 3.13b, we obtain $a_o/(a_o)_{\text{theory}} = 0.887$ and $(a_o)_{\text{theory}} = 6.58$ so that $a_o = 5.8365/\text{rad}$. Using Fig. 3.35, we obtain $(\alpha_\delta)_{C_L}/(\alpha_\delta)_{C_l} = 1.083$. Using the data of Fig. 3.36, we obtain $K_b = 0.55$. From Fig. 3.37b, we obtain $C_{l\delta}/(C_{l\delta})_{\text{theory}} = 0.817$ and, from Fig. 3.37a, $(C_{l\delta})_{\text{theory}} = 3.77/\text{rad}$ so that $C_{l\delta} = 3.08/\text{rad}$. Then, substituting in Eq. (3.98), we obtain $\tau = 0.3143$.

Example 3.6

Estimate the hinge-moment coefficients for a horizontal tail fitted with an elevator and having the following data: $A = 4.0$, $\Lambda_{c/4} = 45$ deg, $\Lambda_{HL} = 45$ deg, $c_b/c_f = 0.09$, $c_f/c = 0.16$, Reynolds number $R_l = 6.1 \times 10^6$, airfoil section NACA 65A006, $\eta_i = 0.25$, $\eta_o = 0.65$, and $t_c/2c_f = 0.08$.

Solution:

The first step is to calculate the balance ratio as given by

$$BR = \sqrt{\left(\frac{c_b}{c_f}\right)^2 - \left(\frac{t_c}{2c_f}\right)^2}$$

$$= 0.0412$$

The next step is to evaluate various parameters appearing in Eq. (3.121).

From Fig. 3.46, we note that $C_{h\alpha_{\text{bal}}}/C'_{h\alpha} = 1$. Using Fig. 3.13c, we obtain $\phi_{TE} = 7$ deg for the given 65A series airfoil with 6% thickness ratio so that $\tan \phi_{TE}/2 = 0.0616$. Then, from Fig. 3.13b, we obtain $a_o/(a_o)_{\text{theory}} = 0.887$ and, from Fig. 3.13a, $(a_o)_{\text{theory}} = 6.58/\text{rad}$ so that $a_o = 5.8365/\text{rad}$. With this and using $c_f/c = 0.16$ from Fig. 3.47a, we obtain $C'_{h\alpha}/(C_{h\alpha})_{\text{theory}} = 0.68$. Using Fig. 3.47b for $t/c = 0.06$, we get $(C_{h\alpha})_{\text{theory}} = -0.39$ so that $C'_{h\alpha} = -0.265/\text{rad}$ and $C_{h\alpha_{\text{bal}}} = -0.265/\text{rad}$.

We have $c_b/c_f = 0.09$ (streamwise). We assume $c'_b/c'_f = 0.09$ (normal to quarter chordline). Similarly, assume $c'_f/c' = 0.16$.

We have $\eta_i = 0.25$ and $\eta_o = 0.65$. Using these values and Fig. 3.48b, we get $(K_\alpha)_{\eta_i} = 1.35$ and $(K_\alpha)_{\eta_o} = 2.45$. Then, using Eq. (3.125), we obtain $K_\alpha = 0.3875$. Furthermore, using Fig. 3.48c, we get $B_2 = 0.76$.

From Fig. 3.48a, for $A = 4.0$, we get $\Delta C_{h\alpha}/a_o B_2 K_\alpha \cos \Lambda_{c/4} = 0.011$. With all these values, we get $\Delta C_{h\alpha} = 0.0134/\text{rad}$. Substitution in Eq. (3.121) gives $C_{h\alpha} = -0.125/\text{rad}$ or $-0.0022/\text{deg}$.

Now let us calculate $C_{h\delta,e}$. From Fig. 3.49, we find $(C_{h\delta,e})_{\text{bal}}/C'_{h\delta,e} = 1$. From Fig. 3.50a, using $c_f/c = 0.16$ and $a_o/(a_o)_{\text{theory}} = 0.887$, we find $C'_{h\delta,e}/(C_{h\delta,e})_{\text{theory}} = 0.91$. Using $t/c = 0.06$ and $c_f/c = 0.16$ from Fig. 3.50b, we find $(C_{h\delta,e})_{\text{theory}} = -0.83$ so that $C'_{h\delta,e} = -0.755/\text{rad}$ and $C_{h\delta,e_{\text{bal}}} = -0.755$.

Using Fig. 3.37b, we obtain $C_{l\delta}/(C_{l\delta})_{\text{theory}} = 0.817$ and, from Fig. 3.37a, $(C_{l\delta})_{\text{theory}} = 3.77$ so that $C_{l\delta} = 3.08/\text{rad}$. We have $\alpha_\delta = -C_{l\delta}/a_o$. Then, with $a_o = 5.8365$, we get $\alpha_\delta = -0.5277$. Using $c'_f/c' = 0.16$ and $A = 4.0$ from Fig. 3.51a, we get $\Delta C_{h\delta,e} = 0.018(C_{l\delta} B_2 K_\delta \cos \Lambda_{c/4} \cos \Lambda_{HL})$.

We have $\eta_i = 0.25$ and $\eta_o = 0.65$. With these values and from Fig. 3.51b, we obtain $(K_\delta)_{\eta_i} = 1.30$ and $(K_\delta)_{\eta_o} = 2.25$. Then, using Eq. (3.130), we obtain $K_\delta = 0.4688$. Further substitution gives $\Delta C_{h\delta,e} = 0.0099/\text{rad}$.

Finally, substituting in Eq. (3.126), we find $C_{h\delta,e} = -0.3501/\text{rad}$ or $-0.0061/\text{deg}$.

3.4 Stability in Maneuvering Flights

The class of flight paths when the load factor exceeds unity is called maneuvers. The load factor n is defined as the ratio of lift to weight.

$$n = \left(\frac{L}{W}\right) \tag{3.171}$$

Therefore, during a maneuver, the lift exceeds the weight and the airplane structure is subjected to higher stresses than those encountered in steady level flight with a load factor of unity. In general, during a maneuver, the airplane experiences accelerations, which makes it necessary for us to consider inertia forces in the analysis. We will study such motions later in the text. However, by introducing some simplifications and ignoring inertia forces, we can extend the methods of static stability and control to study simple maneuvers like the pull-up from a dive in a vertical plane and the coordinated turn in a horizontal plane. Even though not exact, this approach gives us an idea about the level of stability, control requirements, and the stick-force gradients that can be expected in such maneuvers.

We introduce the following assumptions. 1) During the maneuver, a change in the forward speed is small and ignored. In other words, we assume that the airplane is moving at a uniform speed along its maneuver path. 2) The airplane is disturbed only in angle of attack and load factor, and these disturbances are small.

If a maneuver is performed at transonic or low supersonic speeds, the given assumptions may not be justified because even small changes in forward speed can give rise to large variations in aerodynamic forces and moments.

While studying the static longitudinal stability, we considered only one disturbance—angle of attack. Here, we have an additional disturbance in the form of load factor. In the following analysis, we will develop a theory for predicting stability, control requirements, and stick-force gradients for a pull-up from a dive in a vertical plane and the steady turn in a horizontal plane.

3.4.1 Pull-Up in a Vertical Plane

Consider an aircraft to be initially flying in a steady level flight at A as shown in Fig. 3.62. At A, $L = W$ and the load factor n is unity. Let the aircraft climb to point B and enter into a dive (C), and let the pilot effect a pull-out such that at the bottom of the pull-out (D), the aircraft is at the same altitude as it was at A, and the flight path BCDE is approximately semicircular. Thus, the aircraft at D is at the same altitude and forward speed as in A but is operating at a different angle of attack and load factor. In other words, the

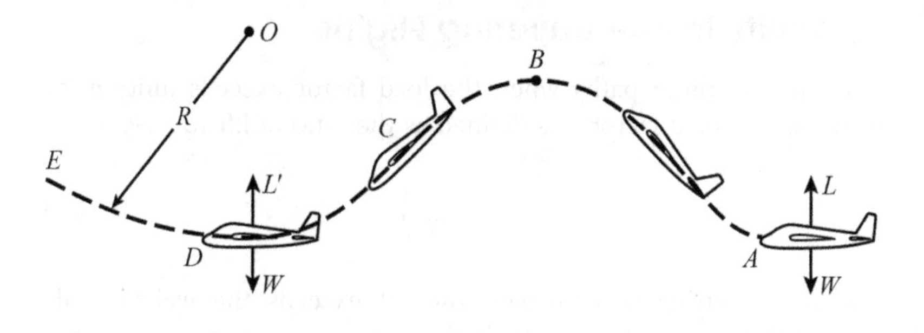

Fig. 3.62 Airplane in a pull-out maneuver.

aircraft at D is disturbed in angle of attack and load factor compared to the steady level flight at A.

An important point here is to observe that, during the pull-up maneuver, the aircraft experiences a steady rate of rotation in vertical plane, which is equivalent to a pitch rate about the y-body axis. We will now show that, on account of this pitch rate, the aircraft will experience higher levels of static stability compared to that in steady level flight and this apparent increase in stability demands additional elevator deflection.

Consider the equilibrium of forces in vertical direction at D. Let L' denote lift at D. Then,

$$L' = nW \tag{3.172}$$

$$= W + \frac{WV^2}{Rg} \tag{3.173}$$

so that

$$n = 1 + \frac{V^2}{Rg} \tag{3.174}$$

and

$$\frac{V}{R} = \frac{(n-1)g}{V} \tag{3.175}$$

Let $\Omega = V/R$ be the angular velocity about the center of the semicircular path O. Then, we have

$$\Omega = \frac{(n-1)g}{V} \tag{3.176}$$

As a result of this angular velocity, the horizontal tail experiences an increase in angle of attack $\Delta\alpha_{i,t}$ as shown in Fig. 3.63. From the geometry, we have

$$\Delta\alpha_{i,t} = \frac{l_t}{R} \tag{3.177}$$

$$= \frac{l_t}{V}\frac{V}{R} \tag{3.178}$$

$$= \frac{\Omega l_t}{V} \tag{3.179}$$

$$= \frac{(n-1)l_t g}{V^2} \tag{3.180}$$

With $\Delta n = n - 1$,

$$\frac{\Delta\alpha_{i,t}}{\Delta n} = \frac{l_t g}{V^2} \tag{3.181}$$

Because both $\Delta\alpha_{i,t}$ and Δn are assumed to be small, we can write

$$\frac{\mathrm{d}\alpha_{i,t}}{\mathrm{d}n} = \frac{l_t g}{V^2} \tag{3.182}$$

At D, $L' = nW$ and at A, $L = W$. The airplane at D is disturbed in angle of attack and load factor compared to the steady level flight at A. These two disturbances together lead to a net increase in lift ΔL so that

$$L' = L + \Delta L \tag{3.183}$$

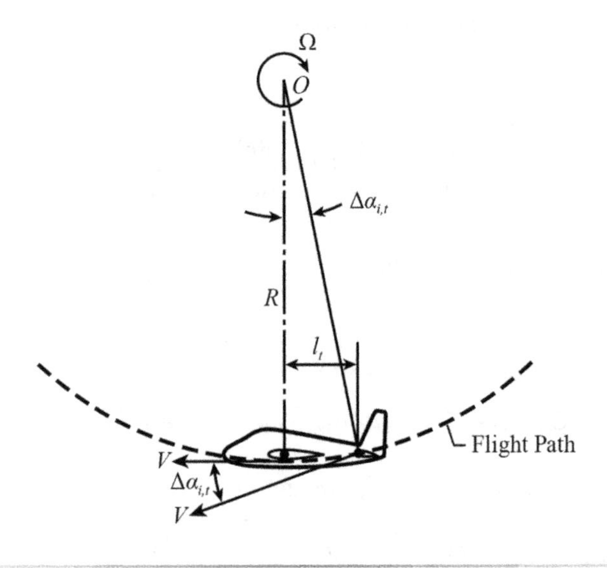

Fig. 3.63 Induced angle of attack at horizontal tail during a pull-out maneuver.

$$n = 1 + \frac{\Delta C_L}{C_L} \tag{3.184}$$

$$n - 1 = \frac{\Delta C_L}{C_L} \tag{3.185}$$

$$\frac{\Delta n}{\Delta C_L} = \frac{1}{C_L} \tag{3.186}$$

or

$$\frac{dn}{dC_L} = \frac{1}{C_L} \tag{3.187}$$

Then we have

$$\frac{d\alpha_{i,t}}{dC_L} = \frac{d\alpha_{i,t}}{d_n}\frac{dn}{dC_L} \tag{3.188}$$

$$= \frac{l_t g}{V^2 C_L} \tag{3.189}$$

$$= \frac{l_t \rho g S}{2W} \tag{3.190}$$

$$= \frac{1}{2\mu_1} \tag{3.191}$$

where $\mu_1 = W/\rho g S l_t$. The parameter μ_1 is often called the longitudinal relative density parameter.

The total angle of attack of the horizontal tail is given by

$$\alpha_t = (\alpha_w - i_w + i_t - \epsilon) + \Delta\alpha_{i,t} \tag{3.192}$$

or

$$\frac{d\alpha_t}{dC_L} = \frac{1}{a_w}\left(1 - \frac{d\epsilon}{d\alpha}\right) + \frac{1}{2\mu_1} \tag{3.193}$$

Stick-Fixed Maneuver Point

We have

$$C_m = C_L\bar{x}_a + C_{mac,w} + C_{mf} - a_t\alpha_t\overline{V}_1\eta_t \tag{3.194}$$

$$= C_L\bar{x}_a + C_{mac,w} + C_{mf} - a_t[\alpha_w - i_w + i_t - \epsilon + \Delta\alpha_{i,t}]\overline{V}_1\eta_t \tag{3.195}$$

Differentiating with respect to C_L, we get

$$\left(\frac{dC_m}{dC_L}\right)_m = \bar{x}_a + \left(\frac{dC_m}{dC_L}\right)_f - \frac{a_t}{a_w}\overline{V}_1\eta_t\left(1 - \frac{d\epsilon}{d\alpha}\right) - \frac{a_t\overline{V}_1\eta t}{2\mu_1} \tag{3.196}$$

The stick-fixed maneuver point N_m is that position of the center of gravity where $(\mathrm{d}C_m/\mathrm{d}C_L)_m = 0$ and is given by

$$N_m = \bar{x}_{ac} - \left(\frac{\mathrm{d}C_m}{\mathrm{d}C_L}\right)_f + \frac{a_t \bar{V}_1 \eta_t}{a_w} = \left[\left(1 - \frac{\mathrm{d}\epsilon}{\mathrm{d}\alpha}\right) + \frac{a_w}{2\mu_1}\right] \tag{3.197}$$

From Eqs. (3.196) and (3.197), we observe that the stability level of the airplane during a pull-up maneuver is higher compared to that in steady level flight and the stick-fixed maneuver point is aft of the stick-fixed neutral point. This increase in stability level, as said before, comes on account of the angular velocity Ω, which causes an increase in the angle of attack of the horizontal tail by $\Delta\alpha_{i,t}$.

The stick-fixed maneuver margin is given by

$$H_m = N_m - \bar{x}_{cg} \tag{3.198}$$

For a stable airplane, H_m is positive.

Elevator Required per g

Because of increased stability level during the maneuver, the magnitude of the elevator deflection required for trim also increases. This increase in elevator deflection is usually expressed by the parameter $(\mathrm{d}\delta_e/\mathrm{d}n)$, which is the elevator deflection per unit increase in load factor above unity.

From Eq. (3.101), we have

$$\left(\frac{\mathrm{d}\delta_e}{\mathrm{d}C_L}\right)_m = -\frac{\left(\frac{\mathrm{d}C_m}{\mathrm{d}C_L}\right)_m}{C_{m\delta}} \tag{3.199}$$

$$= -\frac{\bar{x}_{cg} - N_m}{C_{m\delta}} \tag{3.200}$$

With

$$\left(\frac{\mathrm{d}\delta_e}{\mathrm{d}C_L}\right)_m = \frac{\mathrm{d}\delta_e}{\mathrm{d}n}\frac{\mathrm{d}n}{\mathrm{d}C_L} \tag{3.201}$$

and

$$\frac{\mathrm{d}n}{\mathrm{d}C_L} = \left(\frac{1}{C_L}\right) \tag{3.202}$$

we obtain

$$\frac{\mathrm{d}\delta_e}{\mathrm{d}n} = -\frac{C_L(\bar{x}_{cg} - N_m)}{C_{m\delta}} \tag{3.203}$$

Usually, for a stable airplane, $\bar{x}_{cg} < N_m$ and $C_{m\delta} < 0$ so that we have $\mathrm{d}\delta_e/\mathrm{d}n < 0$. Thus, to pull higher load factors during the maneuver, more up elevator deflection is required compared to that in level flight at the same altitude and forward speed. This incremental elevator deflection can be expressed as

$$\Delta\delta_e = -\frac{C_L(\overline{x}_{cg} - N_m)\Delta n}{C_{m\delta}} \tag{3.204}$$

Stick-Free Maneuver Point

Suppose that the elevator is left free to float during a pull-up maneuver. Then, the horizontal tail angle of attack is given by

$$\alpha_t = (\alpha_w - i_w + i_t - \epsilon + \Delta\alpha_{i,t})\left(1 - \tau\frac{C_{h\alpha}}{C_{h\delta,e}}\right) \tag{3.205}$$

so that

$$\frac{d\alpha_t}{dC_L} = \left[\frac{1}{a_w}\left(1 - \frac{d\epsilon}{d\alpha}\right) + \frac{1}{2\mu_1}\right]\left(1 - \tau\frac{C_{h\alpha}}{C_{h\delta,e}}\right) \tag{3.206}$$

Then,

$$\left(\frac{dC_m}{dC_L}\right)_{m'} = \overline{x}_a + \left(\frac{dC_m}{dC_L}\right)_f - \frac{a_t\overline{V}_1\eta_t}{a_w} \times \left[\left(1 - \frac{d\epsilon}{d\alpha}\right) + \frac{a_w}{2\mu_1}\right]\left(1 - \tau\frac{C_{h\alpha}}{C_{h\delta,e}}\right) \tag{3.207}$$

$$N'_m = \overline{x}_{ac,w} - \left(\frac{dC_m}{dC_L}\right)_f + \frac{a_t\overline{V}_1\eta_t}{a_w} \times \left[\left(1 - \frac{d\epsilon}{d\alpha}\right) + \frac{a_w}{2\mu_1}\right]\left(1 - \tau\frac{C_{h\alpha}}{C_{h\delta,e}}\right) \tag{3.208}$$

We observe that the stick-free maneuver stability is higher than the stick-free stability in level flight and the stick-free maneuver point is aft of the stick-free neutral point. This increase in stick-free stability during the maneuver, as said before, is caused by the angular velocity Ω and the associated increase in tail angle of attack $\Delta\alpha_{i,t}$. The stick-free maneuver margin is given by

$$H'_m = N'_m - \overline{x}_{cg} \tag{3.209}$$

For a stable aircraft, $H'_m > 0$.

Stick-Force Gradient

As we know, the stick force is directly proportional to $\Delta\delta_e = \delta_{e,R} - \delta_{e,f}$, the difference between the required elevator deflection and the floating angle of the elevator. Recall that these two quantities are also related to the stick-fixed and the stick-free stability levels of the airplane, respectively. Assuming $\delta_{e,o} = 0$ and using Eqs. (3.104) and (3.199), we get

$$\delta_{e,R} = -\frac{C_L}{C_{m\delta}}\left(\frac{dC_m}{dC_L}\right)_m \tag{3.210}$$

Assuming $C_{ho} = 0$, the floating angle of the elevator is given by [Eq. (3.120)]

$$\delta_{e,f} = -\frac{C_{h\alpha}}{C_{h\delta,e}}\alpha_s \tag{3.211}$$

$$= -\frac{C_{h\alpha}}{C_{h\delta,e}}\left[\frac{C_L}{a_w}\left(1 - \frac{d\epsilon}{d\alpha}\right) + \frac{\Omega l_t}{V}\right] \tag{3.212}$$

so that

$$\Delta\delta e = \delta_{e,R} - \delta_{e,f} \tag{3.213}$$

$$= -\frac{C_L}{C_{m\delta}}\left(\frac{dC_m}{dC_L}\right)_m + \frac{C_{h\alpha}}{C_{h\delta,e}}\left[\frac{C_L}{a_w}\left(1 - \frac{d\epsilon}{d\alpha}\right) + \frac{\Omega l_t}{V}\right] \tag{3.214}$$

Differentiating with respect to C_L and rearranging, we obtain

$$\frac{d\Delta\delta_e}{dC_L} = -\frac{1}{C_{m\delta}}\left(\frac{dC_m}{dC_L}\right)_{m'} \tag{3.215}$$

In incremental form, we can write Eq. (3.215) as

$$\Delta\delta_e = -\frac{\Delta C_L}{C_{m\delta}}\left(\frac{dC_m}{dC_L}\right)_{m'} \tag{3.216}$$

The corresponding increment in hinge-moment coefficient is given by

$$\Delta C_h = C_{h\delta,e}\Delta\delta e \tag{3.217}$$

$$= -\frac{C_{h\delta,e}\Delta C_L}{C_{m\delta}}\left(\frac{dC_m}{dC_L}\right)_{m'} \tag{3.218}$$

The incremental stick force, i.e., the additional stick force above that required in level flight, is given by

$$\Delta F_s = -G_1\Delta HM \tag{3.219}$$

$$= -G_1 q_t S_e \bar{c}_e \Delta C_h \tag{3.220}$$

$$= G_1 q\eta_t S_e \bar{c}_e \Delta C_L \frac{C_{h\delta,e}}{C_{m\delta}}\left(\frac{dC_m}{dC_L}\right)_{m'} \tag{3.221}$$

where G_1 is the gearing ratio between the stick and the elevator. Using $\Delta C_L = \Delta n C_L$, we obtain the stick force per g as follows:

$$\frac{dF_s}{dn} = G_1\left(\frac{W}{S}\right)\eta_t S_e \bar{c}_e \frac{C_{h\delta,e}}{C_{m\delta}}\left(\frac{dC_m}{dC_L}\right)_{m'} \tag{3.222}$$

Usually, $C_{h\delta,e} < 0$ and $C_{m\delta} < 0$. For a stable airplane, $(dC_m/dC_L)_{m'} < 0$. Then, $dF_s/dn < 0$, i.e., to pull a load factor above unity, for a stable airplane, the pilot has to apply additional negative stick force (pull) as given by Eq. (3.222).

3.4.2 Turning Flight in Horizontal Plane

Consider a level coordinated turning maneuver in a horizontal plane as shown in Fig. 3.64. Let O be the center of the turn and Ω be the angular velocity about a vertical axis passing through O. The angular velocity vector Ω can be resolved along the body y and z axes as

$$\Omega_y = \Omega \sin\phi \tag{3.223}$$

$$\Omega_z = -\Omega \cos\phi \tag{3.224}$$

where $\Omega = V/R$ and R is the radius of turn. The component Ω_y is equivalent to a pitch rate q about the body y axis. This apparent pitching motion gives rise to an induced velocity ql_t at the horizontal tail (see Fig. 3.65) resulting in an increase in the horizontal tail angle of attack.

$$\Delta\alpha_{i,t} = \frac{ql_t}{V} \tag{3.225}$$

$$= \frac{\Omega \sin\phi l_t}{V} \tag{3.226}$$

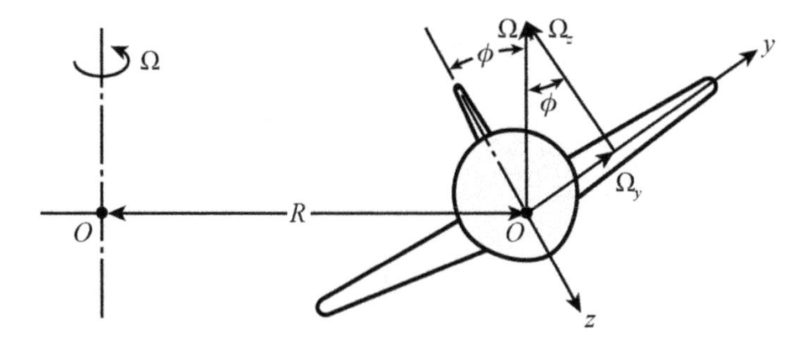

Fig. 3.64 Airplane in a coordinated turn.

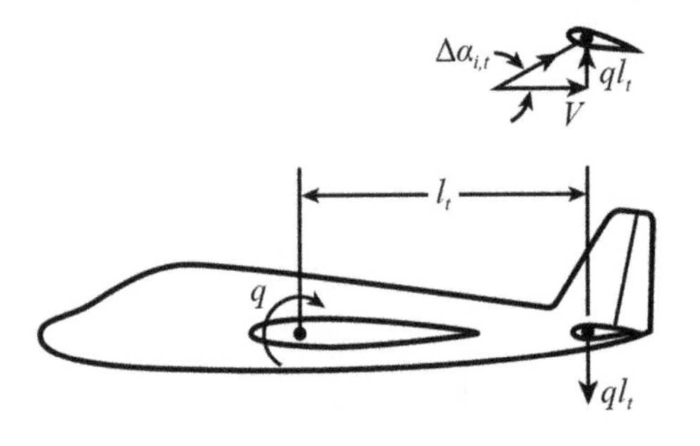

Fig. 3.65 Induced angle of attack at horizontal tail due to pitch rate in a coordinated turn.

From Chapter 2, we have

$$R = \frac{V^2}{g\sqrt{n^2 - 1}} \tag{3.227}$$

$$\sin\phi = \frac{\sqrt{n^2 - 1}}{n} \tag{3.228}$$

$$\Omega = \frac{g\sqrt{n^2 - 1}}{V} \tag{3.229}$$

Then,

$$\Delta\alpha_{i,t} = \frac{l_t g}{V^2}\left(n - \frac{1}{n}\right) \tag{3.230}$$

Proceeding in a similar fashion as we did before for the case of a pull-up in a vertical plane, we obtain

$$\left(\frac{\mathrm{d}\alpha_{i,t}}{\mathrm{d}C_L}\right)_{\text{turn}} = \frac{1}{2\mu_1}\left(1 + \frac{1}{n^2}\right) \tag{3.231}$$

$$\left(\frac{\mathrm{d}C_m}{\mathrm{d}C_L}\right)_{\text{turn}} = \bar{x}_a + \left(\frac{\mathrm{d}C_m}{\mathrm{d}C_L}\right)_f - \frac{a_t\overline{V}_1\eta_t}{a_w}\left[\left(1 - \frac{\mathrm{d}\epsilon}{\mathrm{d}\alpha}\right) + \frac{a_w}{2\mu_1}\left(1 + \frac{1}{n^2}\right)\right] \tag{3.232}$$

$$N_{m,\text{turn}} = \bar{x}_{ac,w} - \left(\frac{\mathrm{d}C_m}{\mathrm{d}C_L}\right)_f + \frac{a_t\overline{V}_1\eta_t}{a_w}\left[\left(1 - \frac{\mathrm{d}\epsilon}{\mathrm{d}\alpha}\right) + \frac{a_w}{2\mu_1}\left(1 + \frac{1}{n^2}\right)\right] \tag{3.233}$$

$$\left(\frac{\mathrm{d}\delta_e}{\mathrm{d}n}\right)_{\text{turn}} = -\frac{C_L\left(\bar{x}_{cg} - N_{m,\text{turn}}\right)}{C_{m\delta}} \tag{3.234}$$

$$N'_{m,\text{turn}} = \bar{x}_{ac,w} - \left(\frac{\mathrm{d}C_m}{\mathrm{d}C_L}\right)_f + \frac{a_t \overline{V}_1 \eta_t}{a_w}\left[\left(1 - \frac{\mathrm{d}\epsilon}{\mathrm{d}\alpha}\right) + \frac{a_w}{2\mu_1}\left(1 + \frac{1}{n^2}\right)\right]$$
$$\times \left(1 - \tau \frac{C_{h\alpha}}{C_{h\delta,e}}\right) \tag{3.235}$$

$$\left(\frac{\mathrm{d}F_s}{\mathrm{d}n}\right)_{\text{turn}} = G_1\left(\frac{W}{S}\right)\eta_t S_e \bar{c}_e \frac{C_{h\delta,e}}{C_{m\delta}}\left(\frac{\mathrm{d}C_m}{\mathrm{d}C_L}\right)_{m',\text{turn}} \tag{3.236}$$

Example 3.7

A trainer aircraft is initially in a steady level flight at an altitude of 5000 m ($\sigma = 0.6$) at a forward speed of 100.0 m/s. The aircraft then climbs to 6000 m at which it enters into a dive and recovers approximately in a semi-circular path so that, at the bottom of the pull-out, it is once again at 5000 m altitude.

Determine 1) the elevator setting above that required for initial level flight, and 2) the stick force per g, based on the following data: wing loading = 1500 N/m^2, $\bar{x}_{ac,w} = 0.24$, $\bar{x}_{cg} = 0.25$, $C_{mf} = 0.15\ C_L$, $a_w = 0.1$ per deg, $a_t = 0.08$ per deg, $\tau = 0.5$, $l_t = 8$ m, $\overline{V}_1 = 0.6$, $\epsilon = 0.35\alpha$, $C_{h\alpha} = -0.003$/deg, $C_{h\delta,e} = -0.006$/deg, $\eta_t = 0.9$, $G_1 = 1.0$ rad/m, $S_e = 2.0$ m^2, and $\bar{c}_e = 0.6$ m.

Solution:

From the information given, we find that $R = 1000$ m. Then,

$$\begin{aligned}
C_L &= \frac{2W}{\rho V^2 S} \\
&= \frac{2 * 1500}{1.225 * 0.6 * 100^2} \\
&= 0.4 \\
\mu_1 &= \frac{W}{\rho g S l_t} \\
&= \frac{1500}{1.225 * 0.6 * 9.81 * 8} \\
&= 25.4842 \\
C_{m\delta} &= -a_t \overline{V}_1 \eta_t \tau \\
&= -0.08 * 0.6 * 0.9 * 0.5 \\
&= -0.0216
\end{aligned}$$

(*Continued*)

Example 3.7 (*Continued*)

$$N_m = \bar{x}_{ac,w} - \left(\frac{dC_m}{dC_L}\right)_f + \frac{a_t \overline{V}_1 \eta_t}{a_w}\left[\left(1 - \frac{d\epsilon}{d\alpha}\right) + \frac{a_w}{2\mu_1}\right]$$

$$= 0.24 - 0.15 + \left(\frac{0.08 * 0.6 * 0.9}{0.1}\right)\left[1 - 0.35 + \frac{0.1 * 57.3}{2 * 25.4842}\right]$$

$$= 0.4193$$

$$\frac{\Delta\delta_e}{\Delta n} = -\frac{C_L(\bar{x}_{cg} - N_m)}{C_{m\delta}}$$

$$= -\frac{0.4(0.25 - 0.4193)}{-0.0216}$$

$$= -3.1352$$

$$\Delta n = \frac{V^2}{Rg}$$

$$= \frac{100^2}{1000 * 9.81}$$

$$= 1.0194$$

$$\Delta\delta_e = -3.1352 * 1.0194$$

$$= -3.1960 \,\text{deg}$$

Now let us calculate the stick force per g as follows:

$$N'_m = \bar{x}_{ac,w} - \left(\frac{dC_m}{dC_L}\right)_f + \frac{a_t \overline{V}_1 \eta_t}{a_w}\left[\left(1 - \frac{d\epsilon}{d\alpha}\right) - \frac{a_w}{2\mu_1}\right]\left(1 - \tau\frac{C_{h\alpha}}{C_{h\delta,e}}\right)$$

$$= 0.24 - 0.15 + \left(\frac{0.08 * 0.6 * 0.9}{0.1}\right)\left(1 - 0.35 - \frac{0.1 * 57.3}{2 * 25.4842}\right)$$

$$\times \left[1 - 0.5\frac{-0.003}{-0.006}\right]$$

$$= 0.3369$$

$$\left(\frac{dC_m}{dC_L}\right)_{m'} = \bar{x}_{cg} - N'_m$$

$$= 0.25 - 0.3369$$

$$= -0.0869$$

$$\frac{dF_s}{dn} = G_1\left(\frac{W}{S}\right)\eta_t S_e \bar{c}_e \frac{C_{h\delta,e}}{C_{m\delta}}\left(\frac{dC_m}{dC_L}\right)_{m'}$$

$$= 1.0 * 1500 * 0.9 * 2.0 * 0.6 * \left(\frac{-0.006}{-0.0216}\right)(-0.0869)$$

$$= 39.105 \,\text{N/g}$$

3.5 Static Directional Stability

In Sec. 3.3, we considered the airplane stability with respect to a disturbance in angle of attack. This disturbance was assumed to be contained in a vertical plane. We assumed that the airplane response in pitch was sufficiently slow so that the pitch rate q could be ignored. In Sec. 3.4, we extended this approach to the study of airplane stability during simple maneuvers. In this section, we will study the airplane stability with respect to a disturbance in sideslip, which is contained in the horizontal plane. Once again, we will assume that the response of the airplane is sufficiently slow so that yaw rate r and roll rate p can be ignored.

The nomenclature used to describe the motion involving the six degrees of freedom is presented in Fig. 3.66. The longitudinal motion of the aircraft involves forward velocity u, vertical velocity w, pitch angle θ, and pitch rate q. Sometimes the longitudinal variables u, w, θ, and q are called symmetric degrees of freedom. The directional motion involves sideslip velocity v, yaw angle ψ, and yaw rate r, and the lateral motion involves bank angle ϕ and roll rate p. However, the lateral and the directional degrees of freedom are always coupled because a sideslip induces both rolling and yawing motions. Similarly, a yawing motion induces both rolling motion and sideslip. In this section, we will study the directional stability of the aircraft with respect to a disturbance in sideslip, and we will study lateral stability in the next section.

Static directional stability is a measure of the aircraft's ability to realign itself along the direction of the resultant wind so that the disturbance in sideslip is effectively eliminated. A disturbance in sideslip could be caused by horizontal gust, wind turbulence, or momentary (small) rudder deflection. Therefore, on encountering a disturbance in the horizontal plane, the aircraft

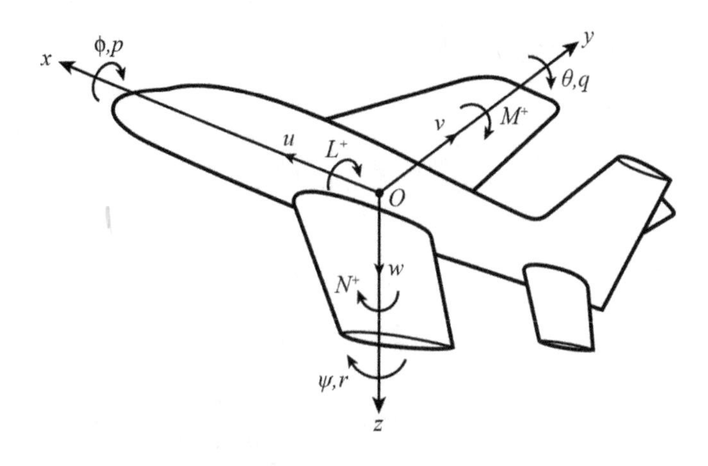

Fig. 3.66 Axis system and nomenclature used in static stability analysis.

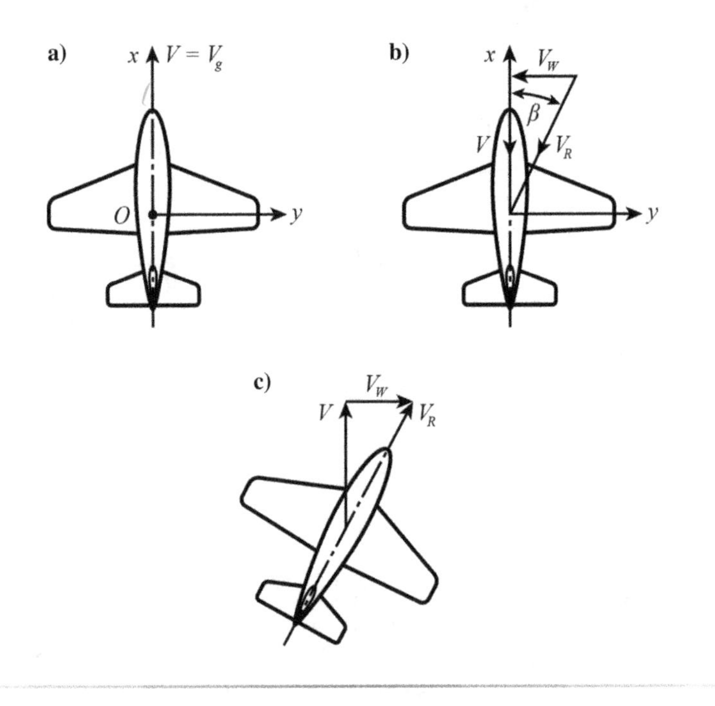

Fig. 3.67 Aircraft orientation in horizontal plane.

orientation in space changes but its heading remains the same as before with respect to the Earth. These concepts are illustrated in Fig. 3.67. The aircraft is in a steady (undisturbed) level flight in Fig. 3.67a, encounters a horizontal gust V_W blowing from starboard side resulting in a sideslip as shown in Fig. 3.67b, and realigns itself with the resultant velocity V_R in Fig. 3.67c. Now, the sideslip is zero but the aircraft orientation in space has changed. Notice that in Fig. 3.67c the aircraft is still moving with the same velocity V with respect to the Earth as before. If the disturbance vanishes, the aircraft orientation will also be restored.

The angles of sideslip and yaw (Fig. 3.68) are two important parameters in the study of directional stability. Both of these angles are measured in a horizontal plane. The angle of sideslip is an aerodynamic angle defined as the angle between the velocity vector and the airplane's plane of symmetry as shown in Fig. 3.68b. The angle of sideslip is usually denoted by β and is given by

$$\sin \beta = \frac{v}{V} \tag{3.237}$$

where v is the sideslip velocity in the y direction and V is the flight velocity.

The usual sign convention is to assume β positive if the airplane sideslips to starboard (right wing leading into sideslip) as shown in Fig. 3.68b and β negative if the airplane sideslips towards port side (left wing leading into sideslip as shown in Fig. 3.68c).

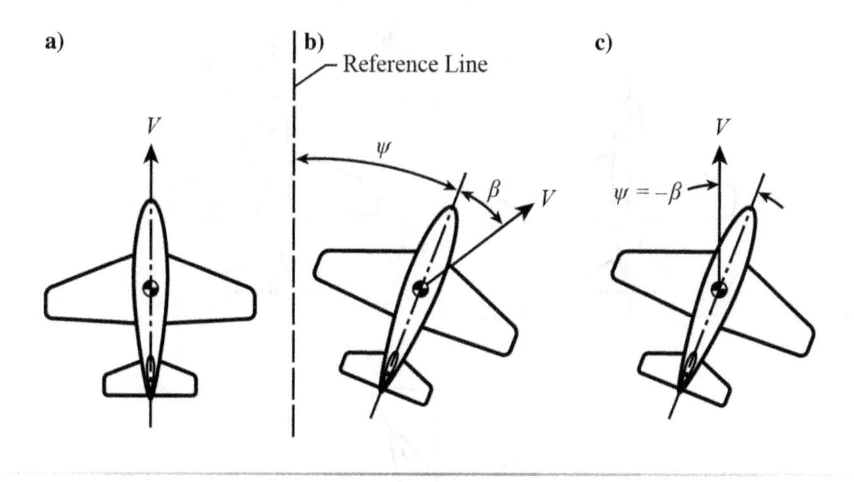

Fig. 3.68 Definitions of angles of yaw and sideslip.

The angle of yaw, usually denoted by ψ, is a kinematic angle and is a measure of the change in the heading or orientation of the aircraft relative to Earth. It is the angle between the airplane's plane of symmetry and a reference plane fixed in space as shown in Fig. 3.68b. Usually, this reference plane is assumed to coincide with the airplane's plane of symmetry when the airplane is in steady level flight before it encounters the disturbance. In principle, the angles of sideslip and yaw are independent of each other. However, in the special case when the direction of motion remains unchanged but the airplane is yawed, the sideslip and yaw are related by the relation $\psi = -\beta$ as shown in Fig. 3.68c. Because the angle of yaw is a kinematic angle, the aerodynamic forces and moments do not depend on ψ except for this special case, which is usually encountered in wind-tunnel testing. On account of this, wind-tunnel test data are sometimes presented as a function of ψ.

3.5.1 Criterion of Directional Stability

An airplane is said to be directionally stable if it has an inherent capability to realign itself into the resultant wind whenever disturbed from steady level flight. Mathematically, this requirement for directional stability can be expressed as follows:

$$N_\beta > 0 \tag{3.238}$$

or, in coefficient form,

$$C_{n\beta} > 0 \tag{3.239}$$

Here, N is the yawing moment and

$$C_n = \frac{N}{qSb} \tag{3.240}$$

$$C_{n\beta} = \frac{\partial C_n}{\partial \beta} \tag{3.241}$$

The concept of directional stability is schematically illustrated in Fig. 3.69.

3.5.2 Evaluation of Static Directional Stability

As done before for the longitudinal static stability, we assume that the $C_{n\beta}$ of the airplane is the sum of individual contributions caused by fuselage, wing, and tail surfaces. A brief discussion on the qualitative effects of power is presented but ignored in the analysis.

Wing Contribution

The wing contribution to directional stability mainly depends on its dihedral and leading-edge sweep. Generally, the magnitude of wing contribution to static directional stability is small. If the wing has no dihedral and not much sweep, its contribution to directional stability can be ignored.

For a swept wing with dihedral, we can assume the total wing contribution to be given by

$$(C_{n\beta})_W = (C_{n\beta})_{\Gamma,W} + (C_{n\beta})_{\Lambda,W} \tag{3.242}$$

We will discuss these effects in the following sections.

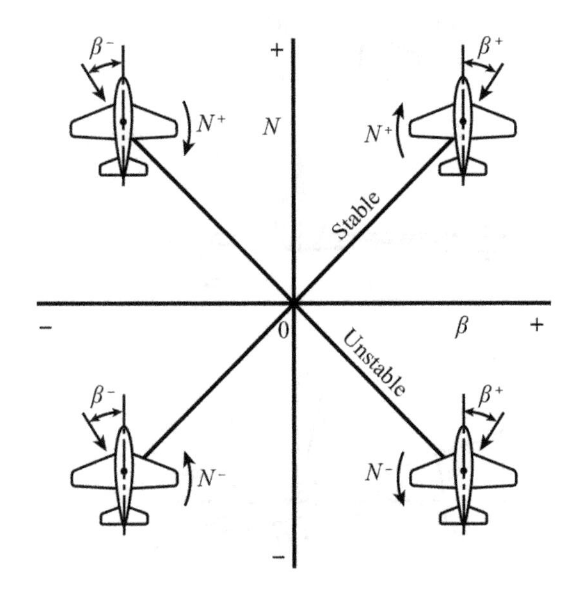

Fig. 3.69 Static directional stability.

Effect of Wing Dihedral

To understand the effect of wing dihedral on static directional stability, consider an unswept, high-aspect ratio rectangular wing with a constant dihedral angle Γ operating at an angle of attack α and sideslip β at a forward velocity V_o as shown in Fig. 3.70. The sideslip velocity $v = V_o\beta$ can be resolved into a spanwise component $V_o\beta \cos\Gamma \simeq V_o\beta$ and a normal component $V_o\beta \sin\Gamma \simeq V_o\beta\Gamma$ as shown in Fig. 3.70b. The spanwise component $V_o\beta$ does not affect the pressure distribution, hence the lift. It only affects the skin-friction drag.

The total component of velocity normal to the plane of wings is given by

$$V_N = V_o(\sin\alpha \pm \beta\sin\Gamma) \tag{3.243}$$

or, assuming α, β, and Γ to be small,

$$V_N = V_o(\alpha \pm \beta\Gamma) \tag{3.244}$$

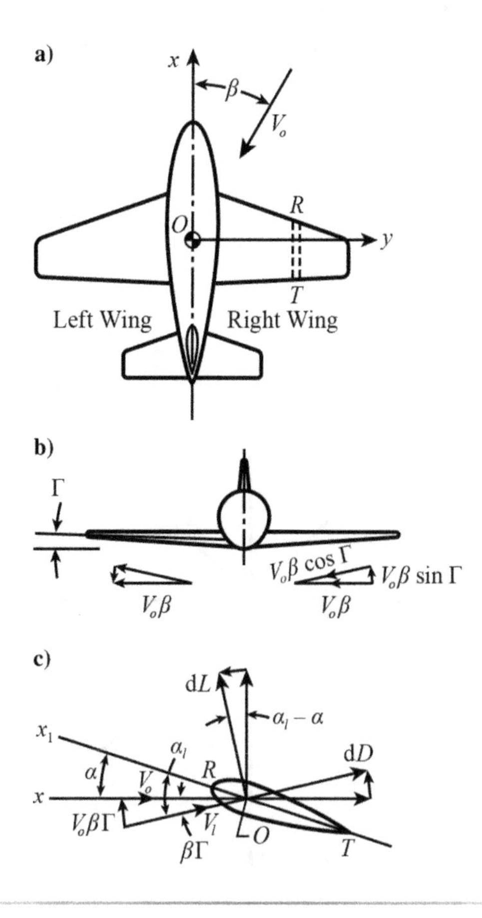

Fig. 3.70 Strip theory analysis of wing dihedral effect.

Note that "+" and "−" signs apply respectively to the right (starboard) and left (port side) wings when sideslip is positive. The chordwise component of velocity is given by

$$V_c = V_o \cos\alpha \tag{3.245}$$

$$= V_o \tag{3.246}$$

The local angle of attack and local dynamic pressure are given by

$$\alpha_l = \frac{V_N}{V_c} \tag{3.247}$$

$$= \alpha \pm \beta\Gamma \tag{3.248}$$

$$q_l = \frac{1}{2}\rho\left(V_N^2 + V_C^2\right) \tag{3.249}$$

$$\simeq \frac{1}{2}\rho V_o^2 \tag{3.250}$$

In the given equations, higher order terms like α^2 and $\alpha\beta$ are ignored. Thus, we observe that for a positive sideslip, the local dynamic pressure on both wings is approximately equal to the freestream dynamic pressure. Furthermore, the leading (starboard) wing experiences an increase in angle of attack and, therefore, an increase in lift and drag coefficients. The port wing experiences opposite effects. As a result of this imbalance in spanwise lift distribution, the wing develops a rolling moment, which we will consider later while studying lateral stability. The imbalance in drag forces gives rise to a yawing moment. For low subsonic speeds, we can approximately estimate this yawing moment using the simple strip theory as discussed in the following.

In the strip theory approach, the wing is divided into a number of spanwise elements or strips. The aerodynamic forces on each strip are calculated assuming that it is a part of a two-dimensional wing having an identical airfoil section as that of the given strip. This concept is illustrated in Fig. 3.71. In other words, the strip theory ignores the downwash (induced angle of attack) variation along the span. This amounts to ignoring the induced drag of the strip and an overestimation of the sectional lift-curve slope. In view of this, the estimations based on strip theory are at best only first approximations. Nevertheless, this approach is quite useful to get an idea of the variables that can have significant influence on the aerodynamic parameter of interest.

Let $c(y)$ be the local chord, $a_o(y)$ be the local sectional lift-curve slope, and V_l be the local resultant velocity of the strip RT on the right wing (see Fig. 3.70c). We assume that the angles α, β, and Γ are small so that their products can be ignored. Here, we resolve the sectional lift ($\mathrm{d}L$) and drag ($\mathrm{d}D$) forces along the stability axis system because all the aerodynamic (stability) derivatives like $C_{n\beta}$ are normally referred to the stability axis

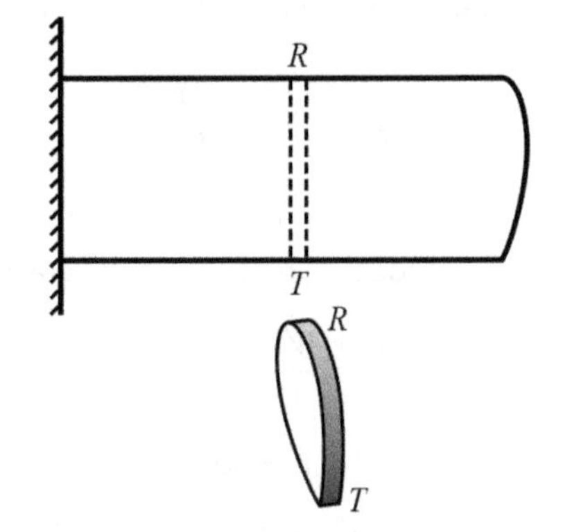

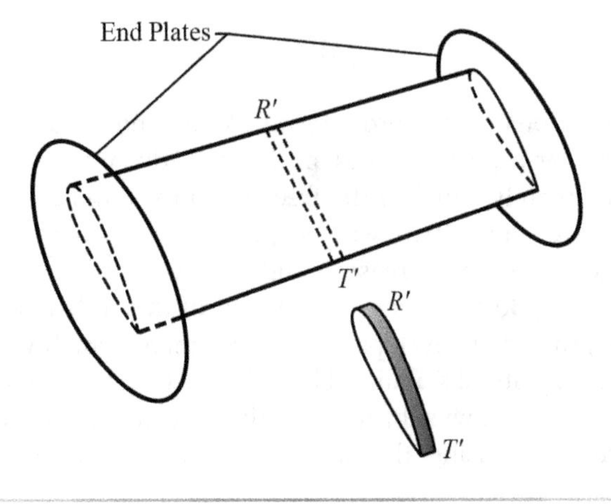

Fig. 3.71 Strip theory.

system (see Chapter 4 for information on various axes systems used in aircraft dynamics). Here, Oxy is the stability axis system with origin at O. The z stability axis is not shown in Fig. 3.70. With this understanding, the component of force along Ox for the right wing strip RT of width dy (Figs. 3.70a and 3.70c) is given by

$$dF = dL \sin(\alpha_l - \alpha) - dD \cos(\alpha_l - \alpha) \tag{3.251}$$

$$= dL(\alpha_l - \alpha) - dD \tag{3.252}$$

With $\alpha_l - \alpha = \beta\Gamma$, we have

$$dF = dL\beta\Gamma - dD \tag{3.253}$$

$$= \frac{1}{2}\rho V_o^2 c(y) dy (C_{l,R}\beta\Gamma - C_{Dl,R}) \tag{3.254}$$

where $C_{l,R}$ is the local lift coefficient of the right wing.

In strip theory, the sectional lift and drag coefficients are assumed to be given by

$$C_l = a_o \alpha_l \tag{3.255}$$

$$C_D = C_{Do,l} + C_{D\alpha,l}\alpha \tag{3.256}$$

where a_o is the sectional (two-dimensional) lift-curve slope, $C_{Do,l}$ is the sectional zero-lift drag coefficient, and $C_{D\alpha,l}$ is the increase in sectional drag coefficient per unit increase in angle of attack. Note that this increase in sectional drag coefficient with angle of attack above $C_{Do,l}$ is caused by an increase in the profile drag coefficient and not caused by the induced drag coefficient. As said earlier, strip theory ignores induced drag.

Then,

$$\mathrm{d}F = \frac{1}{2}\rho V_o^2 c(y)\,\mathrm{d}y\left[C_{l,R}\beta\Gamma - C_{Do,l} - C_{D\alpha,l}(\alpha + \beta\Gamma)\right] \tag{3.257}$$

The yawing moment due to the strip RT on the right wing is given by

$$\mathrm{d}N = -y\mathrm{d}F \tag{3.258}$$

$$= \frac{1}{2}\rho V_o^2\left[-C_{l,R}\beta\Gamma + C_{Do,l} + C_{D\alpha,l}(\alpha + \beta\Gamma)\right]c(y)y\,\mathrm{d}y \tag{3.259}$$

The yawing moment caused by the right (starboard) wing is given by

$$N_R = \frac{1}{2}\rho V_o^2 \int_0^{b/2} [-C_{l,R}\beta\Gamma + C_{Do,l} + C_{D\alpha,l}(\alpha + \beta\Gamma)]c(y)y\,\mathrm{d}y \tag{3.260}$$

Similarly, the yawing moment caused by the left (port) wing is given by

$$N_L = \frac{1}{2}\rho V_o^2 \int_0^{b/2} [-C_{l,L}\beta\Gamma + C_{Do,l} - C_{D\alpha,l}(\alpha - \beta\Gamma)]c(y)y\,\mathrm{d}y \tag{3.261}$$

where $C_{l,L}$ is the local (sectional) lift coefficient of the left wing.

The net or total yawing moment is the sum of the yawing moments due to the right and left wings and is given by

$$N = \frac{1}{2}\rho V_o^2 \int_0^{b/2} [-\beta\Gamma\left(C_{l,R} + C_{l,L}\right) + 2C_{D\alpha,l}\beta\Gamma]c(y)y\mathrm{d}y \tag{3.262}$$

We have

$$C_{l,R} = a_o(\alpha + \beta\Gamma) \tag{3.263}$$

$$C_{l,L} = a_o(\alpha - \beta\Gamma) \tag{3.264}$$

so that

$$C_{l,R} + C_{l,L} = 2a_o\alpha \tag{3.265}$$

$$= 2C_L \tag{3.266}$$

Then,

$$N = -\rho V_o^2 \beta \Gamma \int_0^{b/2} (C_L - C_{D\alpha,l}) c(y) y \, dy \tag{3.267}$$

or, in coefficient form,

$$(C_n)_{\Gamma,W} = -\frac{2\beta\Gamma}{Sb} \int_0^{b/2} (C_L - C_{D\alpha,l}) c(y) y \, dy \tag{3.268}$$

and

$$(C_{n\beta})_{\Gamma,W} = -\frac{2\Gamma}{Sb} \int_0^{b/2} (C_L - C_{D\alpha,l}) c(y) y \, dy \tag{3.269}$$

For a rectangular wing with a constant chord c, the above expression reduces to

$$(C_{n\beta})_{\Gamma,W} = -\frac{\Gamma(C_L - C_{D\alpha,l})}{4} \tag{3.270}$$

Usually for steady level flight conditions, $C_L > C_{D\alpha,l}$ so that $(C_{n\beta})_{\Gamma,W} < 0$, which implies that the wing contribution to directional stability due to dihedral angle is destabilizing. This destabilizing effect increases with lift coefficient and can become significant at high angles of attack. Conversely, the wing unhedral has a stabilizing effect on directional stability.

The simple strip theory analysis, even though very approximate, has given us important information that the wing dihedral has a destabilizing effect on directional stability and this effect is small at low angles of attack but may become significant at high angles of attack or high lift coefficients.

It may be recalled that the strip theory ignores the induced drag effects; hence its predictions will be increasingly in error as the wing aspect ratio decreases. For such cases, the following empirical formula [9] may be used for preliminary estimations at low subsonic speeds:

$$(C_{n\beta})_{\Gamma,W} = -0.075\,\Gamma C_L/\text{rad} \tag{3.271}$$

where the dihedral angle Γ is in radians. For supersonic speeds, no general method is available for estimation of the wing contribution to directional stability due to dihedral effect [1]. According to Datcom [1], this contribution is generally small and can be ignored.

Effect of Sweep

The wing sweep-back has a stabilizing effect on static directional stability. To understand this, let us consider a swept-back wing of sufficiently high aspect ratio and with zero dihedral operating at an angle of attack α and sideslip β and moving at a forward velocity V_o as shown in Fig. 3.72. The velocity components in the spanwise, chordwise (along RT), and normal directions are given by

$$V_s = V_o(\sin\Lambda \cos\alpha \mp \beta \cos\Lambda) \tag{3.272}$$

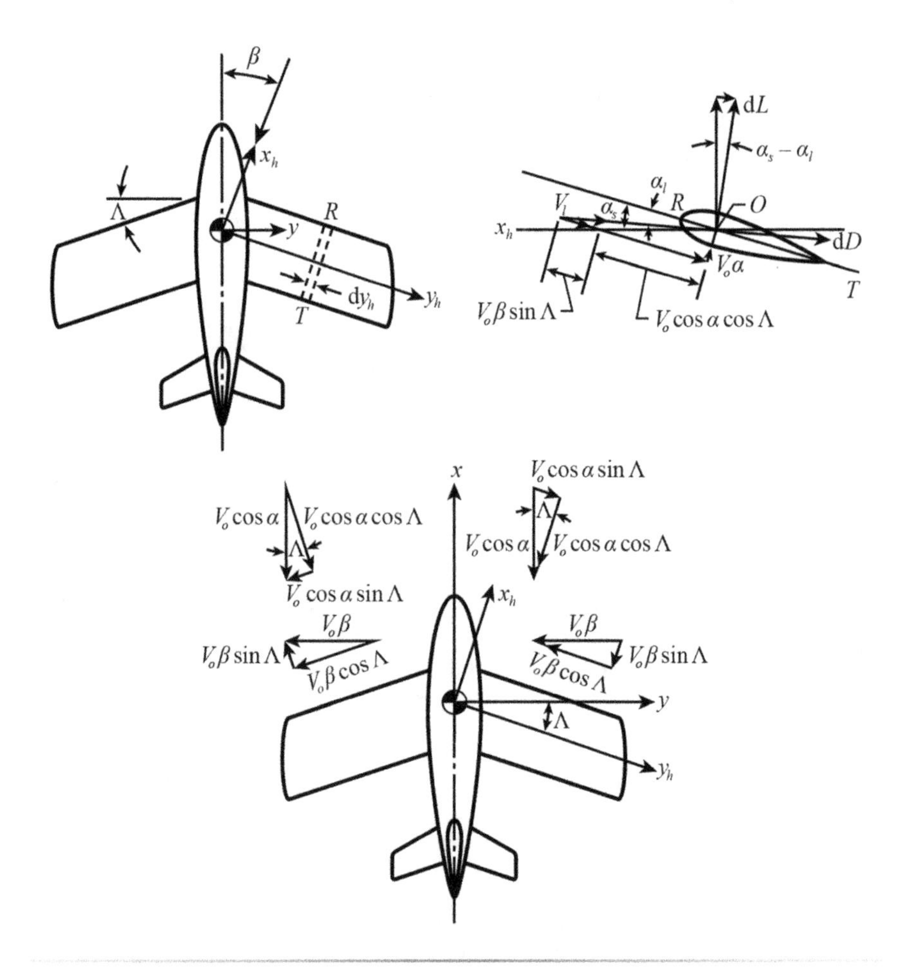

Fig. 3.72 Strip theory analysis of wing sweep effect.

$$V_c = V_o(\cos\Lambda\cos\alpha \pm \beta\sin\Lambda) \tag{3.273}$$

$$= V_o\cos\Lambda(\cos\alpha \pm \beta\tan\Lambda) \tag{3.274}$$

$$V_N = V_o\sin\alpha \tag{3.275}$$

where the first (upper) sign refers to the right wing and the second (lower) sign refers to the left wing when the wing is in positive sideslip.

As before, let us assume that both α and β are small so that higher order terms involving these two parameters can be ignored. With these assumptions, we have

$$V_s = V_o\cos\Lambda(\tan\Lambda \mp \beta) \tag{3.276}$$

$$V_c = V_o\cos\Lambda(1 \pm \beta\tan\Lambda) \tag{3.277}$$

$$V_N = V_o\alpha \tag{3.278}$$

As said before, the spanwise component of velocity does not affect the pressure distribution and hence is ignored in the following calculations. It only adds to skin friction.

The angle of attack of right (starboard) and left (port) wings are given by

$$\alpha_l = \tan\alpha_l = \frac{V_N}{V_c} \tag{3.279}$$

$$= \frac{\alpha}{\cos\Lambda(1 \pm \beta\tan\Lambda)} \tag{3.280}$$

$$= \alpha\sec\Lambda(1 \mp \beta\tan\Lambda) \tag{3.281}$$

For level flight with zero sideslip, the local angle of attack is given by

$$\alpha_s = \alpha\sec\Lambda \tag{3.282}$$

so that

$$\alpha_s - \alpha_l = \pm\alpha\sec\Lambda\beta\tan\Lambda \tag{3.283}$$

The dynamic pressures experienced by the right and left wings are given by

$$q_l = \frac{1}{2}\rho\left(V_c^2 + V_N^2\right) \tag{3.284}$$

$$= \frac{1}{2}\rho V_o^2\cos^2\Lambda(1 \pm \beta\tan\Lambda)^2 \tag{3.285}$$

Here, we have ignored higher terms such as α^2, $\alpha\beta$, or β^2. For low subsonic speeds, we can use the strip theory to estimate the wing contribution to directional stability due to sweep-back as follows [10].

Consider a strip RT of width $\mathrm{d}y_h$ on the right wing. Let y_h denote the spanwise coordinate along the quarter chordline and $c(y_h)$ denote the local chord normal to the wing leading edge. The component of force along the stability axis for the right wing strip RT is given by

$$\mathrm{d}F = -\mathrm{d}L\sin(\alpha_s - \alpha_l) - \mathrm{d}D\cos(\alpha_s - \alpha_l) \tag{3.286}$$

$$= -\mathrm{d}L(\alpha_s - \alpha_l) - \mathrm{d}D \tag{3.287}$$

Substituting for $\alpha_s - \alpha_l$ from Eq. (3.283) we have

$$\mathrm{d}F = -\mathrm{d}L\alpha\sec\Lambda\beta\tan\Lambda - \mathrm{d}D \tag{3.288}$$

$$= -\frac{1}{2}\rho V_o^2\cos^2\Lambda(1 + \beta\tan\Lambda)^2\Bigg[C_{l,R}\alpha\sec\Lambda\beta\tan\Lambda + C_{Do,l} + C_{D\alpha,l}\left(\frac{\alpha\sec\Lambda}{1 + \beta\tan\Lambda}\right)\Bigg]c(y_h)\,\mathrm{d}y_h \tag{3.289}$$

With

$$C_{l,R} = \frac{a_o \alpha \sec \Lambda}{1 + \beta \tan \Lambda} \quad (3.290)$$

where $a_o = a_o(y)$ is the sectional lift-curve slope, we have

$$\begin{aligned} dF = &-\frac{1}{2}\rho V_o^2 \cos^2 \Lambda (1 + \beta \tan \Lambda) \\ &\times \left[a_o \alpha^2 \sec^2 \Lambda \beta \tan \Lambda + C_{Do,l}(1 + \beta \tan \Lambda) + C_{D\alpha,l} \alpha \sec \Lambda\right] \\ &\times c(y_h) dy_h \end{aligned} \quad (3.291)$$

The yawing moment caused by the right wing is given by

$$N_R = -y_h dF \quad (3.292)$$

$$\begin{aligned} &= \frac{1}{2}\rho V_o^2 \cos^2 \Lambda (1 + \beta \tan \Lambda) \int_0^{\frac{b\sec\Lambda}{2}} \left[a_o \alpha^2 \sec^2 \Lambda \beta \tan \Lambda \right. \\ &\left. + C_{Do,l}(1 + \beta \tan \Lambda) + C_{D\alpha,l} \alpha \sec \Lambda\right] c(y_h) y_h \, dy_h \end{aligned} \quad (3.293)$$

Similarly, the yawing moment caused by the left wing is given by

$$\begin{aligned} N_L = &\frac{1}{2}\rho V_o^2 \cos^2 \Lambda (1 - \beta \tan \Lambda) \int_0^{\frac{b\sec\Lambda}{2}} [a_o \alpha^2 \sec^2 \Lambda \beta \tan \Lambda \\ &- C_{Do,l}(1 - \beta \tan \Lambda) - C_{D\alpha,l} \alpha \sec \Lambda] c(y_h) y_h \, dy_h \end{aligned} \quad (3.294)$$

The total yawing moment is given by

$$N = N_R + N_L \quad (3.295)$$

After some simplification, we obtain

$$\begin{aligned} N = &\rho V_o^2 \beta \sin \Lambda \cos \Lambda \\ &\times \int_0^{\frac{b\sec\Lambda}{2}} (a_o \alpha^2 \sec^2 \Lambda + 2C_{Do,l} + C_{D\alpha,l} \alpha \sec \Lambda) c(y_h) y_h \, dy_h \end{aligned} \quad (3.296)$$

In coefficient form,

$$\begin{aligned} (C_n)_{\Lambda,W} = &\left(\frac{2\beta \sin \Lambda}{Sb}\right) \int_0^{\frac{b\sec\Lambda}{2}} (a_o \alpha^2 \sec \Lambda + 2C_{Do,l} \cos \Lambda + C_{D\alpha,l} \alpha) \\ &\times c(y_h) y_h dy_h \end{aligned} \quad (3.297)$$

or, with $C_{L,l} = a_o \alpha \sec \Lambda$ when $\beta = 0$,

$$(C_{n\beta})_{\Lambda,W} = \left(\frac{2\sin\Lambda}{Sb}\right) \int_0^{\frac{b\sec\Lambda}{2}} (C_{L,l}\alpha + 2C_{Do,l}\cos\Lambda + C_{D\alpha,l}\alpha) c(y_h) y_h \, dy_h \tag{3.298}$$

From Eq. (3.298) we observe that wing sweep-back has a stabilizing effect because all the parameters inside the integral are positive, leading to a positive value $C_{n\beta}$. Furthermore, the stabilizing effect increases with angle of attack, and its magnitude depends on the wing leading-edge sweep angle. In a similar way, it can be shown that the forward wing sweep produces a destabilizing effect on static directional stability.

Equation (3.298) is a very crude estimation of $(C_{n\beta})_W$ because the induced drag is ignored in the strip theory approach. As the aspect ratio becomes lower, the induced drag becomes important and the strip theory estimation will be in error. In spite of these shortcomings, the strip theory has helped us to understand two important facts: 1) the sweep-back has a stabilizing effect on static directional stability, and 2) the stabilizing effect increases with angle of attack. It may be noted that the analysis is restricted to the linear angle of attack range.

For more accurate estimation of the wing contribution to static directional stability caused by wing sweep, the following empirical relation [1] can be used for subsonic speeds.

$$\frac{(C_{n\beta})_{\Lambda,W}}{C_L^2} = \frac{1}{4\pi A} - \frac{\tan\Lambda_{c/4}}{\pi A\left(A + 4\cos\Lambda_{c/4}\right)} \times \left(\cos\Lambda_{c/4} - \frac{A}{2} - \frac{A^2}{8\cos\Lambda_{c/4}} - 6\bar{x}_a \frac{\sin\Lambda_{c/4}}{A}\right) \tag{3.299}$$

where $\Lambda_{c/4}$ is the wing quarter chord sweep, A is the wing aspect ratio (theoretical), and $\bar{x}_a$ is the distance between the center of gravity and the wing aerodynamic center in terms of mean aerodynamic chord of the wing. According to our sign convention, $\bar{x}_a > 0$ if the center of gravity is aft of the wing aerodynamic center. The value of $(C_{n\beta})_{\Lambda,W}$ given by Eq. (3.299) is per radian.

The wing quarter chord sweep is given by Eq. (3.47) and is reproduced in the following:

$$\tan\Lambda_{c/4} = \tan\Lambda_{LE} - \left(\frac{c_r - c_t}{2b}\right)$$

For supersonic speeds, no general method is available for estimating the wing contribution to directional stability due to sweep effect. Datcom [1] gives some empirical data for certain wing geometries. The interested reader may refer to Datcom [1] for more information.

Fuselage Contribution

The fuselage contribution to static directional stability is generally destabilizing and is infuenced by wing geometry and wing placement with respect to the fuselage.

The fuselage contribution can be estimated using the following empirical relation [1],

$$\left(C_{n\beta}\right)_{B(W)} = -K_N K_{Rl}\left(\frac{S_{B,S}}{S}\right)\left(\frac{l_f}{b}\right)/\text{deg} \tag{3.300}$$

where K_N is an empirical wing–body interference factor that is a function of the fuselage geometry and the center of gravity position, K_{Rl} is an empirical factor that is a function of the fuselage Reynolds number, based on its length l_f, $S_{B,S}$ is the projected side area of the fuselage, S is the reference wing area, and l_f is the length of the fuselage. According to Datcom [1], Eq. (3.300) is valid for both subsonic and supersonic speeds. The parameters K_N and K_{Rl} can be determined using the data given in Figs. 3.73 and 3.74.

Tail Contribution

The contribution of the horizontal tail, similar to that of the wing, depends on dihedral and sweep. Furthermore, the horizontal tail is usually much smaller in size than the wing. Hence, the contribution of the horizontal tail to static directional stability can be safely ignored.

The vertical tail is perhaps the single largest contributor to static directional stability. Its contribution depends on its moment arm from the center of gravity, surface area, aspect ratio, sweep, and aft fuselage geometry. The aft fuselage and the horizontal tail provide the beneficial endplate effect, which increases its effective aspect ratio, hence the lift-curve slope. The contribution of the vertical tail is also affected by the fuselage sidewash.

For subsonic speeds, the side force developed by the vertical tail when the rudder is held in neutral position (rudder-fixed) is given by Datcom [1]:

$$Y_V = -kq_v a_v(\beta + \sigma)S_v \tag{3.301}$$

or

$$C_{y,V} = -ka_v(\beta + \sigma)\eta_v\left(\frac{S_v}{S}\right) \tag{3.302}$$

and

$$C_{y\beta,V} = -ka_v\left(1 + \frac{\partial\sigma}{\partial\beta}\right)\eta_v\left(\frac{S_v}{S}\right) \tag{3.303}$$

where

$$C_{y,V} = \frac{Y_V}{qS} \tag{3.304}$$

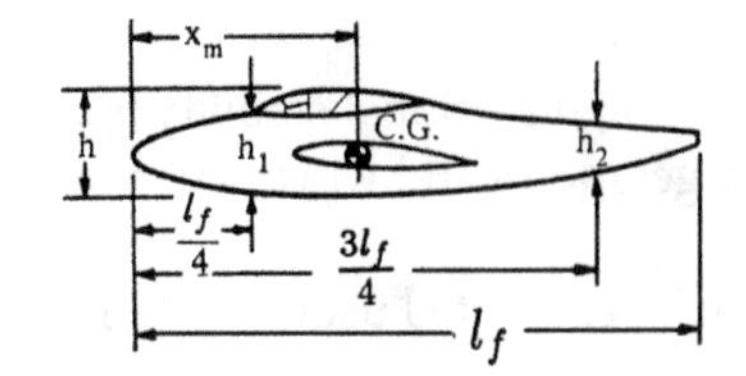

$S_{B,S}$: **Body Side Area**

$b_{f,max}$: **Maximum Body Width**

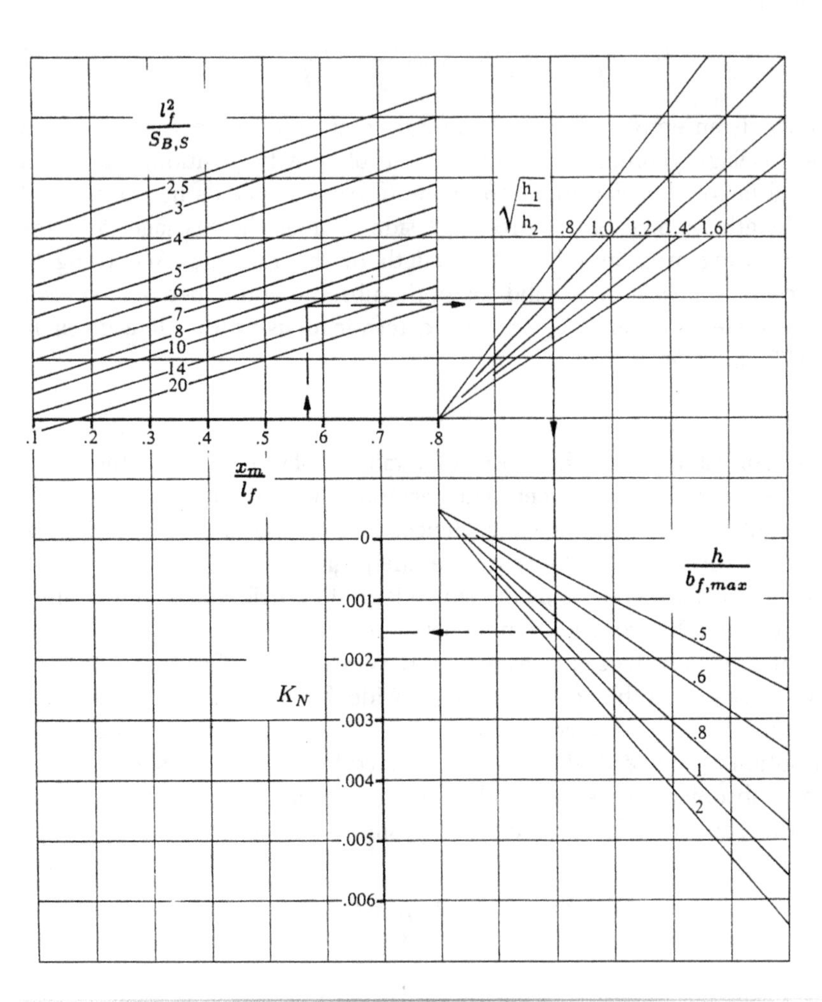

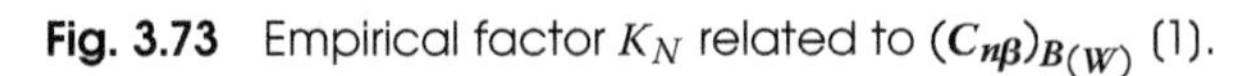

Fig. 3.73 Empirical factor K_N related to $(C_{n\beta})_{B(W)}$ (1).

$$C_{y\beta,V} = \frac{\partial C_{y,V}}{\partial \beta} \tag{3.305}$$

Here Y_V is the side force on the vertical tail.

The parameter k is given in Fig. 3.75 as a function of $b_v/2r_1$, where b_v is the vertical tail span measured up to the fuselage centerline and r_1 is the average radius of the fuselage sections underneath the vertical tail.

Furthermore, S_v is the theoretical vertical tail area (up to the fuselage centerline), σ is the induced sidewash angle at the vertical tail, and η_v is the dynamic pressure ratio at the vertical tail. The sidewash, counterpart of downwash in the horizontal plane, is mainly induced by the forward sections of the fuselage as shown schematically in Fig. 3.76.

The combined sidewash and dynamic pressure ratio term is given by the following empirical relation [1]:

$$\left(1+\frac{\partial\sigma}{\partial\beta}\right)\eta_v = 0.724 + \frac{3.06S_v/S}{1+\cos\Lambda_{c/4}} + \frac{0.4z_w}{d_{f,\max}} + 0.009A \tag{3.306}$$

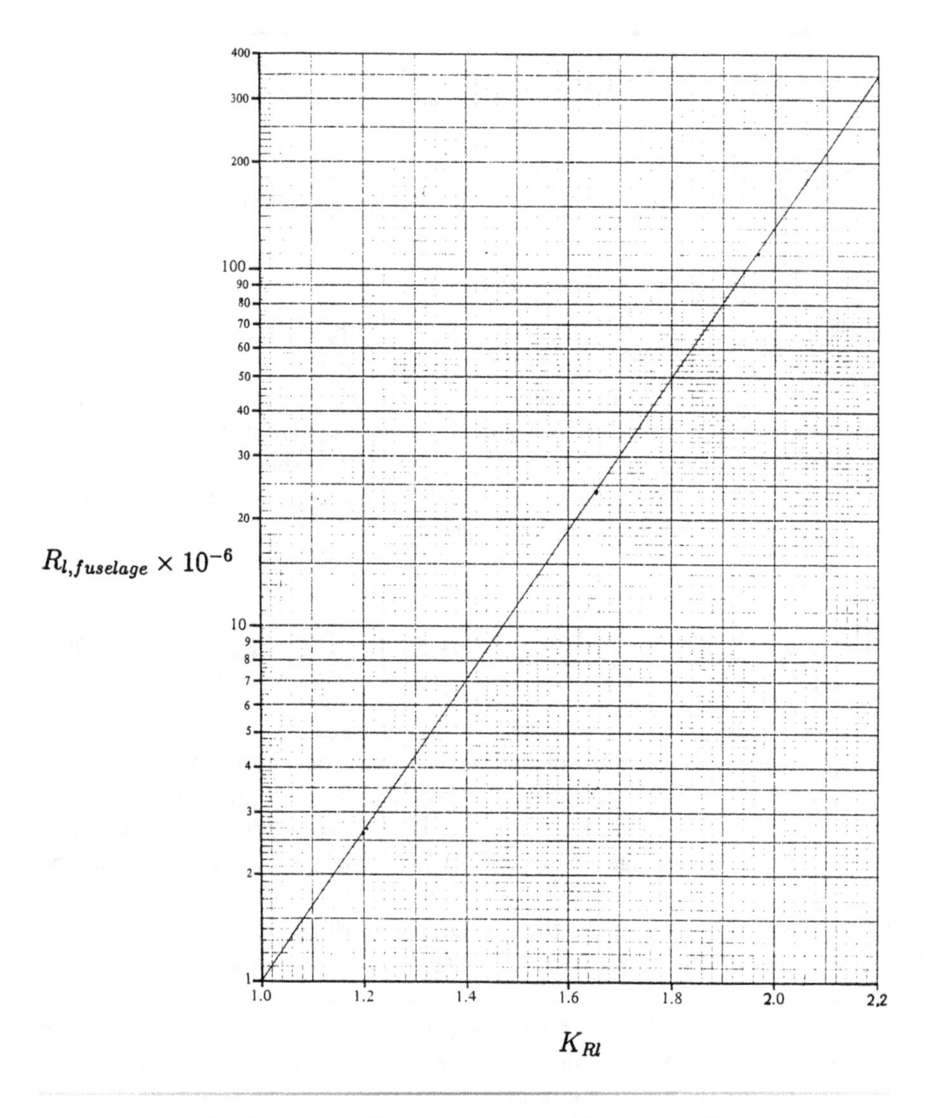

Fig. 3.74 K_{Rl} with fuselage Reynolds number (1).

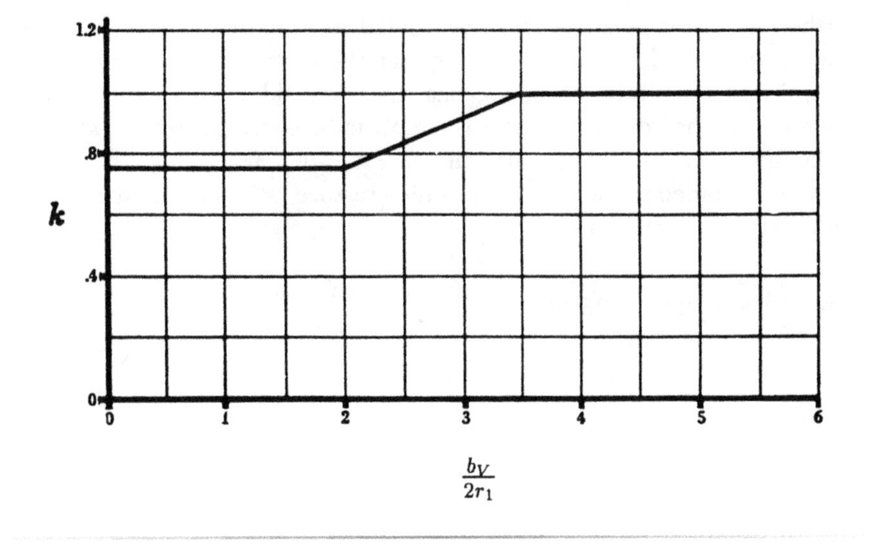

Fig. 3.75 Empirical parameter k as a function of $b_V/2r_1$ (1).

where $d_{f,\max}$ is the maximum fuselage depth, A the theoretical wing aspect ratio, $\Lambda_{c/4}$ is the theoretical wing quarter chord sweep, and z_w is the vertical distance (measured parallel to the z axis) from wing root quarter chord point to the fuselage centerline, positive if the wing is below the fuselage centerline. The lift-curve slope a_v of the vertical tail can be obtained using the methods given in Sec. 3.3 using its effective aspect ratio as given by the following expression [1]:

$$A_{v,\mathrm{eff}} = \left(\frac{A_{v(B)}}{A_v} A_v\right)\left[1 + K_H\left(\frac{A_{v(HB)}}{A_{v(B)}} - 1\right)\right] \tag{3.307}$$

where $A_{v(B)}/A_v$ is the ratio of the vertical tail aspect ratio in the presence of the body to that of the isolated vertical tail as given in Fig. 3.77. The parameters used in Fig. 3.77 are 1) b_v, the vertical tail span measured up to the fuselage centerline; 2) $2r_1$, fuselage average depth in the region of vertical tail; and 3) λ_v, vertical tail taper ratio based on vertical tail surface measured up to the fuselage centerline. A_v is the geometrical aspect ratio of the vertical tail with span and area measured up to the fuselage centerline, $A_{v(HB)}/A_{v(B)}$ is the ratio of vertical tail aspect ratio in the presence of the horizontal tail and body to that of the vertical tail in the presence of the body alone as given in Fig. 3.78, and K_H is a factor accounting for the relative size of the horizontal and vertical tails as given in Fig. 3.79.

The vertical tail contribution to static directional (Fig. 3.80) stability with rudder-fixed is given by

$$(N_V)_{\text{fix}} = Y_V l_v \tag{3.308}$$

$$= kq\eta_v a_v(\beta + \sigma)S_v l_v \tag{3.309}$$

Here, l_v is the distance—measured along the airplane reference line—between the center of gravity and the vertical tail aerodynamic center as shown in Fig. 3.80. In coefficient form,

$$(C_{n,V})_{\text{fix}} = ka_v(\beta + \sigma)\eta_v \overline{V}_2 \tag{3.310}$$

and

$$(C_{n\beta,V})_{\text{fix}} = ka_v\left(1 + \frac{\partial\sigma}{\partial\beta}\right)\eta_v \overline{V}_2 \tag{3.311}$$

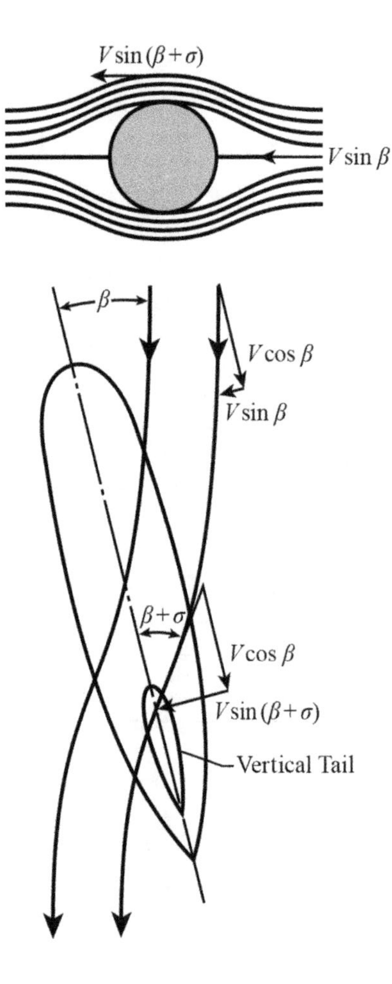

Fig. 3.76 Fuselage sidewash at vertical tail.

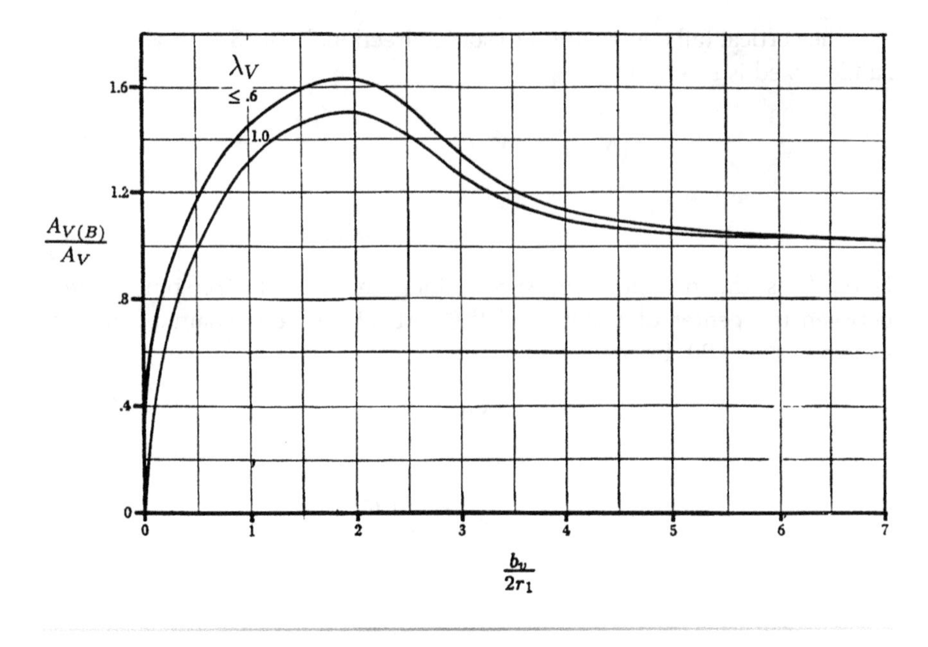

Fig. 3.77 Empirical parameter $A_{v(B)}/A_v$ as a function of $b_v/2r_1$ (1).

where

$$\overline{V}_2 = \frac{S_v l_v}{Sb} \tag{3.312}$$

The parameter $\overline{V}_2$ is often called the vertical tail volume ratio. The suffix "fix" denotes the rudder-fixed condition.

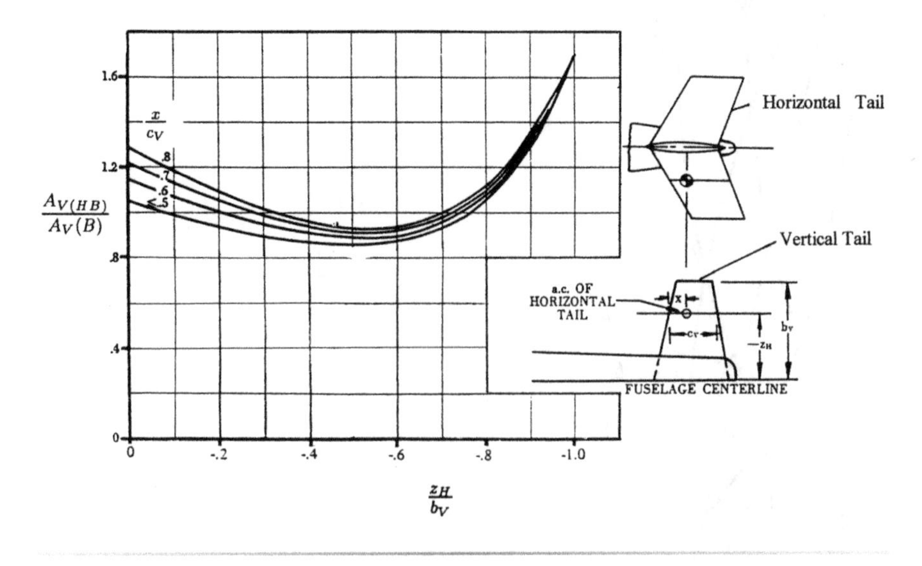

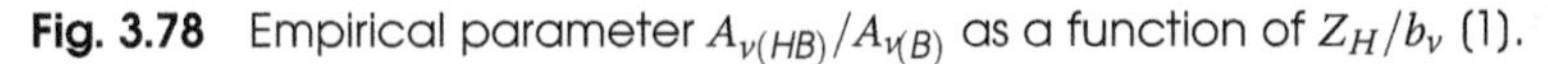

Fig. 3.78 Empirical parameter $A_{v(HB)}/A_{v(B)}$ as a function of Z_H/b_v (1).

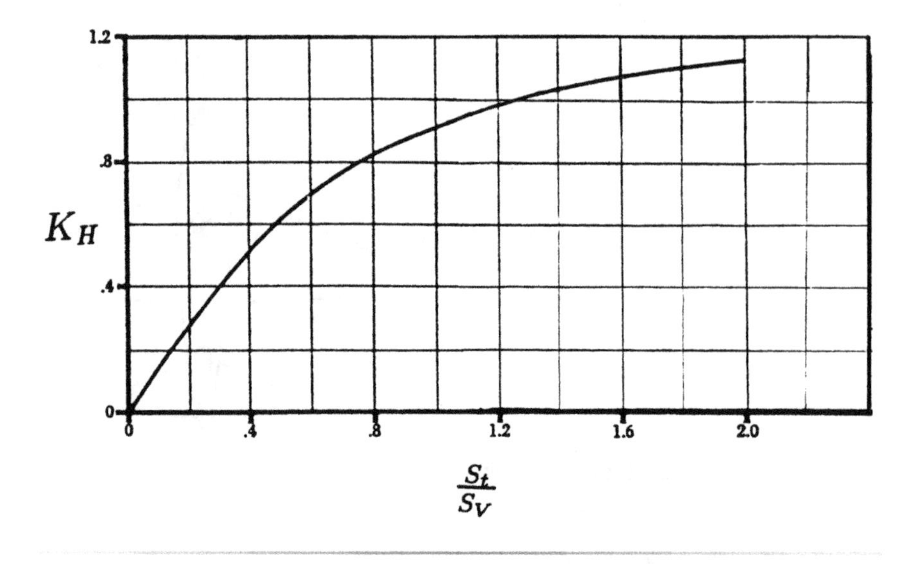

Fig. 3.79 Empirical parameter K_H as a function S_t/S_v (1).

For supersonic speeds, no general method is available for estimating the vertical tail contribution to the directional stability. The flow field around the vertical tail is quite complex because of the presence of shock waves emanating from wing, fuselage, and horizontal tails in addition to the shock waves caused by the vertical tail itself. Datcom [1] gives an approximate engineering method to evaluate the vertical tail contribution for simple configurations. The interested reader may refer to Datcom [1] for more information.

However, to get a crude estimate for preliminary design purposes, Eq. (3.311) may be used with the lift-curve slope a_v of the vertical tail evaluated at supersonic speeds using the methods presented in Sec. 3.3.

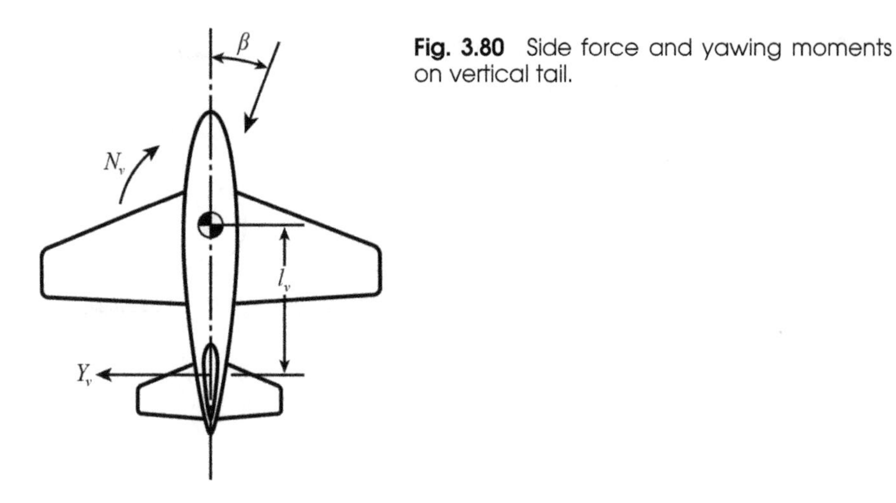

Fig. 3.80 Side force and yawing moments on vertical tail.

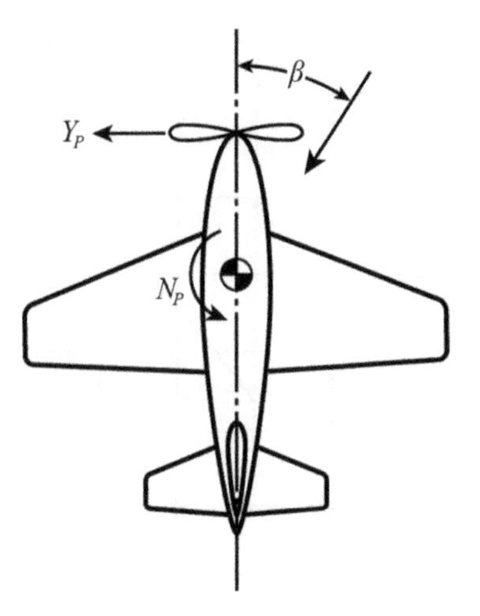

Fig. 3.81 Power effects: propeller airplane.

3.5.3 Effect of Power

For a propeller-driven aircraft, the effect of power on static directional stability consists of two parts: 1) the direct effect due to forces developed by or acting on the propulsion unit, and 2) indirect effect caused by propeller slipstream passing over the wing or tail surfaces.

The direct effect includes thrust developed by the propulsion unit and the side force (drag) acting on the propulsion unit because of sideslip. Because the resultant thrust vector is usually contained in the plane of symmetry, the contribution of the thrust to directional stability can be ignored. The situation, however, is different if there is an engine failure, which we will discuss a little later. The effect of side force depends on the location of the propulsion unit. This effect is destabilizing for a propeller airplane and stabilizing for a pusher airplane as shown Figs. 3.81 and 3.82.

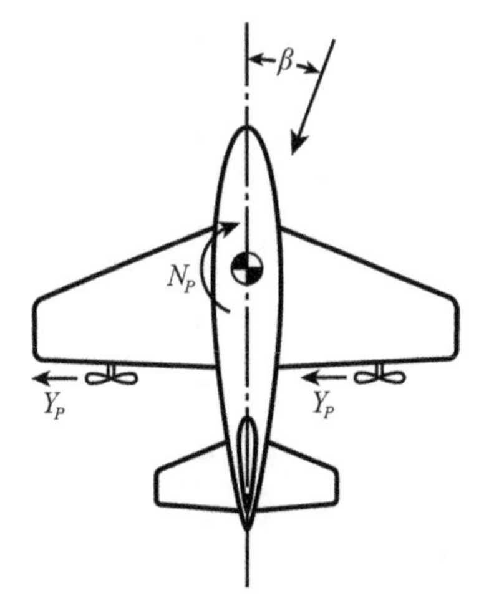

Fig. 3.82 Power effects: pusher airplane.

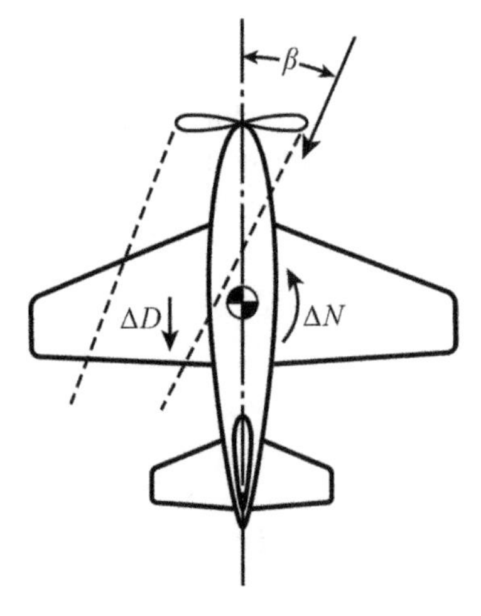

Fig. 3.83 Propeller slipstream effects.

The indirect effects arise mainly because of the slipstream effect on wing as shown in Fig. 3.83. The sections of the wing coming under the influence of propeller slipstream experience higher dynamic pressures leading to higher local lift and drag forces. The asymmetry in lift gives a destabilizing effect in roll. The asymmetry in drag as represented by ΔD produces a destabilizing yawing moment ΔN as shown in Fig. 3.83.

For a jet aircraft, the direct effects due to side forces on the intake are similar to that of a propeller airplane. The indirect effects caused by the jet-induced flow field affect the vertical tails in a manner similar to that shown for horizontal tails in Fig. 3.28.

In general, the evaluation of power effects is quite complex and configuration dependent. For simplicity, we ignore the power effects. The interested reader may refer to Datcom [1] for additional information.

3.5.4 Rudder-Fixed Directional Stability

Ignoring the power effects, the rudder-fixed directional stability of the airplane is given by the sum of the individual contributions as follows:

$$\left(C_{n\beta}\right)_{\text{fix}} = \left(C_{n\beta}\right)_W + \left(C_{n\beta}\right)_{B(W)} + \left(C_{n\beta,V}\right)_{\text{fix}} \tag{3.313}$$

For static directional stability, $(C_{n\beta})_{\text{fix}}$ must be positive over the desired angle of attack and speed range. Generally, a value of $C_{n\beta}$ between 0.0010 and 0.0025 would be considered satisfactory. However, an upper limit on the value of $C_{n\beta}$ may arise from directional control requirements as discussed next.

3.5.5 Directional Control

The rudder is the primary directional control and its effectiveness is measured by the parameter $C_{n\delta r}$, which is equal to the yawing-moment coefficient per unit rudder deflection. The sign convention for the rudder deflection is as follows: the rudder deflection is said to be positive if the rudder is deflected to the left side and that towards the right side is negative as shown in Fig. 3.84. Thus, a positive rudder deflection produces a positive side force and a negative yawing moment, and a negative rudder deflection produces a negative side force and a positive yawing moment. As a result, $C_{n\delta r}$ is usually negative.

Let

$$\tau_2 = \frac{\partial \beta_v}{\partial \delta_r} \tag{3.314}$$

denote the change in the sideslip angle of the vertical tail per unit deflection of the rudder. The parameter τ_2 can be evaluated using the method given in Sec. 3.3.9 using the variable β instead of α. Then, the yawing moment caused by a positive rudder deflection δ_r is given by

$$N_r = -kq\eta_v S_v a_v(\tau_2\delta_r + \sigma)l_v \tag{3.315}$$

or, in coefficient form,

$$C_{nr} = -k\eta_v \overline{V}_2 a_v(\tau_2\delta_r + \sigma) \tag{3.316}$$

so that

$$C_{n\delta r} = -k\eta_v \overline{V}_2 a_v \tau_2 \tag{3.317}$$

where

$$C_{nr} = \frac{N_r}{qSb} \tag{3.318}$$

$$C_{n\delta r} = \frac{\partial C_n}{\partial \delta_r} \tag{3.319}$$

We assume that $\eta_v = \eta_t$. The method to evaluate η_t was presented in Sec. 3.3.

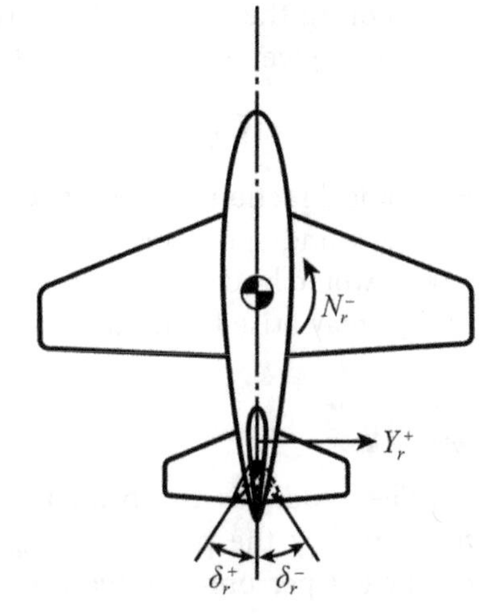

Fig. 3.84 Sign convention for rudder deflection.

The rudder deflection to generate a sideslip β is given by

$$(C_{n\beta})_{\text{fix}}\beta + C_{n\delta r}\delta_r = 0 \tag{3.320}$$

or

$$\delta_r = -\frac{(C_{n\beta})_{\text{fix}}\beta}{C_{n\delta r}} \tag{3.321}$$

From this relation, we observe that the higher the level of static directional stability, the higher the rudder deflection to generate a given sideslip will be. Typically, a value of $C_{n\delta r}$ of -0.001 per deg is usually considered satisfactory.

The rudder effectiveness, like that of any other aerodynamic control surface, is nearly constant at low and moderate sideslip but falls off rapidly at high values of sideslip because of flow separation and stall.

Rudder Requirements

An airplane having an adequate level of static directional stability and symmetric power generally tends to maintain zero sideslip condition and, as such, the deflection of the rudder may not be usually warranted. However, under some critical conditions, it is possible that the static directional stability alone may not be sufficient to maintain zero sideslip, and the operation of the rudder becomes absolutely essential. The rudder should be designed to provide sufficient control authority under such circumstances as discussed in the following sections.

Crosswind Takeoff and Landing

During the ground run, if the aircraft encounters a crosswind, the resultant velocity vector falls out of the airplane's plane of symmetry, producing sideslip. An aircraft with positive directional stability ($C_{n\beta} > 0$) will tend to realign itself with the direction of the resultant wind so that the sideslip is eliminated. Takeoff with this kind of aircraft orientation (Fig. 3.85a) during the ground run can pose safety problems. To prevent this, the rudder should be capable of generating a yawing moment to counter that due to directional stability so that the aircraft sideslips but is properly oriented with respect to the runway (Fig. 3.85b). However, the takeoff or landing performance will be below normal because the sideslipping aircraft experiences a higher drag.

Adverse Yaw

In a level coordinated turn, the aircraft is moving approximately in a circular path in a horizontal plane as shown in Fig. 3.86. The angular velocity Ω about the vertical axis passing through the center of the turn is equal to V/R. During the turn, the outer wing is moving with a higher velocity compared to the inner wing. As a result, the outer wing experiences relatively higher drag, and this imbalance in drag (ΔD) induces a yawing moment that

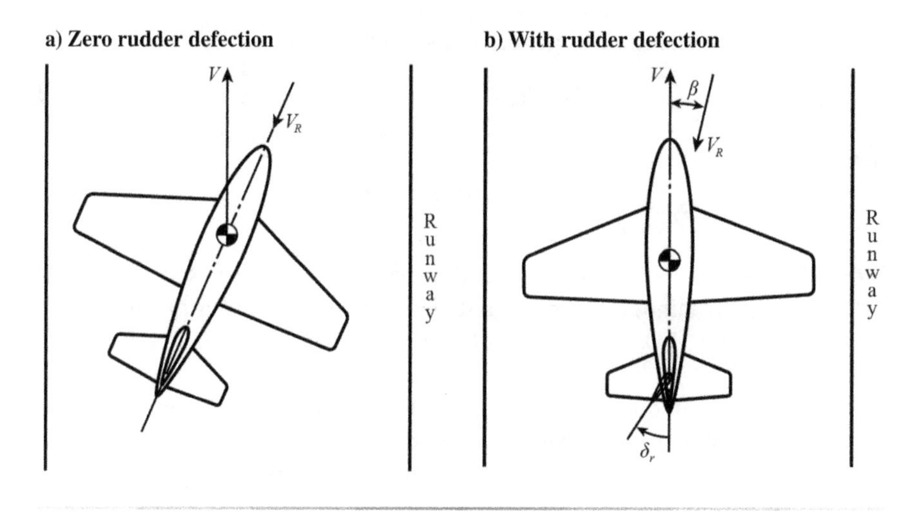

Fig. 3.85 Crosswind takeoff/landing.

tends to turn the nose of the aircraft away from the center of the turn. This phenomenon is known as adverse yaw. Because of this adverse yaw, a rolling motion may also be induced because of the dynamic derivative C_{lr} (which is usually positive) as we will study in Chapters 4 and 6. This rolling motion caused by C_{lr} is called adverse roll because it tends to bank the aircraft away from the direction in which the aircraft is turning. Generally, the magnitude of the adverse yawing moment is small. However, for rapid turns at high angles of attack, it can pose problems, particularly if the adverse roll overpowers the proverse roll because of ailerons causing a roll reversal. Therefore, the rudder should have sufficient control power to prevent the development of adverse yaw during a turn.

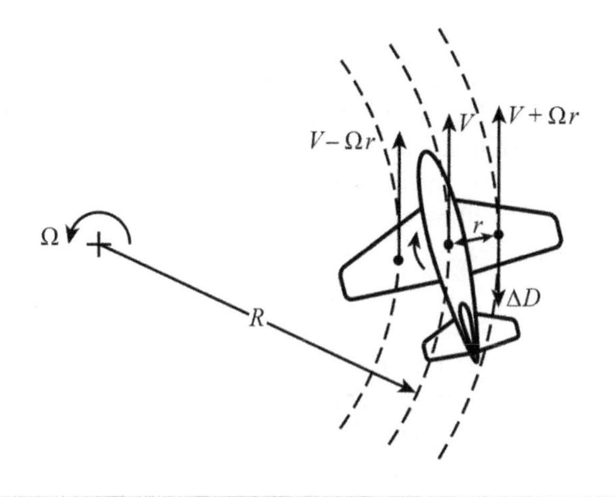

Fig. 3.86 Adverse yaw during a coordinated turn.

Asymmetric Power

On multiengined aircraft, either partial or total failure of one or more engines gives rise to an asymmetric power situation that can generate a significant yawing moment as schematically shown in Fig. 3.87. If this yawing moment is not countered by the rudder, the aircraft will develop a sideslip and, in some cases, may go out of control because of aerodynamic roll–yaw coupling. The rudder must be designed to have sufficient control authority to maintain controlled flight under such conditions.

Spin Recovery

Generally, in spin, the airplane is operating at high angles of attack with wings and horizontal tail surfaces more or less completely stalled. Quite often, the rudder may be the only control that has some effectiveness under these conditions. To break out of an established spin, the rudder must be capable of producing sufficient yawing moment to slow the spin rate and initiate a successful recovery. We will learn more about spin recovery when we discuss the spinning motion in Chapter 7.

3.5.6 Rudder-Free Directional Stability

In sideslipping motion, if the rudder is left free to float (pilot's foot off the rudder pedal), it will assume a condition such that the net hinge moment is zero. This position is called the floating angle of the rudder and is given by

$$\delta_{rf} = -\frac{C_{h\beta}}{C_{h\delta r}}\beta \tag{3.322}$$

where

$$C_{h\beta} = \frac{\partial C_h}{\partial \beta} \tag{3.323}$$

$$C_{h\delta r} = \frac{\partial C_h}{\partial \delta r} \tag{3.324}$$

Fig. 3.87 Asymmetric power effects.

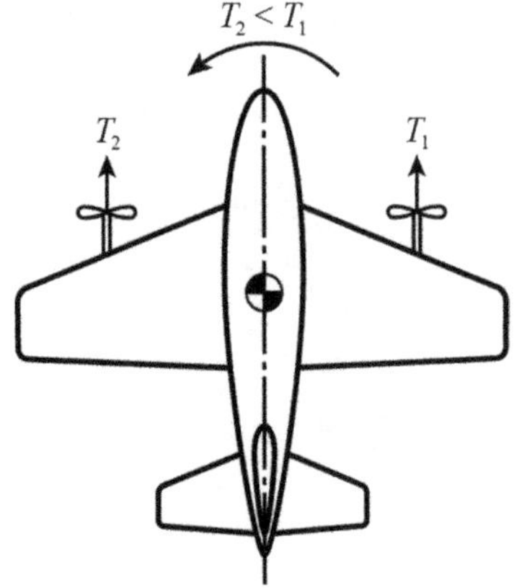

The hinge-moment coefficients $C_{h\beta}$ and $C_{h\delta r}$ can be estimated using the methods given in Sec. 3.3.14 using the variable β instead of α and δr instead of δ_e.

Recall that the elevator hinge moment was assumed positive if it rotated the elevator trailing edge down or in a direction to increase the elevator deflection. In a similar way, we assume rudder hinge moment to be positive if it rotates the rudder such that rudder deflection increases, i.e., it makes the rudder deflect to the left. With this understanding, we observe that a positive sideslip causes a positive hinge moment so that $C_{n\beta} > 0$. Similarly, a positive rudder deflection generates a negative hinge moment so that $C_{h\delta r} < 0$. These concepts are schematically illustrated in Fig. 3.88.

With $C_{h\beta} > 0$ and $C_{h\delta r} < 0$ for positive sideslip, we observe from Eq. (3.322) that the floating angle of the rudder will be positive, i.e., the rudder floats to the left when the aircraft sideslips to starboard and vice versa.

The effective sideslip angle β_V of the vertical tail with rudder free to float is given by

$$\beta_V = \beta + \sigma + \tau_2 \delta_{rf} \tag{3.325}$$

Then,

$$(C_{n,V})_{\text{free}} = k a_v \left[\beta + \sigma - \tau_2 \frac{C_{h\beta}}{C_{h\delta r}} \beta \right] \bar{V}_2 \eta_v \tag{3.326}$$

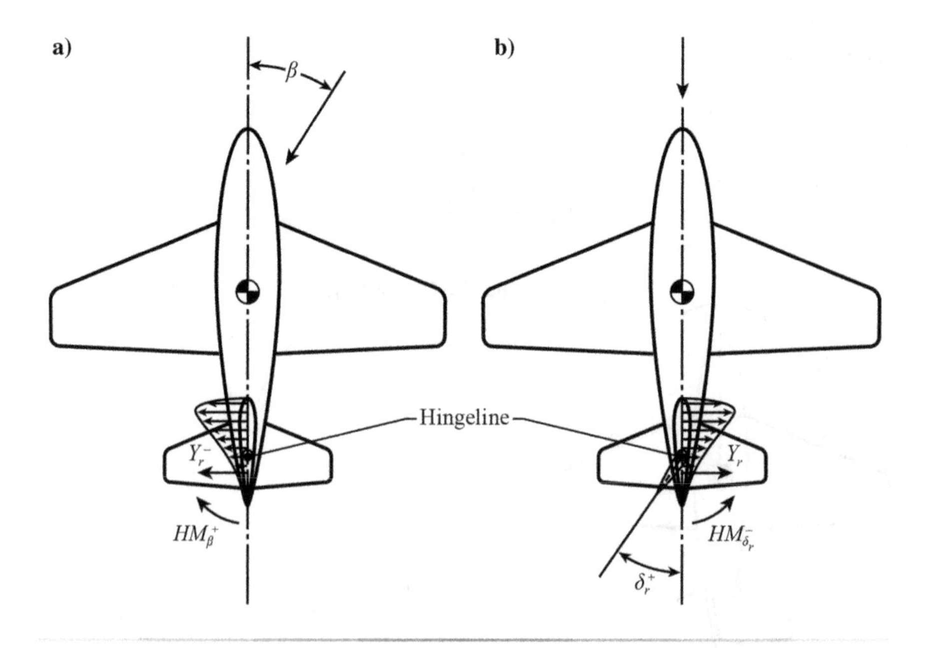

Fig. 3.88 Sign convention for rudder hinge-moment coefficient.

$$\left(C_{n\beta,V}\right)_{\text{free}} = ka_v\left[\left(1+\frac{\partial\sigma}{\partial\beta}\right) - \tau_2\frac{C_{h\beta}}{C_{h\delta r}}\right]\overline{V}_2\eta_v \tag{3.327}$$

Using Eq. (3.317), we get

$$\left(C_{n\beta,V}\right)_{\text{free}} = \left(C_{n\beta,V}\right)_{\text{fix}} + C_{h\beta}\left(\frac{C_{n\delta r}}{C_{h\delta r}}\right) \tag{3.328}$$

and

$$\left(C_{n\beta}\right)_{\text{free}} = \left(C_{n\beta}\right)_W + \left(C_{n\beta}\right)_{B(W)} + \left(C_{n\beta,V}\right)_{\text{free}} \tag{3.329}$$

For positive rudder-free directional stability $(C_{n\beta})_{\text{free}} > 0$.

3.5.7 Pedal Forces

Suppose that the aircraft is required to maintain a given sideslip β. Then, the required rudder deflection is given by

$$\delta r = -\frac{\left(C_{n\beta}\right)_{\text{fix}}\beta}{C_{n\delta r}} \tag{3.330}$$

The pedal forces to maintain this rudder deflection can be calculated as follows.

Assuming that the rudder has a balancing tab similar to the elevator trim tab, we have

$$C_h = C_{h\beta}\beta + C_{h\delta r}\delta r + C_{h\delta_t}\delta_t \tag{3.331}$$

where $C_{h\delta_t} = \partial C_h/\partial\delta_t$. Then, the pedal force F_p is given by

$$F_p = G_2 HM \tag{3.332}$$

$$= G_2 q\eta_v S_r\overline{c}_r\left(C_{h\beta}\beta + C_{h\delta r}\delta r + C_{h\delta_t}\delta_t\right) \tag{3.333}$$

Here, G_2 is the gearing ratio between the pedals and the rudder, S_r is the rudder area, and $\overline{c}_r$ is the rudder mean aerodynamic chord. Substituting for δr from Eq. (3.330), we obtain

$$F_p = G_2 q\eta_v S_r\overline{c}_r\left[C_{h\beta}\beta - C_{h\delta r}\frac{\left(C_{n\beta}\right)_{\text{fix}}\beta}{C_{n\delta r}} + C_{h\delta_t}\delta_t\right] \tag{3.334}$$

Differentiating with respect to β and rearranging, we obtain the pedal-force gradient,

$$\frac{\partial F_p}{\partial\beta} = G_2 q\eta_v S_r\overline{c}_r\left[C_{h\beta} - \left(\frac{C_{h\delta r}}{C_{n\delta r}}\right)\left(C_{n\beta}\right)_{\text{fix}}\right] \tag{3.335}$$

A pedal-force gradient of 5 lb or 22.28 N/deg sideslip at a speed of 150 mph (approximately 240 km/h) is a comfortable minimum value [6]. The FAA [5] specifications require that pedal force vary linearly with sideslip β or that pedal-force gradient be constant up to a sideslip angle of ± 15 deg.

3.5.8 Rudder Lock

The floating angle of the rudder, as given by Eq. (3.322), depends on the hinge-moment parameters $C_{h\beta}$ and $C_{h\delta r}$. As said earlier, with $C_{h\beta} > 0$ and $C_{h\delta r} < 0$, the rudder floating angle δ_{rf} will be positive for positive sideslip. Furthermore, the floating angle increases with sideslip. Schematic variations of the required rudder deflection and floating angle with sideslip are shown in Fig. 3.89. At high sideslip, the floating angle increases beyond the linear rate indicated by Eq. (3.322) because the center of pressure moves aft because of flow separation and stall. This accentuates the floating tendency of the rudder. At one point, the floating angle may catch up with the required rudder deflection. This condition is usually known as rudder lock. Beyond this point, the floating angle may overshoot and opposite pedal forces are required to operate the rudder. Such a situation is undesirable because it may take considerable effort for the pilot to break the rudder lock.

Aerodynamic Balancing to Prevent Rudder Lock

As we have discussed before, aerodynamic balancing helps to alter hinge-moment coefficients and the floating characteristics of a control surface. Therefore, with proper aerodynamic balancing, the floating tendency of the rudder can be adjusted so that the rudder lock phenomenon is avoided.

Dorsal Fin Use to Prevent Rudder Lock

Another method of preventing rudder lock is the use of a device called a dorsal fin. As we know, the stall angle of a given lifting surface increases as the aspect ratio is reduced (see Chapter 1). Extending the chord of inboard sections adds area without extending the span so that the aspect

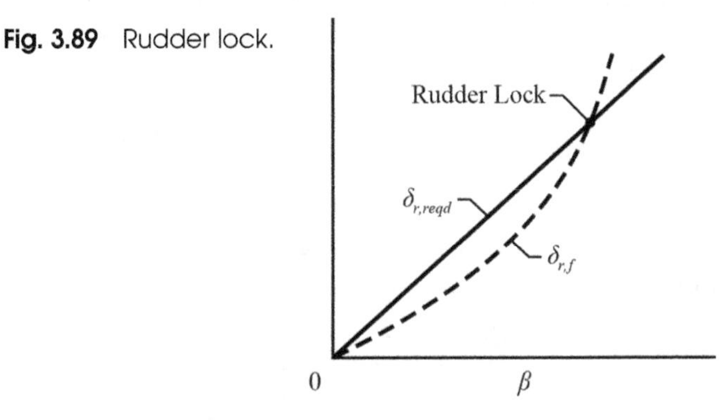

Fig. 3.89 Rudder lock.

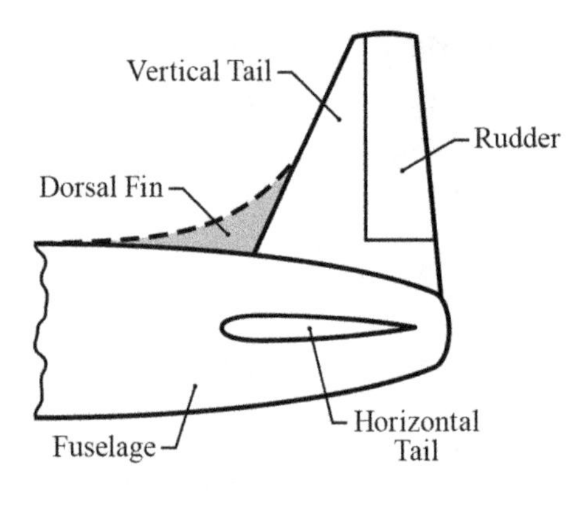

Fig. 3.90 Dorsal fin.

ratio decreases. This form of extension is known as a dorsal fin as shown in Fig. 3.90. Addition of a suitably sized dorsal fin helps to delay the vertical tail stall to higher sideslip (Fig. 3.91a) and minimizes the possibility of rudder lock. Also, the dorsal fin makes the pedal forces vary monotonically with sideslip as indicated in Fig. 3.91b.

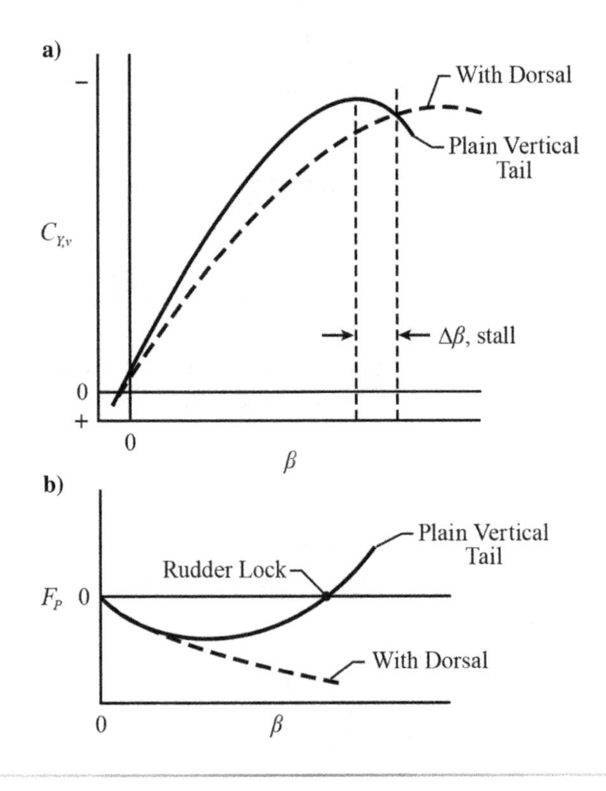

Fig. 3.91 Effect of dorsal fin.

Example 3.8

For the tailless aircraft of Example 3.2, determine the static directional stability parameter $C_{n\beta}$ at $M = 0.7$ and altitude of 8500 m $\sigma = 0.3881$) and at $M = 2.0$ and 18,000 m $\sigma = 0.094$).

Solution:

Let us calculate $C_{n\beta}$ for $M = 0.7$ (subsonic) at 8500 m altitude.

The theoretical aspect ratio of the given wing is 2.8306. Hence, the strip theory cannot be used. We use the empirical relation in Eq. (3.271) to determine approximately the wing contribution due to dihedral as follows:

$$(C_{n\beta})_{\Gamma,W} = -0.075\,\Gamma C_L/\text{rad}$$

where Γ is in radians. We have $\Gamma = 3.5$ deg so that

$$\begin{aligned}(C_{n\beta})_{\Gamma,W} &= -0.075\left(\frac{3.5}{57.3}\right)C_L/\text{rad}\\ &= -0.0046\,C_L/\text{rad}\\ &= -0.0001\,C_L/\deg\end{aligned}$$

The wing contribution caused by sweep is given by Eq. (3.299):

$$\frac{(C_{n\beta})_{\Lambda,W}}{C_L^2} = \frac{1}{4\pi A} - \frac{\tan\Lambda_{c/4}}{\pi A(A + 4\cos\Lambda_{c/4})} \times\left(\cos\Lambda_{c/4} - \frac{A}{2} - \frac{A^2}{8\cos\Lambda_{c/4}} - 6\bar{x}_a\frac{\sin\Lambda_{c/4}}{A}\right)$$

We have $\Lambda_{c/4} = 36.3487$ deg and A (theoretical) $= 2.8306$. Here, $\bar{x}_a$ is the horizontal distance between the wing aerodynamic center of the exposed wing and the center of gravity in terms of mean aerodynamic chord and $\bar{x}_a > 0$ if the center of gravity is aft of the aerodynamic center. From the calculations done for Example 3.2 at $M = 0.7$, we have

$$\left(\frac{x_{ac}}{c_{re}}\right)_W = 0.3367\,M^3 - 0.4810\,M^2 + 0.2214\,M + 0.4918 \quad (0 \le M \le 0.80)$$

For $M = 0.7$, we get

$$\left(\frac{x_{ac}}{c_{re}}\right)_W = 0.5266$$

(*Continued*)

Example 3.8 (*Continued*)

The center of gravity is located 15.9334 m from the leading edge of the fuselage. From Fig. 3.58, we note that the leading edge of the exposed wing is located 14.1275 m from the leading edge of the fuselage. We have $c_{re} = 8.94$ m and $\bar{c} = 6.8072$ m. Then,

$$\frac{x_{cg}}{c_{re}} = \frac{15.9334 - 14.1275}{8.94}$$
$$= 0.2020$$
$$\frac{x_a}{c_{re}} = \left(\frac{x_{cg}}{c_{re}}\right) - \left(\frac{x_{ac}}{c_{re}}\right)_W$$
$$= 0.2020 - 0.5266$$
$$= -0.3246$$
$$\frac{x_a}{\bar{c}} = \left(\frac{x_a}{c_{re}}\right)\left(\frac{c_{re}}{\bar{c}}\right)$$
$$= -0.3246\left(\frac{8.94}{6.8072}\right)$$
$$= -0.4263$$

Substitution of all these values gives

$$\frac{(C_{n\beta})_{\Lambda,W}}{C_L^2} = 0.0461/\text{rad}$$
$$= 0.0008/\text{deg}$$

The combined wing contribution because of dihedral and sweep is then given by

$$(C_{n\beta})_{\text{wing}} = (-0.0001C_L + 0.0008C_L^2)/\text{deg}$$

Using Eq. (3.300), the fuselage contribution is given by

$$(C_{n\beta})_{B(W)} = \left[-K_N K_{Rl}\left(\frac{S_{BS}}{S}\right)\left(\frac{l_f}{b}\right)\right]\Big/\text{deg}$$

where K_N and K_{Rl} are to be obtained from data given in Figs. 3.73 and 3.74. Using data given in Example 3.2, we get $h_1 = 2.7838$ m and $h_2 = 3.048$ m so that $\sqrt{h_1/h_2} = 0.9556$. Furthermore, we have $h = d_{f,\max} = 3.048$ m so that $h/b_{f,\max} = 3.048/3.2715 = 0.9317$, $x_m/l_f = 15.9334/23.2410 = 0.6856$, and $l_f^2/S_{B,S} = 23.2410^2/60.75 = 8.8913$. With these values, we get $K_N = 0.0016$.

(*Continued*)

Example 3.8 (*Continued*)

To find K_{Rl}, we need to find the Reynolds number based on the fuselage length l_f, which is 23.2410 m. At an altitude of 8500 m, the kinematic viscosity is 3.05×10^{-5} m²/s, and the speed of sound is 305.0 m/s (see Appendix A). This gives us a flight velocity of 213.5 m/s and a Reynolds number (Vl_f/ν) of 16.27×10^7. Then we obtain $K_{Rl} = 2.06$ from Fig. 3.74. We have $S_{B,S} = 60.75\ \text{m}^2$, $S = 106.0114\ \text{m}^2$, and $b = 17.3228$ m. Substituting these values in the previous equation, we get

$$(C_{n\beta})_{B(W)} = -0.0025/\text{deg}$$

Now we will evaluate the vertical tail contribution, which is given by

$$(C_{n\beta,V})_{\text{fix}} = ka_v\left(1 + \frac{\partial\sigma}{\partial\beta}\right)\eta_v\overline{V}_2$$

To evaluate a_v, we need to evaluate the effective aspect ratio.

$$A_{v,\text{eff}} = \frac{A_{v(B)}}{A_v}A_v\left[1 + K_H\left(\frac{A_{v(HB)}}{A_{v(B)}} - 1\right)\right]$$

From the data given in Example 3.2, we have $A_v = 1.4869$, $\lambda_v = 0.2951$, $b_v = 5.4864$ m, and $2r_1 = 3.048$ m so that $b_v/2r_1 = 1.8$. Here, $2r_1$ is the average fuselage depth in the region of the vertical tail. Using these data, we obtain $A_{v(B)}/A_v = 1.63$ from Fig. 3.77. Because this vehicle does not have a horizontal tail, $A_{v,HB}/A_{v,B} = 1.0$ so that the effective aspect ratio $A_{v,\text{eff}} = 1.63 * 1.4869 = 2.4236$. Further, we obtain $k = 0.76$ from Fig. 3.75. Using Eq. (3.19), we obtain $\Lambda_{c/2} = 32.3771$ deg. Using Eq. (3.16), we obtain $a_v = 2.8113/\text{rad}$ or $0.0491/\text{deg}$.

Then, we need to evaluate the combined sidewash and dynamic pressure ratio term given by

$$\left(1 + \frac{\partial\sigma}{\partial\beta}\right)\eta_v = 0.724 + 3.06\frac{S_v/S}{1 + \cos\Lambda_{c/4}} + 0.4\frac{z_w}{d_{f,\max}} + 0.009A$$

We have $\Lambda_{c/4} = 39.2488$ deg, $z_w = 1.27$ m, $d_{f,\max} = 3.048$ m, A (theoretical) $= 2.8306$, and $S_v/S = 0.1909$. Substitution gives

$$\left(1 + \frac{\partial\sigma}{\partial\beta}\right)\eta_v = 1.2398$$

We have $l_v = 7.7561$ m. Using this value, we obtain $\overline{V}_2 = 0.0855$. Then, substituting all the required values, we get

$$(C_{n\beta,V})_{\text{fix}} = 0.0040/\text{deg}$$

(*Continued*)

Example 3.8 (*Continued*)

The total value of directional stability parameter at $M = 0.7$ and 8500 m altitude is given by

$$(C_{n\beta})_{\text{fix}} = (C_{n\beta})_W + (C_{n\beta})_{B(W)} + (C_{n\beta,V})_{\text{fix}}$$
$$= -0.0001C_L + 0.0008C_L^2 - 0.0025 + 0.0040$$
$$= 0.0015 - 0.0001C_L + 0.0008C_L^2/\text{deg}$$

Thus, at $M = 0.7$, this aircraft has a positive directional stability.

Now let us evaluate $C_{n\beta}$ for $M = 2.0$ at 18,000 m altitude. In the absence of a reliable method to estimate the wing contribution at supersonic speeds and further noting that wing contribution is generally small, we ignore the wing contribution to directional stability at supersonic speeds.

To evaluate the fuselage contribution $(C_{n\beta})_{B(W)}$, we need to evaluate K_{Rl}, which depends on the Reynolds number. All the other parameters in that expression for fuselage contribution remain the same as calculated for $M = 0.7$. At $M = 2.0$ and altitude of 18,000 m, we find that the Reynolds number is equal to 11.743×10^7. From Fig. 3.74, we get $K_{Rl} = 1.97$. Substituting in the given expression for fuselage contribution, we get $(C_{n\beta})_{B(W)} = -0.00242/\text{deg}$.

Now let us approximately evaluate vertical tail contribution. We have $A_{v,\text{eff}} = 2.4236$, $\lambda_v = 0.2951$, and $\Lambda_{LE,v} = 45$ deg. From Fig. 3.14, we get $\beta(C_{N\alpha})_{\text{theory}} = 3.85/\text{rad}$, where $\beta = \sqrt{M^2 - 1} = 1.732$. This gives $(C_{N\alpha})_{\text{theory}} = 2.2228/\text{rad}$. Note that β here is not sideslip angle.

Now we need to apply the sonic leading correction to the given value of $C_{N\alpha}$ using the data given in Fig. 3.15a. We have $\Delta y = 2.5$ so that $\Delta y\perp = \Delta y/\cos\Lambda_{LE} = 3.5360$. With this, we obtain from Fig. 3.15a, $C_{N\alpha}/(C_{N\alpha})_{\text{theory}} = 0.825$ so that $C_{N\alpha} = 1.8338/\text{rad}$ or $0.0320/\text{deg}$, and $a_v = 0.032/\text{deg}$. All the other values in the expression for vertical tail contribution remain unchanged. Substituting, we obtain

$$\left(C_{n\beta,V}\right)_{\text{fix}} = 0.0026/\text{deg}$$

Then,

$$C_{n\beta} = (C_{n\beta})_{B(W)} + (C_{n\beta,V})_{\text{fix}}$$
$$= -0.0024 + 0.0026$$
$$= 0.0002/\text{deg}$$

We observe that the given aircraft has marginal static directional stability at $M = 2.0$. However, it should be noted that this result is based on a very crude estimation of vertical tail contribution.

Example 3.9

An aircraft is ready for takeoff when it is detected that a crosswind of 8 m/s is blowing across the runway. Determine the rudder angle required to maintain a steady normal heading along the runway at unstick point using the following data.

Wing loading $(W/S) = 2500\,\text{N/m}^2$, span = 25 m, wing area = 70 m^2, unstick velocity = 1.2 V_{stall}, maximum lift coefficient = 1.8, lift-curve slope of the vertical tail = 0.08/deg, $(C_{n\beta})_{\text{fix}} = 0.012/\text{deg}$, vertical tail volume ratio = 0.25, and $\eta_v = 0.9$. Assume that 1 deg of rudder deflection changes the vertical tail incidence by 0.4 deg.

Solution:

Assuming sea-level conditions ($\rho = 1.225\,\text{kg/m}^3$), the stalling velocity is given by

$$V_{\text{stall}} = \sqrt{\frac{2W}{\rho S C_{L,\max}}}$$

$$= \sqrt{\frac{2 * 2500}{1.225 * 1.8}}$$

$$= 47.6191\,\text{m/s}$$

$$V_{\text{unstick}} = 1.2 V_{\text{stall}}$$

$$= 57.1429\,\text{m/s}$$

$$\beta = \tan^{-1} \frac{8}{57.1429}$$

$$= 7.9696\,\text{deg}$$

We have

$$C_{n\delta r} = -k a_v \overline{V}_2 \eta_v \tau_2$$

Assuming $k = 1$, we have

$$C_{n\delta r} = -0.08 * 0.25 * 0.9 * 0.4$$

$$= -0.0072/\text{deg}$$

$$\delta r = -\frac{\beta (C_{n\beta})_{\text{fix}}}{C_{n\delta r}}$$

$$= 13.2827\,\text{deg}$$

The rudder should be deflected to the left by 13.2827 deg.

Example 3.10

A twin jet engine has the following data: thrust per engine = 10,000 N, spanwise distance between the two engines = 10 m, wing area = 50 m^2, wing span = 10 m, rudder effectiveness ($C_{n\delta r}$) = −0.001/deg, and maximum permissible rudder deflection = ±20 deg. Determine the rudder deflection to maintain zero sideslip at 100 m/s in level flight at sea level with one engine completely out.

Solution:

Yawing moment due to asymmetric thrust is given by

$$\begin{aligned} N_T &= \text{Thrust} * \text{distance} \\ &= 10{,}000 * 5.0 \\ &= 50{,}000\ \text{Nm} \end{aligned}$$

The yawing moment due to rudder deflection:

$$N_r = \frac{1}{2}\rho V^2 S b C_{n\delta r}\delta_r$$

For equilibrium,

$$N_T + N_r = 0$$

so that

$$\begin{aligned} \delta_r &= \frac{-2N_T}{\rho V^2 S b C_{n\delta r}} \\ &= \frac{-2 * 50{,}000}{1.225 * 100^2 * 50 * 10(-0.001)} \\ &= 16.32\ \text{deg} \end{aligned}$$

3.6 Lateral Stability

Lateral stability is the inherent capability of the airplane to counter a disturbance in bank. In level flight, both wings are in a horizontal plane and the bank angle is zero. However, because of some disturbance, if the airplane banks—but very slowly so that the roll rate is negligibly small, then there is no aerodynamic mechanism to generate a restoring rolling moment unless sideslip develops. Therefore, the airplane is neutrally stable with respect to a disturbance in bank without sideslip. Fortunately, once banked, the airplane develops a sideslip in the direction of the bank because of a spanwise component of the weight $W \sin \phi$ as shown in Fig. 3.92. As a result of this sideslip, if a restoring rolling moment is

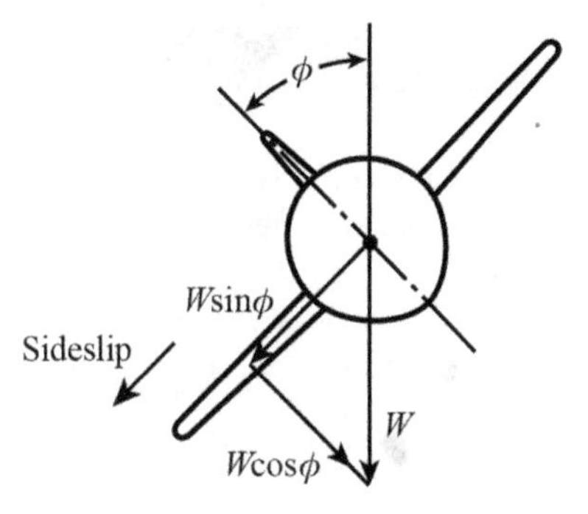

Fig. 3.92 Development of sideslip due to bank.

induced, then the airplane is said to be laterally stable. Once the wings are back in level condition, the disturbances in bank angle and sideslip are eliminated, and the airplane returns to its original, steady level flight condition. On the other hand, if the induced rolling moment causes the bank angle to increase further, generating more and more sideslip, the aircraft is said to be laterally unstable. If the induced rolling moment is zero and the airplane remains constantly banked and keeps on sideslipping, then it is said to be neutrally stable.

The generation of a rolling moment due to sideslip is also called dihedral effect, and an airplane that develops a restoring rolling moment because of sideslip is said to have a positive or stable dihedral effect. Therefore, a laterally stable airplane has a positive dihedral effect and vice versa. Note that the dihedral effect is different from the dihedral angle. The dihedral angle is the angle between the plane of the wing and a horizontal plane, positive if the wingtip is above the wing root. If the wingtip is below the wing root, the dihedral angle is negative (unhedral) (see Fig. 3.93). The dihedral effect, on the other hand, is the rolling moment developed by the airplane because of sideslip and depends on many factors, including the wing dihedral angle.

3.6.1 Criterion for Lateral Stability

From the previous discussion, we observe that, for a laterally stable aircraft, a positive sideslip induces a restoring rolling moment, which according to the usual sign convention, is negative. Similarly, if a stable aircraft sideslips to port, i.e., sideslip is negative, the induced (restoring) rolling moment is positive (see Fig. 3.94). Thus the criterion for lateral stability can be expressed mathematically as

$$L_{\beta} < 0 \tag{3.336}$$

or, in coefficient form,

$$C_{l\beta} < 0 \tag{3.337}$$

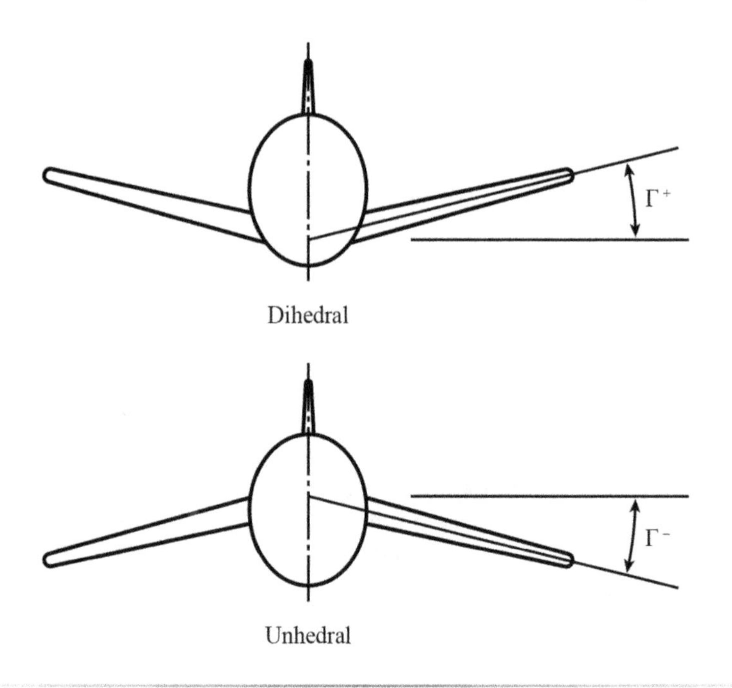

Fig. 3.93 Definition of wing dihedral and unhedral.

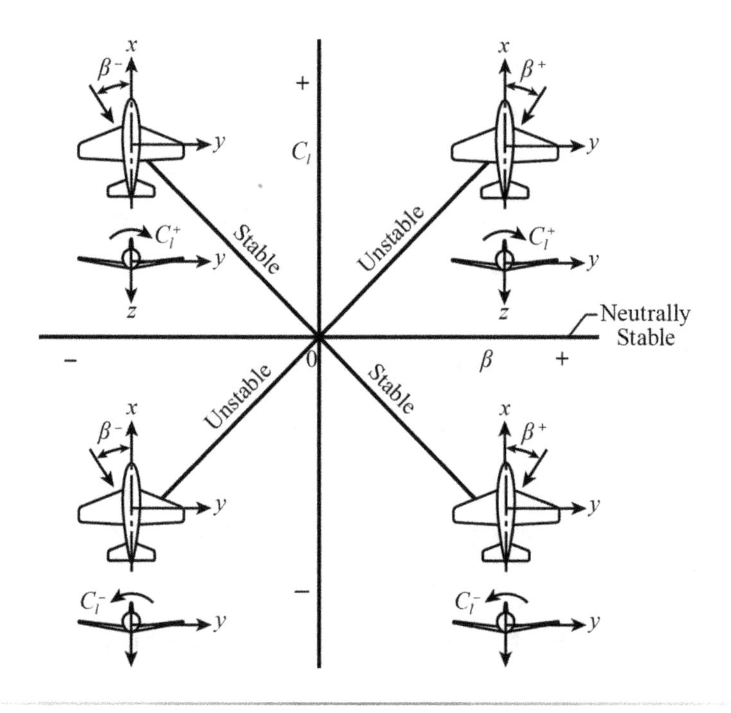

Fig. 3.94 Static lateral stability.

where

$$L_\beta = \frac{\partial L}{\partial \beta} \tag{3.338}$$

$$C_l = \frac{L}{qSb} \tag{3.339}$$

$$C_{l\beta} = \frac{\partial C_l}{\partial \beta} \tag{3.340}$$

Here, L is the rolling moment. (Remember that the symbol L is also used to denote the lift.) Thus, an airplane with $C_{l\beta} < 0$ is said to be laterally stable and that with $C_{l\beta} > 0$ is laterally unstable. An airplane with $C_{l\beta} = 0$ is said to be neutrally stable. An airplane that is neutrally stable or unstable in roll can still be flown but needs constant intervention from the pilot to counter roll disturbances, which can be quite annoying. Usually, such airplanes are made closed-loop stable using feedback control systems, which we will study in Chapters 5 and 6.

3.6.2 Evaluation of Lateral Stability

As before, we assume that the lateral stability of an airplane, as measured by the parameter $C_{l\beta}$, is equal to the sum of individual contributions from the fuselage, wing, and tail surfaces. The effects of power on lateral stability are generally small and ignored.

Fuselage Contribution

The direct contribution of the fuselage to lateral stability is negligible. However, because of its significant interference effect on the wing, it makes an indirect contribution as discussed in the following section under wing contribution.

Wing Contribution

The wing contribution to lateral stability mainly depends on 1) wing–fuselage interference, 2) wing dihedral angle, and 3) wing leading-edge sweep. A brief discussion of these effects is given in the following section.

Wing–Fuselage Interference

This interference effect depends on the location of the wing. A high wing produces a stable contribution, and a low wing produces an unstable or destabilizing contribution.

The flow over a fuselage in sideslip, in principle, is similar to that over a circular cylinder in crossflow as shown in Fig. 3.95. In positive sideslip for a high wing airplane, the inboard sections of the right wing experience a local upwash and an increase in angle of attack, whereas the inboard

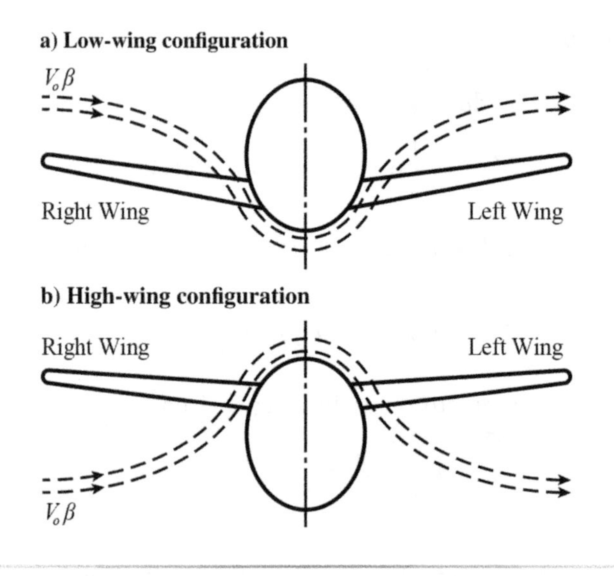

Fig. 3.95 Wing-fuselage interference in sideslip (view from front, looking aft).

sections of the port wing experience a downwash and a decrease in angle of attack. As a result, the lift on the right wing is higher compared to that on the left wing. This imbalance in lift gives rise to a stable or restoring rolling moment for a high-wing configuration (see Fig. 3.95b). In a similar way, we observe that for a low-wing configuration (Fig. 3.95a), the induced rolling moment is destabilizing. If the wing is located in midplane, the interference effects are small, and the induced rolling moment is virtually zero.

Effect of Wing Dihedral

In general, the wing dihedral has a stabilizing effect on lateral stability. To understand how the dihedral influences the lateral stability, let us refer back to Fig. 3.70 and consider an unswept, rectangular wing with a constant dihedral angle Γ operating at an angle of attack α, sideslip β, and a forward velocity V_o. The local angle of attack and local dynamic pressures, as given by Eqs. (3.248) and (3.250), are

$$\alpha_l = \alpha \pm \beta\Gamma \tag{3.341}$$

$$q_l = \frac{1}{2}\rho V_o^2 \tag{3.342}$$

The wing contribution to lateral stability at low subsonic speeds can be estimated using the strip theory as follows.

Let $c(y)$ be the local chord and $a_o(y)$ be the local sectional lift-curve slope. The lift force developed by the elemental strip RT of width dy on the right wing is given by

$$dL = q_l c(y) dy C_{l,l} \tag{3.343}$$

$$= \frac{1}{2}\rho V_o^2 c(y) a_o(y)(\alpha + \beta\Gamma) dy \tag{3.344}$$

The rolling moment due to the elemental strip RT is given by

$$dL = -\frac{1}{2}\rho V_o^2 c(y) a_o(y)(\alpha + \beta\Gamma) y \, dy \tag{3.345}$$

The rolling moment due to the right wing is given by

$$L_R = -\frac{1}{2}\rho V_o^2 \int_0^{b/2} c(y) a_o(y)(\alpha + \beta\Gamma) y \, dy \tag{3.346}$$

Similarly, the rolling moment due to the left wing is given by

$$L_L = \frac{1}{2}\rho V_o^2 \int_0^{b/2} c(y) a_o(y)(\alpha - \beta\Gamma) y \, dy \tag{3.347}$$

The total rolling moment

$$L = -\rho V_o^2 \beta\Gamma \int_0^{b/2} c(y) a_o(y) y \, dy \tag{3.348}$$

or, in coefficient form,

$$C_l = -\frac{2\beta\Gamma}{Sb} \int_0^{b/2} c(y) a_o(y) y \, dy \tag{3.349}$$

and

$$C_{l\beta} = -\frac{2\Gamma}{Sb} \int_0^{b/2} c(y) a_o(y) y \, dy \tag{3.350}$$

For a rectangular wing with a constant chord c and a constant sectional lift-curve slope a_o (constant airfoil section), Eq. (3.350) simplifies to

$$C_{l\beta} = -\frac{a_o \Gamma}{4} \tag{3.351}$$

Thus, the effect of wing dihedral ($\Gamma > 0$) is stabilizing and is directly proportional to the magnitude of wing dihedral angle Γ and the section lift-curve slope. However, having an excessive amount of dihedral may cause discomfort to the passengers because the airplane will be too sensitive in roll to any atmospheric turbulence or gust. Excessive dihedral is also disadvantageous during crosswind takeoff and landing as well as during flights with asymmetric power.

Recall that the strip theory ignores the mutual interference effects between adjacent wing sections, i.e., spanwise variation of downwash and the induced drag are ignored in strip theory.

Effect of Wing Sweep

In general, the sweep-back has a stabilizing effect, and the sweep-forward has the opposite or destabilizing effect. To understand how the wing leading-edge sweep influences the wing contribution to static lateral stability, let us refer back to Fig. 3.72 and consider a swept-back wing with no dihedral and operating at an angle of attack α and sideslip β and moving with a forward velocity V_o.

As before, let us assume that both α and β are small so that higher order terms involving these two parameters can be ignored. Then, from Eqs. (3.280) and (3.285), the effective angle of attack and effective dynamic pressure are given by

$$\alpha = \frac{\alpha}{(1 \pm \beta \tan \Lambda) \cos \Lambda} \tag{3.352}$$

$$q_l = \frac{1}{2} \rho V_o^2 \cos^2 \Lambda^2 (1 \pm \beta \tan \Lambda)^2 \tag{3.353}$$

where the first (upper) sign applies to the right wing and second (lower) sign applies to the left wing when the sideslip is positive.

At low speeds, the contribution of a sufficiently high-aspect ratio swept-back wing to lateral stability can be approximately evaluated using the strip theory as follows.

Consider a strip RT of width dy_h on the right wing. Let y_h denote the spanwise coordinate along the quarter chordline and $c(y_h)$ denote the local chord normal to the wing leading edge. Then the lift on the right wing elemental strip RT is given by

$$\mathrm{d}\,\text{Lift} = q_l c(y_h) \mathrm{d}y_h a_o \alpha_l \tag{3.354}$$

$$= \frac{1}{2} \rho V_o^2 \cos^2 \Lambda (1 + \beta \tan \Lambda)^2 a_o \left[\frac{\alpha}{(1 + \beta \tan \Lambda) \cos \Lambda} \right] c(y_h)\, \mathrm{d}y_h \tag{3.355}$$

$$= \frac{1}{2} \rho V_o^2 \cos \Lambda (1 + \beta \tan \Lambda) a_o \alpha c(y_h)\, \mathrm{d}y_h \tag{3.356}$$

The rolling moment due to strip RT is given by

$$\mathrm{d}L = -\frac{1}{2} \rho V_o^2 \cos \Lambda (1 + \beta \tan \Lambda) a_o \alpha c(y_h) y_h \cos \Lambda\, \mathrm{d}y_h \tag{3.357}$$

The rolling moment due to the right wing is given by

$$L_R = -\frac{1}{2}\rho V_o^2 \cos^2\Lambda(1+\beta\tan\Lambda)\alpha\int_0^{\frac{b}{2}\sec\Lambda} a_o c(y_h)y_h\,dy_h \tag{3.358}$$

Similarly, the rolling moment due to the left wing is given by

$$L_L = \frac{1}{2}\rho V_o^2 \cos^2\Lambda(1-\beta\tan\Lambda)\alpha\int_0^{\frac{b}{2}\sec\Lambda} a_o c(y_h)y_h\,dy_h \tag{3.359}$$

The total rolling moment is given by

$$L = -\rho V_o^2\alpha\beta\cos^2\Lambda\tan\Lambda\int_0^{\frac{b}{2}\sec\Lambda} a_o c(y_h)y_h\,dy_h \tag{3.360}$$

Then,

$$C_l = -\left(\frac{2\alpha\beta\cos\Lambda\sin\Lambda}{Sb}\right)\int_0^{\frac{b}{2}\sec\Lambda} a_o c(y_h)y_h\,dy_h \tag{3.361}$$

$$C_{l\beta} = -\left(\frac{2\alpha\cos\Lambda\sin\Lambda}{Sb}\right)\int_0^{\frac{b}{2}\sec\Lambda} a_o c(y_h)y_h\,dy_h \tag{3.362}$$

For a swept wing in level flight with zero sideslip, the lift is given by

$$\text{Lift} = \rho V_o^2\cos^2\Lambda\int_0^{\frac{b}{2}\sec\Lambda} a_o\alpha\sec\Lambda c(y_h)\,dy_h \tag{3.363}$$

Also,

$$\text{Lift} = \frac{1}{2}\rho V_o^2 S C_L \tag{3.364}$$

Equating Eqs. (3.363) and (3.364),

$$C_L = \frac{2\cos^2\Lambda}{S}\int_0^{\frac{b}{2}\sec\Lambda} a_o\alpha\sec\Lambda c(y_h)\,dy_h \tag{3.365}$$

If the sectional lift-curve slope a_o is constant along the span, then Eq. (3.365) reduces to

$$C_L = a_o\alpha\cos\Lambda \tag{3.366}$$

Substituting this value of C_L in Eq. (3.362),

$$C_{l\beta} = -\left(\frac{2\sin\Lambda C_L}{Sb}\right)\int_0^{\frac{b}{2}\sec\Lambda} c(y_h)y_h\,dy_h \tag{3.367}$$

From Eq. (3.367) we observe that wing sweep-back has a stabilizing effect on static lateral stability. Its magnitude depends on the wing leading-edge sweep angle and the lift coefficient or angle of attack. In a similar way, it can be shown that the forward sweep produces a destabilizing effect on lateral stability.

Estimation of Combined Wing Contribution

For high-aspect ratio swept-wings with dihedral, the combined wing contribution to static lateral stability can be assumed equal to the sum of the individual contributions because of sweep and dihedral, and the strip theory approach may be used. However, it should be noted that the strip theory ignores the mutual interference effects between the adjacent wing sections, i.e., the spanwise variations of downwash and induced drag are not considered in the strip theory analysis.

For more accurate estimation of the wing contribution considering the combined effects of wing–fuselage interference, dihedral, and sweep, the following empirical relation [1] can be used. This relation is valid for untwisted, straight tapered wings of arbitrary aspect ratio and taper ratio. If the wing has a twist, then an additional correction may have to be applied, which is not given here. The interested reader may refer to Datcom [1] to evaluate this correction.

For subsonic speeds [1],

$$(C_{l\beta})_{W(B)} = C_L\left[\left(\frac{C_{l\beta}}{C_L}\right)_{\Lambda_{c/2}} K_{M\Lambda}K_f + \left(\frac{C_{l\beta}}{C_L}\right)_A\right] + \Gamma\left[\frac{C_{l\beta}}{\Gamma}K_{M\Gamma} + \frac{\Delta C_{l\beta}}{\Gamma}\right] + (\Delta C_{l\beta})_{z_w} \tag{3.368}$$

where the dihedral angle Γ is in degrees. The parameter $\left(C_{l\beta}/C_L\right)_{\Lambda c/2}$ is given in Fig. 3.96, $K_{M\Lambda}$ is given in Fig. 3.97, K_f is given in Fig. 3.98, $\left(C_{l\beta}/C_L\right)_A$ is given in Fig. 3.99, $C_{l\beta}/\Gamma$ is given in Fig. 3.100, and $K_{M\Gamma}$ is given in Fig. 3.101. The value of $\left(C_{l\beta}\right)_{WB}$ given by Eq. (3.368) is per degree.

Furthermore,

$$\frac{\Delta C_{l\beta}}{\Gamma} = -0.0005\sqrt{A}\left(\frac{d}{b}\right)^2 \tag{3.369}$$

$$(\Delta C_{l\beta})_{z_w} = \frac{1.2\sqrt{A}}{57.3}\left(\frac{z_w}{b}\right)\left(\frac{2d}{b}\right) \tag{3.370}$$

where A is the theoretical wing aspect ratio, d is the average fuselage diameter at the wing root, b is the wing span, and z_w is vertical distance between the fuselage centerline and wing root quarter chord point, positive if the wing is below the fuselage centerline. The value given by Eq. (3.369) is per degree [2] and that given by the Eq. (3.370) is per degree.

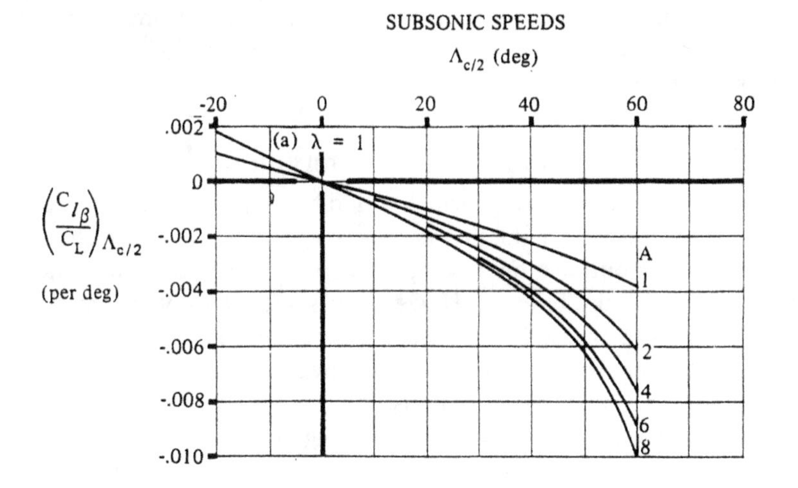

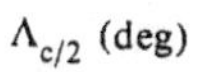

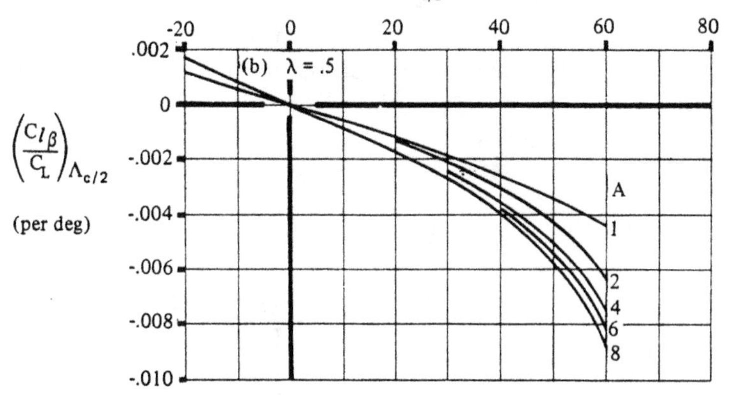

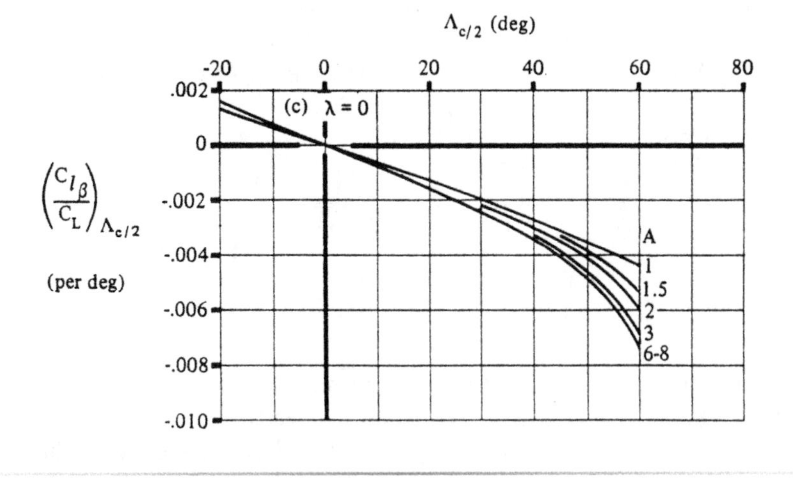

Fig. 3.96 Wing-sweep contribution to $C_{l\beta}$ (1).

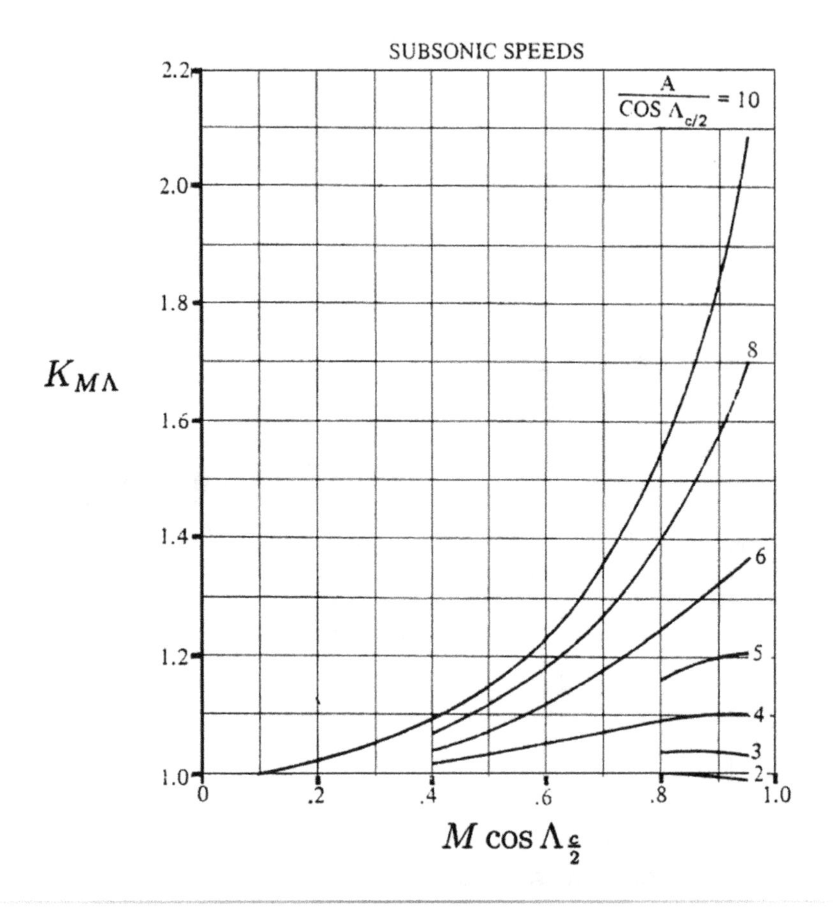

Fig. 3.97 Compressibility correction factor $K_M\Lambda$ (1).

For supersonic speeds, Datcom [1] gives the following empirical relation:

$$(C_{l\beta})_{W(B)} = -0.061C_N\left(\frac{C_{N\alpha}}{57.3}\right)[1+\lambda(1+\Lambda_{LE})]\left(1+\frac{\Lambda_{LE}}{2}\right)$$
$$\times\left(\frac{\tan\Lambda_{LE}}{\sqrt{M^2-1}}\right)\left[\frac{M^2\cos^2\Lambda_{LE}}{A}+\left(\frac{\tan\Lambda_{LE}}{4}\right)^{4/3}\right]$$
$$+\Gamma\left(\frac{C_{l\beta}}{\Gamma}+\frac{\Delta C_{l\beta}}{\Gamma}\right)+(\Delta C_{l\beta})z_w \tag{3.371}$$

where C_N is the normal force coefficient, $C_{N\alpha}$ is the wing normal-force-curve slope (per radian) and can be evaluated using the methods given in Sec. 3.3, λ is the wing taper ratio, Λ_{LE} is the wing leading-edge sweep in radians, A is the wing aspect ratio, and Γ is the wing dihedral in degrees.

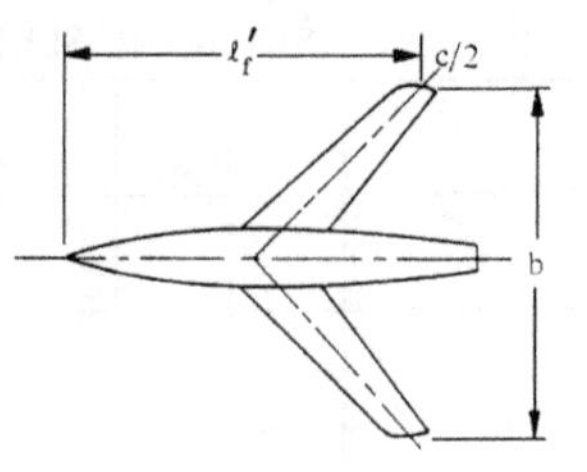

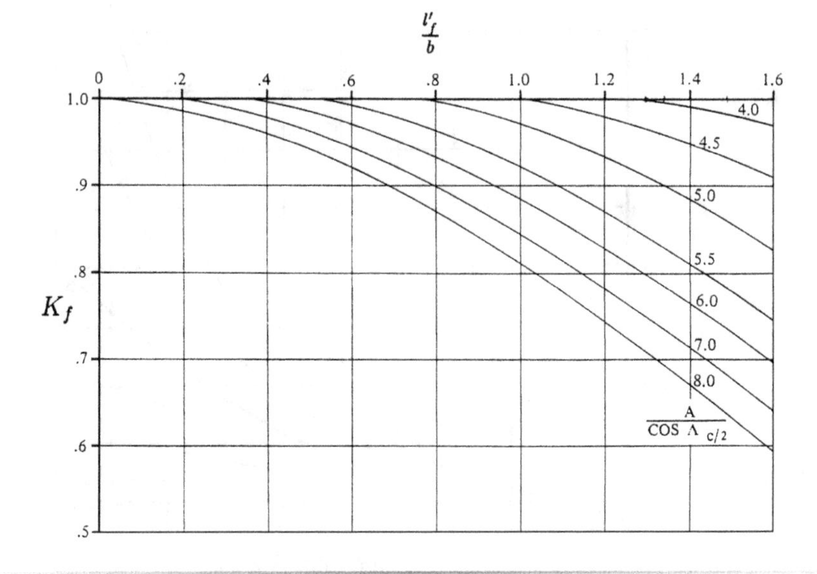

Fig. 3.98 Fuselage correction factor K_f (1).

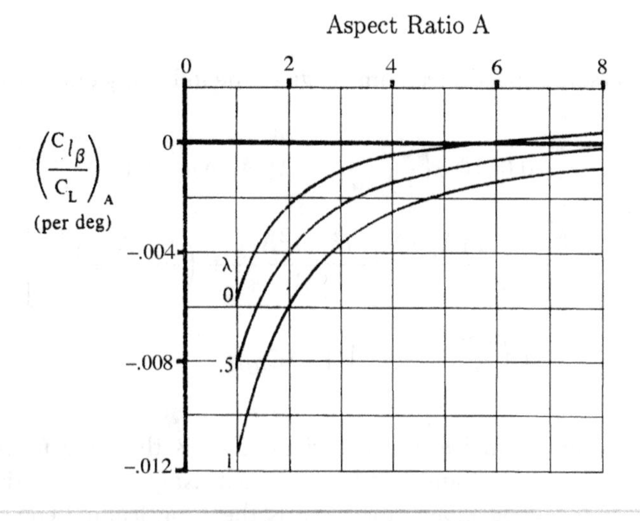

Fig. 3.99 Aspect ratio contribution to $C_{l\beta}$ (1).

The value of $(C_{l\beta})_{W(B)}$ given by Eq. (3.371) is per degree. Furthermore,

$$\frac{C_{l\beta}}{\Gamma} = \left(\frac{2}{57.3^2}\right)\left(\frac{1+2\lambda}{1+3\lambda}\right)C_{lp} \tag{3.372}$$

where C_{lp} is the roll damping parameter assumed to be known. We will learn more about roll damping when we discuss dynamic stability in Chapter 5. The parameters $\Delta C_{l\beta}/\Gamma$ and $\left(\Delta C_{l\beta}\right)_{z_w}$ are given by Eqs. (3.369) and (3.370).

Tail Contribution

The contribution of the horizontal tail to lateral stability is usually small and can be ignored. However, if the horizontal tail is comparable in size to the wing, then the methods given earlier for estimating the wing contribution can be used.

The vertical tail contribution to lateral stability is given by

$$(C_{l\beta})_V = C_{y\beta,V}\left(\frac{z}{b}\right)$$

where

$$z = z_v \cos\alpha - l_v \sin\alpha$$

where z is the rolling moment arm, l_v is distance between the center of gravity and tail aerodynamic center measured along the airplane reference line, and z_v is the distance of the vertical tail aerodynamic center measured normal to the airplane centerline as shown in Fig. 3.102. The parameter $C_{y\beta,V}$ is the slope of the vertical tail side force coefficient with respect to sideslip. As shown in Fig. 3.102, the rolling moment arm z of the vertical tail side force depends on the angle of attack because the rolling moment is with respect to the x-stability axis. As we shall explain later in Chapter 4, the x-stability axis coincides with the projection of the flight velocity vector in the plane of symmetry.

Then,

$$\left(C_{l\beta}\right)_V = C_{y\beta,V}\left(\frac{z_v\cos\alpha - l_v\sin\alpha}{b}\right) \tag{3.373}$$

For subsonic speeds, $C_{y\beta,V}$ is given by Eq. (3.302). Substitution gives

$$\left(C_{l\beta}\right)_V = -ka_v\left(1+\frac{\partial\sigma}{\partial\beta}\right)\eta_v\left(\frac{S_v}{S}\right)\left(\frac{z_v\cos\alpha - l_v\sin\alpha}{b}\right) \tag{3.374}$$

As said before, there is no general method available for estimating the side force on vertical tail at supersonic speeds. Datcom [1] gives an approximate

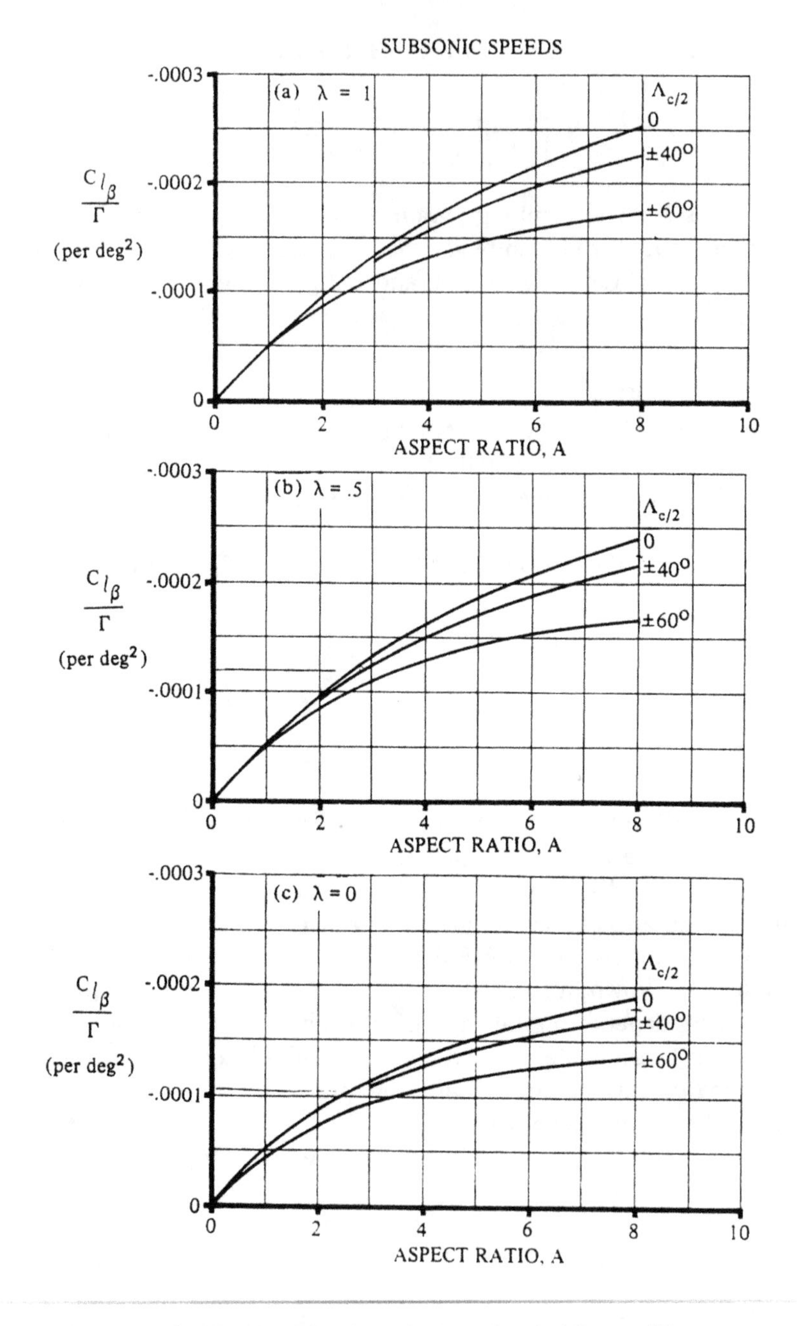

Fig. 3.100 Contribution of wing dihedral to $C_{l\beta}$ (1).

engineering method to evaluate the vertical tail side force at supersonic speeds. The interested reader may refer to Datcom [1] for more information.

Usually, $z_V \cos\alpha > l_V \sin\alpha$, so that $\left(C_{l\beta}\right)_V < 0$. Therefore, the contribution of the vertical tail to lateral stability is generally stabilizing.

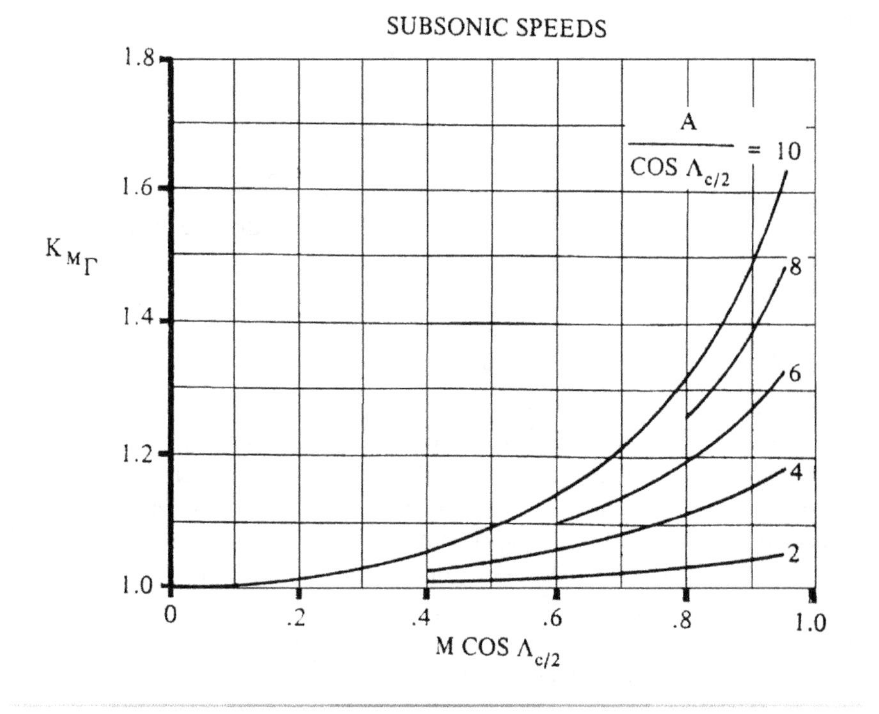

Fig. 3.101 Compressibility correction to wing dihedral effect (1).

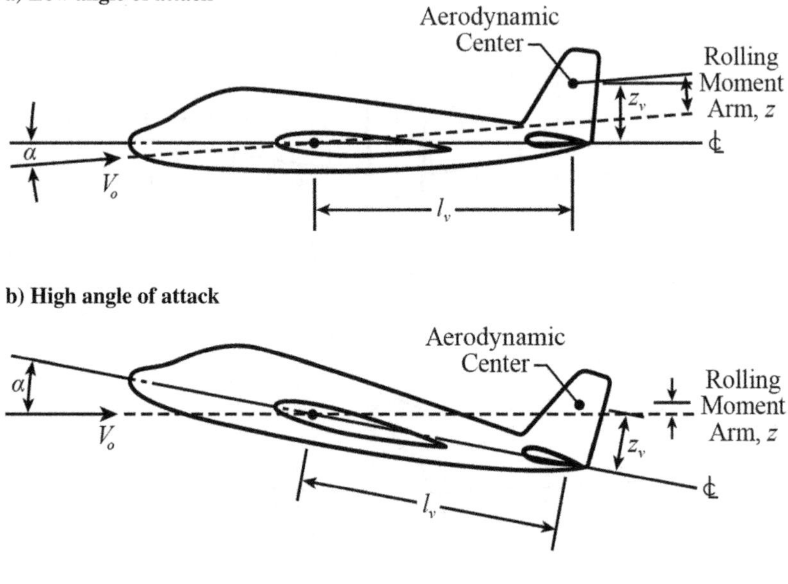

Fig. 3.102 Effect of angle of attack on vertical tail rolling moment arm.

3.6.3 Total Static Lateral Stability

The total static lateral stability of the airplane is given by the sum of the individual contributions as

$$C_{l\beta} = \left[(C_{l\beta})_W \text{ or } (C_{l\beta})_{W(B)}\right] + (C_{l\beta})_V \tag{3.375}$$

For an airplane to be laterally stable, $C_{l\beta} < 0$ over the desired Mach number, angle of attack, and sideslip range.

3.6.4 Lateral Control

On most aircraft, ailerons are usually the primary roll control devices. The ailerons are small flaps located outboard on each wing. The ailerons are deflected differentially to obtain roll control, i.e., one aileron is deflected downward, whereas the other is deflected upward. At positive angles of attack, the lift on the wing with downward deflected aileron increases and that with upward deflected aileron decreases so that a rolling moment is generated in the direction of the upward deflected aileron. The sign convention for the aileron deflection is as follows. We assume that δ_a is positive when the right aileron is deflected downward and the left aileron is deflected

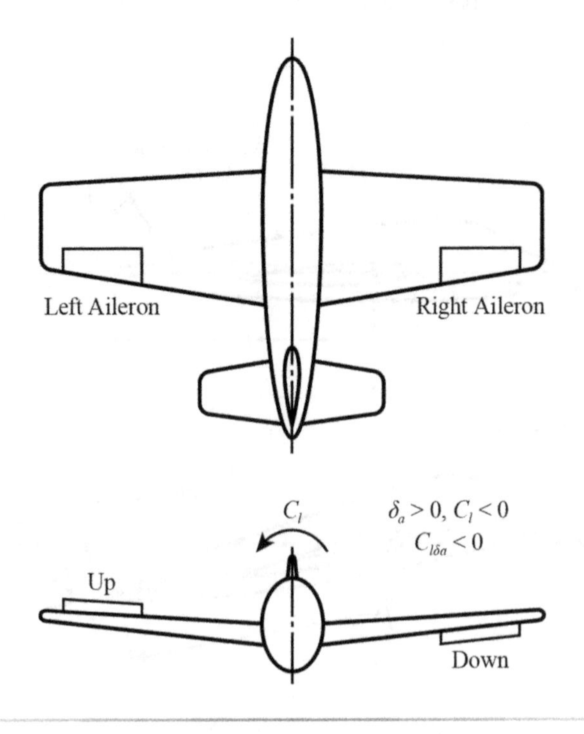

Fig. 3.103 Sign convention for aileron deflection.

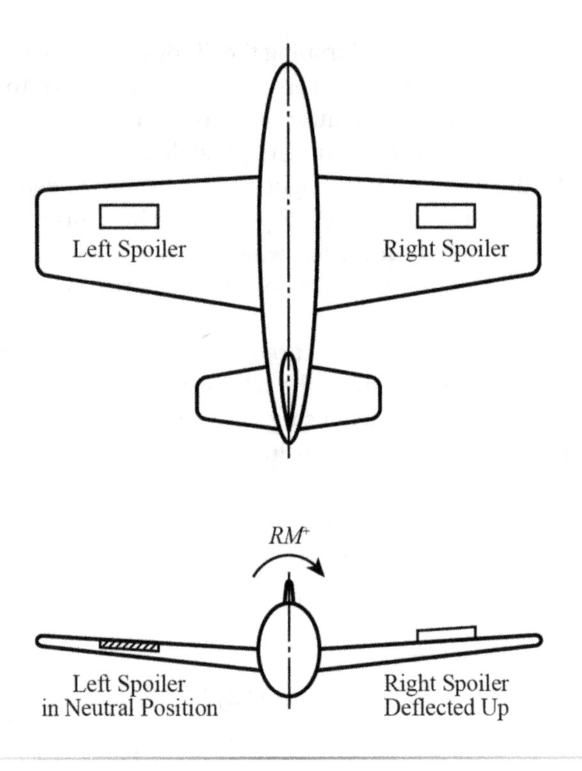

Fig. 3.104 Roll control by spoilers.

upward as shown in Fig. 3.103. This gives a negative rolling moment so that $C_{l\delta a} < 0$. Here, C_l is the rolling moment coefficient defined as

$$C_l = \frac{L}{qSb}, \quad C_{l\delta a} = \frac{\partial C_l}{\delta a} \tag{3.376}$$

and

$$\delta_a = \frac{1}{2}(\delta_{a,R} - \delta_{a,L}) \tag{3.377}$$

where L is the rolling moment, $\delta_{a,R}$ is the right aileron deflection (positive downwards), $\delta_{a,L}$ is the left aileron deflection (positive downwards). For example, if $\delta_{a,R} = +10$ deg (down) and $\delta_{a,L} = -10$ deg (up), then $\delta_a = +10$ deg.

On a tailless aircraft (no horizontal tail), the elevator and ailerons are usually combined into one set of control surfaces called elevons. Differentially deflected, the elevons produce roll control and, symmetrically deflected, they produce pitch control.

Some aircraft also employ spoilers (Fig. 3.104) for enhancing roll control authority. The spoilers are flat platelike devices mounted on the wing upper surface and are mainly used for dumping the lift or increasing the drag during the ground run for landing. Like ailerons, the spoilers have to be deflected differentially for roll control. Because they are mounted on the top surface of the wings, only upward deflection is possible. Therefore, if the aircraft wants to roll to the right, only the right spoiler has to be deflected upward. The left spoiler remains in its neutral position. This causes the right wing to lose lift, whereas the lift on the left wing remains unchanged. This imbalance in lift creates a rolling moment to the right or in the direction of the deflected spoiler.

The aileron effectiveness is measured by the rolling-moment coefficient per unit deflection (differential) of the ailerons and is denoted by $C_{l\delta_a}$ and can be approximately estimated using the strip theory. It may be noted that the strip theory ignores mutual interference effects between adjacent wing sections.

Consider a small strip RT located at a spanwise distance y as shown in Fig. 3.105. The elemental rolling moment due to a differential aileron deflection δ_a is given by

$$\Delta L_{\mathrm{RM}} = -2\Delta L y \tag{3.378}$$

$$= -2qc_a(y)\,\mathrm{d}y\Delta C_l y \tag{3.379}$$

$$= -2qc_a(y)\,\mathrm{d}y a_w \tau_a \delta_a y \tag{3.380}$$

Here, $c_a(y)$ is the local wing chord including the aileron and τ_a is the change in the wing angle of attack per unit aileron deflection. The parameter τ_a can be evaluated using the methods given in Sec. 3.3.9. For ailerons extending from spanwise stations y_i to y_o, the net rolling moment is

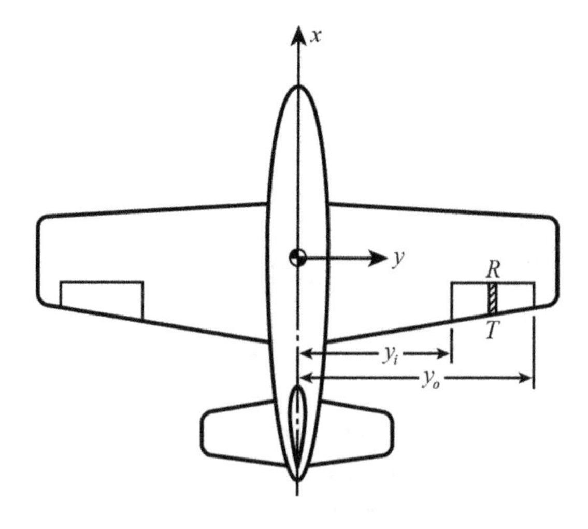

Fig. 3.105 Strip theory estimation of aileron effectiveness.

given by

$$L_{\mathrm{RM}} = -2q \int_{y_i}^{y_o} a_w \tau_a \delta_a c_a(y) y \, \mathrm{d}y \tag{3.381}$$

or

$$C_{l\delta_a} = -\frac{2a_w \tau_a}{Sb} \int_{y_i}^{y_o} c_a(y) y \, \mathrm{d}y \tag{3.382}$$

The angle of attack range in which the ailerons retain their effectiveness depends on the wing planform. As discussed in Chapter 1, for rectangular wings of high aspect ratio, high taper ratio, and low sweep, the ailerons retain their effectiveness all the way up to the wing stall because the stall on such wings originates at the root sections and progresses towards the wing tip.

However, for highly swept-back wings, the stall originates at the wingtip and spreads towards the wing root. As a result, the ailerons lose their effectiveness long before the entire wing stalls. In this respect, the swept-forward wing has a distinct advantage because the stall progresses from root to tip more or less like a rectangular wing.

Example 3.11

For the tailless vehicle configuration of Examples 3.2 and 3.8, obtain the lateral stability derivative $C_{l\beta}$ for $M = 0.7$ and $M = 2.0$.

Solution:

We have for subsonic speeds

$$(C_{l\beta})_{W(B)} = C_L\left[\left(\frac{C_{l\beta}}{C_L}\right)_{\Lambda_{c/2}} K_{M\Lambda}K_f + \left(\frac{C_{l\beta}}{C_L}\right)_A\right] + \Gamma\left[\frac{C_{l\beta}}{\Gamma}K_{M\Gamma} + \frac{\Delta C_{l\beta}}{\Gamma}\right] + (\Delta C_{l\beta})_{z_w}$$

We have $A = 2.8306$, $\Lambda_{c/2} = 25.2485$, and $\lambda = 0.1427$. Using the data given in Fig. 3.96 for $\lambda = 0$ and $\lambda = 0.5$ and interpolating for $\lambda = 0.1427$, we get $(C_{l\beta}/C_L)_{\Lambda_{c/2}} = -0.0020$. For $A/\cos\Lambda_{c/2} = 3.1298$, from Fig. 3.97, we get (approximately) $K_{M\Lambda} = 1.03$. For $l'_f/b = (l_f - 0.5c_t)/b = 21.7174/17.3228 = 1.2537$. From Fig. 3.98, we obtain $K_f = 1.0$ and, from Fig. 3.99, $(C_{l\beta}/C_L)_A = -0.0020$. From Fig. 3.100, we obtain (approximately) $C_{l\beta}/\Gamma = -0.00012$ using interpolation for the given value of λ. $K_{M\Gamma}$ as given by Fig. 3.101 depends on the Mach number. For $M = 0.7$, we obtain $K_{M\Gamma}$. With $d = d_{f,\max} = 3.048$ m, $A = 2.8306$, $z_w = 1.27$ m, and $b = 17.3228$ m

(Continued)

Example 3.11 (*Continued*)

from Eqs. (3.369) and (3.370), we obtain

$$\frac{\Delta C_{l\beta}}{\Gamma} = -0.0000148/\mathrm{deg}^2$$

$$(\Delta C_{l\beta})_{z_w} = 0.00091$$

Substituting, we get

$$(C_{l\beta})_{W(B)} = -0.0041C_L - 0.00014\Gamma + 0.00091$$

With $\Gamma = 3.5$ deg, we get

$$(C_{l\beta})_{W(B)} = -0.0041C_L + 0.0004$$

The vertical tail contribution is given by

$$(C_{l\beta})_V = -ka_v\left(1 + \frac{\partial\sigma}{\partial\beta}\right)\eta_v\left(\frac{S_v}{S}\right)\left(\frac{z_v\cos\alpha - l_v\sin\alpha}{b}\right)$$

From the solution of Example 3.8, we have $k = 0.76$, $a_v = 0.0491/\mathrm{deg}$, and $(1 + [\partial\sigma/\partial\beta])\eta_v = 1.2398$. With $z_v = 3.8290$ m, $l_v = 7.7561$ m, $b = 17.3228$ m, and $S_v/S = 0.1909$ and assuming α to be small, we get

$$(C_{l\beta})_V = -0.76 * 0.0491 * 1.2398 * 0.1909\left(\frac{3.8290 - 7.7561\alpha}{17.3228}\right)$$

$$= -0.0019 + 0.0001\alpha$$

where α is in degrees. Summing the wing and tail contributions, we have

$$C_{l\beta} = \alpha(0.0001 - 0.0041C_{L\alpha}) - 0.0015$$

where α is in degrees. The wing–body lift-curve-slope $C_{L\alpha}$ (per deg) is presented in Fig. 3.60.

At supersonic speeds, we have

$$(C_{l\beta})_{W(B)} = -0.061C_N\left(\frac{C_{N\alpha}}{57.3}\right)[1+\lambda(1+\Lambda_{LE})]\left(1+\frac{\Lambda_{LE}}{2}\right)\left(\frac{\tan\Lambda_{LE}}{\beta}\right)$$
$$\times\left[\frac{M^2\cos^2\Lambda_{LE}}{A}+\left(\frac{\tan\Lambda_{LE}}{4}\right)^{4/3}\right]+\Gamma\left(\frac{C_{l\beta}}{\Gamma}+\frac{\Delta C_{l\beta}}{\Gamma}\right)+(\Delta C_{l\beta})_{z_w}$$

where the wing–body normal force coefficient slope $C_{N\alpha}$ is per radian, Λ_{LE} is in radians, and Γ is in degrees. We have $\Lambda_{LE} = 45$ deg $= 0.7853$ rad and $\lambda = 0.1427$. Using Fig. 3.14, we get $(C_{N\alpha})_{\mathrm{theory}} = 4.0/\mathrm{rad}$. Further, using Fig. 3.15a with $\Delta y_{\perp} = 3.5355$, we get $(C_{N\alpha})/(C_{N\alpha})_{\mathrm{theory}} = 0.825$ so that $C_{N\alpha} = 3.3$ per rad.

Using Eq. (3.372), we obtain $C_{l\beta}/\Gamma = 5.5 * 10^{-4}\, C_{lp}$.

(*Continued*)

Example 3.11 (*Continued*)

Using Datcom [1] for the wing at $1.4 \leq M \leq 4.0$ (see Chapter 4 for more information on C_{lp}), we get

$$C_{lp} = A(-0.0025M^2 + 0.0283M - 0.1154)$$

With $A = 2.8306$ and $M = 2.0$, we get $C_{lp} = -0.1953$. With this, we get $C_{l\beta}/\Gamma = -1.074 * 10^{-4}$.

From calculations performed earlier for $M = 0.7$, we have

$$\frac{\Delta C_{l\beta}}{\Gamma} = -0.0000148/\text{deg}^2$$

$$(\Delta C_{l\beta})_{z_w} = 0.00091$$

Substituting and simplifying, we get

$$\begin{aligned}(C_{l\beta})_{W(B)} &= -0.061C_N\left(\frac{3.3}{57.3}\right)[1 + 0.1427 * (1 + 0.7853)]\left(1 + \frac{0.7853}{2}\right) \\ &\quad \times \left(\frac{1}{\sqrt{3}}\right) + \left[\frac{2^2 * 0.7071^2}{2.8306} + \left(\frac{1}{4}\right)^{4/3}\right] \\ &\quad + 3.5(-1.074 * 10^{-4} - 0.000148) + 0.0091 \\ &= -0.0030C_N + 0.0005\end{aligned}$$

From the solution of Example 3.8, for $M = 2.0$, we have $a_v = 1.8338$/rad, $k = 0.76$, $(1 + [\partial\sigma/\partial\beta])\eta_v = 1.2398$, $S_v/S = 0.1909$, $z_v = 3.8290$ m, $l_v = 7.7561$ m, and $b = 17.3228$ m. Substituting these values in Eq. (3.374), we get

$$(C_{l\beta})_V = -0.0013 + 0.00005\alpha$$

where α is in degrees. Then, summing up wing and tail contributions,

$$C_{l\beta} = \alpha(0.00005 - 0.0030C_{L\alpha}) - 0.0008$$

where α is in degrees and $C_{L\alpha}$ is given in Fig. 3.60. Thus, we observe that this vehicle has adequate lateral stability at subsonic and supersonic speeds. It should be remembered that these results are based on crude approximation of vertical tail contribution to lateral stability at supersonic speeds.

3.7 Summary

In this chapter, we have studied the concepts of static stability and control. The airplane was assumed to be a rigid body and possess a vertical plane of symmetry. The aerodynamic forces and moments were assumed to be linear functions of angle of attack/sideslip and control surface deflections. We assumed that the longitudinal control deflections do not produce lateral-directional forces or moments and vice versa. The effect of power was ignored. Under these assumptions, it was possible to assume that the longitudinal and lateral-directional motions of the airplane are independent of each other, and we could study the associated concepts of static stability and control separately.

The static stability is the inherent capability (open-loop stability) of the airplane to counter a disturbance in angle of attack or sideslip. The stability with respect to a disturbance in angle of attack is called the longitudinal stability. We have two types of stabilities, lateral and directional, with respect to a disturbance in sideslip. Usually, the lateral and directional motions are always aerodynamically coupled. Everything else remaining the same, the location of the center of gravity has a significant influence on the level of the static longitudinal stability. The center of gravity location has some effect on the static directional stability but very little or no influence on the static lateral stability.

The methods of analyses of static longitudinal stability were extended to simple maneuvers like pull-up in a vertical plane and steady turns in a horizontal plane.

For an airplane to be safely flyable, it must be capable of trim and have adequate levels of longitudinal, lateral, and directional stabilities. In this chapter, we have studied how these stability levels are related to various geometric, aerodynamic, and mass properties (center of gravity location). We have also presented methods suitable for preliminary estimation of these characteristics at subsonic and supersonic speeds for typical airplane configurations.

Quite often, to improve the performance, the inherent stability of the airplane is compromised. For safe flyability, such airplanes have to be provided with artificial (closed-loop) stability. This forms the subject matter of automatic flight control systems, which we will study in Chapter 6.

The controllability of the airplane is inversely proportional to the level of stability. The elevator is the primary longitudinal control, and the ailerons and rudder are the primary roll and yaw control devices. We found that the stick or pedal forces associated with the deflection of these control surfaces are directly related to their hinge-moment characteristics. For adequate feel and ease of flying, the stick or pedal forces must lie within normally acceptable limits.

In the next chapters, we will study the dynamic motions of the airplane. The basic concepts we have studied here will be very useful in understanding the more complex subject of airplane dynamics and control.

References

[1] Hoak, D. E. et al., "The USAF Stability and Control Datcom," Air Force Wright Aeronautical Laboratories, TR-83-3048, Oct. 1960 (revised 1978).

[2] Munk, M. M., "Aerodynamic Forces of Airship Hulls," NASA TR 184, 1924.

[3] Multhopp, H., "Aerodynamics of Fuselage," NASA TM-1036, 1942.
[4] Abbott, I. H., and Von Doenhoff, A. E., *Theory of Wing Sections*, Dover, New York, 1958.
[5] *Civil Airworthiness Specifications, Parts 23 and 25, Federal Aviation Regulations*, U.S. Government Printing Office, Washington, DC, 1991.
[6] *British Civil Airworthiness Requirements, Section D*, Air Registration Board England.
[7] Roskam, J., *Airplane Flight Dynamics and Automatic Flight Control, Part I*, Roskam Aviation and Engineering, Lawrence, KS, 1979.
[8] Sova, G., and Divan, P., "Aerodynamic Preliminary Analysis System II," *Part 2, User's Manual*, NASA CR-182077, 1990.
[9] Seckel, E., *Stability and Control of Airplanes and Helicopters*, Academic, New York, 1964.
[10] Pamadi, B. N., and Pai, T. G., "A Note on the Yawing Moment Due to Sideslip for Swept-Back Wings," *Journal of Aircraft*, Vol. 17, No. 5, 1980, pp. 378–380.

Problems

3.1 For an airplane configuration of the type shown in Fig. 3.57, determine the low-speed slope of pitching-moment-coefficient curve using Multhopp's method and the following data: $c_{re} = 3.6$ m, $c_t = 2.0$ m, $\bar{c} = 3.0$ m, $S = 43.5$ m^2, $b = 15$ m, $A = 5.17$, $\Lambda_{c/4} = 3.5$ deg, $l_h = 5.0$ m, $h_H = 0.05b_h$, and $C_{L\alpha,WB} = 0.06$/deg. The geometrical data are given in Table P3.1.

Table P3.1 Geometrical Parameters of the Airplane in Problem 3.1

Section	Δx, m	b_f, m	x_1, m
1	0.70	0.30	4.45
2	0.60	0.50	3.90
3	0.60	0.65	3.30
4	0.60	0.70	2.70
5	1.20	0.80	1.80
6	1.20	0.85	1.20
7	1.20	0.85	0.60
8	1.20	0.80	1.80
9	0.60	0.72	2.70
10	0.60	0.62	3.30
11	0.60	0.50	3.90
12	0.60	0.37	4.50
13	0.60	0.30	5.10
14	0.45	0.20	5.60

3.2 Estimate the lift-curve slope of a wing having a NACA 64009 airfoil section, leading-edge sweep of 45 deg, root chord 3.5 m, tip chord 2.0 m, span 15 m, and Reynolds number 6×10^6 at Mach numbers 0.5 and 2.0.

3.3 Estimate the slope of normal force coefficient and pitching-moment-coefficient curve at $M = 2$ for a cone-cylinder having a semivertex angle of 10 deg, diameter 2.0 m, and length 15 m. Assume that the moment reference point coincides with center of gravity, which is located at 10 m from the leading edge.

3.4 For a generic wing–body configuration shown in Fig. P3.4, determine $C_{L\alpha,WB}$ and $C_{m\alpha,WB}$ at $M = 0.3$ and 2.0. Assume $a_o = 0.085/\text{deg}$ (for low subsonic speeds) and $\Delta y = 2.2$. (Hint: you may ignore the hemispherical nose cap in lift and pitching moment calculations.)

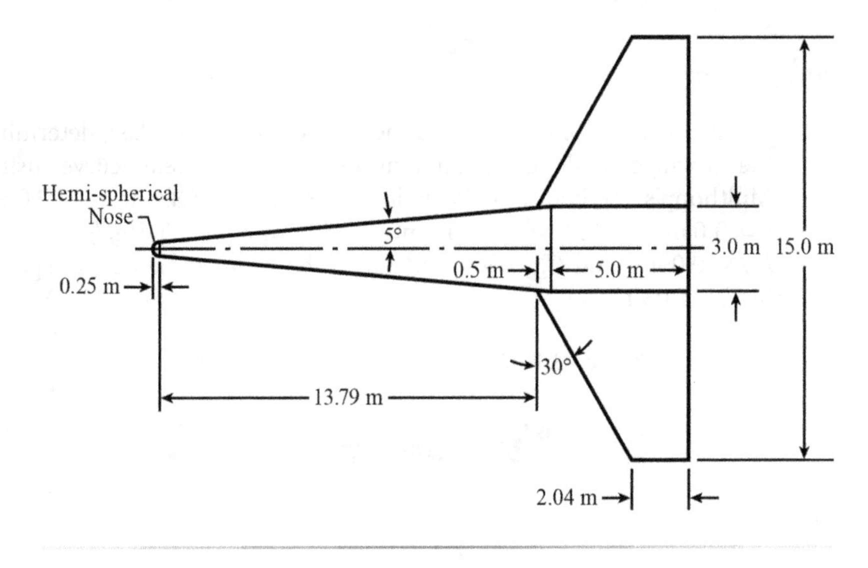

Fig. P3.4 Generic wing–body configuration.

3.5 Estimate the subsonic downwash gradient with respect to angle of attack at the aerodynamic center of a horizontal tail located 2.5 root chords downstream of the wing aerodynamic center and 5% span above the wing root chordline. For the wing, use the data of Exercise 3.2.

3.6 Estimate the dynamic pressure ratio at the horizontal tail of Exercise 3.5 for $M = 0.3$.

3.7 Estimate the low-speed elevator effectiveness τ and hinge-moment coefficients $C_{h\alpha}$ and $C_{h\delta}$ using the following data.

Horizontal tail: root chord 1.2 m, tip chord 0.8 m, span 5 m, leading-edge sweep 30 deg, and NACA 65A009 airfoil section.

Elevator: ratio of chord length ahead of the hingeline to that aft of hingeline 0.085, ratio of chord length aft of the hingeline to the horizontal tail chord 0.165, hingeline sweep 15 deg, and $t_c/2c_f = 0.06$.

Assume that the elevator extends from 20 to 80% semispan of the horizontal tail and the Reynolds number is 10^6.

3.8 Assuming $\delta_{e,o} = i_w = i_t = 0$ and $C_{ho} = 0$, show that

$$\left(\frac{dC_m}{dC_L}\right)_{free} = -\frac{C_{m\delta}(\delta_{e,R} - \delta_{e,f})}{C_L}$$

3.9 The aerodynamic center (a.c.) of wing–fuselage is located at 0.2 mac and the center of gravity is at 0.25 mac. Using the following data, determine the horizontal tail area to give a minimum static margin of 0.08 mac. $C_{m,f} = 0.1C_L$, $a_w = 0.1/\text{deg}$, $a_t = 0.08/\text{deg}$, $\epsilon = 0.45\alpha$, $l_t/\bar{c} = 2.5$, $\eta_t = 0.95$, and $S = 25\ \text{m}^2$. [Answer: $5.5\ \text{m}^2$.]

3.10 A flying wing employs elevons for pitch/roll control. Symmetrically deflected, elevons give pitch control and, when asymmetrically deflected, give roll control. Assuming $C_{m,ac} = -0.02 - 0.01\delta_e$, $W/S = 2890\ \text{N/m}^2$, and a static margin of 7%, determine the symmetric elevon deflection for trim in level flight at 154 m/s. [Answer: −3.3860 deg.]

3.11 An airplane with an all-movable horizontal tail has the following data: $W/S = 1500\ \text{N/m}^2$, $S = 30\ \text{m}^2$, $C_{L,\text{max}} = 1.5$, $\bar{c} = 3\ \text{m}$, $\bar{x}_{ac} = 0.25$, $C_{\text{mac}} = -0.05$, $\alpha_{oL,w} = -2.5$ deg, $i_w = -2.0$ deg, $a_w = 0.1/\text{deg}$, $C_{m,f} = 0.05 + 0.12C_L$, $a_t = 0.08/\text{deg}$, $\epsilon = 0.4\alpha$, $l_t = 2.5\bar{c}$, and $\eta_t = 0.8$. Assuming that the most forward and aft permissible center of gravity (c.g.) locations are $0.20\,\bar{c}$ and $0.35\,\bar{c}$, determine (a) the tail area and (b) tail setting angle. [Answer: (a) $6.87\ \text{m}^2$ and (b) −5.6475 deg.]

3.12 A canard aircraft has the following data: $a_w = 0.10/\text{deg}$, $a_c = 0.08/\text{deg}$, $C_{\text{mac},w} = -0.02$, $\bar{x}_a = -0.5$ (c.g. ahead of wing a.c.), $C_{m,f} = 0.1C_L$, $\bar{c}_c = 0.3\bar{c}$, $C_{\text{mac},c} = -0.015$, and $S_c = 0.3\ S$. Determine the distance between the canard aerodynamic center and the center of gravity in terms of $\bar{c}$ for a static margin of 0.10. Ignore upwash/downwash effects for canard and wing. [Answer: 1.25.]

3.13 An aircraft with a swept-back wing has the following data: wing loading = $2000\ \text{N/m}^2$, wing area = $30\ \text{m}^2$, $\bar{x}_a = -0.15$ (c.g. ahead of wing a.c.), $C_{m,f} = 0.05 + 0.08\ C_L$, $C_{\text{mac},w} = -0.05$, $C_{\text{mac},t} = 0$, $\eta_t = 0.9$, $S_t/S = 0.3$, and $l_t/\bar{c} = 2.5$. (a) Determine the tail-lift coefficient for steady level flight at a speed of 250 m/s and at sea level. (b) If $C_D = 0.025 + 0.05C_L^2$, what is the approximate trim drag penalty? [Answer: −0.0058, 9.46 N.]

3.14 An aircraft light aircraft has the following data: $C_L = 0.085(\alpha + 1.5)$, $i_w = -1$ deg, $i_t = -2$ deg, $C_{\text{mac},w} = -0.012$, $C_{m,f} = 0.05 + 0.10\ C_L$, $a_t = 0.075/\text{deg}$, $\eta_t = 0.9$, $\tau = 0.4$, $\overline{V}_1 = 0.7$, maximum upward elevator deflection equals 15 deg, and $C_{L,\max} = 1.2$. For stick-fixed neutral point at 0.35 mac, plot the elevator deflection vs lift coefficient for various center of gravity locations on either side of the neutral point.

3.15 For the aircraft in Problem 3.14, what is the permissible most forward center of gravity location in terms of mac? [Answer: −0.0164.]

3.16 An aircraft with a static margin of 0.15 has the following data: wing loading = 1800 N/m^2, $a_w = 0.08/\text{deg}$, $a_t = 0.06/\text{deg}$, $C_{\text{mac},w} = -0.04$, $C_{\text{mac},t} = 0$, $\overline{V}_1 = 0.55$, $\alpha_{oLw} = 0$, $\text{d}C_{L,t}/\text{d}\delta_e = 0.025/\text{deg}$, $\epsilon = 0.36\alpha$, $\eta_t = 0.95$, $S_e = 4\ \text{m}^2$, $c_e = 0.5$ m, $i_w = 0$, $i_t = -3$ deg, $G_1 = 1.6$ rad/m $C_{m,f} = 0.05 + 0.1\ C_L$, $C_{h\alpha} = -0.002/\text{deg}$, $C_{h\delta_e} = -0.005/\text{deg}$, and $C_{h\delta_t} = -0.006/\text{deg}$. Assuming level flight at sea-level conditions, determine (a) speed for stick-force trim at zero tab deflection and (b) tab deflection for stick-force trim at a speed of 100 m/s. [Answer: (a) 60.0139 m/s and (b) −3.6044 deg.]

3.17 For the aircraft of Problem 3.16, determine the stick-force gradients for Cases 1 and 2. [Answer: 7.5540 Ns/m and 4.5334 Ns/m.]

3.18 A trainer aircraft has a wing loading of 1850 N/m^2 and a static margin of 0.15 while flying at 8 deg angle of attack at an altitude of 8000 m ($\sigma = 0.4292$). (a) To what altitude should the pilot climb the aircraft so that, at the bottom of a 1.5 g pull-out, he is once again at 8000 m altitude?

For this aircraft, determine (b) the elevator required per g and (c) stick force per g based on the following data: $\epsilon = 0.35\alpha$, $a_w = 0.1/\text{deg}$, $a_t = 0.092/\text{deg}$, $\bar{c} = 2.5$ m, $S_t/S = 0.3$, $l_t = 2.5\ \bar{c}$, $\eta_t = 0.9$, $\tau = 0.35$, $C_{h\alpha} = -0.003/\text{deg}$, $C_{h\delta} = -0.006/\text{deg}$, $S_e = 1.85\ \text{m}^2$, $\bar{c}_e = 0.61$ m, and $G_1 = 1.2$ rad/m. [Answer: (a) 9793.4 m, (b) −6.6605 deg/g, and (c) −65.2776 N/g.]

3.19 An aircraft weighs 66,825 N and has a wing area of 46 m^2 and a tail length of 10.64 m. The center of gravity and wing aerodynamic centers in terms of mean aerodynamic chord are, respectively, at 0.35 and 0.26 from the leading edge of mac. The lift-curve slope of wing and that of horizontal tail are 0.09/deg and 0.07/deg, respectively. The tail volume ratio is 0.6. Assuming $C_{mf} = 0.1C_L$, $\epsilon = 0.3\alpha$, $\eta_t = 0.9$, $\tau = 0.5$, $C_{h\alpha} = -0.003/\text{deg}$, $C_{h\delta} = -0.006/\text{deg}$, $S_e = 1.9\ \text{m}^2$, $\bar{c}_e = 0.55$ m, and $G_1 = 1.2$ rad/m, determine (a) the stick-fixed maneuver margin and (b) the incremental elevator setting for a coordinated turn with 30 deg bank at an altitude of 2200 m ($\sigma = 0.8$) and a lift coefficient of 0.2. (c) What is the stick force per g? [Answer: (a) 0.2374, (b) −0.3886 deg, and (c) −67.9757 N/g.]

3.20 For the airplane configuration shown in Fig. P3.20, the fuselage side area is 40.5 m^2, the maximum fuselage width is 2.5 m, the vertical tail area is 7.0 m^2, and the ratio of rudder chord to vertical tail chord is 0.30. The rudder extends from 1.56 to 3.78 m from the fuselage centerline as shown. Assume that the wing, horizontal tail,

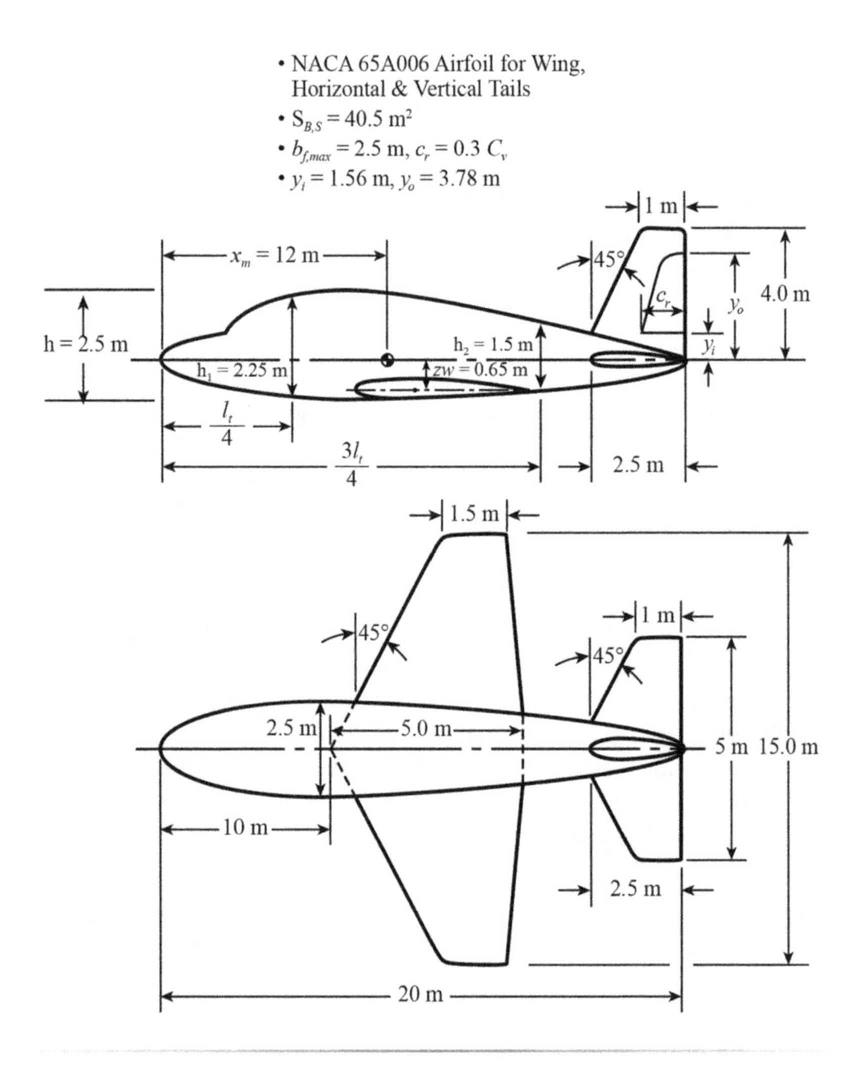

Fig. P3.20 Sketch of an airplane.

and vertical tail have NACA 65A006 airfoil section. Determine the following:

a) The static directional stability parameter $(C_{n\beta})_{\text{fix}}$ at $M = 0.3$ and an altitude of 3000 m.

b) The rudder effectiveness $C_{n\delta r}$ at low subsonic speeds.

c) The rudder hinge-moment coefficients $C_{h\beta}$ and $C_{h\delta r}$ at low subsonic speeds, assuming that the hingeline coincides with the leading edge of the rudder and $\Lambda_{HL} = 20$ deg.

d) The rudder-free directional stability parameter $(C_{n\beta})_{\text{free}}$ at $M = 0.3$.

3.21 Using strip theory, determine the yawing moment developed by a flying wing having a rectangular planform and moving at a forward speed of 75 m/s and a sideslip of 5 m/s at sea level. The wing loading is 2000 N/m^2, aspect ratio is 8, taper ratio is 0.7, sectional lift-curve slope is 0.10/deg, sectional drag coefficient $C_{D,l} = 0.015 + 0.0008\alpha$ deg, dihedral is 5 deg, and wing span is 16 m. [Answer: −1289 Nm.]

3.22 Plot the variation of $(C_{n\beta})_{\Lambda,W}$ with angle of attack using the strip theory for a swept-back wing with the following data: $\Lambda_{LE} = 30$ deg, aspect ratio = 8.0, span = 16.0 m, taper ratio = 1.0, sectional drag coefficient $C_{D,l} = 0.013 + 0.0007\alpha$ deg, sectional lift-curve slope $a_o = 0.1$/deg, sectional stall angle $\alpha_{\text{stall}} = 14$ deg, and $\Gamma = 0$.

3.23 An aircraft has a wing loading of 2850 N/m^2, a wing span of 27 m, a maximum lift coefficient of 1.75, and vertical tail lift-curve slope of 0.082/deg. $(C_{n\beta})_{\text{fix}} = 0.015$/deg, vertical tail volume ratio is 0.2, and the coefficient $k = 0.90$. Assuming that 1 deg of rudder deflection changes the vertical tail sideslip by 0.3 deg and that the maximum rudder deflection is restricted to ±25 deg, determine the maximum crosswind speed that can be permitted for takeoff at sea level. Assume that the unstick velocity is 1.2 times the stall velocity and $\eta_v = 0.92$. [Answer: 7.3598 m/s.]

3.24 A twin jet engine aircraft has a thrust of 20,000 N per engine, and the engines are separated by a spanwise distance of 10 m. The wing area is 60 m^2 and the wing span is 15 m. Assuming that the rudder effectiveness vanishes beyond ±25 deg deflection, determine the minimum rudder effectiveness to hold zero sideslip with one engine losing all the thrust at a forward speed of 75 m/s at sea level. [Answer: −0.0013/deg.]

3.25 For a high-aspect ratio swept-back wing with leading-edge sweep angle A and dihedral Γ, show that

$$(C_{L\beta})_W = -\frac{2\cos^2\Lambda(\Gamma + \alpha\tan\Lambda)}{Sb}\int_0^{\frac{b}{2}\sec\Lambda} a_o c(y_h) y_h\, dy_h$$

3.26 For the airplane in Problem 3.20, determine $C_{l\beta}$ at $M = 0.3$ and an altitude of 3000 m and at $M = 2.0$ and 15,000 m altitude. Assume $\Gamma = 5$ deg for the wing. Also, assuming that $c_a(y) = 0.3c(y)$ and ailerons extend from $y = 0.6s$ to $0.9s$, where s is the semispan, determine the aileron effectiveness at low subsonic speeds.

3.27 For the flying wing in Exercise 3.21, using strip theory, determine the rolling moment.

3.28 For the swept-back wing of Exercise 3.22, using strip theory, plot the variation of $C_{l\beta}$ with angle of attack.

Chapter 4

Equations of Motion and Estimation of Stability Derivatives

4.1 Introduction

In the preceding chapter, we studied static stability and control of airplanes. We assumed that the motion following either an external disturbance such as a wind gust or an internal disturbance like a control input was so slow that the inertia and damping forces/moments could be ignored. Thus, we essentially assumed the airplane to be a static system and studied the stability and control based on the static forces and moments acting on the airplane following a disturbance.

In this chapter, we will study the airplane as a dynamic system and derive equations of motion. We will consider the influence of its mass and inertia on the motion. We will also consider aerodynamic damping effects. However, we will not be considering the aeroelastic effects and, instead, we will assume that the airplane functions like a rigid body.

The foundations of the airplane dynamic stability and response were laid by the pioneering work of Bryan [1]. His formulation was based on two principal assumptions: 1) the instantaneous aerodynamic forces and moments depend only on instantaneous values of the motion variables, and 2) the aerodynamic forces and moments vary linearly with motion variables. This approach of Bryan [1], introduced more than 100 years ago, is used even today in the study of dynamic stability, control, and response of the airplane and forms the basis of the subject matter discussed in this chapter.

To begin with, we will discuss various axes systems used in the study of airplane dynamics and present relations for transforming vectors from one coordinate system into another system. We will then formulate the problem of airplane dynamics and derive equations of motion for six-degree-of-freedom analyses. Because these equations are, in general, coupled and nonlinear, it is difficult to obtain analytical solutions. In view of this, we will assume that the motion following a disturbance is one of small amplitudes in all the disturbed variables. With this assumption and the usual approximation that the airplane has a vertical plane of symmetry, it is possible to linearize and decouple the equations of motion into two sets, one for the longitudinal motion and another for lateral-directional motion. Then we use the method of Bryan [1] and assume that the aerodynamic forces and moments in the disturbed state depend only on the instantaneous values of motion variables and evaluate them using the method of Taylor series expansion. With these approximations, the longitudinal and lateral-directional equations of motion become linear in all the motion variables. The aerodynamic coefficients appearing in the Taylor series expansion are called stability and control derivatives. Finally, we will present engineering methods to evaluate these derivatives for typical airplane configurations.

4.2 Axes Systems

In the formulation of flight dynamic problems, we need to introduce several coordinate systems for specifying the position, velocity, acceleration, forces, and moments acting on the vehicle. Why do we need so many axes systems? Because, we do not have a single axes system that is suitable for specifying all these quantities. Further, the choice of a particular coordinate system in which the equations of motion are written and solved is also a matter of convenience to the analyst. In flight dynamic studies, it is customary to choose the so-called "moving" axes system for the solution of equations of motion as discussed later in this chapter. In the following, we will discuss some of the most commonly used axes systems in flight dynamics and present relations for transforming vectors from one axes system into another.

4.2.1 Inertial Axes System ($Ox_iy_iz_i$)

For every flight dynamic problem, it is necessary to specify an inertial frame of reference because Newton's laws of motion are valid only when the acceleration is measured with respect to an inertial frame. In other words, the acceleration of a body in Newton's second law of motion, $F = ma$, is the acceleration with respect to an inertial frame of reference, which is actually at rest in the universe. While it is a difficult task to find such an inertial reference system, for most of the flight dynamic problems, a nonrotating reference system ($Ox_iy_iz_i$) placed at the center of the Earth (Fig. 4.1) is a reasonably good approximation for an inertial system of reference. In this approximation, the orbital motion of the Earth around the sun is ignored. However, for interplanetary motions like the mission to Mars, the orbital motion of the Earth has to be considered and some other inertial frame of reference such as one centered at the sun may have to be used.

4.2.2 Earth-Fixed Axes System ($Ox_Ey_Ez_E$)

Another axes system that is useful in flight dynamics is an axes system fixed at the center of the Earth and rotating with the Earth. The $Ox_Ey_Ez_E$

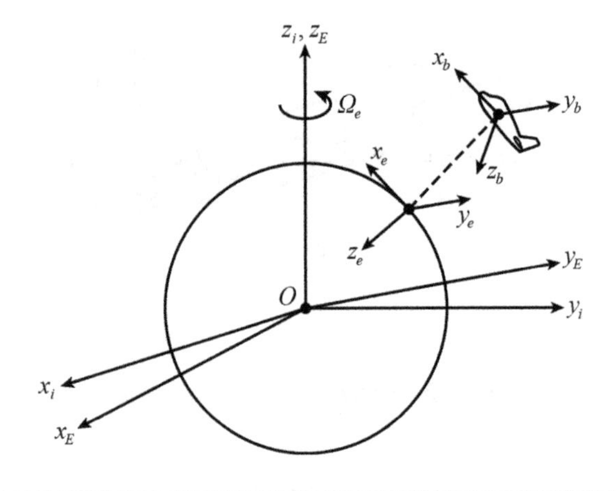

Fig. 4.1 Coordinate systems.

system shown in Fig. 4.1 is such a system. The angular velocity Ω_e of the $Ox_Ey_Ez_E$ system with respect to the $Ox_iy_iz_i$ system is directed along the Oz_i or Oz_E axis. An Earth-fixed reference system is useful in specifying the position and velocity of the vehicle with respect to the rotating Earth.

4.2.3 Navigational System ($Ox_ey_ez_e$)

The origin of this frame of reference $Ox_ey_ez_e$ (Fig. 4.1) is located on the surface of the Earth such that the Oz_e axis is directed towards the center of the spherical Earth. The Ox_e axis usually points towards the local north and Oy_e points towards the local east to form a right-hand system. Usually, the location of the origin of this coordinate system is chosen so that it lies directly beneath the vehicle at $t = 0$. Such a system of reference is useful to define the position and velocity of the vehicle with respect to the launch point.

4.2.4 Body Axes System ($Ox_by_bz_b$)

Any set of axes system fixed to the vehicle and moving with it is called the body axes system (Fig. 4.1). Generally, the mass center or the center of gravity is chosen as the origin of the body axes system. If the vehicle has a plane of symmetry, as most of the aerospace vehicles do, the Ox_bz_b plane coincides with the plane of symmetry (Fig. 4.2). The Ox_b axis usually lies along the longitudinal centerline or zero-lift line of the vehicle and points in the flight direction. The Oy_b axis is perpendicular to the Ox_bz_b plane and points towards the right side or in the direction of the starboard wing.

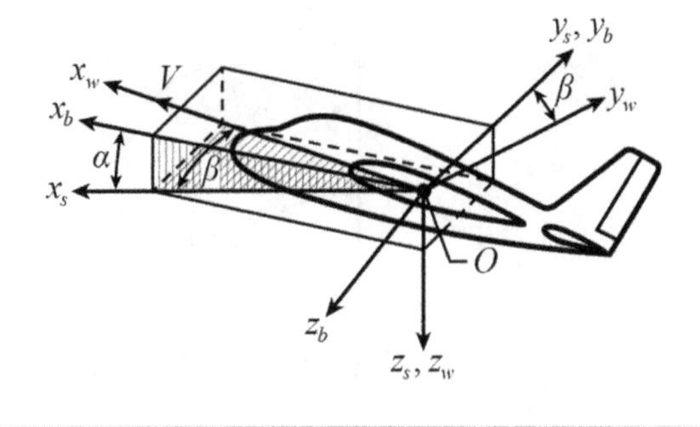

Fig. 4.2 Body-fixed axes systems.

The Oz_b axis lies in the plane of symmetry and points downward so as to form a right-hand system. The body axes system is useful to specify the moments and products of inertia of the body as well as the forces and moments acting on the vehicle.

Stability Axes System ($Ox_sy_sz_s$)

A special case of the body axes system is the so-called stability axes system $Ox_sy_sz_s$ (Fig. 4.2). This axes system is widely used in the study of airplane motion involving small disturbances from a steady reference flight condition. The Ox_s axis lies in the plane of symmetry and, if the reference flight condition is symmetric ($\beta = 0$), points in the opposite direction to the relative wind. If $\beta \neq 0$, then the Ox_s axis is chosen to coincide with the projection of the relative velocity vector in the plane of symmetry. As a result, the component of velocity along Oz_s is zero. The Oy_s axis is normal to the plane of symmetry and points to the right side or the starboard wing and Oz_s points downward to form a right-hand system. The angle between Ox_b and Ox_s or that between the Oz_b and Oz_s axes is usually equal to the angle of attack of the vehicle.

The angle that locates this axes system with respect to the body axes system $Ox_by_bz_b$ is the angle of attack. It may be noted that, if $\beta = 0$, the drag is directed opposite to the Ox_s axis and lift is directed opposite to the Oz_s axis.

Wind Axes System ($Ox_wy_wz_w$)

Another special case of the body axes system is the wind axes system (Fig. 4.2). Here, the Ox_w axis points in the direction opposite to the relative wind and Oz_w lies in the plane of symmetry. The Oy_w axis is normal to the Ox_wz_w plane and points towards the right side of the vehicle.

The angles locating the wind axes system $Ox_wy_wz_w$ with respect to the body axes system $Ox_by_bz_b$ are sideslip β and angle of attack α. Here, the drag and lift are always directed opposite to the Ox_w and Oz_w axes, respectively.

4.2.5 Axes Transformation

There are several methods of performing axes transformations so that a vector given in one reference system can be expressed with respect to another axes system. The methods that are most commonly used in flight dynamics are 1) Euler angles, 2) direction cosine matrices, and 3) quaternions or the Euler four parameter method. In the following, we will discuss each of these three methods.

Euler Angles

In this method, the orientation of a given reference frame relative to another reference system is specified by three angles—ψ, θ, and ϕ, which are called Euler angles. These three angles ψ, θ, and ϕ form three consecutive rotations in that order so that one coordinate axes system is made to coincide with another system. Note that the order of these rotations is extremely important because any other order of rotation would normally result in a different orientation.

Suppose we want to describe the orientation of the $Ox_2y_2z_2$ system with respect to the $Ox_1y_1z_1$ system. We have to perform three consecutive rotations ψ, θ, and ϕ to take the $Ox_1y_1z_1$ system from the given orientation and make it coincide with the $Ox_2y_2z_2$ system (Fig. 4.3). If the $Ox_2y_2z_2$ system happens to be a body-fixed system and $Ox_1y_1z_1$ an Earth-fixed system, then the three Euler angles ψ, θ, and ϕ give the orientation of the aircraft in space with respect to the Earth. Usually, ψ is called the heading or azimuth angle, θ the inclination or pitch angle, and ϕ the bank angle. A continuous display of these three Euler angles on the cockpit instrumentation enables the pilot to have a knowledge of the orientation of the aircraft with respect to the Earth.

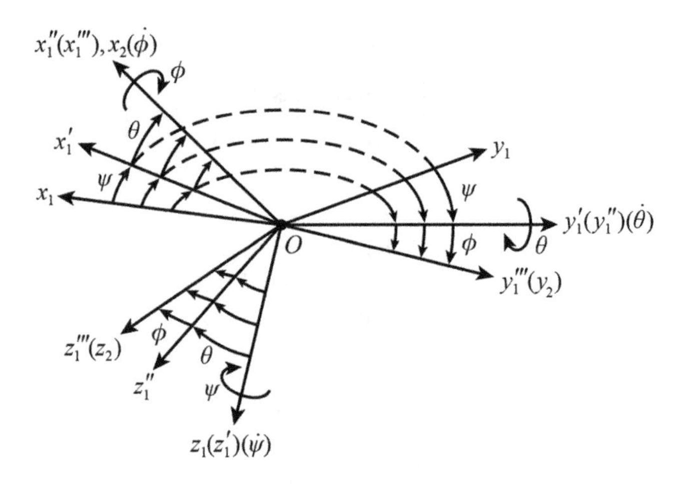

Fig. 4.3 Euler angles.

The sequence of the three Euler rotations is as follows:

1. A rotation ψ about the Oz_1 axis taking Ox_1 to Ox_1' and Oy_1 to Oy_1' such that Ox_1' falls in the plane Ox_2z_2 and Oy_1' falls in the Oy_2z_2 plane.
2. A second rotation θ about Oy_1' taking Ox_1' to Ox_1'' and $Oz_1'$to Oz_1'' so that Ox_1'' coincides with Ox_2 and Oz_1''falls in the Oy_2z_2 plane.
3. A third and final rotation ϕ about $Ox_1''\left(Ox_1'''\text{or } Ox_2\right)$ taking Oy_1'' to $Oy_1'''(Oy_2)$ and Oz_1'' to $Oz_1'''(Oz_2)$.

To avoid ambiguities, the ranges of the Euler angles are limited as follows:

$$-\pi \leq \psi \leq \pi \tag{4.1}$$

$$-\frac{\pi}{2} \leq \theta \leq \frac{\pi}{2} \tag{4.2}$$

$$-\pi \leq \phi \leq \pi \tag{4.3}$$

In view of these restrictions, the Euler angles will have discontinuous (sawtooth) variations for vehicle motions involving continuous rotations. For example, in a steady rolling maneuver, the bank angle ϕ will have a sawtooth variation.

Transformation Matrices Using Euler Angles

1. For first rotation ψ, we have the following relation (see Fig. 4.4a):

$$x_1' = x_1 \cos\psi + y_1 \sin\psi \tag{4.4}$$

$$y_1' = -x_1 \sin\psi + y_1 \cos\psi \tag{4.5}$$

$$z_1' = z_1 \tag{4.6}$$

Or, in matrix form,

$$\begin{bmatrix} x_1' \\ y_1' \\ z_1' \end{bmatrix} = \begin{bmatrix} \cos\psi & \sin\psi & 0 \\ -\sin\psi & \cos\psi & 0 \\ 0 & 0 & 1 \end{bmatrix} \begin{bmatrix} x_1 \\ y_1 \\ z_1 \end{bmatrix} \tag{4.7}$$

Let

$$C = \begin{bmatrix} \cos\psi & \sin\psi & 0 \\ -\sin\psi & \cos\psi & 0 \\ 0 & 0 & 1 \end{bmatrix} \tag{4.8}$$

so that

$$\begin{bmatrix} x_1' \\ y_1' \\ z_1' \end{bmatrix} = C \begin{bmatrix} x_1 \\ y_1 \\ z_1 \end{bmatrix} \tag{4.9}$$

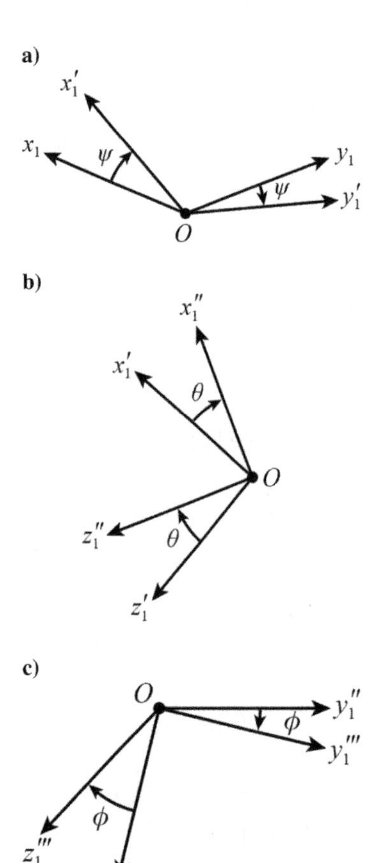

Fig. 4.4 Orientation of various axes during transformation.

2. Next, we perform the rotation θ as shown in Fig. 4.4b about the Oy_1' axis. Then we have

$$x_1'' = x_1' \cos\theta - z_1' \sin\theta \tag{4.10}$$

$$y_1'' = y_1' \tag{4.11}$$

$$z_1'' = x_1' \sin\theta + z_1' \cos\theta \tag{4.12}$$

Or, in matrix form,

$$\begin{bmatrix} x_1'' \\ y_1'' \\ z_1'' \end{bmatrix} = \begin{bmatrix} \cos\theta & 0 & -\sin\theta \\ 0 & 1 & 0 \\ \sin\theta & 0 & \cos\theta \end{bmatrix} \begin{bmatrix} x_1' \\ y_1' \\ z_1' \end{bmatrix} \tag{4.13}$$

Let

$$B = \begin{bmatrix} \cos\theta & 0 & -\sin\theta \\ 0 & 1 & 0 \\ \sin\theta & 0 & \cos\theta \end{bmatrix} \tag{4.14}$$

So that

$$\begin{bmatrix} x_1'' \\ y_1'' \\ z_1'' \end{bmatrix} = B \begin{bmatrix} x_1' \\ y_1' \\ z_1' \end{bmatrix} = BC \begin{bmatrix} x_1 \\ y_1 \\ z_1 \end{bmatrix} \tag{4.15}$$

3. Finally, perform the rotation ϕ about the Ox_1''' axis (see Fig. 4.4c) to obtain

$$x_1''' = x_1'' \tag{4.16}$$

$$y_1''' = y_1'' \cos\phi + z_1'' \sin\phi \tag{4.17}$$

$$z_1''' = -y_1'' \sin\phi + z_1'' \cos\phi \tag{4.18}$$

Or, in matrix form,

$$\begin{bmatrix} x_1''' \\ y_1''' \\ z_1''' \end{bmatrix} = \begin{bmatrix} 1 & 0 & 0 \\ 0 & \cos\phi & \sin\phi \\ 0 & -\sin\phi & \cos\phi \end{bmatrix} \begin{bmatrix} x_1'' \\ y_1'' \\ z_1'' \end{bmatrix} \tag{4.19}$$

Let

$$A = \begin{bmatrix} 1 & 0 & 0 \\ 0 & \cos\phi & \sin\phi \\ 0 & -\sin\phi & \cos\phi \end{bmatrix} \tag{4.20}$$

so that

$$\begin{bmatrix} x_1''' \\ y_1''' \\ z_1''' \end{bmatrix} = \begin{bmatrix} x_2 \\ y_2 \\ z_2 \end{bmatrix} = A \begin{bmatrix} x_1'' \\ y_1'' \\ z_1'' \end{bmatrix} = ABC \begin{bmatrix} x_1 \\ y_1 \\ z_1 \end{bmatrix} \tag{4.21}$$

Let $T_1^2 = ABC$. Here, T_1^2 is the matrix that transforms a vector from the $Ox_1y_1z_1$ system to the $Ox_2y_2z_2$ system. Performing the indicated matrix multiplications, we obtain

$$T_1^2 = \begin{bmatrix} \cos\theta\cos\psi & \cos\theta\sin\psi & -\sin\theta \\ \sin\theta\sin\phi\cos\psi - \sin\psi\cos\phi & \sin\psi\sin\theta\sin\phi + \cos\psi\cos\phi & \sin\phi\cos\theta \\ \sin\theta\cos\phi\cos\psi + \sin\psi\sin\phi & \sin\psi\sin\theta\cos\phi - \cos\psi\sin\phi & \cos\phi\cos\theta \end{bmatrix} \tag{4.22}$$

Then,

$$X_2 = T_1^2 X_1 \tag{4.23}$$

where X denotes a vector expressed in matrix form and its suffix indicates the coordinate system in which the vector X has its components. For example, X_1 is a vector having its components in the $Ox_1y_1z_1$ axes system.

Each transformation matrix A, B, or C is an orthogonal matrix. An important property of an orthogonal matrix is that its transpose is equal to its inverse. The transformation matrix T_1^2, which is the product of the three orthogonal matrices, is also orthogonal. Using this property, we can obtain the transformation matrix T_2^1 which transforms a vector from $Ox_2y_2z_2$ system to $Ox_1y_1z_1$ as

$$T_2^1 = (T_1^2)^{-1} = (T_1^2)' \tag{4.24}$$

$$= \begin{bmatrix} \cos\theta\cos\psi & \sin\theta\sin\phi\cos\psi - \sin\psi\cos\phi & \sin\theta\cos\phi\cos\psi + \sin\psi\sin\phi \\ \cos\theta\sin\psi & \sin\psi\sin\theta\sin\phi + \cos\psi\cos\phi & \sin\psi\sin\theta\cos\phi - \cos\psi\sin\phi \\ -\sin\theta & \sin\phi\cos\theta & \cos\phi\cos\theta \end{bmatrix} \tag{4.25}$$

where superscripts $^{-1}$ and $'$ denote the inverse and transpose of a matrix, respectively.

Transformation of Vectors

Let us consider the transformation of vectors between various coordinate systems as follows.

Inertial to Earth-Fixed System Here we are considering transformation between the $Ox_iy_iz_i$ and $Ox_Ey_Ez_E$ systems. Let the $Ox_Ey_Ez_E$ system coincide with $Ox_iy_iz_i$ system at $t = 0$. For $t > 0$, $\psi = \Omega_e t$, $\theta = \phi = 0$, where Ω_e is the angular velocity of the Earth about Oz_i or Oz_E axis. Then using Eq. (4.22),

$$X_E = T_i^E X_i \tag{4.26}$$

$$= \begin{bmatrix} \cos\Omega_e t & \sin\Omega_e t & 0 \\ -\sin\Omega_e t & \cos\Omega_e t & 0 \\ 0 & 0 & 1 \end{bmatrix} X_i \tag{4.27}$$

The reverse transformation is given by

$$X_i = T_E^i X_E \tag{4.28}$$

where

$$T_E^i = \left(T_i^E\right)^{-1} = \left(T_i^E\right)' \tag{4.29}$$

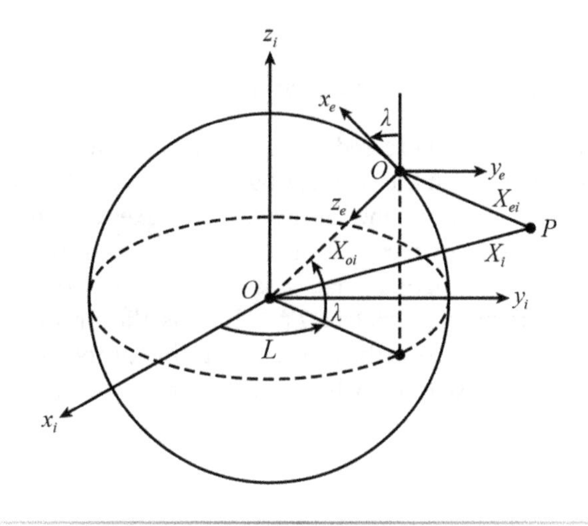

Fig. 4.5 Inertial and navigational coordinate systems.

Inertial System to Navigational System Assume that the navigational system $Ox_ey_ez_e$ is located in the northern hemisphere as shown in Fig. 4.5. Let L denote the longitude and λ the latitude of the origin of the navigational system. The first step is to do a translation of the inertial axes system so that the origin of the inertial axes system coincides with that of the navigational system. Then we have to perform two Euler rotations, $\psi = L$ and then $\theta = -(90 + \lambda)$. With this, the inertial axes system $Ox_iy_iz_i$ coincides with the navigational system $Ox_ey_ez_e$. The third rotation involving ϕ is not necessary. Therefore, $\phi = 0$. Substituting these values of the three Euler angles in Eq. (4.22), we obtain

$$T_i^e = \begin{bmatrix} -\sin\lambda\cos L & -\sin\lambda\sin L & \cos\lambda \\ -\sin L & \cos L & 0 \\ -\cos\lambda\cos L & -\cos\lambda\sin L & -\sin\lambda \end{bmatrix} \tag{4.30}$$

Let the vector X_i denote the position of a particle P in the inertial frame of reference $Ox_iy_iz_i$ as shown in Fig. 4.5. We have

$$X_i = X_{oi} + X_{ei} \tag{4.31}$$

Note that both vectors X_{oi} and X_{ei} are having components in the inertial frame of reference $Ox_iy_iz_i$. Then,

$$X_{ei} = X_i - X_{oi} \tag{4.32}$$

$$X_e = T_i^e X_{ei} \tag{4.33}$$

$$= T_i^e (X_i - X_{oi}) \tag{4.34}$$

Now vector X_e has components in the $Ox_ey_ez_e$ system and $T_e^i = \left(T_i^e\right)'$.

Inertial System to Body Axes System In general, a transformation of this nature will involve all three Euler angles ψ, θ, and ϕ. Equation (4.22) gives this transformation matrix T_i^b as

$$T_i^b = \begin{bmatrix} \cos\theta\cos\psi & \cos\theta\sin\psi & -\sin\theta \\ \sin\theta\sin\phi\cos\psi - \sin\psi\cos\phi & \sin\psi\sin\theta\sin\phi + \cos\psi\cos\phi & \sin\phi\cos\theta \\ \sin\theta\cos\phi\cos\psi + \sin\psi\sin\phi & \sin\psi\sin\theta\cos\phi - \cos\psi\sin\phi & \cos\phi\cos\theta \end{bmatrix} \tag{4.35}$$

The transformation matrix from body to inertial system is given by

$$T_b^i = (T_i^b)' = \begin{bmatrix} \cos\theta\cos\psi & \sin\theta\sin\phi\cos\psi - \sin\psi\cos\phi & \sin\theta\cos\phi\cos\psi + \sin\psi\sin\phi \\ \cos\theta\sin\psi & \sin\psi\sin\theta\sin\phi + \cos\psi\cos\phi & \sin\psi\sin\theta\cos\phi - \cos\psi\sin\phi \\ -\sin\theta & \sin\phi\cos\theta & \cos\phi\cos\theta \end{bmatrix} \tag{4.36}$$

Wind Axes to Body Axes System This transformation involves only two rotations, first $\psi = -\beta$, then $\theta = \alpha$ (Fig. 4.6). Noting that $\phi = 0$, and using Eq. (4.22) we get

$$T_w^b = \begin{bmatrix} \cos\alpha\cos\beta & -\cos\alpha\sin\beta & -\sin\alpha \\ \sin\beta & \cos\beta & 0 \\ \sin\alpha\cos\beta & -\sin\beta\sin\alpha & \cos\alpha \end{bmatrix} \tag{4.37}$$

The velocity vector measured in the wind axes system is given by

$$V_w \begin{bmatrix} U_o \\ 0 \\ 0 \end{bmatrix} \tag{4.38}$$

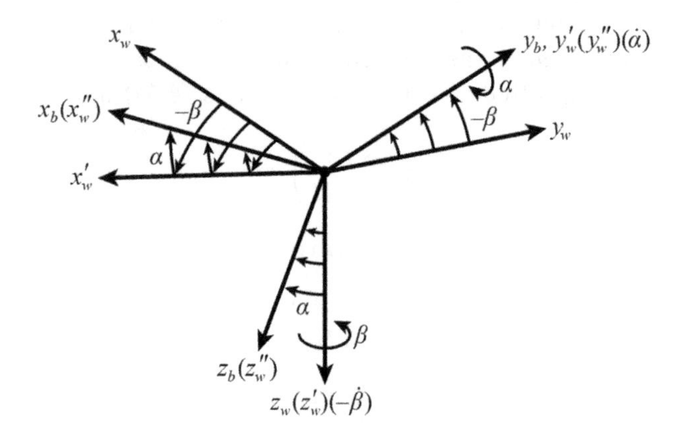

Fig. 4.6 Wind and body axes systems.

where U_o is the flight velocity. In the body axes system,

$$V_b = \begin{bmatrix} U \\ V \\ W \end{bmatrix} = T_w^b V_w$$

$$= \begin{bmatrix} \cos\alpha\cos\beta & -\cos\alpha\sin\beta & -\sin\alpha \\ \sin\beta & \cos\beta & 0 \\ \sin\alpha\cos\beta & -\sin\beta\sin\alpha & \cos\alpha \end{bmatrix} \begin{bmatrix} U_o \\ 0 \\ 0 \end{bmatrix} \tag{4.39}$$

$$= U_o \begin{bmatrix} \cos\alpha\cos\beta \\ \sin\beta \\ \sin\alpha\cos\beta \end{bmatrix} \tag{4.40}$$

so that

$$U = U_o \cos\alpha\cos\beta \tag{4.41}$$

$$V = U_o \sin\beta \tag{4.42}$$

$$W = U_o \sin\alpha\ \cos\beta \tag{4.43}$$

The angle of attack and sideslip are customarily defined as

$$\alpha = \tan^{-1}\left(\frac{W}{U}\right) \tag{4.44}$$

$$\beta = \sin^{-1}\left(\frac{V}{U_o}\right) \tag{4.45}$$

With U, V, and W given by Eqs. (4.41–4.43), we find that the given definitions of α and β hold.

Consider an aircraft model mounted in a wind-tunnel test section. Let σ be the angle between the model longitudinal axis (zero-lift line) and the tunnel axis and ϕ denote the bank angle about the model longitudinal axis.

Using Eq. (4.22) with $\theta = \sigma$ and $\psi = 0$, and Eqs. (4.44) and (4.45), we have

$$\alpha = \tan^{-1}\left(\frac{W}{U}\right)$$

$$= \tan^{-1}(\tan\sigma\ \cos\phi) \tag{4.46}$$

$$\beta = \sin^{-1}\left(\frac{V}{U_o}\right)$$

$$= \sin^{-1}(\sin\sigma\ \sin\phi) \tag{4.47}$$

If $\phi = 90$ deg, then $\alpha = 0$ and $\beta = \sigma$. In other words, when the bank angle is 90 deg, all the angle of attack gets converted to sideslip.

4.2.6 Euler Angle Rates

One of the problems in flight dynamics is to compute the time history of the Euler angles. However, such computations need a knowledge of the Euler angle rates $\dot{\phi}$, $\dot{\theta}$, and $\dot{\psi}$, which are not directly measured or are not available. What are generally available are the angular velocity components p, q, r, which are the body axes components of the angular velocity of the vehicle with respect to an inertial axes system. In other words, we are given p, q, r in a body-fixed system and are asked to find the Euler angle rates $\dot{\psi}$, $\dot{\theta}$, and $\dot{\phi}$. The angular velocity components p, q, r in the body-fixed system may be available either from onboard measurements using rate gyros or may be derived from a solution of the equations of motion.

Let us refer to Fig. 4.3 with an understanding that Ox_1 corresponds to Ox_i, Oy_1 to Oy_i, and Oz_1 to Oz_i. With this understanding, we observe that the angular velocity vector $\dot{\psi}$ is directed along the Oz_i or Oz_i' axis, the angular velocity vector $\dot{\theta}$ along the $Oy_i'\,y_i''$ axis, and the angular velocity vector $\dot{\phi}$ is directed along the Ox_b or Ox_i''' axis. Based on this information, we can determine the relations between body axes rates p, q, r and the Euler angle rates $\dot{\psi}$, $\dot{\theta}$, and $\dot{\phi}$ as follows.

To begin with, consider the $\dot{\psi}$ vector. It has to be transformed from the $Ox_i'y_i'z_i'$ system to the $Ox_by_bz_b$ system, and the corresponding transformation matrix is the matrix product AB so that

$$\dot{\psi}_b = AB\begin{bmatrix} 0 \\ 0 \\ \dot{\psi}_i \end{bmatrix} \tag{4.48}$$

where $\dot{\psi}_b$ denotes the vector $\dot{\psi}$ resolved in the $Ox_by_bz_b$ system.

Next, consider the angular velocity vector $\dot{\theta}$, which is directed along the Oy_i'' axis of the $Ox_i''y_i''z_i''$system (Fig. 4.3). Therefore,

$$\dot{\theta}_b = A\begin{bmatrix} 0 \\ \dot{\theta} \\ 0 \end{bmatrix} \tag{4.49}$$

Finally, we consider the angular velocity vector $\dot{\phi}$, which is directed along the Ox_b axis of the $Ox_by_bz_b$ system (Fig. 4.3) so that

$$\dot{\phi}_b = \begin{bmatrix} \dot{\phi} \\ 0 \\ 0 \end{bmatrix} \tag{4.50}$$

Let the angular velocity vector in body axes system $Ox_by_bz_b$ be denoted by $\vec{\omega}_{i,b}^{b}$. Here, the order of subscripts and superscripts is as follows: the subscripts i, b have the meaning of the body with respect to the inertial system and the superscript b means that the vector has components in the body axes system. An arrow over the symbol denotes a vector. With this understanding, we have

$$\vec{\omega}_{i,b}^{b} = \begin{bmatrix} p \\ q \\ r \end{bmatrix} \tag{4.51}$$

Therefore,

$$\begin{bmatrix} p \\ q \\ r \end{bmatrix} = AB \begin{bmatrix} 0 \\ 0 \\ \dot{\psi}_i \end{bmatrix} + A \begin{bmatrix} 0 \\ \dot{\theta} \\ 0 \end{bmatrix} + \begin{bmatrix} \dot{\phi} \\ 0 \\ 0 \end{bmatrix} \tag{4.52}$$

Substituting for matrices A and B from Eqs. (4.14) and (4.20), we get

$$\begin{aligned} \begin{bmatrix} p \\ q \\ r \end{bmatrix} &= \begin{bmatrix} \cos\theta & 0 & -\sin\theta \\ \sin\phi\sin\theta & \cos\phi & \sin\phi\cos\theta \\ \cos\phi\sin\theta & -\sin\phi & \cos\phi\cos\theta \end{bmatrix} \begin{bmatrix} 0 \\ 0 \\ \dot{\psi} \end{bmatrix} \\ &+ \begin{bmatrix} 1 & 0 & 0 \\ 0 & \cos\phi & \sin\phi \\ 0 & -\sin\phi & \cos\phi \end{bmatrix} \begin{bmatrix} 0 \\ \dot{\theta} \\ 0 \end{bmatrix} + \begin{bmatrix} \dot{\phi} \\ 0 \\ 0 \end{bmatrix} \end{aligned} \tag{4.53}$$

Simplifying, we obtain

$$\begin{bmatrix} p \\ q \\ r \end{bmatrix} = \begin{bmatrix} 1 & 0 & -\sin\theta \\ 0 & \cos\phi & \sin\phi\cos\theta \\ 0 & -\sin\phi & \cos\phi\cos\theta \end{bmatrix} \begin{bmatrix} \dot{\phi} \\ \dot{\theta} \\ \dot{\psi} \end{bmatrix} \tag{4.54}$$

so that

$$p = \dot{\phi} - \dot{\psi}\sin\theta \tag{4.55}$$

$$q = \dot{\theta}\cos\phi + \dot{\psi}\sin\phi\cos\theta \tag{4.56}$$

$$r = \dot{\psi}\cos\phi\cos\theta - \dot{\theta}\sin\phi \tag{4.57}$$

From Eqs. (4.55–4.57), we observe that very-often-used relations such as $p = \dot{\phi}$, $q = \dot{\theta}$, and $r = \dot{\psi}$ are true only when both the pitch angle θ and bank angle ϕ are close to zero.

Let

$$L_{\omega} = \begin{bmatrix} 1 & 0 & -\sin\theta \\ 0 & \cos\phi & \sin\phi\cos\theta \\ 0 & -\sin\phi & \cos\phi\cos\theta \end{bmatrix} \tag{4.58}$$

so that

$$\begin{bmatrix} \dot{\phi} \\ \dot{\theta} \\ \dot{\psi} \end{bmatrix} = L_\omega^{-1} \begin{bmatrix} p \\ q \\ r \end{bmatrix} \tag{4.59}$$

The transformation matrix L_ω is not an orthogonal matrix because the vectors $\dot{\psi}$, $\dot{\theta}$, and $\dot{\phi}$ are not mutually perpendicular. Hence, the inverse is not equal to the transpose. Therefore, we have to compute L_ω^{-1} in the usual way as follows:

$$L_\omega^{-1} = \left(\frac{1}{\Delta(L)}\right) \text{adj}(L_\omega) \tag{4.60}$$

The determinant $\Delta(L)$ of matrix L_ω is given by

$$\Delta(L) = \cos^2 \phi \cos \theta + \sin^2 \phi \cos \theta \tag{4.61}$$

$$= \cos \theta \tag{4.62}$$

The adjoint of matrix L_ω is given by

$$\text{adj}(L_\omega) = \begin{bmatrix} \cos \theta & \sin \theta \sin \phi & \cos \phi \sin \theta \\ 0 & \cos \phi \cos \theta & -\sin \phi \cos \theta \\ 0 & \sin \phi & \cos \phi \end{bmatrix} \tag{4.63}$$

so that

$$L_\omega^{-1} = \begin{bmatrix} 1 & \tan \theta \sin \phi & \cos \phi \tan \theta \\ 0 & \cos \theta & -\sin \phi \\ 0 & \sec \theta \sin \phi & \sec \theta \cos \phi \end{bmatrix} \tag{4.64}$$

Substituting in Eq. (4.59), we get

$$\begin{bmatrix} \dot{\phi} \\ \dot{\theta} \\ \dot{\psi} \end{bmatrix} = \begin{bmatrix} 1 & \tan \theta \sin \phi & \cos \phi \tan \theta \\ 0 & \cos \phi & -\sin \phi \\ 0 & \sec \theta \sin \phi & \sec \theta \sin \phi \end{bmatrix} \begin{bmatrix} p \\ q \\ r \end{bmatrix} \tag{4.65}$$

or

$$\dot{\phi} = p + \tan \theta (q \sin \phi + r \cos \phi) \tag{4.66}$$

$$\dot{\theta} = q \cos \phi - r \sin \phi \tag{4.67}$$

$$\dot{\psi} = \sec \theta (q \sin \phi + r \cos \phi) \tag{4.68}$$

Singularity in Euler Angle Rates

Equations (4.66–4.68) for Euler angle rates become singular at $\theta = \pi/2$. The existence of the singularity can be demonstrated as follows.

Rewrite Eqs. (4.66–4.68),

$$\dot{\phi} = p + \frac{q \sin\theta \sin\phi + r \sin\theta \cos\phi}{\cos\theta} \tag{4.69}$$

$$\dot{\theta} = q \cos\phi - r \sin\phi \tag{4.70}$$

$$\dot{\psi} = \frac{q \sin\phi + r \cos\phi}{\cos\theta} \tag{4.71}$$

For $\theta = \pi/2$, from Eqs. (4.55–4.57),

$$p = \dot{\phi} - \dot{\psi} \tag{4.72}$$

$$q = \dot{\theta} \cos\phi \tag{4.73}$$

$$r = -\dot{\theta} \sin\phi \tag{4.74}$$

Substituting in Eq. (4.71), we have

$$(\dot{\psi})_{\frac{\pi}{2}} = \frac{\dot{\theta} \sin\phi \cos\phi - \dot{\theta} \sin\phi \cos\phi}{\cos\theta} \tag{4.75}$$

$$= \frac{0}{0} \tag{4.76}$$

Similarly,

$$(\dot{\phi})_{\frac{\pi}{2}} = p + \frac{\dot{\theta} \sin\theta \cos\phi \sin\phi - \dot{\theta} \sin\theta \sin\phi \cos\phi}{\cos\theta} \tag{4.77}$$

$$= p + \frac{0}{0} \tag{4.78}$$

Thus, we see that the Euler angle rates $\dot{\psi}$ and $\dot{\phi}$ assume an indeterminate (0/0) form, which means that we have a singularity at $\theta = \pi/2$. We can get around this difficulty using L'Hospital Rule and obtain approximate expressions for Euler angle rates that are valid around $\theta = \pi/2$ as follows [2]:

$$\left(\frac{d\psi}{dt}\right)_{\frac{\pi}{2}} = \lim_{\theta \to \frac{\pi}{2}} \left(\frac{\frac{d}{d\theta}(q \sin\phi + r \cos\phi)}{\frac{d}{d\theta}(\cos\theta)} \right) \tag{4.79}$$

With

$$\frac{d}{d\theta}(\cdot) = \frac{d}{dt}(\cdot) \frac{dt}{d\theta} \tag{4.80}$$

we obtain

$$\left(\frac{d\psi}{dt}\right)_{\frac{\pi}{2}} = \lim_{\theta \to \frac{\pi}{2}} \left[\frac{\dot{q} \sin \phi + q \cos \phi \dot{\phi} + \dot{r} \cos \phi - r \sin \phi \dot{\phi}}{-\dot{\theta} \sin \theta} \right] \tag{4.81}$$

Substituting for q and r from Eqs. (4.73) and (4.74), we obtain

$$\left(\frac{d\psi}{dt}\right)_{\frac{\pi}{2}} = -\frac{1}{\dot{\theta}} (\dot{q} \sin \phi + \dot{\theta}\dot{\phi} + \dot{r} \cos \phi) \tag{4.82}$$

We have from Eq. (4.72),

$$\dot{\phi} = \dot{\psi} + p \tag{4.83}$$

so that

$$(\dot{\phi})_{\frac{\pi}{2}} = p - \frac{1}{\dot{\theta}} (\dot{q} \sin \phi + \dot{\theta}\dot{\phi} + \dot{r} \cos \phi) \tag{4.84}$$

$$= \frac{p}{2} - \frac{1}{2\dot{\theta}} (\dot{q} \sin \phi + \dot{r} \cos \phi) \tag{4.85}$$

From Eq. (4.70), we have

$$(\dot{\theta})_{\frac{\pi}{2}} = q \cos \phi - r \sin \phi$$

Using Eqs. (4.73) and (4.74) we have

$$\tan \phi = -\frac{r}{q}$$

$$\sin \phi = \mp \frac{r}{\sqrt{r^2 + q^2}}$$

$$\cos \phi = \pm \frac{q}{\sqrt{r^2 + q^2}}$$

Then,

$$(\dot{\theta})_{\frac{\pi}{2}} = \pm \sqrt{r^2 + q^2} \tag{4.86}$$

Equations (4.82), (4.85), and (4.86) give the values of Euler angle rates in the neighborhood of $\theta \simeq \pi/2$. Thus, during computations, use Eqs. (4.66–4.68), for all values of θ except in the neighborhood of $|\theta| \simeq \pi/2$. When $|\theta| \to \pi/2$, switch over to Eqs. (4.82), (4.85), and (4.86).

Angular Velocity of a Body in the Navigational System Consider the motion of a spacecraft orbiting the Earth. We want to determine the angular velocity of this orbiting spacecraft with respect to an inertial axes system fixed at the center of the Earth and obtain its components in the navigational system $x_e y_e z_e$.

Suppose the orbit is contained in the plane of the equator (see Fig. 4.7a). Then, the angular velocity of the vehicle with respect to the $x_i y_i z_i$ system is $\hat{k}_i \dot{L}$ where $L = L(t)$ is the longitude at any time t. Here, we assume that the longitude L is measured in the plane of the equator and, from the Ox_i axis, positive counterclockwise. Next, assume that the orbit is contained in a vertical plane $L =$ const. Then, the angular velocity vector is contained in the plane of the equator and has the magnitude $\dot{\lambda}$. Its components along the

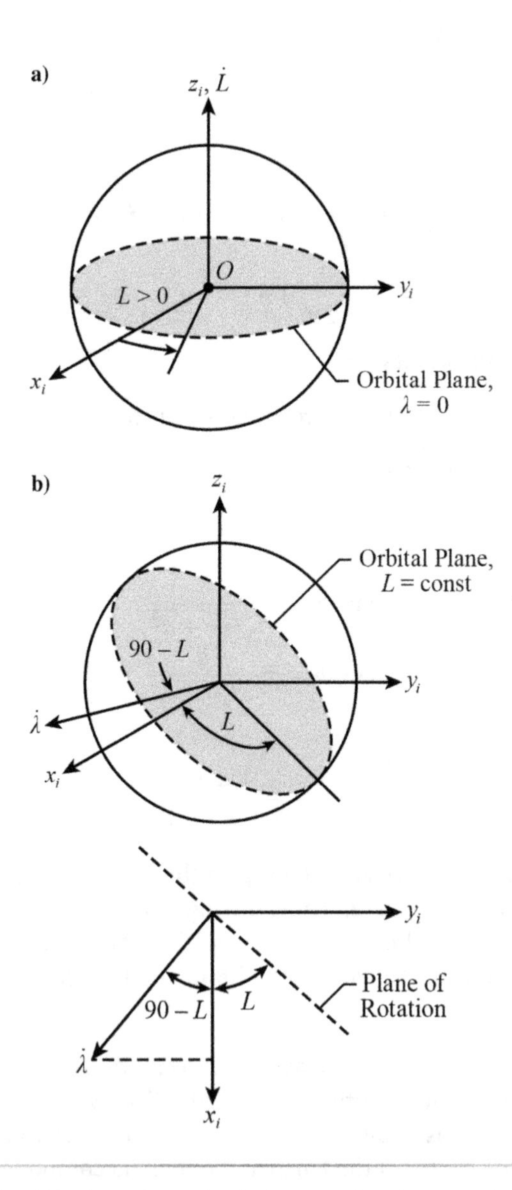

Fig. 4.7 Angular velocity in a navigational system.

Ox_i and Oy_i axes are $\hat{i}_i\dot{\lambda}\sin L$ and $-\hat{j}_i\dot{\lambda}\cos L$. In general, when the orbital plane is inclined, both λ and L vary with time t, and we have

$$\vec{\omega}^i_{i,b} = \begin{bmatrix} \dot{\lambda}\sin L \\ -\dot{\lambda}\cos L \\ \dot{L} \end{bmatrix} \tag{4.87}$$

Here, $\vec{\omega}^i_{i,b}$ is the angular velocity of the spacecraft with respect to the inertial system and has components in the inertial system. As said before, the order of the subscripts and the superscript is as follows: the subscripts i, b have the meaning of the body with respect to the inertial system and the superscript i means that the vector has components in the inertial system.

We have

$$\vec{\omega}^e_{i,b} = T^e_i\vec{\omega}^i_{i,b} \tag{4.88}$$

Using Eq. (4.30) for T^e_i, we obtain

$$\vec{\omega}^e_{i,b} = \begin{bmatrix} -\sin\lambda\cos L & -\sin\lambda\sin L & \cos\lambda \\ -\sin L & \cos L & 0 \\ -\cos\lambda\cos L & -\cos\lambda\sin L & -\sin\lambda \end{bmatrix}\begin{bmatrix} \dot{\lambda}\sin L \\ -\dot{\lambda}\cos L \\ \dot{L} \end{bmatrix} \tag{4.89}$$

$$= \begin{bmatrix} \dot{L}\cos\lambda \\ -\dot{\lambda} \\ -\dot{L}\sin\lambda \end{bmatrix} \tag{4.90}$$

Here, $\vec{\omega}^e_{i,b}$ is the angular velocity of the spacecraft with respect to the inertial system and having components in the navigational system.

Relation between Angular Velocities in Wind and Body Axes Systems

The relative angular velocity vector from the wind axes to the body axes system consists of two components, $-\dot{\beta}$ about the Oz_w axis and $\dot{\alpha}$ about the Oy_b axis (Fig. 4.6). This relative angular velocity vector can be expressed in the wind axes system as

$$\omega^w_{w,b} = \begin{bmatrix} 0 \\ 0 \\ -\dot{\beta} \end{bmatrix} + T^w_b\begin{bmatrix} 0 \\ \dot{\alpha} \\ 0 \end{bmatrix} \tag{4.91}$$

Note that $-\dot{\beta}$ is already in the wind axes system but $\dot{\alpha}$ is about the Oy_b (body) axis and needs to be converted as shown. We can then obtain the angular velocity of the wind axes system relative to the inertial space [12] as

$$\omega_{i,w}^{w} = T_b^w \omega_{i,b}^{b} - \omega_{w,b}^{w} \tag{4.92}$$

$$= T_b^w \begin{bmatrix} p \\ q \\ r \end{bmatrix} + \begin{bmatrix} 0 \\ 0 \\ \dot{\beta} \end{bmatrix} - T_b^w \begin{bmatrix} 0 \\ \dot{\alpha} \\ 0 \end{bmatrix} \tag{4.93}$$

or

$$\begin{bmatrix} p_w \\ q_w \\ r_w \end{bmatrix} = T_b^w \begin{bmatrix} p \\ q - \dot{\alpha} \\ r \end{bmatrix} + \begin{bmatrix} 0 \\ 0 \\ \dot{\beta} \end{bmatrix} \tag{4.94}$$

Using Eq. (4.37) for T_w^b and taking the transpose to obtain T_b^w,

$$\begin{bmatrix} p_w \\ q_w \\ r_w \end{bmatrix} = \begin{bmatrix} \cos\alpha\cos\beta & \sin\beta & \sin\alpha\cos\beta \\ -\cos\alpha\cos\beta & \cos\beta & -\sin\alpha\sin\beta \\ -\sin\alpha & 0 & \cos\alpha \end{bmatrix} \begin{bmatrix} p \\ q - \dot{\alpha} \\ r \end{bmatrix} + \begin{bmatrix} 0 \\ 0 \\ \dot{\beta} \end{bmatrix} \tag{4.95}$$

Simplifying

$$p_w = p\cos\alpha\cos\beta + (q - \dot{\alpha})\sin\beta + r\sin\alpha\cos\beta \tag{4.96}$$
$$q_w = -p\cos\alpha\sin\beta + (q - \dot{\alpha})\cos\beta - r\sin\alpha\sin\beta \tag{4.97}$$
$$r_w = -p\sin\alpha + r\cos\alpha + \dot{\beta} \tag{4.98}$$

Alternately, we can express the relative angular rates $\dot{\alpha}$ and $\dot{\beta}$ in terms of wind axes angular rates as

$$\dot{\alpha} = q - p_w\sin\beta - q_w\cos\beta \tag{4.99}$$
$$\dot{\beta} = r_w + p\sin\alpha - r\cos\alpha \tag{4.100}$$

4.2.7 Method of Direction Cosine Matrix

Consider a vector $\vec{A} = \hat{i}A_x + \hat{j}A_y + \hat{k}A_z$ in the *Oxyz* system. Let δ_1, δ_2, and δ_3 be the angles the vector $\vec{A}$ makes with the *x*, y, and z axes, respectively. Then, we have

$$\cos\delta_1 = \frac{A_x}{|A|} \tag{4.101}$$

$$\cos\delta_2 = \frac{A_y}{|A|} \tag{4.102}$$

$$\cos\delta_3 = \frac{A_z}{|A|} \tag{4.103}$$

so that

$$\cos^2\delta_1 + \cos^2\delta_2 + \cos^2\delta_3 = 1 \tag{4.104}$$

The numbers $\cos\delta_1$, $\cos\delta_2$, and $\cos\delta_3$ are called the direction cosines of the vector $\vec{A}$ with respect to the $Oxyz$ axes system.

Now consider the transformation of a vector from one coordinate system to another using the direction cosines. Suppose we want to transform a vector given in the $Ox_1y_1z_1$ system into $Ox_2y_2z_2$. Let

$$\hat{i}_2 = C_{11}\hat{i}_1 + C_{12}\hat{j}_1 + C_{13}\hat{k}_1 \tag{4.105}$$

$$\hat{j}_2 = C_{21}\hat{i}_1 + C_{22}\hat{j}_1 + C_{23}\hat{k}_1 \tag{4.106}$$

$$\hat{k}_2 = C_{31}\hat{i}_1 + C_{32}\hat{j}_1 + C_{33}\hat{k}_1 \tag{4.107}$$

or, in matrix form,

$$\begin{bmatrix} \hat{i}_2 \\ \hat{j}_2 \\ \hat{k}_2 \end{bmatrix} = \begin{bmatrix} C_{11} & C_{12} & C_{13} \\ C_{21} & C_{22} & C_{23} \\ C_{31} & C_{32} & C_{33} \end{bmatrix} \begin{bmatrix} \hat{i}_1 \\ \hat{j}_1 \\ \hat{k}_1 \end{bmatrix} \tag{4.108}$$

where C_{11}, C_{12}, C_{13} are the direction cosines of the unit vector $\hat{i}_2$ with respect to the $Ox_1y_1z_1$ system and so on. We observe that

$$C_{11} = \hat{i}_2 \cdot \hat{i}_1 \quad C_{12} = \hat{i}_2 \cdot \hat{j}_1 \quad C_{13} = \hat{i}_2 \cdot \hat{k}_1 \tag{4.109}$$

$$C_{21} = \hat{j}_2 \cdot \hat{i}_1 \quad C_{22} = \hat{j}_2 \cdot \hat{j}_1 \quad C_{23} = \hat{j}_2 \cdot \hat{k}_1 \tag{4.110}$$

$$C_{31} = \hat{k}_2 \cdot \hat{i}_1 \quad C_{32} = \hat{k}_2 \cdot \hat{j}_1 \quad C_{33} = \hat{k}_2 \cdot \hat{k}_1 \tag{4.111}$$

where the "·" denotes the scalar product. Thus, knowing all the elements C_{ij}, $i, j = 1, 3$ of the direction cosine matrix, we can transform a vector from the $Ox_1y_1z_1$ to the $Ox_2y_2z_2$ system as follows:

$$\begin{bmatrix} x_2 \\ y_2 \\ z_2 \end{bmatrix} = \begin{bmatrix} C_{11} & C_{12} & C_{13} \\ C_{21} & C_{22} & C_{23} \\ C_{31} & C_{32} & C_{33} \end{bmatrix} \begin{bmatrix} x_1 \\ y_1 \\ z_1 \end{bmatrix} \tag{4.112}$$

Let

$$X_2 = \begin{bmatrix} x_2 \\ y_2 \\ z_2 \end{bmatrix} \tag{4.113}$$

$$X_1 = \begin{bmatrix} x_1 \\ y_1 \\ z_1 \end{bmatrix} \tag{4.114}$$

$$C_1^2 = \begin{bmatrix} C_{11} & C_{12} & C_{13} \\ C_{21} & C_{22} & C_{23} \\ C_{31} & C_{32} & C_{33} \end{bmatrix} \tag{4.115}$$

so that

$$X_2 = C_1^2 X_1 \tag{4.116}$$

or

$$X_1 = \left(C_1^2\right)^{-1} X_2 \tag{4.117}$$

$$= C_2^1 X_2 \tag{4.118}$$

$$= C_2^1 C_1^2 X_1 \tag{4.119}$$

so that

$$C_2^1 C_1^2 = I \tag{4.120}$$

$$\begin{bmatrix} C_{11} & C_{21} & C_{31} \\ C_{12} & C_{22} & C_{32} \\ C_{13} & C_{23} & C_{33} \end{bmatrix} \begin{bmatrix} C_{11} & C_{12} & C_{13} \\ C_{21} & C_{22} & C_{23} \\ C_{31} & C_{32} & C_{33} \end{bmatrix} = \begin{bmatrix} 1 & 0 & 0 \\ 0 & 1 & 0 \\ 0 & 0 & 1 \end{bmatrix} \tag{4.121}$$

where $\left(C_1^2\right)^{-1} = \left(C_1^2\right)' = C_2^1$.

Carrying out the matrix multiplications and equating the corresponding terms on the left- and right-hand sides of Eq. (4.121), we get a total of nine equations. It can be easily verified that three of these equations are redundant, i.e., repeat themselves. In other words, we have only six equations relating nine parameters C_{ij}, $i, j = 1, 3$ as follows:

$$C_{11}^2 + C_{21}^2 + C_{31}^2 = 1 \tag{4.122}$$

$$C_{12}^2 + C_{22}^2 + C_{32}^2 = 1 \tag{4.123}$$

$$C_{13}^2 + C_{23}^2 + C_{33}^2 = 1 \tag{4.124}$$

$$C_{11}C_{12} + C_{21}C_{22} + C_{31}C_{32} = 0 \tag{4.125}$$

$$C_{11}C_{13} + C_{21}C_{23} + C_{31}C_{33} = 0 \tag{4.126}$$

$$C_{12}C_{13} + C_{22}C_{23} + C_{32}C_{33} = 0 \tag{4.127}$$

The fact that the nine parameters C_{ij}, $i, j = 1, 3$ forming the elements of a coordinate transformation matrix have to satisfy six constraint Eqs. (4.122–4.127) implies that only three of them are free. This result should not be a surprise to us because we know that the three Euler angles ψ, θ, and ϕ are necessary and sufficient to perform such coordinate transformations. Therefore, if we introduce more than three parameters, we will have to have that many extra constraint relations among the parameters. These additional constraint relations are often called redundancy relations and are useful to determine any of the missing elements (up to three) from the given direction cosine matrix. We will illustrate this concept at the end of this section with the help of an illustrative example.

Relations Between Euler Angles and Direction Cosines Suppose we are given the Euler angles ψ, θ, and ϕ, which transform a vector given in the $Ox_1y_1z_1$ to the $Ox_2y_2z_2$ system. The transformation matrix based on these Euler angles is given by T_1^2 of Eq. (4.22). Similarly, if we use the method of direction cosine matrix, the transformation matrix is given by C_1^2 of Eq. (4.112). Equating the two matrices, we get the following relations:

$$C_{11} = \cos\theta\cos\psi \tag{4.128}$$

$$C_{12} = \cos\theta\sin\psi \tag{4.129}$$

$$C_{13} = -\sin\theta \tag{4.130}$$

$$C_{21} = \sin\theta\sin\phi\cos\psi - \sin\psi\cos\phi \tag{4.131}$$

$$C_{22} = \sin\psi\sin\theta\sin\phi + \cos\psi\cos\phi \tag{4.132}$$

$$C_{23} = \sin\phi\cos\theta \tag{4.133}$$

$$C_{31} = \sin\theta\cos\phi\cos\psi + \sin\psi\sin\phi \tag{4.134}$$

$$C_{32} = \sin\psi\sin\theta\cos\phi - \cos\psi\sin\phi \tag{4.135}$$

$$C_{33} = \cos\phi\cos\theta \tag{4.136}$$

Therefore, given the elements of a direction cosine matrix, we can obtain the Euler angles using Eqs. (4.128), (4.130), and (4.136) as follows:

$$\theta = \sin^{-1}(-C_{13}) \tag{4.137}$$

$$\phi = \cos^{-1}\left(\frac{C_{33}}{\sqrt{1 - C_{13}^2}}\right)\mathrm{sgn}(C_{23}) \tag{4.138}$$

$$\psi = \cos^{-1}\left(\frac{C_{11}}{\sqrt{1 - C_{13}^2}}\right)\mathrm{sgn}(C_{12}) \tag{4.139}$$

where the sgn() function has the following properties: if the argument is positive, the sign function returns the value of $+1$; if it is negative, it returns -1.

Recall that we have imposed certain restrictions on Euler angles as given by Eqs. (4.1–4.3). Subject to these restrictions, the quadrants in which the Euler angles lie can be determined as follows [2].

Consider the pitch angle θ. If $C_{13} < 0$, then θ lies in the first quadrant or $0 \leq \theta \leq \pi/2$. On the other hand, if $C_{13} > 0$, then θ will lie in the fourth quadrant or $-\pi/2 \leq \theta \leq 0$. Thus, cos θ will always be positive.

Next consider the bank angle ϕ. We have $C_{33} = \cos\phi\cos\theta$. Because cos θ is always positive, the sign of C_{33} is governed by the sign of cos ϕ. Now check the sign of $C_{23} = \sin\phi\cos\theta$. Suppose $C_{33} > 0$ and $C_{23} > 0$. Then ϕ is in the first quadrant. On the other hand, if $C_{33} > 0$ and $C_{23} < 0$, then

ϕ is in the fourth quadrant. If $C_{33} < 0$ and $C_{23} > 0$, then ϕ lies in the second quadrant. If both C_{33} and C_{23} are negative, then ϕ lies in the third quadrant.

Similarly, examining the signs of C_{11} and C_{12}, the quadrant in which the yaw angle ψ falls can be determined. Notice that the sgn() function introduced in Eqs. (4.138) and (4.139) takes care of these considerations.

Updating Direction Cosine Matrix

During a continuous motion, the elements of the direction cosine matrices continuously vary. To determine their variation with time, we need to know the derivative of the direction cosine matrix with respect to time, which can be obtained as follows.

Consider the transformation of unit vectors from body-fixed axes system to inertial axes system using the direction cosine matrix as given by

$$\begin{bmatrix} \hat{i}_i \\ \hat{j}_i \\ \hat{k}_i \end{bmatrix} = \begin{bmatrix} C_{11} & C_{12} & C_{13} \\ C_{21} & C_{22} & C_{23} \\ C_{31} & C_{32} & C_{33} \end{bmatrix} \begin{bmatrix} \hat{i}_b \\ \hat{j}_b \\ \hat{k}_b \end{bmatrix} \tag{4.140}$$

or

$$\begin{bmatrix} \hat{i}_i \\ \hat{j}_j \\ \hat{k}_k \end{bmatrix} = C_b^i \begin{bmatrix} \hat{i}_b \\ \hat{j}_b \\ \hat{k}_b \end{bmatrix} \tag{4.141}$$

Consider the first equation,

$$\hat{i}_i = C_{11}\hat{i}_b + C_{12}\hat{j}_b + C_{13}\hat{k}_b \tag{4.142}$$

$$\frac{d\hat{i}_i}{dt} = \dot{C}_{11}\hat{i}_b + C_{11}\frac{d\hat{i}_b}{dt} + \dot{C}_{12}\hat{j}_b + C_{12}\frac{d\hat{j}_b}{dt} + \dot{C}_{13}\hat{k}_b + C_{13}\frac{d\hat{k}_b}{dt} \tag{4.143}$$

Using the moving axes theorem about which we will learn later in this chapter, we have

$$\frac{d\hat{i}_b}{dt} = \vec{\omega}_{i,b}^b \times \hat{i}_b \tag{4.144}$$

$$\frac{d\hat{j}_b}{dt} = \vec{\omega}_{i,b}^b \times \hat{j}_b \tag{4.145}$$

$$\frac{d\hat{k}_b}{dt} = \vec{\omega}_{i,b}^b \times \hat{k}_b \tag{4.146}$$

where

$$\vec{\omega}_{i,b}^b = \begin{bmatrix} p \\ q \\ r \end{bmatrix}$$

Then,

$$\frac{d\hat{i}_i}{dt} = \dot{C}_{11}\hat{i}_b + \dot{C}_{12}\hat{j}_b + \dot{C}_{13}\hat{k}_b + C_{11}\vec{\omega}^b_{i,b} \times \hat{i}_b + C_{12}\vec{\omega}^b_{i,b} \times \hat{j}_b + C_{13}\vec{\omega}^b_{i,b} \times \hat{k}_b \tag{4.147}$$

$$= \dot{C}_{11}\hat{i}_b + \dot{C}_{12}\hat{j}_b + \dot{C}_{13}\hat{k}_b + \vec{\omega}^b_{i,b} \times \hat{i}_i \tag{4.148}$$

Because $\hat{i}_i$ is a vector of magnitude unity and fixed direction in the inertial system,

$$\frac{d\hat{i}_i}{dt} = \dot{C}_{11}\hat{i}_b + \dot{C}_{12}\hat{j}_b + \dot{C}_{13}\hat{k}_b + \vec{\omega}^b_{i,b} \times \hat{i}_i = 0 \tag{4.149}$$

Similarly,

$$\frac{d\hat{j}_i}{dt} = \dot{C}_{21}\hat{i}_b + \dot{C}_{22}\hat{j}_b + \dot{C}_{23}\hat{k}_b + \vec{\omega}^b_{i,b} \times \hat{j}_i = 0 \tag{4.150}$$

$$\frac{d\hat{k}_i}{dt} = \dot{C}_{31}\hat{i}_b + \dot{C}_{32}\hat{j}_b + \dot{C}_{33}\hat{k}_b + \vec{\omega}^b_{i,b} \times \hat{k}_i = 0 \tag{4.151}$$

Now,

$$\vec{\omega}^b_{i,b} \times \hat{i}_i = \begin{vmatrix} \hat{i}_b & \hat{j}_b & \hat{k}_b \\ p & q & r \\ C_{11} & C_{12} & C_{13} \end{vmatrix} \tag{4.152}$$

$$= \hat{i}_b(C_{13}q - C_{12}r) - \hat{j}_b(C_{13}p - C_{11}r) + \hat{k}_b(C_{12}p - C_{11}q) \tag{4.153}$$

Then,

$$\frac{d\hat{i}_i}{dt} = (\dot{C}_{11} + C_{13}q - C_{12}r)\hat{i}_b + (\dot{C}_{12} + C_{11}r - C_{13}p)\hat{j}_b + (\dot{C}_{13} + C_{12}p - C_{11}q)\hat{k}_b \tag{4.154}$$

$$= 0 \tag{4.155}$$

or

$$\dot{C}_{11} = C_{12}r - C_{13}q \tag{4.156}$$

$$\dot{C}_{12} = C_{13}p - C_{11}r \tag{4.157}$$

$$\dot{C}_{13} = C_{11}q - C_{12}p \tag{4.158}$$

Similarly,

$$\dot{C}_{21} = C_{22}r - C_{23}q \tag{4.159}$$

$$\dot{C}_{22} = C_{23}p - C_{21}r \tag{4.160}$$

$$\dot{C}_{23} = C_{21}q - C_{22}p \quad (4.161)$$
$$\dot{C}_{31} = C_{32}r - C_{33}q \quad (4.162)$$
$$\dot{C}_{32} = C_{33}p - C_{31}r \quad (4.163)$$
$$\dot{C}_{33} = C_{31}q - C_{32}p \quad (4.164)$$

In matrix form,

$$\begin{bmatrix} \dot{C}_{11} & \dot{C}_{12} & \dot{C}_{13} \\ \dot{C}_{21} & \dot{C}_{22} & \dot{C}_{23} \\ \dot{C}_{31} & \dot{C}_{32} & \dot{C}_{33} \end{bmatrix} = \begin{bmatrix} C_{11} & C_{12} & C_{13} \\ C_{21} & C_{22} & C_{23} \\ C_{31} & C_{32} & C_{33} \end{bmatrix} \begin{bmatrix} 0 & -r & q \\ r & 0 & -p \\ -q & p & 0 \end{bmatrix} \quad (4.165)$$

$$\dot{C}_b^i = C_b^i \Omega_{i,b}^b \quad (4.166)$$

where

$$\Omega_{i,b}^b = \begin{bmatrix} 0 & -r & q \\ r & 0 & -p \\ -q & p & 0 \end{bmatrix} \quad (4.167)$$

Here, $\Omega_{i,b}^b$ is called the skew-symmetric form of angular velocity vector $\vec{\omega}_{i,b}^b$. Equation (4.166) is the required relation for propagating the direction cosine matrix C_b^i forward in time given its initial value. Generally, the initial value of C_b^i is not directly given. Instead, the initial values of the Euler angles are given. With this information, the elements of C_b^i can be obtained using Eqs. (4.128–4.136).

The skew-symmetric form of a given vector has the property that premultiplying it to another vector gives the vector cross product of the first vector with the second vector. For example,

$$\Omega_{i,b}^b A = \vec{\omega}_{i,b}^b \times A \quad (4.168)$$

Another property of the skew-symmetric form of angular velocity matrix can be illustrated as follows. Suppose we are given $\Omega_{i,b}^b$ and are asked to find $\Omega_{i,b}^i$, we can do this as follows:

$$C_b^i C_i^b = I \quad (4.169)$$

$$\dot{C}_b^i C_i^b + C_b^i \dot{C}_i^b = 0 \quad (4.170)$$

$$C_b^i \Omega_{i,b}^b C_i^b + C_b^i C_i^b \Omega_{b,i}^i = 0 \quad (4.171)$$

$$C_b^i \Omega_{i,b}^b C_i^b = -C_b^i C_i^b \Omega_{b,i}^i \quad (4.172)$$

We know that

$$\Omega_{i,b}^i = -\Omega_{b,i}^i \quad (4.173)$$

Therefore,

$$C_b^i \Omega_{i,b}^b C_i^b = C_b^i C_i^b \Omega_{i,b}^i \tag{4.174}$$

or

$$\Omega_{i,b}^i = C_b^i \Omega_{i,b}^b C_i^b \tag{4.175}$$

Having found $\Omega_{i,b}^i$, we are now in a position to obtain $\vec{\omega}_{i,b}^i$ using the definition of skew-matrix as given by Eq. (4.167).

Equation (4.175) states that if we have a skew-matrix of angular rates of the b frame relative to the i frame with components in the b frame and we need to find the components of the same vector in the i frame, then we must premultiply it by C_b^i and postmultiply it by C_i^b.

From Eq. (4.166), we have

$$\dot{C}_b^i = C_b^i \Omega_{i,b}^b \tag{4.176}$$

$$\left(\dot{C}_b^i\right)' = \left(C_b^i \Omega_{i,b}^b\right)' \tag{4.177}$$

$$= \left(\Omega_{i,b}^b\right)' \left(C_b^i\right)' \tag{4.178}$$

$$\dot{C}_i^b = -\Omega_{i,b}^b C_i^b \tag{4.179}$$

where we have used the property $(\Omega_{i,b}^b)' = -\Omega_{i,b}^b$. Equation (4.179) is of the form

$$\dot{X} = AX \tag{4.180}$$

The characteristic of Eq. (4.180) is given by

$$\Delta(\lambda I - A) = 0 \tag{4.181}$$

where $\Delta(\cdot)$ denotes the determinant of the argument matrix. Using this definition, the characteristic of Eq. (4.179) is given by

$$\lambda\left(\lambda^2 + p^2 + q^2 + r^2\right) = 0 \tag{4.182}$$

so that

$$\lambda = 0 \tag{4.183}$$

or

$$\lambda = \pm j\sqrt{p^2 + q^2 + r^2} \tag{4.184}$$

$$= \pm j\left|\vec{\omega}_{i,b}^b\right| \tag{4.185}$$

where $j = \sqrt{-1}$. We note that one root is at the origin and the other two are on the imaginary axis. Hence, the system represented by Eq. (4.179) is neutrally stable. Therefore, in numerical computations involving updating of direction cosines, extreme care should be taken because rounding off errors can build up rapidly and blow out the solution.

4.2.8 Quaternions

The method of quaternions presents a practical alternative to the Euler angle rate method and has the advantage that it avoids the singularity that is present in the Euler angle rate equations when the pitch angle θ approaches 90 deg. This method is also known as the Euler four parameter method. In the following, we will present a brief discussion of this method so that the reader is in a position to use this method. Detailed mathematics is avoided to keep the exposition simple. For additional information, the reader may refer to other sources [3, 4].

Fundamental to this method is Euler's theorem that any given frame of axes $Ox_2y_2z_2$ can be made to coincide with another frame $Ox_1y_1z_1$ by a single rotation D about a fixed axis in space making angles A, B, C with $Ox_1y_1z_1$. Then, we can define four parameters, e_0, e_1, e_2, e_3 as follows:

$$e_0 = \cos\frac{D}{2} \tag{4.186}$$

$$e_1 = \cos A \sin\frac{D}{2} \tag{4.187}$$

$$e_2 = \cos B \sin\frac{D}{2} \tag{4.188}$$

$$e_3 = \cos C \sin\frac{D}{2} \tag{4.189}$$

These four parameters are constrained by the following condition:

$$e_0^2 + e_1^2 + e_2^2 + e_3^2 = 1 \tag{4.190}$$

Let $X_2 = T_1^2 X_1$. The transformation matrix T_1^2 is presented in the following without proof. Interested readers may refer elsewhere [4].

$$T_1^2 = \begin{bmatrix} e_0^2 + e_1^2 - e_2^2 - e_3^2 & 2(e_1e_2 + e_0e_3) & 2(e_1e_3 - e_0e_2) \\ 2(e_1e_2 - e_0e_3) & e_0^2 - e_1^2 + e_2^2 - e_3^2 & 2(e_2e_3 + e_0e_1) \\ 2(e_0e_2 + e_1e_3) & 2(e_2e_3 - e_0e_1) & e_0^2 - e_1^2 - e_2^2 + e_3^2 \end{bmatrix} \tag{4.191}$$

With ψ, θ, and ϕ as Euler angles, the transformation matrix T_1^2 is also given by Eq. (4.22). Equating the corresponding elements of the two matrices, it can be shown that [4],

$$e_0 = \cos\frac{\psi}{2}\cos\frac{\theta}{2}\cos\frac{\phi}{2} + \sin\frac{\psi}{2}\sin\frac{\theta}{2}\sin\frac{\phi}{2} \tag{4.192}$$

$$e_1 = \cos\frac{\psi}{2}\cos\frac{\theta}{2}\sin\frac{\phi}{2} - \sin\frac{\psi}{2}\sin\frac{\theta}{2}\cos\frac{\phi}{2} \tag{4.193}$$

$$e_2 = \cos\frac{\psi}{2}\sin\frac{\theta}{2}\cos\frac{\phi}{2} + \sin\frac{\psi}{2}\cos\frac{\theta}{2}\sin\frac{\phi}{2} \tag{4.194}$$

$$e_3 = -\cos\frac{\psi}{2}\sin\frac{\theta}{2}\sin\frac{\phi}{2} + \sin\frac{\psi}{2}\cos\frac{\theta}{2}\cos\frac{\phi}{2} \tag{4.195}$$

According to the method of direction cosines

$$X_2 = C_1^2 X_1 \tag{4.196}$$

where

$$C_1^2 = \begin{bmatrix} C_{11} & C_{12} & C_{13} \\ C_{21} & C_{22} & C_{23} \\ C_{31} & C_{32} & C_{33} \end{bmatrix} \tag{4.197}$$

Then equating the relations (4.22) and (4.197) and using Eqs. (4.192–4.195), the following relations between the elements of the direction cosine matrix, Euler angles, and the four quaternion parameters can be obtained. For detailed mathematical derivations, the interested reader may refer elsewhere [4].

$$C_{11} = \cos\theta\cos\psi = e_0^2 + e_1^2 - e_2^2 - e_3^2 \tag{4.198}$$

$$C_{12} = \cos\theta\sin\psi = 2(e_1e_2 + e_0e_3) \tag{4.199}$$

$$C_{13} = -\sin\theta = 2(e_1e_3 - e_0e_2) \tag{4.200}$$

$$C_{21} = \sin\theta\sin\phi\cos\psi - \sin\psi\cos\phi = 2(e_1e_2 - e_0e_3) \tag{4.201}$$

$$C_{22} = \sin\theta\sin\phi\sin\psi + \cos\psi\cos\phi = e_0^2 - e_1^2 + e_2^2 - e_3^2 \tag{4.202}$$

$$C_{23} = \cos\theta\sin\phi = 2(e_2e_3 + e_0e_1) \tag{4.203}$$

$$C_{31} = \sin\theta\cos\phi\cos\psi + \sin\psi\sin\phi = 2(e_0e_2 + e_1e_3) \tag{4.204}$$

$$C_{32} = \sin\theta\cos\phi\sin\psi - \cos i\sin\phi = 2(e_2e_3 - e_0e_1) \tag{4.205}$$

$$C_{33} = \cos\theta\cos\phi = e_0^2 - e_1^2 - e_2^2 = e_3^2 \tag{4.206}$$

The rate equations for the four quaternion parameters are given by the following relations. For detailed mathematical derivation, the interested reader may refer elsewhere [4].

$$\dot{e}_0 = -\frac{1}{2}(e_1p + e_2q + e_3r) \tag{4.207}$$

$$\dot{e}_1 = \frac{1}{2}(e_0p + e_2r - e_3q) \tag{4.208}$$

$$\dot{e}_2 = \frac{1}{2}(e_0q + e_3p - e_1r) \tag{4.209}$$

$$\dot{e}_3 = \frac{1}{2}(e_0r + e_1q - e_2p) \tag{4.210}$$

Equations (4.207–4.210) along with the specified initial values e_0, e_1, e_2, and e_3 can be used to generate the time history of the four parameters e_0, e_1, e_2, and e_3. The initial values are usually not given directly. What are generally given are the initial values of the three Euler angles. Then using Eqs. (4.192–4.195), the initial values of the four quaternion parameters can be obtained.

An important point in the generation of time history of the four quaternion parameters is that these four parameters are always subject to the constraint given by Eq. (4.190). One way of ensuring that this constraint is always satisfied is to adjoin the constraint to the state Eqs. (4.207–4.210) using a Lagrange-type multiplier λ as follows [3],

$$\dot{e}_0 = -\frac{1}{2}(e_1 p + e_2 q + e_3 r) + \lambda \epsilon e_0 \tag{4.211}$$

$$\dot{e}_1 = \frac{1}{2}(e_0 p + e_2 r - e_3 q) + \lambda \epsilon e_1 \tag{4.212}$$

$$\dot{e}_2 = \frac{1}{2}(e_0 q + e_3 p - e_1 r) + \lambda \epsilon e_2 \tag{4.213}$$

$$\dot{e}_3 = \frac{1}{2}(e_0 r + e_1 q - e_2 p) + \lambda \epsilon e_3 \tag{4.214}$$

where

$$\epsilon = 1 - \left(e_0^2 + e_1^2 + e_2^2 + e_3^2\right) \tag{4.215}$$

Here, λ is a free parameter. Usually, λ is set equal to a small multiple of the integration time step [3].

A forward integration of Eqs. (4.211–4.214) generates the time history of four parameters e_0, e_1, e_2, and e_3. Then, the Euler angles can be obtained as follows. From Eq. (4.200), we have

$$C_{13} = -\sin\theta = 2(e_1 e_3 - e_0 e_2) \tag{4.216}$$

so that

$$\theta = \sin^{-1}[-2(e_1 e_3 - e_0 e_2)] \tag{4.217}$$

Because θ is supposed to be in the range $-\pi/2 \le \theta \le \pi/2$, the angle θ is uniquely determined by Eq. (4.217). Next, from Eq. (4.203) and Eq. (4.206), we have,

$$C_{33} = \cos\theta \cos\phi = e_0^2 - e_1^2 - e_2^2 + e_3^2 \tag{4.218}$$

$$C_{23} = \cos\theta \sin\phi = 2(e_2 e_3 + e_0 e_1) \tag{4.219}$$

so that

$$\phi = \cos^{-1}\left(\frac{C_{33}}{\cos\theta}\right)\text{sgn}(C_{23}) \tag{4.220}$$

$$= \cos^{-1}\left[\frac{e_0^2 - e_1^2 - e_2^2 + e_3^2}{\sqrt{1 - 4(e_1e_3 - e_0e_2)^2}}\right]\text{sgn}[2(e_2e_3 + e_0e_1)] \tag{4.221}$$

where the sgn(·) function has the following properties: if the argument is positive, then the sgn(·) function returns the value of +1. If it is negative, then sgn(·) function returns the value of −1. Equation (4.221) uniquely determines the sign of ϕ. Similarly, using Eqs. (4.198) and (4.199), we get

$$\psi = \cos^{-1}\left[\frac{e_0^2 + e_1^2 - e_2^2 - e_3^2}{\sqrt{1 - 4(e_1e_3 - e_0e_2)^2}}\right]\text{sgn}[2(e_1e_2 + e_0e_3)] \tag{4.222}$$

As said before, we will not encounter the singularity at $\theta = \pi/2$ if we obtain the Euler angles using the method of quaternions. We can summarize the solution sequence of all the three approaches to calculate the Euler angles as follows:

1. Euler angle rates: ψ_0, θ_0, ϕ_0; body rates p, q, r, $\rightarrow$ Euler rates $\dot{\psi}$, $\dot{\theta}$, $\dot{\phi} \rightarrow \psi(t)$, $\theta(t)$, $\phi(t)$.
2. Direction cosine matrices (DCM): ψ_0, θ_0, ϕ_0; DCM at $t = 0$, body rates p, q, r; DCM update equations $\rightarrow$ DCM at time $t \rightarrow \psi(t)$, $\theta(t)$, $\phi(t)$.
3. Quaternions: ψ_0, θ_0, ϕ_0; body rates p, q, r, $\rightarrow$ quaternions e_0, e_1, e_2, and $e_3 \rightarrow \psi(t)$, $\theta(t)$, $\phi(t)$.

Example 4.1

With respect to an Earth-centered inertial frame, a vehicle has the following velocity components:

$$u_i(t) = u_{oi} + a_{xi}t$$
$$v_i(t) = a_{yi}t$$
$$w_i = 0$$

Assuming $u_{oi} = 100$ ft/s, $a_{xi} = 25$ ft/s^2, $a_{yi} = 50$ ft/s^2, determine the position and velocity with reference to the Earth-fixed $x_Ey_Ez_E$ and navigational $x_ey_ez_e$ systems at $t = 50$ s. Assume that, at $t = 0$, the vehicle is located on the equator with $L = 0$.

(Continued)

Example 4.1 (*Continued*)

Solution:

At $t = 0$, we have $x_i = R_e$, $y_i = z_i = 0$. Here, $R_e = 2.0973364 \times 10^7$ ft is the radius of the Earth at the equator. For $t \geq 0$, we have

$$x_i(t) = x_{oi} + \int_0^{t_1} u_i \, \mathrm{d}t$$

$$= R_e + u_{oi} t_1 + a_{xi} \frac{t_1^2}{2}$$

$$y_i(t) = a_{yi} \frac{t_1^2}{2}$$

With $t_1 = 50$ s, $u_{oi} = 100$ ft/s, $a_{xi} = 25$ ft/s^2, and $a_{yi} = 50$ ft/s^2, we get $x_i = R_e + 36{,}250 = 2.1009614 \times 10^7$ ft and $y_i = 62{,}500$ ft. Note that $z_i = 0$.

The angular velocity of the Earth about the Oz_i or Oz_E axis is 1 rev/day or

$$\Omega_e = \frac{2\pi}{24 * 3600}$$

$$= 0.7272 \times 10^{-4} \text{ rad/s}$$

With $t_1 = 50$ s, we have

$$\cos \Omega_e t_1 = 0.999993$$

$$\sin \Omega_e t_1 = 0.003636$$

From Eq. (4.27), we have

$$\begin{bmatrix} x_E \\ y_E \\ z_E \end{bmatrix} = \begin{bmatrix} \cos \Omega_e t_1 & \sin \Omega_e t_1 & 0 \\ -\sin \Omega_e t_1 & \cos \Omega_e t_1 & 0 \\ 0 & 0 & 1 \end{bmatrix} \begin{bmatrix} x_i \\ y_i \\ z_i \end{bmatrix}$$

$$= \begin{bmatrix} 2.1009841 \times 10^7 \\ -13869.95 \\ 0 \end{bmatrix}$$

The altitude is given by

$$h_e = \sqrt{x_E^2 + y_E^2} - R_e$$

Substituting for x_E, y_E, and R_e, we obtain $h_e = 36{,}000.0$ ft.

The origin of the navigational system $Ox_e y_e z_e$ lies directly beneath the vehicle at $t = 0$. Therefore,

$$X_{oi} = \begin{bmatrix} R_e \\ 0 \\ 0 \end{bmatrix}$$

(*Continued*)

Example 4.1 (*Continued*)

With $L \approx 0$, $\lambda = 0$ and using Eq. (4.30),

$$T_i^e = \begin{bmatrix} 0 & 0 & 1 \\ 0 & 1 & 0 \\ -1 & 0 & 0 \end{bmatrix}$$

Using Eq. (4.34)

$$X_e = T_i^e(X_i - X_{oi})$$

or

$$\begin{bmatrix} x_e \\ y_e \\ z_e \end{bmatrix} = \begin{bmatrix} 0 & 0 & 1 \\ 0 & 1 & 0 \\ -1 & 0 & 0 \end{bmatrix} \begin{bmatrix} R_e + 36{,}250 \\ 62{,}500 \\ 0 \end{bmatrix} - \begin{bmatrix} 0 & 0 & 1 \\ 0 & 1 & 0 \\ -1 & 0 & 0 \end{bmatrix} \begin{bmatrix} R_e \\ 0 \\ 0 \end{bmatrix}$$

$$= \begin{bmatrix} 0 \\ 62{,}500 \\ -36{,}250 \end{bmatrix}$$

Example 4.2

Plot the Euler angle time history for an airplane performing a velocity vector roll with 1) $\alpha = 30$ deg, 2) $\alpha = 45$ deg, and 3) $\alpha = 60$ deg. Assume that the angular velocity Ω about the velocity vector in each case is 25 deg/s.

Solution:

The first step is to select a reference frame that remains fixed in space and with respect to which the Euler angles are measured. Let $x_0y_0z_0$ be such a reference frame as shown in Fig. 4.8. The Ox_o axis is assumed to coincide with the velocity vector U_o, the Oy_o axis points to the right in the horizontal plane, and Oz_o points vertically downward to form a right-hand system.

At $t = 0$, the aircraft is assumed to be oriented as shown in Fig. 4.8 so that the initial values of the Euler angles of the body axes system $Ox_by_bz_b$ are $\psi(0) = 0$, $\theta(0) = \alpha$, and $\phi(0) = 0$.

We have

$$p = \Omega \cos \alpha$$

$$q = 0$$

$$r = \Omega \sin \alpha$$

Because the angle of attack remains constant during the velocity vector roll, the values of p, q, r will also remain constant.

(*Continued*)

Example 4.2 (*Continued*)

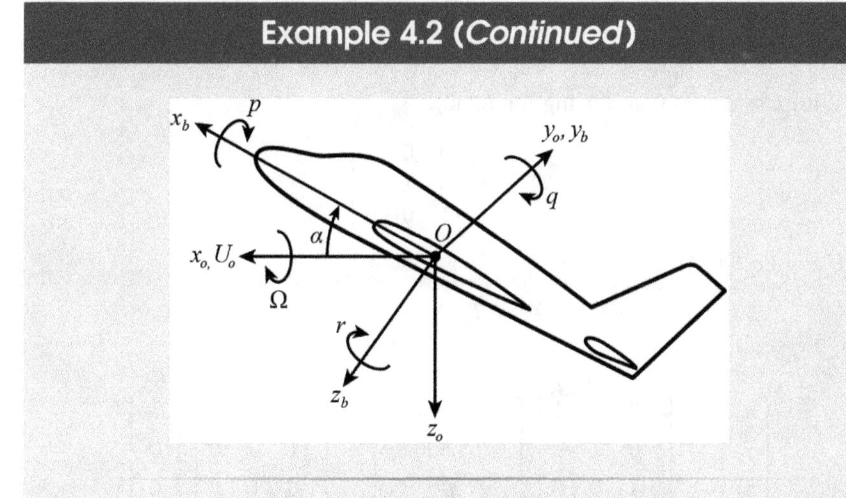

Fig. 4.8 Airplane in a velocity vector roll.

With $q = 0$, the Euler angle rates as given by Eqs. (4.66–4.68) are

$$\dot{\phi} = p + r \tan\theta \cos\phi$$
$$\dot{\theta} = -r \sin\phi$$
$$\dot{\psi} = r \sec\theta \cos\phi$$

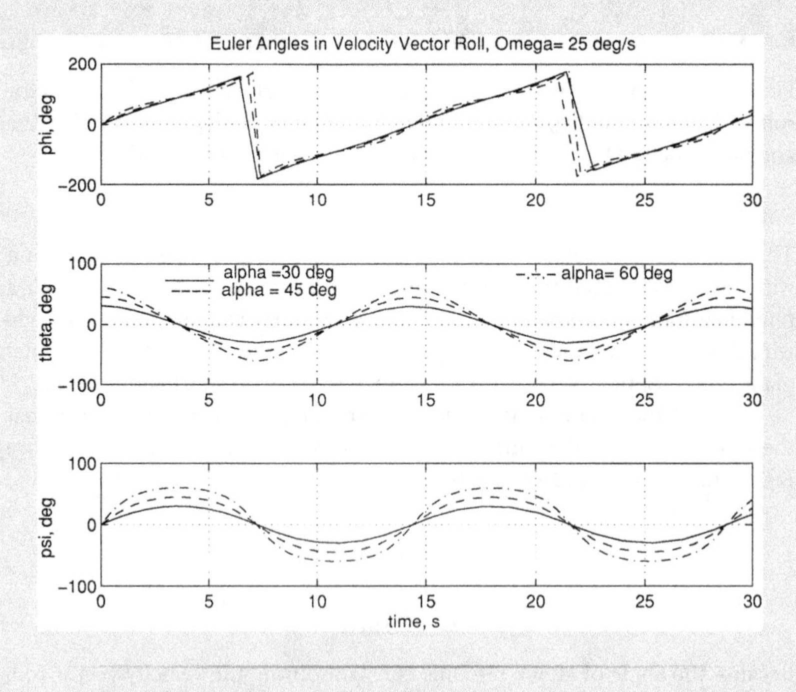

Fig. 4.9 Euler angle time histories in velocity vector rolls.

(*Continued*)

Example 4.2 (*Continued*)

An integration of these equations for the given initial values was performed using MATLAB® [5] routine ODE45, and the results are shown in Fig. 4.9. Here, the roll angle continuously increases with time. In view of the restrictions as given in Eqs. (4.1–4.3), the roll angle ϕ has a sawtooth variation.

Example 4.3

For a spinning airplane as shown in Fig. 4.10, the angular velocity components in the body axes are given by

$$p = \Omega \cos \alpha$$
$$q = 0$$
$$r = \Omega \sin \alpha$$

where Ω is the angular velocity about the spin axis, which is assumed to be vertical and passing through the center of gravity. Plot the Euler angle time history for the first 25 s and two values of $\alpha = 30$ deg and 60 deg. Assume $\Omega = 30$ deg/s.

Solution:

Because $q = 0$, the airplane has zero pitch rate about the body axes. It experiences only rolling and yawing motion about the body axes.

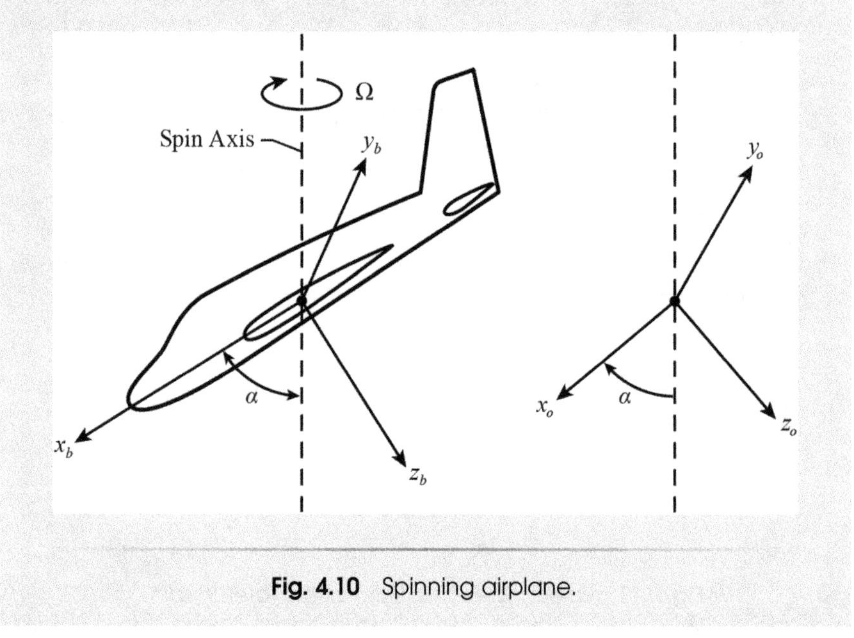

Fig. 4.10 Spinning airplane.

(Continued)

Example 4.3 (*Continued*)

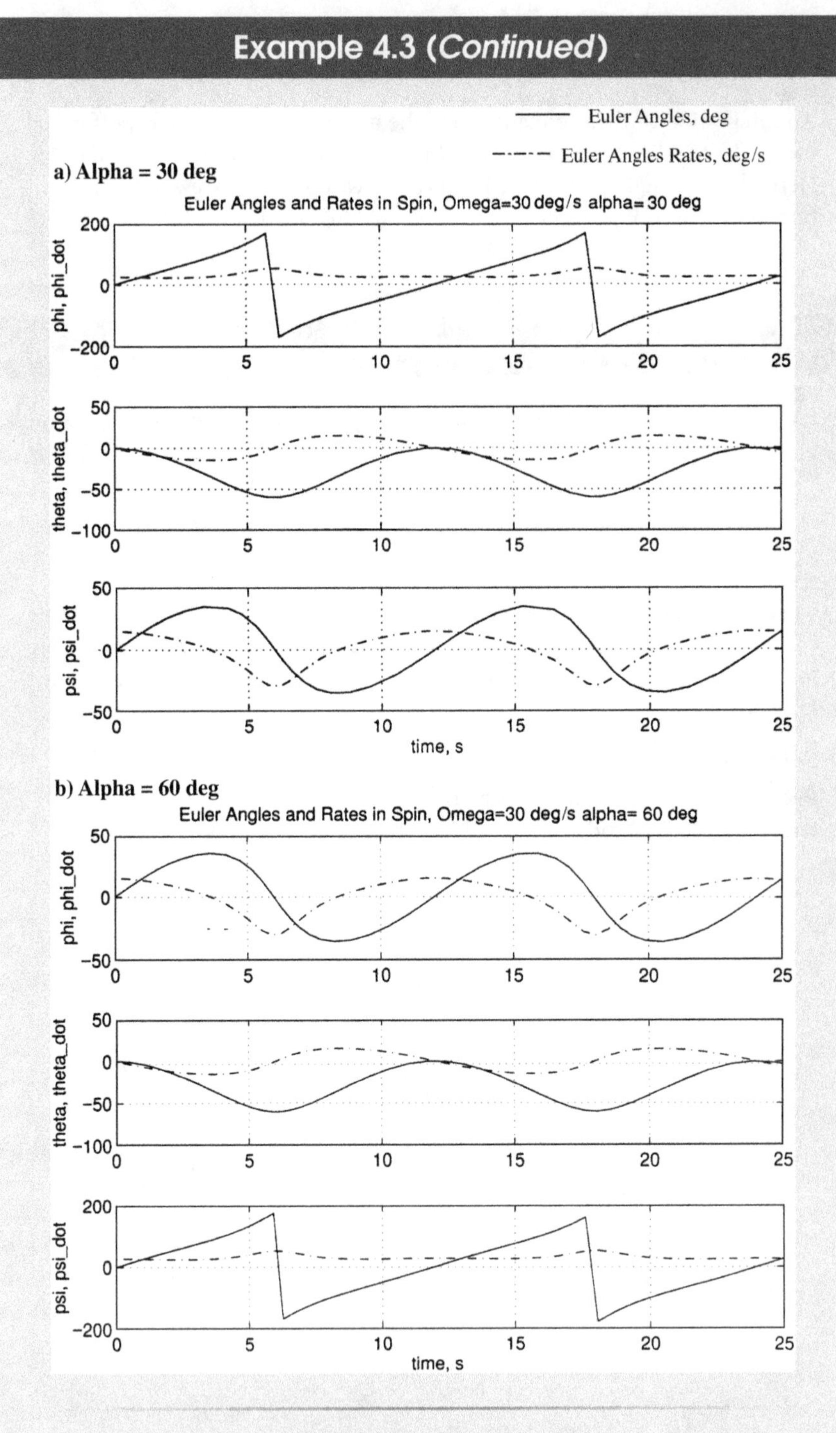

Fig. 4.11 Euler angles and Euler angle rates in spin.

(*Continued*)

Example 4.3 (*Continued*)

We select the reference frame $x_0y_0z_0$ with respect to which the Euler angles are calculated to coincide with the body axes system at $t = 0$. For $t > 0$, the reference $x_0y_0z_0$ remains fixed in space, whereas the body-fixed axes system $x_by_bz_b$ system rotates with the airplane. In view of this, we have $\psi(0) = \theta(0) = \phi(0) = 0$. Note that for each case of $\alpha = 30$ deg and 60 deg, the initial orientation of the $x_0y_0z_0$ axes system is different.

With these initial conditions and using Eqs. (4.66–4.68), the Euler angle time history was obtained as shown in Figs. 4.11a and 4.11b. At low angles of attack, as in the case of $\alpha = 30$ deg, the spinning motion is predominantly the roll about the body axis Ox_b (Fig. 4.11a). On the other hand, at high angles of attack, it will be mostly yawing motion as observed for $\alpha = 60$ deg (Fig. 4.11b).

Example 4.4

An aircraft model is tested in a low-speed wind tunnel at an angle of attack of 20 deg and sideslip of 10 deg. An internal strain gage balance was used to measure the aerodynamic forces acting on the model, which gives components of force in the body axes system. The measurements are $F_x = -21.7$ lb, $F_y = +33.0$ lb, and $F_z = -91$ lb. Determine 1) the transformation matrix T_b^w, and 2) the lift, drag, and side forces acting on the model.

Solution:

The forces F_x, F_y, and F_z measured by the balance are with respect to the body-fixed axes system, whereas the lift, drag, and side force are with respect to the wind axes system. Therefore, we need the transformation matrix T_b^w.

From Eq. (4.37), we have

$$T^b{}_w = \begin{bmatrix} \cos\alpha\cos\beta & -\cos\alpha\sin\beta & -\sin\alpha \\ \sin\beta & \cos\beta & 0 \\ \sin\alpha\cos\beta & -\sin\beta\sin\alpha & \cos\alpha \end{bmatrix}$$

and

$$T_b^w = (T_w^b)'$$

With $\alpha = 20$ deg and $\beta = 10$ deg, we get

$$T_b^w = \begin{bmatrix} 0.9254 & 0.1736 & 0.3368 \\ -0.1632 & 0.9848 & -0.0594 \\ -0.3420 & 0 & 0.9397 \end{bmatrix}$$

(*Continued*)

Example 4.4 (*Continued*)

Then,

$$\begin{bmatrix} F_{xw} \\ F_{yw} \\ F_{zw} \end{bmatrix} = T_b^w \begin{bmatrix} -21.7 \\ 33.0 \\ -91 \end{bmatrix}$$

so that $F_{xw} = -45.0021$ lb, $F_{yw} = 41.4442$ lb, and $F_{zw} = -78.0902$ lb or lift $L = 78.0902$ lb, drag $D = 45.0021$ lb, and side force $Y = 41.4442$ lb.

Example 4.5

For the spinning airplane of Example 4.3, determine the Euler angle history using 1) Euler angles, 2) the method of direction cosines, and 3) quaternions for $\alpha = 30$ deg, 45 deg, and 60 deg.

Solution:

The reference axes system $Ox_0y_0z_0$ with respect to which the Euler angles are measured is assumed to coincide with the body axes system at $t = 0$. Subsequently, the reference axes system $x_0y_0z_0$ remains fixed in space (inertial system for this case). This means that we will have a different reference axes system for each value of α. With these assumptions, $\psi(0) = \theta(0) = \phi(0) = 0$ for each case. We have

$$p = \Omega \cos \alpha$$

$$q = 0$$

$$r = \Omega \sin \alpha$$

Because α is a held constant for each case,$\dot{p} = \dot{q} = \dot{r} = 0$.

1. *Euler angles*: The Euler angle rates are given by Eqs. (4.66–4.68). For this case, these equations reduce to

$$\dot{\phi} = p + r \tan \theta \cos \phi$$

$$\dot{\theta} = -r \sin \phi$$

$$\dot{\psi} = r \sec \theta \cos \phi$$

MATLAB [5] code ODE45 was used for numerical integration of the given Euler angle rate equations. Whenever the value of either ϕ or ψ exceeded 180 deg, it was reset equal to -180 deg.

For $\alpha = 45$ deg, the pitch angle θ approaches -90 deg, and we encounter the singularity present in Euler angle rate equations. To work around

(*Continued*)

Example 4.5 (*Continued*)

this difficulty, when $89.5 \leq \theta \leq 90.5$, we use Eqs. (4.82), (4.85), and (4.86), which assume the following form for this case:

$$(\dot{\phi})_{\pi/2} = \frac{p}{2}$$

$$(\dot{\theta})_{\pi/2} = \pm r$$

$$(\dot{\psi})_{\pi/2} = \frac{-p}{2}$$

2. *Method of direction cosines*: Let X_o and X_b represent vectors in $Ox_oy_oz_o$ and $Ox_by_bz_b$ system respectively. Then,

$$X_o = C_b^o X_b$$

$$= \begin{bmatrix} C_{11} & C_{12} & C_{13} \\ C_{21} & C_{22} & C_{23} \\ C_{31} & C_{32} & C_{33} \end{bmatrix} X_b$$

Using Eqs. (4.128–4.136) with $\psi(0) = \theta(0) = \phi(0) = 0$, we get the initial values of the elements of the direction cosines matrix as follows:

$$\begin{array}{lll} C_{11} = 1 & C_{12} = 0 & C_{13} = 0 \\ C_{21} = 0 & C_{22} = 1 & C_{23} = 0 \\ C_{31} = 0 & C_{32} = 0 & C_{33} = 1 \end{array}$$

The rate equations for updating the elements of the direction cosine matrix elements are given by Eqs. (4.156–4.164), which are reproduced in the following:

$$\dot{C}_{11} = C_{12}r - C_{13}q$$

$$\dot{C}_{12} = C_{13}p - C_{11}r$$

$$\dot{C}_{13} = C_{11}q - C_{12}p$$

$$\dot{C}_{21} = C_{22}r - C_{23}q$$

$$\dot{C}_{22} = C_{23}p - C_{21}r$$

$$\dot{C}_{23} = C_{21}q - C_{22}p$$

$$\dot{C}_{31} = C_{32}r - C_{33}q$$

$$\dot{C}_{32} = C_{33}p - C_{31}r$$

$$\dot{C}_{33} = C_{31}q - C_{32}p$$

All these first-order, coupled ordinary differential equations along with the given initial conditions were integrated using MATLAB [5] code

(*Continued*)

Example 4.5 (*Continued*)

ODE45. Then, knowing the values of C_{ij}, $i = j = 1, 3$, the Euler angles were calculated using the following relations:

$$\theta = \sin^{-1}(-C_{31})$$

$$\phi = \cos^{-1}\left(\frac{C_{33}}{\sqrt{1 - C_{31}^2}}\right)\text{sgn}(C_{32})$$

$$\psi = \cos^{-1}\left(\frac{C_{11}}{\sqrt{1 - C_{31}^2}}\right)\text{sgn}(C_{21})$$

Note that the indices of the elements of the direction cosine matrices in the previous relations for θ, ϕ, and ψ are interchanged from those

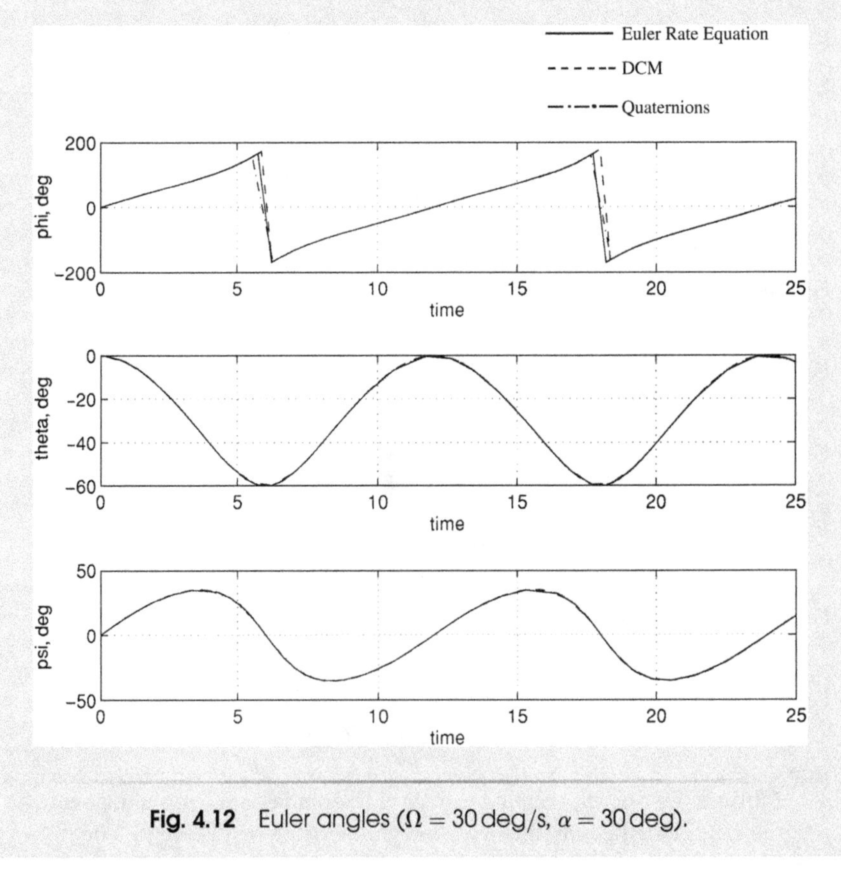

Fig. 4.12 Euler angles ($\Omega = 30$ deg/s, $\alpha = 30$ deg).

(Continued)

Example 4.5 (*Continued*)

given in Eqs. (4.137–4.139). Why? Because the direction cosine matrix in Eq. (4.115) and Eqs. (4.128–4.139) is used for transformation of a vector from inertial system ($Ox_1y_1z_1$) to body axes system ($Ox_2y_2z_2$), whereas C_1^2 in Eq. (4.140) is used for reverse transformation, that is, to transform a vector from body axes system to inertial system. Therefore, we have to interchange the indices to effect a transpose.

3. *Quaternions*: The four Euler parameters are given by Eqs. (4.192–4.195), which are reproduced in the following:

$$e_0 = \cos\frac{\psi}{2}\cos\frac{\theta}{2}\cos\frac{\phi}{2} + \sin\frac{\psi}{2}\sin\frac{\theta}{2}\sin\frac{\phi}{2}$$

$$e_1 = \cos\frac{\psi}{2}\cos\frac{\theta}{2}\sin\frac{\phi}{2} - \sin\frac{\psi}{2}\sin\frac{\theta}{2}\cos\frac{\phi}{2}$$

$$e_2 = \cos\frac{\psi}{2}\sin\frac{\theta}{2}\cos\frac{\phi}{2} + \sin\frac{\psi}{2}\cos\frac{\theta}{2}\sin\frac{\phi}{2}$$

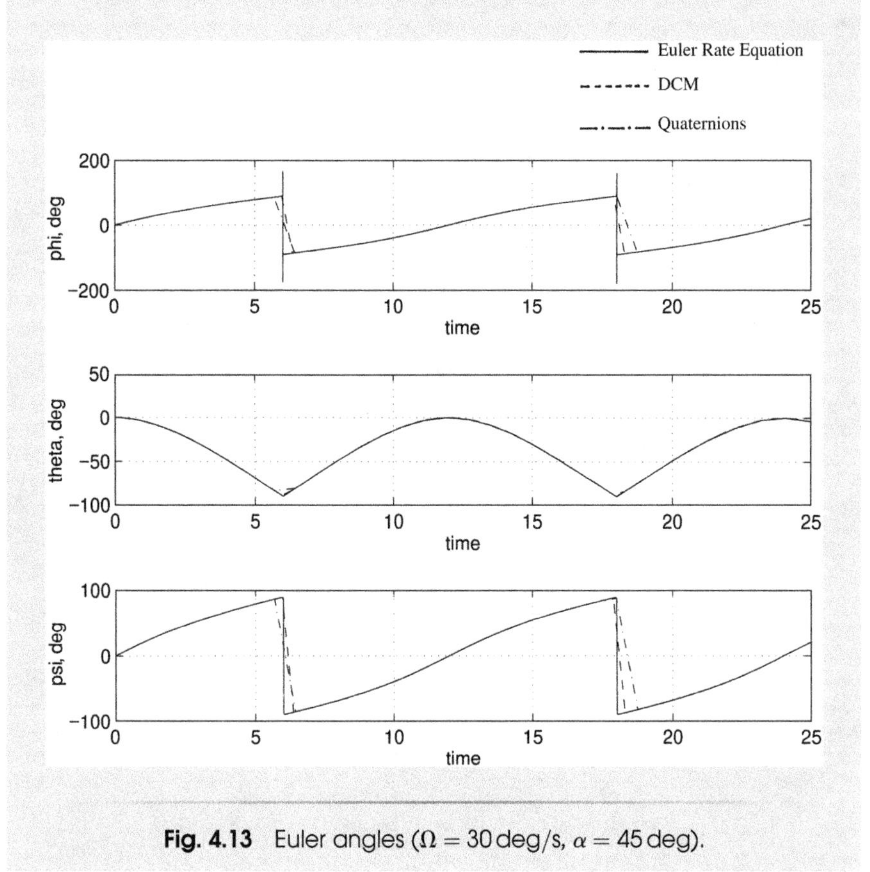

Fig. 4.13 Euler angles ($\Omega = 30$ deg/s, $\alpha = 45$ deg).

(Continued)

Example 4.5 (*Continued*)

$$e_3 = -\cos\frac{\psi}{2}\sin\frac{\theta}{2}\sin\frac{\phi}{2} + \sin\frac{\psi}{2}\cos\frac{\theta}{2}\cos\frac{\phi}{2}$$

Using $\psi(0) = \theta(0) = \phi(0) = 0$, we obtain $e_0 = 1$, $e_1(0) = e_2(0) = e_3(0) = 0$. The rate equations for updating the four quaternion parameters are given by Eqs. (4.211–4.214), which are reproduced in the following:

$$\dot{e}_0 = -\frac{1}{2}(e_1 p + e_2 q + e_3 r) + \lambda \epsilon e_0$$

$$\dot{e}_1 = \frac{1}{2}(e_0 p + e_2 r - e_3 q) + \lambda \epsilon e_1$$

$$\dot{e}_2 = \frac{1}{2}(e_0 q + e_3 p - e_1 r) + \lambda \epsilon e_2$$

$$\dot{e}_3 = \frac{1}{2}(e_0 r + e_1 q - e_2 p) + \lambda \epsilon e_3$$

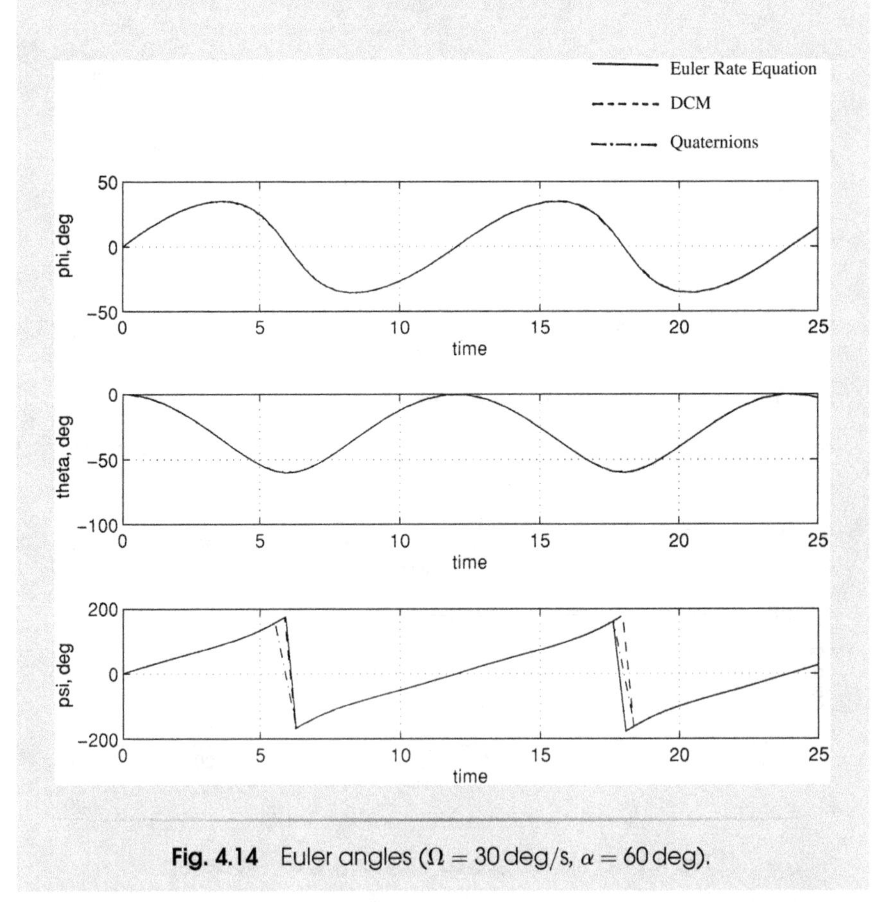

Fig. 4.14 Euler angles ($\Omega = 30$ deg/s, $\alpha = 60$ deg).

(Continued)

Example 4.5 (*Continued*)

where ϵ is given by Eq. (4.215) and is reproduced in the following:

$$\epsilon = 1 - \left(e_0^2 + e_1^2 + e_2^2 + e_3^2\right)$$

We assume $\lambda = 0.0001$ and integrate the given rate equations for four quaternion parameters using MATLAB [5] code ODE45. Then, the Euler angles are obtained using Eqs. (4.217), (4.221), and (4.222).The results are presented in Figs. 4.12–4.14. We observe that all three methods give identical values of Euler angles except around discontinuous switching. The spikes in the roll angle ϕ for $\alpha = 45$ deg (Fig. 4.13) are apparently caused by the fact that Euler angle rates assume large values as θ approaches 90 deg.

Example 4.6

At a certain time during a continuous motion of an airplane, the following direction cosine matrix is recorded. It is suspected that the elements marked *xx* are in error and hence are discarded. Determine the missing elements.

$$C = \begin{bmatrix} 0.8999 & -0.4323 & 0.0578 \\ xx & 0.8665 & -0.2496 \\ xx & xx & 0.9666 \end{bmatrix}$$

Solution:

We observe that the missing elements are C_{21}, C_{31}, and C_{32}. We have nine elements of the direction cosine matrix and six redundancy relations. Therefore, we can determine up to three missing elements at a time. We have

$$\begin{array}{lll} C_{11} = 0.8999 & C_{12} = -0.4323 & C_{13} = 0.0578 \\ C_{21} = xx & C_{22} = 0.8665 & C_{23} = -0.2496 \\ C_{31} = xx & C_{32} = xx & C_{33} = 0.9666 \end{array}$$

Using Eq. (4.127),

$$\begin{aligned} C_{32} &= -\left[\frac{C_{12}C_{13} + C_{22}C_{23}}{C_{33}}\right] \\ &= -\left[\frac{(-0.4323) * 0.0578 + 0.8665 * (-0.2496)}{0.9666}\right] \\ &= -0.2496 \end{aligned}$$

(*Continued*)

Example 4.6 (Continued)

Substituting in Eqs. (4.125) and (4.126), we get

$$0.8999 * (-0.4323) + 0.8665C_{21} + 0.2496C_{31} = 0$$

$$0.8999 * 0.0578 + (-0.2496)C_{21} + 0.9666C_{31} = 0$$

Solving, we get, $C_{21} = 0.4323$ and $C_{31} = 0.0578$. Now to verify our calculations, we have to check whether the remaining three redundancy relations in Eqs. (4.122–4.124) are satisfied. Substituting, we find

$$C_{11}^2 + C_{21}^2 + C_{31}^2 = 1.00004$$

$$C_{12}^2 + C_{22}^2 + C_{32}^2 = 1.000006$$

$$C_{13}^2 + C_{23}^2 + C^2{}_{33} = 0.999996$$

Hence, the calculated values of missing elements are satisfactory.

4.3 Equations of Motion and Concept of Moving Axes System

The equations governing the aircraft motion are based on Newton's laws of motion. We have the force and moment equations in the form

$$F = m\left(\frac{dV}{dt}\right)_i \tag{4.223}$$

$$M = \left(\frac{dH}{dt}\right)_i \tag{4.224}$$

where V is the velocity and H is the angular momentum. The suffix i implies that the acceleration dV/dt and the rate of change of angular momentum dH/dt are supposed to be measured with respect to an inertial frame of reference. We have $H = I_\omega$ where I is the moment of inertia of the body and ω is the angular velocity of the body with respect to an inertial system. Then Eq. (4.224) takes the form

$$M = \left(\frac{dI}{dt}\right)_i \omega + I\left(\frac{d\omega}{dt}\right)_i \tag{4.225}$$

Suppose we choose a space-fixed inertial system of reference to solve the equations of aircraft motion, then we have to compute the angular momentum in this system where the moment of inertia I will continuously vary with time as the aircraft translates and rotates in space. As a result, it will be extremely difficult to solve the equations of motion because, at each time step, we have to evaluate the moment of inertia I and its time derivative $(\mathrm{d}I/\mathrm{d}t)_i$. For the aircraft, I is a dyad with nine components, three principal moments of inertia I_{xx}, I_{yy}, I_{zz}, and six products of inertia I_{xy}, I_{yx}, I_{yz}, I_{zy}, I_{xz}, I_{zx}.

One way of overcoming this difficulty of computing time-varying moments and products of inertia is to use a moving or body axes system, which is fixed to the aircraft all the time and moves with it. Then the moments and products of inertia calculated with respect to this axes system will be constant except for such variations as fuel consumption or control surface deflections that can be easily accounted for.

The introduction of a moving axes system avoids the problem of computing time-varying moments and products of inertia but creates another problem because the accelerations measured in the moving coordinate system are not the accelerations with respect to an inertial frame of reference. Fortunately, this problem is a simpler one to solve, and the following theorem of vector analysis helps us obtain the accelerations with respect to an inertial frame of reference given the accelerations in a moving coordinate system.

The reader may get confused when we talk of aircraft velocity or acceleration in the moving axes system. How can aircraft have velocity or acceleration with respect to an axes system fixed to it and moving with it? Certainly the aircraft (assuming it to be a rigid body) cannot have velocity or acceleration with respect to itself. Then what are these velocity and acceleration in the moving axes system? These are the aircraft velocity and acceleration with respect to an Earth-fixed inertial system but expressed or resolved in the body axes system fixed to the aircraft and moving with it.

4.3.1 Moving Axes Theorem

Let $\vec{A}_b$ be a vector observed in the moving axes system $Ox_by_bz_b$ and $(\mathrm{d}\vec{A}/\mathrm{d}t)_b$ be the time rate of change of $\vec{A}_b$ recorded in the moving axes system. Furthermore, let $\vec{\omega}^b_{i,b}$ be the angular velocity of the moving axes system measured with respect to an inertial system but having components in the body axes system. As said before, the order of the subscripts and superscript is as follows: the subscripts i, b have the meaning of the body axes with respect to the inertial axes system and the superscript b means that the vector $\vec{\omega}$ is resolved or has components in the body axes system. The problem is to determine the time rate of change of a vector $\vec{A}$ in the inertial frame of reference $Ox_iy_iz_i$ given its time rate of change in the body axes system $Ox_by_bz_b$. This can be accomplished using the following theorem known as the moving axes theorem:

$$\left(\frac{\mathrm{d}\vec{A}}{\mathrm{d}t}\right)_i = \left(\frac{\mathrm{d}\vec{A}}{\mathrm{d}t}\right)_b + \vec{\omega}^b_{i,b} \times \vec{A}_b \tag{4.226}$$

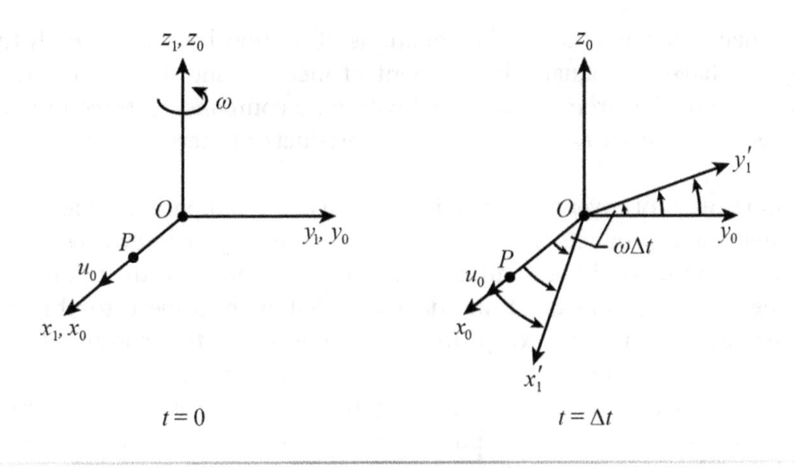

Fig. 4.15 Moving axes theorem.

To understand the concept of moving axes theorem, consider an inertial frame of reference $Ox_0y_0z_0$ and a moving axes system $Ox_1y_1z_1$ as shown in Fig. 4.15.

Let ω be the angular velocity of the $Ox_1y_1z_1$ system with respect to the $Ox_0y_0z_0$ system but having components in the $Ox_1y_1z_1$ system. Assume that at $t = 0$ the $Ox_1y_1z_1$ system coincides with the $Ox_0y_0z_0$ system. Let a particle P move with a constant velocity u_0 along the Ox_0 axis. At $t = \Delta t$, the particle P will still have the same velocity u_0 with respect to $Ox_0y_0z_0$ so that the acceleration measured by an observer stationed at the origin of the $Ox_0y_0z_0$ system is zero.

Now let us find out what an observer stationed at the origin of the moving coordinate system $Ox_1y_1z_1$ has measured. At $t = 0$, he will also record $u_1 = u_0$ and $v_1 = w_1 = 0$. At $t = \Delta t$, he will have $u_1 = u_0 \cos \omega t$, $v_1 = -u_0 \sin \omega t$, and $w_1 = 0$. Thus, according to him, the particle P has the accelerations

$$\dot{u}_1 = \lim_{\Delta t \to 0} \frac{u_0(\cos \omega \Delta t - 1)}{\Delta t} = 0 \tag{4.227}$$

$$\dot{v}_1 = \lim_{\Delta t \to 0} -\frac{u_0 \sin \omega \Delta t - 0}{\Delta t} = -u_0 \omega \tag{4.228}$$

$$\dot{\omega}_1 = 0 \tag{4.229}$$

so that

$$\left(\frac{\mathrm{d}\vec{V}}{\mathrm{d}t}\right)_1 = -\hat{j}_1 u_0 \omega \tag{4.230}$$

Thus, the acceleration measured in the moving coordinate system is different from that recorded in the inertial reference system. According to the moving axes theorem,

$$\left(\frac{d\vec{V}}{dt}\right)_0 = \left(\frac{d\vec{V}}{dt}\right)_1 + \vec{\omega}_{0,1}^1 \times (\vec{V})_1 \tag{4.231}$$

We have

$$(\vec{V})_1 = \hat{i}_1 u_1 + \hat{j}_1 v_1 + \hat{k}_1 w_1 \tag{4.232}$$

$$\vec{\omega}_{0,1}^1 = \hat{k}_1 \omega \tag{4.233}$$

We have $u_1 = u_0$ and $v_1 = w_1 = 0$ so that

$$\vec{\omega}_{0,1}^1 \times (\vec{V})_1 = \hat{j}_1 u_0 \omega \tag{4.234}$$

Substituting in Eq. (4.231), we get

$$\begin{aligned}\left(\frac{d\vec{V}}{dt}\right)_0 &= -\hat{j}_1 u_0 \omega + \hat{j}_i u_0 \omega \\ &= 0\end{aligned} \tag{4.235}$$

Thus, the theorem holds.

4.3.2 Expressions for Velocity and Acceleration

Let us consider the motion of a rigid body as observed in various coordinate systems as shown in Fig. 4.16. Let $x_i y_i z_i$ be a nonrotating reference system fixed at the center of the rotating Earth. Let $x_b y_b z_b$ be a coordinate system fixed to the body and moving with it. Let $x_e y_e z_e$ be a system fixed to the surface of the Earth and located directly below the body at $t = 0$ (navigational system). The $Ox_e y_e z_e$ system rotates with the Earth and has a constant angular velocity Ω_e with respect to the $Ox_i y_i z_i$ system. Here, we assume that $x_i y_i z_i$ serves as an inertial frame of reference.

Let R_e be the radius of the spherical Earth and $\vec{R}_i$ be the position vector of a particle P (mass δm) attached to the rigid body with respect to the inertial system. Let $\vec{R}_o$ be the position vector of the origin of the $Ox_b y_b z_b$ system relative to the Earth-fixed $Ox_e y_e z_e$ system. Let $\vec{r}_b$ be the postion vector of P with respect to the moving coordinate system $Ox_b y_b z_b$. Let $\vec{\omega}_{i,b}^b$ be the angular velocity of the body with respect to the $Ox_i y_i z_i$ system but having components in the $Ox_b y_b z_b$ system.

Then,

$$\vec{R}_i = \vec{R}_e + \vec{R}_o + \vec{r}_b \tag{4.236}$$

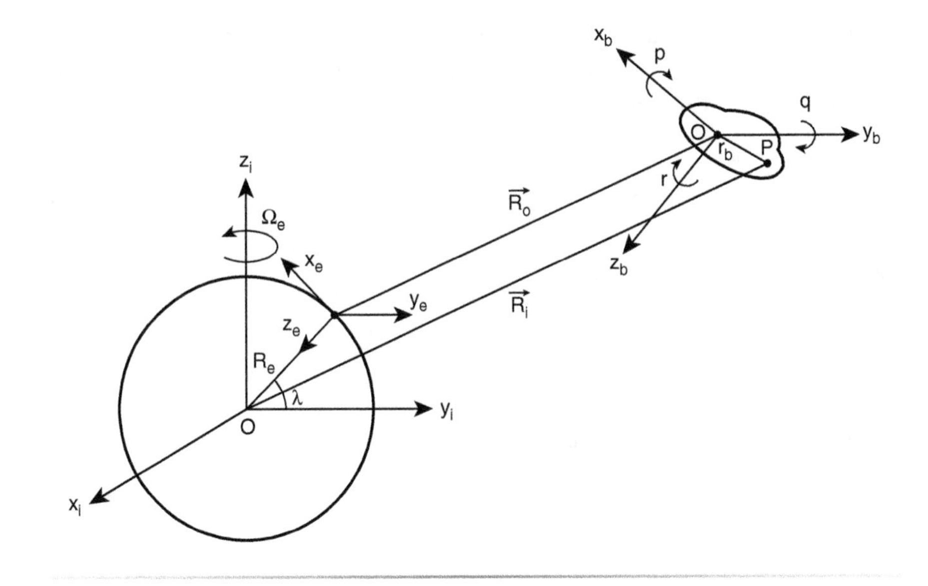

Fig. 4.16 Inertial, navigational, and body axes systems.

$$\left(\frac{\mathrm{d}\vec{R}_i}{\mathrm{d}t}\right)_i = \frac{\mathrm{d}}{\mathrm{d}t}(\vec{R}_e + \vec{R}_o + \vec{r}_b) \tag{4.237}$$

$$= \left(\frac{\mathrm{d}\vec{R}_e}{\mathrm{d}t}\right)_i + \left(\frac{\mathrm{d}\vec{R}_o}{\mathrm{d}t}\right)_i + \left(\frac{\mathrm{d}\vec{r}_b}{\mathrm{d}t}\right)_i \tag{4.238}$$

Using the moving axis theorem shown in Eq. (4.226), we have

$$\left(\frac{\mathrm{d}\vec{R}_e}{\mathrm{d}t}\right)_i = \left(\frac{\mathrm{d}\vec{R}_e}{\mathrm{d}t}\right)_e + \vec{\Omega}_e \times \vec{R}_e \tag{4.239}$$

$$= \vec{\Omega}_e \times \vec{R}_e \tag{4.240}$$

Here, $\vec{\Omega}_e$ is the angular velocity of the Earth-fixed $Ox_ey_ez_e$ system with respect to the $Ox_iy_iz_i$ system but having components in the $Ox_ey_ez_e$ system as given by

$$\vec{\Omega}_e = \hat{i}_e\Omega_e \cos\lambda - \hat{k}_e\Omega_e \sin\lambda \tag{4.241}$$

Here, $\Omega_e = \left|\vec{\Omega}_e\right|$ the magnitude of $\vec{\Omega}_e$. Furthermore,

$$\left(\frac{\mathrm{d}\vec{R}_o}{\mathrm{d}t}\right)_i = \left(\frac{\mathrm{d}\vec{R}_o}{\mathrm{d}t}\right)_e + \vec{\Omega}_e \times \vec{R}_o \tag{4.242}$$

$$= \vec{V}_o^e + \vec{\Omega}_e \times \vec{R}_o \tag{4.243}$$

Here, $\vec{V}_o^e = (\mathrm{d}\vec{R}_o/\mathrm{d}t)_e$ is the velocity of the body with respect to the navigational $(Ox_e y_e z_e)$ system.

$$\left(\frac{\mathrm{d}\vec{r}_b}{\mathrm{d}t}\right)_i = \left(\frac{\mathrm{d}\vec{r}_b}{\mathrm{d}t}\right)_b + \vec{\omega}_{i,b}^b \times \vec{r}_b \tag{4.244}$$

$$= \vec{\omega}_{i,b}^b \times \vec{r}_b \tag{4.245}$$

The first term on the right-hand side of Eq. (4.244) vanishes because we have assumed that the body is rigid. Then,

$$\left(\frac{\mathrm{d}\vec{R}_i}{\mathrm{d}t}\right)_i = \vec{\Omega}_e \times \vec{R}_e + \vec{V}_0^e + \vec{\Omega}_e \times \vec{R}_o + \vec{\omega}_{i,b}^b \times \vec{r}_b \tag{4.246}$$

$$\left(\frac{\mathrm{d}^2\vec{R}_i}{\mathrm{d}t^2}\right)_i = \left[\frac{\mathrm{d}}{\mathrm{d}t}(\vec{\Omega}_e \times \vec{R}_e)\right]_e + \vec{\Omega}_e \times \vec{\Omega}_e \times \vec{R}_e + \left(\frac{\mathrm{d}\vec{V}_o^e}{\mathrm{d}t}\right)_e + \vec{\Omega}_e \times \vec{V}_o^e$$
$$+ \vec{\Omega}_e \times \vec{V}_o^e + \vec{\Omega}_e \times \vec{\Omega}_e \times \vec{R}_o + \left[\frac{\mathrm{d}}{\mathrm{d}t}\left(\vec{\omega}_{i,b}^b \times \vec{r}_b\right)\right]_b + \vec{\omega}_{i,b}^b \times \left(\vec{\omega}_{i,b}^b \times \vec{r}_b\right) \tag{4.247}$$

We note that the first term on the right-hand side is zero because both $\vec{\Omega}_e$ and $\vec{R}_e$ are constant vectors in the $Ox_e y_e z_e$ system. Then,

$$\left(\frac{\mathrm{d}^2\vec{R}_i}{\mathrm{d}t^2}\right)_i = \vec{\Omega}_e \times \left(\vec{\Omega}_e \times \vec{R}_e\right) + (\vec{a}_0)_e + 2\vec{\Omega}_e \times \vec{V}_o^e$$
$$+ \vec{\Omega}_e \times \left(\vec{\Omega}_e \times \vec{R}_o\right) + \left(\frac{\mathrm{d}\vec{\omega}_{i,b}^b}{\mathrm{d}t}\right)_b \times \vec{r}_b + \vec{\omega}_{i,b}^b \times \vec{\omega}_{i,b}^b \times \vec{r}_b \tag{4.248}$$

$$= \vec{\Omega}_e \times \vec{\Omega}_e \times \left(\vec{R}_e + \vec{R}_o\right) + (\vec{a}_0)_e + 2\vec{\Omega}_e \times \vec{V}_o^e + \left(\frac{\mathrm{d}\vec{\omega}_{i,b}^b}{\mathrm{d}t}\right)_b \times \vec{r}_b$$
$$+ \vec{\omega}_{i,b}^b \times \vec{\omega}_{i,b}^b \times \vec{r}_b \tag{4.249}$$

where $(\vec{a}_0)_e = \left(\mathrm{d}\vec{V}_o^e/\mathrm{d}t\right)_e$. Furthermore, we have assumed that the body is rigid so that $(\mathrm{d}\vec{r}_b/\mathrm{d}t)_b = 0$.

In Eq. (4.249), various terms have the following interpretation:

$\vec{\Omega}_e \times \vec{\Omega}_e \times (\vec{R}_e + \vec{R}_o)$: centripetal acceleration at the origin of the body-fixed system due to rotation of the Earth.

$(\vec{a}_o)_e$: acceleration of the body relative to the Earth-fixed system $Ox_e y_e z_e$.

$2\vec{\Omega}_e \times \vec{V}_o^e$: Coriolis acceleration observed in the Earth-fixed system $Ox_e y_e z_e$.

$(\mathrm{d}\vec{\omega}_{i,b}^b/\mathrm{d}t)_b \times \vec{r}_b$: apparent acceleration observed in the body-fixed system due to a time rate of change of the angular velocity of the body.

$\vec{\omega}_{i,b}^b \times \vec{\omega}_{i,b}^b \times \vec{r}_b$: apparent acceleration of the particle due to the angular velocity of the moving axes system and recorded in the body-fixed system.

Force Equations

Multiplying both sides of Eq. (4.249) by δm, mass of the particle, we obtain

$$\delta m \frac{d^2\vec{R}_i}{dt^2} = \delta m \vec{\Omega}_e \times \vec{\Omega}_e \times (\vec{R}_e + \vec{R}_o) + \delta m (\vec{a}_o)_e + 2\delta m \vec{\Omega}_e \times \vec{V}_o^e$$

$$+ \delta m \left(\frac{d\vec{\omega}_{i,b}^b}{dt} \right)_b \times \vec{r}_b + \delta m \vec{\omega}_{i,b}^b \times \left(\vec{\omega}_{i,b}^b \times \vec{r}_b \right) \tag{4.250}$$

$$= \delta \vec{F}_i \tag{4.251}$$

where $\delta \vec{F}_i$ is the net external force acting on the particle P of mass δm. Summing over the entire body,

$$\sum \delta \vec{F}_i = \sum \delta m \vec{\Omega}_e \times \vec{\Omega}_e \times (\vec{R}_e + \vec{R}_o) + \sum \delta m (\vec{a}_o)_e + 2 \sum \delta m \vec{\Omega}_e \times \vec{V}_o^e$$

$$+ \sum \delta m \left(\frac{d\vec{\omega}_{i,b}^b}{dt} \times \vec{r}_b \right) + \sum \delta m \left(\vec{\omega}_{i,b}^b \times \vec{\omega}_{i,b}^b \times \vec{r}_b \right) \tag{4.252}$$

Let

$$\sum \delta m = m \tag{4.253}$$

$$\sum \delta \vec{F}_i = \vec{F} \tag{4.254}$$

We note that

$$r_{cg} = \frac{1}{m} \sum \delta m r_b \tag{4.255}$$

$$m r_{cg} = \sum \delta m r_b \tag{4.256}$$

Then

$$\vec{F} = m \left[\vec{\Omega}_e \times \vec{\Omega}_e \times (\vec{R}_e + \vec{R}_o) + (\vec{a}_o)_e + 2\vec{\Omega}_e \times \vec{V}_o^e + \left(\frac{d\vec{\omega}_{i,b}^b}{dt} \right)_b \times \vec{r}_{cg} \right.$$

$$\left. + \vec{\omega}_{i,b}^b \times \left(\vec{\omega}_{i,b}^b \times \vec{r}_{cg}^b \right) \right] \tag{4.257}$$

Special Cases

When the center of gravity of the vehicle coincides with the origin of the body-fixed coordinate system, $r_{cg} = 0$, then Eq. (4.257) assumes the form

$$\vec{F} = m \left[\vec{\Omega}_e \times \vec{\Omega}_e \times (\vec{R}_e + \vec{R}_o) + (\vec{a}_o)_e + 2\vec{\Omega}_e \times \vec{V}_o^e \right] \tag{4.258}$$

For the low-altitude flight dynamic problems, $|\vec{R}_o| \ll |\vec{R}_e|$, then

$$\vec{F} = m\left[\vec{\Omega}_e \times \vec{\Omega}_e \times \vec{R}_e + (\vec{a}_o)_e + 2\vec{\Omega}_e \times \vec{V}_o^e\right] \tag{4.259}$$

Let us expand the vector products in Eq. (4.259) and derive the equations of motion in the usual Cartesian form. Let λ denote the latitude of the origin of the $Ox_ey_ez_e$ axes system. Then,

$$\vec{\Omega}_e = \hat{i}_e\Omega_e \cos\lambda - \hat{k}_e\Omega_e \sin\lambda \tag{4.260}$$

$$\vec{R}_e = -\hat{k}_e R_e \tag{4.261}$$

$$\vec{V}_o^e = \hat{i}_e V_{ox} + \hat{j}_e V_{oy} + \hat{k}_e V_{oz} \tag{4.262}$$

$$= \hat{i}\dot{x}_e + \hat{j}\dot{y}_e + \hat{k}\dot{z}_e \tag{4.263}$$

$$(\vec{a}_o)_e = \hat{i}\ddot{x}_e + \hat{j}\ddot{y}_e + \hat{k}\ddot{z}_e \tag{4.264}$$

Using the vector cross product rule,

$$\vec{a} \times \vec{b} \times \vec{c} = \vec{b}(\vec{a}\cdot\vec{c}) - \vec{c}(\vec{b}\cdot\vec{a}) \tag{4.265}$$

We have

$$\vec{\Omega}_e \times \vec{\Omega}_e \times \vec{R}_e = \vec{\Omega}_e(\vec{\Omega}_e \cdot \vec{R}_e) - \vec{R}_e(\vec{\Omega}_e \cdot \vec{\Omega}_e) \tag{4.266}$$

$$= (\hat{i}_e\Omega_e\cos\lambda - \hat{k}_e\Omega_e\sin\lambda)(\Omega_e R_e \sin\lambda) - (-\hat{k}_e R_e)\Omega_e^2 \tag{4.267}$$

$$= R_e\Omega_e^2 \cos\lambda(\hat{i}_e \sin\lambda + \hat{k}_e\cos\lambda) \tag{4.268}$$

The vector cross product

$$\vec{\Omega}_e \times \vec{V}_o^e = \hat{i}_e\Omega_e V_{oy}\sin\lambda - \hat{j}_e(\Omega_e V_{oz}\cos\lambda + \Omega_e V_{ox}\sin\lambda) + \hat{k}_e\Omega_e V_{oy}\cos\lambda \tag{4.269}$$

Substituting and rearranging, we obtain the following equations of motion in Cartesian form:

$$R_e\Omega_e^2\cos\lambda\sin\lambda + \ddot{x}_e + 2\Omega_e V_{oy}\sin\lambda = F_{x,e} \tag{4.270}$$

$$\ddot{y}_e - 2\Omega_e(V_{oz}\cos\lambda + V_{ox}\sin\lambda) = F_{y,e} \tag{4.271}$$

$$R_e\Omega_e^2\cos^2\lambda + \ddot{z}_e + 2\Omega_e V_{oy}\cos\lambda = F_{z,e} \tag{4.272}$$

where

$$\vec{F} = \hat{i}_e F_{x,e} + \hat{j}_e F_{y,e} + \hat{k}_e F_{z,e} \tag{4.273}$$

The given system of equations describes the motion of a point mass with the only assumption that the orbital motion of the Earth around the sun is ignored. For the majority of the flight dynamics problems, this assumption is usually satisfactory. However, for interplanetary motions like a mission to the planet Mars or other distant planets, the orbital motion of the Earth has to be considered. The interested reader may refer to a standard text on astrodynamics for this purpose [6].

4.3.3 Force Equations for Aircraft Motion

In the study of aircraft flight dynamics, it is usual to ignore the rotation of the Earth about its own axis. In other words, an Earth-fixed axes system such as the navigational system $x_e y_e z_e$ is assumed to be an inertial frame of reference. In doing so, we ignore the terms $\vec{\Omega}_e \times \vec{\Omega}_e \times \vec{R}_e$ and $2\vec{\Omega}_e \times \vec{V}_o^e$. In other words, we assume $\vec{\omega}_{i,b}^b = \vec{\omega}_{e,b}^b$ and $(\dot{\vec{V}}_o)_i = (\dot{\vec{V}}_o)_e$. The error introduced by these assumptions can be estimated as follows.

Consider an aircraft moving at a flight velocity of 1000 mph (1466.37 ft/s) at sea level (speed of sound $\simeq$1100 ft/s). Let the flight path be contained in the equator ($\lambda = 0$). This flight velocity corresponds to a flight Mach number of approximately 1.34. With $\lambda = 0$ in Eq. (4.268), we have

$$\vec{\Omega}_e \times \vec{\Omega}_e \times \vec{R}_e = \hat{k}_e R_e \vec{\Omega}_e^2 \tag{4.274}$$

$$\left|\vec{\Omega}_e \times \vec{\Omega}_e \times \vec{R}_e\right| = R_e \Omega_e^2 \tag{4.275}$$

With $R_e \simeq 2.097 \times 10^7$ ft and

$$\Omega_e = \frac{1}{24 * 60}\left(\frac{2\pi}{60}\right) \tag{4.276}$$

$$= 7.2722 * 10^{-5}\ \text{rad/s} \tag{4.277}$$

we obtain

$$\left|\vec{\Omega}_e \times \vec{\Omega}_e \times \vec{R}_e\right| = 2.097 \times 10^7 \left(7.2722 * 10^{-5}\right)^2 \tag{4.278}$$

$$= 0.11089\ \text{ft/s}^2 \tag{4.279}$$

Similarly, from Eq. (4.269), with $\lambda = 0$,

$$2\vec{\Omega}_e \times \vec{V}_e = 2\Omega_e\left(-\hat{j}_e V_{oz} + \hat{k}_e V_{oy}\right) \tag{4.280}$$

$$2\left|\vec{\Omega}_e \times \vec{V}_e\right| \simeq 2\Omega_e V_o \tag{4.281}$$

$$\simeq 2(7.2722 * 10^{-5})1466.37 \tag{4.282}$$

$$\simeq 0.2133\ \text{ft/s}^2 \tag{4.283}$$

Thus, the errors caused by ignoring the Earth's rotation about its own axis are relatively small for typical aircraft motions at subsonic and supersonic speeds. However, these errors can become significant if the flight velocity increases.

With these assumptions, Eq. (4.259) reduces to

$$\vec{F} = m(\vec{a}_o)_e \tag{4.284}$$

$$= m(\dot{\vec{V}}_o)_e \tag{4.285}$$

$$= m\left(\frac{\mathrm{d}\vec{V}_o}{\mathrm{d}t}\right)_e \tag{4.286}$$

Using the moving axes theorem, we

$$\left(\frac{\mathrm{d}\vec{V}_o}{\mathrm{d}t}\right)_e = \left(\frac{\mathrm{d}\vec{V}_o}{\mathrm{d}t}\right)_b + \vec{\omega}^b_{e,b} \times \vec{V}_o \tag{4.287}$$

With $(\vec{V}_o)_b = \hat{i}_b U + \hat{j}_b V + \hat{k}_b W$ and $\vec{\omega}^b_{e,b} = \vec{\omega}^b_{i,b} = \hat{i}_b p + \hat{j}_b q + \hat{k}_b r$, we have

$$\left(\frac{\mathrm{d}V_o}{\mathrm{d}t}\right)_b = \hat{i}_b \dot{U} + \hat{j}_b \dot{V} + \hat{k}_b \dot{W} \tag{4.288}$$

$$\vec{\omega}_{e,b} \times \vec{V}_o = \hat{i}_b(qW - Vr) - \hat{j}_b(pW - Ur) + \hat{k}_b(pV - Uq) \tag{4.289}$$

With

$$\vec{F} = \hat{i}_b F_x + \hat{j}_b F_y + \hat{k}_b F_z \tag{4.290}$$

we have the following force equations for aircraft motion in Cartesian form:

$$F_x = m(\dot{U} + qW - rV) \tag{4.291}$$

$$F_y = m(\dot{V} + rU - pW) \tag{4.292}$$

$$F_z = m(\dot{W} + pV - qU) \tag{4.293}$$

Theorem on Angular Momentum

Consider the motion of a particle P of mass δm with respect to the Earth-centered inertial frame of reference $x_i y_i z_i$ (see Fig. 4.16). Let $x_b y_b z_b$ denote the body-fixed axes system and let the origin of the body-axes system be located at the center of gravity of the body. From Newton's first law of motion,

$$\delta\vec{F}_i = \delta m \dot{\vec{V}}_i \tag{4.294}$$

$$= \delta m \ddot{\vec{R}}_i \tag{4.295}$$

$$= \delta m(\ddot{\vec{R}}_o + \ddot{\vec{r}}_b) \tag{4.296}$$

Summing over the entire body,

$$\vec{F}_i = \sum \delta \vec{F}_i \tag{4.297}$$

$$= \sum \delta m \ddot{\vec{R}}_o + \sum \delta m \ddot{\vec{r}}_b \tag{4.298}$$

$$= \sum \delta m \ddot{\vec{R}}_o \tag{4.299}$$

$$= m \dot{\vec{V}}_o \tag{4.300}$$

The second term on the right-hand side of Eq. (4.298) vanishes because we have assumed the body to be rigid. Furthermore, note that all the time derivates are to be taken with respect to the inertial frame of reference $x_i y_i z_i$.

The angular momentum of the particle P with respect to the inertial space $x_i y_i z_i$ is given by

$$\delta h_i = \vec{R}_i \times \delta m \vec{V}_i \tag{4.301}$$

where $\vec{V}_i = \dot{\vec{R}}_i = \left(\dot{\vec{R}}_o + \dot{\vec{r}}_b\right)$.

Now,

$$\frac{\mathrm{d}(\delta \vec{h}_i)}{\mathrm{d}t} = \dot{\vec{R}}_i \times \delta m \vec{V}_i + \vec{R}_i \times \delta m \dot{\vec{V}}_i \tag{4.302}$$

$$= \vec{R}_i \times \delta m \dot{\vec{V}}_i \tag{4.303}$$

$$= \vec{R}_i \times \delta \vec{F}_i \tag{4.304}$$

$$= \delta \vec{G}_i \tag{4.305}$$

Summing over the entire body, we obtain

$$\vec{G}_i = \sum \delta \vec{G}_i \tag{4.306}$$

$$= \sum \vec{R}_i \times \delta \vec{F}_i \tag{4.307}$$

$$= \sum \frac{\mathrm{d}}{\mathrm{d}t} (\delta \vec{h}_i) \tag{4.308}$$

$$= \frac{\mathrm{d}}{\mathrm{d}t} \sum \delta \vec{h}_i \tag{4.309}$$

$$= \frac{\mathrm{d}\vec{H}_i}{\mathrm{d}t} \tag{4.310}$$

where $\vec{H}_i = \sum \delta \vec{h}_i$. Equation (4.310) is a basic relation in mechanics, which states that the time rate of change of angular momentum of a rigid body measured in an inertial space is equal to the net external moment acting on the body with respect to the origin of the inertial frame of reference.

Equation (4.310) can be expressed as

$$\vec{G}_i = \frac{\mathrm{d}\vec{H}_i}{\mathrm{d}t} \tag{4.311}$$

$$\sum \vec{R}_i \times \delta\vec{F}_i = \frac{\mathrm{d}}{\mathrm{d}t}\sum \delta h_i \tag{4.312}$$

$$\sum \left(\vec{R}_o + \vec{r}_b\right) \times \delta\vec{F}_i = \frac{\mathrm{d}}{\mathrm{d}t}\sum \vec{R}_i \times \delta m\vec{V}_i \tag{4.313}$$

$$\vec{R}_o \times \sum \delta\vec{F}_i + \sum \vec{r}_b \times \delta\vec{F}_i = \frac{\mathrm{d}}{\mathrm{d}t}\sum \left(\vec{R}_o + \vec{r}_b\right) \times \delta m\vec{V}_i \tag{4.314}$$

$$= \frac{\mathrm{d}}{\mathrm{d}t}\sum \vec{R}_o \times \delta m\vec{V}_i + \frac{\mathrm{d}}{\mathrm{d}t}\sum \vec{r}_b \times \delta m\vec{V}_i \tag{4.315}$$

$$= \dot{\vec{R}}_o \times \sum \delta m\vec{V}_i + \vec{R}_o \times \sum \delta m\dot{\vec{V}}_i + \frac{\mathrm{d}}{\mathrm{d}t}\sum \vec{r}_b \times \delta m\vec{V}_i \tag{4.316}$$

We note that

$$\dot{\vec{R}}_o \times \sum \delta m\vec{V}_i = \dot{\vec{R}}_o \times \sum \delta m\left(\dot{\vec{R}}_o + \dot{\vec{r}}_b\right) \tag{4.317}$$

$$= \dot{\vec{R}}_o \times \sum \delta m\dot{\vec{R}}_o + \dot{\vec{R}}_o \times \sum \delta m\dot{\vec{r}}_b \tag{4.318}$$

$$= 0 \tag{4.319}$$

The first term on the right-hand side of Eq. (4.318) vanishes because $\dot{\vec{R}}_o \times \dot{\vec{R}}_o = 0$ and the second term vanishes because the body is assumed to be rigid. Furthermore, using Eq. (4.294), we obtain $\vec{R}_o \times \delta\vec{F}_i = \vec{R}_o \times \delta m\dot{\vec{V}}_i$.

With these simplifications, Eq. (4.316) reduces to

$$\sum \vec{r}_b \times \delta\vec{F}_i = \frac{\mathrm{d}}{\mathrm{d}t}\sum \vec{r}_b \times \delta m\vec{V}_i \tag{4.320}$$

Thus, according to Eq. (4.320), the statement $\vec{G}_i = \mathrm{d}\vec{H}_i/\mathrm{d}t$ of Eq. (4.311) is equivalent to the statement that the sum of all the external moments acting on the body about its mass center (or the origin of the body axes system) is equal to the time rate of change of angular momentum with respect to the inertial system.

Let

$$\vec{G}_b = \sum \vec{r}_b \times \delta\vec{F}_i \tag{4.321}$$

$$\vec{H}_b = \sum \vec{r}_b \times \delta m\vec{V}_i \tag{4.322}$$

Then Eq. (4.320) can be expressed as

$$\vec{G}_b = \left(\frac{\mathrm{d}\vec{H}_b}{\mathrm{d}t}\right)_i \tag{4.323}$$

Using the moving axes theorem from Eq. (4.226),

$$\left(\frac{\mathrm{d}\vec{H}_b}{\mathrm{d}t}\right)_i = \left(\frac{\mathrm{d}\vec{H}_b}{\mathrm{d}t}\right)_b + \vec{\omega}_{i,b}^b \times H_b \tag{4.324}$$

As said before, for problems in aircraft dynamics, it is usual to ignore the rotation of the Earth and assume $\vec{\omega}_{i,b}^b = \vec{\omega}_{e,b}^b$.

We have

$$\vec{H}_b = \sum \vec{r}_b \times \delta m \vec{V}_i \tag{4.325}$$

$$= \sum \vec{r}_b \times \delta m \left[\dot{\vec{R}}_o + \dot{\vec{r}}_b\right] \tag{4.326}$$

$$= \sum \vec{r}_b \times \delta m \left[\left(\dot{\vec{R}}_o\right)_b + \vec{\omega}_{i,b}^b \times \vec{R}_o + \vec{\omega}_{i,b}^b \times \vec{r}_b\right] \tag{4.327}$$

$$= \left(\sum \delta m \vec{r}_b\right) \times \left(\dot{\vec{R}}_o\right)_b + \left(\sum \delta m \vec{r}_b\right) \times \vec{\omega}_{i,b}^b \times \vec{R}_o + \sum \left(\delta m \vec{r}_b \times \vec{\omega}_{i,b}^b \times \vec{r}_b\right) \tag{4.328}$$

$$= \sum \delta m \vec{r}_b \times \vec{\omega}_{i,b}^b \times \vec{r}_b \tag{4.329}$$

Note that $\sum \delta m \vec{r}_b = 0$ because we have chosen the origin of the body-fixed axes system to be located at the center of gravity of the body. On account of this, the first two terms on the right-hand side of Eq. (4.328) vanish. Using the vector cross product rule of Eq. (4.265), we obtain

$$\vec{H}_b = \sum \left[\vec{\omega}_{i,b}^b (\vec{r}_b \cdot \vec{r}_b) - \vec{r}_b \left(\vec{\omega}_{i,b}^b \cdot \vec{r}_b\right)\right] \delta m \tag{4.330}$$

We have

$$\vec{\omega}_{i,b}^b = \hat{i}_b p + \hat{j}_b q + \hat{k}_b r \tag{4.331}$$

$$\vec{r}_b = \hat{i}_b x_b + \hat{j}_b y_b + \hat{k}_b z_b \tag{4.332}$$

Here, x_b, y_b, z_b are the coordinates of the particle P of mass δm in the body axes system. Assuming $\vec{H}_b = \hat{i}_b H_{xb} + \hat{j}_b H_{yb} + \hat{k}_b H_{zb}$, Eq. (4.330) can be expressed in Cartesian form as

$$H_{xb} = \sum p \delta m \left(x_b^2 + y_b^2 + z_b^2\right) - \sum \delta m \left(p x_b^2 + q x_b y_b + r x_b z_b\right) \tag{4.333}$$

$$H_{yb} = \sum q \delta m \left(x_b^2 + y_b^2 + z_b^2\right) - \sum \delta m \left(p x_b y_b + q y_b^2 + r y_b z_b\right) \tag{4.334}$$

$$H_{zb} = \sum r \delta m \left(x_b^2 + y_b^2 + z_b^2\right) - \sum \delta m \left(p x_b z_b + q y_b z_b + r z_b^2\right) \tag{4.335}$$

or

$$H_{xb} = pI_x - qI_{xy} - rI_{xz} \tag{4.336}$$

$$H_{yb} = qI_y - rI_{yz} - pI_{yx} \tag{4.337}$$

$$H_{zb} = rI_z - pI_{zx} - qI_{zy} \tag{4.338}$$

where

$$I_x = \sum \delta m \left(y_b^2 + z_b^2\right) \tag{4.339}$$

$$I_y = \sum \delta m \left(x_b^2 + z_b^2\right) \tag{4.340}$$

$$I_z = \sum \delta m \left(x_b^2 + y_b^2\right) \tag{4.341}$$

$$I_{xy} = \sum \delta m x_b y_b \tag{4.342}$$

$$I_{yz} = \sum \delta m y_b z_b \tag{4.343}$$

$$I_{zx} = \sum \delta m x_b z_b \tag{4.344}$$

Note that $I_{xy} = I_{yx}$, $I_{zx} = I_{xz}$, and $I_{yz} = I_{zy}$.

Returning to moment Eq. (4.324), we have

$$\vec{G}_b = \left(\frac{\mathrm{d}\vec{H}_b}{\mathrm{d}t}\right)_b + \vec{\omega}_{i,b}^{\,b} \times \vec{H}_b \tag{4.345}$$

We have

$$\left(\frac{\mathrm{d}\vec{H}_b}{\mathrm{d}t}\right)_b = \hat{i}_b \dot{H}_{xb} + \hat{j}_b \dot{H}_{yb} + \hat{k}_b \dot{H}_{zb} \tag{4.346}$$

$$\dot{H}_{xb} = \dot{p}I_x - \dot{q}I_{yx} - \dot{r}I_{xz} \tag{4.347}$$

$$\dot{H}_{yb} = \dot{q}I_y - \dot{r}I_{yz} - \dot{p}I_{yx} \tag{4.348}$$

$$\dot{H}_{zb} = \dot{r}I_z - \dot{p}I_{zx} - \dot{q}I_{zy} \tag{4.349}$$

and

$$\vec{\omega}_{i,b}^{\,b} \times \vec{H}_b = \begin{vmatrix} \hat{i}_b & \hat{j}_b & \hat{k}_b \\ p & q & r \\ H_{xb} & H_{yb} & H_{zb} \end{vmatrix} \tag{4.350}$$

$$= \hat{i}_b\left(qH_{zb} - rH_{yb}\right) + \hat{j}_b\left(rH_{xb} - pH_{zb}\right) + \hat{k}_b\left(pH_{yb} - qH_{xb}\right) \tag{4.351}$$

In airplane terminology,

$$\vec{G}_b = \vec{i}_b L + \hat{j}_b M + \hat{k}_b N \tag{4.352}$$

where L is the rolling moment, M is the pitching moment, and N is the yawing moment. The use of symbol L to denote the rolling moment may confuse the reader because L is also used to denote lift. This double use of symbol L is so well established in flight-mechanics literature that we have no option but to live with it. Then substituting and simplifying, we obtain the following equations in Cartesian form:

$$L = \dot{p}I_x - \dot{q}I_{yx} - \dot{r}I_{xz} + qH_{zb} - rH_{yb} \tag{4.353}$$

$$= \dot{p}I_x - \dot{q}I_{yx} - \dot{r}I_{xz} + qr(I_z - I_y) - pqI_{zx} + (r^2 - q^2)I_{yz} + prI_{yx} \tag{4.354}$$

Similarly,

$$M = \dot{q}I_y - \dot{r}I_{yz} - \dot{p}I_{yx} + rp(I_x - I_z) - qrI_{xy} + (p^2 - r^2)I_{zx} + pqI_{zy} \tag{4.355}$$

$$N = \dot{r}I_z - \dot{p}I_{zx} - \dot{q}I_{zy} + pq(I_y - I_x) - rpI_{yz} + (q^2 - p^2)I_{xy} + qrI_{xz} \tag{4.356}$$

For an aircraft with vertical plane of symmetry (Ox_bz_b plane), $I_{xy} = I_{yz} = 0$. Then Eqs. (4.354–4.356) assume the following form:

$$L = \dot{p}I_x - I_{xz}(pq + \dot{r}) + qr(I_z - I_y) \tag{4.357}$$

$$M = \dot{q}I_y + rp(I_x - I_z) + (p^2 - r^2)I_{xz} \tag{4.358}$$

$$N = \dot{r}I_z - I_{xz}(\dot{p} - qr) + pq(I_y - I_x) \tag{4.359}$$

The terms on the right-hand side having the form $qr(I_z - I_y)$ are called inertia coupling terms. We will study the problem of inertial coupling in Chapter 7.

The three force Eqs. (4.291–4.293) and the three moment Eqs. (4.357–4.359) constitute a set of six equations needed for the six-degree-of-freedom (6 DOF) simulation of the aircraft dynamics. These six equations govern the motion of an aircraft with respect to an inertial frame of reference fixed to the surface of the flat and nonrotating Earth. In other words, we have ignored the spherical nature of the Earth, its rotation about its own axis as well as its orbital motion around the sun. We have also assumed that the aircraft has a vertical plane of symmetry. Furthermore, we have assumed that the origin of the body axes system is located at the center of gravity of the aircraft. Even after introducing these simplifying assumptions, it is still a formidable task to obtain analytical solutions to these equations because they are a set of coupled, nonlinear, differential equations. However, if it is desired to consider the spherical rotating nature of the Earth, then use the three force equations, Eqs. (4.270–4.272), instead of Eqs. (4.291–4.293) and the three moment equations, Eqs. (4.357–4.359).

Another challenging task is to evaluate the aerodynamic forces and moments as functions of motion variables U, V, W (or α, β), p, q, r, and their time derivatives. However, for the study of stability, response, and automatic control of conventional airplanes, these equations can be further simplified if we assume that the disturbed motion is one of sufficiently small amplitudes in the disturbed variables. The airplane is assumed to be in equilibrium flight condition before encountering a disturbance. With these assumptions, Eqs. (4.291–4.293) and Eqs. (4.357–4.359) can be simplified and separated into two sets of equations, one set of three equations for longitudinal motion and another set of three equations for lateral-directional motion as discussed next.

4.3.4 Equations of Motion with Small Disturbances

For the disturbed motion, we assume

$$U = U_o + \Delta U \quad V = V_o + \Delta V \quad W = W_o + \Delta W \tag{4.360}$$

$$p = p_o + \Delta p \quad q = q_o + \Delta q \quad r = r_o + \Delta r \tag{4.361}$$

$$F_x = F_{xo} + \Delta F_x \quad F_y = F_{yo} + \Delta F_y \quad F_z = F_{zo} + \Delta F_z \tag{4.362}$$

$$L = L_o + \Delta L \quad M = M_o + \Delta M \quad N = N_o + \Delta N \tag{4.363}$$

where the suffix o denotes the steady or equilibrium flight condition. Then force Eqs. (4.291–4.293) and moment Eqs. (4.357–4.359) assume the following form:

$$F_{xo} + \Delta F_x = m\left[\dot{U}_o + \Delta\dot{U} + (q_o + \Delta q)(W_o + \Delta W) - (r_o + \Delta r)(V_o + \Delta V)\right] \tag{4.364}$$

$$F_{yo} + \Delta F_y = m\left[\dot{V}_o + \Delta\dot{V} + (r_o + \Delta r)(U_o + \Delta U) - (p_o + \Delta p)(W_o + \Delta W)\right] \tag{4.365}$$

$$F_{zo} + \Delta F_z = m\left[\dot{W}_o + \Delta\dot{W} + (p_o + \Delta p)(V_o + \Delta V)(q_o + \Delta q)(U_o + \Delta U)\right] \tag{4.366}$$

$$\begin{aligned} L_o + \Delta L = {} & I_x(\dot{p}_o + \Delta\dot{p}) - I_{xz}[(p_o + \Delta p)(q_o + \Delta q) + \dot{r}_o + \Delta\dot{r}] \\ & + (I_z - I_y)(q_o + \Delta q)(r_o + \Delta r) \end{aligned} \tag{4.367}$$

$$\begin{aligned} M_o + \Delta M = {} & I_y(\dot{q}_o + \Delta\dot{q}) + (I_x - I_z)(r_o + \Delta r)(p_o + \Delta p) \\ & + I_{xz}[(p_o + \Delta p)^2 - (r_o + \Delta r)^2] \end{aligned} \tag{4.368}$$

$$\begin{aligned} N_o + \Delta N = {} & I_z(\dot{r}_o + \Delta\dot{r}) - I_{xz}[\dot{p}_o + \Delta\dot{p} - (q_o + \Delta q)(r_o + \Delta r)] \\ & + (I_y - I_x)(p_o + \Delta p)(q_o + \Delta q) \end{aligned} \tag{4.369}$$

We assume that the aircraft is in a steady, unaccelerated flight before it encounters a disturbance, i.e., both the net force and net moment on the aircraft are zero. Furthermore, in the study of airplane dynamics, it is customary to use the stability axes system. Note that the orientation of stability axes system depends on the equlibrium flight condition prior to encountering a disturbance. However, on encountering the disturbance, the airplane is displaced, but the orientation of the stability axes system is assumed to remain fixed in space.

With these assumptions,

$$\dot{U}_o = 0 \quad V_o = W_o = 0 \tag{4.370}$$

$$p_o = q_o = r_o = 0 \tag{4.371}$$

$$F_{xo} = F_{yo} = F_{zo} = 0 \tag{4.372}$$

$$L_o = M_o = N_o = 0 \tag{4.373}$$

For the disturbed flight condition

$$U = U_o + \Delta U \quad V = \Delta V \quad W = \Delta W \tag{4.374}$$

$$p = \Delta p \quad q = \Delta q \quad r = \Delta r \tag{4.375}$$

$$F_x = \Delta F_x \quad F_y = \Delta F_y \quad F_z = \Delta F_z \tag{4.376}$$

$$L = \Delta L \quad M = \Delta M \quad N = \Delta N \tag{4.377}$$

From Eq. (4.375), we note that p, q, and r themselves represent the roll rate, pitch rate, and yaw rate in the disturbed motion. We assume that each of the disturbance velocity components is small in comparison with the reference velocity U_o, and each disturbance angular velocity component is close to zero. Note that the reference value of each angular velocity component is zero. In other words, we assume

$$\frac{\Delta U}{U_o} \ll 1 \quad \frac{\Delta V}{U_o} \ll 1 \quad \frac{\Delta W}{U_o} \ll 1 \tag{4.378}$$

$$p \ll 1 \quad q \ll 1 \quad r \ll 1 \tag{4.379}$$

With these assumptions, second-order terms such as $\Delta q \Delta W$ can be ignored. Then Eqs. (4.364–4.369) reduce to the following form:

$$\Delta F_x = m\Delta\dot{U} \tag{4.380}$$

$$\Delta F_y = m(\Delta\dot{V} + rU_o) \tag{4.381}$$

$$\Delta F_z = m(\Delta\dot{W} - qU_o) \tag{4.382}$$

$$\Delta L = \dot{p}I_x - \dot{r}I_{xz} \tag{4.383}$$

$$\Delta M = \dot{q} I_y \tag{4.384}$$

$$\Delta N = \dot{r} I_z - \dot{p} I_{xz} \tag{4.385}$$

Define

$$u = \frac{\Delta U}{U_o} \quad v = \frac{\Delta V}{U_o} \quad w = \frac{\Delta W}{U_o} \tag{4.386}$$

$$\Delta C_x = \frac{\Delta F_x}{\frac{1}{2}\rho U_o^2 S} \quad \Delta C_y = \frac{\Delta F_y}{\frac{1}{2}\rho U_o^2 S} \quad \Delta C_z = \frac{\Delta F_z}{\frac{1}{2}\rho U_o^2 S} \tag{4.387}$$

$$\Delta C_l = \frac{\Delta L}{\frac{1}{2}\rho U_o^2 S b} \quad \Delta C_m = \frac{\Delta M}{\frac{1}{2}\rho U_o^2 S \bar{c}} \quad \Delta C_n = \frac{\Delta N}{\frac{1}{2}\rho U_0^2 S b} \tag{4.388}$$

Here, U_o is the flight velocity, S is the reference (wing) area, $\bar{c}$ is the mean aerodynamic chord, and b is the wing span. Note that the reference length used for defining the rolling and yawing moment coefficients (C_l and C_n) is span b and that, for pitching moment coefficient C_m is the mean aerodynamic chord $\bar{c}$. Furthermore, we note that

$$\tan\alpha \simeq \Delta\alpha = \frac{W}{U_o} \tag{4.389}$$

$$\sin\beta \simeq \Delta\beta = \frac{V}{U_o} \tag{4.390}$$

so that

$$\Delta\alpha \simeq \frac{\Delta W}{U_o} = w \tag{4.391}$$

$$\Delta\beta \simeq \frac{\Delta V}{U_o} = v \tag{4.392}$$

Then, Eqs. (4.380–4.385) take the following form:

$$\Delta C_x = \frac{m U_o}{\frac{1}{2}\rho U_o^2 S}\dot{u} \tag{4.393}$$

$$\Delta C_y = \frac{m U_o}{\frac{1}{2}\rho U_o^2 S}(\Delta\dot{\beta} + r) \tag{4.394}$$

$$\Delta C_z = \frac{m U_o}{\frac{1}{2}\rho U_o^2 S}(\Delta\dot{\alpha} + q) \tag{4.395}$$

$$\Delta C_l = \frac{1}{\frac{1}{2}\rho U_o^2 S b}(\dot{p} I_x - \dot{r} I_{xz}) \tag{4.396}$$

$$\Delta C_m = \frac{I_y}{\frac{1}{2}\rho U_o^2 S\bar{c}}\dot{q} \tag{4.397}$$

$$\Delta C_n = \frac{1}{\frac{1}{2}\rho U_o^2 Sb}(\dot{r}I_z - \dot{p}I_{xz}) \tag{4.398}$$

4.3.5 Estimation of Aerodynamic Forces and Moments

Our next task is to estimate the aerodynamic coefficients ΔC_x, ΔC_z, ΔC_m, ΔL, ΔM, and ΔN in the disturbed motion. For this purpose, we use Bryan's method [1], which, as said in Sec. 4.1, is based on two principal assumptions: 1) the instantaneous aerodynamic forces and moments depend on the instaneous values of the motion variables, and 2) the aerodynamic forces and moments vary linearly with motion variables. Furthermore, we introduce another important assumption that the longitudinal aerodynamic forces and moment (F_x, F_z, and M) are influenced only by the longitudinal variables u, α, and q. In other words, we assume that the lateral-directional variables β, p, and r do not influence the longitudinal aerodynamic forces F_x and F_z and the pitching moment M. Similarly, we assume that the side force F_y, the rolling moment L, and the yawing moment N depend only on the lateral-directional variables β, p, and r and are not influenced by the longitudinal variables u, α, and q. In other words, we assume that there is no aerodynamic coupling between longitudinal and lateral-directional variables, forces, and moments. These assumptions are usually valid for small angles of attack/sideslip when the aerodynamic coefficients vary linearly with angle of attack/sideslip. At high angles of attack/sideslip, such assumptions are not valid because of flow separation, vortex shedding, and stall. As a result, the aerodynamic coefficients vary nonlinearly with angle of attack/sideslip, and aerodynamic coupling takes place. When this happens, a change in the angle of attack affects side force, rolling, and yawing moments. Similarly, a change in sideslip angle influences lift, drag, and pitching moments. We will study stability and control problems at high angles of attack in Chapter 8.

With these assumptions and remembering that the disturbance variables are assumed to be small, we can use the Taylor series expansion method around the equilibrium level flight condition to obtain the forces and moments in the disturbed state as follows:

$$\Delta C_x = \frac{\partial C_x}{\partial u}u + \frac{\partial C_x}{\partial \alpha}\Delta\alpha + \frac{\partial C_x}{\partial \theta}\Delta\theta + \frac{\partial C_x}{\partial \dot{\alpha}}\Delta\dot{\alpha} + \frac{\partial C_x}{\partial q}q$$
$$+ \frac{\partial C_x}{\partial \delta_e}\Delta\delta_e + \frac{\partial C_x}{\partial \delta_t}\Delta\delta_t + \cdots \tag{4.399}$$

$$\Delta C_y = \frac{\partial C_y}{\partial \beta}\Delta\beta + \frac{\partial C_y}{\partial \phi}\Delta\phi + \frac{\partial C_y}{\partial \dot{\beta}}\Delta\dot{\beta} + \frac{\partial C_y}{\partial p}p + \frac{\partial C_y}{\partial r}r + \frac{\partial C_y}{\partial \delta_a}\Delta\delta_a + \frac{\partial C_y}{\partial \delta_r}\Delta\delta_r + \cdots \quad (4.400)$$

$$\Delta C_z = \frac{\partial C_z}{\partial u}u + \frac{\partial C_z}{\partial \alpha}\Delta\alpha + \frac{\partial C_z}{\partial \theta}\Delta\theta + \frac{\partial C_z}{\partial \dot{\alpha}}\Delta\dot{\alpha} + \frac{\partial C_z}{\partial q}q + \frac{\partial C_z}{\partial \delta_e}\Delta\delta_e + \frac{\partial C_z}{\partial \delta_t}\Delta\delta_t + \cdots \quad (4.401)$$

$$\Delta C_l = \frac{\partial C_l}{\partial \beta}\Delta\beta + \frac{\partial C_l}{\partial \dot{\beta}}\Delta\dot{\beta} + \frac{\partial C_l}{\partial \phi}\Delta\phi + \frac{\partial C_l}{\partial p}p + \frac{\partial C_l}{\partial r}r + \frac{\partial C_l}{\partial \delta_a}\Delta\delta_a + \frac{\partial C_l}{\partial \delta_r}\Delta\delta_r + \cdots \quad (4.402)$$

$$\Delta C_m = \frac{\partial C_m}{\partial u}u + \frac{\partial C_m}{\partial \alpha}\Delta\alpha + \frac{\partial C_m}{\partial \theta}\Delta\theta + \frac{\partial C_m}{\partial \dot{\alpha}}\Delta\dot{\alpha} + \frac{\partial C_m}{\partial p}q + \frac{\partial C_m}{\partial \delta_e}\Delta\delta_e + \frac{\partial C_m}{\partial \delta_t}\Delta\delta_t + \cdots \quad (4.403)$$

$$\Delta C_n = \frac{\partial C_n}{\partial \beta}\Delta\beta + \frac{\partial C_n}{\partial \dot{\beta}}\Delta\dot{\beta} + \frac{\partial C_n}{\partial \phi}\Delta\phi + \frac{\partial C_n}{\partial p}p + \frac{\partial C_n}{\partial r}r + \frac{\partial C_n}{\partial \delta_a}\Delta\delta_a + \frac{\partial C_n}{\partial \delta_r}\Delta\delta_r + \cdots \quad (4.404)$$

In Eqs. (4.399–4.404), δ_e is the elevator/elevon deflection, δ_a is the aileron deflection, δ_r is the rudder deflection, and δ_t is the engine control parameter.

Terms such as $\partial C_x/\partial u$, $\partial C_z/\partial u$, and $\partial C_m/\partial u$ are stability derivatives, and terms such as $\partial C_m/\partial \delta_e$ and $\partial \delta C_m/\partial \delta_t$ are control derivatives. It should be noted that these stability and control derivatives are evaluated at the equilibrium flight condition from which the airplane is supposed to be disturbed. Therefore, it is possible that stability and control derivatives vary from one equilibrium condition to another.

It is important to remember that the Taylor series expansion must include all the motion variables on which the aerodynamic forces and moments are known to depend. If this information is incomplete, estimated forces and moments in the disturbed state will be incorrect. Any predictions based on such incomplete aerodynamic data will also be in error. Therefore, it is the task of the aerodynamicist to understand the physics of the problem, identify all the motion variables on which the aerodynamic forces and moments show dependence, and correctly include all of them in the Taylor series expansion.

In the literature on airplane stability and control, it is customary to use the short-hand notation to denote the stability and control derivatives. For example,

$$C_{xu} = \frac{\partial C_x}{\partial u} \qquad C_{x\alpha} = \frac{\partial C_x}{\partial \alpha} \qquad C_{m\alpha} = \frac{\partial C_m}{\partial \alpha} \tag{4.405}$$

$$C_{y\beta} = \frac{\partial C_y}{\partial \beta} \qquad C_{l\beta} = \frac{\partial C_l}{\partial \beta} \qquad C_{n\beta} = \frac{\partial C_n}{\partial \beta} \tag{4.406}$$

and so on. However, the partial derivatives with respect to variables such as $\dot{\alpha}$, $\dot{\beta}$, p, q, and r are defined somewhat differently as follows:

$$C_{L\dot{\alpha}} = \frac{\partial C_L}{\partial\left(\frac{\dot{\alpha}\bar{c}}{2U_o}\right)} \quad C_{Lq} = \frac{\partial C_L}{\partial\left(\frac{q\bar{c}}{2U_o}\right)} \quad C_{mq} = \frac{\partial C_m}{\partial\left(\frac{q\bar{c}}{2U_o}\right)} \tag{4.407}$$

$$C_{y\dot{\beta}} = \frac{\partial C_y}{\partial\left(\frac{\dot{\beta}b}{2U_o}\right)} \quad C_{l\dot{\beta}} = \frac{\partial C_l}{\partial\left(\frac{\dot{\beta}b}{2U_o}\right)} \quad C_{n\dot{\beta}} = \frac{\partial C_n}{\partial\left(\frac{\dot{\beta}b}{2U_o}\right)} \tag{4.408}$$

$$C_{yp} = \frac{\partial C_y}{\partial\left(\frac{pb}{2U_o}\right)} \quad C_{lp} = \frac{\partial C_l}{\partial\left(\frac{pb}{2U_o}\right)} \quad C_{np} = \frac{\partial C_n}{\partial\left(\frac{pb}{2U_o}\right)} \tag{4.409}$$

$$C_{yr} = \frac{\partial C_y}{\partial\left(\frac{rb}{2U_o}\right)} \quad C_{lr} = \frac{\partial C_l}{\partial\left(\frac{rb}{2U_o}\right)} \quad C_{nr} = \frac{\partial C_n}{\partial\left(\frac{rb}{2U_o}\right)} \tag{4.410}$$

This form of definition of stability derivatives involving time derivatives such as $\dot{\alpha}$, $\dot{\beta}$, and angular velocity components p, q, r is necessary to make these derivatives nondimensional like all the other stability and control derivatives. Furthermore, the nondimensionalization helps us use the results obtained on scaled model tests in wind tunnels for predicting the stability and response of the full-scale airplane. In view of this, it is important to note that $C_{mq} \neq \partial C_m / \partial q$ and so on.

The stability and control derivatives with respect to variables such as $\dot{\alpha}$ and $\dot{\beta}$ are called acceleration derivatives, and those with respect to p, q, and r are called rotary derivatives. Together, they are called dynamic stability-control derivatives.

Using short-hand notation, we can rewrite the expressions for forces and moments in the nondimensional form as follows:

$$\Delta C_x = C_{xu}u + C_{x\alpha}\Delta\alpha + C_{x\theta}\Delta\theta + C_{x\dot{\alpha}}\left(\frac{\Delta\dot{\alpha}\bar{c}}{2U_o}\right) + C_{xq}\left(\frac{q\bar{c}}{2U_o}\right)$$
$$+ C_{x\delta_e}\Delta\delta_e + C_{x\delta_t}\Delta\delta_t + \cdots \tag{4.411}$$

$$\Delta C_y = C_{y\beta}\Delta\beta + C_{y\phi}\Delta\phi + C_{y\dot{\beta}}\left(\frac{\Delta\dot{\beta}b}{2U_o}\right) + C_{yp}\left(\frac{pb}{2U_o}\right) + C_{yr}\left(\frac{rb}{2U_o}\right)$$
$$+ C_{y\delta a}\Delta\delta a + C_{y\delta r}\Delta\delta r + \cdots \tag{4.412}$$

$$\Delta C_z = C_{zu}u + C_{z\alpha}\Delta\alpha + C_{z\theta}\Delta\theta + C_{z\dot{\alpha}}\left(\frac{\Delta\dot{\alpha}\bar{c}}{2U_o}\right) + C_{zq}\left(\frac{q\bar{c}}{2U_o}\right)$$
$$+ C_{z\delta_e}\Delta\delta_e + C_{z\delta_t}\Delta\delta_t + \cdots \tag{4.413}$$

$$\Delta C_l = C_{l\beta}\Delta\beta + C_{l\dot{\beta}}\left(\frac{\Delta\dot{\beta}b}{2U_o}\right) + C_{l\phi}\Delta\phi + C_{lp}\left(\frac{pb}{2U_o}\right) + C_{lr}\left(\frac{rb}{2U_o}\right)$$
$$+ C_{l\delta a}\Delta\delta a + C_{l\delta r}\Delta\delta r + \cdots \tag{4.414}$$

$$\Delta C_m = C_{mu}u + C_{m\alpha}\Delta\alpha + C_{m\theta}\Delta\theta + C_{m\dot{\alpha}}\left(\frac{\Delta\dot{\alpha}\bar{c}}{2U_o}\right) + C_{mp}\left(\frac{q\bar{c}}{2U_o}\right)$$
$$+ C_{m\delta_e}\Delta\delta_e + C_{m\delta_t}\Delta\delta_t + \cdots \tag{4.415}$$

$$\Delta C_n = C_{n\beta}\Delta\beta + C_{n\dot{\beta}}\left(\frac{\Delta\dot{\beta}b}{2U_o}\right) + C_{n\phi}\Delta\phi + C_{np}\left(\frac{pb}{2U_o}\right) + C_{nr}\left(\frac{rb}{2U_o}\right)$$
$$+ C_{n\delta a}\Delta\delta a + C_{n\delta r}\Delta\delta r + \cdots \tag{4.416}$$

It is interesting to observe that, with the assumption of small disturbances and linear uncoupled aerodynamics, Eqs. (4.393), (4.395), and (4.397) with Eqs. (4.411), (4.413), and (4.415) contain only the longitudinal disturbance variables and longitudinal aerodynamic forces and moment. Similarly, Eqs. (4.394), (4.396), and (4.398) with Eqs. (4.412), (4.414), and (4.416) contain only the lateral-directional disturbance variables and lateral-directional aerodynamic force and moments. Thus, we can separate the longitudinal and lateral-directional small disturbance equations of motion into two different sets and solve them separately. In this way, we have reduced a coupled nonlinear six-degree-of-freedom problem to two separate problems of three degrees of freedom, both of which are linear with respect to the disturbance variables. The linearity of the equations makes it easy for us to apply the powerful analytical methods of linear control systems for the study of airplane dynamic stability, control, and response calculations. However, it is necessary to keep in mind that this simplification is based on small disturbances and linear aerodynamics. Obviously, this approach is not valid for complex airplane maneuvers involving large amplitude motion typically occurring at high angles attack and stall/spin.

Longitudinal Equations of Motion for Small Disturbance

Substituting for ΔC_x, ΔC_z, and ΔC_m from Eqs. (4.411), (4.413), and (4.415) in Eqs. (4.393), (4.395), and (4.397), we obtain

$$\left(m_1 \frac{\mathrm{d}}{\mathrm{d}t} - C_{xu}\right)u - \left(C_{x\dot{\alpha}}c_1 \frac{\mathrm{d}}{\mathrm{d}t} + C_{x\alpha}\right)\Delta\alpha - \left(C_{xq}c_1 \frac{\mathrm{d}}{\mathrm{d}t} + C_{x\theta}\right)\Delta\theta$$
$$= C_{x\delta e}\Delta\delta e + C_{x\delta t}\Delta\delta t \tag{4.417}$$

$$-C_{zu}u + \left[\left(m_1 \frac{\mathrm{d}}{\mathrm{d}t} - C_{z\dot{\alpha}}c_1 \frac{\mathrm{d}}{\mathrm{d}t}\right) - C_{z\alpha}\right]\Delta\alpha - \left(m_1 \frac{\mathrm{d}}{\mathrm{d}t} + C_{zq}c_1 \frac{\mathrm{d}}{\mathrm{d}t} + C_{z\theta}\right)\Delta\theta$$
$$= C_{z\delta e}\Delta\delta e + C_{z\delta t}\Delta\delta t \tag{4.418}$$

$$-C_{mu}u - \left(C_{m\dot{\alpha}}c_1 \frac{\mathrm{d}}{\mathrm{d}t} + C_{m\alpha}\right)\Delta\alpha + \frac{\mathrm{d}}{\mathrm{d}t}\left(I_{y1}\frac{\mathrm{d}}{\mathrm{d}t} - C_{mq}c_1\right)\Delta\theta$$
$$= C_{m\delta e}\Delta\delta e + C_{m\delta t}\Delta\delta t \tag{4.419}$$

where

$$c_1 = \frac{\bar{c}}{2U_o} \tag{4.420}$$

$$m_1 = \frac{2m}{\rho U_o S} \tag{4.421}$$

$$I_{y1} = \frac{I_y}{\frac{1}{2}\rho U_o^2 S\,\bar{c}} \tag{4.422}$$

Consider an airplane in a steady (equilibrium) flight as shown in Fig. 4.17a. Assume that the thrust vector is aligned with the flight path. In Fig. 4.17a, $Ox_0y_0z_0$ denotes the stability axes system so that Ox_0 coincides with the velocity vector U_0. We assume that the angle between Ox_0 and the local horizontal is equal to θ_o. The angle between Ox_o and the airplane zero-lift line or the reference axis is equal to angle of attack α. Here, it so happens that θ_o is also equal to the flight path angle γ. This could cause some confusion because this relation $\theta_o = \gamma$ appears to be in contradiction with the usual flight mechanics relation $\gamma = \theta - \alpha$. However, there is no contradiction because θ_o and θ are not the same angles. The angle θ is supposed to be the angle between the local horizontal and the airplane reference axis or the zero-lift line, whereas θ_o is the angle between the local horizontal and the velocity vector or the flight path.

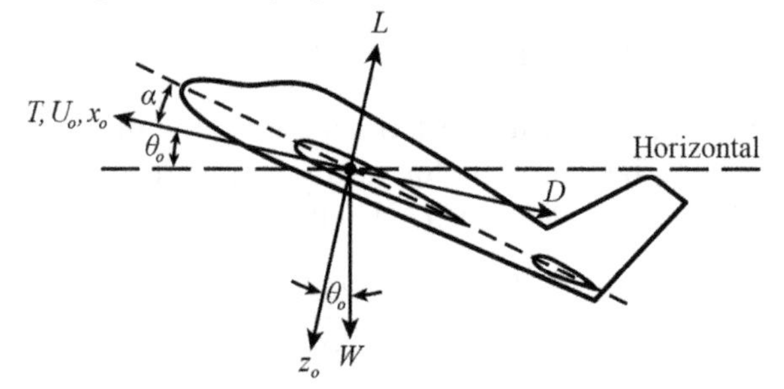

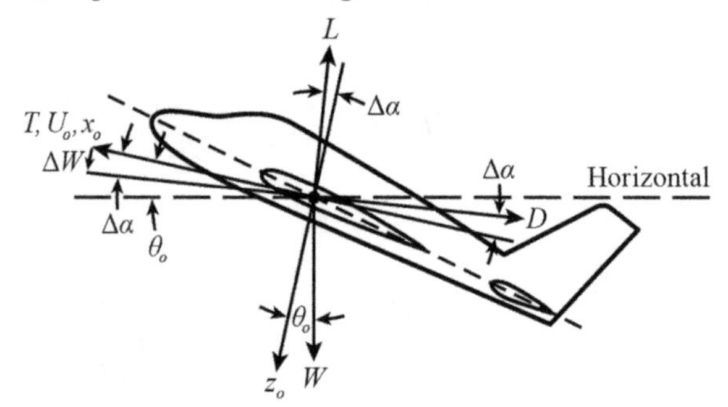

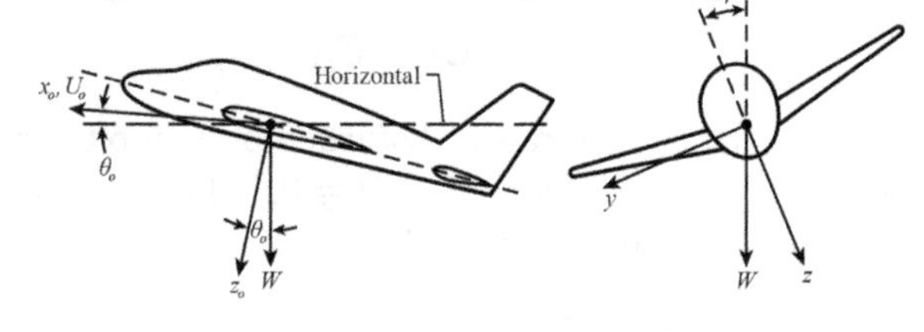

Fig. 4.17 Airplane in steady and disturbed flight

Now let us evaluate various derivatives like C_{xu}, C_{zu}, etc., appearing in Eqs. (4.417–4.419).

The forces acting on the airplane are lift L, drag D, thrust T, and weight W as shown. For steady equilibrium flight,

$$F_{xo} = T - D_o - W \sin\theta_o = 0 \tag{4.423}$$

and, for disturbed flight,

$$F_x = F_{xo} + \Delta F_x \tag{4.424}$$

$$= T - D - W \sin(\theta_o + \Delta\theta) \tag{4.425}$$

$$= T - (D_0 + \Delta D) - W \sin(\theta_o + \Delta\theta) \tag{4.426}$$

so that

$$F_x = \Delta F_x \approx -\Delta D - W \cos \theta_o \Delta\theta \tag{4.427}$$

For simplicity, we ignore thrust variations during the disturbance. Suppose the aircraft is disturbed only in forward speed. We have $U = U_o + \Delta U$, $\Delta\theta = 0$, $\Delta F_x = -\Delta D$ and $\partial U = \partial \Delta U$. Then,

$$\frac{\partial F_x}{\partial U} = \frac{\partial \Delta F_x}{\partial \Delta U} = -\frac{\partial \Delta D}{\partial \Delta U} = -\frac{\partial D}{\partial U} \tag{4.428}$$

$$= -\frac{\partial\left(\frac{1}{2}\rho U^2 S C_D\right)}{\partial U} \tag{4.429}$$

$$= -\rho U S C_D - \frac{1}{2}\rho U^2 S \frac{\partial C_D}{\partial U} \tag{4.430}$$

With $U \simeq U_o$, we have

$$\frac{\partial F_x}{U_o \partial\left(\frac{\Delta U}{U_o}\right)} = -\rho U_o S C_D - \frac{1}{2}\rho U_o^2 S \frac{\partial C_D}{U_o \partial\left(\frac{\Delta U}{U_o}\right)} \tag{4.431}$$

With $u = \Delta U / U_o$ and $C_x = F_x / \frac{1}{2}\rho U_o^2 S$, Eq. (4.431) simplifies to

$$C_{xu} = -2C_D - C_{Du} \tag{4.432}$$

Next, let us suppose that the aircraft is disturbed only in pitch angle so that the new pitch angle is $\theta_o + \Delta\theta$ and $\Delta U = 0$. Then, from Eq. (4.427) $F_x = \Delta F_x = -W \cos \theta_0 \Delta\theta$ and $\partial\theta = \partial\Delta\theta$.

$$\frac{\partial F_x}{\partial \theta} = \frac{\partial \Delta F_x}{\partial \Delta\theta} = -W \cos \theta_o \tag{4.433}$$

or

$$C_{x\theta} = -C_L \cos \theta_o \tag{4.434}$$

Next, consider the forces in the Oz direction. For steady equilibrium flight,

$$F_{zo} = W \cos \theta_o - L_o = 0 \tag{4.435}$$

For the disturbed flight,

$$F_z = F_{zo} + \Delta F_z \tag{4.436}$$

$$= W \cos(\theta_o + \Delta\theta) - L \tag{4.437}$$

$$= W \cos(\theta_o + \Delta\theta) - (L_o + \Delta L) \tag{4.438}$$

Proceeding in a similar fashion as before, we can show that

$$C_{zu} = -2C_L - C_{Lu} \tag{4.439}$$

$$C_{z\theta} = -C_L \sin\theta_o \tag{4.440}$$

Next, suppose the aircraft is disturbed in angle of attack. The forces acting on the airplane in the disturbed state are shown in Fig. 4.17b. We have

$$F_{xo} = -D_o + T - W\sin\theta_o = 0 \tag{4.441}$$

$$F_x = F_{xo} + \Delta F_x \tag{4.442}$$

$$= L\sin\Delta\alpha - D\cos\Delta\alpha + T - W\sin\theta_o \tag{4.443}$$

$$= (L_o + \Delta L)\sin\Delta\alpha - (D_o + \Delta D)\cos\Delta\alpha + T - W\sin\theta_o \tag{4.444}$$

Ignoring higher order terms and assuming cos $\Delta\alpha = 1$ and sin $\Delta\alpha = \Delta\alpha$, We have

$$\Delta F_x = L_o\Delta\alpha - \Delta D \tag{4.445}$$

$$\frac{\partial \Delta F_x}{\partial \Delta\alpha} = \frac{\partial F_x}{\partial \alpha} = L_o - \frac{\partial \Delta D}{\mathrm{d}\Delta\alpha} = L_o - \frac{\partial D}{\partial \alpha} \tag{4.446}$$

or

$$C_{x\alpha} = C_L - C_{D\alpha} \tag{4.447}$$

Similarly,

$$F_{zo} = -L_o + W\cos\theta_o \tag{4.448}$$

$$F_z = F_{zo} + \Delta F_z$$

$$= -(L_o + \Delta L)\cos\Delta\alpha - (D_o + \Delta D)\sin\Delta\alpha + W\cos\theta_o \tag{4.449}$$

$$= -L_o - \Delta L - D_o\Delta\alpha + W\cos\theta_o \tag{4.450}$$

so that

$$\Delta F_z = -\Delta L - D_o\Delta\alpha \tag{4.451}$$

$$\frac{\partial \Delta F_z}{\partial \Delta\alpha} = \frac{\partial F_z}{\partial \alpha} = -\frac{\partial \Delta L}{\partial \Delta\alpha} - D_o \tag{4.452}$$

$$= -\frac{\partial L}{\partial \alpha} - D_o \tag{4.453}$$

or

$$C_{z\alpha} = -C_{L\alpha} - C_D \tag{4.454}$$

It can be shown that

$$C_{x\dot{\alpha}} = C_{L\dot{\alpha}}\Delta\alpha - C_{D\dot{\alpha}} \simeq -C_{D\dot{\alpha}} \tag{4.455}$$

$$C_{xq} = C_{Lq}\Delta\alpha - C_{Dq} \simeq -C_{Dq} \tag{4.456}$$

$$C_{z\dot{\alpha}} = -C_{L\dot{\alpha}} - C_{D\dot{\alpha}}\Delta\alpha \simeq -C_{L\dot{\alpha}} \tag{4.457}$$

$$C_{zq} = -C_{Lq} - C_{Dq}\Delta\alpha \simeq -C_{Lq} \tag{4.458}$$

Usually, $C_{D\dot{\alpha}}$ and C_{Dq} are small and ignored.

Lateral-Directional Equations of Motion for Small Disturbance

Substituting for ΔC_y, ΔC_l, and ΔC_n from Eqs. (4.412), (4.414), and (4.416) in Eqs. (4.394), (4.396), and (4.398), we obtain

$$\left(m_1\frac{\mathrm{d}}{\mathrm{d}t} - b_1C_{y\dot{\beta}}\frac{\mathrm{d}}{\mathrm{d}t} - C_{y\beta}\right)\Delta\beta - \left(b_1C_{yp}\frac{\mathrm{d}}{\mathrm{d}t} + C_{y\phi}\right)\Delta\phi + \left(m_1\frac{\mathrm{d}}{\mathrm{d}t} - b_1C_{yr}\frac{\mathrm{d}}{\mathrm{d}t}\right)\Delta\psi = C_{y\delta r}\Delta\delta_r + C_{y\delta a}\Delta\delta_a \tag{4.459}$$

$$\left(-C_{l\beta} - b_1C_{l\dot{\beta}}\frac{\mathrm{d}}{\mathrm{d}t}\right)\Delta\beta + \left(-b_1C_{lp}\frac{\mathrm{d}}{\mathrm{d}t} + I_{x1}\frac{\mathrm{d}^2}{\mathrm{d}t^2}\right)\Delta\phi + \left(-b_1C_{lr}\frac{\mathrm{d}}{\mathrm{d}t} - I_{xz1}\frac{\mathrm{d}^2}{\mathrm{d}t^2}\right)\Delta\psi = C_{l\delta r}\Delta\delta_r + C_{l\delta a}\Delta\delta_a \tag{4.460}$$

$$\left(-C_{n\beta} - b_1C_{n\dot{\beta}}\frac{\mathrm{d}}{\mathrm{d}t}\right)\Delta\beta + \left(-b_1C_{np}\frac{\mathrm{d}}{\mathrm{d}t} - I_{xz1}\frac{\mathrm{d}^2}{\mathrm{d}t^2}\right)\Delta\phi + \left(-b_1C_{nr}\frac{\mathrm{d}}{\mathrm{d}t} + I_{z1}\frac{\mathrm{d}^2}{\mathrm{d}t^2}\right)\Delta\psi = C_{n\delta r}\Delta\delta_r + C_{n\delta a}\Delta\delta_a \tag{4.461}$$

where

$$\frac{\mathrm{d}\phi}{\mathrm{d}t} = p \tag{4.462}$$

$$\frac{\mathrm{d}\psi}{\mathrm{d}t} = r \tag{4.463}$$

$$b_1 = \frac{b}{2U_o} \tag{4.464}$$

$$I_{x1} = \frac{I_x}{\frac{1}{2}\rho U_o^2 Sb} \tag{4.465}$$

$$I_{z1} = \frac{Iz}{\frac{1}{2}\rho U_o^2 Sb} \tag{4.466}$$

$$I_{xz1} = \frac{I_{xz}}{\frac{1}{2}\rho U_o^2 Sb} \tag{4.467}$$

Now let us evaluate various derivatives like $C_{y\beta}$, $C_{y\phi}$, etc., appearing in Eqs. (4.459–4.461).

Suppose the airplane is disturbed in bank angle $\Delta\phi$ as shown in Fig. 4.17c. We have

$$F_y = F_{yo} + \Delta F_y = W \cos\theta_o \Delta\phi + Y_{\text{aero}} + \Delta Y_{\text{aero}} \tag{4.468}$$

where Y_{aero} denotes the aerodynamic side force. We have $F_{yo} = Y_{\text{aero}} = 0$ so that

$$\Delta F_y = W \cos\theta_o \Delta\phi + \Delta Y_{\text{aero}} \tag{4.469}$$

$$\frac{\partial \Delta F_y}{\partial \Delta\phi} = \frac{\partial F_y}{\partial \phi} = W \cos\theta_o \tag{4.470}$$

or

$$C_{y\phi} = C_L \cos\theta_o \tag{4.471}$$

Similarly, it can be shown that

$$C_{y\beta} = (C_{y\beta})_{\text{aero}} \tag{4.472}$$

$$C_{y\dot{\beta}} = (C_{y\dot{\beta}})_{\text{aero}} \tag{4.473}$$

$$C_{yp} = (C_{yp})_{\text{aero}} \tag{4.474}$$

$$C_{yr} = (C_{yr})_{\text{aero}} \tag{4.475}$$

and so on. Here, it is assumed that the side force Y_{aero} has no dependence on the bank angle ϕ.

It is important to bear in mind that Eqs. (4.417–4.419) and (4.459–4.461) are applicable for small disturbance motion when the airplane is disturbed from a steady, unaccelerated flight with zero sideslip. If this is not the case, then the assumed initial conditions in Eqs. (4.370–4.373) will be different, and we have to derive different sets of equations of motion. This concept will be illustrated with the help of an example later in this section.

Example 4.7

A high-performance aircraft executes a velocity vector roll at a rate of 90 deg/s while flying at 100 m/s and an angle of attack of 30 deg. Determine the acceleration of the aircraft with respect to an Earth-fixed inertial system.

Solution:

We assume that the x axis of the Earth-fixed system points in the direction of the velocity vector, y axis to the right, and z axis vertically downward, to form a right-hand system. Then,

$$p = \Omega \cos \alpha \quad q = 0 \quad r = \Omega \sin \alpha$$

$$U = U_o \cos \alpha \quad V = 0 \quad W = U_o \sin \alpha$$

$$\dot{U} = 0 \quad \dot{V} = 0 \quad \dot{W} = 0$$

The body axes components of the acceleration with respect to the Earth-fixed system are

$$a_x = \dot{U} + qW - rV$$

$$a_y = \dot{V} + rU - pW$$

$$a_z = \dot{W} + pV - qU$$

Substituting, we get

$$a_x = 0$$

$$a_y = \Omega \sin \alpha U_o \cos \alpha - \Omega \cos \alpha U_o \sin \alpha = 0$$

$$a_z = 0$$

Thus, the aircraft does not experience any acceleration with respect to an Earth-fixed axes sytem.

Example 4.8

At $t = 0$, the $Ox_0y_0z_0$ and $Ox_1y_1z_1$ systems coincide, and the $Ox_1y_1z_1$ system has an angular velocity of 1 rad/s with respect to the $Ox_0y_0z_0$ system. The angular velocity vector is directed along the Oz_0 or Oz_1 axis. A particle has a velocity $V_o = \hat{i}_o 50 + \hat{j}_o 25 + \hat{k}_o 0$ as observed in the $Ox_0y_0z_0$ system. What are the velocities and acceleration as measured in the $Ox_1y_1z_1$ system at $t = 5$ s?

(Continued)

Example 4.8 (*Continued*)

Solution:

At $t = 0$, the two coordinate systems coincide. Let $\vec{\Omega} = \hat{k}_o\Omega_1 = \hat{k}_o 1 = \hat{k}_1 1$. Furthermore,

$$\vec{V}_o = \begin{bmatrix} 50 \\ 25 \\ 0 \end{bmatrix} \quad \vec{\omega}_{o,1}^{1} = \begin{bmatrix} 0 \\ 0 \\ \Omega_1 \end{bmatrix} \quad \vec{\omega}_{1,o}^{o} = \begin{bmatrix} 0 \\ 0 \\ -\Omega_1 \end{bmatrix}$$

We have

$$\vec{V}_1 = T_o^1 \vec{V}_o = \begin{bmatrix} \cos\Omega_1 t & \sin\Omega_1 t & 0 \\ -\sin\Omega_1 t & \cos\Omega_1 t & 0 \\ 0 & 0 & 1 \end{bmatrix} \begin{bmatrix} 50 \\ 25 \\ 0 \end{bmatrix}$$

where $\Omega_1 = 1$ rad/s and $t = 5$ s. Substituting and simplifying, we get

$$\vec{V}_1 = \begin{bmatrix} 38.1575 \\ -40.8525 \\ 0 \end{bmatrix}$$

We have

$$\begin{aligned} \left(\frac{d\vec{V}_o}{dt}\right)_1 &= \left(\frac{d\vec{V}_o}{dt}\right)_o + \vec{\omega}_{1,o}^{o} \times \vec{V}_o \\ &= 0 + \begin{vmatrix} \hat{i}_1 & \hat{j}_1 & \hat{k}_1 \\ 0 & 0 & -1 \\ 50 & 25 & 0 \end{vmatrix} \\ &= 25\hat{i}_1 - 50\hat{j}_1 \text{ m/s}^2 \end{aligned}$$

Example 4.9

An aircraft is initially in a dive and, at the bottom of the dive, the pilot effects a steady pull-out with a constant pitch rate q_o rad/s. Obtain the equations of motion for small disturbances during the pull-out.

Solution:

The equations of aircraft motion are

$$F_x = m(\dot{U} + qW - rV)$$
$$F_y = m(\dot{V} + rU - pW)$$

(*Continued*)

Example 4.9 (*Continued*)

$$F_z = m(\dot{W} + pV - qU)$$
$$L = \dot{p}I_x - I_{xz}(pq + \dot{r}) + qr(I_z - I_y)$$
$$M = \dot{q}I_y + rp(I_x - I_z) + (p^2 - r^2)I_{xz}$$
$$N = \dot{r}I_z - I_{xz}(\dot{p} - qr) + pq(I_y - I_x)$$

We have

$$U = U_o + \Delta U \quad V = \Delta V \quad W = \Delta W$$
$$p = \Delta p \quad q = q_o + \Delta q \quad r = \Delta r$$
$$F_x = \Delta F_x \quad F_y = \Delta F_y \quad F_z = \Delta F_z$$
$$L = \Delta L \quad M = \Delta M \quad N = \Delta N$$

Assuming that all the disturbance variables are small (note that q_o is the steady-state pitch rate and is not a disturbance variable) so that we can ignore higher order terms involving products of small disturbance parameters, we get the following equations:

$$\Delta F_x = m(\Delta\dot{U} + q_o \Delta W)$$
$$\Delta F_y = m(\Delta\dot{V} + U_o \Delta r)$$
$$\Delta F_z = m(\Delta\dot{W} - q_o U_o - q_o \Delta U - \Delta q U_o)$$
$$\Delta L = \Delta\dot{p}I_x - I_{xz}(\Delta p q_o + \Delta\dot{r}) + q_o \Delta r(I_z - I_y)$$
$$\Delta M = \Delta\dot{q}I_y$$
$$\Delta N = I_z \Delta\dot{r} - I_{xz}(\Delta\dot{p} - q_o \Delta r) + \Delta p q_o (I_y - I_x)$$

We observe that the longitudinal and lateral-directional equations of motion are now coupled. Here, $q_o = U_o/R$ where R is the radius of curvature. (See Chapter 2 for more information on steady pull-out from a dive in vertical plane.)

Substituting and simplifying, we get the following equations of motion for small disturbances during the pull-out:

$$\left(m_1 \frac{\mathrm{d}}{\mathrm{d}t} - C_{xu}\right)u - \left(C_{x\dot{\alpha}} c_1 \frac{\mathrm{d}}{\mathrm{d}t} + C_{x\alpha} - m_1 q_o\right)\Delta\alpha - \left(C_{xq} c_1 \frac{\mathrm{d}}{\mathrm{d}t} + C_{x\theta}\right)\Delta\theta$$
$$= C_{x\delta e}\Delta\delta e + C_{x\delta t}\Delta\delta t$$

$$(-C_{zu} + m_1 q_o)u + \left[\left(m_1 \frac{\mathrm{d}}{\mathrm{d}t} - C_{z\dot{\alpha}} c_1 \frac{\mathrm{d}}{\mathrm{d}t}\right) - C_{z\alpha}\right]\Delta\alpha - \left(m_1 \frac{\mathrm{d}}{\mathrm{d}t} + C_{zq} c_1 \frac{\mathrm{d}}{\mathrm{d}t}\right)\Delta\theta$$
$$= m_1 q_o + C_{z\delta e}\Delta\delta e + C_{z\delta t}\Delta\delta t$$

(*Continued*)

Example 4.9 (Continued)

$$\left(m_1\frac{d}{dt}-b_1C_{y\dot{\beta}}\frac{d}{dt}-C_{y\beta}\right)\Delta\beta-\left(b_1C_{yp}\frac{d}{dt}+C_{y\phi}\right)\Delta\phi$$
$$+\left(m_1\frac{d}{dt}-b_1C_{yr}\frac{d}{dt}\right)\Delta\psi=C_{y\delta r}\Delta\delta_r+C_{y\delta a}\Delta\delta_a$$

$$\left(-C_{l\beta}-b_1C_{l\dot{\beta}}\frac{d}{dt}\right)\Delta\beta+\left(-b_1C_{lp}\frac{d}{dt}-I_{xz1}q_o\frac{d}{dt}+I_{x1}\frac{d^2}{dt^2}\right)\Delta\phi$$
$$+\left(-b_1C_{lr}\frac{d}{dt}+q_o(I_{z1}-I_{y1})-I_{xz1}\frac{d^2}{dt^2}\right)\Delta\psi=C_{l\delta r}\Delta\delta_r+C_{l\delta a}\Delta\delta_a$$

$$-C_{mu}u-\left(C_{m\dot{\alpha}}c_1\frac{d}{dt}+C_{m\alpha}\right)\Delta\alpha+\frac{d}{dt}\left(I_{y1}\frac{d}{dt}-C_{mq}c_1\right)\Delta\theta$$
$$=C_{m\delta e}\Delta\delta e+C_{m\delta t}\Delta\delta t$$

$$\left(-C_{n\beta}-b_1C_{n\dot{\beta}}\frac{d}{dt}\right)\Delta\beta+\left(-b_1C_{np}\frac{d}{dt}+q_o(I_{y1}-I_{x1})-I_{xz1}\frac{d^2}{dt^2}\right)\Delta\phi$$
$$+\left(-b_1C_{nr}\frac{d}{dt}+I_{xz1}q_o+I_{xz1}\frac{d^2}{dt^2}\right)\Delta\psi=C_{n\delta r}\Delta\delta_r+C_{n\delta a}\Delta\delta_a$$

Now let us evaluate the forces and moments acting on the airplane following a small disturbance during the pull-out. Referring to Fig. 4.18, we have

$$F_{xo}=-D-W\sin\theta_o+T=0$$

$$F_{zo}=-L+W\cos\theta_o-\left(\frac{W}{g}\right)\left(\frac{U_o^2}{R}\right)=0$$

$$M_o=0$$

$$F_{yo}=L_o(RM)=N_o=0$$

For the disturbed flight,

$$F_x=F_{xo}+\Delta F_x=-(D_o+\Delta D)-W\sin(\theta_o+\Delta\theta)+T$$

$$F_z=F_{zo}+\Delta F_z=-(L_o+\Delta L)+W\cos(\theta_o+\Delta\theta)-\left(\frac{W}{g}\right)\left[\frac{(U_o+\Delta U)^2}{R}\right]$$

so that

$$\Delta F_x=-\Delta D-W\cos\theta_o\Delta\theta$$

$$\Delta F_z=-\Delta L-W\sin\theta_o\Delta\theta-\left(\frac{W}{g}\right)\left[\frac{2U_o\Delta U}{R}\right]$$

(Continued)

Example 4.9 (Continued)

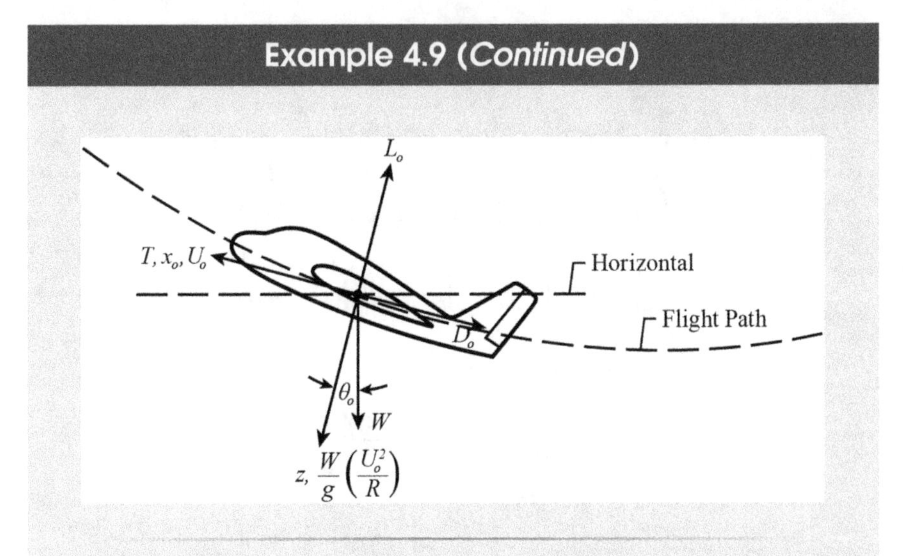

Fig. 4.18 Forces acting on an airplane during a pull-out maneuver.

Proceeding in a similar way as we did for the small disturbance equations of motion for a steady flight in a vertical plane, we obtain

$$C_{xu} = -2C_D - C_{Du}$$

$$C_{x\alpha} = C_L - C_{D\alpha}$$

$$C_{x\theta} = -C_L \cos\theta_o$$

$$C_{x\dot{\alpha}} = -C_{D\dot{\alpha}} = 0$$

$$C_{xq} = -C_{Dq} = 0$$

$$C_{zu} = -2C_L - C_{Lu} - 2m_1 q_o$$

$$C_{z\alpha} = -C_{L\alpha} - C_D$$

$$C_{z\dot{\alpha}} = -C_{L\dot{\alpha}}$$

$$C_{zq} = -C_{Lq}$$

$$C_{z\theta} = -C_L \sin\theta_o$$

The expressions for other lateral-directional stability derivatives essentially remain the same as those derived earlier for steady flight. However, we have to note that all the stability derivatives have to be evaluated for curved flow conditions to simulate the steady pull-out maneuver prior to disturbance. However, if q_o is small, for approximate purposes, the curved flow effects can be ignored, and relations derived for steady flight in a vertical plane can be used with one difference that $C_{m\alpha}$ should be evaluated considering the increase in stability level during the pull-out maneuver.

4.4 Estimation of Stability Derivatives

To solve motion Eqs. (4.417–4.419) and (4.459–4.461) for studying the stability (free response) or the response of the airplane to a given pilot input (forced response), we need to know the values of all the static and dynamic stability derivatives appearing in those equations. These derivatives can be determined either by analytical, semiempirical, computational fluid dynamics (CFD), or experimental methods. The analytical methods based on classical aerodynamic theories can be applied only to idealized wings and bodies. Aircraft configurations of practical interest cannot be analyzed using the classical aerodynamic theories. In view of this difficulty, several empirical/semiempirical methods have been developed over the years for evaluating the stability and control derivatives of aircraft configurations of practical interest. Datcom is one of the most widely used sources for estimating aircraft stability and control derivatives [7]. It is a comprehensive compilation of all the available engineering methods for obtaining the stability and control derivatives of airplane-type configurations.

The modern CFD methods are capable of providing more accurate estimates of static and some dynamic derivatives. But, at present, the level of effort involved precludes their application at preliminary design stages. In view of this, the CFD methods are usually used at a sufficiently later stage in the vehicle design process when the configuration takes its final form.

The experimental methods for obtaining dynamic stability derivatives in ground-based test facilities [8–10] mainly consist of either forced oscillation or free oscillation technique wherein the test model undergoes an oscillatory motion in pitch, roll, or yaw. Another approach that is often used to obtain dynamic stability derivatives in ground-based test facilities is the so-called free flight or semifree flight test technique [11]. Here, the test model is allowed to fly either free or semifree [loosely constrained in some degree(s) of freedom] in the viewing area of the test facility. The model motion is recorded in the form of various accelerometer outputs or high-speed motion picture records. The dynamic stability derivatives are then obtained implicitly by matching the predicted time history of the model with the observed time history. Because the time and effort involved in fabricating and testing aircraft models in the ground-based facilities are considerable, these approaches are also usually used in the final stages of the vehicle design.

On the other hand, the semiempirical or engineering methods provide quick and cost-effective estimates of the static and dynamic stability derivatives, which can be readily used for assessing the vehicle flyability, stability, and control. Also, the predictions based on these engineering methods can be used in the design of flight control and guidance systems. The only disadvantage is that such predictions are less accurate compared to the CFD or the wind-tunnel test data. However, this type of information is very useful in the early stages of the design to quickly evaluate several candidate configurations for their suitability in meeting design goals. As the design matures, these estimates can be replaced by more accurate CFD results or the wind-tunnel test data on high fidelity models as they become available to revise/update the stability and control characteristics as well as the design of flight control and guidance systems.

In this section, we will discuss the engineering methods suitable for evaluating the important stability derivatives appearing in the longitudinal and lateral-directional equations of motion. We will use the simple strip theory wherever applicable. The strip theory analysis is helpful to understand the basic aerodynamic concepts underlying the dynamic derivatives. The semi-empirical methods discussed in this text are taken from Datcom [7]. These relations usually involve several empirical parameters, and Datcom gives numerous charts to evaluate them. It is neither the objective of this text nor it is possible to include all the Datcom [7] information here. Instead, an effort is made to present sufficient information that should be useful for the estimation of static and dynamic stability derivatives of generic aircraft configurations typically used in classroom instruction and study projects. If the reader finds that the data for any fuselage, wing, or tail configuration of interest is not available in this text, please refer to Datcom [7].

4.4.1 Estimation of Longitudinal Stability Derivatives

Estimation of $C_{L\alpha}$ This is the lift-curve slope of the airplane. Essentially, $C_{L\alpha} = (C_{L\alpha})_{WB}$. The methods to evaluate $(C_{L\alpha})_{WB}$ were discussed in Chapter 3.

Estimation of $C_{m\alpha}$ This is the slope of the pitching-moment curve, and essentially $C_{m\alpha} = (C_{m\alpha})_{WB} + (C_{m\alpha})_t$. The methods to evaluate this derivative were discussed in Chapter 3.

Estimation of $C_{D\alpha}$ For linear lift coefficient $C_L = C_{L\alpha}\alpha$ and quadratic drag polar $C_D = C_{Do} + kC_L^2$, the derivative $C_{D\alpha}$ can be expressed as

$$C_{D\alpha} = \left(\frac{\mathrm{d}C_D}{\mathrm{d}C_L}\right)\left(\frac{\mathrm{d}C_L}{\mathrm{d}\alpha}\right) \tag{4.476}$$

$$= 2kC_L C_{L\alpha} \tag{4.477}$$

where $k = 1/\pi Ae$. The planform or Oswald's efficiency parameter e is given by [7]

$$e = \frac{1.1C_{L\alpha}}{RC_{L\alpha} + (1-R)\pi A} \tag{4.478}$$

The parameter R is given by the following expression, which is a curve fit to Datcom data [7]:

$$R = a_1\lambda_1^3 + a_2\lambda_1^2 + a_3\lambda_1 + a_4 \tag{4.479}$$

where $a_1 = 0.0004$, $a_2 = -0.0080$, $a_3 = 0.0501$, $a_4 = 0.8642$, and $\lambda_1 = A\lambda/\cos \Lambda_{LE}$ where A is the aspect ratio, λ is the taper ratio, and Λ_{LE} is the leading-edge sweep of the wing.

Estimation of C_{Du} With $u = \Delta U/U_o$, this derivative can be expressed as

$$\frac{\partial C_D}{\partial u} = M\frac{\partial C_D}{\partial M} \tag{4.480}$$

The derivative $\partial C_D/\partial M$ represents the variation of drag coefficient with Mach number when the angle of attack is held constant. At low subsonic speeds ($M \leq 0.5$), the drag coefficient is practically constant so that $\partial C_D/\partial M = 0$. However, as the flight Mach number approaches the critical Mach number M_{cr}, the drag coefficient starts rising. It assumes a peak value in the transonic Mach number range and starts decreasing as the Mach number becomes supersonic. It tends to assume a steady value at high supersonic or hypersonic Mach numbers. Therefore, if the flight Mach number exceeds 0.5, the derivative C_{Du} should not be ignored. The interested reader may refer to Datcom [7] for information on the methods to evaluate the derivative $\partial C_D/\partial_M$.

Estimation of C_{Lu} This derivative, in a similar fashion to C_{Du}, can be expressed as

$$\frac{\partial C_L}{\partial u} = M\alpha\frac{\partial C_{L\alpha}}{\partial M} \tag{4.481}$$

At low subsonic speeds ($M \leq 0.5$), the lift-curve slope $C_{L\alpha}$ essentially remains constant so that $\partial C_{L\alpha}/\partial M = 0$. However, as the Mach number approaches critical Mach number, the lift-curve slope starts increasing. At some Mach number in the transonic/low supersonic range, it assumes a maximum value and then starts decreasing with further increase in the Mach number. Therefore, this derivative should be considered when $M \geq 0.5$.

The data given in Chapter 3 can be used to obtain $C_{L\alpha}$ as a function of Mach number.

Estimation of C_{mu} This derivative can be expressed as

$$\frac{\partial C_m}{\partial u} = M\alpha\frac{\partial C_{m\alpha}}{\partial M} \tag{4.482}$$

Like the lift-curve slope, the pitching-moment-curve slope also varies with Mach number. Assuming that the airplane center of gravity remains fixed, the variation of $C_{m\alpha}$ with Mach number is caused by the rearward movement of the center of pressure in the transonic and supersonic region. As a result, the aircraft becomes more stable. With further increase in Mach number, especially as it approaches the hypersonic range, the fuselage develops more and more lift, and the center of pressure starts moving forward again. Thus, the derivative C_{mu} assumes significance for $M \geq 0.5$. The data given in Chapter 3 can be used to obtain $C_{m\alpha}$ as a function of Mach number.

Longitudinal Rotary Derivatives

Estimation of C_{Lq} This derivative is a measure of the effect of a steady pitch rate on the lift coefficient. Because of the pitching motion, the effective angle of attack of the fuselage, wing, and tail surfaces will be different from the steady-state value. For a positive pitch rate, the sections of the fuselage and wing that are ahead of the center of gravity experience a reduction in the angle of attack, and those sections aft of center of gravity experience an increase in the local angle of attack. The increase in angle of attack is particularly significant for the aft-located horizontal tail because of its large distance from the center of gravity. For airplanes with short fuselages and a high-aspect ratio wing (small chord and large span) located close to the center of gravity, the contribution of the fuselage and wing to C_{Lq} can be ignored. For such configurations, the contribution to C_{Lq} mainly comes from the horizontal tail and can be estimated as follows.

The increase in the horizontal tail angle of attack (see Fig. 4.19) due to a positive pitch rate q is given by

$$\Delta\alpha_t = \frac{ql_t}{U_o} \tag{4.483}$$

where l_t is the distance between the center of gravity and the aerodynamic center of the horizontal tail and U_o is the flight velocity. Usually, l_t is called the tail length. Note that ql_t is the induced upwash at the aerodynamic center of the horizontal tail. The resulting increase in the tail-lift coefficient is given by

$$\Delta C_{Lt} = a_t\left(\frac{S_t}{S}\right)\eta_t\left(\frac{ql_t}{U_o}\right) \tag{4.484}$$

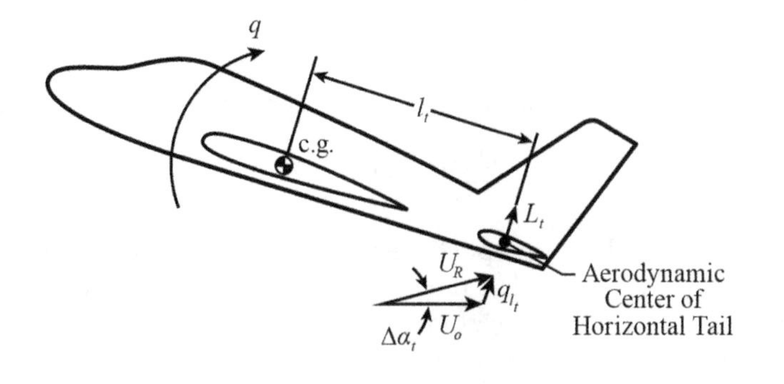

Fig. 4.19 Increase in tail angle of attack during pitching motion.

With $C_{Lq} = \partial C_L/\partial(q\bar{c}/2U_o)$, we obtain

$$(C_{Lq})_t = 2a_t\left(\frac{S_t}{S}\right)\eta_t\left(\frac{l_t}{\bar{c}}\right) \tag{4.485}$$

$$= 2a_t\overline{V}_1\eta_t \tag{4.486}$$

where $\overline{V}_1 = S_t l_t/S\,\bar{c}$.

For configurations with long fuselages and short aspect ratio wings (a large chord and a small span), the contribution of the wing–body combination can become significant so that the total value of C_{Lq} is given by

$$C_{Lq} = (C_{Lq})_{WB} + (C_{Lq})_t \tag{4.487}$$

The contribution of the wing–body combination can be estimated using the following expression [7]:

$$(C_{Lq})_{WB} = [K_{W(B)} + K_{B(W)}]\left(\frac{S_e\bar{c}_e}{S\bar{c}}\right)(C_{Lq})_e + (C_{Lq})_B\left(\frac{S_{B,\max}l_f}{S\bar{c}}\right) \tag{4.488}$$

where $\bar{c}_e$ and $\bar{c}$ are the mean aerodynamic chords of the exposed wing and total (theoretical) wing, respectively; $(C_{Lq})_e$ and $(C_{Lq})_B$ are the contributions of the exposed wing and isolated body, respectively. The wing–body interference parameters $K_{B(W)}$ and $K_{W(B)}$ can be obtained using the data given in Chapter 3. The value of $(C_{Lq})_{WB}$ given by Eq. (4.488) is per radian.

For subsonic speeds [7],

$$(C_{Lq})_e = \left(\frac{1}{2} + 2\xi\right)(C_{L\alpha})_e \tag{4.489}$$

where

$$\xi = \frac{\bar{x}}{\bar{c}_e} \tag{4.490}$$

$$\bar{x} = (x_{ac})_e - x_{cg,le} \tag{4.491}$$

Here, $(x_{ac})_e$ is the distance of the exposed wing aerodynamic center from the leading edge of the root chord, and $x_{cg,le}$ is the distance of the center of gravity from the leading edge of the exposed wing root chord. Both $(x_{ac})_e$ and $x_{cg,le}$ are measured parallel to the exposed wing root chord.

The parameter $\bar{x}$ will be positive if the aerodynamic center of the exposed wing $(x_{ac})_e$ is aft of the center of gravity as shown in the insert at the top of Fig. 4.20. The lift-curve slope of the exposed wing $(C_{L\alpha})_e$ and the aerodynamic center of the exposed wing $(x_{ac})_e$ may be estimated using the procedure discussed in Chapter 3.

For supersonic Mach numbers, Datcom [7] gives

$$(C_{Lq})_e = (C'_{Lq})_e + 2\xi(C_{N\alpha})_e \tag{4.492}$$

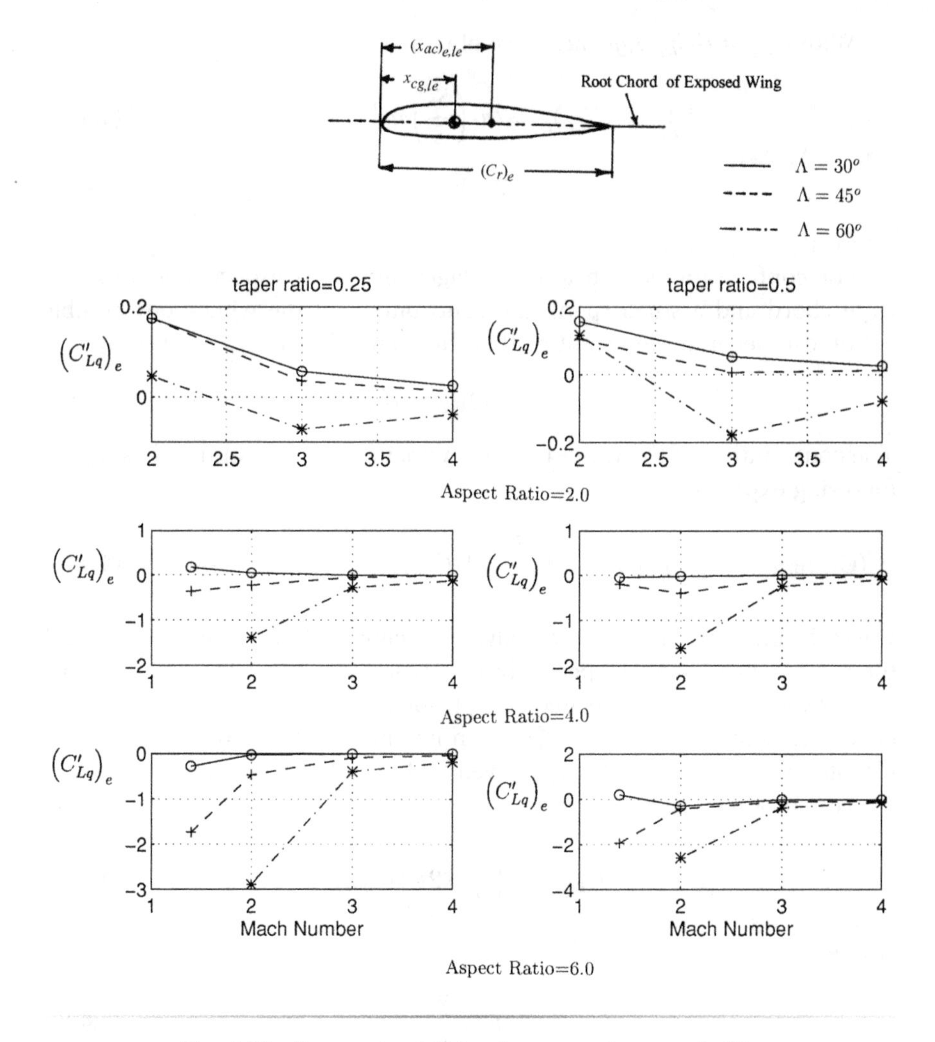

Fig. 4.20 Parameter $(C'_{Lq})_e$ at supersonic speeds (7).

For low supersonic Mach numbers when the leading edge is subsonic or $\beta \cot \Lambda_{LE} \leq 1$, the data [7] is somewhat sparse and is not given here. The interested reader may refer to Datcom [7].

For higher Mach numbers when the leading edge becomes supersonic or $\beta \cot \Lambda \geq 1$, the parameter $(C'_{Lq})_e$ is presented in Fig. 4.20 for typical wing sections. For data on other wing planforms, the reader may refer to Datcom [7].

The body contribution at subsonic speeds is given by [7]

$$(C_{Lq})_B = 2(C'_{L\alpha})_B\left(1 - \frac{x_m}{l_f}\right) \tag{4.493}$$

$$(C'_{L\alpha})_B = (C_{L\alpha})_B \left(\frac{V_B^{2/3}}{S_{B,\max}} \right) \tag{4.494}$$

$$(C_{L\alpha})_B = 2(k_2 - k_1) \left(\frac{S_{B,\max}}{V_B^{2/3}} \right) \tag{4.495}$$

where $x_m = x_{cg}$, the moment reference point (measured from fuselage leading edge), and V_B is the fuselage volume. The fuselage apparent mass coefficient term $(k_2 - k_1)$ can be obtained using the data presented in Fig. 3.6. Note that $(C_{Lq})_B$ is based on $S_{B,\max} l_f$ and $(C_{L\alpha})_B$ is based on $V_B^{2/3}$, where $S_{B,\max}$ is the maximum cross-sectional area of the fuselage, l_f is the total length of the fuselage, and V_B is the volume of the fuselage.

The body contribution at supersonic speeds (based on maximum cross-sectional area and body length) is given by [7]

$$(C_{Lq})_B = 2(C_{N\alpha})_B \left(1 - \frac{x_m}{l_f} \right) \tag{4.496}$$

where the slope of normal force coefficient $(C_{N\alpha})_B$ can be approximately estimated using the data given in Fig. 3.10, which is based on maximum cross-sectional area of the body.

Estimation of C_{mq} This derivative is a measure of the pitching moment induced because of a pitch rate experienced by the aircraft and is known as the damping-in-pitch derivative. This is one of the most important longitudinal dynamic stability derivatives of the aircraft. Generally, the contribution of the aft-located horizontal tail is damping in nature and by far is the largest because of its long moment arm from the center of gravity. The contribution of the wing–body is generally small but can be significant if the fuselage is long and the wing has a small aspect ratio. The contribution of the horizontal tail can be estimated as follows.

As we have seen earlier, because of the pitch rate, the tail angle of attack and the lift coefficient increase. The corresponding increase in pitching-moment coefficient is given by

$$\Delta C_{mt} = -a_t \left(\frac{q l_t}{U_o} \right) \left(\frac{S_t}{S} \right) \eta_t \left(\frac{l_t}{\bar{c}} \right) \tag{4.497}$$

With

$$C_{mq} = \frac{\partial C_m}{\partial \left(\frac{q \bar{c}}{2 U_o} \right)} \tag{4.498}$$

We get

$$(C_{mq})_t = -2a_t\left(\frac{S_t}{S}\right)\eta_t\left(\frac{l_t}{\overline{c}}\right)^2 \tag{4.499}$$

$$= -2a_t\overline{V}_1\eta_t\left(\frac{l_t}{\overline{c}}\right) \tag{4.500}$$

For aircraft configurations with long fuselage and small aspect ratio wings, the contribution of the wing–body combination may become significant. For such configurations, the total value of the damping-in-pitch derivative is given by

$$C_{mq} = (C_{mq})_{WB} + (C_{mq})_t \tag{4.501}$$

The wing–body contribution can be estimated using the following expression [7]:

$$(C_{mq})_{WB} = \left[K_{W(B)} + K_{B(W)}\right]\frac{S_e}{S}\left(\frac{\overline{c}_e}{\overline{c}}\right)^2(C_{mq})_e + (C_{mq})_B\frac{S_{B,\max}}{S}\left(\frac{l_f}{\overline{c}}\right)^2 \Big/ \text{rad} \tag{4.502}$$

where $(C_{mq})_e$ and $(C_{mq})_B$ are the contributions of the exposed wing and body, respectively.

For subsonic speeds [7],

$$(C_{mq})_e = \left[\frac{\dfrac{c_1}{c_3} + c_2}{\dfrac{c_1}{c_4} + 3}\right](C_{mq})_{e,M=0.2} \tag{4.503}$$

where

$$(C_{mq})_{e,M=0.2} = -0.7C_{l\alpha}\cos\Lambda_{c/4}\left[\frac{A(0.5\xi + 2\xi^2)}{c_5} + \left(\frac{c_1}{24c_4}\right) + \frac{1}{8}\right] \tag{4.504}$$

$$c_1 = A^3\tan^2\Lambda_{c/4} \tag{4.505}$$

$$c_2 = \frac{3}{B} \tag{4.506}$$

$$c_3 = AB + 6\cos\Lambda_{c/4} \tag{4.507}$$

$$c_4 = A + 6\cos\Lambda_{c/4} \tag{4.508}$$

$$c_5 = A + 2\cos\Lambda_{c/4} \tag{4.509}$$

$$B = \sqrt{1 - M^2\cos^2\Lambda_{c/4}} \tag{4.510}$$

Here, $A = A_e$, the aspect ratio of the exposed wing and $C_{l\alpha}$ is the sectional or two-dimensional lift-curve slope of the wing and can be evaluated using the methods given in Chapter 3. The parameter ξ is defined in Eq. (4.490). The value of $(C_{mq})_e$ given by Eq. (4.503) is per radian and is based on the exposed wing parameters S_e and c_e^2.

For supersonic speeds, we have [7]

$$(C_{mq})_e = (C'_{mq})_e - \xi(C_{Lq})_e \tag{4.511}$$

where the parameter $(C'_{mq})_e$ is evaluated as follows.

For low supersonic Mach numbers when the wing leading edge operates under subsonic conditions, which happens when the parameter $\beta \cot \Lambda_{LE} \leq 1$, the data [7] are quite limited and are not presented here. The interested reader may refer to Datcom [7].

For higher Mach numbers when the wing leading edge operates under supersonic Mach numbers or $\beta \cot \Lambda_{LE} \geq 1$, the parameter $(C'_{mq})_e$ can be evaluated using the data presented in Fig. 4.21 for typical wing planforms.

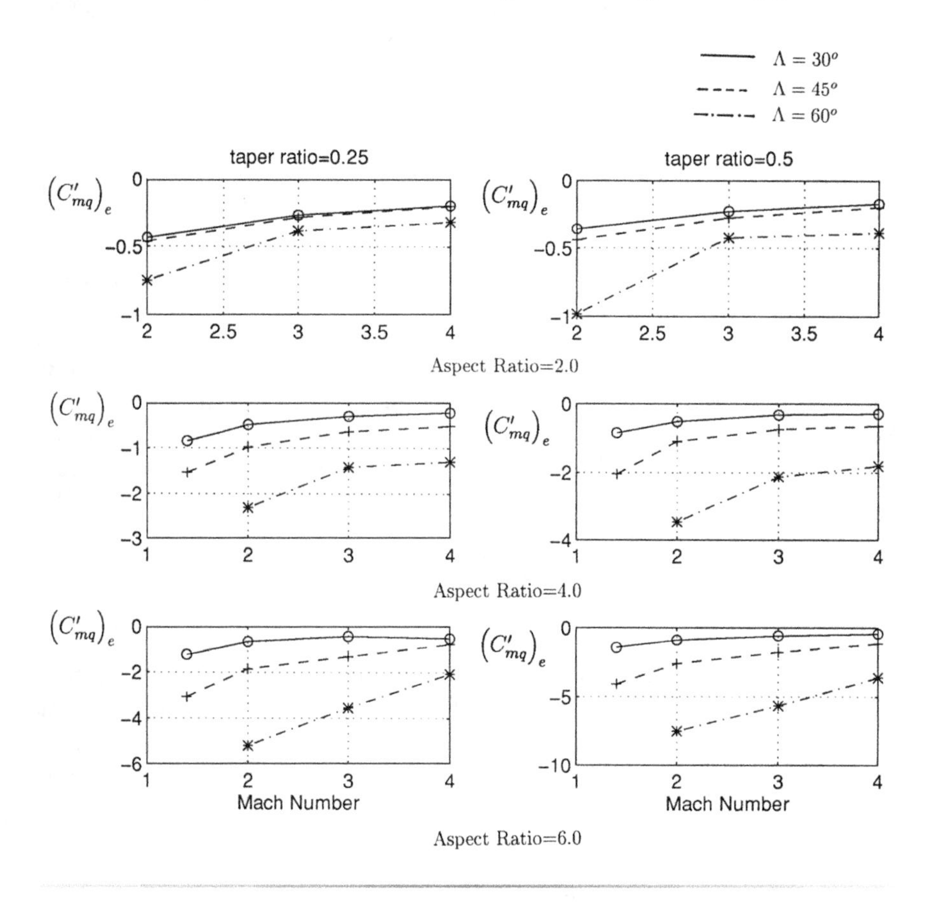

Fig. 4.21 Parameter $(C'_{mq})_e$ at supersonic speeds (7).

For data on wing planforms not covered here, the reader may refer to Datcom [7].

The body contribution at both subsonic and supersonic speeds is given by

$$(C_{mq})_B = 2(C'_{m\alpha})_B \left[\frac{(1 - x_{m1})^2 - V_{B1}(x_{c1} - x_{m1})}{1 - x_{m1} - V_{B1}} \right] \tag{4.512}$$

where

$$x_{m1} = \frac{x_m}{l_f} \qquad x_{c1} = \frac{x_c}{l_f} \qquad V_{B1} = \frac{V_B}{S_{B,\max} l_f} \tag{4.513}$$

$$x_c = \frac{1}{V_B} \int_0^{l_f} S_B(x) x \, \mathrm{d}x \tag{4.514}$$

Here, $x_m = x_{cg}$, the distance of the moment reference point from the leading edge of the fuselage, V_B is the volume of the fuselage, and $S_{B,\max}$ is the maximum cross-sectional area of the fuselage. The value of $(C_{mq})_B$ given by Eq. (4.512) is based on the maximum cross-sectional area and square of the body (fuselage) length [7].

For subsonic speeds,

$$(C'_{m\alpha})_B = (C_{m\alpha})_B \left(\frac{V_B}{S_{B,\max} l_f} \right) \tag{4.515}$$

$$(C_{m\alpha})_B = \frac{2(k_2 - k_1)}{V_B} \int_0^{x_o} \frac{\mathrm{d}S_B(x)}{\mathrm{d}x} (x_m - x) \, \mathrm{d}x \tag{4.516}$$

Here, $k_2 - k_1$ is the fuselage apparent mass coefficient (see Fig. 3.6) and $x_m = x_{cg}$, the distance of the moment reference point from the leading edge of the fuselage, and x_o is the axial location where the fluid flow over the fuselage ceases to be potential. Approximately, this happens at the axial location where the local cross-sectional area $S_B(x)$ assumes a maximum value. In other words, $x = x_o$ when $S_B = S_{B,\max}$.

For supersonic speeds,

$$(C'_{m\alpha})_B = (C_{m\alpha})_B \tag{4.517}$$

However, suitable engineering methods are not available for estimating $(C_{m\alpha})_B$ at supersonic speeds for arbitrary bodies. Approximate values of $(C_{m\alpha})_B$ can be obtained by approximating the given fuselage either as a tangent-ogive or a cone-cylinder, whichever is closer, and using the following relation:

$$(C_{m\alpha})_B = (C_{N\alpha})_B \left(\frac{x_m}{l_f} - \frac{x_{cp}}{l_f} \right) \tag{4.518}$$

where $(C_{N\alpha})_B$ and x_{cp} can be estimated using Fig. 3.10 or 3.11. Note that x_{cp} is the distance from nose to the center of pressure of the body.

Longitudinal Acceleration Derivatives

Estimation of $C_{L\dot{\alpha}}$ This derivative is a measure of the unsteady or time-lag effects in airflow on the lift coefficient when the angle of attack is changing with time, as in the case of an airplane oscillating in pitch. In such situations, the entire flow over the airplane is unsteady. As a result, the aerodynamic coefficients become functions of time. The derivative $C_{L\dot{\alpha}}$ is a measure of the unsteady flow effects on lift coefficient. Obviously, for steady-state flight condition, $C_{L\dot{\alpha}}$ is zero.

The time-lag effects in fluid flow become particularly important in cases involving flow separation or vortex shedding. To understand these concepts, consider an oscillating delta-wing model in a wind tunnel. Let us assume that this delta-wing model has sharp leading edges and the mean angle of attack about which it oscillates is sufficiently high. If this wing model was held at a fixed angle of attack, the vortex strength and vortex breakdown locations will be constant with respect to time. However, on the oscillating wing model, because of flow inertia, the vortex strength and vortex breakdown locations cannot keep pace with angle of attack. As a result, they lag behind, i.e., at any given time, the vortex strength and vortex breakdown locations will not be corresponding to the instantaneous wing angle of attack $\alpha(t)$ but will correspond to $\alpha(t - \Delta t)$, an angle of attack at an earlier time $t - \Delta t$. Similarly, if the flow separates over the upper or lower surface of an oscillating round leading-edge wing, the location of the separation point will lag behind. Because the lift coefficient depends on the vortex strength, vortex breakdown position, or the location of flow separation point, the derivative $C_{L\dot{\alpha}}$ will be nonzero. It may assume a significant numerical value in such cases. On the other hand, if the flow was completely attached with no vortex shedding or flow separation from either the upper or lower surface, then the time-lag effects arise mainly due to flow taking a finite amount of time to adjust from one attached flow condition to the other. Usually, such time lag effects are very small and $C_{L\dot{\alpha}}$ may assume a small numerical value.

For an oscillating airplane, even if the wing flow is completely attached and its time-lag effects are small, the horizontal tail can experience significant time-lag effects. These time-lag effects arise because the wing downwash field takes a certain finite amount of time to reach the aft-located horizontal tail surface. During this time interval, the wing angle attack will have assumed a new value. As a result, the horizontal tail flow field lags behind that of the wing.

The instantaneous angle of attack of the horizontal tail is given by

$$\alpha_t(t) = \alpha_w(t) - i_w + i_t - \epsilon(t) \tag{4.519}$$

Noting that $\alpha_w = \alpha$, the downwash angle $\epsilon(t)$ can be expressed as

$$\epsilon(t) = \frac{d\epsilon}{d\alpha}\alpha(t - \Delta t) \tag{4.520}$$

where Δt is time lag or the time taken by a fluid element to travel from wing surface to the horizontal tail surface.

Using Taylor series expansion and considering only first-order time-lag effects, we have

$$\alpha(t - \Delta t) = \alpha(t) - \dot{\alpha}\Delta t \tag{4.521}$$

Approximately, we can assume $\Delta t = l_t/U_o$. Thus, the larger the tail length l_t is, the longer will be the time delay. With this assumption,

$$\alpha_t(t) = \alpha(t) - i_w + i_t - \frac{d\epsilon}{d\alpha}\left(\alpha(t) - \frac{\dot{\alpha}l_t}{U_o}\right) \tag{4.522}$$

and

$$C_{Lt}(t) = a_t\frac{S_t}{S}\eta_t\left[\alpha(t) - i_w + i_t - \frac{d\epsilon}{d\alpha}\left(\alpha(t) - \frac{\dot{\alpha}l_t}{U_o}\right)\right] \tag{4.523}$$

The derivative $C_{L\dot{\alpha}}$ is defined as

$$C_{L\dot{\alpha}} = \frac{\partial C_L}{\partial\left(\frac{\dot{\alpha}\bar{c}}{2U_o}\right)} \tag{4.524}$$

Differentiating both sides of Eq. (4.523) with respect to α and simplifying,

$$(C_{L\dot{\alpha}})_t = 2a_t\overline{V}_1\eta_t\left(\frac{d\epsilon}{d\alpha}\right) \tag{4.525}$$

The horizontal tail lift-curve slope a_t and the downwash gradient $(d\epsilon/d\alpha)$ at the horizontal tail location can be estimated using the methods discussed in Chapter 3.

From Eq. (4.525), we observe that the tail contribution to acceleration derivative $(C_{L\dot{\alpha}})_t$ depends on the tail volume ratio $\overline{V}_1$ and the downwash velocity gradient $(d\epsilon/d\alpha)$. Recall that $\overline{V}_1$ increases with an increase in tail area and tail length. The downwash gradient depends on the aspect ratio, sweep, and taper ratio of the wing. The higher the wing aspect ratio, the lower the value of the downwash gradient.

For more accurate analysis, it may be desirable to include the contribution of the wing–body combination so that the total value of the derivative $C_{L\dot{\alpha}}$ is given by

$$C_{L\dot{\alpha}} = (C_{L\dot{\alpha}})_{WB} + (C_{L\dot{\alpha}})_t \tag{4.526}$$

The wing–body contribution (based on wing area S and mean aerodynamic chord $\bar{c}$) can be estimated using the following expression [7]:

$$(C_{L\dot{\alpha}})_{WB} = \left[K_{W(B)} + K_{B(W)}\right]\left(\frac{S_e\,\bar{c}_e}{S\,\bar{c}}\right)(C_{L\dot{\alpha}})_e + (C_{L\dot{\alpha}})_B\frac{S_{B,\max}l_f}{S\,\bar{c}} \Big/ \text{rad} \tag{4.527}$$

where $(C_{L\dot{\alpha}})_e$ is the contribution of the exposed wing and $(C_{L\dot{\alpha}})_B$ is that of the isolated body. These two parameters can be determined as follows.

For subsonic speeds [7],

$$(C_{L\dot{\alpha}})_e = 1.5\left(\frac{x_{ac}}{c_r}\right)_e(C_{L\alpha})_e + 3C_L(g)\,\text{rad} \tag{4.528}$$

It may be noted that Eq. (4.528) was derived for a triangular wing [7]. However, no such relation is available for other planforms. In view of this, it is suggested that Eq. (4.528) should be used only to get crude estimates of $C_{L\dot{\alpha}}$ of other wing planforms. The lift-curve slope of the exposed wing $(C_{L\alpha})_e$ can be estimated using the methods discussed in Chapter 3 using exposed wing area, exposed taper ratio, and exposed aspect ratio. The function $C_L(g)$ can be obtained using the following expression, which is curve fit to the Datcom data [7]:

$$C_L(g) = \left(\frac{-\pi A_e}{2\beta^2}\right)(0.0013\tau^4 - 0.0122\tau^3 + 0.0317\tau^2 + 0.0186\tau - 0.0004) \tag{4.529}$$

where $\tau = \beta A_e$, A_e is the exposed wing aspect ratio, and $\beta = \sqrt{1 - M^2}$.

For supersonic speeds, the Datcom [7] procedure for evaluating $(C_{L\dot{\alpha}})_e$ is quite involved and is not presented here. Instead, we have performed calculations using Datcom [7] data for some typical wing planforms and presented these data in Fig. 4.22. It may be observed that the taper ratio has very little influence, whereas the leading-edge sweep and aspect ratio have a large effect on $(C_{L\dot{\alpha}})_e$.

The parameter $(C_{L\dot{\alpha}})_B$ at both subsonic and supersonic speeds (based on $S_{B,max}l_f$) is given by [7]

$$(C_{L\dot{\alpha}})_B = 2(C'_{L\alpha})_B\left(\frac{V_B}{S_{B,\max}l_f}\right) \tag{4.530}$$

For subsonic speeds,

$$(C'_{L\alpha})_B = (C_{L\alpha})_B\left(\frac{V_B^{2/3}}{S_{B,\max}}\right) \tag{4.531}$$

$$(C_{L\alpha})_B = 2(k_2 - k_1)\left(\frac{S_{B,\max}}{V_B^{2/3}}\right) \tag{4.532}$$

where the fuselage apparent mass coefficient $(k_2 - k_1)$ can be obtained using the data presented in Fig. 4.6.

For supersonic speeds,

$$(C'_{L\alpha})_B = (C_{N\alpha})_B \tag{4.533}$$

where $(C_{N\alpha})_B$ can be estimated by approximating the given fuselage either as a tangent-ogive or a cone-cylinder, whichever is closer, and using Fig. 3.10 or 3.11. Note that the value of $C_{N\alpha}$ given in Figs. 3.10 and 3.11 is based on $S_{B,\max}$.

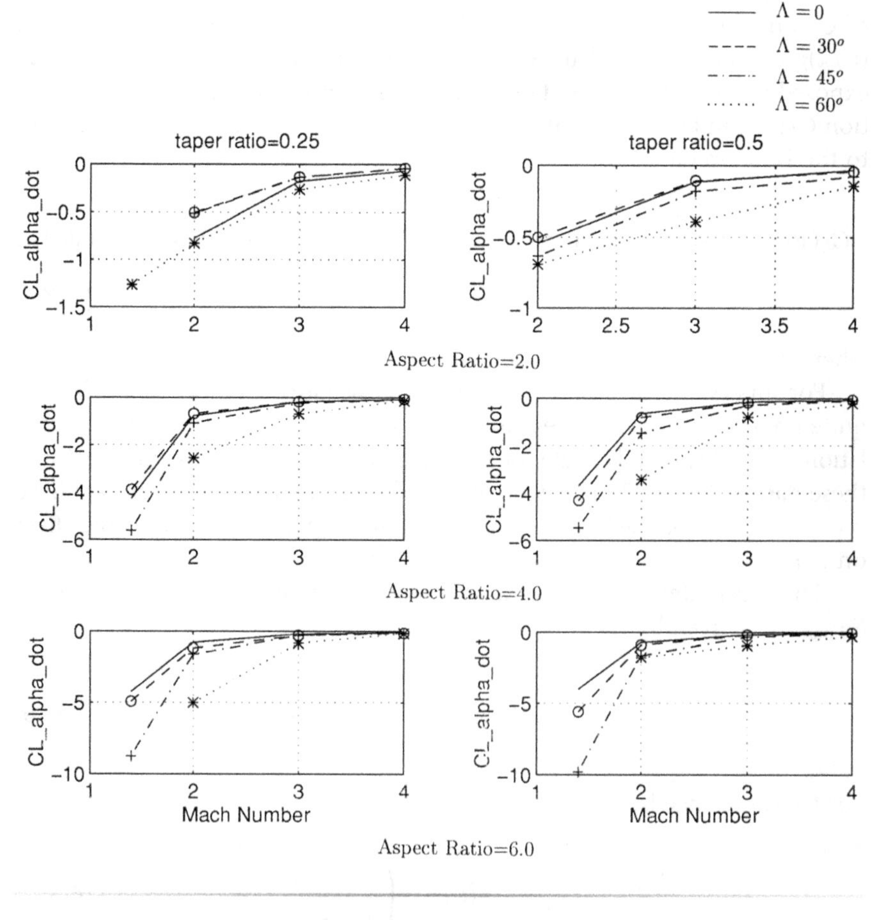

Fig. 4.22 Parameter $(C_{L\dot{\alpha}})_e$ for supersonic speeds (7).

Estimation of $C_{m\dot{\alpha}}$ This derivative is a measure of the time-lag effects on the pitching moment when the aircraft experiences a change in the angle of attack with respect to time. As said before in our discussion on $C_{L\dot{\alpha}}$, the horizontal tail contribution is usually the most significant one. From Eq. (4.523), we have

$$C_{Lt}(t) = a_t\left(\frac{S_t}{S}\right)\eta_t\left[\alpha(t) - i_w + i_t - \frac{d\epsilon}{d\alpha}\left(\alpha(t) - \frac{\dot{\alpha}l_t}{U_o}\right)\right] \tag{4.534}$$

Taking moments about the center of gravity,

$$C_{mt}(t) = -a_t\left(\frac{S_t l_t}{S\bar{c}}\right)\eta_t\left[\alpha(t) - i_w + i_t - \frac{d\epsilon}{d\alpha}\left(\alpha_w(t) - \frac{\dot{\alpha}l_t}{U_o}\right)\right] \tag{4.535}$$

The derivative $C_{m\dot{\alpha}}$ is defined as

$$C_{m\dot{\alpha}} = \frac{\partial C_m}{\partial\left(\frac{\dot{\alpha}\bar{c}}{2U_o}\right)} \tag{4.536}$$

Differentiating Eq. (4.535) with respect to $\dot{\alpha}$ and simplifying,

$$(C_{m\dot{\alpha}})_t = -2a_t\overline{V}_1\eta_t\left(\frac{d\epsilon}{d\alpha}\right)\left(\frac{l_t}{\bar{c}}\right) \tag{4.537}$$

For more accurate analysis, the contribution of the wing–body combination can also be included. The total value of $C_{m\dot{\alpha}}$ for the complete airplane is given by

$$C_{m\dot{\alpha}} = (C_{m\dot{\alpha}})_{WB} + (C_{m\dot{\alpha}})_t \tag{4.538}$$

The wing–body contribution can be estimated using the following expression [7]:

$$(C_{m\dot{\alpha}})_{WB} = \left[K_{W(B)} + K_{B(W)}\right]\left(\frac{S_e\bar{c}_e^2}{S\bar{c}^2}\right)(C_{m\dot{\alpha}})_e + (C_{m\dot{\alpha}})_B\frac{S_{B,\max}l_f^2}{S\bar{c}^2}\Big/\text{rad} \tag{4.539}$$

where $(C_{m\dot{\alpha}})_e$ is the contribution of the exposed wing and $(C_{m\dot{\alpha}})_B$ is that of the isolated body.

For both subsonic and supersonic speeds [7],

$$(C_{m\dot{\alpha}})_e = (C''_{m\dot{\alpha}})_e + \left(\frac{x_{cg,le}}{\bar{c}_e}\right)(C_{L\dot{\alpha}})_e\Big/\text{rad} \tag{4.540}$$

where $x_{cg,le}$ is the distance of the center of gravity from the leading edge of the exposed wing, positive aft (see the insert at the top of Fig. 4.20), and $(C_{L\dot{\alpha}})_e$ is to be evaluated as discussed earlier.

For subsonic speeds,

$$(C''_{m\dot{\alpha}})_e = -\left(\frac{81}{32}\right)\left(\frac{x_{ac}}{c_r}\right)_e^2 (C_{L\alpha})_e + \frac{9}{2} C_{mo}(g) \Big/ \text{rad} \tag{4.541}$$

It may be noted that Eq. (4.541) was derived for a triangular wing [7]. Such relations are not available for wings of other planform. Therefore, it is suggested that Eq. (4.541) may be used only to get crude estimates for wings of other planform. The function $C_{mo}(g)$ can obtained using the following expression, which is a curve fit to Datcom data [7]:

$$C_{mo}(g) = \left(\frac{\pi A_e}{2\beta^2}\right)(0.0008\tau^4 - 0.0075\tau^3 + 0.0185\tau^2 + 0.0128\tau - 0.0003) \tag{4.542}$$

where $\tau = \beta A_e$, A_e is the exposed wing aspect ratio, and $\beta = \sqrt{1 - M^2}$.

For supersonic speeds [7], the procedure of evaluating $(C''_{m\dot{\alpha}})_e$ is quite involved and is not discussed here. Instead, using Datcom [7] data, we have performed some calculations and presented the data on $(C''_{m\dot{\alpha}})_e$ for some typical wing planforms in Fig. 4.23. As noted in the case of $(C''_{L\dot{\alpha}})_e$, the wing leading-edge sweep and wing aspect ratio have more significant influence, and the wing taper ratio has very little effect on $(C''_{m\dot{\alpha}})_e$.

The parameter $(C_{m\dot{\alpha}})_B$ can be evaluated as follows [7].

For both subsonic and supersonic speeds,

$$(C_{m\dot{\alpha}})_B = 2(C'_{m\alpha})_B \left[\frac{x_{c1} - x_{m1}}{1 - x_{m1} - V_{B1}}\right]\left(\frac{V_B}{S_{B,\max} l_f}\right) \tag{4.543}$$

where x_{c1}, x_{m1}, and V_{B1} are defined in Eqs. (4.513) and (4.514). The parameter $(C'_{m\alpha})_B$ can be determined using Eqs. (4.515) and (4.517).

4.4.2 Lateral-Directional Derivatives

Rotary Derivatives

Estimation of C_{yp} This derivative is a measure of the side force induced due to a roll rate experienced by the airplane. Generally, the contribution of the fuselage and horizontal tail are very small and can be ignored. The major contributions come from the wing and the vertical tail so that

$$C_{yp} = (C_{yp})_W + (C_{yp})_V \tag{4.544}$$

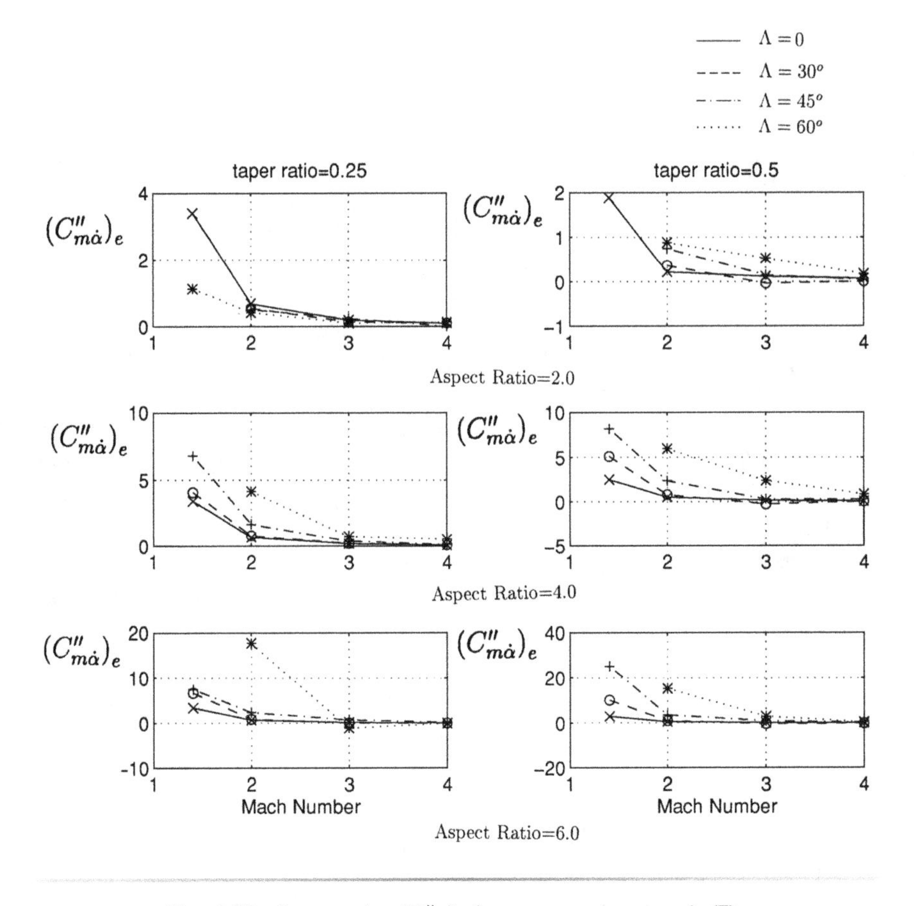

Fig. 4.23 Parameter $(C''_{m\dot{\alpha}})_e$ for supersonic speeds (7).

Consider an airplane in rolling motion with $p > 0$. Suppose we ignore the effect of wing sidewash, then the isolated vertical tail experiences an approximate increase in local sideslip

$$(\Delta\beta)_V = \frac{pz_v}{U_o}$$

and

$$(C_{yp})_V = \frac{z_v}{b} C_{y\beta,V}$$

However, according to Datcom [7], in the presence of wing sidewash, the contribution of the vertical tail is modified and is given by the following expression which is based on the approximation that $(C_{yp})_V$ vanishes at zero angle of attack.

$$(C_{yp})_V = \frac{2(z - z_v)}{b} C_{y\beta,V} \tag{4.545}$$

where

$$z = z_v \cos\alpha - l_v \sin\alpha \tag{4.546}$$

and

$$C_{y\beta,V} = -ka_v\left(1 + \frac{\partial\sigma}{\partial\beta}\right)\eta_v \frac{S_v}{S} \tag{4.547}$$

Here, k is an empirical parameter (Fig. 3.75), z_v is the vertical distance between the aerodynamic center of the vertical tail and the center of gravity measured perpendicular to the fuselage centerline, l_v is the corresponding horizontal distance measured parallel to the fuselage centerline (Fig. 3.102), a_v is the vertical tail lift-curve slope, and $(1 + [\partial\sigma/\partial\beta])$ is the fuselage sidewash parameter. The methods to evaluate a_v and $(1 + [\partial\sigma/\partial\beta])$ were discussed in Chapter 3. Note that Eq. (4.547) is applicable only for subsonic speeds [7]. For supersonic speeds, no general method is available for the estimation of $(C_{yp})_V$ [7].

The wing contribution $(C_{yp})_W$ may be obtained for subsonic speeds as follows [7].

$$(C_{yp})_W = K\left(\frac{C_{yp}}{C_L}\right)_{C_L=0,M} C_L + (\Delta C_{yp})_\Gamma \Big/ \text{rad} \tag{4.548}$$

where

$$K = \frac{1 - a_{w1}}{1 - a_{w2}} \qquad a_{w1} = \frac{(C_{L\alpha})_e}{\pi A e} \qquad a_{w2} = e a_{w1} \tag{4.549}$$

$$\left(\frac{C_{yp}}{C_L}\right)_{C_L=0,M} = \frac{(A + B\cos\Lambda_{c/4})(AB + \cos\Lambda_{c/4})}{(AB + 4\cos\Lambda_{c/4})(A + \cos\Lambda_{c/4})}\left(\frac{C_{yp}}{C_L}\right)_{C_L=0,M=0} \tag{4.550}$$

$$B = \sqrt{1 - M^2\cos^2\Lambda_{c/4}} \tag{4.551}$$

Here, $A = A_e$ the exposed wing aspect ratio. The planform efficiency factor e can be evaluated using Eq. (4.478). Data for $(C_{yp}/C_L)_{C_L=0,M=0}$ for some typical wing planforms are presented in Fig. 4.24.

Furthermore,

$$(\Delta C_{yp})_\Gamma = 3\sin\Gamma\left[1 - \frac{4z}{b}\sin\Gamma\right](C_{lp})_{\Gamma=0,C_L=0} \Big/ \text{rad} \tag{4.552}$$

where

$$(C_{lp})_{\Gamma=0,C_L=0} = \left(\frac{\beta C_{lp}}{k}\right)_{C_L=0}\frac{k}{\beta} \tag{4.553}$$

$$k = \frac{a_o}{2\pi} \tag{4.554}$$

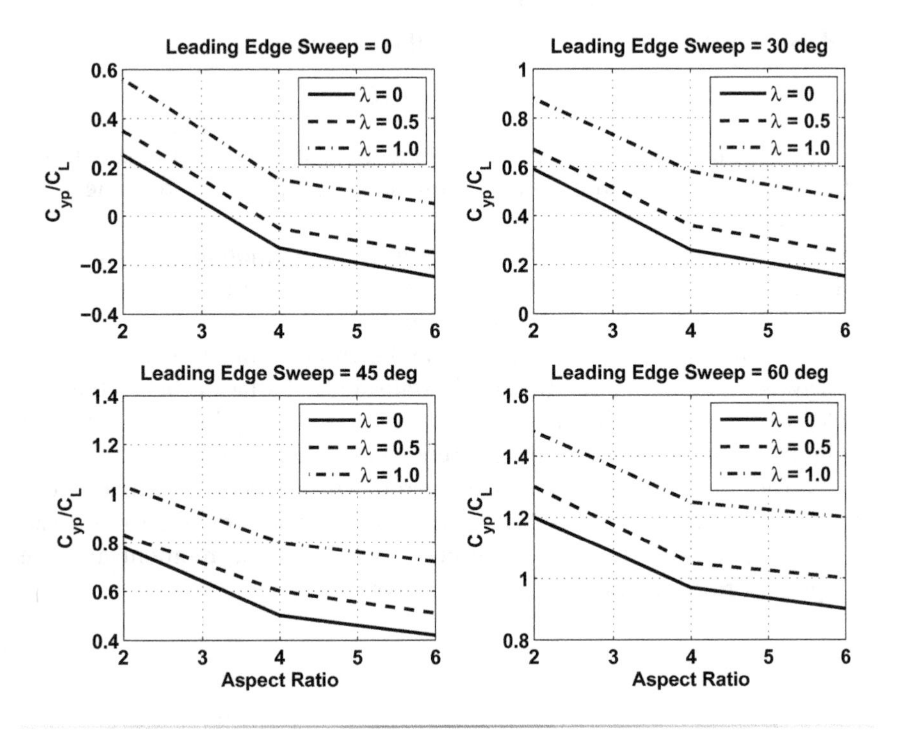

Fig. 4.24 Parameter $(C_{yp}/C_L)_{C_L=0,M=0}$ speeds (7).

Here, $\beta = \sqrt{1 - M^2}$, Γ is the wing dihedral in deg, a_o is the sectional or two dimensional lift-curve slope of the wing at low subsonic speeds and can be evaluated using the methods presented previously in Chapter 3.

Values of $(\beta C_{lp}/k)_{C_L=0}$ for typical wing planforms are presented in Fig. 4.25.

For supersonic speeds, no general method is available to estimate $(C_{yp})_W$. Using the information given in Datcom [7], calculations were performed for typical wing planforms and the results in terms of $(C_{yp}/\alpha)_W$ per rad^2 are presented in Fig. 4.26.

Estimation of C_{lp} This derivative is a measure of the rolling moment induced due to a roll rate experienced by the aircraft and is called the damping-in-roll derivative. This is one of the most important lateral-directional dynamic derivatives. The major contribution comes from the wing and the vertical tail, and the contributions of the fuselage and horizontal tail are usually small and can be ignored. However, the contribution of the horizontal tail can become significant if it is comparable in size to the wing. In that case, the same approach as that used for the wing can be used to estimate its contribution. With these assumptions,

$$C_{lp} = (C_{lp})_W + (C_{lp})_V \tag{4.555}$$

For aircraft with high-aspect ratio rectangular wings, an approximate estimation of the wing contribution at low subsonic speeds can be done using the strip theory as follows.

Consider a rectangular wing in a uniform rolling motion with a positive roll rate p about the Ox axis as shown in Fig. 4.27a. Because of this rolling motion, the local angle of attack of wing sections on the down-going or right wing increases (upwash) and that on the up-going or left wing decreases (downwash). Assuming that the steady-state angle of attack is below the stall angle, we observe that the lift developed by the right wing increases and that of the left wing decreases. This difference in lift gives rise to a restoring or a negative rolling moment and is the aerodynamic mechanism that generates the damping-in-roll because of the wing. Using strip theory, we can approximately estimate the magnitude of this restoring moment and the damping-in-roll derivative $(C_{lp})_W$ as follows.

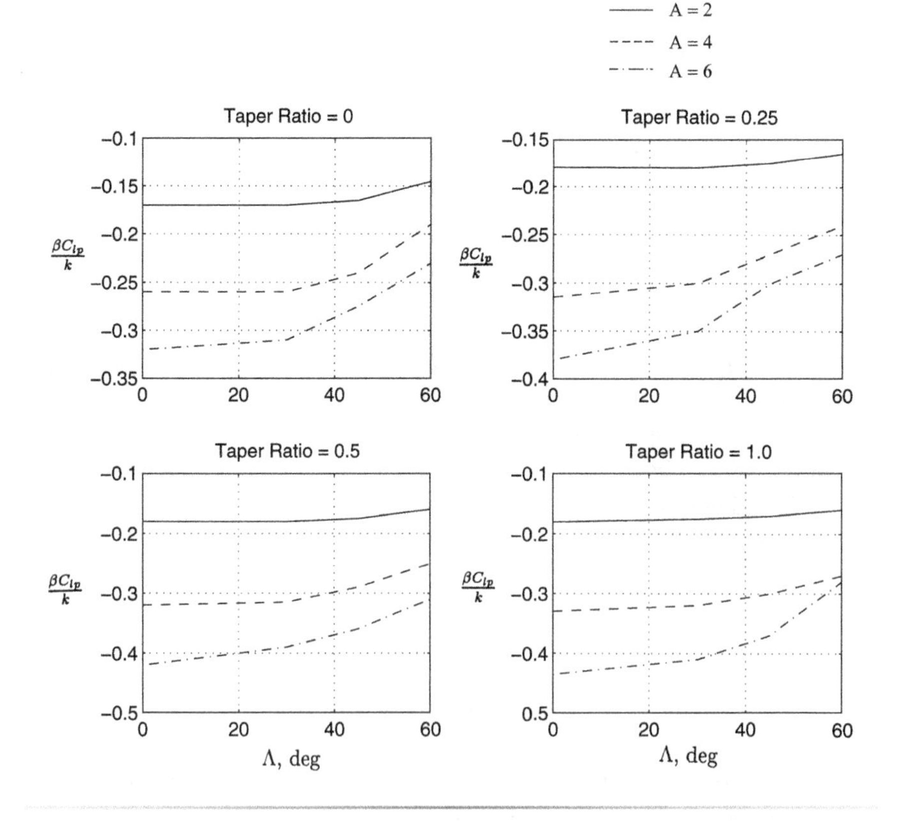

Fig. 4.25 Parameter $(\beta C_{lp}/k)_{C_L} = 0$ at subsonic speeds (7).

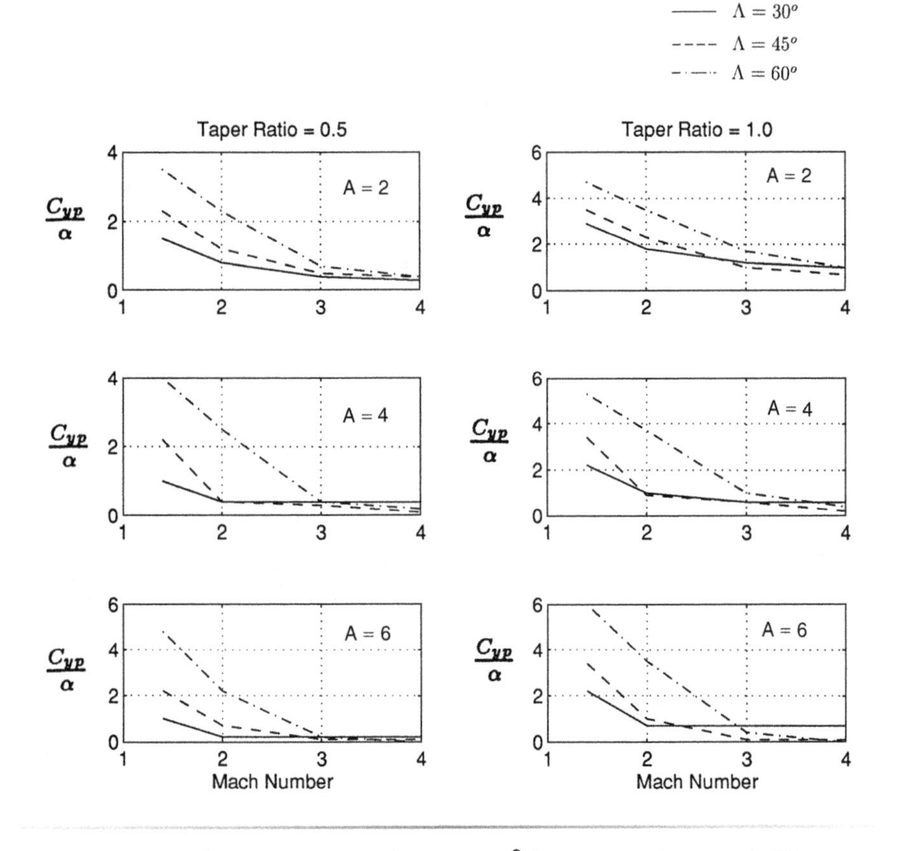

Fig. 4.26 Parameter $(C_{yp}/\alpha)_W$ per rad^2 for supersonic speeds (7).

The increase in the local angle of attack of an elemental wing strip RT of width dy at a spanwise distance y on the right wing is given by

$$\tan \alpha_p = \frac{py}{U_o} \tag{4.556}$$

where U_o is the flight velocity. Assuming that the roll rate p is small, we obtain

$$\alpha_p = \frac{py}{U_o} \tag{4.557}$$

so that the local angle of attack of the strip RT

$$\alpha_l(y) = \alpha + \alpha_p = \alpha + \frac{py}{U_o} \tag{4.558}$$

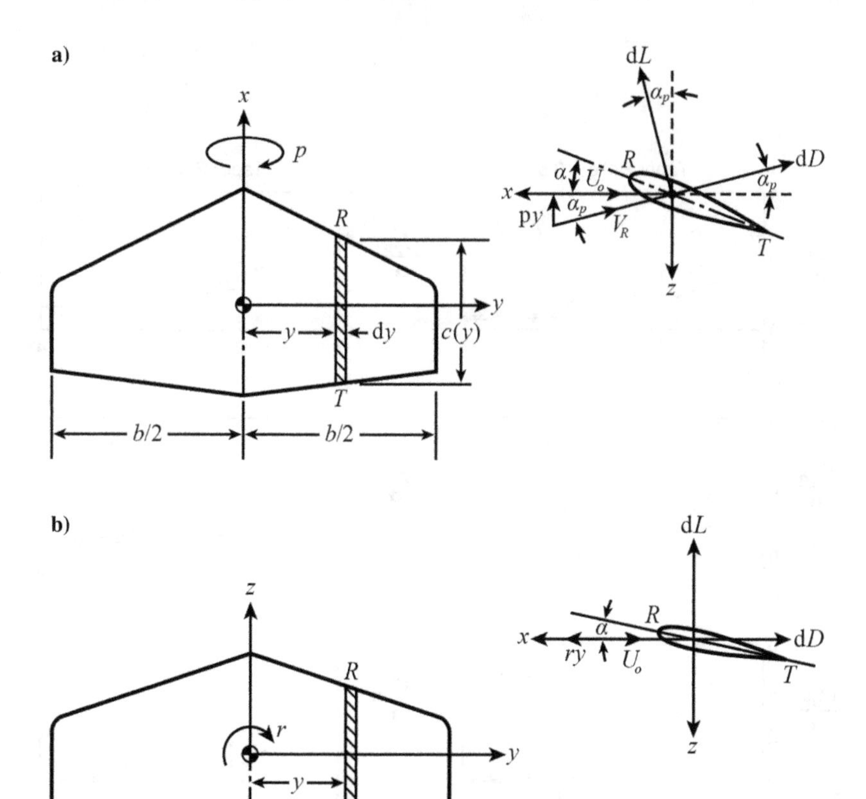

Fig. 4.27 Wing in rolling and yawing motion.

The resultant velocity at the strip RT is given by

$$V_R = \sqrt{U_o^2 + (py)^2} \tag{4.559}$$

$$\simeq U_o \tag{4.560}$$

Let $c(y)$ be the local chord and $a_o(y)$ be the local sectional lift-curve slope of the wing. The lift on the elemental strip RT is given by

$$\mathrm{d}L = \frac{1}{2}\rho U_o^2 a_o(y) \alpha_l c(y)\,\mathrm{d}y \tag{4.561}$$

$$= \frac{1}{2}\rho U_o^2 a_o(y)\left[\alpha + \frac{py}{U_o}\right] c(y)\,\mathrm{d}y \tag{4.562}$$

The drag on the elemental strip RT is given by

$$dD = \frac{1}{2}\rho U_o^2 C_{D,l} c(y)\,dy \tag{4.563}$$

where

$$C_{D,l} = C_{Do,l} + C_{D\alpha,l}\alpha \tag{4.564}$$

Here, $C_{Do,l}$ is the sectional zero-lift drag coefficient (profile drag coefficient when $C_L = 0$), and $C_{D\alpha,l}$ is the incremental sectional profile drag coefficient per unit angle of attack. Note that the strip theory ignores the induced drag.

The normal force on the elemental strip RT is given by

$$dZ = -dL\cos\alpha_p - dD\sin\alpha_p \tag{5.565}$$

$$= -dL - dD\alpha_p \tag{5.566}$$

$$= -\frac{1}{2}\rho U_o^2\left[a_o(y)\alpha + [a_o(y) + C_{D,l}]\left(\frac{py}{U_o}\right)\right]c(y)\,dy \tag{5.567}$$

The rolling moment due to the elemental strip RT is given by

$$dRM = -\frac{1}{2}\rho U_o^2\left[a_o(y)\alpha + [a_o(y) + C_{D,l}]\left(\frac{py}{U_o}\right)\right]c(y)y\,dy \tag{4.568}$$

The total rolling moment due to the right wing is given by

$$RM_R = -\frac{1}{2}\rho U_o^2\int_0^{\frac{b}{2}}\left[a_o(y)\alpha + [a_o(y) + C_{D,l}]\left(\frac{py}{U_o}\right)\right]c(y)y\,dy \tag{4.569}$$

Similarly, the rolling moment due to the left wing (change $+y$ to $-y$) is given by

$$RM_L = -\frac{1}{2}\rho U_o^2\int_0^{\frac{b}{2}}\left[a_o(y)\alpha - [a_o(y) + C_{D,l}]\left(\frac{py}{U_o}\right)\right]c(y)y\,dy \tag{4.569}$$

The total rolling moment

$$RM = -\rho U_o^2\left(\frac{p}{U_o}\right)\int_0^{\frac{b}{2}}[a_o(y) + C_{D,l}]c(y)y^2\,dy \tag{4.571}$$

We have

$$C_l = \frac{RM}{\frac{1}{2}\rho U_o^2 Sb} \tag{4.572}$$

$$C_{lp} = \frac{\partial C_l}{\partial\left(\frac{p\bar{b}}{2U_o}\right)} \tag{4.573}$$

Then,

$$(C_{lp})_W = \frac{-4}{Sb^2}\int_0^{\frac{b}{2}} [a_o(y) + C_{D,l}]c(y)y^2\, dy \tag{4.574}$$

For an untwisted rectangular wing with a constant chord and an identical airfoil section along the span, Eq. (4.574) reduces to the following form:

$$(C_{lp})_W = -\frac{1}{6}(a_o + C_{D,l}) \tag{4.575}$$

Usually, $|a_o(y)| > C_{D,l}$ so that the sign of C_{lp} depends on the sign of $a_o(y)$. Therefore, for angles of attack below stall where the lift-curve slope a_o is positive, $(C_{lp})_W < 0$, i.e., the wing provides a positive damping or simply damping effect. For angles of attack above the stall angle, the lift coefficient decreases with an increase in the angle of attack so that a_o becomes negative and $(C_{lp})_W > 0$. This loss of damping-in-roll is one of the main causes of autorotation of the wings in poststall region and aircraft spin entry, which we will discuss in Chapter 7.

The given result based on strip theory is very approximate because the strip theory ignores the induced drag and mutual interference effect between adjacent wing sections. As a result, the strip theory prediction becomes increasingly erroneous as the wing aspect ratio becomes smaller. For such configurations, the following Datcom [7] relation can be used:

$$(C_{lp})_W = \left(\frac{\beta C_{lp}}{k}\right)_{C_L=0}\left(\frac{k}{\beta}\right)\left(\frac{(C_{lp})_\Gamma}{(C_{lp})_{\Gamma=0}}\right)\Big/\text{rad} \tag{4.576}$$

In Eq. (4.576), it is assumed that the angle of attack is in the linear range or $C_L = a_w\alpha$ and the effect of drag force on the rolling moment is ignored. Parameter $(C_{lp})_\Gamma/(C_{lp})_{\Gamma=0}$ is given by the following relation [7]:

$$\frac{(C_{lp})_\Gamma}{(C_{lp})_{\Gamma=0}} = (1 - 2z'\sin\Gamma + 3z'^2\sin^2\Gamma)/\text{rad} \tag{4.577}$$

where

$$z' = \frac{2z_\omega}{b} \tag{4.578}$$

Here, z_W is the vertical distance between the center of gravity and the wing root chord, positive for center of gravity above the root chord.

The data to estimate $(\beta C_{lp}/k)_{C_L=0}$ are presented in Fig. 4.25 for typical wing planforms. These data were calculated for $M = 0$ and hence, are applicable only for low subsonic speeds. For high subsonic speeds, the interested reader may refer to Datcom [7].

The vertical tail contribution, $(C_{lp})_V$ is given by

$$(C_{lp})_V = \left|2\left(\frac{z}{b}\right)\left(\frac{z - z_v}{b}\right)\right| C_{y\beta,V} \tag{4.579}$$

where

$$z = z_v \cos \alpha - l_v \sin \alpha \tag{4.580}$$

Here, z_v is the vertical distance between the aerodynamic center of the vertical tail and the center of gravity and is measured perpendicular to the fuselage centerline, l_v is the horizontal distance between the aerodynamic center of the vertical tail and the center of gravity and is measured parallel to the fuselage centerline. The parameter $C_{y\beta,V}$ can be obtained using Eq. (4.547).

For supersonic speeds, no general method suitable for engineering purposes is available for estimating the contributions of the wing and the vertical tail to damping-in-roll derivative. Datcom [7] presents data for some selected wing planforms. Interested readers may refer to Datcom [7].

Estimation of C_{np} This derivative is a measure of the yawing moment induced due to a roll rate experienced by the aircraft. The contributions of the fuselage and horizontal tail to C_{np} are usually small and can be ignored. The contributions mainly come from the wing and the vertical tail so that

$$C_{np} = (C_{np})_W + (C_{np})_V \tag{4.581}$$

For low subsonic speeds, an approximate estimation of the wing contribution can be done using the strip theory as follows.

Consider once again the strip RT on right wing (Fig. 4.27a). The force in the Ox direction is given by

$$dF = dL \sin \alpha_p - dD \cos \alpha_p \tag{4.582}$$

$$= \frac{1}{2}\rho U_o^2 [a_o(y)(\alpha + \alpha_p)\alpha_p - (C_{Do,l} + C_{D\alpha,l}(\alpha + \alpha_p))] c(y)\, dy \tag{4.583}$$

$$= \frac{1}{2}\rho U_o^2 [-C_{D,l} + (a_o(y)\alpha - C_{D\alpha,l})\alpha_p] c(y)\, dy \tag{4.584}$$

Here, we have assumed α_p is small and ignored higher-order terms like α_p^2. Then

$$C_{D,l} = C_{Do,l} + C_{D\alpha,l}\alpha \tag{4.585}$$

Substituting $\alpha_p = py/U_o$, the yawing moment developed by the elemental strip RT is given by

$$dYM = -\frac{1}{2}\rho U_o^2 \left[-C_{D,l} + \left[a_o(y)\alpha - C_{D\alpha,l}\right]\frac{py}{U_o}\right] c(y) y\, dy \tag{4.586}$$

The total yawing moment due to the right wing is given by

$$YM_R = -\frac{1}{2}\rho U_o^2 \int_0^{\frac{b}{2}} \left[-C_{D,l} + [a_o(y)\alpha - C_{D\alpha,l}] \frac{py}{U_o} \right] c(y) y \, dy \tag{4.587}$$

Similarly, the yawing moment developed by the left wing (change $+y$ to $-y$) is given by

$$YM_L = \frac{1}{2}\rho U_o^2 \int_0^{\frac{b}{2}} \left[-C_{D,l} - [a_o(y)\alpha - C_{D\alpha,l}] \frac{py}{U_o} \right] c(y) y \, dy \tag{4.588}$$

The total or net yawing moment, which is the sum of the right and left wing yawing moments, is given by

$$YM = -\rho U_o^2 \int_0^{\frac{b}{2}} [a_o(y)\alpha - C_{D\alpha,l}] \left(\frac{py}{U_o} \right) c(y) y \, dy \tag{4.589}$$

We have

$$C_n = \frac{YM}{\frac{1}{2}\rho U_o^2 Sb} \tag{4.590}$$

$$C_{np} = \frac{\partial C_n}{\partial \left(\frac{pb}{2U_o} \right)} \tag{4.591}$$

so that

$$(C_{np})_W = \frac{-4}{Sb^2} \int_0^{\frac{b}{2}} [a_o(y)\alpha - C_{D\alpha,l}] c(y) y^2 \, dy \tag{4.592}$$

For an untwisted rectangular wing with a constant chord and an identical airfoil section along the span, Eq. (4.592) reduces to

$$(C_{np})_W = -\frac{1}{6}(C_L - C_{D\alpha})/\text{rad} \tag{4.593}$$

Usually, for an airplane operating at an angle of attack below the stall, $C_L > C_{D\alpha}$ and the wing contribution $(C_{np})_W$ is negative. In other words, if the aircraft has a positive roll rate, then the wing develops a yawing moment that tends to yaw the aircraft to the left.

It may be recalled once again that the strip theory ignores the induced drag effects and the mutual interference between adjacent wing sections. Hence, the given strip theory prediction of $(C_{np})_W$ is quite approximate and will be in error if the wing aspect ratio is small.

For more accurate estimations, the following formula can be used for subsonic speeds [7]:

$$(C_{np})_W = C_{lp}\tan\alpha(K-1) + K\left(\frac{C_{np}}{C_L}\right)_{C_L=0,M} C_L \Big/ \text{rad} \tag{4.594}$$

where the parameter K is given by Eq. (4.549) and

$$\left(\frac{C_{np}}{C_L}\right)_{C_L=0,M} = \left(\frac{A+4\cos\Lambda_{c/4}}{AB+4\cos\Lambda_{c/4}}\right)\left[\frac{AB+0.5(AB+\cos\Lambda_{c/4})\tan^2\Lambda_{c/4}}{A+0.5(A+\cos\Lambda_{c/4})\tan^2\Lambda_{c/4}}\right] \times \left(\frac{C_{np}}{C_L}\right)_{C_L=M=0} \Big/ \text{rad} \tag{4.595}$$

where

$$B = \sqrt{1 - M^2\cos^2\Lambda_{c/4}} \tag{4.596}$$

$$\left(\frac{C_{np}}{C_L}\right)_{C_L=M=0} = -\left[\frac{A+6(A+\cos\Lambda_{c/4})\left(\dfrac{\xi\tan\Lambda_{c/4}}{A}+\dfrac{\tan^2\Lambda_{c/4}}{12}\right)}{6(A+4\cos\Lambda_{c/4})}\right] \tag{4.597}$$

Here, A is the exposed aspect ratio (A_e) and ξ is defined in Eq. (4.490).

The vertical tail contribution is given by

$$(C_{np})_V = -\left(\frac{2}{b}\right)(l_v\cos\alpha + z_v\sin\alpha)\left(\frac{z-z_v}{b}\right)C_{y\beta,V} \Big/ \text{rad} \tag{4.598}$$

where $C_{y\beta,V}$ can be obtained using Eq. (4.547). The parameters l_v, z_v, and z are defined in Eq. (4.546).

For supersonic speeds, no general method suitable for engineering purposes is available for estimation of $(C_{np})_W$ and $(C_{np})_V$. Datcom [7] gives some information and the interested reader may refer to it.

Estimation of C_{yr} This derivative is a measure of the side force induced due to yaw rate experienced by the aircraft. Generally, the contributions of the wing, fuselage, and horizontal tail are quite small and can be ignored. The only meaningful contribution comes from vertical tail, which can be estimated using the following formula [7]:

$$(C_{yr})_V = -\frac{2}{b}(l_v\cos\alpha + z_v\sin\alpha)C_{y\beta,V}/\text{rad} \tag{4.599}$$

where $C_{y\beta,V}$ is given by Eq. (4.547).

For supersonic speeds, no general method is available for the estimation of $(C_{yr})_V$ [7].

Estimation of C_{lr} This derivative is a measure of the rolling moment induced due to yaw rate experienced by the aircraft. Generally, the contributions of the fuselage and the horizontal tail surfaces are small and can be ignored. The contribution mainly comes from the wing and vertical tail surfaces so that

$$C_{lr} = (C_{lr})_W + (C_{lr})_V \tag{4.600}$$

For approximate purposes, we can use the strip theory to estimate the wing contribution at low subsonic speeds as follows.

Because of yaw rate, the relative velocity experienced by the wing sections varies along the span. Suppose the aircraft is experiencing a positive yaw rate. Then the sections on the right wing experience a decrease in the relative velocity, whereas those on the left wing experience an increase in the relative velocity. As a result, the lift on right wing is smaller compared to that on the left wing. This gives rise to a positive rolling moment. Thus, conceptually we see that $(C_{lr})_W > 0$.

The relative velocity experienced by the elemental strip RT (Fig. 4.27b) on the right wing is given by

$$V(y) = U_o - ry \tag{4.601}$$

so that

$$q_l = \frac{1}{2}(U_o - ry)^2 \tag{4.602}$$

$$\simeq \frac{1}{2}(U_o^2 - 2rU_o y) \tag{4.603}$$

The lift acting on the elemental strip RT is given by

$$\mathrm{d}L = \frac{1}{2}\rho(U_o^2 - 2rU_o y)a_o(y)\alpha c(y)\,\mathrm{d}y \tag{4.604}$$

Then the rolling moment due to the elemental strip RT is given by

$$\mathrm{d}RM = -\frac{1}{2}\rho(U_o^2 - 2rU_o y)a_o(y)\alpha c(y)y\,\mathrm{d}y \tag{4.605}$$

The rolling moment due to the right wing is given by

$$RM_R = -\frac{1}{2}\rho\int_0^{\frac{b}{2}}(U_o^2 - 2rU_o y)a_o(y)\alpha c(y)y\,\mathrm{d}y \tag{4.606}$$

Similarly, the rolling moment due to the left wing (change y to $-y$ is given by

$$RM_L = \frac{1}{2}\rho \int_0^{\frac{b}{2}} (U_o^2 + 2rU_o y) a_o(y) \alpha c(y) y \, dy \tag{4.607}$$

The total or net rolling moment, which is the sum of the right and left wing rolling moments,

$$RM = 2\rho U_o r \int_0^{\frac{b}{2}} a_o(y) \alpha c(y) y^2 \, dy \tag{4.608}$$

We have

$$C_l = \frac{RM}{\frac{1}{2}\rho U_o^2 S b} \tag{4.609}$$

$$C_{lr} = \frac{\partial C_l}{\partial \left(\frac{rb}{2U_o} \right)} \tag{4.610}$$

so that

$$(C_{lr})_W = \frac{8}{Sb^2} \int_0^{\frac{b}{2}} a_o(y) \alpha c(y) y^2 \, dy \tag{4.611}$$

For an untwisted rectangular wing of constant chord and constant airfoil section all along the span, Eq. (4.611) reduces to

$$(C_{lr})_W = \frac{C_L}{3} \tag{4.612}$$

Thus, for positive angles of attack, $(C_{lr})_W > 0$, which implies that if an aircraft experiences a positive yaw rate then the wing induces a positive rolling moment that tends to raise the left wing and lower the right wing.

The strip theory estimate ignores the induced drag effects and mutual interference between adjacent wing sections. Thus, the strip theory predictions will be in error for wings with small aspect ratio. For more accurate prediction of $(C_{lr})_W$ at subsonic speeds, the following Datcom [7] formula can be used:

$$(C_{lr})_W = C_L \left(\frac{C_{lr}}{C_L} \right)_{C_L = 0, M} + \left(\frac{\Delta C_{lr}}{\Gamma} \right) \Gamma \Big/ \text{rad} \tag{4.613}$$

where

$$\left(\frac{C_{lr}}{C_L} \right)_{C_L = 0, M} = \frac{\text{Num}}{\text{Den}} \left(\frac{C_{lr}}{C_L} \right)_{C_L = 0, M = 0} \tag{4.614}$$

$$\text{Num} = 1 + \frac{A(1 - B^2)}{2B\left(AB + 2\cos\Lambda_{c/4}\right)} + \left(\frac{AB + 2\cos\Lambda_{c/4}}{AB + 4\cos\Lambda_{c/4}}\right)\left(\frac{\tan^2\Lambda_{c/4}}{8}\right) \tag{4.615}$$

$$\text{Den} = 1 + \left(\frac{A + 2\cos\Lambda_{c/4}}{A + 4\cos\Lambda_{c/4}}\right)\left(\frac{\tan^2\Lambda_{c/4}}{8}\right) \tag{4.616}$$

The data to evaluate the parameter $(C_{lr}/C_L)_{C_L=0,M=0}$ are presented in Fig. 4.28. Furthermore,

$$\left(\frac{\Delta C_{lr}}{\Gamma}\right) = \left(\frac{1}{12}\right)\left(\frac{\pi A\sin\Lambda_{c/4}}{A + 4\cos\Lambda_{c/4}}\right) \Big/ \text{rad}^2 \tag{4.617}$$

Here, $A = A_e$, the exposed wing aspect ratio. The vertical tail contribution is given by [7]

$$(C_{lr})_V = \frac{-2}{b^2}(l_v\cos\alpha + z_v\sin\alpha)(z_v\cos\alpha - l_v\sin\alpha)C_{y\beta,V} \tag{4.618}$$

The value of $(C_{lr})_V$ given by Eq. (4.618) is per radian.

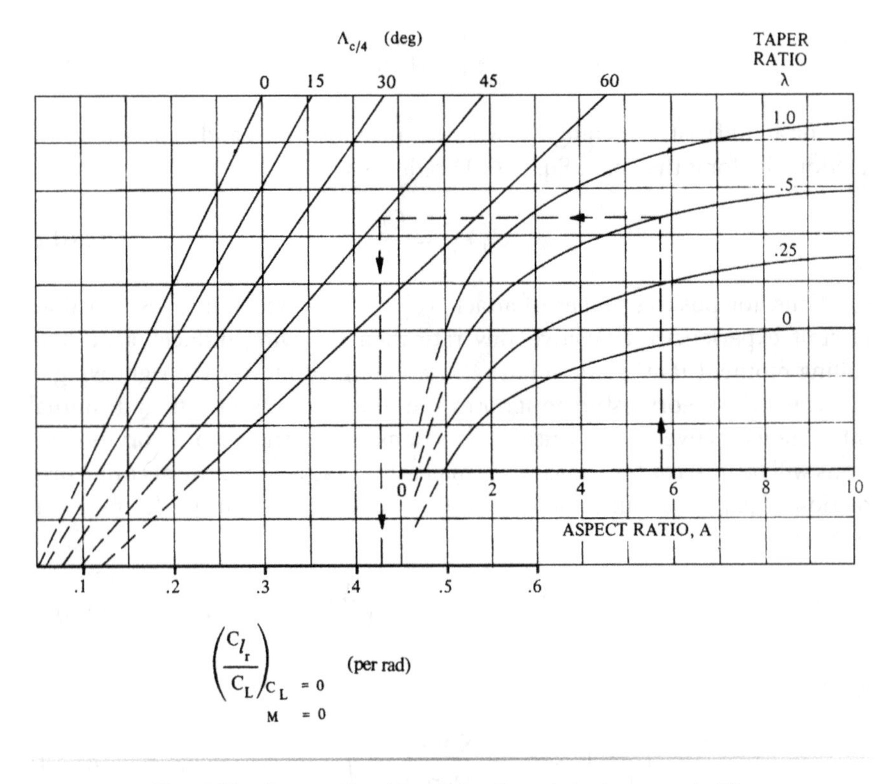

Fig. 4.28 Parameter $(C_{lr}/C_L)_W$ for subsonic speeds [7].

For supersonic speeds, no general method suitable for engineering purposes is available [7].

Estimation of C_{nr} This derivative is a measure of the yawing moment induced due to the yaw rate experienced by the aircraft and is known as the damping-in-yaw derivative. This is one of the most important lateral-directional dynamic stability derivatives. Generally, the contributions of the fuselage and horizontal tail are small and can be ignored so that

$$C_{nr} = (C_{nr})_W + (C_{nr})_V \tag{4.619}$$

The vertical tail contribution is given by

$$(C_{nr})_V = \frac{2}{b^2}(l_v \cos\alpha + z_v \sin\alpha)^2 C_{y\beta,V} \tag{4.620}$$

The value of $(C_{nr})_V$ given by Eq. (4.620) is per radian.

An approximate estimate of $(C_{nr})_W$ can be obtained using the strip theory as follows. Consider once again the elemental strip RT on the right wing of an aircraft undergoing yawing motion with a positive yaw rate r (Fig. 4.27b). As a result of this yawing motion, the relative velocity experienced by the wing sections varies along the span, but the angle of attack can be assumed to remain approximately constant. The relative air velocity experienced by the left wing sections is higher compared to that of the right wing sections. Consequently, the drag of the left wing is higher than that of the right wing, leading to a negative or restoring yawing moment. Thus, conceptually $(C_{nr})_W < 0$ or wing provides a damping effect in yaw.

We have

$$dF = -dD \tag{4.621}$$

$$= -\frac{1}{2}\rho(U_o^2 - 2U_o ry)C_{D,l}c(y)\,dy \tag{4.622}$$

The yawing moment developed by the elemental strip RT is given by

$$dYM = \frac{1}{2}\rho(U_o^2 - 2U_o ry)C_{D,l}c(y)y\,dy \tag{4.623}$$

The total yawing moment due to the right wing is given by

$$YM_R = \frac{1}{2}\rho\int_0^{\frac{b}{2}}(U_o^2 - 2U_o ry)C_{D,l}c(y)y\,dy \tag{4.624}$$

Similarly, the yawing moment due to the left wing (change y to $-y$) is given by

$$YM_L = -\frac{1}{2}\rho\int_0^{\frac{b}{2}} (U_o^2 + 2U_o ry)\, C_{D,l}c(y)y\, dy \tag{4.625}$$

so that the total yawing moment is given by

$$YM = -2\rho U_o r\int_0^{\frac{b}{2}} C_{D,l}c(y)y^2\, dy \tag{4.626}$$

We have

$$C_n = \frac{YM}{\frac{1}{2}\rho U_o^2 Sb} \tag{4.627}$$

$$C_{nr} = \frac{\partial C_n}{\partial\left(\frac{rb}{2U_o}\right)} \tag{4.628}$$

so that

$$(C_{nr})_W = \frac{-8}{Sb^2}\int_0^{\frac{b}{2}} C_{D,l}c(y)y^2\, dy \tag{4.629}$$

For an untwisted rectangular wing of constant chord and an identical airfoil section along the span, Eq. (4.629) reduces to

$$(C_{nr})_W = -\frac{1}{3}C_{D,l} \tag{4.630}$$

The strip theory estimate is based on neglecting the induced drag and mutual interference effects between adjacent wing sections. Hence, this prediction will be in error for low-aspect ratio wings.

For more accurate analysis, the following Datcom [7] formula may be used:

$$(C_{nr})_W = \left(\frac{C_{nr}}{C_L^2}\right)C_L^2 + \left(\frac{C_{nr}}{C_{Do}}\right)C_{Do}\Big/\text{rad} \tag{4.631}$$

The data to evaluate (C_{nr}/C_L^2) and (C_{nr}/C_{Do}) are presented in Figs. 4.29a and 4.29b.

For supersonic speeds, no general method suitable for engineering purposes is available to evaluate $(C_{nr})_W$.

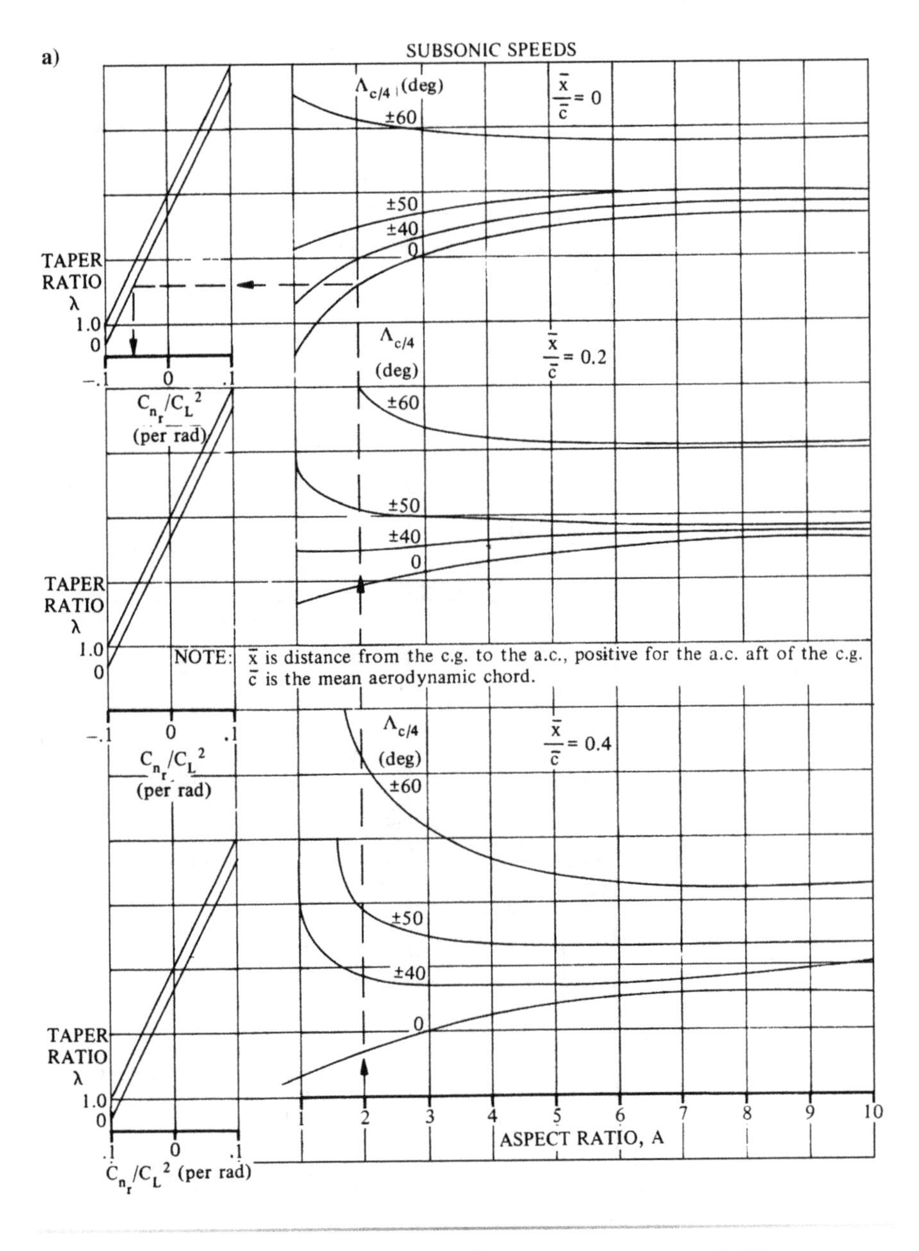

Fig. 4.29a Parameter $(C_{nr}/C_L^2)_W$ for subsonic speeds (7).

Lateral-Directional Acceleration Derivatives

Estimation of $C_{y\dot{\beta}}$ This derivative is a measure of the unsteady effects or the effects of time rate of change of sideslip on the side-force coefficient. For subsonic speeds, Datcom [7] presents some methods that are discussed here. At low angles of attack, the contributions of the wing and body are

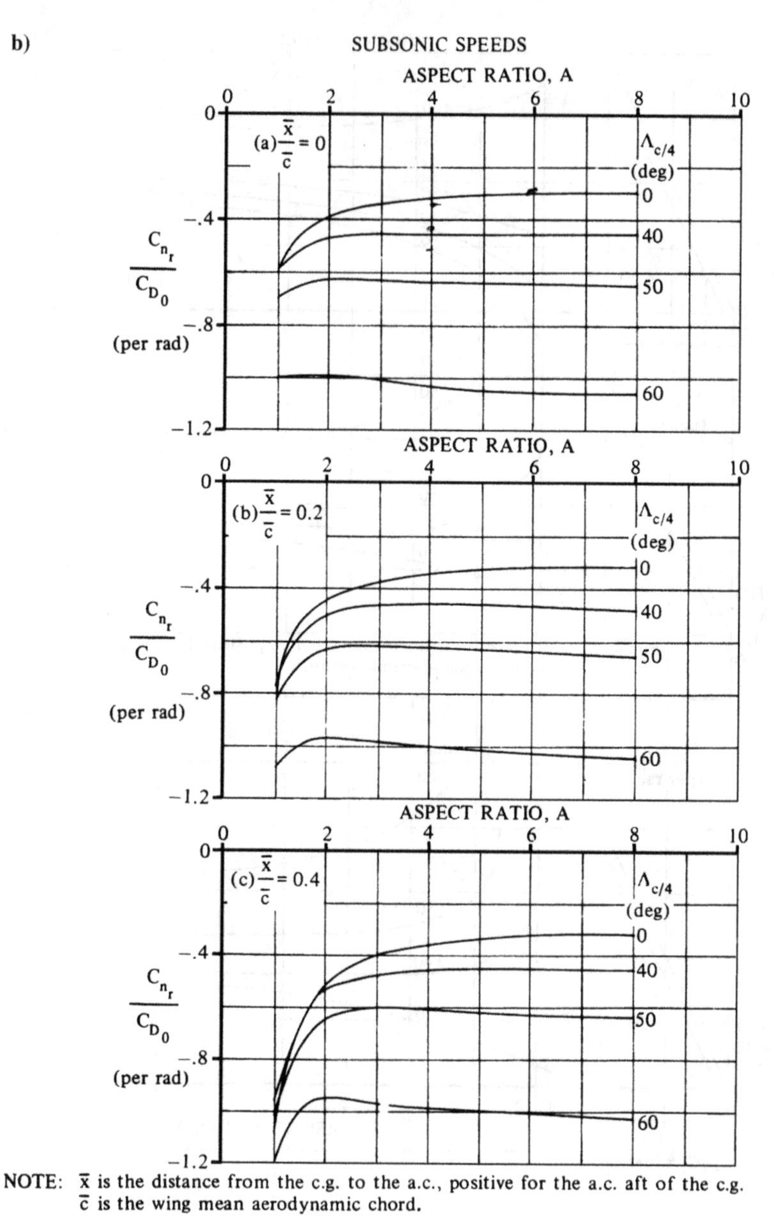

Fig. 4.29b Parameter $(C_{nr}/C_{Do})_W$ for subsonic speeds (7).

small and can be ignored. The largest contribution comes from the vertical tail, which can be estimated using the following expression [7]:

$$(C_{y\dot{\beta}})_V = 2a_V\sigma_\beta\left(\frac{S_V}{S}\right)\left[\frac{l_V\cos\alpha + z_V\sin\alpha}{b}\right] \tag{4.632}$$

where a_v is the lift-curve slope of the vertical tail and can be estimated using the methods discussed in Chapter 3. From Datcom [7],

$$\sigma_\beta = \sigma_{\beta\alpha}\alpha + \sigma_{\beta\Gamma}\Gamma + \sigma_{\beta,WB} \tag{4.633}$$

where α is the angle of attack in degrees and Γ is the dihedral angle in degrees. Here, we have assumed that the wing has zero twist. Here, $\sigma_{\beta\alpha}$ and $\sigma_{\beta\Gamma}$ are per deg.

The term $\sigma_{\beta\alpha}$ gives the variation of the sidewash with angle of attack, $\sigma_{\beta\Gamma}$ represents the influence of dihedral angle Γ on the sidewash, and $\sigma_{\beta,wB}$ represents the influence of the wing–body interference effect on the sidewash. The wing–body interference effects depend on the location of the wing with respect to the fuselage, i.e., whether it is a low, middle, or high wing configuration.

Data to estimate $\sigma_{\beta\alpha}$, $\sigma_{\beta\Gamma}$, and $\sigma_{\beta,WB}$ for typical airplane configurations and at subsonic speeds (M = 0.2 and 0.8) are presented in Figs. 4.30–4.32. No data is available for supersonic speeds [7]. The results presented in Fig. 4.31 for $\sigma_{\beta\Gamma}$ are applicable for $0 \leq \lambda \leq 1$. The data on $\sigma_{\beta,WB}$ given in Fig. 4.32 are

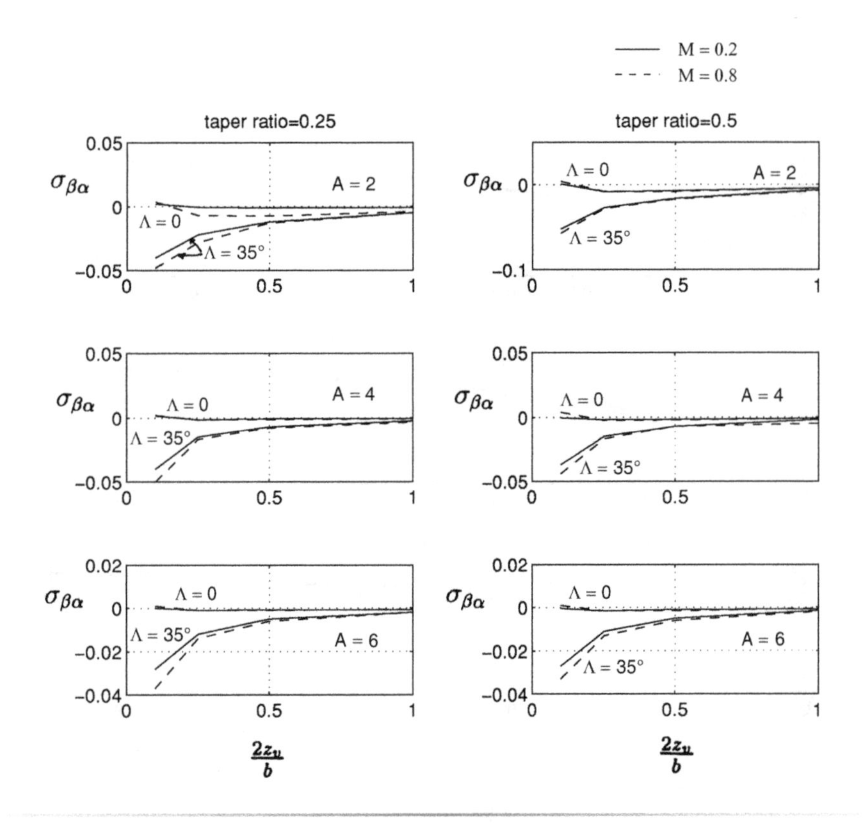

Fig. 4.30 Parameter $\sigma_{\beta\alpha}$ (per deg) at subsonic speeds (7).

for only one value of the ratio of body diameter to wing semispan of 0.12 and apply for $0 \leq \lambda \leq 1$ and $0 \leq \Lambda \leq 35$ deg. For other configurations not covered here, the reader may refer to Datcom [7].

Estimation of $C_{l\dot{\beta}}$ and $C_{n\dot{\beta}}$ For subsonic speeds, the major contribution comes from vertical tail and is given by

$$(C_{l\dot{\beta}})_V = (C_{y\dot{\beta}})_V \left[\frac{z_v \cos\alpha - l_v \sin\alpha}{b} \right] \tag{4.634}$$

$$(C_{n\dot{\beta}})_V = -(C_{y\dot{\beta}})_V \left[\frac{l_v \cos\alpha + z_v \sin\alpha}{b} \right] \tag{4.635}$$

At present, no engineering level methods are available for the estimation of $C_{l\dot{\beta}}$ and $C_{n\dot{\beta}}$ at supersonic speeds [7].

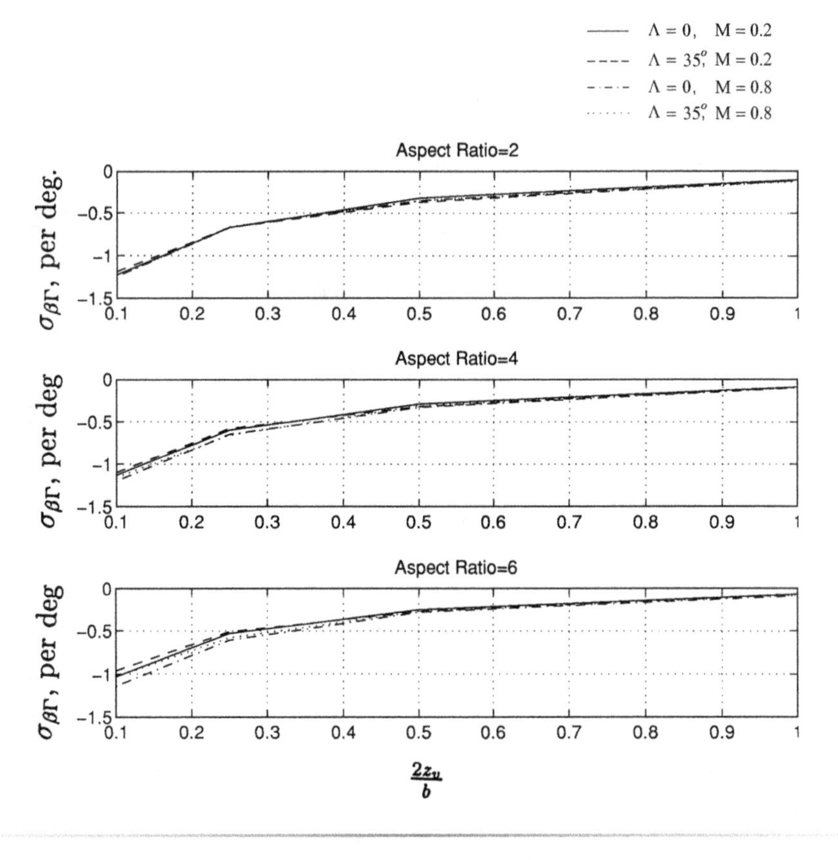

Fig. 4.31 Parameter $\sigma_{\beta\Gamma}$ at subsonic speeds (7).

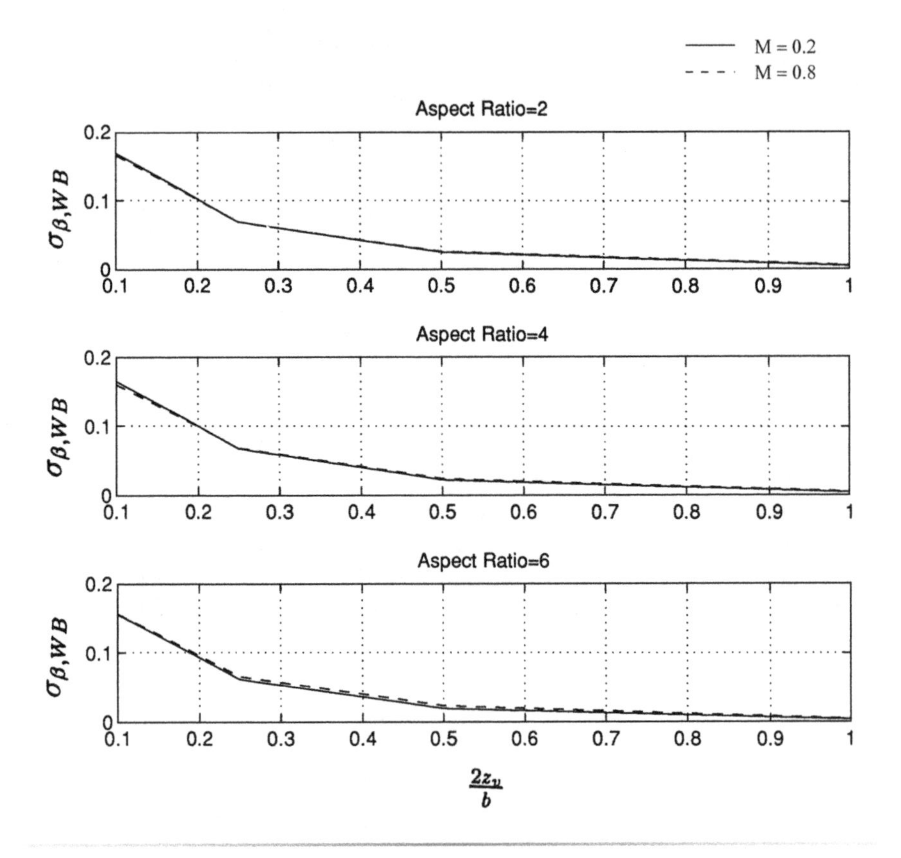

Fig. 4.32 Parameter $(\sigma_\beta)_{WB}$ at subsonic speeds (7).

Example 4.10

For the tailless aircraft of Examples 3.2 and 3.8, estimate the dynamic stability parameters at subsonic and supersonic speeds.

Solution:

Longitudinal dynamic stability derivatives

C_{Lq}: Because the given wing has a low aspect ratio ($A_e = 2.6893$), we use the Datcom [7] method to evaluate the wing–body contribution, which is given by Eq. (4.489). We have $S_e = 73.6282\ \text{m}^2$, $S = 106.0114\ \text{m}^2$, $k_2 - k_1 = 0.89$, $S_{B,\max} = 8.3193\ \text{m}^2$, and $l_f = 23.2410$ m. For both subsonic and supersonic speeds, $K_{W(B)} = 1.1607$. For subsonic speeds, $K_{B(W)} = 0.2627$ and, at supersonic speeds, $K_{B(W)} = 0.0011M^2 - 0.1044M + 0.3274$.

The lift-curve slope of the exposed wing $C_{L\alpha,e}$ and the aerodynamic center of the exposed wing $(x_{ac}/cr)_e$ were estimated as explained in Example 3.8.

(Continued)

Example 4.10 (*Continued*)

For supersonic speeds, C_{Lq} of Eq. (4.492) was evaluated using Datcom data [7].

The lift-curve slope of the body $(C_{L\alpha})_B$ was evaluated at subsonic and supersonic speeds using the methods discussed in Chapter 3.

Using all these values, C_{Lq} was calculated at subsonic and supersonic Mach numbers as shown in Fig. 4.33a. In Figs. 4.33–4.37, the circles indicate the values of Mach number at which the calculations were performed.

C_{mq}: Most of the steps involved in calculating this derivative are similar to those in the calculation of C_{Lq}. The parameter (C''_{mq}) was evaluated using Datcom data [7]. The calculated values of C_{mq} at subsonic and supersonic speeds are shown in Fig. 4.33b.

$C_{L\dot{\alpha}}$: Because the given configuration is tailless, we have $C_{L\dot{\alpha}} = (C_{L\dot{\alpha}})_{WB}$, which is given by Eq. (4.527). The term $(C_{L\dot{\alpha}})_e$ was evaluated at subsonic speeds using Eq. (4.528) and Datcom data [7] at supersonic speeds. The body contribution was evaluated using Eq. (4.530). The calculated values of $C_{L\dot{\alpha}}$ are shown in Fig. 4.34a.

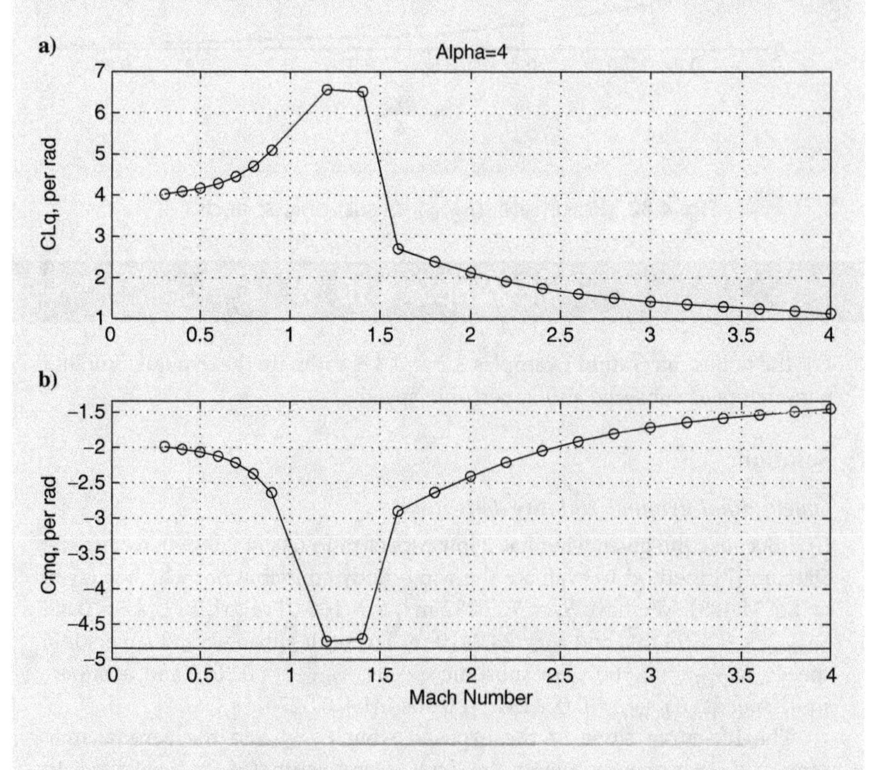

Fig. 4.33 C_{Lq} and C_{mq} for the aircraft.

(*Continued*)

Example 4.10 (*Continued*)

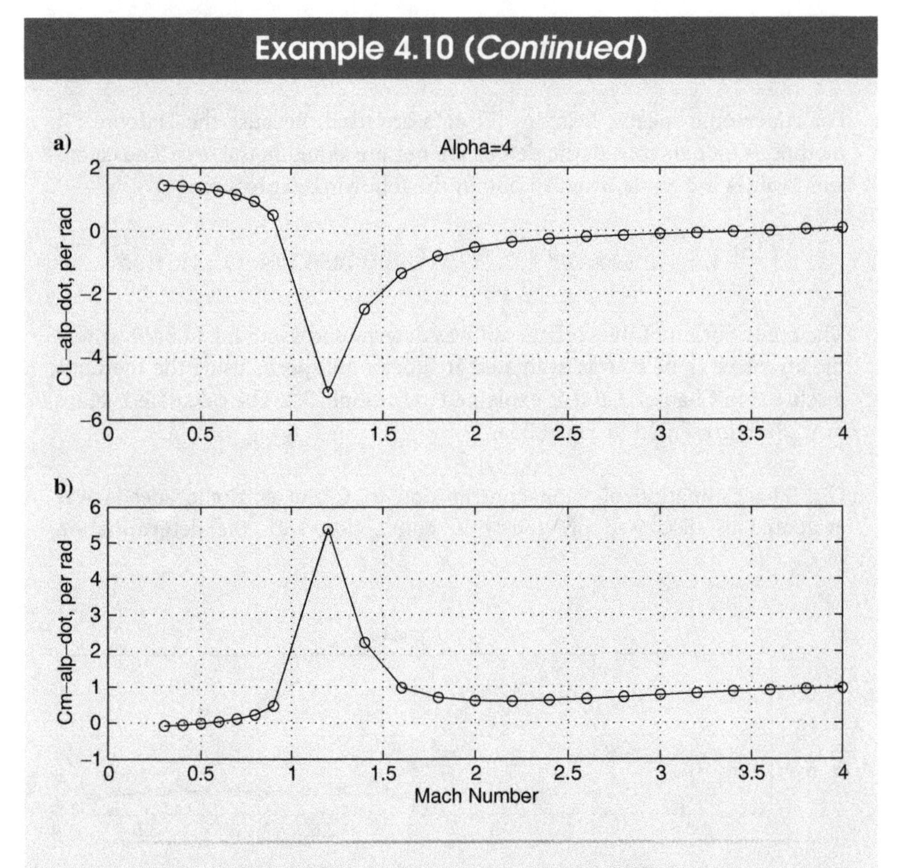

Fig. 4.34 $C_{L\dot{\alpha}}$ and $C_{m\dot{\alpha}}$ for the aircraft.

$C_{m\dot{\alpha}}$: We have $C_{m\dot{\alpha}} = (C_{m\dot{\alpha}})_{WB}$, which is given by Eq. (4.539). The term $(C_{m\dot{\alpha}})_e$ was evaluated at subsonic and supersonic Mach numbers using Eq. (4.540) and Datcom data [7]. The body contribution was evaluated using Eq. (4.543). The calculated values of $C_{m\dot{\alpha}}$ are shown in Fig. 4.34b.

Lateral-directional dynamic derivatives

C_{yp}: We have $C_{yp} = (C_{yp})_W + (C_{yp})_V$. For subsonic speeds, the wing contribution was evaluated using Eq. (4.548). The parameter k appearing in Eq. (4.554) was chosen equal to 0.8. The planform efficiency parameter e was evaluated using Eq. (4.478) based on exposed aspect ratio A_e. We have $\Gamma = 3.5$ deg, $z_w = 1.27$ m, and span $b = 17.3228$ m. For $\Lambda_{LE} = 45$ deg, $\lambda = 0.25$, and $A_e = 2.6893$ and at $M = 0.3$, we obtain from Fig. 4.25, $(\beta C_{lp}/k)_{C_L=0} = -0.225$. Similarly, the values at other Mach numbers are obtained and the data are curve fitted to obtain the following expression:

$$\left(\frac{\beta C_{lp}}{k}\right)_{C_L=0}\left(\frac{k}{\beta}\right) = 0.3708\,M^3 - 0.6662\,M^2 + 0.3128\,M - 0.2325/\text{rad}$$

(*Continued*)

Example 4.10 (*Continued*)

For supersonic speeds, Datcom [7] data are used. Because the Datcom [7] method is quite involved, the details are not presented in the text. The calculated values are curve fitted to obtain the following expression:

$$\left(\frac{C_{yp}}{\alpha}\right) = -0.3806\,M^3 + 3.422\,M^2 - 10.1458\,M + 10.1346/\text{rad}$$

The contribution of the vertical tail was determined using Eq. (4.545), where the lift-curve slope a_v was evaluated at supersonic speeds using the methods discussed in Chapter 3 and as explained in Example 3.8. The calculated values of C_{yp} are presented in Fig. 4.35a.

C_{lp}: The estimation of wing contribution to C_{lp} at subsonic speeds was evaluated as discussed previously in connection with the determination of C_{yp}.

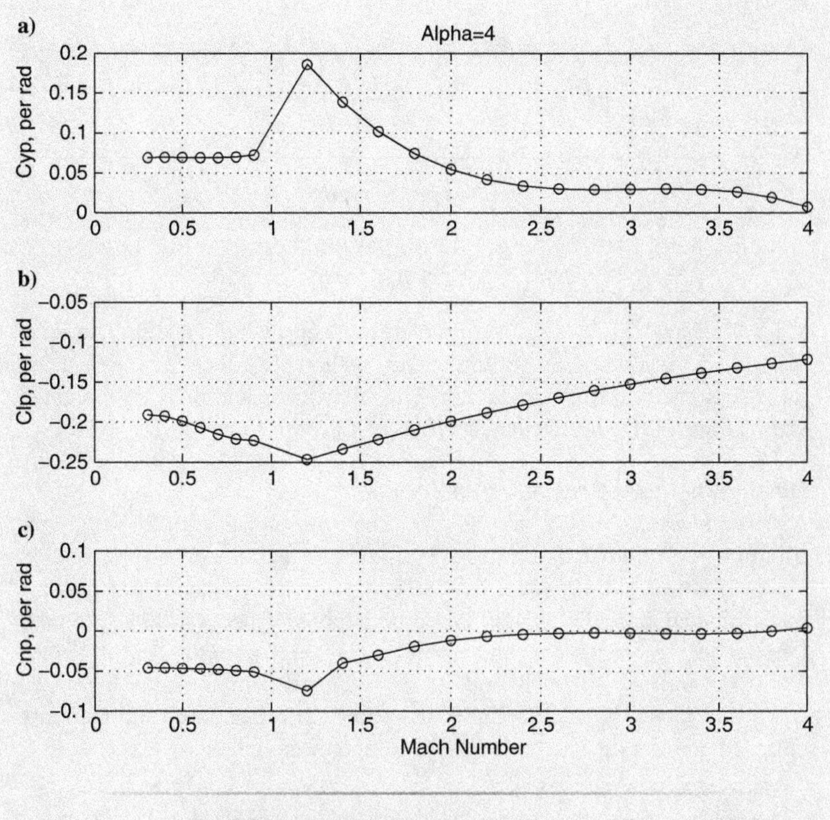

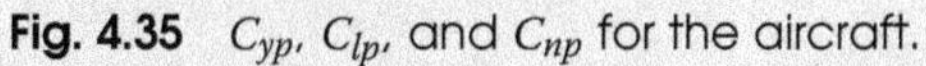

Fig. 4.35 C_{yp}, C_{lp}, and C_{np} for the aircraft.

(Continued)

Example 4.10 (*Continued*)

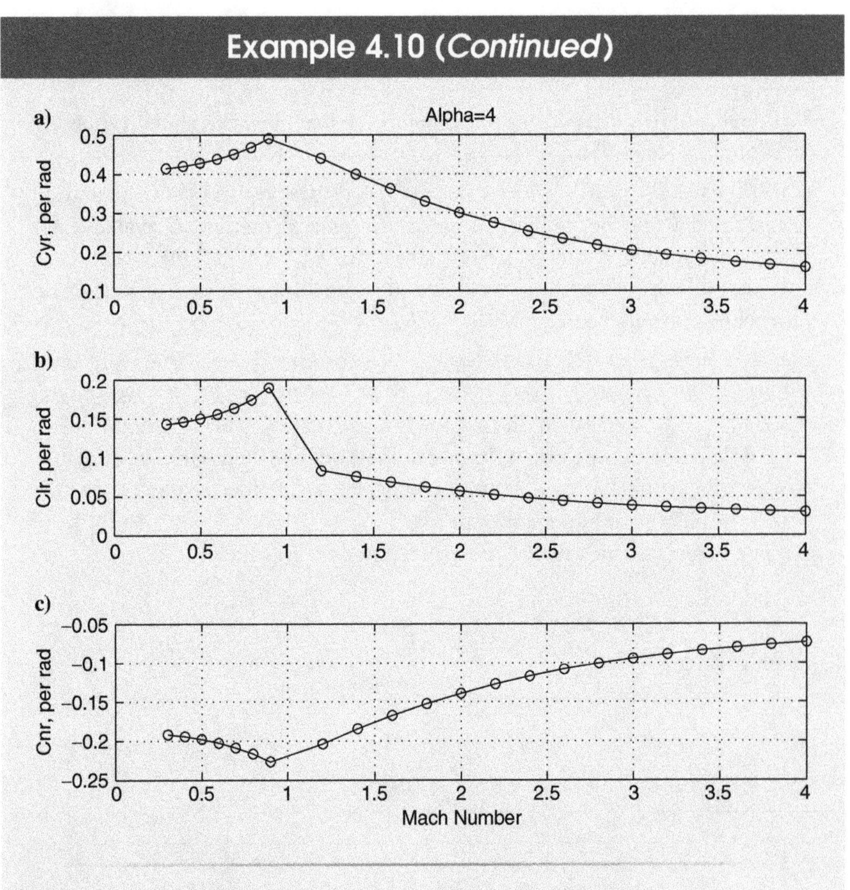

Fig. 4.36 C_{yr}, C_{lr}, and C_{nr} for the aircraft.

For supersonic speeds, Datcom [7] data are used to calculate $(C_{lp})_W$. This method is not discussed in the text because it is quite involved. The calculated values are curve fitted to obtain the following expression:

$$(C_{lp})_W = (-0.0025\,M^2 + 0.0283\,M - 0.1154)\,A/\text{rad}$$

where A is the theoretical wing aspect ratio and, for this case, $A = 2.8306$.

The vertical tail contribution at subsonic speeds is evaluated using Eq. (4.579) with $z_v = 3.8290$ m, $l_v = 7.7561$ m, and $b = 17.3228$ m. We have already explained the procedure of calculating $C_{y\beta,v}$.

The calculated values of C_{lp} at subsonic and supersonic speeds are presented in Fig. 4.35b.

C_{np}: We have $C_{np} = (C_{np})_W + (C_{np})_V$. At subsonic speeds, the wing contribution is evaluated using Eq. (4.594). The vertical tail contribution is evaluated using Eq. (4.598). At supersonic speeds, Datcom [7] data are used.

(Continued)

Example 4.10 (*Continued*)

The details of this method are not given in the text. The calculated values are presented in Fig. 4.35c.

C_{yr}: We have $C_{yr} = (C_{yr})_V$. At subsonic speeds, the vertical tail contribution is evaluated using Eq. (4.599). Because no general method is available for supersonic speeds, to get a crude estimate of C_{yr}, Eq. (4.599) is also used at supersonic speeds with a_v evaluated at supersonic speeds. The calculated values are presented in Fig. 4.36a.

C_{lr}: We have $C_{lr} = (C_{lr})_W + (C_{lr})_V$. At subsonic speeds, the wing contribution is evaluated using Eq. (4.613). From Fig. 4.28, we obtain $(C_{lr}/C_L)_{C_L=0,M=0} = 0.30$. With this, we can calculate C_{lr} for subsonic speeds.

For supersonic speeds, no general method of calculation is available. However, a crude estimate of the vertical tail is obtained using Eq. (4.618) with a_v evaluated at supersonic speeds.

The calculated values of C_{lr} are presented in Fig. 4.36b.

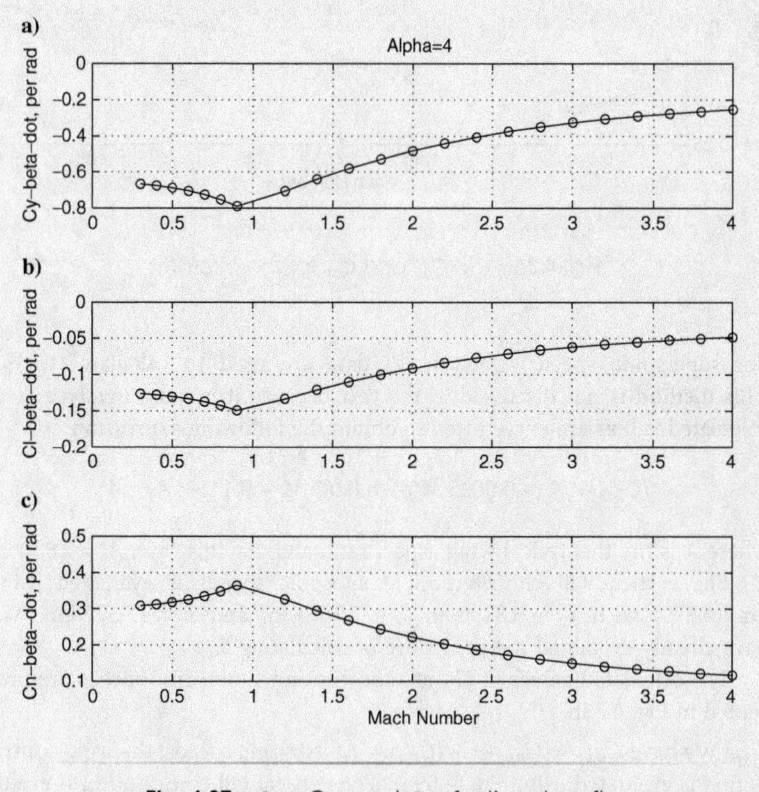

Fig. 4.37 $C_{y\dot{\beta}}$, $C_{l\dot{\beta}}$, and $C_{n\dot{\beta}}$ for the aircraft.

(Continued)

Example 4.10 (*Continued*)

C_{nr}: We have $C_{nr} = (C_{nr})_W + (C_{nr})_V$. The wing contribution was estimated using Datcom data [7] with $C_{Do} = 0.0025$. The vertical tail contribution at subsonic speeds is estimated using Eq. (4.620). No general method is available for supersonic speeds. However, using Eq. (4.620), crude estimates for vertical tail contribution are obtained at supersonic speeds. The calculated values are presented in Fig. 4.36c.

Lateral-directional acceleration derivatives

The derivatives $C_{y\dot{\beta}}$, $C_{l\dot{\beta}}$, and $C_{n\dot{\beta}}$ were obtained using Eqs. (4.632), (4.634), and (4.635). The calculated values are plotted in Fig. 4.37. Because no methods are available for supersonic speeds, the subsonic methods were used with a_v calculated at supersonic speeds. In view of this crude approximation, these results are to be used only for academic purposes.

Example 4.11

An aircraft with a wing loading of 2000 N/m^2 is in a steady level flight with a velocity of 100 m/s. The drag polar of the aircraft is given by $C_D = 0.025 + 0.05C_L^2$. Assuming $\rho = 1.25$ kg/m^3, $C_L = 0.08\alpha$ deg, $l_t = 3\bar{c}$, $S_t/S = 0.35$, $\epsilon = 0.35\alpha$ deg, $a_t = 0.06\alpha_t$ deg, and $\eta_t = 0.9$, determine the stability derivatives $C_{x\alpha}$, C_{xu}, $C_{z\alpha}$, C_{zu}, $C_{z\dot{\alpha}}$, C_{mq}, and $C_{m\dot{\alpha}}$.

Solution:

We have

$$C_L = \frac{2\left(\frac{W}{S}\right)}{\rho U_o^2} = \frac{2 * 2000}{1.25 * 100^2}$$

$$= 0.32$$

$$\overline{V}_1 = \frac{S_t l_t}{S\bar{c}}$$

$$= 0.35 * 3.0 = 1.05$$

Then

$$C_D = C_{Do} + kC_L^2 = 0.025 + 0.05 * 0.32^2 = 0.03012$$

$$C_{D\alpha} = 2kC_L C_{L\alpha} = 2 * 0.05 * 0.32(0.08 * 57.3) = 0.1467/\text{rad}$$

$$C_{xu} = -2C_D - C_{Du} = -2 * 0.03012 = -0.06024$$

$$C_{x\alpha} = C_L - C_{D\alpha} = 0.32 - 0.1467 = 0.1733/\text{rad}$$

(Continued)

Example 4.11 (*Continued*)

$$C_{zu} = -2C_L - C_{Lu} = -2 * 0.32 = -0.64$$

$$C_{z\alpha} = -C_{L\alpha} - C_D = -0.08 * 57.3 - 0.03012 = -4.614/\text{rad}$$

$$\begin{aligned} C_{z\dot{\alpha}} &= -C_{L\dot{\alpha}} \\ &= -2a_t\overline{V}_1\eta_t\left(\frac{\mathrm{d}\epsilon}{\mathrm{d}\alpha}\right) \\ &= -2(0.06 * 57.3)1.05 * 0.9 * 0.3 \\ &= -1.9494/\text{rad} \end{aligned}$$

$$\begin{aligned} C_{mq} &= -2a_t\overline{V}_1\eta_t\left(\frac{l_t}{\bar{c}}\right) \\ &= -2(0.06 * 57.3)1.05 * 0.9 * 3.0 \\ &= -19.4935/\text{rad} \end{aligned}$$

$$\begin{aligned} C_{m\dot{\alpha}} &= -2a_t\overline{V}_1\eta_t\left(\frac{\mathrm{d}\epsilon}{\mathrm{d}\alpha}\right)\left(\frac{l_t}{\bar{c}}\right) \\ &= -2(0.06 * 57.3)1.05 * 0.9 * 0.3 * 3.0 \\ &= -5.8480/\text{rad} \end{aligned}$$

In these calculations, we have assumed $C_{Du} = C_{Lu} = 0$.

Example 4.12

For the generic wing–body configuration shown in Fig. 4.38, estimate the stability derivatives C_{Lq}, C_{mq}, $C_{L\dot{\alpha}}$, and $C_{m\dot{\alpha}}$ for a flight Mach number of 0.10 at sea level and based on the following data.

Exposed wing area $S_e = 31.0\ \text{m}^2$, total (theoretical) wing area $S = 35.94\ \text{m}^2$, exposed aspect ratio $A_e = 3.22$, theoretical aspect ratio $A = 3.27$, exposed span $b_e = 10$ m, leading-edge sweep $\Lambda_{LE} = 45$ deg, exposed root chord $c_{re} = 5.6$ m, tip chord $c_t = 0.6$ m, total (tip-to-tip) span $b = 10.85$ m, theoretical root chord 6.025 m, maximum fuselage cross-sectional area $S_{B,\max} = 0.5675\ \text{m}^2$, fuselage length $l_f, = 10$ m, maximum fuselage width $b_{f,\max} = 0.85$ m, fuselage apparent mass coefficient $k_2 - k_1 = 0.95$, and distance of center of gravity from fuselage nose $x_{cg} = 5.0$ m. Assume that the cross section of the fuselage is circular and the variation of the width/diameter along the fuselage axis is shown in Table 4.1.

(*Continued*)

Example 4.12 (*Continued*)

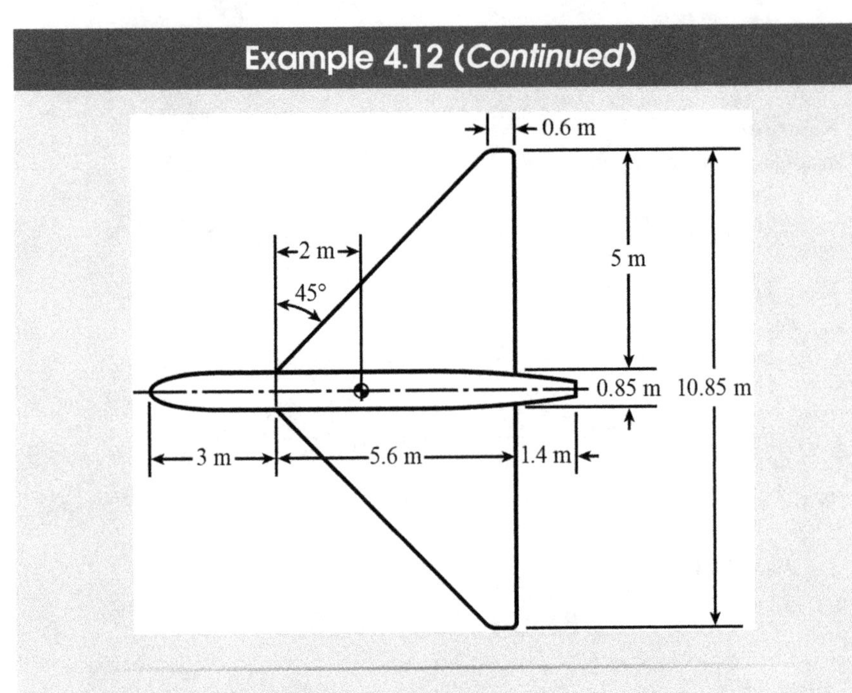

Fig. 4.38 Generic wing-body.

Table 4.1 Fuselage Geometrical Data

x, m	b_f, m
0	0.05
0.25	0.40
0.50	0.50
1.0	0.60
2.0	0.70
3.0	0.80
4.0	0.825
5.0	0.85
6.0	0.85
7.0	0.825
8.0	0.80
9.0	0.65
9.5	0.50
10.0	0.30

(*Continued*)

Example 4.12 (*Continued*)

Solution:

We have from Eq. (4.488)

$$(C_{Lq})_{WB} = [K_{W(B)} + K_{B(W)}]\left(\frac{S_e\bar{c}_e}{S\bar{c}}\right)(C_{Lq})_e + (C_{Lq})_B\left(\frac{S_{B,\max}l_f}{S\bar{c}}\right)$$

From Eqs. (3.27) and (3.28), we have

$$K_{WB} = 0.1714\left(\frac{b_{f,\max}}{b}\right)^2 + 0.8326\left(\frac{b_{f,\max}}{b}\right) + 0.9974$$

$$K_{BW} = 0.7810\left(\frac{b_{f,\max}}{b}\right)^2 + 1.1976\left(\frac{b_{f,\max}}{b}\right) + 0.0088$$

With $b_{f,\max} = 0.85$ m and $b = 10.85$ m, we obtain $K_{WB} = 1.0636$ and $K_{BW} = 0.1074$.

We have Eq. (4.489)

$$(C_{Lq})_e = \left(\frac{1}{2} + 2\xi\right)(C_{L\alpha})_e$$

where

$$\xi = \frac{\bar{x}}{\bar{c}_e}$$

$$\bar{x} = (x_{ac})_e - x_{cg,le}$$

The exposed taper ratio $\lambda_e = 0.6/5.6 = 0.1071$. The data on x_{ac} given in Fig. 3.19 is for $\lambda = 0$ and 0.2. Using these two sources of data with $\beta = \sqrt{1 - M^2} = 0.995$, we find that the mean value of $(x_{ac}/c_r)_e = 0.535$ for $\lambda_e = 0.1071$. Then, we get $(x_{ac})_e = 0.535 * 5.6 = 2.9960$. We have $\bar{x} = (x_{ac})_e - (x_{cg})_{le} = 2.996 - 2.0 = 0.996$ m. The mean aerodynamic chord is given by Eq. (1.25) as

$$\bar{c} = \left(\frac{2c_r}{3}\right)\left(\frac{1 + \lambda + \lambda^2}{1 + \lambda}\right)$$

Substituting $\lambda_e = 0.1071$ and $c_{re} = 5.6$ m, we get $\bar{c}_e = 4.1058$ m and $\xi = 0.2426$. Similarly, with theoretical root chord $c_r = 6.025$ m and theoretical taper ratio $\lambda = 0.09958$, we get the theoretical mean aerodynamic chord $\bar{c} = 4.0529$ m.

The subsonic lift-curve slope of a straight tapered wing is given by the following expression [Eq. (3.16)]:

$$C_{L\alpha} = \frac{2\pi A}{2 + \sqrt{\frac{A^2\beta^2}{k^2}\left(1 + \frac{\tan^2\Lambda_{c/2}}{\beta^2}\right) + 4}}$$

(*Continued*)

Example 4.12 (Continued)

where $\beta = \sqrt{1 - M^2}$ and the midchord sweep angle $\Lambda_{c/2}$ is given by the following expression [Eq. (3.19)]:

$$\tan \Lambda_{c/2} = \tan \Lambda_{LE} - \left(\frac{c_r - c_t}{b}\right)$$

With $\Lambda_{LE} = 45$ deg, $c_{re} = 5.6$ m, $c_t = 0.6$ m for the exposed wing, we get $\tan\ \Lambda_{c/2} = 0.5$. Then, with $A_e = 3.22$ and assuming $k = 0.8$, we find $(C_{L\alpha})_e = 2.9287/\text{rad}$. With these calculations, we get $(C_{Lq})_e = \ 2.8853/\text{rad}$.

Now let us evaluate the body contribution $(C_{Lq})_B$. We have Eqs. (4.493–4.495)

$$(C_{Lq})_B = 2(C'_{L\alpha})_B\left(1 - \frac{x_m}{l_f}\right)$$

$$(C'_{L\alpha})_B = (C_{L\alpha})_B\left(\frac{V_B^{2/3}}{S_{B,\max}}\right)$$

$$(C_{L\alpha})_B = 2(k_2 - k_1)\left(\frac{S_{B,\max}}{V_B^{2/3}}\right)$$

so that

$$(C'_{L\alpha})_B = 2(k_2 - k_1) = 2 * 0.95 = 1.9/\text{rad}$$

With $x_m = x_{cg} = 5.0$ m and $l_f = 10$ m, we get $(C_{Lq})_B = 1.9/\text{rad}$. With these calculations, we obtain $(C_{Lq})_{WB} = 3.0263/\text{rad}$.

From Eq. (4.502), we have

$$(C_{mq})_{WB} = [K_{W(B)} + K_{B(W)}]\frac{S_e}{S}\left(\frac{\bar{c}_e}{\bar{c}}\right)^2(C_{mq})_e + (C_{mq})_B\frac{S_{B,\max}}{S}\left(\frac{l_f}{\bar{c}}\right)^2 \Big/ \text{rad}$$

The contribution of the exposed wing $(C_{mq})_e$ is given by Eq. (4.503) as

$$(C_{mq})_e = \left[\frac{\frac{c_1}{c_3} + c_2}{\frac{c_1}{c_4} + 3}\right](C_{mq})_{e,M} = 0.2$$

Using Eq. (3.47), we have

$$\tan \Lambda_{c/4} = \tan \Lambda_{LE} - \left(\frac{c_r - c_t}{2b}\right)$$

(*Continued*)

Example 4.12 (Continued)

$$B = \sqrt{1 - M^2 \cos^2 \Lambda_{c/4}}$$

$$c_1 = A^3 \tan^2 \Lambda_{c/4}$$

$$c_2 = \frac{3}{B}$$

$$c_3 = AB + 6 \cos \Lambda_{c/4}$$

$$c_4 = A + 6 \cos \Lambda_{c/4}$$

$$c_5 = A + 2 \cos \Lambda_{c/4}$$

Substituting, we get $\Lambda_{c/4} = 36.8698$ deg, $B = 0.9968$, $c_1 = 18.7759$, $c_2 = 3.0096$, $c_3 = 8.0099$, $c_4 = 8.0202$, and $c_5 = 4.8201$. Furthermore, using Eq. (4.504)

$$(C_{mq})_{e,M=0.2} = -0.7 C_{l\alpha} \cos \Lambda_{c/4} \left[\frac{A(0.5\xi + 2\xi^2)}{c_5} + \left(\frac{c_1}{24 c_4} \right) + \frac{1}{8} \right]$$

We assume $C_{l\alpha} = 2\pi k = 2\pi * 0.8 = 5.0265/\text{rad}$. With these values we get $(C_{mq})_{e,M=0.2} = -1.0759/\text{rad}$ and then $(C_{mq})_e = -1.0784/\text{rad}$.

The body contribution is given by Eqs. (4.512–4.514) as

$$(C_{mq})_B = 2(C'_{m\alpha})_B \left[\frac{(1 - x_{m1})^2 - V_{B1}(x_{c1} - x_{m1})}{1 - x_{m1} - V_{B1}} \right]$$

Here,

$$x_{m1} = \frac{x_m}{l_f} \quad x_{c1} = \frac{x_c}{l_f} \quad V_{B1} = \frac{V_B}{S_{B,\max} l_f}$$

$$V_B = \int_0^{l_f} S_B(x)\, dx$$

$$x_c = \frac{1}{V_B} \int_0^{l_f} S_B(x) x\, dx$$

Furthermore, using Eqs. (4.515) and (4.516)

$$(C'_{m\alpha})_B = (C_{m\alpha})_B \left(\frac{V_B}{S_{B,\max} l_f} \right)$$

$$(C_{m\alpha})_B = \frac{2(k_2 - k_1)}{V_B} \int_0^{x_o} \frac{dS_B(x)}{dx} (x_m - x)\, dx$$

Here, $x_m = x_{cg} = 5.0$ m. From Table 4.1, we find that $x_o = 6.0$ m. Performing the required calculations, we get $V_B = 4.2762\ \text{m}^3$, $x_c = 5.1510$ m, $x_{m1} = 0.5$,

(Continued)

Example 4.12 (*Continued*)

$x_{c1} = 0.5151$, $V_{B1} = 0.7535$, $(C_{m\alpha})_B = 0.9105/\text{rad}$, $(C'_{m\alpha})_B = 0.6860$, and $(C_{mq})_B = -1.2820/\text{rad}$. With these, we get $(C_{mq})_{WB} = -1.2410/\text{rad}$.

Now let us calculate the acceleration derivatives $C_{L\dot{\alpha}}$ and $C_{m\dot{\alpha}}$ for the given wing–body configuration. Using Eqs. (4.527–4.529), we have

$$(C_{L\dot{\alpha}})_{WB} = [K_{W(B)} + K_{B(W)}]\left(\frac{S_e \bar{c}_e}{S\bar{c}}\right)(C_{L\dot{\alpha}})_e + (C_{L\dot{\alpha}})_B \frac{S_{B,\max} l_f}{S\bar{c}} \Big/ \text{rad}$$

where

$$(C_{L\dot{\alpha}})_e = 1.5\left(\frac{x_{ac}}{c_r}\right)_e (C_{L\alpha})_e + 3C_L(g)/\text{rad}$$

$$C_L(g) = \left(-\frac{\pi A_e}{2\beta^2}\right)(0.0013\tau^4 - 0.0122\tau^3 + 0.0317\tau^2 + 0.0186\tau - 0.0004)$$

$$\beta = \sqrt{1 - M^2}$$

$$\tau = \beta A_e$$

Substituting we obtain $\tau = 3.22$, $C_L(g) = -0.6148$, and $(C_{L\dot{\alpha}})_e = 0.5059/\text{rad}$.

Using Eqs. (4.530–4.532), the body contribution is given by

$$(C_{L\dot{\alpha}})_B = 2(C'_{L\alpha})_B\left(\frac{V_B}{S_{B,\max} l_f}\right)$$

We obtain $(C_{L\dot{\alpha}})_B = 2.8643/\text{rad}$. With these values, we obtain $(C_{L\dot{\alpha}})_{WB} = 0.6292/\text{rad}$.

Finally using Eqs. (4.539–4.543),

$$(C_{m\dot{\alpha}})_{WB} = [K_{W(B)} + K_{B(W)}]\left(\frac{S_e \bar{c}_e^2}{S\bar{c}^2}\right)(C_{m\dot{\alpha}})_e + (C_{m\dot{\alpha}})_B \frac{S_{B,\max} l_f^2}{S\bar{c}^2} \Big/ \text{rad}$$

where

$$(C_{m\dot{\alpha}})_e = (C''_{m\dot{\alpha}})_e + \left(\frac{x_{cg,le}}{\bar{c}}\right)(C_{L\dot{\alpha}})_e$$

$$(C''_{m\dot{\alpha}})_e = -\left(\frac{81}{32}\right)\left(\frac{x_{ac}}{c_r}\right)_e^2 (C_{L\alpha})_e + \frac{9}{2} C_{mo}(g)$$

$$C_{mo}(g) = \left(\frac{\pi A_e}{2\beta^2}\right)(0.0008\tau^4 - 0.0075\tau^3 + 0.0185\tau^2 + 0.0128\tau - 0.0003)$$

(*Continued*)

Example 4.12 (Continued)

Substituting, we obtain $C_{mo}(g) = 0.3487$, $(C''_{m\dot{\alpha}})_e = -0.5528$, and $(C_{m\dot{\alpha}})_e = -0.3064/\text{rad}$.

$$(C_{m\dot{\alpha}})_B = 2(C'_{m\dot{\alpha}})_B \left[\frac{x_{c1} - x_{m1}}{1 - x_{m1} - V_{B1}}\right]\left(\frac{V_B}{S\bar{c}}\right)$$

We have $(C'_{m\dot{\alpha}})_B = 0.6860$. Substitution gives $(C_{m\dot{\alpha}})_B = -0.06117/\text{rad}$. With these values, we get $(C_{m\dot{\alpha}})_{WB} = -0.3235/\text{rad}$.

Example 4.13

An aircraft wing has the following characteristics: leading-edge sweep = 0, aspect ratio = 6, taper ratio = 1.0, sectional lift-curve slope = 0.1/deg, zero-lift sectional drag coefficient $C_{Do} = 0.022$, and increase in sectional profile drag coefficient per deg angle of attack $C_{D\alpha,l} = 0.001/\text{deg}$.

Assuming that it is a midwing configuration, determine C_{lp}, C_{np}, C_{lr}, and C_{nr} using strip theory for $\alpha = 5$ deg and M = 0.20. Compare your results with those obtained using Datcom methods.

Solution:

The strip theory [Eq. (4.575)] gives

$$(C_{lp})_W = -\frac{1}{6}(a_o + C_{D,l})$$

We have $a_o = 0.1 * 57.3 = 5.73/\text{rad}$ and

$$\begin{aligned} C_{D,l} &= C_{Do,l} + C_{D\alpha,l}\alpha \\ &= 0.022 + 0.001 * 5 = 0.027 \end{aligned}$$

Substituting, we get the strip theory value of $(C_{lp})_W = -0.9595/\text{rad}$.

Now let us determine the value of $(C_{lp})_W$ using the Datcom [7] method. We have from Eq. (4.576)

$$(C_{lp})_W = \left(\frac{\beta C_{lp}}{k}\right)_{C_L=0}\left(\frac{k}{\beta}\right)\left(\frac{(C_{lp})\Gamma}{(C_{lp})_{\Gamma=0}}\right) / \text{rad}$$

We assume $k = 1.0$ and $\beta = \sqrt{1 - 0.2^2} = 0.9797$. From Fig. 4.25, $(\beta C_{lp}/k)_{C_L=0} = -0.43$ for $\Lambda_{LE} = 0$, $\lambda = 1.0$, and $A = 6.0$. Because $\Gamma = 0$, we find $[(C_{Lp})_\Gamma/(C_{lp})_{\Gamma=0}] = 1.0$. With these values, we get

(Continued)

Example 4.13 (Continued)

$(C_{Lp})_W = -0.430/\text{rad}$. Therefore, the strip theory result differs from the Datcom value considerably, by as much as 100%.

Now let us estimate $(C_{np})_W$. The strip theory [Eq. (4.593)] gives

$$(C_{np})_W = -\frac{1}{6}(C_L - C_{D\alpha})/\text{rad}$$

Substituting the required values, we get $(C_{np})_W = -0.08317/\text{rad}$.

Using Eqs. (4.594–4.597)

$$(C_{np})_W = C_{lp} \tan\alpha(K-1) + K\left(\frac{C_{np}}{C_L}\right)_{C_L=0,M} C_L \Big/ \text{rad}$$

$$\left(\frac{C_{np}}{C_L}\right)_{C_L=0,M} = \left(\frac{A + 4\cos\Lambda_{c/4}}{AB + 4\cos\Lambda_{c/4}}\right)\left[\frac{AB + 0.5(AB + \cos\Lambda_{c/4})\tan^2\Lambda_{c/4}}{A + 0.5(A + \cos\Lambda_{c/4})\tan^2\Lambda_{c/4}}\right] \times \left(\frac{C_{np}}{C_L}\right)_{C_L=0} \Big/ \text{rad}$$

$$\left(\frac{C_{np}}{C_L}\right)_{C_L=0} = -\left[\frac{A + 6(A + \cos\Lambda_{c/4})\left(\frac{\bar{x}}{\bar{c}}\frac{\tan\Lambda_{c/4}}{A} + \frac{\tan^2\Lambda_{c/4}}{12}\right)}{A + 4\cos\Lambda_{c/4}}\right]$$

$$B = \sqrt{1 - M^2\cos^2\Lambda_{c/4}}$$

$$K = \frac{1 - a_{w1}}{1 - a_{w2}} \quad a_{w1} = \frac{(C_{L\alpha})_e}{\pi Ae} \quad a_{w2} = \frac{(C_{L\alpha})_e}{\pi A}$$

Using Eqs. (4.478) and (4.479)

$$e = \frac{1.1 C_{L\alpha}}{RC_{L\alpha} + (1 - R)\pi A}$$

$$R = a_1\lambda_1^3 + a_2\lambda_1^2 + a_3\lambda_1 + a_4$$

$$a_1 = 0.0004 \quad a_2 = -0.0080 \quad a_3 = 0.0501 \quad a_4 = 0.8642$$

$$\lambda_1 = \frac{A\lambda}{\cos\Lambda_{LE}}$$

Using Eq. (1.52), we have

$$C_{L\alpha,e} = \frac{a_o}{1 + \frac{a_o}{\pi A}}$$

Substituting $a_o = 0.1 * 57.3 = 5.73/\text{rad}$ and $A = 6$, we get $C_{L\alpha,e} = 0.0767/\text{deg}$ or $4.3943/\text{rad}$. With this, $C_L = C_{L\alpha,e}\alpha = 0.3834$. Further using $\Lambda_{c/4} = 0$ and $\lambda = 1.0$, we get $R = 0.9632$, $e = 0.9812$, and $K = 0.9942$.

(Continued)

Example 4.13 (*Continued*)

We have $B = \sqrt{1 - M^2 \cos^2 \Lambda_{c/4}} \simeq 1$. Substituting, we get $(C_{np}/C_L)_{C_L=0,M} = (C_{np}/C_L)_{C_L=0} = -0.6$ and $(C_{np})_W = -0.2285$. Therefore, the strip theory prediction of -0.08317/rad is in error by as much as 63%.

Now let us calculate $(C_{lr})_W$ and $(C_{nr})_W$. The strip theory [Eqs. (4.612) and (4.630)] gives

$$(C_{lr})_W = \frac{C_L}{3} = \frac{0.1 * 5}{3} = 0.1667/\text{rad}$$

$$(C_{nr})_W = \frac{-C_{D,l}}{3} = -\frac{C_{Do} + C_{D\alpha,l} * \alpha}{3} = -\frac{0.022 + 0.0011 * 5}{3}$$
$$= -0.0092/\text{rad}$$

Now let us estimate $(C_{lr})_W$ and $(C_{nr})_W$ using Eqs. (4.613–4.617) as

$$(C_{lr})_W = C_L \left(\frac{C_{lr}}{C_L}\right)_{C_L=0,M} + \left(\frac{\Delta C_{lr}}{\Gamma}\right)\Gamma/\text{rad}$$

$$\left(\frac{C_{lr}}{C_L}\right)_{C_L=0,M} = \frac{\text{Num}}{\text{Den}}\left(\frac{C_{lr}}{C_L}\right)_{C_L=0,M=0}$$

$$\text{Num} = 1 + \frac{A(1 - B^2)}{2B(AB + 2\cos\Lambda_{c/4})}\left(\frac{AB + 2\cos\Lambda_{c/4}}{AB + 4\cos\Lambda_{c/4}}\right)\left(\frac{\tan^2\Lambda_{c/4}}{8}\right)$$

$$\text{Den} = 1 + \left(\frac{A + 2\cos\Lambda_{c/4}}{A + 4\cos\Lambda_{c/4}}\right)\left(\frac{\tan^2\Lambda_{c/4}}{8}\right)$$

Here, we have $C_L = 0.0767 * 5 = 0.3834$, $\Gamma = 0$, $\Lambda_{LE} = \Lambda_{c/4} = 0$, $A = 6$, and $M = 0.2$. Using Fig. 4.28, we get $(C_{lr}/C_L)_{C_L=M=0} = 0.27$. Substituting and simplifying, we get $(C_{lr})_W = 0.1051$/rad. Thus, the strip theory prediction of 0.1667 differs by as much as 50% compared to the Datcom [7] result.

Using Eq. (4.631), we have

$$(C_{nr})_W = \left(\frac{C_{nr}}{C_L^2}\right)C_L^2 + \left(\frac{C_{nr}}{C_{Do}}\right)C_{Do}$$

Assuming $\xi = 0$ from Fig. 4.29, we get $(C_{nr}/C_L^2) = -0.02$ and $(C_{nr}/C_{Do}) = -0.30$. We have $C_L = 0.3834$ and $C_{Do} = 0.022$. Substituting, we get $(C_{nr})_W = -0.0095$/rad. In this case, the strip theory does well and its result of -0.009/rad differs from the Datcom[7] result by just about 5%.

Example 4.14

Estimate the low-speed vertical tail contributions $(C_{lp})_V$, $(C_{np})_V$, $(C_{lr})_V$, and $(C_{nr})_V$, at an angle of attack of 5 deg using the following data: lift-curve slope $a_v = 0.07/\text{deg}$, leading-edge sweep $\Lambda_{LE} = 0$, vertical tail length $l_v = 9.0$ m, wing mean aerodynamic chord $\bar{c} = 3.0$ m, wing span $b = 15$ m, $z_v = 0.9$ m, ratio of vertical tail area to wing area $S_v/S = 0.20$, wing aspect ratio $A = 6$, and wing taper ratio $\lambda = 0.5$.

Solution:

From Eqs. (3.303) and (3.306), we have

$$C_{y\beta,V} = -ka_v\left(1 + \frac{\partial\sigma}{\partial\beta}\right)\eta_v\frac{S_v}{S}$$

$$\left(1 + \frac{\partial\sigma}{\partial\beta}\right)\eta_v = 0.724 + \frac{3.06S_v/S}{1 + \cos\Lambda_{c/4}} + \frac{0.4z_w}{d_{f,\max}} + 0.009A$$

Using Eq. (4.579)

$$(C_{lp})_V = \left|2\left(\frac{z}{b}\right)\left(\frac{z - z_v}{b}\right)\right|C_{y\beta,V}$$

$$z = z_v\cos\alpha - l_v\sin\alpha$$

Using Eq. (4.598)

$$(C_{np})_V = -\left(\frac{2}{b}\right)(l_v\cos\alpha + z_v\sin\alpha)\left(\frac{z - z_v}{b}\right)C_{y\beta,V}/\text{rad}$$

Using Eq. (4.618)

$$(C_{lr})_V = \frac{-2}{b^2}(l_v\cos\alpha + z_v\sin\alpha)(z_v\cos\alpha - l_v\sin\alpha)C_{y\beta,V}$$

From Eq. (4.620)

$$(C_{nr})_V = \frac{2}{b^2}(l_v\cos\alpha + z_v\sin\alpha)^2C_{y\beta,V}$$

We assume $k = 0.80$ and that we have a midwing configuration so that $z_w = 0$. Substituting in the given expressions, we obtain, $(1 + [\partial\sigma/\partial\beta])\eta_v = 1.39$, $C_{y\beta,V} = -0.8920/\text{rad}$, $z = 0.12\,\text{m}$, $(C_{lp})_V = -0.000742/\text{rad}$, $(C_{np})_V = -0.05615/\text{rad}$, $(C_{lr})_V = 0.00804/\text{rad}$, and $(C_{nr})_V = -0.07171/\text{rad}$.

4.5 Summary

In this chapter we have studied various axes systems used in airplane dynamics and discussed various methods of transforming vectors from one coordinate system into another. We also studied the methods of calculating time history of Euler angles. The method based on using Euler angle rates has a singularity when the pitch angle approaches 90 deg. However, the direction cosine method and the quaternions do not encounter this problem. We then formulated the problem of airplane dynamics making use of a moving coordinate system to avoid the problem of computing time-varying moments and products of inertia but had to deal with more complex acceleration and angular momentum terms, which render the equations of motion coupled and nonlinear. We then introduced the concept of small disturbances, which enabled us to linearize and decouple the equations of motion into two categories, one set of three equations for longitudinal motion and another set of three equations for lateral-directional motion. The advantage of this approach is that the two motions can be studied independent of each other. We then used Bryan's method and assumed that the aerodynamic forces and moments depend linearly on the instantaneous values of motion variables. We used the method of Taylor series expansion to estimate the aerodynamic forces and moments in the disturbed state. We also discussed engineering methods to evaluate the stability and control derivatives appearing in the Taylor series expansions.

Our next task is to solve the equations of longitudinal and lateral-directional motions to study the stability and response of the airplane to control inputs. The linearity of the equations enables us to use the powerful methods of linear control systems. In Chapter 5, we will discuss basic principles of linear system theory and design and then, in Chapter 6, discuss their application for the study of dynamic stability and response of the aircraft as well as the design of autopilots and stability augmentation systems.

References

[1] Bryan, G. H., *Stability in Aviation,* Macmillan, London, 1911.

[2] Regan, F., *Reentry Dynamics,* AIAA Education Series, AIAA, New York, 1984.

[3] Rolfe, J. M., and Staples, K. J., *Flight Simulation,* Cambridge Univ. Press, New York, 1986.

[4] Robinson, A. C., "On the Use of Quaternions in Simulation of Rigid-Body Motion," Wright Air Development Center, WAEDC TR 58-17, Wright–Patterson Air Force Base, Dec. 1958.

[5] *MATLAB High Performance Numeric Computation and Visualization Software,* The Math Works, Natick, MA, Oct. 1992.

[6] Bate, R. R., Mueller, D. D., and White, J. E., *Fundamentals of Astrodynamics,* Dover, New York, 1971.

[7] Hoak, D. E., et al., "The USAF Stability and Control Datcom," Air Force Wright Aeronautical Laboratories, TR-83-3048, Oct. 1960 (revised 1978).

[8] Schuler, C. J., Ward, L. K., and Hodapp, A. E., "Techniques for Measurement of Dynamic Stability Derivatives in Ground Test Facilities," AGARDograph 121, Oct. 1967.

[9] "Dynamic Stability Parameters," AGARD, LS-114, May 1981.

[10] Orlick-Rueckmann, K. J., "Dynamic Stability Parameters," *AGARD,* CP-235, 1978.

[11] Dayman, B., Jr., "Free Flight Testing in High Speed Wind Tunnels," AGARDograph 113, 1966.

[12] Etkin, B., *Dynamics of Atmospheric Flight,* Dover, New York, 2005.

Problems

4.1 At $t = 0$, a launch vehicle takes off from the surface of the Earth at the equator with longitude $L = 0$. At $t = 50$ s, the vehicle has a longitude of 10 deg, latitude of 0, and an altitude of 50,000 ft. The velocity components measured with respect to the Earth-fixed $Ox_Ey_Ez_E$ system are $u_E = 1500$ ft/s, $v_E = 3000$ ft/s, and $w_E = 0$. Assuming that the Earth-fixed system $Ox_Ey_Ez_E$ coincides with the inertial system at $t = 0$, determine the position and velocity of the vehicle in the inertial reference system at $t = 50$ s.

4.2 An aircraft is undergoing a steady rotation with angular velocity components in body axes system $p = 10$ deg/s, $q = 2$ deg/s, and $r = 5$ deg/s. At a certain instant of time, the Euler angles are $\psi = -30$ deg, $\theta = 10$ deg, and $\phi = 15$ deg. Determine the corresponding Euler angle rates.

4.3 An aircraft executes a velocity vector roll at an angle of attack of 40 deg with an angular velocity of 30 deg/s about the wind axis. Plot the time history of Euler angles for first 20 s.

4.4 Using the method of direction cosines and quaternions, compute the time history of Euler angles for the aircraft in Example 4.3 and compare your results.

4.5 An aircraft is flying at an angle of attack of 10 deg and sideslip of 5 deg. The onboard accelerometers record $a_{xb} = 10$ ft/s^2, $a_{yb} = 5$ ft/s^2, and $a_{zb} = -5$ ft/s^2. Determine the components of accelerations in the wind axes system.

4.6 An aircraft model is tested in a low-speed wind tunnel at an angle of attack of 25 deg and sideslip of -5 deg. The internal strain gage balance records $F_x = -25$ lb, $F_y = -3$ lb, and $F_z = -75$ lb with respect to the model axes system. Determine lift, drag, and side force acting on the model.

4.7 An airplane is flying at a velocity of 150 ft/s. At an angle of attack of 12 deg and sideslip of 2 deg, the instantaneous angular velocities in pitch, roll, and yaw measured by onboard rate gyros are 10 deg/s, 5 deg/s, and 10 deg/s, respectively. If the instantaneous values of $\dot{\alpha}$ and $\dot{\beta}$ are 3 deg/s and -5 deg/s, respectively, determine the wind axis angular velocity components p_w, q_w, and r_w.

4.8 Given $p = 10$ deg/s, $q = 5$ deg/s, $r = 10$ deg/s, $\psi = 30$ deg, $\theta = 25$ deg, and $\phi = -10$ deg, determine $\Omega_{i,b}^{b}$ and $\Omega_{i,b}^{i}$.

4.9 An aircraft is in a spin at an angle of attack of 60 deg. Assuming that the spin rate about a vertical axis through the center of gravity is 40 deg/s, plot the time history of Euler angles using (a) method of Euler angles, (b) the method of direction cosines, and (c) the quaternions.

4.10 Determine the missing elements (marked xx) of the following direction cosine matrices:

a) $$\mathrm{DCM}_1 = \begin{bmatrix} 0.1587 & xx & 0.4858 \\ 0.8595 & -0.1218 & xx \\ xx & 0.4963 & 0.7195 \end{bmatrix}$$

b) $$\mathrm{DCM}_2 = \begin{bmatrix} xx & 0.5283 & 0.0888 \\ -0.5253 & xx & 0.3033 \\ 0.0888 & xx & 0.9487 \end{bmatrix}$$

4.11 An acrobatic aircraft flying at 150 m/s and a 30-deg angle of attack executes a body axis roll at a rate of 150 deg/s. Determine the accelerations measured by onboard accelerometers.

4.12 An airplane flying at an angle of attack below stall angle is in a steady roll at a rate p_0 deg/s. Derive the equations of motion for small disturbance motion.

4.13 An aircraft weighs 50,000 N and is in a steady level flight at 150 m/s at sea level ($\rho = 1.225\ \mathrm{kg/m^3}$). The drag polar is given by $C_D = 0.018 + 0.24C_L^2$. The lift-curve slope of the wings is 0.095/deg, the wing area is 25 m^2 and the wing mean aerodynamic chord is 2.5 m. The lift-curve slope of the horizontal tail is 0.06/deg, the horizontal tail area is 6 m^2, the tail length $l_t = 6$ m, and $\epsilon = 0.35\alpha$. Assuming a tail efficiency of 0.9, estimate the stability derivatives C_{xu}, C_{zu}, $C_{x\alpha}$, $C_{z\alpha}$, C_{zq}, $C_{z\dot{\alpha}}$, C_{mq}, and $C_{m\dot{\alpha}}$. [Answer: $C_{xu} = -0.0370$, $C_{zu} = -0.2902$, $C_{x\alpha} = 0.1072$, $C_{z\alpha} = -0.1135$, $C_{zq} = -3.5645$, $C_{z\dot{\alpha}} = -1.2476$, $C_{mq} = -8.5548$, and $C_{m\dot{\alpha}} = -2.9942$. All values are per radian.]

4.14 For the wing–body of Example 4.12, estimate the stability derivatives C_{Lq}, C_{mq}, $C_{L\dot{\alpha}}$, and $C_{m\dot{\alpha}}$ at a Mach number of 0.4.

4.15 For an aircraft wing with leading-edge sweep 30 deg, aspect ratio 4, sectional lift-curve slope 0.1/deg, dihedral angle 3 deg, taper ratio 0.5, $C_{DO} = 0.021$, $C_{D\alpha,l} = 0.0012$/deg, and span 10 m, estimate the stability derivatives C_{lp}, C_{np}, C_{lr}, and C_{nr} at M $= 0.3$ and $\alpha = 8$ deg using strip theory and compare your results with those obtained using Datcom methods. Assume that the center of gravity (moment reference point) is located 0.2 c_r ahead of the aerodynamic center and the wing root chord is located 1 m below the center of gravity ($z_W = 1.0$ m).

4.16 Estimate the vertical tail contribution to C_{lp}, C_{np}, C_{lr}, C_{nr}, $C_{y\dot{\beta}}$, $C_{l\dot{\beta}}$, and $C_{n\dot{\beta}}$ for an aircraft with following data: wing leading edge sweep = 30 deg, wing mean aerodynamic chord $\bar{c} = 2.5$ m, wing dihedral = 3 deg, wing aspect ratio = 4, wing taper ratio = 0.5, wing span = 12 m, vertical tail length = 2.5 $\bar{c}$, $a_v = 0.08/\text{deg}$, $z_v = 0.85$ m, and ratio of vertical tail area to wing area = 0.25. Assume vertical tail parameter $k = 0.80$, $z_W = 0$, and that the aircraft is operating at an angle of attack of 6 deg and M = 0.10. [Answer: $C_{lp} = -0.0019$, $C_{np} = -0.0615$, $C_{lr} = 0.0179$, $C_{nr} = -0.5888$, $C_{y\dot{\beta}} = -3.2971$, $C_{l\dot{\beta}} = -0.0527$, and $C_{n\dot{\beta}} = 1.7322$. All values are per radian.]

Chapter 5 Linear Systems, Theory, and Design: A Brief Review

5.1 Introduction

Generally, a dynamical system is characterized by a differential equation that gives a relation between the input and the output of that system. A dynamical system may be linear or nonlinear. It is said to be linear if the differential equation that characterizes the system is linear.

A differential equation is linear if the coefficients are constants or functions of only the independent variable and not that of the dependent variable. The most important property of the linear systems is the applicability of the principle of superposition, i.e., if $y_1(t)$ and $y_2(t)$ are two solutions to inputs $r_1(t)$ and $r_2(t)$, then the solution to the new input $r(t) = c_1 r_1(t) + c_2 r_2(t)$ is given by $y(t) = c_1 y_1(t) + c_2 y_2(t)$. This feature enables us to build system response to any complex input function by expressing it as a sum of several simple input functions.

A system is said to be nonlinear if the differential equation that characterizes it is nonlinear. A differential equation is nonlinear if it contains products or powers of the dependent variable or its derivatives. Nonlinear differential equations, in general, are quite difficult to solve. Furthermore, the property of superposition does not hold for nonlinear systems.

A control system may be an open- or closed-loop system. An open-loop system is one in which the output has no effect on the input. In other words, in an open-loop system the output is not fed back for comparison with the input for regulation. An open-loop control can be used in practice if the relation between the output and the input is precisely known and the system is not subject to internal parameter variations or external disturbances. A closed-loop system is one in which the output is measured and is fed back to the input for comparison and system regulation. An advantage of the closed-loop or feedback system is that the system response will be relatively insensitive to internal parameter variations or external disturbances. For open-loop systems, stability is not a major concern. However, it is of major concern for a closed-loop system because a closed-loop system may tend to overcorrect itself and in that process develop instability.

In this chapter, we will review the basic principles of linear time-invariant systems and their representation in the transfer function form using Laplace transform. We will also discuss system response to standard inputs such as unit-step function and derive expressions for steady-state errors. Furthermore, we will briefly discuss the frequency response and stability of closed-loop systems and the design of compensators. Finally, we will give a brief exposure to modern state-space analyses and design methods.

5.2 Laplace Transform

For linear systems, the application of Laplace transform enables us to express the given differential equation in an algebraic form that greatly simplifies the analyses of control systems. Obviously this type of simplification is not possible for nonlinear systems. In view of this, one often introduces what is called an equivalent linear system. Such a linearized system is valid for only a limited range of parameter values. The linearization process may have to be repeated several times to cover the entire range of parameter values of interest.

In this section, we will briefly review the main results on Laplace transform that are useful in the analyses and design of aircraft control systems. For more information, the reader may refer to the standard texts on linear systems [1–3].

The Laplace transform of a function $f(t)$ is defined as [2]

$$L[f(t)] = \bar{f}(s) = \int_0^\infty f(t)e^{-st}\,\mathrm{d}t \tag{5.1}$$

where s is a complex variable, equal to $\sigma + j\omega$. Quite often, s is also called the Laplace variable. We assume that $f(t)$ satisfies all the conditions for the existence of its Laplace transform.

The inverse of a Laplace transform is defined as [2]

$$f(t) = L^{-1}[\bar{f}(s)] = \frac{1}{2\pi j}\int_{\sigma - j\omega}^{\sigma + j\omega} \bar{f}(s)e^{st}\,\mathrm{d}s, \quad t > 0 \tag{5.2}$$

A table of useful Laplace and inverse Laplace transforms is presented in Appendix B. Some of the important theorems on Laplace transform are summarized in the following.

1. ***Translated function.*** The Laplace transform of the translated function $f(t - \alpha)$ (Fig. 5.1), where $f(t - \alpha) = 0$ for $0 < t < \alpha$ can be obtained as follows.

   We have

$$L[f(\tau)] = \bar{f}(s) = \int_0^\infty f(\tau)e^{-s\tau}\,\mathrm{d}\tau \tag{5.3}$$

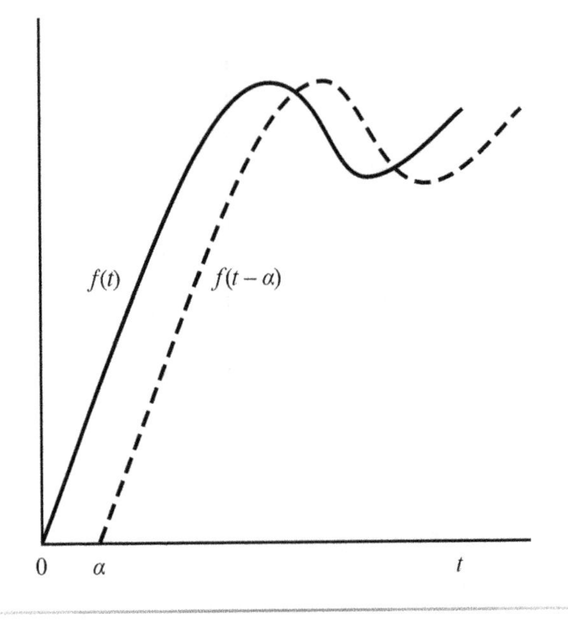

Fig. 5.1 Translated function.

Let $\tau = t - \alpha$. Then,

$$\bar{f}(s) = \int_0^\infty f(t-\alpha)\, e^{-s(t-\alpha)}\, \mathrm{d}t = e^{\alpha s} \int_0^\infty f(t-\alpha) e^{-st}\, \mathrm{d}t$$
$$= e^{\alpha s} L[f(t-\alpha)] \tag{5.4}$$

Therefore,

$$L[f(t-\alpha)] = e^{-\alpha s}\bar{f}(s) \tag{5.5}$$

This theorem states that the translation of the time function $f(t)$ by α corresponds to a multiplication of its transform by $e^{-\alpha s}$.

Using this theorem, one can obtain the Laplace transform of a pulse function as follows. A pulse function (Fig. 5.2a) is given by

$$\begin{aligned} f(t) &= A \quad 0 \le t \le t_0 \\ &= 0 \quad t < 0, \quad t_0 < t \end{aligned} \tag{5.6}$$

The given pulse function can be expressed as a sum of two-step functions, each of height A, one positive step function beginning at $t = 0$ and the other negative step function at $t = t_0$. Thus,

$$f(t) = Al(t) - Al(t - t_0) \tag{5.7}$$

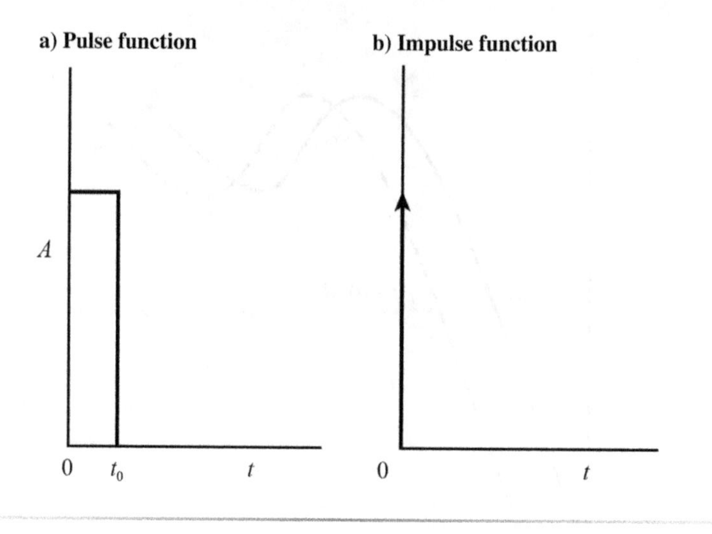

Fig. 5.2 Pulse and impulse functions.

where $l(t)$ and $l(t - t_0)$ are the unit step functions originating at $t = 0$ and $t = t_0$, respectively. The Laplace transform of a unit step function is given by

$$L[f(t)] = \int_0^\infty l(t)e^{-st}\,dt = \int_0^\infty 1e^{-st}dt = \frac{1}{s} \tag{5.8}$$

Then,

$$L[f(t)] = L[Al(t)] - L[Al(t - t_0)] = \frac{A}{s}(1 - e^{-st_0}) \tag{5.9}$$

The impulse function is a special limiting case of a pulse function. Consider the pulse function given by

$$f(t) = \lim_{t_0 \to 0}\left(\frac{A}{t_0}\right) \quad 0 \leq t \leq t_0 \tag{5.10}$$

$$= 0 \quad t < 0 \quad t_0 < t \tag{5.11}$$

The height of the impulse is A/t_0 and its duration is t_0. Therefore, the area under the impulse is equal to A. As the duration t_0 approaches zero, the height of the pulse approaches infinity giving us an impulse (Fig. 5.2b). Note that, even though the height of the impulse tends to infinity, the area remains finite.

The Laplace transform of an impulse function is given by

$$L[f(t)] = \lim_{t_0 \to 0} \frac{A}{t_0 s}\left(1 - e^{-st_0}\right) = \lim_{t_0 \to 0} \frac{\dfrac{d}{dt_0}\left[A\left(1 - e^{-st_0}\right)\right]}{\dfrac{d}{dt_0}(t_0 s)} = A \tag{5.12}$$

An impulse function of infinite magnitude and zero duration is a mathematical fiction. However, if the magnitude of a pulse input is very large and its duration is very small, then we can approximate it as an impulse input. An impulse function whose area is equal to unity is called a unit-impulse function or Dirac delta function. The Laplace transform of a unit-impulse function is unity. A unit-impulse function occurring at $t = t_1$ is usually denoted by $\delta(t - t_1)$, which has the following properties:

$$\delta(t - t_1) = 0 \quad t \neq t_1 \tag{5.13}$$

$$= \infty \quad t = t_1 \tag{5.14}$$

$$\int_{-\infty}^{\infty} \delta(t - t_1)\,dt = 1 \tag{5.15}$$

The concept of the unit-impulse is very useful in differentiating discontinuous functions. For example,

$$\delta(t) = \frac{d}{dt} l(t) \tag{5.16}$$

where $\delta(t)$ and $l(t)$ are the unit-impulse and unit-step functions, respectively, and both occur at the origin. Thus, integrating a unit-impulse function $\delta(t)$, we get a unit-step function $l(t)$. The concept of impulse function helps us represent functions involving multiple discontinuities. Such a representation will have that many impulse functions as the number of discontinuities and the magnitude of each impulse function will be equal to the magnitude of the corresponding discontinuity.

2. ***Multiplication of $f(t)$ by $e^{-\alpha t}$.*** The Laplace transform of the function $e^{-\alpha t} f(t)$ is given by

$$L\left[e^{-\alpha t} f(t)\right] = \int_0^{\infty} f(t) e^{-\alpha t} e^{-st}\,dt = \bar{f}(s + \alpha) \tag{5.17}$$

Thus, multiplying the function $f(t)$ by $e^{-\alpha t}$ has the effect of replacing the Laplace variable s by $s + \alpha$. Here, α may be real or complex.

3. ***Change of time scale.*** Suppose the time t is changed to t/α, then

$$L\left[f\left(\frac{t}{\alpha}\right)\right] = \int_0^{\infty} f\left(\frac{t}{\alpha}\right) e^{-st}\,dt \tag{5.18}$$

Let $t_1 = t/\alpha$ and $s_1 = \alpha s$ so that $s_1 t_1 = st$. Then,

$$L\left[f\left(\frac{t}{\alpha}\right)\right] = \int_0^{\infty} f(t_1)\, e^{-s_1 t_1}\, \mathrm{d}(\alpha t_1) = \alpha \bar{f}(s_1) = \alpha \bar{f}(\alpha s) \tag{5.19}$$

As an example, consider $f(t) = e^{-t}$ so that $f(t/4) = e^{-0.25t}$. We have $\alpha = 4$ and $s_1 = 4s$. Then,

$$L[f(t)] = \bar{f}(s) = \frac{1}{s+1} \tag{5.20}$$

and

$$L\left[f\left(\frac{t}{4}\right)\right] = \alpha \bar{f}(\alpha s) = \frac{4}{4s+1} \tag{5.21}$$

4. ***Differentiation.*** The Laplace transform of the derivative of a function is given by

$$L\left[\frac{\mathrm{d}}{\mathrm{d}t} f(t)\right] = s\bar{f}(s) - f(0) \tag{5.22}$$

To prove this theorem, we proceed as follows:

$$\int_0^{\infty} f(t) e^{-st}\, \mathrm{d}t = f(t) \frac{e^{-st}}{-s}\bigg|_{t=0}^{t=\infty} - \int_0^{\infty} \frac{\mathrm{d}f(t)}{\mathrm{d}t} \frac{e^{-st}}{-s}\, \mathrm{d}t \tag{5.23}$$

Then,

$$\bar{f}(s) = \frac{f(0)}{s} + \frac{1}{s} L\left[\frac{\mathrm{d}f(t)}{\mathrm{d}t}\right] \tag{5.24}$$

so that

$$L\left[\frac{\mathrm{d}f(t)}{\mathrm{d}t}\right] = s\bar{f}(s) - f(0) \tag{5.25}$$

In a similar fashion, we can obtain the Laplace transforms of higher order derivatives of $f(t)$. For example,

$$L\left[\frac{\mathrm{d}^2 f(t)}{\mathrm{d}t^2}\right] = s^2 \bar{f}(s) - sf(0) - \dot{f}(0) \tag{5.26}$$

where $\dot{f}(0)$ is the value of $\mathrm{d}f(t)/\mathrm{d}t$ at $t = 0$.

5. ***Final value theorem.*** This theorem gives

$$f(\infty) = \lim_{t \to \infty} f(t) = \lim_{s \to 0} \left[s\bar{f}(s)\right] \tag{5.27}$$

To prove this theorem, take the limit as s approaches zero in Eq. (5.25).

$$\lim_{s\to 0}\int_0^\infty \left[\frac{df(t)}{dt}\right] e^{-st}\,dt = \lim_{s\to 0}\left[s\bar{f}(s)\right] - f(0) \tag{5.28}$$

Because $e^{-st} = 1$ as $s \to 0$, we have

$$\int_0^\infty \left[\frac{df(t)}{dt}\right] dt = f(\infty) - f(0) = \lim_{s\to 0}\left[s\bar{f}(s)\right] - f(0) \tag{5.29}$$

Hence,

$$f(\infty) = \lim_{s\to 0}\left[s\bar{f}(s)\right] \tag{5.30}$$

This theorem is very useful in determining the steady-state value of a given function using its Laplace transform.

6. ***Initial value theorem.*** The initial value of a function $f(t)$ is given by

$$f(0_+) = \lim_{s\to\infty} s\bar{f}(s) \tag{5.31}$$

To prove this theorem, consider the Laplace transform of $df(t)/dt$ and take the limit as $s \to \infty$ in Eq. (5.25).

$$\lim_{s\to\infty}\int_0^\infty \left[\frac{df(t)}{dt}\right] e^{-st}\,dt = \lim_{s\to\infty}\left[s\bar{f}(s) - f(0)\right] \tag{5.32}$$

As s approaches infinity, e^{-st} approaches zero. Hence,

$$f(0) = \lim_{s\to\infty}\left[s\bar{f}(s)\right] \tag{5.33}$$

7. ***Integration theorem.*** The Laplace transform of the integral of $f(t)$ is given by

$$L\left[\int f(t)\,dt\right] = \frac{\bar{f}(s)}{s} + \frac{f^{-1}(0)}{s} \tag{5.34}$$

where $f^{-1} = \int f(t)\,dt$ evaluated $t = 0$.

To prove this theorem, we proceed as follows:

$$L\left[\int f(t)\,dt\right] = \int_0^\infty \left[\int f(t)\,dt\right] e^{-st}\,dt \tag{5.35}$$

$$= \left[\int f(t)\,dt\right] \frac{e^{-st}}{-s}\bigg|_0^\infty + \frac{1}{s}\int_0^\infty f(t)e^{-st}\,dt \tag{5.36}$$

$$= \frac{1}{s}\int f(t)\,dt\big|_{t=0} + \frac{1}{s}\int_0^\infty f(t)e^{-st}\,dt \tag{5.37}$$

$$= \frac{f^{-1}(0)}{s} + \frac{\bar{f}(s)}{s} \tag{5.38}$$

Hence, the theorem is proved.

8. ***Convolution integral.*** The integral of the form $\int_0^t f_1(t-\tau)f_2(\tau)\,d\tau$ is called the convolution integral and is frequently encountered in the study of control systems.

   The Laplace transform of the convolution integral is given by

$$L\left[\int_0^t f_1(t-\tau)f_2(\tau)\,d\tau\right] = \bar{f}_1(s)\bar{f}_2(s) \tag{5.39}$$

For the proof of this theorem, the reader may refer elsewhere [1, 3].

5.3 Transfer Function

Let us consider a linear system represented by the following differential equation:

$$\ddot{y} + a\dot{y} + by = kr(t) \tag{5.40}$$

where a is the damping constant, b is frequency parameter, and $r(t)$ is the input function. We assume $\dot{y}(0) = y(0) = 0$.

Taking the Laplace transform of both sides and using the initial conditions $\dot{y}(0) = y(0) = 0$,

$$s^2\bar{y}(s) + as\bar{y}(s) + b\bar{y}(s) = k\bar{r}(s) \tag{5.41}$$

$$\frac{\bar{y}(s)}{\bar{r}(s)} = \frac{k}{s^2 + as + b} \tag{5.42}$$

Let

$$G(s) = \frac{\bar{y}(s)}{\bar{r}(s)} \tag{5.43}$$

so that

$$G(s) = \frac{k}{s^2 + as + b} \tag{5.44}$$

Here, $G(s)$ is the ratio of the Laplace transform of the output to the Laplace transform of the input and is called the transfer function of the given system. If the input is a unit-impulse function whose Laplace transform is unity, then the transfer function is equal to the Laplace transform of the output. In other words, by measuring the output for a unit-impulse function input, one can deduce the information on the system transfer function.

The points in the s-plane where the function $G(s)$ equals zero are called zeros of $G(s)$ and the points where $G(s)$ approaches infinity are called poles of $G(s)$. As an example, consider the transfer function

$$G(s) = \frac{k(s + z_1)(s + z_2)}{(s + p_1)(s + p_2)(s + p_3)}$$

Here, the transfer function $G(s)$ has two zeros at $s = -z_1$ and $-z_2$, and three poles at $s = -p_1$, $-p_2$, and $-p_3$. If z_1 or z_2 is equal to zero, then the corresponding zero is located at the origin. Similarly, p_1, p_2, or p_3 is zero, then the corresponding pole is located at the origin.

5.4 System Response

The system response depends on the order of the system. The order of the system refers to the order of the differential equation representing the physical system or the degree of the denominator of the corresponding transfer function. For example, the system represented by the following second-order differential equation

$$m\ddot{x} + c\dot{x} + kx = u(t) \tag{5.45}$$

is a second-order system. If

$$G_1(s) = \frac{k}{(s + a)} \tag{5.46}$$

$$G_2(s) = \frac{k}{s^2 + as + b} \tag{5.47}$$

then $G_1(s)$ is a first-order system and $G_2(s)$ is a second-order system.

Generally, the output or response of a system consists of two parts: 1) the natural or free response, and 2) forced response. In the following, we discuss the unit-step response of typical first- and second-order systems.

5.4.1 Response of First-Order Systems

Consider a typical first-order system

$$G(s) = \frac{s+b}{s+a} \tag{5.48}$$

The response of this system to a unit-step function whose Laplace transform $1/s$ is given by

$$\bar{y}(s) = G(s)\bar{r}(s) \tag{5.49}$$

$$= \left(\frac{s+b}{s+a}\right)\left(\frac{1}{s}\right) \tag{5.50}$$

It is convenient to use the method of partial fractions to factor the right-hand side. Let

$$\frac{s+b}{s(s+a)} = \frac{A}{s} + \frac{B}{s+a} \tag{5.51}$$

Multiply throughout by $s(s+a)$ so that

$$s + b = A(s+a) + Bs \tag{5.52}$$

This identity is supposed to hold for all values of s. Therefore, with $s = -a$, we get $B = (a-b)/a$ and, with $s = 0$, we get $A = b/a$. Then,

$$\bar{y}(s) = \frac{b}{as} + \frac{(a-b)}{a(s+a)} \tag{5.53}$$

Taking the inverse Laplace transform,

$$y(t) = \frac{b}{a} + \left(\frac{a-b}{a}\right)e^{-at} \tag{5.54}$$

We note that $y(0) = 1$ and $y(\infty) = b/a$.

The first term on the right-hand side represents forced response and the second term represents the natural response. The forced response is also known as steady-state response, and the natural response is also known as transient response. Observe that the pole at $s = 0$ corresponding to the input unit-step function generates the forced response. The transient response is generated by the system pole at $s = -a$ and is of the form e^{-at}. Thus, the farther to the left the pole is located on the negative real axis, the faster the transient response will decay to zero. On the other hand, if this pole is located on the positive real axis, then the response will be of diverging nature because the output will increase steadily with time.

The zeros of the system and the input function influence the amplitude of both the steady-state and the transient response. In this case, we have only one system zero at $s = -b$, and there is no zero because of the input function.

The effect of this system zero on the amplitude of both the transient and steady-state response can be seen in Eq. (5.54).

The quantity $1/a$ is called the time constant of the given first-order system represented in Eq. (5.48). The time constant is a measure of the speed with which a system responds to an external input. The lower the value of the time constant (or higher the value of a), the faster will be the system response. Sometimes, a is also called the exponential decay frequency. For $t = 1/a$, $y(t) = 0.63$ times its final rise above the initial value. In other words, at time equal to one time constant, the output rises to 63% of its steady-state value above the initial value.

The rise time T_r is the time for the output to increase from 0.1 to 0.9 times its final or steady-state value. However, it may be noted that some authors define it as the time for the output to rise from 0.1 to 100% of the final value. However, this alternative definition is not used in this text.

For a first-order system,

$$T_r = \frac{2.2}{a} \tag{5.55}$$

The settling time T_s is defined as the time required for the output to reach, for the first occurrence, within 2% of its final or steady-state value. For the first-order system, this value is approximately given by

$$T_s = \frac{4}{a} \tag{5.56}$$

5.4.2 Response of Second-Order Systems

Now let us consider a second-order system given by

$$G(s) = \frac{b}{s^2 + as + b} \tag{5.57}$$

The response to an input $r(t)$ whose Laplace transform is $\bar{r}(s)$

$$\bar{y}(s) = G(s)\bar{r}(s) = \frac{b\bar{r}(s)}{s^2 + as + b} \tag{5.58}$$

If the input is a unit step function with $\bar{r}(s) = 1/s$, then

$$\bar{y}(s) = \frac{b}{s(s^2 + as + b)} \tag{5.59}$$

A second-order system has two poles. In general, the response of a second-order system can be any one of the four types of responses as shown in Fig. 5.3. Suppose the system poles, which depend on the values of a and b, are both real and negative as shown in Fig. 5.3a; then the corresponding response is a steady rise, without any overshoot, to the final value. This type of response is called an "overdamped" response. If the

× Poles of $G(s) = \dfrac{b}{s^2 + as + b}$

a) Overdamped system

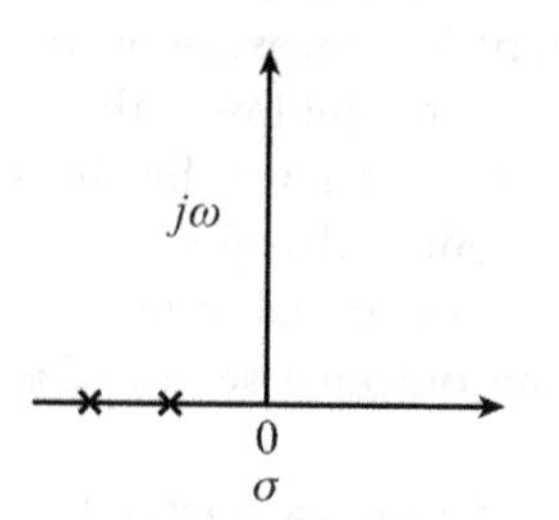

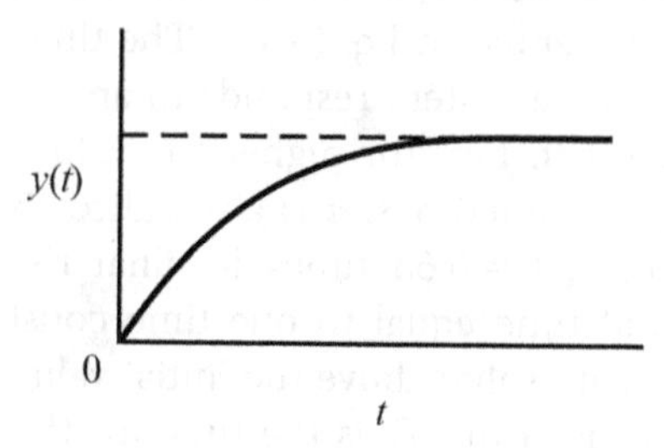

b) Oscillatory response

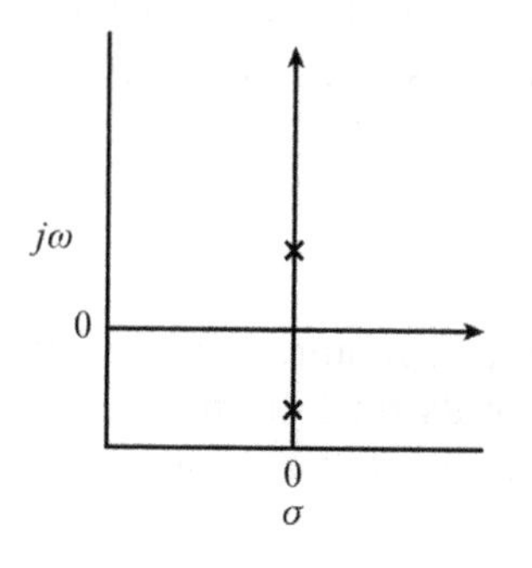

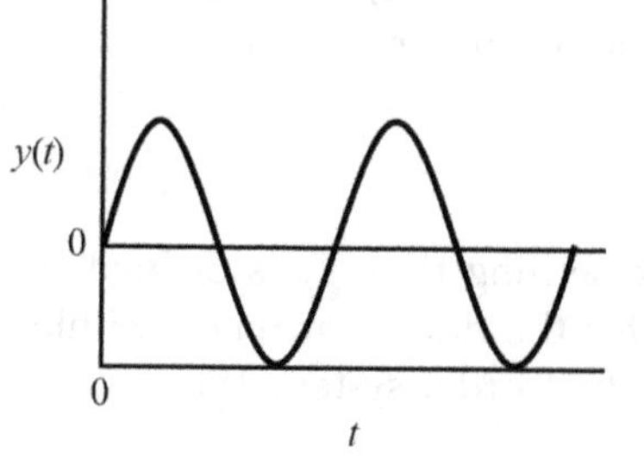

c) Underdamped response

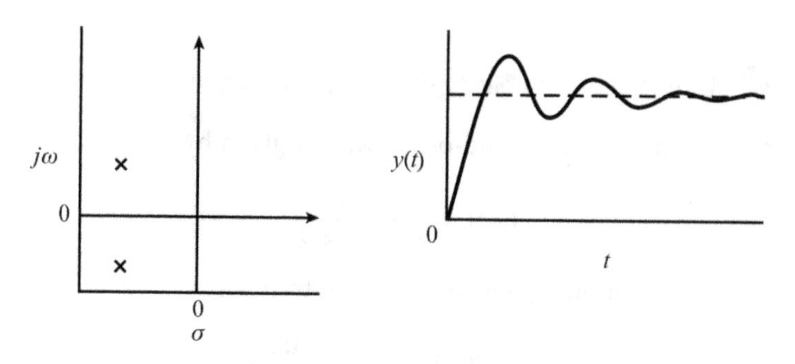

d) Critically damped response

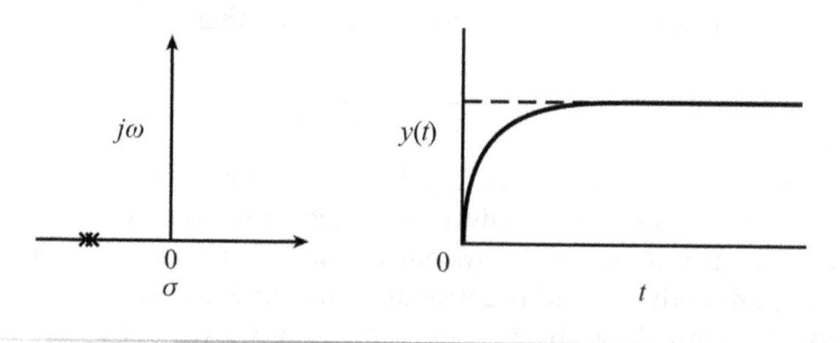

Fig. 5.3 Second-order system response.

poles are purely imaginary, then the response is a constant amplitude sinusoid that will continue forever because there is no damping in the system (Fig. 5.3b). This type of response is called "oscillatory response." The frequency of this undamped oscillation is called the natural frequency of the system. If the system poles are a pair of complex conjugate numbers with negative real parts, then the transient response will be oscillatory and is characterized by overshoots as shown in Fig. 5.3c. This type of response is called "underdamped response" and the frequency of this oscillation is called the exponential decay frequency or the damped frequency. If the poles are real, negative, and equal to each other, then the response is said to be critically damped as shown in Fig. 5.3d.

The transfer function of the second-order system given by Eq. (5.57) can be expressed in the standard form as follows:

$$G(s) = \frac{\omega_n^2}{s^2 + 2\zeta\omega_n s + \omega_n^2} \tag{5.60}$$

where

$$\omega_n = \sqrt{b} \tag{5.61}$$

$$\zeta = \frac{a}{2\omega_n} = \frac{a}{2\sqrt{b}} \tag{5.62}$$

Here, ω_n is the natural frequency of the system, and ζ is the damping ratio of the system. The damping ratio is defined as the ratio of the existing damping to that required for critical damping. For $\zeta > 1$, the second-order system has two real, negative, and unequal roots, and the system has an overdamped response as in Fig. 5.3a. When $\zeta = 1$, the two real negative roots become equal, and the motion associated with this case is called critically damped motion as shown in Fig. 5.3d. When $\zeta < 1.0$, the second-order system has a pair of complex roots with negative real parts, and the system displays a damped oscillatory motion as in Fig. 5.3c. Thus, the condition $\zeta = 1$ represents the boundary between the overdamped exponential motion and the damped oscillatory motion.

The damping ratio ζ and the natural frequency ω_n are two important parameters that characterize a second-order system. The response of a second-order system depends on the values of these two parameters. We can express the poles of the second-order system in Eq. (5.60) in terms of natural frequency ω_n and damping ratio ζ as

$$s_{1,2} = -\zeta\omega_n \pm j\omega_n\sqrt{1-\zeta^2} \tag{5.63}$$

Let

$$\sigma_d = \zeta\omega_n \quad \omega_d = \omega_n\sqrt{1-\zeta^2} \tag{5.64}$$

so that

$$s_{1,2} = -\sigma_d \pm j\omega_d \tag{5.65}$$

Here, σ_d is called the damping parameter and ω_d the damped frequency.

The unit-step input response of a second-order system of Eq. (5.60) is given by

$$\bar{y}(s) = \frac{\omega_n^2}{s\left(s^2 + 2\zeta\omega_n s + \omega_n^2\right)} \tag{5.66}$$

$$= \frac{k_1}{s} + \frac{k_2 s + k_3}{s^2 + 2\zeta\omega_n s + \omega_n^2} \tag{5.67}$$

where k_1, k_2, and k_3 are constants. Expanding the partial fractions, taking the inverse Laplace transforms, and simplifying, we obtain

$$y(t) = 1 - \frac{1}{\sqrt{1-\zeta^2}} e^{-\zeta\omega_n t} \cos(\omega_d t - \phi) \tag{5.68}$$

$$= 1 - \frac{1}{\sqrt{1-\zeta^2}} e^{-\zeta\omega_n t} \cos\left(\omega_n t\sqrt{1-\zeta^2} - \phi\right) \tag{5.69}$$

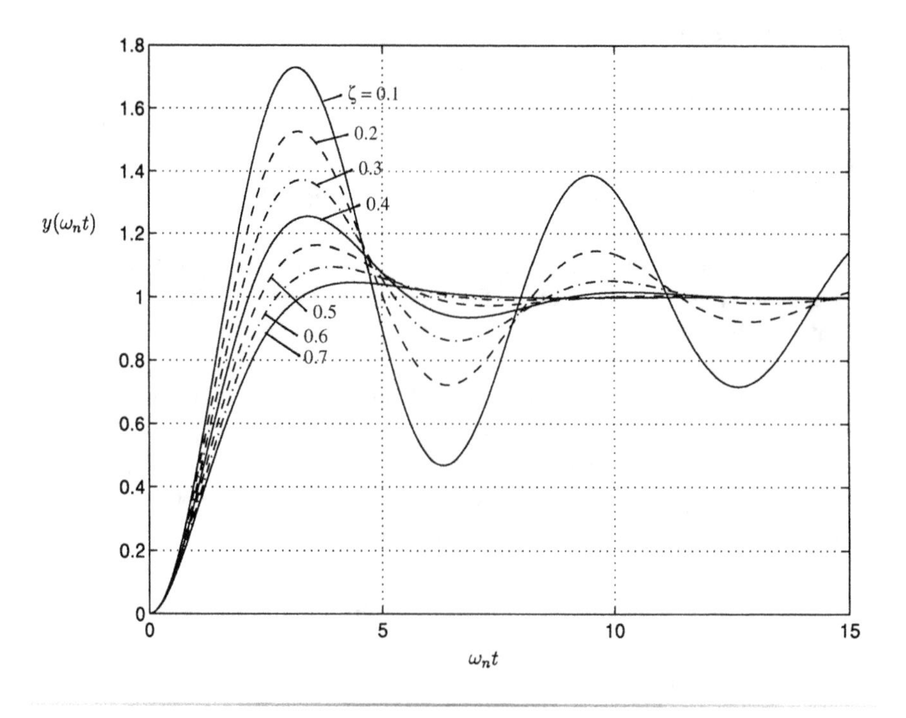

Fig. 5.4 Typical second-order system response.

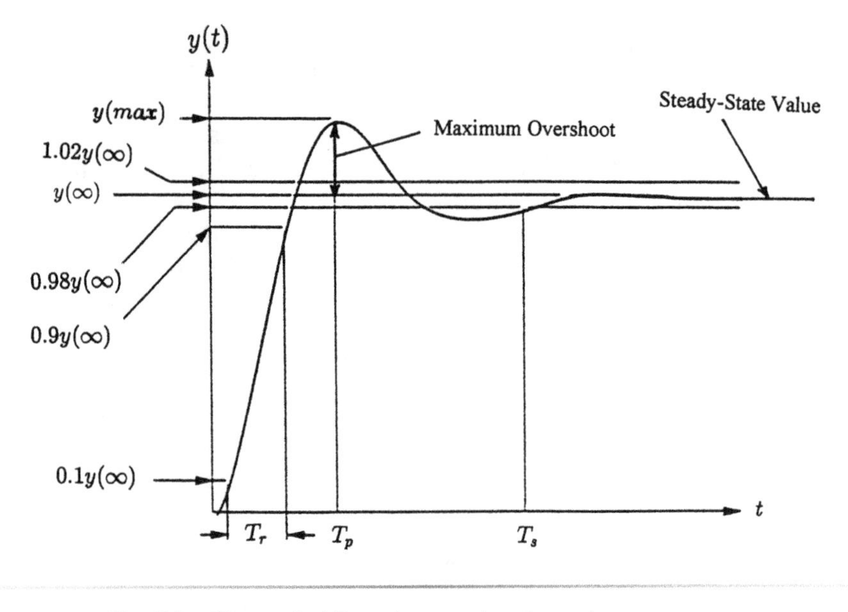

Fig. 5.5 Characteristics of second-order system response.

Here, ϕ is called the phase angle and is given by

$$\phi = \tan^{-1}\frac{\zeta}{\sqrt{1-\zeta^2}} \tag{5.70}$$

If $\phi > 0$, then the output is said to lag the input and if $\phi < 0$, then the output leads the input.

The typical response for various values of the damping parameter ζ are shown in Fig. 5.4. Because the time appears as a product $\omega_n t$ in Eq. (5.69), it is convenient to plot $y(\omega_n t)$ vs $\omega_n t$, which has the effect of normalizing the time with respect to the system natural frequency ω_n. We observe that the lower the value of ζ, the more oscillatory is the response and the larger is the overshoot.

The parameters that characterize the response of a second-order system (see Fig. 5.5) are as follows:

1. ***Peak time*** T_p. It is the time required to reach the first or maximum peak $y(\max)$.
2. ***Percent overshoot*** O_s. It is the maximum overshoot above the final or steady-state value and expressed as a percentage of steady-state value $y(\infty)$.
3. ***Settling time*** T_s. It is the time required for the transient response to come and stay within $\pm 2\%$ of the steady-state value.
4. ***Rise time*** T_r. This is the time required for the response to rise from 0.1 to 0.9 of the final or steady-state value at its first occurrence.

Notice that the settling time and rise time are basically the same as those defined for first-order systems. The given definitions are general in nature and as such apply to systems of any order.

In the following, we will derive analytical expressions for T_p, O_s, and T_s for second-order systems. However, for the rise time T_r, it is not possible to obtain a simple analytical expression.

The peak time T_p can be obtained by differentiating Eq. (5.69) with respect to time t and finding the first zero crossing for $t \geq 0$ as

$$T_p = \frac{\pi}{\omega_d} \tag{5.71}$$

$$= \frac{\pi}{\omega_n \sqrt{1 - \zeta^2}} \tag{5.72}$$

The percent overshoot O_s is given by

$$O_s = \frac{y(\max) - y(\infty)}{y(\infty)} \times 100 \tag{5.73}$$

where $y(\max)$ is the value of $y(t)$ at $t = T_p$. For the unit-step input, $y(\infty) = 1$. Then,

$$O_s = e^{\frac{-\zeta\pi}{\sqrt{1-\zeta^2}}} \times 100 \tag{5.74}$$

This relation states that the percent overshoot depends uniquely on the damping ratio ζ. The value of the damping ratio corresponding to a given percent overshoot is given by

$$\zeta = \frac{-\ell n(O_s/100)}{\sqrt{\pi^2 + \ell n^2(O_s/100)}} \tag{5.75}$$

The settling time T_s is the value of time t when the amplitude of the damped response comes within ± 0.02 for the first occurrence. Using Eq. (5.69),

$$\frac{1}{\sqrt{1 - \zeta^2}} e^{-\zeta \omega_n T_s} = 0.02 \tag{5.76}$$

Solving, we get

$$T_s = \frac{-\ell n\left(0.02\sqrt{1 - \zeta^2}\right)}{\zeta \omega_n} \tag{5.77}$$

However, this expression is somewhat complex for frequent use. Instead, the following simple approximation is used to evaluate T_s. The numerator of Eq. (5.77) varies from 3.91 to 4.74 as ζ varies from 0 to 0.9. For typical underdamped second-order systems, the numerator is usually close to 4. In view of this, the following approximation is often used:

$$T_s = \frac{4}{\zeta\omega_n} \tag{5.78}$$

5.4.3 Minimum and Nonminimum Phase Systems

If all the poles and zeros of a system lay in the left half of the s-plane, then the given system is called a minimum phase system. If a system has at least one pole or one zero in the right half of the s-plane, then such a system is called a nonminimum phase system. A characteristic property of a nonminimum phase system is that the transient response may start out in the opposite direction to the input but comes back eventually in the same direction as the input.

For the first-order system given by Eq. (5.48), if $b \leq 0$, the system becomes a nonminimum phase system. The steady-state value will be negative, whereas the response starts out in the positive direction with an initial value equal to $+1.0$.

5.5 Steady-State Errors of Unity Feedback Systems

Ideally, control systems are designed so that the output follows the reference input all the time. In other words, it is desired that the steady-state value of the output be equal to the value of the reference input as closely as possible. However, it may not always be possible to achieve this goal and, in reality, the steady-state value of the output differs from the value of the reference input.

The steady-state error is the difference between the steady-state value of the output and the reference input. One important point to bear in mind is that the units of reference input $\bar{r}(s)$ and output $\bar{y}(s)$ must match. If not, appropriate conversion must be made before they are compared. Usually, unit-step, unit-ramp, or parabolic functions are used as test inputs to determine the steady-state error. In the following, we will derive expressions for steady-state error for unity feedback systems as shown in Fig. 5.6. It may be noted that any given nonunity feedback system (Fig. 5.7) can be expressed as an equivalent unity feedback system by adding and subtracting a unity feedback loop as shown in Fig. 5.8a and obtaining an equivalent unity feedback system as shown in Fig. 5.8b.

Let $e(t)$ be the error signal that is the difference between the output and the input. For steady-state error to be zero, $e(t) \to 0$ as $t \to \infty$.

We have

$$e(t) = r(t) - y(t) \tag{5.79}$$

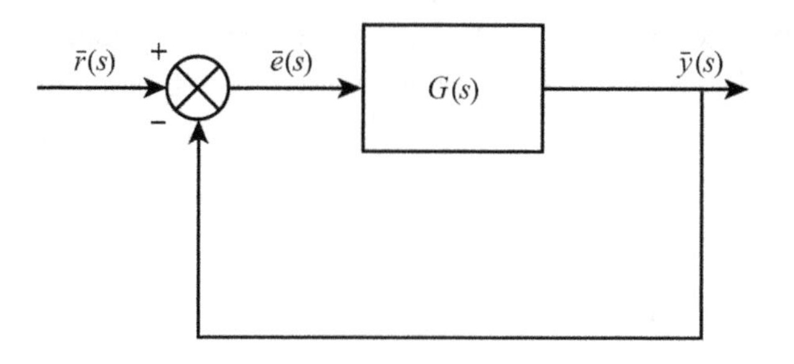

Fig. 5.6 Unity feedback system.

Taking Laplace transforms,

$$\bar{e}(s) = \bar{r}(s) - \bar{y}(s) \tag{5.80}$$

$$= \bar{r}(s) - G(s)\bar{e}(s) \tag{5.81}$$

$$= \frac{\bar{r}(s)}{1 + G(s)} \tag{5.82}$$

Using the final value theorem in Eq. (5.30), we can obtain the steady-state error $e(t)$ as follows:

$$e(\infty) = \lim_{s \to 0} [s\bar{e}(s)] \tag{5.83}$$

$$= \lim_{s \to 0} \left[\frac{s\bar{r}(s)}{1 + G(s)} \right] \tag{5.84}$$

The steady-state error to a unit-step function $\bar{r}(s) = 1/s$ is given by

$$e(\infty) = \frac{1}{1 + \lim_{s \to 0} G(s)} \tag{5.85}$$

In other words, for steady-state error to a unit-step function to be zero, $\lim_{s \to 0} G(s) = \infty$.

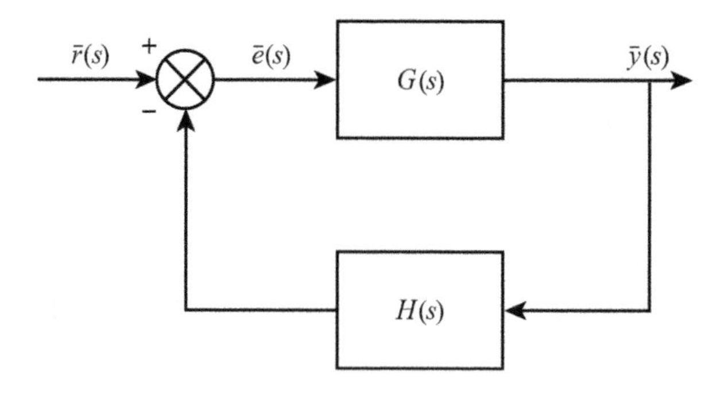

Fig. 5.7 Nonunity feedback system.

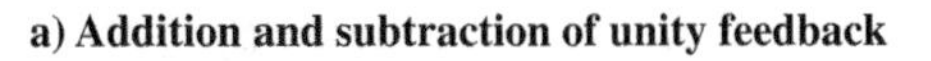

a) Addition and subtraction of unity feedback

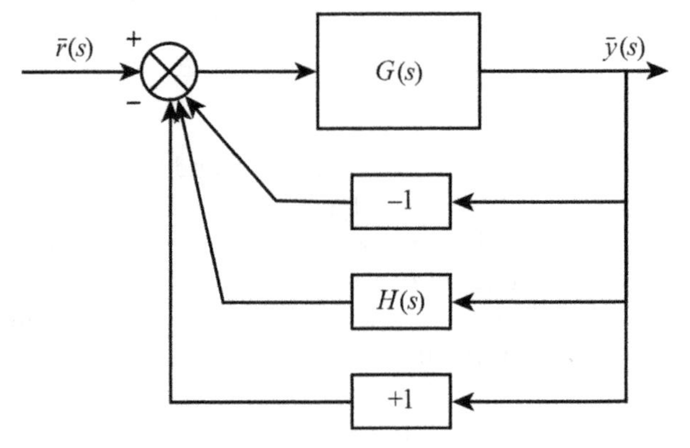

b) Equivalent unity feedback system

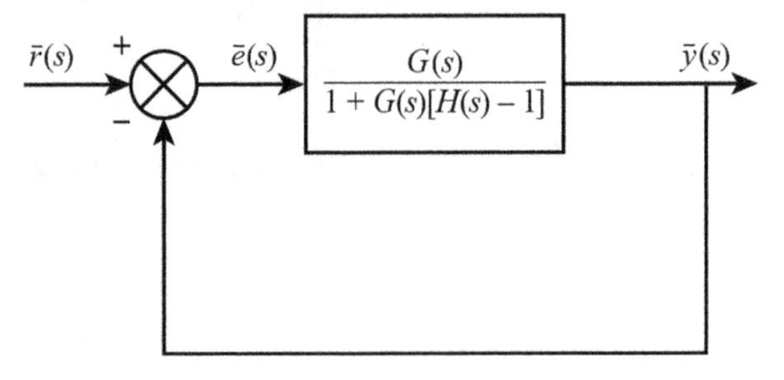

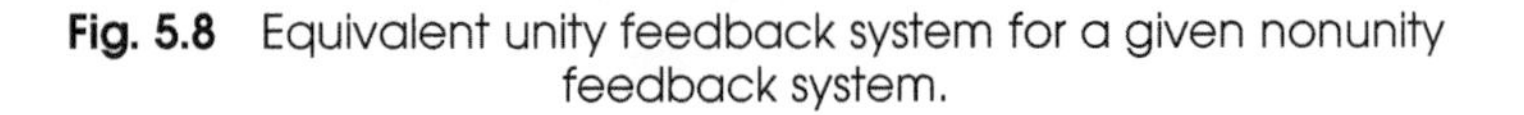
Fig. 5.8 Equivalent unity feedback system for a given nonunity feedback system.

Generally, we have

$$G(s) = \frac{(s + z_1)(s + z_2)\cdots(s + z_m)}{s^q(s + p_1)(s + p_2)\cdots(s + p_n)} \tag{5.86}$$

For rational transfer functions, $n \geq m$, i.e., the number of poles exceeds the number of zeros. The value of the index q designates the type of the system. For example, if $q = 0$, the system is said to be a type "0" system and, for a type "0" system,

$$\lim_{s \to 0} G(s) = \frac{z_1 z_2 \cdots z_m}{p_1 p_2 \cdots p_n} \tag{5.87}$$

which is finite. Hence, the steady-state error for a type "0" system to a unit-step input is nonzero and is given by

$$e(\infty) = \frac{1}{1 + K_p} \tag{5.88}$$

Here, K_p is called the position constant and is given by

$$K_p = \lim_{s \to 0} G(s) = \frac{z_1 z_2 \cdots z_m}{p_1 p_2 \cdots p_n} \tag{5.89}$$

For type "1" or higher systems ($q \geq 1$), $K_p = \infty$, and the steady-state error to a unit-step input approaches zero. A type "1" system is said to have one integrator in the forward path. In other words, the integer value of q corresponds to the number of integrators in the forward path.

It can be shown that the steady-state error to a unit-ramp function $r(t) = t$ or $\bar{r}(s) = 1/s^2$ is given by

$$e(\infty) = \frac{1}{K_v} \tag{5.90}$$

where the velocity error coefficient K_v is given by

$$K_v = \lim_{s \to 0} [sG(s)] \tag{5.91}$$

Thus, if the steady-state error for a unit-ramp function is to vanish, the velocity constant K_v must be very large, which implies that we must have $q \geq 2$. In other words, we must have at least two integrators in the forward path. Thus, for a type "0" system, the steady-state error for a unit-ramp function is infinity; for a type "1" system, it is finite; and for systems of type "2" or higher, it is zero.

Similarly, the steady-state error to a unit-parabolic input function, $r(t) = t^2$ or $\bar{r}(s) = 1/s^3$, can be obtained as

$$e(\infty) = \frac{1}{K_a} \tag{5.92}$$

where

$$K_a = \lim_{s \to 0} \left[s^2 G(s)\right] \tag{5.93}$$

We observe that for a given system to have zero steady-state error to a unit parabolic input, we must have at least have three integrators in the forward path. Therefore, the steady-state error to unit parabolic input of type "0" and type "1" systems is infinite; for type "2" systems, it is finite; and for systems of type "3" or higher, it is zero.

5.6 Frequency Response

In steady-state, a sinusoidal input to a linear system generates a sinusoidal response (output) of the same frequency. However, the magnitude and phase angles of the response are generally different from those of the input and also vary with the frequency of the applied input. In the following, we will determine the steady-state response (magnitude and phase angle) to sinusoidal inputs [2].

In general, a sinusoid input function can be represented as

$$r(t) = A\cos\omega t + B\sin\omega t \tag{5.94}$$

$$= M_i\cos(\omega t + \phi_i) \tag{5.95}$$

Here, M_i and ϕ_i are the magnitude and phase angle of the input sinusoid function and are given by the following expressions:

$$M_i = \sqrt{A^2 + B^2} \tag{5.96}$$

$$\phi_i = -\tan^{-1}\frac{B}{A} \tag{5.97}$$

We can also express a sinusoid function in phasor notation as

$$r(t) = M_i\angle\phi_i \tag{5.98}$$

Taking the Laplace transform of Eq. (5.94), we get

$$\bar{r}(s) = \left(\frac{As + B\omega}{s^2 + \omega^2}\right) \tag{5.99}$$

The response to a sinusoidal input is given by

$$\bar{y}(s) = \left(\frac{As + B\omega}{s^2 + \omega^2}\right)G(s) \tag{5.100}$$

$$= \left(\frac{As + B\omega}{(s + j\omega)(s - j\omega)}\right)G(s) \tag{5.101}$$

$$= \frac{k_1}{s + j\omega} + \frac{k_2}{s - j\omega} + \cdots \tag{5.102}$$

Because we are interested in only the steady-state response, we have ignored the terms corresponding to $G(s)$, which generate the transient response. Recall that the steady-state response comes from the poles because of input function, which in this case are at $s = \pm j\omega$. Using partial fraction method, we get

$$k_1 = \left[\frac{As + B\omega}{s - j\omega}G(s)\right]_{s=-j\omega} \tag{5.103}$$

$$= \frac{A + jB}{2}G(-j\omega) \tag{5.104}$$

Because $G(j\omega)$ is a complex number, we can write $G(j\omega) = M_g e^{j\phi_g}$ or $G(-j\omega) = M_g e^{-j\phi_g}$. Let $A + jB = M_i e^{-j\phi i}$. Then,

$$k_1 = \frac{1}{2} M_i e^{-j\phi_i} M_g e^{-j\phi_g} \tag{5.105}$$

$$= \frac{M_i M_g}{2} e^{-j(\phi_i + \phi_g)} \tag{5.106}$$

Similarly,

$$k_2 = \frac{M_i M_g}{2} e^{j(\phi_i + \phi_g)} \tag{5.107}$$

$$= k_1^* \tag{5.108}$$

where $*$ denotes the complex conjugate. Then,

$$\bar{y}_\infty(s) = \frac{M_i M_g}{2} \left[\frac{e^{-j(\phi_i + \phi_g)}}{(s + j\omega)} + \frac{e^{j(\phi_i + \phi_g)}}{(s - j\omega)} \right] \tag{5.109}$$

where the suffix ∞ denotes the steady-state value ($t \to \infty$) and M_g and ϕ_g are the magnitude and phase angle of the transfer function $G(s)$ (with $s = j\omega$) and are given by

$$M_g = |G(j\omega)| \tag{5.110}$$

$$\angle\phi_g = \angle G(j\omega) \tag{5.111}$$

Taking the inverse Laplace transforms in Eq. (5.109), we get

$$y_\infty(t) = \frac{M_i M_g}{2} \left[e^{-j(\phi_i + \phi_g + \omega t)} + e^{j(\phi_i + \phi_g + \omega t)} \right] \tag{5.112}$$

$$= M_i M_g \cos(\phi_i + \phi_g + \omega t) \tag{5.113}$$

or, in phasor notation,

$$M_\infty \angle\phi_\infty = M_i M_g \angle(\phi_i + \phi_g) \tag{5.114}$$

Thus, at any frequency, the magnitude of the steady-state output is the product of the magnitude of the input and the magnitude of the transfer function. The phase of the steady-state output is the sum of the phase of the input and the phase of the transfer function. Therefore, if we want to know how the magnitude and phase of the system response vary with frequency of a given sinusoidal input, it is sufficient to know the variation of M_g and ϕ_g with frequency because M_i and ϕ_i are supposed to be known.

This process of determining the variation of M_g and ϕ_g with frequency is called the frequency response of the system. In other words, the frequency response of a system whose transfer function is $G(s)$, $s = j\omega$ is nothing but the variation of M_g and ϕ_g with frequency ω.

One of the most widely used methods of obtaining the frequency response of a transfer function is the Bode plot. It consists of two parts: the magnitude plot in decibels where one decibel of $M = 20\log_{10} M$ and the phase plot in degrees, both plotted against frequency ω, which is usually expressed in radians/second.

Generally, the Bode plot is drawn for open-loop transfer function $G(s)$. Furthermore, if the transfer function contains a variable gain k, then the Bode plot is made for $k = 1$. For any other value of k, the corresponding Bode plot can be easily obtained by shifting the entire Bode plot by $20\log_{10} k$. The plot shifts upward if $k > 0$ and downward if $k < 0$. We will illustrate the method of drawing a Bode plot with the help of Example 5.1.

Example 5.1

Draw the Bode plot for a system given by

$$G(s) = \frac{k(s+3)}{s(s+1)(s+2)}$$

Solution:

The first step is to assume $k = 1$ and rewrite the given transfer function in the following form:

$$G(s) = \frac{\frac{3}{2}\left(\frac{s}{3}+1\right)}{s(s+1)\left(\frac{s}{2}+1\right)}$$

$$= \frac{3}{2} G_1(s)G_2(s)G_3(s)G_4(s)$$

where

$$G_1(s) = \left(\frac{s}{3}+1\right)$$

$$G_2(s) = \frac{1}{s}$$

$$G_3(s) = \frac{1}{s+1}$$

$$G_4(s) = \frac{1}{\left(\frac{s}{2}+1\right)}$$

(*Continued*)

Example 5.1 (*Continued*)

Substituting $s = j\omega$, taking logarithms on both sides, multiplying by 20 to convert to decibels, and taking absolute values, we get

$$20\log_{10}|G(j\omega)| = 20\log_{10}\frac{3}{2} + 20\log_{10}|G_1(j\omega)| + 20\log_{10}|G_2(j\omega)| + 20\log_{10}|G_3(j\omega)| + 20\log_{10}|G_4(j\omega)|$$

Now let us consider the magnitude plot of each one of the terms on the right-hand side separately. The magnitude plot of the term $20\log_{10}\frac{3}{2} = 3.5218$ for all values of the frequency ω. For the second term,

$$20\log_{10}|G_1(j\omega)| = 20\log_{10}\left|\left(\frac{j\omega}{3}+1\right)\right| = 20\log_{10}\left|\sqrt{\frac{\omega^2}{9}+1}\right|$$

For smaller values of ω (low-frequency approximation), we assume

$$20\log_{10}|G_1(j\omega)| = 20\log_{10}1 = 0$$

and, for higher values of ω (high-frequency approximation), we assume

$$20\log_{10}|G_1(j\omega)| = 20\log_{10}\frac{\omega}{3}$$

Thus, for $\omega = 3$ rad/s, $20\log_{10}|G_1(j\omega)| = 0$ and, for $\omega = 30$ rad/s, $20\log_{10} \times |G_1(j\omega)| = 20$ db. The slope of the high-frequency approximation of this term is +20 db/decade. Here, one decade means a tenfold increase in frequency.

The frequency at which the low-frequency approximation intersects the high-frequency approximation is called the corner frequency. For the magnitude plot of $G_1(j\omega)$, the corner frequency is 3 rad/s. The low-frequency approximation holds for frequencies that are below the corner frequency, and the high-frequency approximation holds for frequencies that are above the corner frequency. Proceeding in a similar way, we find that the magnitude plots of $G_3(j\omega)$ and $G_4(j\omega)$ have corner frequencies of 1.0 and 2.0 rad/s, respectively. The magnitude plot of $G_2(j\omega)$ is a straight line and hence has no corner frequency. Each of the magnitude plots of $G_2(j\omega)$, $G_3(j\omega)$, and $G_4(j\omega)$ have a slope of –20 db/decade. The component magnitude plots of $G_1(j\omega)$, $G_2(j\omega)$, $G_3(j\omega)$, and $G_4(j\omega)$ are shown in Fig. 5.9.

The phase plot can be drawn using the relation

$$\angle G(j\omega) = \angle\frac{3}{2} + \angle G_1(j\omega) + \angle G_2(j\omega) + \angle G_3(j\omega) + \angle G_4(j\omega)$$

(*Continued*)

Example 5.1 (Continued)

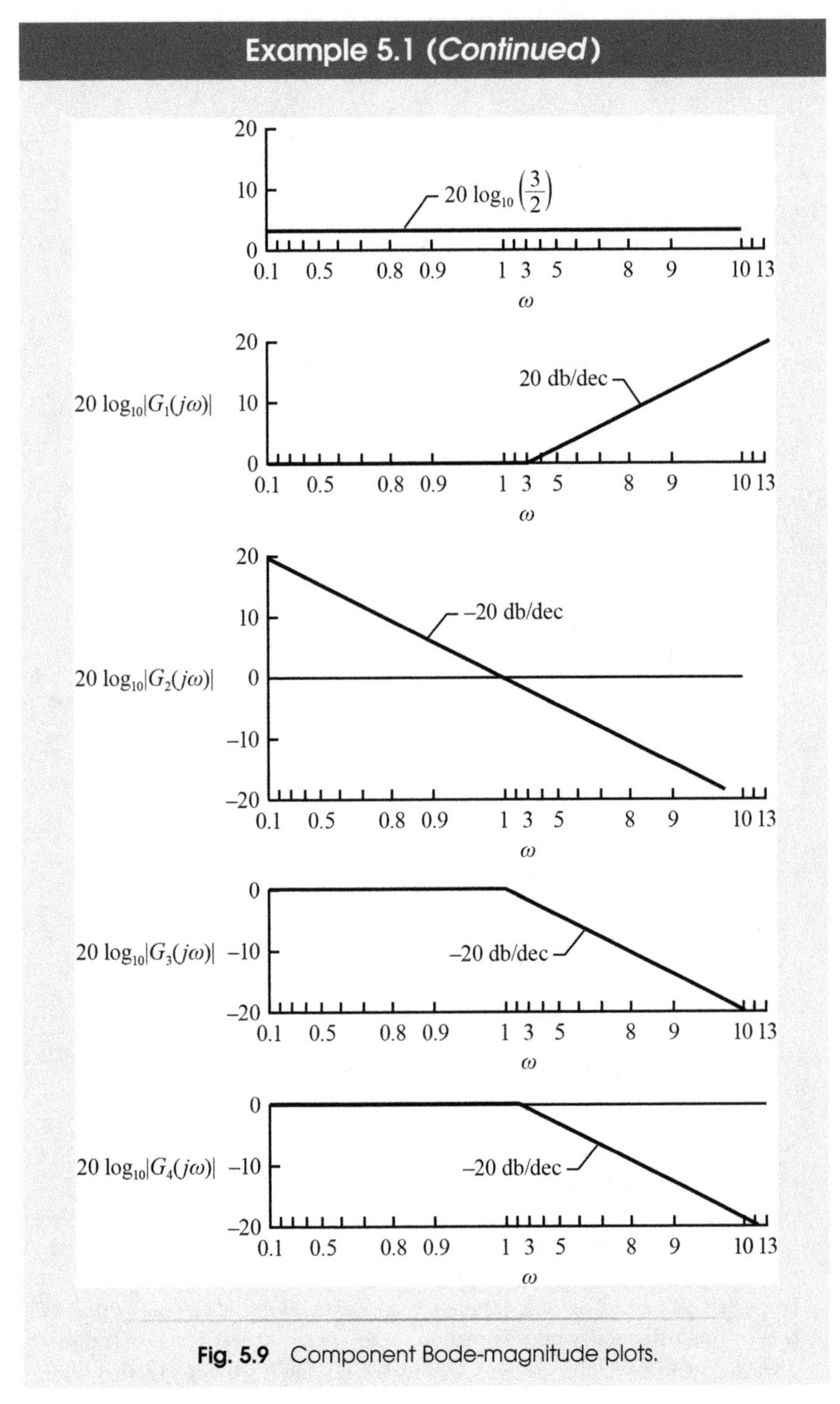

Fig. 5.9 Component Bode-magnitude plots.

(*Continued*)

Example 5.1 (*Continued*)

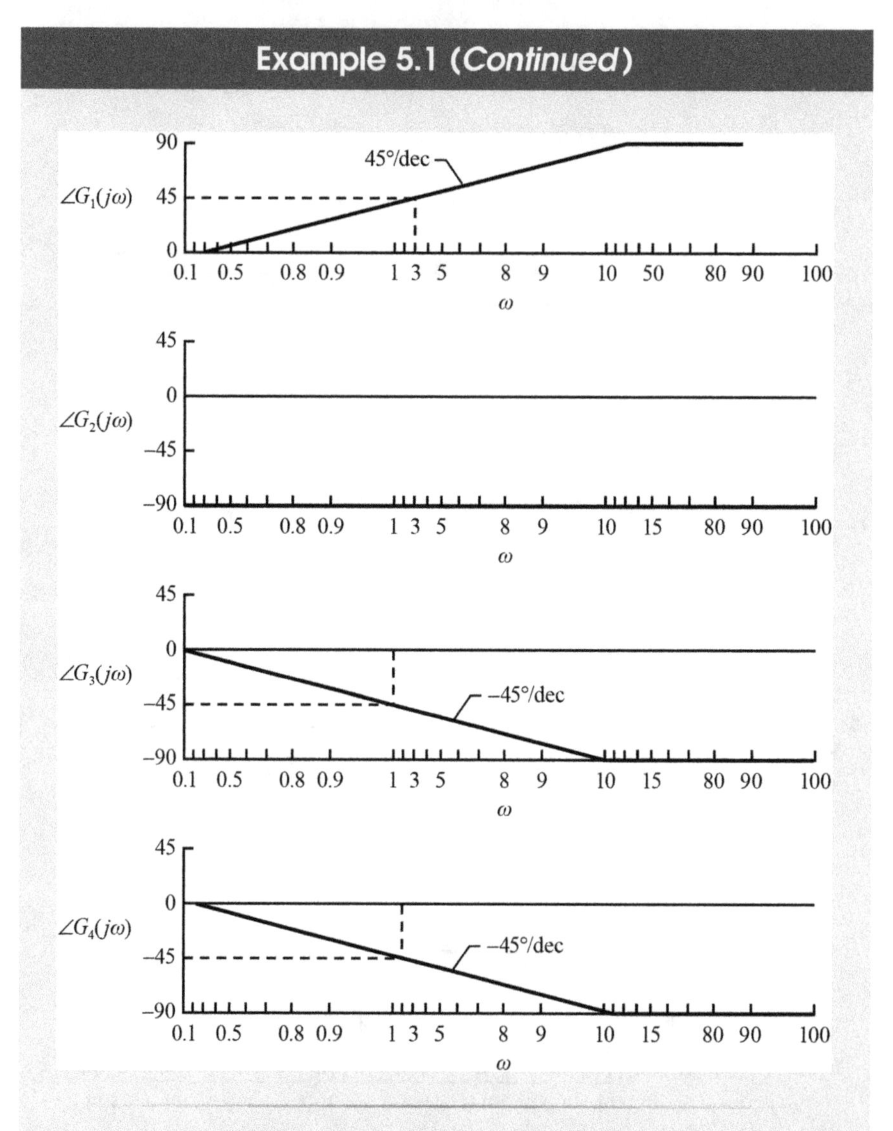

Fig. 5.10 Component Bode-phase plots.

Note that $\angle 3/2 = 0$. As before, let us consider the terms on the right-hand side one by one. The phase of the second term is given by

$$\angle G_1(j\omega) = \angle\left(\frac{j\omega}{3} + 1\right) = \tan^{-1}\left(\frac{\omega}{3}\right)$$

For small values of ω (low-frequency approximation), $\angle G_1(j\omega) = 0$. As $\omega \to \infty$ (high-frequency approximation), $\angle G_1(j\omega) = 90$ deg. For $\omega = 3\,\text{rad/s}$, $\angle G_1(j\omega) = 45$ deg and, for $\omega = 30\,\text{rad/s}$, $\angle G_1(j\omega) \simeq 90$ deg so that the

(*Continued*)

Example 5.1 (*Continued*)

slope of high-frequency approximation is 45 deg/decade. We assume that the low-frequency approximation holds for frequencies that are one decade below the frequency at which the phase angle is 45 deg. For $\angle G_1(j\omega)$, this value is 0.3 rad/s.

With these approximations, the phase plot $\angle G_1(j\omega)$ and those of other terms are shown in Fig. 5.10.

The combined magnitude and phase plots, which are the sum of component plots, are shown in Fig. 5.11.

MATLAB® is a convenient tool for control system analysis and design. We assume that the reader has access to this or a software with similar

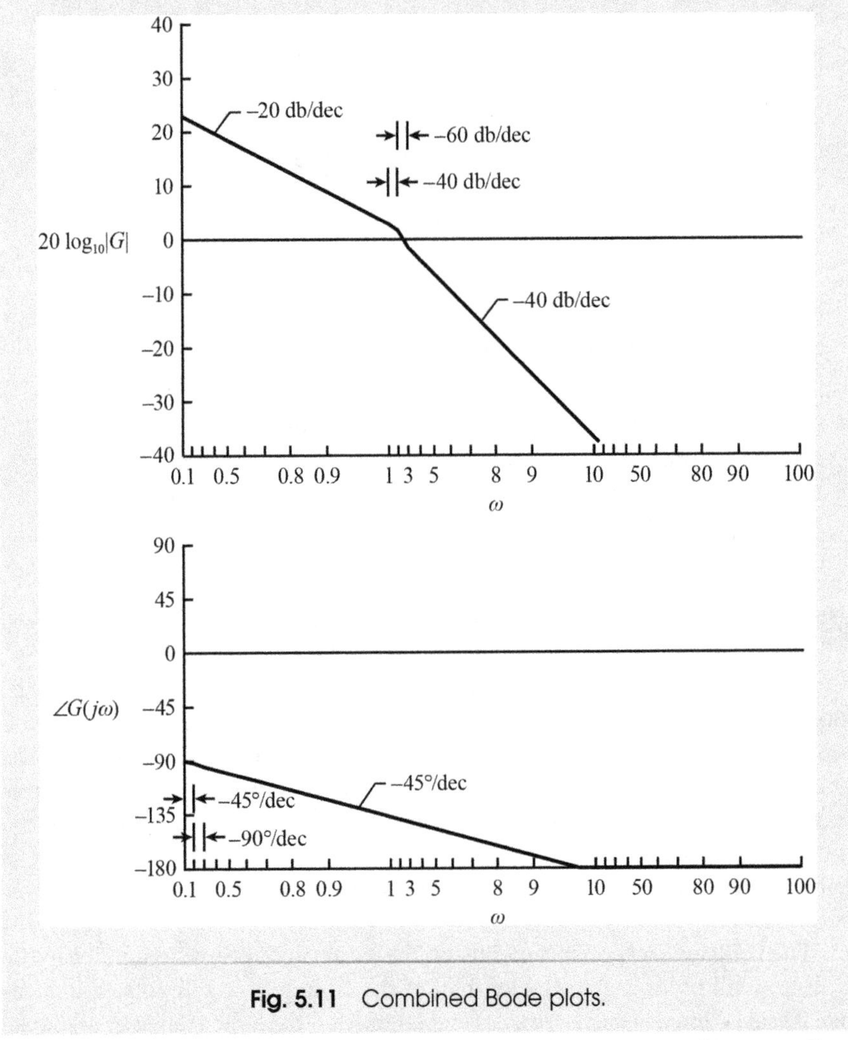

Fig. 5.11 Combined Bode plots.

(*Continued*)

Example 5.1 (Continued)

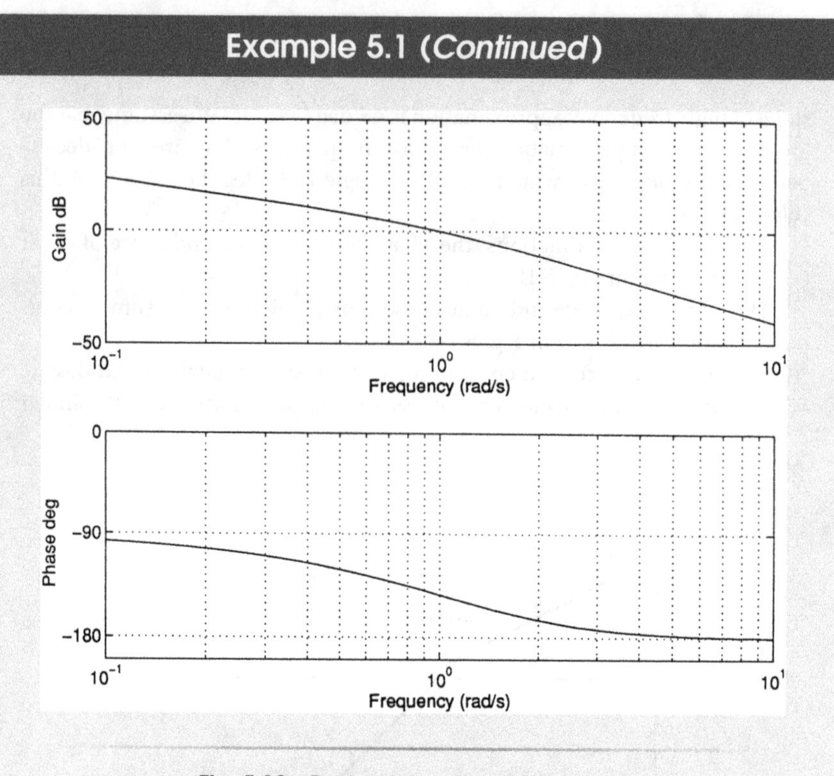

Fig. 5.12 Bode plots using MATLAB (4).

capabilities. Using MATLAB [4], the magnitude and phase plots of the given transfer function are drawn as shown in Fig. 5.12. It is interesting to observe that the approximate method that involves the concept of corner frequencies comes close to the more accurate plots given by MATLAB [4].

5.7 Stability of Closed-Loop Systems

One of the most important requirements for a control system is stability. A linear, time-invariant system is said to be stable if a bounded input produces a bounded output. In other words, for a stable system the output for a bounded input should reach a steady state. If the input is zero, then the transient or free response must decay or go to zero as the time approaches infinity. Therefore, if the output to a bounded input is not bounded and the free response does not decay, the system is said to be unstable.

The transient response depends on the location of the system poles in the s-plane. If all poles are on the left half of the s-plane—i.e., all poles are negative if real or have negative real parts if complex—then the transient response is one of exponential decay or damped oscillation, and the system is stable.

On the other hand, if any one or more of the system poles are located on the right half of the s-plane—i.e., are positive if real or have positive real parts if complex—then the transient response is one of exponential divergence or an oscillatory motion with ever-increasing amplitude. Such a system is said to be unstable. Thus, a stable system has all poles located in the left half of the s-plane, and an unstable system has one or more of its poles located in the right half of the s-plane. If some or all the system poles are located on the imaginary axis, the transient response will consist of pure oscillatory motion in which the amplitude of oscillation neither increases nor decreases. Such a system is said to be neutrally stable.

The task of determining the stability of an open-loop system is simple and straightforward because the open-loop poles are known. However, it is not so straightforward for the closed-loop systems because the closed-loop poles, which are the roots of the characteristic equation $1 + kGH = 0$, are not known. Moreover, the exercise of finding the closed-loop poles has to be repeated many times if a system parameter like the gain k is a variable. It is a simple task to determine the roots of the characteristic equation $1 + kGH = 0$ if this expression is a polynomial in s of degree lower than three. For fourth- or higher degree polynomials, the analytical determination of the roots is not a simple task. In the following, we will discuss methods that help us determine the stability of the closed-loop systems without actually solving the characteristic equation $1 + kGH = 0$. These are 1) Routh's stability criterion, 2) the root-locus method, and 3) Nyquist stability criterion.

5.7.1 Routh's Stability Criterion

Routh's stability criterion helps us determine whether any of the closed-loop poles are positive if real or have positive real parts if complex without actually solving the closed-loop characteristic equation. The procedure is as follows.

1. Express the characteristic polynomial in the following form:

$$a_0 s^n + a_1 s^{n-1} + a_2 s^{n-2} + \cdots + a_{n-1} s + a_n = 0 \tag{5.115}$$

   where the coefficients $a_0, a_1, \ldots, a_n$ are real quantities. We assume that $a_n \neq 0$ so that any zero root is removed.
2. Examine the value of each coefficient. If any coefficient is zero or negative when at least one other coefficient is positive, then Routh's criterion states that there will be at least one root of the characteristic polynomial that is imaginary or has a positive real part. In such a case, the system is not stable. Therefore, for stability, all the coefficients must be positive or must have the same sign. This forms the necessary condition for stability.

3. To check whether the sufficiency condition is satisfied, form the Routh's array as follows:

$$
\begin{array}{lccccc}
s^n & a_0 & a_2 & a_4 & a_6 & \cdot \\
s^{n-1} & a_1 & a_3 & a_5 & a_7 & \cdot \\
s^{n-2} & b_1 & b_2 & b_3 & b_4 & \cdot \\
s^{n-3} & c_1 & c_2 & c_3 & c_4 & \cdot \\
s^{n-4} & d_1 & d_2 & d_3 & d_4 & \cdot \\
\cdot & \cdot & \cdot & \cdot & \cdot & \cdot \\
\cdot & \cdot & \cdot & \cdot & \cdot & \cdot \\
\cdot & \cdot & \cdot & \cdot & \cdot & \cdot \\
\cdot & \cdot & \cdot & \cdot & \cdot & \cdot \\
s^2 & e_1 & e_2 & & & \\
s^1 & f_1 & & & & \\
s^0 & g_1 & & & &
\end{array}
\tag{5.116}
$$

where

$$b_1 = \frac{a_1 a_2 - a_0 a_3}{a_1}, \quad b_2 = \frac{a_1 a_4 - a_0 a_5}{a_1}, \quad b_3 = \frac{a_1 a_6 - a_0 a_7}{a_1}, \ldots \tag{5.117}$$

$$c_1 = \frac{b_1 a_3 - a_1 b_2}{b_1}, \quad c_2 = \frac{b_1 a_5 - a_1 b_3}{b_1}, \quad c_3 = \frac{b_1 a_7 - a_1 b_4}{b_1}, \ldots \tag{5.118}$$

and

$$d_1 = \frac{c_1 b_2 - b_1 c_2}{c_1}, \quad d_2 = \frac{c_1 b_3 - b_1 c_3}{c_1} \tag{5.119}$$

$$\ldots\ldots \tag{5.120}$$

This process is continued until nth row has been completed. The complete array of coefficients is triangular. Note that the evaluation of b_i, c_i, and d_i, etc., is continued until the remaining ones are zero. For example, for a fourth-degree polynomial in s, $a_5 = a_6 = \cdots = 0$ so that $b_3 = b_4 = \cdots = 0$, $c_2 = c_3 = \cdots = 0$, $d_2 = d_3 = \cdots = 0$, $e_2 = e_3 = \cdots = 0$, and $f_2 = f_3 = \cdots = 0$.

Routh's stability criterion states that the number of roots of the characteristic polynomial with positive real parts is equal to the number of changes in sign of the coefficients of the first column of Routh's array. It is important to note that the exact values of these coefficients need not be known; instead only the signs are required. Thus, the sufficiency condition for a closed-loop system to be stable is that all the elements of the first column of Routh's array must be positive or must have the same sign.

To summarize, the necessary and sufficient condition for the stability of a closed-loop system is that all the coefficients of the characteristic polynomial and the elements of the first column of Routh's array must be positive or must have the same sign.

If any of the coefficients of the characteristic polynomial or any element of the first column in the Routh's array is zero, then replace that term by a very small positive number ϵ and proceed as usual with the evaluation of the rest of the elements of Routh's array.

For a fourth-order polynomial, the Routh's stability criterion reduces to the following:

1. All coefficients a_0, a_1, a_2, a_3, and a_4 must be positive.
2. The Routh's discriminant $(a_1a_2 - a_0a_3)a_3 - a_1^2a_4$ must be positive.

Example 5.2

Using Routh's criterion, determine the stability of the system represented by the following characteristic polynomial:

$$s^4 + 2s^3 + 5s^2 + 2s + 2 = 0$$

Solution:

This is a fourth-degree polynomial in s. We have $a_0 = 1$, $a_1 = 2$, $a_2 = 5$, $a_3 = 2$, $a_4 = 2$, and $a_5 = 0$. Because all the coefficients are positive and none of the coefficients a_0 to a_4 is zero, the necessary condition for stability is satisfied. To examine whether the sufficiency condition is satisfied, we form Routh's array as follows:

$$\begin{array}{llll} s^4: & 1 & 5 & 2 \\ s^3: & 2 & 2 & 0 \\ s^2: & 4 & 2 & 0 \\ s^1: & 1 & 0 & \\ s^0: & 2 & & \end{array}$$

We observe that all the elements of the first column of this table are positive; hence the sufficiency condition is also satisfied. Hence, the characteristic polynomial has no positive real root or a complex root with positive real part and the given system is stable.

Example 5.3

Given the characteristic polynomial

$$s^4 + 3s^3 + 2s^2 + 4s + 1 = 0$$

Examine the stability of the system using Routh's stability criterion.

(*Continued*)

Example 5.3 (*Continued*)

Solution:

Because all of the coefficients of this fourth-degree polynomial are positive, the necessary condition is satisfied. To see whether the sufficiency condition is satisfied, form Routh's array as follows:

$$
\begin{array}{llll}
s^4: & 1 & 2 & 1 \\
s^3: & 3 & 4 & 0 \\
s^2: & \frac{2}{3} & 1 & 0 \\
s^1: & -\frac{1}{2} & 0 & \\
s^0: & 1 & &
\end{array}
$$

There are two sign changes in the first column starting with the row corresponding to s^2. Hence, there will be two roots that are either positive or have positive real parts and the given system is unstable.

Example 5.4

For the system whose characteristic polynomial is given by

$$s^4 + 2s^2 + 5s + 2 = 0$$

Examine the stability of the system using Routh's criterion.

Solution:

Notice that the s^3 term is missing. Hence, we rewrite the given polynomial as follows:

$$s^4 + \epsilon s^3 + 2s^2 + 5s + 2 = 0$$

where ϵ is a small positive number, say 0.0001. With this, we observe that the necessary condition is satisfied. To see whether the sufficiency condition is satisfied, we form Routh's array as follows:

$$
\begin{array}{llll}
s^4: & 1 & 2 & 2 \\
s^3: & \epsilon & 5 & 0 \\
s^2: & -50{,}000 & 2 & 0 \\
s^1: & 5 & 0 & \\
s^0: & 2 & &
\end{array}
$$

(*Continued*)

Example 5.4 (*Continued*)

There are two sign changes. Hence, there will be two roots that are either positive or have positive real parts. Hence, the given system is unstable. Note that if ϵ appears in any expression, we have to evaluate the value of that expression by taking the limit as ϵ tends to zero.

5.7.2 Root-Locus Method

The root-locus is a powerful method of determining the nature of transient response and stability of closed-loop control systems without actually solving the characteristic equation to determine the closed-loop poles. It is a graphical method and is particularly well suited for application to those problems where any parameter or the loop-gain is a variable.

Consider a closed-loop system as shown in Fig. 5.13. The closed-loop transfer function is given by

$$T(s) = \frac{kG(s)}{1 + kG(s)H(s)} \tag{5.121}$$

The equation

$$1 + kG(s)H(s) = 0 \tag{5.122}$$

is known as the characteristic equation of the given closed-loop system. The roots of this equation are also called the eigenvalues of the closed-loop system. In other words, the poles of $T(s)$ are the eigenvalues of the closed-loop system. The root-locus is a plot of the variation of roots

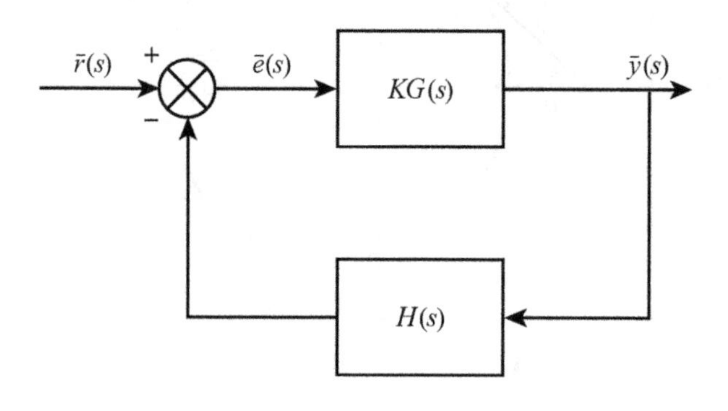

Fig. 5.13 Feedback control system.

of the characteristic equation of the closed-loop system as the parameter k is varied from zero to infinity. Using Eq. (5.122), we can deduce the following two conditions for a given point to lie on the root-locus [1–3].

1. ***Magnitude condition.*** If a given point s is to lie on the root-locus, we must have

$$|kG(s)H(s)| = 1 \tag{5.123}$$

2. ***Phase condition.*** For a given point to lie on the root-locus, we must have

$$\angle kG(s)H(s) = (2n+1)180 \tag{5.124}$$

where $n = 0, \pm 1, \pm 2, \ldots$ Note that the expression on the right-hand side of Eq. (5.124) is an odd multiple of 180 deg with either positive or negative sign. These two conditions form the basis of sketching the root-locus as the parameter k varies from zero to infinity.

To understand the meaning of Eqs. (5.123) and (5.124), let us refer to Fig. 5.14. Suppose P is to be a point on the root-locus; then according to the magnitude condition

$$k = \frac{1}{|G(s)H(s)|} = \frac{\prod |(s+p_i)|}{\prod |(s+z_i)|} = \frac{\prod l_{i,p}}{\prod l_{i,z}} \tag{5.125}$$

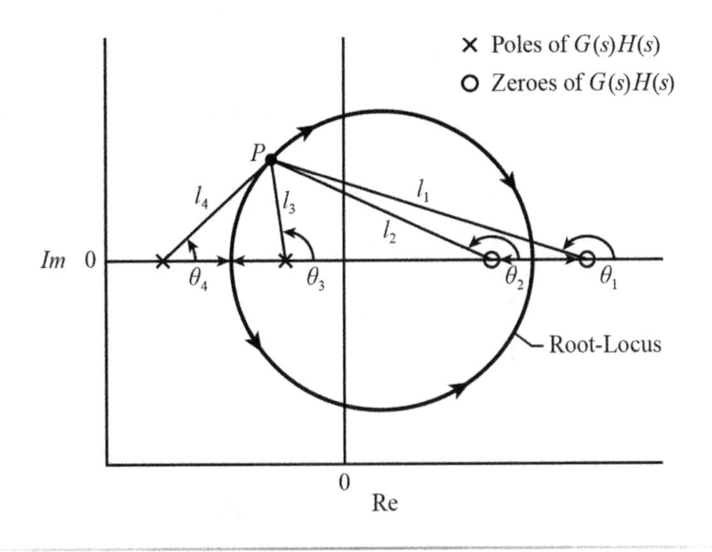

Fig. 5.14 Magnitudes and angles of vectors for a point on the root-locus.

where $|(s+p_i)| = l_{i,p}, i = 1,\ldots,n,$ and $|(s+z_i)| = l_{i,z}, i = 1,\ldots,m,$ are the magnitudes of the vectors drawn from each of the n poles and m zeros to the given point on the root-locus. As said before, for rational transfer functions, $n \geq m$. The phase (angle) condition is given by

$$\angle kG(s)H(s) = \sum \angle(s+z_i) - \sum \angle(s+p_i) = \sum \theta_{zi} - \sum \theta_{pi} = (2n+1)180 \tag{5.126}$$

where $\angle(s+z_i) = \theta_{z_i}, i = 1,\ldots,m$ and $\angle(s+p_i) = \theta_{p_i}, i = 1,\ldots,n$ are the angles measured from the positive real axis to the vectors drawn to point P from each of the m zeros and n poles. For the root-locus shown in Fig. 5.14, the magnitude and phase conditions are

$$k = \frac{l_3 l_4}{l_1 l_2}, \quad \theta_1 + \theta_2 - \theta_3 - \theta_4 = (2n+1)180 \tag{5.127}$$

where $n = 0, \pm 1, \pm 2, \ldots.$

Rules for Sketching Root-Locus

1. ***Number of branches of root-locus.*** Note that each of the closed-loop poles moves in the s-plane as the parameter k varies. Therefore, the number of branches of the root-locus is equal to the number of closed-loop poles.
2. ***Symmetry.*** For all physical systems, the coefficients of the characteristic equation are real. As a result, if any of its roots are complex, then they occur in pairs as complex conjugates. All the real roots lie on either the positive or negative real axes. Hence, the root-locus of a physical system is always symmetric with respect to the real axis.
3. ***Real axis segments.*** Whether a given segment of the real axis forms a part of the root-locus depends on the angle condition shown in Eq. (5.124), i.e., the algebraic sum of the angles subtended at that point because all the poles and zeros must be equal to an odd multiple of 180 deg. The net angle contribution of the complex poles or zeros is zero because they always occur as complex conjugate pairs. Furthermore, the angle contribution of a real axis pole or zero located to the right of a point on the real axis is zero. The angle contribution to a point on the root-locus comes only from those real axis poles and zeros that are located on the left side and is equal to –180 deg for poles and +180 deg for zeros. Because the sum of all such contributions has to be an odd multiple of 180 deg, it is clear that only that part of the real axis segment forms a branch of the root-locus that lies to the left of odd number of poles and/or zeros.

4. ***Starting and ending points of root-locus.*** To understand where the root-locus begins and where it ends as the parameter k is varied from zero to infinity, let

$$G(s) = \frac{N_g(s)}{D_g(s)} \tag{5.128}$$

$$H(s) = \frac{N_h(s)}{D_h(s)} \tag{5.129}$$

Note that $N_g = 0$ and $D_g = 0$ give us, respectively, the zeros and poles of the open-loop transfer function $G(s)$. Similarly, $N_h(s) = 0$ and $D_h(s) = 0$ give, respectively, the zeros and poles of $H(s)$. Then,

$$T(s) = \frac{kN_g(s)D_h(s)}{D_g(s)D_h(s) + kN_g(s)N_h(s)} \tag{5.130}$$

When the parameter $k \to 0$, the closed-loop transfer function $T(s)$ can be approximated as

$$T(s) = \frac{kN_g(s)D_h(s)}{D_g(s)D_h(s)} \tag{5.131}$$

that is, when $k \to 0$, the poles of $T(s)$ coincide with the combined open-loop poles of $G(s)$ and $H(s)$. Therefore, the root-locus starts at the open-loop poles of the system.

When $k \to \infty$, we have

$$T(s) = \frac{N_g(s)D_h(s)}{N_g(s)N_h(s)} \tag{5.132}$$

That is, when $k \to \infty$, the poles of $T(s)$ approach the combined zeros of $G(s)$ and $H(s)$. In other words, the root-locus ends at the open-loop zeros of the system. Summarizing, the root-locus starts at the open-loop poles and ends at the open-loop zeros. This statement implies that the system should have equal number of poles and zeros, which is true if we assume that the missing zeros and poles are located at infinity. To understand this point, consider

$$G(s) = \frac{k}{s(s+3)(s+5)} \tag{5.133}$$

We have three poles at $s = 0, -3, -5$ and no finite zeros. Therefore, the missing zeros are located at $s = \infty$.

As $s \to \infty$, $G(s) = 1/s^3 = 0$, i.e., $G(s)$ has three zeros at $s = \infty$. Consider

$$G(s) = s \tag{5.134}$$

This system has a zero at $s = 0$ and a pole at infinity because, as $s \to \infty$, $G(s) \to \infty$. Similarly, $G(s) = 1/s$ has a zero at infinity because $G(s) \to 0$ as $s \to \infty$.

5. ***Asymptotes.*** The asymptotes give the behavior of the root-locus as the parameter k approaches infinity. The point of intersection of the asymptotes with the real axis σ_o (see Fig. 5.15a) and the slopes of the asymptotes M at this point are given by

$$\sigma_o = \frac{\sum \text{poles} - \sum \text{zeros}}{n_p - n_z} \tag{5.135}$$

$$M = \tan \frac{(2n+1)\pi}{n_p - n_z} \tag{5.136}$$

where n_p and n_z are the number of open-loop poles and zeros, respectively, and $n = 0, \pm 1, \pm 2, \ldots$. The running index n gives the slopes of the asymptotes that form the branches of the root-locus as $k \to \infty$.

Imaginary Axis Crossing Another characteristic feature that is of interest in the root-locus method is the point where the root-locus crosses the imaginary axis because the system stability changes at this point. If the imaginary axis crossing is from right to left of the s-plane, the closed-loop becomes stable as the gain is increased. If it is from left to right, then the closed-loop system becomes unstable on increasing the gain.

The point(s) where the root-locus crosses the imaginary axis can be determined by 1) using the Routh's criterion and finding the values of the gain k that give all the zeros in any one row of the Routh's table or 2) substituting $s = j\omega$ in the characteristic equation, setting both real and imaginary parts to zero and solving for the gain k and frequency ω. We will illustrate this second procedure in the following example.

Example 5.5

Sketch the root-locus for the unity feedback system with

$$G(s) = \frac{k(s+4)}{s(s+1)(s+2)(s+5)}$$

Solution:

We have four poles at $s = 0, -1, -2, -5$ and only one finite zero at $s = -4$. Therefore, the other three missing zeros are at infinity. We have four branches of the root-locus. Furthermore, the root-locus will be symmetrical with respect to the real axis. That segment of the real axis forms a part of the

(Continued)

Example 5.5 (*Continued*)

root-locus, which lies to the left of the odd number of poles and/or zeros. Thus, the real axis segment between the poles at 0 and -1 and between the pole at -2 and zero at -4 and all the real axis that is to the left of the pole at -5 forms the branches of the root-locus.

We should have three asymptotes corresponding to three branches of the root-locus, which seek zeros at infinity. We have

$$\begin{aligned}\sigma_o &= \frac{\sum \text{poles} - \sum \text{zeros}}{n_p - n_z} \\ &= \frac{(0-1-2-5)-(-4)}{4-1} \\ &= -\frac{4}{3}\end{aligned}$$

$$\begin{aligned}M &= \tan\frac{(2n+1)\pi}{n_p - n_z} \\ &= \tan\frac{(2n+1)\pi}{3} \\ &= \tan\frac{\pi}{3} \qquad n = 0 \\ &= \tan\pi \qquad n = 1 \\ &= \tan\frac{5\pi}{3} \qquad n = 2\end{aligned}$$

With this information, the root-locus can be sketched as shown in Fig. 5.15a. The root-locus crosses the imaginary axis from the left half to the right half of the s-plane, i.e., the closed-loop system becomes unstable as the value of the gain k is increased beyond this point.

The value of the gain k and frequency ω where the root-locus crosses the imaginary axes can be obtained as follows.

The characteristic equation [Eq. (5.122)] is

$$s^4 + 8s^3 + 17s^2 + s(10+k) + 4k = 0$$

Substituting $s = j\omega$, we obtain

$$\omega^4 - 17\omega^2 + 4k + j(-8\omega^3 + \omega[10+k]) = 0$$

Equating real and imaginary parts to zero, we get $k = 8.4856$ and $\omega = \pm 1.5201$.

MATLAB [4] is a convenient tool for plotting the root-locus. The MATLAB command RLOCUS sketches the root-locus, and the command

(*Continued*)

Example 5.5 (*Continued*)

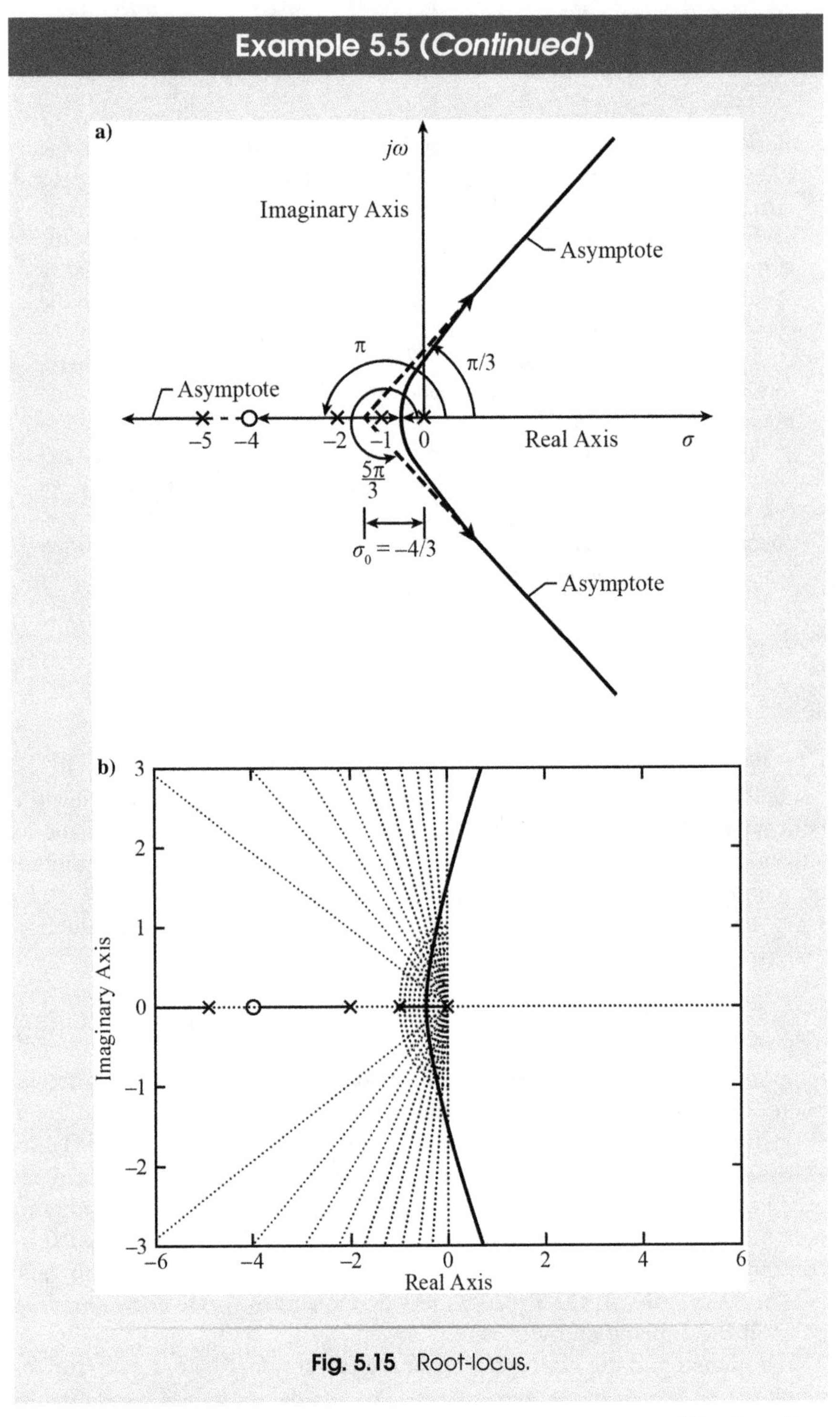

Fig. 5.15 Root-locus.

(Continued)

Example 5.5 (*Continued*)

RLOCFIND enables us to find the value of the gain k and the location of the closed-loop poles corresponding to any point on the root-locus. Using RLOCFIND, we find that $k = 8.8$ and $\omega = 1.55$ when the root-locus crosses the imaginary axis. These values are in good agreement with the analytical values. Furthermore, the corresponding locations of the closed-loop poles are -5.15, -2.88, and $0 \pm j1.55$. Note that some subjective element is involved in using the Matlab cursor to pick a desired point on the root-locus. In view of this, the values of the gain k and the locations of the closed-loop poles are likely to vary a little bit from person to person. Therefore, all such numbers given in this book should be treated as guidelines and not as absolute numbers.

The root-locus obtained using MATLAB [4] is shown in Fig. 5.15b.

We can also find other information using MATLAB [4]. For example, we can find the value of the gain k so that the closed-loop system is stable and operates with a damping ratio ζ of 0.4. Using RLOCFIND, we obtain $k = 2.0$ and closed-loop poles $p = -5.04$, -2.4, and $-0.3 \pm j0.8$.

5.7.3 Nyquist Stability Criterion

Concept of Mapping

Before we discuss the Nyquist stability criterion, let us briefly review the concept of mapping. Suppose we are given a contour A in the s-plane as shown in Fig. 5.16a and a function $F(s) = s^2 + 2s + 1$. Consider a point P on contour A in the s-plane, and let the coordinates of point P be $s = 4 + j3$. If we substitute this complex number into the given function $F(s)$, we get another complex number:

$$F(s) = (4 + j3)^2 + 2(4 + j3) + 1 = 16 + j30 \tag{5.137}$$

Suppose we plot the real and imaginary parts of this number in another plane, called the F-plane; we get point P_1 as shown in Fig. 5.16b. Point P_1 in the F-plane is said to be the image of point P in the s-plane. Here, $F(s)$ is called the mapping function. In a similar way, we can map all other points on contour A to corresponding points in the F-plane and obtain contour B. Then contour A in the s-plane is said to be mapped to contour B in the F-plane. We assume that the mapping is one to one, i.e., for every point in the s-plane, there is one and only one corresponding point in the F-plane and vice versa.

To understand the concept of mapping further, let $F(s) = s - z_1$ and let point $s = z_1$, which is the zero of $F(s)$, lie outside contour A as shown in Fig. 5.17a. Instead of using the coordinates of point P, let us use the vector

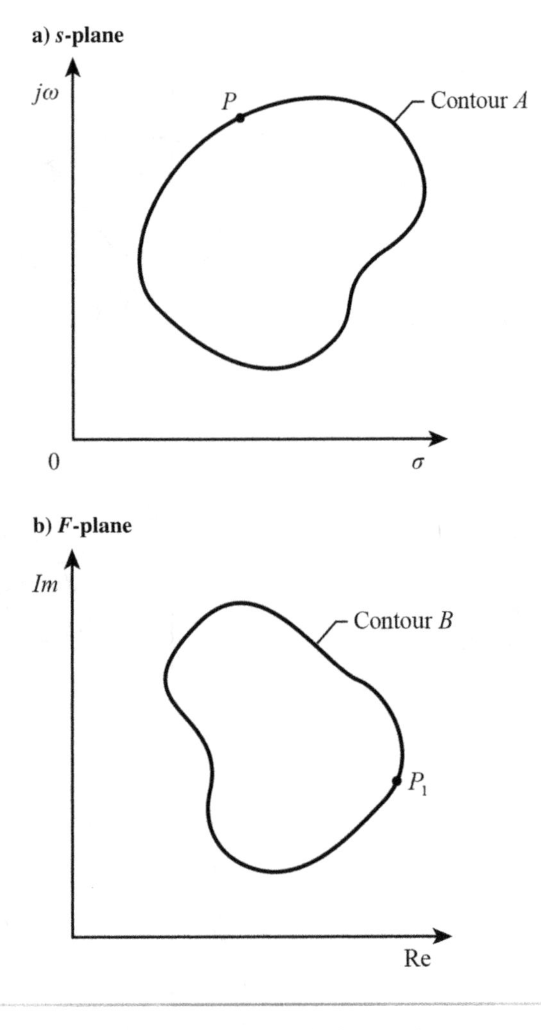

Fig. 5.16 Concept of mapping.

approach. Note that any complex number like $s - z_1$ can be represented by a vector drawn from the zero of function $s = z_1$ to point s. Every point P on contour A is associated with a vector V. Let V' be the image vector in the F-plane. For this case, $|V'| = |V|$ and $\angle V' = \angle V$. Now as we move clockwise along contour A, the magnitude and phase of the vector V vary. The phase oscillates between the two limiting values ϕ_1 and ϕ_2. In this case, a clockwise movement along contour A corresponds to a clockwise movement along image contour B in the F-plane.

Now let $F(s) = 1/(s - p_1)$ and let the pole $s = p_1$ lie outside contour A as shown in Fig. 5.17b. For this case, $|V'| = 1/|V|$ and $\angle V' = -\angle V$. As a result, contour A in the first quadrant maps to contour B in the fourth quadrant. Observe that a clockwise movement along contour A in the s-plane

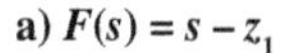

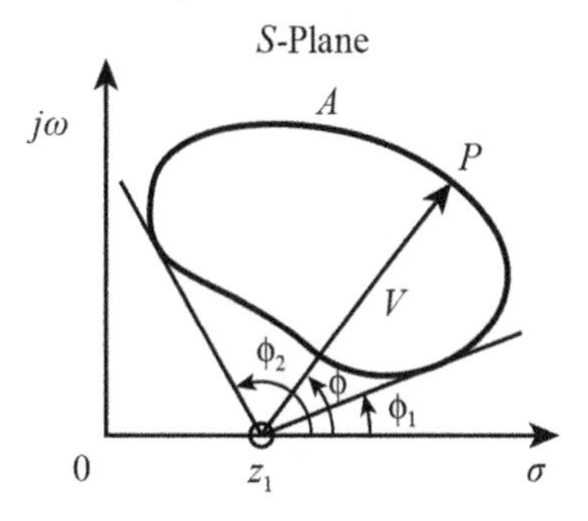

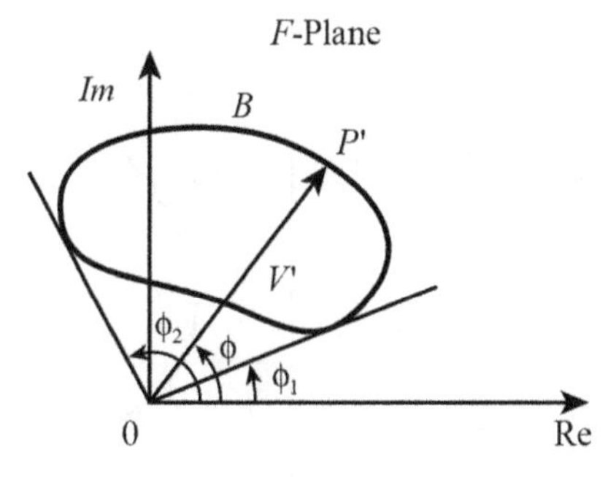

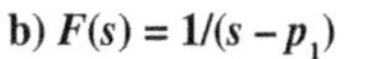

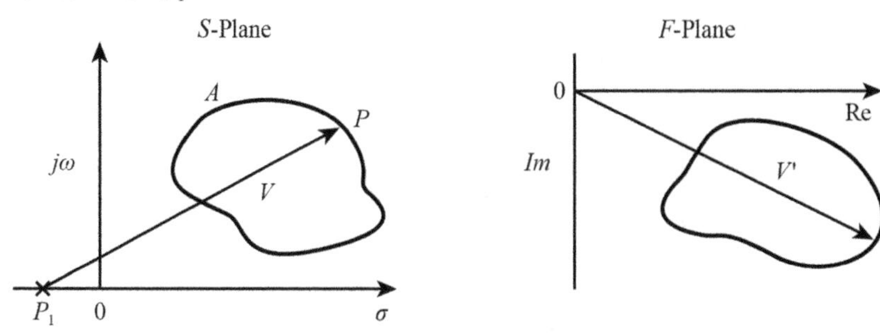

c) $F(s) = s - z_1$

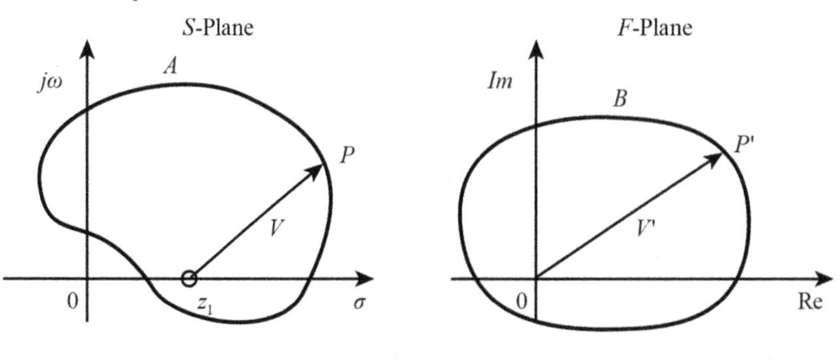

Fig. 5.17 Contour mapping.

corresponds to a counterclockwise movement along contour B in the F-plane because the phase angle in the F-plane is negative of that in the s-plane.

Suppose the zero of $F(s) = s - z_1$ lies inside contour A as shown in Fig. 5.17c. Then the vector V' makes one complete clockwise rotation of 360 deg in the F-plane so that contour B encloses the origin. Similarly, if we have a pole of the mapping function $F(s) = 1/(s - p_1)$ that lies inside contour A; then image contour B in the F-plane makes one complete

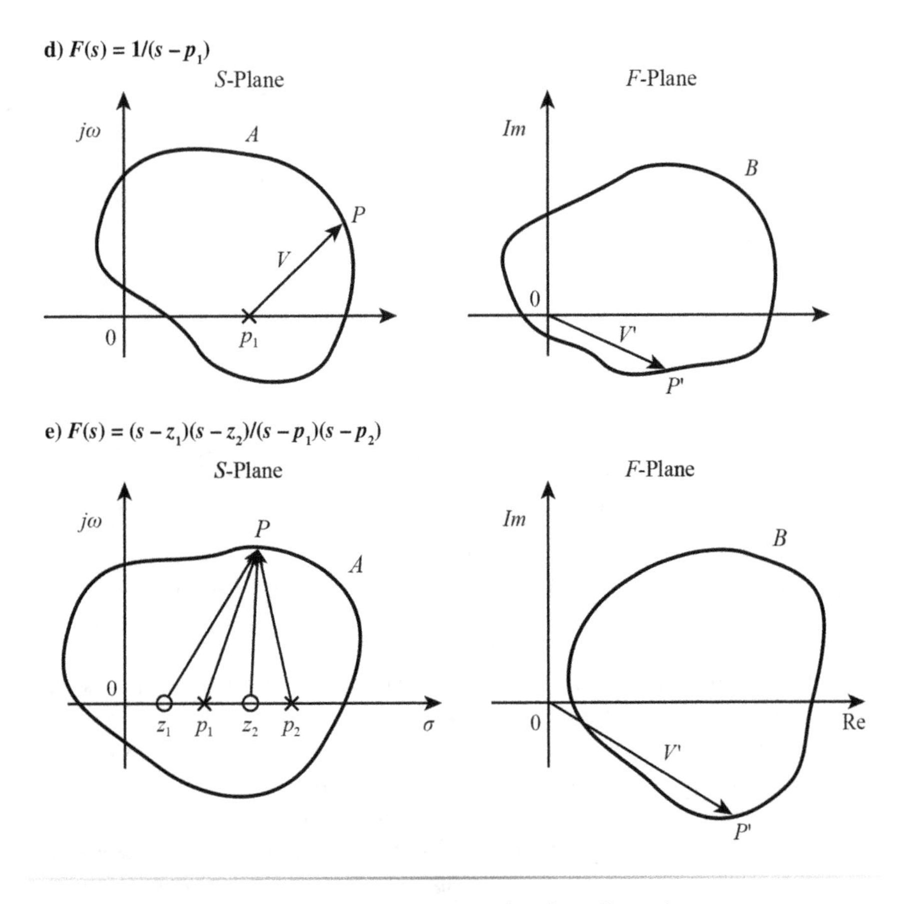

Fig. 5.17 Contour mapping (*continued*).

counterclockwise rotation and encloses the origin as shown in Fig. 5.17d. If contour A encloses an equal number of poles and zeros of the mapping function $F(s)$, then clockwise encirclement of the origin due to the zeros cancels the counterclockwise encirclement due to poles, and image contour B does not enclose the origin as shown in Fig. 5.17e.

Nyquist Plot

Suppose contour A in the s-plane is a semicircle of infinite radius covering the entire right half of the s-plane, then the corresponding image contour in the F-plane is said to be the Nyquist plot of right half of the s-plane through the given mapping function $F(s)$. If we have zeros and/or poles in the right half of the s-plane, then image contour B in the F-plane will encircle the origin n times where $n = n_z - n_p$ and n_z and n_p are, respectively, the number of zeros and poles of the mapping function $F(s)$ located in the right half of the s-plane. If $n > 0$, then we will have n clockwise encirclements and, if $n < 0$, we will have that many counter clockwise encirclements of the origin.

Nyquist Criterion of Stability

The transfer function of a closed-loop system is given by

$$T(s) = \frac{G(s)}{1 + G(s)H(s)} \tag{5.138}$$

$$= \frac{N_g D_h}{D_g D_h + N_g N_h} \tag{5.139}$$

where $G(s) = N_g/D_g$ and $H(s) = N_h/D_h$.

The poles of the closed-loop transfer function $T(s)$ are generally not known and have to be determined actually by solving the closed-loop characteristic equation $1 + G(s)H(s) = 0$. Note that the poles of $T(s)$ are the zeros of $1 + G(s)H(s)$. The closed-loop system will be unstable if any of the poles of $T(S)$ are located in the right half of the s-plane. The usefulness of the Nyquist stability criterion is that it enables us to know whether any of the poles of $T(s)$ are located in the right half of the s-plane without actually solving the closed-loop characteristic equation $1 + G(s)H(s) = 0$. In this way, it gives us an idea whether the given closed-loop system is stable or not without actually knowing the location of the closed-loop poles. This information is very useful in evaluating the stability of a closed-loop system as a certain system parameter, say the gain k, is varied.

Suppose we make a Nyquist plot of the function $F(s) = 1 + G(s)H(s)$. Note that the zeros of this function $F(s)$ are the poles of the closed-loop transfer function $T(s)$ and the poles of $F(s)$ are the combined poles of the open-loop transfer function $G(s)H(s)$, which are known. Then, the Nyquist criterion for stability centers around the determination of the parameter, $N = P - Z$, where N is the number of encirclements of the origin in the F-plane, P is the number of poles of $F(s)$ located in the right half of the s-plane, and Z is the number of zeros of $F(s)$ that are located in the right half of the s-plane. Note that a positive value of N corresponds to counterclockwise encirclement of the origin and a negative value to clockwise encirclement. Here, P is known but Z is not known. Therefore, unless we have a method to determine Z, we cannot sketch a Nyquist plot and determine the system stability. As said before, we do not have an easy method of finding Z.

Suppose we use the function $G(s)H(s)$ as the mapping function instead of $1 + G(s)H(s)$ because all the poles and zeros of the function $G(s)H(s)$ are known. The resulting Nyquist plot is the same as that of $1 + G(s)H(s)$ except that it is displaced by one unit to the left of the origin. Then, instead of counting the encirclement of the origin, we can count the encirclement of point -1. Everything else remains the same, and we can now use the Nyquist plot to determine the system stability. With this modification, the Nyquist criterion for the stability of a closed-loop system can be restated as follows.

If a contour A in the s-plane that covers the entire right half of the s-plane is mapped to the F-plane with the mapping function $F(s) = G(s)H(s)$, then the number of closed-loop poles Z that lie in the right half of the s-plane equals the number of open-loop poles P that are in the right half of the s-plane minus the number of counterclockwise rotations N of the Nyquist plot around point -1 in the F-plane, i.e., $Z = P - N$. For stability of the closed-loop system, Z must be equal to zero. The sign convention for N is as follows: For counterclockwise rotation, N is positive, and, for clockwise rotation it is negative.

To understand the Nyquist criterion, let us study two cases shown in Fig. 5.18. Let us assume that somehow we know the zeros of $1 + G(s)H(s)$, which are poles of the closed-loop transfer function $T(s)$. The poles of $1 + G(s)H(s)$ are the combined poles of the open-loop transfer function $G(s)H(s)$ and are known. Let the open circles denote the zeros of

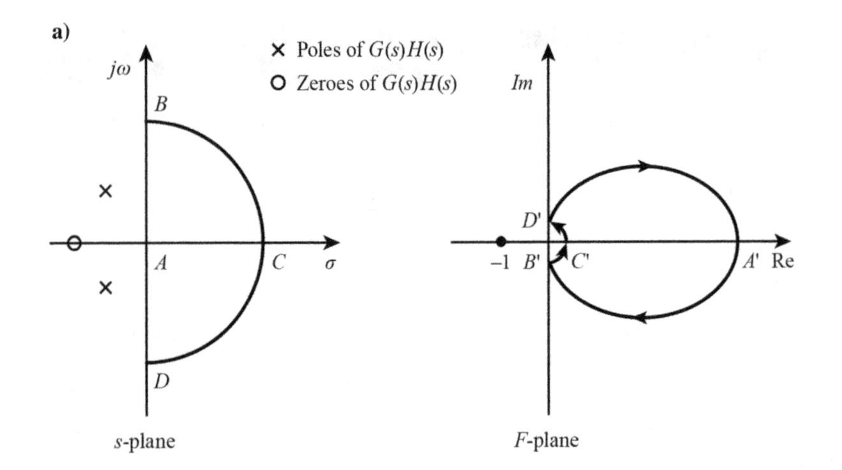

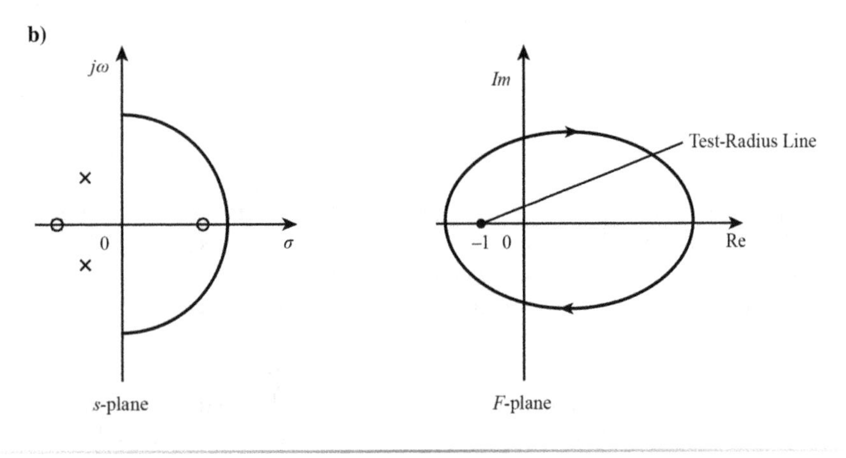

Fig. 5.18 Nyquist plots for mapping function $1 + G(s)H(s)$.

$1 + G(s)H(s)$ and cross the poles of $G(s)H(s)$. For Fig. 5.18a, there are no poles or zeros of $1 + G(s)H(s)$ in the right half of the s-plane, i.e., $P = 0$ and $Z = 0$. Hence, the Nyquist plot will not encircle point -1 in the F-plane as shown in Fig. 5.18b. For this case, $N = P - Z = 0$ and the system is stable. For Fig. 5.18b, we have one zero of $1 + G(s)H(s)$ located in the right half of the s-plane (unstable system). Therefore, $Z = 1$. Furthermore, $P = 0$ because there are no poles of $1 + G(s)H(s)$ located in right half of the s-plane. Hence, according to the Nyquist criterion, $N = P - Z = -1$, i.e., the Nyquist plot in the F-plane will encircle point -1 once in the clockwise direction as shown in Fig. 5.19b.

The number of encirclements can be conveniently determined by drawing a radial line from point -1 and counting the number of intersections with the Nyquist plot as shown in Fig. 5.18b. However, the reader should keep in mind that, in a given problem, the locations of closed-loop poles are not known as assumed in this discussion.

Example 5.6

Draw the Nyquist plot for a unity feedback system with

$$G(s) = \frac{5(s+2)}{(s+1)(s+3)}$$

Solution:

Here, $H(s) = 1$ so that $G(s)H(s) = G(s)$. The first step is to select some points along the Nyquist contour in the s-plane as shown in Fig. 5.19a. Consider an arbitrary point P. Let V_1, V_2, and V_3 be the vectors drawn to point P from the zeros and poles as shown. Then, the magnitude and phase of the image vector in the F-plane are given by

$$|V'_P| = \frac{5|V_1|}{|V_2||V_3|}$$

$$\angle V'_P = \angle V_1 - \angle V_2 - \angle V_3$$

If point P coincides with A, $|V_1| = 2$, $|V_2| = 1$, and $|V_3| = 3$ so that $|V'_A| = 10/3$. The phase angles $\angle V_1 = \angle V_2 = \angle V_3 = 0$ so that $\angle V'_A = 0$. Thus point A in the s-plane maps to point A' on the real axis in the F-plane with abscissa equal to $10/3$. In a similar fashion, we find the magnitude and phase angles at other image points such as $|V'_B| = 0$, $\angle V'_B = -90$ deg; $|V'_c| = 0$, $\angle V'_C = 0$; and $|V'_D| = 0$, $\angle V'_D = 90$ deg.

Based on this information, the Nyquist plot can be sketched as shown in Fig. 5.19b. Observe that the Nyquist plot does not encircle the origin but just

(*Continued*)

Example 5.6 (*Continued*)

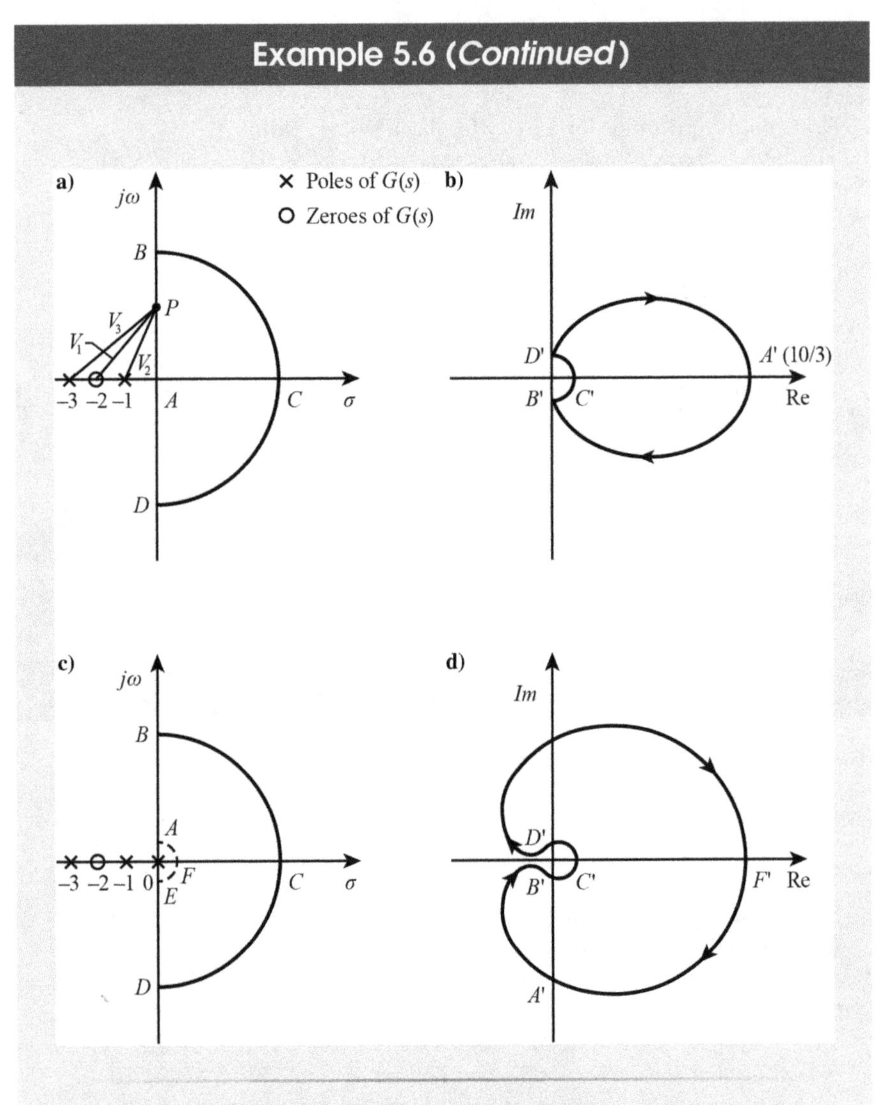

Fig. 5.19 Nyquist plots for Examples 5.6 and 5.7.

goes around it in a semicircle of "zero" radius. Why? Because we do not have any poles or zeros of $G(s)$ in the right half of the s-plane. Further, as we move clockwise in the s-plane starting at point A, we move in the counterclockwise direction in the F-plane from A'.

In this example we didn't have any poles of the mapping function on the imaginary axis. If we did, then we have to draw semicircles of infinitesimally small radii around each one to prevent a breakdown of the mapping procedure at these points. We illustrate the procedure of drawing such Nyquist plots with the help of Example 5.7.

Example 5.7

Draw the Nyquist plot for a unity feedback system with

$$G(s) = \frac{5(s+2)}{s(s+1)(s+3)}$$

Solution:

Here, we have $H(s) = 1$ and a pole ($s = 0$) at the origin. As said previously, we draw a semicircle of infinitesimally small radius around it as shown in Fig. 5.19c. The magnitudes and phase angles of the image points $A' - F'$ are as follows: $|V'_A| = \infty$, $\angle V'_A = -90$ deg; $|V'_B| = 0$, $\angle V'_B = -180$ deg; $|V'_C| = 0$, $\angle V'_C = 0$; $|V'_D| = 0$, $\angle V'_D = 180$ deg; $|V'_E| = \infty$, $\angle V'_E = 90$ deg; and, $|V'_F| = \infty$, $\angle V'_F = 0$.

The Nyquist plot is shown in Fig. 5.19d. Observe that the small circle of "zero" radius encircles the origin in the F-plane in the counterclockwise direction because the phase angle changes from -180 deg at B' to $+180$ deg at D'.

Example 5.8

Determine the stability of a unity feedback system given by

$$G(s) = \frac{k(s+2)}{(s-2)(s-3)}$$

Solution:

Here, $H(s) = 1$. Furthermore, assume that the gain k is a variable. First, we use MATLAB [4] to draw the root-locus as shown in Fig. 5.20a. We observe that the root-locus starts in the right half of the s-plane, implying that the closed-loop system is unstable for small values of the gain k and, for $k \geq 4.9420$, the root-locus crosses over to the left half of the s-plane, indicating that the closed-loop system becomes stable for $k \geq 4.9420$.

Now using MATLAB [4], let us draw the Nyquist plot as shown in Fig. 5.20b assuming $k = 1$. We observe that the Nyquist plot in the F-plane does not encircle point -1, which means that $N = 0$. Instead, it intersects the negative real axis at $s = -0.2$. We have $P = 2$ because the poles $s = 2$ and $s = 3$ of $G(s)$ are located in the right half of the s-plane. According to Nyquist criterion, we get $Z = P - N = 2$. In other words, the Nyquist criterion predicts that there are two poles of the closed-loop transfer function $T(s)$ located in the right half of the s-plane and, therefore, the system is unstable. From the root-locus of Fig. 5.20a, we find this to be true.

Suppose we increase the gain k beyond unity. Then the Nyquist plot will expand and eventually touch the critical point -1. When this happens, the value of k is equal to $1/0.2 = 5$, which is quite close to that predicted

(*Continued*)

Example 5.8 (*Continued*)

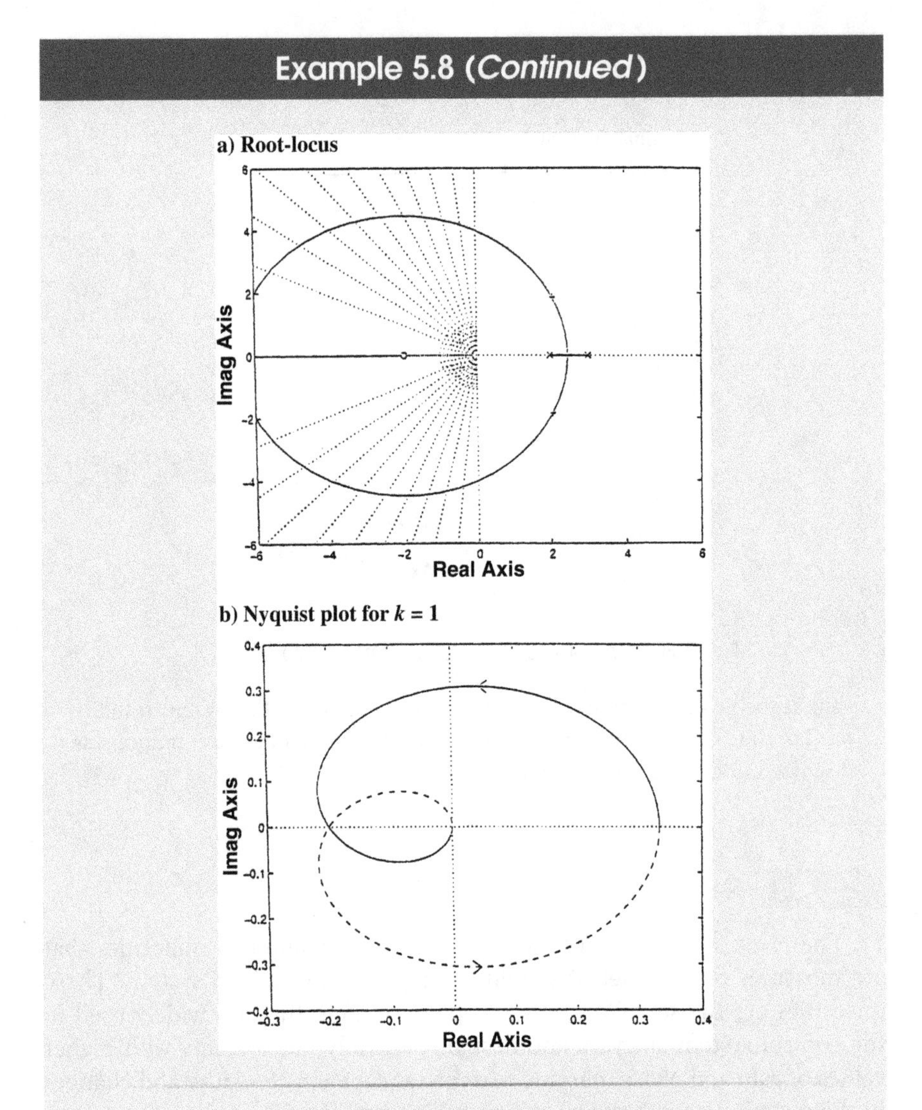

Fig. 5.20 Root-locus and Nyquist plots.

by the root-locus method. For higher values of gain k, the Nyquist plot will expand further and will encircle the critical point -1 twice in a counter-clockwise direction as shown in Fig. 5.20c for $k = 6$. We then have $N = 2$ and $Z = P - N = 2 - 2 = 0$, which indicates that the closed-loop system has become stable.

This example has illustrated an important concept that the stability of closed-loop systems depends on the value of the gain. Feedback systems that are unstable for low values of gain can become stable for higher values of gain, and those that are stable for low values of gain can become unstable for higher values of gain. The Nyquist criterion can be used to determine the

(*Continued*)

Example 5.8 (*Continued*)

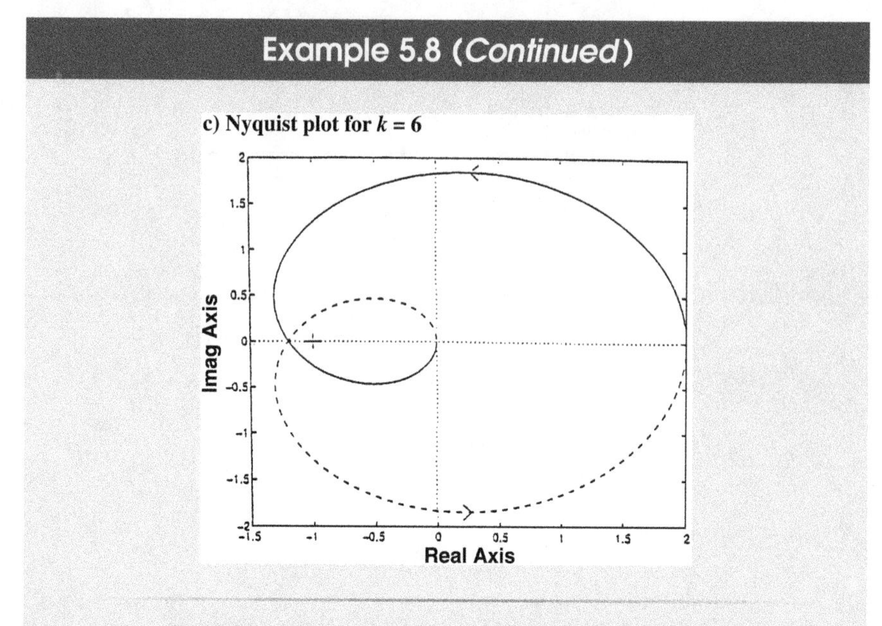

Fig. 5.20 Root-locus and Nyquist plots (*continued*).

gain at the crossover point. This kind of dependence of the system stability on the value of the gain leads to the concepts of gain and phase margins as discussed in the next subsection.

5.7.4 Gain and Phase Margins

The Nyquist stability criterion enables us to define two quantities that are measures of the level of stability of a given closed-loop system. These quantities are the so-called gain and phase margin that are widely used in the control system analyses and design. Generally, the systems with higher values of gain and phase margins have better capability to withstand changes in the system parameters before becoming unstable.

The concepts of gain margin and phase margin are illustrated in Fig. 5.21 based on using Nyquist mapping of only the positive imaginary axis. Why only the positive imaginary axis? Because, the part of the Nyquist plot that corresponds to the semicircle of infinite radius in the s-plane is usually mapped as a circle(s) of "zero" radius around the origin in the F-plane and hence is not needed in evaluating the gain and phase margins. Furthermore, the mapping of the positive imaginary axis is equivalent to studying the frequency response of the system because the Nyquist diagram of the positive imaginary axis is a polar plot of the magnitude vs phase of the open-loop transfer function with frequency as an implicit variable.

The gain margin G_M is the reciprocal of the magnitude $|G(j\omega)|$ at the phase crossover frequency. The phase crossover frequency is the frequency

ω_1 at which the phase angle of the open-loop transfer function $G(j\omega)$ is -180 deg. The gain margin is given by

$$G_M = \frac{1}{|G(j\omega_1)|} \tag{5.140}$$

If $a = |G(j\omega_1)|$ (see Fig. 5.21), then

$$G_M = \frac{1}{a} \tag{5.141}$$

The gain margin is usually expressed in decibels as

$$G_M(\text{db}) = 20\log_{10}\left(\frac{1}{a}\right) \tag{5.142}$$

$$= -20\log_{10} a \tag{5.143}$$

The gain margin expressed in decibels is positive if $a < 1$ (Fig. 21a) and is negative if $a > 1$ as shown in Fig. 5.21b. A positive gain margin (in decibels) means that the system is stable, and a negative gain margin (in decibels) means that the system is unstable. For a stable minimum phase system, the value of the gain margin indicates how much the open-loop gain can be increased before the closed-loop system becomes unstable. For example, a gain margin of 30 db implies that the open-loop gain can be increased by a factor of 31.6228 before the closed-loop system becomes unstable. On the other hand, if the gain margin is -30 db, then the closed-loop system is already unstable, and the gain has to be reduced by a factor of 31.6228 to make the closed-loop system stable.

The phase margin is defined as the amount of additional phase lag at the gain crossover frequency that can be introduced in the open-loop system to make the closed-loop system unstable. The gain crossover frequency ω_2 is

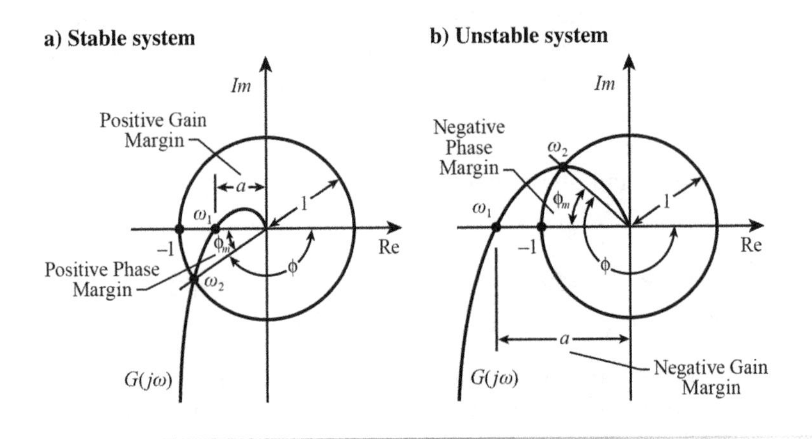

Fig. 5.21 Gain and phase margins.

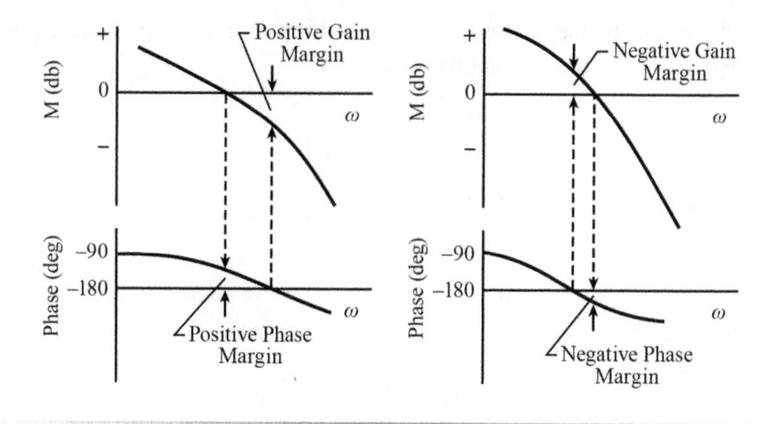

Fig. 5.22 Gain and phase margins using Bode plots.

that frequency when the magnitude of the open-loop transfer function $G(j\omega)$ is unity, i.e., when the Nyquist plot of $G(j\omega)$ intersects the unit circle as shown in Fig. 5.21.

The phase margin is usually denoted by ϕ_M and is expressed in degrees and is given by

$$\phi_M = 180 + \phi \tag{5.144}$$

where $\phi = \angle G(j\omega_2)$ is the open-loop phase angle at the gain crossover frequency ω_2 as shown in Fig. 5.21. Note that the value of the phase angle ϕ is negative in Fig. 5.21 because it is measured in the clockwise direction. Thus, for a stable system (Fig. 5.21a), the phase margin is positive because $|\phi| < 180$ deg, and for an unstable system the phase margin is negative because $|\phi| > 180$ deg as indicated in Fig. 5.21b. For example, a phase margin of 30 deg indicates that the open-loop phase lag can be increased further by 30 deg before making the system unstable. On the other hand, a phase margin of -30 deg indicates that the system is already unstable, and the open-loop phase lag has to be reduced by 30 deg to make the system stable. It is important to bear in mind that the Nyquist plots shown in Fig. 5.21 are drawn for unity gain.

The Bode plot also offers an alternative and a convenient method to estimate the gain and phase margins as illustrated in Fig. 5.22.

5.8 Relations Between Time-Domain and Frequency-Domain Parameters

Generally, the performance requirements for control systems are specified in terms of time-domain parameters like rise time T_r, settling time T_s, time for peak amplitude T_p, and percent overshoot O_s. In the following, we present some relations between these time-domain parameters and frequency-domain parameters. These relations will be useful in the analyses and design of control systems using frequency-domain methods.

Consider a second-order system whose open-loop and unity feedback closed-loop transfer functions are given by

$$G(s) = \frac{\omega_n^2}{s(s + 2\zeta\omega_n)} \tag{5.145}$$

$$T(s) = \frac{\omega_n^2}{s^2 + 2\zeta\omega_n s + \omega_n^2} \tag{5.146}$$

Let M denote the magnitude of the closed-loop frequency response. Then,

$$M = |T(j\omega)| = \frac{\omega_n^2}{\sqrt{(\omega_n^2 - \omega^2)^2 + 4\zeta^2\omega_n^2\omega^2}} \tag{5.147}$$

A typical plot of M vs ω is shown in Fig. 5.23.

To determine the peak amplitude M_p and the corresponding frequency ω_p, take the squares of both sides of Eq. (5.147), differentiate with respect

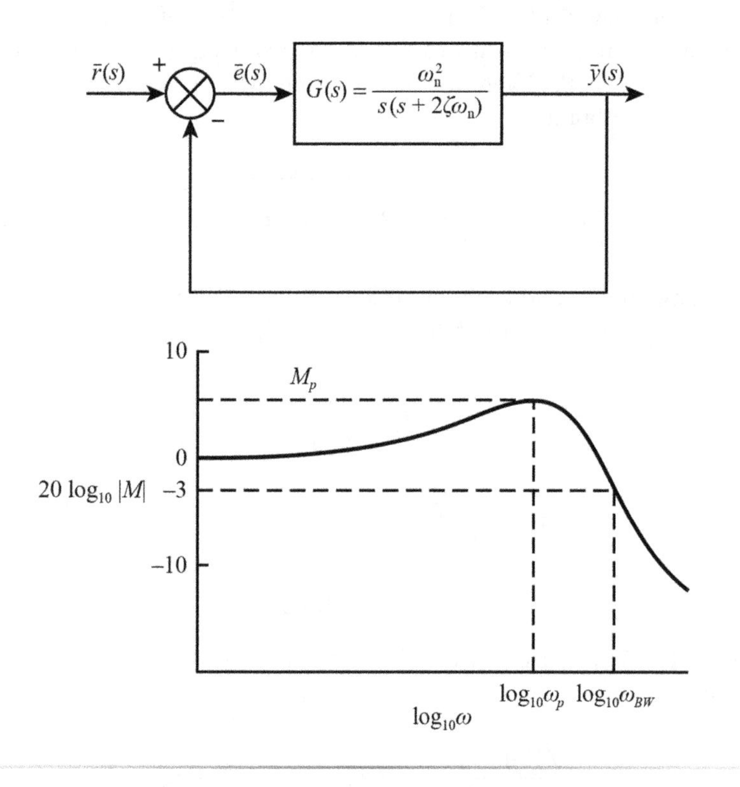

Fig. 5.23 Frequency response parameters.

to ω, and equate the resulting expression to zero to obtain

$$M_p = \frac{1}{2\zeta\sqrt{1-\zeta^2}} \tag{5.148}$$

$$\omega_p = \omega_n\sqrt{1-2\zeta^2} \tag{5.149}$$

Equations (5.148) and (5.149) show that M_p and ω_p of the closed-loop frequency response are directly related to the damping ratio ζ. The percent overshoot O_s and damping ratio ζ are related through Eq. (5.75), which is reproduced here in the following:

$$\zeta = \frac{-\ell n(O_s/100)}{\sqrt{\pi^2 + \ell n^2(O_s/100)}} \tag{5.150}$$

Using these relations, given the value of M_p, we can determine the percent overshoot O_s of the given closed-loop system and vice versa.

The bandwidth ω_{BW} is another important characteristic of the closed-loop frequency response. The bandwidth is defined as that frequency at which the magnitude M drops to 0.707 or $1/\sqrt{2}$ of its value at $\omega = 0$. This is also equivalent to a drop by 3 db. From Eq. (5.147), we find that $M = 1$ when $\omega = 0$. Therefore substituting $M = 1/\sqrt{2}$ and $\omega = \omega_{BW}$ in Eq. (5.147), we obtain

$$\omega_{BW} = \omega_n\sqrt{(1-2\zeta^2) + \sqrt{4\zeta^4 - 4\zeta^2 + 2}} \tag{5.151}$$

The settling time T_s and time for peak amplitude T_p given by Eqs. (5.78) and (5.72) are reproduced in the following:

$$T_s = \frac{4}{\zeta\omega_n} \tag{5.152}$$

$$T_P = \frac{\pi}{\omega_d} \tag{5.153}$$

$$= \frac{\pi}{\omega_n\sqrt{1-\zeta^2}} \tag{5.154}$$

Using Eq. (5.151), we can rewrite these relations in terms of bandwidth ω_{BW} as follows:

$$T_s = \left(\frac{4}{\omega_{BW}\zeta}\right)\sqrt{(1-2\zeta^2) + \sqrt{4\zeta^4 - 4\zeta^2 + 2}} \tag{5.155}$$

$$T_p = \left(\frac{\pi}{\omega_{BW}\sqrt{1-\zeta^2}}\right)\sqrt{(1-2\zeta^2)+\sqrt{4\zeta^4-4\zeta^2+2}} \tag{5.156}$$

Equations (5.155) and (5.156) give relations between the time-domain parameters T_s and T_p and the frequency response parameter ω_{BW} for second-order systems. These relations contain the two basic system parameters, ζ and ω_n.

Another important parameter of the frequency-domain design method is the phase margin ϕ_M. A relation between phase margin ϕ_M and the damping parameter ζ can be obtained as follows.

Let $\omega = \omega_1$ when the magnitude of the open-loop frequency response is unity, i.e., $|G(j\omega)| = 1$, or

$$|Gj(\omega_1)| = \frac{\omega_n^2}{\left|\left(-\omega_n^2 + j2\zeta\omega_n\omega_1\right)\right|} = 1 \tag{5.157}$$

so that

$$\omega_1 = \omega_n\sqrt{-2\zeta^2+\sqrt{1+4\zeta^4}} \tag{5.158}$$

The phase angle of $G(j\omega)$ at $\omega = \omega_1$ is given by

$$\angle G(j\omega_1) = -90 - \tan^{-1}\frac{\omega_1}{2\zeta\omega_n} \tag{5.159}$$

$$= -90 - \tan^{-1}\left(\frac{\sqrt{-2\zeta^2+\sqrt{1+4\zeta^4}}}{2\zeta}\right) \tag{5.160}$$

The phase margin is given by

$$\phi_M = 180 + \angle G(j\omega_1) = 90 - \tan^{-1}\left(\frac{\sqrt{-2\zeta^2+\sqrt{1+4\zeta^4}}}{2\zeta}\right) \tag{5.161}$$

$$= \tan^{-1}\left(\frac{2\zeta}{\sqrt{-2\zeta^2+\sqrt{1+4\zeta^4}}}\right) \tag{5.162}$$

The variation of ϕ_M with ζ is shown in Fig. 5.24.

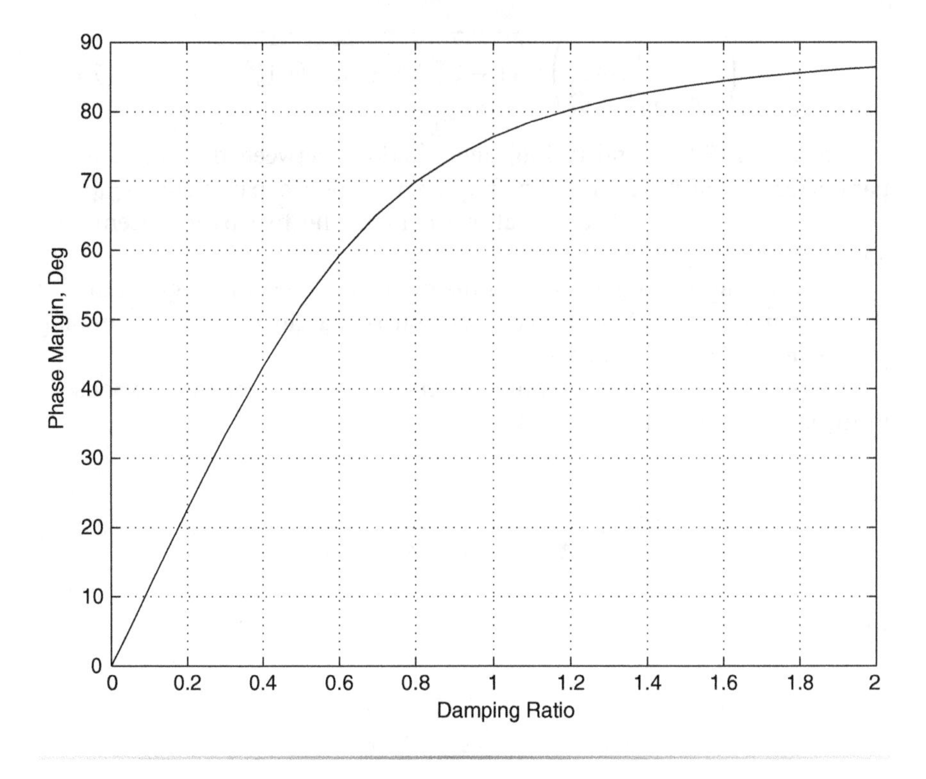

Fig. 5.24 Relation between phase margin and damping ratio.

5.9 Design of Compensators

The response characteristics of a given system depend on the internal physical nature of the system and may not meet the specified performance requirements. A simple modification of the plant dynamics is an obvious first choice to meet the performance specifications. However, such a thing may not be possible in practice because the plant can be quite complex so that it may not be easy to affect the necessary modifications. In such cases, the adjustment of the gain is the next obvious step. However, in many cases, this alone may not be sufficient, and one may have to redesign the entire plant. Such a process can be quite expensive and time consuming. A simpler alternative is to introduce a compensator into the system that compensates for the deficiencies of the original plant so that the overall system, including the compensator, meets the specified performance requirements.

A compensator is also called a controller. Compensators that employ pure integration to improve steady-state error or pure differentiation to speed up the transient response are called ideal compensators. However, a disadvantage of ideal compensators is that their implementation requires active networks like operational amplifiers. It is possible to construct passive networks involving resistors, inductors, and capacitors and to achieve performances close to those of ideal compensators. Such compensators are called either lead or lag or lead-lag or lag-lead compensators depending on their type. We will not be dealing with the issues concerning hardware implemention of the compensator designs discussed here. The interested reader may refer elsewhere [1–3].

In this section, we will discuss the design of compensators to obtain the specified transient response or steady-state error or both for single-input-single-output systems. Basically, there are two design methods: 1) the root-locus method, and 2) the frequency-response method. The frequency-response method has the advantage that the explicit knowledge of the plant transfer function is not needed. All that is needed is the plant frequency response. However, the main disadvantage of the frequency-response method is that the quantities one deals with are not directly related to the time-response parameters, which are specified as design requirements. Hence, the design becomes more of trial and error, and the number of iterations depends on the knowledge and experience of the designer. On the other hand, the root-locus method has a clear advantage in that the quantities it deals with are directly related to the design requirements. Furthermore, the correlation of the root-locus with time response is quite good. Also, the effect of changing compensator parameters can be easily observed by studying the root-locus. With the availability of tools like MATLAB [4], the root-locus method becomes very attractive for control system design. However, a disadvantage of the root-locus method is that it becomes more complex as the order of the system increases.

Here, we will use the root-locus method for compensator design of single-input-single-output systems. Readers interested in using frequency-domain methods may refer elsewhere [1, 2]. For multi-input-multi-output systems, the modern state-space methods are quite convenient. We will discuss these approaches in Sec. 5.10.

5.9.1 Proportional–Integral Compensator

To understand the basic principles of designing an integral compensator, consider a type "0" system with unity feedback as shown in Fig. 5.25a. The root-locus for this system is sketched in Fig. 5.25b. Let us assume that the system is operating at point A, having the desired transient response. Recall that the transient response is characterized by the settling time T_s, time for peak amplitude T_p, and the rise time T_r, all of which depend on the damping ratio ζ and frequency ω_n. At point A, the closed-loop poles are a pair of complex roots and one real root. The steady-state error of this system is equal to $\epsilon(\infty) = 1/(1 + K_p)$, where $K_p = k/p_1 p_2 p_3$. Because K_p is finite, the steady-state error is nonzero.

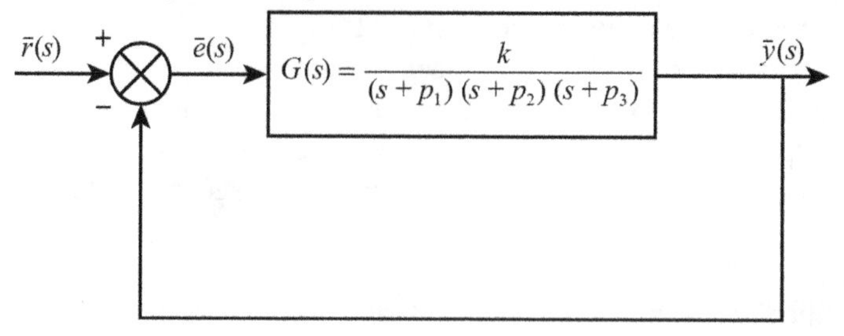

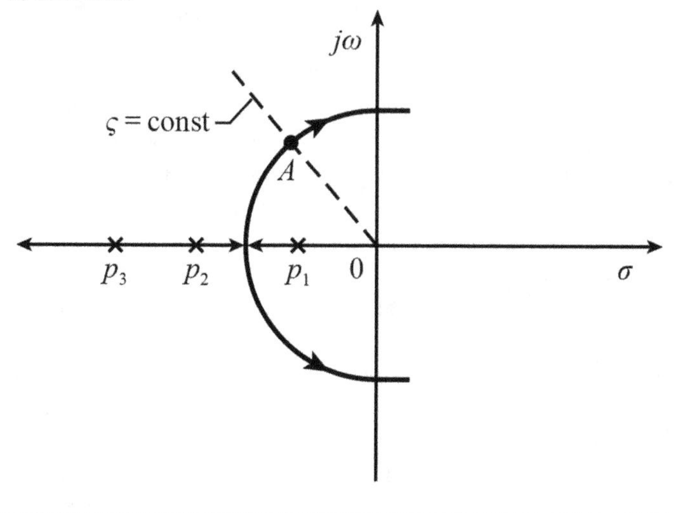

Fig. 5.25 System operating at the desired point A.

To drive the steady-state error to zero, let us make it a type "1" system by adding a pure integrator in the forward path as shown in Fig. 5.26a. This amounts to adding a pole at the origin as shown in Fig. 5.26b. The root-locus of the entire system is now changed and does not go through point A as shown in Fig. 5.26b. In other words, the steady-state error is driven to zero, but the transient response has changed. To solve this problem, add a compensator zero at $s = -z_c$, which is close to the origin so that this zero almost cancels the compensator pole. Such a compensator is called a proportional–integral (PI) compensator. Now, the root-locus with PI compensation (Fig. 5.26c) is nearly the same as that of the basic uncompensated system (Fig. 5.25b). Therefore, the transient response will remain unaffected while the steady-state error is driven to zero.

The transfer function of a PI compensator is given by

$$G_c(s) = \frac{s + z_c}{s} \tag{5.163}$$

a) Integral compensator

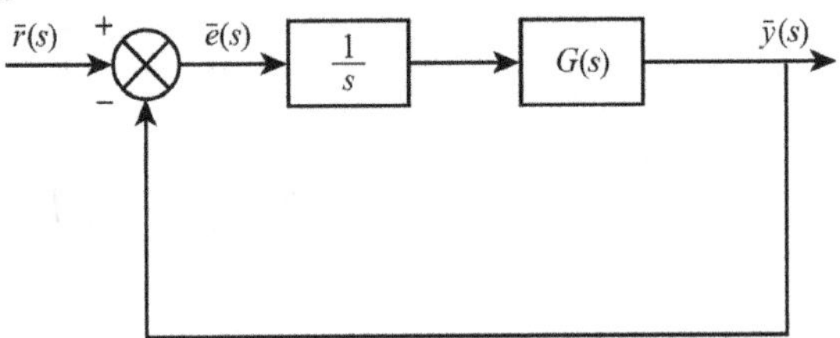

b) Root-locus of pure integral compensator **c) Root-locus of PI compensator**

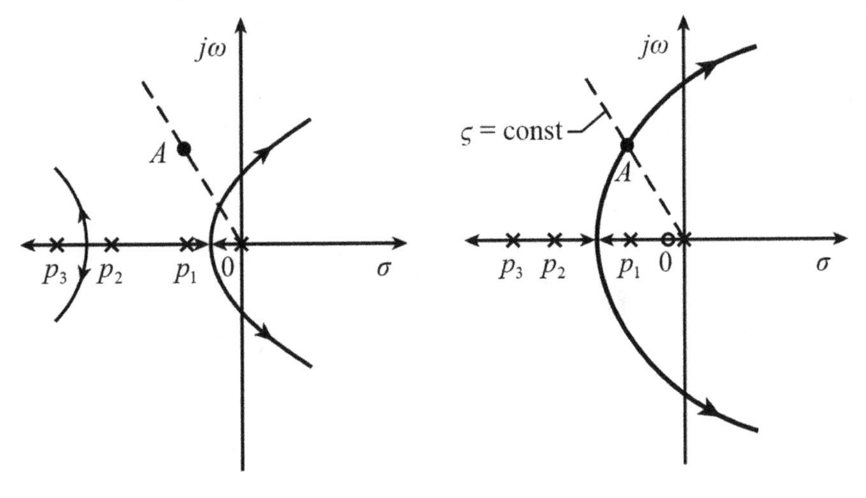

Fig. 5.26 Proportional–integral compensator.

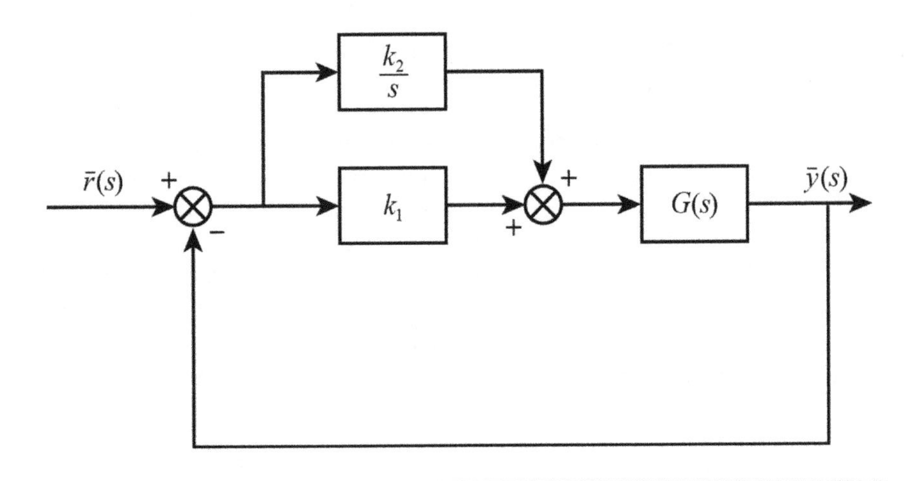

Fig. 5.27 Implementation of a PI compensator.

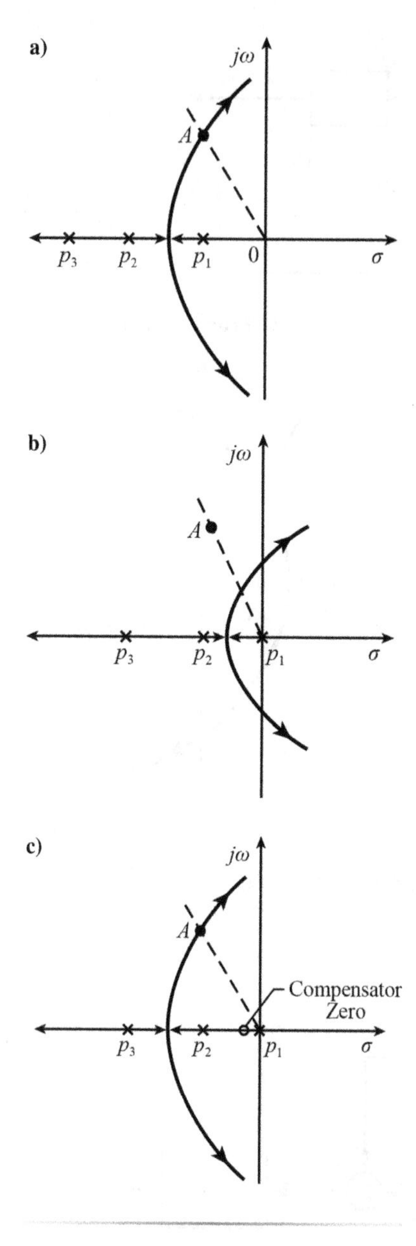

Fig. 5.28 Proportional-derivative compensation.

The schematic implementation of a PI compensator is shown in Fig. 5.27 with

$$G_c(s) = k_1 + \left(\frac{k_2}{s}\right) \tag{5.164}$$

$$= k_1\left[1 + \frac{k_2}{k_1}\left(\frac{1}{s}\right)\right] \tag{5.165}$$

In this implementation, $k_1 = 1$ and $k_2 = z_c$. Note that the first term on the right-hand side is the "proportional" part and the second term is the "integral" part.

5.9.2 Proportional-Derivative Compensator

Generally, the derivative compensation is used when a simple gain adjustment alone cannot give the desired transient response of the closed-loop system. This concept is illustrated in Fig. 5.28. In Fig. 5.28a, a simple gain adjustment is sufficient because the root-locus passes through the desired operating point A. However, in Fig. 5.28b, the root-locus cannot pass through the desired operating point A for any value of gain k. The addition of a compensator zero close to the origin modifies the root-locus so that it is made to pass through point A as shown in Fig. 5.28c. Such a compensator that produces a zero in the forward path is called a proportional–derivative (PD) compensator.

The transfer function of a PD compensator is of the form

$$G_c(s) = s + z_c \tag{5.166}$$

which is essentially the sum of differentiator s and gain z_c.

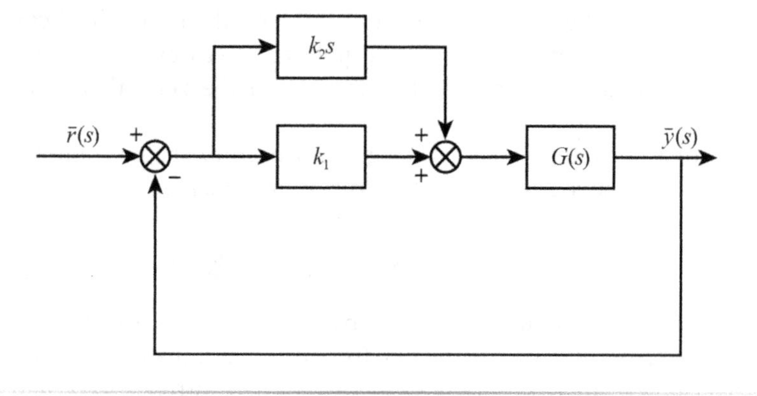

Fig. 5.29 Implementation of PD compensator.

The implementation of a PD compensator is schematically shown in Fig. 5.29. The transfer function of such a compensator can also be written in the following form:

$$G_c(s) = k_1 + k_2 s = k_2\left(\frac{k_2}{k_1} + s\right) \tag{5.167}$$

The first term on the right-hand side is the "proportional" part, and the second term is the "derivative" part. Gains k_1 and k_2 are design variables to be determined so that the system attains the specified transient response.

5.9.3 Lead/Lag Compensator

The transfer function of a lead/lag compensator is of the form

$$G_c(s) = \frac{s + z_c}{s + p_c} \tag{5.168}$$

A schematic diagram of a lead/lag compensator is shown in Fig. 5.30. By a suitable choice of the locations of pole and zero, we can have either a lag or lead compensator. For example, if the zero is close to the origin ($z_c \simeq 0$)

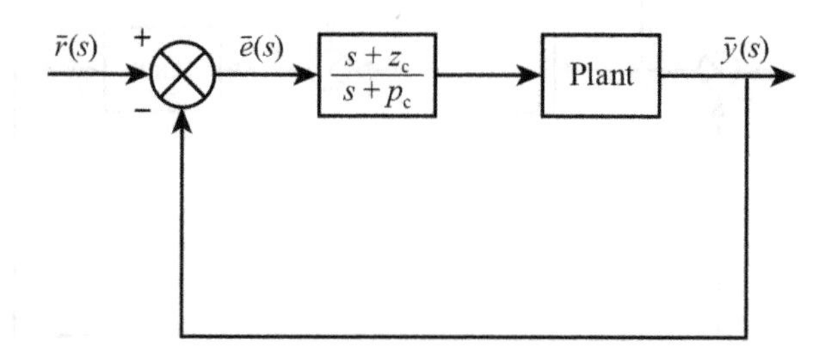

Fig. 5.30 Lead/lag compensator.

and the pole is farther away to the left of the zero, then it is a lead compensator. On the other hand, if both the pole and zero are located close to the origin with the pole located to the right of the zero, then it is a lag compensator.

Note that for both the lead and lag compensators, the pole and the zero are supposed to be located in the left half of the s-plane.

5.9.4 Proportional–Integral and Derivative Controller

Suppose we want to improve the transient response as well as reduce the steady-state error, then we use the PID (proportional–integral and derivative) controller. There are two ways of designing a PID controller: 1) design a PD controller first to improve the transient response and then add a PI controller to improve the steady-state error or 2) design a PI controller first and then add a PD controller. Both methods are iterative because one affects the other.

The schematic diagram of a PID controller is shown in Fig. 5.31.

5.9.5 Feedback Compensation

The desired transient response can also be obtained by feedback compensation. With a proper choice of the feedback-loop transfer function, the root-locus can be modified to obtain the desired transient response. This method offers an added advantage that the parts of the system can be isolated for improvement in transient response prior to closing the major loop. This approach is also equivalent to relocating the open-loop poles of the system so that the root-locus is reshaped to obtain the desired closed-loop poles.

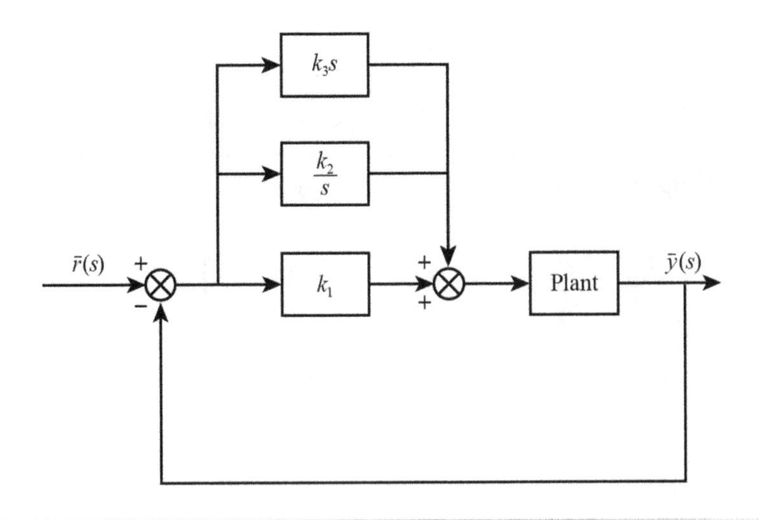

Fig. 5.31 Proportional–integral and derivative controller.

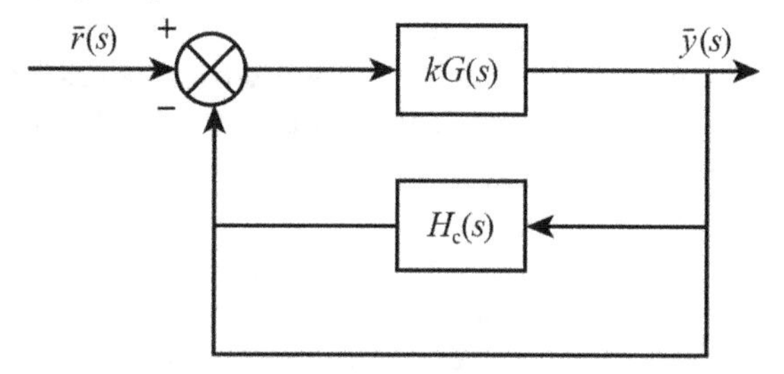

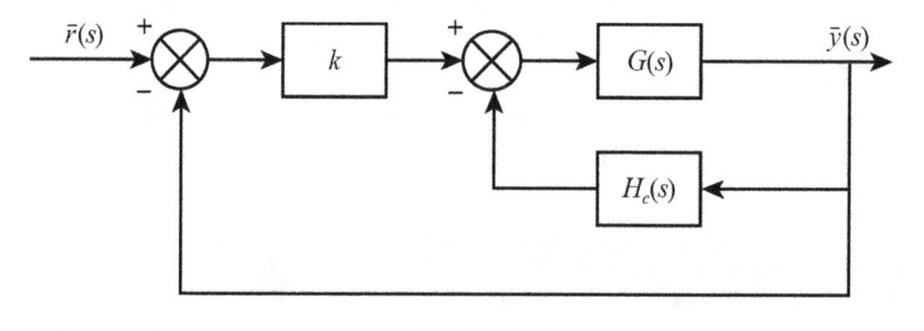

Fig. 5.32 Feedback compensation.

Feedback compensation can be accomplished in two ways: 1) major-loop compensation and 2) minor-loop compensation as schematically shown in Fig. 5.32.

Major-Loop Compensation

Let $H_c(s) = k_h s$ be the transfer function of the major loop. This form of transfer function is representative of a tachometer or a rate gyro whose input is a displacement and output is the time rate of change of displacement or velocity. For example, if the input is bank angle, then the output will be roll rate.

The closed-loop transfer function is given by

$$T(s) = \frac{kG(s)}{1 + kk_h G(s)\left(s + \frac{1}{k_h}\right)} \tag{5.169}$$

The characteristic equation of this major-loop feedback compensated system is given by

$$1 + kk_h G(s)\left(s + \frac{1}{k_h}\right) = 0 \tag{5.170}$$

Thus, in principle, the major-loop feedback compensation introduces a zero into the system at $s = -1/k_h$ so that the root-locus is reshaped to pass through the desired operating point. By varying the parameter k_h, we can vary the gain as well as the location of this zero. Even though this concept is similar to the PD compensation, there is a difference. The compensator zero in the case of a PD compensator is an open-loop zero, whereas the zero introduced in major-loop feedback compensation is not an open-loop zero.

Minor-Loop Compensation

With $H_c(s) = k_h s$, the open-loop transfer function of the minor loop is $G_c(s)k_h s$. Thus, the addition of a zero at the origin of the minor-loop root-locus considerably speeds up the response of the minor loop and also has an effect on the overall system performance. Once gain k_h is adjusted to obtain the desired performance of the minor loop, the outer loop is closed, and gain k is adjusted to obtain the specified overall system performance. This method of compensation is usually used in aircraft control systems to improve the response to individual degrees of freedom like pitch, roll, or yaw before closing the outer loop as we will discuss in Chapter 6.

Example 5.9

For the system shown in Fig. 5.33, 1) design a PI compensator to reduce the steady-state error to zero, and 2) a lag compensator to reduce the steady-state error by a factor of 10 for a step input without affecting the transient response. Assume that the system is required to operate with a damping ratio of $\zeta = 0.2$.

Solution:

We have

$$G(s) = \frac{k}{(s+2)(s+5)(s+12)}$$

The first step is to draw the root-locus of the basic (uncompensated) system and determine the value of the gain for operation at $\zeta = 0.2$. Using MATLAB [4],

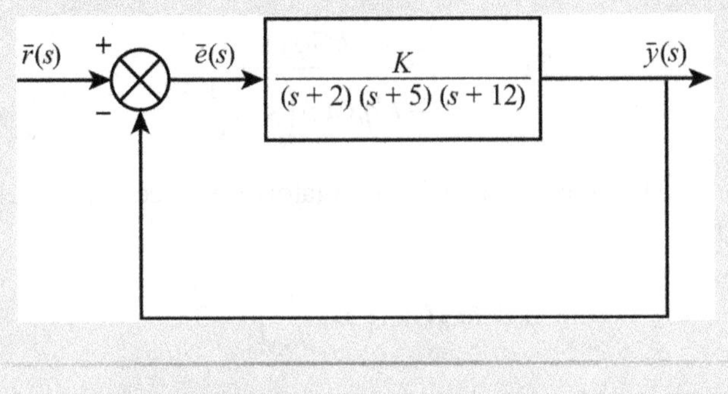

Fig. 5.33 Control system.

(Continued)

Example 5.9 (*Continued*)

the root-locus is drawn as shown in Fig. 5.34a. For operation with $\zeta = 0.2$, the values of the gain and closed-loop pole locations are $k = 679.086$ and $p = -16.2416, -1.3792 \pm j6.8773$.

The position constant K_p and the steady-state error $e(\infty)$ are given by

$$K_p = \frac{k}{p_1 p_2 p_3} = \frac{679.086}{2 * 5 * 12} = 5.6590$$

$$e(\infty) = \frac{1}{1 + k_p} = \frac{1}{1 + 5.6590} = 0.1502$$

With this, we get the steady-state value of the output (for a unit-step input), $y(\infty) = 1 - e(\infty) = 0.8498$.

Design of the PI Compensator

The PI compensator is characterized by a pole at the origin and a zero close to it. Let us choose the zero at $s = -0.05$. With this, the open-loop transfer function of the PI-compensated system is

$$G_c(s) = \frac{k(s + 0.05)}{s(s + 2)(s + 5)(s + 12)}$$

Now let us determine the value of gain k so that the PI-compensated system operates at a damping ratio of 0.2, while the steady-state error is driven to zero. Using MATLAB [4], we draw the root-locus for the PI-compensated system as shown in Fig. 5.34b. For operating with $\zeta = 0.2$, we obtain $k = 673.175$ and $p = -16.2118, -1.3728 \pm j6.8408$, and -0.0426.

Comparing these results with those obtained earlier for the basic system, we observe that the dominant second-order complex poles that determine the transient response are virtually unchanged because the pole at $s = -0.0426$ almost cancels with the zero at $s = -0.05$. The pole at $s = -16.2118$ is so far away on the left-hand side of the s-plane that its influence is negligible. In view of this, the system will essentially behave like a second-order system with dominant poles at $-1.3728 \pm j6.8408$.

Design of the Lag Compensator

We have to design the lag compensator to achieve a reduction in the steady-state error by a factor of 10, i.e.,

$$e(\infty) = \frac{0.1502}{10} = 0.01502$$

(Continued)

Example 5.9 (*Continued*)

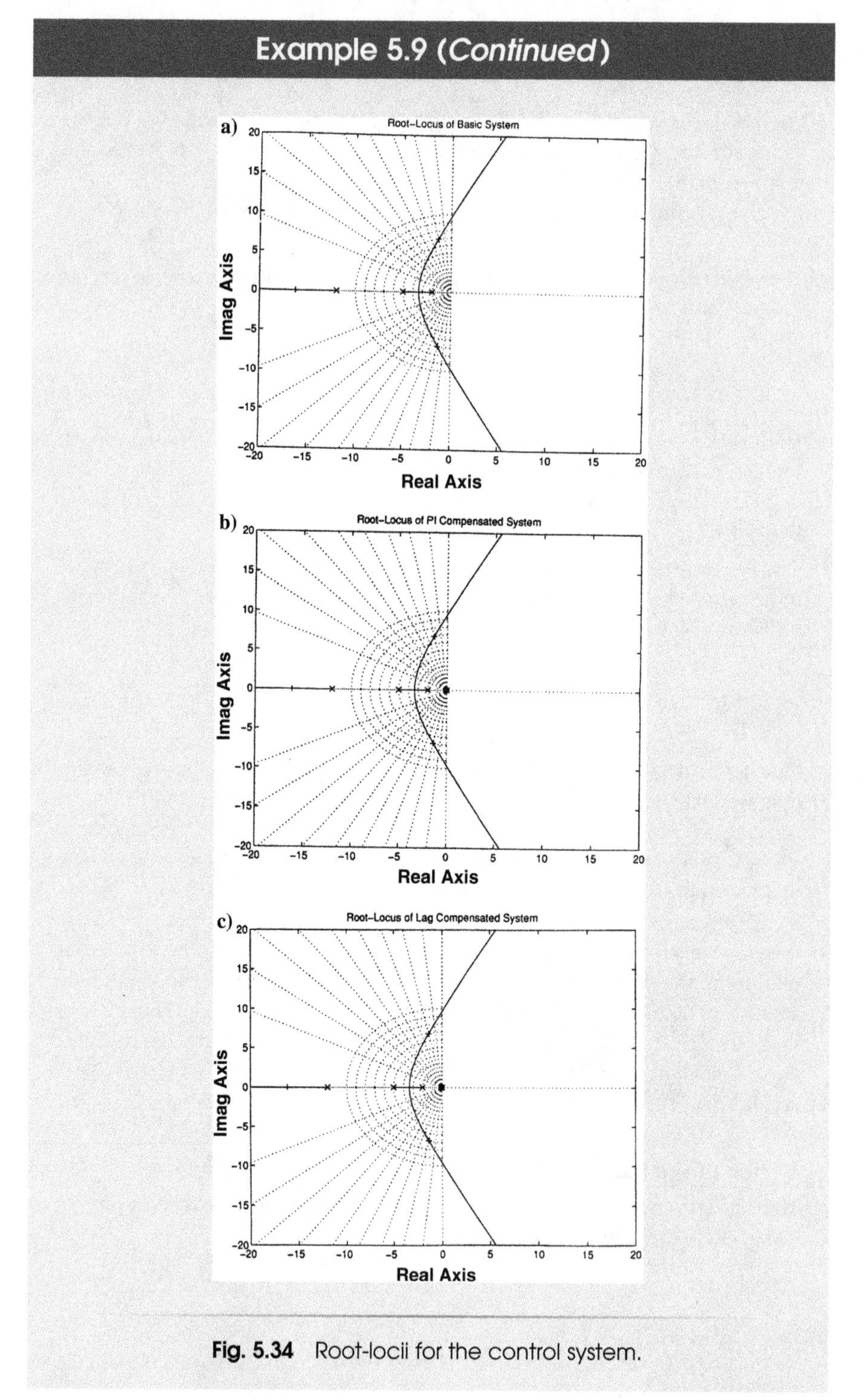

Fig. 5.34 Root-locii for the control system.

(*Continued*)

Example 5.9 (*Continued*)

Then,

$$K_p = \frac{1 - e(\infty)}{e(\infty)} = \frac{1 - 0.01502}{0.01502} = 64.7895$$

For the lag-compensated system,

$$K_p = \frac{z_c k}{p_c * 2 * 5 * 12}$$

Here, we have three unknowns k, z_c, and p_c and one relation as given. To begin with, let us assume $k = 679.086$ (uncompensated gain). Then, we can choose one of the two remaining unknowns arbitrarily. Let us choose $z_c = 0.05$ so that we get $p_c = 0.0044$. The open-loop transfer function of the lag-compensated system is given by

$$G(s) = \frac{k(s + 0.05)}{(s + 0.0044)(s + 2)(s + 5)(s + 12)}$$

Now we can draw the root-locus as shown in Fig. 5.34c and determine the value of the gain and closed-loop poles for operating with $\zeta = 0.2$. We get $k = 670.3530$ and $p = -16.2019$, $-1.3796 \pm j6.8304$, and -0.0433.

Thus, the pole locations are similar to those observed for the PI compensator.

With this new value of gain $k = 670.3530$, the steady-state error is slightly changed. We have

$$\begin{aligned} K_p &= \frac{z_c k}{p_c * 2 * 5 * 12} \\ &= \frac{0.05 * 670.3530}{0.0044 * 2 * 5 * 12} \\ &= 63.4804 \end{aligned}$$

$$\begin{aligned} e(\infty) &= \frac{1}{1 + K_p} \\ &= \frac{1}{1 + 63.4804} \\ &= 0.0155 \end{aligned}$$

which is close to the target value of 0.01502. Hence, we need not repeat the design process.

(*Continued*)

Example 5.9 (*Continued*)

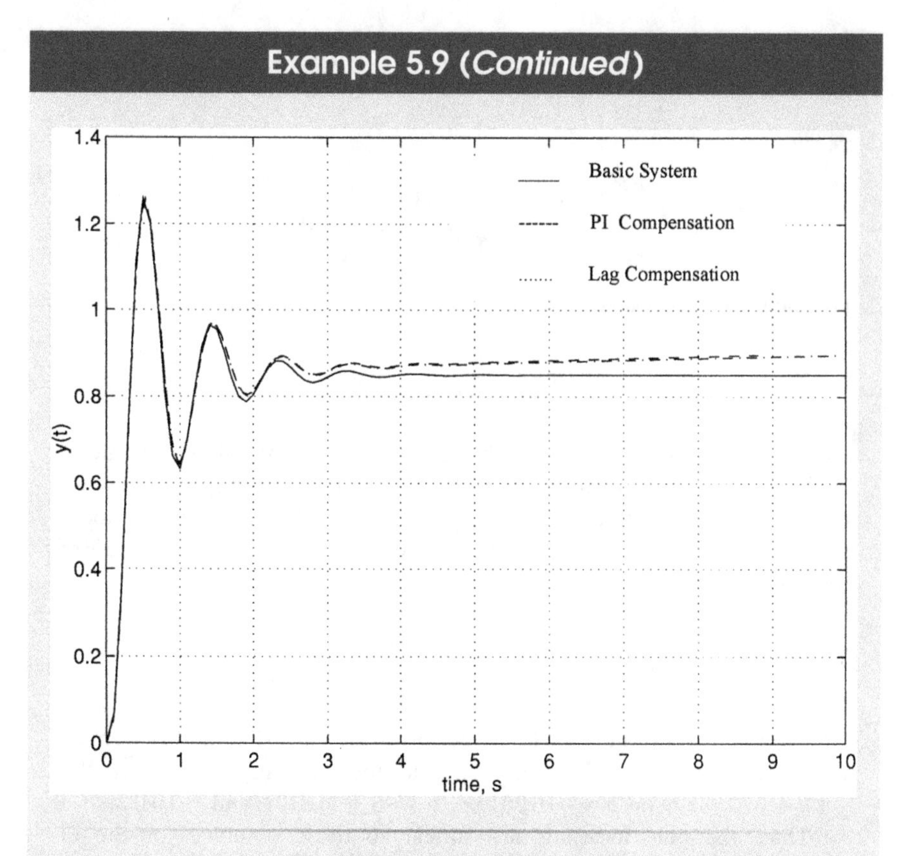

Fig. 5.35 Unit-step responses.

Now to test the designs of PI- and lag-compensated systems, we have obtained unit-step responses of the basic, PI-, and lag-compensated systems as shown in Fig. 5.35. We observe that the design objectives are met. The transient response of the PI- and lag-compensated systems are almost identical to that of the basic system. Furthermore, as t assumes large values, the steady-state error for the PI compensator approaches zero and that for the lag compensator approaches the target value of 0.01502.

Example 5.10

For the following system, design 1) a PD compensator, and 2) a lead-lag compensator so that the peak time is reduced by a factor of 3, while the percent overshoot remains unchanged at 25.38%.

$$G(s) = \frac{k}{s(s+3)(s+5)}$$

(Continued)

Example 5.10 (*Continued*)

Solution:

PD Compensator

The first step is to draw the root-locus of the basic system as shown in Fig. 5.36a. Using Eq. (5.75), we find the damping ratio that corresponds to 25.38% overshoot is equal to 0.4. For the basic (uncompensated) system, the value of the gain that corresponds to $\zeta = 0.4$ is equal to 28.3825, and the closed-loop pole locations are $-0.8299 \pm j1.9426$ and -6.3402. Because the third pole located at -6.3402 is farther from the second-order

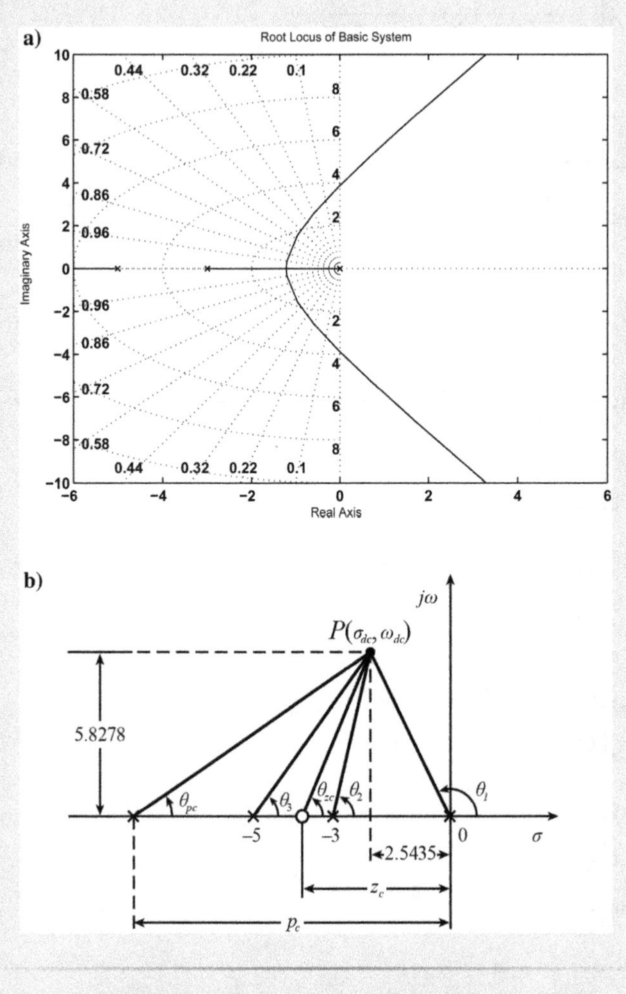

Fig. 5.36 Root-locii for the control system.

(*Continued*)

Example 5.10 (*Continued*)

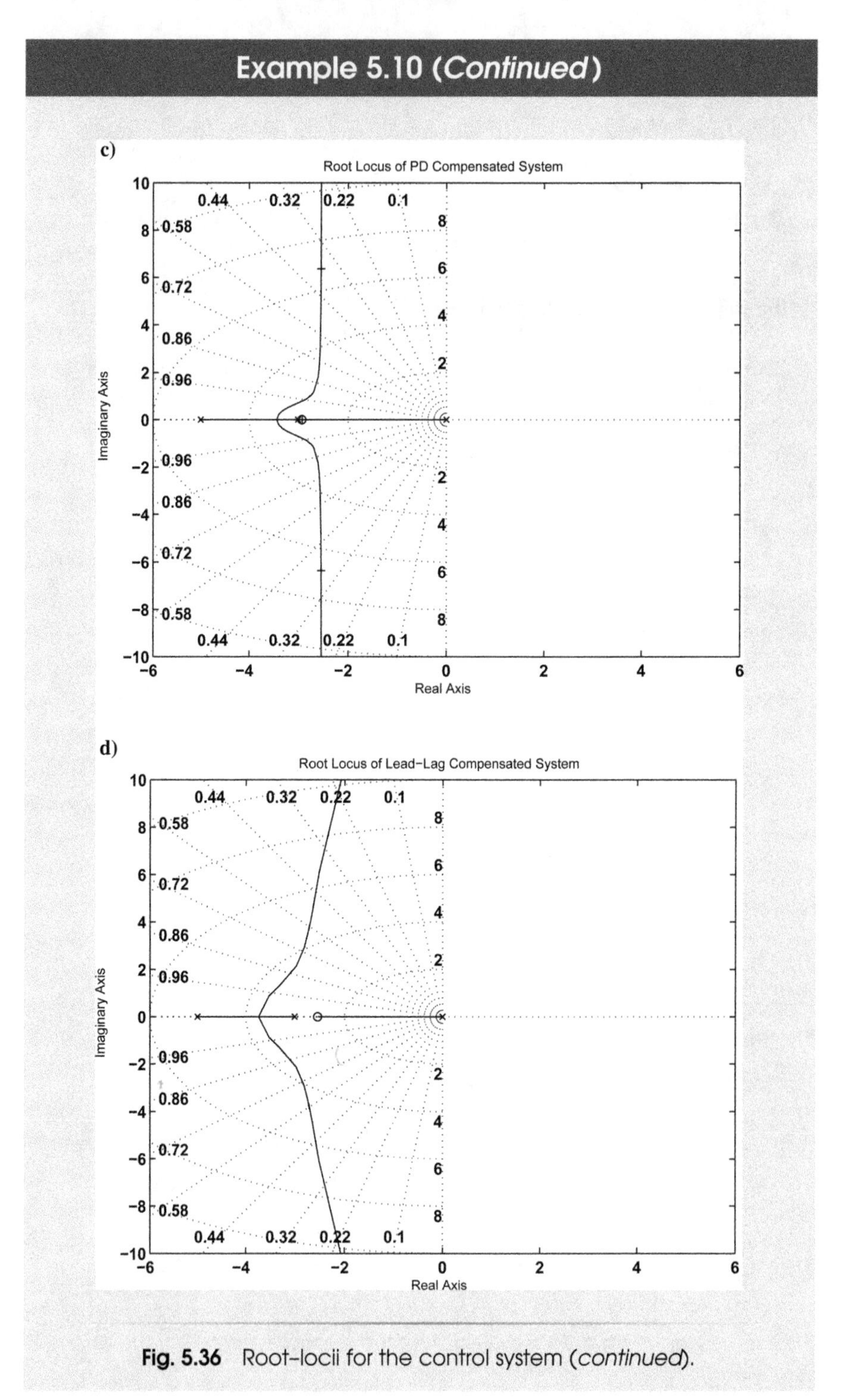

Fig. 5.36 Root-locii for the control system (*continued*).

(*Continued*)

Example 5.10 (*Continued*)

poles, we can use the second-order approximation. With this assumption, the times for peak amplitude for the basic system T_p and that for the compensated system T_{pc} are given by

$$T_p = \frac{\pi}{\omega_d}$$

$$= \frac{\pi}{1.9426} = 1.6172\,\text{s}$$

$$T_{pc} = \frac{T_p}{3} = 0.5391\,\text{s}$$

Then,

$$\omega_{dc} = \frac{\pi}{T_{pc}}$$

$$= \frac{\pi}{0.5391}$$

$$= 5.8278$$

$$\omega_{nc} = \frac{\omega_{dc}}{\sqrt{1-\zeta^2}}$$

$$= \frac{5.8278}{\sqrt{1-0.4^2}}$$

$$= 6.3587$$

$$\sigma_{dc} = -\zeta\omega_{nc}$$

$$= -0.4 * 6.3587$$

$$= -2.5435$$

Here, σ_{dc} and ω_{dc} are the real and imaginary parts for the dominant second-order poles.

The transfer function of the PD compensator is given by

$$G_c(s) = s + z_c$$

Now we have to determine the location of the compensating zero z_c so that the root-locus passes through the point $(\sigma_{dc}, \omega_{dc})$. The value of z_c is determined by the angle condition of Eq. (5.124), which in this case leads to

$$\theta_{zc} - (\theta_1 + \theta_2 + \theta_3) = (2n+1)180$$

Referring to Fig. 5.36b (ignoring the pole at $s = p_c$), we find $\theta_1 = 113.5736$ deg, $\theta_2 = 85.5274$ deg, and $\theta_3 = 67.1488$ deg. Choosing $n = -1$,

(*Continued*)

Example 5.10 (*Continued*)

we obtain $\theta_{zc} = 86.2498$ deg and $z_c = 2.9261$. Then, the transfer function of the PD-compensated system is given by

$$G_c(s) = \frac{k(s + 2.9261)}{s(s+3)(s+5)}$$

Then we draw the root-locus using MATLAB [4] as shown in Fig. 5.36c and obtain $k = 40.2971$ and $p = -2.5435 \pm j5.8317$ and -2.9130 for operating at $\zeta = 0.4$.

Lead-Lag Compensator

The transfer function of the lead compensator is given by Eq. (5.168) as

$$G_c(s) = \frac{s + z_c}{s + p_c}$$

We have to find the locations of the zero z_c and the pole p_c on the real axis so that the design objectives are met.

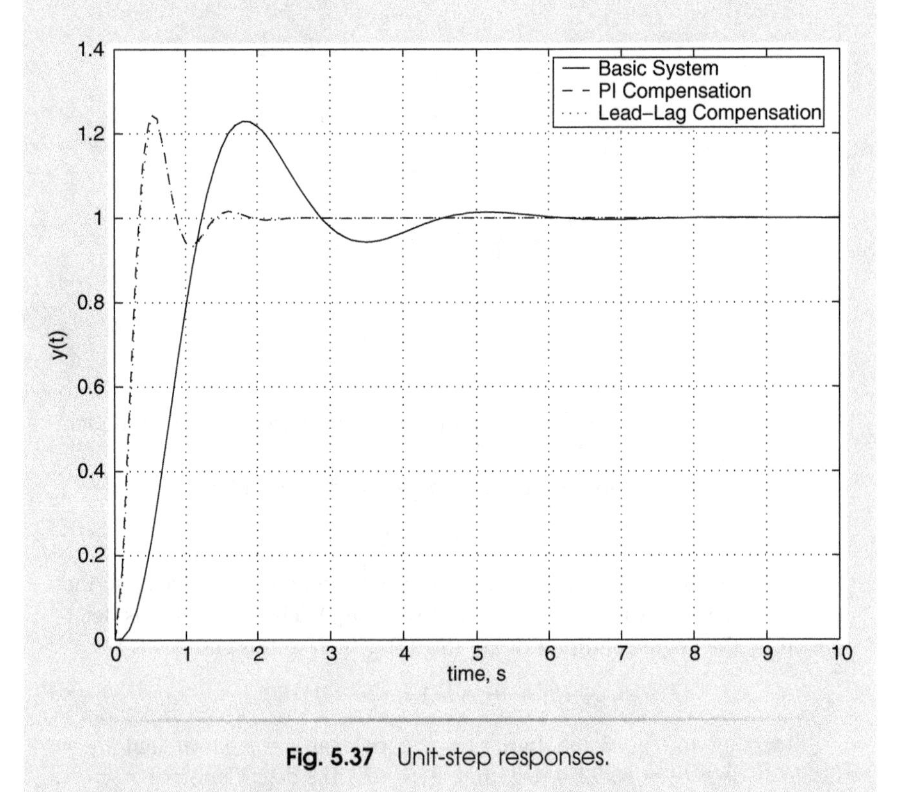

Fig. 5.37 Unit-step responses.

(Continued)

Example 5.10 (*Continued*)

From the analysis of PD compensator as above, we know that the net angle contribution due to the zero at $s = -z_C$ and pole at $s = -p_C$ must be equal to 86.2498 deg. Let us assume that the angle contribution due to zero at $s = -z_C$ is $\theta_{zc} = 90$ deg so that $z_C = 2.5435$. Then, the angle contribution due to the pole at $s = -p_C$ is $\theta_p = 3.7502$ deg so that $p_C = 88.9$. Then, the transfer function of the lead-compensated system is given by

$$G_C(s) = \frac{k(s + 2.5435)}{s(s + 3)(s + 5)(s + 88.90)}$$

Now we draw the root-locus of the lead-lag compensated system as shown in Fig. 5.36d (the root-locus around the origin is shown in this figure) and select the operating point for $\zeta = 0.4$. We get $k = 3451.4$ and closed-loop poles at -89.3603, $-2.5405 \pm j5.7879$, and -2.4588.

Let us verify the designs by performing the simulation, i.e., we determine the unit-step response using MATLAB [4]. The results are shown in Fig. 5.37. We observe that the peak amplitudes (hence the percent overshoot O_s) for all three cases are nearly equal. For the basic system, $T_p = 1.80$ s and, for PD- and lead-compensated systems, $T_p \simeq 0.6$ s. Thus, the design objectives have been realized.

Example 5.11

Design a PID controller for a unity feedback system with

$$G(s) = \frac{k(s + 15)}{(s + 1)(s + 3)(s + 9)}$$

to operate at a peak time, which is 50% of the basic system and has a zero steady-state error for a unit-step input while continuing to operate at a damping ratio of 0.3.

Solution:

The approach we take here is to design the PD controller first and then add a PI controller.

Draw the root-locus for the basic system using MATLAB [4] as shown in Fig. 5.38a and pick the point on the root-locus corresponding to $\zeta = 0.3$. We get $k = 14.2869$ and $p = -10.089$, $-1.4555 \pm j4.6689$. Using this

(*Continued*)

Example 5.11 (*Continued*)

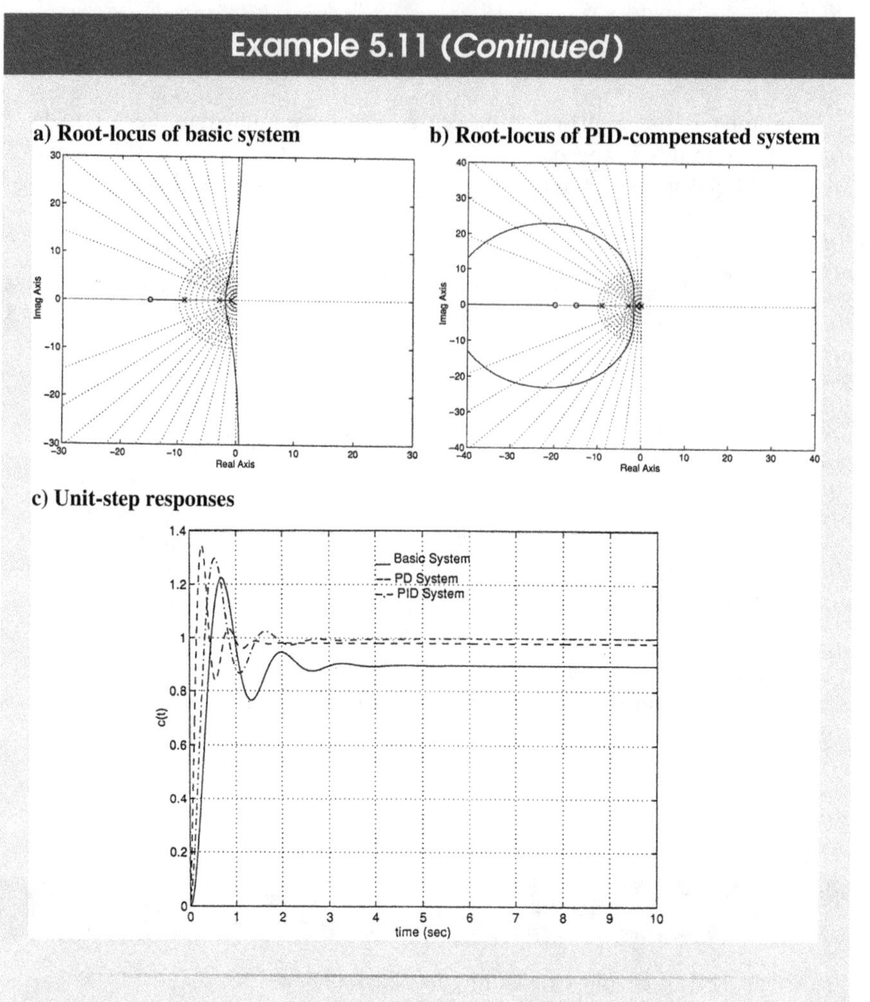

Fig. 5.38 PID controller.

information, we get $T_p = 0.6729$ s. Then, for the compensated system, $T_{pc} = T_p/2 = 0.3365$ s, which corresponds to $\omega_{dc} = 9.3378$, $\omega_{nc} = \omega_{dc}/\sqrt{1-\zeta^2} = 9.7887$, and $\sigma_{dc} = \zeta\omega_{nc} = 2.9366$.

Now we calculate the angle contributions. Proceeding as before in Example 5.10, we get $\theta_1 = 101.1851$ due to pole at $s = -1$, $\theta_2 = 89.6355$, due to pole at $s = -3$ and $\theta_3 = 39.0601$ due to zero at $s = -15$, and $\theta_4 = 58.2290$ due to pole at $s = -9$. As a result, $\theta_{zc} = 29.9885$ giving $z_c = 19.9014$. With this, the transfer function of the PD-compensated system is given by

$$G(s) = \frac{k(s+15)(s+19.9014)}{(s+1)(s+3)(s+9)}$$

(Continued)

Example 5.11 (*Continued*)

Next, we add the PI controller. Select a pole at $s = 0$ and a zero at $s = -0.5$ so that the transfer function of the PID controller is given by

$$G(s) = \frac{k(s + 0.5)(s + 15)(s + 19.9014)}{s(s + 1)(s + 3)(s + 9)}$$

Now we draw the root-locus of the PID system as shown in Fig. 5.38b and pick the point corresponding to $\zeta = 0.3$. We get $k = 6.1112$ and closed-loop poles at $-3.6846 \pm j12.2343$, -11.2460, and -0.4969.

The unit-step responses of the basic, PD-, and PID-compensated systems are shown in Fig. 5.38c. It may be observed that the PID-compensated system meets the design requirements.

Example 5.12

For the system of Example 5.10, design a major-loop feedback to achieve the same performance.

Solution:

We have found in Example 5.10 that $\theta_{zc} = 86.2498$. With this, we obtain the equivalent pole location, $z_c = 2.9261$, and $k_h = 1/z_c = 0.3418$. We then plot the root-locus using MATLAB and obtain the value of the gain as 40.2971, which is equal to kk_h so that $k = 117.8967$. The reader may verify that this response is identical to that of the PD controller of Example 5.10.

Example 5.13

For the control system shown in Fig. 5.39a, determine gain k_h so that the minor loop operates with a damping ratio of 0.707 and the complete system has a damping ratio of 0.4.

Solution:

Consider the minor loop. We draw the root-locus using MATLAB [4] as shown in Fig. 5.39b and pick the point on the root-locus corresponding to $\zeta = 0.707$. Then, we get $k_h = 17.2510$ and $p = 0, -4.0 \pm j4.0313$.

(*Continued*)

Example 5.13 (*Continued*)

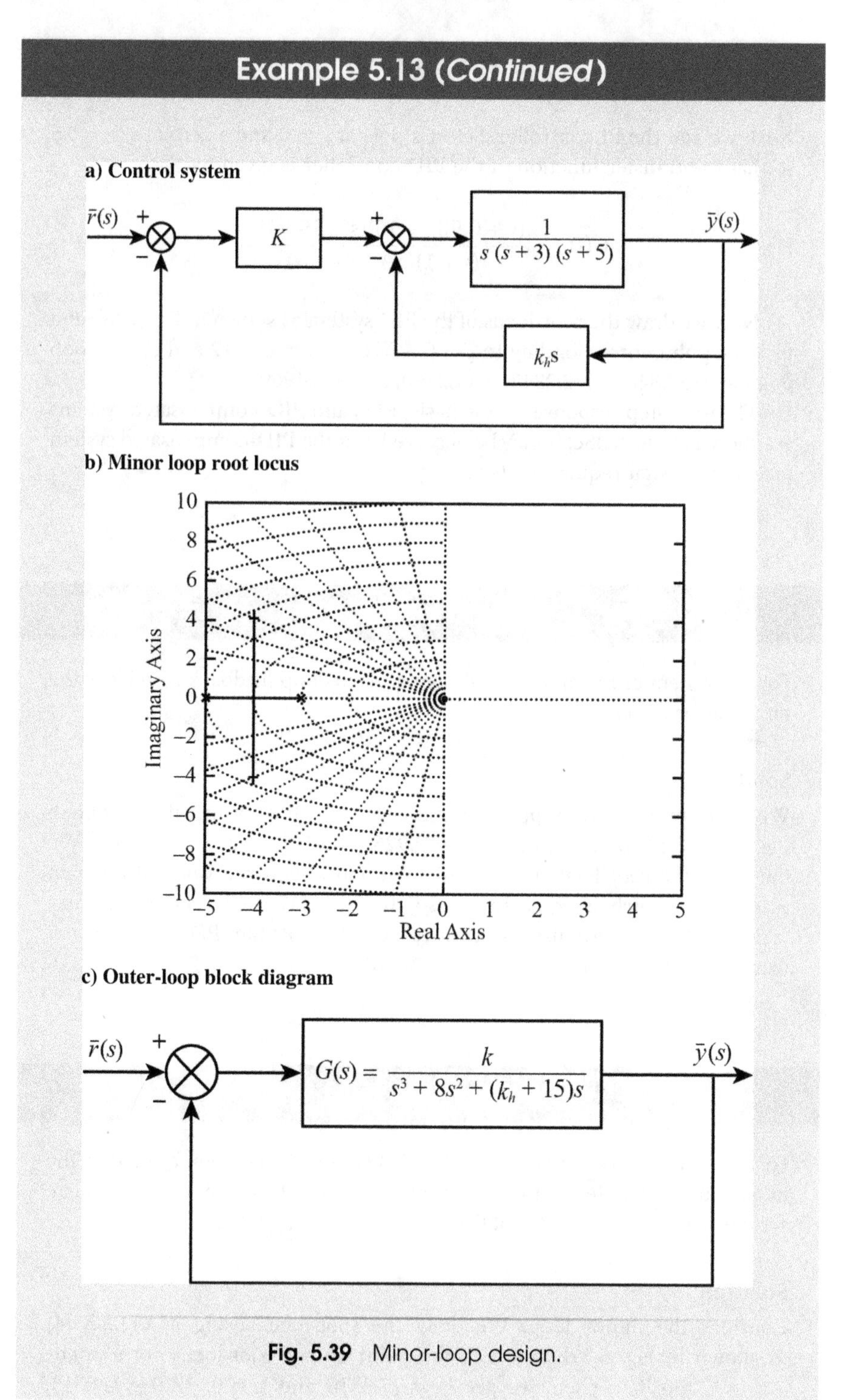

Fig. 5.39 Minor-loop design.

(*Continued*)

Example 5.13 (*Continued*)

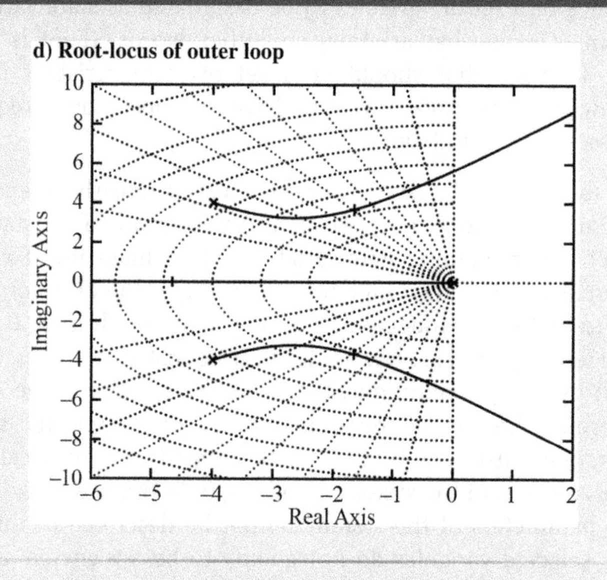

Fig. 5.39 Minor-loop design (*continued*).

Having designed the minor loop and knowing the value of k_h, we can simplify the system block diagram as shown in Fig. 5.39c. We have

$$G_1(s) = \frac{k}{s^3 + 8s^2 + (k_h + 15)s}$$

For this system, we draw the root-locus as shown in Fig. 5.39d and obtain $k = 77.5786$ and $p = -4.6521, -1.6740 \pm j3.7248$ for operating at $\zeta = 0.4$. This completes the design.

5.10 State-Space Analysis and Design

The classical method of analyses discussed in the previous sections is called the frequency-domain technique because it is based on system representation in the form of transfer function. The main advantage of this approach is that the governing differential equation of the system is replaced by an algebraic transfer function. However, a disadvantage of this classical method is that it is limited to linear time-invariant systems with zero initial conditions. The modern state-space approach is more general in nature because it can be used to represent nonlinear, time-varying systems with nonzero initial conditions. The state-space method can also handle multi-input-multi-output systems in a compact manner. Furthermore, the state-space approach becomes very attractive because it is based on matrix algebra, and powerful matrix analyses tools like MATLAB [4] are commercially available.

5.10.1 Concept of State Variable

The choice of a set of variables to be designated as state variables for a given system is somewhat arbitrary. In other words, there is no unique method of defining what should be a set of state variables for a given system. However, the state variables have to meet some requirements, which can be stated as follows.

1. The variables selected as state variables must be linearly independent, i.e., it should not be possible to express any one or more of the state variables in terms of the remaining state variables. Mathematically, if x is an n dimensional vector with components x_i, $i = 1, n$, then the components x_i are said to be linearly independent if $\alpha_i x_i \neq 0$, for all $\alpha_i \neq 0$ and all $x_i \neq 0$. Here, α_i, $i = 1, \ldots, n$ are arbitrary constants.
2. Given all the initial conditions, the input for $t \geq 0$, and the solution of the governing differential equation in terms of the selected state variables, one must be able to describe uniquely any physical parameter (state and output) of the system for all $t \geq 0$. In other words, if any of the physical parameters of the system cannot be described in this manner, then the selected variables do not qualify to be designated as state variables.

5.10.2 State-Space Representation

A state vector is a vector whose elements are the state variables satisfying the requirements just discussed. If n is the dimension of a state vector, then the state-space is an n dimensional space whose axes are the state variables. For example, if x_1, x_2, and x_3 are the elements of the state vector x, then $n = 3$, and the state-space is a three-dimensional space with x_1, x_2, and x_3 as three axes.

The state equation is a set of n simultaneous first-order differential equations involving n state variables and m inputs. Usually, $m < n$. The output equation is a set of algebraic equations that relate the outputs of the system to the state variables.

For example,

$$\dot{x} = Ax + Bu \tag{5.171}$$

$$y = Cx + Du \tag{5.172}$$

where

$$x = \begin{bmatrix} x_1 \\ x_2 \\ x_3 \\ \cdot \\ \cdot \\ x_n \end{bmatrix} \tag{5.173}$$

$$A = \begin{bmatrix} a_{11} & a_{12} & a_{13} & \cdot & a_{1n} \\ a_{21} & a_{22} & a_{23} & \cdot & a_{2n} \\ a_{31} & a_{32} & a_{33} & \cdot & a_{3n} \\ \cdot & \cdot & \cdot & \cdot & \cdot \\ \cdot & \cdot & \cdot & \cdot & \cdot \\ a_{n1} & a_{n2} & a_{n3} & \cdot & a_{nn} \end{bmatrix} \quad B = \begin{bmatrix} b_{11} & b_{12} & b_{13} & \cdot & b_{1m} \\ b_{21} & b_{22} & b_{23} & \cdot & b_{2m} \\ b_{31} & b_{32} & b_{33} & \cdot & b_{3m} \\ \cdot & \cdot & \cdot & \cdot & \cdot \\ \cdot & \cdot & \cdot & \cdot & \cdot \\ b_{n1} & b_{n2} & b_{n3} & \cdot & b_{nm} \end{bmatrix} \quad u = \begin{bmatrix} u_1 \\ u_2 \\ u_3 \\ \cdot \\ \cdot \\ u_m \end{bmatrix} \tag{5.174}$$

$$y = \begin{bmatrix} y_1 \\ y_2 \\ y_3 \\ \cdot \\ \cdot \\ y_q \end{bmatrix} \quad C = \begin{bmatrix} c_{11} & c_{12} & c_{13} & \cdot & c_{1n} \\ c_{21} & c_{22} & c_{23} & \cdot & c_{2n} \\ c_{31} & c_{32} & c_{33} & \cdot & c_{3n} \\ \cdot & \cdot & \cdot & \cdot & \cdot \\ \cdot & \cdot & \cdot & \cdot & \cdot \\ c_{q1} & c_{q2} & c_{q3} & \cdot & c_{qn} \end{bmatrix} \quad D = \begin{bmatrix} d_{11} & d_{12} & d_{13} & \cdot & d_{1m} \\ d_{21} & d_{22} & d_{23} & \cdot & d_{2m} \\ d_{31} & d_{32} & d_{33} & \cdot & d_{3m} \\ \cdot & \cdot & \cdot & \cdot & \cdot \\ \cdot & \cdot & \cdot & \cdot & \cdot \\ d_{q1} & d_{q2} & d_{q3} & \cdot & d_{qm} \end{bmatrix} \tag{5.175}$$

is a state-space representation of an nth order system. Here, the order of the system is equal to the number of simultaneous first-order differential equations.

In Eqs. (5.171) and (5.172), x is the state vector of dimension n, A is the system matrix of dimension $n \times n$, B is an $n \times m$ input coupling matrix, u is an $m \times 1$ input vector, y is the output vector of dimension $q \times 1$, C is a $q \times n$ output matrix, and D is a $q \times m$ feed forward matrix. The term feed forward is used when a part of the input directly appears at the output. A schematic diagram of the state-space representation is shown in Fig. 5.40.

5.10.3 State Transition Matrix

Consider the homogeneous part ($u = 0$) of state Eqs. (5.171) and (5.172) as given by

$$\dot{x} = Ax \tag{5.176}$$

$$y = Cx \tag{5.177}$$

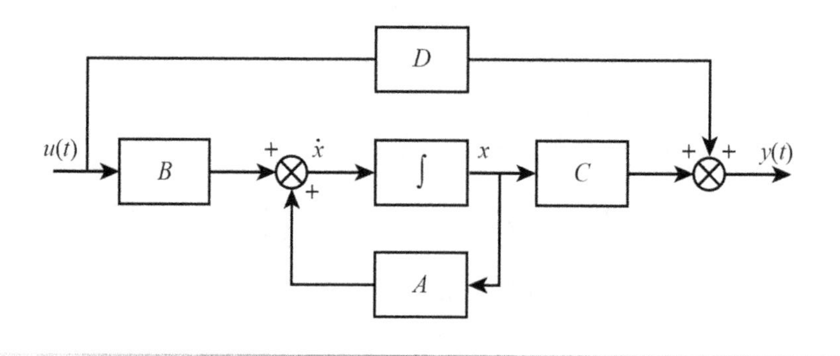

Fig. 5.40 State-space representation.

The state transition matrix $\Phi(t)$ is defined as a matrix that satisfies the equation

$$x(t) = \Phi(t)x(0) \tag{5.178}$$

In other words, given the initial conditions $x(0)$, the state transition matrix enables us to predict the state vector at $t \geq 0$.

Substituting for $x(t)$ from Eq. (5.178) into state Eq. (5.176), we obtain

$$\dot{\Phi}(t)x(0) = A\Phi(t)x(0) \tag{5.179}$$

$$[\dot{\Phi}(t) - A\Phi(t)]x(0) = 0 \tag{5.180}$$

If this identity is to hold for all arbitrary values of $x(0)$, we must have

$$\dot{\Phi}(t) - A\Phi(t) = 0 \tag{5.181}$$

This shows that the state transition matrix $\Phi(t)$ is a solution to the homogeneous state Eq. (5.176).

Determination of State Transition Matrix

Take the Laplace transformation of Eq. (5.176),

$$s\bar{x}(s) - x(0) = A\bar{x}(s) \tag{5.182}$$

$$\bar{x}(s) = (sI - A)^{-1}x(0) \tag{5.183}$$

Here, I is an identity matrix. An identity matrix is a square matrix with all diagonal elements equal to unity and all the rest of the elements equal to zero. We assume that $(sI - A)^{-1}$ exists, i.e., $(sI - A)$ is nonsingular. Then,

$$x(t) = L^{-1}[(sI - A)^{-1}]x(0) \tag{5.184}$$

for $t \geq 0$. Comparing Eqs. (5.184) and (5.178), we get

$$\Phi(t) = L^{-1}[(sI - A)^{-1}] \tag{5.185}$$

Let

$$x(t) = e^{At}x(0) \tag{5.186}$$

The matrix exponential is given by

$$e^{At} = I + At + \frac{A^2t^2}{2!} + \cdots + \frac{A^nt^n}{n!} + \cdots \tag{5.187}$$

We note that Eq. (5.186) satisfies the homogeneous state Eq. (5.176). Hence,

$$\Phi(t) = e^{At} = I + At + \frac{A^2t^2}{2!} + \cdots + \frac{A^nt^n}{n!} + \cdots \tag{5.188}$$

Using this, the solution of the complete nonhomogeneous state Eq. (5.171) can be expressed as [1, 3]

$$x(t) = \Phi(t)x(0) + \int_0^t \Phi(t-\tau)Bu(\tau)\,\mathrm{d}\tau \tag{5.189}$$

and the output

$$y(t) = C\left[\Phi(t)x(0) + \int_0^t \Phi(t-\tau)Bu(\tau)\,\mathrm{d}\tau\right] + Du \tag{5.190}$$

The integral in Eq. (5.189) is the convolution integral, which was introduced earlier in Eq. (5.39). The first term on the right-hand side of Eq. (5.189) represents the solution to the homogeneous part of the state equation and gives the free (transient) response. The second term represents the forced response and is independent of the initial conditions $x(0)$.

Properties of State Transition Matrix

The state transition matrix $\Phi(t)$ has the following properties. The proof of these identities is left as an exercise to the reader.

$$\Phi(0) = I \tag{5.191}$$

$$\Phi^{-1}(t) = \Phi(-t) \tag{5.192}$$

$$\Phi(t_2 - t_1)(t_1 - t_0) = \Phi(t_2 - t_0) \tag{5.193}$$

$$\Phi[(t)]^k = \Phi(kt) \tag{5.194}$$

Characteristic Equation

Given a square matrix A, the equation

$$\Delta(\lambda) = |\lambda I - A| = 0 \tag{5.195}$$

is called the characteristic equation of matrix A. Here, $|\cdot|$ denotes the determinant of the argument (square) matrix.

Eigenvalues and Eigenvectors

The roots of characteristic Eq. (5.195) are called the eigenvalues of matrix A and are usually denoted by λ_i, $i = 1, \ldots, n$, where n is the number of rows or columns of matrix A. As an example, let

$$A = \begin{bmatrix} 1 & 1 \\ 4 & 1 \end{bmatrix} \tag{5.196}$$

so that

$$\lambda I - A = \begin{bmatrix} \lambda - 1 & -1 \\ -4 & \lambda - 1 \end{bmatrix} \tag{5.197}$$

and

$$\Delta(\lambda) = |\lambda I - A| = \lambda^2 - 2\lambda - 3 = 0 \tag{5.198}$$

Solving, we get the eigenvalues $\lambda_1 = 3$ and $\lambda_2 = -1$.

An important property of the eigenvalues is that they remain invariant under any linear transformation. A direct consequence of this property is that the closed-loop characteristic equation also remains invariant under any linear transformation. To make this point clear, suppose we are given a linear, time-invariant system $x = Ax + Bu$ and we transform this system using a linear transformation $x = Pz$ so that the transformed system is $\dot{z} = P^{-1}APz + P^{-1}Bu$. Then, the eigenvalues of matrix A and those of $P^{-1}AP$ are identical. Interested readers may verify this statement by working out the details.

If λ_i is an eigenvalue of the square matrix A, then any vector x that satisfies the equation

$$Ax = \lambda_i x \tag{5.199}$$

is called the eigenvector corresponding to the eigenvalue λ_i. In other words, every eigenvalue will have an associated eigenvector. Returning to the previous example,

$$\begin{bmatrix} 1 & 1 \\ 4 & 1 \end{bmatrix} \begin{bmatrix} x_1 \\ x_2 \end{bmatrix} = (-1) \begin{bmatrix} x_1 \\ x_2 \end{bmatrix} \tag{5.200}$$

or

$$2x_1 + x_2 = 0 \tag{5.201}$$

$$4x_1 + 2x_2 = 0 \tag{5.202}$$

Note that the two equations are identical, stating that the eigenvector is not unique and depends on our choice of one of the two variables. Let $x_2 = 1$ so that $x_1 = -1/2$. The eigenvector corresponding to $\lambda = -1$ is $[-1/2 \ 1]^T$. Here, the superscript "T" denotes the matrix transpose. It can be shown that the eigenvector obtained by choosing any other value for x_2 would be a scalar multiple of this eigenvector. Similarly, the eigenvector corresponding to $\lambda_2 = 3$ is $[1 \ \ 2]^T$.

To understand the physical meaning of eigenvalues and eigenvectors, consider a system with two first-order, coupled linear differential equations

$$\dot{x}_1 = x_1 + x_2 \tag{5.203}$$

$$\dot{x}_2 = 4x_1 + x_2 \tag{5.204}$$

Assume that the solution is given by $x_1(t) = u_1 e^{\lambda t}$ and $x_2(t) = u_2 e^{\lambda t}$. Substituting, we get

$$\lambda u_1 e^{\lambda t} = u_1 e^{\lambda t} + u_2 e^{\lambda t} \tag{5.205}$$

$$\lambda u_2 e^{\lambda t} = 4u_1 e^{\lambda t} + u_2 e^{\lambda t} \tag{5.206}$$

Because $e^{\lambda t} \geq 0$ for all $t \geq 0$, we can write

$$\lambda u_1 = u_1 + u_2 \tag{5.207}$$

$$\lambda u_2 = 4u_1 + u_2 \tag{5.208}$$

or

$$\begin{bmatrix} 1 & 1 \\ 4 & 1 \end{bmatrix} \begin{bmatrix} u_1 \\ u_2 \end{bmatrix} = \lambda \begin{bmatrix} u_1 \\ u_2 \end{bmatrix} \tag{5.209}$$

which is of the form $Au = \lambda u$. This equation is of the same form as Eq. (5.199) with λ as the eigenvalue of matrix A. Furthermore, we recognize that the matrix A in Eq. (5.209) is the same as matrix A in Eq. (5.196), which has the eigenvalues of -1 and 3 and eigenvectors of $[-1/2 \ \ 1]^T$ and $[1 \ \ 2]^T$. This means, with $\lambda = -1$, $u_1 = -1/2$, and $u_2 = 1$, we get the first solution as

$$x_1 = -\frac{1}{2}e^{-t} \quad x_2 = e^{-t} \tag{5.210}$$

and, with $\lambda = 3$, $u_1 = 1$, and $u_2 = 2$, we get the second solution as

$$x_1 = e^{3t} \quad x_2 = 2e^{3t} \tag{5.211}$$

Therefore, the general solution is given by

$$x_1 = -\frac{1}{2}e^{-t} + e^{3t} \quad x_2 = e^{-t} + 2e^{3t} \tag{5.212}$$

Thus, we observe that the eigenvalues determine the nature of the transient response and the eigenvectors determine the amplitude of the transient response.

5.10.4 Controllability and Observability

The concept of controllability is linked to the question of whether the input u affects or controls the variation of each one of the state variables x_i. If it does, then we say that the given system is controllable. On the other hand, if any of the state variables are not influenced by the input u, then the system is said to be uncontrollable. Alternatively, if we can take the system from a given initial state $x(0)$ to a specified final state $x(t_f)$

using the available control u, then the system is said to be controllable. If not, the system is uncontrollable.

The given nth order linear, time-invariant system

$$\dot{x} = Ax + Bu \tag{5.213}$$

$$y = Cx \tag{5.214}$$

is said to be controllable if matrix

$$Q_c = [B \quad AB \quad A^2B \quad \cdots \quad A^{n-1}B] \tag{5.215}$$

is nonsingular or has full rank n. Matrix Q_c is called the controllability matrix [1–3].

The concept of observability is related to the question whether each one of the state variables affects or controls the variation of output y. If the answer to this question is yes, then the system is said to be observable. If not, the system is unobservable. Thus, for an unobservable system, the input does not affect some or all of the output variables.

A given linear, time-invariant system in the state-space form is said to be observable if matrix

$$Q_0 = \begin{bmatrix} C \\ CA \\ CA^2 \\ \cdot\cdot \\ \cdot\cdot \\ CA^{n-1} \end{bmatrix} \tag{5.216}$$

is nonsingular or has the full rank n. Matrix Q_0 is called the observability matrix [1–3].

5.10.5 Phase-Variable Form

Let us suppose that we have a state-space representation in the form

$$\dot{x} = A_p x + B_p u \tag{5.217}$$

where

$$x = \begin{bmatrix} x_1 \\ x_2 \\ x_3 \\ \cdot \\ \cdot \\ x_n \end{bmatrix} \tag{5.218}$$

$$A_p = \begin{bmatrix} 0 & 1 & 0 & 0 & \cdot & 0 \\ 0 & 0 & 1 & 0 & \cdot & 0 \\ 0 & 0 & 0 & 1 & \cdot & 0 \\ \cdot & \cdot & \cdot & \cdot & \cdot & \cdot \\ \cdot & \cdot & \cdot & \cdot & \cdot & \cdot \\ -a_0 & -a_1 & -a_2 & \cdot & \cdot & -a_{n-1} \end{bmatrix} \tag{5.219}$$

$$B_p = \begin{bmatrix} 0 \\ 0 \\ 0 \\ \cdot \\ \cdot \\ 1 \end{bmatrix} \tag{5.220}$$

Equation (5.217) with matrices A_p and B_p given by Eqs. (5.219) and (5.220) is called the phase-variable form of state Eq. (5.171). The advantage of the phase-variable form is that there is a minimum amount of coupling between the state variables. For example, given all the initial conditions, $x_1(0)$, $x_2(0), \ldots, x_n(0)$, we can first solve $\dot{x}_1(0) = x_2(0)$ and obtain $x_1(\Delta t)$ by an integration over a small time step Δt. Similarly, we can get $x_2(\Delta t)$, $x_3(\Delta t), \ldots, x_n(\Delta t)$ by successively solving the other equations.

The phase-variable form of representation has another advantage that the elements of the last row constitute the coefficients of the characteristic equation as follows:

$$\Delta(s) = s^n + a_{n-1}s^{n-1} + \cdots + a_1 s + a_0 = 0 \tag{5.221}$$

This property is useful in the design of compensators using the pole-placement method, which we will discuss a little later.

5.10.6 Conversion of Differential Equations to Phase-Variable Form

Let the given dynamical system be represented by the following linear differential equation:

$$\frac{\mathrm{d}^n y}{\mathrm{d}t^n} + a_{n-1}\frac{\mathrm{d}^{n-1} y}{\mathrm{d}t^{n-1}} + \cdots + a_1\frac{\mathrm{d}y}{\mathrm{d}t} + a_0 y = u(t) \tag{5.222}$$

Let us select a set of state variables such that each subsequent state variable is defined as the derivative of the previous state variable. That is,

$$x_1 = y \qquad x_2 = \dot{y} = \dot{x}_1, \ldots, x_n = \frac{\mathrm{d}^{n-1} y}{\mathrm{d}t^{n-1}} = \dot{x}_{n-1} \tag{5.223}$$

Then

$$\dot{x}_1 = x_2 \tag{5.224}$$

$$\dot{x}_2 = x_3 \tag{5.225}$$

$$\cdots$$

$$\cdots$$

$$\cdots$$

$$\dot{x}_n = -a_0 x_1 - a_1 x_2 - a_2 x_3 - \cdots - a_{n-1} x_n + u(t) \tag{5.226}$$

Or, in matrix form,

$$\dot{x} = A_p x + B_p u \tag{5.227}$$

where

$$x = \begin{bmatrix} x_1 \\ x_2 \\ x_3 \\ \cdot \\ \cdot \\ x_n \end{bmatrix} \tag{5.228}$$

$$A_p = \begin{bmatrix} 0 & 1 & 0 & 0 & \cdot & 0 \\ 0 & 0 & 1 & 0 & \cdot & 0 \\ 0 & 0 & 0 & 1 & \cdot & 0 \\ \cdot & \cdot & \cdot & \cdot & \cdot & \cdot \\ \cdot & \cdot & \cdot & \cdot & \cdot & \cdot \\ -a_0 & -a_1 & -a_2 & \cdot & \cdot & -a_{n-1} \end{bmatrix} \tag{5.229}$$

$$B_p = \begin{bmatrix} 0 \\ 0 \\ 0 \\ \cdot \\ \cdot \\ 1 \end{bmatrix} \tag{5.230}$$

The output is given by

$$y = Cx = \begin{bmatrix} 1 & 0 & \cdots & 0 \end{bmatrix} \begin{bmatrix} x_1 \\ x_2 \\ x_3 \\ \cdot \\ \cdot \\ x_n \end{bmatrix} \tag{5.231}$$

Thus, by choosing each successive state variable to be the derivative of the previous one, we are able to express the given differential equation in the state-space, phase-variable form.

5.10.7 Conversion of General State-Space Representation to Phase-Variable Form

A system given in a general state-space form can be expressed in the phase-variable form if the system is controllable. Let

$$\dot{x} = Ax + Bu \tag{5.232}$$

$$y = Cx \tag{5.233}$$

be the given plant, which is not in phase-variable form. We assume that this system is controllable. Furthermore, let us assume that there exists a matrix P, which is defined as

$$z = Px \tag{5.234}$$

which transforms the given system into phase-variable form,

$$\begin{bmatrix} \dot{z}_1 \\ \dot{z}_2 \\ \dot{z}_3 \\ \cdot \\ \cdot \\ \dot{z}_n \end{bmatrix} = \begin{bmatrix} 1 & 0 & 0 & \cdot & \cdot & 0 \\ 0 & 1 & 0 & \cdot & \cdot & 0 \\ 0 & 0 & 1 & \cdot & \cdot & 0 \\ \cdot & \cdot & \cdot & \cdot & \cdot & 0 \\ \cdot & \cdot & \cdot & \cdot & \cdot & 0 \\ -a_0 & -a_1 & \cdot & \cdot & \cdot & -a_{n-1} \end{bmatrix} \begin{bmatrix} z_1 \\ z_2 \\ z_3 \\ \cdot \\ \cdot \\ z_n \end{bmatrix} + \begin{bmatrix} 0 \\ 0 \\ 0 \\ \cdot \\ \cdot \\ 1 \end{bmatrix} u \tag{5.235}$$

The transformation matrix P has the form

$$P = \begin{bmatrix} p_{11} & p_{12} & \cdot & \cdot & \cdot & p_{1n} \\ p_{21} & p_{22} & \cdot & \cdot & \cdot & p_{2n} \\ p_{31} & p_{32} & \cdot & \cdot & \cdot & p_{3n} \\ \cdot & \cdot & \cdot & \cdot & \cdot & \cdot \\ \cdot & \cdot & \cdot & \cdot & \cdot & \cdot \\ p_{n1} & p_{n2} & \cdot & \cdot & \cdot & p_{nn} \end{bmatrix} = \begin{bmatrix} P_1 \\ P_2 \\ P_3 \\ \cdot \\ \cdot \\ P_n \end{bmatrix} \tag{5.236}$$

We have

$$z_1 = P_1 x \tag{5.237}$$

so that

$$\dot{z}_1 = P_1 A x + P_1 B u \tag{5.238}$$

Because this transformed equation is supposed to be in phase-variable form, we must have $\dot{z}_1 = z_2$. This gives $z_2 = P_1 A x$ and $P_1 B = 0$. From Eq. (5.234), we have $z_2 = P_2 x$. Therefore, $P_2 = P_1 A$. Continuing this further, we find that $\dot{z}_2 = P_1 A^2 x$, $P_1 AB = 0$, $P_3 = P_1 A^2, \ldots, \dot{z}_{n-1} = P_1 A^{n-1} x$, $P_1 A^{n-2} B = 0$, and $P_{n-1} = P_1 A^{n-2}$. Finally, we have

$$\dot{z}_n = P_1 A^n x + P_1 A^{n-1} B u \tag{5.239}$$

Comparing this with Eq. (5.235), we find that $P_1 A^{n-1} B = 1$. With this, we can construct the following matrix:

$$[P_1 B \quad P_1 AB \quad P_1 A^2 B \quad \cdot \quad P_1 A^{n-1} B] = [0 \quad 0 \quad 0 \quad \cdot \quad 1] \tag{5.240}$$

or

$$P_1 = [0 \quad 0 \quad 0 \quad \cdot \quad 1][B \quad AB \quad A^2 B \quad \cdot \quad A^{n-1} B]^{-1} \tag{5.241}$$

$$= [0 \quad 0 \quad 0 \quad \cdot \quad 1] Q_c^{-1} \tag{5.242}$$

Because we have assumed that the given system is controllable, the controllability matrix Q_c is nonsingular and Q_c^{-1} exists. Once P_1 is known, then P_2, $P_3, \ldots, P_n$ can be calculated using the relations derived as shown. Then, using Eqs. (5.234) and (5.232), we obtain the phase-variable form of the given system is

$$\dot{z} = (PAP^{-1}) z + (PB) u \tag{5.243}$$

5.10.8 Conversion of Transfer Function Form to Phase-Variable Form

Suppose the relation between the input and output of a system is given in the form of a transfer function; we can convert this to state-space phase-variable representation using a number of different approaches. Here, we will discuss a method based on decomposition of the transfer function. To illustrate the method, consider the system shown in Fig. 5.41a.

Let the open-loop transfer function of a system be given by

$$G(s) = \frac{k(s^2 + a_1 s + a_2)}{s^3 + b_1 s^2 + b_2 s + b_3} \tag{5.244}$$

The first step is to decompose the given system into two blocks, one for the denominator with transfer function $G_1(s)$ and the other for the

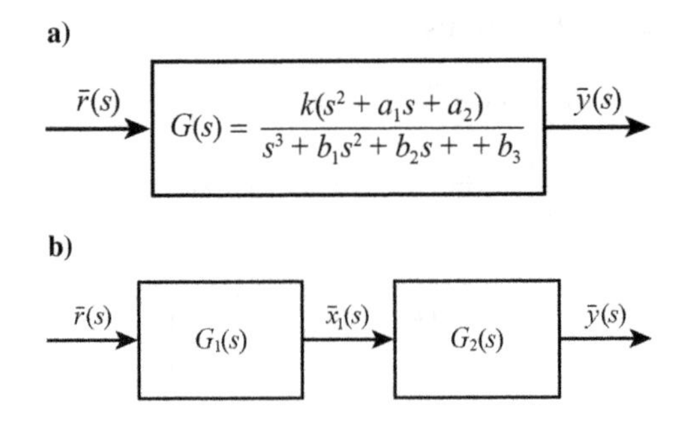

Fig. 5.41 Decomposition of a given control system in phase-variable form.

numerator with transfer function $G_2(s)$ as shown in Fig. 5.41b. Let the output of the first block be denoted as $\bar{x}_1(s)$. Then, for the first block,

$$G_1(s) = \frac{\bar{x}_1(s)}{\bar{r}(s)} = \frac{k}{s^3 + b_1 s^2 + b_2 s + b_3} \tag{5.245}$$

so that

$$\bar{x}_1(s)(s^3 + b_1 s^2 + b_2 s + b_3) = k\bar{r}(s) \tag{5.246}$$

Taking the inverse Laplace transforms,

$$\frac{d^3 x_1}{dt^3} + b_1 \frac{d^2 x_1}{dt^2} + b_2 \frac{dx_1}{dt} + b_3 x_1 = kr(t) \tag{5.247}$$

Let $x_2 = \dot{x}_1$ and $x_3 = \dot{x}_2 = \ddot{x}_1$ so that

$$\dot{x}_1 = x_2 \tag{5.248}$$

$$\dot{x}_2 = x_3 \tag{5.249}$$

$$\dot{x}_3 = -b_1 x_3 - b_2 x_2 - b_3 x_1 + kr(t) \tag{5.250}$$

In matrix form,

$$\begin{bmatrix} \dot{x}_1 \\ \dot{x}_2 \\ \dot{x}_3 \end{bmatrix} = \begin{bmatrix} 0 & 1 & 0 \\ 0 & 0 & 1 \\ -b_3 & -b_2 & -b_1 \end{bmatrix} \begin{bmatrix} x_1 \\ x_2 \\ x_3 \end{bmatrix} + \begin{bmatrix} 0 \\ 0 \\ 1 \end{bmatrix} kr(t) \tag{5.251}$$

which is the required phase-variable representation.

Consider the second block. This block gives the output matrix. We have

$$\bar{y}(s) = (s^2 + a_1 s + a_2)\bar{x}_1(s) \tag{5.252}$$

Taking the inverse Laplace transform, we get

$$y(t) = \ddot{x}_1 + a_1\dot{x}_1 + a_2x_1 \tag{5.253}$$

$$= x_3 + a_1x_2 + a_2x_1 \tag{5.254}$$

$$= \begin{bmatrix} a_2 & a_1 & 1 \end{bmatrix} \begin{bmatrix} x_1 \\ x_2 \\ x_3 \end{bmatrix} \tag{5.255}$$

5.10.9 Pole-Placement Method

The root-locus method discussed earlier is essentially a pole-placement method in frequency-domain analyses. The term pole refers to the poles of the closed-loop transfer function. When we consider higher order systems greater than two, the classical PD or PI type of controllers will not be able to place all the poles as desired because there are only two free variables [k_1 and k_2, see Eqs. (5.165) and (5.167)] at our disposal in PD or PI controllers. Therefore, for higher order systems, the pole-placement method of the state-space approach becomes very attractive because it can place all the closed-loop poles arbitrarily, but subject to the conditions that all the states are available for feedback and the given plant satisfies the controllability condition. In such a case, the pole-placement method is also called full-state feedback method.

Consider the system represented in state-space form as given by

$$\dot{x} = Ax + Bu \tag{5.256}$$

$$y = Cx \tag{5.257}$$

With full-state feedback, $u = r(t) - Kx$, where $r(t)$ is the $m \times 1$ input vector and K is an $m \times n$ matrix of feedback gains. Then,

$$\dot{x} = (A - BK)x + Br(t) \tag{5.258}$$

$$y = Cx \tag{5.259}$$

The block diagram implementation of the full-state feedback control system is shown in Fig. 5.42.

For simplicity, consider a system with single input ($m = 1$), which means K is of dimension $1 \times n$. Then, the design procedure is as follows.

1. Represent the given plant in phase-variable form.

$$\dot{z} = A_pz + B_pu \tag{5.260}$$

where $z = Px$ and P is the required transformation matrix, $A_p = PAP^{-1}$, $B_p = PB$.

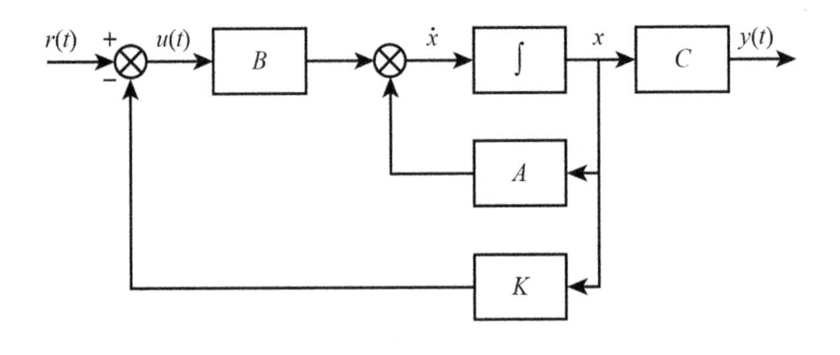

Fig. 5.42 Pole-placement method.

2. Feed back each state variable to the input of the plant with gains k_i so that

$$A_p - B_pK = \begin{bmatrix} 0 & 1 & 0 & 0 & . & 0 \\ 0 & 0 & 1 & 0 & . & 0 \\ 0 & 0 & 0 & 1 & . & 0 \\ . & . & . & . & . & . \\ . & . & . & . & . & . \\ -(a_0+k_1) & -(a_1+k_2) & . & . & . & -(a_{n-1}+k_n) \end{bmatrix} \tag{5.261}$$

3. Write down the characteristic equation for the plant as follows:

$$\left|sI-(A_p-B_pK)\right| = s^n + (a_{n-1}+k_n)s^{n-1} + \cdots + (a_1+k_2)s + (a_0+k_1) = 0 \tag{5.262}$$

4. Decide on all the closed-loop pole locations that give the desired system response, i.e., let the desired characteristic equation be given by

$$s^n + d_{n-1}s^{n-1} + \cdots + d_1 s + d_0 = 0 \tag{5.263}$$

5. Equate coefficients of the two characteristic equations

$$d_{n-1} = a_{n-1} + k_n, \ldots, d_1 = a_1 + k_2,\; d_0 = a_0 + k_1 \tag{5.264}$$

so that

$$k_1 = d_0 - a_0,\; k_2 = d_1 - a_1, \ldots, k_n = d_{n-1} - a_{n-1} \tag{5.265}$$

and

$$K = [k_1 \quad k_2 \quad \cdot \quad \cdot \quad k_n] \tag{5.266}$$

The full-state feedback law in the transformed z-space is given by

$$u = -Kz + r(t) \tag{5.267}$$

or, in the original state-space,

$$u = -KPx + r(t) \tag{5.268}$$

so that the given system with full-state feedback is given by

$$\dot{x} = Ax + Bu \tag{5.269}$$

$$= (A - BKP)x + Br(t) \tag{5.270}$$

6. Perform a simulation to verify the design.

The advantages of expressing the given plant in the phase-variable form is that equations for gains k_i are uncoupled and k_i can be easily obtained as given in Eq. (5.265). However, if the plant is not controllable, then it is not possible to represent it in the phase-variable form. For such a case, the described design procedure remains same except for the fact the equations for k_i will be coupled. Then gains k_i have to be obtained by solving the n coupled algebraic equations.

5.10.10 Dual Phase-Variable Form

The state-space representation, which is in the form

$$\dot{x} = Ax + Bu \tag{5.271}$$

where

$$x = \begin{bmatrix} x_1 \\ x_2 \\ x_3 \\ \cdot \\ \cdot \\ x_n \end{bmatrix} \quad A = \begin{bmatrix} -a_{n-1} & 1 & 0 & 0 & \cdot & 0 \\ -a_{n-2} & 0 & 1 & 0 & \cdot & 0 \\ \cdot & \cdot & \cdot & \cdot & \cdot & \cdot \\ -a_2 & 0 & 0 & \cdot & 1 & 0 \\ -a_1 & \cdot & \cdot & \cdot & \cdot & 1 \\ -a_0 & \cdot & \cdot & \cdot & \cdot & \cdot \end{bmatrix} \quad B = \begin{bmatrix} 0 \\ 0 \\ 0 \\ \cdot \\ \cdot \\ 1 \end{bmatrix} \tag{5.272}$$

is said to be in dual phase-variable form. Similar to the phase-variable form, the elements of the first column of matrix A in dual phase-variable form constitute the coefficients of the characteristic equation as follows:

$$s^n + a_{n-1}s^{n-1} + a_{n-2}s^{n-2} + \cdots + a_1 s + a_0 = 0 \tag{5.273}$$

Furthermore, this form of representation of the system matrix A is very useful in the design of state observers, which we will be discussing a little later.

5.10.11 Conversion of Transfer Function Form to Dual Phase-Variable Form

We will illustrate this method with the help of an example. Consider once again the system (see Fig. 5.41) given by

$$G(s) = \frac{k(s^2 + a_1 s + a_2)}{s^3 + b_1 s^2 + b_2 s + b_3} \tag{5.274}$$

Rewrite this in the following form:

$$G(s) = \frac{\frac{k}{s} + \frac{a_1 k}{s^2} + \frac{a_2 k}{s^3}}{1 + \frac{b_1}{s} + \frac{b_2}{s^2} + \frac{b_3}{s^3}} \tag{5.275}$$

$$= \frac{\bar{y}(s)}{\bar{r}(s)} \tag{5.276}$$

or,

$$\bar{y}(s)\left(1 + \frac{b_1}{s} + \frac{b_2}{s^2} + \frac{b_3}{s^3}\right) = \bar{r}(s)\left(\frac{k}{s} + \frac{a_1 k}{s^2} + \frac{a_2 k}{s^3}\right) \tag{5.277}$$

Then,

$$\bar{y}(s) = \frac{1}{s}\left[-b_1\bar{y}(s) + k\bar{r}(s) + \frac{1}{s}\left([\bar{r}(s)ka_1 - b_2\bar{y}(s)]. + \frac{1}{s}[ka_2\bar{r}(s) - b_3\bar{y}(s)]\right)\right] \tag{5.278}$$

Let

$$s\bar{x}_3(s) = ka_2\bar{r}(s) - b_3\bar{y}(s) \tag{5.279}$$

$$s\bar{x}_2(s) = ka_1\bar{r}(s) - b_2\bar{y}(s) + \bar{x}_3(s) \tag{5.280}$$

$$s\bar{x}_1(s) = -b_1\bar{y}(s) - \bar{x}_2(s) + k\bar{r}(s) \tag{5.281}$$

so that

$$\bar{y}(s) = \bar{x}_1(s) \tag{5.282}$$

Taking the inverse Laplace transforms, we obtain the desired dual phase-variable form as follows:

$$\begin{bmatrix} \dot{x}_1 \\ \dot{x}_2 \\ \dot{x}_3 \end{bmatrix} = \begin{bmatrix} -b_1 & -1 & 0 \\ -b_2 & 0 & 1 \\ -b_3 & 0 & 0 \end{bmatrix} \begin{bmatrix} x_1 \\ x_2 \\ x_3 \end{bmatrix} + \begin{bmatrix} 1 \\ a_1 \\ a_2 \end{bmatrix} kr(t) \tag{5.283}$$

and

$$y(t) = x_1(t) \tag{5.284}$$

5.10.12 Observer Design

The pole-placement design method requires that all the state variables are accurately measured and are available for feedback. If this requirement is met and the system is controllable, then a complete control over all the eigenvalues is possible. A problem arises if some or all of the states are not actually measured or are not available for state feedback. An obvious solution would be to add more sensors that can measure the missing states. However, this approach may not always be feasible and often can be quite expensive. The other option is to estimate the unavailable states using a subsystem called a state observer. An observer that estimates all the states, including those that are actually measured, is called a full-state observer, and one that estimates only those states that are not measured is called a reduced-state observer. Here, we will discuss the procedure for the design of a full-state observer.

The design of an observer is based on the knowledge of a mathematical model of the plant, input(s), and output(s). The basic idea is to make the estimated states as close to the actual states as possible, but the problem is that all the actual states are not available for comparison. However, we do know the output of the given plant, and we can compare it with the estimated output of the observer. The design objective is then to drive the error between the actual and estimated outputs to zero as rapidly as possible so that, in the limit, the estimated states approach the actual states. The schematic diagram of such a full-state observer is shown in Fig. 5.43.

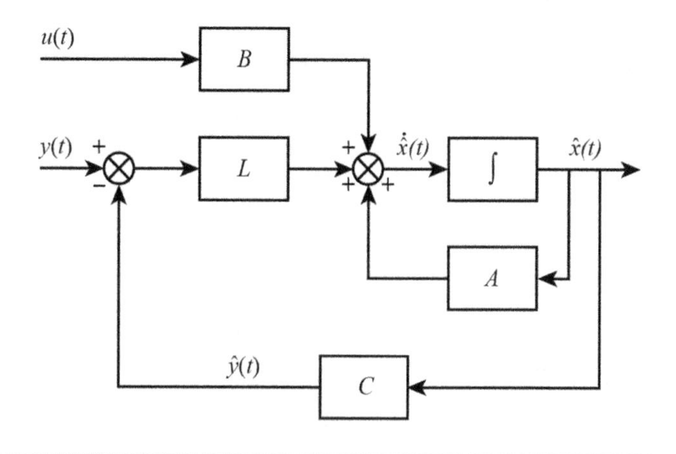

Fig. 5.43 Full-state observer.

Suppose the dynamics of the plant and output are given by

$$\dot{x} = Ax + Bu \tag{5.285}$$

$$y = Cx \tag{5.286}$$

Let the observer dynamics and the output be given by

$$\dot{\hat{x}} = A\hat{x} + Bu + L(y - \hat{y}) \tag{5.287}$$

$$\hat{y} = C\hat{x} \tag{5.288}$$

so that

$$y - \hat{y} = C(x - \hat{x}) \tag{5.289}$$

Here, $\hat{x}$ is the estimated state vector, L is the observer gain matrix and $\hat{y}$ is the estimated output. Let $e_x = x - \hat{x}$ be the error between the actual states and the estimated states and $e_y = y - \hat{y}$ be the error between the measured and estimated outputs. Then,

$$\dot{e}_x = \dot{x} - \dot{\hat{x}} = A(x - \hat{x}) - L(y - \hat{y}) \tag{5.290}$$

$$= (A - LC)(x - \hat{x}) = (A - LC)e_x \tag{5.291}$$

$$e_y = C(x - \hat{x}) = Ce_x \tag{5.292}$$

The objective of the design is to choose the observer gain matrix L such that the errors e_x and e_y approach zero as rapidly as possible. In other words, we choose the observer gain matrix L so that the closed-loop eigenvalues produce the desired transient response of the observer.

Plant Given in Dual Phase-Variable Form

Suppose that the given plant A is in dual phase-variable form

$$A = \begin{bmatrix} -a_{n-1} & 1 & 0 & 0 & \cdot & 0 \\ -a_{n-2} & 0 & 1 & 0 & \cdot & 0 \\ -a_{n-3} & 0 & 0 & 1 & \cdot & 0 \\ \cdot & \cdot & \cdot & \cdot & \cdot & \cdot \\ -a_1 & \cdot & \cdot & \cdot & \cdot & 1 \\ -a_0 & 0 & 0 & 0 & \cdot & 0 \end{bmatrix} \tag{5.293}$$

Then, let us assume that matrix $A - LC$ has the following form

$$A - LC = \begin{bmatrix} -(a_{n-1} + l_1) & 1 & 0 & 0 & \cdot & 0 \\ -(a_{n-2} + l_2) & 0 & 1 & 0 & \cdot & 0 \\ -(a_{n-3} + l_3) & 0 & 0 & 1 & \cdot & 0 \\ \cdot & \cdot & \cdot & \cdot & \cdot & \cdot \\ -(a_1 + l_n) & \cdot & \cdot & \cdot & \cdot & 1 \\ -(a_0 + l_n) & 0 & 0 & 0 & \cdot & 0 \end{bmatrix} \tag{5.294}$$

The characteristic equation of the observer system is then given by

$$s^n + (a_{n-1} + l_1)s^{n-1} + (a_{n-2} + l_2)s^{n-2} + \cdots + (a_0 + l_n) = 0 \quad (5.295)$$

Let the characteristic equation that gives the desired transient response of the observer be given by

$$s^n + d_{n-1}s^{n-1} + d_{n-2}s^{n-2} + \cdots + d_0 = 0 \quad (5.296)$$

Equating the coefficients of like powers of s, we obtain

$$l_1 = d_{n-1} - a_{n-1}, \quad l_2 = d_{n-2} - a_{n-2}, \ \ldots, l_n = d_0 - a_0 \quad (5.297)$$

Here, $l_1, l_2, \ldots, l_n$ are the elements of the feedback gain matrix to achieve the desired performance of the full-state observer.

Plant Not Given in Dual Phase-Variable Form

Suppose the given plant

$$\dot{x} = Ax + Bu \quad (5.298)$$

$$y = Cx \quad (5.299)$$

is not in dual phase-variable form. Then we assume that there exists a matrix P where $x = Pz$ that transforms this plant into a dual phase-variable form as given by

$$\dot{z} = A_z z + B_z u \quad (5.300)$$

$$y = C_z z \quad (5.301)$$

where $A_z = P^{-1}AP$, $B_z = P^{-1}B$, and $C_z = CP$. The observability matrix of the original system is given by

$$Q_{0,x} = \begin{bmatrix} C \\ CA \\ CA^2 \\ \cdot \\ \cdot \\ CA^{n-1} \end{bmatrix} \quad (5.302)$$

and that of the transformed system in phase-variable form is given by

$$Q_{0,z} = \begin{bmatrix} C \\ CA \\ CA^2 \\ \cdot \\ \cdot \\ CA^{n-1} \end{bmatrix} P \quad (5.303)$$

so that

$$P = Q_{0,x}^{-1} Q_{0,z} \tag{5.304}$$

Let

$$e_z = z - \hat{z} \tag{5.305}$$

Then,

$$\dot{e}_z = (A_z - L_z C_z) e_z \tag{5.306}$$

$$y - \hat{y} = C_z e_z \tag{5.307}$$

where $\hat{z}$ and $\hat{y}$ are the estimated state and the output vectors in the transformed system. Thus, the design procedure is as follows: 1) design the observer in the transformed space and obtain the gain matrix L_z, and 2) transform back to the original system to get the corresponding gain matrix L_x as follows.

We have $z = P^{-1}x$, $\hat{z} = P^{-1}\hat{x}$, so that $e_z = z - \hat{z} = P^{-1}(x - \hat{x}) = P^{-1}e_x$ and $\dot{e}_z = P^{-1}\dot{e}_x$. Then, substitution in Eq. (5.306) gives

$$\dot{e}_x = (A - PL_z C) e_x \tag{5.308}$$

Also, Eq. (5.291) with $L = L_x$ gives

$$\dot{e}_x = (A - L_x C) e_x \tag{5.309}$$

Therefore,

$$L_x = PL_z \tag{5.310}$$

Here, L_x is the gain matrix of the full-state observer corresponding to the original system.

Now one question that remains to be answered is how to construct the transformation matrix P. We will discuss one method that is based on knowing the characteristic equation or the eigenvalues of the given system. This approach is feasible because we have software tools like MATLAB that are available. We will illustrate this method in Example 5.16.

Example 5.14

Given the plant,

$$\dot{x} = \begin{bmatrix} 1 & 2 & 1 \\ 3 & 5 & 2 \\ 4 & 0 & 3 \end{bmatrix} \begin{bmatrix} x_1 \\ x_2 \\ x_3 \end{bmatrix} + \begin{bmatrix} 0 \\ 1 \\ 2 \end{bmatrix} u$$

Convert it to the phase-variable form.

(*Continued*)

Example 5.14 (*Continued*)

Solution:

For the given system, the controllability matrix is given by [Eq. (5.215)]

$$Q_c = [B \quad AB \quad A^2B] = \begin{bmatrix} 0 & 4 & 28 \\ 1 & 9 & 69 \\ 2 & 6 & 34 \end{bmatrix}$$

The rank of this matrix is three because all the columns are linearly independent. The inverse of this matrix exists and is given by

$$Q_c^{-1} = \begin{bmatrix} -1.35 & 0.4 & 0.3 \\ 1.30 & -0.7 & 0.35 \\ -0.15 & 0.10 & -0.05 \end{bmatrix}$$

Then using Eq. (5.241),

$$\begin{aligned} P_1 &= [0 \quad 0 \quad 1][B \quad AB \quad A^2B]^{-1} \\ &= [0 \quad 0 \quad 1]\begin{bmatrix} -1.35 & 0.4 & 0.3 \\ 1.30 & -0.7 & 0.35 \\ -0.15 & 0.10 & -0.05 \end{bmatrix} \\ &= [-0.15 \quad 0.10 \quad -0.05] \end{aligned}$$

Similarly,

$$\begin{aligned} P_2 &= P_1 A \\ &= [-0.05 \quad 0.20 \quad -0.10] \\ P_3 &= P_1 A^2 \\ &= [0.15 \quad 0.90 \quad 0.05] \end{aligned}$$

so that

$$P = \begin{bmatrix} -0.15 & 0.10 & -0.05 \\ -0.05 & 0.20 & -0.10 \\ 0.15 & 0.90 & 0.05 \end{bmatrix}$$

$$PAP^{-1} = \begin{bmatrix} 0 & 1 & 0 \\ 0 & 0 & 1 \\ -7 & -13 & 9 \end{bmatrix}$$

$$PB = \begin{bmatrix} 0 \\ 0 \\ 1 \end{bmatrix}$$

(*Continued*)

Example 5.14 (*Continued*)

Then, the phase-variable form of the given system is given by

$$\begin{bmatrix} \dot{z}_1 \\ \dot{z}_2 \\ \dot{z}_3 \end{bmatrix} = \begin{bmatrix} 0 & 1 & 0 \\ 0 & 0 & 1 \\ -7 & -13 & 9 \end{bmatrix} \begin{bmatrix} z_1 \\ z_2 \\ z_3 \end{bmatrix} + \begin{bmatrix} 0 \\ 0 \\ 1 \end{bmatrix} u$$

where $z = Px$.

Example 5.15

For the plant,

$$G(s) = \frac{25(s+2)}{(s+1)(s+3)(s+5)}$$

1. Represent the plant in phase-variable, state-space form.
2. Design a phase-variable, full-state feedback controller to yield 15% overshoot with a settling time of 1 s.

Solution:

For 15% overshoot, from Eq. (5.75), we find that $\zeta = 0.5169$. With $T_s = 1.0$ and using Eq. (5.78), $\omega_n = 4/\zeta = 7.7384$. Further, using Eq. (5.64), $\omega_d = \omega_n\sqrt{1-\zeta^2} = 6.6245$, and $\sigma = \zeta\omega_n = 4.0$. Therefore, the dominant poles are at $-4 \pm j6.6245$. Because the given system is a third-order system, we must choose one more pole. Let this pole be located at –2.1 so that it nearly cancels the zero at –2 justifying the second-order approximation.

Therefore, the characteristic equation that gives the desired response is given by

$$(s+4-j6.6245)(s+4+j6.6245)(s+2.1) = 0$$

or

$$s^3 + 10.1s^2 + 76.6s + 125.7543 = 0$$

The next step is to express the given plant in state-space, phase-variable form. For this purpose, let us decompose the transfer function into two blocks (one for the numerator and another for the denominator) in cascade as schematically shown earlier in Fig. 5.41b.

(Continued)

Example 5.15 (*Continued*)

For the first block,

$$G_1(s) = \frac{\bar{x}_1(s)}{\bar{r}(s)} = \frac{1}{(s+1)(s+3)(s+5)} = \frac{1}{s^3 + 9s^2 + 23s + 15}$$

or

$$(s^3 + 9s^2 + 23s + 15)\bar{x}_1(s) = \bar{r}(s)$$

Taking the inverse Laplace transform,

$$\frac{d^3x_1}{dt^3} + \frac{9d^2x_1}{dt^2} + \frac{23dx_1}{dt} + 15x_1 = r(t)$$

Let $\dot{x}_1 = x_2$ and $\dot{x}_2 = x_3$ so that

$$\begin{bmatrix} \dot{x}_1 \\ \dot{x}_2 \\ \dot{x}_3 \end{bmatrix} = \begin{bmatrix} 0 & 1 & 0 \\ 0 & 0 & 1 \\ -15 & -23 & -9 \end{bmatrix} \begin{bmatrix} x_1 \\ x_2 \\ x_3 \end{bmatrix} + \begin{bmatrix} 0 \\ 0 \\ 1 \end{bmatrix} r(t)$$

Consider the second block. Following a similar procedure,

$$y(t) = [50 \quad 25 \quad 0] \begin{bmatrix} x_1 \\ x_2 \\ x_3 \end{bmatrix}$$

The phase-variable form with full-state feedback system is given by

$$\begin{bmatrix} \dot{x}_1 \\ \dot{x}_2 \\ \dot{x}_3 \end{bmatrix} = \begin{bmatrix} 0 & 1 & 0 \\ 0 & 0 & 1 \\ -(15+k_1) & -(23+k_2) & -(9+k_3) \end{bmatrix} \begin{bmatrix} x_1 \\ x_2 \\ x_3 \end{bmatrix} + \begin{bmatrix} 0 \\ 0 \\ 1 \end{bmatrix} r(t)$$

The characteristic equation of the given system is given by

$$s^3 + (9+k_3)s^2 + (23+k_2)s + (15+k_1) = 0$$

Comparing the coefficients of the given characteristic equation with that of the desired characteristic equation, we get $k_1 = 110.7543$, $k_2 = 53.6$, and $k_3 = 1.10$.

(*Continued*)

Example 5.15 (*Continued*)

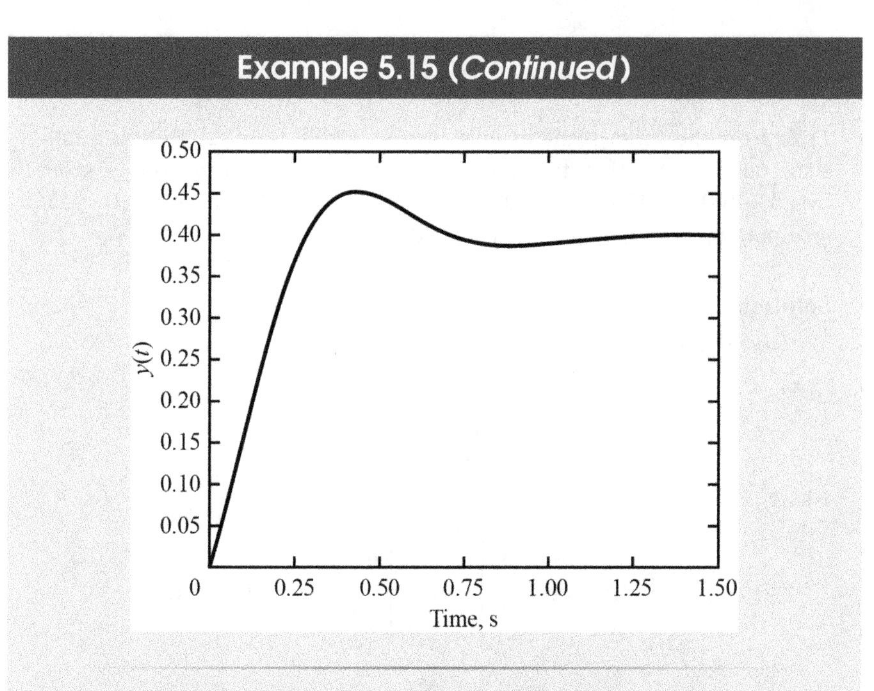

Fig. 5.44 Unit-step response of full-state feedback design.

Thus, the given system with full-state feedback is given by

$$\begin{bmatrix} \dot{x}_1 \\ \dot{x}_2 \\ \dot{x}_3 \end{bmatrix} = \begin{bmatrix} 0 & 1 & 0 \\ 0 & 0 & 1 \\ -125.7543 & -76.6 & -10.10 \end{bmatrix} \begin{bmatrix} x_1 \\ x_2 \\ x_3 \end{bmatrix} + \begin{bmatrix} 0 \\ 0 \\ 1 \end{bmatrix} r(t)$$

Now, we verify the design by simulating a response to a unit-step input using MATLAB [4]. The results of the simulation are shown in Fig. 5.44. It can be observed that the design requirements have been met.

Example 5.16

Given the system

$$\begin{bmatrix} \dot{x}_1 \\ \dot{x}_2 \\ \dot{x}_3 \end{bmatrix} = \begin{bmatrix} -5 & 2 & 0 \\ 1 & -3 & 1 \\ 0 & 1 & -1 \end{bmatrix} \begin{bmatrix} x_1 \\ x_2 \\ x_3 \end{bmatrix} + \begin{bmatrix} 0 \\ 0 \\ 1 \end{bmatrix} (t)$$

$$y = \begin{bmatrix} 1 & 0 & 0 \end{bmatrix} \begin{bmatrix} x_1 \\ x_2 \\ x_3 \end{bmatrix}$$

(*Continued*)

Example 5.16 (*Continued*)

1) Express the given plant in dual phase-variable form. 2) Design a full-state observer so that the closed-loop characteristic equation is given by $s^3 + 100s^2 + 1500s + 40{,}000 = 0.3$) Verify the design for $u(t) = 25t$ assuming $x_1(0) = 5$, $x_2(0) = 1.5$, and $x_3(0) = 0.25$.

Solution:

We have the given system

$$\dot{x} = Ax + Bu$$

$$y = Cx$$

where

$$A = \begin{bmatrix} -5 & 2 & 0 \\ 1 & -3 & 1 \\ 0 & 1 & -1 \end{bmatrix} \quad B = \begin{bmatrix} 0 \\ 0 \\ 1 \end{bmatrix} \quad C = [1 \quad 0 \quad 0]$$

It is convenient to use MATLAB [4] to compute the observability matrix, which for this system is found to be

$$Q_{0,x} = \begin{bmatrix} 1 & 0 & 0 \\ -5 & 2 & 0 \\ 27 & -16 & 2 \end{bmatrix}$$

Using MATLAB [4], we find that this system has full rank of three; hence the system is observable. Furthermore,

$$Q_{0,x}^{-1} = \begin{bmatrix} 1 & 0 & 0 \\ 2.5 & 0.5 & 0 \\ 6.5 & 4.0 & 0.5 \end{bmatrix}$$

Next step is to express the given plant in the dual phase-variable form. We know that the elements of the first column of the matrix in dual phase-variable form are the coefficients of the characteristic equation. Conversely, given the coefficients of the characteristic equation, we can directly write down the dual phase-variable form of matrix A.

As we know, the characteristic equation is given by

$$\Delta(s) = |sI - A| = 0$$

However, expanding the determinant in the characteristic equation is simple if it is of an order lower than three. However, for determinants of orders

(*Continued*)

Example 5.16 (*Continued*)

greater than three, this is a tedious job. For this purpose, MATLAB [4] comes out very handy. Using MATLAB [4], we get the characteristic equation

$$s^3 + 9s^2 + 20s + 8 = 0$$

Then, the dual phase-variable form of the given system is

$$A_z = \begin{bmatrix} -9 & 1 & 0 \\ -20 & 0 & 1 \\ -8 & 0 & 0 \end{bmatrix}$$

Furthermore, we assume that

$$C_z = [1 \quad 0 \quad 0]$$

Then,

$$Q_{0,z} = \begin{bmatrix} C_z \\ C_z A_z \\ C_z A_z^2 \end{bmatrix} = \begin{bmatrix} 1 & 0 & 0 \\ -9 & 1 & 0 \\ 61 & -9 & 1 \end{bmatrix}$$

and

$$P = Q_{0,x}^{-1} Q_{0,z} = \begin{bmatrix} 1 & 0 & 0 \\ -2 & 0.5 & 0 \\ 1 & -0.5 & 0.5 \end{bmatrix}$$

Now let us design the full-state observer for the system transformed in dual phase-variable form. We have

$$A_z - L_z C_z = \begin{bmatrix} -(9 + l_1) & 1 & 0 \\ -(20 + l_2) & 0 & 1 \\ -(8 + l_3) & 0 & 0 \end{bmatrix}$$

The characteristic equation of the observer is given by

$$s^3 + (9 + l_1)s^2 + (20 + l_2)s + 8 + l_3 = 0$$

The desired characteristic equation is

$$s^3 + 100s^2 + 1500s + 40{,}000 = 0$$

(Continued)

Example 5.16 (*Continued*)

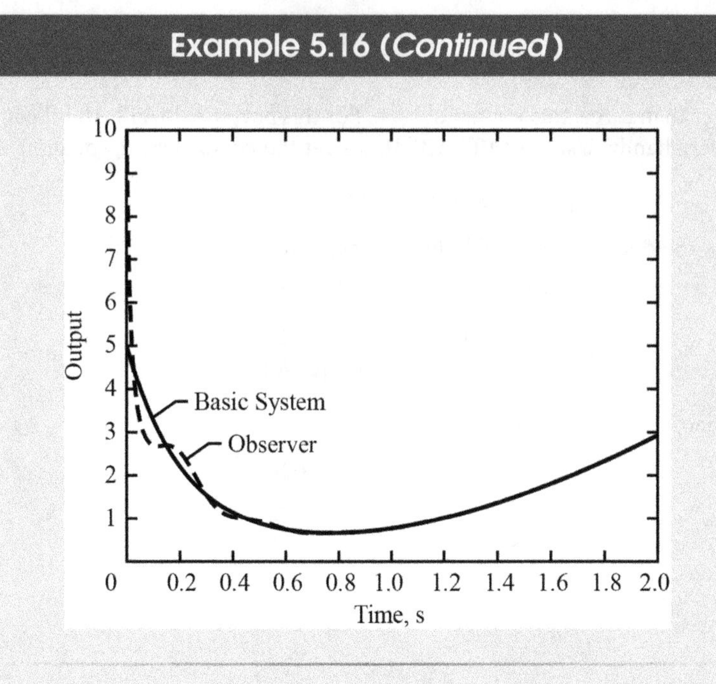

Fig. 5.45 Response of full-state observer.

Comparing the coefficients of like powers of s, we get $l_1 = 91$, $l_2 = 1480$, and $l_3 = 39{,}992$. Then, the gain matrix corresponding to the given system is obtained by transforming back as

$$L_x = PL_z = \begin{bmatrix} 91 \\ 558 \\ 19{,}347 \end{bmatrix}$$

The performance of the observer for the given initial conditions of $x_1(0) = 5$, $x_2(0) = 1.5$, and $x_3(0) = 0.25$ and to the input $r(t) = 25t$ is shown in Fig. 5.45 using MATLAB [4]. We note that the output of the observer catches up with that of the basic system very quickly as expected.

5.11 Summary

In this chapter, we have reviewed the basic principles of linear systems and illustrated the theory with a number of solved examples. This background will be useful in the study of aircraft dynamics and control. We will derive longitudinal and lateral-directional transfer functions and study the free response of the aircraft. We will also use the design methods we have learned here for the design of stability augmentation systems and automatic flight control systems of the aircraft to obtain the desired handling qualities. It was not possible to go into all the details of linear systems theory. Readers interested in getting more detailed information on control systems may refer elsewhere [1–3].

References

[1] Ogata, K., *Modern Control Engineering*, Prentice–Hall, Englewood Cliffs, NJ, 1986.
[2] Norman, S., and Nise, N. S., *Control System Engineering*, Benjamin/Cummings, 1992.
[3] Kuo, B., *Automatic Control Systems*, 4th ed., Prentice–Hall, Englewood Cliffs, NJ, 1982.
[4] *Pro-MATLAB for Sun Workstations*, The MathWorks, Natick, MA, Jan. 1990.

Problems

5.1 Sketch the Bode plot for the open-loop systems with

a) $G(s) = \dfrac{50}{s(s+5)}$

b) $G(s) = \dfrac{25(s+2)}{(s+3)(s+5)(s+7)}$

5.2 Using Routh's criterion, examine the stability of the closed-loop system with a characteristic polynomial given by

a) $s^4 + 5s^3 + 3s^2 + s + 2 = 0$

b) $s^4 + 2s^3 + 0.001s^2 + 3s + 4 = 0$

c) $s^4 + 4s^3 + 7s + 2 = 0$

[Answer: (a) Two sign changes, unstable; (b) two sign changes, unstable; and (c) two sign changes, unstable.]

5.3 Sketch the root-locus for a unity feedback system with

a) $G(s) = \dfrac{k(s+3)}{s(s+2)(s+4)}$

b) $G(s) = \dfrac{k(s+1)}{s(s+2)(s+3)(s+7)}$

Determine the value of gain k when the closed-loop system in (b) becomes unstable.

5.4 Sketch the root-locus for a unity feedback system with

$$G(s) = \frac{k(s+3)(s+7)}{s(s+1)}$$

Find the value of gain k so that the closed-loop system is stable and is operating with a damping ratio of 0.7.

5.5 For the following unity feedback systems, sketch the Nyquist plot.

a) $G(s) = \dfrac{25}{(s+1)(s+3)}$

b) $G(s) = \dfrac{k(s+2)}{(s-3)(s-4)}$

For the system in (b), determine the value of the gain for which the closed-loop system becomes unstable.

5.6 Using the Nyquist plot, determine the gain and phase margins of the system given by

$$G(s) = \frac{k(s-2)(s-3)}{s(s+5)(s+9)}$$

5.7 For the system in Problem 5.6, use Bode plots to obtain the gain and phase margins.

5.8 For a unity feedback system with

$$G(s) = \frac{k}{(s+2)(s+4)}$$

design a PI controller to reduce the steady-state error to zero for a unit-step input. Assume that the system is operating with a damping ratio of 0.6. Plot the unit-step response to verify your design. Compare the values of T_s and T_p for the basic and compensated systems.

5.9 For the system given in Problem 5.8, design a lag compensator to reduce the steady-state error by a factor of 15.

5.10 For the unity feedback system given by

$$G(s) = \frac{k}{(s+1)(s+2)(s+3)}$$

a) Determine the value of gain k for 15% overshoot. Determine the corresponding values of T_s and T_p.

b) Design a PD controller for reducing T_p by a factor of 2 and T_s by 50%, while operating at 15% overshoot in both cases.

5.11 Given the unity feedback system with

$$G(s) = \frac{k}{s(s+3)(s+5)}$$

determine the value of gain k for the system to operate with a damping ratio of 0.51. Find the corresponding locations of closed-loop poles. If a lead compensator is to be designed to reduce the time for peak amplitude T_p by 50%, with compensator zero placed at -2.5, find the compensator pole location. How does the performance of the compensated system compare with that of the basic system?

5.12 For the unity feedback system with

$$G(s) = \frac{k}{(s+1)(s+3)}$$

a) show that the system cannot be made to operate with time for peak amplitude of 2.0 s and 25.38% overshoot by simple gain adjustment and (b) design a suitable compensator to achieve this performance.

5.13 For the unity feedback system with

$$G(s) = \frac{k}{(s+1)(s+2)(s+7)}$$

design a PID controller that will give time for peak amplitude of 1.2 s and 15% overshoot with zero steady-state error for a unit-step input. Plot the unit-step response for the basic and PID-compensated systems.

5.14 For the system shown in Fig. P5.14, determine the values of gains k_h, k, and the minor loop damping ratio so that the entire closed-loop system has 15% overshoot.

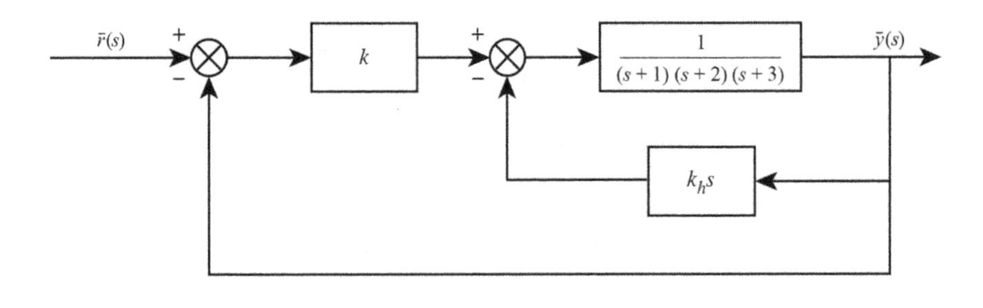

Fig. P5.14 Control system for Problems 5.14 and 5.15.

5.15 Determine the rate gyro gain k_h for the system shown in Fig. P5.14 so that the major-loop compensated system operates at one-half the settling time compared to the basic system while continuing to have the same 15% overshoot.

5.16 Given the linear time-invariant system

$$\dot{x}(t) = Ax(t) + Bu(t)$$

find (a) eigenvalues of matrix A, (b) the state transition matrix $\Phi(t)$, and (c) state vector $x(t)$ for

$$A = \begin{bmatrix} 0 & 1 \\ -1 & 2 \end{bmatrix} \quad B = \begin{bmatrix} 0 \\ 1 \end{bmatrix} \quad x(0) = \begin{bmatrix} 1 \\ 2 \end{bmatrix}$$

5.17 Given the state equation

$$\begin{bmatrix} \dot{x}_1 \\ \dot{x}_2 \\ \dot{x}_3 \end{bmatrix} = \begin{bmatrix} 1 & 2 & 1 \\ 3 & 5 & 2 \\ 4 & 1 & 3 \end{bmatrix} \begin{bmatrix} x_1 \\ x_2 \\ x_3 \end{bmatrix} + \begin{bmatrix} 1 \\ 2 \\ 3 \end{bmatrix} u(t)$$

$$y = \begin{bmatrix} 1 & 0 & 0 \end{bmatrix} \begin{bmatrix} x_1 \\ x_2 \\ x_3 \end{bmatrix}$$

Can this system be transformed into phase-variable form? If so, find the transformation $z = Px$ so that the transformed system $\dot{z} = A_z z + B_z u(t)$, $y = C_z z$ is in phase-variable form.

5.18 Represent the following system in state-space, phase-variable form:

$$\frac{d^3x}{dt^3} + \frac{2d^2x}{dt^2} + \frac{3dx}{dt} + 5x = u(t)$$

5.19 Given the state equation

$$\begin{bmatrix} \dot{x}_1 \\ \dot{x}_2 \\ \dot{x}_3 \end{bmatrix} = \begin{bmatrix} 1 & 2 & -1 \\ 1 & 2 & -3 \\ 1 & 0 & 2 \end{bmatrix} \begin{bmatrix} x_1 \\ x_2 \\ x_3 \end{bmatrix} + \begin{bmatrix} 0 \\ 0 \\ 1 \end{bmatrix} u(t)$$

$$y = \begin{bmatrix} 1 & 0 & 0 \end{bmatrix} \begin{bmatrix} x_1 \\ x_2 \\ x_3 \end{bmatrix}$$

Can this system be transformed to dual-phase variable form? If so, find the transformation $z = Px$ such that the transformed system $\dot{z} = A_z z + B_z u(t)$, $y = C_z z$ is in dual phase-variable form.

5.20 Design a phase-variable, full-state feedback controller for the plant given by

$$G(s) = \frac{10(s + 0.8)}{(s + 2)(s + 3)(s + 5)}$$

to yield a 15% overshoot with a settling time of 0.8 s.

5.21 Design a full-state feedback observer for the plant

$$G(s) = \frac{1}{s(s + 3)(s + 6)}$$

so that the closed-loop characteristic equation of the observer system is given by

$$s^3 + 90s^2 + 2000s + 10{,}000 = 0$$

Chapter 6 Airplane Response and Closed-Loop Control

6.1 Introduction

In Chapter 4, we studied various coordinate systems used in airplane dynamics and derived equations of motion applicable for small disturbances. We also discussed the methods of evaluating various stability and control derivatives appearing in those equations. Under the assumption of small disturbance, the equations of motion could be grouped into two sets of three equations each: one set for longitudinal motion and another set for lateral-directional motion of the aircraft. This kind of decoupling enables us to study separately the longitudinal and lateral-directional response and closed-loop control. In Chapter 5, we studied the basic principles of linear system theory and design of closed-loop control systems.

In this chapter we will discuss the solution of the small-disturbance equations to determine the airplane response. The airplane response depends on the initial conditions and the input time history. The response to a given set of initial conditions with zero input is called the natural or free response. The free response is indicative of the transient behavior or the dynamic stability of the system. The initial conditions are equivalent to suddenly imposed disturbances. For example, the response with $u(0) = 0$, $\Delta\alpha(0) = 5$ deg, and $q = \Delta\theta = 0$ is equivalent to the response for a suddenly imposed vertical gust that momentarily increases the airplane angle of attack by 5 deg.

The forced response is the solution of equations of motion with zero initial conditions and a given input time history. The forced response is indicative of an airplane's steady-state behavior. The common input test functions used to obtain the forced response are the unit-step and impulse functions. For example, airplane longitudinal response to a unit-step function describes the motion of the airplane to a sudden unit deflection of the elevator. The steady-state solution gives the corresponding steady-state values of forward velocity, angle of attack, attitude, and pitch rate.

Finally, we will discuss the closed-loop control of the airplane to obtain desired level of handling qualities. The closed-loop control systems used for obtaining the specified level of free response are called stability augmentation systems, and those used to establish and hold the desired flight conditions are called autopilots. In this chapter, we will discuss some of the important stability augmentation systems and autopilots.

6.2 Longitudinal Response

In this section we will discuss the longitudinal response of the airplane. We will discuss two types of responses: 1) the free response and 2) the forced response. The free response corresponds to the solution with a given set of initial conditions and zero input and is indicative of the dynamic stability of the system. The forced response corresponds to the solution with zero initial conditions and a given input time history. For free response, we assume that the elevator is held fixed (stick-fixed), and, for forced response, we assume that the elevator is moved in a specified manner to a new position and subsequently held fixed at that position.

The longitudinal equations of motion for elevator control are given by Eqs. (4.417–4.419) as

$$\left(m_1\frac{\mathrm{d}}{\mathrm{d}t}-C_{xu}\right)u-\left(C_{x\dot{\alpha}}c_1\frac{\mathrm{d}}{\mathrm{d}t}+C_{x\alpha}\right)\Delta\alpha-\left(C_{xq}c_1\frac{\mathrm{d}}{\mathrm{d}t}+C_{x\theta}\right)\Delta\theta=C_{x\delta e}\Delta\delta e \tag{6.1}$$

$$-C_{zu}u+\left[\left(m_1\frac{\mathrm{d}}{\mathrm{d}t}-C_{z\dot{\alpha}}c_1\frac{\mathrm{d}}{\mathrm{d}t}\right)-C_{z\alpha}\right]\Delta\alpha-\left(m_1\frac{\mathrm{d}}{\mathrm{d}t}+C_{zq}c_1\frac{\mathrm{d}}{\mathrm{d}t}+C_{z\theta}\right)\Delta\theta$$
$$=C_{z\delta e}\Delta\delta e \tag{6.2}$$

$$-C_{mu}u-\left(C_{m\dot{\alpha}}c_1\frac{\mathrm{d}}{\mathrm{d}t}+C_{m\alpha}\right)\Delta\alpha+\frac{\mathrm{d}}{\mathrm{d}t}\left(I_{y1}\frac{\mathrm{d}}{\mathrm{d}t}-C_{mq}c_1\right)\Delta\theta=C_{m\delta e}\Delta\delta e \tag{6.3}$$

For the study of airplane response, it is convenient to express Eqs. (6.1–6.3) in the state-space form as follows:

$$\frac{du}{dt}=\frac{1}{m_1}\left[(C_{xu}+\xi_1C_{zu})u+(C_{x\alpha}+\xi_1C_{z\alpha})\Delta\alpha+\left[C_{xq}c_1+\xi_1\left(m_1+C_{zq}c_1\right)\right]q\right.$$
$$\left.+(C_{x\theta}+\xi_1C_{z\theta})\Delta\theta+\left(C_{x\delta_e}+\xi_1C_{z\delta_e}\right)\Delta\delta_e\right] \tag{6.4}$$

$$\frac{\mathrm{d}\Delta\alpha}{\mathrm{d}t}=\frac{1}{(m_1-C_{z\dot{\alpha}}c_1)}\left[C_{zu}u+C_{z\alpha}\Delta\alpha+(m_1+C_{zq}c_1)q+C_{z\theta}\Delta\theta+C_{z\delta e}\Delta\delta e\right] \tag{6.5}$$

$$\frac{\mathrm{d}q}{\mathrm{d}t}=\frac{1}{I_{y1}}\left[(C_{mu}+\xi_2C_{zu})u+(C_{m\alpha}+\xi_2C_{z\alpha})\Delta\alpha+\left[C_{mq}c_1+\xi_2\left(m_1+C_{zq}c_1\right)\right]q\right.$$
$$\left.+\xi_2C_{z\theta}\Delta\theta+\left(C_{m\delta_e}+\xi_2C_{z\delta_e}\right)\Delta\delta_e\right] \tag{6.6}$$

$$\frac{\mathrm{d}\Delta\theta}{\mathrm{d}t}=q \tag{6.7}$$

where

$$\xi_1 = \frac{C_{x\dot{\alpha}}c_1}{m_1 - C_{z\dot{\alpha}}c_1} \tag{6.8}$$

$$\xi_2 = \frac{C_{m\dot{\alpha}}c_1}{m_1 - C_{z\dot{\alpha}}c_1} \tag{6.9}$$

Let

$$x_1 = u \qquad x_2 = \Delta\alpha \qquad x_3 = q \qquad x_4 = \Delta\theta \tag{6.10}$$

Then, Eqs. (6.4–6.7) can be expressed in the state-space form as

$$\dot{x} = Ax + Bu_c \tag{6.11}$$

where

$$x = \begin{bmatrix} x_1 \\ x_2 \\ x_3 \\ x_4 \end{bmatrix} \quad A = \begin{bmatrix} a_{11} & a_{12} & a_{13} & a_{14} \\ a_{21} & a_{22} & a_{23} & a_{24} \\ a_{31} & a_{32} & a_{33} & a_{34} \\ a_{41} & a_{42} & a_{43} & a_{44} \end{bmatrix} \quad B = \begin{bmatrix} b_1 \\ b_2 \\ b_3 \\ b_4 \end{bmatrix} \quad u_c = \delta_e \tag{6.12}$$

and

$$a_{11} = \frac{C_{xu} + \xi_1 C_{zu}}{m_1} \qquad a_{12} = \frac{C_{x\alpha} + \xi_1 C_{z\alpha}}{m_1}$$

$$a_{13} = \frac{C_{xq}c_1 + \xi_1(m_1 + C_{zq}c_1)}{m_1} \qquad a_{14} = \frac{C_{x\theta} + \xi_1 C_{z\theta}}{m_1}$$

$$a_{21} = \frac{C_{zu}}{m_1 - C_{z\dot{\alpha}}c_1} \qquad a_{22} = \frac{C_{z\alpha}}{m_1 - C_{z\dot{\alpha}}c_1}$$

$$a_{23} = \frac{m_1 + C_{zq}c_1}{m_1 - C_{z\dot{\alpha}}c_1} \qquad a_{24} = \frac{C_{z\theta}}{m_1 - c_1 C_{z\dot{\alpha}}}$$

$$a_{31} = \frac{C_{mu} + \xi_2 C_{zu}}{I_{y1}} \qquad a_{32} = \frac{C_{m\alpha} + \xi_2 C_{z\alpha}}{I_{y1}}$$

$$a_{33} = \frac{C_{mq}c_1 + \xi_2(m_1 + C_{zq}c_1)}{I_{y1}} \qquad a_{34} = \frac{\xi_2 C_{z\theta}}{I_{y1}}$$

$$a_{41} = 0 \quad a_{42} = 0 \quad a_{43} = 1 \quad a_{44} = 0$$

$$b_1 = \frac{C_{x\delta_e} + \xi_1 C_{z\delta_e}}{m_1} \qquad b_2 = \frac{C_{z\delta_e}}{m_1 - c_1 C_{z\dot{\alpha}}}$$

$$b_3 = \frac{C_{m\delta_e} + \xi_2 C_{z\delta_e}}{I_{y1}} \qquad b_4 = 0$$

$$m_1 = \frac{2m}{\rho U_o S} \qquad c_1 = \frac{\bar{c}}{2U_o}$$

$$I_{y1} = \frac{I_y}{\frac{1}{2}\rho U_o^2 S \bar{c}}$$

For free response, $u_c = 0$ so that

$$\dot{x} = Ax \tag{6.13}$$

A solution to Eq. (6.13) can be obtained in the usual way by assuming

$$x = x_o e^{\lambda t} \tag{6.14}$$

so that

$$\dot{x} = x_o \lambda e^{\lambda t} \tag{6.15}$$

and

$$x_o \lambda e^{\lambda t} - A x_o e^{\lambda t} = 0 \tag{6.16}$$

or

$$(\lambda I - A)x_o = 0 \tag{6.17}$$

where I is the identity matrix

$$I = \begin{bmatrix} 1 & 0 & 0 & 0 \\ 0 & 1 & 0 & 0 \\ 0 & 0 & 1 & 0 \\ 0 & 0 & 0 & 1 \end{bmatrix} \tag{6.18}$$

For nontrivial solutions, the determinant of $(\lambda I - A)$ must be zero

$$|\lambda I - A| = 0 \tag{6.19}$$

An expansion of the determinant in Eq. (6.19) results in a fourth-order algebraic equation of the form

$$A_\delta \lambda^4 + B_\delta \lambda^3 + C_\delta \lambda^2 + D_\delta \lambda + E_\delta = 0 \tag{6.20}$$

where

$$A_\delta = m_1 I_{y1}(m_1 - C_{z\dot{\alpha}} c_1) \tag{6.21}$$

$$B_\delta = m_1(-I_{y1}C_{z\alpha} - C_{mq}c_1[m_1 - C_{z\dot{\alpha}}c_1] - C_{m\dot{\alpha}}c_1[m_1 + C_{zq}c_1]) \\ -C_{xu}I_{y1}(m_1 - C_{z\dot{\alpha}}c_1) - C_{x\dot{\alpha}}c_1C_{zu}I_{y1} \tag{6.22}$$

$$C_\delta = m_1(C_{z\alpha}C_{mq}c_1 - C_{m\alpha}[m_1 + C_{zq}c_1] - C_{z\theta}C_{m\dot{\alpha}}c_1) \\ + C_{xu}(I_{y1}C_{z\alpha} + C_{mq}c_1[m_1 - C_{z\dot{\alpha}}c_1] + C_{m\dot{\alpha}}c_1[m_1 + C_{zq}c_1]) \\ - C_{x\alpha}C_{zu}I_{y1} + C_{x\dot{\alpha}}c_1(C_{zu}C_{mq}c_1 - C_{mu}[m_1 + C_{zq}c_1]) \\ - C_{xq}c_1(C_{zu}C_{m\dot{\alpha}}c_1 + C_{mu}[m_1 - C_{z\dot{\alpha}}c_1]) \tag{6.23}$$

$$D_\delta = -C_{xu}(C_{z\alpha}C_{mq}c_1 - C_{m\alpha}[m_1 + C_{zq}c_1] - C_{z\theta}C_{m\dot{\alpha}}c_1) \\ - m_1C_{m\alpha}C_{z\theta} + C_{x\alpha}(C_{zu}C_{mq}c_1 - C_{mu}[m_1 + C_{zq}c_1]) \\ - C_{x\dot{\alpha}}c_1C_{mu}C_{z\theta} - C_{xq}c_1(C_{zu}C_{m\alpha} - C_{z\alpha}C_{mu}) \\ - C_{x\theta}(C_{zu}C_{m\dot{\alpha}}c_1 + C_{mu}[m_1 - C_{z\dot{\alpha}}c_1]) \tag{6.24}$$

$$E_\delta = C_{xu}C_{m\alpha}C_{z\theta} - C_{x\alpha}C_{mu}C_{z\theta} - C_{x\theta}(C_{zu}C_{m\alpha} - C_{z\alpha}C_{mu}) \tag{6.25}$$

Equation (6.20) is also called the characteristic equation of the longitudinal motion and gives four values of the root λ. The nature of the free response can be any combination of the typical free responses shown in Fig. 6.1. A steady convergence corresponds to a real negative root, a steady divergence if the real root is positive, an oscillatory motion of constant amplitude if the real part of the complex root is zero, a damped oscillatory motion if the real part of the complex root is negative, and a divergent oscillatory motion if the real part of the complex root is positive. Thus for dynamic stability, the roots of the characteristic equation must be negative if real or must have negative real parts if complex.

For a conventional, statically stable airplane ($C_{m\alpha} < 0$), the longitudinal characteristic equation usually has a pair of complex conjugate roots of the form

$$\lambda_{1,2} = -r_1 \pm js_1 \tag{6.26}$$

$$\lambda_{3,4} = -r_2 \pm js_2 \tag{6.27}$$

so that

$$\Delta_{\text{long}}(\lambda) = (-r_1 + js_1)(-r_1 - js_1)(-r_2 + js_2)(-r_2 - js_2) \tag{6.28}$$

The roots of the characteristic equation are also called the eigenvalues of the system represented by matrix A. To determine eigenvalues λ_i, $i = 1, 4$, we can either solve the fourth-order characteristic equation or obtain them directly using the methods of matrix algebra. However, either of these tasks can be easily performed using commercially available matrix software tools like MATLAB [1].

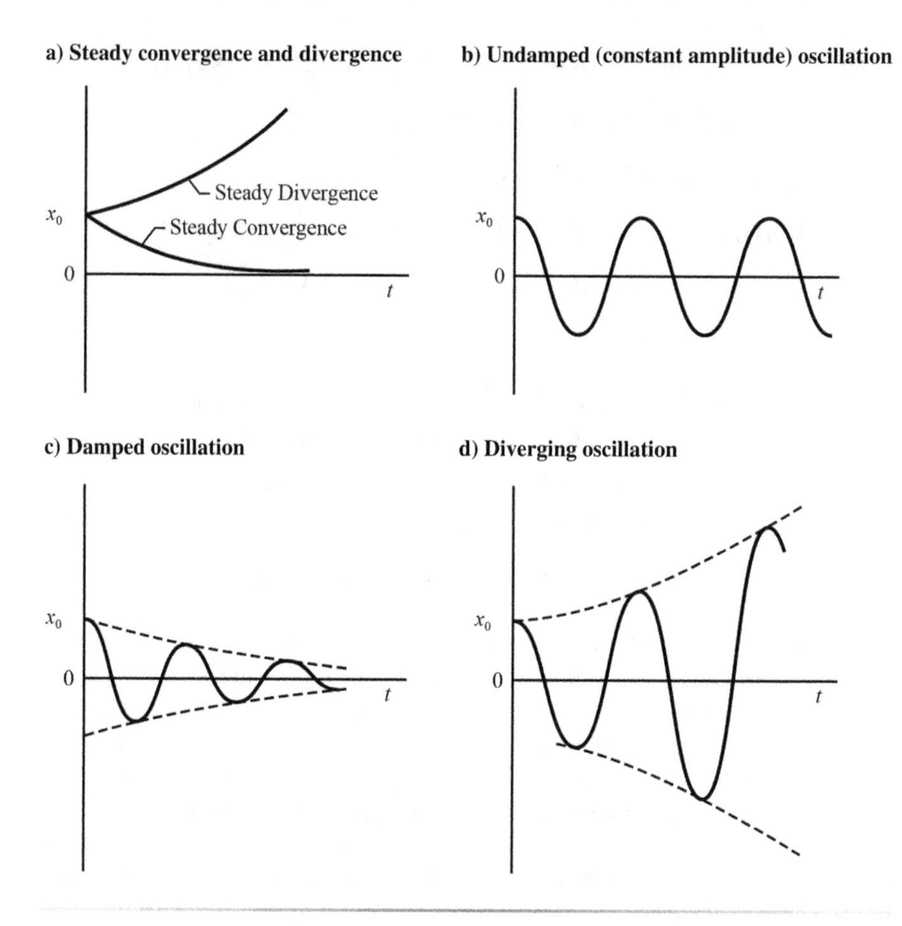

Fig. 6.1 Typical dynamic motions.

To illustrate the nature of free response and the concept of airplane dynamic stability, let us determine the response of a general aviation airplane [2], which is shown in Fig. 6.2. The mass and aerodynamic properties of this airplane are as follows [2].

Wing area $S = 16.094\ \mathrm{m}^2$, weight $W = 12{,}232.6$ N, wing span $b = 10.1803$ m, wing mean aerodynamic chord $\bar{c} = 1.7374$ m, distance of center of gravity from wing leading edge in terms of mean aerodynamic chord $\bar{x}_{cg} = 0.295$, $I_x = 1420.8973\ \mathrm{kg \cdot m^2}$, $I_y = 4067.454\ \mathrm{kg \cdot m^2}$, $I_z = 4786.0375\ \mathrm{kg \cdot m^2}$, and $I_{xy} = I_{yz} = I_{zx} = 0$. $C_L = 0.41$, $C_D = 0.05$, $C_{L\alpha} = 4.44$, $C_{L\dot{\alpha}} = 0$, $C_{D\dot{\alpha}} = 0$, $C_{LM} = 0$, $C_{L\delta_e} = 0.355$, $C_{m\delta_e} = -0.9230$, $C_{D\alpha} = 0.33$, $C_{DM} = 0$, $C_{D\delta_e} = 0$, $C_{m\alpha} = -0.683$, $C_{m\dot{\alpha}} = -4.36$, $C_{mM} = 0$, $C_{mq} = -9.96$, $M = 0.158$, and $\rho = 1.225\ \mathrm{kg/m^3}$. All the derivatives are per radian.

We assume that the airplane is in level flight with $\theta_o = 0$ before it encounters any disturbance. This assumption gives us $C_{x\theta} = -C_L$ and

$C_{z\theta} = 0$. In [2], the data on C_{Lq} is not given. For the purpose of this text, we assume $C_{Lq} = 3.8$ per radian.

Recall from Chapter 4 that $C_{xu} = -2C_D - C_{Du}$, $C_{x\alpha} = C_L - C_{D\alpha}$, $C_{x\theta} = -C_L \cos\theta$, $C_{z\theta} = -C_L \sin\theta$, $C_{x\dot{\alpha}} = -C_{D\dot{\alpha}}$, $C_{xq} = -C_{Dq}$, $C_{z\alpha} = -C_{L\alpha} - C_D$, $C_{zu} = -2C_L - C_{Lu}$, $C_{z\theta} = -C_L \sin\theta$, $C_{x\delta_e} = -C_{D\delta_e}$, and $C_{z\delta_e} = -C_{L\delta_e}$.

Substituting the required quantities in Eq. (6.12), we get

$$A = \begin{bmatrix} -0.0453 & 0.0363 & 0 & -0.1859 \\ -0.3717 & -2.0354 & 0.9723 & -0 \\ 0.3398 & -7.0301 & -2.9767 & 0 \\ 0 & 0 & 1 & 0 \end{bmatrix} \tag{6.29}$$

$$B = \begin{bmatrix} 0 \\ -0.1609 \\ -11.8674 \\ 0 \end{bmatrix} \tag{6.30}$$

Note that the units in matrix B for $C_{x\delta_e}$, $C_{z\delta_e}$, and $C_{m\delta_e}$ are per radian. Using MATLAB [1], we obtain the eigenvalues of matrix A as

$$\lambda_{1,2} = -2.5118 \pm j2.5706 \tag{6.31}$$

$$\lambda_{3,4} = -0.0169 \pm j0.2174 \tag{6.32}$$

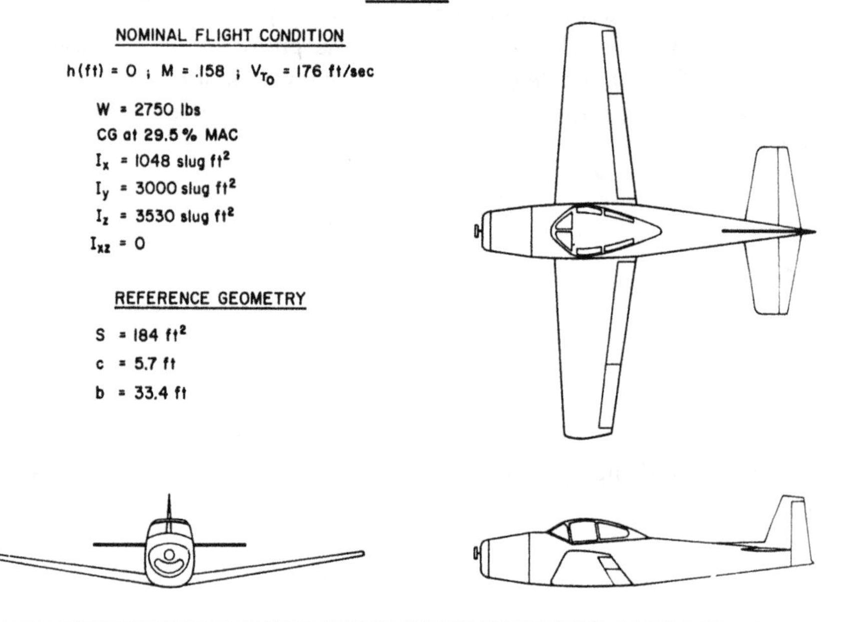

NAVION
NOMINAL FLIGHT CONDITION
h(ft) = 0 ; M = .158 ; V_{T_0} = 176 ft/sec
W = 2750 lbs
CG at 29.5 % MAC
I_x = 1048 slug ft²
I_y = 3000 slug ft²
I_z = 3530 slug ft²
I_{xz} = 0
REFERENCE GEOMETRY
S = 184 ft²
c = 5.7 ft
b = 33.4 ft

Fig. 6.2 Three-view drawing of the general aviation airplane [2].

Because both the complex roots have negative real parts, the free response of this airplane is of stable nature. It consists of two decaying oscillatory motions, which are superposed one on another. Thus, the general aviation airplane considered here is dynamically stable.

An alternative way of examining the dynamic stability of the airplane without actually solving for the roots of the characteristic polynomial or the eigenvalues of matrix A is to make use of the Routh's stability criterion we discussed in Chapter 5. Substituting in Eq. (6.20), the characteristic equation for the general aviation airplane is given by

$$0.3739\lambda^4 + 1.8907\lambda^3 + 4.9103\lambda^2 + 0.2526\lambda + 0.2296 = 0 \tag{6.33}$$

The necessary condition for stability is that all the coefficients of the characteristic equation must be positive or have the same sign. This condition is satisfied here. To examine whether the sufficiency condition is satisfied, we form Routh's array as follows:

$$\begin{array}{llll} s^4: & 0.3739 & 4.9103 & 0.2296 \\ s^3: & 1.8907 & 0.2526 & 0 \\ s^2: & 4.8603 & 0.2296 & \\ s^1: & 0.1633 & & \\ s^0: & 0.2296 & & \end{array}$$

We observe that all the elements of the first column of Routh's array are positive or have the same sign; hence the sufficiency condition is also satisfied. Hence, the characteristic polynomial has no positive real root or a complex root with positive real part. Therefore, the given system is stable as we have noted from direct determination of eigenvalues of matrix A.

The standard form of the characteristic equation of a second-order system is

$$\lambda^2 + 2\zeta\omega_n\lambda + \omega_n^2 = 0 \tag{6.34}$$

Let

$$\lambda = -r \pm js \tag{6.35}$$

$$= -\zeta\omega_n \pm j\omega_n\sqrt{1-\zeta^2} \tag{6.36}$$

where ζ is the damping ratio and ω_n is the natural frequency of the system. Then,

$$\zeta = \frac{r}{\sqrt{r^2+s^2}} \tag{6.37}$$

$$\omega_n = \sqrt{r^2+s^2} \tag{6.38}$$

The period T and time for the amplitude t_a to become either half or double the initial amplitude are given by

$$T = \frac{2\pi}{\omega_n \sqrt{1-\zeta^2}} \tag{6.39}$$

$$t_a = \frac{0.6931}{|r|} \tag{6.40}$$

Here, t_a is the time for half amplitude if r is positive, and it is the time for the amplitude to double if r is negative.

For the general aviation airplane, we get $\zeta_{1,2} = 0.6989$ and $\omega_{1,2} = 3.5940$, corresponding to $\lambda_{1,2}$ and $\zeta_{3,4} = 0.0775$, and $\omega_{3,4} = 0.2181$, corresponding to $\lambda_{3,4}$. The motion corresponding to $\lambda_{1,2}$ is heavily damped and is of higher frequency or a shorter period. The other motion corresponding to $\lambda_{3,4}$ is lightly damped and is of lower frequency or longer period. These values of $\lambda_{1,2}$ and $\lambda_{3,4}$ are typical of conventional, statically stable airplanes. The high-frequency, heavily damped oscillatory motion is called the short-period mode, and the lightly damped, long-period oscillatory mode is known as phugoid or long-period mode. For the general aviation airplane, we find that the periods of short-period and phugoid modes are 2.4442 and 28.9015 s, respectively. The corresponding values of the time for half amplitude are 0.2759 and 44.018 s. Because the short-period mode is fast and heavily damped, it is just felt as a bump by the pilot or the passengers. The pilot does not have to take any action to kill this mode. The phugoid mode is very lightly damped and usually persists for a long time. It can be quite annoying if left to die by itself. Because the period is quite long, the pilot can easily operate the longitudinal control (elevator) and kill the phugoid mode.

Physically, the motion of the airplane during the phugoid motion can be depicted as shown in Fig. 6.3. Beginning at the bottom of one cycle, we

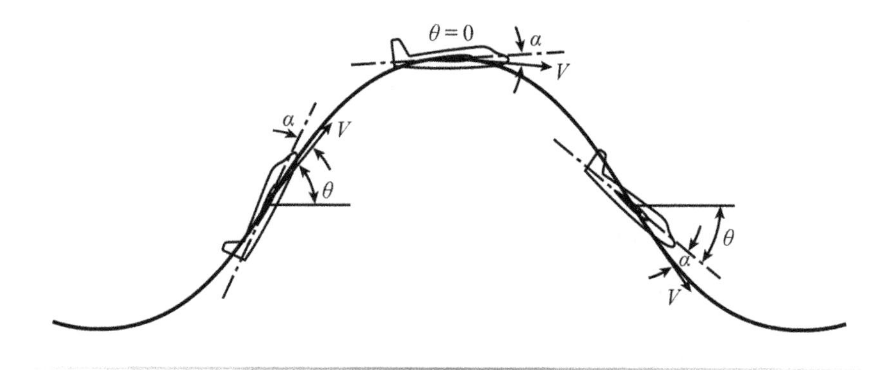

Fig. 6.3 Physical motion of an airplane during phugoid motion.

observe that the pitch angle increases and the airplane gains altitude and loses forward speed. During phugoid motion, the angle of attack remains constant so that a drop in forward speed amounts to a decrease in lift and flattening of the pitch attitude. As a result, at the top of the cycle, the pitch angle goes to zero. Beyond this point, the airplane begins to lose altitude, the pitch angle goes negative, and the air speed increases. At the bottom of the cycle, the pitch attitude is nearly level, the air speed is at its maximum, and the cycle repeats once again.

In view of the fact that the longitudinal response of statically stable airplanes consists of two distinct oscillatory motions, it is customary to introduce the short-period and phugoid approximations as follows.

6.2.1 Short-Period Approximation

Of the two oscillatory modes, the short-period mode is the more heavily damped oscillatory motion with a higher frequency. This oscillatory motion lasts just for a few seconds, usually fewer than 10 s, during which the angle of attack, pitch angle, and pitch rate vary rapidly and the forward speed nearly remains constant. Therefore, we can assume $u = \dot{u} = 0$ during short-period oscillation. With this assumption, we can ignore the force equation in the x direction. Then, the other two equations assume the following form:

$$\left(m_1 \frac{\mathrm{d}}{\mathrm{d}t} - C_{z\dot{\alpha}} c_1 \frac{\mathrm{d}}{\mathrm{d}t} - C_{z\alpha}\right)\Delta\alpha - \left(m_1 \frac{\mathrm{d}}{\mathrm{d}t} + C_{zq} c_1 \frac{\mathrm{d}}{\mathrm{d}t} + C_{z\theta}\right)\Delta\theta = C_{z\delta e}\Delta\delta e \tag{6.41}$$

$$-\left(C_{m\dot{\alpha}} c_1 \frac{\mathrm{d}}{\mathrm{d}t} + C_{m\alpha}\right)\Delta\alpha + \frac{\mathrm{d}}{\mathrm{d}t}\left(I_{y1} \frac{\mathrm{d}}{\mathrm{d}t} - C_{mq} c_1\right)\Delta\theta = C_{m\delta_e}\Delta\delta_e \tag{6.42}$$

Let

$$x_1 = \Delta\alpha \tag{6.43}$$

$$x_2 = q = \frac{\mathrm{d}\Delta\theta}{\mathrm{d}t} \tag{6.44}$$

$$x_3 = \Delta\theta \tag{6.45}$$

Then rearranging, we have

$$\begin{bmatrix} \dot{x}_1 \\ \dot{x}_2 \\ \dot{x}_3 \end{bmatrix} = \begin{bmatrix} a_{11} & a_{12} & a_{13} \\ a_{21} & a_{22} & a_{23} \\ a_{31} & a_{32} & a_{33} \end{bmatrix} \begin{bmatrix} x_1 \\ x_2 \\ x_3 \end{bmatrix} + \begin{bmatrix} b_1 \\ b_2 \\ b_3 \end{bmatrix} \Delta\delta_e \tag{6.46}$$

Let

$$A_{sp} = \begin{bmatrix} a_{11} & a_{12} & a_{13} \\ a_{21} & a_{22} & a_{23} \\ a_{31} & a_{32} & a_{33} \end{bmatrix} \tag{6.47}$$

$$B_{sp} = \begin{bmatrix} b_1 \\ b_2 \\ b_3 \end{bmatrix} \tag{6.48}$$

where

$$a_{11} = \frac{C_{z\alpha}}{m_1 - C_{z\dot{\alpha}}c_1} \quad a_{12} = \frac{m_1 + C_{zq}c_1}{m_1 - C_{z\dot{\alpha}}c_1}$$

$$a_{13} = \frac{C_{z\theta}}{m_1 - C_{z\dot{\alpha}}c_1} \quad a_{21} = \left(\frac{1}{I_{y1}}\right)\left(C_{m\alpha} + \frac{C_{m\dot{\alpha}}c_1C_{z\alpha}}{m_1 - C_{z\dot{\alpha}}c_1}\right)$$

$$a_{22} = \left(\frac{1}{I_{y1}}\right)\left[C_{mq}c_1 + \left(\frac{C_{m\dot{\alpha}}c_1}{m_1 - C_{z\dot{\alpha}}c_1}\right)(m_1 + c_1C_{zq})\right]$$

$$a_{23} = \frac{C_{m\dot{\alpha}}c_1C_{z\theta}}{I_{y1}(m_1 - C_{z\dot{\alpha}}c_1)} \quad a_{31} = 0$$

$$a_{32} = 1 \quad a_{33} = 0$$

$$b_1 = \frac{C_{z\delta_e}}{m_1 - C_{z\dot{\alpha}}c_1} \quad b_2 = \left(\frac{1}{I_{y1}}\right)\left(C_{m\delta_e} + \frac{C_{m\dot{\alpha}}c_1C_{z\delta_e}}{m_1 - C_{z\dot{\alpha}}c_1}\right)$$

$$b_3 = 0$$

We have

$$C_{z\theta} = -C_L \sin\theta_o \tag{6.49}$$

For equilibrium level flight, $\theta_o = 0$, so that $C_{z\theta} \simeq 0$, $a_{13} = 0$, and $a_{23} = 0$. With this assumption, Eq. (6.46) reduces to the following form:

$$\begin{bmatrix} \dot{x}_1 \\ \dot{x}_2 \end{bmatrix} = \begin{bmatrix} a_{11} & a_{12} \\ a_{21} & a_{22} \end{bmatrix}\begin{bmatrix} x_1 \\ x_2 \end{bmatrix} + \begin{bmatrix} b_1 \\ b_2 \end{bmatrix}\Delta\delta_e \tag{6.50}$$

To get an idea of the physical parameters that have a major influence on the short-period mode, we have to introduce some more simplifications. We assume $C_{zq} = C_{z\dot{\alpha}} = 0$ because they are usually small. With these assumptions, it can be shown that the characteristic equation of the short-period mode is given by

$$\lambda^2 + B\lambda + C = 0 \tag{6.51}$$

where

$$B = -\left(\frac{C_{z\alpha}}{m_1} + \frac{c_1}{I_{y1}}\left(C_{mq} + C_{m\dot{\alpha}}\right)\right) \tag{6.52}$$

$$C = \frac{C_{z\alpha}c_1C_{mq}}{m_1I_{y1}} - \frac{C_{m\alpha}}{I_{y1}} \tag{6.53}$$

Comparing this equation with the standard form of a characteristic equation of a second-order system [Eq. (6.34)], we get

$$\omega_n = \sqrt{C} \tag{6.54}$$

$$= \sqrt{\frac{C_{z\alpha}c_1C_{mq}}{m_1I_{y1}} - \frac{C_{m\alpha}}{I_{y1}}} \tag{6.55}$$

$$\zeta = \frac{B}{2\omega_n} \tag{6.56}$$

$$= \frac{-\left(\frac{C_{z\alpha}}{m_1} + \frac{c_1}{I_{y1}}\left(C_{mq} + C_{m\dot{\alpha}}\right)\right)}{2\sqrt{\frac{C_{z\alpha}c_1C_{mq}}{m_1I_{y1}} - \frac{C_{m\alpha}}{I_{y1}}}} \tag{6.57}$$

From this analysis we observe that 1) the frequency of the short-period mode depends directly on the magnitude of the static stability parameter $C_{m\alpha}$ and 2) damping of the short-period mode directly depends on the damping-in-pitch derivative C_{mq} and the acceleration derivative $C_{m\dot{\alpha}}$. We know that the static stability parameter $C_{m\alpha}$ is directly related to the center of gravity position. As the center of gravity moves forward, $C_{m\alpha}$ decreases (becomes more stable), and the frequency of the short-period mode also increases. Conversely, if the center of gravity moves aft, $C_{m\alpha}$ increases (becomes less stable), and the frequency of the short-period mode decreases. At one point, as we see later, when $C_{m\alpha} \geq 0$, the short-period mode breaks up into two exponential modes.

The damping of the short-period mode improves with an increase in the stable values of C_{mq} and $C_{m\dot{\alpha}}$. Recall that the major contribution to C_{mq} and $C_{m\dot{\alpha}}$ comes from the horizontal tail. The higher the tail-volume ratio, the larger will be the horizontal tail contribution to C_{mq} and $C_{m\dot{\alpha}}$ and the higher will be the short-period damping ratio.

6.2.2 Phugoid Approximation

Because the disturbance in angle of attack quickly decays to zero during the short-period oscillation and subsequently remains close to zero, we assume $\Delta\alpha = \Delta\dot{\alpha} = 0$ for the phugoid motion. Furthermore, we assume that the pitching motion is quite slow so that pitch acceleration can be ignored, i.e., $\dot{q} = \ddot{\theta} = 0$. In view of this, the pitching moment Eq. (6.3) can be ignored. Then, Eqs. (6.1) and (6.2) reduce to the following form:

$$m_1 \frac{du}{dt} = C_{xu}u + C_{x\theta}\Delta\theta + C_{xq}c_1\left(\frac{d\Delta\theta}{dt}\right) + C_{x\delta_e}\Delta\delta_e \tag{6.58}$$

$$\left(m_1 + C_{zq}c_1\right)\frac{d\Delta\theta}{dt} = -C_{zu}u - C_{z\theta}\Delta\theta - C_{z\delta_e}\Delta\delta_e \tag{6.59}$$

Rearranging in a form suitable for state-space representation, we obtain

$$\frac{du}{dt} = \left(\frac{C_{xu} - \xi_3 C_{zu}}{m_1}\right)u + \left(\frac{C_{x\theta} - \xi_3 C_{z\theta}}{m_1}\right)\Delta\theta + \left(\frac{C_{x\delta_e} - \xi_3 C_{z\delta_e}}{m_1}\right)\Delta\delta_e \tag{6.60}$$

$$\frac{d\Delta\theta}{dt} = \left(\frac{-C_{zu}}{m_1 + C_{zq}c_1}\right)u + \left(\frac{-C_{z\theta}}{m_1 + C_{zq}c_1}\right)\Delta\theta + \left(\frac{-C_{z\delta_e}}{m_1 + C_{zq}c_1}\right)\Delta\delta_e \tag{6.61}$$

where

$$\xi_3 = \frac{C_{xq}c_1}{m_1 + C_{zq}c_1} \tag{6.62}$$

Let

$$x_1 = u \tag{6.63}$$

$$x_2 = \Delta\theta \tag{6.64}$$

$$x = \begin{bmatrix} x_1 \\ x_2 \end{bmatrix} \tag{6.65}$$

$$u = \Delta\delta_e \tag{6.66}$$

so that

$$\begin{bmatrix} \dot{x}_1 \\ \dot{x}_2 \end{bmatrix} = \begin{bmatrix} a_{11} & a_{12} \\ a_{21} & a_{22} \end{bmatrix}\begin{bmatrix} x_1 \\ x_2 \end{bmatrix} + \begin{bmatrix} b_1 \\ b_2 \end{bmatrix} u \tag{6.67}$$

Let

$$A_{ph} = \begin{bmatrix} a_{11} & a_{12} \\ a_{21} & a_{22} \end{bmatrix} \tag{6.68}$$

$$B_{ph} = \begin{bmatrix} b_1 \\ b_2 \end{bmatrix} \tag{6.69}$$

where

$$a_{11} = \left(\frac{C_{xu} - \xi_3 C_{zu}}{m_1} \right) \quad a_{12} = \left(\frac{C_{x\theta} - \xi_3 C_{z\theta}}{m_1} \right)$$

$$a_{21} = \left(\frac{-C_{zu}}{m_1 + C_{zq} c_1} \right) \quad a_{22} = \left(\frac{-C_{z\theta}}{m_1 + C_{zq} c_1} \right)$$

$$b_1 = \left(\frac{C_{x\delta_e} - \xi_3 C_{z\delta_e}}{m_1} \right) \quad b_2 = \left(\frac{-C_{z\delta_e}}{m_1 + C_{zq} c_1} \right)$$

Equations (6.60) and (6.61) are approximate equations for the phugoid or long-period mode and are applicable for conventional statically stable airplanes. These equations may not be applicable if the airplane is statically unstable or incorporates relaxed static stability concepts wherein the aircraft is rendered statically unstable to improve the performance. For such airplanes, as we will discuss later, the conventional short-period and phugoid modes do not exist.

To get an idea of the physical parameters that have major influence on the frequency and damping of the phugoid mode, we need to introduce some additional simplifications. As said before, $C_{z\theta} \simeq 0$. Furthermore, assume that C_{xq} and C_{zq} are small and can be ignored. With these assumptions, it can be shown that the characteristic equation of the phugoid mode is given by

$$\lambda^2 - \frac{C_{xu}}{m_1} \lambda + \frac{C_{x\theta} C_{zu}}{m_1^2} = 0 \tag{6.70}$$

Comparing this equation with that of the standard second-order system [Eq. (6.34)], we obtain

$$\omega_n = \frac{1}{m_1} \sqrt{C_{x\theta} C_{zu}} \tag{6.71}$$

$$\zeta = -\frac{C_{xu}}{2 m_1 \omega_n} \tag{6.72}$$

We have

$$C_{xu} = -2C_D - C_{Du} \tag{6.73}$$

$$C_{x\theta} = -C_L \cos\theta_o \tag{6.74}$$

$$C_{zu} = -2C_L - C_{Lu} \tag{6.75}$$

For low subsonic speeds, $C_{Du} \simeq C_{Lu} \simeq 0$. For level flight, $\theta_o = 0$ so that

$$C_{xu} = -2C_D \tag{6.76}$$

$$C_{x\theta} = -C_L \tag{6.77}$$

$$C_{zu} = -2C_L \tag{6.78}$$

Furthermore, using $C_L = 2W/\rho U_o^2 S$, we obtain

$$\omega_n = \frac{\sqrt{2}g}{U_o} \tag{6.79}$$

$$T = \frac{\sqrt{2}\pi U_o}{g} \tag{6.80}$$

$$\zeta = \frac{1}{\sqrt{2}}\left(\frac{C_D}{C_L}\right) \tag{6.81}$$

$$= \frac{1}{\sqrt{2}E} \tag{6.82}$$

Thus, we observe that the damping of the phugoid mode is inversely proportional to the aerodynamic efficiency E. Because E varies with angle of attack, the phugoid damping will vary with angle of attack, becoming minimum when the airplane flies at that angle of attack when $E = E_{max}$ or $C_L = \sqrt{C_{Do}/k}$ [see Eqs. (2.11) and (2.20)]. We also note that the period of the phugoid motion increases with forward speed.

6.2.3 Accuracy of Short-Period and Phugoid Approximations

It is instructive to assess the accuracy of the short-period and the phugoid approximations. For this purpose, let us consider the general aviation airplane once again as follows.

Substituting the values of mass, inertia, and aerodynamic parameters in Eqs. (6.46) and (6.67), we get

$$A_{sp} = \begin{bmatrix} -2.0354 & 0.9723 \\ -7.0301 & -2.9767 \end{bmatrix} \tag{6.83}$$

$$B_{sp} = \begin{bmatrix} -0.1609 \\ -11.8674 \end{bmatrix} \tag{6.84}$$

$$A_{ph} = \begin{bmatrix} -0.0453 & -0.183 \\ 0.3823 & 0 \end{bmatrix} \tag{6.85}$$

$$B_{ph} = \begin{bmatrix} 0 \\ 0.1655 \end{bmatrix} \tag{6.86}$$

Note that control derviatives are per radian. Using MATLAB [1], we obtain the eigenvalues for the short-period and phugoid approximations as

$$\lambda_{sp} = -2.5060 \pm j2.5717 \tag{6.87}$$

$$\lambda_{ph} = -0.0227 \pm j0.2656 \tag{6.88}$$

Comparing these eigenvalues with those of the complete fourth-order system obtained earlier [Eqs. (6.31) and (6.32)], we observe that the short-period approximation is quite satisfactory for the prediction of free response of the general aviation aircraft, whereas the damping and frequency of

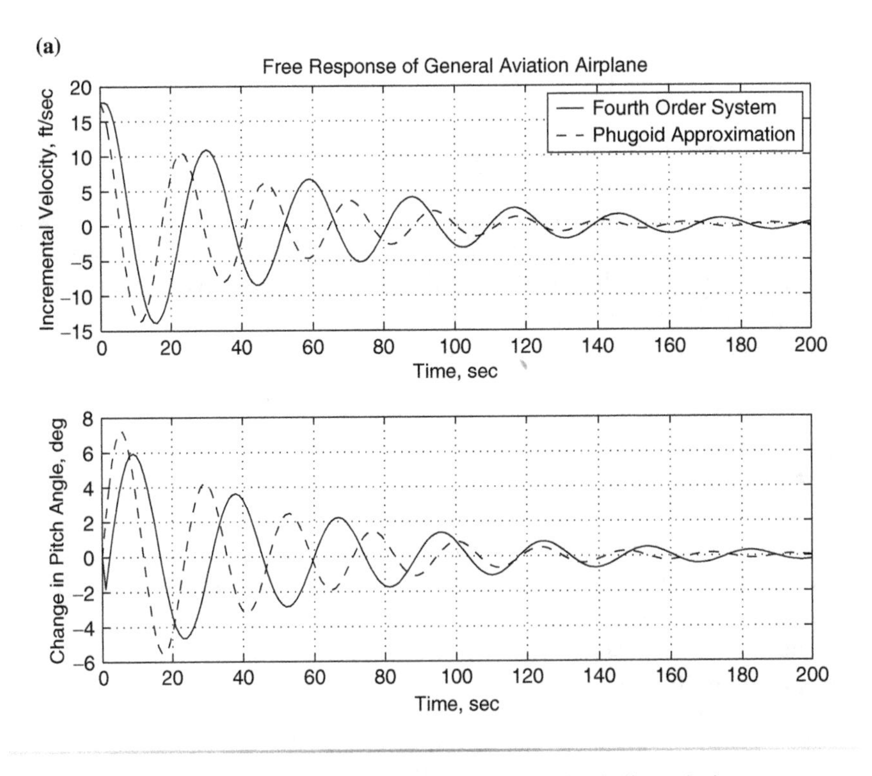

Fig. 6.4a Longitudinal response of general aviation airplane.

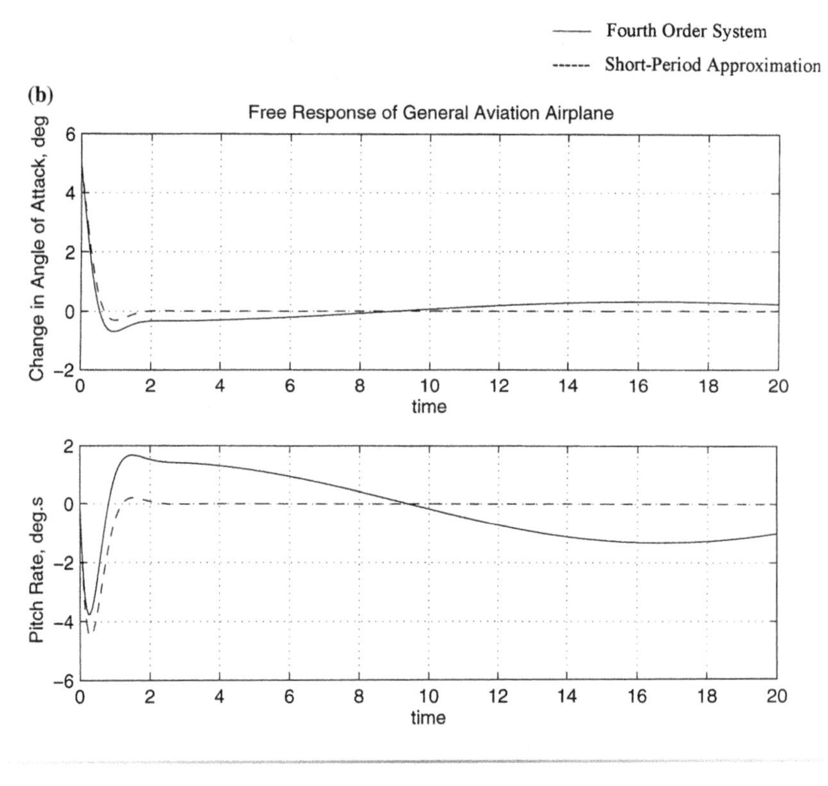

Fig. 6.4b Longitudinal response of general aviation airplane (*continued*).

phugoid motion are in error. In general, for typical statically stable aircraft, this type of comparison is usually obtained.

The free responses of the general aviation airplane based on complete fourth-order system and based on short-period and phugoid approximations with assumed initial conditions $\Delta\alpha = 5$ deg and $u = 0.1$ are shown in Figs. 6.4a and 6.4b. MATLAB [1] was used for these calculations. We observe that the disturbances in angle of attack and pitch rate q decay quickly and come close to zero within 3–5 s, whereas the disturbances in forward velocity and pitch angle persist for a long time and decay slowly. In other words, disturbances in angle of attack and pitch rate decay rapidly during the short-period mode. During the phugoid mode, which continues after the decay of the short-period mode, the angle of attack nearly remains constant, and the pitch rate is close to zero.

Next, let us examine the accuracy of the short-period and the phugoid approximations for forced response. For this purpose, we have computed the unit-step response of the general aviation airplane, and the results are presented in Figs. 6.5–6.7. It is interesting to observe that the unit step responses for 1-deg elevator input differ considerably. The short-period response agrees well with that of the complete fourth-order system initially, but the steady-state values differ. The short-period approximation results in a

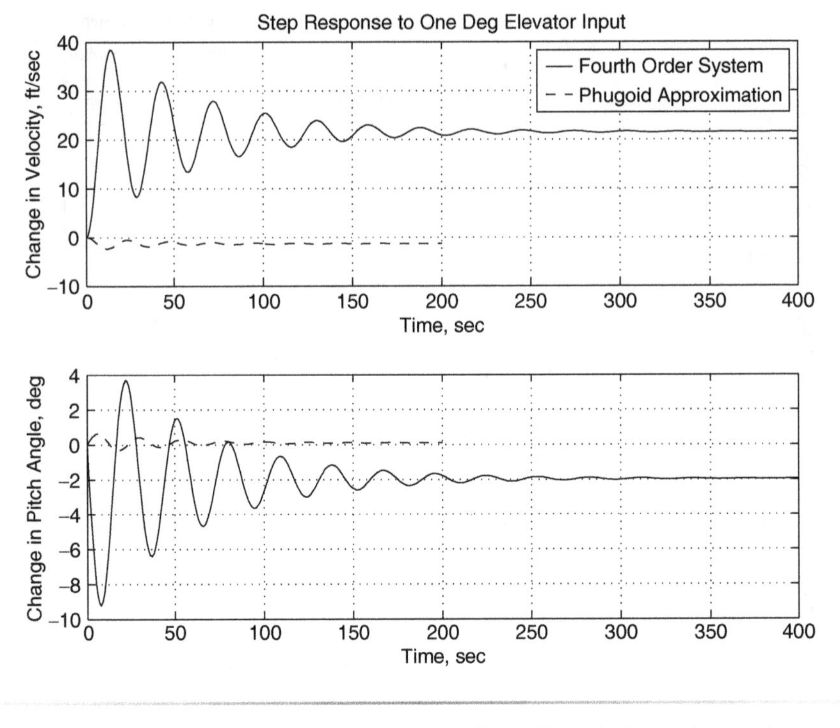

Fig. 6.5 Unit-step response of general aviation airplane: forward velocity and pitch angle.

quick decay of the oscillatory motion, whereas the complete system continues to oscillate for a much longer time. The step response based on phugoid approximation differs considerably from that for the complete system. There is a significant difference in the transient as well as steady-state values. Thus, for the general aviation airplane, the forced response based on short-period approximation is somewhat satisfactory, whereas the phugoid approximation is not adequate at all. Therefore, it is always a good practice to check any control law design based on such approximations by a simulation of the complete system using the same control law. If the two simulations are in fair agreement, then the design is satisfactory. If not, the exercise has to be repeated using the complete system instead of a reduced-order system based on either short-period or phugoid approximations.

6.2.4 Effect of Static Stability on Longitudinal Response

When the center of gravity moves aft, the static stability level of the airplane decreases. When it goes aft of the stick-fixed neutral point, the aircraft becomes statically unstable. When this happens, in the s-plane, the short-period and phugoid roots move towards the real axis as shown in

Fig. 6.8. Here, *AA* and *BB* denote the short-period and phugoid roots for a stable location of the center of gravity. When the center of gravity moves forward, the roots move onto the real axis, and the conventional short-period and phugoid approximations break down. The right-moving branch of the short-period mode meets the left-moving branch of the phugoid mode and a new oscillatory mode emerges. This mode has the short-period-like damping and phugoid-like frequency. This mode is sometimes known as the third oscillatory mode. At the same time, the right-moving branch of the phugoid mode crosses the imaginary axis and moves into the right half of the s-plane. This indicates that the airplane will exhibit an exponential instability in pitch.

The condition when the real root crosses the imaginary axis and moves into the right half of the s-plane can be determined as follows.

According to Routh's criterion, the necessary condition for stability is that all coefficients of the characteristic polynomial must be positive or have the same sign. When any one of the coefficients becomes negative while the rest are positive, then the necessary condition is violated and the system becomes unstable. By examining the expression of all the coefficients of the characteristic polynomial, we observe that the coefficient that is most likely to change sign due to a change in static stability level is the coefficient E_δ as given by

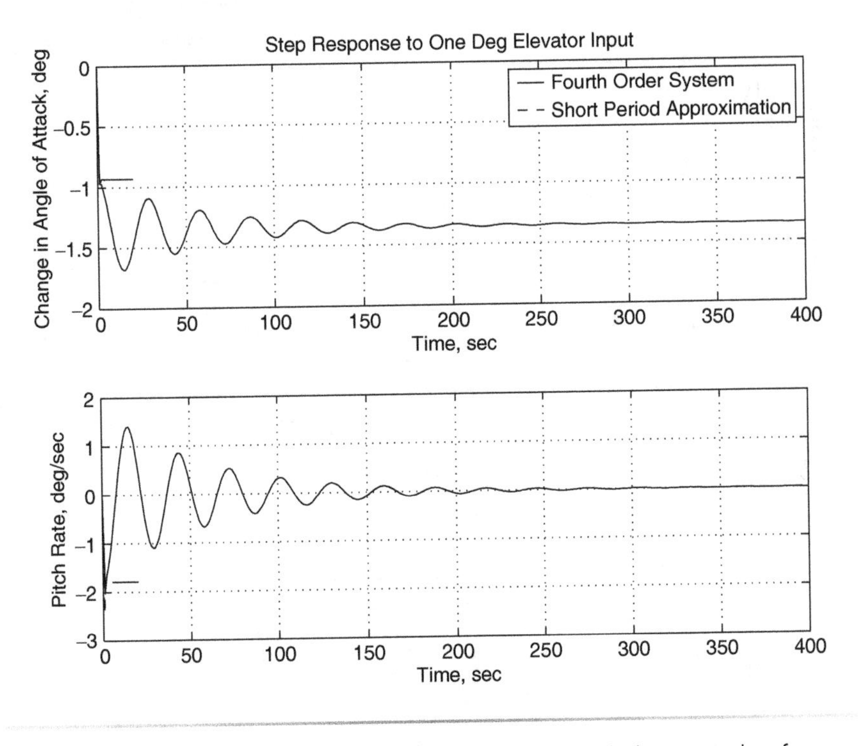

Fig. 6.6 Unit-step response of general aviation airplane: angle of attack and pitch rate.

Eq. (6.25). Therefore, the condition for the root-locus to cross the imaginary axis is

$$E_\delta = C_{z\theta}(C_{xu}C_{m\alpha} - C_{x\alpha}C_{mu}) - C_{x\theta}(C_{zu}C_{m\alpha} - C_{z\alpha}C_{mu}) \\ = 0 \tag{6.89}$$

or

$$C_{m\alpha} = \frac{C_{mu}(C_{x\alpha}C_{z\theta} - C_{x\theta}C_{z\alpha})}{C_{xu}C_{z\theta} - C_{x\theta}C_{zu}} \tag{6.90}$$

We can express the derivative C_{mu} as

$$C_{mu} = \frac{\partial C_m}{\partial\left(\dfrac{\Delta U}{U_o}\right)} \tag{6.91}$$

$$= M\frac{\partial C_m}{\partial M} \tag{6.92}$$

$$= M\alpha\frac{\partial C_{m\alpha}}{\partial M} \tag{6.93}$$

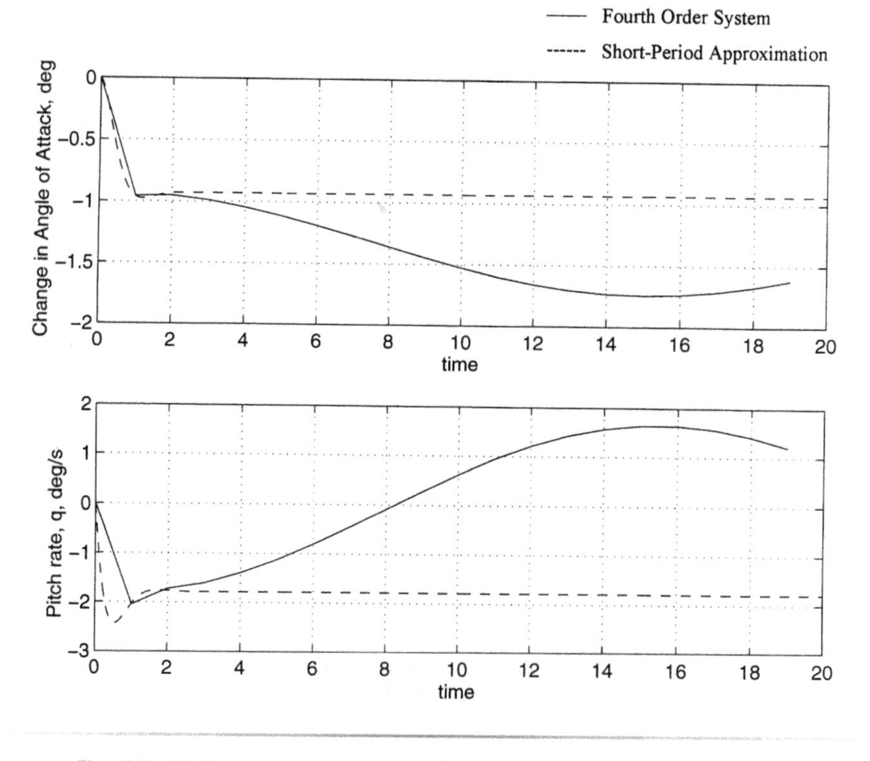

Fig. 6.7 Unit-step response of general aviation airplane: angle of attack and pitch rate for first 20 s.

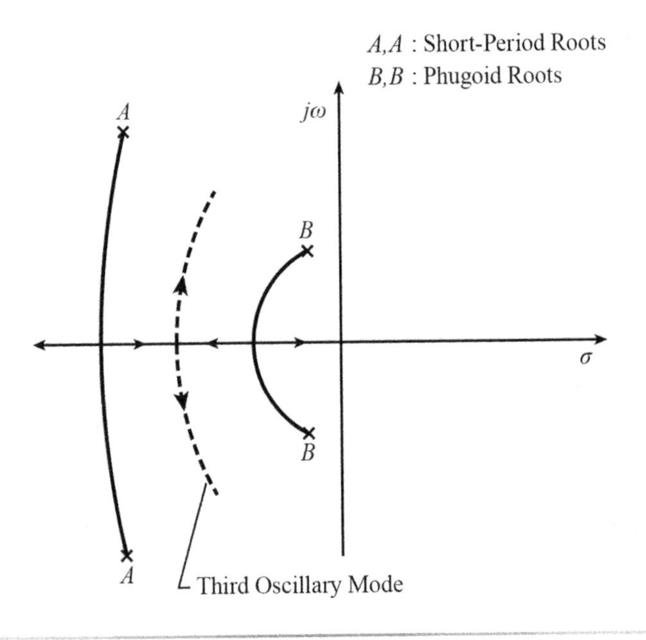

Fig. 6.8 Effect of center of gravity location on longitudinal modes.

Furthermore, we have

$$C_{x\alpha} = C_L - C_{D\alpha} \tag{6.94}$$

$$C_{x\theta} = -C_L \cos\theta_o \tag{6.95}$$

$$C_{z\theta} = -C_L \sin\theta_o \tag{6.96}$$

$$C_{xu} = -2C_D - C_{Du} \qquad C_{zu} = -2C_L - C_{Lu} \tag{6.97}$$

$$C_{z\alpha} = -C_{L\alpha} - C_D \tag{6.98}$$

However, if we assume $\theta_o = 0$ (equilibrium level flight) so that $C_{z\theta} = 0$, then we have

$$C_{m\alpha} = \frac{C_{z\alpha}C_{mu}}{C_{zu}} \tag{6.99}$$

Let us ignore the compressibility effects, i.e., we assume that the flight Mach number is below 0.5, so that $C_{mu} = 0$. Then, for such cases, the given condition reduces to

$$C_{m\alpha} = 0 \tag{6.100}$$

From Eq. (3.85), we have

$$C_{m\alpha} = \left(\bar{x}_{cg} - N_o\right) C_{L\alpha} \tag{6.101}$$

so that

$$\bar{x}_{cg} = N_o \tag{6.102}$$

Thus, when the center of gravity coincides with the stick-fixed neutral point, the root-locus crosses the imaginary axis. For further aft locations of the center of gravity, the root-locus moves into the right half of the s-plane, and the aircraft becomes dynamically unstable as well. To have an idea of the nature of free response of a statically unstable airplane, let us consider the general aviation airplane once again with the assumption that the center of gravity is moved aft to make it unstable in pitch. We know that the center of gravity has a direct effect on the damping terms C_{mq} and $C_{m\dot{\alpha}}$ through horizontal tail terms. To account for these effects, let $C_{m\alpha} = 0.07/\text{rad}$ and both $C_{m\dot{\alpha}}$ and C_{mq} be reduced by 50%.

The eigenvalues of this statically unstable version of the general aviation airplane were calculated as

$$\lambda_1 = -3.1303 \tag{6.103}$$

$$\lambda_{2,3} = -0.2965 \pm j0.2062 \tag{6.104}$$

$$\lambda_4 = 0.1542 \tag{6.105}$$

Notice that the pair of complex roots $\lambda_{2,3}$ has a short-period-like damping and phugoid-like frequency. This is the so-called third oscillatory mode.

The free response to an initial disturbance in angle of attack of 5 deg is shown in Fig. 6.9. We note that the disturbance in angle of attack decays initially because of the damping effect provided by the negative real root and to some extent by the negative real part of the complex root. However, the angle of attack starts building up subsequently due to the real positive root λ_4. Along with this, pitch rate also builds up as shown. For such airplanes, it is desirable to have an automatic stability augmentation system that can take care of this problem so that the airplane has the conventional short-period and phugoid modes in the closed-loop sense. We will discuss the design of such a pitch augmentation system later in this chapter.

6.2.5 Longitudinal Transfer Functions

The transfer function gives a relation between the input and the output of a control system in the Laplace domain. It forms the basis of

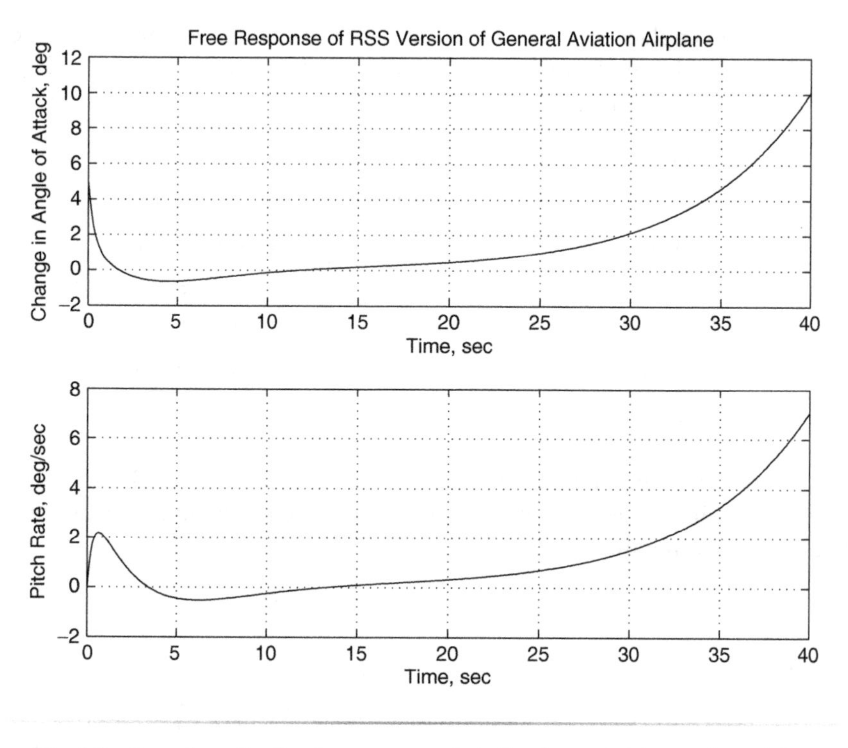

Fig. 6.9 Free (longitudinal) response of statically unstable version of the general aviation airplane.

analysis and design using frequency-domain methods. Here, we will derive longitudinal transfer functions for the elevator input for the complete fourth-order system as well as those based on short-period and phugoid approximations.

Transfer Functions of a Complete System

Assuming that an elevator is the only longitudinal control and taking the Laplace transform of Eqs. (6.1–6.3), we obtain

$$(m_1 s - C_{xu})\bar{u}(s) - (C_{x\alpha} + C_{x\dot{\alpha}}c_1 s)\Delta\bar{\alpha}(s) - \big(C_{xq}c_1 s + C_{x\theta}\big)\Delta\bar{\theta}(s) \\ = C_{x\delta_e}\Delta\bar{\delta}_e(s) \tag{6.106}$$

$$-C_{zu}\bar{u}(s) + (m_1 s - C_{z\dot{\alpha}}c_1 s - C_{z\alpha})\Delta\bar{\alpha}(s) - \big(m_1 s + C_{zq}c_1 s + C_{z\theta}\big)\Delta\bar{\theta}(s) \\ = C_{z\delta_e}\Delta\bar{\delta}_e(s) \tag{6.107}$$

$$-C_{mu}\bar{u}(s) - (C_{m\dot{\alpha}}c_1 s + C_{m\alpha})\Delta\bar{\alpha}(s) + s\big(I_{y1}s - C_{mq}c_1\big)\Delta\bar{\theta}(s) = C_{m\delta e}\Delta\bar{\delta}_e(s) \tag{6.108}$$

Here, s is the Laplace variable, and a bar over the symbol denotes its Laplace transform. Dividing throughout by $\Delta\bar{\delta}_e(s)$ and using Cramer's rule (see Appendix C), we obtain

$$\frac{\bar{u}(s)}{\Delta\bar{\delta}_e(s)} = \frac{\begin{vmatrix} C_{x\delta_e} & -(C_{x\alpha} + C_{x\dot{\alpha}}c_1 s) & -(C_{xq}c_1 s + C_{x\theta}) \\ C_{z\delta_e} & (m_1 s - C_{z\dot{\alpha}}c_1 s - C_{z\alpha}) & -(m_1 s + C_{zq}c_1 s + C_{z\theta}) \\ C_{m\delta_e} & -(C_{m\dot{\alpha}}c_1 s + C_{m\alpha}) & s(I_{y1}s - C_{mq}c_1) \end{vmatrix}}{\begin{vmatrix} (m_1 s - C_{xu}) & -(C_{x\alpha} + C_{x\dot{\alpha}}c_1 s) & -(C_{xq}c_1 s + C_{x\theta}) \\ -C_{zu} & (m_1 s - C_{z\dot{\alpha}}c_1 s - C_{z\alpha}) & -(m_1 s + C_{zq}c_1 s + C_{z\theta}) \\ -C_{mu} & -(C_{m\dot{\alpha}}c_1 s + C_{m\alpha}) & s(I_{y1}s - C_{mq}c_1) \end{vmatrix}} \tag{6.109}$$

Let $N_{\bar{u}}$ denote the determinant in the numerator and $\Delta_{\text{long}}(s)$ denote the determinant in the denominator of Eq. (6.109). Then, expanding the determinant in the numerator, we obtain

$$N_{\bar{u}} = A_{\bar{u}}s^3 + B_{\bar{u}}s^2 + C_{\bar{u}}s + D_{\bar{u}} \tag{6.110}$$

where

$$A_{\bar{u}} = I_{y1}\left(C_{x\delta_e}m_1 - C_{x\delta_e}C_{z\dot{\alpha}}c_1 + C_{x\dot{\alpha}}c_1 C_{z\delta_e}\right)$$

$$\begin{aligned} B_{\bar{u}} = &-C_{x\delta_e}\left(C_{z\alpha}I_{y1} + m_1 C_{mq}c_1 - C_{z\dot{\alpha}}C_{mq}c_1^2 + m_1 C_{m\dot{\alpha}}c_1 + C_{zq}C_{m\dot{\alpha}}c_1^2\right) \\ &-C_{z\delta_e}C_{x\dot{\alpha}}c_1^2 C_{mq} + C_{z\delta_e}I_{y1}C_{x\alpha} + C_{x\dot{\alpha}}c_1 C_{m\delta_e}(m_1 + C_{zq}c_1) \\ &+C_{xq}c_1^2 C_{z\delta_e}C_{m\dot{\alpha}} + C_{xq}c_1 C_{m\delta_e}(m_1 - C_{z\dot{\alpha}}c_1) \end{aligned}$$

$$\begin{aligned} C_{\bar{u}} = &\, C_{x\delta_e}\left(C_{z\alpha}C_{mq}c_1 - m_1 C_{m\alpha} - C_{zq}c_1 C_{m\alpha} - C_{z\theta}C_{m\dot{\alpha}}c_1\right) \\ &- C_{mq}C_{z\delta_e}C_{x\alpha}c_1 + C_{x\alpha}C_{m\delta_e}(m_1 + C_{zq}c_1) \\ &+ C_{x\dot{\alpha}}c_1 C_{z\theta}C_{m\delta_e} + C_{xq}c_1 C_{z\delta_e}C_{m\alpha} \\ &- C_{xq}c_1 C_{m\delta_e}C_{z\alpha} + C_{x\theta}C_{z\delta_e}C_{m\dot{\alpha}}c_1 \\ &+ C_{x\theta}C_{m\delta_e}(m_1 - C_{z\dot{\alpha}}c_1) \end{aligned}$$

$$D_{\bar{u}} = C_{z\theta}\left(C_{x\alpha}C_{m\delta_e} - C_{m\alpha}C_{x\delta_e}\right) + C_{x\theta}\left(C_{z\delta_e}C_{m\alpha} - C_{m\delta_e}C_{z\alpha}\right)$$

Furthermore, expanding the determinant in the denominator, we obtain the longitudinal characteristic polynomial

$$\Delta_{\text{long}}(s) = A_\delta s^4 + B_\delta s^3 + C_\delta s^2 + D_\delta s + E_\delta \tag{6.111}$$

where the coefficients A_δ, B_δ, C_δ, D_δ, and E_δ are identical to those given in Eqs. (6.21–6.25) and are reproduced here in the following:

$$A_\delta = m_1 I_{y1}(m_1 - C_{z\dot{\alpha}}c_1) \tag{6.112}$$

$$B_\delta = m_1\left(-I_{y1}C_{z\alpha} - C_{mq}c_1[m_1 - C_{z\dot{\alpha}}c_1] - C_{m\dot{\alpha}}c_1[m_1 + C_{zq}c_1]\right) - C_{xu}I_{y1}(m_1 - C_{z\dot{\alpha}}c_1) - C_{x\dot{\alpha}}c_1C_{zu}I_{y1} \tag{6.113}$$

$$C_\delta = m_1\left(C_{z\alpha}C_{mq}c_1 - C_{m\alpha}[m_1 + C_{zq}c_1] - C_{z\theta}C_{m\dot{\alpha}}c_1\right) + C_{xu}\left(I_{y1}C_{z\alpha} + C_{mq}c_1[m_1 - C_{z\dot{\alpha}}c_1] + C_{m\dot{\alpha}}c_1[m_1 + C_{zq}c_1]\right) - C_{x\alpha}C_{zu}I_{y1} + C_{x\dot{\alpha}}c_1\left(C_{zu}C_{mq}c_1 - C_{mu}[m_1 + C_{zq}c_1]\right) - C_{xq}c_1(C_{zu}C_{m\dot{\alpha}}c_1 + C_{mu}[m_1 - C_{z\dot{\alpha}}c_1]) \tag{6.114}$$

$$D_\delta = -C_{xu}\left(C_{z\alpha}C_{mq}c_1 - C_{m\alpha}[m_1 + C_{zq}c_1] - C_{z\theta}C_{m\dot{\alpha}}c_1\right) - m_1C_{m\alpha}C_{z\theta} + C_{x\alpha}\left(C_{zu}C_{mq}c_1 - C_{mu}[m_1 + C_{zq}c_1]\right) - C_{x\dot{\alpha}}c_1C_{mu}C_{z\theta} - C_{xq}c_1(C_{zu}C_{m\alpha} - C_{z\alpha}C_{mu}) - C_{x\theta}(C_{zu}C_{m\dot{\alpha}}c_1 + C_{mu}[m_1 - C_{z\dot{\alpha}}c_1]) \tag{6.115}$$

$$E_\delta = C_{xu}C_{m\alpha}C_{z\theta} - C_{x\alpha}C_{mu}C_{z\theta} - C_{x\theta}(C_{zu}C_{m\alpha} - C_{z\alpha}C_{mu}) \tag{6.116}$$

The transfer function for angle of attack is given by

$$\frac{\Delta\bar{\alpha}(s)}{\Delta\bar{\delta}_e(s)} = \frac{\begin{vmatrix} m_1s - C_{xu} & C_{x\delta_e} & -(C_{xq}c_1s + C_{x\theta}) \\ -C_{zu} & C_{z\delta_e} & -(m_1s + C_{zq}c_1s + C_{z\theta}) \\ -C_{mu} & C_{m\delta_e} & s(I_{y1}s - C_{mq}c_1) \end{vmatrix}}{\Delta_{\text{long}}(s)} \tag{6.117}$$

Expanding the determinant in the numerator, we obtain

$$N_{\bar{\alpha}} = A_{\bar{\alpha}}s^3 + B_{\bar{\alpha}}s^2 + C_{\bar{\alpha}}s + D_{\bar{\alpha}} \tag{6.118}$$

where

$$A_{\bar{\alpha}} = m_1I_{y1}C_{z\delta_e} \tag{6.119}$$

$$B_{\bar{\alpha}} = m_1\left(-C_{z\delta_e}C_{mq}c_1 + C_{m\delta_e}m_1 + C_{m\delta_e}C_{zq}c_1\right) - C_{xu}C_{z\delta_e}I_{y1} + C_{x\delta_e}C_{zu}I_{y1} \tag{6.120}$$

$$C_{\bar{\alpha}} = C_{xu}\left(C_{z\delta_e}C_{mq}c_1 - C_{m\delta_e}m_1 - C_{m\delta_e}C_{zq}c_1\right) + m_1C_{m\delta_e}C_{z\theta} - C_{x\delta_e}C_{zu}C_{mq}c_1 + m_1C_{x\delta_e}C_{mu} + C_{x\delta_e}C_{mu}C_{zq}c_1 + C_{xq}c_1C_{zu}C_{m\delta_e} - C_{xq}c_1C_{mu}C_{z\delta_e} \tag{6.121}$$

$$D_{\bar{\alpha}} = C_{x\theta}\left(C_{zu}C_{m\delta_e} - C_{mu}C_{z\delta_e}\right) + C_{z\theta}\left(C_{x\delta_e}C_{mu} - C_{m\delta_e}C_{xu}\right) \tag{6.122}$$

The transfer function for pitch angle is given by

$$\frac{\Delta\bar{\theta}(s)}{\Delta_{\bar{\delta}_e}(s)} = \frac{\begin{vmatrix} m_1 s - C_{xu} & -(C_{x\alpha} + C_{x\dot{\alpha}} c_1 s) & C_{x\delta_e} \\ -C_{zu} & (m_1 s - C_{z\dot{\alpha}} c_1 s - C_{z\alpha}) & C_{z\delta_e} \\ -C_{mu} & -(C_{m\dot{\alpha}} c_1 s + C_{m\alpha}) & C_{m\delta_e} \end{vmatrix}}{\Delta_{\text{long}}(s)} \tag{6.123}$$

Expanding the determinant in the numerator, we obtain

$$N_\theta = A_{\bar{\theta}} s^2 + B_{\bar{\theta}} s + C_{\bar{\theta}} \tag{6.124}$$

where

$$A_{\bar{\theta}} = m_1 \left(m_1 C_{m\delta_e} - C_{z\dot{\alpha}} c_1 C_{m\delta_e} + C_{z\delta_e} C_{m\dot{\alpha}} c_1 \right) \tag{6.125}$$

$$\begin{aligned} B_{\bar{\theta}} = {} & m_1 \left(C_{z\delta_e} C_{m\alpha} - C_{z\alpha} C_{m\delta_e} \right) - C_{xu} \left(m_1 C_{m\delta_e} - C_{z\dot{\alpha}} c_1 C_{m\delta_e} + C_{z\delta_e} C_{m\dot{\alpha}} c_1 \right) \\ & + c_1 C_{x\dot{\alpha}} \left(-C_{zu} C_{m\delta_e} + C_{mu} C_{z\delta_e} \right) + C_{x\delta_e} C_{zu} C_{m\dot{\alpha}} c_1 \\ & + C_{x\delta_e} C_{mu} m_1 - C_{x\delta_e} C_{mu} C_{z\dot{\alpha}} c_1 \end{aligned} \tag{6.126}$$

$$\begin{aligned} C_{\bar{\theta}} = {} & -C_{xu} \left(C_{z\delta_e} C_{m\alpha} - C_{z\alpha} C_{m\delta_e} \right) + C_{x\delta_e} (C_{m\alpha} C_{zu} - C_{mu} C_{z\alpha}) \\ & + C_{x\alpha} \left(-C_{zu} C_{m\delta_e} + C_{mu} C_{z\delta_e} \right) \end{aligned} \tag{6.127}$$

We have $q = \dot{\theta} = \Delta\dot{\theta}$ or $\bar{q}(s) = s\bar{\theta}(s)$ so that

$$\frac{\bar{q}(s)}{\Delta\bar{\delta}_e(s)} = \frac{s\Delta\bar{\theta}(s)}{\Delta\bar{\delta}_e(s)} \tag{6.128}$$

Transfer Function for Short-Period Approximation

Taking the Laplace transform of Eqs. (6.41) and (6.42), we have

$$\begin{aligned} & (m_1 s - C_{z\dot{\alpha}} c_1 s - C_{z\alpha}) \Delta\bar{\alpha}(s) - \left[s \left(m_1 + C_{zq} c_1 \right) + C_{z\theta} \right] \Delta\bar{\theta}(s) \\ & \quad = C_{z\delta_e} \Delta\bar{\delta}_e(s) \end{aligned} \tag{6.129}$$

$$-(C_{m\alpha} + C_{m\dot{\alpha}} c_1 s) \Delta\bar{\alpha}(s) + \left(I_{y1} s^2 - C_{mq} c_1 s \right) \Delta\bar{\theta}(s) = C_{m\delta_e} \Delta\bar{\delta}_e(s) \tag{6.130}$$

Dividing throughout by $\Delta\bar{\delta}_e(s)$ and using Cramer's rule, we obtain the transfer function for angle of attack.

$$\frac{\Delta\bar{\alpha}(s)}{\Delta\bar{\delta}_e(s)} = \frac{\begin{vmatrix} C_{z\delta_e} & -\left[s \left(m_1 + C_{zq} c_1 \right) + C_{z\theta} \right] \\ C_{m\delta_e} & s \left(I_{y1} s - C_{mq} c_1 \right) \end{vmatrix}}{\begin{vmatrix} s(m_1 - C_{z\dot{\alpha}}) - C_{z\alpha} & -\left[s \left(m_1 + C_{zq} c_1 \right) + C_{z\theta} \right] \\ -(C_{m\dot{\alpha}} c_1 s + C_{m\alpha}) & s \left(I_{y1} s - C_{mq} c_1 \right) \end{vmatrix}} \tag{6.131}$$

Let $N_{\bar{\alpha},\mathrm{spo}}$ and Δ_{spo} denote the determinants in the numerator and the denominator of Eq. (6.131). Expanding these determinants, we obtain

$$N_{\bar{\alpha},\mathrm{spo}} = A_{\bar{\alpha},\mathrm{spo}}s^2 + B_{\bar{\alpha},\mathrm{spo}}s + C_{\bar{\alpha},\mathrm{spo}} \tag{6.132}$$

where

$$A_{\bar{\alpha},\mathrm{spo}} = C_{z\delta_e}I_{y1} \tag{6.133}$$

$$B_{\bar{\alpha},\mathrm{spo}} = C_{m\delta_e}(m_1 + C_{zq}c_1) - C_{z\delta_e}C_{mq}c_1 \tag{6.134}$$

$$C_{\bar{\alpha},\mathrm{spo}} = C_{m\delta_e}C_{z\theta} \tag{6.135}$$

and

$$\Delta_{\mathrm{spo}} = \begin{vmatrix} s(m_1 - C_{z\dot{\alpha}}) - C_{z\alpha} & -[s(m_1 + C_{zq}c_1) + C_{z\theta}] \\ -(C_{m\dot{\alpha}}c_1 s + C_{m\alpha}) & s(I_{y1}s - C_{mq}c_1) \end{vmatrix} \tag{6.136}$$

$$= a_1 s^3 + a_2 s^2 + a_3 s + a_4 \tag{6.137}$$

where

$$a_1 = (m_1 - C_{z\dot{\alpha}}c_1)I_{y1} \tag{6.138}$$

$$a_2 = -[C_{z\alpha}I_{y1} + C_{mq}c_1(m_1 - C_{z\dot{\alpha}}c_1) + (m_1 + C_{zq}c_1)C_{m\dot{\alpha}}c_1] \tag{6.139}$$

$$a_3 = C_{z\alpha}C_{mq}c_1 - [(m_1 + C_{zq}c_1)C_{m\alpha} + C_{z\theta}C_{m\dot{\alpha}}c_1] \tag{6.140}$$

$$a_4 = -C_{z\theta}C_{m\alpha} \tag{6.141}$$

The transfer function for pitch angle is given by

$$\frac{\Delta\bar{\theta}(s)}{\Delta\bar{\delta}_e(s)} = \frac{\begin{vmatrix} s(m_1 - C_{z\dot{\alpha}}c_1) - C_{z\alpha} & C_{z\delta_e} \\ -(C_{m\alpha} + C_{m\dot{\alpha}}c_1 s) & C_{m\delta_e} \end{vmatrix}}{\Delta_{\mathrm{spo}}(s)} \tag{6.142}$$

Expanding the numerator in Eq. (6.142), we get

$$N_{\bar{\theta},\mathrm{spo}} = A_{\bar{\theta},\mathrm{spo}}s + B_{\bar{\theta},\mathrm{spo}} \tag{6.143}$$

where

$$A_{\bar{\theta},\mathrm{spo}} = C_{m\delta_e}(m_1 - C_{z\dot{\alpha}}c_1) + C_{m\dot{\alpha}}c_1 C_{z\delta_e} \tag{6.144}$$

$$B_{\bar{\theta},\mathrm{spo}} = C_{m\alpha}C_{z\delta_e} - C_{z\alpha}C_{m\delta_e} \tag{6.145}$$

We will be making use of these transfer functions in the design of flight control systems to be discussed later in this chapter.

Transfer Function for a Phugoid Approximation

Taking the Laplace transform of Eqs. (6.58) and (6.59) and rearranging, we get

$$(m_1 s - C_{xu})\bar{u}(s) - (C_{x\theta} + C_{xq}c_1 s)\Delta\bar{\theta}(s) = C_{x\delta_e}\Delta\bar{\delta}_e(s) \tag{6.146}$$

$$-C_{zu}\bar{u}(s) - \left[C_{z\theta} + s(m_1 + C_{zq}c_1)\right]\Delta\bar{\theta}(s) = C_{z\delta_e}\Delta\bar{\delta}_e(s) \tag{6.147}$$

Dividing throughout by $\Delta\bar{\delta}_e(s)$ and using Cramer's rule (Appendix C), the transfer function for forward velocity can be obtained as

$$\frac{\bar{u}(s)}{\Delta\bar{\delta}_e(s)} = \frac{\begin{vmatrix} C_{x\delta_e} & -(C_{x\theta} + C_{xq}c_1 s) \\ C_{z\delta_e} & -\left[C_{z\theta} + s(m_1 + C_{zq}c_1)\right] \end{vmatrix}}{\begin{vmatrix} m_1 s - C_{xu} & -(C_{x\theta} + C_{xq}c_1 s) \\ -C_{zu} & -\left[C_{z\theta} + s(m_1 + C_{zq}c_1)\right] \end{vmatrix}} \tag{6.148}$$

Let $N_{\bar{u},\mathrm{lpo}}$ and Δ_{lpo} denote the determinants in the numerator and denominator of Eq. (6.148). Expanding these determinants, we get

$$N_{\bar{u},\mathrm{lpo}} = A_{\bar{u},\mathrm{lpo}}s + B_{\bar{u},\mathrm{lpo}} \tag{6.149}$$

$$A_{\bar{u},\mathrm{lpo}} = -\left[C_{x\delta_e}(m_1 + C_{zq}c_1) - C_{z\delta_e}C_{xq}c_1\right] \tag{6.150}$$

$$B_{\bar{u},\mathrm{lpo}} = C_{z\delta_e}C_{x\theta} - C_{x\delta_e}C_{z\theta} \tag{6.151}$$

and

$$\Delta_{\mathrm{lpo}} = A_{\delta,\mathrm{lpo}}s^2 + B_{\delta,\mathrm{lpo}}s + C_{\delta,\mathrm{lpo}} \tag{6.152}$$

$$A_{\delta,\mathrm{lpo}} = -m_1(m_1 + C_{zq}c_1) \tag{6.153}$$

$$B_{\delta,\mathrm{lpo}} = -m_1 C_{z\theta} + C_{xu}(m_1 + C_{zq}c_1) - C_{zu}C_{xq}c_1 \tag{6.154}$$

$$C_{\delta,\mathrm{lpo}} = C_{xu}C_{z\theta} - C_{zu}C_{x\theta} \tag{6.155}$$

The transfer function for pitch angle is given by

$$\frac{\bar{\theta}(s)}{\Delta\bar{\delta}_e(s)} = \frac{\begin{vmatrix} m_1 s - C_{xu} & C_{x\delta_e} \\ -C_{zu} & C_{z\delta_e} \end{vmatrix}}{\Delta_{\mathrm{spo}}} \tag{6.156}$$

Expanding the determinant in the numerator of Eq. (6.156), we obtain

$$N_{\bar{\theta},\mathrm{lpo}} = A_{\theta,\mathrm{lpo}}s + B_{\theta,\mathrm{lpo}} \tag{6.157}$$

$$A_{\bar{\theta},\mathrm{lpo}} = m_1 C_{z\delta_e} \tag{6.158}$$

$$B_{\bar{\theta},\mathrm{lpo}} = C_{x\delta_e}C_{zu} - C_{xu}C_{z\delta_e} \tag{6.159}$$

For the general aviation airplane, substitution of various parameters in Eqs. (6.109), (6.123), and (6.117) gives the following longitudinal transfer functions for the complete (fourth-order) system:

$$\frac{\bar{u}(s)}{\Delta\bar{\delta}_e} = \frac{-0.0022\,s^2 + 0.6617\,s + 1.5997}{0.3739\,s^4 + 1.8907\,s^3 + 4.9103\,s^2 + 0.2526\,s + 0.2296} \tag{6.160}$$

$$\frac{\Delta\bar{\theta}(s)}{\Delta\bar{\delta}_e} = \frac{-4.4367\,s^2 - 8.8084\,s - 0.4507}{0.3739\,s^4 + 1.8907\,s^3 + 4.9103\,s^2 + 0.2526\,s + 0.2296} \tag{6.161}$$

$$\frac{\Delta\bar{\alpha}(s)}{\Delta\bar{\delta}_e} = \frac{-0.0602\,s^3 - 4.4954\,s^2 - 0.2037\,s - 0.3103}{0.3739\,s^4 + 1.8907\,s^3 + 4.9103\,s^2 + 0.2526\,s + 0.2296} \tag{6.162}$$

Similarly, the transfer functions of the general aviation airplane based on short-period approximation are given by

$$\frac{\Delta\bar{\alpha}(s)}{\Delta\bar{\delta}_e} = \frac{-0.0273\,s^2 - 2.0366\,s}{0.1695\,s^3 + 0.8494\,s^2 + 2.1851\,s} \tag{6.163}$$

$$\frac{\Delta\bar{\theta}(s)}{\Delta\bar{\delta}_e} = \frac{-2.0112\,s - 3.9018}{0.1695\,s^3 + 0.8494\,s^2 + 2.1851\,s} \tag{6.164}$$

$$\frac{\bar{q}(s)}{\Delta\bar{\delta}_e} = \frac{-2.0112\,s^2 - 3.9018\,s}{0.1695\,s^3 + 0.8494\,s^2 + 2.1851\,s} \tag{6.165}$$

The transfer functions of the general aviation airplane based on phugoid approximation [substitute in Eqs. (6.148) and (6.156)] are given by

$$\frac{\bar{u}(s)}{\Delta\bar{\delta}_e} = \frac{0.1455}{-4.7314\,s^2 - 0.2145\,s - 0.3362} \tag{6.166}$$

$$\frac{\Delta\bar{\theta}(s)}{\Delta\bar{\delta}_e} = \frac{-0.7831\,s - 0.0355}{-4.7314\,s^2 - 0.2145\,s - 0.3362} \tag{6.167}$$

We will be making use of these transfer functions for the design of stability augmentation systems and autopilots later in this chapter. Note that we have used control derivatives per radian.

6.2.6 Longitudinal Frequency Response

The frequency response of the airplane is of interest because the pilot input to a control surface can be considered as a signal with a frequency content. Generally, the bandwidth of a human pilot is about 4 rad/s [3]. Therefore, it can be expected that a human pilot can control flight path variables that respond to input frequencies within this range. The autopilots have a higher bandwidth. A typical value of autopilot bandwidth is about 20–25 rad/s [3].

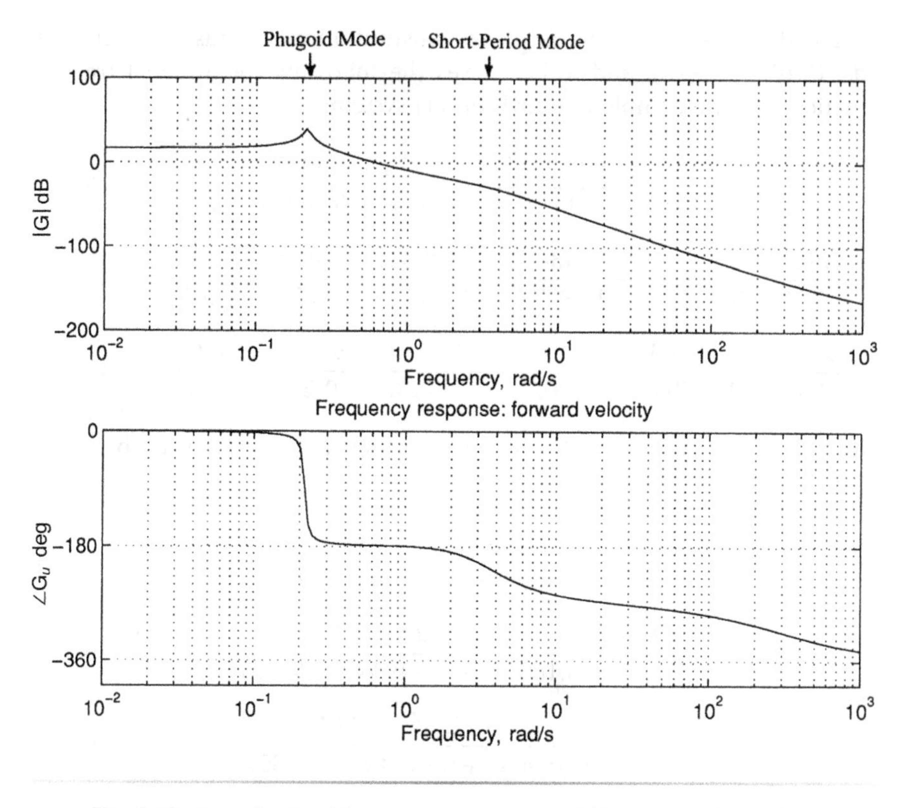

Fig. 6.10 Longitudinal frequency response of the general aviation airplane: forward velocity.

To illustrate the nature of longitudinal frequency response of conventional, statically stable airplanes, we have presented the longitudinal frequency response of the general aviation airplane based on complete fourth-order transfer function in Figs. 6.10–6.12. Here, G_u, G_α, and G_θ are the longitudinal transfer functions of the general aviation airplane given in Eqs. (6.160–6.162). From the magnitude plot of Fig. 6.10, we observe that $|G_u|$ attains a peak value around the phugoid frequency ω_{ph} and then drops off rapidly. The magnitude crossover frequency (the frequency where the magnitude is 0 db) for $|G_u|$ is about 0.5 rad/s. This implies that elevator input with frequencies beyond 0.5 rad/s have little or no effect on the forward speed. For example at $\omega = 3.6$ rad/s, which corresponds to the short-period frequency ω_{spo}, $20 \log_{10} |G_u| = -40$ db or $|G_u| = 0.01$, i.e., just 1% change in the forward speed for a unit deflection of the elevator with a frequency of 3.6 rad/s.

Thus, we understand that forward speed is influenced only by slow elevator inputs and rapid elevator movements have little or no effect on the forward speed. In other words, elevator input is not the best way to quickly change the forward speed.

The phase angle $\angle G_u$ is approximately zero at lower values and then suddenly drops off to -180 around $\omega = 0.21$ rad/s, which corresponds to the phugoid mode. This type of behavior of $\angle G_u$ is characteristic of lightly damped systems. As said earlier, this implies that only a slow movement of the elevator results in a change of forward speed without any phase lag. For $\omega > \omega_{spo}$, the phase lag increases, which is characteristic of heavily damped systems.

The frequency response of G_α displays the influence of frequency both at short-period and phugoid frequencies. The $|G_\alpha|$ remains nearly constant at low values of frequency but displays somewhat complex behavior around ω_{ph}. Such a behavior is due to a pole-zero cancellation (close proximity of pole and zero) in the transfer function G_α. The frequency response of G_θ also displays a similar and complex behavior around the phugoid frequency that is characteristic of lightly damped systems. Furthermore, around the short-period frequency, the phase angle of G_θ is close to 90 deg, which is a characteristic feature of second-order systems. We may recall that, for a second-order system, when $\omega = \omega_n$, the phase angle is 90 deg for all values of ζ. The magnitude crossover frequency for G_α and G_θ is about 2.0 rad/s. Therefore, a human pilot with a bandwidth of 4 rad/s can have a good control over the angle of attack and pitch angle.

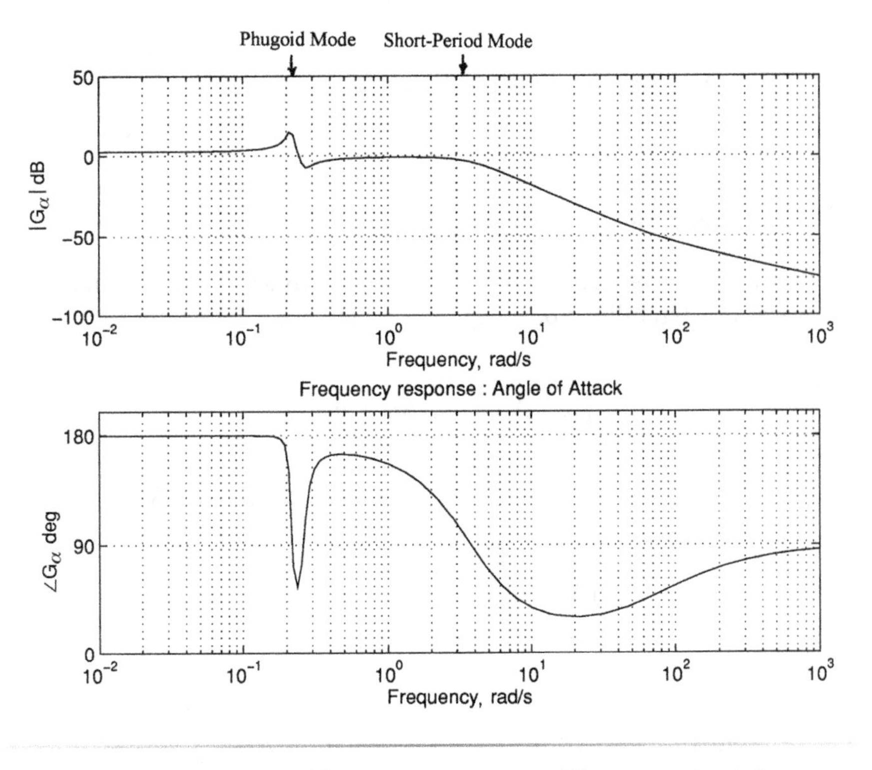

Fig. 6.11 Longitudinal frequency response of the general aviation airplane: angle of attack.

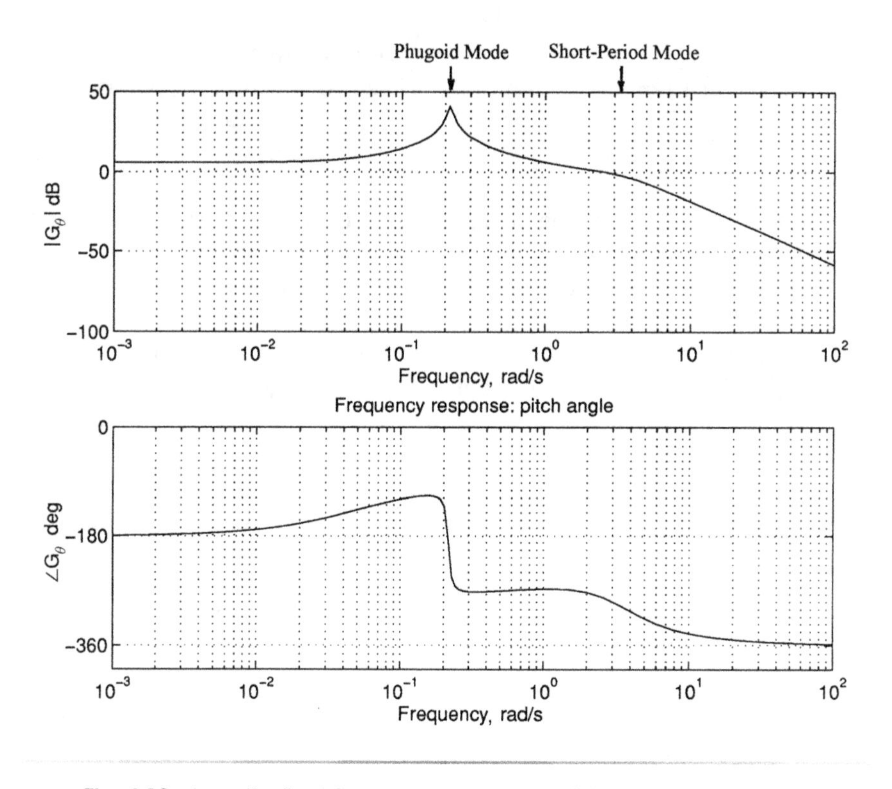

Fig. 6.12 Longitudinal frequency response of the general aviation airplane: pitch angle.

6.3 Lateral-Directional Response

Equations (4.459–4.461),which are based on the assumption of small disturbances in lateral-directional degrees of freedom, can be expressed as

$$\frac{d\beta}{dt} = \left(\frac{1}{m_1 - b_1 C_{y\dot{\beta}}}\right)\Big[C_{y\beta}\Delta\beta + C_{y\phi}\Delta\phi + b_1 C_{yp}p - (m_1 - b_1 C_{yr})r + C_{y\delta_a}\Delta\delta_a + C_{y\delta_r}\Delta\delta_r\Big] \tag{6.168}$$

$$\dot{p} = \frac{1}{I_{x1}}(C_{l\beta}\Delta\beta + C_{l\dot{\beta}}b_1\Delta\dot{\beta} + b_1 C_{lp}p + C_{lr}rb_1 + I_{xz1}\dot{r} + C_{l\delta_a}\Delta\delta_a + C_{l\delta_r}\Delta\delta_r) \tag{6.169}$$

$$\dot{r} = \frac{1}{I_{z1}}(C_{n\beta}\Delta\beta + C_{n\dot{\beta}}b_1\Delta\dot{\beta} + b_1 C_{np}p + C_{nr}rb_1 + I_{xz1}\dot{p} + C_{n\delta_a}\Delta\delta_a + C_{n\delta_r}\Delta\delta_r) \tag{6.170}$$

where $b_1 = b/2U_o$ and

$$I_{x1} = \frac{I_x}{\frac{1}{2}\rho U_o^2 Sb} \tag{6.171}$$

$$I_{z1} = \frac{I_z}{\frac{1}{2}\rho U_o^2 Sb} \tag{6.172}$$

$$I_{xz1} = \frac{I_{xz}}{\frac{1}{2}\rho U_o^2 Sb} \tag{6.173}$$

Note that some of the other variables appearing in Eqs. (6.168–6.170) are defined in Eqs. (4.464–4.467). However, Eqs. (6.169) and (6.170) are not in the standard state-space form because $\dot{\beta}$, $\dot{p}$, and $\dot{r}$ terms appear on their right-hand sides. To get around this problem, we proceed as follows. From Eqs. (4.396) and (4.398), we have

$$\Delta C_l = \dot{p} I_{x1} - I_{xz1}\dot{r} \tag{6.174}$$

$$\Delta C_n = \dot{r} I_{z1} - I_{xz1}\dot{p} \tag{6.175}$$

Multiply Eq. (6.174) by I_{z1} and Eq. (6.175) by I_{xz1}, add the two resulting equations, and simplify to obtain

$$\dot{p} = I'_{z1}\Delta C_l + I'_{xz1}\Delta C_n \tag{6.176}$$

$$\dot{r} = I'_{xz1}\Delta C_l + I'_{x1}\Delta C_n \tag{6.177}$$

where

$$I'_{x1} = \frac{I_{x1}}{I_{x1}I_{z1} - I_{xz1}^2} \tag{6.178}$$

$$I'_{z1} = \frac{I_{z1}}{I_{x1}I_{z1} - I_{xz1}^2} \tag{6.179}$$

$$I'_{xz1} = \frac{I_{xz1}}{I_{x1}I_{z1} - I_{xz1}^2} \tag{6.180}$$

Then,

$$\begin{aligned}\dot{p} &= (C_{l\beta}I'_{z1} + C_{n\beta}I'_{xz1})\Delta\beta + (C_{l\dot{\beta}}b_1 I'_{z1} + C_{n\dot{\beta}}b_1 I'_{xz1})\Delta\dot{\beta} \\ &\quad + (C_{lp}b_1 I'_{z1} + C_{np}b_1 I'_{xz1})p + (C_{lr}b_1 I'_{z1} + C_{nr}b_1 I'_{xz1})r \\ &\quad + (C_{l\delta_a}I'_{z1} + C_{n\delta_a}I'_{xz1})\Delta\delta_a + (C_{l\delta_r}I'_{z1} + C_{n\delta_r}I'_{xz1})\Delta\delta_r\end{aligned} \tag{6.181}$$

$$\dot{r} = (C_{n\beta}I'_{x1} + C_{l\beta}I'_{xz1})\Delta\beta + (C_{n\dot{\beta}}b_1I'_{x1} + C_{l\dot{\beta}}b_1I'_{xz1})\Delta\dot{\beta}$$
$$+ (C_{np}b_1I'_{x1} + C_{lp}b_1I'_{xz1})p + (C_{nr}b_1I'_{x1} + C_{lr}b_1I'_{xz1})r$$
$$+ (C_{n\delta_a}I'_{x1} + C_{l\delta_a}I'_{xz1})\Delta\delta_a + (C_{n\delta_r}I'_{x1} + C_{l\delta_r}I'_{xz1})\Delta\delta_r \qquad (6.182)$$

Now we have to substitute for $\dot{\beta}$ from Eq. (6.168) in Eqs. (6.181) and (6.182) and do some simplification. With this, Eqs. (6.168), (6.181), and (6.182) can be expressed in the standard state-space form as follows:

Let

$$x = \begin{bmatrix} x_1 \\ x_2 \\ x_3 \\ x_4 \\ x_5 \end{bmatrix} = \begin{bmatrix} \Delta\beta \\ \Delta\phi \\ p \\ \Delta\psi \\ r \end{bmatrix} \qquad (6.183)$$

$$u = \begin{bmatrix} \Delta\delta_a \\ \Delta\delta_r \end{bmatrix} \qquad (6.184)$$

so that

$$\dot{x} = Ax + Bu \qquad (6.185)$$

where

$$A = \begin{bmatrix} a_{11} & a_{12} & a_{13} & a_{14} & a_{15} \\ a_{21} & a_{22} & a_{23} & a_{24} & a_{25} \\ a_{31} & a_{32} & a_{33} & a_{34} & a_{35} \\ a_{41} & a_{42} & a_{43} & a_{44} & a_{45} \\ a_{51} & a_{52} & a_{53} & a_{54} & a_{55} \end{bmatrix} \qquad (6.186)$$

$$B = \begin{bmatrix} b_{11} & b_{12} \\ b_{21} & b_{22} \\ b_{31} & b_{32} \\ b_{41} & b_{42} \end{bmatrix} \qquad (6.187)$$

and

$$a_{11} = \frac{C_{y\beta}}{m_1 - b_1C_{y\dot{\beta}}} \quad a_{12} = \frac{C_{y\phi}}{m_1 - b_1C_{y\dot{\beta}}}$$

$$a_{13} = \frac{C_{yp}b_1}{m_1 - b_1C_{y\dot{\beta}}} \quad a_{14} = 0 \quad a_{15} = -\left(\frac{m_1 - b_1C_{yr}}{m_1 - b_1C_{y\dot{\beta}}}\right)$$

$$a_{21} = 0 \quad a_{22} = 0$$

$$a_{23} = 1 \quad a_{24} = 0 \quad a_{25} = 0$$

$$a_{31} = C_{l\beta}I'_{z1} + C_{n\beta}I'_{xz1} + \xi_1 b_1 a_{11} \quad a_{32} = \xi_1 b_1 a_{12}$$

$$a_{33} = C_{lp}b_1 I'_{z1} + C_{np}I'_{xz1}b_1 + \xi_1 b_1 a_{13} \quad a_{34} = 0$$

$$a_{35} = C_{lr}b_1 I'_{z1} + C_{nr}I'_{xz1}b_1 + \xi_1 b_1 a_{15}$$

$$a_{41} = a_{42} = a_{43} = a_{44} = 0 \quad a_{45} = 1$$

$$a_{51} = I'_{x1}C_{n\beta} + I'_{xz1}C_{l\beta} + b_1 \xi_2 a_{11}$$

$$a_{52} = \xi_2 b_1 a_{12} \quad a_{53} = b_1(C_{np}I'_{x1} + C_{lp}I'_{xz1} + \xi_2 a_{13})$$

$$a_{54} = 0 \quad a_{55} = b_1(I'_{x1}C_{nr} + I'_{xz1}C_{lr} + \xi_2 a_{15})$$

$$b_{11} = \frac{C_{y\delta_a}}{(m_1 - b_1 C_{y\dot{\beta}})} \quad b_{12} = \frac{C_{y\delta_r}}{(m_1 - b_1 C_{y\dot{\beta}})}$$

$$b_{21} = 0 \quad b_{22} = 0$$

$$b_{31} = C_{l\delta_a}I'_{z1} + C_{n\delta_a}I'_{xz1} + \xi_1 b_1 b_{11} \quad b_{32} = C_{l\delta_r}I'_{z1} + C_{n\delta_r}I'_{xz1} + \xi_1 b_1 b_{12}$$

$$b_{41} = 0 \quad b_{42} = 0$$

$$b_{51} = C_{n\delta_a}I'_{x1} + C_{l\delta_a}I'_{xz1} + \xi_2 b_1 b_{11} \quad b_{52} = C_{n\delta_r}I'_{x1} + C_{l\delta_r}I'_{xz1} + \xi_2 b_1 b_{12}$$

where

$$\xi_1 = I'_{z1}C_{l\dot{\beta}} + I'_{xz1}C_{n\dot{\beta}}$$

$$\xi_2 = I'_{x1}C_{n\dot{\beta}} + I'_{xz1}C_{l\dot{\beta}}$$

For free response, $u_c = 0$ so that the previous equation reduces to

$$\dot{x} = Ax \tag{6.188}$$

As said earlier, for nontrivial solution, the determinant of $(\lambda I - A)$ must be zero

$$|\lambda I - A| = 0 \tag{6.189}$$

An expansion of the determinant results in a fifth-order algebraic equation of the form

$$A_{\delta,\text{lat}}\lambda^5 + B_{\delta,\text{lat}}\lambda^4 + C_{\delta,\text{lat}}\lambda^3 + D_{\delta,\text{lat}}\lambda^2 + E_{\delta,\text{lat}}\lambda = 0 \tag{6.190}$$

where

$$A_{\delta_{lat}} = (m_1 - b_1 C_{y\dot{\beta}})(I_{x1} I_{z1} - I_{xz1}^2) \tag{6.191}$$

$$\begin{aligned} B_{\delta_{lat}} &= (-C_{y\beta})(I_{x1} I_{z1} - I_{xz1}^2) - (m_1 - b_1 C_{y\dot{\beta}})(I_{x1} C_{nr} b_1 + I_{z1} C_{lp} b_1 \\ &+ I_{xz1} C_{lr} b_1 + I_{xz1} C_{np} b_1) - b_1 C_{yp}(C_{l\dot{\beta}} b_1 I_{z1} + C_{n\dot{\beta}} b_1 I_{xz1}) \\ &+ (m_1 - b_1 C_{yr})(C_{l\dot{\beta}} b_1 I_{xz1} + I_{x1} C_{n\dot{\beta}} b_1) \end{aligned} \tag{6.192}$$

$$\begin{aligned} C_{\delta_{lat}} &= b_1^2 (m_1 - b_1 C_{y\dot{\beta}})(C_{lp} C_{nr} - C_{lr} C_{np}) + b_1 C_{y\beta}(I_{x1} C_{nr} + I_{z1} C_{lp} \\ &+ I_{xz1} C_{lr} + I_{xz1} C_{np} - C_{y\phi} b_1 (C_{l\dot{\beta}} I_{z1} + C_{n\dot{\beta}} I_{xz1}) + b_1 C_{yp} \\ &\times (C_{l\dot{\beta}} C_{nr} b_1^2 - C_{l\beta} I_{z1} - I_{xz1} C_{n\beta} - C_{lr} C_{n\dot{\beta}} b_1^2) + (m_1 - b_1 C_{yr}) \\ &\times (C_{l\beta} I_{xz1} + C_{l\dot{\beta}} C_{np} b_1^2 - C_{lp} C_{n\dot{\beta}} b_1^2 + I_{x1} C_{n\beta}) \end{aligned} \tag{6.193}$$

$$\begin{aligned} D_{\delta_{lat}} &= -C_{y\beta} b_1^2 (C_{lp} C_{nr} - C_{lr} C_{np}) + C_{y\phi}(C_{l\dot{\beta}} C_{nr} b_1^2 - C_{l\beta} I_{z1} \\ &- I_{xz1} C_{n\beta} - C_{lr} C_{n\dot{\beta}} b_1^2) + b_1^2 C_{yp}(C_{l\beta} C_{nr} - C_{lr} C_{n\beta}) \\ &+ b_1 (m_1 - b_1 C_{yr})(C_{l\beta} C_{np} - C_{lp} C_{n\beta}) \end{aligned} \tag{6.194}$$

$$E_{\delta_{lat}} = C_{y\phi} b_1 (C_{l\beta} C_{nr} - C_{n\beta} C_{lr}) \tag{6.195}$$

Equation (6.190) is also called the lateral-directional characteristic equation and gives five values of the root λ. One of them is the zero root corresponding to neutral stability to a disturbance only in angle of yaw. For conventional airplanes, the other roots are two real roots and a pair of complex conjugate roots. One real root is large and negative and corresponds to the heavily damped roll subsidence mode. The other real root is small and may be positive or negative, and the response associated with this root is called the spiral mode. If this real root is positive, then the spiral mode is slowly divergent. The motion associated with the pair of complex roots is called the Dutch-roll oscillation. The damping ratio and frequency of the Dutch roll depend on the type of airplane.

To illustrate the nature of lateral-directional response, let us consider the general aviation airplane once again. In addition to the data given earlier, we have the following data for the lateral-directional aerodynamic coefficients of this airplane [2]: $C_{y\beta} = -0.564$, $C_{y\delta_a} = 0$, $C_{y\delta_r} = 0.157$, $C_{l\beta} = -0.074$, $C_{lp} = -0.410$, $C_{lr} = 0.107$, $C_{l\delta_a} = -0.1342$, $C_{l\delta_r} = 0.0118$, $C_{n\beta} = 0.0701$, $C_{np} = -0.0575$, $C_{nr} = -0.125$, $C_{n\delta_a} = 0.00346$, $C_{n\delta_r} = -0.0717$, $C_{y\dot{\beta}} = 0$, $C_{l\dot{\beta}} = 0$, $C_{n\dot{\beta}} = 0$, $C_{yp} = 0$, and $C_{yr} = 0$. Note that all the derivatives are per radian. Furthermore, as before, we assume $M = 0.158$ and $\rho = 1.225\ \mathrm{kg/m^3}$.

Substituting these values and other parameters for the general aviation airplane in Eqs. (6.186) and (6.187) and assuming that the airplane is in equilibrium level flight ($\theta_o = 0$) before encountering a lateral-directional disturbance, we obtain

$$A = \begin{bmatrix} -0.2557 & 0.1820 & 0 & 0 & -1.0000 \\ 0 & 0 & 1.000 & 0 & 0 \\ -16.1572 & 0 & -8.4481 & 0 & 2.2048 \\ 0 & 0 & 0 & 0 & 1.0000 \\ 4.5440 & 0 & -0.3517 & 0 & -0.7647 \end{bmatrix} \tag{6.196}$$

$$B = \begin{bmatrix} 0 & 0.0712 \\ 0 & 0 \\ -29.3013 & 2.5764 \\ 0 & 0 \\ 0.2243 & -4.6477 \end{bmatrix} \tag{6.197}$$

Note that control derivatives are per radian.

Using MATLAB [1], we obtain the eigenvalues of matrix A in Eq. (6.196) as

$$\lambda_1 = 0 \quad \text{(Neutral Stability in Yaw)} \tag{6.198}$$

$$\lambda_2 = -8.4804 \quad \text{(Roll Subsidence)} \tag{6.199}$$

$$\lambda_3 = -0.0087 \quad \text{(Spiral Mode)} \tag{6.200}$$

$$\lambda_{4,5} = -0.4897 \pm j2.3468 \quad \text{(Dutch Roll)} \tag{6.201}$$

The damping ratio ζ, natural frequency ω_n, period T, and time for half amplitude t_a for the Dutch-roll oscillation are obtained using Eqs. (6.35–6.40) as follows:

$$\zeta\omega_n = 0.4897 \tag{6.202}$$

$$\omega_n\sqrt{1-\zeta^2} = 2.3468 \tag{6.203}$$

$$T = \frac{2\pi}{\omega_n\sqrt{1-\zeta^2}} \tag{6.204}$$

$$t_a = \frac{0.69}{\zeta\omega_n} \tag{6.205}$$

so that $\zeta = 0.2043$, $\omega_n = 2.3973$ rad/s, $T = 2.6773$ s, and $t_a = 1.4090$ s.

The free response of the general aviation airplane to various assumed initial conditions is shown in Figs. 6.13–6.16. From Fig. 6.13, we observe that the response to an initial disturbance in sideslip of 5 deg exhibits oscillatory motion due to the Dutch-roll roots. The disturbance in sideslip decays

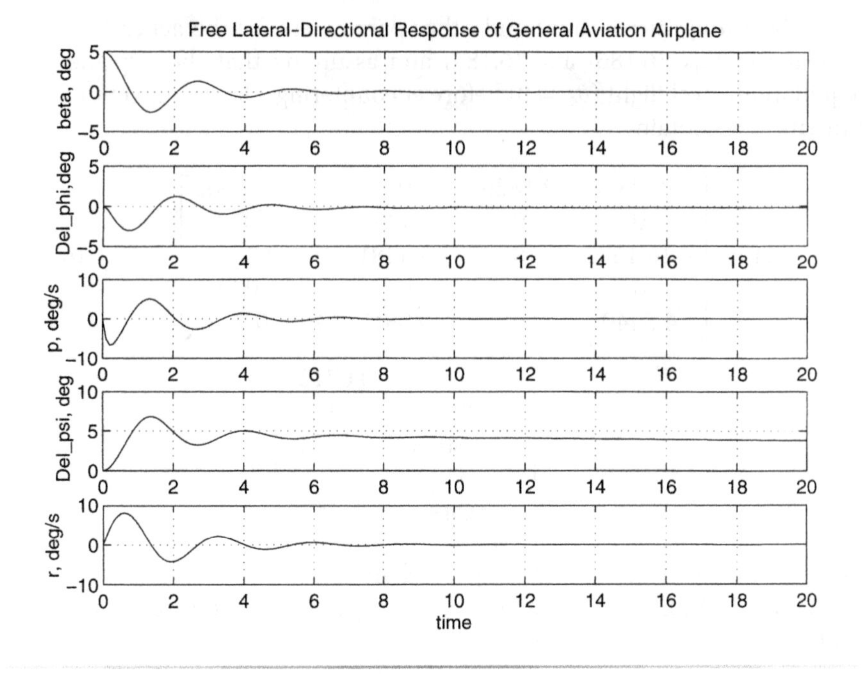

Fig. 6.13 Free (lateral/directional) response of the general aviation airplane to a disturbance in sideslip.

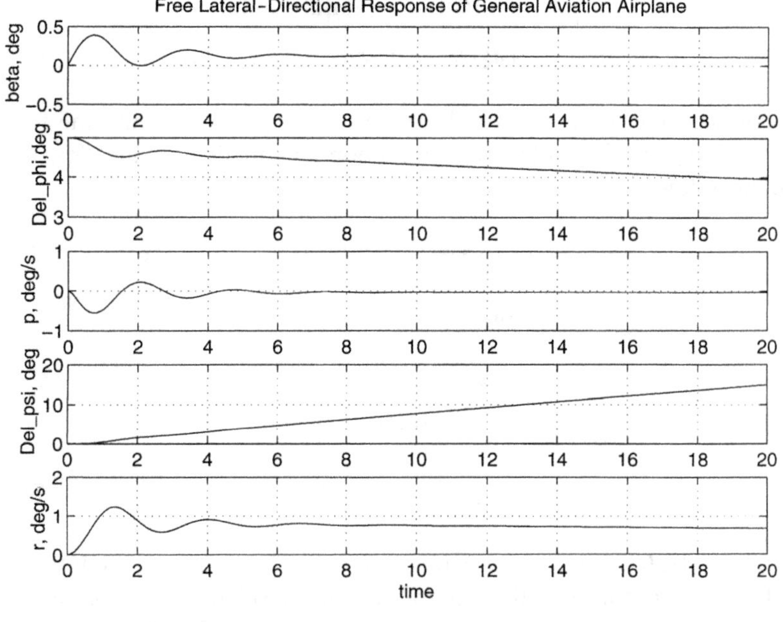

Fig. 6.14 Free (lateral/directional) response of the general aviation airplane to a disturbance in bank angle.

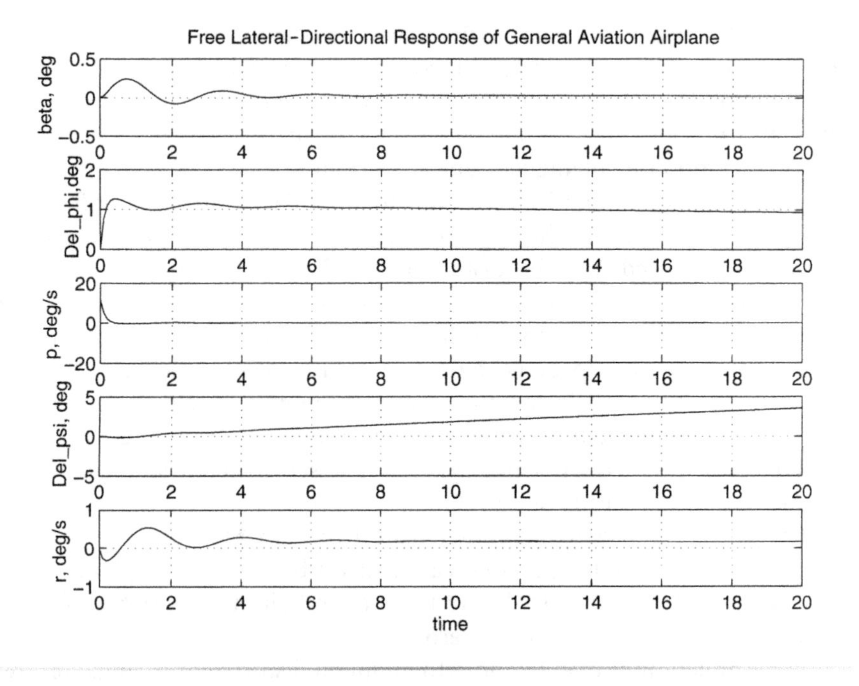

Fig. 6.15 Free (lateral/directional) response of the general aviation airplane to a disturbance in roll rate.

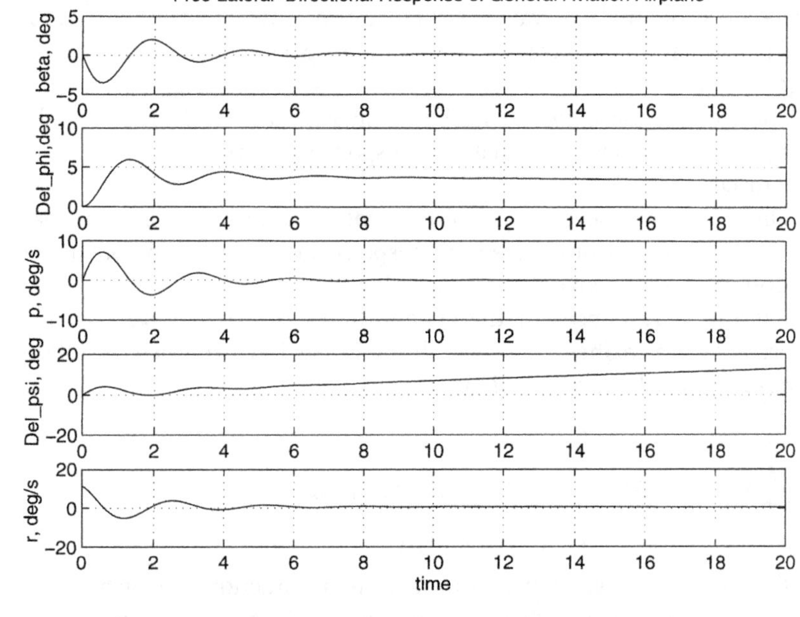

Fig. 6.16 Free (lateral/directional) response of the general aviation airplane to a disturbance in yaw rate.

to zero within two or three oscillations lasting for about 5 to 6 s. The disturbance in sideslip induces rolling and yawing motions, which also decay along with the sideslip. However, the angle of yaw does not go to zero. Instead, it assumes a nonzero steady-state value. This is due to the zero root, which makes the motion involving yaw angle neutrally stable.

Figure 6.14 shows the free response to an initial disturbance of 5 deg in bank angle. From a time-history plot for large values of time (not shown here), it is observed that the disturbance in bank angle took as much as 500 s to decay to zero, whereas the roll rate vanishes rapidly. The response following the heavily damped roll subsidence mode is the Dutch roll during which the sideslip, bank angle, and yaw rate display oscillatory behavior. Once the Dutch roll decays, the subsequent motion is a slow convergence and is called spiral mode. During the spiral mode, the bank angle and yaw rates decay to zero slowly. Once again, it is observed that the angle of yaw assumes a nonzero steady-state value.

The free response to an initial disturbance in roll rate of 0.2 rad/s (Fig. 6.15) is similar to the previous case of disturbance in roll angle. The disturbance in roll rate quickly decays to zero, and all other variables, except angle of yaw, gradually approach zero.

The simulation of free response to an initial disturbance in angle of yaw of 5 deg (not shown here) shows that a disturbance involving only angle of yaw does not induce any sideslip, banking, or roll, nor increase the yaw rate. The disturbance in angle of yaw remains constant and does not decay at all. This is due to the fact that no aerodynamic force or moment depends on the angle of yaw. Mathematically, the associated root is zero, and the airplane is neutrally stable with respect to a disturbance in the angle of yaw.

The free response to a disturbance in yaw rate of 0.2 rad/s is shown in Fig. 6.16. It is observed that this response is very similar to the other cases discussed.

The free response to any arbitrary combination of intial disturbances can be constructed from these basic responses using the principle of superposition because the given system is a linear system.

These free responses of the general aviation airplane are typical of a majority of the airplanes. In view of this, it is usual to introduce the following lateral-directional approximations.

6.3.1 Lateral-Directional Approximations

Roll-Subsidence Approximation

The motion immediately following a lateral-directional disturbance is the heavily damped roll subsidence mode during which the airplane motion is predominantly rolling about the x-body axis, and, during this process, other variables vary very slowly so that we can assume $\Delta\beta = \Delta\dot{\beta} = \Delta\psi =$

$r = \dot{r} = 0$. With this assumption, the side force and yawing moment equations can be ignored. Furthermore, the rolling moment equation [Eq. (6.169)] assumes the following form:

$$I_{x1}\dot{p} - C_{lp}b_1 p = C_{l\delta_a}\Delta\delta_a + C_{l\delta_r}\Delta\delta_r \tag{6.206}$$

For free response, $\Delta\delta_a = \Delta\delta_r = 0$ so that

$$I_{x1}\dot{p} - C_{lp}b_1 p = 0 \tag{6.207}$$

or

$$\tau\dot{p} + p = 0 \tag{6.208}$$

where

$$\tau = -\frac{I_{x1}}{C_{lp}b_1} \tag{6.209}$$

The parameter τ is the time constant for the rolling motion of the airplane. Assuming $p = p_o e^{\lambda_r t}$ and substituting in Eq. (6.208), we get

$$\lambda_r = -\frac{1}{\tau} \tag{6.210}$$

$$= \frac{C_{lp}b_1}{I_{x1}} \tag{6.211}$$

Usually, for angles of attack below stall $C_{lp} < 0$. Therefore, λ_r is real. Usually, λ_r has a large negative value. As a result, the roll subsidence mode is well damped and is hardly felt by the pilot or passengers.

Dutch-Roll Approximation

The oscillatory motion following the heavily damped roll subsidence mode is the lateral-directional oscillatory motion known as the Dutch roll. The aircraft trajectory in a Dutch-roll mode is schematically shown in Fig. 6.17. It involves mainly sideslip and yawing motions. However, it is possible that there is some rolling motion due to aerodynamic roll–yaw coupling as we have observed earlier for the general aviation airplane. If the rolling motion is small and can be neglected ($\Delta\phi = p = 0$), we can ignore the rolling moment equation. Then, Eqs. (6.168) and (6.170) reduce to the following form:

$$(m_1 - b_1 C_{y\dot{\beta}})\frac{\mathrm{d}\beta}{\mathrm{d}t} = C_{y\beta}\Delta\beta - (m_1 - b_1 C_{yr})r + C_{y\delta_a}\Delta\delta_a + C_{y\delta_r}\Delta\delta_r \tag{6.212}$$

$$I_{z1}\dot{r} = C_{n\beta}\Delta\beta + C_{n\dot{\beta}}b_1\Delta\dot{\beta} + C_{nr}b_1 r + C_{n\delta_a}\Delta\delta_a + C_{n\delta_r}\Delta\delta_r \tag{6.213}$$

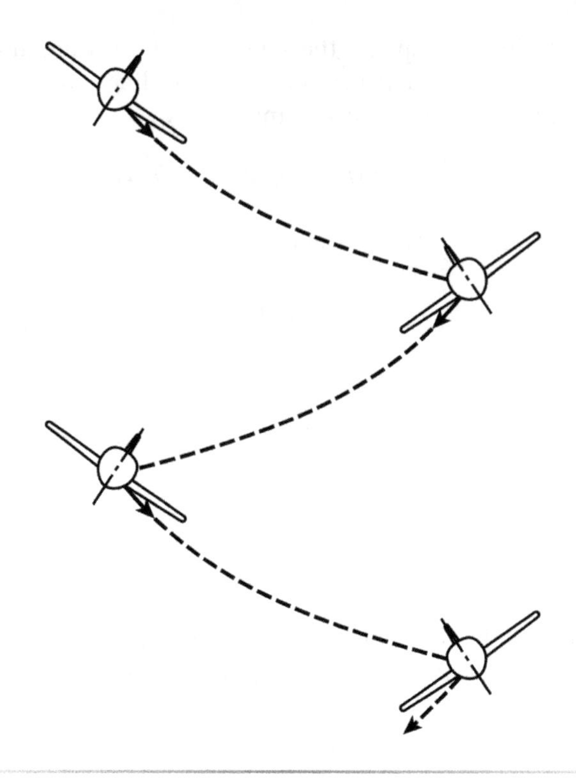

Fig. 6.17 Physical motion of the airplane during a Dutch roll.

Substituting for $\Delta\dot{\beta}$ from Eq. (6.212) in Eq. (6.213) and simplifying, we get these two equations in state-space form

$$\begin{bmatrix} \Delta\dot{\beta} \\ \dot{r} \end{bmatrix} = \begin{bmatrix} a_{11} & a_{12} \\ a_{21} & a_{22} \end{bmatrix} \begin{bmatrix} \Delta\beta \\ r \end{bmatrix} + \begin{bmatrix} b_{11} & b_{12} \\ b_{21} & b_{22} \end{bmatrix} \begin{bmatrix} \Delta\delta_a \\ \Delta\delta_r \end{bmatrix} \tag{6.214}$$

where

$$a_{11} = \frac{C_{y\beta}}{m_1 - C_{y\dot{\beta}} b_1} \tag{6.215}$$

$$a_{12} = -\frac{m_1 - b_1 C_{yr}}{m_1 - C_{y\dot{\beta}} b_1} \tag{6.216}$$

$$a_{21} = \frac{1}{I_{z1}} \left(C_{n\beta} + \frac{C_{n\dot{\beta}} b_1 C_{y\beta}}{m_1 - C_{y\dot{\beta}} b_1} \right) \tag{6.217}$$

$$a_{22} = \frac{1}{I_{z1}} \left(\frac{-C_{n\dot{\beta}} b_1 (m_1 - b_1 C_{yr})}{m_1 - C_{y\dot{\beta}} b_1} + b_1 C_{nr} \right) \tag{6.218}$$

and

$$b_{11} = \frac{C_{y\delta_a}}{m_1 - C_{y\dot{\beta}} b_1}$$

$$b_{12} = \frac{C_{y\delta_r}}{m_1 - C_{y\dot{\beta}} b_1}$$

$$b_{21} = \frac{1}{I_{z1}} \left(C_{n\delta_a} + \frac{C_{n\dot{\beta}} b_1 C_{y\delta_a}}{m_1 - C_{y\dot{\beta}} b_1} \right)$$

$$b_{22} = \frac{1}{I_{z1}} \left(C_{n\delta_r} + \frac{C_{n\dot{\beta}} b_1 C_{y\delta_r}}{m_1 - C_{y\dot{\beta}} b_1} \right)$$

For free response, $\Delta\delta_a = \Delta\delta_r = 0$.

To have an understanding of the physical parameters that have a major effect on the damping ratio and frequency of the Dutch-roll oscillation, let us introduce some additional simplifications. Usually $C_{y\dot{\beta}}$ and $C_{n\dot{\beta}}$ are small so that we can assume $C_{y\dot{\beta}} = C_{n\dot{\beta}} = 0$. With this, it can be shown that the characteristic equation corresponding to the system given by Eq. (6.214) is given by

$$\lambda^2 + B\lambda + C = 0 \tag{6.219}$$

where

$$B = -\left(\frac{C_{y\beta}}{m_1} + \frac{b_1 C_{nr}}{I_{z1}} \right) \tag{6.220}$$

$$C = \left(\frac{1}{m_1 I_{z1}} \right) \left[C_{y\beta} C_{nr} b_1 + C_{n\beta} (m_1 - b_1 C_{yr}) \right] \tag{6.221}$$

Comparing Eq. (6.219) with the standard second-order Eq. (6.34), we obtain

$$\omega_n = \sqrt{C} = \sqrt{\left(\frac{1}{m_1 I_{z1}} \right) \left[C_{y\beta} C_{nr} b_1 + C_{n\beta} (m_1 - b_1 C_{yr}) \right]} \tag{6.222}$$

$$\zeta = \frac{B}{2\omega_n} = -\left(\frac{1}{2\omega_n} \right) \left(\frac{C_{y\beta}}{m_1} + \frac{b_1 C_{nr}}{I_{z1}} \right) \tag{6.223}$$

Usually, $C_{y\beta} < 0$ (side force due to sideslip), $C_{n\beta} > 0$ (static directional stability parameter), $C_{yr} > 0$ (side force due to yaw rate), and $C_{nr} < 0$ (damping in yaw). Generally, C_{yr} is small so that $m_1 > b_1 C_{yr}$. The major contribution to the frequency term comes from $C_{n\beta}$, and that for damping ratio comes from C_{nr}.

Spiral Mode

For the general aviation airplane, we have observed that, during the slow convergence corresponding to the small negative real root, the sideslip varies very slowly so that $\Delta\dot{\beta} \simeq 0$. Hence, the side force equation can be ignored. Furthermore, the roll rate is practically zero during this slow spiral motion so that the net rolling moment must be zero. In addition, we ignore the contribution due to the product of inertia term I_{xz1}. With these simplifying assumptions, the rolling moment and yawing moment Eqs. (6.169) and (6.170) reduce to

$$C_{l\beta}\Delta\beta + C_{lr}b_1 r + C_{l\delta a}\Delta\delta_a + C_{l\delta r}\Delta\delta_r = 0 \tag{6.224}$$

$$I_{z1}\dot{r} = C_{n\beta}\Delta\beta + C_{nr}rb_1 + C_{n\delta_a}\Delta\delta_a + C_{n\delta_r}\Delta\delta_r \tag{6.225}$$

For free response, $\Delta\delta_a = \Delta\delta_r = 0$.

Rearranging Eqs. (6.224) and (6.225), we get

$$\dot{r} = \frac{(-C_{n\beta}C_{lr} + C_{nr}C_{l\beta})b_1 r}{I_{z1}C_{l\beta}} \tag{6.226}$$

We can obtain an analytical solution to this equation. Let $r = r_o e^{\lambda_{sp}t}$ so that $\dot{r} = r_o\lambda_{sp}e^{\lambda_{sp}t}$. Then,

$$r_o\lambda_{sp}e^{\lambda_{sp}t} = \frac{(-C_{n\beta}C_{lr} + C_{nr}C_{l\beta})}{I_{z1}C_{l\beta}} b_1 r_o e^{\lambda_{sp}t} \tag{6.227}$$

or

$$\lambda_{sp} = \frac{b_1(C_{l\beta}C_{nr} - C_{lr}C_{n\beta})}{I_{z1}C_{l\beta}} \tag{6.228}$$

Usually, $C_{l\beta} < 0$ (stable dihedral effect) and $C_{nr} < 0$ (positive damping in yaw) so that the term $C_{l\beta}C_{nr} > 0$. Also, $C_{n\beta} > 0$ (directionally stable) and $C_{lr} > 0$ (positive roll due to positive yaw) so that $C_{lr}C_{n\beta} > 0$. Therefore, when $|C_{l\beta}C_{nr}| > |C_{lr}C_{n\beta}|$, $\lambda_{sp} < 0$, and the spiral mode will be stable. Therefore, for spiral stability, we must have $C_{l\beta}C_{nr} > C_{lr}C_{n\beta}$. In other words, the airplane should have a relatively higher level of lateral stability or dihedral effect compared to the static directional stability. If not, it is possible that the spiral mode becomes unstable. However, it should be remembered that this inference is based on spiral approximation and not on the complete fifth-order lateral-directional system.

The physical motion of an aircraft experiencing a spiral divergence can be approximately described as follows.

On encountering a gust that raises one wing relative to the other, the aircraft banks, the nose drops ($L \cos \phi < W$), and the aircraft develops sideslip in the direction of the lower wing due to $L \sin \phi$ component. Let us assume that the left wing is raised, the right wing is dropped, and the aircraft sideslips to the right. Because the aircraft has a low level of lateral stability or only a small stable value of $C_{l\beta}$, it cannot come to the wing level condition immediately. Instead, because of the strong directional stability ($C_{n\beta} > 0$), it yaws to the right to orient the nose in the direction of the relative wind. In this process it develops a positive yaw rate and, because of this, it further rolls to the right due to C_{lr} because C_{lr} is usually positive. This rolling motion increases the sideslip further. Because the lateral stability $C_{l\beta}$ is insufficient, the condition worsens, with the bank angle continuously increasing and the nose dropping further. The net result is that the aircraft is losing altitude, gaining air speed, and banking more and more to the right with an ever-increasing turn rate. The aircraft is essentially in a tightening spiral motion. What distinguishes the spiral motion from a spin is that the angle of attack is below the stall and control surfaces are still effective.

To avoid the spiral divergence, we have to improve spiral stability of the aircraft while keeping a check on the level of directional stability. This can be done by increasing the dihedral effect (for example increase wing dihedral) and simultaneously keep the static directional stability level to an acceptable minimum.

6.3.2 Accuracy of Lateral-Directional Approximations

Generally, the accuracy of these approximations depends on type of the airplane. It may so happen that for one aircraft these approximations may be satisfactory but for the other they can be considerably in error.

To have an idea of the accuracy of lateral-directional approximations, let us refer to Table 6.1, which gives various values of the lateral-directional roots based on the complete fifth-order system, roll subsidence, Dutch roll, and spiral mode approximations for the general aviation airplane.

Table 6.1 Lateral-Directional Roots of the General Aviation Airplane

Mode	Approximation	Complete system	Comment
Roll subsidence, λ_r	-8.4481	-8.4804	Excellent
Dutch roll, λ_{dr}	$-0.5102 \pm j2.1164$	-0.4897 ± 2.3468	Good
Spiral mode	-0.1446	-0.0087	Poor

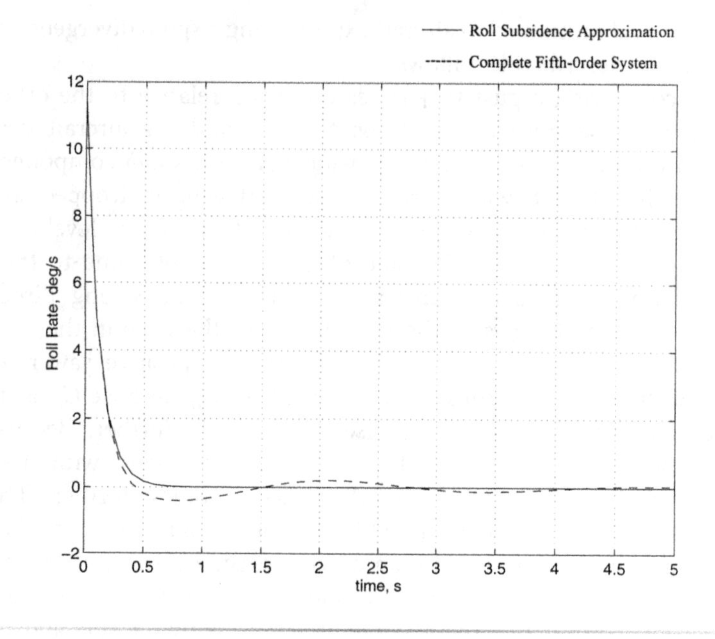

Fig. 6.18 Free response (lateral-directional) of the general aviation airplane to an initial disturbance in roll rate of 0.2 rad/s (11.5 deg/s).

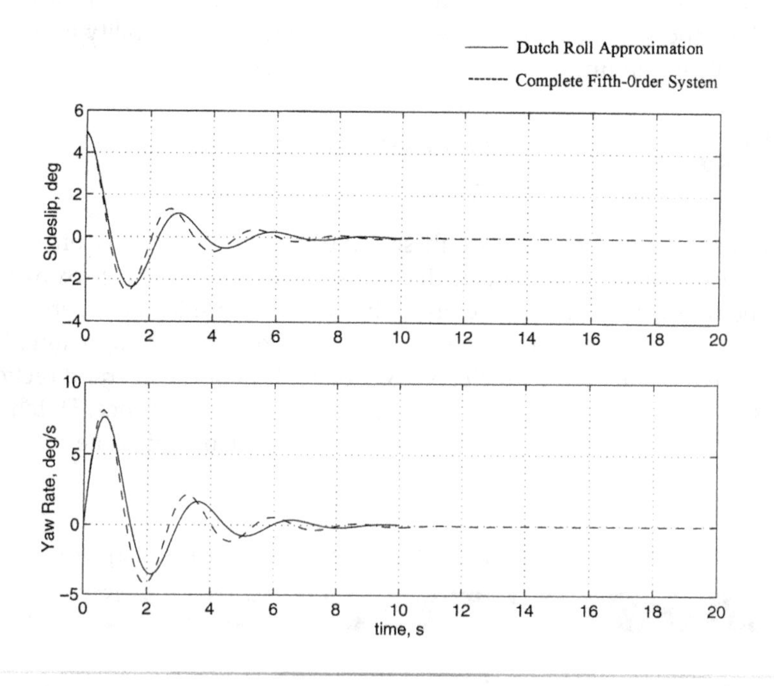

Fig. 6.19 Free response (lateral-directional) of the general aviation airplane to an initial sideslip disturbance of 5 deg.

We observe that the roll subsidence and Dutch-roll approximations are satisfactory but the spiral approximation is in poor agreement with the complete fifth-order lateral-directional system.

The computed free responses based on the complete fifth-order system and lateral-directional approximations for the general aviation airplane are shown in Figs. 6.18–6.20. It may be observed that the motion corresponding to spiral root differs considerably.

The unit-step responses to the rudder input based on complete fifth-order system and lateral-directional approximations for the first 10 seconds are shown in Figs. 6.21–6.25. It is observed that the transient behaviors are in fair agreement but it is evident that the steady-state values differ considerably. The reason for large differences in steady-state values is that these lateral-directional approximations are primarily aimed at predicting the free response or the dynamic stability of the lateral-directional motion of the airplane. Therefore, they do a reasonably good job of predicting the poles but not the zeros of the lateral-directional transfer function. We know that the steady-state values also depend on the zeros (see Chapter 5). Because the zeros of the approximate transfer functions are not the same as the zeros of the complete fifth-order transfer function, the steady-state values differ significantly.

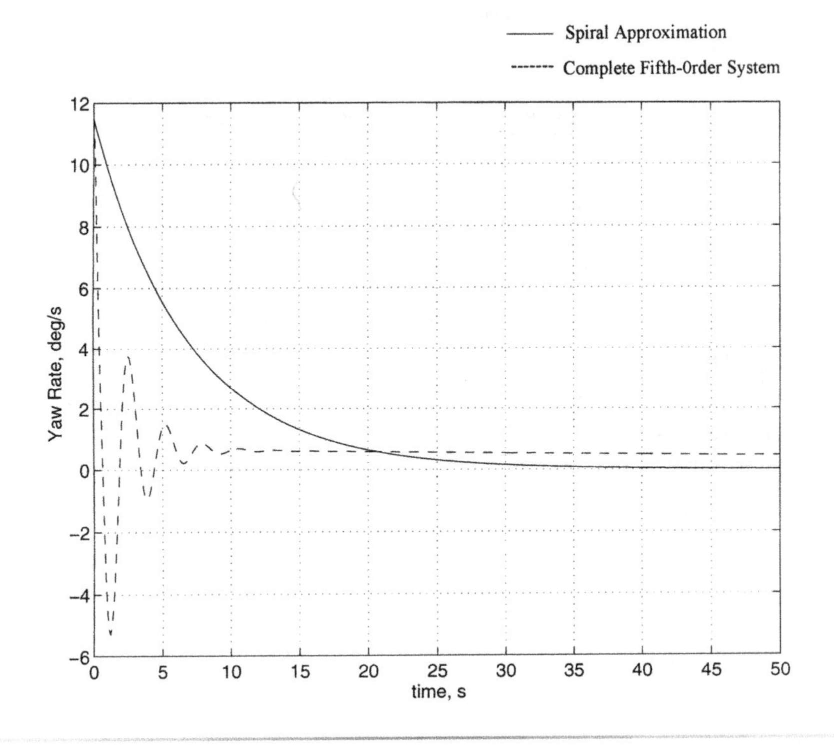

Fig. 6.20 Free response (lateral-directional) of the general aviation airplane to an initial yaw rate disturbance of 0.2 rad/s (11.5 deg/s).

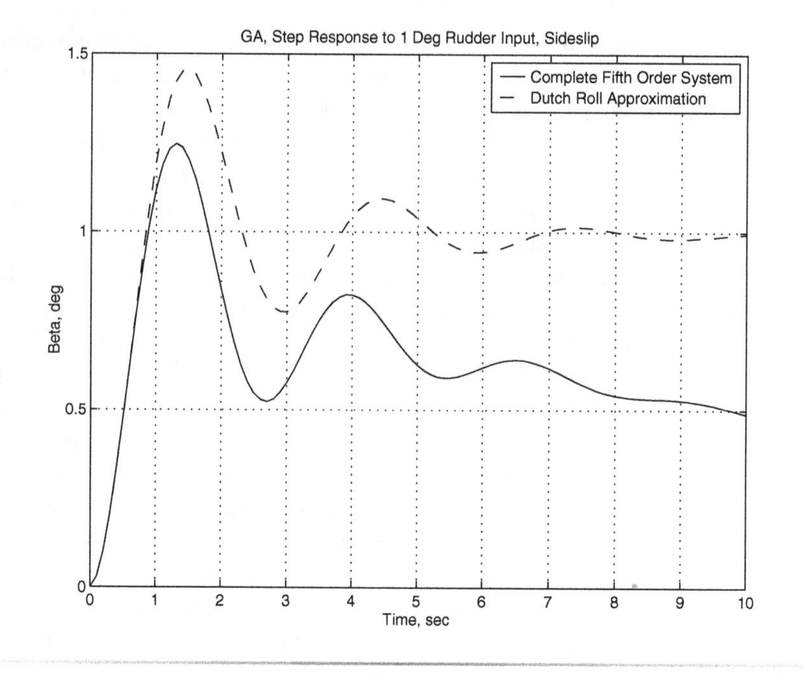

Fig. 6.21 Unit-step response (lateral-directional) of the general aviation airplane: sideslip.

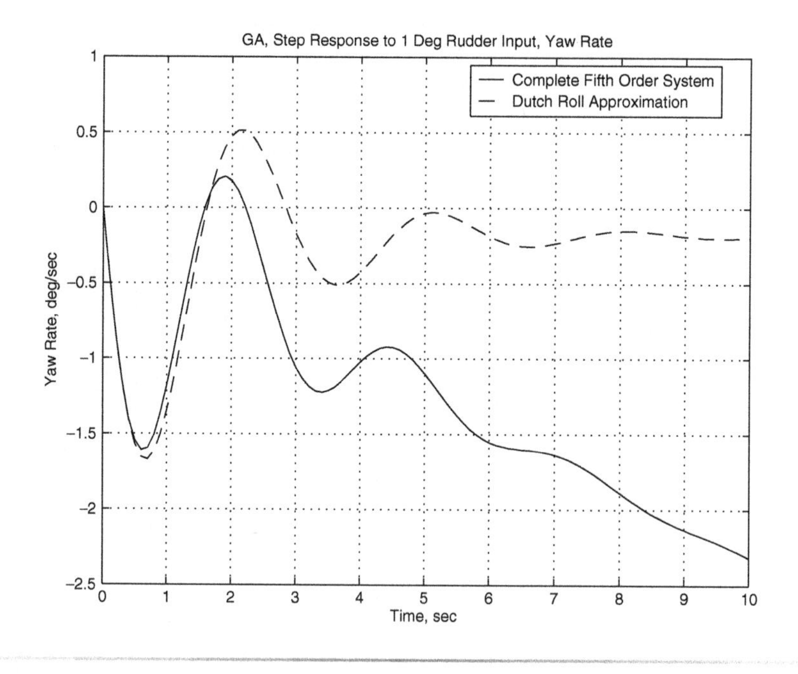

Fig. 6.22 Unit-step response (lateral-directional) of the general aviation airplane: yaw rate.

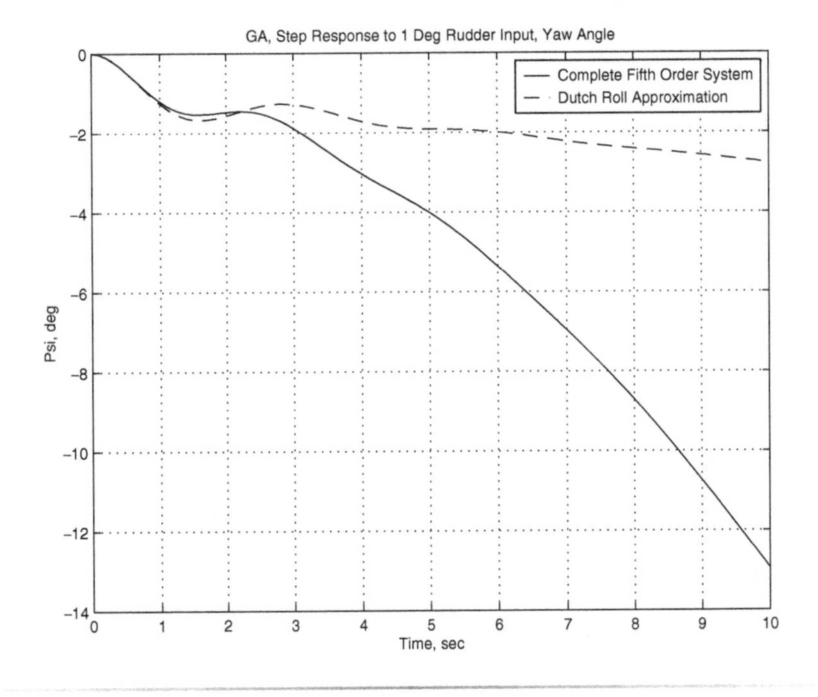

Fig. 6.23 Unit-step response (lateral-directional) of the general aviation airplane: angle of yaw.

6.3.3 Lateral-Directional Transfer Functions

Equations (6.168–6.170) for lateral-directional motion based on the assumption of small disturbance can be expressed as

$$\left(m_1\frac{\mathrm{d}}{\mathrm{d}t} - b_1C_{y\dot{\beta}}\frac{\mathrm{d}}{\mathrm{d}t} - C_{y\beta}\right)\Delta\beta - \left(b_1C_{yp}\frac{\mathrm{d}}{\mathrm{d}t} + C_{y\phi}\right)\Delta\phi$$
$$+ \left(m_1 - b_1C_{yr}\right)\frac{\mathrm{d}\Delta\psi}{\mathrm{d}t} = C_{y\delta_a}\Delta\delta_a + C_{y\delta_r}\Delta\delta_r \tag{6.229}$$

$$\left(-C_{l\beta} - b_1C_{l\dot{\beta}}\frac{\mathrm{d}}{\mathrm{d}t}\right)\Delta\beta + \left(-C_{lp}b_1\frac{\mathrm{d}}{\mathrm{d}t} + I_{x1}\frac{\mathrm{d}^2}{\mathrm{d}t^2}\right)\Delta\phi$$
$$- \left(C_{lr}b_1\frac{\mathrm{d}}{\mathrm{d}t} + I_{xz1}\frac{\mathrm{d}^2}{\mathrm{d}t^2}\right)\Delta\psi = C_{l\delta_a}\Delta\delta_a + C_{l\delta_r}\Delta\delta_r \tag{6.230}$$

$$\left(-C_{n\beta} - b_1C_{n\dot{\beta}}\frac{\mathrm{d}}{\mathrm{d}t}\right)\Delta\beta + \left(-C_{np}b_1\frac{\mathrm{d}}{\mathrm{d}t} - I_{xz1}\frac{\mathrm{d}^2}{\mathrm{d}t^2}\right)\Delta\phi$$
$$+ \left(-C_{nr}b_1\frac{\mathrm{d}}{\mathrm{d}t} + I_{z1}\frac{\mathrm{d}^2}{\mathrm{d}t^2}\right)\Delta\psi = C_{n\delta_a}\Delta\delta_a + C_{n\delta_r}\Delta\delta_r \tag{6.231}$$

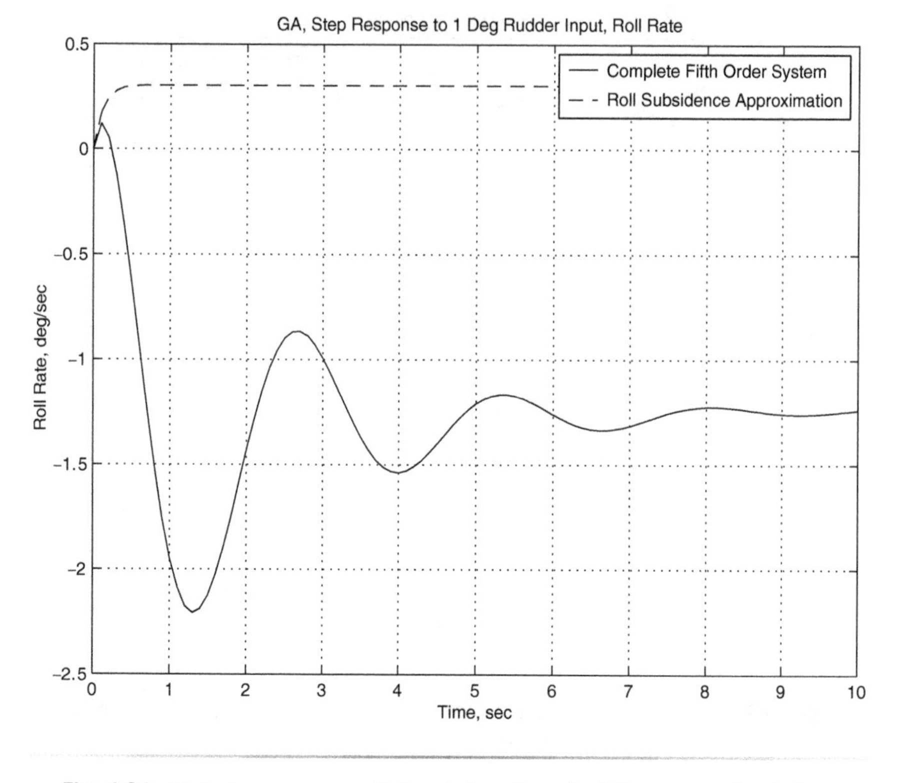

Fig. 6.24 Unit-step response (lateral-directional) of the general aviation airplane: roll rate.

In the following, we will derive the transfer function for aileron deflection assuming that the rudder is held fixed, i.e., $\Delta\delta_r = 0$. Similarly, to derive transfer functions for rudder deflection, assume $\delta_a = 0$. Using these transfer functions, the airplane response to a given aileron input can be obtained. To obtain the airplane response to rudder inputs, we need the transfer functions for rudder deflection, which can be obtained by replacing $\Delta\delta_a$ by $\Delta\delta_r$, $C_{y\delta_a}$ by $C_{y\delta_r}$, $C_{l\delta_a}$ by $C_{l\delta_r}$, and $C_{n\delta_a}$ by $C_{n\delta_r}$.

The airplane response to combined aileron and rudder deflections can be obtained by a linear addition of the two responses because the system is assumed to be linear.

Taking the Laplace transform of Eqs. (6.229–6.231), we obtain

$$(m_1 s - b_1 C_{y\dot{\beta}} s - C_{y\beta})\Delta\bar{\beta}(s) - (b_1 C_{yp} s + C_{y\phi})\Delta\bar{\phi}(s) + (m_1 - b_1 C_{yr}) s \Delta\bar{\psi}(s) = C_{y\delta_a}\Delta\bar{\delta}_a(s) \tag{6.232}$$

$$(-C_{l\beta} - b_1 C_{l\dot{\beta}} s)\Delta\bar{\beta}(s) + (-C_{lp} b_1 s + I_{x1} s^2)\Delta\bar{\phi}(s) - (C_{lr} b_1 s + I_{xz1} s^2)\Delta\bar{\psi}(s) = C_{l\delta_a}\Delta\bar{\delta}_a(s) \tag{6.233}$$

$$
\begin{aligned}
&(-C_{n\beta} - b_1 C_{n\dot{\beta}} s)\Delta\bar{\beta}(s) + (-C_{np} b_1 s - I_{xz1} s^2)\Delta\bar{\phi}(s) \\
&\quad + (-C_{nr} b_1 s + I_{z1} s^2)\Delta\bar{\psi}(s) = C_{n\bar{\delta}_a}\Delta\bar{\delta}_a(s)
\end{aligned} \tag{6.234}
$$

Dividing throughout by $\Delta\bar{\delta}_a(s)$ and using Cramer's rule, we obtain

$$
\frac{\Delta\bar{\beta}(s)}{\Delta\bar{\delta}_a(s)} = \frac{\begin{vmatrix} C_{y\delta_a} & -(b_1 C_{yp} s + C_{y\phi}) & s(m_1 - b_1 C_{yr}) \\ C_{l\delta_a} & -C_{lp} b_1 s + I_{x1} s^2 & -(C_{lr} b_1 s + I_{xz1} s^2) \\ C_{n\delta_a} & -C_{np} b_1 s - I_{xz1} s^2 & -C_{nr} b_1 s + I_{z1} s^2 \end{vmatrix}}{\begin{vmatrix} m_1 s - b_1 C_{y\dot{\beta}} s - C_{y\beta} & -(b_1 C_{yp} s + C_{y\phi}) & s(m_1 - b_1 C_{yr}) \\ -C_{l\beta} - b_1 C_{l\dot{\beta}} s & -C_{lp} b_1 s + I_{x1} s^2 & -(C_{lr} b_1 s + I_{xz1} s^2) \\ -C_{n\beta} - b_1 C_{n\dot{\beta}} s & -C_{np} b_1 s + I_{xz1} s^2 & -C_{nr} b_1 s + I_{z1} s^2 \end{vmatrix}} \tag{6.235}
$$

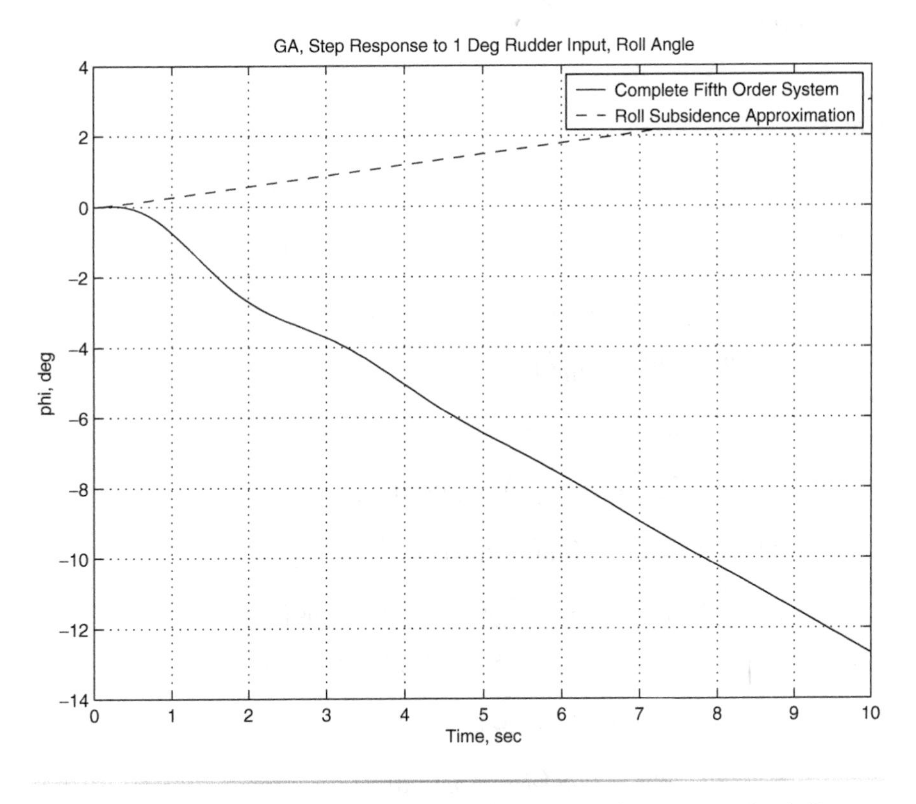

Fig. 6.25 Unit-step response (lateral-directional) of the general aviation airplane: roll angle.

Let $N_{\bar{\beta},\delta_a}$ denote the determinant in the numerator and Δ_{lat} denote the determinant in the denominator. Expanding the determinant $N_{\bar{\beta},\delta_a}$, we obtain

$$N_{\bar{\beta},\delta_a} = A_{\beta,\delta_a}s^4 + B_{\beta,\delta_a}s^3 + C_{\beta,\delta_a}s^2 + D_{\beta,\delta_a}s \tag{6.236}$$

where

$$A_{\beta,\delta_a} = C_{y\delta_a}(I_{x1}I_{z1} - I_{xz1}^2) \tag{6.237}$$

$$B_{\beta,\delta_a} = -b_1 C_{y\delta_a}(I_{x1}C_{nr} + C_{lp}I_{z1} + C_{lr}I_{xz1} + C_{np}I_{xz1}) + b_1 C_{yp}(I_{z1}C_{l\delta_a} + C_{n\delta_a}I_{xz1}) - (m_1 - b_1 C_{yr})(C_{l\delta_a}I_{xz1} + C_{n\delta_a}I_{x1}) \tag{6.238}$$

$$C_{\beta,\delta_a} = b_1^2 C_{y\delta_a}(C_{lp}C_{nr} - C_{lr}C_{np}) + C_{y\phi}(C_{l\delta_a}I_{z1} + C_{n\delta_a}I_{xz1}) - b_1^2 C_{yp}(C_{l\delta_a}C_{nr} - C_{n\delta_a}C_{lr}) + b_1(m_1 - b_1 C_{yr})(C_{n\delta_a}C_{lp} - C_{l\delta_a}C_{np}) \tag{6.239}$$

$$D_{\beta,\delta_a} = -b_1 C_{y\phi}(C_{l\delta_a}C_{nr} - C_{n\delta_a}C_{lr}) \tag{6.240}$$

Expanding the determinant Δ_{lat}, we get a fifth-order polynomial in s, which is of the form

$$\Delta_{\text{lat}} = A_{\delta_{\text{lat}}}s^5 + B_{\delta_{\text{lat}}}s^4 + C_{\delta_{\text{lat}}}s^3 + D_{\delta_{\text{lat}}}s^2 + E_{\delta_{\text{lat}}}s \tag{6.241}$$

where

$$A_{\delta_{\text{lat}}} = (m_1 - b_1 C_{y\dot{\beta}})(I_{x1}I_{z1} - I_{xz1}^2) \tag{6.242}$$

$$B_{\delta_{\text{lat}}} = (-C_{y\beta})(I_{x1}I_{z1} - I_{xz1}^2) - (m_1 - b_1 C_{y\dot{\beta}})(I_{x1}C_{nr}b_1 + I_{z1}C_{lp}b_1 + I_{xz1}C_{lr}b_1 + I_{xz1}C_{np}b_1) - b_1 C_{yp}(C_{l\dot{\beta}}b_1 I_{z1} + C_{n\dot{\beta}}b_1 I_{xz1}) + (m_1 - b_1 C_{yr})(C_{l\dot{\beta}}b_1 I_{xz1} + I_{x1}C_{n\dot{\beta}}b_1) \tag{6.243}$$

$$C_{\delta_{\text{lat}}} = b_1^2(m_1 - b_1 C_{y\dot{\beta}})(C_{lp}C_{nr} - C_{lr}C_{np}) + b_1 C_{y\beta}(I_{x1}C_{nr} + I_{z1}C_{lp} + I_{xz1}C_{lr} + I_{xz1}C_{np}) - C_{y\phi}b_1(C_{l\dot{\beta}}I_{z1} + C_{n\dot{\beta}}I_{xz1}) + b_1 C_{yp}(C_{l\dot{\beta}}C_{nr}b_1^2 - C_{l\beta}I_{z1} - I_{xz1}C_{n\beta} - C_{lr}C_{n\dot{\beta}}b_1^2) + (m_1 - b_1 C_{yr})(C_{l\beta}I_{xz1} + C_{l\dot{\beta}}C_{np}b_1^2 - C_{lp}C_{n\dot{\beta}}b_1^2 + I_{x1}C_{n\beta}) \tag{6.244}$$

$$D_{\delta_{\text{lat}}} = -C_{y\beta}b_1^2(C_{lp}C_{nr} - C_{lr}C_{np}) + C_{y\phi}(C_{l\dot{\beta}}C_{nr}b_1^2 - C_{l\beta}I_{z1} - I_{xz1}C_{n\beta} - C_{lr}C_{n\dot{\beta}}b_1^2) + b_1^2 C_{yp}(C_{l\beta}C_{nr} - C_{lr}C_{n\beta}) + b_1(m_1 - b_1 C_{yr})(C_{l\beta}C_{np} - C_{lp}C_{n\beta}) \tag{6.245}$$

$$E_{\delta_{\text{lat}}} = C_{y\phi}b_1(C_{l\beta}C_{nr} - C_{n\beta}C_{lr}) \tag{6.246}$$

The transfer function for angle of yaw is given by

$$\frac{\Delta\bar{\psi}(s)}{\Delta\bar{\delta}_a(s)} = \frac{\begin{vmatrix} m_1 s - b_1 C_{y\dot{\beta}} s - C_{y\beta} & -(b_1 C_{yp} s + C_{y\phi}) & C_{y\delta_a} \\ -(C_{l\beta} + b_1 C_{l\dot{\beta}} s) & -C_{lp} b_1 s + I_{x1} s^2 & C_{l\delta_a} \\ -(C_{n\beta} + b_1 C_{n\dot{\beta}} s) & -C_{np} b_1 s - I_{xz1} s^2 & C_{n\delta_a} \end{vmatrix}}{\Delta_{\text{lat}}} \tag{6.247}$$

Let $N_{\bar{\psi},\delta_a}$ denote the determinant in the numerator. Expanding, we get

$$N_{\bar{\psi},\delta_a} = A_{\bar{\psi}} s^3 + B_{\bar{\psi}} s^2 + C_{\bar{\psi}} s + D_{\bar{\psi}} \tag{6.248}$$

where

$$A_{\bar{\psi}} = (m_1 - b_1 C_{y\dot{\beta}})(I_{x1} C_{n\delta_a} + I_{zx1} C_{l\delta_a}) + C_{y\delta_a}(I_{zx1} C_{l\dot{\beta}} b_1 + I_{x1} C_{n\dot{\beta}} b_1) \tag{6.249}$$

$$\begin{aligned} B_{\bar{\psi}} &= -C_{y\beta}(I_{xz1} C_{n\delta_a} + I_{xz1} C_{l\delta_a}) + (m_1 - b_1 C_{y\dot{\beta}})(C_{l\delta_a} C_{np} b_1 - C_{n\delta_a} C_{lp} b_1) \\ &+ C_{yp} b_1(-C_{n\delta_a} C_{l\dot{\beta}} b_1 + C_{l\delta_a} C_{n\dot{\beta}} b_1) + C_{y\delta_a}(I_{xz1} C_{l\beta} + C_{l\dot{\beta}} C_{np} b_1^2 \\ &- C_{n\dot{\beta}} C_{lp} b_1^2 + I_{x1} C_{n\beta}) \end{aligned} \tag{6.250}$$

$$\begin{aligned} C_{\bar{\psi}} &= -C_{y\beta} b_1(C_{l\delta_a} C_{np} - C_{n\delta_a} C_{lp}) + b_1 C_{yp}(C_{n\beta} C_{l\delta_a} - C_{l\beta} C_{n\delta_a}) \\ &+ C_{y\phi}(-C_{n\delta_a} C_{l\dot{\beta}} b_1 + C_{l\delta_a} C_{n\dot{\beta}} b_1) + b_1 C_{y\delta_a}(C_{l\beta} C_{np} - C_{lp} C_{n\beta}) \end{aligned} \tag{6.251}$$

$$D_{\bar{\psi}} = C_{y\phi}(C_{n\beta} C_{l\delta_a} - C_{l\beta} C_{n\delta_a}) \tag{6.252}$$

The transfer function for the bank angle is given by

$$\frac{\Delta\bar{\phi}(s)}{\Delta\bar{\delta}_a(s)} = \frac{\begin{vmatrix} (m_1 s - b_1 C_{y\dot{\beta}} s - C_{y\beta}) & C_{y\delta_a} & m_1 s - b_1 C_{yr} s \\ -(C_{l\beta} + b_1 C_{l\dot{\beta}} s) & C_{l\delta_a} & -(C_{lr} b_1 s + I_{xz1} s^2) \\ -(C_{n\beta} + b_1 C_{n\dot{\beta}} s) & C_{n\delta_a} & -(C_{nr} b_1 s + I_{z1} s^2) \end{vmatrix}}{\Delta_{\text{lat}}} \tag{6.253}$$

Let $N_{\bar{\phi},\delta_a}$ denote the determinant in the numerator. Expanding, we get

$$N_{\bar{\phi},\delta_a} = \mathrm{A}_{\bar{\phi}} s^3 + B_{\bar{\phi}} s^2 + C_{\bar{\phi}} s \tag{6.254}$$

where

$$A_{\bar{\phi}} = (m_1 - b_1 C_{y\dot{\beta}})(I_{xz1} C_{n\delta_a} + I_{z1} C_{l\delta_a}) + C_{y\delta_a}(I_{xz1} C_{n\dot{\beta}} b_1 + I_{z1} C_{l\dot{\beta}} b_1) \tag{6.255}$$

$$B_{\bar{\phi}} = -C_{y\beta}(I_{xz1}C_{n\delta_a} + I_{z1}C_{l\delta_a}) + (m_1 - b_1C_{y\dot{\beta}})(C_{n\delta_a}C_{lr} + C_{l\delta_a}C_{nr})b_1$$
$$- C_{y\delta_a}(C_{l\dot{\beta}}C_{nr}b_1^2 - C_{l\beta}I_{z1} - C_{n\beta}I_{xz1} - C_{n\dot{\beta}}C_{lr}b_1^2)$$
$$+ (m_1 - b_1C_{yr})(C_{l\delta_a}C_{n\dot{\beta}}b_1 - C_{n\delta_a}C_{l\dot{\beta}}b_1) \quad (6.256)$$

$$C_{\bar{\phi}} = (m_1 - b_1C_{yr})(C_{l\delta_a}C_{n\beta} - C_{n\delta_a}C_{l\beta}) + b_1C_{y\beta}(C_{l\delta_a}C_{nr} - C_{n\delta_a}C_{lr})$$
$$+ b_1C_{y\delta_a}(C_{n\beta}C_{lr} - C_{l\beta}C_{nr}) \quad (6.257)$$

Substituting the mass and aerodynamic characteristics of the general aviation air-plane, we get the transfer functions of the general aviation airplane as follows:

$$\frac{\Delta\bar{\beta}(s)}{\Delta\bar{\delta}_a(s)} = \frac{-3.4958\times10^{-5}s^3 - 0.0027s^2 - 6.2143\times10^{-4}s}{1.5586\times10^{-4}s^5 + 0.0015s^4 + 0.0022s^3 + 0.0076s^2 + 6.6266\times10^{-5}s} \quad (6.258)$$

$$\frac{\Delta\bar{\psi}(s)}{\Delta\bar{\delta}_a(s)} = \frac{3.4958\times10^{-5}s^3 + 0.0019s^2 + 4.8622\times10^{-4}s - 0.0037}{1.5586\times10^{-4}s^5 + 0.0015s^4 + 0.0022s^3 + 0.0076s^2 + 6.6266\times10^{-5}s} \quad (6.259)$$

$$\frac{\Delta\bar{\phi}(s)}{\Delta\bar{\delta}_a(s)} = \frac{-0.0046s^3 - 0.0046s^2 - 0.0211s}{1.5586\times10^{-4}s^5 + 0.0015s^4 + 0.0022s^3 + 0.0076s^2 + 6.6266\times10^{-5}s} \quad (6.260)$$

We note that

$$\frac{\Delta\bar{r}(s)}{\Delta\bar{\delta}_a(s)} = \frac{s\Delta\bar{\psi}(s)}{\Delta\bar{\delta}_a(s)} \quad (6.261)$$

$$\frac{\Delta\bar{p}(s)}{\Delta\bar{\delta}_a(s)} = \frac{s\Delta\bar{\phi}(s)}{\Delta\bar{\delta}_a(s)} \quad (6.262)$$

The transfer functions for the general aviation airplane (Fig. 6.2) for the rudder input are as follows:

$$\frac{\Delta\bar{\beta}(s)}{\Delta\bar{\delta}_r(s)} = \frac{1.1093\times10^{-5}s^3 + 8.2661\times10^{-4}s^2 + 0.0064s - 2.3475\times10^{-4}}{1.5586\times10^{-4}s^5 + 0.0015s^4 + 0.0022s^3 + 0.0076s^2 + 6.6266\times10^{-5}s} \quad (6.263)$$

$$\frac{\Delta\bar{\psi}(s)}{\Delta\bar{\delta}_r(s)} = \frac{-7.2441\times10^{-4}\,s^3 - 0.0064s^2 - 0.0011s - 0.0018}{1.5586\times10^{-4}\,s^5 + 0.0015\,s^4 + 0.0022\,s^3 + 0.0076\,s^2 + 6.6266\times10^{-5}\,s} \tag{6.264}$$

$$\frac{\Delta\bar{\phi}(s)}{\Delta\bar{\delta}_r(s)} = \frac{4.0157\times10^{-4}s^3 - 0.0014\,s^2 - 0.0102\,s}{1.5586\times10^{-4}\,s^5 + 0.0015\,s^4 + 0.0022\,s^3 + 0.0076\,s^2 + 6.6266\times10^{-5}\,s} \tag{6.265}$$

Note that control derivatives are per radian.

Approximate Lateral-Directional Transfer Functions

Roll Subsidence

Taking the Laplace transform of Eq. (6.206) for aileron input with rudder held fixed ($\Delta\delta_r = 0$) and rearranging the terms, we get

$$\frac{\Delta\bar{\phi}(s)}{\Delta\bar{\delta}_a(s)} = \frac{C_{l\delta_a}}{s(I_{x1}s - C_{lp}b_1)} \tag{6.266}$$

so that

$$\frac{\bar{p}(s)}{\Delta\bar{\delta}_a(s)} = \frac{s\Delta\bar{\phi}(s)}{\Delta\bar{\delta}_a(s)} \tag{6.267}$$

Transfer Functions for Dutch-Roll Approximation

Taking the Laplace transform of Eqs. (6.212) and (6.213) and using Cramer's rule (Appendix C), we obtain the following transfer functions for aileron deflections:

$$\frac{\Delta\bar{\beta}(s)}{\Delta\bar{\delta}_a(s)} = \frac{N_{\bar{\beta}}}{\Delta_{\text{lat},dr}} \tag{6.268}$$

$$\frac{\Delta\bar{\psi}(s)}{\Delta\bar{\delta}_a(s)} = \frac{N_{\bar{\psi}}}{\Delta_{\text{lat},dr}} \tag{6.269}$$

where

$$N_{\bar{\beta}} = I_{z1}C_{y\delta_a}s^2 + (b_1C_{n\delta_a}C_{yr} - C_{y\delta_a}C_{nr}b_1 - m_1C_{n\delta_a})s \tag{6.270}$$

$$\begin{aligned}\Delta_{\text{lat},dr} = {} & I_{z1}(m_1 - b_1C_{y\dot{\beta}})s^3 - [C_{y\beta}I_{z1} + C_{nr}b_1(m_1 - b_1C_{y\dot{\beta}}) \\ & - C_{n\dot{\beta}}b_1(m_1 - b_1C_{yr})]s^2 + [C_{y\beta}C_{nr}b_1 + C_{n\beta}(m_1 - b_1C_{yr})]s\end{aligned} \tag{6.271}$$

$$N_{\bar{\psi}} = [C_{n\delta_a}(m_1 - b_1C_{y\dot{\beta}}) + C_{y\delta_a}C_{n\dot{\beta}}b_1]s + [C_{y\delta_a}C_{n\beta} - C_{n\delta_a}C_{y\beta}] \tag{6.272}$$

Spiral Approximation

Taking the Laplace transform of Eqs. (6.224) and (6.225) and using Cramer's rule (Appendix C), we obtain the following transfer functions for aileron deflections:

$$\frac{\Delta\bar{\beta}(s)}{\Delta\bar{\delta}_a(s)} = \frac{A_{\bar{\beta}}s^2 + B_{\bar{\beta}}s}{A_\delta s^2 + B_\delta s} \tag{6.273}$$

$$\frac{\Delta\bar{\psi}(s)}{\Delta\bar{\delta}_a(s)} = \frac{A_{\bar{\psi}}}{A_\delta s^2 + B_\delta s} \tag{6.274}$$

where

$$A_{\bar{\beta}} = C_{l\delta_a}I_{z1} \tag{6.275}$$

$$B_{\bar{\beta}} = b_1(C_{n\delta_a}C_{lr} - C_{l\delta_a}C_{nr}) \tag{6.276}$$

$$A_\delta = -C_{l\beta}I_{z1} \tag{6.277}$$

$$B_\delta = b_1(C_{l\beta}C_{nr} - C_{n\beta}C_{lr}) \tag{6.278}$$

$$A_{\bar{\psi}} = C_{n\beta}C_{l\delta_a} - C_{l\beta}C_{n\delta_a} \tag{6.279}$$

Substituting the mass and aerodynamic characteristics of the general aviation airplane, we get the approximate transfer functions of the general aviation airplane for aileron and rudder inputs (per radian) as follows. Note that we replace $\Delta\delta_a$ by $\Delta\delta_r$, $C_{y\delta_a}$ by $C_{y\delta_r}$, and so on.

For the roll subsidence mode,

$$\frac{\Delta\bar{\phi}(s)}{\Delta\bar{\delta}_a(s)} = \frac{0.1342}{s(0.0046\,s + 0.0387)} \tag{6.280}$$

$$\frac{\Delta\bar{\phi}(s)}{\Delta\bar{\delta}_r(s)} = \frac{0.0118}{s(0.0046\,s + 0.0387)} \tag{6.281}$$

For the Dutch-roll approximation,

$$\frac{\Delta\bar{\beta}(s)}{\Delta\bar{\delta}_a(s)} = \frac{-0.0076\,s}{0.0340\,s^3 + 0.0347\,s^2 + 0.1613\,s} \tag{6.282}$$

$$\frac{\Delta\bar{\psi}(s)}{\Delta\bar{\delta}_a(s)} = \frac{0.0076\,s + 0.002}{0.0340\,s^3 + 0.0347\,s^2 + 0.1613\,s} \tag{6.283}$$

$$\frac{\Delta\bar{\beta}(s)}{\Delta\bar{\delta}_r(s)} = \frac{0.0024\,s^2 + 0.16\,s}{0.0340\,s^3 + 0.0347\,s^2 + 0.1613\,s} \tag{6.284}$$

$$\frac{\Delta\bar{\psi}(s)}{\Delta\bar{\delta}_r(s)} = \frac{-0.1582\,s - 0.0294}{0.0340\,s^3 + 0.0347\,s^2 + 0.1613\,s} \tag{6.285}$$

For the spiral approximation,

$$\frac{\Delta\bar{\beta}(s)}{\Delta\bar{\delta}_a(s)} = \frac{-0.0021\,s^2 - 0.0016\,s}{0.0011\,s^2 + 1.6508 \times 10^{-4}\,s} \tag{6.286}$$

$$\frac{\Delta\bar{\psi}(s)}{\Delta\bar{\delta}_a(s)} = \frac{-0.0092\,s}{0.0011\,s^2 + 1.6508 \times 10^{-4}\,s} \tag{6.287}$$

$$\frac{\Delta\bar{\beta}(s)}{\Delta\bar{\delta}_r(s)} = \frac{1.8204 \times 10^{-4}\,s^2 - 5.4861 \times 10^{-4}\,s}{0.0011\,s^2 + 1.6508 \times 10^{-4}\,s} \tag{6.288}$$

$$\frac{\Delta\bar{\psi}(s)}{\Delta\bar{\delta}_r(s)} = \frac{-0.0045}{0.0011\,s^2 + 1.6508 \times 10^{-4}\,s} \tag{6.289}$$

We note that

$$\bar{p}(s) = s\Delta\bar{\phi}(s) \tag{6.290}$$

$$\bar{r}(s) = s\Delta\bar{\psi}(s) \tag{6.291}$$

Using these transfer functions of the general aviation airplane and the final value theorem [see Eq. (5.30)], the steady-state values of the lateral-directional variables $\Delta\beta$, $\Delta\phi$, and p for a given input such as a step input can be obtained. However, this is left as an exercise to the reader.

6.3.4 Lateral-Directional Frequency Response

The lateral-directional frequency response to aileron and rudder inputs is of interest because the pilot input to these two control surfaces can be considered as a signal with a frequency content. The bandwidth of the human pilot is about 4 rad/s [3]. Therefore, it is reasonable to expect that a human pilot can control any lateral-directional flight path variable using aileron/rudder inputs within this range of frequency. The autopilots have much higher bandwidths, up to 20–25 rad/s [3].

As an example of the lateral-directional frequency response, we have presented the frequency responses for sideslip, roll angle, roll rate, yaw angle, and yaw rate to aileron input in Figs. 6.26–6.30 for the general aviation airplane. Here, G_β, G_ψ, G_ϕ, G_p, and G_r are the lateral-directional transfer functions of the general aviation airplane given in Eqs. (6.258–6.262). These frequency-response (Bode magnitude and phase) plots were drawn using MATLAB [1]. Similar frequency responses can be obtained for rudder input using the transfer functions given in Eqs. (6.263–6.265).

For the general aviation airplane, the Dutch-roll frequency is 2.3973 rad/s, and the damping ratio is 0.2043. Thus, the Dutch roll is lightly damped. In view of this, the magnitude responses of all the variables display a peak value around this frequency.

The magnitude crossover frequency for sideslip is around 0.09 rad/s, which means that the aileron inputs of frequencies of 0.09 rad/s and

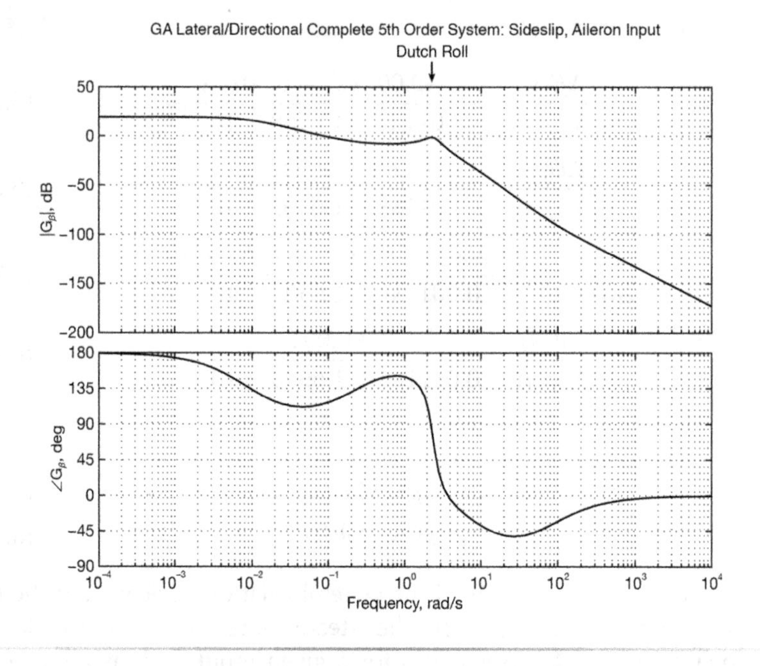

Fig. 6.26 Frequency response of the general aviation airplane: sideslip to aileron input.

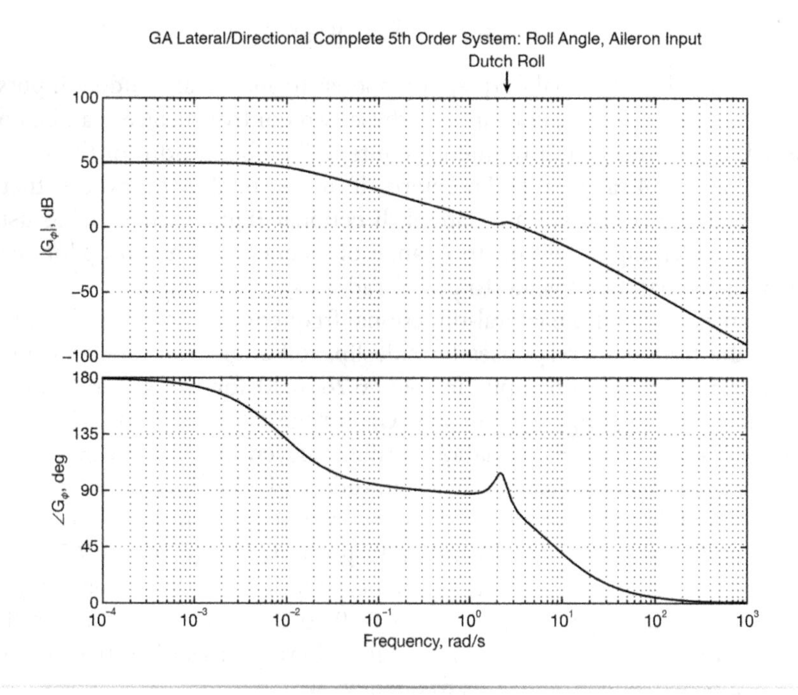

Fig. 6.27 Frequency response of the general aviation airplane: roll angle to aileron input.

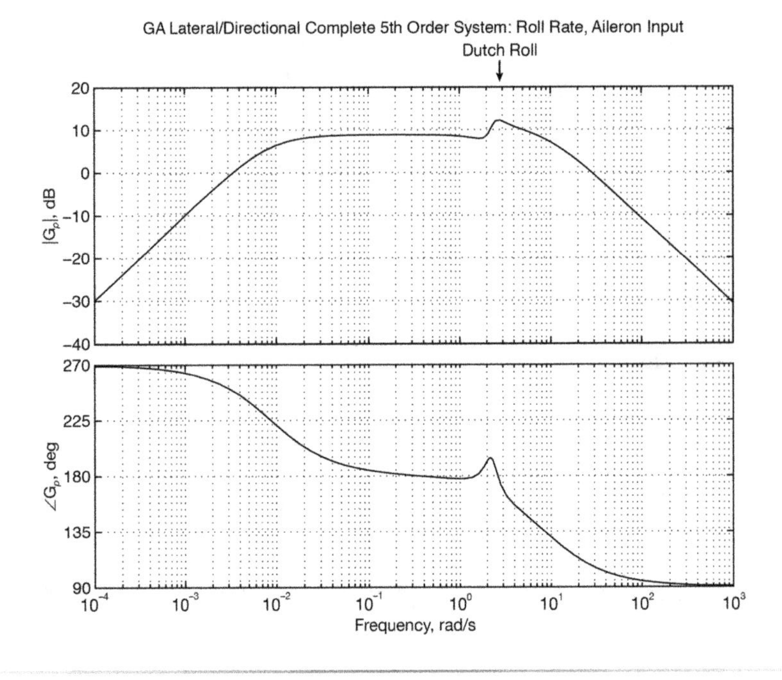

Fig. 6.28 Frequency response of the general aviation airplane: roll rate to aileron input.

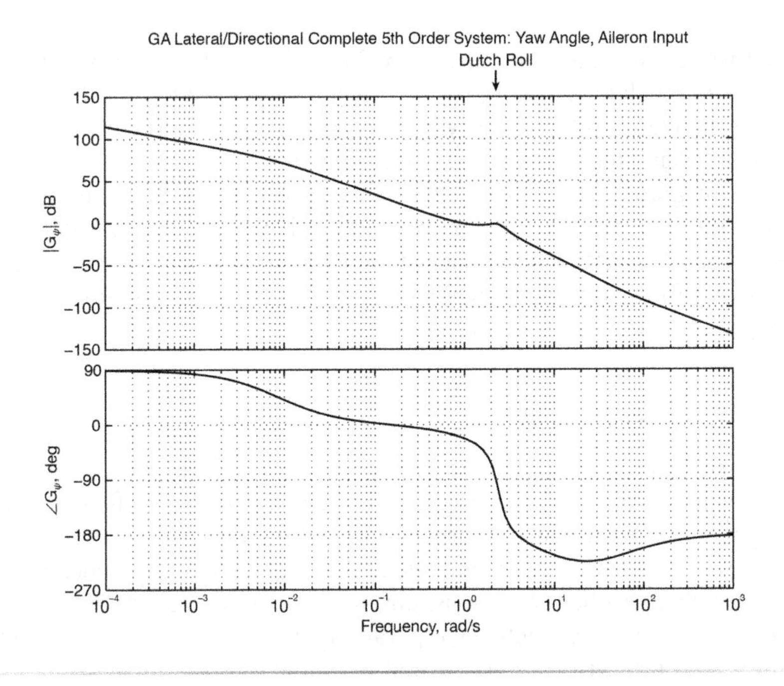

Fig. 6.29 Frequency response of the general aviation airplane: yaw angle to aileron input.

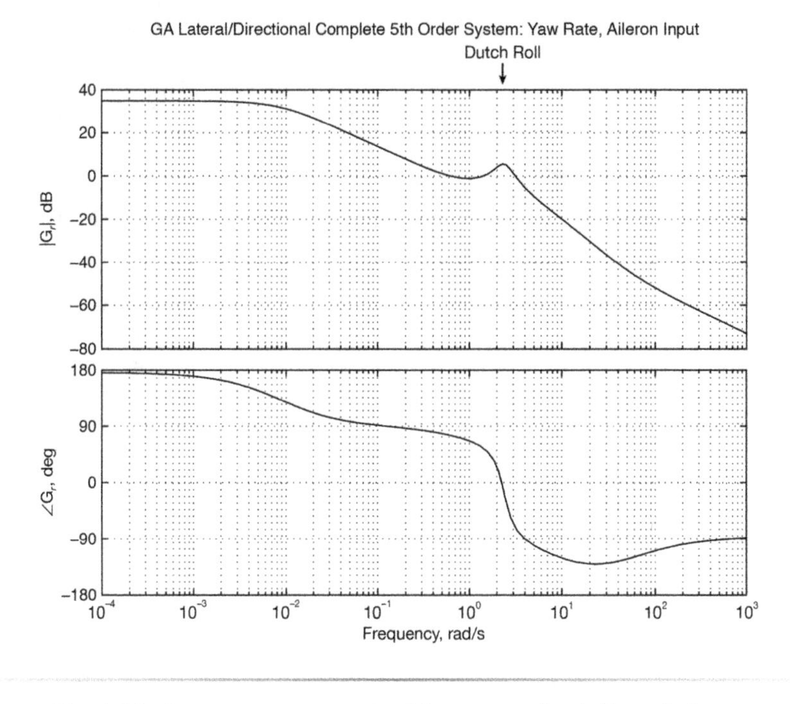

Fig. 6.30 Frequency response of the general aviation airplane: yaw rate to aileron input.

higher will have little or no effect on the sideslip. Only aileron inputs of frequencies much lower than 0.09 rad/s will have some effect on the sideslip. Around the Dutch-roll frequency, the phase angle of $|G_\beta|$ is close to 90 deg, which is a characteristic of lightly damped systems.

The magnitude crossover frequency for roll angle is about 4 rad/s. Hence, the human pilot can have good roll control using the aileron inputs. We observe that the magnitude of the crossover frequency for yaw rate is quite low and is about 0.5 rad/s. Thus, it is not possible to have good yaw control using the aileron inputs.

6.4 Flying Qualities

The subject of flying qualities attempts to quantify the ease or difficulty with which a pilot can fly an airplane in a given category of flight. Besides the dynamic behavior of the airplane, pilot opinion is also influenced by such factors like stick force gradients, design and layout of the cockpit controls, visibility, weather conditions, physical and emotional condition of the pilot, etc. In view of such widely varying factors that can affect pilot opinion, it is usual to conduct several tests and ask a number of pilots to fly the given airplane for each phase of flight and then take an average of all the test results. The Cooper–Harper scale is a systematic method of quantifying these test results. It assigns three levels to describe the flying qualities of an airplane as follows [4,5].

Level I Flying qualities clearly adequate for the mission flight phase.

Level II Flying qualities adequate to accomplish the mission flight phase when some increase in pilot workload or degradation in mission effectiveness exists or both.

Level III Flying qualities such that the mission can be controlled safely, but pilot workload is excessive or mission effectiveness is inadequate or both.

Category A flight phases can be terminated safely, and category B and C flight phases can be completed. The flight phases are defined as follows.

Nonterminal Flight Phases

Category A Nonterminal flight phases that require rapid maneuvering, precision tracking, or precise flight-path control. Included in the category are air-to-air combat, ground attack, weapon delivery/launch, aerial recovery, reconnaissance, in-flight refueling (receiver), terrain following, antisubmarine search, and close-formation flying.

Category B Nonterminal flight phases that are normally accomplished using gradual maneuvers and without precision tracking, although accurate flight-path control may be required. Included in the category are climb, cruise, loiter, in-flight refueling (tanker), descent, emergency descent, emergency deceleration, and aerial delivery.

Terminal Flight Phases

Category C Terminal flight phases are normally accomplished using gradual maneuvers and usually require accurate flight-path control. Included in this category are takeoff, catapult takeoff, approach, waveoff/go-around, and landing.

The aircraft are classified as follows.

Class I Small, light airplanes, such as light utility, primary trainer, and light observation craft.

Class II Medium-weight, low-to-medium maneuverability airplanes, such as heavy utility/search and rescue, light or medium transport/cargo/tanker, reconnaissance, tactical bomber, heavy attack, and trainer for class II.

Class III Large, heavy, low-to-medium maneuverability airplanes, such as heavy transport/cargo/tanker, heavy bomber, and trainer for class III.

Class IV High-maneuverability airplanes, such as fighter/interceptor, attack, tactical reconnaissance, observation, and trainer for class IV.

Extensive research and development activities have been conducted by various government agencies and the aviation industry to relate the

Table 6.2 Damping Ratios for Short-Period Mode

	Cat A and C $\zeta_{sp,\min}$	Cat A and C $\zeta_{sp,\max}$	Cat B $\zeta_{sp,\min}$	Cat B $\zeta_{sp,\max}$
Level I	0.35	1.30	0.30	2.0
Level II	0.25	2.0	0.20	2.0
Level III	0.15	—	0.15	—

Cooper–Harper flying qualities to physical parameters like the frequency and damping ratio of the oscillatory motions of the aircraft. The military specifications MIL-F-8785C [6] (used also for civilian aircraft) are as follows.

6.4.1 Longitudinal Flying Qualities

Phugoid Mode (4,5)

$$\text{Level I} = \zeta > 0.04$$

$$\text{Level II} = \zeta > 0$$

$$\text{Level III} = T_2 > 55\,\text{s}$$

At level III, the aircraft is assumed to have an unstable (divergent) phugoid mode, and T_2 denotes the time required for the amplitude to double the initial value.

Short-Period Mode The damping ratios (ζ) of the short-period for mode [4,5] various levels of flying qualities for category A, B, and C flight phases are given in Table 6.2.

The requirements on the natural frequencies (ω_n) of the short-period mode [4,5] for category A, B, and C flight phases according to MIL-F-8785C [6] should lie within the limits as given in Table 6.3.

In Table 6.3, n is the load factor as given by

$$n = \frac{L}{W} \tag{6.292}$$

$$= \frac{\frac{1}{2}\rho U_o^2 S C_L}{W} \tag{6.293}$$

Table 6.3 Limits on $\omega_{n,sp}^2/(n/\alpha)$

	Cat A, min	Cat A, max	Cat B, min	Cat B, max	Cat C, min	Cat C, max
Level I	0.28	3.6	0.085	3.6	0.16	3.6
Level II	0.16	10.0	0.038	10.0	0.096	10.0
Level III	0.16	—	0.038	—	0.096	—

Table 6.4 Time Constants for Roll Subsidence Mode

Class	Category	Level I	Level II	Level III
I, IV	A	1.0	1.4	10.0
II, III	A	1.4	3.0	10.0
All	B	1.4	3.0	10.0
I, IV	C	1.0	1.4	10.0
II, III	C	1.4	3.0	10.0

For linear range of angle of attack,

$$\frac{n}{\alpha} = \left(\frac{1}{2W}\right)\rho U_o^2 S C_{L\alpha} \tag{6.294}$$

Thus, the ratio n/α depends on the flight velocity and flight altitude.

As an example, let us calculate the limiting values of the short-period frequency for the general aviation airplane level I for category B phase of flight at sea level. We have $U_o = Ma = 0.158 * 340.592 = 53.8135$ m/s, $\rho = 1.225$ kg/m^3, $S = 16.7225$ m^2, $C_{L\alpha} = 4.44$ per radian, and $W =$ 12,232.6 N. Substituting, we get $n/\alpha = 10.6689$. Therefore, for the general aviation airplane, the limits on the short-period frequency for level I category B phase of flight are $0.085 \leq \omega_{n,sp}^2/(n/\alpha) \leq 3.6$ or $0.9523 \leq \omega_n \leq 6.1974$ rad/s.

6.4.2 Lateral-Directional Flying Qualities

Roll Mode The maximum allowable values of the roll-subsidence mode time constant [4] are shown in Table 6.4.

Table 6.5 Dutch-Roll Flying Qualities

Level	Category	Class	Min ζ[a]	Min $\zeta\omega_n$,[a] rad/s	Min ω_n, rad/s
I	A	I, IV	0.19	0.35	1.0
I	A	II, III	0.19	0.35	0.4
I	B	All	0.08	0.15	1.0
I	C	I, II-C,[b] IV	0.08	0.15	1.0
I	C	II-L,[b] III	0.08	0.15	0.4
II	All	All	0.02	0.05	0.4
III	All	All	0.02	—	0.4

[a]The governing damping requirement is that yielding the larger value of ζ.
[b]Letters C and L denote carrier-based and land-based aircraft.

Table 6.6 Time to Double Amplitude for Spiral Mode

Class	Category	Level I	Level II	Level III
I, IV	A	12 s	12 s	4 s
I, IV	B and C	20 s	12 s	4 s
II and III	All	20 s	12 s	4 s

Dutch Roll The Dutch-roll flying qualities [4] are specified in Table 6.5. The natural frequency (ω_n) and damping ratio (ζ) should be generally better or exceed the minimum values given in Table 6.5.

Spiral Mode The spiral mode flying qualities [4] are given in Table 6.6 in terms of minimum time to double the amplitude.

6.5 Closed-Loop Flight Control

In the previous sections, we have discussed longitudinal and lateral-directional free and forced responses of the aircraft. These responses can be classified into two broad categories: 1) those involving mainly rotational degrees of freedom such as short-period mode, roll-subsidence mode, and Dutch-roll oscillation and 2) those involving flight path changes such as phugoid and spiral modes. The responses belonging to first category are of high frequency. Therefore, it is essential that they be adequately damped. If not, it will be difficult for the human pilot to take corrective action especially if the frequencies of the unstable modes are higher than the bandwidth of the human pilot, which is about 4 rad/s. For such cases, it is necessary to provide an automatic control system that ensures that these modes are adequately damped in the closed-loop sense. Such a control system is called a stability augmentation system.

There is another class of flight control systems called autopilots. The autopilots are used for establishing and maintaining desired flight conditions. Once the autopilot is engaged, it continues to hold that flight condition without further intervention from the pilot.

In the following, we will discuss some of the typical stability augmentation systems and autopilots.

6.5.1 Pitch Stabilization System

The pitch stabilization system is used when the short-period mode is unstable. Typically, modern high-performance aircraft experience these problems because such aircraft are intentionally made statically unstable for better performance. As we have seen earlier for the relaxed static stability version of the general aviation airplane, such aircraft experience exponential instability in pitch. Here, we will discuss the design of a pitch stabilization system to make the aircraft closed-loop stable in pitch.

In Fig. 6.31a, the schematic diagram of a typical pitch stabilization system is shown and, in Fig. 6.31b, its block diagram implementation is presented. The elevator servo is approximated as a first-order lag with a break frequency $a = 10$ rad/s. We have an inner loop with angle of attack feedback and an outer loop with pitch rate feedback. The purpose of the inner-loop alpha feedback is to make the aircraft statically stable. The outer-loop pitch rate feedback enables us to select a suitable value of the damping ratio so that the aircraft has the desired level of longitudinal handling qualities in the closed-loop sense.

To illustrate the design procedure, we consider the statically unstable version of the general aviation airplane. Here, we consider the transfer function of the complete (fourth-order) system because the short-period approximation is not valid. For the purpose of this illustration, we assume $C_{m\alpha} = 0.07$/rad and C_{mq} and $C_{m\dot{\alpha}}$ at 50% of the baseline values. All the other stability and control derivatives remain unchanged. With this assumption, we get

$$\frac{\Delta\bar{\alpha}(s)}{\Delta\bar{\delta}_e(s)} = \frac{-(0.0602\,s^3 + 4.4326\,s^2 + 0.2008\,s + 0.3103)}{0.3739\,s^4 + 1.3343\,s^3 + 0.5280\,s^2 + 0.0381\,s - 0.0235} \tag{6.295}$$

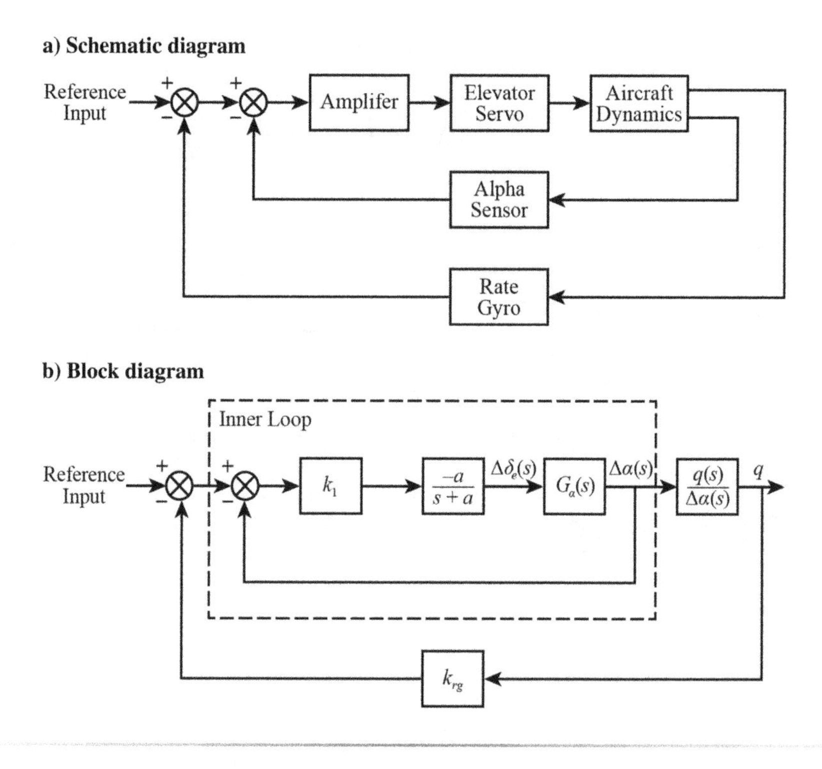

Fig. 6.31 Pitch stabilization system.

$$\frac{\Delta\bar{q}(s)}{\Delta\bar{\delta}(s)} = \frac{-s(4.4642\,s^2 + 9.3994\,s + 0.4775)}{0.3739\,s^4 + 1.3343\,s^3 + 0.5280\,s^2 + 0.0381\,s - 0.0235} \tag{6.296}$$

In Fig. 6.31b, G_α denotes the transfer function given by Eq. (6.295). Because G_α has a negative sign, we choose a negative sign for the transfer function of the elevator servo so that the loop transfer function is positive, and we can use a negative feedback as usual.

The first step is to design the inner loop. The block diagram of the inner loop is shown in Fig. 6.32a. We draw the root-locus of the inner loop and select a point on the root-locus corresponding to $\zeta = 0.8$ as shown in Fig. 6.33 (complete plot in Fig. 6.33a and close-up in Fig. 6.33b). We get $k_1 = 0.2351$. Next, we proceed to the design of the outer loop. The block diagram of the outer loop is shown in Fig. 6.32b. Here, T_α represents the closed-loop transfer function of the inner loop and is given by

$$T_\alpha = \frac{\dfrac{-k_1 a G_\alpha}{s+a}}{1 - \dfrac{k_1 a G_\alpha}{s+a}} \tag{6.297}$$

The transfer function for the pitch rate to angle of attack is obtained as follows:

$$\frac{\bar{q}(s)}{\Delta\bar{\alpha}(s)} = \left(\frac{\bar{q}(s)}{\Delta\bar{\delta}_e(s)}\right)\left(\frac{\Delta\bar{\delta}_e(s)}{\Delta\bar{\alpha}(s)}\right) \tag{6.298}$$

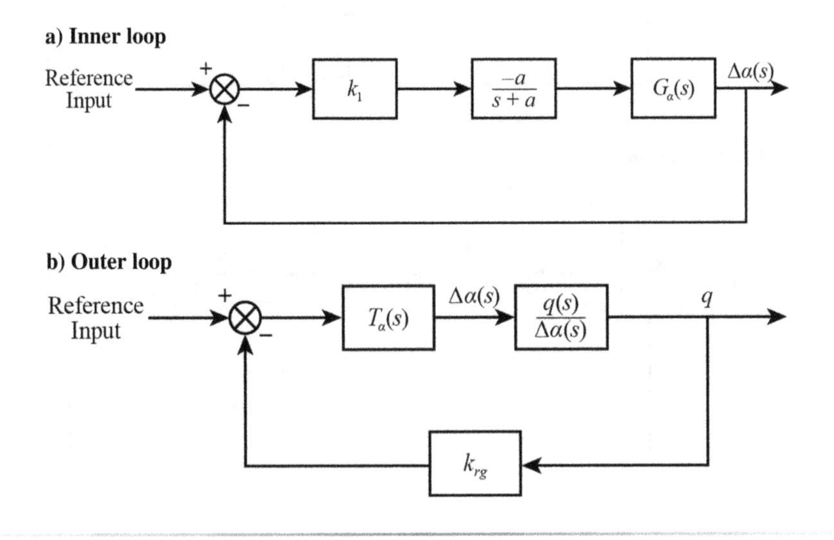

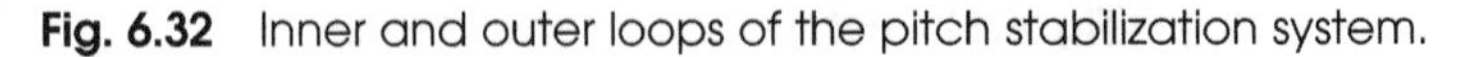
Fig. 6.32 Inner and outer loops of the pitch stabilization system.

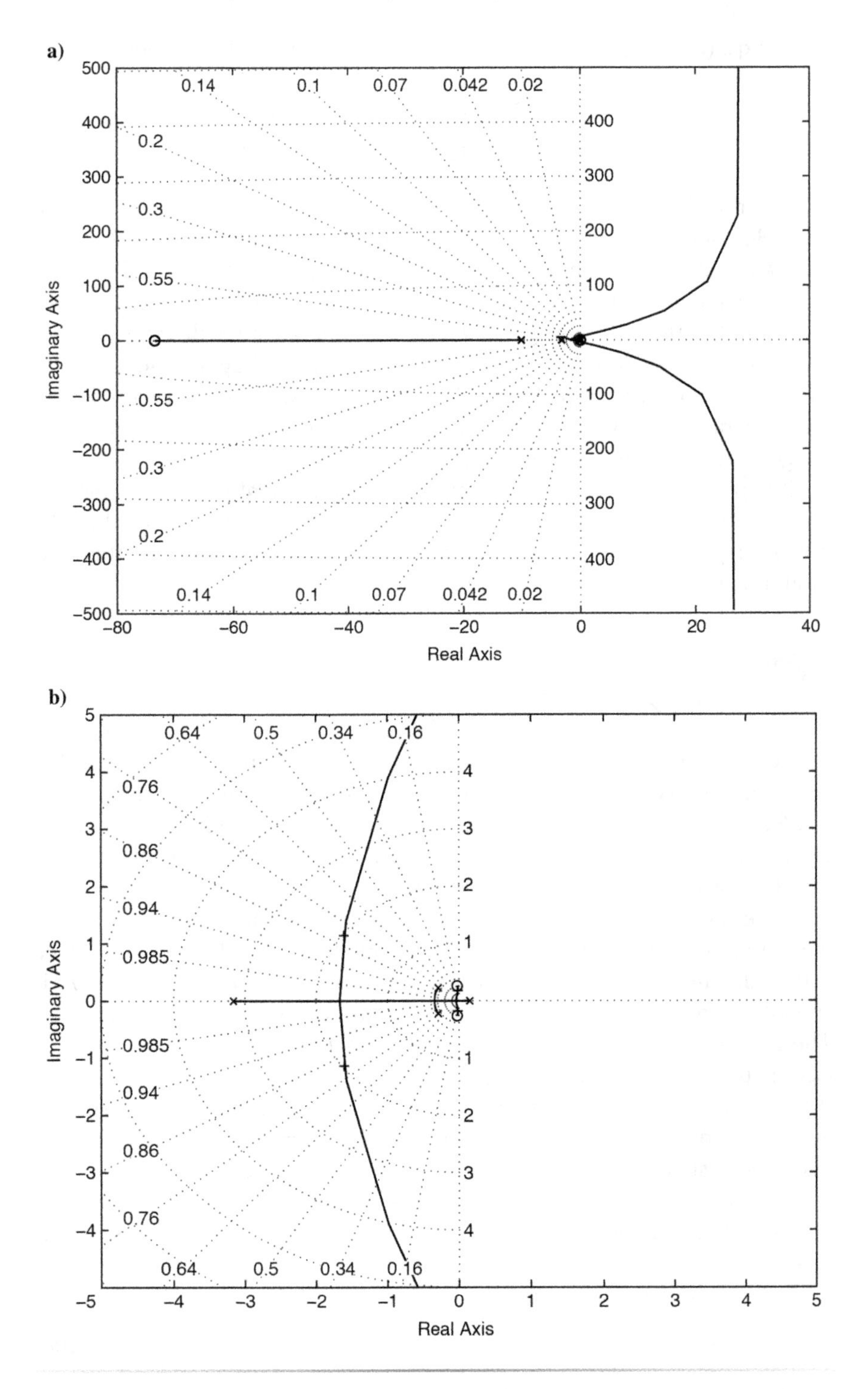

Fig. 6.33 Root-locii of the inner loop of the pitch stabilization system of the general aviation airplane.

Using Eqs. (6.295) and (6.296) and the relation $\bar{q}(s) = s\bar{\theta}(s)$, we get

$$\frac{\bar{q}(s)}{\bar{\alpha}(s)} = \frac{s(4.4642\, s^2 + 9.3994\, s + 0.4775)}{0.0602\, s^3 + 4.4326\, s^2 + 0.2008\, s + 0.3103} \tag{6.299}$$

The root-locus of the outer loop is shown in Fig. 6.34 (complete plot in Fig. 6.34a and close-up in Fig. 6.34b). We pick a point for $\zeta = 0.9$, which gives $k_{rg} = 0.5695$. We have the closed-loop poles at -8.1257, $-2.7057 \pm j1.2806$ (short-period), and $-0.0220 \pm j0.1329$ (phugoid). For these closed-loop pole locations, we get $\zeta = 0.9$, $\omega_n = 2.9935$ rad/s for the short-period mode and $\zeta = 0.1633$, $\omega_n = 0.1347$ rad/s for the phugoid mode. Thus, with alpha and pitch rate feedback, the relaxed static stability version of the general aviation airplane has the conventional short-period and phugoid modes with level I flying qualities.

It may be noted that there is some subjective element involved in picking a point on the root-locus in Matlab using the cursor. In view of this, the values of gain and locations of the closed-loop poles given in this text should be used as guidelines and not as absolute numbers.

6.5.2 Full-State Feedback Design for Longitudinal Stability Augmentation System

To illustrate the design procedure of a full-state feedback stability augmentation system, let us consider the general aviation airplane (see Fig. 6.2) once again. Here, we assume that all four longitudinal states u, $\Delta\alpha$, $\Delta\theta$, and q are accurately measured and are available for feedback. The air speed is usually measured by a pitot-static sensor, the pitch angle and pitch rates are measured by positional and rate gyros, and the angle of attack is measured by a vane-type sensor. Of these four state variables, angle of attack is the difficult one to measure because the local flow around the airplane is altered due to wing–body upwash in front and downwash behind the wing. The alpha sensor is usually mounted on the wingtips or at the nose of the fuselage. Thus, it is subjected to the upwash field. Hence, it is necessary to correct the measured values of the angle of attack for the induced upwash/downwash effects. For our purpose, we will assume that such corrections are done and that all four states are accurately measured and are available for feedback.

Let us assume that the desired poles or the roots of the longitudinal characteristic equations that give the specified handling qualities are as follows.

For the short-period mode, let

$$\lambda_{1,2} = -4.8 \pm j2.16 \tag{6.300}$$

which corresponds to a damping ratio of 0.9119 and a natural frequency of 5.2636 rad/s for the short-period mode. As we have seen earlier, these

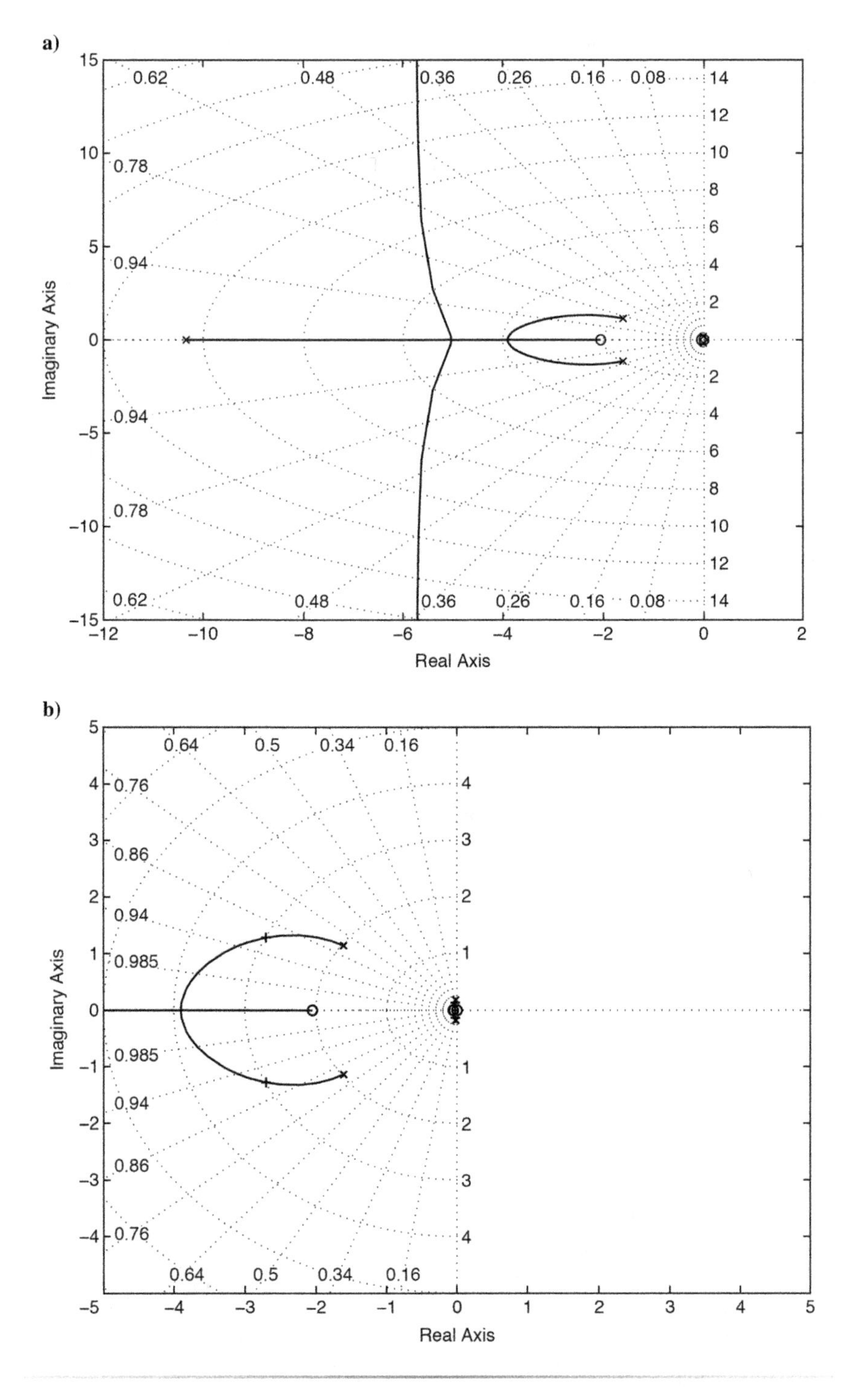

Fig. 6.34 Root-locii of the outer loop of the pitch stabilization system of the general aviation airplane.

values of damping ratio and natural frequency give us level I short-period flying qualities.

For the phugoid mode, let

$$\lambda_{3,4} = -0.04 \pm j0.1960 \tag{6.301}$$

which corresponds to a damping ratio of 0.20 and a natural frequency of 0.20 rad/s, giving level I phugoid flying qualities.

For the values of the roots given by Eqs. (6.300) and (6.301), the desired characteristic equation is given by

$$s^4 + 9.68\,s^3 + 28.5136\,s^2 + 2.6006\,s + 1.1087 \tag{6.302}$$

The objective is to design a full-state feedback law so that the closed-loop characteristic equation is identical to the desired characteristic equation.

We have the given system in the state-space form

$$\dot{x} = Ax + Bu_c \tag{6.303}$$

where

$$x = \begin{bmatrix} u \\ \Delta\alpha \\ q \\ \Delta\theta \end{bmatrix} \tag{6.304}$$

and matrices A and B given by Eqs. (6.29) and (6.30) are

$$A = \begin{bmatrix} -0.0453 & 0.0363 & 0 & -0.1859 \\ -0.3717 & -2.0354 & 0.9723 & 0 \\ 0.3398 & -7.0301 & -2.9767 & 0 \\ 0 & 0 & 1 & 0 \end{bmatrix} \tag{6.305}$$

$$B = \begin{bmatrix} 0 \\ -0.1609 \\ -11.8674 \\ 0 \end{bmatrix} \tag{6.306}$$

The design procedure for the state feedback was explained in Sec. 5.10.9. The first step is to express the given plant in the phase-variable form. Because the given plant is not in the phase-variable form, we have to do a transformation. The advantage is that the elements of the last row of a matrix in phase-variable form are the coefficients of the characteristic equation. Then, we do the full-state feedback design in the transformed space, and then, finally, we do a reverse transformation to obtain the state feedback law in the original state-space.

We express the given plant in the phase-variable form as follows:

$$\dot{z} = A_p z + B_p u_c \tag{6.307}$$

where z is a new state vector (4×1) and matrices A_p and B_p are in phase-variable form and are defined as

$$z = Px \tag{6.308}$$

$$A_p = PAP^{-1} \tag{6.309}$$

$$B_p = PB \tag{6.310}$$

The transformation matrix P is constructed as follows.

1. Obtain the state-controllability matrix Q_c given by

$$Q_c = [B \;\; AB \;\; A^2B \;\; A^3B] \tag{6.311}$$

2. Form matrices P_1, P_2, P_3, and P_4 as follows:

$$P_1 = [0 \;\; 0 \;\; 0 \;\; 1]Q_c^{-1} \tag{6.312}$$

$$P_2 = P_1 A \tag{6.313}$$

$$P_3 = P_1 A^2 \tag{6.314}$$

$$P_4 = P_1 A^3 \tag{6.315}$$

$$P = \begin{bmatrix} P_1 \\ P_2 \\ P_3 \\ P_4 \end{bmatrix} \tag{6.316}$$

Then

$$A_p = PAP^{-1} \tag{6.317}$$

$$B_p = PB \tag{6.318}$$

For the general aviation airplane, substitution gives

$$Q_c = \begin{bmatrix} 0 & -0.0058 & 1.7994 & -4.7437 \\ -0.1609 & -11.2112 & 58.2684 & -148.1528 \\ -11.8674 & 36.4568 & -29.7073 & -320.5912 \\ 0 & -11.8674 & 36.4568 & -29.7073 \end{bmatrix} \tag{6.319}$$

$$Q_c^{-1} = \begin{bmatrix} 0.0597 & 0.3006 & -0.0883 & -0.5554 \\ 3.0533 & -0.1042 & 0.0014 & 0.0171 \\ 1.1855 & -0.0490 & 0.0007 & 0.0477 \\ 0.2352 & -0.0184 & 0.0003 & 0.0181 \end{bmatrix} \tag{6.320}$$

$$P = \begin{bmatrix} 0.2352 & -0.0184 & 0.0003 & 0.0181 \\ -0.0037 & 0.0443 & -0.0006 & 0.0437 \\ -0.0165 & -0.0861 & 0.0012 & 0.0007 \\ 0.0332 & 0.1665 & -0.0865 & 0.0031 \end{bmatrix} \tag{6.321}$$

$$A_p = \begin{bmatrix} 0 & 1 & 0 & 0 \\ 0 & 0 & 1 & 0 \\ 0 & 0 & 0 & 1 \\ -0.6143 & -0.6754 & -13.1347 & -5.0574 \end{bmatrix} \tag{6.322}$$

$$B_p = \begin{bmatrix} 0 \\ 0 \\ 0 \\ 1 \end{bmatrix} \tag{6.323}$$

The input with full-state feedback in the transformed system is

$$u_c = r(t) - Kz \tag{6.324}$$

where K is given by

$$K = [k_1 \;\; k_2 \;\; k_3 \;\; k_4] \tag{6.325}$$

Then

$$\dot{z} = (A_p - B_p K)z + B_p r(t) \tag{6.326}$$

and

$$\begin{bmatrix} \dot{z}_1 \\ \dot{z}_2 \\ \dot{z}_3 \\ \dot{z}_4 \end{bmatrix} = \begin{bmatrix} 0 & 1 & 0 & 0 \\ 0 & 0 & 1 & 0 \\ 0 & 0 & 0 & 1 \\ -(0.6143 + k_1) & -(0.6754 + k_2) & -(13.1347 + k_3) & -(5.0574 + k_4) \end{bmatrix} \times \begin{bmatrix} z_1 \\ z_2 \\ z_3 \\ z_4 \end{bmatrix} + \begin{bmatrix} 0 \\ 0 \\ 0 \\ 1 \end{bmatrix} r(t) \tag{6.327}$$

The characteristic equation of this transformed phase-variable form is

$$s^4 + (5.0574 + k_4)\, s^3 + (13.1347 + k_3)\, s^2 + (0.6754 + k_2)\, s + (0.6143 + k_1) = 0 \tag{6.328}$$

Comparing the coefficient of Eqs. (6.302) and (6.328), we obtain

$$k_1 = 0.4944 \quad k_2 = 1.9252 \quad k_3 = 15.3789 \quad k_4 = 4.6226 \tag{6.329}$$

Then next step is to do a reverse transformation back to the original state-space. Substituting for Z, A_p, and B_p in Eq. (6.307) and rearranging, we obtain

$$\dot{x} = (A - BKP)x + Br(t) \tag{6.330}$$

$$= Ax + B[r(t) - KPx] \tag{6.331}$$

This gives us the desired full-state feedback control law as

$$u_c = r(t) - KPx \tag{6.332}$$

To verify the design, we have performed a simulation of the free response [$r(t) = 0$] using MATLAB [1] for the initial conditions $\Delta\alpha = 5$ deg, $u = 0$, $\theta = q = 0$. The results are shown in Fig. 6.35. We observe that the system with full-state feedback law performs as expected.

6.5.3 Yaw Damper

The purpose of the yaw damper is to generate a yawing moment that opposes any yaw rate that builds up from an unstable Dutch-roll mode. However, this type of control augmentation has one drawback. It will

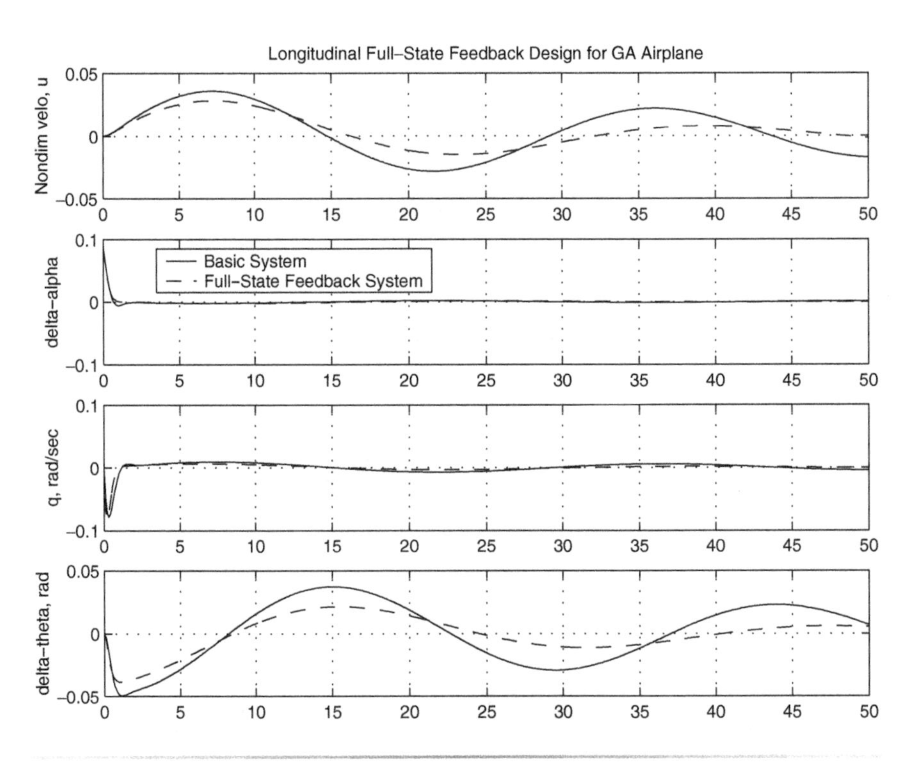

Fig. 6.35 Longitudinal response of the full-state feedback controller for the general aviation airplane.

oppose any steady-state yaw rate the pilot wants to intentionally generate, for example, during a steady coordinated level turn. In this case the pilot will have to fight the yaw damper system. One way of avoiding this problem is to use a wash-out circuit in the feedback loop. The wash-out circuit will filter out all the steady-state components so that the yaw damper system will not respond to steady-state yaw rates.

A simple yaw damper with yaw rate feedback is shown in Fig. 6.36. To illustrate the design procedure, consider the general aviation airplane and use the Dutch-roll transfer function to rudder input. From Eq. (6.285),

$$\frac{\Delta\bar{\psi}(s)}{\Delta\bar{\delta}_r(s)} = \frac{-(0.1582\,s + 0.0294)}{s(0.0340\,s^2 + 0.0347\,s + 0.1613)} \tag{6.333}$$

so that

$$\frac{\bar{r}(s)}{\Delta\bar{\delta}_r(s)} = \frac{s\Delta\bar{\psi}(s)}{\Delta\bar{\delta}_r(s)} \tag{6.334}$$

$$= \frac{-(0.1582\,s + 0.0294)}{0.0340\,s^2 + 0.0347\,s + 0.1613} \tag{6.335}$$

We assume that the rudder servo is a first-order system with $a = 10$ so that the time constant $1/a = 0.10$. As said before, we use the negative sign in the numerator of the rudder servo transfer function because we have a negative sign in the numerator of the yaw-rate-to-rudder-input transfer function as given by Eq. (6.335). With this sign adjustment, we can use the negative feedback as shown in Fig. 6.36b.

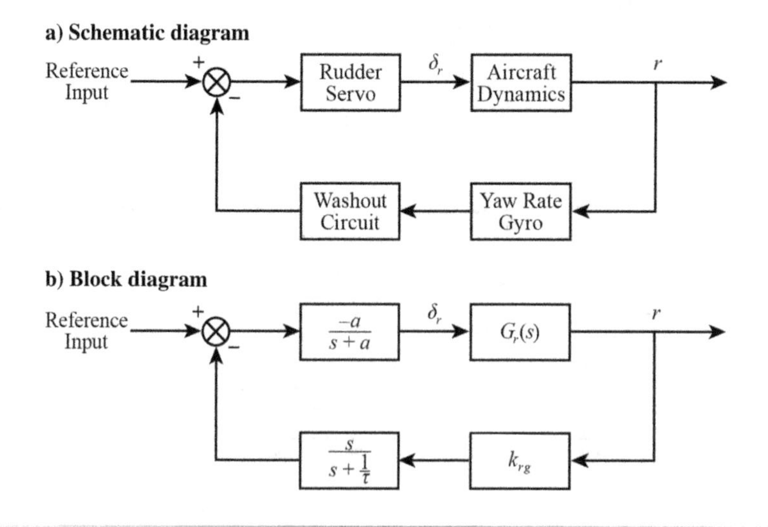

Fig. 6.36 Yaw damper.

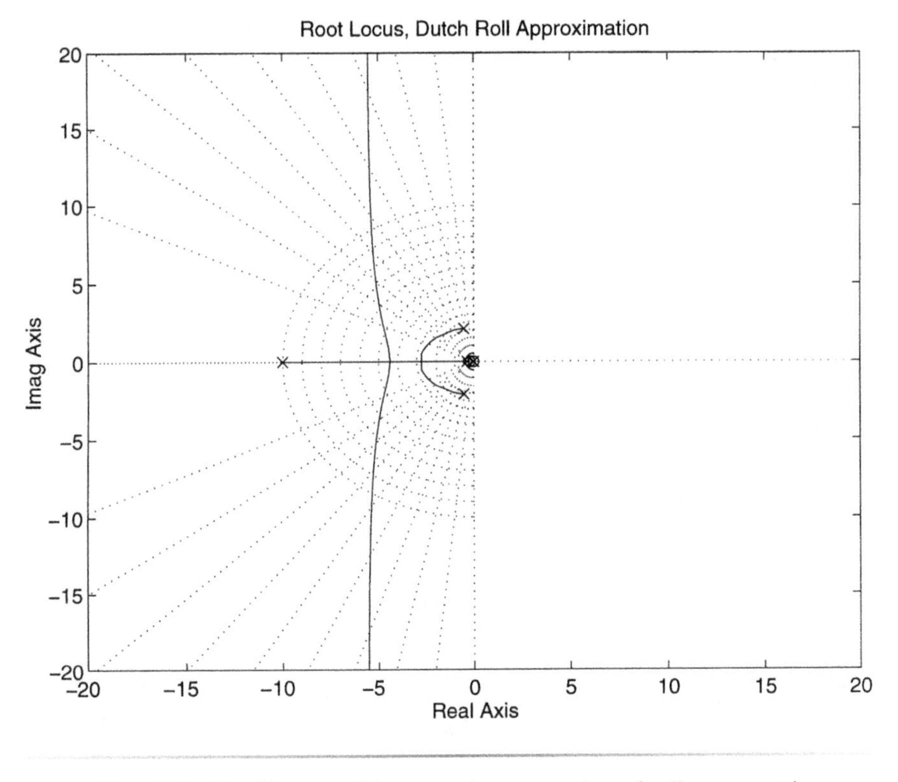

Fig. 6.37 Root-locus of the yaw damper system for the general aviation airplane.

We assume $\tau = 0.3$, so that the transfer function of the wash-out circuit is given by

$$G_c = \frac{s}{s + 0.3333} \tag{6.336}$$

Now we draw the root-locus as shown in Figs. 6.37 and 6.38. The root-locus of Fig. 6.38 is a close-up of the root-locus around the origin. We select a point on root-locus to have $\zeta = 0.8$ for the Dutch-roll mode. We get $k_{rg} = 0.4228$ and the location of the closed-loop poles at -6.9705, $-2.0108 \pm j1.5140$, -0.3619. This gives us a damping ratio of 0.8 and a natural frequency of 2.5170 rad/s, giving us level I Dutch-roll flying qualities for the general aviation airplane, which is a class I airplane.

We can now verify the design by simulating the response to a unit-step function (one-rad rudder input) as shown in Fig. 6.39a. We observe that the closed-loop system with yaw damper is performing satisfactory. To check the design further, we perform a simulation for the complete fifth-order system as shown in Fig. 6.39b. We observe that the design is satisfactory. The difference in the steady-state behaviors is due to the approximations involved in deriving Dutch-roll transfer functions as we have discussed earlier.

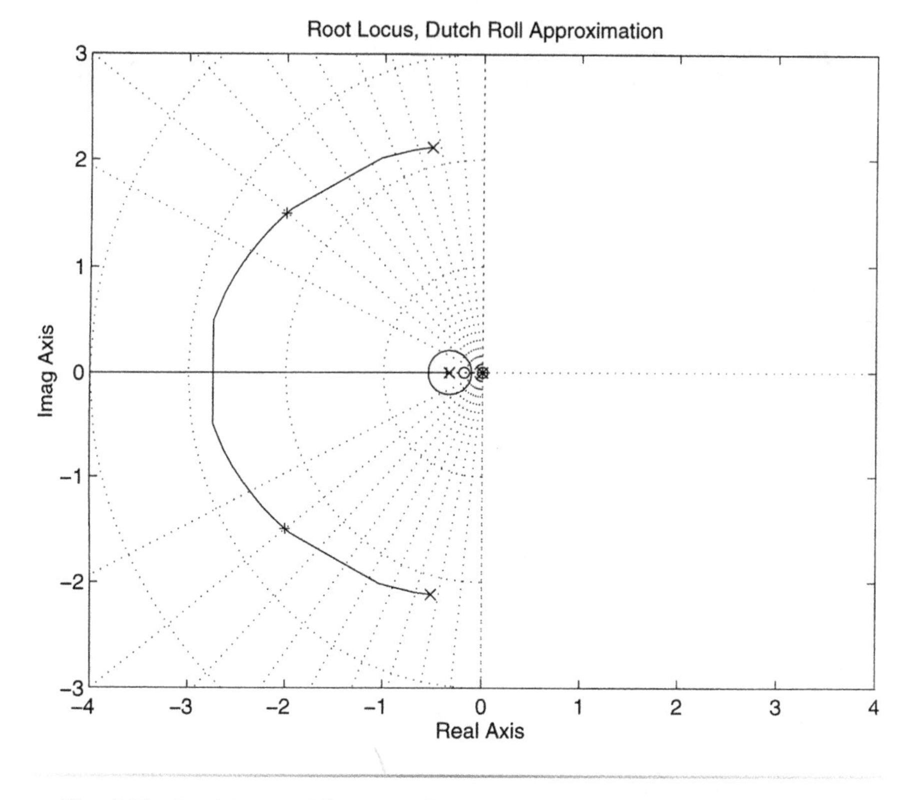

Fig. 6.38 Root-locus of the yaw damper system for the general aviation airplane (close-up around the origin).

6.5.4 Full-State Feedback Design for Lateral-Directional Stability Augmentation System

In this section we will design a full-state feedback control law for a lateral-directional stability augmentation system. Such a flight control system needs that all five state variables β, ϕ, p, ψ, and r be accurately measured and be available for feedback. The roll and yaw angles can be accurately measured by positional gyros and the roll and yaw rates by the rate gyros. The sideslip angle is perhaps the most difficult one to measure. A vane-type instrument is usually used to measure the sideslip but, like angle of attack sensor, it is also subject to positional errors due to fuselage sidewash. Another method used to obtain sideslip is the integration of accelerometer outputs. However, this is also subject to measurement noise. In any case, it is essential that the sideslip be accurately estimated and be available for feedback. For our purpose, we will assume that all five state variables are measured accurately and are available for feedback.

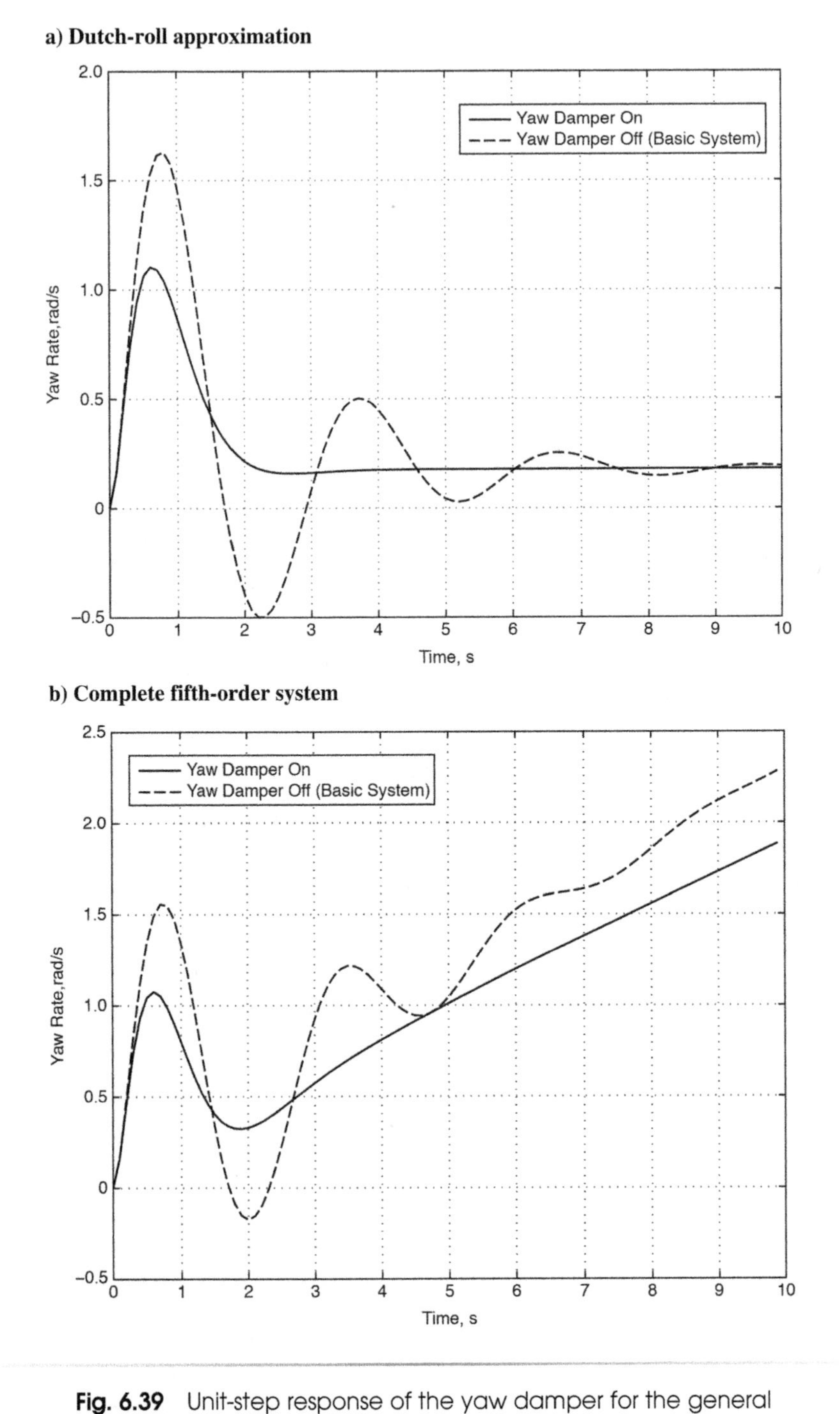

Fig. 6.39 Unit-step response of the yaw damper for the general aviation airplane.

Let us consider the general aviation airplane once again with only rudder input and assume that the desired poles of the lateral-directional closed-loop characteristic equations are as follows.

For Dutch-roll mode, let

$$\lambda_{dr} = -1.2 \pm j2.75 \tag{6.337}$$

so that we have a damping ratio of 0.4 and a natural frequency ω_n of 3.0 rad/s, corresponding to level I Dutch-roll flying qualities.

For roll subsidence and spiral mode, let

$$\lambda_r = -8.50 \tag{6.338}$$

$$\lambda_{sp} = -0.0080 \tag{6.339}$$

The other root is the zero root, $\lambda = 0$.

We have a roll-time constant of 0.1176, which gives level I flying qualities for the roll-subsidence mode. The spiral mode is damped and, therefore, it also has level I flying qualities. Thus, overall, the assumed pole locations give us level I lateral-directional flying qualities.

For these values of five roots, the desired lateral-directional characteristic equation is given by

$$s\left(s^4 + 10.9080\,s^3 + 29.4897\,s^2 + 76.7565\,s + 0.6122\right) = 0 \tag{6.340}$$

The objective is to design a full-state feedback law so that the closed-loop lateral-directional characteristic equation of the general aviation airplane is identical to the desired characteristic equation.

From Eqs. (6.196) and (6.197), we have

$$A = \begin{bmatrix} -0.2557 & 0.1820 & 0 & 0 & -1.0000 \\ 0 & 0 & 1.0 & 0 & 0 \\ -16.1572 & 0 & -8.4481 & 0 & 2.2048 \\ 0 & 0 & 0 & 0 & 1.0 \\ 4.5440 & 0 & -0.3517 & 0 & -0.7647 \end{bmatrix} \tag{6.341}$$

$$B = \begin{bmatrix} 0 & 0.0712 \\ 0 & 0 \\ -29.3013 & 2.5764 \\ 0 & 0 \\ 0.2243 & -4.6477 \end{bmatrix} \tag{6.342}$$

Note that for rudder input, $B = B(:, 2)$. Because the given plant is not in phase-variable form, we have to do a transformation and express it in the phase-variable form (see Sec. 5.10.7). Then, do the full-state feedback design in this new transformed space. Finally, do a reverse transformation to obtain the full-state feedback law in the original state-space.

The phase-variable form is given by

$$\dot{z} = A_p z + B_p u_c \tag{6.343}$$

where z is the new state vector and matrices A_p and B_p are in the phase-variable form

$$z = Px \tag{6.344}$$

$$A_p = PAP^{-1} \tag{6.345}$$

$$B_p = PB \tag{6.346}$$

The state vector x was defined in Eq. (6.183). The transformation matrix P is constructed as follows.

1. Obtain the lateral-directional controllability matrix Q_c given by

$$Q_c = [B \quad AB \quad A^2B \quad A^3B \quad A^4B] \tag{6.347}$$

2. Form matrices P_1 to P_5 as follows:

$$P_1 = [0 \quad 0 \quad 0 \quad 0 \quad 1]Q_c^{-1} \tag{6.348}$$

$$P_2 = P_1 A \tag{6.349}$$

$$P_3 = P_1 A^2 \tag{6.350}$$

$$P_4 = P_1 A^3 \tag{6.351}$$

$$P_5 = P_1 A^4 \tag{6.352}$$

$$P = \begin{bmatrix} P_1 \\ P_2 \\ P_3 \\ P_4 \\ P_5 \end{bmatrix} \tag{6.353}$$

Then the phase-variable forms A_p and B_p are given by

$$A_p = PAP^{-1} \tag{6.354}$$

$$B_p = PB \tag{6.355}$$

For the general aviation airplane, the lateral-directional controllability matrix is given by

$$Q_c = \begin{bmatrix} 0 & 5.0 & -4.0 & -36.0 & 162.0 \\ 0 & 3.0 & -33.0 & 212.0 & -1664.0 \\ 3.0 & -33.0 & 212.0 & -1664.0 & 14{,}376.0 \\ 0 & -5.0 & 3.0 & 30.0 & -115.0 \\ -5.0 & 3.0 & 30.0 & -115.0 & 511.0 \end{bmatrix} \tag{6.356}$$

so that

$$Q_c^{-1} = \begin{bmatrix} 0.0000 & -0.6953 & -0.0717 & -0.0369 & -0.2549 \\ -3.8467 & 0.3386 & 0.0499 & -4.2354 & -0.0312 \\ -1.1859 & 0.0187 & 0.0063 & -1.2253 & -0.0147 \\ -0.7785 & 0.0926 & 0.0131 & -0.8208 & -0.0046 \\ -0.0815 & 0.0113 & 0.0016 & -0.0867 & -0.0003 \end{bmatrix} \tag{6.357}$$

$$P = \begin{bmatrix} -0.0815 & 0.0113 & 0.0016 & -0.0867 & -0.0003 \\ -0.0070 & -0.0148 & -0.0022 & 0 & -0.0014 \\ 0.0320 & -0.0013 & 0.0046 & 0 & 0.0031 \\ -0.0692 & 0.0058 & -0.0415 & 0 & -0.0241 \\ 0.5791 & -0.0126 & 0.3650 & 0 & -0.0039 \end{bmatrix} \tag{6.358}$$

$$A_p = \begin{bmatrix} 0.0000 & 1.0000 & 0.0000 & 0.0000 & 0.0000 \\ 0.0000 & 0.0000 & 1.0000 & 0.0000 & 0.0000 \\ 0.0000 & 0.0000 & 0.0000 & 1.0000 & 0.0000 \\ 0.0000 & 0.0000 & 0.0000 & 0.0000 & 1.0000 \\ 0.0000 & -0.4253 & -48.8614 & -14.1354 & -9.4685 \end{bmatrix} \tag{6.359}$$

$$B_p = \begin{bmatrix} 0 \\ 0 \\ 0 \\ 0 \\ 1 \end{bmatrix} \tag{6.360}$$

The input with full-state feedback in the transformed system is

$$u_c = r(t) - Kz \tag{6.361}$$

where K is given by

$$K = [k_1 \quad k_2 \quad k_3 \quad k_4 \quad k_5] \tag{6.362}$$

then,

$$\dot{z} = (A_p - B_pK)z + B_pr(t) \tag{6.363}$$

Substituting, we get

$$\begin{bmatrix} \dot{z}_1 \\ \dot{z}_2 \\ \dot{z}_3 \\ \dot{z}_4 \\ \dot{z}_5 \end{bmatrix} = \begin{bmatrix} 0 & 1 & 0 & 0 & 0 \\ 0 & 0 & 1 & 0 & 0 \\ 0 & 0 & 0 & 1 & 0 \\ 0 & 0 & 0 & 0 & 1 \\ k_1 & -(0.4253 + k_2) & -(48.8614 + k_3) & -(14.1354 + k_4) & -(9.4685 + k_5) \end{bmatrix} \times \begin{bmatrix} z_1 \\ z_2 \\ z_3 \\ z_4 \\ z_5 \end{bmatrix} + \begin{bmatrix} 0 \\ 0 \\ 0 \\ 0 \\ 1 \end{bmatrix} r(t) \tag{6.364}$$

The characteristic equation of this transformed phase-variable form is

$$s[s^4 + (9.4685 + k_5)s^3 + (14.1354 + k_4)s^2 + (48.8614 + k_3)s + (0.4253 + k_2) + k_1] = 0 \tag{6.365}$$

Comparing the coefficient of Eqs. (6.340) and (6.365), we obtain $k_1 = 0$, $k_2 = 0.1869$, $k_3 = 27.8950$, $k_4 = 15.3543$, and $k_5 = 1.4395$.

Then next step is to transform back to the original state-space. Substituting for z, A_p, and B_p in Eq. (6.363), we obtain

$$\dot{x} = (A - BK\,P)x + Br(t) \tag{6.366}$$

$$= Ax + B[r(t) - K\,Px] \tag{6.367}$$

This gives us the desired state feedback control law as

$$u_c = r(t) - K\,Px \tag{6.368}$$

where the feedback gain matrix K is given by Eq. (6.362) with $k_1 \ldots k_5$ obtained as described and the transformation matrix P is given by Eq. (6.358).

To verify the design, we obtain the free response $[r(t) = 0]$ using MATLAB [1] for the initial conditions $\Delta\beta = 5$ deg, $\Delta\phi = p = \Delta\psi = r = 0$ as shown in Fig. 6.40. We observe that the system with full-state feedback law performs as expected.

6.5.5 Autopilots

Autopilots come in various types and degrees of sophistication and perform different functions. Autopilots are commonly used for maintaining pitch attitude, heading, altitude, Mach number, prescribed glide slope during approach, and landing.

The principal element of an autopilot is a sensing element called gyro (gyroscope). It may be a positional gyro or a rate gyro. Whenever there is a deviation from the preselected condition, the gyro senses the change and generates electrical signals, which are amplified and are used to drive the servo devices that move the appropriate control surfaces in the desired manner.

Pitch Displacement Autopilot

The function of the pitch displacement autopilot is to maintain the given pitch attitude of the aircraft. A schematic diagram of a typical pitch displacement autopilot is shown in Fig. 6.41a, and the block diagram implementation is shown in Fig. 6.41b. The vertical gyro is replaced by a summer, and the elevator servo is represented by a first-order lag with a time constant $\tau = 1/a$. The parameter a characterizes the response time of the servo and is usually in the range of 10 to 20. In this design, we will assume $a = 10$ so that $\tau = 0.1$. As said before in Sec. 6.5.1, we choose a negative sign for the elevator servo transfer function.

Let us consider the general aviation airplane and design a pitch displacement autopilot for it. We will use the short-period approximation because it

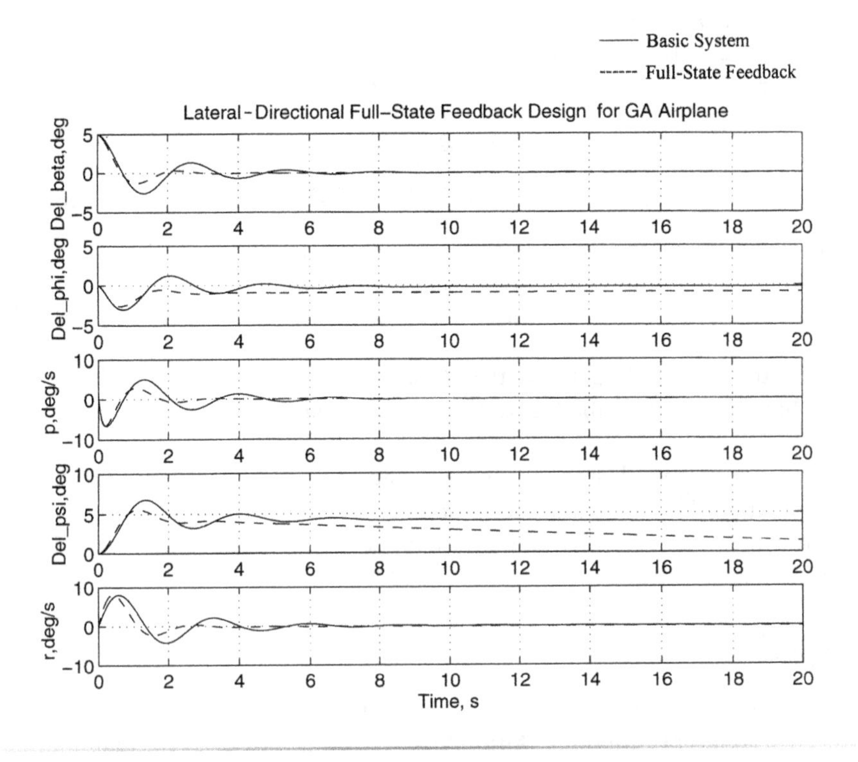

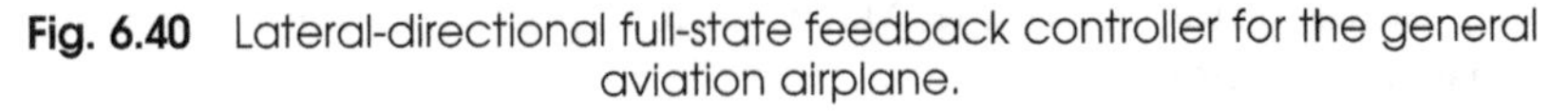

Fig. 6.40 Lateral-directional full-state feedback controller for the general aviation airplane.

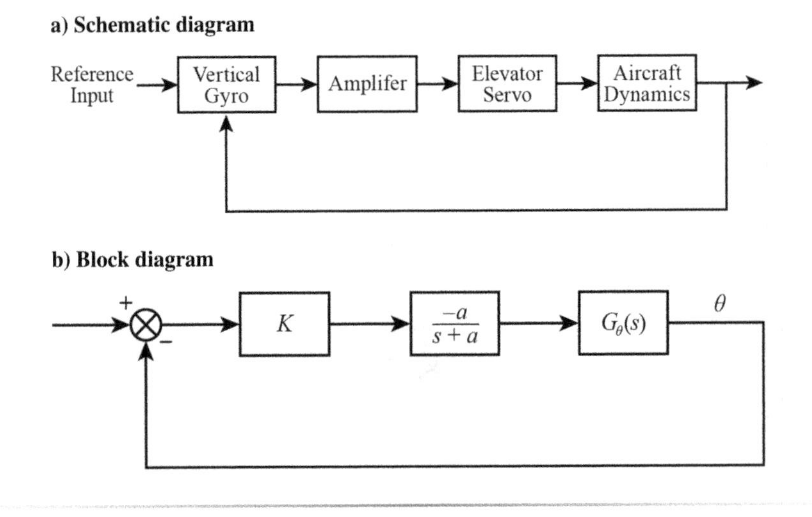

Fig. 6.41 Pitch displacement autopilot.

is found to be adequate for predicting the free and forced longitudinal responses. The short-period transfer function of the general aviation airplane for the pitch angle is given by Eq. (6.164)

$$\frac{\Delta\bar{\theta}(s)}{\Delta\bar{\delta}_e} = \frac{-2.0112\,s - 3.9018}{s(0.1695\,s^2 + 0.8494\,s + 2.1851)} \tag{6.369}$$

The first step is to draw the root-locus of the system with pitch angle feedback. We use MATLAB [1] to draw the root-locus as shown in Fig. 6.42. We observe that the aircraft (open-loop system) has a damping ratio in pitch of about 0.69 and a natural frequency of 2.6 rad/s, which still gives us level I handling qualities. However, as the amplifier gain increases, the closed-loop damping ratio reduces, and the aircraft becomes dynamically unstable. Therefore, we have to select a small value of the gain. We choose $k = 0.3335$, which gives a damping ratio of $\zeta = 0.53$ and still gives us level I flying qualities. With this value of the gain, the closed-loop poles are at -10.4636, $-1.9940 \pm j3.305$, and -0.5597.

To verify the design, we obtain a unit-step (one-rad elevator input) response using MATLAB [1] as shown in Fig. 6.43. The steady-state value of the output (pitch angle in rad) is unity indicating that the steady-state error is zero. Thus, the design is found to be satisfactory.

Pitch Rate Feedback In the just discussed design of pitch displacement autopilot for the general aviation airplane, the open-loop system had adequate damping in pitch; hence, on closing the loop, we had a small margin to select the amplifier gain. However, this may not always be the case.

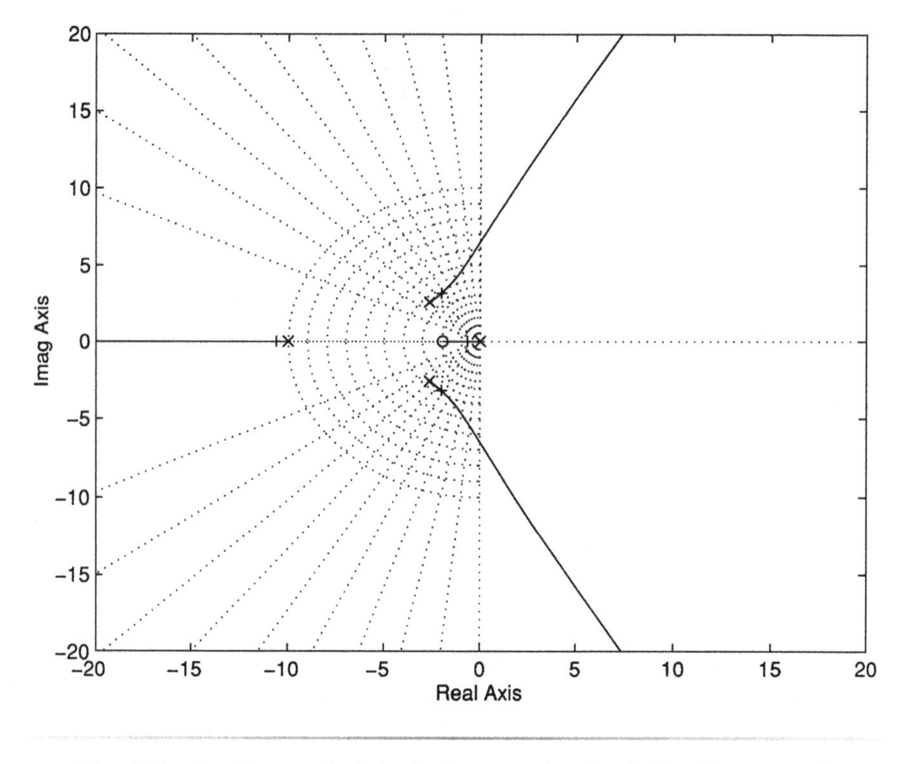

Fig. 6.42 Root-locus of pitch displacement autopilot for the general aviation airplane.

Several aircraft have much lower open-loop damping ratios in pitch so that, on closing the loop, they tend to become unstable even for very small values of the amplifier gain. In such cases, we have to add an inner-loop pitch rate feedback as shown in Fig. 6.44 and select the gain of the rate gyro so that the overall system has adequate damping. To illustrate this design procedure, we will consider a business jet [4] (see Exercise 6.1) whose short-period transfer function is given by

$$\frac{\Delta\bar{\theta}(s)}{\Delta\bar{\delta}_e(s)} = \frac{d_1 s + d_2}{s(a_1 s^2 + a_2 s + a_3)} \tag{6.370}$$

$$\frac{\bar{q}(s)}{\Delta\bar{\delta}_e(s)} = \frac{s\Delta\bar{\theta}(s)}{\Delta\bar{\delta}_e(s)} \tag{6.371}$$

$$= \frac{d_1 s + d_2}{a_1 s^2 + a_2 s + a_3} \tag{6.372}$$

where $d_1 = -6.6246$, $d_2 = -3.8069$, $a_1 = 3.1536$, $a_2 = 4.1624$, and $a_3 = 7.5662$.

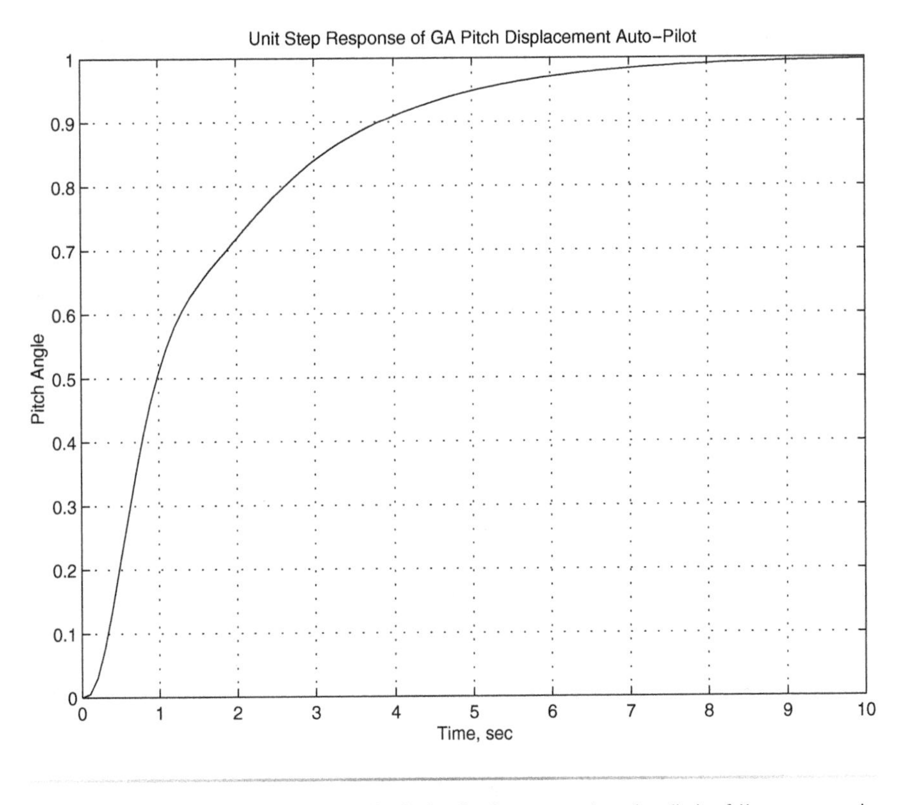

Fig. 6.43 Unit-step response of pitch displacement autopilot of the general aviation airplane.

The first step is to draw the root-locus of the inner loop as shown in Fig. 6.45 and select a suitable value for the rate gyro gain k_{rg}. We select a point on the root-locus for $\zeta = 0.9$ so that the system will have an adequate damping ratio when the outer loop is closed. For $\zeta = 0.9$, we get $k_{rg} = 0.8322$, and the locations of the closed-loop poles for the inner loop are -7.5275, $-1.8959 \pm j0.9623$.

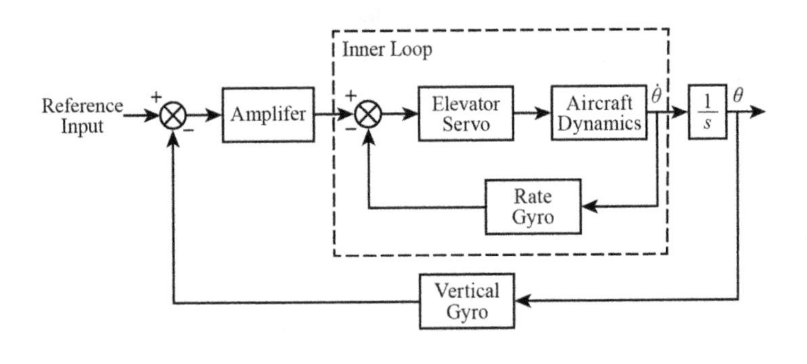

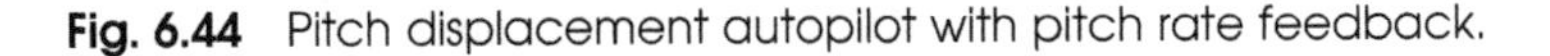
Fig. 6.44 Pitch displacement autopilot with pitch rate feedback.

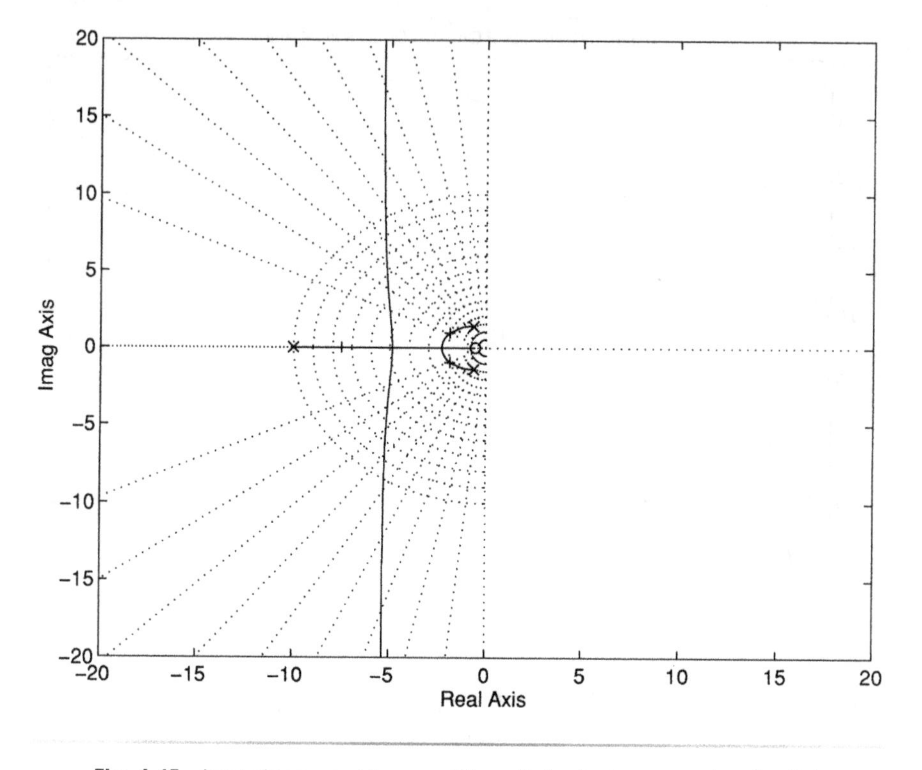

Fig. 6.45 Inner-loop root-locus of the pitch displacement autopilot for the business jet.

The block diagram of the outer loop is shown in Fig. 6.46. Here, G_{eq} is the equivalent transfer function of the inner loop, which is given by

$$G_{eq} = \frac{-10(d_1 s + d_2)}{a_1 s^3 + (a_2 + 10a_1)s^2 + s(a_3 + 10a_2 - 10k_{rg}d_1) + 10(a_3 - k_{rg}d_2)} \tag{6.373}$$

We draw the outer-loop root-locus using MATLAB [1] as shown in Fig. 6.47. We select a point on the root-locus to have a damping ratio of $\zeta = 0.7$ as shown, which is satisfactory to have level I handling qualities. For this value of $\zeta = 0.7$, we get $k_1 = 0.753$ and closed-loop poles at -7.9213, $-1.5942 \pm j1.7139$, and -0.2095. The pole at -7.9213 is due to the elevator

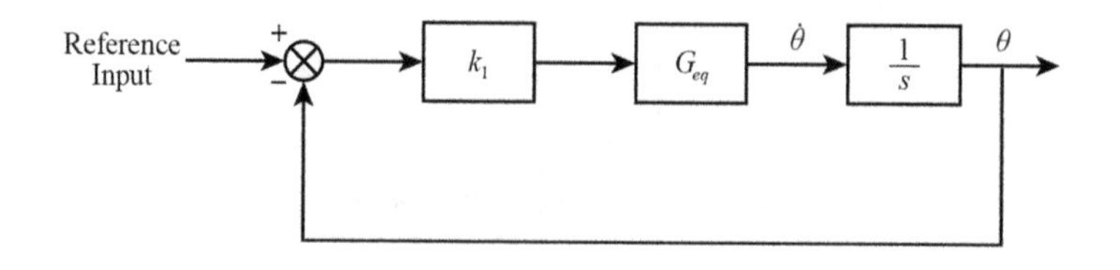

Fig. 6.46 Outer loop of the pitch displacement autopilot.

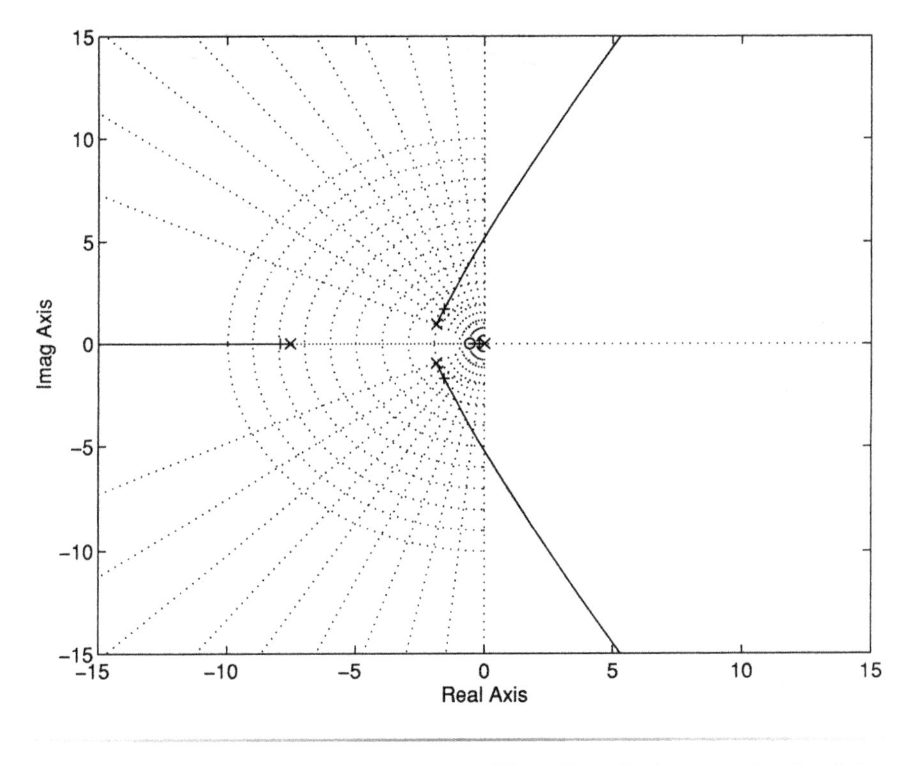

Fig. 6.47 Root-locus of the outer loop of the pitch displacement autopilot for the business jet.

servo and that at -0.2095 is the system pole due to the integrator ($s = 0$) in the forward path. To check the design, we perform a simulation to unit-step (one-deg elevator) input as shown in Fig. 6.48. We noted that the system response was satisfactory and the steady state error approached zero for $t > 30$ s.

Altitude-Hold Autopilot Altitude-hold autopilots are generally used by commercial airplanes during the cruise flight when the airplane is in the cruise mode. Usually, with such a system, the airplane is manually operated during the climb and descent portions of the flight. Once the aircraft has reached the desired cruise altitude, the altitude-hold autopilot is engaged. The autopilot will then maintain that altitude, making whatever corrections necessary when updrafts or downdrafts tend to cause the aircraft to gain or lose altitude.

Some autopilots have the feature that enables the pilot to enter a given flight altitude in advance. In this case, the autopilot flies the aircraft, making it climb or descend to attain the desired altitude. On reaching that altitude, the autopilot automatically levels the aircraft and, afterwards, it will maintain that altitude until told to do otherwise.

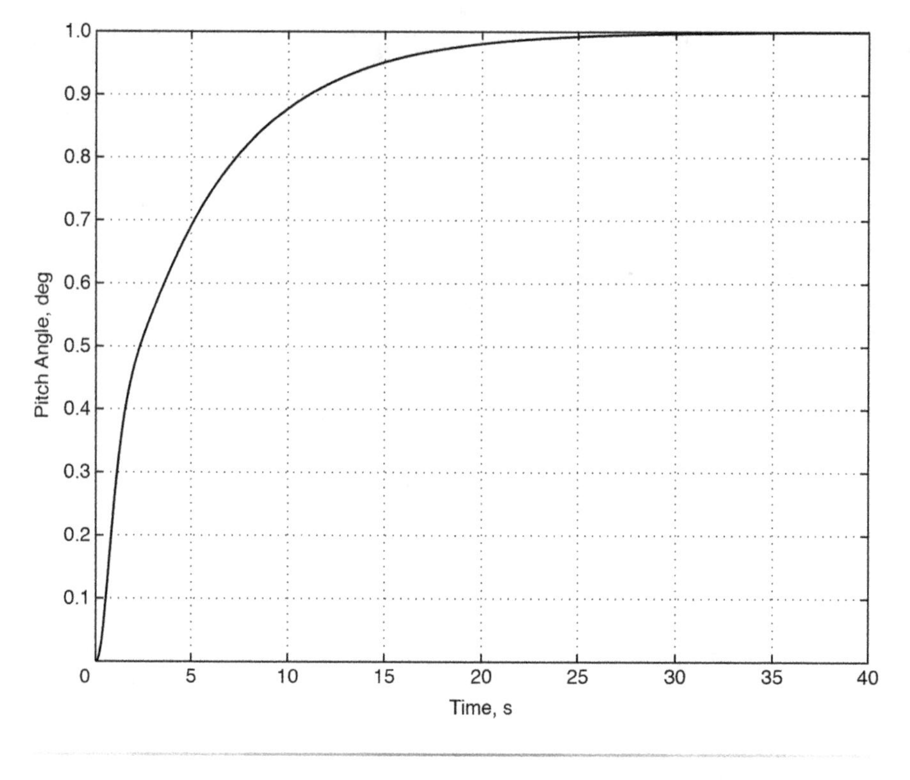

Fig. 6.48 Unit-step response of the pitch displacement autopilot for the business jet.

The schematic diagram of a typical altitude-hold autopilot is shown in Fig. 6.49a. The corresponding block diagram is shown in Fig. 6.49b. We have a lag compensator in the forward path, which is required to retain stability on closing the outer loop. The transfer function of the lag compensator is assumed as

$$G_c = \frac{k_3(s + 0.8)}{s + 8} \tag{6.374}$$

The transfer function for the altitude-to-elevator input can be obtained as follows:

$$\dot{h} = U_o \sin \gamma \tag{6.375}$$

$$\gamma = \theta - \alpha \tag{6.376}$$

where U_o is the flight velocity and γ is the flight path angle, which is assumed to be small. Then,

$$s\bar{h}(s) = U_o \bar{\gamma}(s) \tag{6.377}$$

$$\frac{\bar{h}(s)}{\Delta\bar{\delta}_e(s)} = \frac{U_o\bar{\gamma}(s)}{s\Delta\bar{\delta}_e(s)} \tag{6.378}$$

$$= \frac{U_o[\bar{\theta}(s) - \bar{\alpha}(s)]}{s\Delta\bar{\delta}_e(s)} \tag{6.379}$$

To illustrate the design procedure, let us consider the general aviation airplane once again and design an altitude-hold autopilot for it. Using Eqs. (6.163) and (6.164), we get

$$\frac{\bar{h}(s)}{\Delta\bar{\delta}_e(s)} = \frac{U_o\left(-b_1s^2 + (d_1 - b_2)s + (d_2 - b_3)\right)}{s(a_1s^3 + a_2s^2 + a_3s)} \tag{6.380}$$

where $b_1 = -0.0273$, $b_2 = -2.0366$, $b_3 = 0$, $a_1 = 0.1695$, $a_2 = 0.8494$, $a_3 = 2.1851$, $d_1 = -2.0112$, and $d_2 = -3.9018$.

At first, let us assume that there is no pitch rate feedback. The corresponding root-locus of the altitude-hold autopilot with only altitude feedback is shown in Fig. 6.50. We observe that, as soon as the outer loop is closed, the system becomes unstable for any positive value of gain k_3. Therefore, we need to add the inner loop with pitch rate feedback as shown in Fig. 6.51. The transfer function G_q is given by

$$G_q = \frac{\bar{q}(s)}{\Delta\bar{\delta}_e(s)} = \frac{d_1s + d_2}{a_1s^2 + a_2s + a_3} \tag{6.381}$$

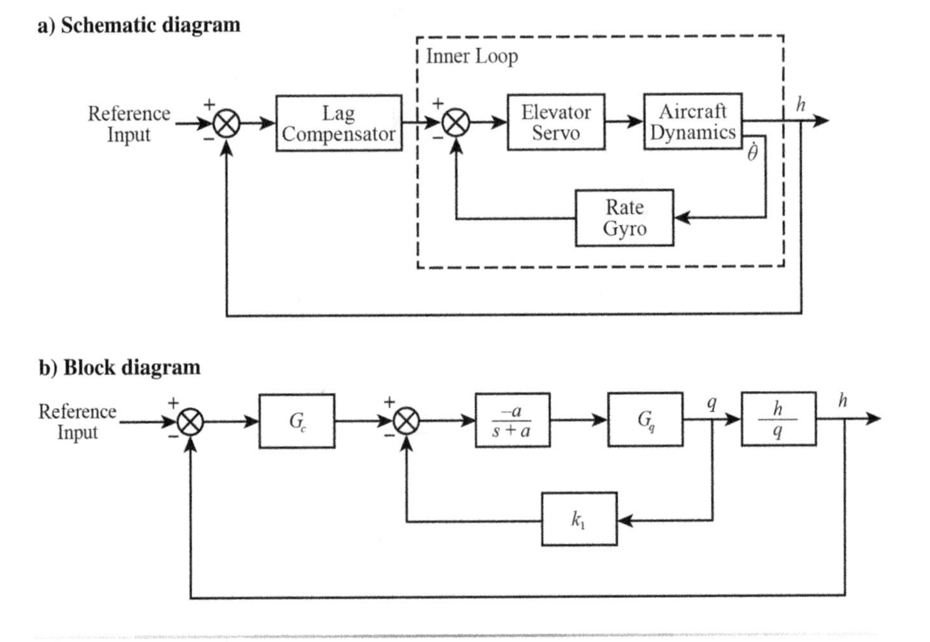

Fig. 6.49 Altitude-hold autopilot.

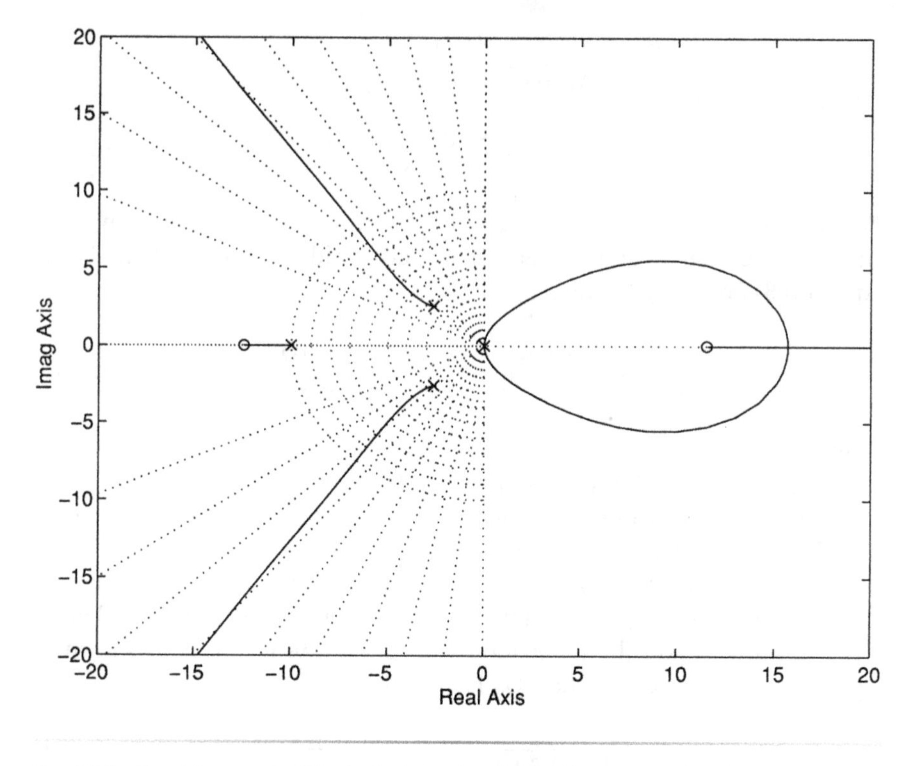

Fig. 6.50 Root-locus of altitude-hold autopilot for the general aviation airplane.

Now we plot the root-locus of the inner loop as shown in Fig. 6.52 and pick a point on the root-locus corresponding to $\zeta = 0.8$, which gives the gain of rate gyro $k_1 = 0.2296$. This is a small improvement in the damping ratio compared to the basic system (open loop) damping of 0.7 but may be sufficient to make the system stable on closing the outer loop.

The next step is to design the outer loop (see Fig. 6.53). The transfer functions G_{eq} and $\bar{h}(s)/\bar{q}(s)$ are given by the following expressions:

$$G_{eq} = \frac{-10(d_1 s + d_2)}{a_1 s^3 + (a_2 + 10a_1)s^2 + (a_3 + 10a_2 - 10k_1 d_1)s + 10(a_3 - k_1 d_2)} \tag{6.382}$$

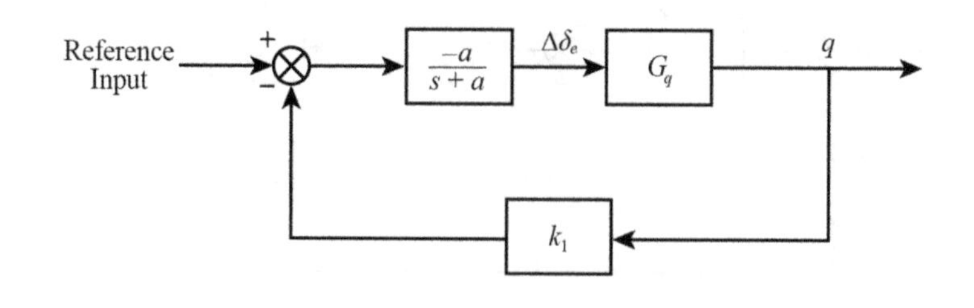

Fig. 6.51 Inner loop of the altitude-hold autopilot.

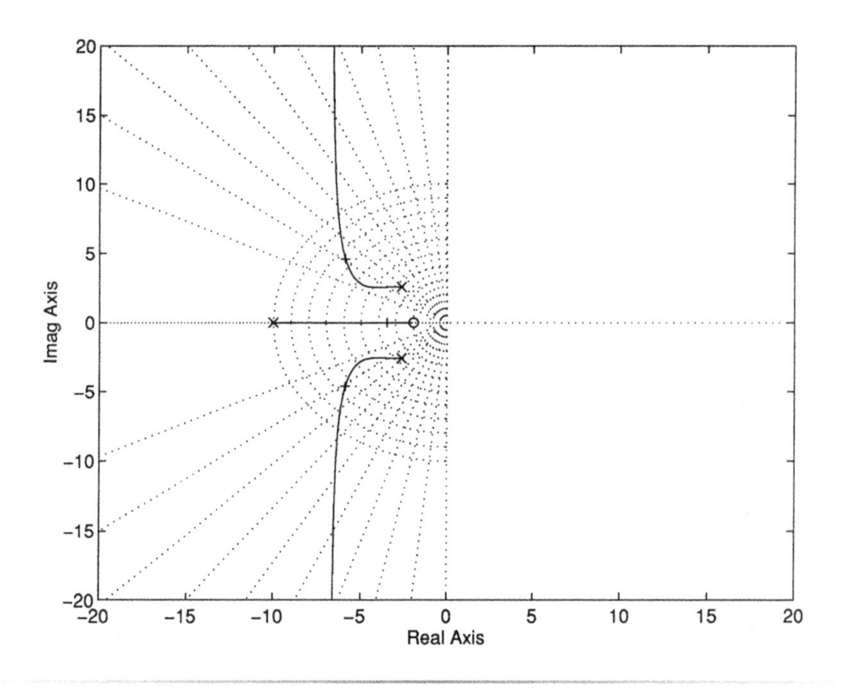

Fig. 6.52 Root-locus of the inner loop of the altitude-hold autopilot for the general aviation airplane.

$$\frac{\bar{h}(s)}{\bar{q}(s)} = \left(\frac{U_o}{s^2}\right)\left(\frac{-b_1 s^2 + (d_1 - b_2)s + d_2 - b_3}{d_1 s + d_2}\right) \tag{6.383}$$

The lag-compensator pole ($s = -8$) is sufficiently away from the origin so that it has very little effect on the short-period dynamics. The root-locus of the outer loop is shown in Fig. 6.54. We observe that we have to confine ourselves to very low values of compensator gain k_3 to keep the system stable. A higher value of k_3 will make the system unstable. The effect of the compensator is obvious. It has pulled to the left the branches of the root-locus that originate from the two poles at the origin and migrate to the right half of the plane. This provides a small range of stable values for

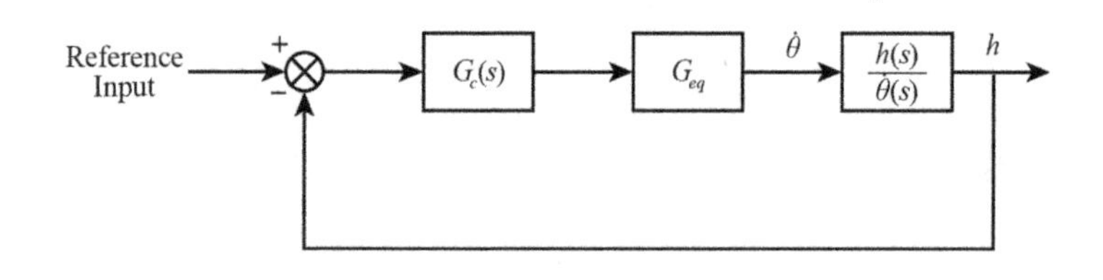

Fig. 6.53 Outer loop of the altitude-hold autopilot.

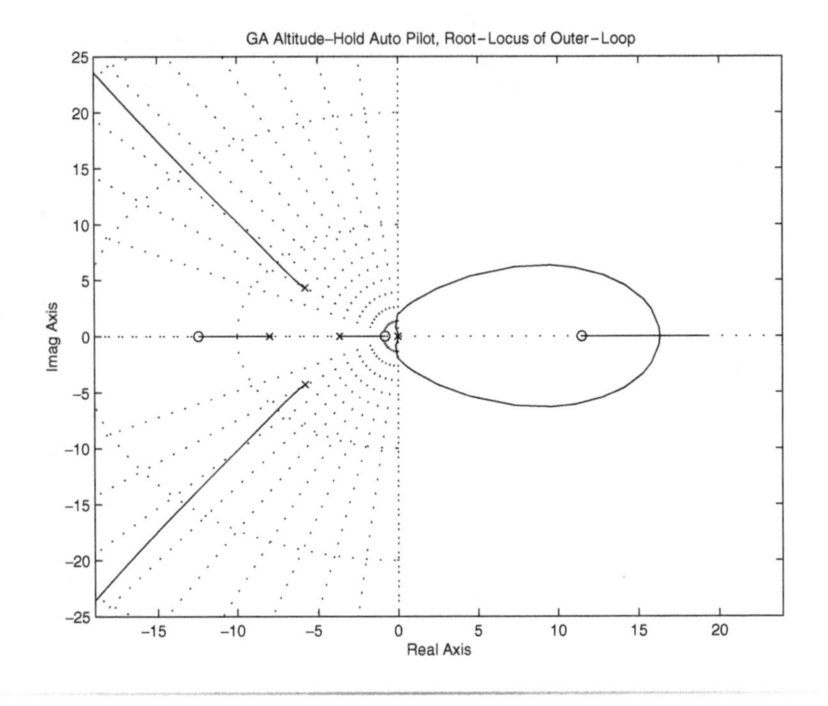

Fig. 6.54 Root-locus of the outer loop of the pitch displacement autopilot for the general aviation airplane.

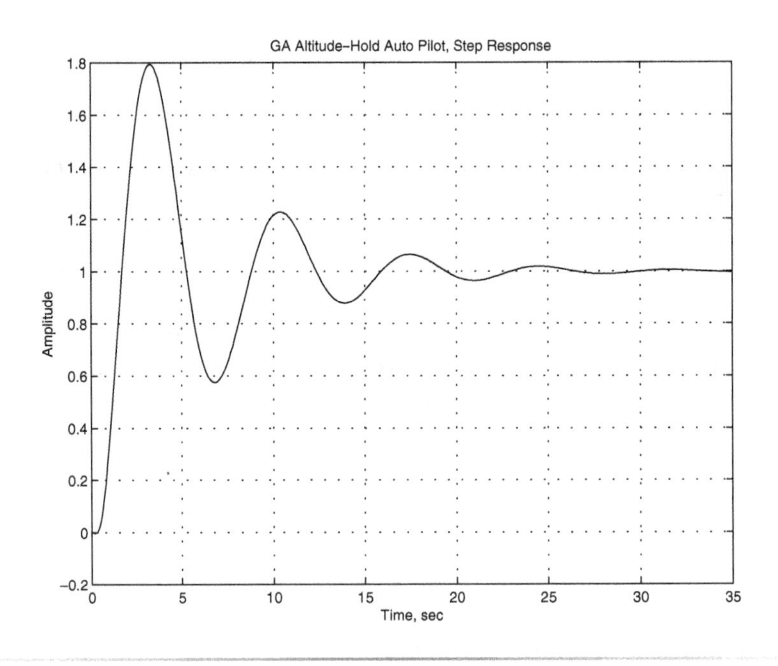

Fig. 6.55 Unit-step response of the pitch displacement autopilot for the general aviation airplane.

the complex conjugate poles. With $k_3 = 0.0214$, we get the closed-loop poles at -8.4427, $-6.1427 \pm j4.6397$, $-0.1536 \pm j0.7955$, and -2.2242.

To verify the design, we simulate the response of the system to a unit-step elevator input as shown in Fig. 6.55. The large overshoot is due to the low value of the damping ratio. Except for this, the system performance is satisfactory.

6.6 Summary

In this chapter, we have studied the longitudinal and lateral-directional responses of the airplane. The longitudinal response of a statically stable airplane consists of two distinct modes, the short-period mode and the phugoid mode. The lateral-direction motion comprises of roll-subsidence mode, the Dutch-roll oscillation, and the spiral modes. Accordingly, we obtained simplified transfer functions and state-space models to study these modes individually. These simplified transfer functions were found to be of considerable help in designing the flight control systems.

For a statically unstable airplane, the usual short-period and phugoid modes do not exist. Instead, we have two exponential modes and an oscillatory mode. The oscillatory mode is called the third oscillatory mode that has short-period-like damping and phugoid-like frequency. Because of static instability, at least one of the two exponential modes will be unstable.

The handling quality requirements form the basis of the design of stability augmentation systems and autopilots. The longitudinal flying quality requirements were presented in terms of damping ratios and frequencies of the short-period and phugoid modes. Similarly, the lateral-directional flying qualities were presented in terms of damping ratios and frequency for Dutch-roll mode and in terms of time for half or double the amplitude for roll-subsidence and spiral modes.

We have illustrated the basic design procedures for stability augmentation systems and autopilots. It was not possible to cover all types of stability augmentation systems and autopilots that are in use on modern airplanes. The interested reader may refer elsewhere [5,7] for additional information.

References

[1] *Pro-MATLAB for Sun Workstations*, The MathWorks, Natick, MA, Jan. 1990.
[2] Teper, G. L., "Aircraft Stability and Control Data," NASA CR-96008, April 1969.
[3] Roskam, J., *Airplane Flight Dynamics and Automatic Flight Control, Part II*, Roskam Aviation and Engineering Corp., Lawrence, KS, 1979.
[4] Nelson, R. C., *Flight Stability and Automatic Control*, McGraw-Hill, New York, 1989.
[5] Stevens, B. L., and Lewis, F. L., *Aircraft Control and Simulation*, Wiley, New York, 1994.

[6] MIL-F-8785C, "Flying Qualities of Piloted Airplanes," *U.S. Department of Defense Military Specifications*, Nov. 5, 1980.
[7] Blakelock, J. H., *Automatic Control of Aircraft and Missiles*, 2nd ed., Wiley, New York, 1991.

Problems

6.1 The mass properties and aerodynamic characteristics of a business jet airplane [4] are as follows.

Mass properties: weight $W = 169{,}921.24$ N, wing area $S = 50.3983\ \text{m}^2$, wing span $b = 16.383$ m, mean aerodynamic chord $\bar{c} = 3.3315$ m, distance between center of gravity and aerodynamic center in terms of mean aerodynamic chord $\bar{x}_{cg} = 0.25$, $I_x = 161{,}032.43\ \text{kgm}^2$, $I_y = 184{,}211.19\ \text{kgm}^2$, $I_z = 330{,}142.72\ \text{kgm}^2$, $I_{xy} = I_{yz} = 0$, and $I_{xz} = 6861.7038\ \text{kgm}^2$.

Longitudinal aerodynamic characteristics: $C_L = 0.737$, $C_D = 0.095$, $C_{L\alpha} = 5.0$, $C_{D\alpha} = 0.75$, $C_{m\alpha} = -0.8$, $C_{L\dot{\alpha}} = 0$, $C_{D\dot{\alpha}} = 0$, $C_{m\dot{\alpha}} = -3.0$, $C_{LM} = 0$, $C_{DM} = 0$, $C_{mM} = -0.05$, $C_{L\delta_e} = 0.4$, $C_{D\delta_e} = 0$, $C_{m\delta_e} = -0.81$, $C_{Lq} = 0$, $C_{Dq} = 0$, and $C_{mq} = -8.0$.

Lateral-directional aerodynamic characteristics: $C_{y\beta} = -0.720$, $C_{l\beta} = -0.103$, $C_{n\beta} = 0.137$, $C_{y\delta_a} = 0$, $C_{l\delta_a} = -0.054$, $C_{n\delta_a} = 0.0075$, $C_{y\delta_r} = 0.175$, $C_{l\delta_r} = 0.029$, $C_{n\delta_r} = -0.063$, $C_{yp} = 0$, $C_{lp} = -0.37$, $C_{np} = -0.14$, $C_{yr} = 0$, $C_{lr} = 0.11$, $C_{nr} = -0.16$, $C_{y\dot{\beta}} = 0$, $C_{l\dot{\beta}} = 0$, and $C_{n\dot{\beta}} = 0$.

Note that all derivatives are per radian.

For flight at Mach 0.2, determine the following:

a) The state-space representations, eigenvalues, free responses, and unit-step responses for the longitudinal and lateral-directional dynamics based on a complete system as well as longitudinal (short-period and phugoid) and lateral-directional (roll-subsidence, Dutch-roll, and spiral) approximations.

b) Transfer functions, roots of characteristic equations, steady-state values for unit-step inputs for longitudinal and lateral-directional variables based on a complete system as well as longitudinal (short-period and phugoid) and lateral-directional (roll-subsidence, Dutch-roll, and spiral) approximations. Compare these results with those obtained in (a).

c) The longitudinal (elevator input) and lateral-directional (aileron and rudder inputs) frequency responses based on a complete system.

6.2 The longitudinal characteristic equation of an airplane is given by

$$s^4 + 1.8\,s^3 + 3.5\,s^2 + 0.25\,s + 0.05 = 0$$

Examine the longitudinal stability of the airplane using Routh's criterion.

6.3 The longitudinal transfer functions of an airplane based on the short-period approximations are given by

$$G_\theta(s) = \frac{-(5.25\,s + 3.15)}{s(2.25\,s^2 + 3.15\,s + 6.53)}$$

$$G_\alpha(s) = \frac{-(0.04\,s^2 + 4\,s)}{s(2.25\,s^2 + 3.15\,s + 6.53)}$$

Design (a) a pitch displacement autopilot to give level I handling qualities and (b) an altitude-hold autopilot for $M = 0.2$ at sea level to give acceptable handling qualities.

6.4 A light airplane has the following longitudinal transfer functions:

$$G_\alpha(s) = \frac{-(0.2\,s^3 + 10\,s^2 + s + 1.5)}{(s^4 + 4\,s^3 + 2\,s^2 + 0.15\,s - 0.1)}$$

$$G_\theta(s) = \frac{-(10\,s^2 + 20\,s + 1.5)}{(s^4 + 4\,s^3 + 2\,s^2 + 0.15\,s - 0.1)}$$

Determine the longitudinal eigenvalues. Is the airplane longitudinally stable? If not, design a suitable pitch augmentation system so that the airplane has conventional short-period and phugoid modes in the closed-loop sense.

6.5 The transfer function of an airplane for yaw rate to rudder deflection is given by

$$\frac{\bar{r}(s)}{\Delta\bar{\delta}_r(s)} = \frac{-(4.5\,s + 0.75)}{(s^2 + 1.1\,s + 4.2)}$$

Design a yaw damper to give level I Dutch-roll flying qualities.

6.6 The longitudinal dynamics of an airplane in state-space form is given by

$$\begin{bmatrix} \dot{u} \\ \Delta\dot{\alpha} \\ \dot{q} \\ \Delta\dot{\theta} \end{bmatrix} = \begin{bmatrix} -0.01 & 0.05 & 0 & -0.2 \\ -0.50 & -2.5 & 1.0 & -0.05 \\ 0.5 & -5.0 & -3.5 & 0.05 \\ 0 & 0 & 1 & 0 \end{bmatrix} \begin{bmatrix} u \\ \Delta\alpha \\ q \\ \Delta\theta \end{bmatrix} + \begin{bmatrix} 0 \\ -0.5 \\ -7.5 \\ 0 \end{bmatrix} \Delta\delta_e$$

Design a full-state feedback controller to give a natural frequency of 5.0 rad/s and damping ratio of 0.8 for the short-period mode and a natural frequency of 0.15 rad/s and damping ratio of 0.15 for the phugoid mode.

6.7 The lateral-directional dynamics of an airplane in state-space form is given by

$$\begin{bmatrix} \Delta\dot{\beta} \\ \Delta\dot{\phi} \\ \dot{p} \\ \Delta\dot{\psi} \\ \dot{r} \end{bmatrix} = \begin{bmatrix} -0.2 & 0.2 & 0 & 0 & -1.0 \\ 0 & 0 & 1.0 & 0 & 0 \\ -15.0 & 0 & -8.0 & 0 & 2.0 \\ 0 & 0 & 0 & 0 & 1 \\ 3.5 & 0 & -0.5 & 0 & -0.20 \end{bmatrix} \begin{bmatrix} \Delta\beta \\ \Delta\phi \\ p \\ \Delta\psi \\ r \end{bmatrix} + \begin{bmatrix} 0 & 0.05 \\ 0 & 0 \\ 25.0 & 2.0 \\ 0 & 0 \\ -0.15 & -4.0 \end{bmatrix} \begin{bmatrix} \Delta\delta_a \\ \Delta\delta_r \end{bmatrix}$$

Design a full-state feedback controller to place the closed-loop eigenvalues at 0, $-0.6 \pm j1.9079$, -0.005, and -8.0.

Chapter 7 Inertia Coupling and Spin

7.1 Introduction

In the preceding chapters, we have studied the airplane stability, dynamics, and control with the assumption that the longitudinal and lateral-directional motions of the airplane could be decoupled and studied separately. With these assumptions, the problem of airplane dynamics and control was linearized so that the analyses methods of linear systems could be used. For example, the response of the airplane to a 2-deg elevator input would be exactly double that for a 1-deg elevator input. Also, there was no cross coupling between the longitudinal and lateral-directional degrees of freedom. In other words, operating the rudder or the ailerons would not generate any pitching motion or a change in the forward speed or a change in the angle of attack. Similarly, moving the elevator would not generate any sideslip, rolling, or yawing motion. As a consequence, one had to solve the problem for only one set of control inputs and, using those solutions, the response to any other combination of control inputs could be quickly deduced. Furthermore, the response to combined control inputs was the sum of the responses to individual control inputs. In other words, the flight dynamicist had to solve the problem only once. Then, he or she had it solved for all other cases. What could be simpler?

However, this type of simple approach cannot be used for problems in which the longitudinal and lateral motions are coupled. Such coupling occurs because of either inertial cross coupling or aerodynamic nonlinearities. The examples of the first category are the inertia or roll coupling problems and the spinning motion. We will study these two problems in this chapter. The cross coupling between longitudinal and lateral motions due to aerodynamic nonlinearities and large amplitude motions occurs in flight at high angles of attack close to or exceeding the stalling angle. We will study these problems in Chapter 8 for airplanes and in Chapter 9 for unmanned aerial vehicles.

7.2 Inertia Coupling

The problem of inertia coupling was totally unknown to the aeronautical engineer until the closing years of World War II. During a demonstration flight in late 1944, the German fighter aircraft Heinkel 162 disintegrated in a fast high-speed rolling maneuver. The Heinkel 162 was a small single-engine jet fighter. It so happened that a number of cameras recorded this event, which helped a close reconstruction of the entire episode. However, with the knowledge and expertise available then, no one could come up with a convincing explanation as to why the Heinkel 162 aircraft behaved that way. The second example was that of the British Fairey Delta aircraft during an air show at Farnborough, England. This aircraft had an impressive roll rate capability of 500 deg/s. Halfway through the roll, the pilot lost control, or more appropriately, the airplane took over. Luckily, this incidence did not result in a fatality or severe damage to the aircraft. The Fairey Delta aircraft carried extensive flight test instrumentation, which recorded the event. Once again, this incident offered a frustrating puzzle because no calculated response even with wild assumptions about the misbehaved aerodynamics could come near the observed motion of the Fairey Delta aircraft. At the same time, the very same approach gave perfectly good results when the roll rate was small.

These two examples demonstrate vividly the essential features of the problem now known as inertia coupling or roll coupling. It has been widely recognized that inertia coupling will occur on aircraft that have long slender fuselages and short aspect ratio wings and have most of their mass concentrated in the fuselage. Such aircraft have a low value of moment of inertia about the longitudinal (x) axis and fairly large values of moments of inertia about y- and z-axes. Furthermore, it is also understood that a loss of either longitudinal (pitch) stability or the directional stability compounds this problem further. This type of mass/inertia distribution and stability characteristics are typical of modern high-speed fighter aircraft.

A fundamental analysis of inertia coupling was presented by Philips in 1948 [1]. He examined the stability of a steadily rolling aircraft. He assumed that all the disturbance variables except the steady roll rate were small. He also ignored the damping in pitch and yaw. Based on this analysis, he demonstrated that a steadily rolling airplane deficient in pitch stability experiences a divergence in pitch, whereas one deficient in directional stability experiences a divergence in yaw or sideslip. Even though this simple analysis did not actually consider the real rolling maneuvers like that of the Heinkel 162 or Fairey Delta aircraft, it helped the flight dynamicist identify the root cause of the problem. Since then, the problem of inertia coupling has received considerable attention from many authors who have performed more rigorous analyses of this problem. Interested readers may refer to the literature for more information on this subject [2–6].

In this section, we will first present a brief physical explanation of the problem of inertia coupling. Then, we will develop the theory of stability of a steadily rolling aircraft and derive Philip's criteria for divergence in pitch or yaw.

7.2.1 Yaw and Pitch Divergence in Rolling Maneuvers

To understand the basic physical principles of inertia coupling, let us assume that the mass of the airplane is concentrated in four distinct lumps at the extremities of the fuselage and wings as shown in Fig. 7.1. The two masses M_1 and M_2 at either end of the fuselage represent the inertia in pitch I_y, and the two masses M_3 and M_4 at the wingtips represent the roll inertia I_x. All the four masses acting together contribute to the inertia in yaw I_z.

Now let us consider what happens when the aircraft starts rolling about its longitudinal (x) axis as shown in Fig. 7.1a. The fuselage masses M_1 and M_2 representing the inertia in pitch do not develop any centrifugal reaction, whereas the wing masses representing the roll inertia develop equal and opposite centrifugal reactions that will try to tear the wings apart. Because they cancel each other, these forces do not cause any inertia coupling but concern the structural engineer who has to provide enough strength and rigidity to the airframe to deal with this situation.

Now let us assume that, for some reason, the rolling aircraft is disturbed in yaw from an equilibrium condition. For this case, all four masses are subjected to centrifugal reactions as shown in Fig. 7.1b. The fuselage masses M_1 and M_2 representing the inertia in pitch will attempt to further increase the yaw, and the wing masses M_3 and M_4 representing the inertia in roll will try to reduce yaw and restore the aircraft to its original equilibrium condition. Whether the net result is stabilizing or destabilizing depends on the relative magnitudes of these two centrifugal reactions. If the inertia in pitch exceeds the inertia in roll, which is usually the case for most of the modern combat aircraft, then the destabilizing couple will dominate. Whether the aircraft will experience a divergence in yaw because of this destabilizing inertia-induced yawing couple depends on the level of static directional stability. If the restoring yawing moment due to directional stability overpowers the inertia-induced destabilizing yawing moment, the aircraft returns to its equilibrium condition. If not, the aircraft will experience a divergence in yaw or sideslip. This phenomenon is known as the roll–yaw coupling.

Now let us consider another situation where the aircraft is rolling about the velocity vector as shown in Fig. 7.1c. Such a rolling motion is usually preferred at high angles of attack to avoid the sideslip buildup. In this case, only the fuselage masses representing pitch inertia will contribute to the centrifugal reaction, which produces a destabilizing couple in pitch. If this destabilizing inertia-induced pitching moment is greater than the restoring moment due to static longitudinal stability, the aircraft will experience a divergence in pitch. In other words, an air-craft deficient in static longitudinal stability is prone to divergence in pitch due to inertia coupling during velocity vector rolls.

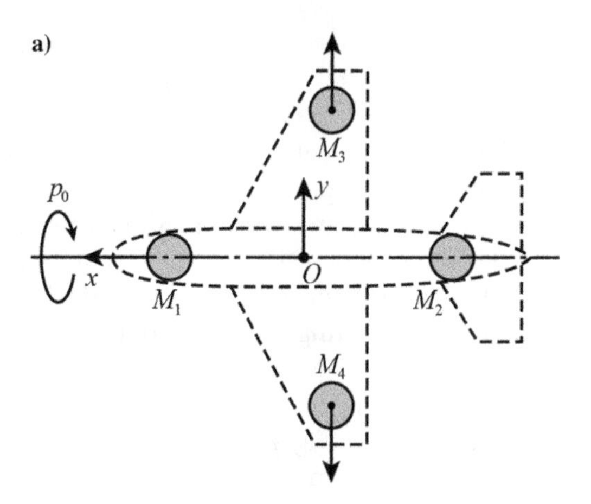

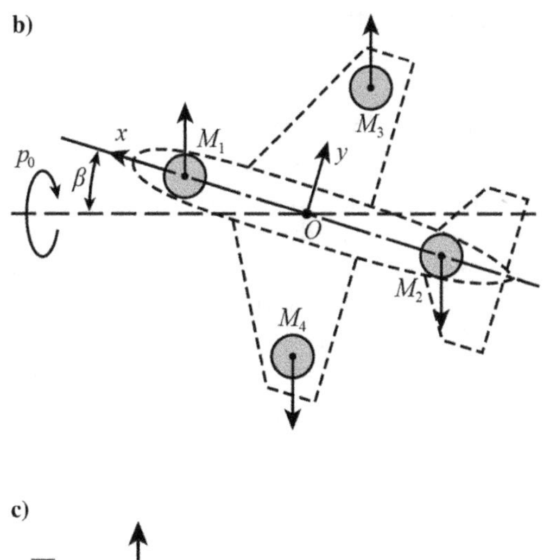

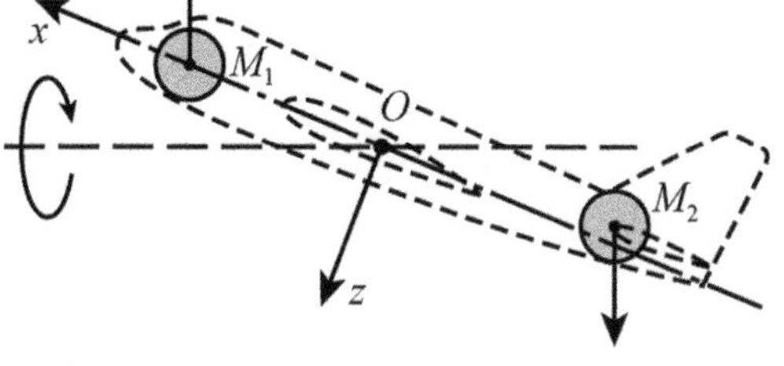

Fig. 7.1 Inertia coupling effects.

The pitch inertia increases with the length of the fuselage, thereby increasing the magnitude of the destabilizing inertia-induced pitching couple. Furthermore, to improve performance, modern fighter aircraft compromise static stability in pitch. These two factors make modern combat aircraft vulnerable to inertia cross coupling problems.

7.2.2 Equations of Motion of a Steadily Rolling Aircraft

We have the following force and moment equations [Eqs. (4.291–4.293, 4.357–4.359)],

$$F_x = m(\dot{U} + qW - rV) \tag{7.1}$$

$$F_y = m(\dot{V} + Ur - pW) \tag{7.2}$$

$$F_z = m(\dot{W} + pV - Uq) \tag{7.3}$$

$$L = \dot{p}I_x - I_{xz}(pq + \dot{r}) + qr(I_z - I_y) \tag{7.4}$$

$$M = \dot{q}I_y + rp(I_x - I_z) + (p^2 - r^2)I_{xz} \tag{7.5}$$

$$N = \dot{r}I_z - I_{xz}(\dot{p} - qr) + pq(I_y - I_x) \tag{7.6}$$

Equations (7.1–7.6) are the complete six-degree-of-freedom equations for airplane motion. In general, it is difficult to obtain analytical solutions to these equations. In the preceding chapters, we assumed that all the flight-path variables in the disturbed motion are small. With this assumption, we obtained two sets of linear decoupled equations, one set of three equations for longitudinal motion and another set of three equations for lateral-directional motion. Here, for the stability of a steadily rolling airplane, we will proceed along similar lines assuming that all the disturbance variables are small. Note that the nonzero steady-state roll rate p_0 is not restricted to be small. However, we assume that p_0 is constant so that $\dot{p} = 0$. We need to introduce some more assumptions to simplify the problem further. We will assume that the product of inertia I_{xz} is small so that it can be ignored and further assume that the flight velocity is constant. With the assumptions of constant flight velocity and constant roll rate, Eqs. (7.1) and (7.4) reduce to two algebraic equations. These equations can be used to derive control settings, which is not the main issue here. Hence, we ignore these two equations from our analysis. With these assumptions, we can express the remaining four equations as (see Sec. 4.3.4)

$$\Delta F_y = mU_0(\Delta\dot{\beta} + r - p_0\Delta\alpha) \tag{7.7}$$

$$\Delta F_z = mU_0(\Delta\dot{\alpha} + p_0\Delta\beta - q) \tag{7.8}$$

$$\Delta M = \dot{q}I_y + rp_0(I_x - I_z) \tag{7.9}$$

$$\Delta N = \dot{r}I_z + p_0q(I_y - I_x) \tag{7.10}$$

Notice that the longitudinal and lateral-directional motions of the aircraft are now coupled because of the nonzero steady-state rolling velocity p_0. However, Eqs. (7.7–7.10) are still linear because we have retained the assumption of small disturbances in all the disturbance variables. For a nonrolling aircraft ($p_0 = 0$), these equations reduce to two sets of uncoupled, linear equations, one set of two equations for the longitudinal motion (short-period oscillation) and another set of two equations for the lateral-directional motion (Dutch-roll oscillation) as we have studied in Chapter 6.

Divide both sides of Eqs. (7.7) and (7.8) by qS, those of Eq. (7.9) by $qS\bar{c}$ and those of Eq. (7.10) by qSb so that we can express them in the nondimensional form as follows:

$$\Delta C_y = m_1(\Delta\dot{\beta} + r - p_0\Delta\alpha) \tag{7.11}$$

$$\Delta C_z = m_1(\Delta\dot{\alpha} + p_0\Delta\beta - q) \tag{7.12}$$

$$\Delta C_m = \dot{q}I_{y1} + rp_0(I_{x1} - I_{z1}) \tag{7.13}$$

$$\Delta C_n = \dot{r}I'_{z1} + p_0q(I'_{y1} - I'_{x1}) \tag{7.14}$$

where

$$m_1 = \frac{2m}{\rho U_0 S} \tag{7.15}$$

$$I_{x1} = \frac{I_x}{qS\bar{c}} \tag{7.16}$$

$$I_{y1} = \frac{I_y}{qS\bar{c}} \tag{7.17}$$

$$I_{z1} = \frac{I_z}{qS\bar{c}} \tag{7.18}$$

$$I'_{x1} = \frac{I_x}{qSb} \tag{7.19}$$

$$I'_{y1} = \frac{I_y}{qSb} \tag{7.20}$$

$$I'_{z1} = \frac{I_z}{qSb} \tag{7.21}$$

Here, q is the freestream dynamic pressure, S is the wing (reference) area, $\bar{c}$ is the mean aerodynamic chord, and b is the wing span. It may be noted that some of the given nomenclature involving the moments of inertia terms differ from those introduced in Chapter 4.

Let

$$\Delta C_y = C_{y\beta}\Delta\beta \tag{7.22}$$

$$\Delta C_z = C_{z\alpha}\Delta\alpha \tag{7.23}$$

$$\Delta C_m = C_{m\alpha}\Delta\alpha + C_{mq}\left(\frac{q\bar{c}}{2U_0}\right) \tag{7.24}$$

$$\Delta C_n = C_{n\beta}\Delta\beta + C_{nr}\left(\frac{rb}{2U_0}\right) \tag{7.25}$$

For simplicity, we have ignored the acceleration derivatives like $C_{y\dot{\beta}}$, $C_{m\dot{\alpha}}$, and $C_{n\dot{\beta}}$ and other rotary derivatives like C_{yp}, C_{yr}, and C_{np}. With these assumptions, Eqs. (7.11–7.14) can be expressed as follows:

$$m_1 p_0 \Delta\alpha + \left(C_{y\beta} - m_1\frac{\mathrm{d}}{\mathrm{d}t}\right)\Delta\beta - m_1 r = 0 \tag{7.26}$$

$$\left(C_{z\alpha} - m_1\frac{\mathrm{d}}{\mathrm{d}t}\right)\Delta\alpha - m_1 p_0\Delta\beta + m_1 q = 0 \tag{7.27}$$

$$C_{m\alpha}\Delta\alpha + \left(C_{mq}c_1 - I_{y1}\frac{\mathrm{d}}{\mathrm{d}t}\right)q - (I_{x1} - I_{z1})\,p_0 r = 0 \tag{7.28}$$

$$C_{n\beta}\Delta\beta - (I'_{y1} - I'_{x1})\,p_0 q + \left(C_{nr}b_1 - I'_{z1}\frac{\mathrm{d}}{\mathrm{d}t}\right)r = 0 \tag{7.29}$$

where

$$c_1 = \frac{\bar{c}}{2U_0} \tag{7.30}$$

$$b_1 = \frac{b}{2U_0} \tag{7.31}$$

Taking Laplace transforms, we get

$$m_1 p_0 \Delta\bar{\alpha}(s) + (C_{y\beta} - m_1 s)\Delta\bar{\beta}(s) - m_1\bar{r}(s) = 0 \tag{7.32}$$

$$(C_{z\alpha} - m_1 s)\Delta\bar{\alpha}(s) - m_1 p_0\Delta\bar{\beta}(s) + m_1\bar{q}(s) = 0 \tag{7.33}$$

$$C_{m\alpha}\Delta\bar{\alpha}(s) + (C_{mq}c_1 - I_{y1}s)\bar{q}(s) - (I_{x1} - I_{z1})\,p_0\bar{r}(s) = 0 \tag{7.34}$$

$$C_{n\beta}\Delta\bar{\beta}(s) + (I'_{y1} - I'_{x1})\,p_0\bar{q}(s) + (C_{nr}b_1 - I'_{z1}s)\bar{r}(s) = 0 \tag{7.35}$$

For a nontrivial solution of these equations, we must have

$$\begin{vmatrix} m_1 p_0 & (C_{y\beta} - m_1 s) & 0 & -m_1 \\ (C_{z\alpha} - m_1 s) & -m_1 p_0 & m_1 & 0 \\ C_{m\alpha} & 0 & (C_{mq}c_1 - I_{y1}s) & -(I_{x1} - I_{z1})\,p_0 \\ 0 & C_{n\beta} & -(I'_{y1} - I'_{x1})\,p_0 & (C_{nr}b_1 - I'_{z1}s) \end{vmatrix} = 0 \tag{7.36}$$

Expanding the determinant, we get the characteristic equation

$$A_1 s^4 + B_1 s^3 + C_1 s^2 + D_1 s + E_1 = 0 \tag{7.37}$$

where

$$A_1 = m_1^2 I_{y1} I'_{z1} \tag{7.38}$$

$$B_1 = -m_1^2(I_{y1} C_{nr} b_1 + I'_{z1} C_{mq} c_1) - m_1 I_{y1} I'_{z1} C_{z\alpha} - m_1 C_{y\beta} I_{y1} I'_{z1} \tag{7.39}$$

$$\begin{aligned} C_1 = {} & m_1^2 p_0^2 I_{y1} I'_{z1} + C_{y\beta}[m_1(I_{y1} C_{nr} b_1 + I'_{z1} C_{mq} c_1) + I_{y1} I'_{z1} C_{z\alpha}] \\ & - m_1[m_1 p_0^2(I_{x1} - I_{z1})(I'_{y1} - I'_{x1}) - C_{z\alpha}(I_{y1} C_{nr} b_1 + I'_{z1} C_{mq} c_1) \\ & - m_1 C_{mq} c_1 C_{nr} b_1 + m_1 C_{m\alpha} I'_{z1}] + m_1^2 I_{y1} C_{n\beta} \end{aligned} \tag{7.40}$$

$$\begin{aligned} D_1 = {} & -m_1^2 p_0^2(I_{y1} C_{nr} b_1 + I'_{z1} C_{mq} c_1) + C_{y\beta}[-C_{z\alpha}(I_{y1} C_{nr} b_1 + I'_{z1} C_{mq} c_1) \\ & + m_1^2 p_0^2(I_{x1} - I_{z1})(I'_{y1} - I'_{x1}) - m_1 C_{mq} c_1 C_{nr} b_1 + m_1 C_{m\alpha} I'_{z1}] \\ & - m_1[C_{z\alpha} C_{mq} c_1 C_{nr} b_1 - m_1 C_{m\alpha} C_{nr} b_1 - (I_{x1} - I_{z1})(I'_{y1} - I'_{x1}) C_{z\alpha} p_0^2] \\ & - m_1 C_{n\beta}(m_1 C_{mq} c_1 + C_{z\alpha} I_{y1}) \end{aligned} \tag{7.41}$$

$$\begin{aligned} E_1 = {} & m_1^2(I_{z1} - I_{x1})(I'_{y1} - I'_{x1}) p_0^4 - p_0^2[-m_1^2 C_{mq} c_1 C_{nr} b_1 + m_1^2 C_{n\beta}(I_{z1} - I_{x1}) \\ & - C_{y\beta}(I_{z1} - I_{x1})(I'_{y1} - I'_{x1}) C_{z\alpha} - m_1^2 C_{m\alpha}(I'_{y1} - I'_{x1})] \\ & + C_{y\beta}(C_{z\alpha} C_{mq} c_1 C_{nr} b_1 - m_1 C_{m\alpha} C_{nr} b_1) \\ & + m_1 C_{n\beta} C_{z\alpha} C_{mq} c_1 - m_1^2 C_{m\alpha} C_{n\beta} \end{aligned} \tag{7.42}$$

Special Case, $p_0 = 0$ It is interesting to note that in Eqs. (7.38–7.42), wherever the roll rate p_0 appears, it has only the powers of 2 and 4. For very small values of roll rates $p_0^2 \simeq p_0^4 \simeq 0$. Thus, assuming $p_0 \simeq 0$, Eq. (7.36) reduces to

$$\begin{vmatrix} 0 & (C_{y\beta} - m_1 s) & 0 & -m_1 \\ (C_{z\alpha} - m_1 s) & 0 & m_1 & 0 \\ C_{m\alpha} & 0 & (C_{mq} c_1 - I_{y1} s) & 0 \\ 0 & C_{n\beta} & 0 & (C_{nr} b_1 - I'_{z1} s) \end{vmatrix} = 0 \tag{7.43}$$

Expanding this determinant, we get

$$\begin{aligned} & [-(C_{z\alpha} - m_1 s)(C_{mq} c_1 - I_{y1} s) + m_1 C_{m\alpha}] \\ & \times [(C_{y\beta} - m_1 s)(C_{nr} b_1 - I'_{z1} s) + m_1 C_{n\beta}] = 0 \end{aligned} \tag{7.44}$$

Expanding the first term, we get

$$s^2 - \left(\frac{C_{z\alpha}}{m_1} + \frac{C_{mq}c_1}{I_{y1}}\right)s + \left(\frac{C_{z\alpha}C_{mq}c_1 - m_1 C_{m\alpha}}{m_1 I_{y1}}\right) = 0 \tag{7.45}$$

Comparing this equation with Eq. (6.51), we recognize that the two equations are identical if we note that we have ignored the damping term $C_{m\dot{\alpha}}$. Thus, the first term in Eq. (7.44) represents the conventional short-period oscillation of the aircraft. Similarly, the second term reduces to

$$s^2 - \left(\frac{m_1 C_{nr} b_1 + C_{y\beta} I'_{z1}}{m_1 I'_{z1}}\right)s + \left(\frac{m_1 C_{n\beta} + C_{y\beta} C_{nr} b_1}{m_1 I'_{z1}}\right) = 0 \tag{7.46}$$

This equation is identical with Eq. (6.219) for the Dutch-roll approximation if we note that we have ignored the C_{yr} term.

Thus, for $p_0 = 0$, the characteristic Eq. (7.37) has a pair of complex roots, one pair for the short-period oscillation and another pair for the Dutch-roll oscillation. We have already studied these two cases in Chapter 6.

Stability Criterion for a Steadily Rolling Aircraft Now, consider the case of a steadily rolling airplane with $p_0 \neq 0$. According to Routh's necessary condition for stability, all the coefficients of the characteristic Eq. (7.37) must be positive or must have the same sign. If any one of the coefficients is negative while at least one other coefficient is positive, the system is likely to become unstable. Normally, the coefficients A_1, B_1, C_1, and D_1 are likely to be positive. The only coefficient that is most likely to become negative is E_1. However, the expression for E_1 as given by Eq. (7.42) is still too complicated to derive a criterion for the stability of a steadily rolling aircraft. Therefore, we need to introduce some more simplifications. As done by Philips [1], we assume that $C_{mq} = C_{nr} = C_{y\beta} = 0$. With these assumptions, Eq. (7.42) reduces to

$$\begin{aligned} E_1 = m_1^2[&(I_{z1} - I_{x1})(I'_{y1} - I'_{x1})p_0^4 - p_0^2(C_{n\beta}(I_{z1} - I_{x1}) \\ &- C_{m\alpha}(I'_{y1} - I'_{x1})) - C_{m\alpha}C_{n\beta}] \end{aligned} \tag{7.47}$$

From Eq. (6.55), we have

$$\omega_\theta = \sqrt{\frac{C_{z\alpha}c_1 C_{mq}}{m_1 I_{y1}} - \frac{C_{m\alpha}}{I_{y1}}} \tag{7.48}$$

Here, ω_θ denotes the natural frequency of the short-period mode. With $C_{mq} = 0$, this equation reduces to

$$\omega_\theta = \sqrt{\frac{-C_{m\alpha}}{I_{y1}}} \tag{7.49}$$

or

$$-C_{m\alpha} = \omega_{\theta}^2 I_{y1} \tag{7.50}$$

Similarly, using Eq. (6.222) with $C_{nr} = C_{y\beta} = C_{yr} = 0$, we get

$$C_{n\beta} = \omega_{\psi}^2 I'_{z1} \tag{7.51}$$

Here, ω_{ψ} denotes the natural frequency of the Dutch-roll mode. Using Eqs. (7.50) and (7.51), Eq. (7.47) can be expressed as

$$E_1 = I_{y1} I'_{z1} m_1^2 \left[\left(\frac{I_{z1} - I_{x1}}{I_{y1}} \right) p_0^2 - \omega_{\theta}^2 \right] \left[\left(\frac{I'_{y1} - I'_{x1}}{I'_{z1}} \right) p_0^2 - \omega_{\psi}^2 \right] \tag{7.52}$$

$$= I_{y1} I'_{z1} m_1^2 \left[\left(\frac{I_z - I_x}{I_y} \right) p_0^2 - \omega_{\theta}^2 \right] \left[\left(\frac{I_y - I_x}{I_z} \right) p_0^2 - \omega_{\psi}^2 \right] \tag{7.53}$$

Thus, for $E_1 < 0$ (instability), one of the two terms in the square brackets must be negative. Both terms should not become negative at the same time. Thus, we have two conditions for instability. The first condition is given by

$$\left(\frac{I_z - I_x}{I_y} \right) p_0^2 > \omega_{\theta}^2 \tag{7.54}$$

$$\left(\frac{I_y - I_x}{I_z} \right) p_0^2 < \omega_{\psi}^2 \tag{7.55}$$

and the second condition is given by

$$\left(\frac{I_z - I_x}{I_y} \right) p_0^2 < \omega_{\theta}^2 \tag{7.56}$$

$$\left(\frac{I_y - I_x}{I_z} \right) p_0^2 > \omega_{\psi}^2 \tag{7.57}$$

The first condition is equivalent to

$$\left(\frac{\omega_{\theta}^2 I_y}{I_z - I_x} \right) < p_0^2 < \left(\frac{\omega_{\psi}^2 I_z}{I_y - I_x} \right) \tag{7.58}$$

or

$$\left(\frac{\omega_{\theta}}{p_0} \right)^2 < \left(\frac{I_z - I_x}{I_y} \right) \tag{7.59}$$

and

$$\left(\frac{\omega_\psi}{p_0}\right)^2 > \left(\frac{I_y - I_x}{I_z}\right) \tag{7.60}$$

Suppose we plot $(\omega_\theta/p_0)^2$ vs $(\omega_\psi/p_\theta)^2$ as shown in Fig. 7.2; then the first condition corresponds to the shaded region I of pitch divergence. Here, the aircraft experiences pitch divergence because it has a smaller level of longitudinal stability as indicated by Eq. (7.59) and a larger level of directional stability as indicated by Eq. (7.60).

Similarly, the second condition given by Eqs. (7.56) and (7.57) can be expressed as

$$\left(\frac{\omega_\psi}{p_0}\right)^2 < \left(\frac{I_y - I_x}{I_z}\right) \tag{7.61}$$

and

$$\left(\frac{\omega_\theta}{p_0}\right)^2 > \left(\frac{I_z - I_x}{I_y}\right) \tag{7.62}$$

which corresponds to the shaded region II of yaw divergence as shown in Fig. 7.2. In this case, the aircraft has a relatively large longitudinal stability as indicated by Eq. (7.62) and a small directional stability as indicated by Eq. (7.61). Thus, if a point lies in region I, then the aircraft experiences a

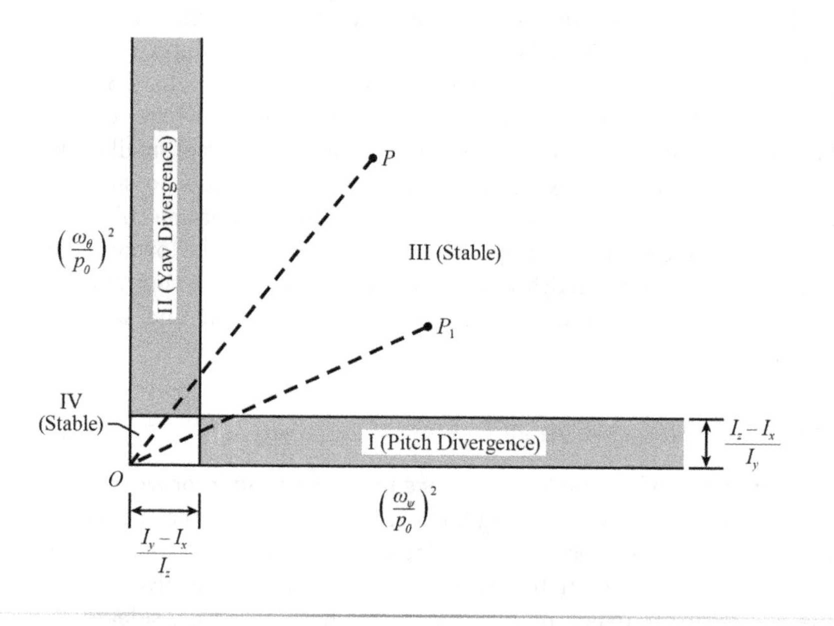

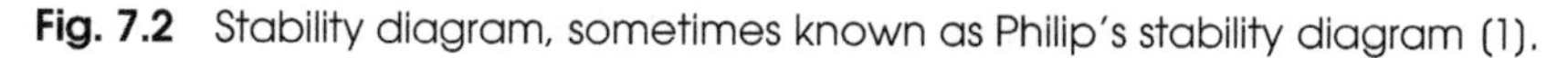

Fig. 7.2 Stability diagram, sometimes known as Philip's stability diagram (1).

divergence in pitch and, if it lies in region II, it will diverge in yaw or sideslip.

We have two regions, III and IV, where the aircraft is stable. Region III corresponds to those cases where the rolling velocity p_0 is small, and region IV corresponds to large rolling velocities. Let us consider what happens when a given aircraft starts rolling at a steady rate. Initially, let us assume that the roll rate is small so that this condition corresponds to the point P in region III. Note that the slope of the line connecting point P to the origin is equal to the ratio $(\omega_\theta/\omega_\psi)^2$. As the rate of roll increases, we approach the origin along the straight line OP. Because this aircraft has a higher short-period frequency compared to the Dutch-roll frequency, at some value of roll rate p_0, the straight line OP will intersect region II, indicating that the aircraft will diverge in yaw or sideslip. On the other hand, if it has a higher Dutch-roll frequency compared to the short-period frequency, it will experience a divergence in pitch as shown by the line OP_1 intersecting region I.

The narrow region around the origin in which the aircraft becomes stable at extremely high roll rates corresponds to the case of spin stabilization. This is the case of artillery shells and bullets, which are known to be flying through the atmosphere at high velocities and at high roll rates without experiencing any inertia coupling problems. Generally, a body becomes spin stabilized about an axis of least inertia. Usually, for modern combat aircraft, the axis of least inertia happens to be the x-body axis about which the aircraft is rolling. So as the roll rate increases, the aircraft becomes spin stabilized.

In the analysis just discussed, for simplicity, we have ignored acceleration and damping derivatives. With the inclusion of these terms, the stability boundaries are modified as shown by the solid lines in Fig. 7.3. Thus, we observe that the narrow region around the origin where the aircraft becomes spin stabilized now connects to the stable region of small rolling velocities. By properly choosing the values of ω_θ and ω_ψ, it is possible to avoid the instabilities in pitch and yaw as shown by the straight line OP. As the roll rate increases, the aircraft simply becomes spin stabilized. However, we must remember that this analysis is based on several simplifying assumptions pertaining to the airplane's mass/inertia distribution and aerodynamic characteristics.

7.2.3 Prevention of Inertia Coupling

For an aircraft prone to divergence in yaw, we can improve its resistance to inertia coupling by increasing the level of static directional stability $C_{n\beta}$ or by increasing the inertia in roll I_x relative to the inertia in pitch I_y. Note that the width of the region II shrinks as we increase I_x relative to I_y. These methods call for major modifications and have to be considered early in the design cycle. As we know, the vertical tail is the major contributor to

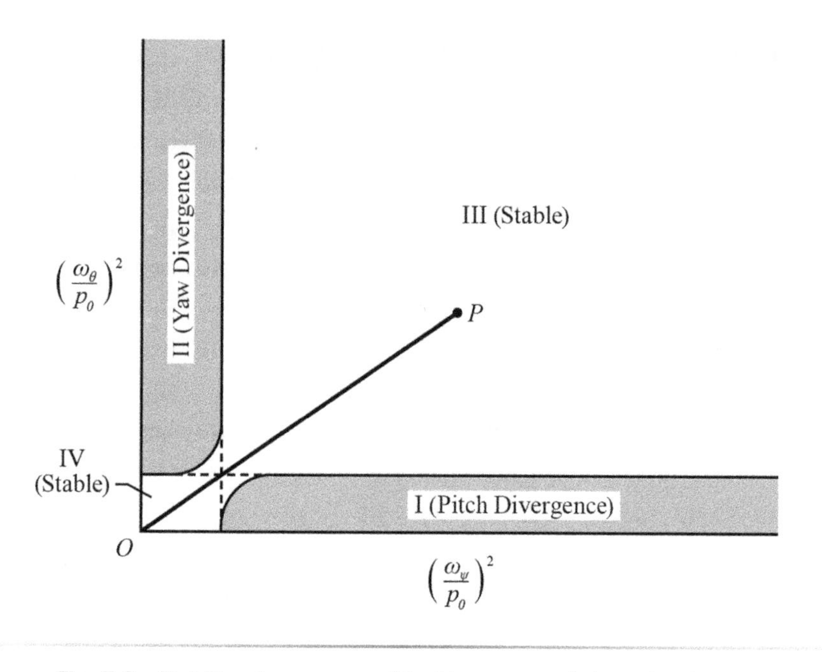

Fig. 7.3 Stability diagram modified because of damping terms.

the static directional stability parameter $C_{n\beta}$. Therefore, to increase $C_{n\beta}$, we have to increase the vertical tail area and vertical tail arm. However, this may cause some problems in spiral stability. Thus, a tradeoff is involved in deciding how much directional stability can be permitted on a given aircraft. The second option calls for distributing more mass into the wings such as placing fuel tanks in the wings. Also, we can have a larger wing span or a higher aspect ratio. This explains why the earlier aircraft that had a short bulky fuselage and high aspect ratio wings did not experience inertia coupling problems.

Similarly, for an aircraft prone to pitch divergence, we can increase its resistance to inertia coupling by increasing the static longitudinal stability level $C_{m\alpha}$ or increasing the inertia in roll I_x relative to the inertia in yaw I_z. The main factors affecting the longitudinal stability parameter $C_{m\alpha}$ are the horizontal tail area, horizontal tail moment arm, and the location of the center of gravity. The designer has to make a proper selection of these variables. However, the second option is of very limited value in this case because we cannot increase I_x without increasing I_z.

Another option is to use a feedback control system, which will increase both the damping in pitch and damping in yaw. Referring to Fig. 7.3, we observe that this will result in widening the area, which connects regions III and IV. However, we will not be going into the design of such a feedback control system. The interested reader may refer elsewhere [7] for additional information on this subject.

Example 7.1

To illustrate the theory discussed, we consider the following aircraft [6]. The data given in [6] is in the FPS (foot-pound-second) system, which has been converted to SI units as given in the following: mass $= 1.0872 \times 10^4$ kg, $I_x = 1.4881 \times 10^4$ kgm^2, $I_y = 7.7417 \times 10^4$ kgm^2, $I_z = 8.7850 \times 10^4$ kgm^2, wing area $S = 35.0233$ m^2, mean aerodynamic chord $\bar{c} = 3.442$ m, wing span $b = 11.1557$ m, $C_{nr} = -0.095$, $C_{mq} = -3.5$, $C_{m\alpha} = -0.36$, $C_{n\beta} = 0.057$, $C_{y\beta} = -0.28$, $C_{L\alpha} = 3.85$, and $C_{lp} = -0.255$. All the derivatives are per radian.

For flight at a velocity of 210.6168 m/s and a dynamic pressure of 9432.4 Nm^2, examine the stability of this aircraft in steady rolling maneuvers.

Solution:

We have

$$\left(\frac{I_z - I_x}{I_y}\right) = \frac{8.7850 \times 10^4 - 1.4881 \times 10^4}{7.7417 \times 10^4} = 0.9425$$

$$\left(\frac{I_y - I_x}{I_z}\right) = \frac{7.7417 \times 10^4 - 1.4881 \times 10^4}{8.7850 \times 10^4} = 0.7118$$

The pitch divergence and yaw divergence boundaries are shown in Fig. 7.4. We have

$$\begin{aligned} I_{y1} &= \frac{I_y}{qS\bar{c}} \\ &= \frac{7.7417 \times 10^4}{9432.4 \times 35.0233 \times 3.442} \\ &= 0.06808 \end{aligned}$$

$$\begin{aligned} \omega_\theta &= \sqrt{\frac{-C_{m\alpha}}{I_{y1}}} \\ &= \sqrt{\frac{0.36}{0.6808}} \\ &= 2.299 \text{ rad/s} \end{aligned}$$

$$\begin{aligned} I'_{z1} &= \frac{I_z}{qSb} \\ &= \frac{8.7850 \times 10^4}{9432.4 \times 35.0233 \times 11.1557} \\ &= 0.0238 \end{aligned}$$

(Continued)

Example 7.1 (*Continued*)

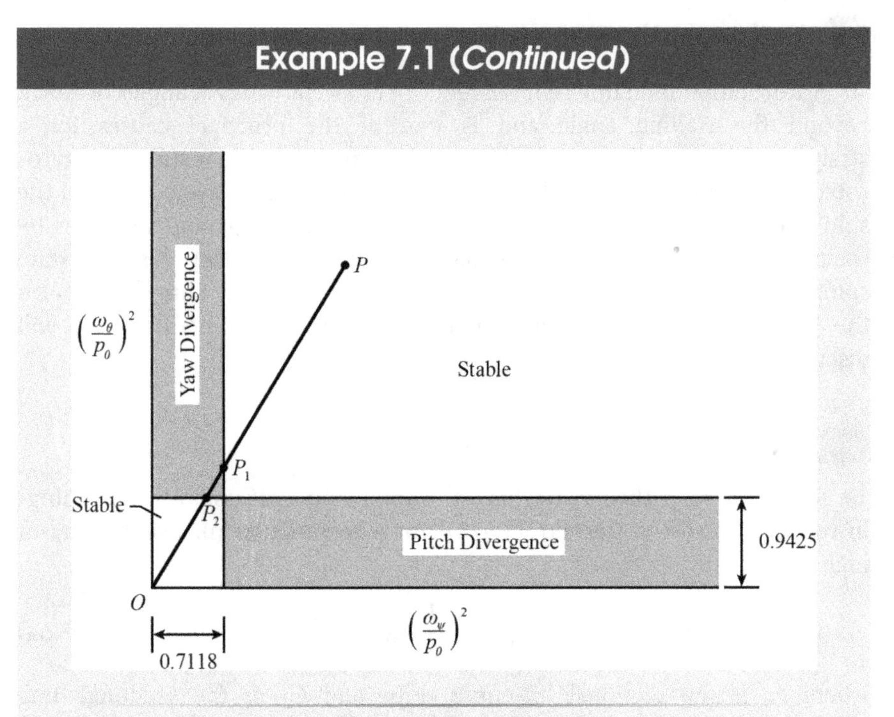

Fig. 7.4 Stability diagram.

$$\omega_\psi = \sqrt{\frac{C_{n\beta}}{I'_{z1}}}$$

$$= \sqrt{\frac{0.057}{0.0238}}$$

$$= 1.5476 \text{ rad/s}$$

The slope of the line *OP* for the given aircraft is given by

$$\left(\frac{\omega_\theta}{\omega_\psi}\right)^2 = 2.2068$$

As the roll rate increases, we move towards the origin along the line *OP*. When the roll rate increases to that corresponding to point P_1, the aircraft experiences yaw divergence. However, when the roll rate is further increased to that corresponding to P_2, the aircraft becomes stable again (spin stabilized). The roll rates corresponding to P_1 and P_2 are, respectively, equal to 1.8344 and 2.3680 rad/s. Thus, the aircraft experiences yaw divergence due to inertial coupling when

$$1.8344 \text{ rad/s} < p_0 < 2.3680 \text{ rad/s}$$

7.3 Autorotation of Wings and Fuselages

Autorotation is an inherent tendency of unswept wings at angles of attack beyond the stalling angle and is one of the principal causes for a straight-wing, propeller-driven light airplane to enter into a spin. The autorotative tendency of the fuselage also contributes to the development of the spin. However, the autorotative tendencies of the wings and fuselages by themselves are not sufficient to make an airplane develop a steady-state spin. Whether or not an airplane develops a steady-state spin depends on the balance between the aerodynamic and inertial moments as we will discuss later in this chapter.

7.3.1 Autorotation of Wings

Based on strip theory, we have the following expression for damping-in-roll derivative C_{lp} [see Eq. (4.575)] for a rectangular (unswept) wing of high aspect ratio:

$$(C_{lp})_W = -\frac{1}{6}(a_0 + C_D) \tag{7.63}$$

where a_0 is the sectional lift-curve slope and C_D is the sectional drag coefficient. For flight at low angles of attack, a_0 is positive so that $C_{lp} < 0$. For angles of attack above stall, $a_0 < 0$ and, if $|a_0| > C_D$, then the damping-in-roll derivative changes sign and becomes positive ($C_{lp} > 0$). When this happens, the wing becomes unstable in roll. Thus, for instability in roll,

$$a_0 + C_D < 0 \tag{7.64}$$

Equation (7.63) is based on the assumption that the angle of attack is small and is in the linear range. This approximation is satisfactory for most of the airfoils whose stalling angles are in the range of 10–15 deg. However, if the stalling angles are higher, then the criterion for instability in roll is given by [8]

$$\frac{\partial C_R}{\partial \alpha} < 0 \tag{7.65}$$

where C_R is the resultant force coefficient given by

$$C_R = \sqrt{C_L^2 + C_D^2} \tag{7.66}$$

At high angles of attack exceeding the stalling angle, the resultant force is approximately normal to the chordline so that

$$C_L = C_R \cos \alpha \tag{7.67}$$

$$C_D = C_R \sin \alpha \tag{7.68}$$

Suppose we mount an unswept (rectangular) wing on a single-degree-of-freedom, free-to-roll apparatus having frictionless bearings and place it in the test section of a low-speed wind tunnel at an angle of attack below stall. When it is disturbed from its equilibrium position, the disturbance in roll will quickly die out because $C_{lp} < 0$, and the wing will immediately return to its equilibrium position. However, if the angle of attack is above the stalling angle, the disturbance in roll will increase because $C_{lp} > 0$. In other words, the wing is unstable in roll and starts autorotating. The rate of roll will increase initially but eventually reach a steady value. The steady-state roll rate is called the autorotational speed.

The autorotational characteristics of a wing depend on the nature of the variation of lift and drag coefficients with an angle of attack beyond stall. In Fig. 7.5, schematic variations of lift and drag coefficients of an airfoil with angle of attack are shown. Generally, it is possible to identify five regions. In region I, where $\alpha < \alpha_{\text{stall}}$, the damping in roll derivative C_{lp} is negative, and the airfoil is stable in roll. Region II, with $C_{lp} > 0$, is one of spontaneous autorotation because even a slight disturbance will initiate autorotation. In region III, even though the lift-curve slope is negative, the airfoil is stable again because the magnitude of the lift-curve slope is smaller than the drag coefficient. In other words, the damping effect due to drag is sufficient to make the airfoil stable in roll. In region IV, once again the airfoil exhibits autorotative tendency but only to large disturbances in roll so that the angle of attack of the down-going wing falls in region IV and that of the up-going wing in region II. To distinguish this autorotational tendency from the spontaneous autorotative tendency of region II, region IV is called one of latent autorotation. In region V, the airfoil is stable again because of large values of drag coefficient.

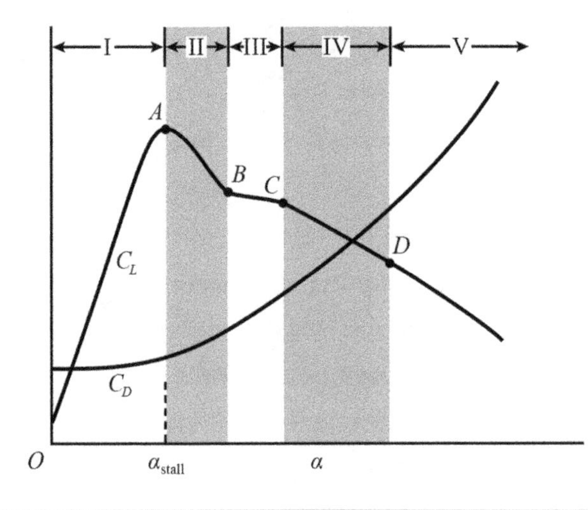

Fig. 7.5 Airfoil lift and drag coefficients with angle of attack.

7.3.2 Autorotation of Fuselages

The autorotation of a fuselage depends on its cross-sectional shape. Generally, fuselages with circular cross sections or cross sections with round bottoms are resistant to autorotation. On the other hand, cross sections with flat bottoms are prone to autorotation.

To understand the aerodynamics associated with autorotation of the fuselage, consider a noncircular cylindrical model mounted on a single-degree-of-freedom, free-to-roll apparatus that is held at an angle of attack in an airstream of uniform velocity U_0 as shown in Fig. 7.6. We assume that the model is pivoted at its center of gravity and is constrained to rotate about the velocity vector U_0. Let the origin of the body-fixed coordinate system coincide with the center of gravity.

Suppose that the model is imparted a disturbance that makes it rotate in a clockwise direction (viewed from aft end) with an angular velocity Ω. The crossflow angle ϕ at an axial location x (station AA or BB) is given by

$$\phi = \tan^{-1}\left(\frac{\Omega x}{U_0}\right) \tag{7.69}$$

We note that, for the cross sections ahead of the center of gravity, the crossflow angle is positive and, for the sections aft of the center of gravity, the crossflow angle is negative.

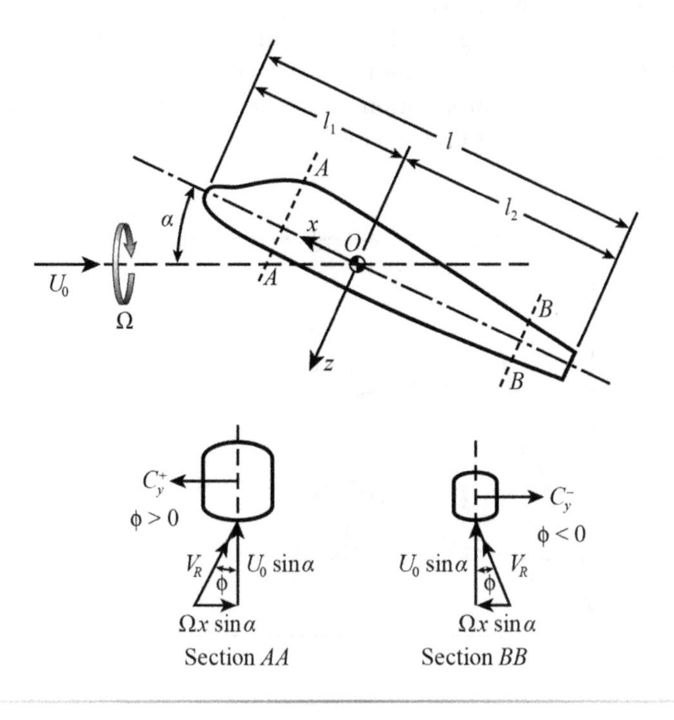

Fig. 7.6 Autorotating fuselage.

The local dynamic pressure at the axial location x is given by

$$q_l = \frac{1}{2}\rho U_0^2\left[1 + \left(\frac{\Omega x \sin\alpha}{U_0}\right)^2\right] \tag{7.70}$$

For simplicity, we assume that the side force coefficient C_y of the fuselage cross sections depends only on the crossflow angle ϕ.

The moment developed by section AA about the axis of rotation (velocity vector) is given by

$$\Delta N_\Omega = \frac{1}{2}\rho U_0^2\left[1 + \left(\frac{\Omega x \sin\alpha}{U_0}\right)^2\right] C_y(\phi) b_0 x \sin\alpha \, dx \tag{7.71}$$

or, in coefficient form,

$$\Delta C_\Omega = \frac{1}{l^2}\left[1 + \left(\frac{\Omega x \sin\alpha}{U_0}\right)^2\right] C_y(\phi) x \sin\alpha \, dx \tag{7.72}$$

where

$$\Delta C_\Omega = \frac{\Delta N_\Omega}{\frac{1}{2}\rho U_0^2 b_0 l^2} \tag{7.73}$$

Here, b_0 is the width of the body, l is the total length of the body, and $x \sin\alpha$ is the moment arm.

Given the variation of C_y with crossflow angle ϕ, an integration of Eq. (7.72) gives the variation of the moment coefficient C_Ω as a function of the angular velocity Ω. The schematic variation of C_Ω is shown in Fig. 7.7. It is possible that this variation is of two types. For type I variation, the yawing moment coefficient is initially positive and at some point crosses zero and becomes negative for higher values of the angular velocity Ω. Let $\Omega = \Omega_c$ when $C_\Omega = 0$. We observe that for $0 < \Omega < \Omega_c$, $C_\Omega > 0$, which implies that the induced yawing moment will assist the imparted disturbance, and, for $\Omega > \Omega_c$, $C_\Omega < 0$, the induced yawing moment will oppose the rotation. Hence, the angular velocity Ω_c when $C_\Omega = 0$ is the steady-state or equilibrium autorotational speed. If the variation of C_Ω curve is of type II, then the body does not autorotate. It is resistant to autorotation.

The side force variations that can give type I and type II variations of the yawing moment are shown in Fig. 7.8. The type A variation of side force has a positive value of side force for low values of the crossflow angle, and the side force becomes negative for higher values of the crossflow angle. This type of side force variation generates autorotative (yawing) moments on sections close to the center of gravity and damping (yawing) moments for sections away from the center of gravity. Equilibrium or steady autorotation speed is reached when the autorotative moments balance the damping moments

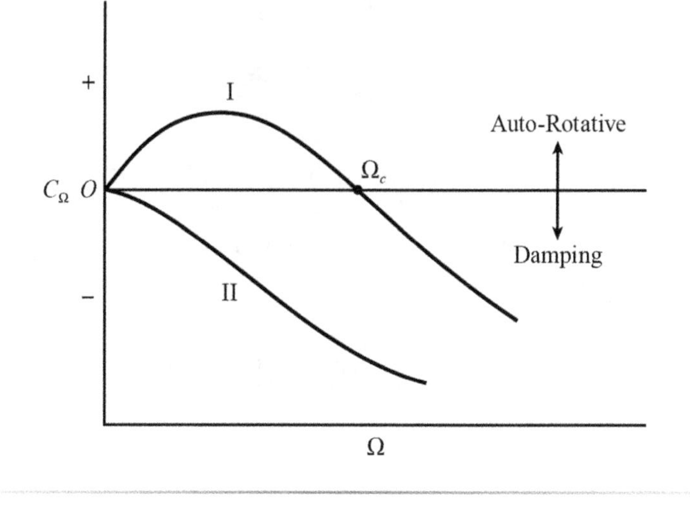

Fig. 7.7 Yawing moment coefficient of a rotating fuselage.

so that the net yawing moment on the body is zero. For type B, the side force coefficient is always negative, leading to damping (yawing) moments at all cross sections. This corresponds to type II variation of C_Ω vs Ω as shown in Fig. 7.7.

The cross-sectional shapes having rounded bottom surfaces (Fig. 7.9) generally have type B side force coefficient variation, and those having flat

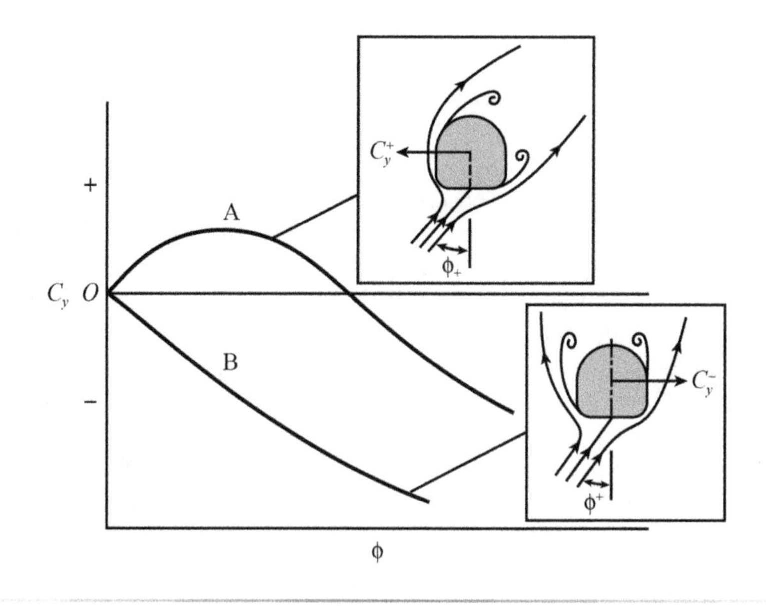

Fig. 7.8 Side force variation in crossflow.

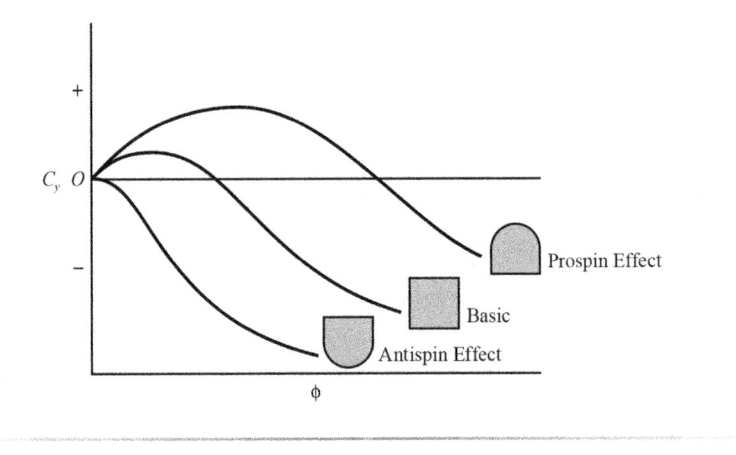

Fig. 7.9 Effect of fuselage cross-sectional shape on spin behavior.

bottom surfaces have type A side force coefficient variation. Rounding the top surface while the bottom surface is kept flat enhances the autorotational tendency [22].

The Reynolds number is known to have a significant influence on the side force characteristics of noncircular cylinders [9]. For certain cross-sectional shapes, the side force variation changes from type A to type B as the Reynolds number is increased. Thus, it is quite possible that a spin-tunnel model may indicate that the configuration is prospin, whereas full-scale airplane may exhibit an antispin behavior.

7.4 Airplane Spin

Stall/spin problems have been encountered since the very beginning of aviation. The very low altitudes at which early aircraft were flown precluded the progression of stall to a fully developed spin prior to the ground impact. As a result, it was not possible to clearly ascertain the true causes of these early crashes.

The stall–spin is one of the major causes of light airplane accidents even today. Such an accident is characterized by an inadvertent stall and a spin entry at an altitude that is too low to effect a successful recovery [10, 11]. There are three approaches to preventing an inadvertent stall departure [12]: 1) pilot training, 2) stall warning, and 3) increased spin resistance. Although the Federal Aviation Administration (FAA) places significant importance on pilot training, the demonstration of competency in spin recovery does not form a part of the private or commercial pilot licensing in the United States [20]. Stall warning systems have helped improve the safety of light general aviation airplanes, but the pilot is still required to take some form of corrective action to prevent the airplane from becoming uncontrollable. Improving the inherent spin resistance offers great potential to make the configuration spinproof and improve the safety of general aviation airplanes.

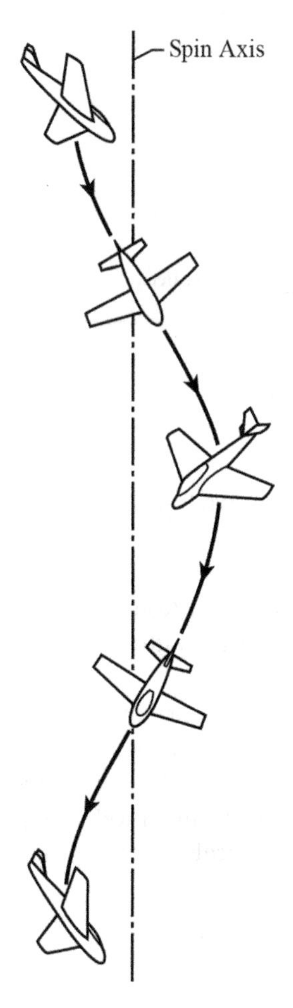

Fig. 7.10 Spinning airplane.

The straight-wing, light propeller-driven airplanes usually have good longitudinal (pitch) and directional (yaw) stability at stall. The critical aerodynamic characteristic of such airplanes at stall is the autorotative tendency. Also, such airplanes experience an asymmetric stall. In other words, both wings do not stall at the same time, or one wing stalls earlier than the other. One of the possible causes for an asymmetric stall is the propeller sidewash effect. Following anasymmetric stall, the nose drops and the airplane rolls in the direction of the fallen wing and continues to roll owing to the autorotative tendency of the stalled wings. In this process, yawing motion develops due to the aerodynamic roll–yaw coupling. (Note that we are not talking of the inertial roll–yaw coupling here.) Such a motion of the stalled airplane involving combined pitch, roll, and yaw is often called poststall gyration. As the yaw rate builds up, the nose rises, the flight path steepens, and the aircraft starts losing height. The airplane is now in a spin. In spin, the airplane descends vertically downward in a helical path as schematically shown in Fig. 7.10. If a steady-state spin develops, the descent velocity, pitch, roll, and yaw rates attain constant values.

The radius of the helix or the spin radius is usually of the order of one-half of the wing span. However, as the spin becomes flatter, the spin radius decreases further and, in the limiting case, the spin axis may pass through the airplane center of gravity.

Whether or not an airplane develops a steady-state spin depends on the balance between the inertia couples and the aerodynamic moments. If such a balance cannot be achieved, the spinning motion remains oscillatory. Sometimes, this type of spin is also called incipient spin.

In contrast to a straight-wing light airplane, modern aircraft with highly swept wings and long, slender fuselages experience loss of longitudinal and directional stabilities as well as directional control at stall. As a result, the motion following a stall is predominantly in yaw involving directional instability and divergence that may lead to a spin entry.

The spin is not of any tactical value for a military aircraft. Also, it serves no useful purpose in civil aviation. The spin is different from a spiral dive.

What distinguishes a spin from the spiral dive is the low air speed and operation of the wings well beyond the stalling angle. In a spiral dive, the airplane is not stalled and the air speed is continuously increasing. Further, the spin axis is much closer to the airplane center of gravity than the axis of the spiral dive. The airplane has a much higher yaw rate in a spin compared to that in a spiral dive. Typical yaw rates for straight-wing light general aviation airplanes are about 120–150 deg/s in a developed spin.

The spin may be in either direction, i.e., to the right or to the left. However, both the right and left spins may not be identical because of the effect of rotating engine masses or the propellers. The spin may be due to an inadvertent stall or could be intentionally initiated by stalling the airplane and applying prospin controls. Initially, the pitch, roll, and yaw rates oscillate, but these oscillations decay if the airplane establishes a steady spin, in which case all the parameters tend to assume constant values.

The spin may be erect or inverted. In erect spin, the roll (as seen visually by the pilot) and yaw (as indicated by the turn indicator needle) are in the same direction. In an inverted spin, the two are in opposite directions.

The characteristics of the steady-state spin differ between the types of airplanes. No two airplanes spin in the same way. Even airplanes of the same type may spin differently. Also, the spin is affected to a considerable degree by the amount of control deflections used to initiate a spin entry.

Basically, there are two types of erect spins. The first is steep spin in which the angle of attack is around 30 deg so that the nose is approximately 60 deg below the horizontal plane. The other is flat spin in which the angle of attack is 60 deg or more. The attitudes of the airplane in steep and flat spins are shown in Fig. 7.11. Steep spin is the slower of the two. In flat spin, the rate of rotation is considerably higher. The recovery from the steep spin is relatively easier, whereas recovery from a flat spin may be quite difficult. Usually, a steep spin preceeds a flat spin.

The airplane spin is not very amenable for theoretical analysis because of nonlinear, inertial cross coupling between the longitudinal and lateral degrees of freedom. Furthermore, the aerodynamics of the spinning airplane are extremely complex because of extensive flow separation over the wing and tail surfaces. Owing to these difficulties, most of the investigations of airplane spin have been traditionally based on experimental studies using static and dynamic testing in wind tunnels, rotary balance tests, flying model tests in vertical spin tunnels, outdoor radio-controlled model tests, and full-scale airplane flight tests [11]. Even with this kind of experimental aerodynamic database, it is not always possible to obtain a good correlation between the observed and the predicted spin behavior.

However, if we restrict ourselves to the study of the developed or steady-state spin, it is possible to get some physical understanding of various factors that influence the spinning motion and use the semiempirical approach based on strip theory to get approximate estimates of the aerodynamic coefficients of a steadily spinning airplane [13–15]. In a steady-state spin, the

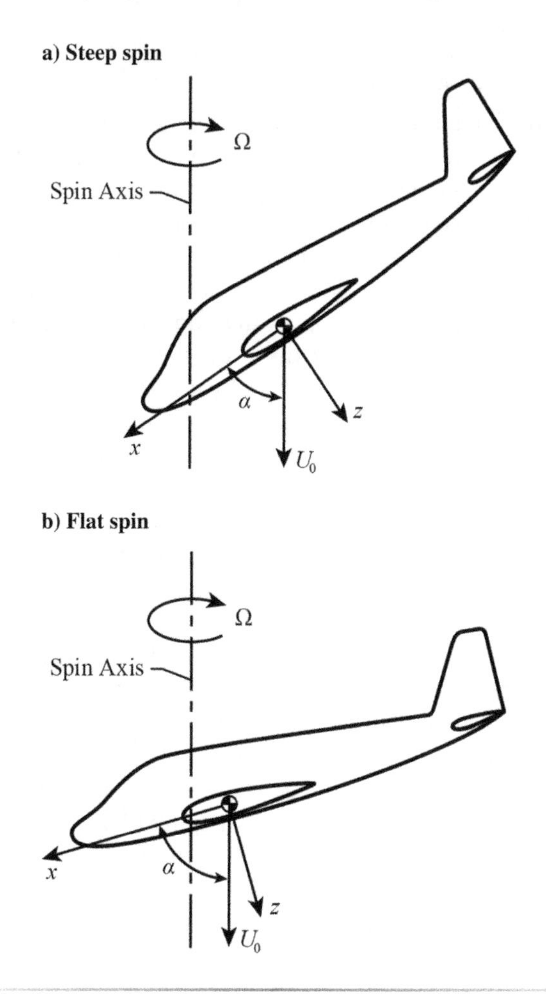

Fig. 7.11 Airplane attitudes in spin.

balance of forces is necessary, but this fact alone does not guarantee that the steady-state spin will be established. It is the balance between the inertial and aerodynamic moments that is of crucial importance to the establishment of a steady-state spin. Even for the case of steady-state spin, it is not easy to analytically determine the equilibrium spin modes because: 1) a simple and comprehensive analytical aerodynamic model of the spinning airplane is not available and, 2) the equations of motion are coupled and nonlinear. An example of determining the steady-state spin modes using the static and dynamic wind-tunnel test data may be found elsewhere [16].

In this chapter, we will discuss various factors that infiuence the balance between inertia and aerodynamic moments and also discuss various methods of spin recovery. Finally, we will discuss some of the recent methods for improving the spin resistance of airplanes.

7.5 Equations of Motion for Steady-State Spin

The equations governing the spinning motion are the complete six-degree-of-freedom equations [Eqs. (7.1–7.6)]. For simplicity, we ignore the product of inertia term I_{xz}. With this assumption, Eqs. (7.1–7.6) assume the following form:

$$F_x = m(\dot{U} + qW - rV) \tag{7.74}$$

$$F_y = m(\dot{V} + Ur - pW) \tag{7.75}$$

$$F_z = m(\dot{W} + pV - Uq) \tag{7.76}$$

$$L = \dot{p}I_x + qr(I_z - I_y) \tag{7.77}$$

$$M = \dot{q}I_y + rp(I_x - I_z) \tag{7.78}$$

$$N = \dot{r}I_z + pq(I_y - I_x) \tag{7.79}$$

In a steady-state spin, the spin axis is nearly vertical, and the center of gravity of the airplane moves downward in a helical path around the spin axis with a constant velocity. Let U_0 denote the velocity of descent. Resolving the velocity vector U_0 in the body axes system, we have

$$U = U_0 \cos\alpha \tag{7.80}$$

$$W = U_0 \sin\alpha \tag{7.81}$$

where α is the angle of attack. Because the descent velocity is in a vertical direction, α is the angle between the chordline and the vertical as shown in Fig. 7.12. Note that the spin axis is also vertical.

In a steady-state spin, when viewed from the top, the airplane's center of gravity appears to be moving in a circular path with a constant angular velocity. Let Ω denote this constant angular velocity. Let us assume that the airplane is spinning to its right, i.e., it is rotating in a clockwise direction when viewed from the top. Because of this angular velocity, the airplane will have a component of velocity along the y-body axis. At the center of gravity, the velocity component along the y-body axis is given by

$$V = -\Omega R \tag{7.82}$$

where R is the radius of the helix or the spin radius. As said earlier, the spin radius is usually about one-half of the wing span for steep spins and still smaller for flat spins.

Because in a steady-state spin U, V, and W are constants

$$\dot{U} = \dot{V} = \dot{W} = 0 \tag{7.83}$$

In spin, the angle of attack is well above the stalling angle. At such angles of attack, the resultant aerodynamic force is approximately normal to the wing

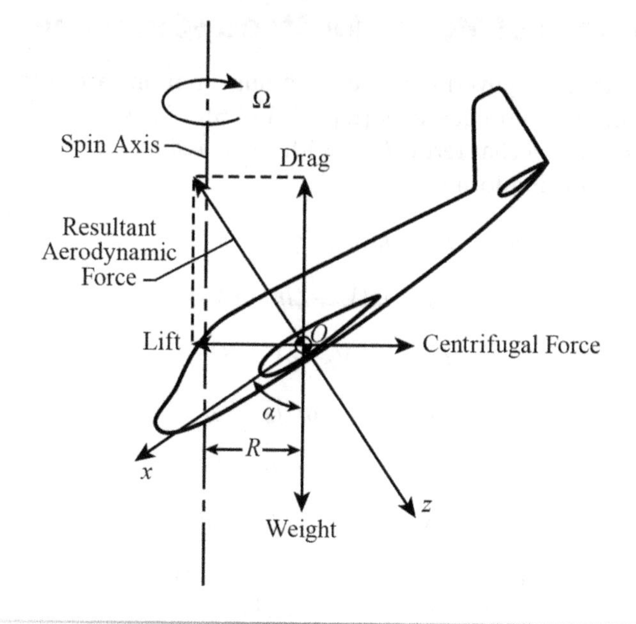

Fig. 7.12 Forces acting on an airplane in steady-state spin.

chordline. Note that the lift acts in the horizontal plane and drag is directed in the vertical plane opposite to the gravity as shown in Fig. 7.12.

The angular velocity vector Ω can be resolved along the x- and z-body axes (see Fig. 7.13) as

$$p = \Omega \cos \alpha \tag{7.84}$$

$$r = \Omega \sin \alpha \tag{7.85}$$

If the wings are in the horizontal plane,

$$q = 0 \tag{7.86}$$

Because of the helical motion, the spinning airplane experiences a sideslip. For example, in a positive spin (spin to the right), the sideslip is towards the left or port wing, which is also the leading wing. In right spin, the right or starboard wing is the trailing wing. As shown in Fig. 7.14, the sideslip angle is related to the helix angle by the following relation:

$$\beta = -\gamma \tag{7.87}$$

where the helix angle γ is given by

$$\gamma = \tan^{-1}\left(\frac{\Omega R}{U_0}\right) \tag{7.88}$$

Now consider a more general case where the wings are tilted out of the horizontal plane. Let θ_y denote the wing tilt. We assume that θ_y is

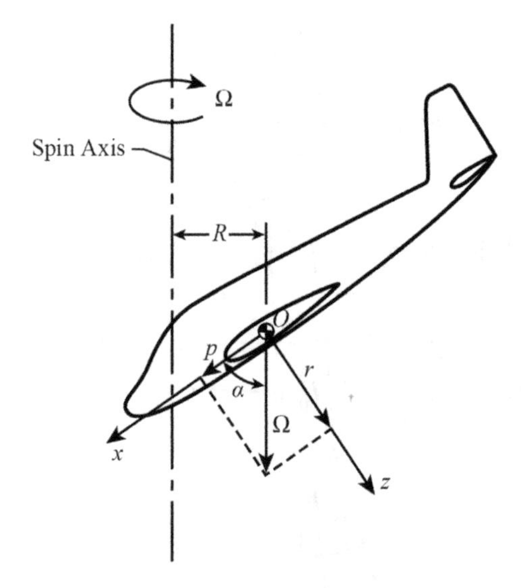

Fig. 7.13 Angular velocity components in spin.

positive when the right wing is tilted down and the left wing is raised with respect to the horizontal plane as shown in Fig. 7.15. The sideslip is now given by

$$\beta = \theta_y - \gamma \tag{7.89}$$

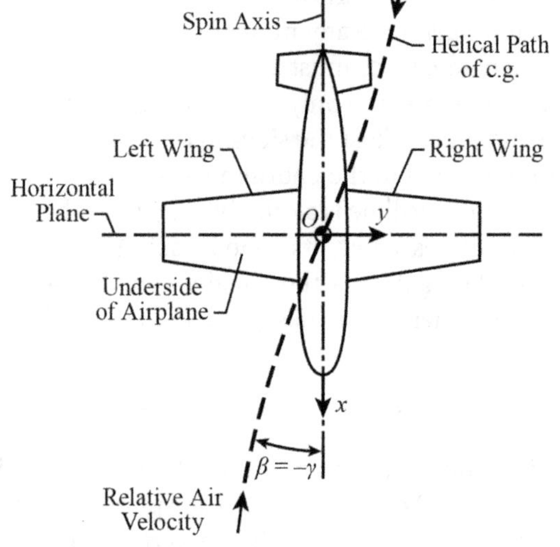

Fig. 7.14 Sideslip in spin.

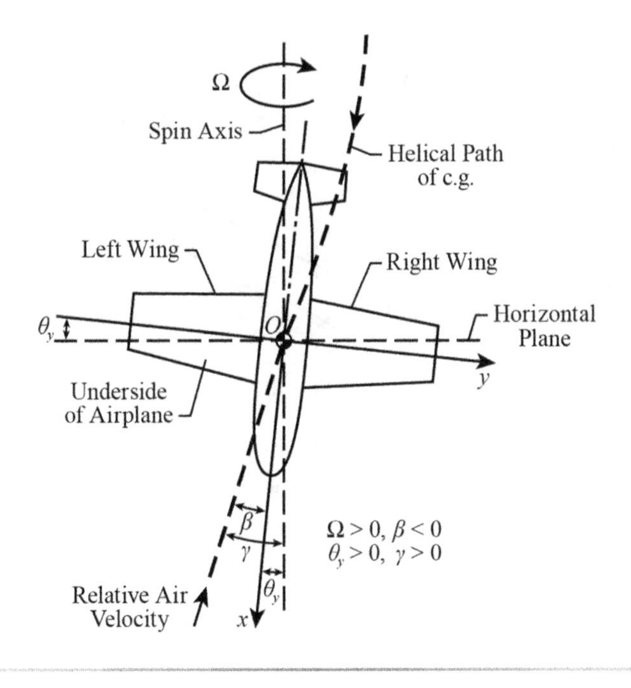

Fig. 7.15 Effect of wing tilt in spin.

We observe that a positive wing tilt θ_y reduces the sideslip. When $\theta_y = \gamma$, the sideslip is zero.

The sideslip plays an important role in the balance of moments. Usually, a certain amount of sideslip is always necessary to achieve a balance of all three components of the moment. Because the centrifugal force acting on all the components of the airplane is directed radially outward and passes through the spin axis, it cannot generate any moment about the spin axis. Therefore, the resultant aerodynamic force must also pass through the spin axis as indicated in Fig. 7.12. Therefore, the only way in which an airplane can have the right amount of wing tilt to adjust the sideslip to the required value is through a rotation about the normal to the chordline as shown in Fig. 7.16. Note that the normal to the wing chordline is along the negative z-body axis. Hence, the airplane essentially rotates about its z-body axis to generate the required amount of wing tilt. Let χ denote the angle by which the aircraft is rotated about the z-body axis. Then the angles χ, θ_y, and α are related by the following expression [17]:

$$\sin\theta_y = -\cos\alpha\sin\chi \tag{7.90}$$

To get a physical understanding of this relation, consider the two extreme cases of $\alpha = 0$ and $\alpha = 90$ deg. At $\alpha = 0$, the spinning motion is all rolling because the x-body axis coincides with the vertical spin axis. The z-body axis is now in the horizontal plane. So the rotation χ about the z-body axis is numerically equal to θ_y.

For $\alpha = 90$ deg, the airplane is in a flat spin, and the spinning motion is all yaw about the z-body axis, which now coincides with the spin axis. Thus, any amount of rotation about the z-body axis does not give any wing tilt because the wings are in the horizontal plane for all values of χ. Hence, the $\theta_y = 0$ for $\alpha = 90$ deg.

The angular velocity vector Ω now has the following components in the body axes system:

$$p = \Omega \cos\alpha \cos\chi \tag{7.91}$$

$$q = -\Omega \cos\alpha \sin\chi \tag{7.92}$$

$$r = \Omega \sin\alpha \tag{7.93}$$

7.5.1 Balance of Forces in Steady-State Spin

With $\dot{U} = \dot{V} = \dot{W} = 0$, Eqs. (7.74–7.76) reduce to

$$F_x = m(qW - rV) \tag{7.94}$$

$$F_y = m(Ur - pW) \tag{7.95}$$

$$F_z = m(pV - Uq) \tag{7.96}$$

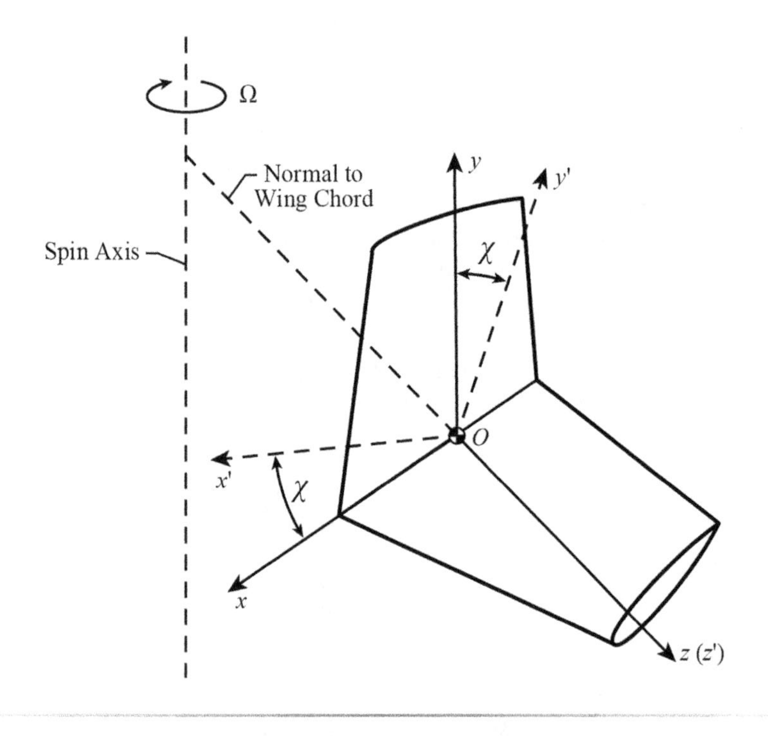

Fig. 7.16 Rotation about normal-to-wing chord.

Substituting for U, V, W from Eqs. (7.80–7.82) for p, q, r from Eqs. (7.91–7.93) and ignoring χ ($\cos\chi = 1$, $\sin\chi = 0$), we get

$$F_x = m\Omega^2 R \sin\alpha \tag{7.97}$$

$$F_y = 0 \tag{7.98}$$

$$F_z = -m\Omega^2 R \cos\alpha \tag{7.99}$$

Ignoring power effects and resolving the aerodynamic and gravity forces acting on the airplane (Fig. 7.12),

$$F_x = L\sin\alpha - D\cos\alpha + W\cos\alpha = m\Omega^2 R\sin\alpha \tag{7.100}$$

$$F_z = -L\cos\alpha - D\sin\alpha + W\sin\alpha = -m\Omega^2 R\cos\alpha \tag{7.101}$$

Multiply Eq. (7.100) by $\cos\alpha$ and Eq. (7.101) by $\sin\alpha$ and add the two equations to obtain

$$D = W \tag{7.102}$$

With this, substitution in either Eq. (7.100) or Eq. (7.101) gives

$$L = m\Omega^2 R \tag{7.103}$$

It so happens that we could have arrived at these simple relations directly by looking at Fig. 7.12. However, in this process of deriving these results using equations of motion, we have obtained some understanding of the kinematics of the spinning motion.

With

$$D = \frac{1}{2}\rho U_0^2 S C_D \tag{7.104}$$

$$L = \frac{1}{2}\rho U_0^2 S C_L \tag{7.105}$$

we get

$$U_0 = \sqrt{\frac{2W}{\rho S C_D}} \tag{7.106}$$

$$R = \left(\frac{1}{2m\Omega^2}\right)\rho U_0^2 S C_L \tag{7.107}$$

From Eqs. (7.67) and (7.68), we have

$$C_L = C_R \cos\alpha \tag{7.108}$$

$$C_D = C_R \sin\alpha \tag{7.109}$$

Suppose we could determine the angle of attack and spin rate from some other criteria, then we could use Eqs. (7.106) and (7.107) to determine the descent velocity and spin radius.

7.5.2 Balance of Moments

For balance of moments, the sum of all the moments must be zero, regardless of the axes system chosen. Suppose we consider the airplane motion with respect to the spin axis; then the centrifugal forces acting on all the components of the airplane acting radially outward from the spin axis do not produce any moment about the spin axis. Therefore, the net aerodynamic moment about the spin axis must also be zero. This means that the resultant aerodynamic force must also pass through the spin axis. In other words, to determine the equilibrium spin modes, we simply have to find the combination of angle of attack, sideslip, and spin rate at which the resultant aerodynamic force passes through the spin axis. However, the scenario is not this simple. Usually, the aerodynamic data does not contain the magnitude, direction, and point of action of the resultant aerodynamic force. Instead, we usually have the aerodynamic data in the form of lift, drag, side force, pitching, rolling, and yawing moment coefficients in the stability axes system. Using this information, we have to compute the balance of pitching, rolling, and yawing moments about the body axes system to determine the equilibrium spin modes.

For steady-state spin, $\dot{p} = \dot{q} = \dot{r} = 0$. With this, moment Eqs. (7.77–7.79) take the following form:

$$L = qr(I_z - I_y) \tag{7.110}$$

$$M = rp(I_x - I_z) \tag{7.111}$$

$$N = pq(I_y - I_x) \tag{7.112}$$

Here, L, M, and N denote the net external rolling, pitching, and yawing moments acting on the airplane during the steady-state spin. Because we have ignored the power effects, the only external moments acting on the airplane are the aerodynamic moments. The right-hand side of Eqs. (7.110–7.112) represent the moments due to inertia cross coupling effects. Substituting for p, q, and r from Eqs. (7.91–7.93) in Eqs. (7.110–7.112) take the following form:

$$L = -\frac{\Omega^2}{2}\sin 2\alpha \sin\chi(I_z - I_y) \tag{7.113}$$

$$M = \frac{\Omega^2}{2}\sin 2\alpha \cos\chi(I_x - I_z) \tag{7.114}$$

$$N = -\frac{\Omega^2}{2}\cos^2\alpha \sin 2\chi(I_y - I_x) \tag{7.115}$$

Equations (7.113–7.115) can be written as

$$L + L_i = 0 \tag{7.116}$$

$$M + M_i = 0 \tag{7.117}$$

$$N + N_i = 0 \tag{7.118}$$

where L_i is the inertia-rolling moment, M_i is the inertia-pitching moment, and N_i is the inertia-yawing moment as given by

$$L_i = qr(I_y - I_z) \tag{7.119}$$

$$= \frac{\Omega^2}{2}\sin 2\alpha \sin\chi(I_z - I_y) \tag{7.120}$$

$$M_i = rp(I_z - I_x) \tag{7.121}$$

$$= \frac{\Omega^2}{2}\sin 2\alpha \cos\chi(I_z - I_x) \tag{7.122}$$

$$N_i = pq(I_x - I_y) \tag{7.123}$$

$$= \frac{\Omega^2}{2}\cos^2\alpha \sin 2\chi(I_y - I_x) \tag{7.124}$$

Balance of Pitching Moments We have

$$M_i = (I_z - I_x)rp \tag{7.125}$$

$$= \frac{\Omega^2}{2}\sin 2\alpha \cos\chi(I_z - I_x) \tag{7.126}$$

To have a physical understanding of how the inertia-pitching moment arises, let us assume that the masses M_1 and M_2 located at the fuselage extremities represent inertia in yaw I_z and masses M_3 and M_4 located at the wingtips represent the inertia in roll I_x as shown in Fig. 7.17. In a right spin, the inertia-induced pitching couple due to M_1 and M_2 is noseup so as to flatten the spin attitude, whereas that due to M_3 and M_4 is nosedown so

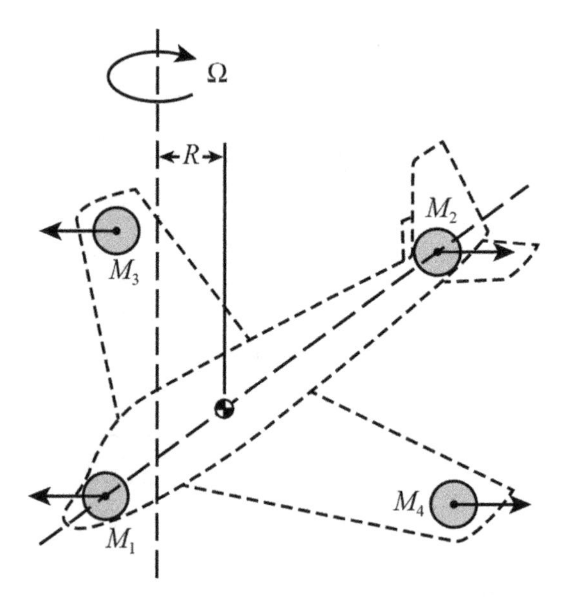

Fig. 7.17 Inertia-pitching moment.

as to steepen the spin attitude. The net inertial pitching couple is the difference between these two opposing contributions.

An alternative way of understanding the inertia-induced pitching moment is to use the gyroscopic analogy. Let us imagine that the inertias in pitch, roll, and yaw are represented by three gyroscopes located at the airplane's center of gravity as shown in Fig. 7.18. The gyroscope with inertia I_x is aligned along the x-body axis and is assumed to be rotating with an angular velocity p. Similarly, the other two gyroscopes with I_y and I_z are aligned along y- and z-body axes and are assumed to be rotating with angular velocities of q and r, respectively.

Now if the gyroscope in Fig. 7.18c is disturbed because of an external torque that imparts it a rolling velocity p, then, according to the gyroscopic principles, it will precess in a direction perpendicular to a plane containing both the vectors r and p. Hence, in this case, the angular velocity vector of precession will be along positive y-axis (Fig. 7.19), which corresponds to a noseup pitching motion. Similarly, if the gyroscope in Fig. 7.18a is disturbed by a torque that imparts it a yawing velocity r, it will pitch in a nosedown direction. Thus, the net gyroscopic action will be the difference between these two induced pitching motions.

Generally, $I_z > I_x$ so that the inertia pitching couple is positive or noseup that tends to flatten the spin attitude. For equilibrium, the aerodynamic pitching moment must be negative or nosedown. At angles of attack well above stall, the nosedown aerodynamic pitching moment mainly comes from the horizontal tail with wing making very little contribution.

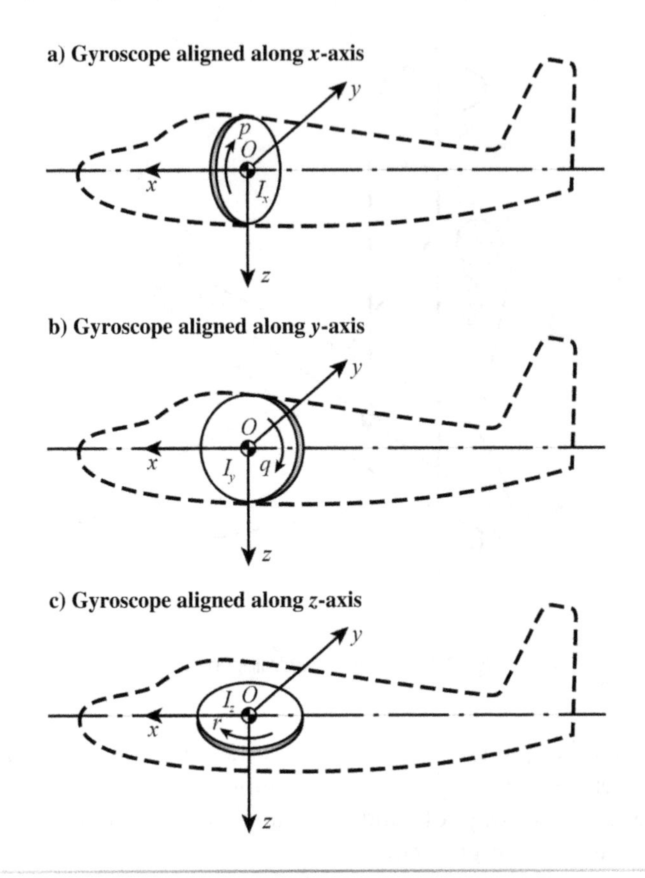

Fig. 7.18 Gyroscopic representation of a spinning airplane.

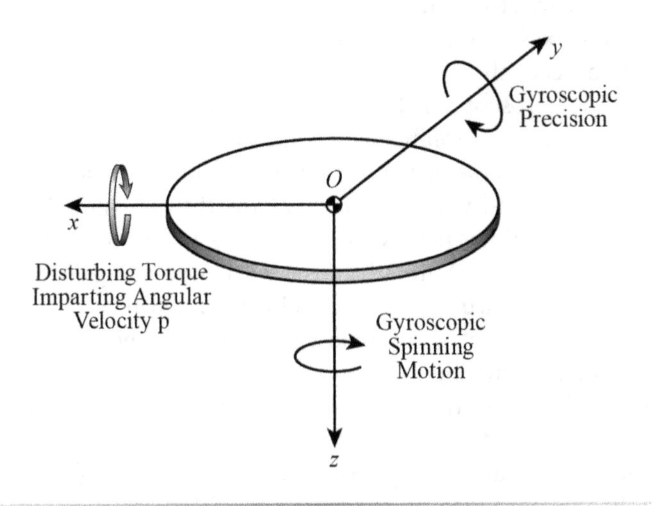

Fig. 7.19 Gyroscopic precision.

Balance of Rolling Moments As we have seen earlier, unswept wings have a strong autorotative tendency at stall. As a result, for angles of attack beyond the stalling angle, the aerodynamic rolling moment for airplanes having unswept wings is generally prospin. Another factor that contributes to the development of aerodynamic rolling moment is the dihedral effect $C_{l\beta}$. However, beyond the stalling angle, the magnitude of $C_{l\beta}$ is quite small. Therefore, for airplanes with unswept wings, a major contribution to rolling moment comes from the wings, which produce autorotative or prospin rolling moments for $\alpha \geq \alpha_{\text{stall}}$. Therefore, for such airplanes the inertia-rolling moment must be antispin to achieve the required balance of rolling moments. The inertia-rolling moment is given by

$$L_i = (I_y - I_z)qr \tag{7.127}$$

$$= \left(\frac{\Omega^2}{2}\right)\sin 2\alpha \sin\chi(I_z - I_y) \tag{7.128}$$

Thus, if $\chi = 0$, the inertial-rolling moment is zero. Because the wing tilt directly depends on angle χ, we observe that a wing tilt one way or the other is necessary to achieve the rolling-moment balance.

A physical explanation using the lumped wing or fuselage masses is difficult in this case. The gyroscopic analogy comes in handy. If the gyroscope in Fig. 7.18b with moment of inertia I_y and rotating about the y-axis with an angular velocity q is disturbed with yaw rate r, it will roll to the right. Similarly, the gyroscope in Fig. 7.18c with moment of inertia I_z aligned along the z-body axis and rotating with an angular velocity r disturbed by a pitching velocity q will roll to the left. The net rolling motion will be the difference between the two induced motions. Usually, $I_z > I_y$ so that the inertia-induced rolling moment is negative or antispin.

Balance of Yawing Moments We have

$$N_i = (I_x - I_y)pq \tag{7.129}$$

$$= \left(\frac{\Omega^2}{2}\right)\cos^2\alpha \sin 2\chi(I_y - I_x) \tag{7.130}$$

As in the case of balance of rolling moments, we observe that the wing tilt plays an important role in the balance of yawing moments.

The concept of lumped fuselage and wing masses can be used to understand the inertia yawing moment as shown in Fig. 7.20 for an aircraft in positive or right spin. In Fig. 7.20a, the wing tilt is negative (right wing above the horizontal, $\theta_y < 0$, $\chi > 0$), and we observe that the fuselage masses M_1 and M_2 representing inertia in pitch I_y produce a prospin yawing moment and

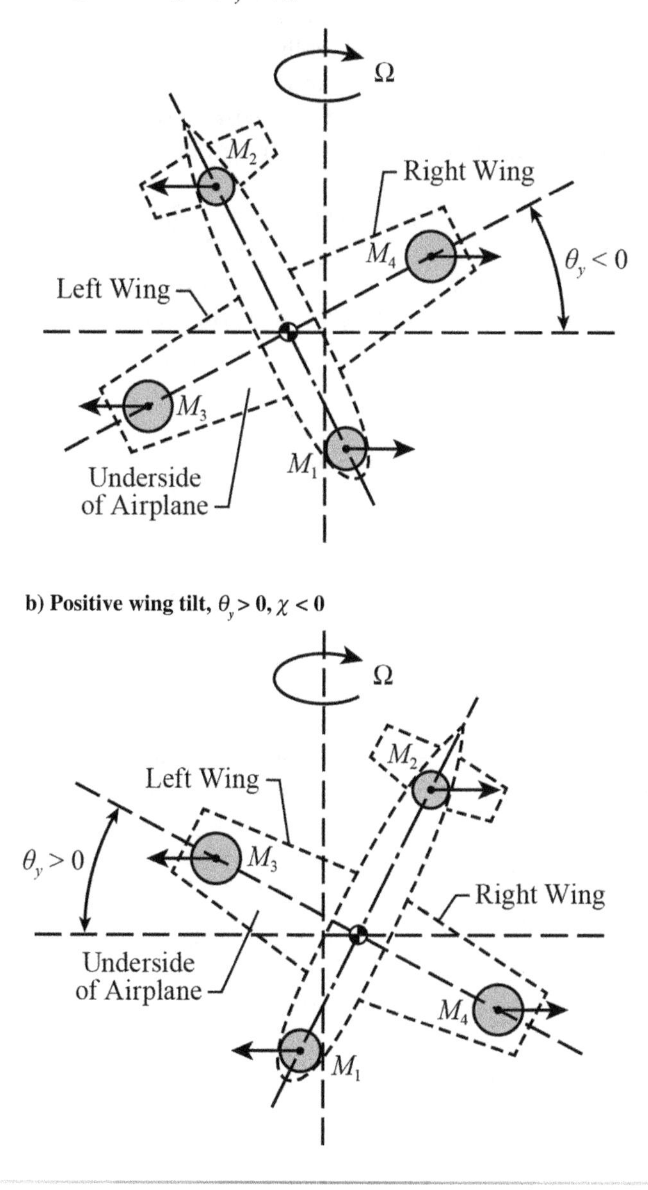

Fig. 7.20 Inertia-yawing moment in spin.

the wing masses M_3 and M_4 representing inertia in roll I_x produce an antispin yawing moment. The net inertia-induced yawing moment is the difference of these two contributions. If $I_y > I_x$, then the fuselage masses dominate and the induced yawing moment is prospin. On the other hand, if $I_x > I_y$, the wing masses will dominate and the inertia-yawing couple will be antispin.

If the wing tilt is the other way, i.e., $\theta_y > 0$ or $\chi < 0$ as shown in Fig. 7.20b, then the nature of each contribution changes. The wing masses M_3 and M_4 produce prospin yawing moment, whereas fuselage masses produce antispin yawing moment. If $I_y > I_x$, then the inertia yawing couple will be antispin and, if $I_x > I_y$, it will be prospin.

This simple concept of lumped masses helps us understand how important the wing tilt is in the establishment of equilibrium spin modes. Generally, in a right spin, the majority of airplanes have a small positive wing tilt ($\theta_y > 0$, $\chi < 0$) and a small outward sideslip (sideslip to port wing, $\beta < 0$).

The contribution to aerodynamic yawing moment mainly comes from the fuselage and vertical tail. The cross-sectional shape of the fuselage has a strong influence on the fuselage contribution [18, 19]. In particular, the aft cross sections have a stronger influence owing to their large moment arm with respect to the center of gravity. The cross sections with flat-bottom surfaces contribute to prospin yawing moment, whereas those with rounded-bottom surfaces generate antispin yawing moments.

The effectiveness of the vertical tail and the rudder in spin depend on the extent of shielding of these components due to the wake of the horizontal tail as shown in Fig. 7.21. A deflection of the elevator also has an effect on the shielding of the vertical tail and rudder surfaces. A downward deflection of the elevator increases the shielded area, whereas an upward deflection alleviates the shielding effect.

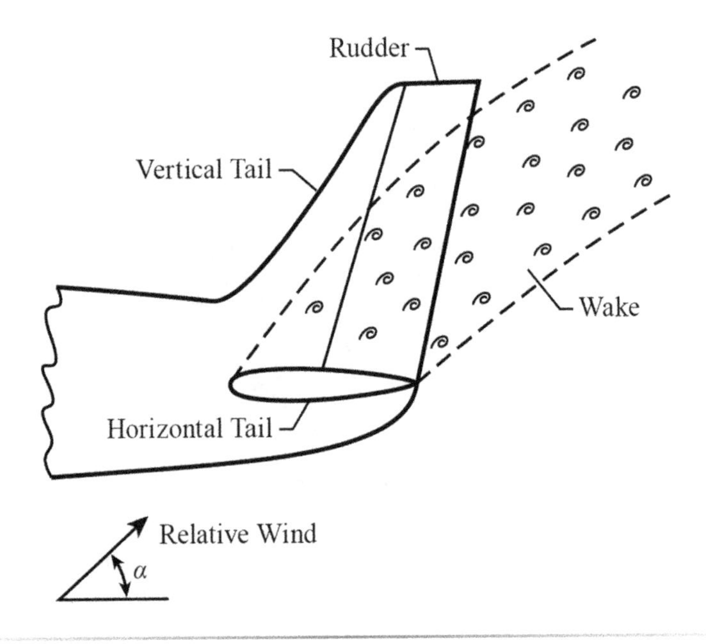

Fig. 7.21 Shielding effect on vertical tail and rudder.

The nature and extent of shielding of the vertical tail and rudder surfaces also depend on the relative placement of the horizontal tail with respect to the vertical tail as shown in Fig. 7.22. In Fig. 7.22a the aft-body part A and

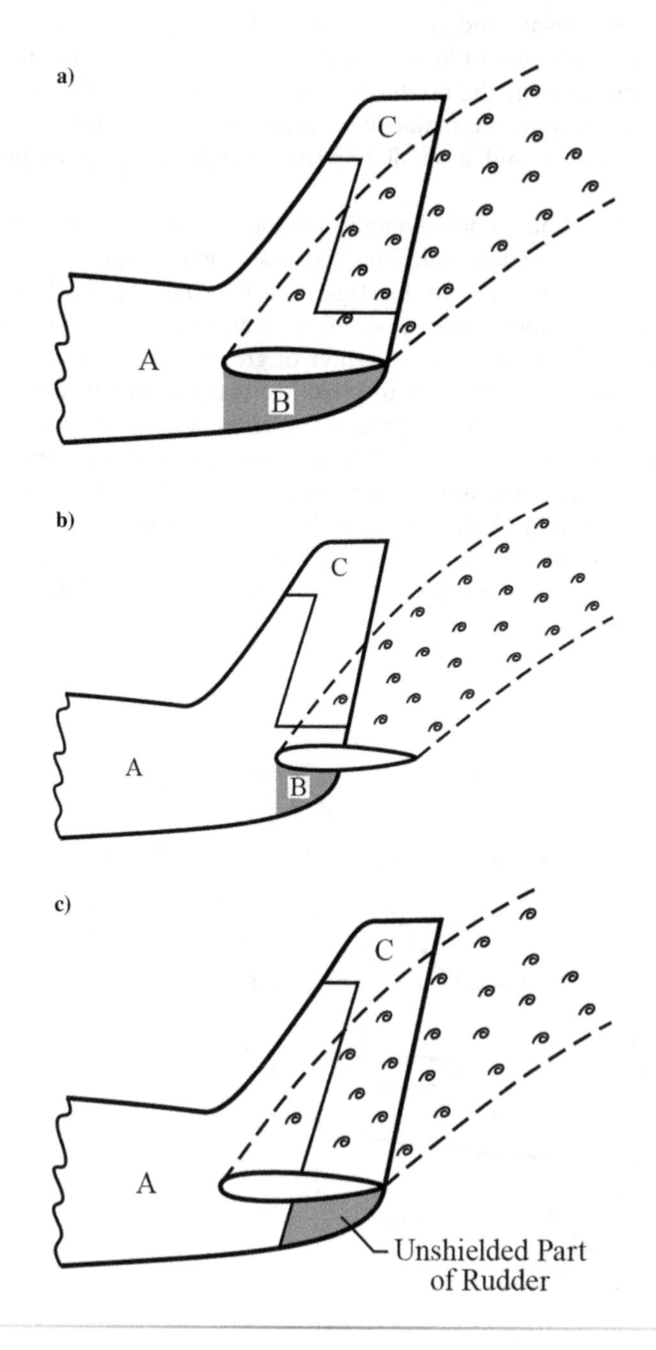

Fig. 7.22 Effect of vertical tail design on shielding effect.

the part B of the body below the horizontal tail contribute to the yawing moment developed by the fuselage. The unshielded part C of the rudder contributes to the rudder effectiveness in generating yawing moments. We observe that the tail design of Fig. 7.22a results in substantial shielding of the rudder. By moving the horizontal tail further aft as in Fig. 7.22b, the shielding of the vertical tail and the rudder can be reduced. Also, extending the rudder below the horizontal tail as shown in Fig. 7.22c improves rudder effectiveness.

7.6 Spin Recovery

The FAR Part 23 [20] requires that, for a normal recovery, a single-engine airplane be capable of recovering from a one-turn spin or from a 3-s spin, whichever takes longer, in no more than an additional turn using normal recovery controls.

The standard method of spin recovery as given in so-called "NACA Recovery Procedure" for conventional single-engine light airplanes [21, 22] is as follows: move the rudder briskly against the spin, followed by forward stick about one-half turn later while maintaining neutral ailerons. It has been found that this recovery technique is effective in majority of the test cases NASA has evaluated over the years on a variety of single-engine light general aviation airplanes [11].

The role of ailerons as recovery controls depends on the mass and inertia distribution of the airplane. For an airplane that has most of its mass concentrated into a long fuselage and has relatively lighter wings with a short span ($I_y/I_x > 1$), then the aileron applied in the direction of the spin (or direction of rolling) tends to assist the recovery. The longer the fuselage and greater the weight concentrated in nose and tail, the larger will be the numerical value of the ratio I_y/I_x and greater will be the effect of in-spin ailerons in aiding the recovery. For an airplane that has heavy wings and a large wing span giving $I_y/I_x < 1$, then out-spin ailerons or ailerons applied against the direction of spin (or the direction of roll) is most beneficial. Such a situation exists for airplanes fitted with wing tip tanks. Once again, the smaller the numerical value of the ratio I_y/I_x, the greater will be the effect of out-spin ailerons in aiding the recovery process.

Application of the opposite rudder slows down the spin rate and steepens the spin attitude. While the spin rate is decreasing because of the opposite rudder, the elevator becomes increasingly effective in assisting the recovery. However, an opposite sequence of operation, i.e., one in which the elevator movement preceeds the rudder movement, is decidedly objectionable and may result in an aggravated spin, often resulting in higher rate of rotation and, for some configurations, it may even result in the development of flat spin.

For some airplanes, it may be difficult to recover from a flat spin using normal recovery techniques. Such airplanes often make use of a spin-chute to effect a recovery from flat spin (Fig. 7.23). The opening of the spin-chute generates a nosedown pitching moment in addition to slowing the yaw rate. When the yaw rate decreases, application of forward stick (elevator up) results in a successful recovery.

A typical time history of a spin entry, developed spin, and recovery is shown in Fig. 7.24 [23]. The prospin controls for this aircraft consist of full-aft stick, full-rudder deflection in the direction of spin, and full-aileron deflection against the spin. As shown in Fig. 7.24, full-aft stick is gradually applied from 10 to 17 s. At about 17 s, full right rudder ($\delta_r < 0$) for right spin and antispin aileron ($\delta_a > 0$, right aileron down and left aileron up to roll to left) are applied. For the next 5–6 s, the angle of attack and yaw rate build up. Around 25 s, the airplane has entered into a right spin, but it takes about two turns to establish a steady-state spin as indicated by a steady value of angle of attack around 52 to 55 deg and a steady value of yaw rate around 110 deg/s. After six complete turns, recovery controls

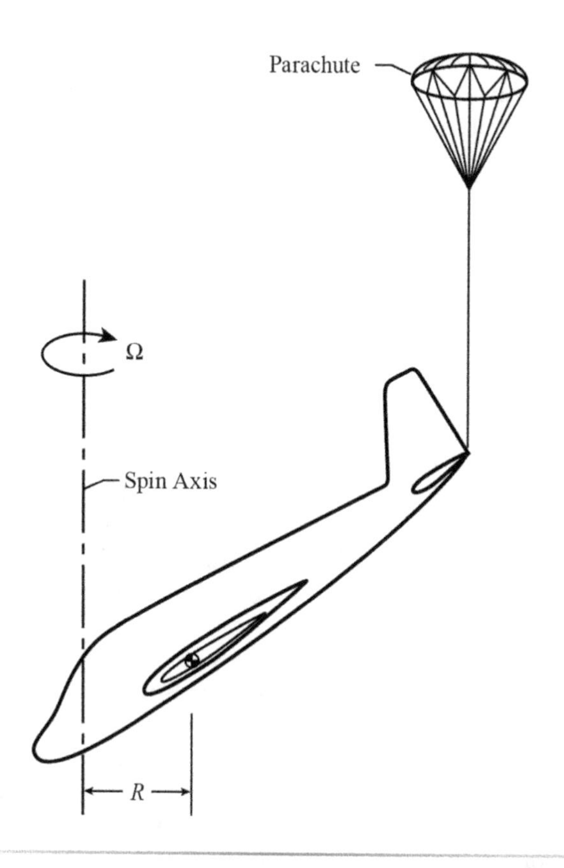

Fig. 7.23 Parachute deployment for spin recovery.

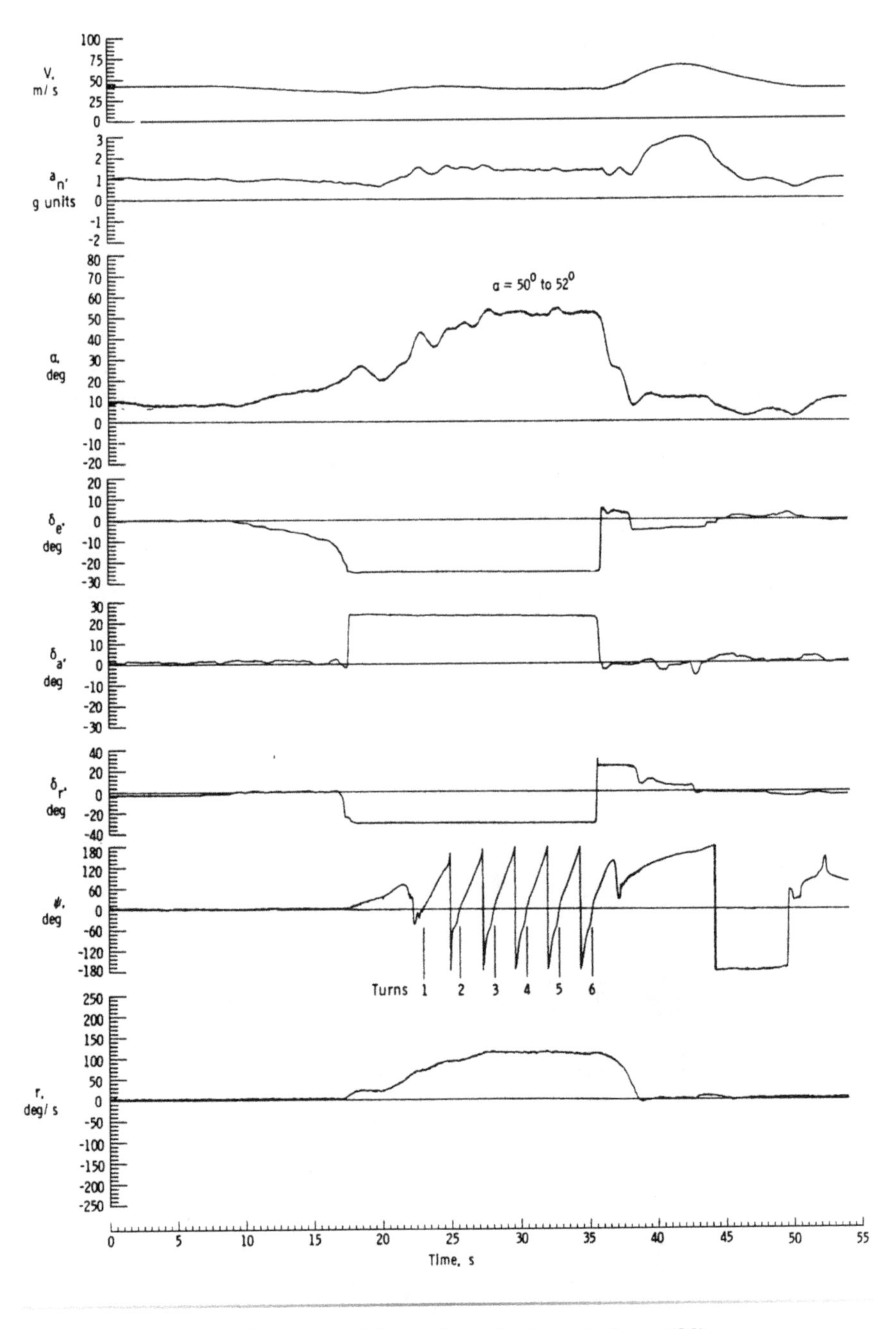

Fig. 7.24 Time history of a spinning airplane (23).

(opposite rudder, stick forward, and ailerons neutral) are applied. The aircraft recovers successfully and returns to normal steady-level flight within next 5–10 s.

7.7 Geometrical Modifications to Improve Spin Resistance

7.7.1 Modification to Wing

The autorotational characteristics of the wing have a large impact on the stall–spin behavior of the airplane. Eliminating or reducing the autorotative tendency of the wings can greatly improve the spin resistance of the airplane.

One modification to the wing that has been successfully tested by NASA on a number of general aviation airplanes is the reshaping of the wing leading edge of the outer wing panels. The modification consists of a chord extension of about 3%, a drooped-nose airfoil design, and an abrupt discontinuity between the undrooped inner wing and drooped outer wing [23–25]. An application of this method to a general aviation airplane is shown in Fig. 7.25. The basic airplane (Fig. 7.25a) had an untwisted wing with a NACA 64(2)-415 airfoil section. The modification extended from 57–95% semispan locations as shown in Fig. 7.25b. The variation of the resultant

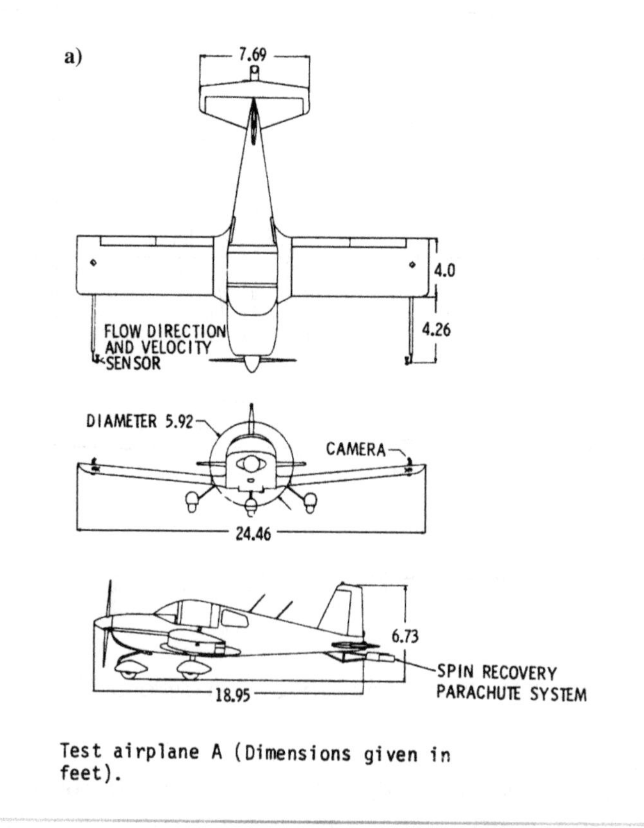

Test airplane A (Dimensions given in feet).

Fig. 7.25 An example of wing leading-edge modification (23, 24) (*continued*).

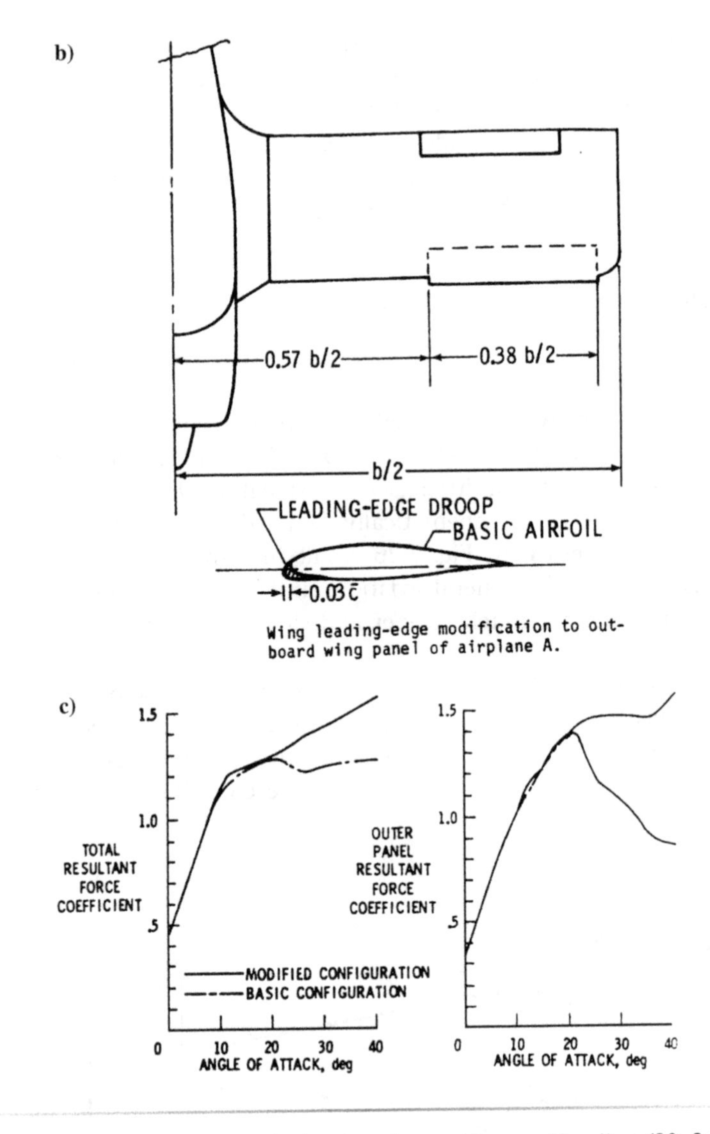

Fig. 7.25 An example of wing leading-edge modification (23, 24).

force coefficient of the basic airplane and that with the modification are shown in Fig. 7.25c. We observe that the basic airplane exhibits the usual drop in the resultant force coefficient at stall (around $\alpha = 20$ deg) that leads to instability in roll and autorotation, whereas the modified configuration does not exhibit such tendency for the range of angle of attack shown in Fig. 7.25c.

The basic aircraft exhibited two spin modes, one at $\alpha = 46$ deg and another at $\alpha = 61$ deg. However, the modified airplane was highly resistant to enter the spin [25].

The aerodynamic phenomenon responsible for this beneficial effect is a strong vortex-type flow that originates at the leading edge discontinuity and tails over the upper surface of the wing functioning like an aerodynamic fence that prevents the spreading of the stalled inboard flow towards the wingtips. As a result, the drooped outboard wing panels remain effective and generate lift to very high angles of attack [11].

7.7.2 Fuselage Modifications

The aft sections of the fuselage have a relatively stronger influence on the autorotational/spin behavior of the airplane. Hence, modifications to aft fuselage sections will be more effective in controlling the spin behavior. Examples of such modifications tested by NASA [19] include rounding of the bottom surface of the aft fuselage and installation of strakes on the underside of the aft fuselage as schematically depicted in Fig. 7.26. Modifications similar to those shown in Fig. 7.26 result in significant improvement in spin resistance of light general aviation airplanes. The spins exhibited by modified airplanes are much steeper and slower compared to the baseline airplanes [19].

The aerodynamic flow mechanisms giving rise to these beneficial effects are discussed elsewhere [26, 27]. Essentially, these modifications alter the sectional side force coefficient characteristics from prospin type A to antispin type B (see Fig. 7.8). As a result, the fuselage contribution becomes more damping in nature.

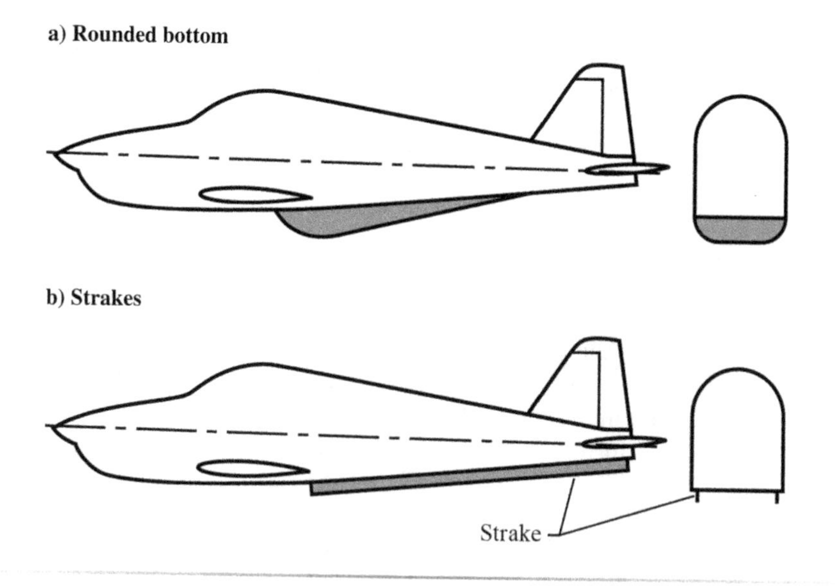

Fig. 7.26 Examples of modification of fuselage geometry.

Example 7.2

A light single-engine, general aviation airplane has the following data: gross weight = 10,915 N, wing span = 9.9822 m, wing area = 13.53 m^2, mean aerodynamic chord = 1.34 m, $I_x = 2304\ \text{kg/m}^2$, $I_y = 2602\ \text{kg/m}^2$, and $I_z = 4336\ \text{kg/m}^2$.

This aircraft has a steady spin mode at an angle of attack of 40 deg. The aircraft takes about 3 min per turn and spins with right wing 5 deg below the horizontal plane. Assuming that the resultant force coefficient at 40-deg angle of attack is equal to 1.2, determine 1) velocity of descent, 2) spin radius as a fraction of wing semispan, 3) angular velocity components in the body axes system, and 4) aerodynamic moments acting on the airplane during the steady-state spin. Assume sea-level conditions.

Solution:

We have

$$\sin\theta_y = -\cos\alpha\sin\chi$$

With $\theta_y = 5$ deg and $\alpha = 40$ deg, we get $\chi = -6.5326$ deg.

The angular velocity components are given by

$$p = \Omega\cos\alpha\cos\chi$$

$$q = -\Omega\cos\alpha\sin\chi$$

$$r = \Omega\sin\alpha$$

We have

$$\Omega = \frac{360}{3}$$

$$= 120\ \text{deg/s}$$

$$= 2.0942\ \text{rad/s}$$

Substituting, we get

$$p = 91.3285\ \text{deg/s}$$

$$q = 10.4578\ \text{deg/s}$$

$$r = 77.1298\ \text{deg/s}$$

We have $C_R = 1.2$. The lift and drag coefficients are given by

$$C_L = C_R\cos\alpha$$

$$C_D = C_R\sin\alpha$$

Substituting, we get $C_L = 0.9193$ and $C_D = 0.7713$.

(Continued)

Example 7.2 (*Continued*)

We have $W = 10{,}915$ N, $\rho = 1.225$ kg/m^3, and $S = 13.53$ m^2. The velocity of descent is given by

$$U_0 = \sqrt{\frac{2W}{\rho S C_D}}$$

Substituting, we get $U_0 = 41.3236$ m/s. The spin radius is given by

$$R = \left(\frac{1}{2m\Omega^2}\right)\rho U_0^2 S C_L$$

where $m = W/g$. Substitution gives $R = 2.6664$ m or, in terms of wing semispan,

$$\frac{2R}{b} = \frac{2 \times 2.6664}{9.9822}$$
$$= 0.5342$$

Thus, we see that the spin radius is close to one-half of wing semispan.

The inertia moments are given by

$$M_i = rp(I_z - I_x)$$
$$L_i = qr(I_y - I_z)$$
$$N_i = pq(I_x - I_y)$$

In these equations, the values of p, q, and r are in rad/s. Substitution gives

$$M_i = 4359.5633 \text{ Nm}$$
$$L_i = -425.9929 \text{ Nm}$$
$$N_i = -86.6869 \text{ Nm}$$

For steady-state spin, the sum of aerodynamic and inertial moments is zero. Therefore,

$$M_{\text{aero}} = -4359.5633 \text{ Nm}$$
$$L_{\text{aero}} = 425.9929 \text{ Nm}$$
$$N_{\text{aero}} = 86.6869 \text{ Nm}$$

Converting to coefficient form, we get

$$C_{m,\text{aero}} = \frac{M_{\text{aero}}}{qS\bar{c}}$$
$$= -0.2299$$

(*Continued*)

Example 7.2 (Continued)

$$C_{l,\text{aero}} = \frac{L_{\text{aero}}}{qSb} = 0.003$$

$$C_{n,\text{aero}} = \frac{N_{\text{aero}}}{qSb} = 0.0006$$

7.8 Summary

In this chapter, we have discussed two classes of flight where the longitudinal and lateral-directional motions of the airplane are strongly coupled. The first case was the inertia coupling or roll coupling problem where a steadily rolling aircraft experiences instability as the roll rate is increased. We observed that an aircraft having a higher level of directional (yaw) stability compared to longitudinal (pitch) stability experiences a divergence in pitch, whereas an aircraft deficient in directional stability is susceptible to a yaw divergence. The roll rates at which an aircraft experiences either of these two instabilities depend on its inertia characterisitics. As the roll rate is further increased, the aircraft becomes stable again, but this time it is spin stabilized about its longitudinal body axis.

The other class of problems we considered is the airplane spin. The autorotational tendency of the unswept wings at stall is the driving mechanism for light general aviation airplanes to enter into a spin. Whether the steady-state spin is established or not depends on the balance between the aerodynamic and inertia moments. If such a balance is not achieved, the spin remains oscillatory. The wing tilt plays an important role in the establishment of a steady-state spin. The magnitude and sign of the inertia-rolling and yawing moments depend on the wing tilt.

Also, the wing tilt affects the sideslip, thereby affecting the aerodynamic rolling and yawing moments. The application of opposite rudder followed by a forward stick is the standard recovery procedure for most of the light general aviation airplanes. However, the use of ailerons for recovery depends on the mass and inertia characteristics of the airplane. On some airplanes, prospin ailerons assist recovery, whereas on others it may be the opposite. Modifying the outboard leading edge of the wing considerably helps improve the spin resistance. Also, some modifications to aft fuselage geometry such as rounding the bottom surface or installing strakes increases the damping effect of the fuselage in spin.

References

[1] Philips, W. H., "Effect of Steady Rolling on Longitudinal and Lateral Stability," NACA TN 1627, June 1948.
[2] Pinsker, W. J. G., "The Theory and Practice of Inertia Coupling," *Journal of the Royal Aeronautical Society*, Vol. 73, Aug. 1973, pp. 695–702.
[3] Gates, O. B. Jr., and Woodling, C. H., "A Theoretical Analysis of the Effects of Engine Angular Momentum on Longitudinal and Directional Stability in Steady Rolling Maneuvers," NACA TN 4249, April 1958.
[4] Hacker, T., and Oprisiu, C., "A Discussion on the Roll Coupling Problem," *Progress in Aerospace Sciences*, Vol. 15, 1974, pp. 151–180.
[5] Schy, A. A., and Hannah, M. E., "Prediction of Jump Phenomena in Roll-Coupled Maneuvers of Airplanes," *Journal of Aircraft*, Vol. 14, No. 4, April 1977, pp. 375–382.
[6] Sternfield, L., "A Simplified Method for Approximating the Transient Motion in Angles of Attack and Sideslip During Constant Rolling Maneuver," NACA Rept. 1344, 1958.
[7] Blakelock, J. H., *Automatic Control of Aircraft and Missiles*, 2nd ed., Wiley, New York, 1991.
[8] Knight, M., "Wind Tunnel Tests on the Autorotation and the Flat Spin," NACA Rept. 273, 1927.
[9] Polhamus, E. C., "Effect of Flow Incidence and Reynolds Number on Low Speed Aerodynamic Characteristics of Several Noncircular Cylinders with Application to Directional Stability and Spinning," NASA TR R-29, 1959.
[10] Anderson, S. B., "Historical Overview of Stall/Spin Characteristics of General Aviation Aircraft," AIAA Paper 78-1551, Aug. 1978.
[11] Chambers, J. R., and Paul Stough, H., III., "Summary of NASA Stall/Spin Research for General Aviation Configurations," AIAA Paper 86-2597, 1986.
[12] DiCarlo, D. J., Stough, H. P. III., Glover, K. E., Brown, P. W., and Patton, J. M. Jr., "Development of Spin Resistance Criteria for General Aviation Airplanes," AIAA Paper 86-9812, 1986.
[13] Gates, S. B., and Bryant, L. W., "The Spinning of Aeroplanes," Aeronautical Research Committee Reports and Memoranda (ARC R & M) No. 1001, 1926.
[14] Pamadi, B. N., and Taylor, L. W. Jr., "A Semi-Empirical Method for Estimation of Aerodynamic Characteristics of Steadily Spinning Light Airplanes," NASA TM 4009, Dec. 1987.
[15] Pamadi, B. N., and Taylor, L. W. Jr., "Estimation of Aerodynamic Characteristics of Steadily Spinning Light Airplanes," *Journal of Aircraft*, Vol. 21, No. 12, 1984, pp. 943–954.
[16] Adams, W. M. Jr., "Analytical Prediction of Airplane Equilibrium Spin Characteristics," NASA TN D-6926, Nov. 1972.
[17] Dickinson, B., *Aircraft Stability and Control for Pilots and Engineers*, Sir Isaac Pitman and Sons, London, 1968.
[18] Beaurain, L., "General Study of Light Airplane Spin, Aft Geometry, Part I," NASA TTF-17446, June 1977.
[19] Burk, S. M., Bowman, J. S., and White, W. L., "Spin-Tunnel Investigation of the Spinning Characteristics of Typical Single-Engine General Aviation Airplane Designs, I-Low Wing Model A, Effect of Tail Configurations," NASA TP 1009, 1977.
[20] *Airworthiness Standards*, "Normal, Utility and Acrobatic Category Airplanes," *Federal Aviation Regulations*, Vol. III, Part 23, FAA, June 1974.
[21] Neihouse, A. I., and Klinar, W. J., "Status of Spin Research for Recent Airplane Designs," NASA TR R-57, 1960 (supersedes NACA RM L57F12).
[22] Bowman, J. S., "Summary of Spin Technology as Related to Light General Aviation Airplanes," NASA TND-7575, Dec. 1971.
[23] Staff of Langley Research Center, "Exploratory Study of the Effects of Wing-Leading Edge Modifications on the Stall/Spin Behavior of a Light General Aviation Airplane," NASA TP 2011, June 1982.

[24] DiCarlo, D. J., Glover, K. E., Stewart, E. C., and Stough, H. P., "Discontinuous Wing Leading Edge to Enhance Spin Resistance," *Journal of Aircraft*, Vol. 22, No. 4, 1985, pp. 283–288.
[25] Stough, H. P., III., Jordan, F. L. Jr., DiCarlo, D. J., and Glover, K. E., "Leading-Edge Design for Improved Spin Resistance of Wings Incorporating Conventional and Advanced Airfoils," Society of Automotive Engineers, SAE Aerospace Technology Conference and Exposition, Paper 830720, Long Beach, CA, Oct. 14–17, 1985.
[26] Pamadi, B. N., and Pordal, H. S., "Effect of Strakes on the Autorotational Characteristics of Noncircular Cylinders," *Journal of Aircraft*, Vol. 24, No. 2, 1987, pp. 84–97.
[27] Pamadi, B. N., Jambunathan, V., and Rahman, A., "Control of Autorotational Characteristics of Light-Airplane Fuselages," *Journal of Aircraft*, Vol. 25, No. 8, 1988, pp. 695–701.

Problems

7.1 If the aircraft in Example 7.1 is flying at 175 m/s at an altitude of 8000 m, determine the range of roll rates for divergence in yaw or pitch. Assume $C_{n\beta} = 0.045$ per radian. Except for this change, all the other parameters remain the same.

7.2 The aircraft of solved Example 7.2 has another steady-state spin mode at $\alpha = 55$ deg and takes 2.2 s per turn. Assuming that the aircraft spins with its right wing 3.5 deg below the horizontal plane, determine (a) descent velocity, (b) spin radius in terms of wing semispan, (c) angular velocity components in body-fixed axes system, and (d) inertia and aerodynamic coefficients acting on the airplane.

Assume that the aircraft is operating at an altitude of 6000 m and the resultant force coefficient at $\alpha = 55$ deg is 1.30.

Chapter 8 Stability and Control Problems at High Angles of Attack

8.1 Introduction

In previous chapters, we considered the airplane stability and control in the linear range of angle of attack based on the assumption of small disturbances in all the motion variables. Under this assumption, the longitudinal and lateral-directional motions could be decoupled and analyzed separately. Such an approach is not possible when the cross coupling between longitudinal and lateral dynamics becomes important. This type of coupling occurs because of inertia terms in the equations of motion or nonlinearities in aerodynamics. In Chapter 7, we discussed two examples of inertia cross coupling, inertia coupling during rapid rolls and airplane spin. In this chapter, we will study the stability and control problems caused by cross coupling between longitudinal and lateral-directional motions on account of nonlinearities in aerodynamics and large amplitude motions, which typically occur during maneuvers at high angles of attack approaching or exceeding the stalling angle.

8.2 A Brief Historical Sketch

Flight at high angles of attack, or even at angles of attack close to the stall angle, was recognized to be disastrous from the very early days of aviation. The Wright Brothers had experienced perhaps the first military accident in 1908 during a demonstration flight because of an inadvertent stall and loss of control. During this early period and up to the end of World War II, several accidents occurred because of inadvertent stall, loss of control, and spin. These events led airplane designers to focus their attention to the study of spinning problems with an objective to develop successful spin recovery techniques. Such an approach continues even today in civil aviation for the design and operation of light general aviation airplanes, with increased emphasis on improving spin resistance and spin avoidance as we discussed in Chapter 7. However, for military airplanes, the design philosophy has undergone a revolutionary change.

In the years following World War II and through the 1950s, military airplane designers, especially in the United States, emphasized the importance of high speed and high altitude flying [1]. Such capabilities were essential for an interceptor aircraft. The scenario envisioned was one involving an aerial combat beyond the visual range (BVR) consisting of missile launches between enemy aircraft that would never come within the visual range (WVR). Because of such belief, the prevailing requirement for achieving air superiority in close combat in WVR disappeared. The F-104 (Star Fighter) and the F-105 designs were based on this philosophy. The belief that flight at high angles of attack must be avoided at all costs still persisted, and airplanes were equipped with artificial limiters like stick pushers to limit the angle of attack.

A turnaround in the thinking concerning the importance of flight at high angles of attack occurred in the late 1960s and 1970s. The Southeast Asian conflict and a large number of other fatal accidents caused by inadvertent stall, loss of control, and spins contributed to the realization that, rather than concentrating on spin recovery, attention must be devoted to the prevention of departure from controlled flight at high angles of attack. Furthermore, it dawned on the military planners and airplane designers in the 1970s that the development of such technologies like electronic countermeasures and radar jammers severely limited a positive identification and acquisition of enemy aircraft in BVR. They also realized that close encounters in WVR would be inevitable in future aerial combats. This kind of changed thinking led to a resurgence of interest in high angle of attack aerodynamics and control technology and to the development of such fighter aircraft like the U.S. F-15 and the F-16, the Israel Lavi, the Swedish Grippen, the French Rafale and the Mirage III, the European Fighter Aircraft, and the Soviet MiG 29 and the MiG 31.

The line of thinking in the area of high angles of attack took another turn in the late 1980s and early 1990s when it was further realized that the maximum sustained turn rate can no longer be improved by conventional methods. The longer the turn lasts, the more time an opponent has to set up an attack. On the other hand, the capability of controlled flight at high angles of attack well beyond the stall angle of attack offers distinct and virtually unlimited tactical advantage over the adversary in WVR encounters. The Herbst maneuver (see Chapter 2) is one such example. Thus, the technical requirements for flight at high angles of attack became directly opposite of those advocated in the 1960s. Whereas in the 1960s, the high-angle-of-attack regime was regarded as extremely hazardous and to be avoided at all cost, the current requirements for fighter aircraft demand a high degree of agility and unlimited capability at high angles of attack. With the advent of highly lethal, all-aspect air-to-air missiles, the capability to get a first shot has become extremely important in air combat. The F-15 STOL and Maneuver Technology Demonstrator (S/MTD), the F/A-18, and experimental programs like the X-29 and the X-31 are examples of fighter aircraft designed and developed with this philosophy.

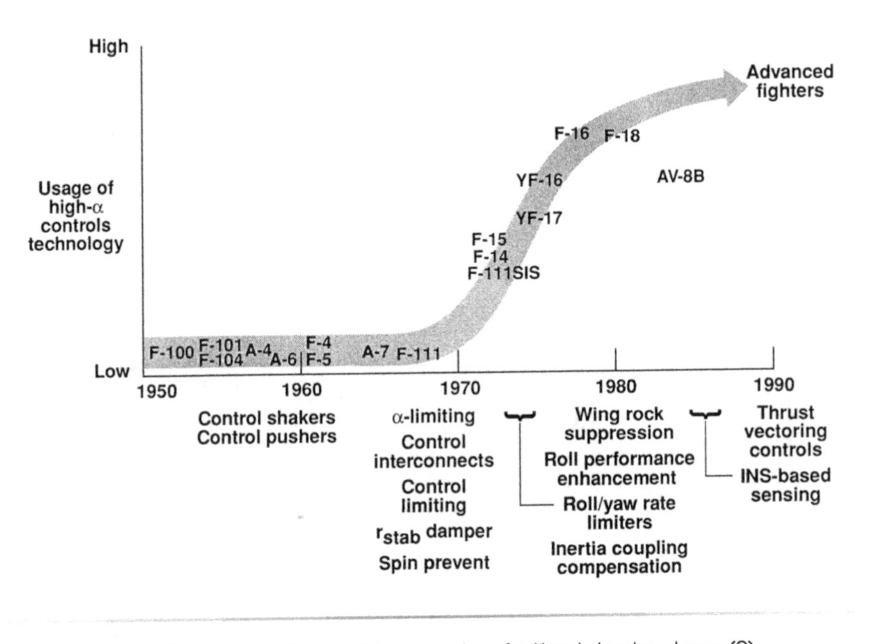

Fig. 8.1 Application of high-angle-of-attack technology (2).

A summary of usage of high-angle-of-attack controls technology in the design and evolution of combat aircraft in the United States is schematically presented in Fig. 8.1 [2].

8.3 Brief Overview of High-Alpha Problems

Typically, a modern fighter aircraft features a long, pointed slender fuselage and thin, sharp leading-edge, highly swept delta wings. Furthermore, they also feature wing strakes, or leading-edge extensions (LEX), and close-coupled canards. Such aircraft configurations attain their maximum lift coefficients in the angle of attack range of 25–35 deg. The high angle of attack problems encountered by these types of aircraft are schematically illustrated in Fig. 8.2 [3].

As the angle of attack is increased, the aircraft experiences the onset of buffeting. The buffeting is characterized by heavy fluctuations in aerodynamic forces over the horizontal tail because of the separated, turbulent wing and fuselage wake passing over it. In this angle of attack range, the aircraft may also experience pitch-up. With further increase in angle of attack, the aircraft is likely to experience such problems as wing rock, roll reversal, directional instability, or directional divergence.

Wing rock is an oscillatory motion predominantly in roll about the body axis. Wing rock can be highly annoying to the pilot and may pose serious limitation to the combat effectiveness. In addition, the wing rock can be a safety problem during landing.

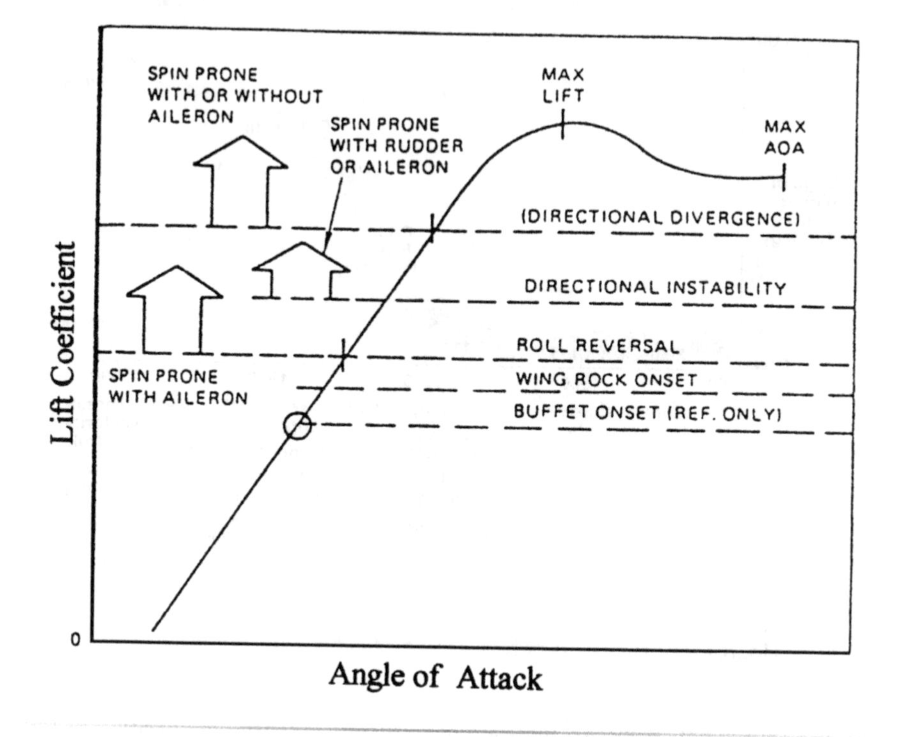

Fig. 8.2 Various limitations on the use of maximum lift coefficient (3). (Courtesy AGARD.)

The roll reversal phenomenon is one in which the aircraft rolls in the opposite direction to aileron input. The roll due to adverse yaw overpowers the proverse roll due to ailerons.

The directional instability/directional departure may cause the aircraft to depart from controlled flight and enter into a spin. Whether the aircraft can develop a steady spin depends on the balance between the inertial and aerodynamic moments as we discussed in Chapter 7.

In this chapter, we will study the problems of wing rock, roll reversal, directional instability/departure, and the control concepts that are pertinent to flight at high angles of attack. However, before we do so, let us develop some basic understanding of the high-angle-of-attack aerodynamics of delta wings and slender forebodies.

8.4 Delta Wings at High Angles of Attack

For a highly swept, thin, sharp leading-edge delta wing held at high angles of attack, the boundary layer cannot negotiate the large adverse pressure gradients formed at the wing leading edge. As a result, the flow separates right at the leading edge, forming a fixed line of flow separation all along the leading edge of the wing. The separated flow rolls up into a pair of spiral vortices with concentrated vortex cores on the lee side as shown in Fig. 8.3 [4]. The leading-edge geometry has very little effect on such vortex flow fields formed on the lee side as long as the leading edges are sufficiently thin and sharp. The external flow reattaches to the surface of the wing, forming a primary reattachment line as shown in Fig. 8.3b. As this reattached flow moves in the spanwise direction, it encounters an adverse pressure gradient and separates once again, forming another vortex called the secondary vortex. The secondary vortex is much smaller in size and has a sense of rotation opposite that of the primary vortex. The flow in the region between the two primary reattachment lines on either side of the centerline is the attached flow, which moves in the freestream direction. Evidently, this flow is not influenced by the primary or secondary vortices as illustrated in Fig. 8.3b.

The suction formed over the lee side of the wing under the influence of the vortices enhances the lift and causes the lift coefficient to vary nonlinearly with angle of attack as shown in Fig. 8.4. This part of the total lift, which is in

a) Schematic diagram

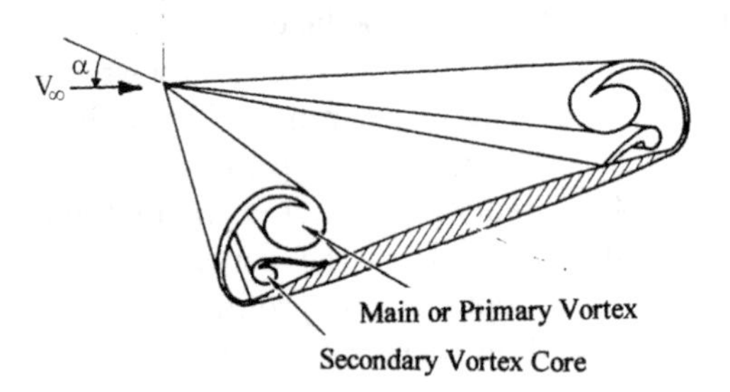

b) Details of flow pattern

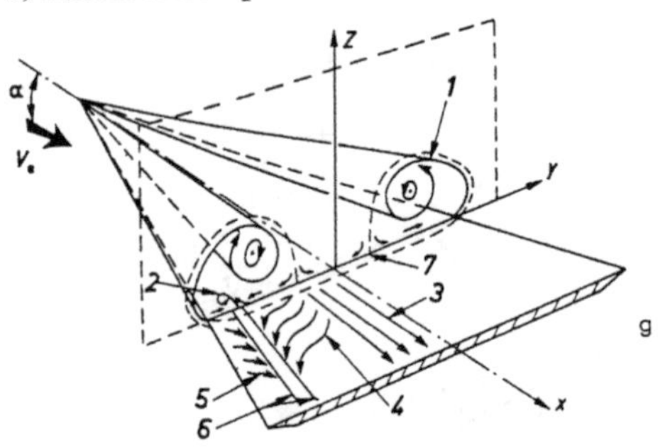

Schematic Flow Diagram:

1) Primary or main vortex
2) Secondary vortex
3) Central zone without vortex
4) Zone induced by the main vortex
5) Zone induced by the secondary vortex
6) Accumulation zone (air bubble or coating)
7) Reattachment line

Fig. 8.3 Vortex flow over sharp leading-edge delta wing at high angles of attack (4). (Courtesy AGARD.)

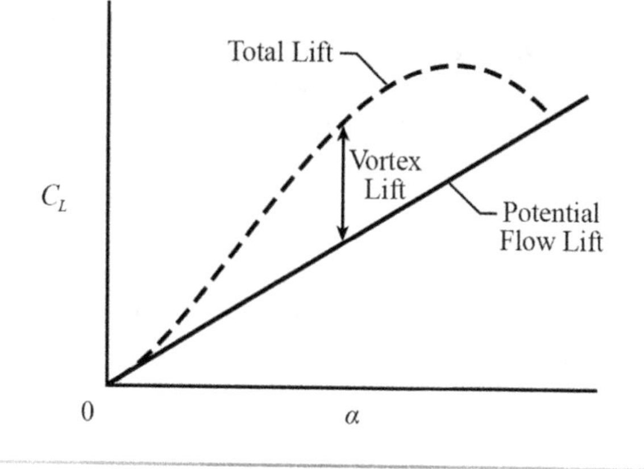

Fig. 8.4 Lift coefficient with angle of attack for slender delta wings.

excess over the potential or the attached flow lift, is called vortex lift. The vortex lift increases with increase in angle of attack until the vortex breakdown occurs on the lee side of the wing. However, this beneficial vortex lift is accompanied by a large increase in the drag coefficient because the leading-edge suction or thrust, omnipresent on the typical low-speed, round leading-edge wings, is lost because of flow separation. Researchers have tried several combinations of twist, camber, and sweep to recover the leading-edge thrust as much as possible to reduce the drag coefficient and increase the lift-to-drag ratio. One such example is the concept of the vortex flap that attempts to achieve an acceptable compromise between the beneficial vortex lift and the adverse drag increase at high angles of attack. As shown in Fig. 8.5a [5], the leading-edge vortex flap tilts its lift

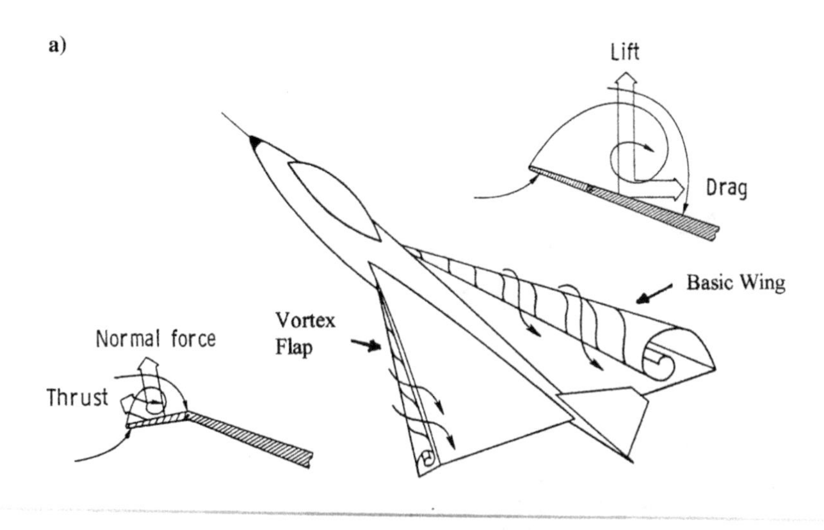

Fig. 8.5a Vortex flap concept (5).

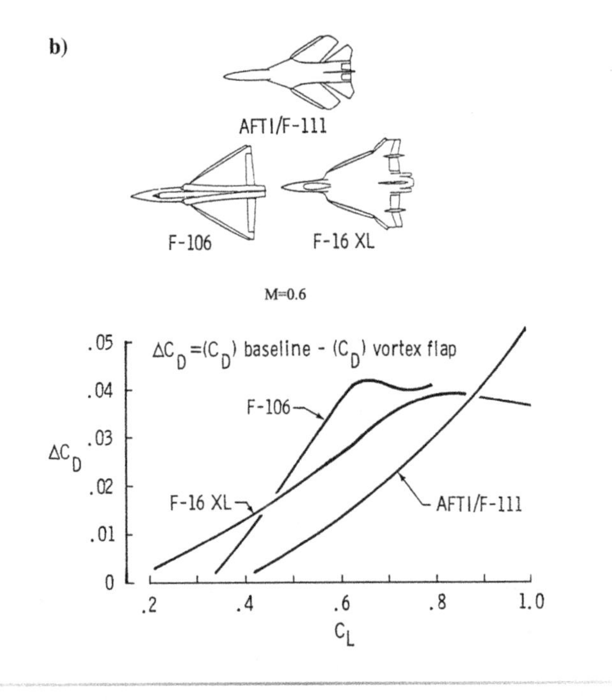

Fig. 8.5b Subsonic drag reduction using vortex flaps (6).

vector slightly forward to produce a "thrust" component. Vortex flaps have been incorporated on the F-106, the F-16XL, and the AFTI/F-111 aircraft. An example of drag reduction due to a vortex flap is shown in Fig. 8.5b [6].

As the angle of attack increases, the amount of vorticity fed into the vortex core increases. The vortices become stronger and physically lift themselves above the wing surface. For highly swept delta wings, the vortices assume asymmetric configuration as depicted in Fig. 8.6. The angle of attack for the onset of asymmetry depends on the leading-edge sweep [7].

8.4.1 Vortex Breakdown and Stall

Another interesting phenomenon associated with the leading-edge vortices is the vortex breakdown. The breakdown, or bursting as it is commonly called, refers to a sudden and rather dramatic structural change, which usually results in the turbulent dissipation of the vortex. Vortex bursting is characterized by a sudden deceleration of the axial flow in the vortex core, formation of a small recirculatory flow region, a decrease in the circumferential velocity, and an increase in the size of the vortex.

As shown in Fig. 8.7, two types of vortex breakdown are commonly observed on delta wings at high angles of attack, bubble type and spiral type. The bubble type or "axisymmetric" mode of vortex breakdown is characterized by a stagnation point on the swirl axis, followed by an

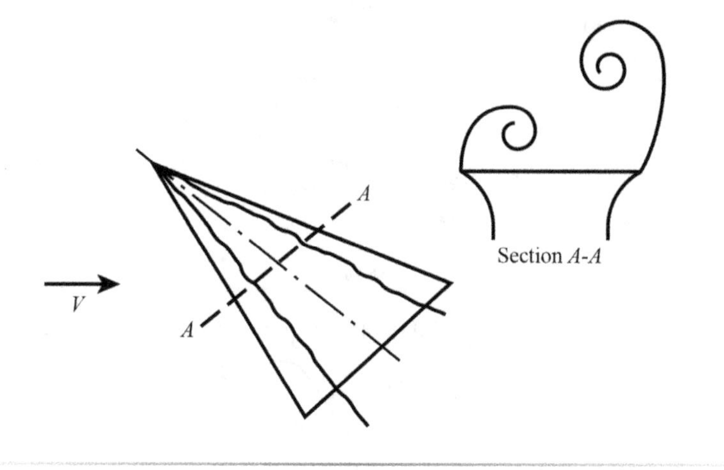

Fig. 8.6 Vortex asymmetry over slender delta wings at high angles of attack.

oval-shaped recirculation bubble. The spiral mode of breakdown is characterized by a rapid deceleration of the core flow, followed by an abrupt kink. At the kink, the core flow takes the form of a spiral, which persists for one or two turns before breaking up into a large-scale turbulence. While the bubble-type breakdown has been observed on delta wings in low Reynolds number water-tunnel tests, it is the spiral type that is more routinely observed in wind-tunnel experiments [8].

The stall of a thin, highly swept, sharp-edged, delta wing (such a wing is also called slender delta wing) is different from the stall of a conventional

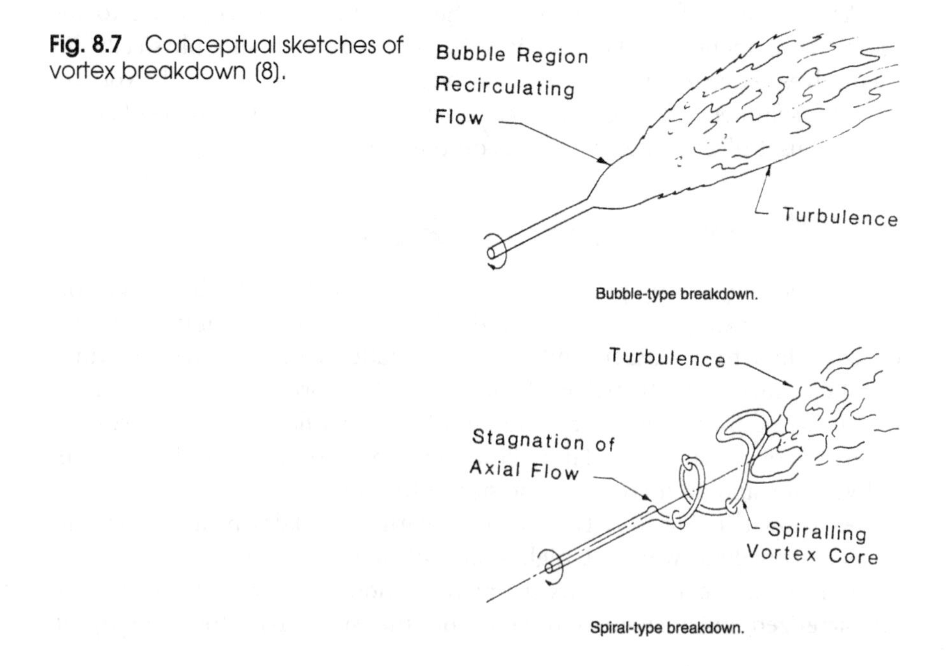

Fig. 8.7 Conceptual sketches of vortex breakdown (8).

round leading-edge airfoil or a finite wing, which is largely influenced by the boundary-layer and flow separation characteristics (see Chapter 1). The stall of a slender delta wing essentially depends on the location of the vortex breakdown point on the wing. The variation of the vortex breakdown point on the wing with angle of attack is shown in Fig. 8.8.

The variations of lift coefficient with angle of attack for slender delta wings of leading edge sweep angles ranging from 55 deg to 82 deg are shown in Fig. 8.9. Also indicated in Fig. 8.9 are the values of angles of attack when the vortex breakdown point crosses the trailing edge. We observe that a slender delta wing of high leading-edge sweep, 75 deg or higher, attains its maximum lift and stalls approximately at the angle of attack when the vortex breakdown point crosses the trailing edge. For delta wings of lower leading-edge sweeps, the stall occurs after the vortex breakdown point crosses the wing trailing edge and starts moving towards the wing apex. Also, there is a break in the lift-curve slope at the angle of attack when the vortex breakdown point reaches the trailing edge as observed in Fig. 8.9b. With further increase in angle of attack, the vortex breakdown point moves towards the wing apex and, at some point, the wing stalls. For example, the 70-deg delta wing stalls at $\alpha = 33$ deg when the vortex breakdown point is located approximately 35% chord upstream of the trailing edge (Fig. 8.8). In general, as the leading-edge sweep decreases, the stall occurs when the vortex breakdown point is located more closer to the leading edge.

The slender delta wing experiences the so-called pitch-up phenomenon (Fig. 8.10) at high angles of attack because of the progressive loss of vortex

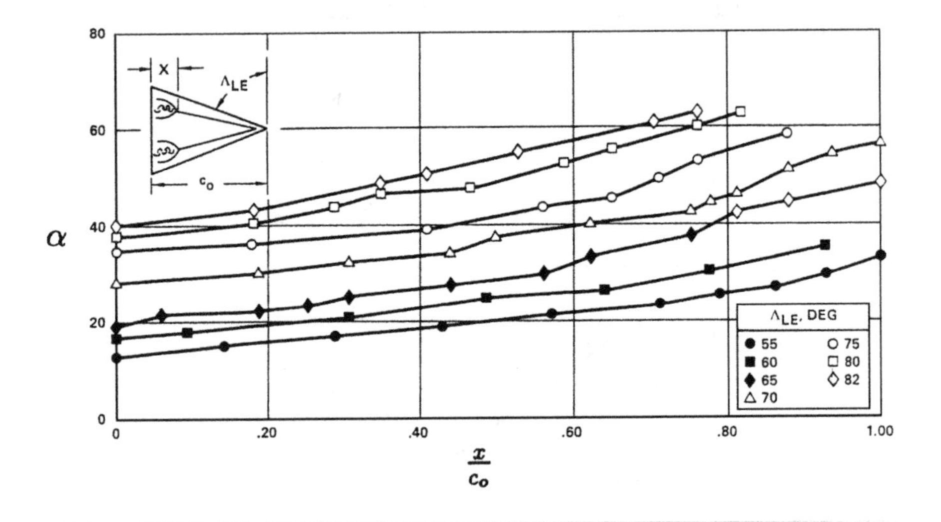

Fig. 8.8 Location of vortex breakdown point for slender delta wings (9). (Courtesy AGARD.)

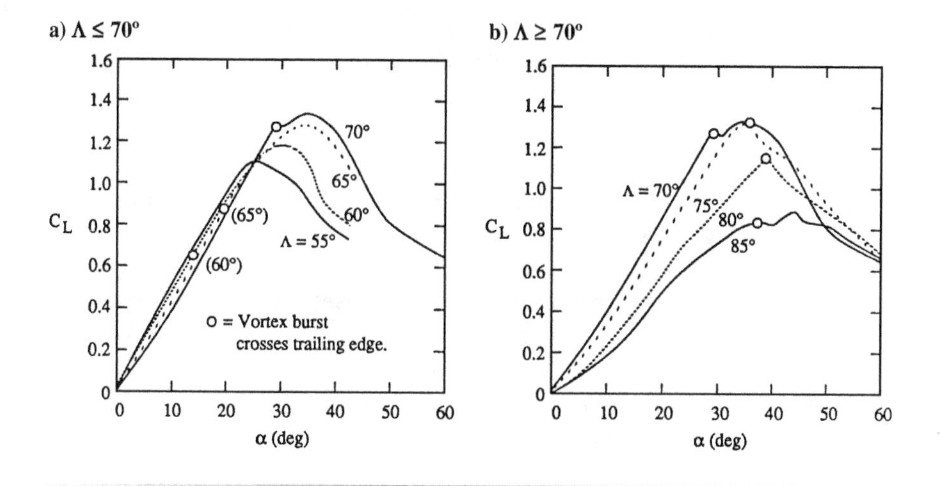

Fig. 8.9 Effect of vortex breakdown location on the lift coefficient of delta wings (10, 11).

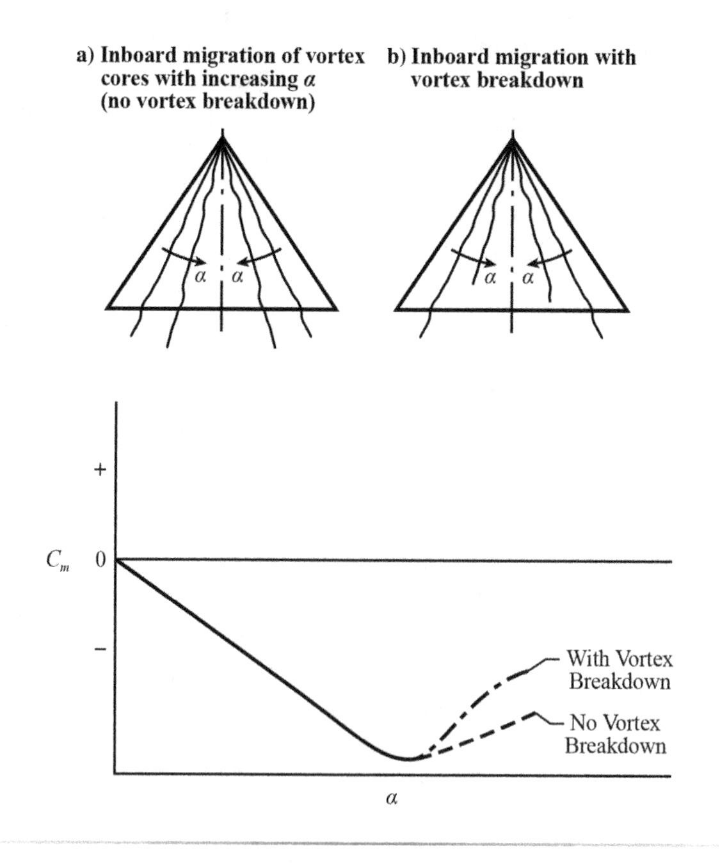

Fig. 8.10 Pitch-up of slender delta wings.

lift from the wingtip regions as the leading-edge vortices migrate inboard. This pitch-up problem may be further accentuated by the vortex breakdown and movement of the vortex breakdown point over the wing surface at high angles of attack.

Effects of Sideslip The sideslip has a strong influence on the vortex breakdown phenomenon. Assume that, for a certain delta wing held at some angle of attack, the vortex breakdown occurs at the trailing edge as shown in Fig. 8.11a. If this wing is in sideslip, then, on the windward side, the vortex breakdown point progresses towards the wing apex, whereas on the lee side it moves aft and away from the trailing edge as shown in Fig. 8.11b. In other words, the vortex on the windward side becomes more unstable and that on the lee side becomes more stable. It should be noted that, by vortex stability, we mean the fluid dynamic stability of the vortex motion and not the aircraft stability. In essence, the effective sweep of the windward side decreases and that of the lee side increases as illustrated in Fig. 8.12. This asymmetric vortex burst due to sideslip also generates an asymmetric flow field around the vertical tail. On the wing, the lift on the windward side decreases and that on the lee side slightly increases. As a result, the delta wing experiences a destabilizing effect in roll as indicated by the increase in the lateral stability parameter $C_{l\beta}$ (Fig. 8.13).

This destabilizing effect in roll increases with angle of attack, and the maximum reduction in lateral stability level occurs approximately at the same angle of attack at which the burst asymmetry as measured by $\Delta x/c$ also reaches a maximum value. However, with further increase in angle of attack, the burst asymmetry and the associated destabilizing effect decrease, and there is some improvement in lateral stability as shown in Fig. 8.13.

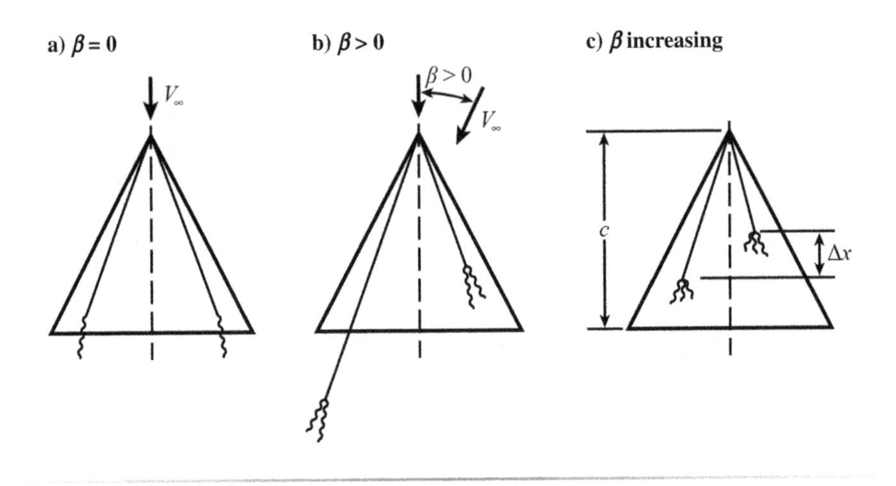

Fig. 8.11 Effect of sideslip on vortex breakdown.

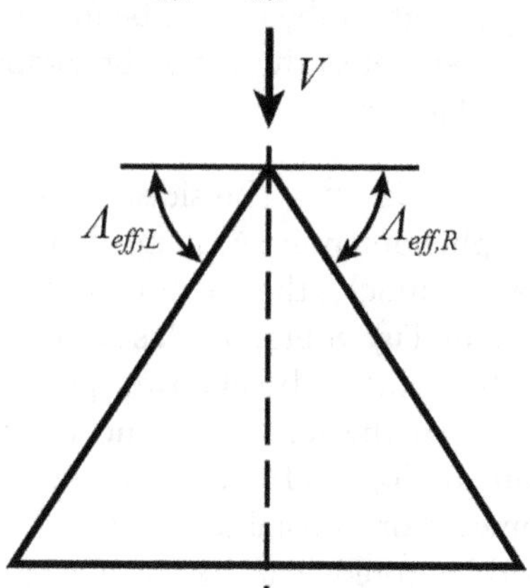

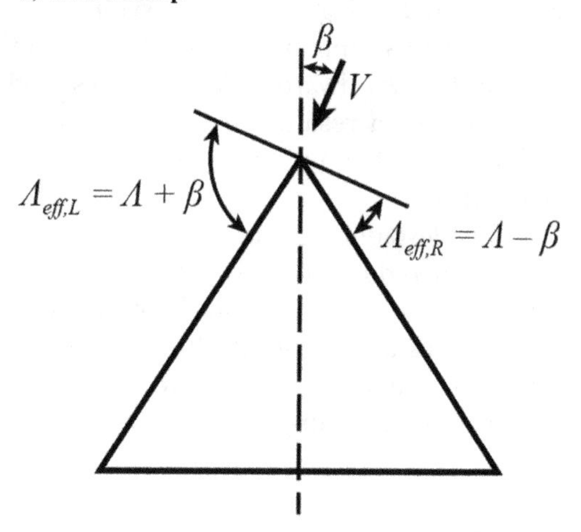

Fig. 8.12 Effective sweep for delta wings in sideslip.

Effects of Downstream Obstacle The vortex breakdown location over the wing is also susceptible to downstream disturbances. An obstacle placed downstream of the wing trailing edge has a strong influence on the vortex breakdown location [12] as shown schematically in Fig. 8.14. Such a sensitivity of the vortex breakdown to downstream obstacles demands extreme care and caution in conducting wind-tunnel tests on sting mounted slender delta wings at high angles of attack.

Several fighter aircraft like the F-15 and the F/A-18 feature twin vertical tail surfaces for enhancing the directional stability at high angles of attack. The designer has to carefully select the location of the twin vertical tails because an improper placement can promote premature vortex bursting, limit the maximum lift coefficient, and influence the longitudinal/lateral-directional stability parameters [3, 12].

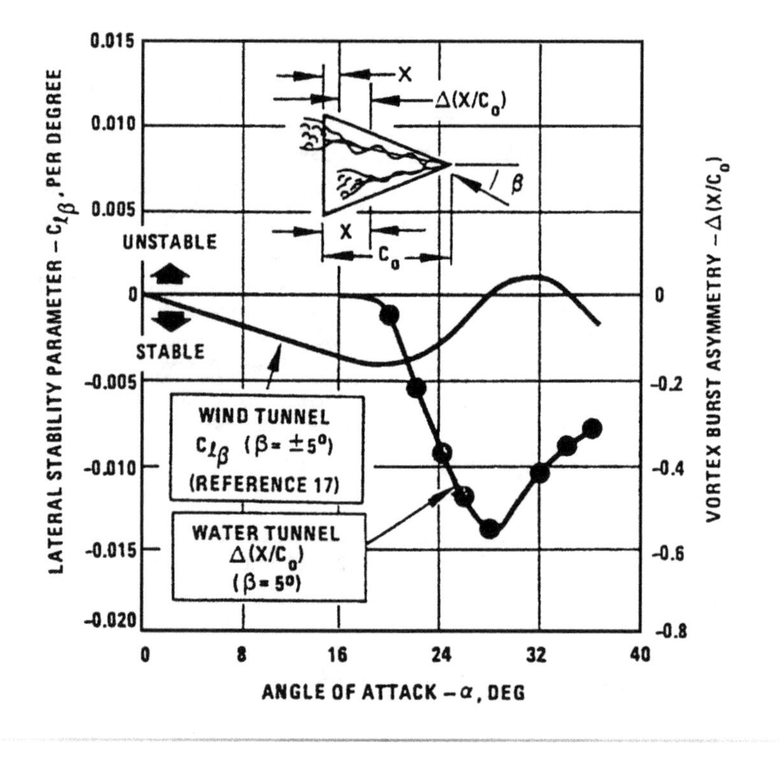

Fig. 8.13 Effect of asymmetric vortex breakdown on lateral stability at high angles of attack (12). (Courtesy AGARD.)

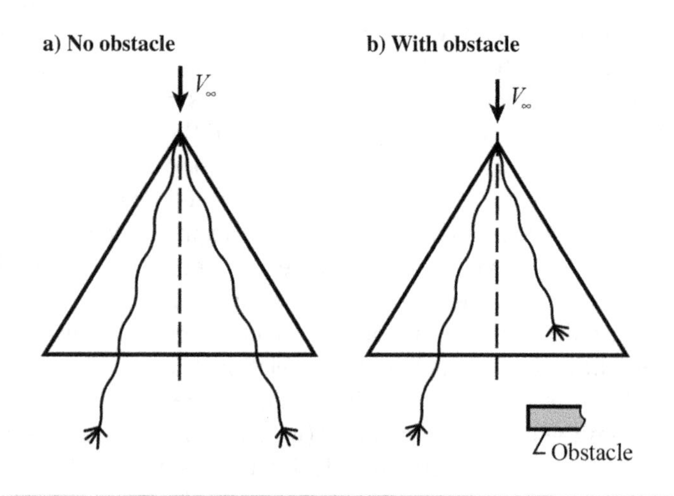

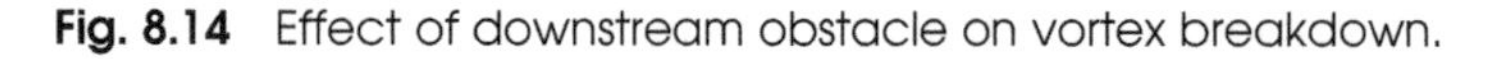

Fig. 8.14 Effect of downstream obstacle on vortex breakdown.

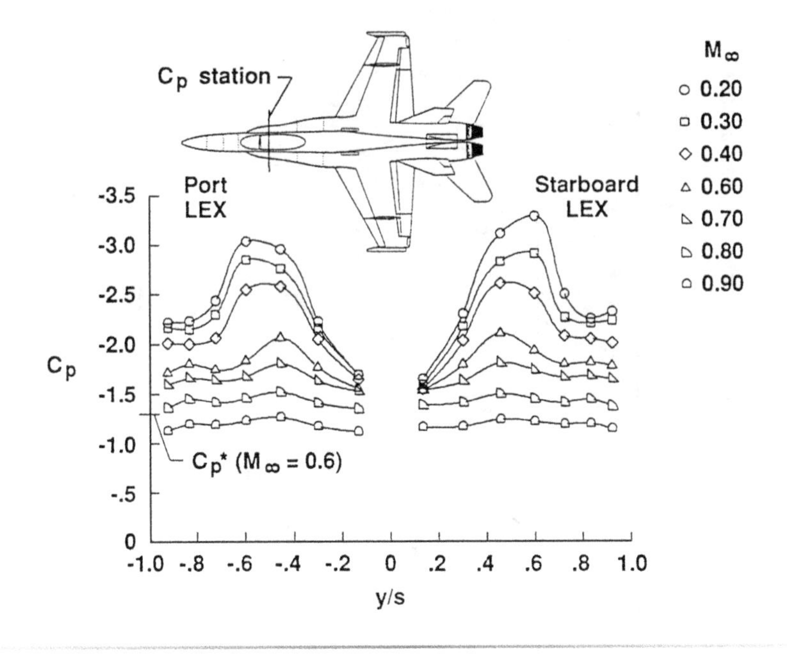

Fig. 8.15 Effect of Mach number on vortex flow (14).

Effects of Reynolds Number Generally, the separated vortex flows where the primary separation line is fixed at the sharp leading edge are relatively insensitive to the flow Reynolds number. This is true in a broader sense because the main structure of the vortex flow field on the lee side of the delta wing is not influenced by the flow Reynolds number. However, the Reynolds number can have some secondary effects, i.e., it can influence the flow transition in the boundary layer of the reattached flow on the lee side. Beneath the primary vortex, the upper surface boundary-layer is moving outboard and aft. As noted earlier, this boundary-layer flow encounters an adverse pressure gradient outboard and separates and rolls up to form a secondary vortex of opposite sense (see Fig. 8.3). If the boundary layer undergoes a transition before separation, the separation will be delayed, and the secondary vortex will be much smaller in size compared to that formed with laminar separation. Furthermore, the displacement effect of the secondary vortex on the primary vortex will be reduced, allowing the latter to move outboard and downward towards the wing surface.

It is suggested [13] that, to ignore the effects of Reynolds number on separated vortex flows over delta wings, the relation $L \gg x/\sqrt{R_x}$ must be satisfied. Here, L is the typical lateral dimension of the large-scale vortex at a distance x from the origin of the vortex (wing apex), and R_x is the local Reynolds number based on freestream velocity and distance x. This relation is based on the assumption that the structure of the vortex is determined mainly by vorticity convection and not by vorticity diffusion. Another way of interpreting this relation is that the typical lateral dimension of the vortex should be much greater than the local boundary-layer thickness. Recall that the local laminar boundary-layer thickness is inversely proportional to the square root of the local Reynolds number.

Effect of Mach Number In general, the vortex lift decreases with increase in Mach number. To illustrate this point, a typical spanwise pressure distribution over a delta wing is shown in Fig. 8.15. It may be observed that the footprint of the wing vortex as characterized by the suction peak is quite prominent at low Mach numbers and gradually disappears as the Mach number is increased.

8.4.2 Control of Vortex Breakdown

Several attempts have been made to control the vortex breakdown phenomenon using various methods, including blowing on the upper surface. Essentially, such methods aim in postponing the vortex breakdown to higher angles of attack and soften its impact on the aircraft stability and control. Typical examples of upper surface blowing are shown in Fig. 8.16.

a) Distributed leading-edge blowing

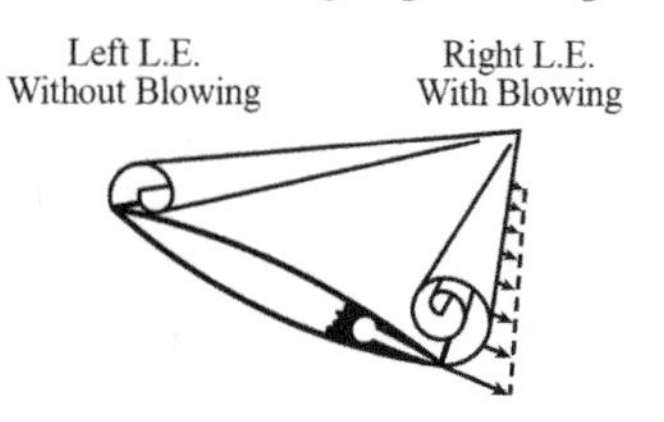

b) Concentrated leading-edge blowing

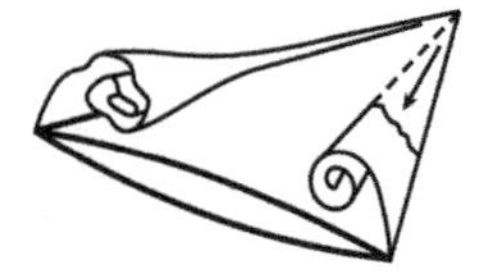

c) Effect of blowing on lift coeffcient

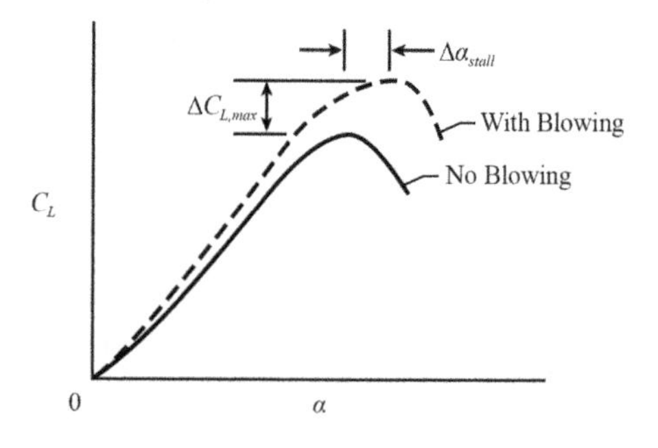

Fig. 8.16 Pneumatic vortex control of delta wings (15).

The ejection of high-energy air in the spanwise direction (Fig. 8.16a) energizes the vortex core and delays the vortex breakdown to higher angles of attack. Similarly, concentrated blowing from the wing apex (Fig. 8.16b) in the axial direction of the core also energizes the vortex core and delays the vortex breakdown to higher angles of attack. The benefits of upper surface blowing on vortex lift are schematically shown in Fig. 8.16c.

8.5 Leading-Edge Extensions

A significant enhancement of aerodynamic characteristics in terms of an increase in the maximum lift and reduction in the drag of delta wings at high angles of attack can be obtained using leading-edge extensions (LEX). Usually, the LEX has a higher sweep angle to minimize the supersonic wave drag, and the (main) wing has a lower sweep to improve subsonic lift-to-drag ratio. A forward-placed, close-coupled canard also functions in a similar way. The schematic arrangement of these surfaces is shown in Fig. 8.17.

To understand how such devices enhance the maximum lift and reduce the drag at high angles of attack, consider the flow field over a delta wing with LEX as shown in Fig. 8.18. As expected, the flow separation occurs over the LEX, and separated flow rolls up to form a pair of spiral vortices. The core of each of the LEX vortices stretches downstream. There is a kink in the wing leading edge at the junction of the LEX and the wing. At this kink, a second vortex, which is the main or wing vortex, originates and also stretches downstream. With increasing downstream distance, the wing vortex moves inboard and upward. The LEX vortex moves outboard and downward towards the wing surface. Downstream of the leading-edge kink, the LEX vortex is no longer fed by the vorticity shed from the leading edge. Therefore, the strength of the LEX vortex aft of the leading-edge kink can be assumed to remain more or less constant if viscous dissipation is ignored. On the other hand, the wing vortex is continuously fed by the vorticity shed from the leading edge and, therefore, its strength will increase until the wing trailing edge is reached. The LEX and wing vortices interact with one another, and the degree of the interaction

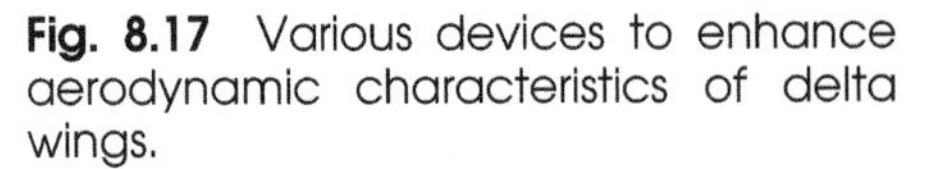
Fig. 8.17 Various devices to enhance aerodynamic characteristics of delta wings.

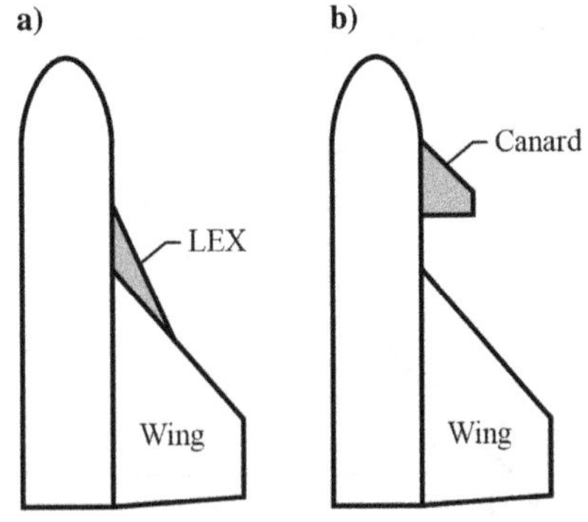

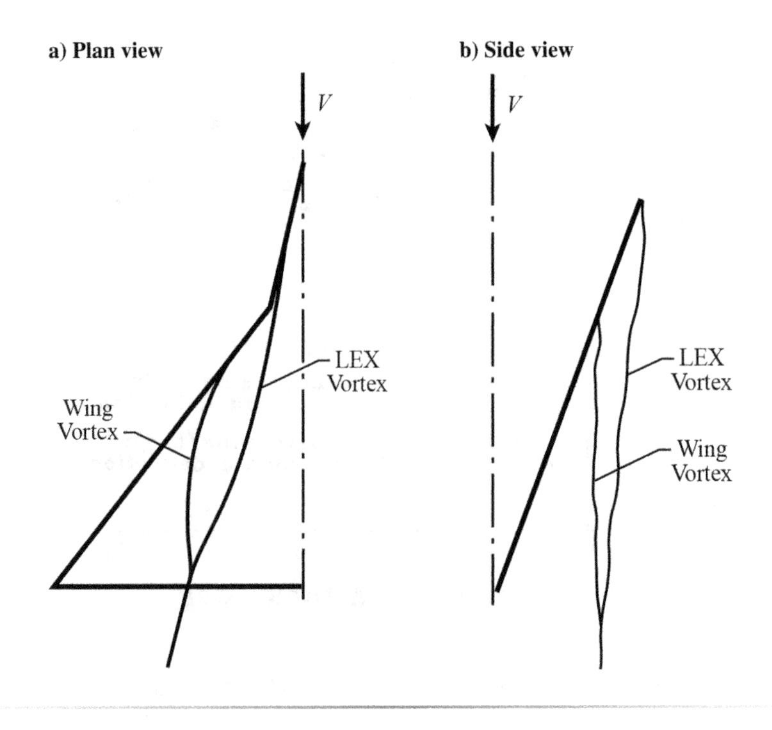

Fig. 8.18 Vortex interactions.

depends on the relative strengths of the two vortices and the distance between them. Eventually, the two vortices merge to form a single vortex stretching in the downstream direction as shown in Fig. 8.18.

The LEX vortex induces downwash over the inboard wing section and upwash/sidewash over the outboard wing sections. The induced downwash over the inboard wing sections reduces the local angle of attack, hence the lift on the inboard wing sections. In other words, the addition of the LEX simulates a positive camber effect for the inboard wing sections. The upwash over the outboard wing sections increases the local angle of attack and leads to increased lift in that region, simulating a local negative camber effect. The net effect of this interaction is to delay the occurrence of the vortex breakdown to higher angles of attack in comparison to the wing alone case. As said before, a forward-placed, close-coupled canard also functions in a similar fashion.

The F-16 and the F/A-18 aircraft make use of the LEX for enhancement of their aerodynamic characteristics at high angles of attack. The Swedish (SAAB) Viggen is perhaps the first successfully operating combat aircraft to feature a forward-placed, close-coupled canard. For this aircraft, it is reported that, with the deployment of close-coupled canards [16], the lift on the final approach during landing increased by as much as 65% over a simple delta wing. Several other aircraft such as the Israel Lavi, the Swedish Grippen, the French Rafale and the Mirage III, the Russian Su-35,

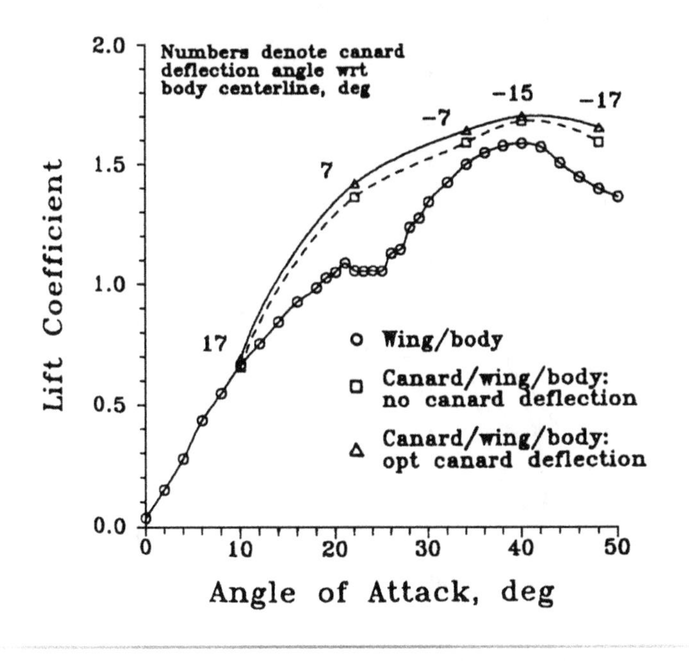

Fig. 8.19 Example of optimal canard deflections for lift enhancement of delta wings (17).

and the European Fighter Aircraft (EFA) employ a variety of close-coupled canard configurations for enhancement of aerodynamic lift and drag characteristics. An exception is the X-31 aircraft, which uses a long-coupled canard for the purpose of pitch recovery at high angles of attack rather than for enhanced lift. An example of lift enhancement using optimal canard deflection at various angles of attack is presented in Fig. 8.19.

The sideslip has an adverse effect on lateral-directional stability characteristics of delta wings equipped with LEX or leading-edge strakes or close-coupled canards [3, 12]. The adverse effect could be more pronounced compared to plain delta wings. In other words, the very same fluid dynamic phenomenon that gives rise to lift augmentation and drag reduction at high angles of attack, zero sideslip conditions, leads to a deterioration in lateral-directional stability and control characteristics if sideslip is experienced at high angles of attack. Therefore, a compromise may have to be made in the design of slender delta wings with LEX/strakes or forward-located, close-coupled canards between the benefits in lift and drag and the degradation in lateral-directional stability at high angles of attack.

8.6 Forebodies at High Angles of Attack

One of the most puzzling aerodynamic problems encountered in recent years is the side force developed by an axially symmetric slender body, typical of the forebody of a modern fighter aircraft in symmetrical flow at high angles of attack. The magnitude of the yawing moment required to counter the yawing moment due to this side force can far exceed the rudder capability as schematically shown in Fig. 8.20. If this forebody-induced yawing moment is of unstable nature, it can have a strong influence on the aircraft stability and control at high angles of attack. The vortices shed from the forebody at high incidence are primarily responsible for this phenomenon through a direct effect on the forebody pressure distribution and through secondary effects involving interaction with other vortices shed from the wings and the LEX.

At low and moderate angles of attack, the forebody vortex system consists of a pair of a steady symmetric vortices as shown in Fig. 8.21. As the angle of attack is increased, the vortex pattern becomes asymmetric. The asymmetric vortex pattern is of "bistable" nature, i.e., it has two stable patterns that are mirror images of each other. For any reason, if one asymmetric vortex pattern is disturbed, it switches to the other asymmetric pattern. The vortices will essentially stay stable in the new asymmetric pattern until disturbed again. In its natural form, the microasymmetries in the forebody geometry or slight flow asymmetries prompt the vortex system to assume any one of the two bistable patterns.

With angle of attack increasing further and approaching 90 deg, the organized pair of asymmetric vortex patterns disappears and the flow assumes a pattern typical of classical time-dependent Kármán vortex shedding from a circular-cylinder in crossflow.

The side force on the forebody arises in that angle of attack range when the vortices are asymmetric and the wake flow is steady. The side force vanishes as the angle of attack increases towards 90 deg because of the establishment of classical unsteady, Kármán vortex shedding.

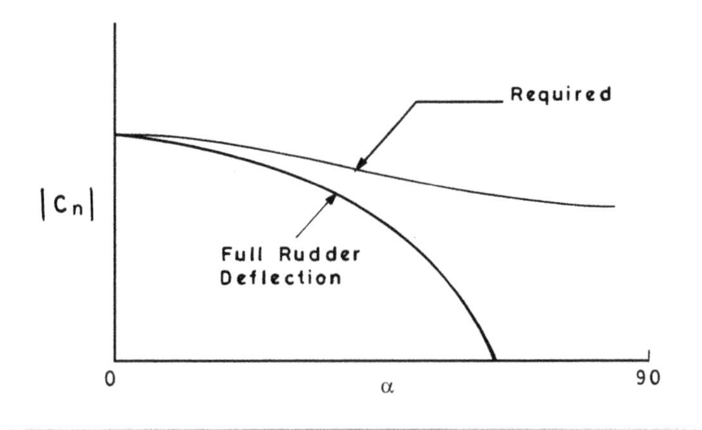

Fig. 8.20 Available and required yawing moment at high angles of attack.

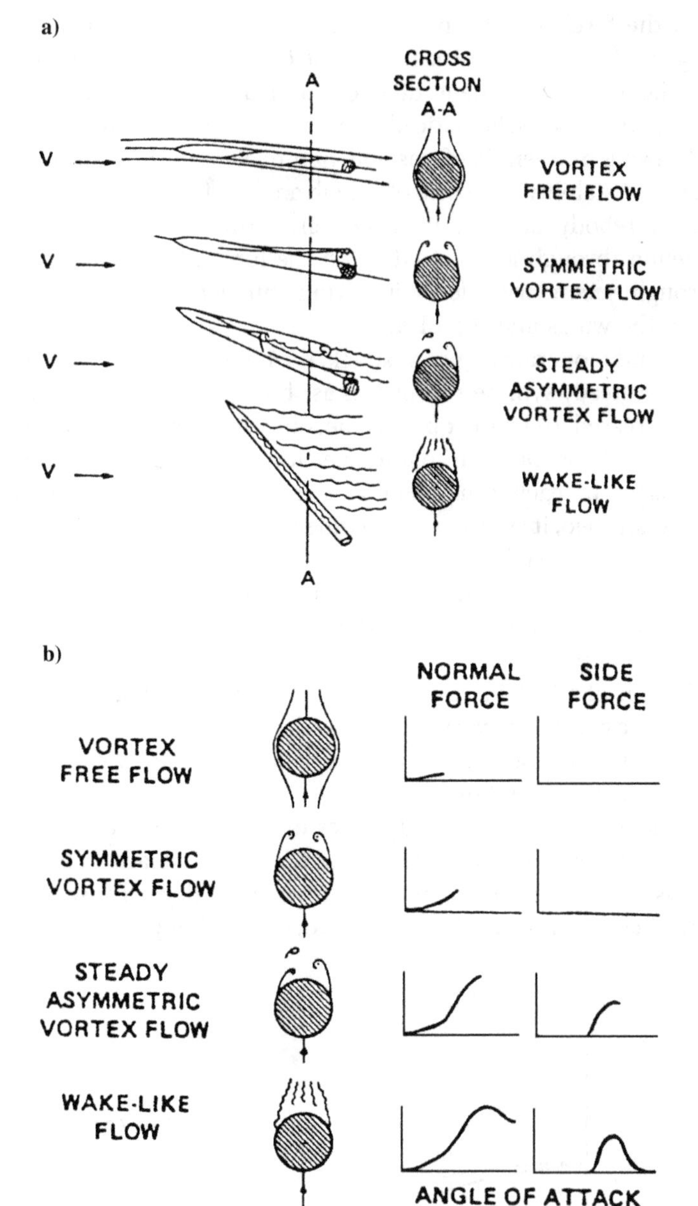

Fig. 8.21 Flow pattern and side force over slender bodies at high angles of attack (7).

8.6.1 Influence of Reynolds Number

The forebody side force is found to decrease drastically in the transitional Reynolds number range. A typical result for an ogive-cylinder of fineness ratio of 2 at $\alpha = 55$ deg is shown in Fig. 8.22. It is interesting to observe that, in the transitional Reynolds number range, the side force decreases and, at high Reynolds numbers when the flow is turbulent, the magnitude of side force increases back to a value close to that at low Reynolds numbers.

8.6.2 Effect of Forebody Geometry

A properly designed forebody can significantly reduce the magnitude of the side force and the associated yawing moment. Sometimes, a proper design of the forebody can even enhance the directional stability at high angles of attack. One of the primary geometric parameters that influences the forebody aerodynamics is the forebody fineness ratio, which is defined as the forebody length l divided by its maximum diameter D. The angle of attack at which the separated vortex flow becomes asymmetric is shown in Fig. 8.23. We observe that, as the forebody fineness ratio increases, the angle of attack for the vortex asymmetry decreases rapidly.

Another geometrical parameter that has a strong influence on the vortex flow field is the forebody cross-sectional shape. A forebody with circular cross section loses directional stability beyond 20 deg of angle of attack as shown in Fig. 8.24. However, as the cross section becomes

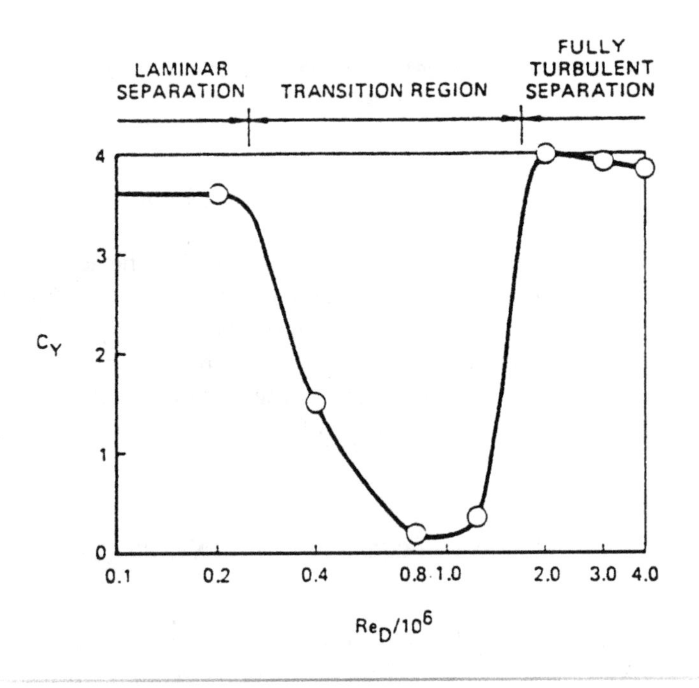

Fig. 8.22 Effect of Reynolds number on maximum side force of an ogive-cylinder at $\alpha = 55$ deg (18).

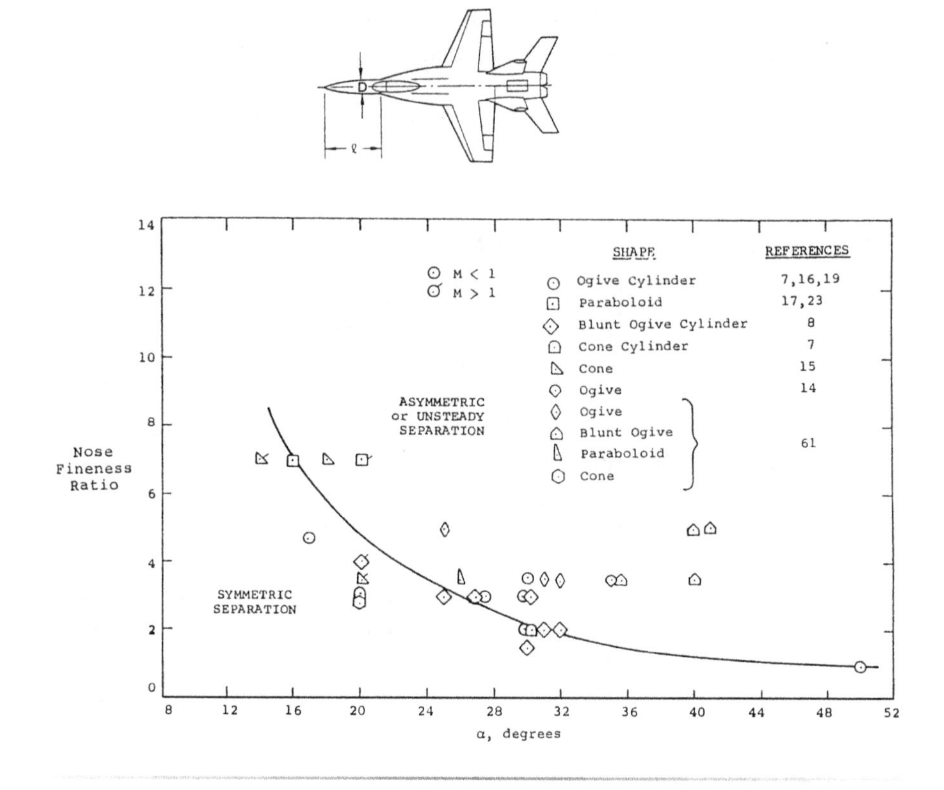

Fig. 8.23 Onset of asymmetric flow separation on slender bodies of revolution (19).

flatter, the directional stability level improves significantly as shown in Fig. 8.25.

The nose bluntness has a similar effect in reducing the adverse effect of aerodynamic asymmetries at high angles of attack. As the bluntness is increased, the onset of aerodynamic asymmetries is postponed to higher angles of attack. Another geometric modification that can improve the directional stability at high angles of attack is the addition of nose strakes [3]. However, it is possible that the effect of nose strakes may not always be beneficial.

8.6.3 Interaction of Forebody and Wing Vortex Flow Fields

The vortices emanating from the forebody interact with those emanating from the LEX/wing and canard. This interaction becomes stronger at high angles of attack and often leads to severe problems in stability and control at high angles of attack. The proximity of the LEX to the forebody causes the forebody vortices to be "sucked down" towards the wing upper surface. The magnitude of this interaction depends on the forebody geometry and LEX area and planform shape. This interaction of the forebody and LEX/wing vortices affects the lateral-directional stability levels, especially at angles of attack close to the stalling angle, and leads to nonlinear variations in $C_{l\beta}$ and $C_{n\beta}$. Whether the interaction is favorable or adverse depends on the forebody geometry, the LEX/wing geometry, and the values of angle of attack and sideslip. A favorable interaction can delay the wing stall and improve the lateral-directional stability at high angles of attack. An adverse interaction results in opposite effects.

The forebody, wing/LEX vortex system further interacts with vertical tail surfaces. A significant beneficial effect can be derived in terms of directional stability by properly placing the vertical tail to take full benefit of the energy contained in the vortex flow field emanating from the forebody, wing/LEX system. An example of this is the reported increase in directional stability and control on the YF-17 and the F-5G aircraft configurations by moving the vertical tail(s) forward on the body and closer to the wing/LEX vortex flow field [3, 12]. However, much of the beneficial effects may be lost at

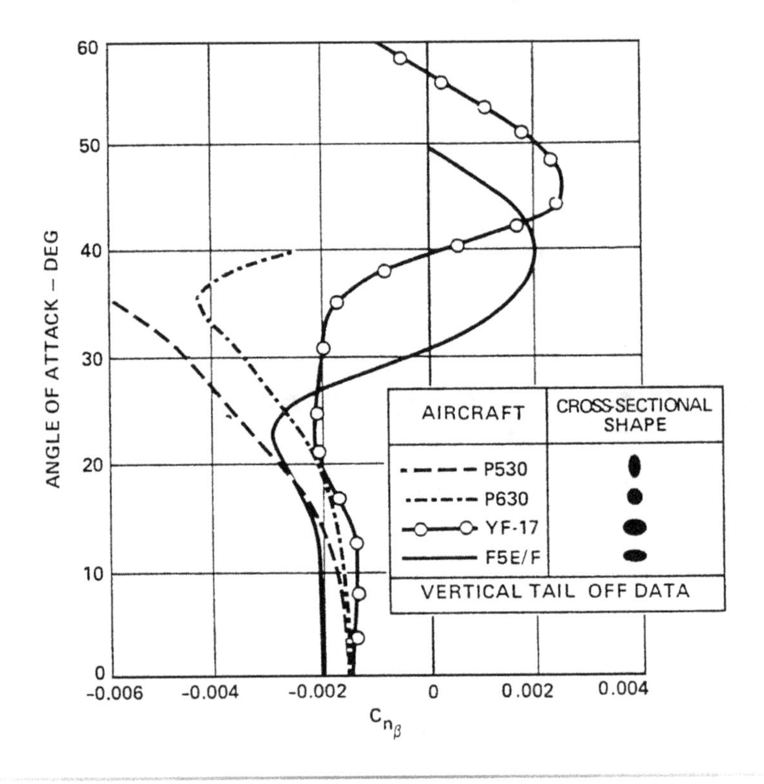

Fig. 8.24 Effect of nose shape on static directional stability at high angles of attack (12, 20). (Courtesy AGARD.)

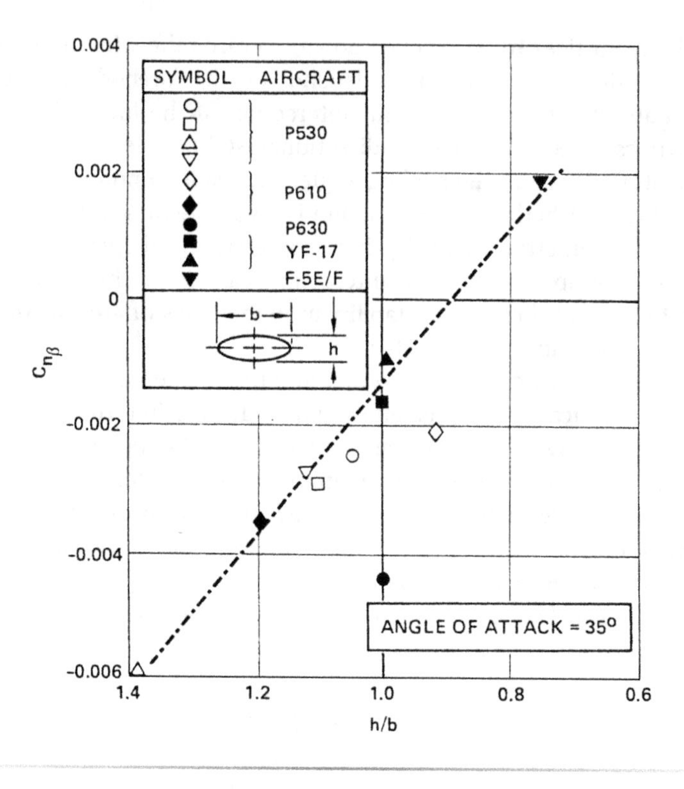

Fig. 8.25 Effect of cross-sectional shape on static directional stability (12, 20). (Courtesy AGARD.)

higher sideslip because of the sensitivity of the vortex trajectory and vortex stability to sideslip as discussed previously.

8.7 Relation Between Angle of Attack, Sideslip, and Roll Angle

Before we discuss the high-angle-of-attack stability and control problems, it is necessary to understand that, at high angles of attack, the rolling motion about the body axis reduces the effective angle of attack and builds up the sideslip as given by the following equations:

$$\alpha = \tan^{-1}(\tan\sigma\cos\phi) \tag{8.1}$$

$$\beta = \sin^{-1}(\sin\sigma\sin\phi) \tag{8.2}$$

where σ is the angle between the x-body (roll) axis and the velocity vector and ϕ is the roll angle about the x-body axis (see Fig. 8.26). For $\phi = 0$, $\alpha = \sigma$, and $\beta = 0$; and for $\phi = 90$, $\alpha = 0$, and $\beta = \sigma$. Thus, at extreme roll angles, all the angle of attack gets converted to sideslip.

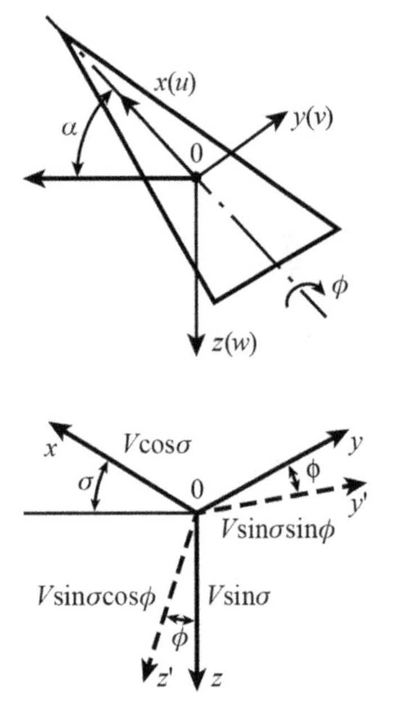

Fig. 8.26 Delta wing at combined high angle of attack and roll angle.

8.8 Wing Rock

Wing rock is a limit cycle oscillation predominantly in roll about the body axis. The wing rock is exhibited by a variety of aircraft configurations. Some examples are the F-5, the F-16, the X-29, and the X-31. The principal source of wing rock is the wing itself. The thin, low-aspect ratio, highly swept delta wings are prone to develop wing rock at high angles of attack. However, aircraft configurations not having highly swept delta wings but featuring fuselages with long, slender forebodies are also known to exhibit wing rock. This latter type of wing rock is called forebody-induced wing rock and is known to occur even when the wing is removed. It may also be noted that thin, sharp-edged rectangular wings of very low aspect ratio, usually less than 0.5, also exhibit wing rock, which is generated by the dynamic motion of the side edge vortices. Because aircraft rarely have such low-aspect ratio rectangular wings, this type of wing rock is not discussed here. The interested reader may refer elsewhere [21].

A loss of damping in roll at high angles of attack makes a configuration susceptible to wing rock but does not necessarily generate a sustained wing rock. An example of loss of roll damping at high angles of attack is presented in Fig. 8.27 for the F-5 and the X-29 aircraft. It is interesting to observe that these two aircraft have similar forebodies but radically different wing planforms, yet both lose damping in roll at high angles of attack.

To generate a sustained wing rock motion, some additional aerodynamic cause is necessary. The various causes that have been identified so far are 1) nonlinearities in lateral-directional aerodynamic characteristics,

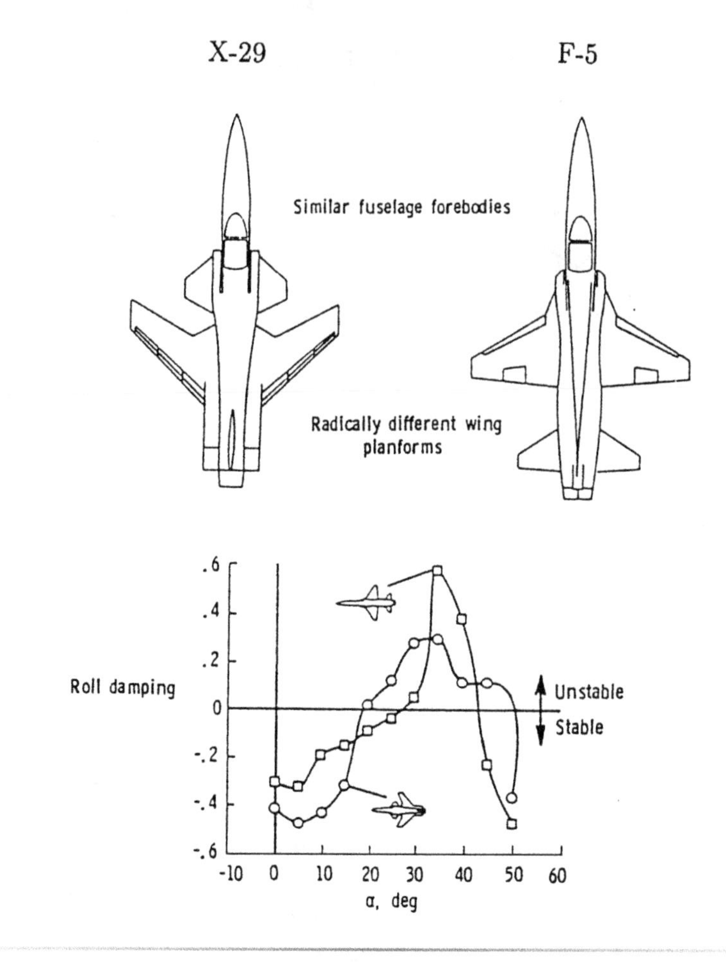

Fig. 8.27 Loss of damping in roll at high angles of attack (22).

2) aerodynamic hysteresis in the variation of rolling moment with sideslip/roll angle, and 3) nonlinear variation of roll damping such that negative damping (destabilizing) exists at low values of sideslip/roll angle and positive damping (stabilizing) at higher values of sideslip/roll angle.

The nonlinearities in aerodynamic characteristics of modern fighter aircraft with highly swept, thin delta wings and long, pointed slender forebodies arise because of the complex vortex flow fields existing at high angles of attack as discussed earlier. As an example, the damping and cross derivatives of a wing–body configuration are shown in Fig. 8.28a. These data illustrate how drastically and how suddenly the dynamic derivatives vary with angle of attack. In such situations, the use of traditional stability derivative concept, at best, can give only a locally linearized description of the aerodynamic forces and moments. For more accurate estimation, advanced nonlinear mathematical modeling is necessary.

Another example of nonlinearities in aerodynamic coefficients is presented in Fig. 8.28b for a fighter aircraft at low speeds. The damping-in-yaw derivative exhibits a very sudden and very large unstable peak at angle of attack of about 60 deg, while the two dynamic cross derivatives exhibit an equally large and very sudden variation. It is interesting to note that the angle of attack at which the peaks occur is very much independent of wing sweep and the presence or the absence of the vertical tail. This suggests that the forebody vortices are responsible for these effects.

The aerodynamic hysteresis, in general, arises because of the convective time lag. An example of the convective time lag is the familiar downwash lag at the tail, which we have studied in Chapter 3. Because of this convective time lag in downwash, the airplane develops damping in pitch. For highly swept delta wings, one possible fluid flow mechanism that may cause hysteresis in rolling moment (Fig. 8.29) is the time lag in the vortex strength, vortex location, and location of vortex breakdown point [24]. On an oscillating delta wing, these parameters will be out of phase with roll angle and hence differ

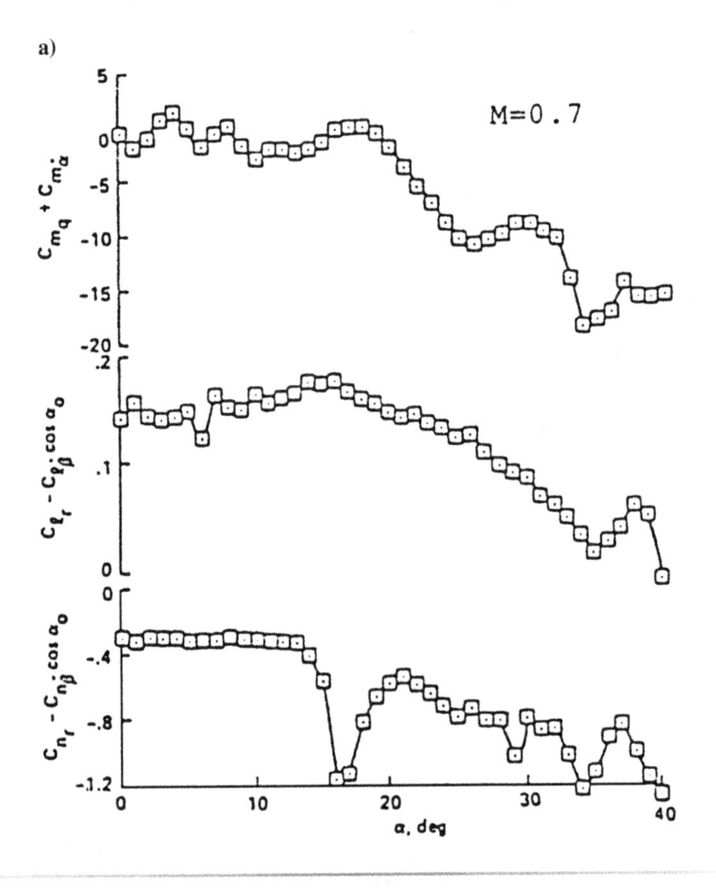

Fig. 8.28a Damping derivatives of a wing-body combination (23).

b)

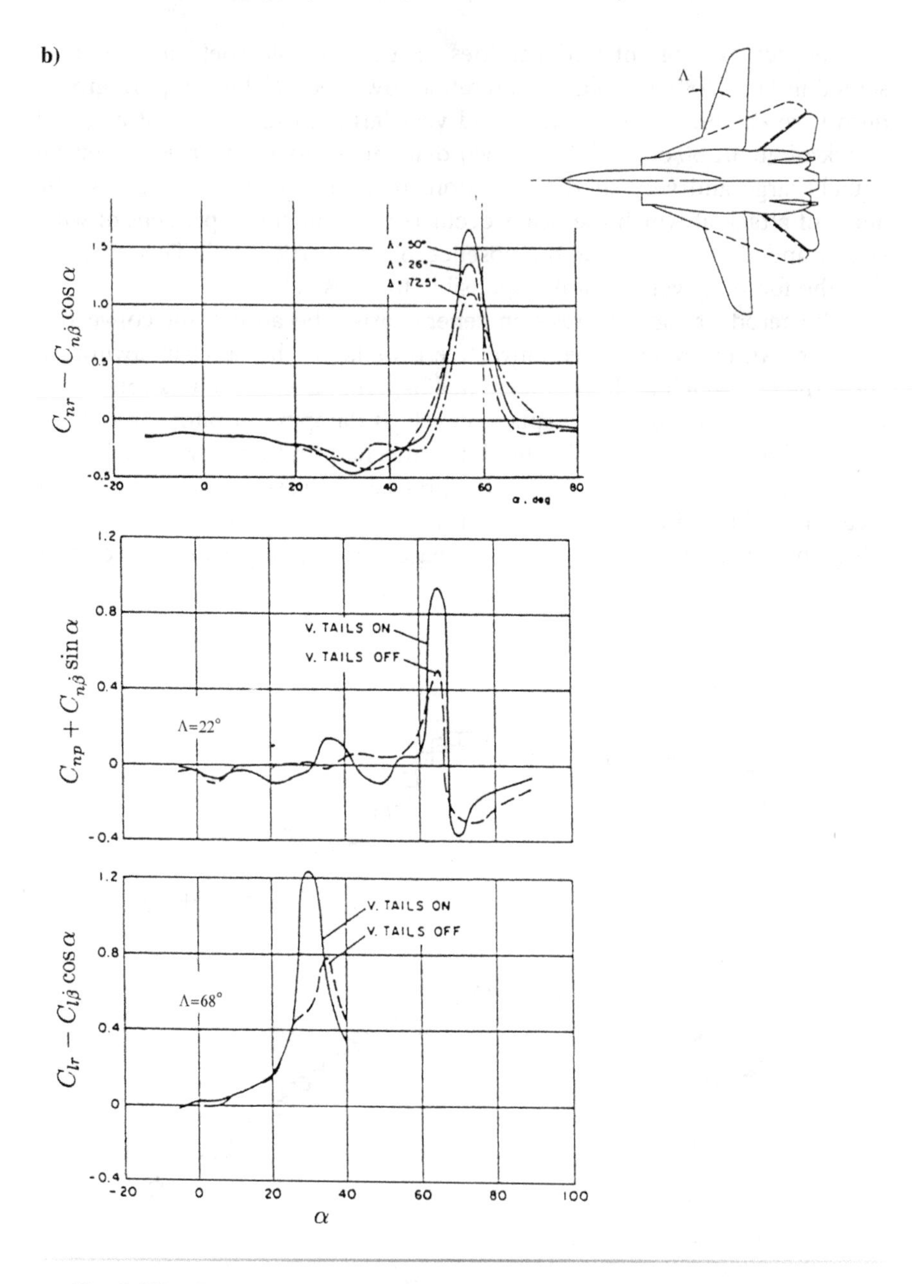

Fig. 8.28b Damping derivatives of a generic fighter configuration (23).

from their static counterparts. We will learn more about this phenomenon later in this chapter. In view of such complex flow physics, the quasi-static approach fails to make a satisfactory prediction of wing rock. The quasi-static approach is based on the assumption that the instantaneous value of an aerodynamic force or moment coefficient during a dynamic state is equal to its static value corresponding to the instantaneous values of motion variables. Such an approach may be justified when the time lag effects are small. The time lag effects depend on the frequency and amplitude of the oscillatory motion.

The study of wing rock caused by the nonlinearities in aerodynamic parameters and hysteresis is not addressed here. The interested reader may refer elsewhere [25, 26].

8.8.1 Wing Rock of a Delta Wing Model

To understand the wing rock generated by the wing, let us study the wing rock of a single-degree-of-freedom, free-to-roll delta wing model. It is interesting to note that only those delta wings whose leading-edge sweep is 75 deg or more exhibit wing rock. The angle of attack for the onset of wing rock of such a wing is below that for the vortex breakdown to occur at the trailing edge. Therefore, it is generally accepted that only thin, highly swept (slender) delta wings exhibit wing rock, and vortex breakdown is not the cause for the onset of wing rock of such wing models. As an example, let us consider the wing rock of a thin, sharp-edged 80-deg delta wing model, which has received considerable attention in the literature.

At high angles of attack, the 80-deg delta wing model loses roll damping. On receiving a disturbance, the free-to-roll model starts oscillating, and the amplitude of oscillation builds up rapidly as shown in Fig. 8.30. Even a small disturbance in the form of wind-tunnel turbulence or flow unsteadiness is usually sufficient to initiate the rocking motion. In view of this, such wing rock is often called a self-induced wing rock.

It is interesting to observe that the limit cycle amplitude is reached within few oscillations. The maximum amplitude and the frequency of wing rock vary with angle of attack as shown in Fig. 8.31. For the 80-deg delta wing model, the angle of attack for the onset of wing rock is around 20 deg and,

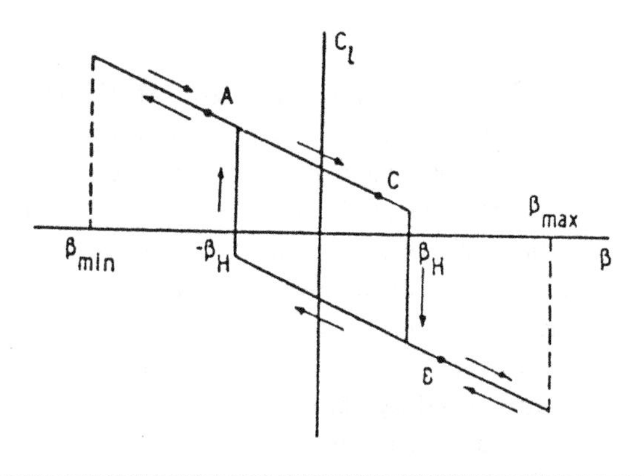

Fig. 8.29 Hysteresis in rolling moment (24).

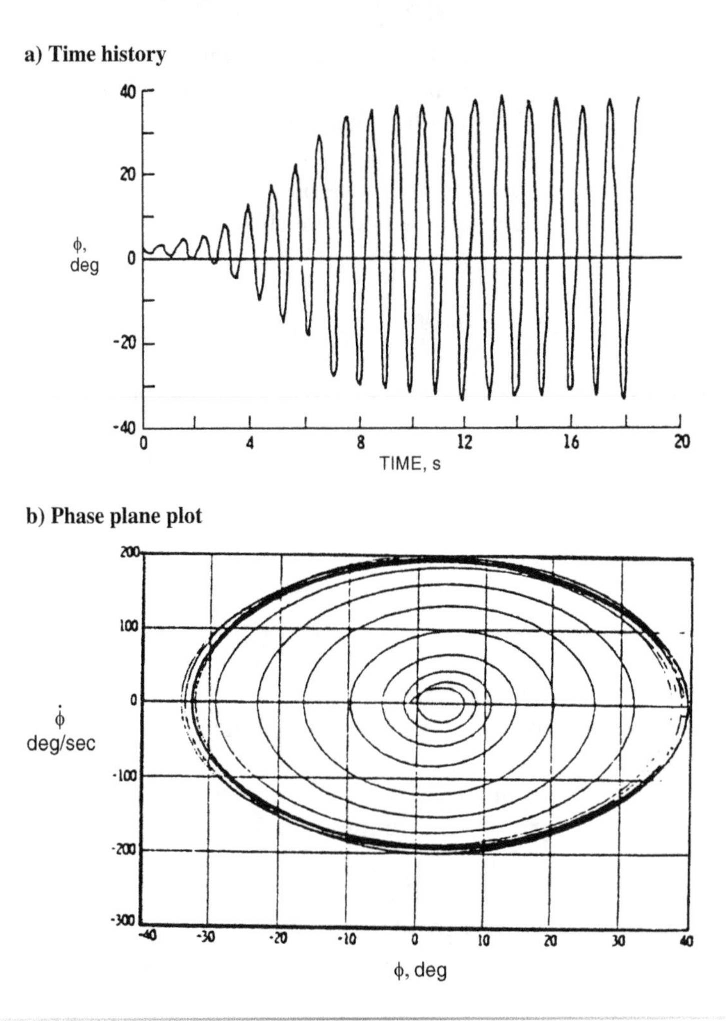

Fig. 8.30 Typical wing rock buildup of 80-deg delta wing (24).

below this value of angle of attack, any imparted disturbance will be damped out. In Fig. 8.32, the angle of attack boundaries of vortex asymmetry and vortex breakdown for the 80-deg delta wing are shown. From Figs. 8.31 and 8.32, we observe that the angle of attack of 20 deg for the onset of wing rock is below the angle of attack for the vortex asymmetry. Also, it is below that for the vortex breakdown occurring at the trailing edge.

There are some variations in the values of angle of attack for the onset of wing rock, the peak amplitude, and the angle of attack at which the peak amplitude occurs as measured by various researchers. The main reason for these variations is believed to be the bearing friction in the free-to-roll apparatus. If the bearing friction is high, the wing rock starts at a higher value of angle of attack because a larger destabilizing aerodynamic moment is necessary to set the model rolling. On the other hand, if such a wing were in free flight (no bearing friction), it will exhibit wing rock at a much lower angle of attack than measured in the ground-based, free-to-roll tests. For the data presented in Fig. 8.31, the peak wing rock amplitude occurs around $\alpha = 25$ deg and, for $\alpha > 50$ deg, the orderly periodic large amplitude motion characteristic of wing rock degenerates into an unsteady, low-amplitude, random oscillation.

8.8.2 Flow Mechanism

To gain some understanding of the physical causes of the wing rock of an 80-deg delta wing, let us refer to the available experimental data. From Fig. 8.33a, we observe that, for the range of angle of attack when the wing rock occurs, the static rolling moment varies in a stable manner with sideslip, i.e., $C_{l\beta} < 0$. This is to be expected because, for oscillatory motion in roll, the delta wing must be statically stable in roll. At high angles of attack, the roll damping parameter C_{lp} varies nonlinearly with roll angle. It is positive (destabilizing) at low values of roll angle and negative (stabilizing) at high values of roll angle as shown in Fig. 8.33b. The roll angle $\overline{\phi}$ is the mean roll angle about which the forced oscillation roll damping data are obtained. As stated [24], the roll damping parameter obtained by forced oscillation tests on the 80-deg delta wing is not sensitive to the oscillation frequency. This implies that hysteresis is not the cause of wing rock of the 80-deg delta wing. The hysteresis, if present, would make the roll damping depend on frequency of oscillations. In view of this, the principal cause of the wing rock of the single-degree-of-freedom, free-to-roll, 80-deg delta wing model is believed to be the nonlinear variation of roll damping with roll angle or sideslip.

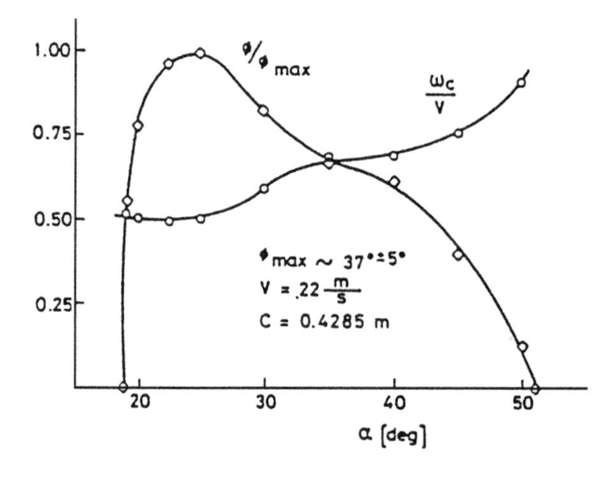

Fig. 8.31 Wing rock amplitude and frequency with angle of attack for 80-deg delta wing (27).

The normal force, even though not having a direct effect on the wing rock motion, has an interesting variation during the wing rock as shown in Fig. 8.34. The frequency of the normal force is almost twice that of the roll angle, whereas the side force has the same frequency and phase as the roll angle. According to the static data [27], the normal force coefficient should vary between 1.28 ($\phi = 0$) and 0.8 ($\phi = \phi_{max} = 28$ deg). However, the mean value of the dynamic normal force is around 0.65 (Fig. 8.34b), which is far less than the expected average value of 1.04 for the static range of 0.8–1.28.

The physical mechanism responsible for the complex variations of aerodynamic characteristics during wing rock is not very clear. However, to get an idea of the flow physics, let us take a look at the histogram of C_l vs ϕ and try to correlate it with available flow visualization data.

The histogram of C_l vs ϕ for one limit cycle of wing rock [27] is shown in Fig. 8.35. Also shown in this figure is the corresponding histogram with

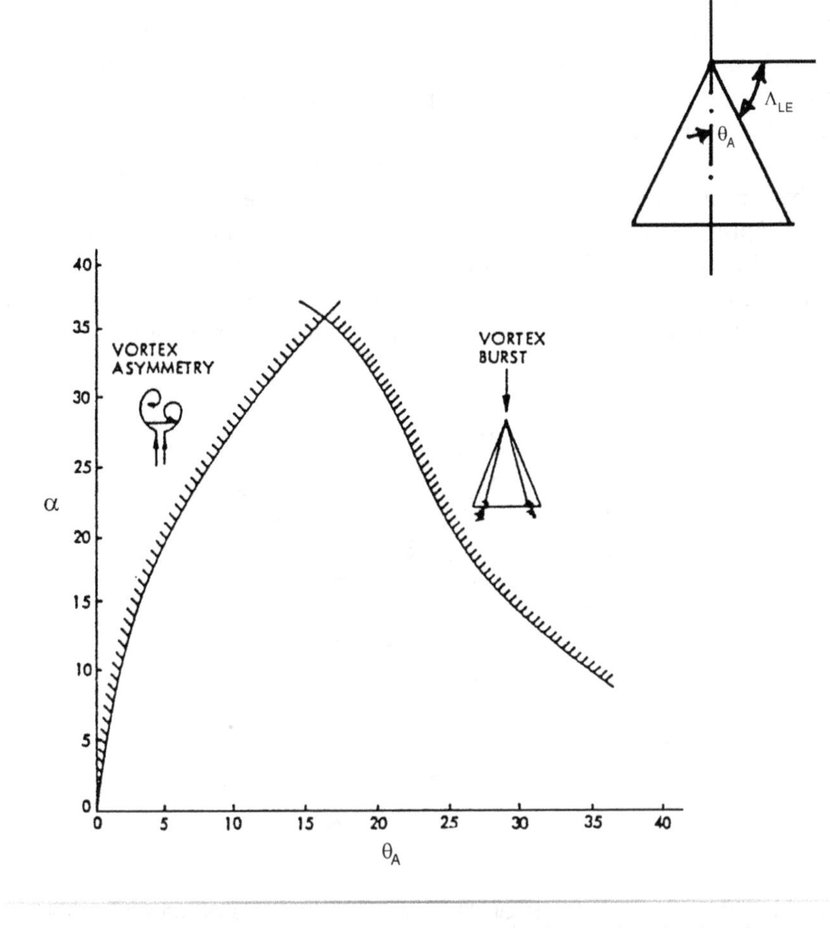

Fig. 8.32 Boundaries for vortex asymmetry and vortex burst.

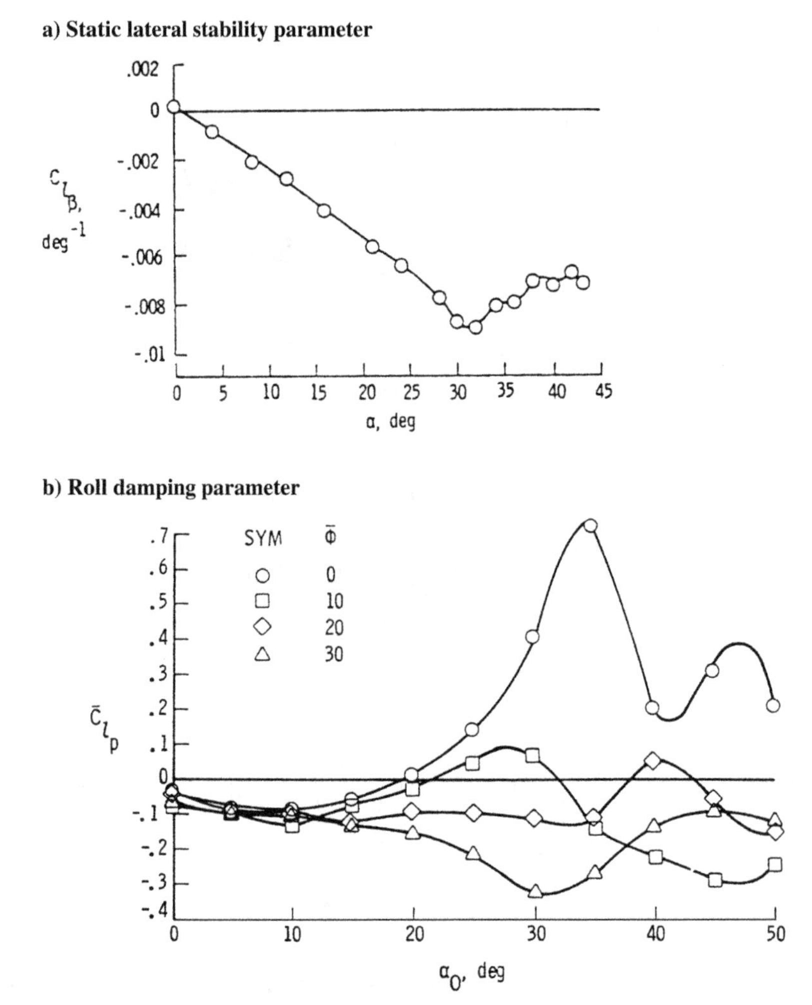

Fig. 8.33 Lateral stability and roll damping parameters with angle of attack (24). (Courtesy AGARD.)

sideslip as the independent variable. The sideslip angle was evaluated using Eq. (8.2). The time-varying rolling moment coefficient $C_l(t)$ presented in Fig. 8.35 was obtained from the measured roll angle time history as follows:

$$C_l(t) = \frac{I_x \ddot{\phi}}{qSb} \tag{8.3}$$

Usually, $\ddot{\phi}$ is determined using numerical differentiation or curve fitting to the measured roll angle or roll rate time history.

The histogram of Fig. 8.35 indicates that the inner loop encloses the area in a clockwise sense (positive) and the two outer loops enclose the area in a

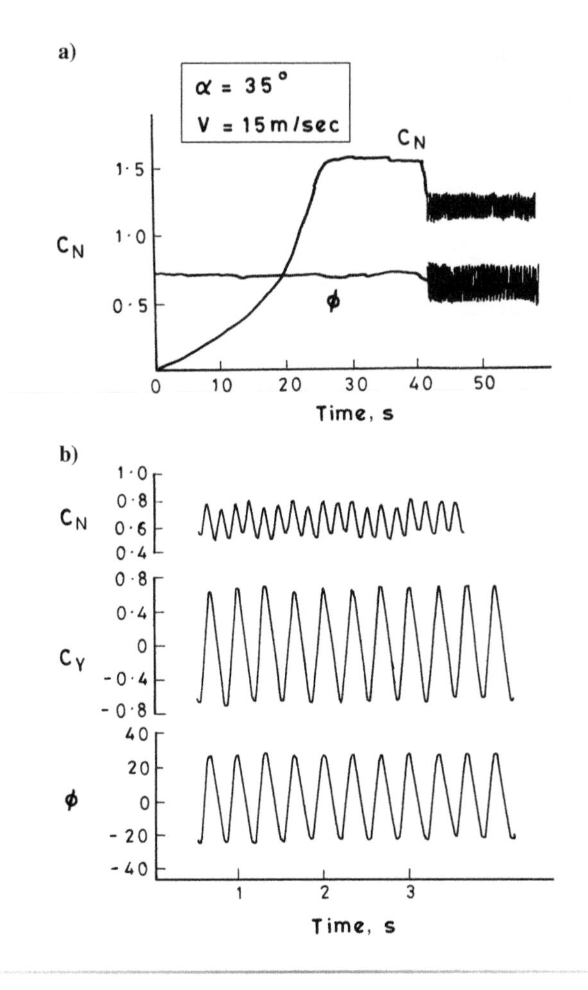

Fig. 8.34 Normal force and side force coefficients during wing rock of 80-deg delta wing (27).

counterclockwise sense (negative) as the loop is traversed in the direction of increasing time. The positive area corresponds to the addition of energy to the system, and the negative area implies extraction of energy (dissipation) from the system. Therefore, the positive loop for $-20 \text{ deg} \leq \phi \leq 20 \text{ deg}$ is destabilizing, and the outer lobes for $20 \text{ deg} \leq \phi \leq 55 \text{ deg}$ are stabilizing. If the net area is positive, the amplitude of the oscillatory motion increases and leads to a divergence. On the other hand, if the net area is negative, the oscillatory motion decays gradually (damped oscillation). If the net area is zero, it is a case of constant amplitude or limit cycle motion. It may be observed that the net area of the histogram shown in Fig. 8.35 is close to zero, confirming that the wing rock is an example of limit cycle oscillation.

Now, let us study the available flow visualization data on the delta wing model during a wing rock cycle. The static vortex locations are shown in

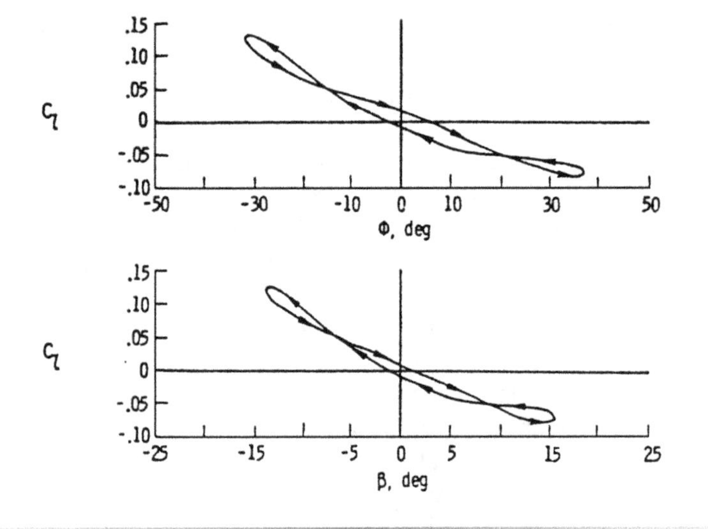

Fig. 8.35 Histogram of rolling moment coefficient (29). (Courtesy AGARD.)

Fig. 8.36 for various roll angles corresponding to the roll angles recorded in a typical wing rock cycle. The peak amplitude of the roll angle for this wing rock cycle is 45 deg. The numbers above the open circle symbols are the values of roll angle ϕ. Starting from $\phi = 0$, as the model is rolled in the positive direction (right wing down and left wing up), the right vortex moves

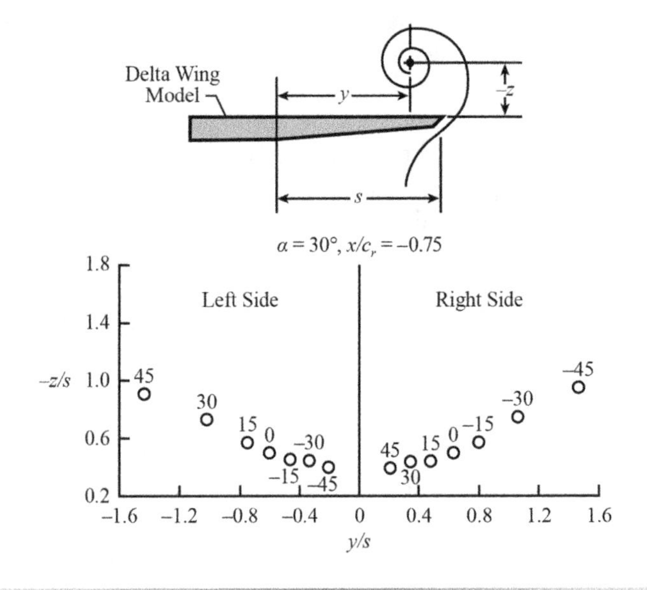

Fig. 8.36 Static vortex locations over 80-deg delta wing at various roll angles for $\alpha = 30$ deg (30).

inboard and vertically closer to the wing surface, while the left vortex moves outboard and further away from the wing surface. The picture is other way when the model rolls in the negative direction. This alternating vortex asymmetry is a characteristic feature of the wing rock of the 80-deg delta wing. As said earlier, the angle of attack for the onset of wing rock is below that for vortex asymmetry. Therefore, what actually happens is that, once the free-to-roll delta wing model starts oscillating because of loss of roll damping, the vortex asymmetry gets established and leads to a sustained, limit-cycle oscillatory motion.

When the model rolls to the right, the vortex on the right wing (down going) moves closer to the wing surface, leading to an increase in the lift, whereas that on the left wing moves away from the wing surface, leading to a decrease in the lift. This leads to a stable variation in the static rolling moment with sideslip/roll angle as observed in Fig. 8.33a. The closer the vortex is to the wing upper surface, the more the lift on that part of the wing is because the suction is higher.

The dynamic vortex positions are presented pictorially in Fig. 8.37a. It is evident that the dynamic vortex positions differ appreciably from the corresponding static positions. In addition, the dynamic vortex position for a given roll angle depends on the direction in which the wing is rolling. For example, consider all the vortex locations when $\phi = +15$ deg. The vortex locations are quite different depending on whether the model is rolling in the positive or negative direction, i.e., whether $\dot{\phi} > 0$ or $\dot{\phi} < 0$. A larger vertical displacement occurs when the wing is rolling in a direction away from the vortex, and a lower displacement occurs when the wing is moving towards the vortex. From Fig. 8.37b, we observe that the time lag exists only in the normal direction and not in the spanwise direction.

To quantify the effects of the normal time lag, let $\Delta z = z_r - z_l$ denote normal asymmetry in vortex location as shown in Fig. 8.38. Let us assume that the strength of both of the vortices is identical. Then, $\Delta z > 0$ gives rise to a positive rolling moment and vice versa. Thus, when ϕ is increasing or $\dot{\phi} > 0$, if $\Delta z > 0$, it leads to a destabilizing effect. In other words, $\partial\dot{\phi}/\partial\Delta z > 0$ corresponds to a destabilizing effect and $\partial\dot{\phi}/\partial\Delta z < 0$ to a stable effect.

In Fig. 8.39, the static and dynamic variations of Δz during a wing rock cycle are presented. We observe that $\partial\dot{\phi}/\partial\Delta z < 0$ for 20 deg $\leq \phi \leq$ 45 deg during positive half cycle and for -45 deg $\leq \phi \leq -20$ deg during the negative half cycle. For all of the rest of the cycle, $\partial\dot{\phi}/\partial\Delta z > 0$. Therefore, the net effect because of normal vortex asymmetry is predominantly destabilizing. Therefore, for limit-cycle wing rock, this destabilizing contribution must be balanced by a stabilizing contribution from a source other than the positional asymmetry of the vortices. It is possible that this stabilizing contribution could come from the time lag in vortex strengths. Even though not much information is available on this subject, some preliminary measurements have supported this line of thinking [31].

8.8.3 Effects of Vortex Breakdown

As we have mentioned previously, for the 80-deg delta wing model, the angle of attack for the onset of wing rock is well below the value of angle of attack at which the vortex breakdown is further away from the trailing edge. Therefore, for wing rock at such values of angles of attack, it is possible that the vortex breakdown may not occur over the wing during a complete wing rock cycle. However, as the angle of attack is increased, the vortex breakdown will eventually occur over the wing surface during wing rock. In this section, we will discuss the static and dynamic (time lag) effects of vortex breakdown on the wing rock.

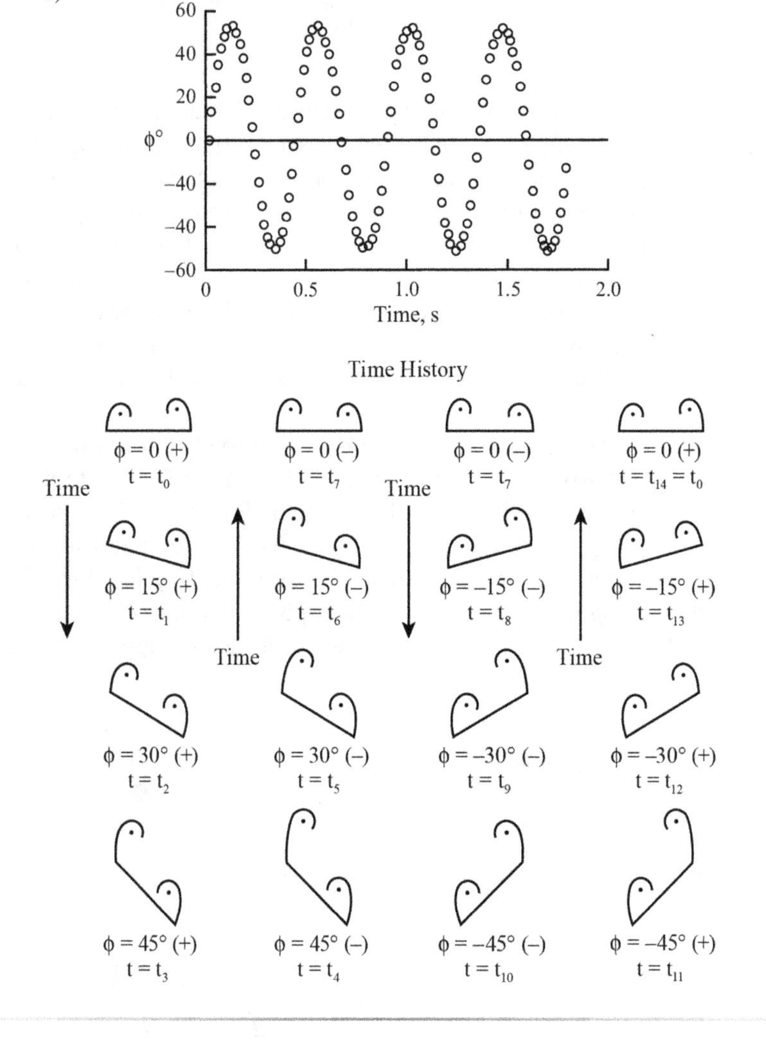

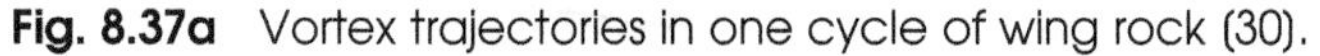

Fig. 8.37a Vortex trajectories in one cycle of wing rock (30).

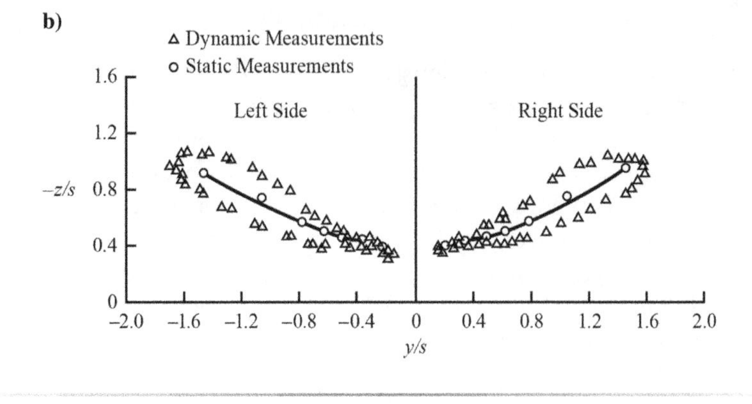

Fig. 8.37b Static and dynamic vortex locations for $\alpha = 30$ deg, $\pi/C_r = -0.75$ (30).

Let $\Delta x = x_l - x_r$ denote the asymmetry in chordwise location of the vortex breakdown point (Fig. 8.40). Let us assume that the strengths of the right and left vortices are identical. We observe that $\Delta x > 0$ corresponds to a positive rolling moment and $\Delta x < 0$ to a negative rolling moment. Therefore, statically the vortex breakdown effect will be destabilizing if $\partial\phi/\partial\Delta x > 0$. The static variation of Δx for roll angles corresponding to a typical wing rock cycle at an angle of attack of 40 deg is shown in Fig. 8.40. We observe from Fig. 8.40 that the static asymmetry in Δx is destabilizing. In other words, this static asymmetry in vortex breakdown would lead to autorotation.

Dynamically, the effect of asymmetry in chordwise location of vortex breakdown point would be destabilizing if $\partial\dot{\phi}/\partial\Delta x > 0$ and stabilizing if $\partial\dot{\phi}/\partial\Delta x < 0$. From Fig. 8.40, we observe that, for the positive half cycle when the roll angle is increasing ($\dot{\phi} > 0$), $\Delta x < 0$ so that the effect is damping. Similarly, for the remainder of positive half cycle when the roll angle is decreasing ($\dot{\phi} < 0$), $\Delta x > 0$, which gives damping effect. Similarly, we find that, for negative half cycle, the effect is also damping. As a result,

Fig. 8.38 Normal vortex asymmetry on a rolling delta wing (31).

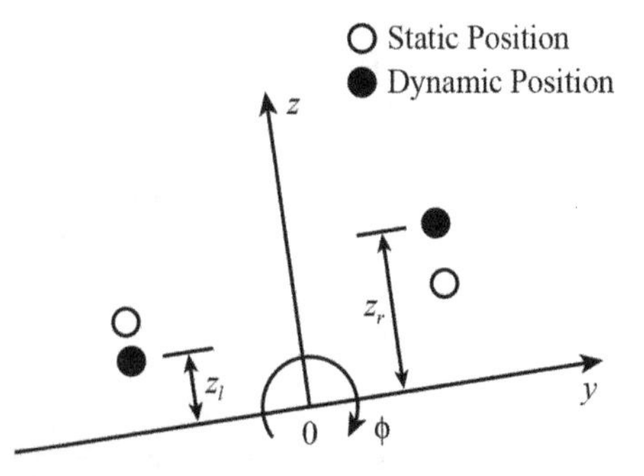

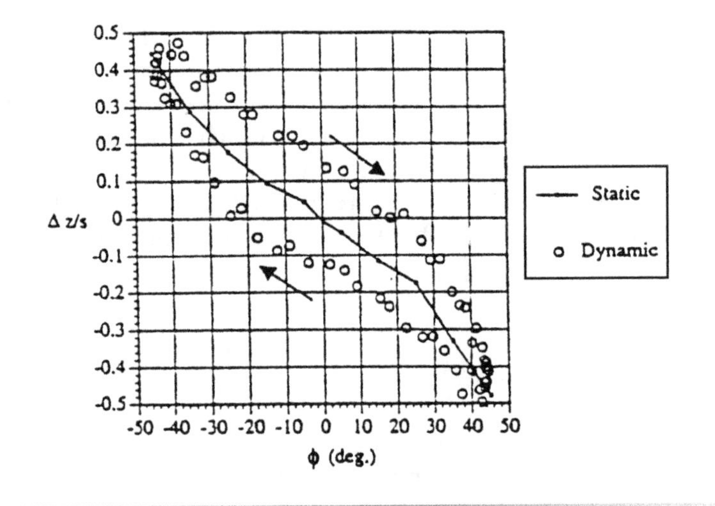

Fig. 8.39 Normal vortex asymmetry during wing rock cycle (31).

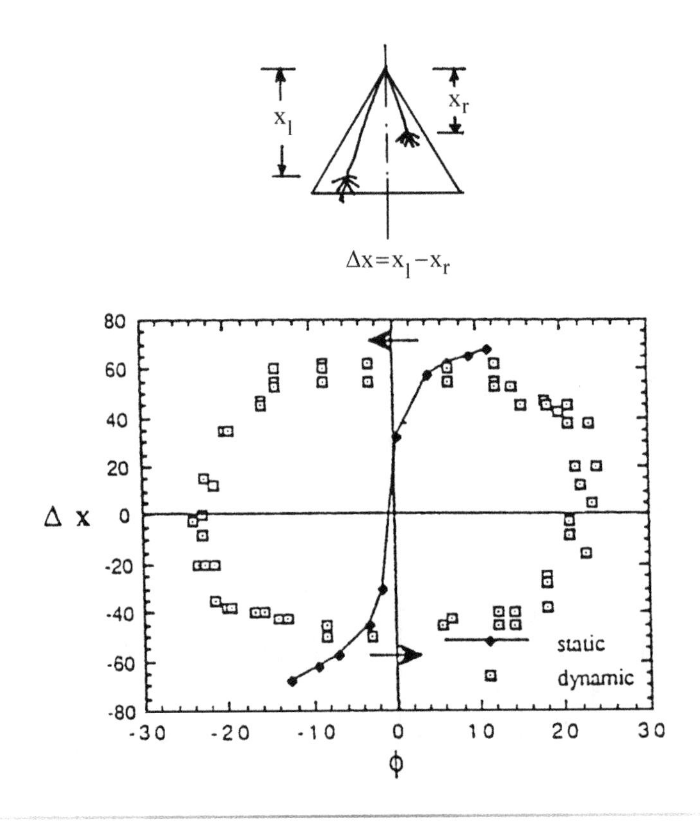

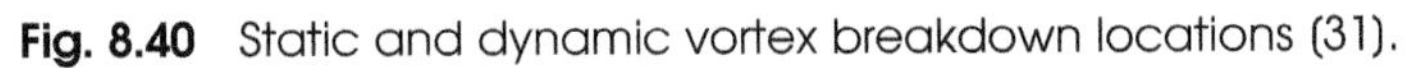
Fig. 8.40 Static and dynamic vortex breakdown locations (31).

the hysteresis loop is counterclockwise or encloses a negative area. This implies that the dynamic effect of vortex breakdown is of stabilizing nature. In other words, the occurrence of vortex breakdown over the wing during a wing rock cycle has a damping effect, which is contrary to the destabilizing effect predicted by static considerations alone.

This damping effect of the vortex breakdown is perhaps one of the factors that limits the wing rock amplitude and causes a degeneration of the wing rock beyond a 55-deg angle of attack into a random, small-amplitude oscillation.

8.8.4 Effect of Sideslip

The effect of sideslip on the wing rock of slender delta wings like the 80-deg delta wing is one of damping in nature. In other words, if sideslip is increased, the wing rock amplitude decreases, and eventually it degenerates into a random, small-amplitude oscillation [24].

8.9 Roll Attractor of Delta Wings

Roll attractor is another interesting phenomenon exhibited by thin, sharp-edged, highly swept delta wings. The roll attractor is the steady-state

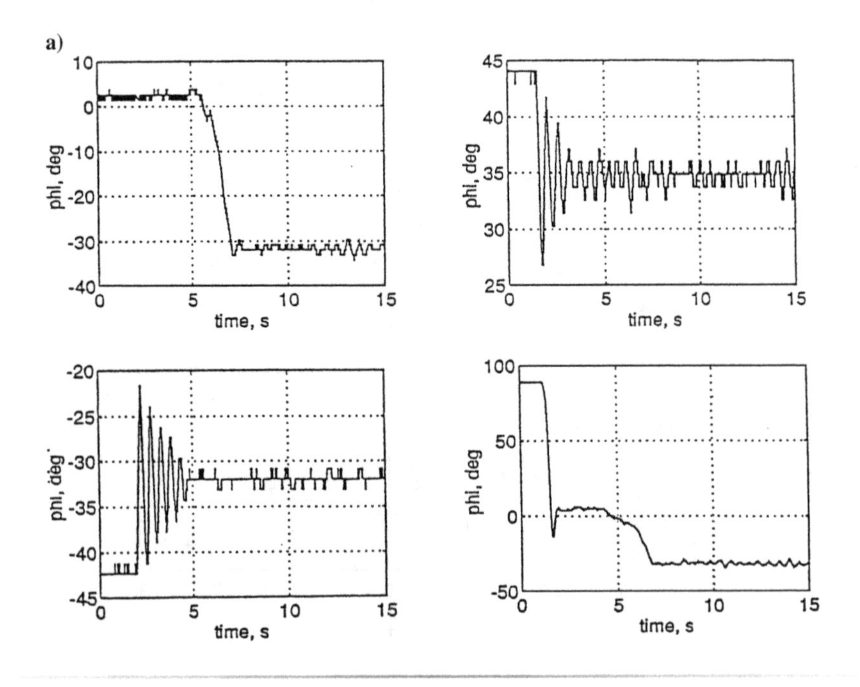

Fig. 8.41a Free-to-roll time histories of 60-deg delta wing (32). (Courtesy ICAS.)

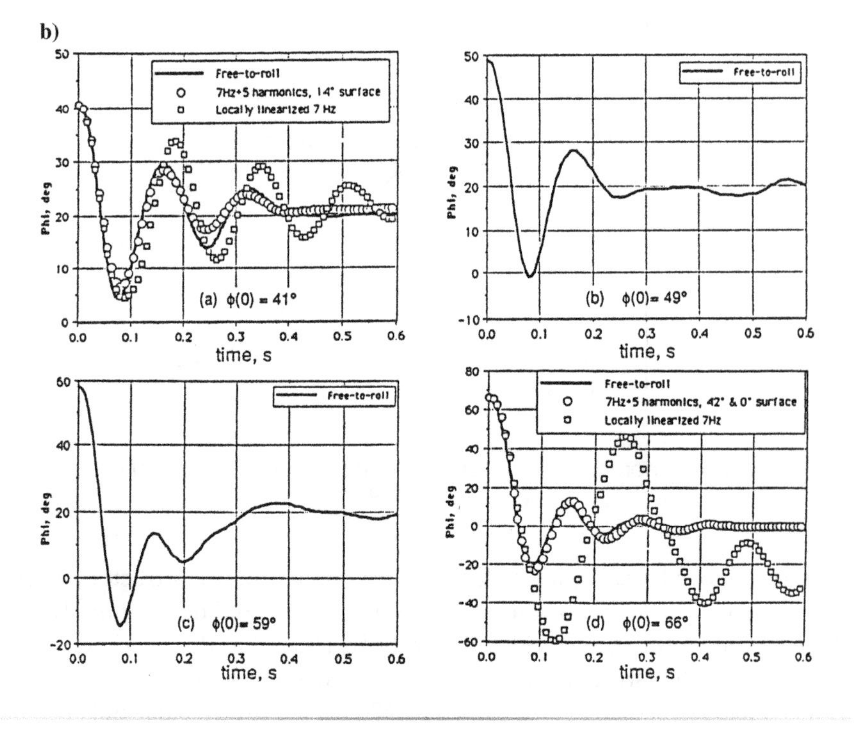

Fig. 8.41b Free-to-roll time histories of 65-deg delta wing (33). (Courtesy AGARD.)

roll angle attained by a free-to-roll model. The leading-edge sweep of delta wings, which exhibits the existence of multiple-roll attractors is high but generally below the value that leads to wing rock. Typical time histories of free-to-roll delta wings with leading-edge sweeps of 60 and 65 deg at $\alpha = 30$ deg are shown in Fig. 8.41 [32, 33]. The 65-deg delta wing has three roll attractors at 0 and ± 21 deg [33], whereas the 60-deg delta wing has two attractors at 34.5 deg and -31 deg [32]. The usual zero attractor is not observed for the 60-deg delta wing at $\alpha = 30$ deg. The asymmetry in numerical values of the attractors for the 60-deg delta wing is attributed to the slight asymmetry in the free-to-roll model as stated elsewhere [32]. The typical variation of static rolling moment with roll angle for the 65-deg delta wing at $\alpha = 30$ deg is shown in Fig. 8.42. It is evident that the existence of roll attractors is caused by the stable zero crossing of the rolling-moment curve at $\phi = 0$ and ± 21 deg. The zero crossing at $\phi = \pm 7$ deg is unstable. The multiple-zero crossings of the rolling moment occur because of the complex variation of vortex breakdown point over the 65-deg delta wing [33]. However, a clear physical understanding of this phenomenon is still not available at present.

The sideslip has an interesting effect on the roll attractor of the 60-deg delta wing as shown in Fig. 8.43. With the increase of sideslip, the two

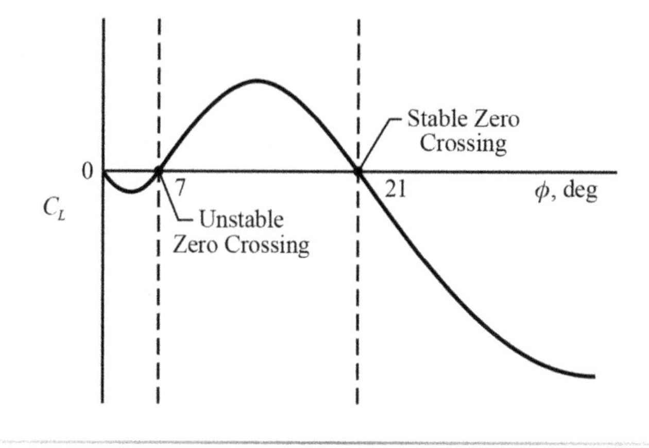

Fig. 8.42 Rolling moment of 65-deg delta wing at 30-deg angle of attack (33). (Courtesy AGARD.)

attractors move toward each other and, at $\beta = 20$, the two attractors merge into one, i.e., for $\beta = 20$, the attractor is at $\phi = -45$ deg. Similarly, for $\beta = -20$, the attractor can be expected to be located at $\phi = +45$ deg.

Another interesting effect of the sideslip is the development of wing rock as indicated by the hatched line in Fig. 8.43 and the time history of Fig. 8.44.

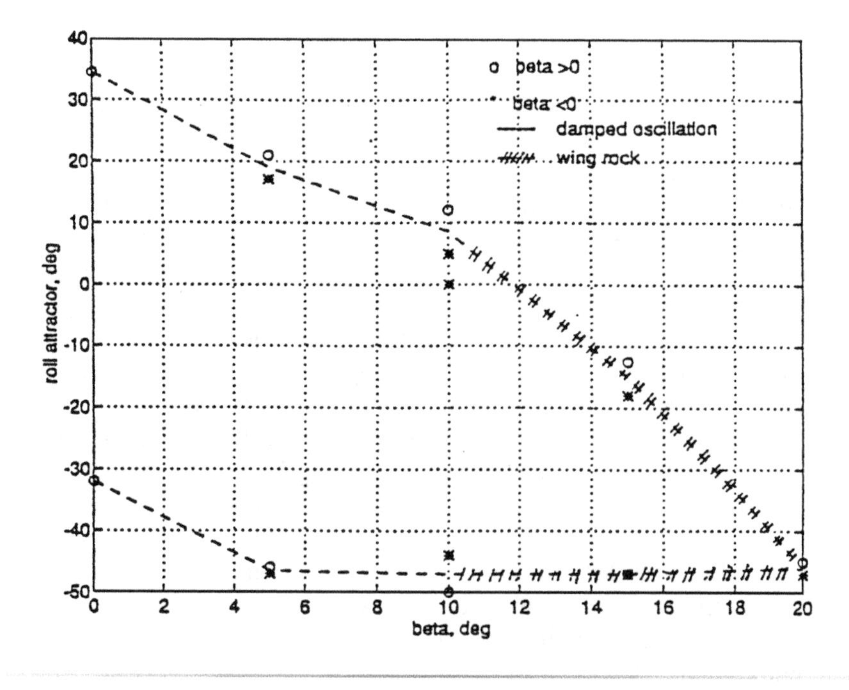

Fig. 8.43 Effect of sideslip on roll attractors of 60-deg delta wing at $\alpha = 30$ deg and $\beta = -15$ deg (32). (Courtesy ICAS.)

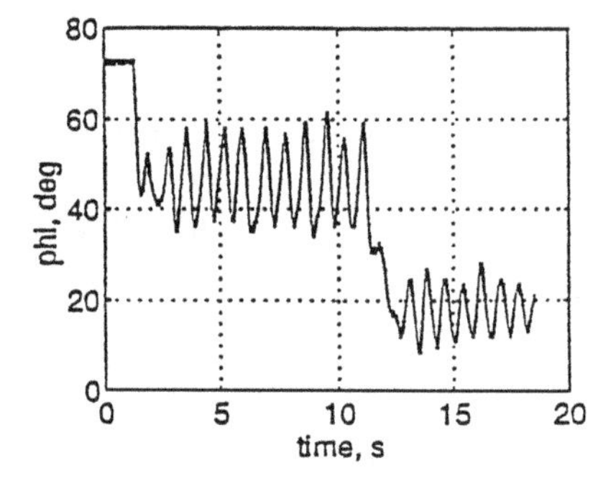

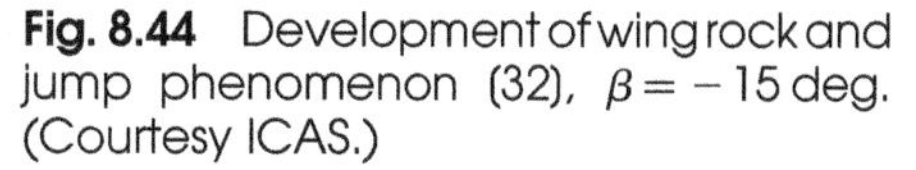
Fig. 8.44 Development of wing rock and jump phenomenon (32), $\beta = -15$ deg. (Courtesy ICAS.)

As can be observed, the 60-deg delta wing develops wing rock about one roll attractor, undergoes some wing rock cycles, and then jumps to another attractor and continues wing rock. At present, no clear physical explanation exists for such complex behavior.

An interesting consequence of the existence of multiple-roll attractors is that, in the presence of a long slender fuselage, it can lead to a wing rock with nonzero mean roll angle [34]. The wing rock is apparently induced by the forebody, and the existence of the multiple, nonzero attractors causes the wing rock to develop around one of the nonzero attractors.

8.10 Forebody-Induced Wing Rock

As stated earlier, the vortices shed from the slender, pointed fuselage forebodies at high angles of attack significantly affect the static stability characteristics of fighter aircraft. At high angles of attack, the forebody vortices assume asymmetric, bistable configuration similar to that observed on slender delta wings. Furthermore, the forebody vortices interact with LEX/wing vortices and produce nonlinearities in static aerodynamic coefficients. Therefore, it is natural to expect that the forebody has a significant effect on the dynamic motions like wing rock. In this section, we will study how the forebody vortices affect the wing rock.

Consider a generic fighter aircraft model shown in Fig. 8.45. This generic aircraft model has a swept-back wing with a leading-edge sweep of 26 deg and a slender, pointed forebody. A typical, single-degree-of-freedom, free-to-roll time history at an angle of attack of 35 deg is shown in Fig. 8.46a with corresponding phase plane plot in Fig. 8.46b. As stated before, the isolated delta wing with 26-deg sweep is not likely to generate the wing rock. Therefore, the wing rock of this generic fighter model must be generated by the forebody vortex flow field. This type of wing rock is called forebody-induced wing rock. The fact that the limit cycle condition is reached within three or four cycles indicates that the forebody-induced wing rock can be much more violent compared to that due to the slender-delta wing.

This generic aircraft model exhibited wing rock when fitted with different wings of varying aspect ratio and sweep. The wing rock was observed even on a model without the wing. This implies that the geometry of the wing has virtually no effect on the forebody-induced wing rock. For the configuration without the wing, the horizontal and vertical tail surfaces coming under the influence of the forebody vortex system produce the necessary rolling moments. In other words, the wing and tail surfaces, coming under the influence of forebody vortices, merely serve as aerodynamic surfaces to generate the necessary rolling moments to sustain the wing rock.

8.10.1 Effect of Forebody Geometry

The fact that the forebody-induced wing rock is caused by the forebody vortices suggests that the forebody geometry must have a significant effect on the forebody-induced wing rock. The generic aircraft model of Fig. 8.45 was tested with four different forebody cross-sectional shapes, which were horizontal ellipse, circle, vertical ellipse, and triangle. In Fig. 8.47, the static lateral and directional stability derivatives for these four different cross-sectional shapes are presented. It is interesting to observe that the forebody with horizontal ellipse cross section has the highest levels of static directional and static lateral stabilities yet has the maximum wing rock amplitude as

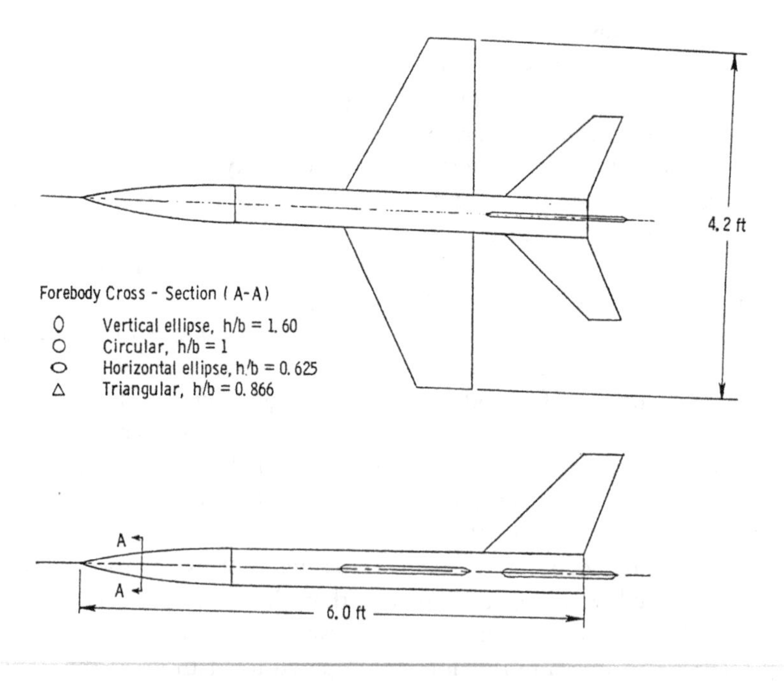

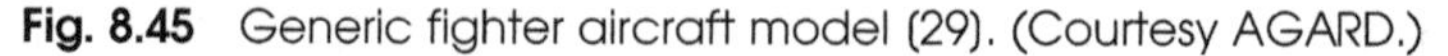

Fig. 8.45 Generic fighter aircraft model (29). (Courtesy AGARD.)

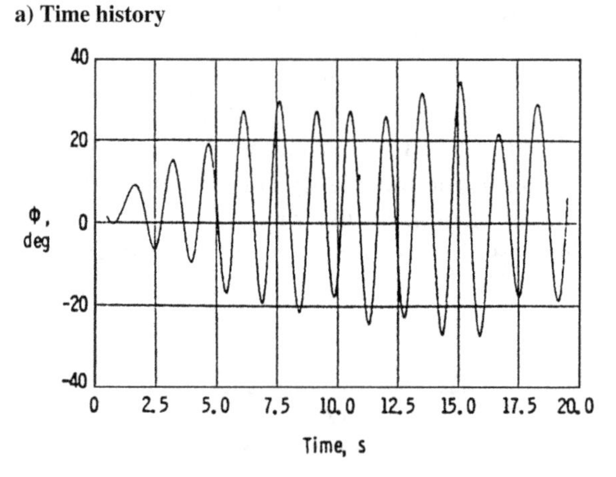

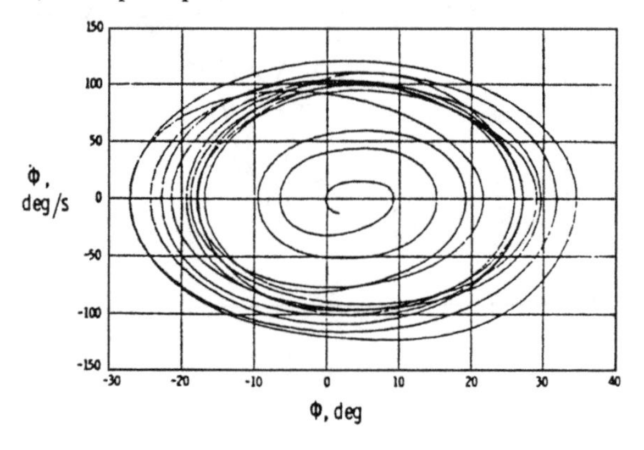

Fig. 8.46 Wing rock characteristics of generic aircraft model (29). (Courtesy AGARD.)

shown in Fig. 8.48. The vertical elliptical cross section has the lowest levels of static directional and lateral stabilities yet displays a wing rock of modest amplitude.

8.10.2 Physical Flow Mechanism for Forebody-Induced Wing Rock

As we know, the loss of damping in roll at high angles of attack makes a configuration susceptible to wing rock but does not necessarily generate the wing rock. To generate and sustain the wing rock in single-degree-of-freedom, free-to-roll tests, the additional requirement is that the configuration must be statically stable, and the roll damping must vary so that it is positive (undamping) at low roll angles and negative (damping) at higher roll angles. However, for the forebody-induced wing rock, the physical flow mechanism that can cause this type of variation of roll damping is not very clear. Ericson has forwarded a conceptual hypothesis based on the well-known Magnus effect observed on circular cylinders in crossflow and Swanson's experimental results [35] on rotating circular cylinders in crossflow. Ericson refers to this as the moving wall effect hypothesis [36].

Swanson's experimental results on a rotating circular cylinder are presented in Fig. 8.49. Ericson constructs his moving wall effect hypothesis

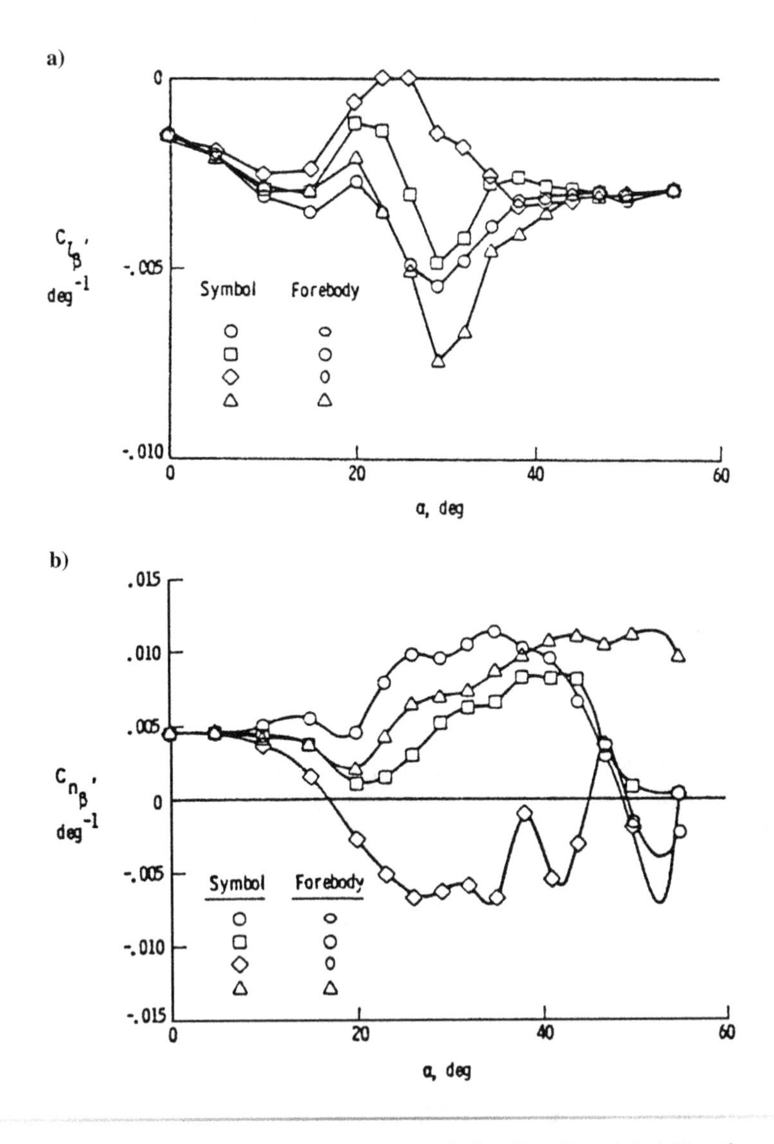

Fig. 8.47 Effect of cross-sectional shape on static directional stability of generic fighter (29). (Courtesy AGARD.)

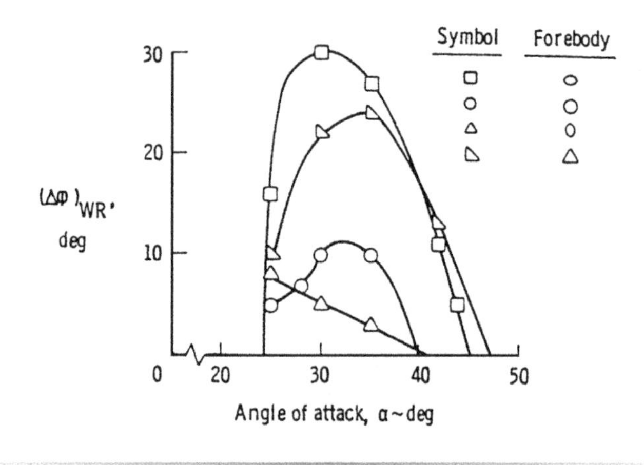

Fig. 8.48 Effect of forebody cross-sectional shape on wing rock of generic aircraft model (29). (Courtesy AGARD.)

as a conceptual explanation to Swanson's measurements and then uses these ideas to develop a physical explanation for the forebody-induced wing rock.

Ericson's Moving Wall Effect Hypothesis Consider the fluid flow about a circular-cylinder. The flow pattern over a stationary cylinder depends on the Reynolds number. Hence, it is natural to expect that, for a rotating cylinder, the moving wall effect also depends on the Reynolds number. In addition, the moving wall effect depends on another parameter, the nondimensional rotation rate, $p = U_W/U_\infty$, where $U_W = \omega d/2$ is the tangential velocity of the cylinder surface, ω is the angular velocity, and U_∞ is the free-stream velocity. Analogous to the critical Reynolds number, we have a critical value of the nondimensional rotation rate p_{cr} associated with the given rotating cylinder. When $p = p_{cr}$, the slope of the Magnus lift changes sign, and the Magnus lift starts building up in the opposite direction as observed in Fig. 8.49. The value of p_{cr} depends on the Reynolds number. For group 1 (laminar flow), p_{cr} varies from 0.52 at a Reynolds number of 3.58×10^4 to 0.3 at 12.8×10^6. For group 2 (transitional flow, Reynolds number from 15.2×10^4 to 36.5×10^4), p_{cr} is around 0.2 and for group 3 (turbulent flow, Reynolds number from 42×10^4 to 50.1×10^4), and p_{cr} is in the range 0.1–0.2. Thus, the lower the Reynolds number, the higher the critical rotation rate. In the following, we will discuss the moving wall effect for various flow regimes.

Laminar Flow Here, we consider the effect of rotation rate when the flow is laminar, corresponding to the data of group 1 of Fig. 8.49. For a stationary cylinder ($p = 0$, Fig. 8.50a), the flow separates ahead of the lateral meridian point or the maximum thickness point, forming a large wake characterized by the classical Kármán vortex shedding. When the rotation rate $p < p_{cr}$, the Magnus effect comes mainly from the downstream moving wall effect on the top side of the rotating cylinder as shown in Fig. 8.50b. On the top side, the flow separation point is moved in the downstream direction, whereas, on the bottom surface, the flow separation is still of laminar type. This movement of the separation point on the top surface takes place because of the moving wall effect and not because of flow transition. On the bottom side, the moving wall effect does not do much to alter the location of the flow separation point. As a result, the suction on the top side increases relative to that

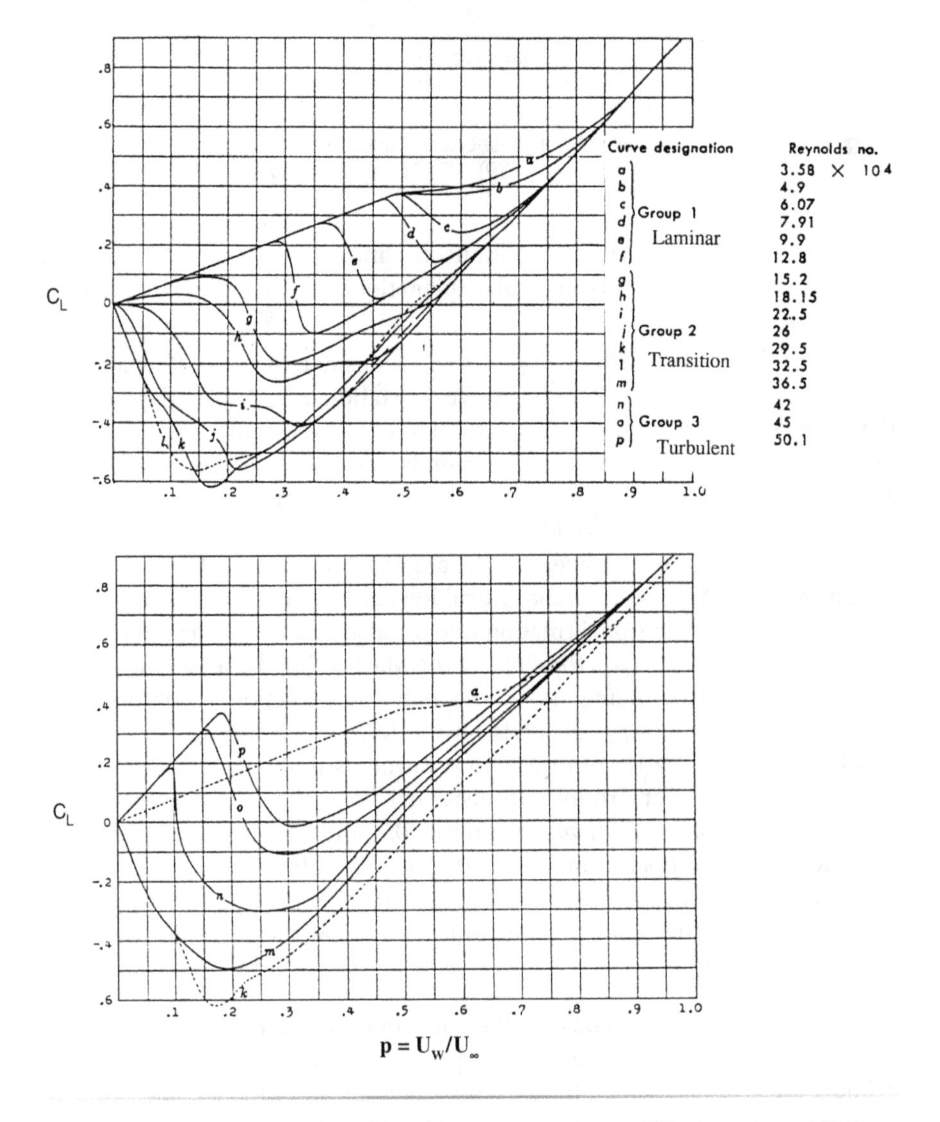

Fig. 8.49 Magnus lift on a rotating circular cylinder (35). (Courtesy ASME, *Journal of Basic Engineering.*)

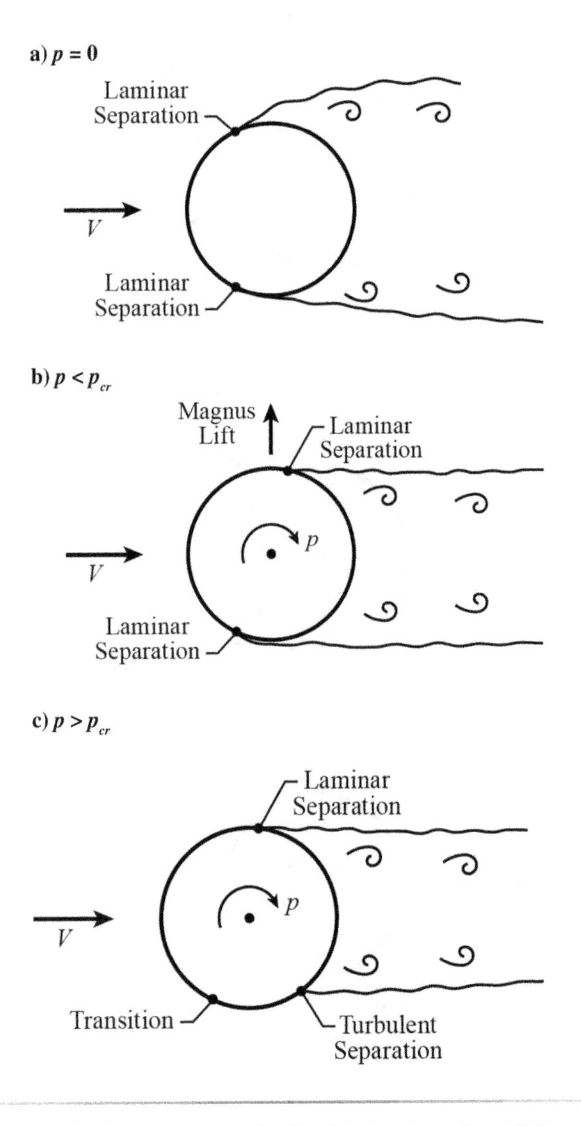

Fig. 8.50 Moving wall effect in laminar flow (36).

on the bottom side, and a positive Magnus lift is generated as shown in Fig. 8.49 for the data of group 1 (laminar flow).

When $p \geq p_{cr}$, the moving wall effect influences the flow pattern on the bottom side. It causes the transition to occur prior to separation and, therefore, moves the separation point downstream (Fig. 8.50c). On the top side, the moving wall effect remains the same as said for the case of $p < p_{cr}$. However, the net result is that the Magnus lift starts dropping and assumes negative values for some cases as indicated by the data for group 1 in Fig. 8.49.

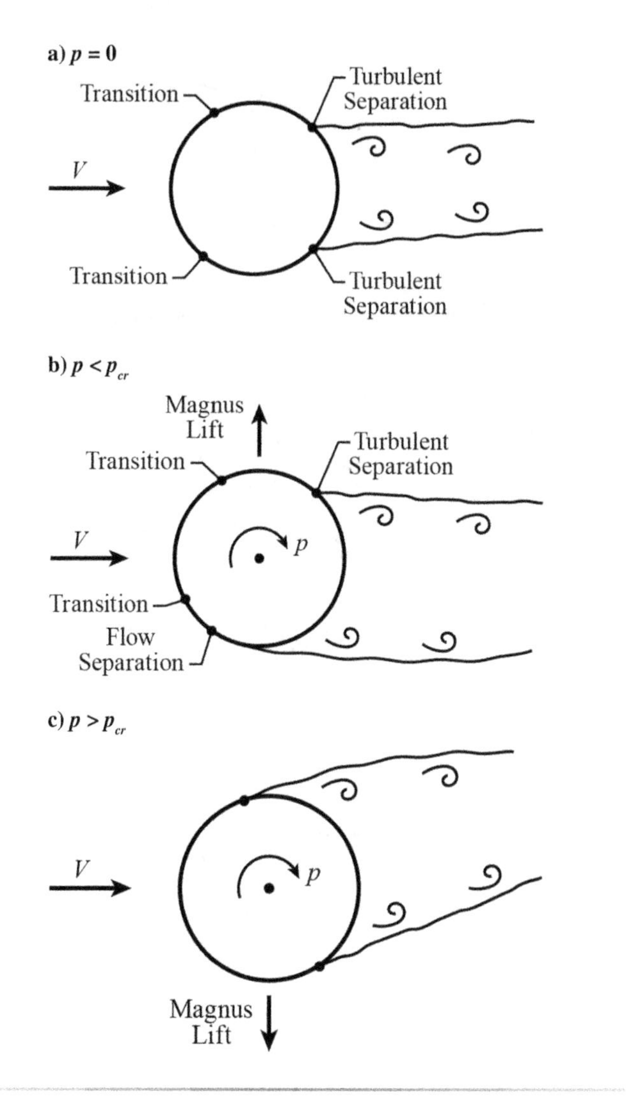

Fig. 8.51 Moving wall effect in turbulent flow (36).

Turbulent Flow Consider the effect of rotation rate when the flow is turbulent corresponding to the data of group 3 in Fig. 8.49. For a stationary cylinder, when the Reynolds number exceeds the critical Reynolds number, the flow over the circular cylinder is of the supercritical type, which is characterized by the flow transition ahead of the lateral meridian point and a turbulent separation downstream of the lateral meridian point as shown in Fig. 8.51a. When $p < p_{cr}$, the moving wall effect does not do much to alter the supercritical-type separation on the top side. However, on the bottom side, the moving wall effect causes the flow separation point to move upstream towards the subcritical location as shown in Fig. 8.51b. The net effect is to generate a positive Magnus lift similar to that observed for laminar flow, but the magnitude of the Magnus lift is much higher.

When $p > p_{cr}$, the downstream moving wall effect on the top side causes a delay in the transition to such an extent that separation occurs before transition, thereby moving the separation point towards the subcritical (laminar-type) location. On the bottom side, the moving wall effect does not do much to alter the supercritical flow separation. As a result, the Magnus lift starts building up in the opposite direction as indicated in Fig. 8.51c.

Transition Reynolds Numbers This case corresponds to the data of group 2 in Fig. 8.49. Let us first consider the flow pattern for $p = 0$. On a stationary circular cylinder, at transition Reynolds numbers, the flow is characterized by the presence of a separation bubble as schematically shown in Fig. 8.52a. The flow separates at P_1 just ahead of the lateral meridian, and this separation is of the subcritical or laminar type. A transition occurs in the lifted shear layer at P_2, causing the flow to become turbulent. This turbulent flow reattaches to the surface of the cylinder at P_3, forming a bubble. The reattached flow eventually separates at P_4. An important feature of this type of flow pattern is that the separation point P_4 is further downstream of the

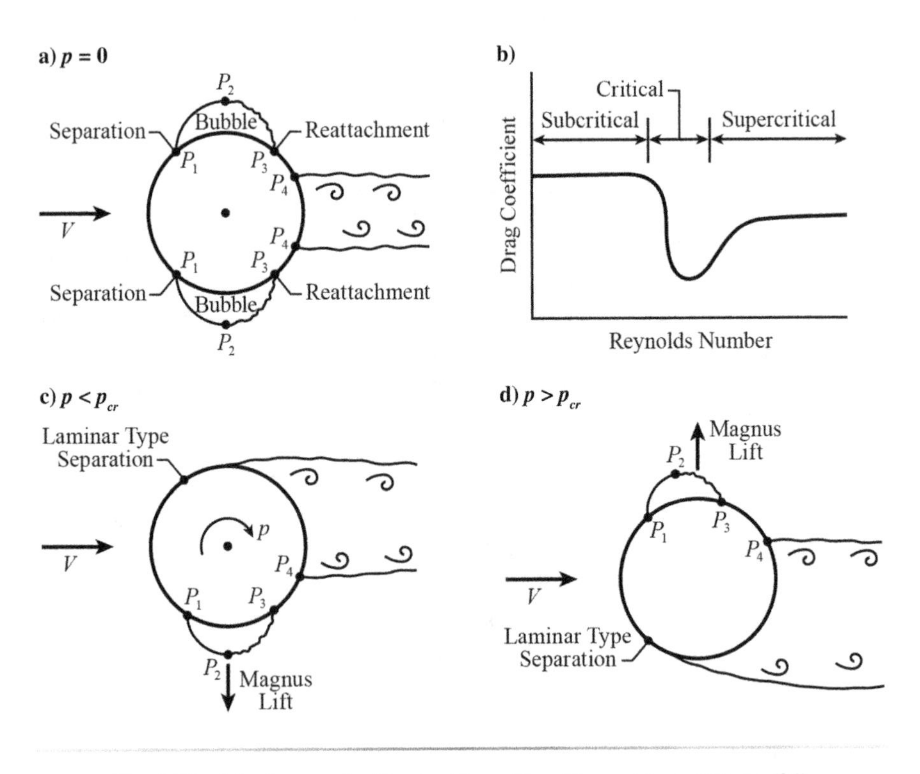

Fig. 8.52 Moving wall effect at transition Reynolds numbers (36).

supercritical (turbulent-type) separation point. As a consequence of this type of flow pattern, the drag coefficient assumes minimum value, forming a drag bucket as shown schematically in Fig. 8.52b. When the freestream Reynolds number increases further, the transition point gradually moves ahead in the bubble so that, at one point, it jumps ahead of the bubble and wipes it out completely, establishing the supercritical flow pattern of the type shown earlier in Fig. 8.51a.

Now consider a rotating cylinder ($p \neq 0$) when the freestream Reynolds number is in the transition range. The data in Fig. 8.49 for group 2 give the Magnus lift for this range of Reynolds numbers. We observe that the Magnus lift is negative for $p < p_{cr}$, decreases further as p increases and assumes a minimum value in the neighborhood of $p = p_{cr} = 0.2$, and starts rising for higher values of p. For $p \geq 0.55$, it becomes positive.

For $p \leq p_{cr}$, the upstream moving wall effect on the bottom side causes a forward movement in the transition point in the lifted shear layer. On the top side, the downstream moving wall effect causes a delay in the transition to such an extent that the boundary layer may not reattach to the cylinder surface. In other words, the flow separation essentially is of subcritical type on the top side. As a result, the Magnus lift is negative. For $p > p_{cr}$, the transition takes place in the separated shear layer, and the flow reattaches to the top surface, forming a bubble. On the bottom side, the upstream moving wall effect establishes a subcritical flow pattern. As a result, the Magnus lift assumes positive values.

Ericson's Conceptual Flow Mechanism for Forebody-Induced Wing Rock One of the main assumptions in Ericson's hypothesis for the forebody-induced wing rock is that the freestream Reynolds number is in the critical range [37]. Thus, according to Ericson, the forebody-induced wing rock is not likely to occur if the Reynolds number is in the subcritical or supercritical range. Furthermore, he assumes that the wing and tail surfaces provide the downstream surfaces needed to generate the necessary rolling moments. With these assumptions, an explanation to the development of a vortex-asymmetry-switching mechanism that is necessary for the occurrence of forebody-induced wing rock is offered as follows [37].

When disturbed at high angles of attack, the free-to-roll wing–body model starts oscillating because of a loss of damping in roll. Let this happen at $t = t_1$ and let us assume that the model starts rolling in the clockwise direction as shown in Fig. 8.53a. Once the model starts rolling, the moving wall effect comes into existence. As shown in Fig. 8.53a, the adverse moving wall effect promotes transition and a turbulent reattachment on the right side. On the left side, the proverse moving wall effect does not do much to alter the subcritical-type flow separation. As a result, an asymmetric vortex pattern is formed as shown. However, it takes a certain time δt for this vortex pattern to reach the wing. During this interval of time, the model continues to roll in the clockwise direction. At $t = t_2 = t_1 + \delta t$, the asymmetric

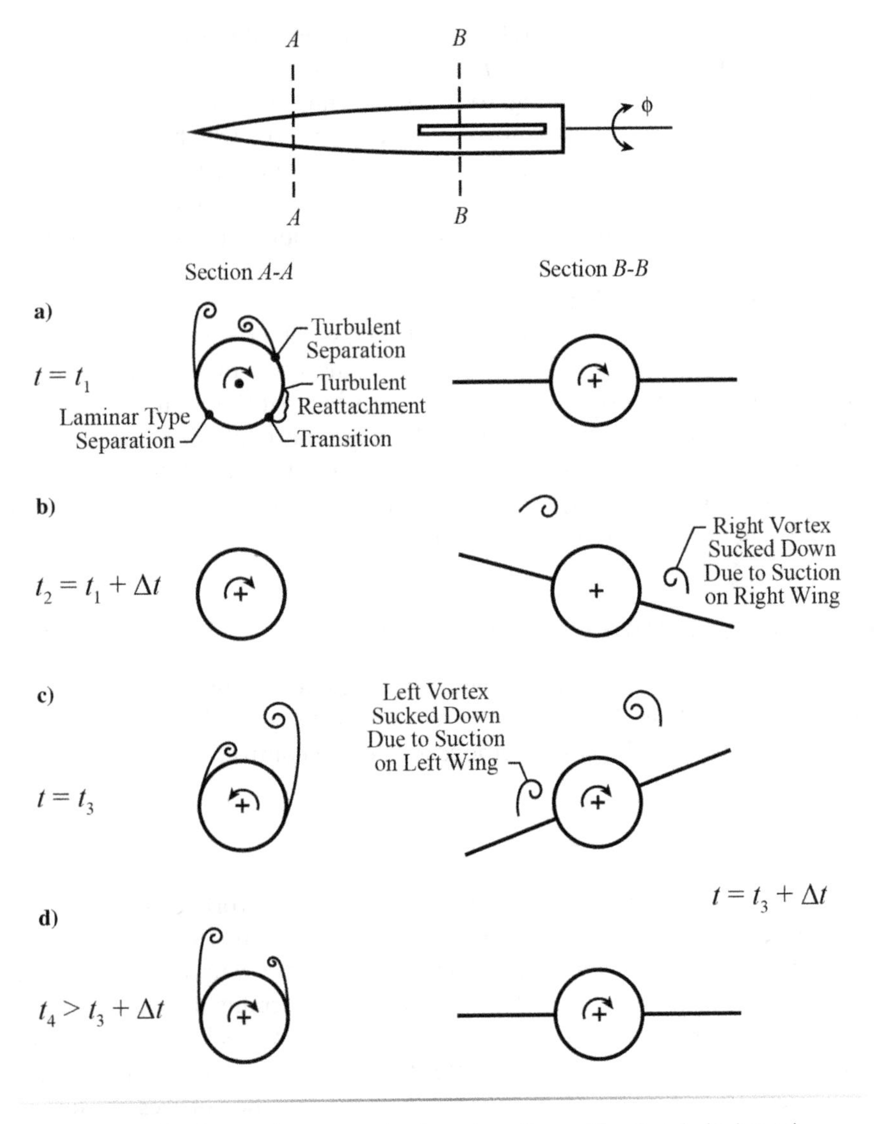

Fig. 8.53 Ericson's conceptual explanation of forebody-induced wing rock (37).

vortex pattern reaches the wing and establishes a type of pressure distribution that begins to generate a rolling motion in the counterclockwise direction. With this, the moving wall effect switches sides. The adverse moving wall effect on the left side causes a transition and turbulent reattachment with the result that the vortex on this side is closer to the body. On the right side, the proverse moving wall effect leads to laminar-type (subcritical) flow separation and leads to a vortex located further away from the body as shown in Fig. 8.53c. As before, because of the time delay of δt, the model will continue to roll in the counterclockwise direction until this new asymmetric vortex reaches the wing at $t = t_3 + \Delta t$. At this point, the rolling motion reverses, starting a new wing rock cycle. Thus, the moving wall effect generates an alternating asymmetric vortex pattern, which produces the necessary aerodynamic spring or the statically stabilizing effect. Because of the time delay effect, this statically stabilizing effect becomes dynamically destabilizing. For additional information, the reader may refer elsewhere [36, 37].

The flow over the slender forebody of a fighter aircraft at high angles of attack is much more complex than the classical crossflow over a two-dimensional circular cylinder, which forms the basis of Ericson's moving wall hypothesis. In view of this, it is quite possible that the actual mechanism causing the forebody-induced wing rock on aircraft configurations is different from that assumed in Ericson's hypothesis. In spite of this, Ericson's hypothesis provides a physical insight into the possible flow mechanism causing the forebody-induced wing rock.

8.11 Suppression of Wing Rock

Several attempts have been made in recent years to find solutions to the wing rock problem. These approaches can be broadly classified into two groups: passive aerodynamic methods and active control.

8.11.1 Passive Aerodynamic Methods

As we have seen, wing rock is caused by wing vortices or forebody vortices or by an interaction between these two vortices. The loss of damping in roll at high angles of attack initiates the wing rock, and some form of vortex-asymmetry-switching mechanism sustains it as a limit cycle oscillation. Therefore, the key to suppressing the wing rock by passive aerodynamic methods is to manipulate the vortices so that the vortex-asymmetry-switching mechanism is suppressed and the vortices are forced to assume a symmetric disposition.

The basic process that leads to the formation of leading-edge vortices on the lee side of a delta wing is the flow separation at the sharp leading edges. Therefore, a direct manipulation of this process holds the key to suppressing wing rock. One such example is the concept of spanwise tangential jet blowing along the leading edges as shown schematically in Fig. 8.54. An application of this concept for suppression of wing rock is shown in Fig. 8.55 [38]. The test model used [38] is a 60-deg delta wing with a LEX of 78-deg sweep. The wing rock of this model is primarily induced by the LEX vortices. The Coanda effect induced by the jet energizes the boundary layer in the vicinity of the leading edge and, thereby, delays the flow separation. This process leads to smaller and better organized vortices. The

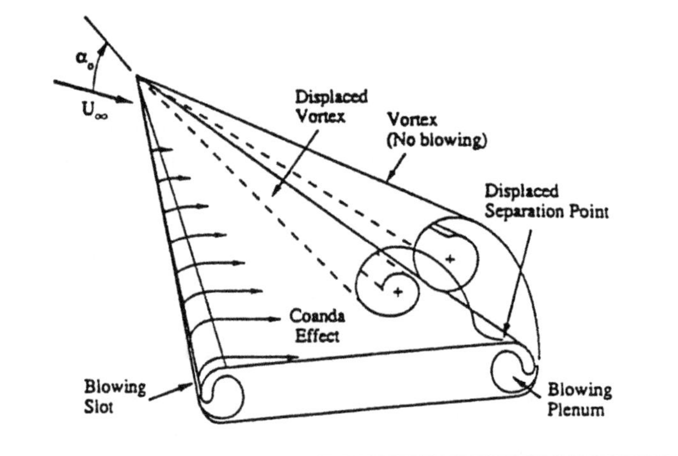

Fig. 8.54 Tangential leading-edge spanwise blowing concept (38).

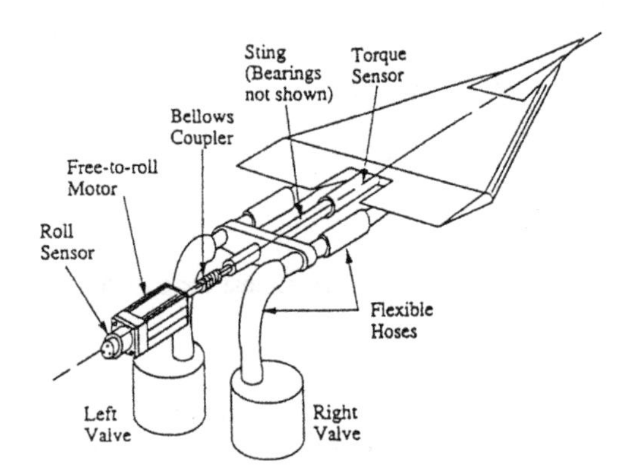

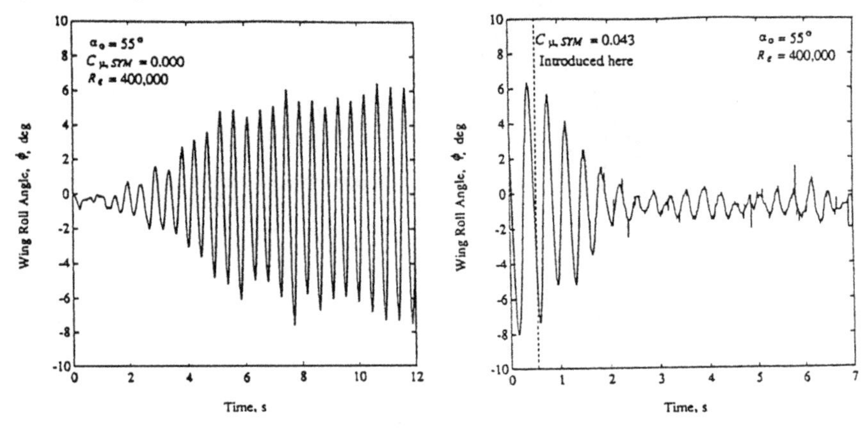

Fig. 8.55 Suppression of wing rock with tangential leading-edge blowing (38).

symmetrical blowing on both wings leads to symmetrically disposed vortices. As a result, the wing regains roll damping, and wing rock is suppressed as shown in Fig. 8.55b.

An example of suppression of forebody-induced wing rock using the blowing concept [37] is presented schematically in Fig. 8.56. The model tested was a 60-deg delta wing with a conical nose, and the angle of attack was 45 deg. The model exhibited wing rock with an amplitude of about 22 deg as shown in Fig. 8.57a. Note that at $\alpha = 45$ deg and $\beta = 0$, the wing rock of this model is induced by the forebody because an isolated 60-deg delta wing is not likely to exhibit wing rock. At high angles of attack, such a wing exhibits the roll attractor phenomenon as described earlier.

The blowing slots were located on either side of the conical nose as shown in Fig. 8.56. The wing rock was suppressed by asymmetrical (one-sided, right or left) tangential aft blowing as shown in Fig. 8.57b. However, the disadvantage of one-sided blowing is that it is accompanied by significant amounts of undesirable side force and yawing moments. To avoid this, it is preferable to employ simultaneous symmetric, tangential aft blowing from both sides. However, this approach of symmetrical blowing was not much effective in suppressing the wing rock [39].

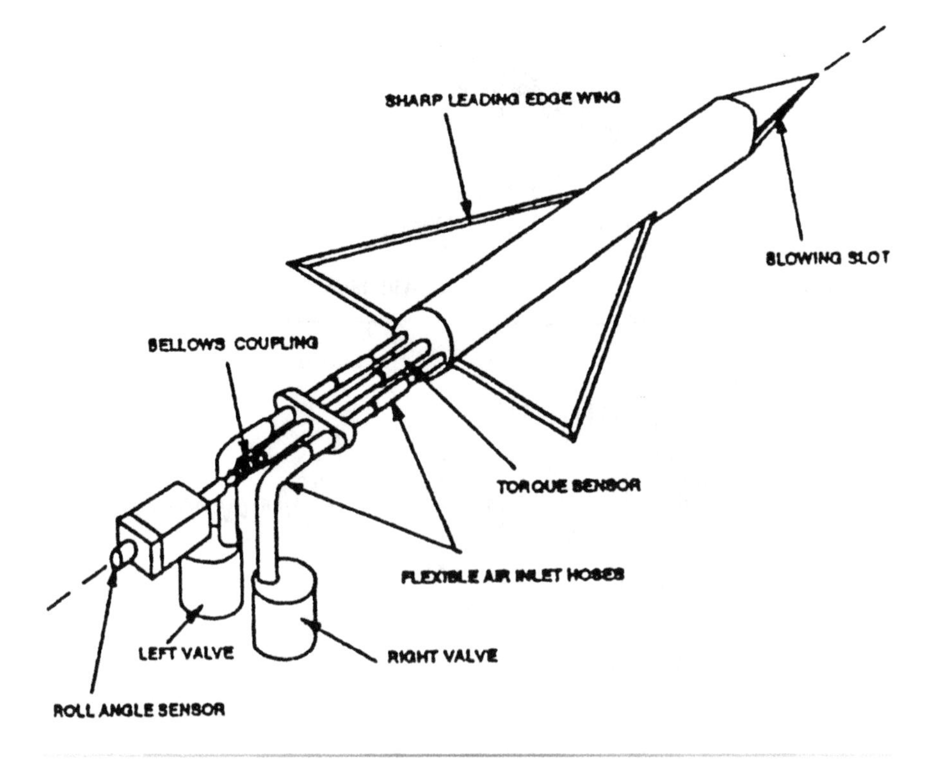

Fig. 8.56 Schematic arrangement of forebody blowing (39).

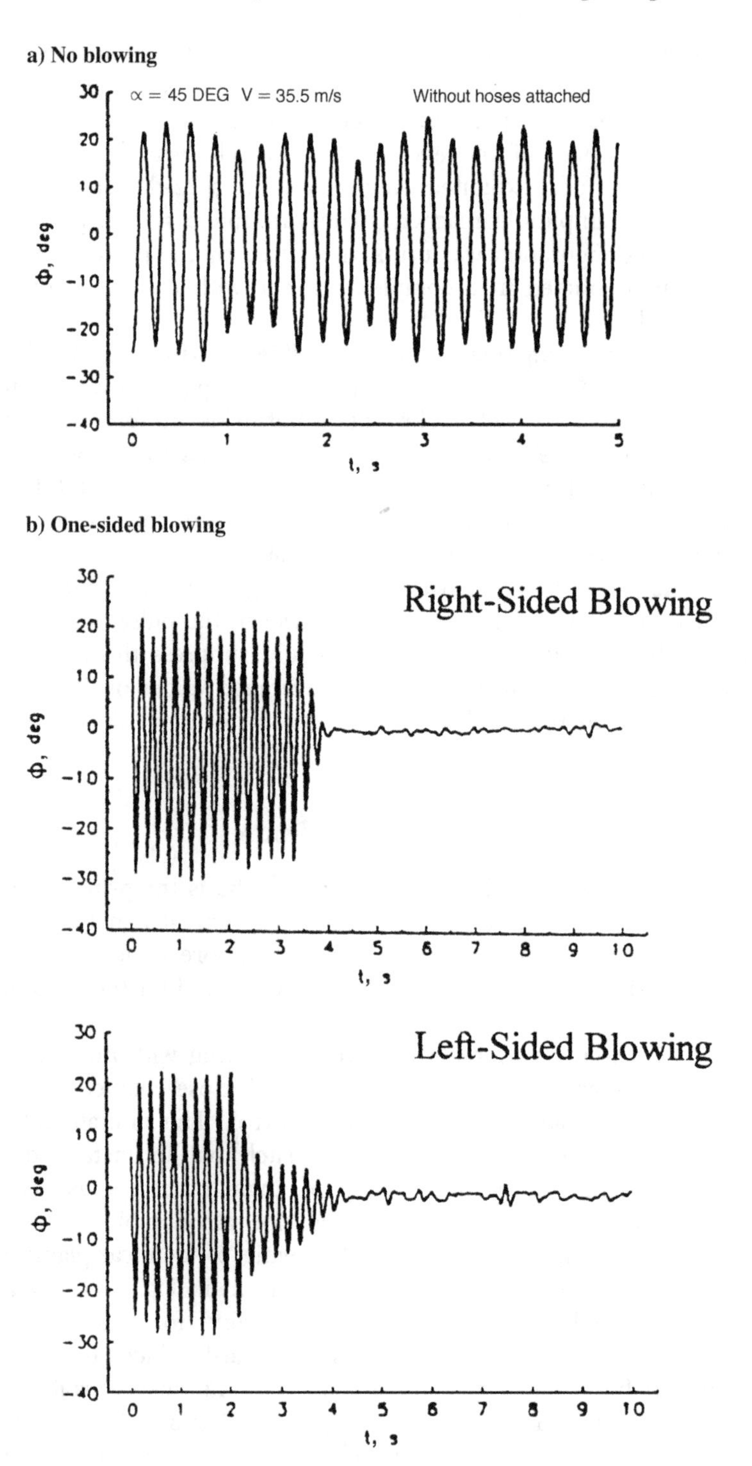

Fig. 8.57 Wing rock suppression with one-sided forebody blowing (39).

Another interesting example of the application of the blowing concept to the suppression of forebody-induced wing rock is discussed in [40]. The model tested was a generic aircraft configuration with a circular-cross-section forebody of fineness ratio 6 and a delta wing of 78-deg sweep (Fig. 8.58). This model exhibited wing rock above a 22-deg angle of attack. The wing rock amplitude reached 40 deg around $\alpha = 30$ deg. The wing rock for this configuration was primarily caused by the powerful forebody vortices that interacted in a complex manner with the wing vortices as described [40].

To suppress the wing rock, a tangential, aft jet blowing was done from the slots located on the leeward side near the forebody tip [40]. It was interesting to note that steady, symmetric blowing on both sides to force a symmetric disposition of the forebody vortices did not suppress the wing rock. This observation is similar to that described earlier [39] for the 60-deg delta wing model with a conical nose. Instead, a steady one-sided blowing suppressed the wing rock, but the model assumed a nonzero equilibrium roll angle. As said earlier, the asymmetric blowing was accompanied by undesirable side force and yawing moments. However, the pulsed blowing from left and right slots at a high frequency could suppress the wing rock as shown in Fig. 8.58. In this way, the time-averaged values of the side force and yawing moments could be kept down to a minimum for this configuration.

8.11.2 Active Control of Wing Rock

Because the loss of roll damping at high alpha is the primary cause of instability in roll and the development of wing rock, it is natural to think of a stability augmentation system in roll to suppress the occurrence of wing rock during flight. Such an approach is discussed for the X-29 aircraft in [41].

The X-29 aircraft features a forward-swept wing with more than 32% negative static margin. It uses a forward-placed, close-coupled canard and an aft-located fuselage strake for pitch control. The aircraft exhibits a decrease in roll damping above $\alpha = 15$ deg such that an unstable damping in roll is encountered beyond 20 deg of angle of attack as shown in Fig. 8.59. This loss of roll damping is not attributable to the wing because the tests on an isolated flat plate wing of identical planform did not exhibit such a loss of roll damping as shown in Fig. 8.59. Apparently the powerful forebody vortices were responsible for this loss of roll damping at high angles of attack. The use of canard reduced the severity of this problem but could not eliminate it [41]. This loss of roll damping coupled with a high yaw-to-roll inertia ($I_z/I_x = 10$) made the configuration susceptible to wing rock. The wing rock amplitude increased with angle of attack and, at $\alpha = 35$ deg, the peak amplitude exceeded 25 deg. An interesting aspect of this study [42] was that the wing rock was not fundamentally

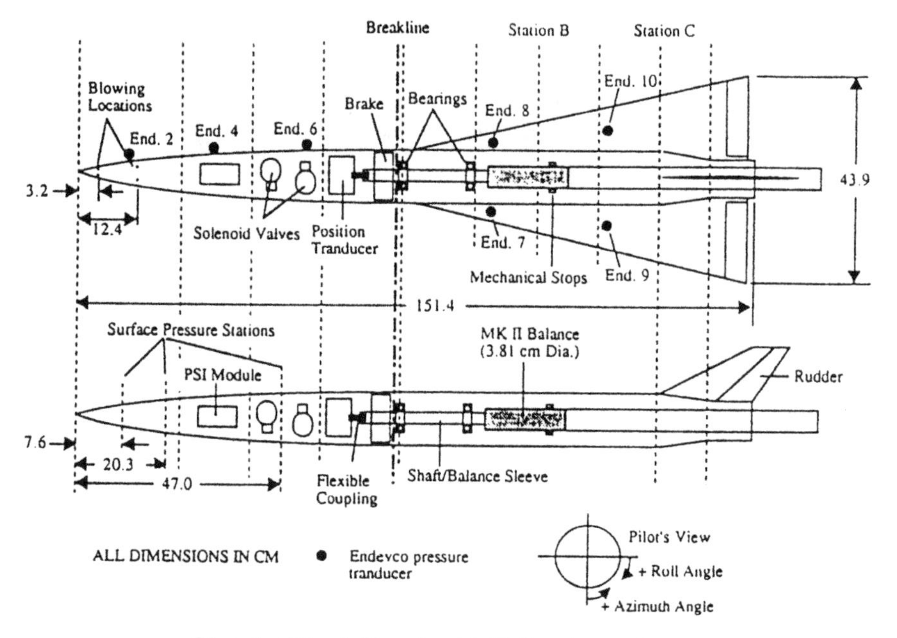

Schematic of the wind-tunnel model. All dimensions are in centimeters.

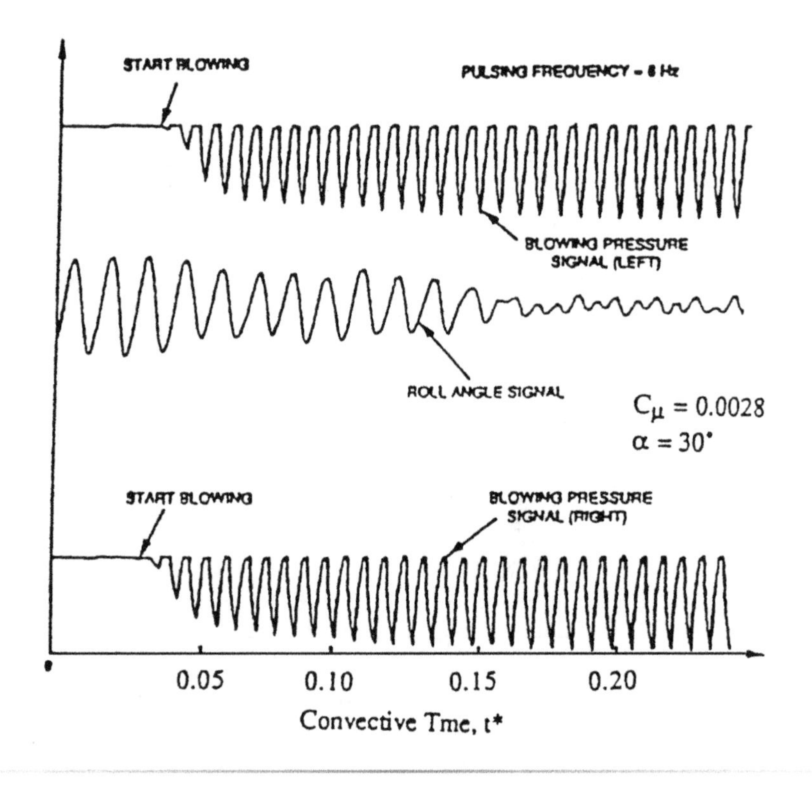

Fig. 8.58 Wing rock suppression using pulsed, tangential forebody blowing (40).

altered by the individual removal of either the canard or the wings or the vertical tail.

The use of a stability augmentation system (SAS) with roll rate feedback effectively suppresses the wing rock as shown in Fig. 8.60. With SAS off at 7 s, the configuration rapidly develops wing rock. The traces of roll and yaw rates indicate that the motion is predominantly a roll about the body axis. Turning the SAS on at 18 s effectively suppressed the wing rock.

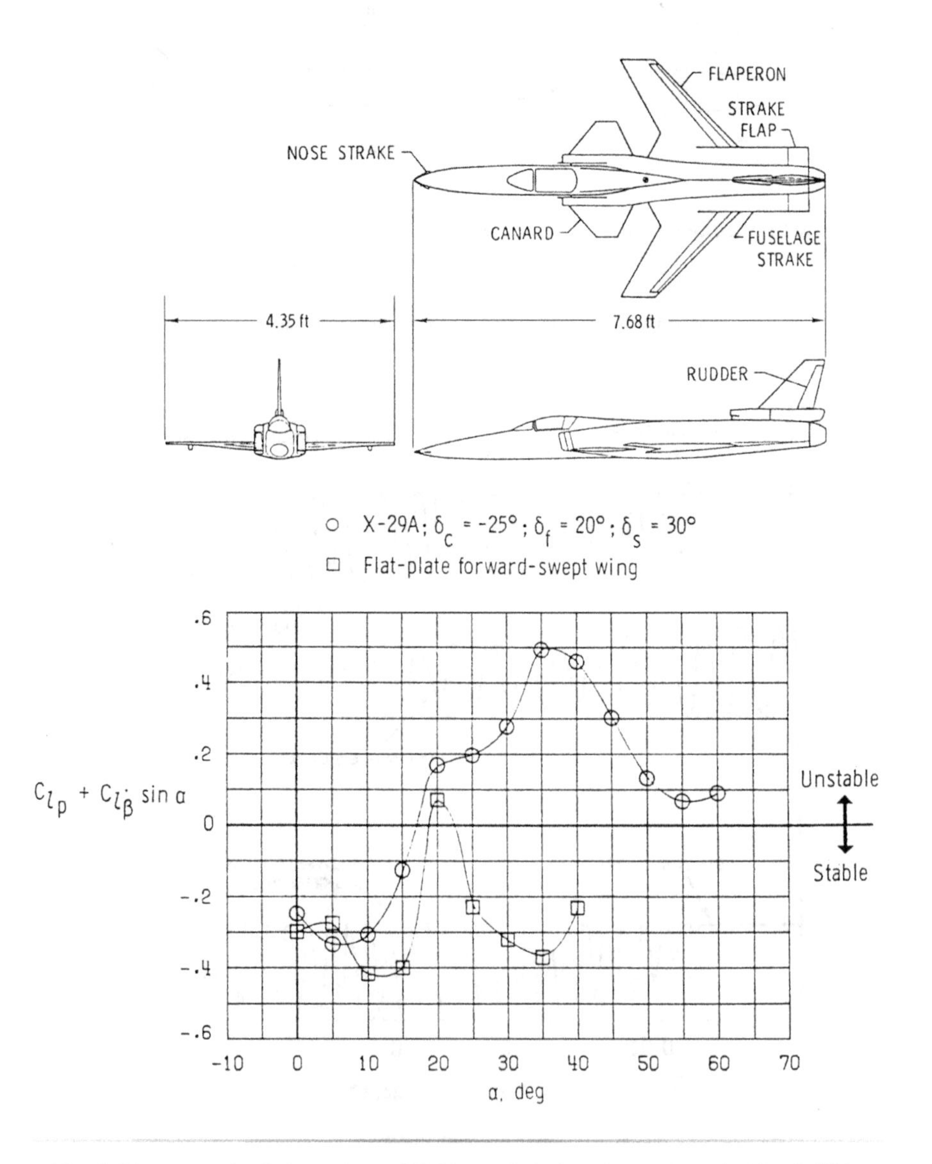

Fig. 8.59 Loss of roll damping of X-29 model at high angles of attack (41).

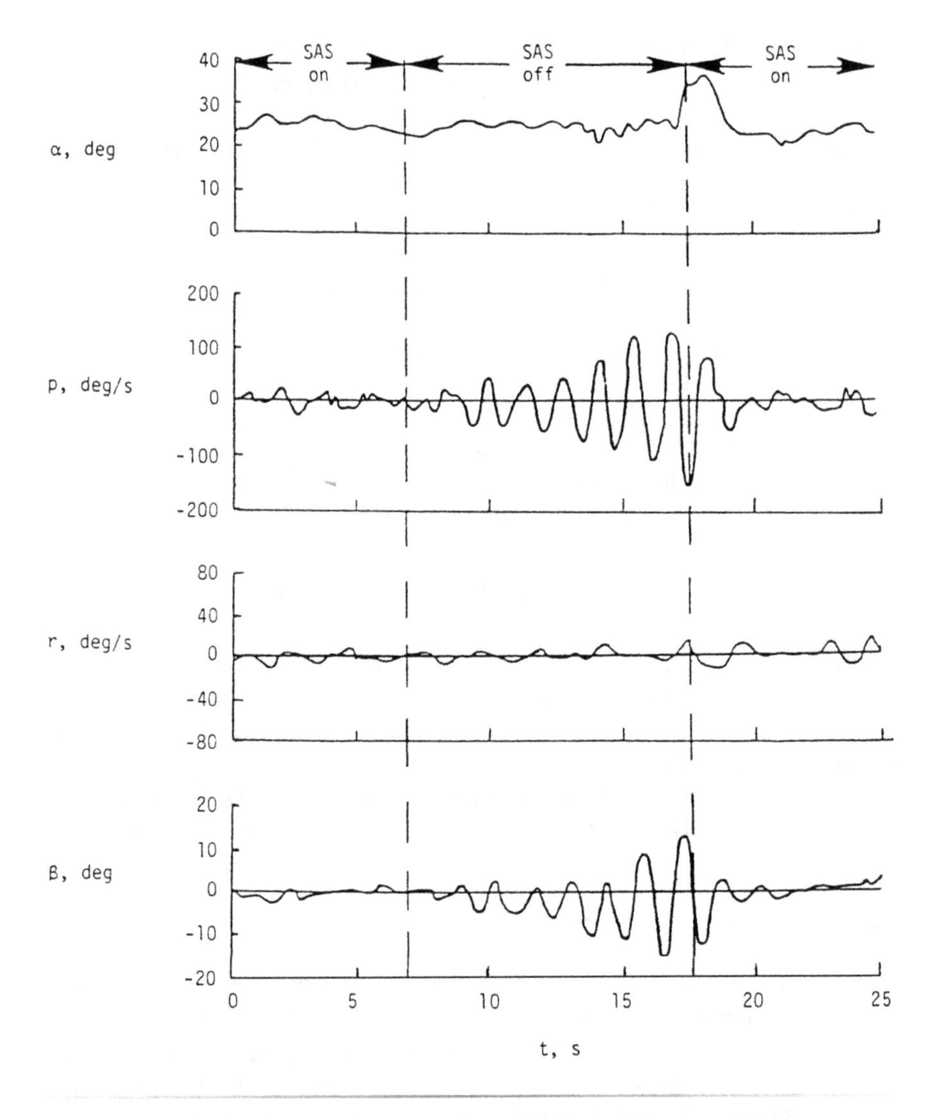

Fig. 8.60 Wing rock suppression of X-29 model using stability augmentation system (41).

8.12 Roll Reversal and Yaw Departure

8.12.1 Roll Reversal

Generally, the ailerons are used for producing rolling motion. However, at high angles of attack, much of the aileron effectiveness may be lost. In addition, the aileron deflection at high angles of attack may produce significant adverse yaw, which may lead to the so-called roll reversal phenomenon. As a result, the aircraft rolls in the opposite direction to aileron input. The angle of attack at which the roll reversal occurs can be approximately determined using the following criterion, called lateral control departure parameter or LCDP [43]. The LCDP is derived from simplified equations for yawing and rolling motions and is defined as follows.

Ailerons Alone for Roll The LCDP is given by

$$\mathrm{LCDP} = C_{n\beta} - \frac{C_{n\delta_a}}{C_{l\delta_a}} C_{l\beta} \tag{8.4}$$

The LCDP for this case of ailerons alone is also called aileron alone departure parameter (AADP).

Aileron Plus Rudder for Roll On some aircraft, the ailerons and rudder are interconnected for proper roll–yaw coordination during velocity vector rolls at high angles of attack. For such cases, the LCDP is given by

$$\mathrm{LCDP} = C_{n\beta} - C_{l\beta} \frac{(C_{n\delta_a} + kC_{n\delta_r})}{(C_{l\delta_a} + kC_{l\delta_r})} \tag{8.5}$$

where $k = \delta_r/\delta_a$, the ratio of rudder deflection to aileron defiection.

For normal response, i.e., the aircraft to roll in the proper direction to aileron input, LCDP > 0. If LCDP < 0, then the aircraft rolls in the opposite direction to aileron input. If the LCDP is negative with a large magnitude, then it is possible that the aircraft may also depart in yaw, and spin entry is likely. When LCDP is zero, the aircraft refuses to roll for an aileron input.

8.12.2 Yaw Departure

Many attempts have been made to derive simple expressions for predicting the yaw departure at high angles of attack. One expression that is commonly used in aircraft dynamics is the so-called $C_{n\beta\,\mathrm{DYN}}$ criterion. The expression for $C_{n\beta\,\mathrm{DYN}}$ is derived from the open-loop lateral-directional quartic equation [43–45] and is given by

$$C_{n\beta\,\mathrm{DYN}} = C_{n\beta} \cos\alpha - \frac{I_z}{I_x} C_{l\beta} \sin\alpha \tag{8.6}$$

For directional stability at high angles of attack, $C_{n\beta\,\mathrm{DYN}}$ must be positive. This is the equivalent of the static stability criterion $C_{n\beta} > 0$ used for low angles of attack. If $C_{n\beta\,\mathrm{DYN}}$ is negative, the aircraft may experience yaw departure, and the Dutch-roll mode may become unstable [44]. Modern combat aircraft usually have $I_z > I_x$. A loss of static directional stability at high angles of attack makes such configurations highly susceptible to yaw departure. One way of improving resistance to yaw departure is to increase the dihedral effect. However, a high dihedral effect makes an aircraft very sensitive in roll to gusts and control inputs due to aileron or rudder.

The yaw departure to aileron and rudder control inputs can be predicted using the following approach called β-plus-δ axis stability indicator [43]:

1. Calculate $\alpha_{-\beta}$ and α_{δ} given by the following expressions:

$$\alpha_{-\beta} = \alpha - \tan^{-1}\left[\frac{C_{n\beta} I_x}{C_{l\beta} I_z}\right] \tag{8.7}$$

$$\alpha_{\delta} = \alpha - \tan^{-1}\left[\frac{C_{n\delta_a} I_x}{C_{l\delta_a} I_z}\right] \tag{8.8}$$

2. Plot the graphs $\alpha_{-\beta}$ vs α and α_{δ} vs α.

For stability at high angles of attack, $\alpha_{-\beta} > 0$ and $\alpha_{-\beta} > \alpha_{\delta}$. In other words, the curve $\alpha_{-\beta}$ must lie above the α axis and above the α_{δ} curve.

There is a direct relationship between β-plus-δ axis stability indicator and the $C_{n\beta \mathrm{DYN}}$ and LCDP criteria. When $\alpha_{-\beta} = 0$, we find that $C_{n\beta \mathrm{DYN}} = 0$. Furthermore, if $\alpha_{-\beta} = \alpha_{\delta}$, LCDP = 0.

8.12.3 Application of Departure Criteria

Let us first consider the application of $C_{n\beta \mathrm{DYN}}$ and LCDP criteria for predicting the aircraft departure/spin susceptibility at high angles of attack. This exercise consists of plotting the values of $C_{n\beta \mathrm{DYN}}$ and LCDP for selected altitude and Mach number on a typical chart as shown in Fig. 8.61. This chart

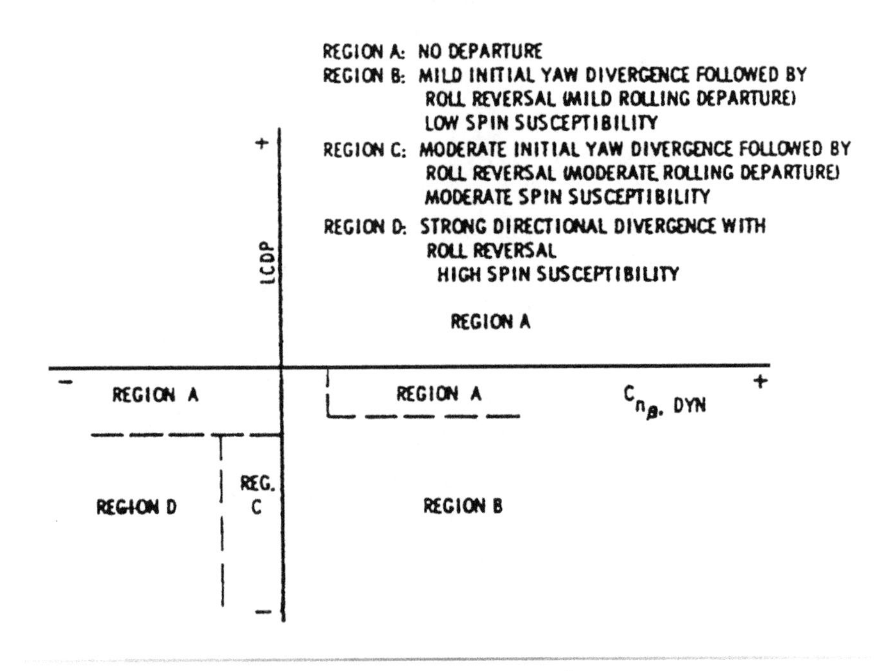

Fig. 8.61 Departure and spin susceptibility criteria (43).

has four regions, A, B, C, and D. In region A, the aircraft has good handling characteristics, and the departure is not likely to occur. In region B, the aircraft may experience a mild divergence in yaw followed by a roll reversal. In region C, the aircraft is likely to experience moderate initial yaw divergence, followed by a roll reversal, and this tendency is likely to become worse in region D, where a strong directional divergence coupled with roll reversal may occur, leading to a spin entry.

These definitions of various regions A–D are derived by the application of $C_{n\beta\mathrm{DYN}}$ and LCDP criteria to various aircraft [43]. Examples of this application for the F-8, the F-102, the F-106, and the SAAB 37 are shown in Fig. 8.62. The flight tests on the F-8 aircraft model indicate that beyond 26-deg angle of attack, the departure is likely to occur. Full-scale flight tests confirm this and further indicate that the aircraft experience a mild to moderate rolling departure. The motion immediately following departure (from controlled flight) is primarily a rolling motion. From Fig. 8.62, we note that the points for the F-8 aircraft at $\alpha = 25$ deg and $\alpha = 30$ deg fall in region B. Therefore, the observed departure of the F-8 aircraft is correctly predicted by the $C_{n\beta\mathrm{DYN}}$ and LCDP criteria.

Model tests on the F-102 indicate that the aircraft experiences a directional divergence above 28-deg angle of attack, and the severity of this divergence increases with further increase in the angle of attack. From Fig. 8.62, we observe that this behavior is correctly predicted because the point for $\alpha = 30$ deg falls in the overlapping area of regions C and D.

The full-scale F-106 aircraft indicates a departure followed by poststall gyrations for angles of attack between 34 and 38 deg. The correlation

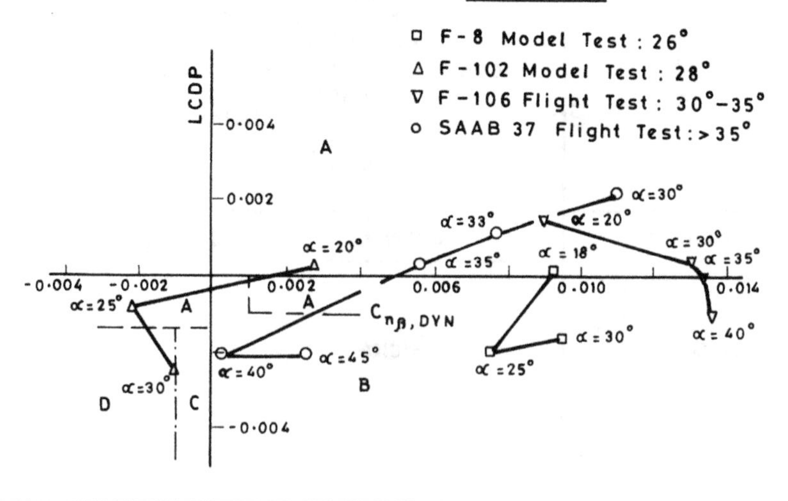

Fig. 8.62 Application of $C_{n\beta\mathrm{DYN}}$ and LCDP criteria (43).

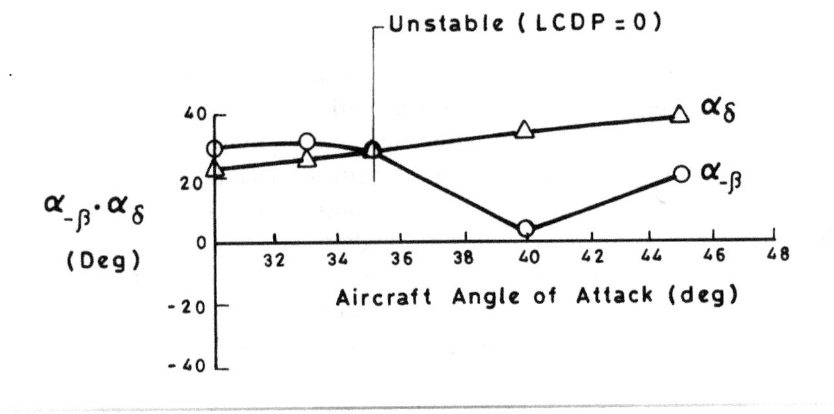

Fig. 8.63 β-plus-δ stability indicator, SAAB 37 airplane (43).

shown in Fig. 8.62 appears to be quite satisfactory as the points between 33- and 40-deg angle of attack fall in region *B*.

The SAAB 37 aircraft has good flying qualities up to $\alpha = 35$ deg. For higher angles of attack, the aircraft experiences a departure to lateral control inputs and exhibits spin susceptibility. Application of β-plus-δ axis stability indicator to the SAAB 37 is shown in Fig. 8.63. Up to $\alpha = 35$ deg, $\alpha_{-\beta} > 0$ and $\alpha_{-\beta} > \alpha_\delta$, indicating good flying qualities. At $\alpha = 35$ deg, $\alpha_{-\beta} = \alpha_\delta$, which implies that LCDP = 0. Beyond $\alpha = 35$ deg, $\alpha_{-\beta} < \alpha_\delta$ and LCDP becomes negative, indicating departure to lateral control inputs as experienced by the aircraft.

In summary, both $C_{n\beta\,\mathrm{DYN}}$–LCDP and β-plus-δ axes stability indicators correlate well with experimental data on model and full-scale flight tests. Therefore, these criteria can be used with some degree of confidence to predict the departure characteristics during preliminary design stages. Although both criteria are equivalent, the $C_{n\beta\,\mathrm{DYN}}$–LCDP criterion is better suited for practical application.

8.13 Control Concepts at High Angles of Attack

To achieve maneuverability at high angles of attack/poststall conditions, it is desirable that the fighter aircraft possess a high level of control authority about all three axes and more particularly in yaw and pitch throughout the maneuver envelope. To illustrate this point, let us consider the Herbst maneuver discussed in Chapter 2, which consists of three phases as follows.

1. Pitch-up to a high angle of attack.
2. Roll about the velocity vector.
3. Pitch-down back to the conventional flight regime.

The control requirements for each of these three phases are discussed in the following.

8.13.1 Pitch-Up

The first phase of a typical high angle of attack, poststall maneuver is a pitch-up to a high angle of attack orientation. To generate the required pitch-up as rapidly as possible, sufficient pitching moment capability must be available throughout the poststall maneuver envelope. The effectiveness of the conventional aerodynamic pitch control surfaces, as measured by the pitch acceleration, decreases with angle of attack and is far below the required level as schematically shown in Fig. 8.64 for the F-15 aircraft. This decrease in longitudinal control effectiveness occurs because of the immersion of control surfaces in the low-energy stalled flow at high angles of attack.

8.13.2 Velocity Vector Roll

Suppose an aircraft executes a pure rolling motion about the x-body axis as shown in Fig. 8.65a. From Eq. (8.2), with $\sigma = \alpha$, we have

$$\sin\beta = \sin\alpha \sin\phi \tag{8.9}$$

For small values of β and ϕ, we have

$$\beta = \phi \sin\alpha \tag{8.10}$$

so that

$$\dot{\beta} = p \sin\alpha \tag{8.11}$$

Thus, if the angle of attack is small, a pure rolling motion about the x-body axis is not likely to generate any significant sideslip excursions. However,

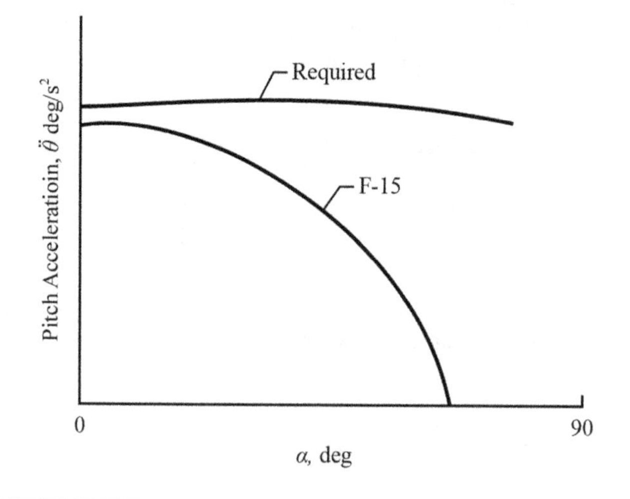

Fig. 8.64 Pitch control requirements at high angles of attack (46).

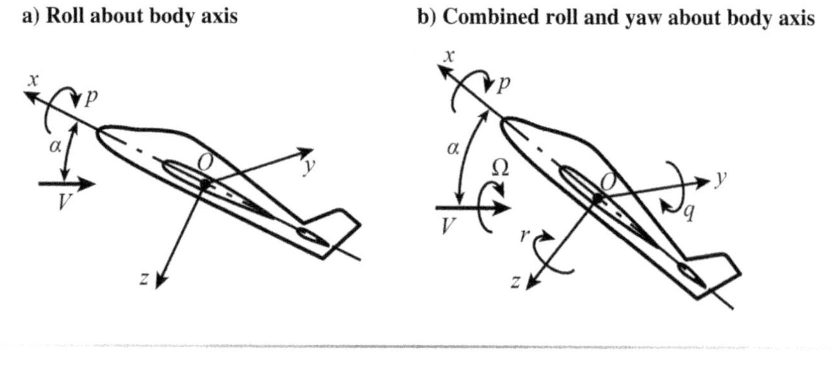

Fig. 8.65 Rolling motion at high angles of attack.

at high angles of attack, body axis roll may produce significantly large and undesirable sideslip excursions. To avoid this adverse sideslip buildup, the aircraft must roll about the velocity vector as shown in Fig. 8.65b. To understand the concept of velocity vector roll, consider an aircraft that simultaneously rolls and yaws about the body axes with rates p and r, respectively. For this case, the rate of sideslip buildup is given by

$$\dot{\beta} = p \sin \alpha - r \cos \alpha \tag{8.12}$$

To avoid sideslip buildup, we must have $\dot{\beta} = 0$. Then, Eq. (8.12) gives

$$r = p \tan \alpha \tag{8.13}$$

In other words, to produce a body axis roll rate of p while suppressing the buildup of adverse sideslip, the aircraft must simultaneously yaw about the z-body axis at a rate, $r = p \tan \alpha$. This coordinated rolling and yawing about the body axes is equivalent to the velocity vector roll at a rate Ω, which is given by

$$\Omega = p \cos \alpha + r \sin \alpha \tag{8.14}$$

$$= \frac{p}{\cos \alpha} \tag{8.15}$$

Quite often, Ω is also called the stability axis roll rate.

From Eq. (8.13), we observe that the proportion of body axes yaw rate in a velocity vector roll increases as the angle of attack increases. This places a severe requirement on yaw control during velocity vector rolls at high angles of attack at a time when the conventional aerodynamic controls like rudder become ineffective because of shielding by the forebody and wing wakes. With current fighter aircraft, the rudder capability degrades rapidly as the angle of attack approaches the stall angle due to the rudder becoming immersed in the low-energy stalled wake shed from the wing and fuselage.

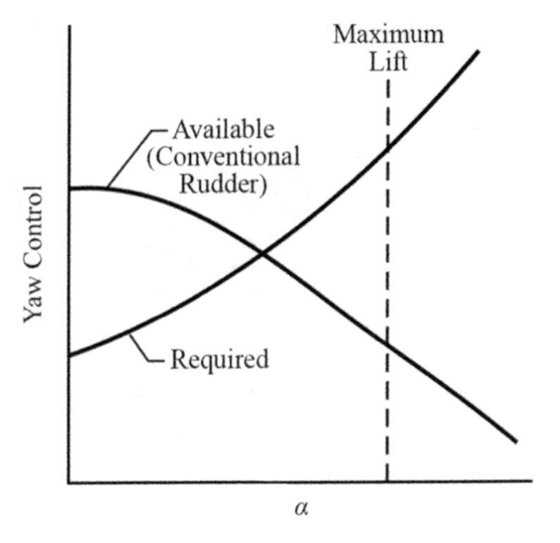

Fig. 8.66 Required and available yaw control at high angles of attack (47).

A typical example of the level of available and required yaw control is shown in Fig. 8.66, which depicts that the required yaw control far exceeds the available amount even below the stall angle. This deficiency in yaw control inherently limits the high-alpha roll rate capability, hence the maneuver effectiveness of the aircraft.

8.13.3 Pitch-Down

A combined rolling and yawing motion gives rise to an inertia-induced pitching moment and an equivalent pitching acceleration [see Eq. (7.5) with $I_{xz} = 0$], which is given by

$$\dot{q} = \frac{pr(I_z - I_x)}{I_y} \tag{8.16}$$

For velocity vector roll, $r = p \tan \alpha$ so that

$$\dot{q} = \frac{p^2 \tan \alpha (I_z - I_x)}{I_y} \tag{8.17}$$

For a typical combat aircraft of current generation, $I_z > I_x$ so that the inertia-induced pitch acceleration or the inertia-induced pitching moment is positive (noseup). If sufficient nosedown control authority is not available to counter this inertia-induced noseup pitching moment, the aircraft may experience departure from controlled flight (divergence in pitch as discussed in Chapter 7) and get into a deep stall trim point while executing a velocity vector roll [46].

The deep stall [48] is one of the most significant flight dynamic problems associated with nonlinear variation of pitching-moment coefficient with angle of attack as illustrated in Fig. 8.67. We have two stable trim points: one is the conventional trim point at low angles of attack below the stall, and the other is a poststall trim point, which is called the deep-stall trim point. This kind of variation of the pitching moment with angle of attack is caused by various factors [48] such as 1) progression of the stall over the wing, starting at the wingtips; 2) inboard movement of the wing and LEX vortices; 3) bursting of the forebody, LEX, and main wing vortices; and 4) immersion of the tail surfaces in the wake of the forebody, LEX, and the main wing surfaces. Recovery from deep stall can be quite difficult or even impossible because control surfaces may be totally ineffective because of immersion in the stalled wake.

The F-16 aircraft has a deep-stall trim point around $\alpha = 50$ deg. In view of the limited nosedown pitch authority, the maximum roll capability of the F-16 is limited, and an alpha limiter is also used to prevent the aircraft from getting locked up in a deep-stall trim point during velocity vector rolls [46].

Therefore, for high angles of attack/poststall maneuverability, sufficient control authority about all three axes must be available. Usually, sufficient roll control authority is always available. The limitations usually arise in yaw and pitch control authority as we have mentioned earlier. These limitations plus any susceptibility of the aircraft to pitch divergence on account of inertial coupling may force the use of an alpha limiter. However, in this process the combat effectiveness of the aircraft is reduced.

To overcome the discussed deficiencies in yaw and pitch control at high angles of attack, combat aircraft designers have explored a variety of control concepts. Some of these concepts are schematically shown in Fig. 8.68. However, the two concepts that have received considerable attention are forebody vortex control and propulsive or thrust vectoring control (TVC).

8.13.4 Forebody Vortex Control

The purpose of forebody vortex control is to systematically manipulate the forebody vortices to generate controlled yawing moments at high

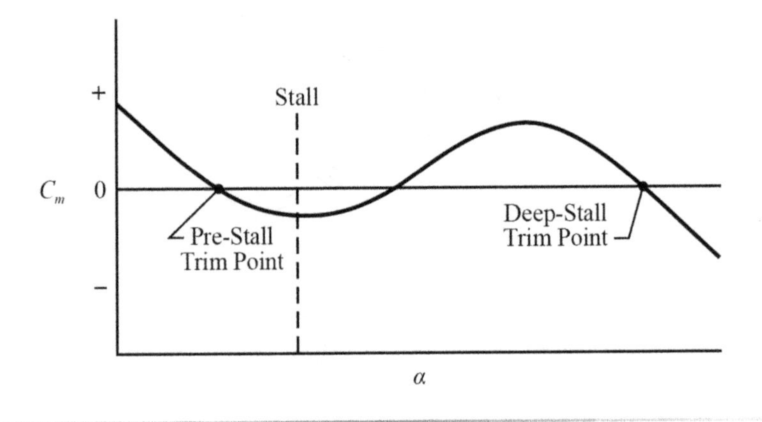

Fig. 8.67 Deep stall.

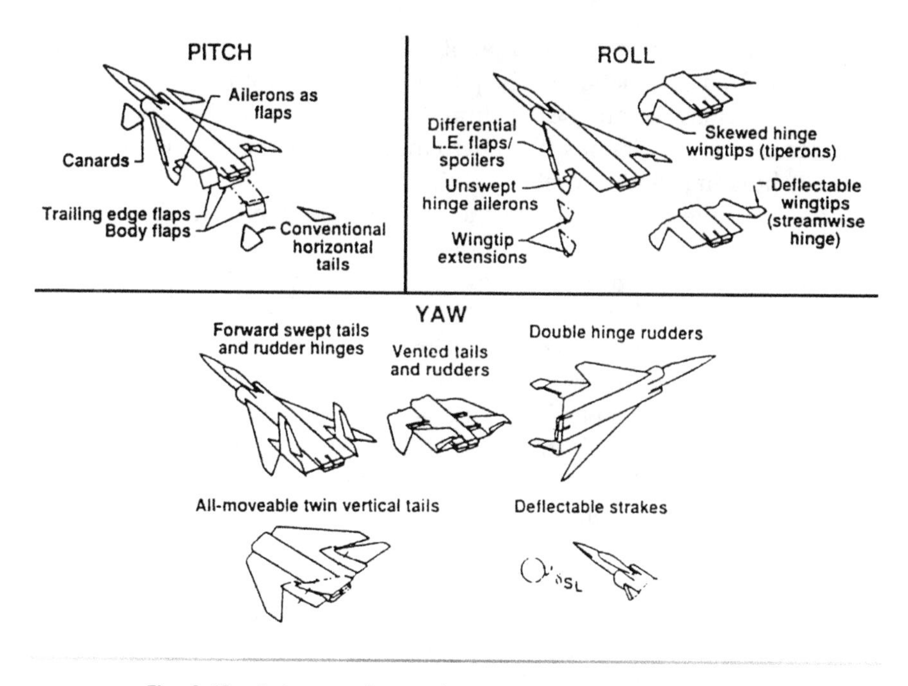

Fig. 8.68 Advanced aerodynamic control concepts (2).

angles of attack when the conventional rudder loses its effectiveness. The forebody vortex control also offers the potential to improve the directional stability at high angles of attack, augment roll–yaw damping, and prevent departure/spin entry.

The important forebody vortex control concepts that have received attention are actuated forebody strakes and forebody blowing and suction. These concepts are schematically illustrated in Fig. 8.69.

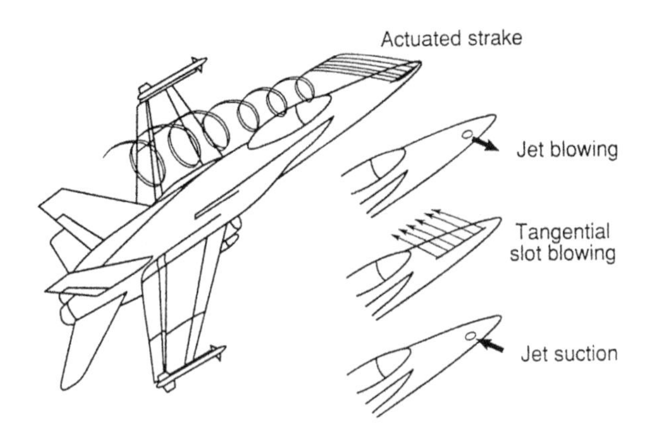

Fig. 8.69 Forebody vortex control concepts (2).

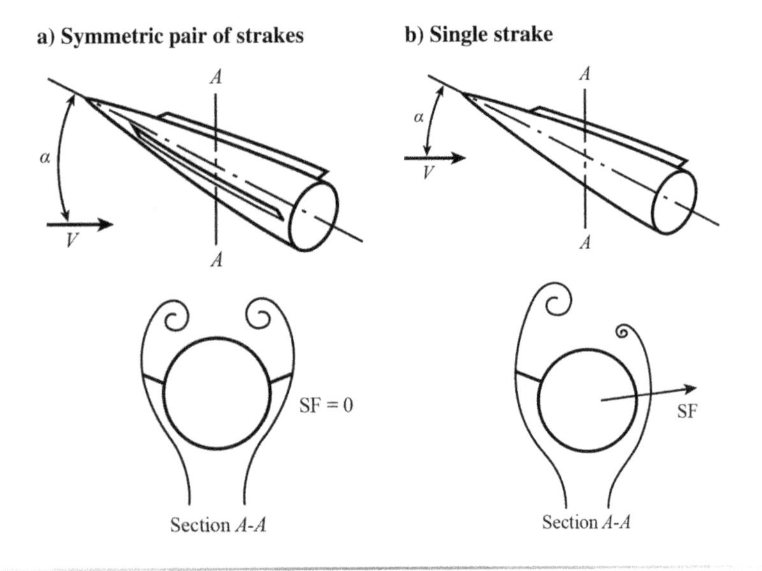

Fig. 8.70 Concepts of forebody strakes.

Actuated Forebody Strakes The use of a fixed symmetric pair of forebody strakes produces a pair of symmetric vortices suppressing the naturally occurring vortex asymmetry at high angles of attack and eliminates the associated side force as shown in Fig. 8.70a. On the other hand, the deployment of a single strake can systematically manipulate the forebody vortex system to produce a side force with large moment arm in a controlled manner for generating the much needed yaw control at high angles of attack as shown in Fig. 8.70b.

The magnitude of the side force depends on the axial and circumferential location of the strakes and height of the strakes. For example, a strake located at a given axial location is found to produce the highest levels of yaw control if located circumferentially at $\phi = 120$ deg as shown in Fig. 8.71. It may be noted that the value of ϕ for maximum strake effectiveness also depends on the forebody cross-sectional shape, fineness ratio, angle of attack, and sideslip. As shown in Fig. 8.71, the conventional rudder begins to lose its effectiveness (as measured by ΔC_n) around $\alpha = 20$ deg and, at just about this angle of attack, the forebody strake starts to develop side force/yawing moment. The yawing moment coefficient due to the forebody strake around $\alpha = 50$ deg is more than twice the maximum yawing moment produced by the rudder at its peak effectiveness. The strake effectiveness is higher if located (axially) closer to the apex because it is here that the forebody vortex system originates.

The strake height has a prominent effect on the direction rather than the magnitude of the side force. If the strake height is below a certain critical value, the strake acts like a boundary-layer tripping device and causes a

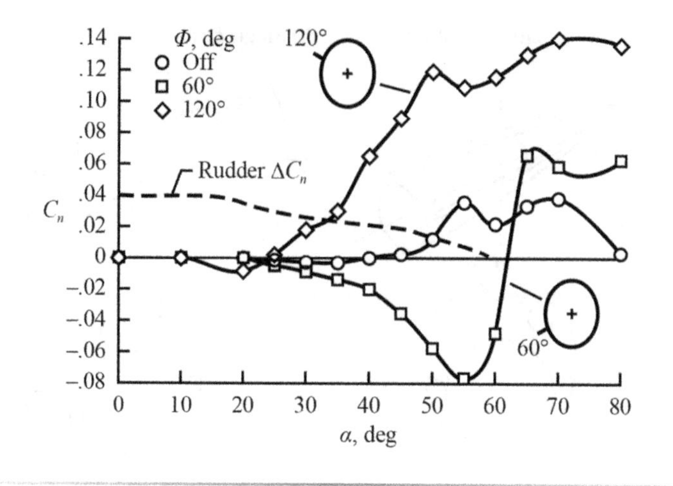

Fig. 8.71 Effects of radial location of forebody strakes (47).

transition from laminar to turbulent flow in the boundary layer. The turbulent boundary layer adheres to the body surface to a much greater extent and delays the flow separation. As a result, the vortex on the side of the strake lies closer to the body. On the other side where there is no strake, the flow pattern essentially remains the same as the basic case with the vortex positioned further away from the body surface. As a result, the suction is relatively higher on the side of the strake, and a side force is produced in the direction of the deployed strake (Fig. 8.72a). It is to be noted that this type of flow pattern is likely to be sensitive to flow Reynolds numbers. On the other hand, if the strake height is above the critical value, it promotes flow separation on its side and produces a forebody side force in the direction opposite

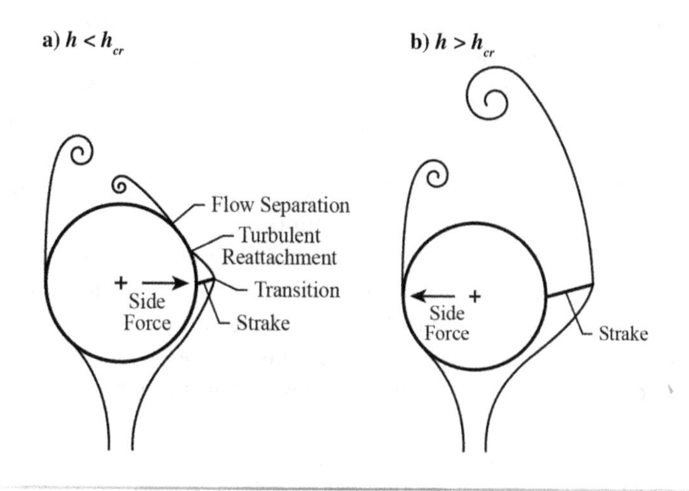

Fig. 8.72 Effect of strake height on the forebody flow pattern (49).

to the deployed strake as shown in Fig. 8.72b. Because the flow separation point is almost fixed at the strake, the direction of the side force is always fixed and not affected by the flow Reynolds number. Therefore, to ensure a fixed direction of the side force, it is desirable to use a sufficiently large strake height so that the strake always acts as a flow separator on its side and produces a side force in the (fixed) opposite direction. For the example considered in Fig. 8.72, the critical strake height is around 5 percent of the forebody diameter [49].

The forebody strakes can either be fixed or be of actuated type as shown in Fig. 8.73. The actuated forebody strakes have the advantage that they can be retracted and remain conformal with the forebody shape at low angles of attack when not required. With this concept, strakes on either side of the forebody would deflect individually from their conformal positions depending on the desired direction of yaw control. An important advantage of the forebody strakes is that they produce very little rolling or pitching moment when deployed as yaw control devices. These favorable characteristics make the forebody strakes a promising yaw control concept for poststall angles of attack. The conformal forebody strakes have been flight tested on F-18 aircraft [49].

The basic fluid flow mechanism responsible for generating such large side forces and yawing moments is schematically shown in Fig. 8.74. The basic forebody operating at high angles of attack has a pair of small counter-rotating vortices. However, the forebody with a single strake of sufficiently large height (exceeding the critical height) has a much larger and stronger strake vortex, which is displaced far above the strake. This type of fluid flow pattern gives rise to the pressure distribution shown in Fig. 8.74 for various circumferential locations (ϕ) of the left strake. On the left half, even though the vortex is stronger, the pressure is higher compared to the basic case owing to earlier flow separation induced by the strake. On the

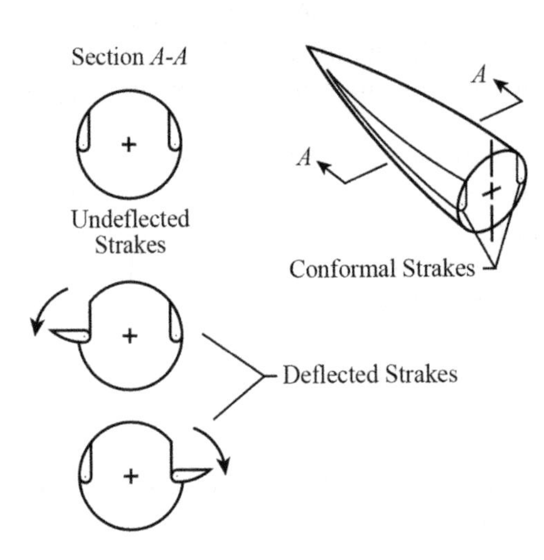

Fig. 8.73 Actuated forebody strakes (47).

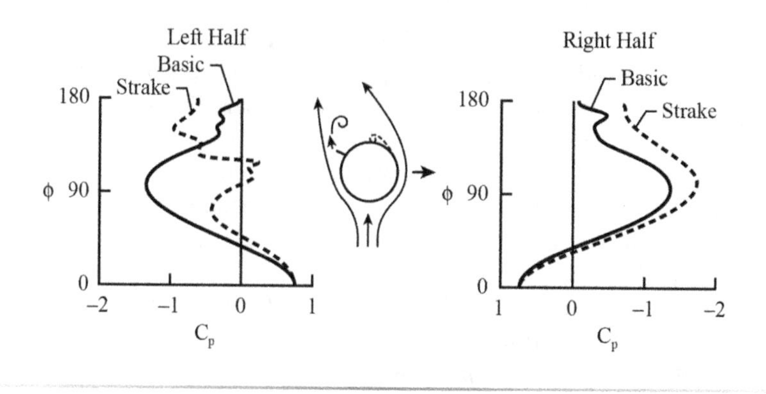

Fig. 8.74 Fluid flow mechanism associated with forebody strakes (47).

right half (no strake), flow pattern and the pressure distribution are essentially similar to the basic case. Therefore, the net effect of the strake is that the suction on the side of the strake is reduced, and the forebody develops a side force to the right as shown. Here, the forebody strake acts like a spoiler. The magnitude of this side force and the associated yawing moment depends on the pressure differentials, which in turn depend on the axial and circumferential location of the strake.

One possible limitation to the deployment of the forebody strakes on aircraft is that they may cause an adverse interference on the operation of the radar, which is usually housed inside the forebody. In this regard, the pneumatic methods of forebody blowing and suction offer better alternatives as discussed next.

Forebody Blowing The jet blowing is perhaps one of the earliest concepts that was explored for forebody vortex control [49]. The blowing is usually accomplished using single or multiple nozzles or through slots located at various places on the forebody. The direction of jet blowing is either forward (against the airflow) or aft (along the flow) or slot (tangential to the body surface) as illustrated in Fig. 8.75. The optimal blowing configuration for a given aircraft usually depends on the forebody geometry, angle of attack, and sideslip.

An example of the effect of forebody jet blowing for the F-16 aircraft is shown in Fig. 8.76a. The nozzles on either side are fixed at fuselage station (FS) 5 and blow the jets aft along the forebody surface. The direction of the side force is opposite to the blowing side, and the magnitude of the side force increases with the blowing parameter C_μ, which is defined as

$$C_\mu = \frac{\dot{m}_j v_j}{q_\infty S} \tag{8.18}$$

where $\dot{m}_j$ is the jet mass flow rate, v_j is the jet velocity, q_∞ is the freestream dynamic pressure, and S is the reference area.

The physical flow mechanism associated with the tangential aft blowing concept is schematically shown in Fig. 8.76b. The basic principle behind the jet blowing concept is the jet entrainment effect. The jet induces an inward velocity field in its proximity and continuously entrains the surrounding fluid as it blows aft (downstream). As a result, the vortex on the blowing side moves closer to the body surface. Consequently, the other vortex on the nonblowing side gets pushed away from the body surface, giving rise to a side force in the direction of blowing.

An example of slot blowing for the F-16 aircraft is presented in Fig. 8.77. The blowing on the right causes a yawing moment to the right and vice versa. The slot extends from FS 5 to FS 39 [50]. The slot blowing operates on the principle of circulation control. As shown in Fig. 8.77b, the blowing energizes the boundary-layer flow near the surface so that the flow separation is delayed. On the nonblowing side, the flow pattern remains more or less the same as in the basic case. As a result, the vortex on the blowing side is located closer to the body. On the other side, the vortex

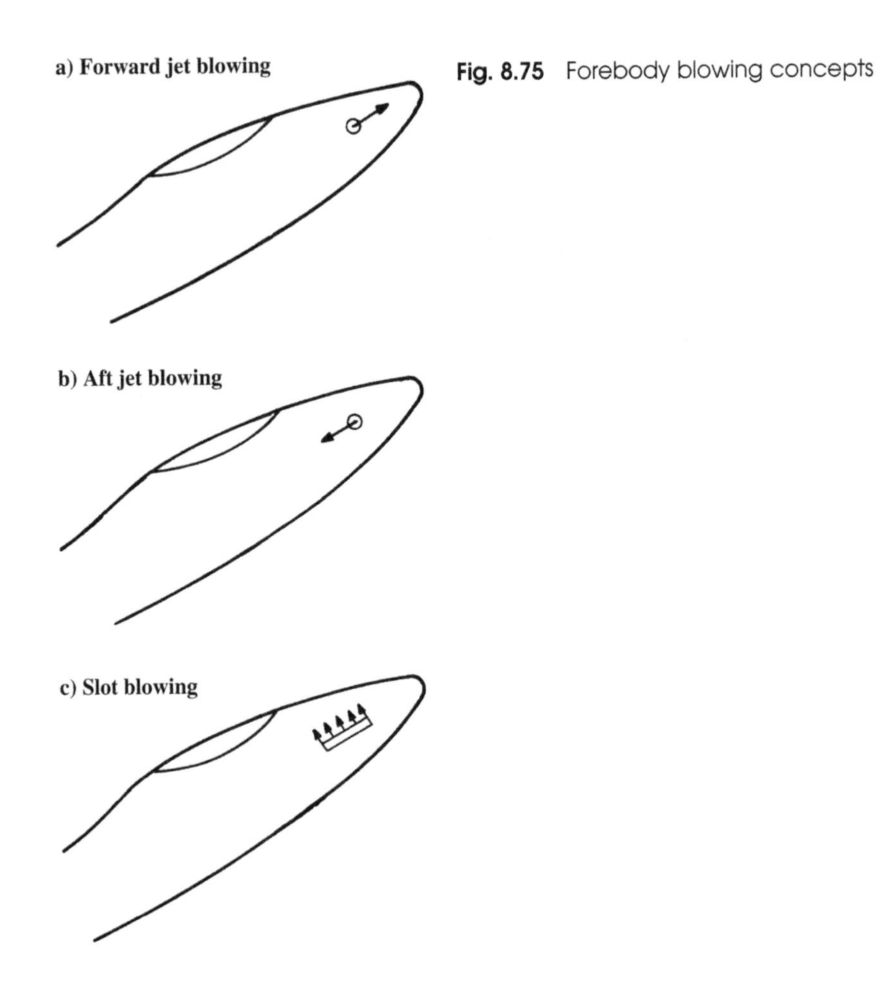

Fig. 8.75 Forebody blowing concepts

a) C_n vs α, $\beta = 0$

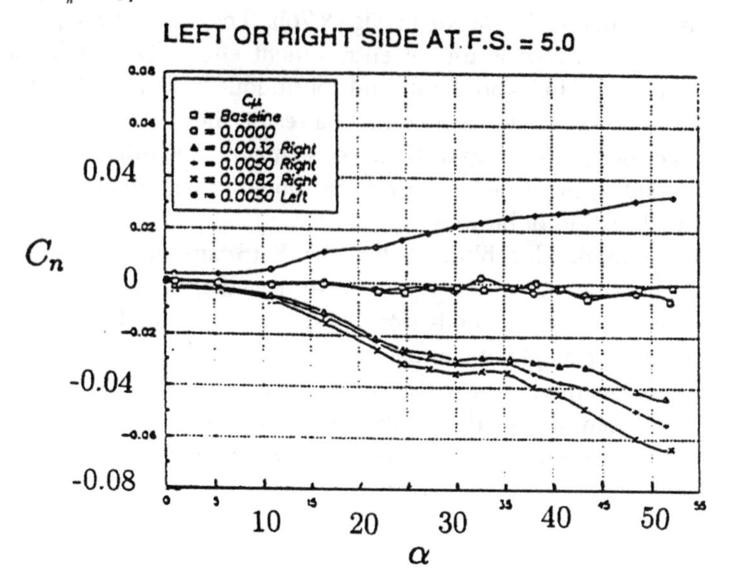

b) Flow near the forebody apex

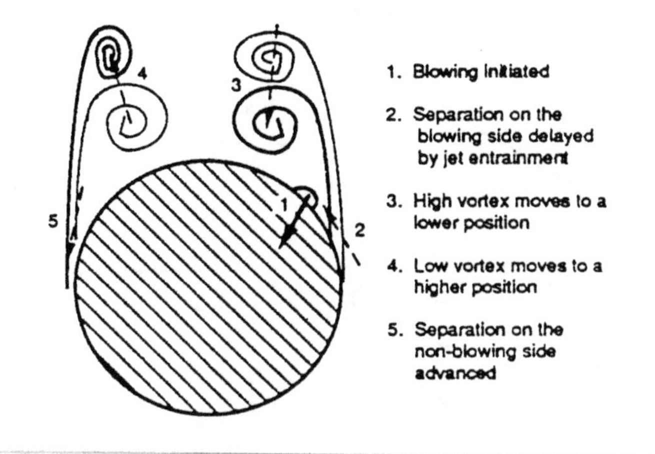

Fig. 8.76 Nozzle jet blowing for forebody vortex control on the F-16 aircraft (49, 50).

gets pushed further away, giving rise to a side force directed towards the blowing side.

Forebody Suction An example of the application of forebody suction for vortex control is shown in Fig. 8.78a. The vehicle configuration to which the forebody suction was applied is the High Incidence Research Model (HIRM) of the Royal Aircraft Establishment (RAE), U.K. The suction ports are located near the apex of the nose.

The nonzero side force when the suction is off is due to the naturally occurring vortex asymmetry. The effect of suction on the left is to produce a negative yawing moment (towards the suction side) and vice versa.

The forebody suction, like forebody blowing, works on the principle of circulation control, wherein the boundary-layer flow is pulled towards the surface to keep it attached to a greater extent and delay the flow separation. As a result, a side force develops towards the suction side.

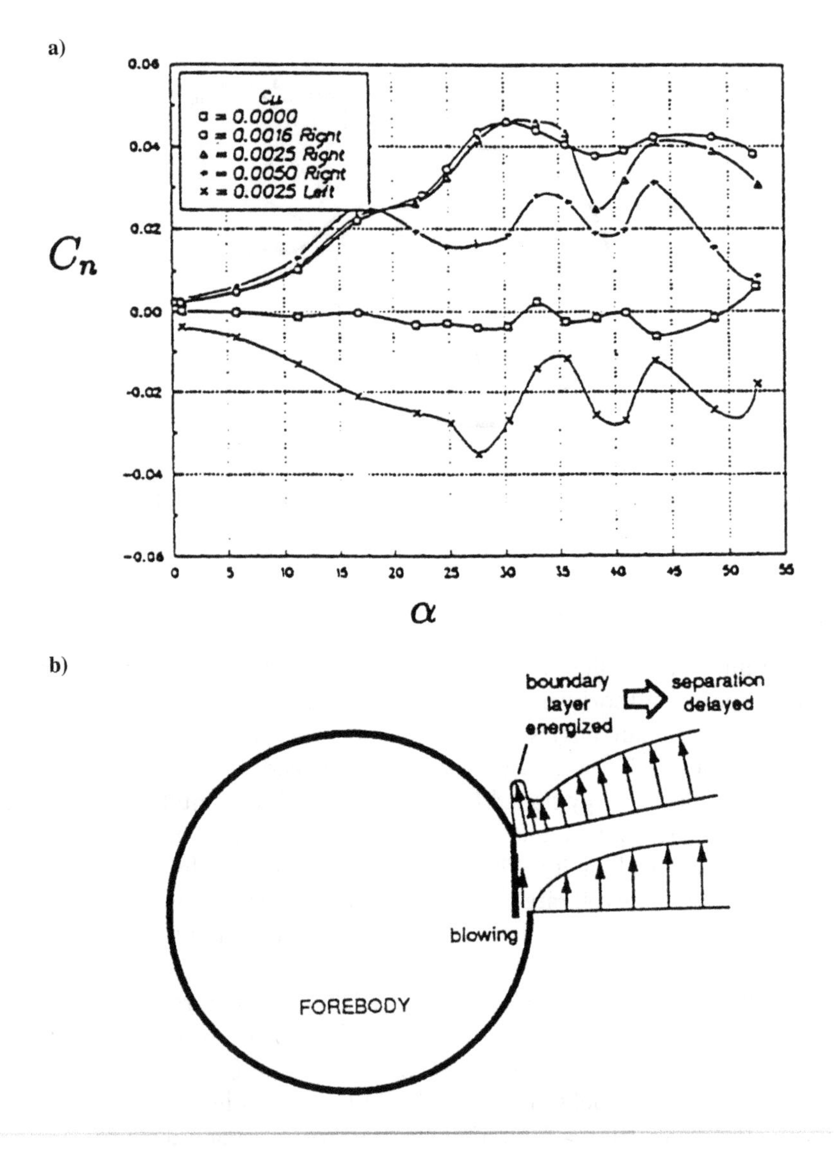

Fig. 8.77 Slot blowing for forebody vortex control on the F-16 aircraft (49, 50).

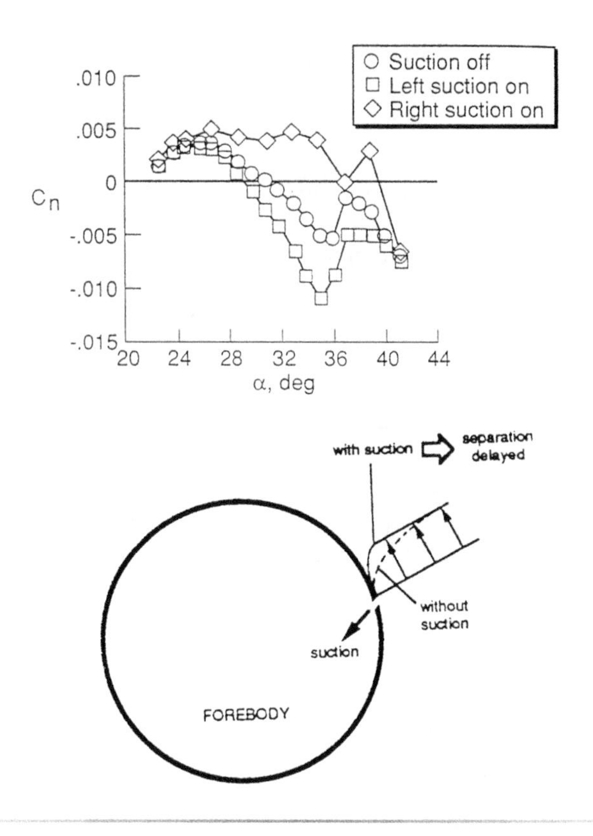

Fig. 8.78 Forebody vortex control using suction for RAE HIRM1 aircraft (51).

8.13.5 Thrust Vector Control (TVC)

The interest in using TVC has arisen mainly from the requirement to increase control authority in the low-speed, low dynamic pressure conditions experienced at high angles of attack when the traditional aerodynamic control surfaces will have lost most of their effectiveness because of their immersion in low-energy stalled flow. The application of multiaxis TVC provides increasing control authority at high angles of attack, leading to improved low-speed agility. The TVC can eliminate the limitations on the maximum attainable roll rate capability imposed because of insufficient aerodynamic control authority at high angles of attack. Some typical multiaxis TVC concepts are shown in Fig. 8.79.

The primary benefit of TVC is the availability of control power at high angles of attack, low dynamic pressure conditions where the conventional aerodynamic control will have lost most of its effectiveness. However, another significant benefit offered by the TVC is that it leads to smaller aerodynamic control surfaces. Traditionally, the size of an aerodynamic control surface is determined from control effectiveness at low-speed considerations.

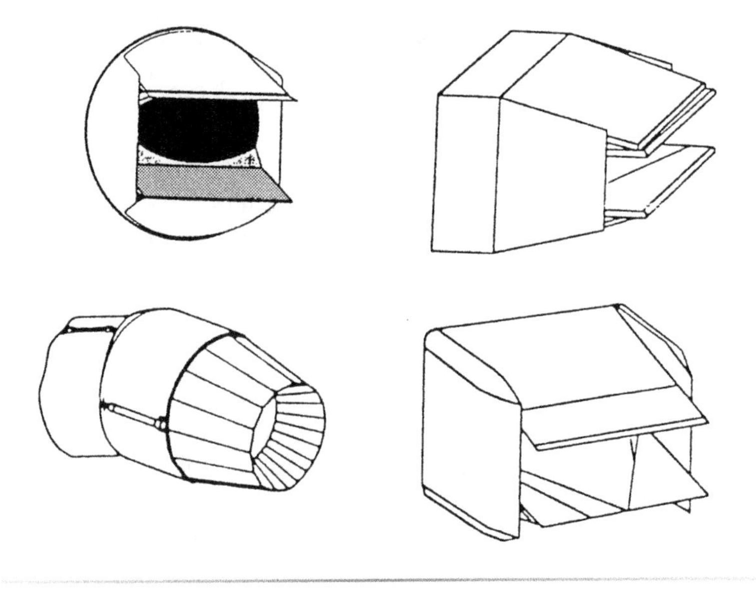

Fig. 8.79 Multiaxis thrust vectoring concepts (4).

The fact that the TVC can provide the necessary control authority at low-speed or low dynamic pressure conditions makes it unnecessary to size the aerodynamic surfaces from low-speed considerations. However, at high-speed or high dynamic pressure conditions, the aerodynamic control surfaces become effective, and much smaller surfaces can be used compared to those determined from low-speed considerations. The use of small-size aerodynamic control surfaces leads to lower structural weight, reduced trim drag, and improved overall performance.

An ideal scenario would be one in which the aerodynamic controls and the TVC are smoothly blended as shown in Fig. 8.80, the TVC providing most of the control power at high-angle-of-attack/low dynamic pressure

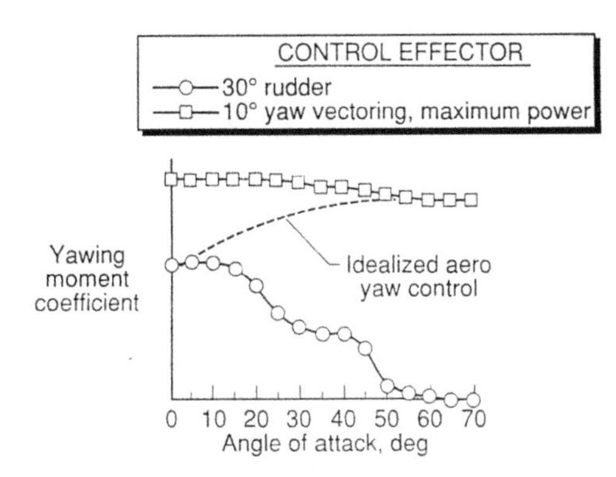

Fig. 8.80 Example of smooth blending of aerodynamic and thrust vector controls (2).

conditions and the aerodynamic surfaces taking over at low-angle-of-attack/high dynamic pressure conditions. As shown in Fig. 8.80, 10 deg of thrust vectoring in the yaw plane on a typical modern fighter aircraft matches or exceeds the effectiveness of the ideal control at low-speed conditions [29]. In view of the enormous potential of TVC concept, extensive research was conducted by NASA using the F/A-18 aircraft and by NASA/Boeing using the tailless X-36 research airplane as test beds for evaluating the benefits of TVC technologies for future combat aircraft designs.

8.14 Summary

Interest in unlimited high-angle-of-attack/poststall maneuver capability has led to extensive research and technology development in high-angle-of-attack aerodynamics, stability, and control. Modern combat aircraft feature thin, highly swept delta wings and long, slender fuselages. The airflow over such configurations at high angles of attack is quite complex and is characterized by extensive separated flow regions dominated by the forebody and LEX/wing vortices and their mutual interactions. The existence of such complex flow fields leads to aerodynamic nonlinearities, loss of lateral-directional stability and control, roll reversal, departure from controlled flight, and unsteady motions like wing rock. The control authority of conventional aerodynamic surfaces decreases very rapidly as the angle of attack increases beyond the stalling angle. A successful design and operation of the modern combat aircraft require good understanding of the aerodynamic stability and control problems at high angles of attack. In this chapter, we have attempted to review and highlight some of the main aspects of these problems. It is not possible to review every aspect of the experimental and computational studies of vortex flows over slender bodies and delta wings at high angles of attack. We have addressed only those aerodynamic aspects that are relevant to the stability and control problems of aircraft at high angles of attack.

References

[1] Chambers, J. R., "High-Angle-of-Attack Technology, Progress and Challenges," NASA CP-3149, Part I, Vol. I, 1990, pp. 13–22.
[2] Nguyen, L. T., Arbuckle, P. D., and Gera, J., "Progress in Controls Technology for High-Angle-of-Attack Applications," NASA CP-3149, Part I, Vol. I, pp. 117–156.
[3] Skow, A. M., and Titiriga, A., Jr., "A Survey of Analytical and Experimental Techniques to Predict Aircraft Dynamic Characteristics at High-Angle-of-Attack," CP-235, AGARD, May 1978.
[4] Werle, H., "Flow Visualization Technique for the Study of High Incidence Aerodynamics," LS-121, AGARD, Paper 3, 1982.
[5] Johnson, T. D., Jr., and Huffman, J. K., "Experimental Study of Vortex Flaps on a Delta Wing Sweep Series at High Angles of Attack," NASA CP-2416, Vol. II, 1985, pp. 169–184.
[6] Campbell, J. F., and Osborn, R. F., "Leading-Edge Vortex Research, Some Nonplanar Concepts and Current Challenges," NASA CP-2416, 1992, pp. 31–64.

[7] Hunt, B. L., "Asymmetric Vortex Forces and Wakes on Slender Bodies," AIAA Paper 82-1336, Aug. 1982.
[8] Payne, F. M., and Nelson, R. C., "An Experimental Study of Vortex Breakdown on a Delta Wing," NASA CP-2416, 1992, pp. 135–161.
[9] Skow, A. M., Titiriga, A., Jr., and Moore, W. A., "Forebody–Wing Vortex Interactions and their Influence on Departure and Spin Resistance," CP-247, AGARD, Paper No. 6, 1979.
[10] Wentz, W. H., and Kohlman, D. L., "Vortex Breakdown on Slender Sharp-Edged Wings," *Journal of Aircraft*, Vol. 3, March 1971.
[11] Roos, F. W., and Kegelman, J. T., "Leading Edge Vortex Flow Fields," NASA CP-3149, Part I, 1990, pp. 157–172.
[12] Skow, A. M., and Erickson, G. E., "Modern Fighter Design for High-Angle-of-Attack Maneuvering," LS-121, AGARD, Paper 4, 1982.
[13] Thompson, D. H., "A Visualisation Study of the Vortex Flow Around Double-Delta Wings," Australian Dept. of Defence, Aerodynamics Rept. 165, Melbourne, Australia, Aug. 1985.
[14] Hall, R. M., Erickson, G. E., Banks, D. W., and Fisher, D. F., "Advances in High-Alpha Experimental Aerodynamics: Ground Test and Flight," NASA CP-3149, Part I, Vol. I, 1990, pp. 69–87.
[15] Rao, D. M., "Pneumatic Concept for Tip-Stall Control of Cranked Arrow Wings," *Journal of Aircraft*, Vol. 31, No. 6, Nov.–Dec. 1994, pp. 1380–1386.
[16] Behrbohm, H., "Basic Low Speed Aerodynamics of the Short Coupled Canard Configuration of Small Aspect Ratio," Svenska Aeroplan Aktiebolag, SAAB TN 60, Linkoping, Sweden, July 1965.
[17] Howard, R. M., and Kersh, J. M., Jr., "Effect of Canard Deflection on Enhanced Lift for Close-Coupled Canard Conflguration," AIAA Paper 91-3222, Sept. 1991.
[18] Lamont, P. J., "Pressure Measurements on an Ogive-Cylinder at High Angles of Attack with Laminar, Transitional and Turbulent Separation," AIAA Paper 80-1556, Jan. 1980.
[19] Mendenhall, M. R., and Nielsen, J. N., "Effect of Symmetrical Vortex Shedding On the Longitudinal Aerodynamic Characteristics of Wing-Body-Tail Combinations," NASA CR-2473, 1975.
[20] Headley, J. W., "Analysis of Wind Tunnel Data Pertaining to High-Angle-of-Attack Aerodynamics," Vol. I—Technical Discussion and Analysis of Results, Air Force Flight Dynamics Laboratory, AFFDL-TR-78-94, Wright-Patterson Air Force Base, OH, July 1978.
[21] Levin, D., and Katz, J., "Self Induced Roll Oscillations of Low Aspect Ratio Rectangular Wings," AIAA Paper 90-2811, 1990.
[22] Chambers, J. R., "High-Angle-of-Attack Aerodynamics: Lessons Learned," AIAA Paper 86-1774, 1986.
[23] Orlick-Rueckmann, W. J., "Aerodynamic Aspects of Aircraft Dynamics at High Angles of Attack," *Journal of Aircraft*, Vol. 20, No. 9, 1983, pp. 737–752.
[24] Nguyen, L. T., Yip, L. P., and Chambers, J. R., "Self Induced Wing Rock of Slender Delta Wings," AIAA Paper 81-1883, Aug. 1981.
[25] Ross, A. J., "Investigation of Nonlinear Motion Experienced on a Slender-Wing Research Aircraft," *Journal of Aircraft*, Vol. 9, No. 9, 1972, pp. 625–631.
[26] Schmidt, L. V., "Wing Rock due to Aerodynamic Hysteresis," *Journal of Aircraft*, Vol. 16, 1979, pp. 129–133.
[27] Levin, D., and Katz, J., "Dynamic Load Measurements with Delta Wings Undergoing Self Induced Roll Oscillations," *Journal of Aircraft*, Vol. 21, No. 1, 1984, pp. 30–36.
[28] Ericson, L. E., "The Fluid Mechanics of Slender Wing Rock," *Journal of Aircraft*, Vol. 21, No. 5, 1984, pp. 322–328.
[29] Nguyen, L. T., Whipple, R. D., and Brandon, J. M., "Recent Experiences of Unsteady Aerodynamic Effects on Aircraft Flight Dynamics at High-Angles-of-Attack," Paper 28, CP 386, AGARD, 1985.

[30] Jun, Y. W., and Nelson, R. C., "Leading-Edge Vortex Dynamics on a Slender Oscillating Wing," *Journal of Aircraft*, Vol. 25, No. 9, 1988, pp. 815–819.
[31] Arena, A. S., and Nelson, R. C., "Experimental Investigations on Limit Cycle Wing Rock of Slender Delta Wings," *Journal of Aircraft*, Vol. 31, No. 5, 1994, pp. 1148–1155.
[32] Pamadi, B. N., Rao, D. M., and Niranjana, T., "Roll Attractor of Delta Wings at High Angles of Attack," ICAS Paper 94-7.3.2, Sept. 1994.
[33] Hanff, E. S., and Ericson, L. E., "Multiple Roll Attractors of a Delta Wing at High Angles of Attack," Paper 31, CP-494, AGARD, 1990.
[34] Pamadi, B. N., Rao, D. M., and Niranjana, T., "Wing Rock and Roll Attractor of Delta Wings at High Angles of Attack," AIAA Paper 94-0807, Jan. 1994.
[35] Swanson, W. M., "Magnus Effect: A Summary of Investigation to Date," *ASME Journal of Basic Engineering*, 1961, pp. 461–470.
[36] Ericson, L. E., "Moving Wall Effect in Unsteady Flow," *Journal of Aircraft*, Vol. 25, No. 11, 1988, pp. 977–990.
[37] Ericson, L. E., "Wing Rock Generated by Forebody Vortices," AIAA Paper 87-0268, 1987.
[38] Wong, G. S., Rock, S. M., Wood, N. J., and Roberts, L., "Active Control of Wing Rock Using Tangential Leading-Edge Blowing," *Journal of Aircraft*, Vol. 31, No. 3, 1994, pp. 659–665.
[39] Celik, Z. Z., Pedrico, N., and Roberts, L., "The Control of Wing Rock by Forebody Blowing," AIAA Paper 93-3685, Aug. 1993.
[40] Ng, T. T., Suarez, C. J., Kramer, B. R., Ong, L. Y., Ayers, B., and Malcolm, G. N., "Forebody Vortex Control for Wing Rock Supression," *Journal of Aircraft*, Vol. 31, No. 2, 1994, pp. 298–305.
[41] Murri, D. G., Nguyen, L. T., and Grafton, S. B., "Wind-Tunnel Free-Flight Investigation of a Model of a Forward-Swept-Wing Fighter Configuration," NASA TP 2230, Feb. 1984.
[42] Gilbert, W. P., Nguyen, L. T., and Gera, J., "Control Research in the NASA High-Alpha Technology Program," *AGARD Fluid Dynamics Panel Symposium on Aerodynamics of Combat Aircraft Control and Ground Effects*, Madrid, Spain, Oct. 2–5, 1989.
[43] Weissman, R., "Status of Design Criteria for Predicting Departure Characteristics and Spin Susceptibility," *Journal of Aircraft*, Vol. 12, No. 12, 1975, pp. 989–993.
[44] Chambers, J. R., and Anglin, E. L., "Analysis of Lateral-Directional Stability Characteristics of a Twin-Jet Fighter Airplane at High Angles of Attack," NASA TN D-5361, 1969.
[45] Moul, M. T., and Paulson, J. W., "Dynamic Lateral Behavior of High Performance Aircraft," NASA RM L5E16, 1958.
[46] Bitten, R., "Operational Benefits of Thrust Vectoring (TVC)," NASA CP-3149, Part 2, Vol. I, 1990, pp. 587–601.
[47] Murri, D. G., Bierdron, R. T., Erickson, G. E., and Jordan, F. L., Jr., "Development of Actuated Forebody Strake Controls for the F-18 High Alpha Research Vehicle," NASA CP-3149, Part I, Vol. 1, 1990, pp. 335–380.
[48] Chambers, J. R., and Grafton, S. B., "Aerodynamic Characteristics of Aiplanes at High Angles of Attack," NASA TM 74097, 1977.
[49] Malcolm, G. N., "Forebody Vortex Control—A Progress Review," AIAA Paper 93-3540-CP, 1993.
[50] Ng, T. T., and Malcolm, G. N., "Aerodynamic Control Using Forebody Vortex Control," NASA CP-3149, Part I, Vol. I, 1990, pp. 507–532.
[51] White, R. E., "Effect of Suction on High Angle-of-Attack Directional Control Characteristics of Isolated Forebodies," NASA CP-3149, Part I, Vol. 1, 1990, pp. 533–556.

Chapter 9 Stability, Control Issues, and Challenges of UAVs

9.1 Introduction

Unmanned air vehicles (UAVs) historically have been used for military intelligence, surveillance, reconnaissance, and attack as well as for a variety of civil applications such as traffic surveillance, area mapping, and forest fire detection. UAVs are also used for aircraft development and testing because they offer significant cost and safety benefits compared to the development and testing of full-scale aircraft. Unmanned vehicles that are small in size and weight are known as micro air vehicles (MAVs). MAVs are typically 30 cm or less with gross weights ranging from few grams to kilograms [1]. These operate at very low altitudes and speeds often in an urban or even indoor environment. Various types of unmanned air vehicles, their roles, and their capabilities are discussed in [1–3].

UAVs provide a clear and distinct advantage over piloted aircraft in certain combat operations because the usual restrictions necessary for piloted aircraft do not apply and there is no pilot to put in harm's way [4, 5]. Such vehicles used for aerial combat are called unmanned combat aerial vehicles (UCAVs). UCAVs can be highly maneuverable because they are not subject to the typical structural/operational constraints of a piloted combat aircraft.

Examples of UAVs are presented in Figs. 9.1 and 9.2. The Pioneer (Fig. 9.1) has been operated by the U.S. Navy since 1985 [1]. It operates from sea level to 4.56 km (15,000 ft) altitude with a cruise speed of about 65 kts and has a range of about 185 km. Its gross takeoff weight is about 450 kg. The Predator (Fig. 9.2) operated by the U.S. Air Force [1], weighs about 953 kg (2100 lb) and operates from sea level to 7.9 km (26,000 ft) altitude with a cruise speed of 70–90 kts, a range of about 770 miles (1238 km), and an endurance of over 24 hr. The NASA-Boeing X-48 Blended Wing Body (Fig. 9.3) is an example of the application of the UAV concept for flight dynamics and flight controls research [6]. The X-45A (Fig. 9.4) is an example of the UCAV concept [5]. The empty weight is around 3629 kg (8000 lb), with a fuel capacity of 1224 kg (2700 lb), and a payload of about 405 kg (1500 lb). Dragonfly [7] is a MAV (Fig. 9.5) with a wingspan of about 30 cm. It weighs around 186 g, has a cruise speed of 12 m/s, an endurance of about 30 min, and carries a video camera as its payload.

Fig. 9.1 Pioneer UAV (1).

Fig. 9.2 Predator UAV (1).

Fig. 9.3 NASA-Boeing BWB concept (6).

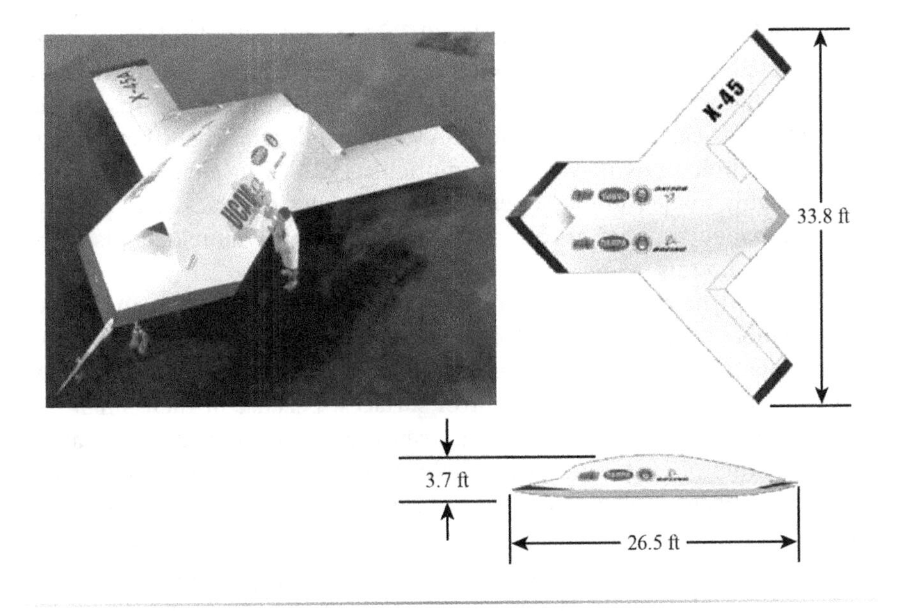

Fig. 9.4 X-45A UCAV concept (5).

In this chapter, the stability, control issues, and challenges related to UAVs, UCAVs, and MAVs will be discussed. A detailed study of these issues is beyond the scope of this chapter, but for additional information, the interested reader may refer to the references provided herein.

9.2 Stability, Control Issues, and Challenges

The aircraft stability and control theory/design discussed in previous chapters was based on rigid body dynamics and a linear variation of aerodynamic force/moment coefficients with the angle of attack/sideslip. The aircraft response was characterized by longitudinal short period and phugoid

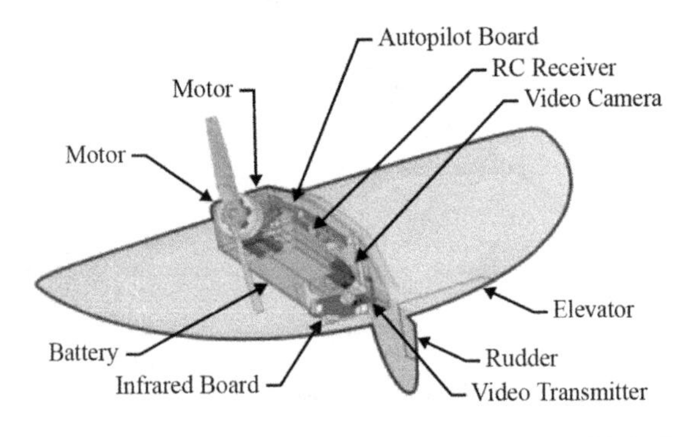

Fig. 9.5 Dragonfly Micro Air Vehicle (7).

modes and lateral/directional Dutch rolls, spiral, and roll subsidence modes. Various stability augmentation systems and autopilots were studied based on the application of the linear control system theory/design. In general, these methods can also be used for UAV control design when they operate within the limitations of these assumptions. UCAVs in general, however, can be highly maneuverable, operate at high angles of attack, and perform extreme maneuvers that are usually not attempted by human pilots flying the combat aircraft.

The aerodynamics, stability, and control issues of small UAVs and MAVs are even more challenging because they are small in size, have low mass, have low moments of inertia, and experience low Reynolds number flows. Some MAVs have flexible wings and control surfaces. Because of their small size and very low flight speeds, MAVS are sensitive to wind gusts that may be the same order of magnitude as the flight speed itself. Another area that presents a significant challenge for MAV design and operation is the nonavailability of established flying or handling quality specifications. The existing flying qualities were developed for the design and operation of piloted aircraft. Similar information is not yet available for the design and operation of UAVs, UCAVs, and MAVs. These issues will be briefly addressed in the following subsections.

9.2.1 Low Reynolds Number Aerodynamics

Figure 9.6 shows the Reynolds number variation for flying objects in nature ranging from small insects to a large aircraft, like a Boeing 747 [8].

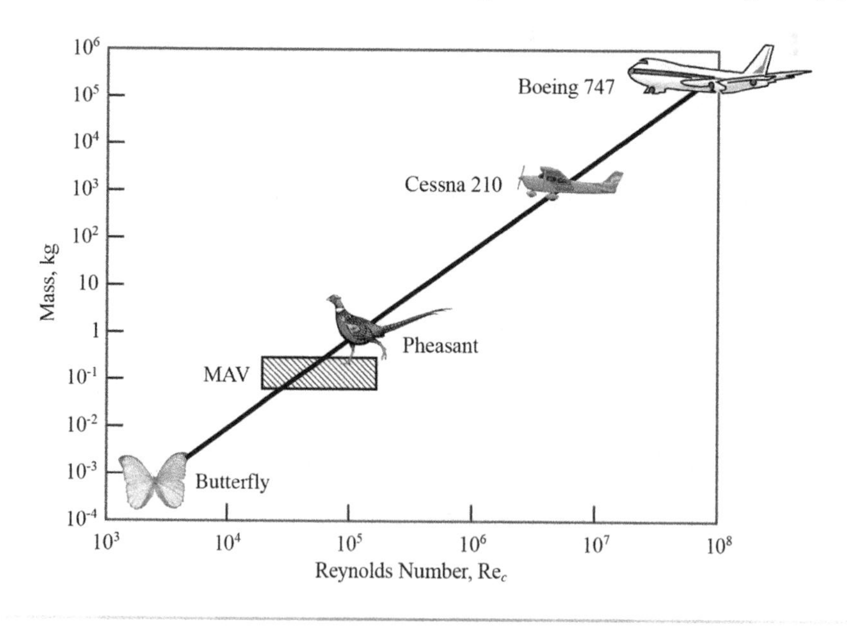

Fig. 9.6 Reynolds number (based on wing chord) variation with vehicle size/weight (8).

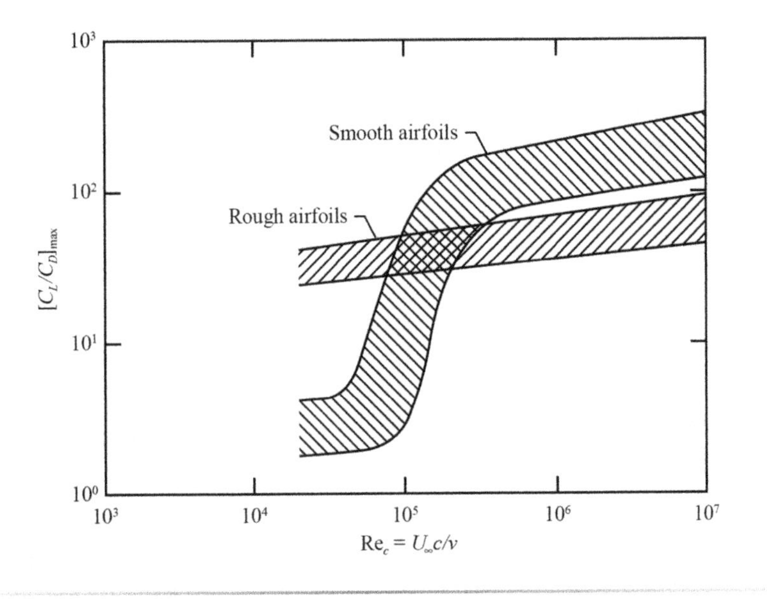

Fig. 9.7 Aerodynamic efficiency with Reynolds number (based on wing chord) (8).

The airfoil lift-to-drag ratio (aerodynamic efficiency), which is a measure of airfoil effectiveness, depends on the surface finish and on the Reynolds number, as shown in Fig. 9.7. Observe that, for smooth airfoils, the lift-to-drag ratio is quite low at low Reynolds numbers but increases significantly with the Reynolds number, particularly in the range of $10^4 - 10^6$ where small UAVs and MAVs usually operate. For example, the Predator drone, with a wing root chord of 3 ft 6 in. (1.067 m), operating at 26,000 ft (7.9 km) altitude and with a speed of 90 knots (indicated air speed), has a Reynolds number of about 2.7×10^6. However, for UAVs much smaller than the Predator and MAVs, the Reynolds number based on wing chord falls in the range $10^4 - 10^5$. This range of Reynolds numbers is usually referred to as the low Reynolds number range. For this range of Reynolds number range, complicated flow phenomena occur within the boundary layer [8]. This involves initial laminar flow separation, transition in the separated shear layer, reattachment forming a bubble, and, possibly, an eventual turbulent separation, all of which can occur within a short distance over the upper surface of the airfoil. Depending on the airfoil shape, surface finish, and the Reynolds number, either a short bubble or long bubble may be formed that leads to different types of stall patterns [8]. This type of flow pattern is analogous to that of the circular cylinder in a cross flow at critical Reynolds numbers discussed in Chapter 1.

The challenge for the designers is to develop aerodynamic models needed for the flight control design of small UAVs and MAVs operating at low Reynolds numbers. The aircraft aerodynamic data available in the literature generally apply for Reynolds numbers that are orders of magnitude higher than those applicable for small UAVs and MAVs. In view of this, the aerodynamic data based on engineering methods like DATCOM discussed in Chapter 4 may not be suitable for application to small UAVs and MAVs operating at low Reynolds numbers.

9.2.2 Unsteady Aerodynamic Effects

Small UAVs, particularly MAVs, are potentially very sensitive to atmospheric disturbances while operating at low altitudes and in the vicinity of the ground or complex urban environments (Fig. 9.8). For example, a MAV responding to atmospheric gusts that are of the order of its flight speed may experience severe unsteady effects. Figure 9.9 shows the flow patterns over the lee side (top surface) of an airfoil in a water tunnel, pitching at a rate of 35 deg/s from 0 to 45 deg and subsequently held fixed [9]. This study was an attempt to simulate the approach and landing of a MAV in a wind gust. The test Reynolds number based on the wing chord was around 6.0×10^5. At a 24-deg angle of attack, a leading edge vortex is formed and, by about a 30-deg angle of attack, it becomes prominent. The leading edge vortex keeps growing until the pitching motion stops at a 45-deg angle of attack. Then notice in the last photograph that a massive separated flow region is formed over the top surface leading to the formation of an organized, alternating Karman vortex street downstream of the airfoil. The aerodynamic force and moment coefficients would then depend on the vortex

Fig. 9.8 MAV operating in urban environment.

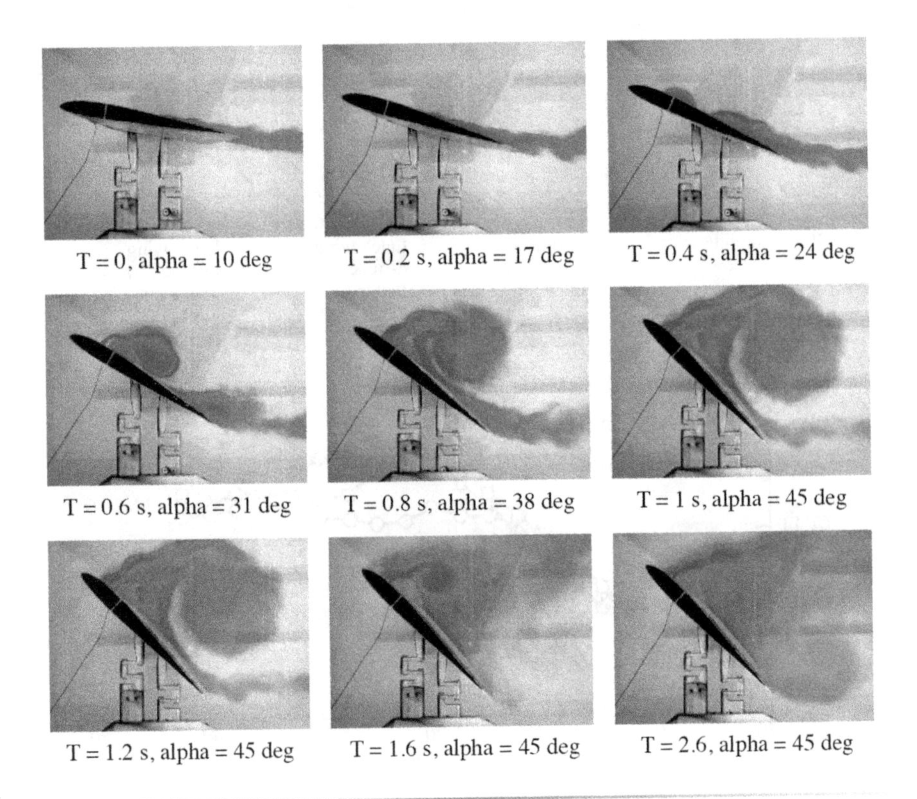

Fig. 9.9 Flow field on the lee side of a pitching airfoil (9).

shedding frequency and the Strouhal number in a manner similar to the case of the circular cylinder in a cross flow discussed in Chapter 1. The challenge for the aerodynamicist is to understand such low Reynolds numbers' unsteady flow effects and develop aerodynamic force/moment coefficient models suitable for the flight control design of these vehicles.

9.2.3 Vehicle Flexibility

These effects arise due to the deformation of the vehicle surface as a consequence of external loading on the vehicles, which are mainly aerodynamic in nature. Another issue/challenge is that of control and structure interaction, which is even more complicated. Traditionally, vehicle flexibility is an undesirable trait in terms of structural design considerations. However, pilots remotely flying flexible MAVs in a wind gust environment often report that the flexible MAVs are relatively easier to fly than rigid ones because gust energy is partially absorbed by the flexible structure of the vehicle. Therefore, vehicle movement because of gust is reduced resulting in a smaller impact on the vehicle motion. Also, the onset of stall may be delayed, as shown in Fig. 9.10, for a MAV wing with various levels of

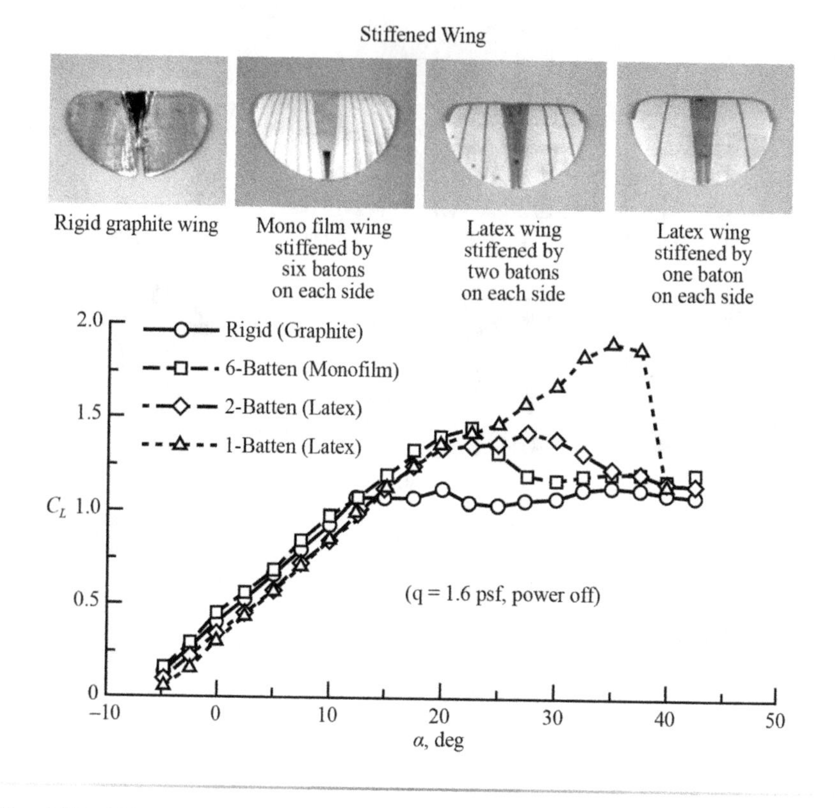

Fig. 9.10 Lift coefficient for a MAV wing with different levels of flexibility (10).

flexibility [10]. The rigid wing stalls around a 20-deg angle of attack, which progressively increases towards 40 deg as the wing is made more flexible using fewer stiffeners (batons).

9.2.4 Flight Control

The methods of linear control system theory and design discussed in Chapter 5 generally apply to piloted aircraft at small angles of attack (below stall) and are considered as linear systems. However, most UCAVs and MAVs are typically nonlinear systems. For flight control design using a linear system approach, the nonlinear system must be linearized at several points along its expected operational envelope so there will be a different linear system at each one of those operating points. Then, a linear closed-loop control system can be designed to determine the feedback gains at each one of those operating points. This approach is called gain scheduling [11]. If the operating envelope covers multiple parameters such as Mach or angle of attack/sideslip, then multidimensional interpolation must be performed to obtain gains at all the intermediate operating points in the flight envelope. Using this approach, essentially a set of linear controllers that collectively operate like a nonlinear controller is designed.

Figure 9.11 shows an example of nonlinear aerodynamic stability and control characteristics. This figure shows the ratios of longitudinal and lateral/directional static stability and control parameters for the X-45A UCAV [5]. Notice the nonlinear variations, particularly the abrupt unstable break in pitching moments above 10-deg angles of attack. According to [5], designing flight control systems becomes increasingly difficult when these ratios are greater than two. Therefore, the airspeeds for the X-45A's landing and takeoff were increased to keep the angles of attack below 10 deg and to keep this ratio below two. For flight control design, the plant was linearized at various points in the flight envelope to determine gain scheduling to cover the complete flight envelope [5].

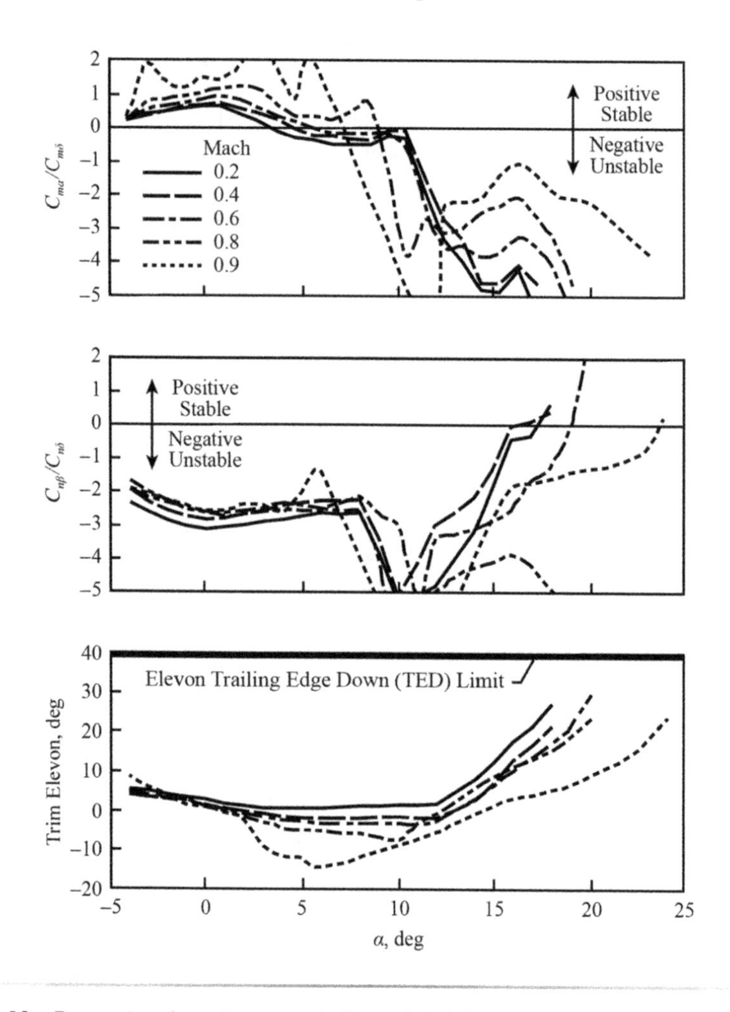

Fig. 9.11 Example of nonlinear variation of stability and control characteristics, X-45A UCAV concept (5).

An alternative method that does not require gain scheduling is the dynamic inversion technique [11]. The dynamic inversion technique can deal directly with known system nonlinearities by effectively canceling them using feedback linearization. For dynamic inversion to work, the plant characteristics must be well known and all states available for feedback similar to the linear state space pole placement technique discussed in Chapter 5. The objective of the dynamic inversion controller is to generate inputs to the plant such that the system output matches the desired output with minimal error (tracking). Mathematical details for the derivation of the dynamic inversion controller are discussed in [11]. In situations where the plant characteristics are not well known, other approaches such as incremental dynamic inversion can be used [12].

Figure 9.12 shows a simple schematic representation of the application of the dynamic inversion technique for the flight control of a flexible MAV [13]. The system inputs are the commanded flight path angle γ_C, wind axis heading angle χ_C, and total flight speed V_{tc}. The desired input vector $(\dot{y}_d)$ to the nonlinear dynamic inversion controller consists of three elements—roll rate acceleration $\dot{p}$, pitch rate acceleration $\dot{q}$, and yaw rate acceleration $\dot{r}$. The inner loop dynamic inversion controller generates symmetric and antisymmetric elevon commands so that the plant output $\dot{y}$ tracks the desired output $(\dot{y}_d)$ with minimal error. Reference 13 provides additional details of this design. Other examples of application of dynamic inversion technique for the UAV flight control design are reported in [14, 15].

MAVs are usually flown by a remote pilot utilizing the image relayed to a ground station by an onboard optical or infrared video-camera. This method of control proves to be very difficult and inefficient at times depending on the nature of the environment. The angle of view provided by the miniaturized video cameras often limits remote pilots' perception of the space, making it very difficult for the pilot to assess the operational scenario. For efficient operation under a changing environment, it is desirable that MAVs be capable of adaptive, autonomous, and reconfigurable flight control. This scenario applies to UAVs and UCAVs for efficient and autonomous operation in a changing environment. Generally, UAVs have been designed with very limited envelopes to try and keep within predictable flight

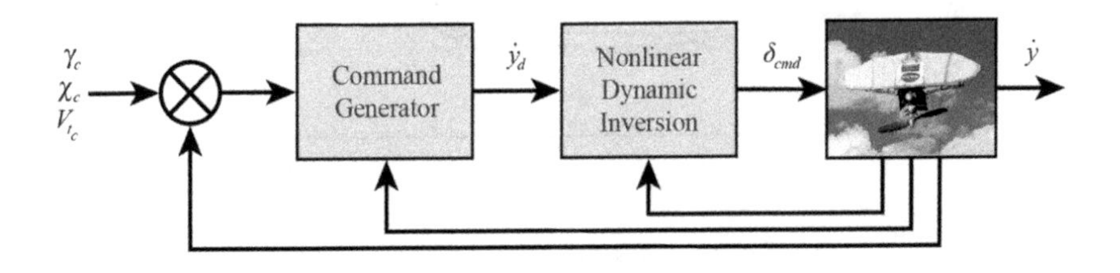

Fig. 9.12 Nonlinear dynamic inversion controller (12).

envelopes. This will be a problem when "panic" type maneuvers are allowed to "see-and-avoid" a last-minute pop-up target, start doing real air combat instead of being a stable platform to launch weapons, or encounter other large upset conditions resulting in excursions into very nonlinear aerodynamic stability and control regimes. In such scenarios, UAVs require sophisticated, intelligent, autonomous flight control to maintain vehicle stability and control as well as to recover from unstable and possibly unsteady flight conditions.

Several studies [16–28] are reported in the literature regarding intelligent, fault-tolerant control systems capable of maintaining vehicle stability and control under changing vehicle and environmental conditions. The majority of these studies are focused on intelligent control where the vehicle learns online, reconfigures itself, and adapts to changing environments. Using an online, real-time parameter identification technique, the vehicle updates its plant and actuator models and evaluates the available control authority. Then, it reconfigures the flight control system to accomplish the remainder of its mission. All these processes happen online and in real time as the vehicle continues its mission.

Artificial neural networks are suited for intelligent flight control [17]. They mimic the input–output relations and learning properties of biological neural systems. They are intrinsically nonlinear and can address problems in multiple dimensions. The artificial neural networks can be considered generalized spline functions that identify efficient input–output mappings from observations. For control system applications, neural networks perform functions analogous to the gain scheduling of a linear system. Note that artificial neural networks do not require gain scheduling. Artificial neural networks can be used for online adaptation or reconfigurable control following a failure resulting in unexpected variations in vehicle aerodynamics, mass properties or propulsion characteristics, or dealing with external wind gusts that can significantly impact vehicle performance at low speeds. Reference 28 reports an example of the application of artificial neural networks for UAV flight control.

9.2.5 Flying Qualities

Although UAVs have been in use for several decades and operate over the national airspace, there are not yet clearly defined flying qualities suitable for the design and operation of UAVs. Detailed requirements for piloted aircraft specifications include longitudinal and lateral-directional stability and control; takeoff, landing, recovery from stalls, spins, and out-of-control flight characteristics of the primary and secondary control systems; and response to atmospheric gusts. The existing flying qualities applicable for piloted aircraft were developed over the past several decades and are not directly applicable to UAVs. As in the case of piloted aircraft, the intent of UAV flying qualities is supposed to guarantee a safe, controlled, and operationally effective flight vehicle. Chapter 6 provided some aspects of flying qualities pertaining to the design of stability augmentation systems and autopilots for general aviation aircraft.

UCAVs are not subject to restrictions like piloted combat aircraft. Application of piloted flying qualities criteria to UCAVs would add unnecessary cost and may increase the weight of the vehicle. Additionally, without a pilot the same requirements may not be appropriate for UAV flights resulting in lower stability margins than desired or even resulting in loss of control. Another issue and challenge is that UAVs do not fit into the class of aircraft for which the current flying qualities are defined. A new class for aircraft flying qualities to suit these smaller and lighter vehicles must be developed.

One of the biggest challenges to developing UAV flying qualities is the lack of a database related to UAV missions, flight tests, lessons learned, and so on. Recently, some studies have addressed these issues [29–31]. These studies essentially focus on adapting the current piloted flying qualities to UAVs where applicable. For example, Foster and Bowman have proposed that the longitudinal short period mode frequency upper/lower limits for level I, II, and III categories (see Table 6.6) be dynamically scaled by the square root of a suitably chosen representative length scale [31]. For selected UAV models, they estimated dynamic stability parameters using empirical methods, conducted flight tests, and obtained flight test data. They demonstrated that the dynamic frequency scaling approach correlated well with the flight test data and was consistent with the feedback they received from pilots who flew them remotely.

9.3 Summary

In this chapter, the aerodynamics, stability, control, and autopilot design issues and challenges related to UAVs have been outlined briefly. These issues become more complex as the vehicles become smaller in size and weight and become more flexible. These vehicles usually operate at low Reynolds numbers and often experience severe unsteady effects as a result. For efficient operation, it is desirable that these vehicles be capable of intelligent, fault-tolerant, autonomous, and reconfigurable control. These issues are very complex, and a detailed discussion of these issues is beyond the scope of this chapter. However, to make the reader aware of these issues and challenges, an effort has been made to highlight technical problems and present a brief discussion on the progress made in the literature to address those problems.

References

[1] "Earth Observation and the Role of UAVs: A Capabilities Assessment," Version 1.1, Civil UAV Assessment Team, Aug. 2006; http://www.nasa.gov/centers/dryden/research/civuav/index.html.

[2] Francis, M. S., "The Role of Unmanned Air Vehicles in Advancing System-of-Systems (SOS) Technologies and Capabilities," AIAA Paper 2003-2739.
[3] Tsach, S., Peled, A., Penn, D., Keshales, B., and Guedj, R., "Development Trends for Next Generation UAV Systems," AIAA Paper 2007-2762.
[4] Woolvin, S. J., "UCAV Configuration and Performance Trade-Offs," AIAA Paper 2006-1264.
[5] Davidson, R. W., "Flight Control Design and Test of the Joint Unmanned Air Combat System (J-UCAS) X-45A," AIAA Paper 2004-6557.
[6] Risch, T., Cosentino, G., and Regan, C. D., "X-48B Flight-Test Progress Overview," AIAA Paper 2009-934.
[7] Krashanitsa, R., Platanitis, G., Silin, B., and Shkarayev, S., "Aerodynamics and Controls Design for Autonomous Micro Air Vehicles," AIAA Paper 2006-6639.
[8] Gad-el-Hak, M., "Micro Air Vehicles: Can They be Controlled Better?," *Journal of Aircraft*, Vol. 38, No. 3. May–June 2001, pp. 419–429.
[9] Ol, M., Parker, G., Abate, G., and Evers, J., "Flight Controls and Performance Challenges for MAVs in Complex Environments," AIAA Paper 2008-6508.
[10] Ifju, P. G., Jenkins, D. A., Ettinger, S., Lian, Y., Shyy, W., and Waszak, M., "Flexible-Wing-Based Micro Air Vehicles," AIAA Paper 2002-0705.
[11] Stevens, B. L., and Lewis, F. L., *Aircraft Control and Simulation*, 2nd ed., John Wiley & Sons Inc., 2003.
[12] Bacon, B. J., and Ostroff, A. J., "Reconfigurable Flight Control Using Nonlinear Dynamic Inversion With A Special Accelerometer Implementation," AIAA Paper 2000-4565.
[13] Waszak, M. R., Davidson, J. B., and Ifju, P. G., "Simulation and Flight Control of an Aeroelastic Fixed Wing Micro Air Vehicle," AIAA 2002-4875.
[14] Singh, S. P., and Padhi, R., "Automatic Landing of UAV using Dynamic Inversion," AIAA Paper 2008-082.
[15] Shin, Y., Calise, A. J., and Motter, M. A., "Adaptive Autopilot Design for an Unmanned Aerial Vehicle," AIAA Paper 2005-6166.
[16] Schmitz, Janardan, V., and Balakrishnan, S. N., "Implementation of Nonlinear Reconfigurable Controllers for Autonomous Unmanned Vehicles," AIAA Paper 2005–348.
[17] Stengel, R. F., "Towards Intelligent Flight Control," *IEEE Transactions on Systems, Man and Cybernetics*, Vol. 23, No. 6, Nov–Dec 1993.
[18] Ward, D. G., Sharma, M. M., and Richards, N. D., "Intelligent Control of Unmanned Air Vehicles: Program Summary and Representative Results," AIAA Paper 2003-6641.
[19] Zhou, Quing-Li, Zhang, Y., "Reconfigurable Control Allocation Technology Using Weighted Least Squares for Nonlinear System in Unmanned Aerial Vehicle," AIAA Paper 2010-1138.
[20] Ortiz, J. E., and Klenke, R. H., "Low Cost Autopilot System for Unmanned Aerial Vehicles," AIAA Paper 2011-1410.
[21] Jin, F., Hiroshi, T., and Shigueru, S., "Development of Small Unmanned Aerial Vehicle and Flight Control Design," AIAA Paper 2007-6501.
[22] Kordes, T., Buschmann, S., Winkler, S., and Vorsmann, H-W. S., "Progress in the Development of the Fully Autonomous MAV "CAROLO," AIAA Paper 2003-6547.
[23] Poinsot, D., Berard, C., Krashanista, R., and Shkarayev, S., "Investigation of Flight Dynamics and Automatic Controls for Hovering Micro Air Vehicles," AIAA Paper 2008-6332.
[24] Stewart, K., Abate, G., and Evers, J., "Flight Mechanics and Control Issues of Micro Air Vehicles," AIAA Paper 2006-6638.
[25] Krashanista, R., Platantinis, G., Silin, B., and Shkarayev, S., "Aerodynamics and Controls Design for Autonomous Micro Air Vehicles," AIAA Paper 2006-6639.
[26] Pines, D. J., and Bohorquez, F., "Challenges Facing Future Micro-Air-Vehicle Development," *Journal of Aircraft*, Vol. 43, No. 2, March–April 2006, pp. 290–305.
[27] Arning, R. K., and Sassen, S., "Flight Control of Micro Air Vehicles," AIAA Paper 2004-4911.

[28] Ippolito, C., Yeh, Yoo-Hsiu, and Kaneshige, J., "Neural Adaptive Flight Control Testing of an Unmanned Experimental Aerial Vehicle," AIAA Paper 2007-2827.
[29] McFarlene, C., Richardson, T. S., and Jones, C. D. C., "Unmanned Aerial Vehicle Flying Qualities," AIAA Paper 2008-7156.
[30] Holmberg, J., King, D. J., and Leonard, J. R., "Flying Qualities Specification & Design Standards for Unmanned Air Vehicles," AIAA Paper 2008-6555.
[31] Foster, T. M., and Bowman, W. J., "Dynamic Stability and Handling Qualities of Small Unmanned Aerial Vehicles," AIAA Paper 2005-1023.

Appendix A Standard Atmospheres

U.S. 1976 standard atmospheres

Altitude, m	Temperature, K	Pressure, N/m^2	Density, kg/m^3	Speed of sound, m/s	Kinematic viscosity, m^2/s
0.00	288.15	101,326.04	1.2250	340.29	0.146070E-04
200.00	286.85	98,946.34	1.2017	339.53	0.148384E-04
400.00	285.55	96,612.11	1.1786	338.76	0.150748E-04
600.00	284.25	94,322.67	1.1560	337.98	0.153160E-04
800.00	282.95	92,077.29	1.1336	337.21	0.155621E-04
1,000.00	281.65	89,875.47	1.1116	336.43	0.158131E-04
1,200.00	280.35	87,716.47	1.0900	335.66	0.160690E-04
1,400.00	279.05	85,599.63	1.0686	334.88	0.163304E-04
1,600.00	277.75	83,524.38	1.0476	334.10	0.165969E-04
1,800.00	276.45	81,490.05	1.0269	333.31	0.168690E-04
2,000.00	275.15	79,496.02	1.0065	332.53	0.171476E-04
2,200.00	273.85	77,541.70	0.9864	331.74	0.174316E-04
2,400.00	272.55	75,626.41	0.9666	330.95	0.177210E-04
2,600.00	271.25	73,749.67	0.9472	330.16	0.180167E-04
2,800.00	269.95	71,910.85	0.9280	329.37	0.183190E-04
3,000.00	268.65	70,109.26	0.9091	328.58	0.186280E-04
3,200.00	267.35	68,344.40	0.8905	327.78	0.189437E-04
3,400.00	266.05	66,615.75	0.8723	326.98	0.192660E-04
3,600.00	264.75	64,922.60	0.8543	326.18	0.195950E-04
3,800.00	263.45	63,264.51	0.8366	325.38	0.199311E-04
4,000.00	262.15	61,640.88	0.8191	324.58	0.202748E-04
4,200.00	260.85	60,051.13	0.8020	323.77	0.206260E-04
4,400.00	259.55	58,494.75	0.7851	322.96	0.209846E-04
4,600.00	258.25	56,971.19	0.7685	322.16	0.213516E-04
4,800.00	256.95	55,479.94	0.7522	321.34	0.217270E-04
5,000.00	255.65	54,020.46	0.7361	320.53	0.221102E-04
5,200.00	254.35	52,592.22	0.7203	319.71	0.225020E-04
5,400.00	253.05	51,194.70	0.7048	318.90	0.229030E-04
5,600.00	251.75	49,827.42	0.6895	318.08	0.233135E-04

(Continued)

U.S. 1976 standard atmospheres (*Continued*)

Altitude, m	Temperature, K	Pressure, N/m^2	Density, kg/m^3	Speed of sound, m/s	Kinematic viscosity, m^2/s
5,800.00	250.45	48,489.85	0.6745	317.25	0.237326E-04
6,000.00	249.15	47,181.51	0.6597	316.43	0.241610E-04
6,200.00	247.85	45,901.91	0.6452	315.60	0.246004E-04
6,400.00	246.55	44,650.54	0.6309	314.77	0.250499E-04
6,600.00	245.25	43,426.93	0.6169	313.94	0.255090E-04
6,800.00	243.95	42,230.65	0.6031	313.11	0.259788E-04
7,000.00	242.65	41,061.17	0.5895	312.27	0.264601E-04
7,200.00	241.35	39,918.07	0.5762	311.44	0.269530E-04
7,400.00	240.05	38,800.86	0.5631	310.60	0.274573E-04
7,600.00	238.75	37,709.12	0.5502	309.75	0.279739E-04
7,800.00	237.45	36,642.38	0.5376	308.91	0.285030E-04
8,000.00	236.15	35,600.18	0.5252	308.06	0.290445E-04
8,200.00	234.85	34,582.15	0.5130	307.21	0.295989E-04
8,400.00	233.55	33,587.81	0.5010	306.36	0.301670E-04
8,600.00	232.25	32,616.74	0.4892	305.51	0.307495E-04
8,800.00	230.95	31,668.55	0.4777	304.65	0.313460E-04
9,000.00	229.65	30,742.78	0.4663	303.79	0.319570E-04
9,200.00	228.35	29,839.07	0.4552	302.93	0.325838E-04
9,400.00	227.05	28,956.98	0.4443	302.07	0.332260E-04
9,600.00	225.75	28,096.14	0.4336	301.20	0.338840E-04
9,800.00	224.45	27,256.12	0.4230	300.33	0.345591E-04
10,000.00	223.15	26,436.56	0.4127	299.46	0.352511E-04
10,200.00	221.85	25,637.08	0.4026	298.59	0.359610E-04
10,400.00	220.55	24,857.28	0.3926	297.71	0.366900E-04
10,600.00	219.25	24,096.81	0.3829	296.83	0.374333E-04
10,800.00	217.95	23,355.26	0.3733	295.95	0.382020E-04
11,000.00	216.65	22,632.33	0.3639	295.07	0.390536E-04
11,200.00	216.65	21,929.70	0.3526	295.07	0.401554E-04
11,400.00	216.65	21,248.88	0.3417	295.07	0.414740E-04
11,600.00	216.65	20,589.20	0.3311	295.07	0.428038E-04
11,800.00	216.65	19,950.00	0.3208	295.07	0.441611E-04
12,000.00	216.65	19,330.64	0.3108	295.07	0.455740E-04
12,200.00	216.65	18,730.51	0.3012	295.07	0.470290E-04
12,400.00	216.65	18,149.02	0.2918	295.07	0.485291E-04
12,600.00	216.65	17,585.57	0.2828	295.07	0.500780E-04

(*Continued*)

U.S. 1976 standard atmospheres (*Continued*)

Altitude, m	Temperature, K	Pressure, N/m^2	Density, kg/m^3	Speed of sound, m/s	Kinematic viscosity, m^2/s
12,800.00	216.65	17,039.62	0.2740	295.07	0.516761E-04
13,000.00	216.65	16,510.61	0.2655	295.07	0.533248E-04
13,200.00	216.65	15,998.03	0.2572	295.07	0.550260E-04
13,400.00	216.65	15,501.37	0.2493	295.07	0.567814E-04
13,600.00	216.65	15,020.12	0.2415	295.07	0.585928E-04
13,800.00	216.65	14,553.82	0.2340	295.07	0.604620E-04
14,000.00	216.65	14,101.98	0.2268	295.07	0.623907E-04
14,200.00	216.65	13,664.18	0.2197	295.07	0.643808E-04
14,400.00	216.65	13,239.97	0.2129	295.07	0.664340E-04
14,600.00	216.65	12,828.93	0.2063	295.07	0.685525E-04
14,800.00	216.65	12,430.65	0.1999	295.07	0.707388E-04
15,000.00	216.65	12,044.73	0.1937	295.07	0.729950E-04
15,200.00	216.65	11,670.80	0.1877	295.07	0.753227E-04
15,400.00	216.65	11,308.48	0.1818	295.07	0.777241E-04
15,600.00	216.65	10,957.40	0.1762	295.07	0.802020E-04
15,800.00	216.65	10,617.22	0.1707	295.07	0.827592E-04
16,000.00	216.65	10,287.60	0.1654	295.07	0.853973E-04
16,200.00	216.65	9,968.22	0.1603	295.07	0.881190E-04
16,400.00	216.65	9,658.75	0.1553	295.07	0.909281E-04
16,600.00	216.65	9,358.89	0.1505	295.07	0.938264E-04
16,800.00	216.65	9,068.34	0.1458	295.07	0.968160E-04
17,000.00	216.65	8,786.81	0.1413	295.07	0.999006E-04
17,200.00	216.65	8,514.02	0.1369	295.07	0.103085E-03
17,400.00	216.65	8,249.70	0.1327	295.07	0.106370E-03
17,600.00	216.65	7,993.59	0.1285	295.07	0.109757E-03
17,800.00	216.65	7,745.42	0.1245	295.07	0.113252E-03
18,000.00	216.65	7,504.96	0.1207	295.07	0.116860E-03
18,200.00	216.65	7,271.97	0.1169	295.07	0.120584E-03
18,400.00	216.65	7,046.20	0.1133	295.07	0.124426E-03
18,600.00	216.65	6,827.45	0.1098	295.07	0.128390E-03
18,800.00	216.65	6,615.49	0.1064	295.07	0.132478E-03
19,000.00	216.65	6,410.11	0.1031	295.07	0.136694E-03

Ref: *U.S. Standard Atmosphere: 1976*, National Oceanic and Atmospheric Administration, NASA, and U.S. Air Force, Washington, DC, 1976.

Appendix B Table of Laplace Transforms

$f(t)$	$F(s)$
unit impulse $\delta(t)$	1
unit step $l(t)$	$\frac{1}{s}$
t	$\frac{1}{s^2}$
$t^n = (n = 1, 2, 3 \ldots)$	$\frac{n!}{s^{n+1}}$
e^{-at}	$\frac{1}{s+a}$
te^{-at}	$\frac{1}{(s+a)^2}$
$\frac{t^2}{2}e^{-at}$	$\frac{1}{(s+a)^3}$
$1 - e^{-at}$	$\frac{a}{s(s+a)}$
$\frac{1}{(b-a)}(e^{-at} - e^{-bt})$	$\frac{1}{(s+a)(s+b)}$
$\frac{1}{(b-a)}(be^{-at} - ae^{-bt})$	$\frac{s}{(s+a)(s+b)}$
$\sin \omega t$	$\frac{\omega}{(s^2+\omega^2)}$
$\cos \omega t$	$\frac{s}{(s^2+\omega^2)}$
$e^{-at} \sin \omega t$	$\frac{\omega}{(s+a)^2+\omega^2}$
$e^{-at} \cos \omega t$	$\frac{s+a}{(s+a)^2+\omega^2}$
$\frac{1}{a^2}(at - 1 + e^{-at})$	$\frac{1}{s^2(s+a)}$
$\frac{\omega_n}{\sqrt{1-\zeta^2}} e^{-\zeta\omega_n t} \sin \omega_n t\sqrt{1-\zeta^2}$	$\frac{\omega_n^2}{(s^2+2\zeta\omega_n s+\omega_n)^2}$
$\frac{-1}{\sqrt{1-\zeta^2}} e^{-\zeta\omega_n t} \sin(\omega_n t\sqrt{1-\zeta^2} - \phi)$	$\frac{s}{(s^2+2\zeta\omega_n s+\omega_n)^2}$
$\phi = \tan^{-1}\left(\frac{\sqrt{1-\zeta^2}}{\zeta}\right)$	

Appendix C Cramer's Rule

To illustrate the use of Cramer's rule for solving a set of simultaneous, linear, algebraic equations, consider the following set of three equations with three unknowns, x_1, x_2, and x_3.

$$a_{11}x_1 + a_{12}x_2 + a_{13}x_3 = b_1$$
$$a_{21}x_1 + a_{22}x_2 + a_{23}x_3 = b_2$$
$$a_{31}x_1 + a_{32}x_2 + a_{33}x_3 = b_3$$

Using Cramer's rule, we can solve for each of these three unknowns as follows:

$$x_1 = \frac{D_1}{D}$$
$$x_2 = \frac{D_2}{D}$$
$$x_3 = \frac{D_3}{D}$$

where

$$D = \begin{vmatrix} a_{11} & a_{12} & a_{13} \\ a_{21} & a_{22} & a_{23} \\ a_{31} & a_{32} & a_{33} \end{vmatrix}$$

$$D_1 = \begin{vmatrix} b_1 & a_{12} & a_{13} \\ b_2 & a_{22} & a_{23} \\ b_3 & a_{32} & a_{33} \end{vmatrix}$$

$$D_2 = \begin{vmatrix} a_{11} & b_1 & a_{13} \\ a_{21} & b_2 & a_{23} \\ a_{31} & b_3 & a_{33} \end{vmatrix}$$

$$D_3 = \begin{vmatrix} a_{11} & a_{12} & b_1 \\ a_{21} & a_{22} & b_2 \\ a_{31} & a_{32} & b_3 \end{vmatrix}$$

In general, if we have k simultaneous, linear algebraic equations for k unknowns, x_i, $i = 1, 2, \ldots, k$, then we can write

$$x_k = \frac{D_k}{D}$$

where the determinant D_k is formed by replacing the kth column of the determinant D with the elements $b_1, b_2, \ldots, b_k$ from the right-hand side.

Appendix D Conversion of U.S. Customary Units to SI Units

Physical Quantity	Multiply U.S. Customary Unit	by Conversion Factor	to obtain SI Units
Length	ft	0.3048	m
Velocity	ft/sec	0.3048	m/sec
	mph	0.44704	m/sec
Acceleration	ft/sec^2	0.3048	m/sec^2
Pressure	lbf/ft^2	47.88026	N/m^2
Mass	slugs	14.59390	kg
Moment of Inertia	$slugs\text{-}ft^2$	1.355818	$kg\text{-}m^2$
Force	lbf	4.448222	N
Moment	ft-lbf	1.355818	N-m
Area	ft^2	0.092903	m^2
Mass Flow	slugs/sec	14.59390	kg/sec
Density	$slugs/ft^3$	515.3788	kg/m^3
Kinematic Viscosity	ft^2/sec	0.092903	m^2/sec
Coefficient of Viscosity	slugs/ft-sec	47.880258	$N\text{-}sec/m^2$

Appendix E Solved Examples

	Example	Page		Example	Page
Chapter 1	1.1	64	Chapter 4	4.3	375
	1.2	65		4.4	377
	1.3	66		4.5	378
	1.4	66		4.6	383
	1.5	67		4.7	412
	1.6	68		4.8	412
Chapter 2	2.1	84		4.9	413
	2.2	86		4.10	453
	2.3	96		4.11	459
	2.4	97		4.12	460
	2.5	111		4.13	466
	2.6	112		4.14	469
	2.7	125	Chapter 5	5.1	497
	2.8	128		5.2	505
	2.9	149		5.3	505
	2.10	150		5.4	506
	2.11	165		5.5	511
Chapter 3	3.1	249		5.6	520
	3.2	253		5.7	522
	3.3	259		5.8	522
	3.4	261		5.9	538
	3.5	263		5.10	542
	3.6	264		5.11	547
	3.7	274		5.12	549
	3.8	306		5.13	549
	3.9	310		5.14	571
	3.10	311		5.15	573
	3.11	329		5.16	575
Chapter 4	4.1	371	Chapter 7	7.1	694
	4.2	373		7.2	725

Summary of Equations of Motion, Stability Derivatives, and Transfer Functions

This appendix lists the basic equations of motion, stability derivatives, equations of motion in state-space form, and longitudinal and lateral/directional transfer functions applicable for aircraft motion involving small disturbance for convenience and ready reference. Detailed information on the theory and mathematical derivation are presented in Chapters 4 and 6.

F.1 Equations of Motion

$$F_x = m(\dot{U} + qW - rV)$$

$$F_y = m(\dot{V} + rU - pW)$$

$$F_z = m(\dot{W} + pV - qU)$$

$$L = \dot{p}I_x - \dot{q}I_{yx} - \dot{r}I_{xz} + qr(I_z - I_y) - pqI_{zx} + (r^2 - q^2)I_{yz} + prI_{yx}$$

$$M = \dot{q}I_y - \dot{r}I_{yz} - \dot{p}I_{yx} + rp(I_x - I_z) - qrI_{xy} + (p^2 - r^2)I_{zx} + pqI_{zy}$$

$$N = \dot{r}I_z - \dot{p}I_{zx} - \dot{q}I_{zy} + pq(I_y - I_x) - rpI_{yz} + (q^2 - p^2)I_{xy} + qrI_{xz}$$

For an aircraft with a vertical plane of symmetry (Ox_bz_b plane), $I_{xy} = I_{yz} = 0$

$$L = \dot{p}I_x - I_{xz}(pq + \dot{r}) + qr(I_z - I_y)$$

$$M = \dot{q}I_y + rp(I_x - I_z) + (p^2 - r^2)I_{xz}$$

$$N = \dot{r}I_z - I_{xz}(\dot{p} - qr) + pq(I_y - I_x)$$

F.1.1 Longitudinal Equations of Motion for Small Disturbance

$$\left(m_1\frac{\mathrm{d}}{\mathrm{d}t} - C_{xu}\right)u - \left(C_{x\dot{\alpha}}c_1\frac{\mathrm{d}}{\mathrm{d}t} + C_{x\alpha}\right)\Delta\alpha - \left(C_{xq}c_1\frac{\mathrm{d}}{\mathrm{d}t} + C_{x\theta}\right)\Delta\theta$$

$$= C_{x\delta e}\Delta\delta e + C_{x\delta t}\Delta\delta t$$

$$-C_{zu}u+\left[\left(m_1\frac{\mathrm{d}}{\mathrm{d}t}-C_{z\dot{\alpha}}c_1\frac{\mathrm{d}}{\mathrm{d}t}\right)-C_{z\alpha}\right]\Delta\alpha-\left(m_1\frac{\mathrm{d}}{\mathrm{d}t}+C_{zq}c_1\frac{\mathrm{d}}{\mathrm{d}t}+C_{z\theta}\right)\Delta\theta$$
$$=C_{z\delta e}\Delta\delta e+C_{z\delta t}\Delta\delta t$$

$$-C_{mu}u-\left(C_{m\dot{\alpha}}c_1\frac{\mathrm{d}}{\mathrm{d}t}+C_{m\alpha}\right)\Delta\alpha+\frac{\mathrm{d}}{\mathrm{d}t}\left(I_{y1}\frac{\mathrm{d}}{\mathrm{d}t}-C_{mq}c_1\right)\Delta\theta$$
$$=C_{m\delta e}\Delta\delta e+C_{m\delta t}\Delta\delta t$$

where

$$c_1=\frac{\bar{c}}{2U_o}$$

$$m_1=\frac{2m}{\rho U_o S}$$

$$I_{y1}=\frac{I_y}{\frac{1}{2}\rho U_o^2 S\bar{c}}$$

Longitudinal Stability Derivatives

$$C_{xu}=-2C_D-C_{Du}$$

$$C_{x\theta}=-C_L\cos\theta_o$$

$$C_{zu}=-2C_L-C_{Lu}$$

$$C_{z\theta}=-C_L\sin\theta_o$$

$$C_{x\alpha}=C_L-C_{D\alpha}$$

$$C_{z\alpha}=-C_{L\alpha}-C_D$$

$$C_{x\dot{\alpha}}=C_{L\dot{\alpha}}\Delta\alpha-C_{D\dot{\alpha}}\simeq-C_{D\dot{\alpha}}$$

$$C_{xq}=C_{Lq}\Delta\alpha-C_{Dq}\simeq-C_{Dq}$$

$$C_{z\dot{\alpha}}=-C_{L\dot{\alpha}}-C_{D\dot{\alpha}}\Delta\alpha\simeq-C_{L\dot{\alpha}}$$

$$C_{zq}=-C_{Lq}-C_{Dq}\Delta\alpha\simeq-C_{Lq}$$

Usually, stability derivatives like C_{xu} and C_{zu} are reported as nondimensional; static aerodynamic derivatives like $C_{L\alpha}$ and $C_{D\alpha}$, acceleration derivatives like $C_{L\dot{\alpha}}$ and $C_{m\dot{\alpha}}$, and rotary derivatives like C_{Lq}, C_{Dq} and C_{mq} are reported per radian.

F.1.2 Lateral-Directional Equations of Motion for Small Disturbance

$$\left(m_1\frac{d}{dt}-b_1C_{y\dot{\beta}}\frac{d}{dt}-C_{y\beta}\right)\Delta\beta-\left(b_1C_{yp}\frac{d}{dt}+C_{y\phi}\right)\Delta\phi$$
$$+\left(m_1\frac{d}{dt}-b_1C_{yr}\frac{d}{dt}\right)\Delta\psi=C_{y\delta r}\Delta\delta_r+C_{y\delta a}\Delta\delta_a$$

$$\left(-C_{l\beta}-b_1C_{l\dot{\beta}}\frac{d}{dt}\right)\Delta\beta+\left(-b_1C_{lp}\frac{d}{dt}+I_{x1}\frac{d^2}{dt^2}\right)\Delta\phi$$
$$+\left(-b_1C_{lr}\frac{d}{dt}-I_{xz1}\frac{d^2}{dt^2}\right)\Delta\psi=C_{l\delta r}\Delta\delta_r+C_{l\delta a}\Delta\delta_a$$

$$\left(-C_{n\beta}-b_1C_{n\dot{\beta}}\frac{d}{dt}\right)\Delta\beta+\left(-b_1C_{np}\frac{d}{dt}-I_{xz1}\frac{d^2}{dt^2}\right)\Delta\phi$$
$$+\left(-b_1C_{nr}\frac{d}{dt}+I_{xz1}\frac{d^2}{dt^2}\right)\Delta\psi=C_{n\delta r}\Delta\delta_r+C_{n\delta a}\Delta\delta_a$$

where

$$b_1=\frac{b}{2U_o}$$
$$I_{x1}=\frac{I_x}{\frac{1}{2}\rho U_o^2Sb}$$
$$I_{z1}=\frac{I_z}{\frac{1}{2}\rho U_o^2Sb}$$
$$I_{xz1}=\frac{I_{xz}}{\frac{1}{2}\rho U_o^2Sb}$$

Lateral Stability Derivatives

$$C_{y\beta}=(C_{y\beta})_{\text{aero}}$$

$$C_{y\dot{\beta}}=(C_{y\dot{\beta}})_{\text{aero}}$$

$$C_{yp}=(C_{yp})_{\text{aero}}$$

$$C_{yr}=(C_{yr})_{\text{aero}}$$

Usually, stability derivatives like $C_{y\beta}$, acceleration derivatives like $C_{y\dot{\beta}}$, and rotary derivatives like C_{yp}, C_{lp} and C_{nr} are reported per radian.

F.2 Longitudinal Equations in State-Space Form

F.2.1 Complete Fourth Order System

$$\dot{x} = Ax + Bu_c$$

$$x = \begin{bmatrix} u \\ \Delta\alpha \\ q \\ \Delta\theta \end{bmatrix} \qquad A = \begin{bmatrix} a_{11} & a_{12} & a_{13} & a_{14} \\ a_{21} & a_{22} & a_{23} & a_{24} \\ a_{31} & a_{32} & a_{33} & a_{34} \\ a_{41} & a_{42} & a_{43} & a_{44} \end{bmatrix} \qquad B = \begin{bmatrix} b_1 \\ b_2 \\ b_3 \\ b_4 \end{bmatrix}$$

$$a_{11} = \frac{C_{xu} + \xi_1 C_{zu}}{m_1} \qquad a_{12} = \frac{C_{x\alpha} + \xi_1 C_{z\alpha}}{m_1}$$

$$a_{13} = \frac{C_{xq}c_1 + \xi_1(m_1 + C_{zq}c_1)}{m_1} \qquad a_{14} = \frac{C_{x\theta} + \xi_1 C_{z\theta}}{m_1}$$

$$a_{21} = \frac{C_{zu}}{m_1 - C_{z\dot{\alpha}}c_1} \qquad a_{22} = \frac{C_{z\alpha}}{m_1 - C_{z\dot{\alpha}}c_1}$$

$$a_{23} = \frac{m_1 + C_{zq}c_1}{m_1 - C_{z\dot{\alpha}}c_1} \qquad a_{24} = \frac{C_{z\theta}}{m_1 - c_1 C_{z\dot{\alpha}}}$$

$$a_{31} = \frac{C_{mu} + \xi_2 C_{zu}}{I_{y1}} \qquad a_{32} = \frac{C_{m\alpha} + \xi_2 C_{z\alpha}}{I_{y1}}$$

$$a_{33} = \frac{C_{mq}c_1 + \xi_2(m_1 + C_{zq}c_1)}{I_{y1}} \qquad a_{34} = \frac{\xi_2 C_{z\theta}}{I_{y1}}$$

$$a_{41} = 0 \qquad a_{42} = 0 \qquad a_{43} = 1 \qquad a_{44} = 0$$

$$b_1 = \frac{C_{x\delta_e} + \xi_1 C_{z\delta_e}}{m_1} \qquad b_2 = \frac{C_{z\delta_e}}{m_1 - c_1 C_{z\dot{\alpha}}}$$

$$b_3 = \frac{C_{m\delta_e} + \xi_2 C_{z\delta_e}}{I_{y1}} \qquad b_4 = 0$$

$$\xi_1 = \frac{C_{x\dot{\alpha}}c_1}{m_1 - C_{z\dot{\alpha}}c_1}$$

$$\xi_2 = \frac{C_{m\dot{\alpha}}c_1}{m_1 - C_{z\dot{\alpha}}c_1}$$

$$m_1 = \frac{2m}{\rho U_o S} \qquad c_1 = \frac{\bar{c}}{2U_o}$$

$$I_{y1} = \frac{I_y}{\frac{1}{2}\rho U_o^2 S\bar{c}}$$

The longitudinal characteristic equation:

$$A_\delta\lambda^4 + B_\delta\lambda^3 + C_\delta\lambda^2 + D_\delta\lambda + E_\delta = 0$$

$$A_\delta = m_1 I_{y1}(m_1 - C_{z\dot{\alpha}}c_1)$$

$$B_\delta = m_1(-I_{y1}C_{z\alpha} - C_{mq}c_1[m_1 - C_{z\dot{\alpha}}c_1] - C_{m\dot{\alpha}}c_1[m_1 + C_{zq}c_1])$$
$$- C_{xu}I_{y1}(m_1 - C_{z\dot{\alpha}}c_1) - C_{x\dot{\alpha}}c_1 C_{zu}I_{y1}$$

$$C_\delta = m_1(C_{z\alpha}C_{mq}c_1 - C_{m\alpha}[m_1 + C_{zq}c_1] - C_{z\theta}C_{m\dot{\alpha}}c_1)$$
$$+ C_{xu}(I_{y1}C_{z\alpha} + C_{mq}c_1[m_1 - C_{z\dot{\alpha}}c_1] + C_{m\dot{\alpha}}c_1)[m_1 + C_{zq}c_1]$$
$$- C_{x\alpha}C_{zu}I_{y1} + C_{x\dot{\alpha}}c_1(C_{zu}C_{mq}c_1 - C_{mu}[m_1 + C_{zq}c_1])$$
$$- C_{xq}c_1(C_{zu}C_{m\dot{\alpha}}c_1 + C_{mu}[m_1 - C_{z\dot{\alpha}}c_1])$$

$$D_\delta = -C_{xu}(C_{z\alpha}C_{mq}c_1 - C_{m\alpha}[m_1 + C_{zq}c_1] - C_{z\theta}C_{m\dot{\alpha}}c_1)$$
$$- m_1 C_{m\alpha}C_{z\theta} + C_{x\alpha}(C_{zu}C_{mq}c_1 - C_{mu}[m_1 + C_{zq}c_1])$$
$$- C_{x\dot{\alpha}}c_1 C_{mu}C_{z\theta} - C_{xq}c_1(C_{zu}C_{m\alpha} - C_{z\alpha}C_{mu})$$
$$- C_{x\theta}(C_{zu}C_{m\dot{\alpha}}c_1 + C_{mu}[m_1 - C_{z\dot{\alpha}}c_1])$$

$$E_\delta = C_{xu}C_{m\alpha}C_{z\theta} - C_{x\alpha}C_{mu}C_{z\theta} - C_{x\theta}(C_{zu}C_{m\alpha} - C_{z\alpha}C_{mu})$$

F.2.2 Short-Period Approximation

$$\dot{x} = A_{sp}x + B_{sp}u_c$$

$$x = \begin{bmatrix} \Delta\alpha \\ q \\ \Delta\theta \end{bmatrix} \qquad A = \begin{bmatrix} a_{11} & a_{12} & a_{13} \\ a_{21} & a_{22} & a_{23} \\ a_{31} & a_{32} & a_{33} \end{bmatrix} \qquad B = \begin{bmatrix} b_1 \\ b_2 \\ b_3 \end{bmatrix} \qquad u_c = \Delta\delta_e$$

$$a_{11} = \frac{C_{z\alpha}}{m_1 - C_{z\dot{\alpha}}c_1} \qquad a_{12} = \frac{m_1 + C_{zq}c_1}{m_1 - C_{z\dot{\alpha}}c_1}$$

$$a_{13} = \frac{C_{z\theta}}{m_1 - C_{z\dot{\alpha}}c_1} \qquad a_{21} = \left(\frac{1}{I_{y1}}\right)\left(C_{m\alpha} + \frac{C_{m\dot{\alpha}}c_1 C_{z\alpha}}{m_1 - C_{z\dot{\alpha}}c_1}\right)$$

$$a_{22} = \left(\frac{1}{I_{y1}}\right)\left[C_{mq}c_1 + \left(\frac{C_{m\dot{\alpha}}c_1}{m_1 - C_{z\dot{\alpha}}c_1}\right)(m_1 + c_1 C_{zq})\right]$$

$$a_{23} = \frac{C_{m\dot{\alpha}}c_1 C_{z\theta}}{I_{y1}(m_1 - C_{z\dot{\alpha}}c_1)}$$

$$a_{31} = 0 \qquad a_{32} = 1 \qquad a_{33} = 0$$

$$b_1 = \frac{C_{z\delta_e}}{m_1 - C_{z\dot{\alpha}}c_1} \qquad b_2 = \left(\frac{1}{I_{y1}}\right)\left(C_{m\delta_e} + \frac{C_{m\dot{\alpha}}c_1 C_{z\delta_e}}{m_1 - C_{z\dot{\alpha}}c_1}\right)$$

$$b_3 = 0$$

F.2.3 Phugoid Approximation

$$\dot{x} = A_{ph}x + B_{ph}u_c$$

$$x = \begin{bmatrix} u \\ \Delta\theta \end{bmatrix} \qquad A_{ph} = \begin{bmatrix} a_{11} & a_{12} \\ a_{21} & a_{22} \end{bmatrix} \qquad B_{ph} = \begin{bmatrix} b_1 \\ b_2 \end{bmatrix} \qquad u_c = \Delta\delta_e$$

$$a_{11} = \left(\frac{C_{xu} - \xi_3 C_{zu}}{m_1}\right) \qquad a_{12} = \left(\frac{C_{x\theta} - \xi_3 C_{z\theta}}{m_1}\right)$$

$$a_{21} = \left(\frac{-C_{zu}}{m_1 + C_{zq}c_1}\right) \qquad a_{22} = \left(\frac{-C_{z\theta}}{m_1 + C_{zq}c_1}\right)$$

$$b_1 = \left(\frac{C_{x\delta_e} - \xi_3 C_{z\delta_e}}{m_1}\right) \qquad b_2 = \left(\frac{-C_{z\delta_e}}{m_1 + C_{zq}c_1}\right)$$

$$\xi_3 = \frac{C_{xq}c_1}{m_1 + C_{zq}c_1}$$

F.3 Longitudinal Transfer Functions

F.3.1 Complete Fourth Order System

Transfer function for forward velocity:

$$\frac{\bar{u}(s)}{\Delta\bar{\delta}_e(s)} = \frac{N_{\bar{u}}}{\Delta_{\text{long}}(s)}$$

$$N_{\bar{u}} = A_{\bar{u}}s^3 + B_{\bar{u}}s^2 + C_{\bar{u}}s + D_{\bar{u}}$$

$$A_{\bar{u}} = I_{y1}\left(C_{x\delta_e}m_1 - C_{x\delta_e}C_{z\dot{\alpha}}c_1 + C_{x\dot{\alpha}}c_1C_{z\delta_e}\right)$$

$$\begin{aligned} B_{\bar{u}} = {} & -C_{x\delta_e}\left(C_{z\alpha}I_{y1} + m_1C_{mq}c_1 - C_{z\dot{\alpha}}C_{mq}c_1^2 + m_1C_{m\dot{\alpha}}c_1 + C_{zq}C_{m\dot{\alpha}}c_1^2\right) \\ & - C_{z\delta_e}C_{x\dot{\alpha}}c_1^2C_{mq} + C_{z\delta_e}I_{y1}C_{x\alpha} + C_{x\dot{\alpha}}c_1C_{m\delta_e}(m_1 + C_{zq}c_1) \\ & + C_{xq}c_1^2C_{z\delta_e}C_{m\dot{\alpha}} + C_{xq}c_1C_{m\delta_e}(m_1 - C_{z\dot{\alpha}}c_1) \end{aligned}$$

$$\begin{aligned} C_{\bar{u}} = {} & C_{x\delta_e}(C_{z\alpha}C_{mq}c_1 - m_1C_{m\alpha} - C_{zq}c_1C_{m\alpha} - C_{z\theta}C_{m\dot{\alpha}}c_1) \\ & - C_{mq}C_{z\delta_e}C_{x\alpha}c_1 + C_{x\alpha}C_{m\delta_e}(m_1 + C_{zq}c_1) \\ & + C_{x\dot{\alpha}}c_1C_{z\theta}C_{m\delta_e} + C_{xq}c_1C_{z\delta_e}C_{m\alpha} - C_{xq}c_1C_{m\delta_e}C_{z\alpha} \\ & + C_{x\theta}C_{z\delta_e}C_{m\dot{\alpha}}c_1 + C_{x\theta}C_{m\delta_e}(m_1 - C_{z\dot{\alpha}}c_1) \end{aligned}$$

$$D_{\bar{u}} = C_{z\theta}(C_{x\alpha}C_{m\delta_e} - C_{m\alpha}C_{x\delta_e}) + C_{x\theta}(C_{z\delta_e}C_{m\alpha} - C_{m\delta_e}C_{z\alpha})$$

$$\Delta_{\text{long}}(s) = A_\delta s^4 + B_\delta s^3 + C_\delta s^2 + D_\delta s + E_\delta$$

The transfer function for angle of attack:

$$\frac{\Delta\bar{\alpha}(s)}{\Delta\bar{\delta}_e(s)} = \frac{N_{\bar{\alpha}}}{\Delta_{\text{long}}(s)}$$

$$N_{\bar{\alpha}} = A_{\alpha}s^3 + B_{\bar{\alpha}}s^2 + C_{\bar{\alpha}}s + D_{\bar{\alpha}}$$

$$A_{\bar{\alpha}} = m_1 I_{y1} C_{z\delta_e}$$

$$B_{\bar{\alpha}} = m_1\left(-C_{z\delta_e}C_{mq}c_1 + C_{m\delta_e}m_1 + C_{m\delta_e}C_{zq}c_1\right) - C_{xu}C_{z\delta_e}I_{y1} + C_{x\delta_e}C_{zu}I_{y1}$$

$$C_{\bar{\alpha}} = C_{xu}\left(C_{z\delta_e}C_{mq}c_1 - C_{m\delta_e}m_1 - C_{m\delta_e}C_{zq}c_1\right) + m_1 C_{m\delta_e}C_{z\theta} - C_{x\delta_e}C_{zu}C_{mq}c_1 + m_1 C_{x\delta_e}C_{mu} + C_{x\delta_e}C_{mu}C_{zq}c_1 + C_{xq}c_1 C_{zu}C_{m\delta_e} - C_{xq}c_1 C_{mu}C_{z\delta_e}$$

$$D_{\bar{\alpha}} = C_{x\theta}\left(C_{zu}C_{m\delta_e} - C_{mu}C_{z\delta_e}\right) + C_{z\theta}\left(C_{x\delta_e}C_{mu} - C_{m\delta_e}C_{xu}\right)$$

The transfer function for pitch angle:

$$\frac{\Delta\bar{\theta}(s)}{\Delta\bar{\delta}_e(s)} = \frac{N_{\bar{\theta}}}{\Delta_{\text{long}}(s)}$$

$$N_{\theta} = A_{\bar{\theta}}s^2 + B_{\bar{\theta}}s + C_{\bar{\theta}}$$

$$A_{\bar{\theta}} = m_1\left(m_1 C_{m\delta_e} - C_{z\dot{\alpha}}c_1 C_{m\delta_e} + C_{z\delta_e}C_{m\dot{\alpha}}c_1\right)$$

$$B_{\bar{\theta}} = m_1\left(C_{z\delta_e}C_{m\alpha} - C_{z\alpha}C_{m\delta_e}\right) - C_{xu}\left(m_1 C_{m\delta_e} - C_{z\dot{\alpha}}c_1 C_{m\delta_e} + C_{z\delta_e}C_{m\dot{\alpha}}c_1\right) + c_1 C_{x\dot{\alpha}}\left(-C_{zu}C_{m\delta_e} + C_{mu}C_{z\delta_e}\right) + C_{x\delta_e}C_{zu}C_{m\dot{\alpha}}c_1 + C_{x\delta_e}C_{mu}m_1 - C_{x\delta_e}C_{mu}C_{z\dot{\alpha}}c_1$$

$$C_{\bar{\theta}} = -C_{xu}\left(C_{z\delta_e}C_{m\alpha} - C_{z\alpha}C_{m\delta_e}\right) + C_{x\delta_e}(C_{m\alpha}C_{zu} - C_{mu}C_{z\alpha}) + C_{x\alpha}\left(-C_{zu}C_{m\delta_e} + C_{mu}C_{z\delta_e}\right)$$

The transfer function for pitch rate:

$$\frac{\bar{q}(s)}{\Delta\bar{\delta}_e(s)} = \frac{s\Delta\bar{\theta}(s)}{\Delta\bar{\delta}_e(s)}$$

F.3.2 Transfer Functions for Short-Period Approximation

The transfer function for angle of attack:

$$\frac{\Delta\bar{\alpha}(s)}{\Delta\bar{\delta}_e(s)} = \frac{N_{\bar{\alpha},\text{spo}}}{\Delta_{\text{spo}}(s)}$$

$$N_{\overline{\alpha},\text{spo}} = A_{\overline{\alpha},\text{spo}} s^2 + B_{\overline{\alpha},\text{spo}} s + C_{\overline{\alpha},\text{spo}}$$
$$A_{\overline{\alpha},\text{spo}} = C_{z\delta_e} I_{y1}$$
$$B_{\overline{\alpha},\text{spo}} = C_{m\delta_e}(m_1 + C_{zq}c_1) - C_{z\delta_e} C_{mq} c_1$$
$$C_{\overline{\alpha},\text{spo}} = C_{m\delta_e} C_{z\theta}$$
$$\Delta_{\text{spo}} = a_1 s^3 + a_2 s^2 + a_3 s + a_4$$
$$a_1 = (m_1 - C_{z\dot{\alpha}} c_1) I_{y1}$$
$$a_2 = -[C_{z\alpha} I_{y1} + C_{mq} c_1 (m_1 - C_{z\dot{\alpha}} c_1) + (m_1 + C_{zq} c_1) C_{m\dot{\alpha}} c_1]$$
$$a_3 = C_{z\alpha} C_{mq} c_1 - [(m_1 + C_{zq} c_1) C_{m\alpha} + C_{z\theta} C_{m\dot{\alpha}} c_1]$$
$$a_4 = -C_{z\theta} C_{m\alpha}$$

The transfer function for pitch angle:

$$\frac{\Delta\overline{\theta}(s)}{\Delta\overline{\delta}_e(s)} = \frac{N_{\overline{\theta},\text{spo}}}{\Delta_{\text{spo}}(s)}$$
$$N_{\overline{\theta},\text{spo}} = A_{\overline{\theta},\text{spo}} s + B_{\overline{\theta},\text{spo}}$$
$$A_{\overline{\theta},\text{spo}} = C_{m\delta_e}(m_1 - C_{z\dot{\alpha}} c_1) + C_{m\dot{\alpha}} c_1 C_{z\delta_e}$$
$$B_{\overline{\theta},\text{spo}} = C_{m\alpha} C_{z\delta_e} - C_{z\alpha} C_{m\delta_e}$$

F.3.3 Transfer Functions for a Phugoid Approximation

The transfer function for forward velocity:

$$\frac{\Delta\overline{u}(s)}{\Delta\overline{\delta}_e(s)} = \frac{N_{\overline{u},\text{lpo}}}{\Delta_{\text{lpo}}(s)}$$
$$N_{\overline{u},\text{lpo}} = A_{\overline{u},\text{lpo}} s + B_{\overline{u},\text{lpo}}$$
$$A_{\overline{u},\text{lpo}} = -\left[C_{x\delta_e}(m_1 + C_{zq} c_1) - C_{z\delta_e} C_{xq} c_1\right]$$
$$B_{\overline{u},\text{lpo}} = C_{z\delta_e} C_{x\theta} - C_{x\delta_e} C_{z\theta}$$

and

$$\Delta_{\text{lpo}} = A_{\delta,\text{lpo}} s^2 + B_{\delta,\text{lpo}} s + C_{\delta,\text{lpo}}$$
$$A_{\delta,\text{lpo}} = -m_1(m_1 + C_{zq} c_1)$$
$$B_{\delta,\text{lpo}} = -m_1 C_{z\theta} + C_{xu}(m_1 + C_{zq} c_1) - C_{zu} C_{xq} c_1$$
$$C_{\delta,\text{lpo}} = C_{xu} C_{z\theta} - C_{zu} C_{x\theta}$$

The transfer function for pitch angle:

$$\frac{\Delta\overline{\theta}(s)}{\Delta\overline{\delta}_e(s)} = \frac{N_{\overline{\theta},\text{lpo}}}{\Delta_{\text{lpo}}(s)}$$

$$N_{\bar{\theta},\text{lpo}} = A_{\bar{\theta},\text{lpo}} s + B_{\bar{\theta},\text{lpo}}$$

$$A_{\bar{\theta},\text{lpo}} = m_1 C_{z\delta_e}$$

$$B_{\bar{\theta},\text{lpo}} = C_{x\delta_e} C_{zu} - C_{xu} C_{z\delta_e}$$

F.4 Lateral-Directional Equations in State-Space Form

F.4.1 Complete Fifth Order System

$$\dot{x} = Ax + Bu_c$$

$$x = \begin{bmatrix} \Delta\beta \\ \Delta\phi \\ p \\ \Delta\psi \\ r \end{bmatrix} \quad A = \begin{bmatrix} a_{11} & a_{12} & a_{13} & a_{14} & a_{15} \\ a_{21} & a_{22} & a_{23} & a_{24} & a_{25} \\ a_{31} & a_{32} & a_{33} & a_{34} & a_{35} \\ a_{41} & a_{42} & a_{43} & a_{44} & a_{45} \\ a_{51} & a_{52} & a_{53} & a_{54} & a_{55} \end{bmatrix}$$

$$B = \begin{bmatrix} b_{11} & b_{12} \\ b_{21} & b_{22} \\ b_{31} & b_{32} \\ b_{41} & b_{42} \end{bmatrix} \qquad u_c = \begin{bmatrix} \Delta\delta_a \\ \Delta\delta_r \end{bmatrix}$$

$$a_{11} = \frac{C_{y\beta}}{m_1 - b_1 C_{y\dot{\beta}}} \qquad a_{12} = \frac{C_{y\phi}}{m_1 - b_1 C_{y\dot{\beta}}}$$

$$a_{13} = \frac{C_{yp} b_1}{m_1 - b_1 C_{y\dot{\beta}}} \qquad a_{14} = 0 \qquad a_{15} = -\left(\frac{m_1 - b_1 C_{yr}}{m_1 - b_1 C_{y\dot{\beta}}}\right)$$

$$a_{21} = 0 \qquad a_{22} = 0$$

$$a_{23} = 1 \qquad a_{24} = 0 \qquad a_{25} = 0$$

$$a_{31} = C_{l\beta} I'_{z1} + C_{n\beta} I'_{xz1} + \xi_1 b_1 a_{11} \qquad a_{32} = \xi_1 b_1 a_{12}$$

$$a_{33} = C_{lp} b_1 I'_{z1} + C_{np} I'_{xz1} b_1 + \xi_1 b_1 a_{13} \qquad a_{34} = 0$$

$$a_{35} = C_{lr} b_1 I'_{z1} + C_{nr} I'_{xz1} b_1 + \xi_1 b_1 a_{15}$$

$$a_{41} = a_{42} = a_{43} = a_{44} = 0 \qquad a_{45} = 1$$

$$a_{51} = I'_{x1} C_{n\beta} + I'_{xz1} C_{l\beta} + b_1 \xi_2 a_{11}$$

$$a_{52} = \xi_2 b_1 a_{12} \qquad a_{53} = b_1 \left(C_{np} I'_{x1} + C_{lp} I'_{xz1} + \xi_2 a_{13} \right)$$

$$a_{54} = 0 \qquad a_{55} = b_1 \left(I'_{x1} C_{nr} + I'_{xz1} C_{lr} + \xi_2 a_{15} \right)$$

$$b_{11} = \frac{C_{y\delta_a}}{(m_1 - b_1 C_{y\dot{\beta}})} \qquad b_{12} = \frac{C_{y\delta_r}}{(m_1 - b_1 C_{y\dot{\beta}})}$$

$$b_{21} = 0 \qquad b_{22} = 0$$

$$b_{31} = C_{l\delta_a} I'_{z1} + C_{n\delta_a} I'_{xz1} + \xi_1 b_1 b_{11} \qquad b_{32} = C_{l\delta_r} I'_{z1} + C_{n\delta_r} I'_{xz1} + \xi_1 b_1 b_{12}$$

$$b_{41} = 0 \qquad b_{42} = 0$$

$$b_{51} = C_{n\delta_a} I'_{x1} + C_{l\delta_a} I'_{xz1} + \xi_2 b_1 b_{11} \qquad b_{52} = C_{n\delta_r} I'_{x1} + C_{l\delta_r} I'_{xz1} + \xi_2 b_1 b_{12}$$

$$\xi_1 = I'_{z1} C_{l\dot{\beta}} + I'_{xz1} C_{n\dot{\beta}} \qquad \xi_2 = I'_{x1} C_{n\dot{\beta}} + I'_{xz1} C_{l\dot{\beta}}$$

F.4.2 Roll Subsidence Approximation

$$I_{x1}\dot{p} - C_{lp} b_1 p = C_{l\delta_a}\Delta\delta_a + C_{l\delta_r}\Delta\delta_r$$

F.4.3 Dutch-Roll Approximation

$$\begin{bmatrix} \Delta\dot{\beta} \\ \dot{r} \end{bmatrix} = \begin{bmatrix} a_{11} & a_{12} \\ a_{21} & a_{22} \end{bmatrix} \begin{bmatrix} \Delta\beta \\ r \end{bmatrix} + \begin{bmatrix} b_{11} & b_{12} \\ b_{21} & b_{22} \end{bmatrix} \begin{bmatrix} \Delta\delta_a \\ \Delta\delta_r \end{bmatrix}$$

$$a_{11} = \frac{C_{y\beta}}{m_1 - C_{y\dot{\beta}} b_1}$$

$$a_{12} = -\frac{m_1 - b_1 C_{yr}}{m_1 - C_{y\dot{\beta}} b_1}$$

$$a_{21} = \frac{1}{I_{z1}}\left(C_{n\beta} + \frac{C_{n\dot{\beta}} b_1 C_{y\beta}}{m_1 - C_{y\dot{\beta}} b_1} \right)$$

$$a_{22} = \frac{1}{I_{z1}}\left(\frac{-C_{n\dot{\beta}} b_1 (m_1 - b_1 C_{yr})}{m_1 - C_{y\dot{\beta}} b_1} + b_1 C_{nr} \right)$$

$$b_{11} = \frac{C_{y\delta_a}}{m_1 - C_{y\dot{\beta}} b_1}$$

$$b_{12} = \frac{C_{y\delta_r}}{m_1 - C_{y\dot{\beta}} b_1}$$

$$b_{21} = \frac{1}{I_{z1}}\left(C_{n\delta_a} + \frac{C_{n\dot{\beta}} b_1 C_{y\delta_a}}{m_1 - C_{y\dot{\beta}} b_1} \right)$$

$$b_{22} = \frac{1}{I_{z1}}\left(C_{n\delta_r} + \frac{C_{n\dot{\beta}} b_1 C_{y\delta_r}}{m_1 - C_{y\dot{\beta}} b_1} \right)$$

F.4.4 Spiral Mode Approximation

$$C_{l\beta}\Delta\beta + C_{lr}b_1 r + C_{l\delta a}\Delta\delta_a + C_{l\delta r}\Delta\delta_r = 0$$

$$I_{z1}\dot{r} = C_{n\beta}\Delta\beta + C_{nr}rb_1 + C_{n\delta_a}\Delta\delta_a + C_{n\delta_r}\Delta\delta_r$$

F.5 Lateral-Directional Transfer Functions

F.5.1 Complete Fifth Order System

The transfer functions for aileron deflection are presented here. To derive transfer functions for rudder deflections, replace $\Delta\bar{\delta}_a(s)$ with $\Delta\bar{\delta}_r(s)$. The following is the sideslip transfer function for aileron deflection:

$$\frac{\Delta\bar{\beta}(s)}{\Delta\bar{\delta}_a(s)} = \frac{N_{\beta,\bar{\delta}_a}}{\Delta_{\text{lat}}(s)}$$

$$N_{\bar{\beta},\delta_a} = A_{\beta,\delta_a}s^4 + B_{\beta,\delta_a}s^3 + C_{\beta,\delta_a}s^2 + D_{\beta,\delta_a}s$$

$$A_{\beta,\delta_a} = C_{y\delta_a}\left(I_{x1}I_{z1} - I_{xz1}^2\right)$$

$$\begin{aligned}B_{\beta,\delta_a} &= -b_1 C_{y\delta_a}(I_{x1}C_{nr} + C_{lp}I_{z1} + C_{lr}I_{xz1} + C_{np}I_{xz1})\\ &\quad + b_1 C_{yp}\left(I_{z1}C_{l\delta_a} + C_{n\delta_a}I_{xz1}\right) - (m_1 - b_1 C_{yr})\left(C_{l\delta_a}I_{xz1} + C_{n\delta_a}I_{x1}\right)\end{aligned}$$

$$\begin{aligned}C_{\beta,\delta_a} &= b_1^2 C_{y\delta_a}(C_{lp}C_{nr} - C_{lr}C_{np}) + C_{y\phi}(C_{l\delta_a}I_{z1} + C_{n\delta_a}I_{xz1})\\ &\quad - b_1^2 C_{yp}\left(C_{l\delta_a}C_{nr} - C_{n\delta_a}C_{lr}\right) + b_1(m_1 - b_1 C_{yr})\left(C_{n\delta_a}C_{lp} - C_{l\delta_a}C_{np}\right)\end{aligned}$$

$$D_{\beta,\delta_a} = -b_1 C_{y\phi}\left(C_{l\delta_a}C_{nr} - C_{n\delta_a}C_{lr}\right)$$

The determinant Δ_{lat}:

$$\Delta_{\text{lat}}(s) = A_{\delta_{\text{lat}}}s^5 + B_{\delta_{\text{lat}}}s^4 + C_{\delta_{\text{lat}}}s^3 + D_{\delta_{\text{lat}}}s^2 + E_{\delta_{\text{lat}}}s$$

$$A_{\delta_{\text{lat}}} = (m_1 - b_1 C_{y\dot{\beta}})\left(I_{x1}I_{z1} - I_{xz1}^2\right)$$

$$\begin{aligned}B_{\delta_{\text{lat}}} &= (-C_{y\beta})\left(I_{x1}I_{z1} - I_{xz1}^2\right) - (m_1 - b_1 C_{y\dot{\beta}})(I_{x1}C_{nr}b_1 + I_{z1}C_{lp}b_1\\ &\quad + I_{xz1}C_{lr}b_1 + I_{xz1}C_{np}b_1) - b_1 C_{yp}(C_{l\dot{\beta}}b_1 I_{z1} + C_{n\dot{\beta}}b_1 I_{xz1})\\ &\quad + (m_1 - b_1 C_{yr})(C_{l\dot{\beta}}b_1 I_{xz1} + I_{x1}C_{n\dot{\beta}}b_1)\end{aligned}$$

$$\begin{aligned}C_{\delta_{\text{lat}}} &= b_1^2(m_1 - b_1 C_{y\dot{\beta}})(C_{lp}C_{nr} - C_{lr}C_{np}) + b_1 C_{y\beta}(I_{x1}C_{nr} + I_{z1}C_{lp}\\ &\quad + I_{xz1}C_{lr} + I_{xz1}C_{np}) - C_{y\phi}b_1(C_{l\dot{\beta}}I_{z1} + C_{n\dot{\beta}}I_{xz1})\\ &\quad + b_1 C_{yp}(C_{l\dot{\beta}}C_{nr}b_1^2 - C_{l\beta}I_{z1} - I_{xz1}C_{n\beta} - C_{lr}C_{n\dot{\beta}}b_1^2)\\ &\quad + (m_1 - b_1 C_{yr})(C_{l\beta}I_{xz1} + C_{l\dot{\beta}}C_{np}b_1^2 - C_{lp}C_{n\dot{\beta}}b_1^2 + I_{x1}C_{n\beta})\end{aligned}$$

$$D_{\delta_{\text{lat}}} = -C_{y\beta}b_1^2(C_{lp}C_{nr} - C_{lr}C_{nr}) + C_{y\phi}(C_{l\dot{\beta}}C_{nr}b_1^2 - C_{l\beta}I_{z1} - I_{xz1}C_{n\beta}$$
$$-C_{lr}C_{n\dot{\beta}}b_1^2) + b_1^2C_{yp}(C_{l\beta}C_{nr} - C_{lr}C_{n\beta}) + b_1(m_1 - b_1C_{yr})$$
$$\times (C_{l\beta}C_{np} - C_{lp}C_{n\beta})$$
$$E_{\delta_{\text{lat}}} = C_{y\phi}b_1(C_{l\beta}C_{nr} - C_{n\beta}C_{lr})$$

The transfer function for angle of yaw for aileron deflection:

$$\frac{\Delta\bar{\psi}(s)}{\Delta\bar{\delta}_a(s)} = \frac{N_{\psi,\bar{\delta}_a}}{\Delta_{\text{lat}}(s)}$$

$$N_{\bar{\psi},\delta_a} = A_{\bar{\psi}}s^3 + B_{\bar{\psi}}s^2 + C_{\bar{\psi}}s + D_{\bar{\psi}}$$

$$A_{\bar{\psi}} = (m_1 - b_1C_{y\dot{\beta}})(I_{x1}C_{n\delta_a} + I_{xz1}C_{l\delta_a}) + C_{y\delta_a}(I_{xz1}C_{l\dot{\beta}}b_1 + I_{x1}C_{n\dot{\beta}}b_1)$$

$$B_{\bar{\psi}} = -C_{y\beta}(I_{xz1}C_{n\delta_a} + I_{xz1}C_{l\delta_a}) + (m_1 - b_1C_{y\dot{\beta}})\left(C_{l\delta_a}C_{np}b_1 - C_{n\delta_a}C_{lp}b_1\right)$$
$$+ C_{yp}b_1(-C_{n\delta_a}C_{l\dot{\beta}}b_1 + C_{l\delta_a}C_{n\dot{\beta}}b_1) + C_{y\delta_a}(I_{xz1}C_{l\beta} + C_{l\dot{\beta}}C_{np}b_1^2$$
$$- C_{n\dot{\beta}}C_{lp}b_1^2 + I_{x1}C_{n\beta})$$

$$C_{\bar{\psi}} = -C_{y\beta}b_1\left(C_{l\delta_a}C_{np} - C_{n\delta_a}C_{lp}\right) + b_1C_{yp}(C_{n\beta}C_{l\delta_a} - C_{l\beta}C_{n\delta_a})$$
$$+ C_{y\phi}(-C_{n\delta_a} + C_{l\dot{\beta}}b_1 + C_{l\delta_a}C_{n\dot{\beta}}b_1) + b_1C_{y\delta_a}(C_{l\beta}C_{np} - C_{lp}C_{n\beta})$$
$$D_{\bar{\psi}} = C_{y\phi}(C_{n\beta}C_{l\delta_a} - C_{l\beta}C_{n\delta_a})$$

The transfer function for the bank angle is given by

$$\frac{\Delta\bar{\phi}(s)}{\Delta\bar{\delta}_a(s)} = \frac{N_{\phi,\bar{\delta}_a}}{\Delta_{\text{lat}}(s)}$$

$$N_{\bar{\phi},\delta_a} = A_{\bar{\phi}}s^2 + B_{\bar{\phi}}s + C_{\bar{\phi}}$$

$$A_{\bar{\phi}} = (m_1 - b_1C_{y\dot{\beta}})(I_{xz1}C_{n\delta_a} + I_{z1}C_{l\delta_a}) + C_{y\delta_a}(I_{xz1}C_{n\dot{\beta}}b_1 + I_{z1}C_{l\dot{\beta}}b_1)$$

$$B_{\bar{\phi}} = -C_{y\beta}(I_{xz1}C_{n\delta_a} + I_{z1}C_{l\delta_a}) + (m_1 - b_1C_{y\dot{\beta}})(C_{n\delta_a}C_{lr} - C_{l\delta_a}C_{nr})b_1$$
$$- C_{y\delta_a}(C_{l\dot{\beta}}C_{nr}b_1^2 - C_{l\beta}I_{z1} - C_{n\beta}I_{xz1} - C_{n\dot{\beta}}C_{lr}b_1^2)$$
$$+ (m_1 - b_1C_{yr})(C_{l\delta_a}C_{n\dot{\beta}}b_1 - C_{n\delta_a}C_{l\dot{\beta}}b_1)$$

$$C_{\bar{\phi}} = (m_1 - b_1C_{yr})(C_{l\delta_a}C_{n\beta} - C_{n\delta_a}C_{l\beta}) + b_1C_{y\beta}(C_{l\delta_a}C_{nr} - C_{n\delta_a}C_{lr})$$
$$+ b_1C_{y\delta_a}(C_{n\beta}C_{lr} - C_{l\beta}C_{nr})$$

F.5.2 Roll Subsidence Approximation

$$\frac{\Delta\bar{\phi}(s)}{\Delta\bar{\delta}_a(s)} = \frac{C_{l\delta_a}}{s(I_{x1}s - C_{lp}b_1)}$$

$$\frac{\bar{p}(s)}{\Delta\bar{\delta}_a(s)} = \frac{s\Delta\bar{\phi}(s)}{\Delta\bar{\delta}_a(s)}$$

F.5.3 Dutch-Roll Approximation

$$\frac{\Delta\bar{\beta}(s)}{\Delta\bar{\delta}_a(s)} = \frac{N_{\bar{\beta}}}{\Delta_{\text{lat},dr}(s)}$$

$$\frac{\Delta\bar{\psi}(s)}{\Delta\bar{\delta}_a(s)} = \frac{N_{\bar{\psi}}}{\Delta_{\text{lat},dr}(s)}$$

$$N_{\bar{\beta}} = I_{z1}C_{y\delta_a}s^2 + (b_1C_{n\delta_a}C_{yr} - C_{y\delta_a}C_{nr}b_1 - m_1C_{n\delta_a})s$$

$$\Delta_{\text{lat},dr}(s) = I_{z1}(m_1 - b_1C_{y\dot{\beta}})s^3 - [C_{y\beta}I_{z1} + C_{nr}b_1(m_1 - b_1C_{y\dot{\beta}}) - C_{n\dot{\beta}}b_1(m_1 - b_1C_{yr})]s^2 + [C_{y\beta}C_{nr}b_1 + C_{n\beta}(m_1 - b_1C_{yr})]s$$

$$N_{\bar{\psi}} = [C_{n\delta_a}(m_1 - b_1C_{y\dot{\beta}}) + C_{y\delta_a}C_{n\dot{\beta}}b_1]s + (C_{y\delta_a}C_{n\beta} - C_{n\delta_a}C_{y\beta})$$

F.5.4 Spiral Mode Approximation

$$\frac{\Delta\bar{\beta}(s)}{\Delta\bar{\delta}_a(s)} = \frac{A_{\bar{\beta}}s^2 + B_{\bar{\beta}}s}{A_\delta s^2 + B_\delta s}$$

$$\frac{\Delta\bar{\psi}(s)}{\Delta\bar{\delta}_a(s)} = \frac{A_{\bar{\psi}}}{A_\delta s^2 + B_\delta s}$$

$$A_{\bar{\beta}} = C_{l\delta_a}I_{z1}$$

$$B_{\bar{\beta}} = b_1\left(C_{n\delta_a}C_{lr} - C_{l\delta_a}C_{nr}\right)$$

$$A_\delta = -C_{l\beta}I_{z1}$$

$$B_\delta = b_1(C_{l\beta}C_{nr} - C_{n\beta}C_{lr})$$

$$A_{\bar{\psi}} = C_{n\beta}C_{l\delta_a} - C_{l\beta}C_{n\delta_a}$$

BIBLIOGRAPHY

Anderson, J. D., *Introduction to Flight,* McGraw-Hill, New York, 1989.

Babister, A. W., *Aircraft Stability and Control,* Pergamon, New York, 1961.

Blakelock, J. H., *Automatic Control of Aircraft and Missiles,* 2nd ed., Wiley, New York, 1991.

Dickinson, B., *Aircraft Stability and Control for Pilots and Engineers,* Sir Isaac Pitman and Sons, London, 1968.

Etkin, B., *Dynamics of Atmospheric Flight,* Dover, New York, 2005.

Gerlach, O. H., *Lecture Notes on Stability and Control,* Parts 1 and 2, Dept. of Aerospace Engineering, Delft Univ., Delft, Netherlands, 1983.

Hale, F. J., *Aircraft Performance, Selection, and Design,* Wiley, New York, 1984.

McCormick, B. W., *Aerodynamics, Aeronautics, and Flight Mechanics,* Wiley, New York, 1979.

McRuer, D., Askenas, I., and Graham, D., *Aircraft Dynamics and Automatic Control,* Princeton Univ., Princeton, NJ, 1973.

Miele, A., *Flight Mechanics, Vol. I, Theory of Flight Paths,* Addison-Wesley, Reading, MA, 1962.

Nelson, R. C., *Flight Stability and Automatic Control,* McGraw-Hill, New York, 1989.

Perkins, C. D., and Hage, R. E., *Airplane Performance, Stability and Control,* 10th Printing, Wiley, New York, 1965.

Raymer, D. P., *Aircraft Design: A Conceptual Approach,* AIAA Education Series, AIAA, Washington, DC, 1992.

Roskam, J., *Airplane Flight Dynamics and Automatic Flight Control, Part I,* Roskam Aviation and Engineering, Lawrence, KS, 1979.

INDEX

Page numbers followed by *f* indicate figures. Page numbers followed by **t** indicate tables.

SUPPORTING MATERIALS

A complete listing of titles in the AIAA Education Series is available from AIAA's electronic library, Aerospace Research Central (ARC), at arc.aiaa.org. Visit ARC frequently to stay abreast of product changes, corrections, special offers, and new publications.

AIAA is committed to devoting resources to the education of both practicing and future aerospace professionals. In 1996, the AIAA Foundation was founded. Its programs enhance scientific literacy and advance the arts and sciences of aerospace. For more information, please visit www.aiaafoundation.org.